MW00463937

Springer Finance

Springer Finance

Springer Finance is a programme of books aimed at students, academics and practitioners working on increasingly technical approaches to the analysis of financial markets. It aims to cover a variety of topics, not only mathematical finance but foreign exchanges, term structure, risk management, portfolio theory, equity derivatives, and financial economics.

Ammann M., Credit Risk Valuation: Methods, Models, and Application (2001)
Back K., A Course in Derivative Securities: Introduction to Theory and Computation (2005)
Barucci E., Financial Markets Theory. Equilibrium, Efficiency and Information (2003)
Bielecki T.R. and Rutkowski M., Credit Risk: Modeling, Valuation and Hedging (2002)
Bingham N.H. and Kiesel R., Risk-Neutral Valuation: Pricing and Hedging of Financial Derivatives (1998, 2nd ed. 2004)
Brigo D. and Mercurio F., Interest Rate Models: Theory and Practice (2001, 2nd ed. 2006)
Buff R., Uncertain Volatility Models-Theory and Application (2002)
Carmona R.A. and Tehranchi M.R., Interest Rate Models: an Infinite Dimensional Stochastic Analysis Perspective (2006)
Dana R.A. and Jeanblanc M., Financial Markets in Continuous Time (2002)
Deboeck G. and Kohonen T. (Editors), Visual Explorations in Finance with Self-Organizing Maps (1998)
Delbaen F. and Schachermayer W., The Mathematics of Arbitrage (2005)
Elliott R.J. and Kopp P.E., Mathematics of Financial Markets (1999, 2nd ed. 2005)
Fengler M.R., Semiparametric Modeling of Implied Volatility (2005)
Geman H., Madan D., Pliska S.R. and Vorst T. (Editors), Mathematical Finance–Bachelier Congress 2000 (2001)
Gundlach M., Lehrbass F. (Editors), CreditRisk$^+$ in the Banking Industry (2004)
Kellerhals B.P., Asset Pricing (2004)
Külpmann M., Irrational Exuberance Reconsidered (2004)
Kwok Y.-K., Mathematical Models of Financial Derivatives (1998)
Malliavin P. and Thalmaier A., Stochastic Calculus of Variations in Mathematical Finance (2005)
Meucci A., Risk and Asset Allocation (2005)
Pelsser A., Efficient Methods for Valuing Interest Rate Derivatives (2000)
Prigent J.-L., Weak Convergence of Financial Markets (2003)
Schmid B., Credit Risk Pricing Models (2004)
Shreve S.E., Stochastic Calculus for Finance I (2004)
Shreve S.E., Stochastic Calculus for Finance II (2004)
Yor M., Exponential Functionals of Brownian Motion and Related Processes (2001)
Zagst R., Interest-Rate Management (2002)
Zhu Y.-L., Wu X., Chern I.-L., Derivative Securities and Difference Methods (2004)
Ziegler A., Incomplete Information and Heterogeneous Beliefs in Continuous-time Finance (2003)
Ziegler A., A Game Theory Analysis of Options (2004)

Damiano Brigo · Fabio Mercurio

Interest Rate Models – Theory and Practice

With Smile, Inflation and Credit

With 124 Figures and 131 Tables

Damiano Brigo
Head of Credit Models
Banca IMI, San Paolo-IMI Group
Corso Matteotti 6
20121 Milano, Italy
and
Fixed Income Professor
Bocconi University, Milano, Italy
E-mail: damiano.brigo@gmail.com

Fabio Mercurio
Head of Financial Modelling
Banca IMI, San Paolo-IMI Group
Corso Matteotti 6
20121 Milano, Italy
E-mail: fabio.mercurio@bancaimi.it

Mathematics Subject Classification (2000): 60H10, 60H35, 62P05, 65C05, 65C20, 90A09

JEL Classification: G12, G13, E43

Library of Congress Control Number: 2006929545

ISBN-10 3-540-22149-2 2nd ed. Springer Berlin Heidelberg New York
ISBN-13 978-3-540-22149-4 2nd ed. Springer Berlin Heidelberg New York

ISBN 3-540-41772-9 1st ed. Springer-Verlag Berlin Heidelberg New York

Springer is a part of Springer Science+Business Media
springer.com

Printed in Germany

Cover design: *design & production*, Heidelberg
Typesetting: by the authors using a Springer LaTeX macro package
Production: LE-TeX Jelonek, Schmidt & Vöckler GbR, Leipzig

Printed on acid-free paper 41/3100YL - 5 4 3 2 1 0

To Our Families

Preface

"Professor Brigo, will there be any new quotes in the second edition?"
"Yes... for example this one!"

A student at a London training course, following a similar question by a Hong Kong student to Massimo Morini, 2003.

"I would have written you a shorter letter, but I didn't have the time"
Benjamin Franklin

MOTIVATION.... five years later.

...I'm sure he's got a perfectly good reason... for taking so long...
Emily, "Corpse Bride", Tim Burton (2005).

Welcome onboard the second edition of this book on interest rate models, to all old and new readers. We immediately say this second edition is actually almost a new book, with four hundred fifty and more new pages on smile modeling, calibration, inflation, credit derivatives and counterparty risk.

As explained in the preface of the first edition, the idea of writing this book on interest-rate modeling crossed our minds in early summer 1999. We both thought of different versions before, but it was in Banca IMI that this challenging project began materially, if not spiritually (more details are given in the trivia Appendix G). At the time we were given the task of studying and developing financial models for the pricing and hedging of a broad range of derivatives, and we were involved in medium/long-term projects.

The first years in Banca IMI saw us writing a lot of reports and material on our activity in the bank, to the point that much of those studies ended up in the first edition of the book, printed in 2001.

In the first edition preface we described motivation, explained what kind of theory and practice we were going to address, illustrated the aim and readership of the book, together with its structure and other considerations. We do so again now, clearly updating what we wrote in 2001.

Why a book on interest rate models, and why this new edition?

"Sorry I took so long to respond, Plastic Man. I'd like to formally declare my return to active duty, my friends... This is J'onn J'onzz activating full telepathic link. Counter offensive has begun". JLA 38, DC Comics (2000).

In years where every month a new book on financial modeling or on mathematical finance comes out, one of the first questions inevitably is: why one more, and why one on interest-rate modeling in particular?

The answer springs directly from our job experience as mathematicians working as quantitative analysts in financial institutions. Indeed, one of the major challenges any financial engineer has to cope with is the practical implementation of mathematical models for pricing derivative securities.

When pricing market financial products, one has to address a number of theoretical and practical issues that are often neglected in the classical, general basic theory: the choice of a satisfactory model, the derivation of specific analytical formulas and approximations, the calibration of the selected model to a set of market data, the implementation of efficient routines for speeding up the whole calibration procedure, and so on. In other words, the general understanding of the theoretical paradigms in which specific models operate does not lead to their complete understanding and immediate implementation and use for concrete pricing. This is an area that is rarely covered by books on mathematical finance.

Undoubtedly, there exist excellent books covering the basic theoretical paradigms, but they do not provide enough instructions and insights for tackling concrete pricing problems. We therefore thought of writing this book in order to cover this gap between theory and practice.

The first version of the book achieved this task in several respects. However, the market is rapidly evolving. New areas such as smile modeling, inflation, hybrid products, counterparty risk and credit derivatives have become fundamental in recent years. New bridges are required to cross the gap between theory and practice in these recent areas.

The Gap between Theory and Practice

But Lo! Siddârtha turned/ Eyes gleaming with divine tears to the sky,/ Eyes lit with heavenly pity to the earth;/ From sky to earth he looked, from earth to sky,/ As if his spirit sought in lonely flight/ Some far-off vision, linking this and that,/ Lost - past - but searchable, but seen, but known.
From "The Light of Asia", Sir Edwin Arnold (1879).

A gap, indeed. And a fundamental one. The interplay between theory and practice has proved to be an extremely fruitful ingredient in the progress of science and modeling in particular. We believe that practice can help to appreciate theory, thus generating a feedback that is one of the most important and intriguing aspects of modeling and more generally of scientific investigation.

If theory becomes deaf to the feedback of practice or vice versa, great opportunities can be missed. It may be a pity to restrict one's interest only to extremely abstract problems that have little relevance for those scientists or quantitative analysts working in "real life".

Now, it is obvious that everyone working in the field owes a lot to the basic fundamental theory from which such extremely abstract problems stem.

It would be foolish to deny the importance of a well developed and consistent theory as a fundamental support for any practical work involving mathematical models. Indeed, practice that is deaf to theory or that employs a sloppy mathematical apparatus is quite dangerous.

However, besides the extremely abstract refinement of the basic paradigms, which are certainly worth studying but that interest mostly an academic audience, there are other fundamental and more specific aspects of the theory that are often neglected in books and in the literature, and that interest a larger audience.

Is This Book about Theory? What kind of Theory?

> *"Our paper became a monograph. When we had completed the details, we rewrote everything so that no one could tell how we came upon our ideas or why. This is the standard in mathematics."*
>
> David Berlinski, "Black Mischief" (1988).

In the book, we are not dealing with the fundamental no-arbitrage paradigms with great detail. We resume and adopt the basic well-established theory of Harrison and Pliska, and avoid the debate on the several possible definitions of no-arbitrage and on their mutual relationships. Indeed, we will raise problems that can be faced in the basic framework above. Insisting on the subtle aspects and developments of no-arbitrage theory more than is necessary would take space from the other theory we need to address in the book and that is more important for our purposes.

Besides, there already exist several books dealing with the most abstract theory of no-arbitrage. On the theory that we deal with, on the contrary, there exist only few books, although in recent years the trend has been improving. What is this theory? For a flavor of it, let us select a few questions at random:

- How can the market interest-rate curves be defined in mathematical terms?
- What kind of interest rates does one select when writing the dynamics? Instantaneous spot rates? Forward rates? Forward swap rates?
- What is a sufficiently general framework for expressing no-arbitrage in interest-rate modeling?
- Are there payoffs that do not require the interest-rate curve dynamics to be valued? If so, what are these payoffs?
- Is there a definition of volatility (and of its term structures) in terms of interest-rate dynamics that is consistent with market practice?
- What kinds of diffusion coefficients in the rate dynamics are compatible with different qualitative evolutions of the term structure of volatilities over time?
- How is "humped volatility shape" translated in mathematical terms and what kind of mathematical models allow for it?

- What is the most convenient probability measure under which one can price a specific product, and how can one derive concretely the related interest-rate dynamics?
- Are different market models of interest-rate dynamics compatible?
- What does it mean to calibrate a model to the market in terms of the chosen mathematical model? Is this always possible? Or is there a degree of approximation involved?
- Does terminal correlation among rates depend on instantaneous volatilities or only on instantaneous correlations? Can we analyze this dependence?
- What is the volatility smile, how can it be expressed in terms of mathematical models and of forward-rate dynamics in particular?
- Is there a diffusion dynamics consistent with the quoting mechanism of the swaptions volatility smile in the market?
- What is the link between dynamics of rates and their distributions?
- What kind of model is more apt to model correlated interest-rate curves of different currencies, and how does one compute the related dynamics under the relevant probability measures?
- When does a model imply the Markov property for the short rate and why is this important?
- What is inflation and what is its link with classical interest-rate modeling?
- How does one calibrate an inflation model?
- Is the time of default of a counterparty predictable or not?
- Is it possible to value payoffs under an equivalent pricing measure in presence of default?
- Why are Poisson and Cox processes so suited to default modeling?
- What are the mathematical analogies between interest-rate models and credit-derivatives models? For what kind of mathematical models do these analogies stand?
- Does counterparty risk render a payoff dynamics-dependent even if without counterparty risk the payoff valuation is model-independent?
- What kind of mathematical models may account for possible jump features in the stochastic processes needed in credit spread modeling?
- Is there a general way to model dependence across default times, and across market variables more generally, going beyond linear correlation? What are the limits of these generalizations, in case?
- ......

We could go on for a while with questions of this kind. Our point is, however, that the theory dealt with in a book on interest-rate models should consider this kind of question.

We sympathize with anyone who has gone to a bookstore (or perhaps to a library) looking for answers to some of the above questions with little success. We have done the same, several times, and we were able to find only limited material and few reference works, although in the last few years the situation has improved. We hope the second edition of this book will cement the steps forward taken with the first edition.

We also sympathize with the reader who has just finished his studies or with the academic who is trying a life-change to work in industry or who is considering some close cooperation with market participants. Being used to precise statements and rigorous theory, this person might find answers to the above questions expressed in contradictory or unclear mathematical language. This is something else we too have been through, and we are trying not to disappoint in this respect either.

Is This Book about Practice? What kind of Practice?

If we don't do the work, the words don't mean anything. Reading a book or listening to a talk isn't enough by itself.
Charlotte Joko Beck, "Nothing Special: Living Zen", Harper Collins, 1995.

We try to answer some questions on practice that are again overlooked in most of the existing books in mathematical finance, and on interest-rate models in particular. Again, here are some typical questions selected at random:

- What are accrual conventions and how do they impact on the definition of rates?
- Can you give a few examples of how time is measured in connection with some aspects of contracts? What are "day-count conventions"?
- What is the interpretation of most liquid market contracts such as caps and swaptions? What is their main purpose?
- What kind of data structures are observed in the market? Are all data equally significant?
- How is a specific model calibrated to market data in practice? Is a joint calibration to different market structures always possible or even desirable?
- What are the dangers of calibrating a model to data that are not equally important, or reliable, or updated with poor frequency?
- What are the requirements of a trader as far as a calibration results are concerned?
- How can one handle path-dependent or early-exercise products numerically? And products with both features simultaneously?
- What numerical methods can be used for implementing a model that is not analytically tractable? How are trees built for specific models? Can instantaneous correlation be a problem when building a tree in practice?
- What kind of products are suited to evaluation through Monte Carlo simulation? How can Monte Carlo simulation be applied in practice? Under which probability measure is it convenient to simulate? How can we reduce the variance of the simulation, especially in presence of default indicators?

- Is there a model flexible enough to be calibrated to the market smile for caps?
- How is the swaptions smile quoted? Is it possible to "arbitrage" the swaption smile against the cap smile?
- What typical qualitative shapes of the volatility term structure are observed in the market?
- What is the impact of the parameters of a chosen model on the market volatility structures that are relevant to the trader?
- What is the accuracy of analytical approximations derived for swaptions volatilities and terminal correlations?
- Is it possible to relate CMS convexity adjustments to swaption smiles?
- Does there exist an interest-rate model that can be considered "central" nowadays, in practice? What do traders think about it?
- How can we express mathematically the payoffs of some typical market products?
- How do you handle in practice products depending on more than one interest-rate curve at the same time?
- How do you calibrate an inflation model in practice, and to what quotes?
- What is the importance of stochastic volatility in inflation modeling?
- How can we handle hybrid structures? What are the key aspects to take into account?
- What are typical volatility sizes in the credit market? Are these sizes motivating different models?
- What's the impact of interest-rate credit-spread correlation on the valuation of credit derivatives?
- Is counterparty risk impacting interest-rate payoffs in a relevant way?
- Are models with jumps easy to calibrate to credit spread data?
- Is there a way to imply correlation across default times of different names from market quotes? What models are more apt at doing so?
- ......

Again, we could go on for a while, and it is hard to find a single book answering these questions with a rigorous theoretical background. Also, answering some of these questions (and others that are similar in spirit) motivates new theoretical developments, maintaining the fundamental feedback between theory and practice we hinted at above.

AIMS, READERSHIP AND BOOK STRUCTURE

> *"And these people are sitting up there seriously discussing intelligent stars and trips through time to years that sound like telephone numbers. Why am I here?"* Huntress/Helena Bertinelli, DC One Million (1999).

Contrary to what happens in other derivatives areas, interest-rate modeling is a branch of mathematical finance where no general model has been yet accepted as "standard" for the whole sector, although the LIBOR market model is emerging as a possible candidate for this role. Indeed, there exist market standard models for both main interest-rate derivatives "submarkets", namely the caps and swaptions markets. However, such models are theoretically incompatible and cannot be used jointly to price other interest-rate derivatives.

Because of this lack of a standard, the choice of a model for pricing and hedging interest-rate derivatives has to be dealt with carefully. In this book, therefore, we do not just concentrate on a specific model leaving all implementation issues aside. We instead develop several types of models and show how to use them in practice for pricing a number of specific products.

The main models are illustrated in different aspects ranging from theoretical formulation to a possible implementation on a computer, always keeping in mind the concrete questions one has to cope with. We also stress that different models are suited to different situations and products, pointing out that there does not exist a single model that is uniformly better than all the others.

Thus our aim in writing this book is two-fold. First, we would like to help quantitative analysts and advanced traders handle interest-rate derivatives with a sound theoretical apparatus. We try explicitly to explain which models can be used in practice for some major concrete problems. Secondly, we would also like to help academics develop a feeling for the practical problems in the market that can be solved with the use of relatively advanced tools of mathematics and stochastic calculus in particular. Advanced undergraduate students, graduate students and researchers should benefit as well, from seeing how some sophisticated mathematics can be used in concrete financial problems.

The Prerequisites

The prerequisites are some basic knowledge of stochastic calculus and the theory of stochastic differential equations and Poisson processes in particular. The main tools from stochastic calculus are Ito's formula, Girsanov's theorem, and a few basic facts on Poisson processes, which are, however, briefly reviewed in Appendix C.

The Book is Structured in Eight Parts

The first part of the book reviews some basic concepts and definitions and briefly explains the fundamental theory of no-arbitrage and its implications as far as pricing derivatives is concerned.

In the second part the first models appear. We review some of the basic short-rate models, both one- and two-dimensional, and then hint at forward-rate models, introducing the so called Heath-Jarrow-Morton framework.

In the third part we introduce the "modern" models, the so-called market models, describing their distributional properties, discussing their analytical tractability and proposing numerical procedures for approximating the interest-rate dynamics and for testing analytical approximations. We will make extensive use of the "change-of-numeraire" technique, which is explained in detail in a initial section. This third part contains a lot of new material with respect to the earlier 2001 edition. In particular, the correlation study and the cascade calibration of the LIBOR market model have been considerably enriched, including the work leading to the Master's and PhD theses of Massimo Morini.

The fourth part is largely new, and is entirely devoted to smile modeling, with a parade of models that are studied in detail and applied to the caps and swaptions markets.

The fifth part is devoted to concrete applications. We in fact list a series of market financial products that are usually traded over the counter and for which there exists no uniquely consolidated pricing model. We consider some typical interest-rate derivatives dividing them into two classes: i) derivatives depending on a single interest-rate curve; ii) derivatives depending on two interest-rate curves.

Part Six is new and we introduce and study inflation derivatives and related models to price them.

Part Seven is new as well and concerns credit derivatives and counterparty risk, and besides introducing the payoffs and the models we explain the analogies between credit models and interest-rate models.

Appendices

Part Eight regroups our appendices, where we have also moved the "other interest rate models" and the "equity payoffs under stochastic rates" sections, which were separate chapters in the first edition. We updated the appendix on stochastic calculus with Poisson processes and updated the "Talking to the Traders" appendix with conversations on the new parts of the book.

We also added an appendix with trivia and frequently asked questions such as "who's who of the two authors", "what does the cover represent", "what about all these quotes" etc.

It is sometimes said that no one ever reads appendices. This book ends with eight appendices, and the last one is an interview with a quantitative trader, which should be interesting enough to convince the reader to have a look at the appendices, for a change.

FINAL WORD AND ACKNOWLEDGMENTS

Whether our treatment of the theory fulfills the targets we have set ourselves, is for the reader to judge. A disclaimer is necessary though. Assembling a book in the middle of the "battlefield" that is any trading room, while quite stimulating, leaves little space for planned organization. Indeed, the book is not homogeneous, some topics are more developed than others.

We have tried to follow a logical path in assembling the final manuscript, but we are aware that the book is not optimal in respect of homogeneity and linearity of exposition. Hopefully, the explicit contribution of our work will emerge over these inevitable little misalignments.

Acknowledgments

A book is always the product not only of its authors, but also of their colleagues, of the environment where the authors work, of the encouragements and critique gathered from conferences, referee reports for journal publications, conversations after seminars, university lectures, training courses, summer and winter schools, e-mail correspondence, and many analogous events. While we cannot do justice to all the above, we thank explicitly our recently acquired colleagues Andrea "Fifty levels of backtrack and I'm not from Vulcan" Pallavicini, who joined us in the last year with both analytical and numerical impressive skills, and Roberto "market-and-modeling-super-speed" Torresetti, one of the founders of the financial engineering department, who came back after a tour through Chicago and London, enhancing our activity with market understanding and immediate and eclectic grasp of modeling issues.

Some of the most important contributions, physically included in this book, especially this second edition, come from the "next generation" of quants and PhD students. Here is a roll call:

- Aurélien Alfonsi (PhD in Paris and Banca IMI trainee, Credit Derivatives with Damiano);
- Cristina Capitani (Banca IMI trainee, LIBOR model calibration, with Damiano);
- Laurent Cousot (PhD student in NY and Banca IMI trainee, Credit Derivatives with Damiano);
- Naoufel El-Bachir (PhD student in Reading and Banca IMI trainee, Credit Derivatives with Damiano);
- Eymen Errais (PhD student at Stanford and Banca IMI trainee, Credit Derivatives with Damiano and Smile Modeling with Fabio);
- Jan Liinev (PhD in Ghent, LIBOR / Swap models distance with differential geometric methods, with Damiano);
- Dmitri Lvov (PhD in Reading and Banca IMI trainee, Bermudan Swaption Pricing and Hedging with the LFM, with Fabio);
- Massimo Masetti (PhD in Bergamo, currently working for a major bank in London, Counterparty Risk and Credit Derivatives with Damiano);
- Nicola Moreni (PhD in Paris, and Banca IMI trainee, currently our colleague, Inflation Modeling with Fabio);

- Giulio Sartorelli (PhD in Pisa, Banca IMI trainee and currently our colleague, Short-Rate and Smile Modeling with Fabio);
- Marco Tarenghi (Banca IMI trainee and our former colleague, Credit derivatives and counterparty risk with Damiano);

Special mention is due to Massimo Morini, Damiano's PhD student in Milan, who almost learned the first edition by heart. His copy is the most battered and travel-worn we have ever seen; Massimo is virtually a co-author of this second edition, having contributed largely to the new parts of Chapters 6 and 7, and having developed recent and promising results on smile calibration and credit derivatives market models that we have not been in time to include here. Massimo also taught lectures and training courses based on the book all around the world, helping us whenever we were too busy to travel abroad.

As before we are grateful to our colleagues Gianvittorio "Tree and Optimization Master" Mauri and Francesco "Monte Carlo" Rapisarda, for their help and continuous interaction concerning both modeling and concrete implementations on computers. Francesco also helped by proofreading the manuscript of the first edition and by suggesting modifications.

The feedback from the trading desks (interest-rate-derivatives and credit derivatives) has been fundamental, first in the figures of Antonio Castagna and Luca Mengoni (first edition) and then Andrea Curotti (now back in London), Stefano De Nuccio (now with our competitors), Luca Dominici, Roberto Paolelli, and Federico Veronesi. They have stimulated many developments with their objections, requirements and discussions. Their feeling for market behavior has guided us in cases where mere mathematics and textbook finance could not help us that much. Antonio has also helped us with stimulating discussions on inflation modeling and general pricing and hedging issues. As before, this book has been made possible also by the farsightedness of our head Aleardo Adotti, who allowed us to work on the frontiers of mathematical finance inside a bank.

Hundreds of e-mails in the last years have reached us, suggesting improvements, asking questions, and pointing out errors. Again, it would be impossible to thank all single readers who contacted us, so we say here a big collective "Thank-you" to all our past readers. All mistakes that are left are again, needless to say, ours.

It has been tough to remain mentally sane in these last years, especially when completing this almost-one-thousand pages book. So the list of "external" acknowledgments has lengthened since the last time.

Damiano is grateful to the next generation above, in particular to the ones he is working/has worked with most: his modeling colleagues Andrea and Roberto, and then Aurélien, Eymen, Jan and Massimo Masetti (who held the "fortress" all alone in a difficult moment); Marco and Massimo Morini are gratefully mentioned also for the Boston MIT -Miami-Key West-Cape Canaveral "tournee" of late 2004. Umberto Cherubini has been lots of fun with the "Japanese experiences" of 1999-2004 and many professional suggestions. Gratitude goes also to Suzuki "Freccetta" SV650, to the Lake of Como (Lario) and the Dolomites (Dolomiti), to Venice, Damiano's birthplace, a dream still going after all these years, to the Venice carnival for the tons of fun with the "Difensori della Terra" costume players, including Fabrizio "Spidey", Roberto "Cap", Roberto "Ben" and Graziano "Thor" among many others; to Diego and Bojana, they know why, to Chiara and Marco Salcoacci (and the newly arrived Carlo!), possibly the nicest persons on Earth, to Lucia and Massimo (Hayao Miyazaki is the greatest!), and to the many on-line young friends at ComicUS and DCForum. Damiano's gratitude goes finally to his young fiancée, who in the best tradition of comic-books and being quite shy asked to maintain a secret identity here, and especially to his whole family past and present, for continued affection, support and encouragement, in particular to Annamaria, Francesco, Paolo, Dina and Mino.

Fabio is grateful to his colleagues Gianvittorio Mauri, Andrea Pallavicini, Francesco Rapisarda and Giulio Sartorelli for their invaluable contribution in the modeling, pricing and hedging of the bank's derivatives. Their skilled efficiency has allowed (and still allows) him to devote himself also to more speculative matters. Special thanks then go to his friends, and especially to the "ammiragliato" (admiralty) group, for the fun they have planning their missions around a table in "trattorie" near Treviso, to Antonio, Jacopo and Raffaele for the great time they spend together in Milan and travelling all over the world, to Chiara and Eleonora for their precious advices and sincere affection, to his pastoral friends for their spiritual support, and last, but not least, to his family for continued affection and support.

Finally, our ultimate gratitude is towards transcendence and is always impossible to express with words. We just say that we are grateful for the Word of the Gospel and the Silence of Zen.

A Special Final Word for Young Readers and Beginners

It looked insanely complicated, and this was one of the reasons why the snug plastic cover it fitted into had the words "Don't Panic" printed on it in large friendly letters.
Douglas Adams (1952 - 2001).

We close this long preface with a particular thought and encouragement for young readers. Clearly, if you are a professional or academic experienced in interest-rate modeling, we believe you will not be scared by a first quick look at the table of contents and at the chapters.

However, even at a first glance when flipping through the book, some young readers might feel discouraged by the variety of models, by the difference in approaches, by the book size, and might indeed acquire the impression of a chaotic sequence of models that arose in mathematical finance without a particular order or purpose. Yet, we assure you that this subject is interesting, relevant, and that it can (and should) be fun, however "clichéd" this may sound to you. We have tried at times to be colloquial in the book, in an attempt to avoid writing a book on formal mathematical finance from A to Zzzzzzzzzz... (where have you heard this one before?).

We are trying to avoid the two apparent extremes of either scaring or boring our readers. Thus you will find at times opinions from market participants, guided tours, intuition and discussion on things as they are seen in the market. We would like you to give it at least a try. So, if you are one of the above young readers, and be you a student or a practitioner, we suggest you take it easy. This book might be able to help you a little in entering this exciting field of research. This is why we close this preface with the by-now classic recommendations...

.. a brief hiss of air as the green plasma seals around him and begins to photosynthesize oxygen, and then the dead silence of space. A silence as big as everything. [...] Cool green plasma flows over his skin, maintaining his temperature, siphoning off sweat, monitoring muscle tone, repelling micro-meteorites. He thinks green thoughts. And his thoughts become things. Working the ring is like giving up cigarettes.
He feels like a "sixty-a-day" man.
Grant Morrison on Green Lantern (Kyle Rayner)'s ring, *JLA*, 1997

"May fear and dread not conquer me". Majjhima Nikaya VIII.6

"Do not let your hearts be troubled and do not be afraid". St. John XIV.27

Martian manhunter: *"...All is lost...."*
Batman: *"I don't believe that for a second. What should I expect to feel?"*
M: *"Despair. Cosmic despair. Telepathic contact with Superman is only possible through the Mageddon mind-field that holds him in thrall. It broadcasts on the lowest psychic frequencies...horror...shame...fear...anger..."*
B: *"Okay, okay. Despair is fine. I can handle despair and so can you."*
Grant Morrison, JLA: World War Three, 2000, DC Comics.

"Non abbiate paura!" [Don't be afraid!]. Karol Wojtyla (1920- 2005)

"For a moment I was afraid." "For no reason".
Irma [Kati Outinen] and M [Markku Peltola], "The Man without a Past", Aki Kaurismaki (2002).

Venice and Milan, May 4, 2006

Damiano Brigo and Fabio Mercurio

DESCRIPTION OF CONTENTS BY CHAPTER

We herewith provide a detailed description of the contents of each chapter, highlighting the updates for the new edition.

Part I: BASIC DEFINITIONS AND NO ARBITRAGE

Chapter 1: Definitions and Notation. The chapter is devoted to standard definitions and concepts in the interest-rate world, mainly from a static point of view. We define several interest-rate curves, such as the LIBOR, swap, forward-LIBOR and forward-swap curves, and the zero-coupon curve.

We explain the different possible choices of rates in the market. Some fundamental products, whose evaluation depends only on the initially given curves and not on volatilities, such as bonds and interest-rate swaps, are introduced. A quick and informal account of fundamental derivatives depending on volatility such as caps and swaptions is also presented, mainly for motivating the following developments.

Chapter 2: No-Arbitrage Pricing and Numeraire Change. The chapter introduces the theoretical issues a model should deal with, namely the no-arbitrage condition and the change of numeraire technique. The change of numeraire is reviewed as a general and powerful theoretical tool that can be used in several situations, and indeed will often be used in the book.

We remark how the standard Black models for either the cap or swaption markets, the two main markets of interest-rate derivatives, can be given a rigorous interpretation via suitable numeraires, as we will do later on in Chapter 6.

We finally hint at products involving more than one interest-rate curve at the same time, typically quanto-like products, and illustrate the no-arbitrage condition in this case.

Part II: FROM SHORT RATE MODELS TO HJM

Chapter 3: One-Factor Short-Rate Models. In this chapter, we begin to consider the dynamics of interest rates. The chapter is devoted to the short-rate world. In this context, one models the instantaneous spot interest rate via a possibly multi-dimensional driving diffusion process depending on some parameters. The whole yield-curve evolution is then characterized by the driving diffusion.

If the diffusion is one-dimensional, with this approach one is directly modeling the short rate, and the model is said to be "one-factor". In this chapter, we focus on such models, leaving the development of the multi-dimensional (two-dimensional in particular) case to the next chapter.

As far as the dynamics of one-factor models is concerned, we observe the following. Since the short rate represents at each instant the initial point of the yield curve, one-factor short-rate models assume the evolution of the whole yield curve to be completely determined by the evolution of its initial point. This is clearly a dangerous assumption, especially when pricing products depending on the correlation between different rates of the yield curve at a certain time (this limitation is explicitly pointed out in the guided tour of the subsequent chapter).

We then illustrate the no-arbitrage condition for one-factor models and the fundamental notion of market price of risk connecting the objective world, where rates are observed, and the risk-neutral world, where expectations leading to prices occur. We also show how choosing particular forms for the market price of risk can lead to models to which one can apply both econometric techniques (in the objective world) and calibration to market prices (risk-neutral world). We briefly hint at this kind of approach and subsequently leave the econometric part, focusing on the market calibration.

A short-rate model is usually calibrated to some initial structures in the market, typically the initial yield curve, the caps volatility surface, the swaptions volatility surface, and possibly other products, thus determining the model parameters. We introduce the historical one-factor time-homogeneous models of Vasicek, Cox Ingersoll Ross (CIR), Dothan, and the Exponential Vasicek (EV) model. We hint at the fact that such models used to be calibrated only to the initial yield curve, without taking into account market volatility structures, and that the calibration can be very poor in many situations.

We then move to extensions of the above one-factor models to models including "time-varying coefficients", or described by inhomogeneous diffusions. In such a case, calibration to the initial yield curve can be made perfect, and the remaining model parameters can be used to calibrate the volatility structures. We examine classic one-factor extensions of this kind such as Hull and White's extended Vasicek (HW) model, classic extensions of the CIR model, Black and Karasinski's (BK) extended EV model and a few more.

We discuss the volatility structures that are relevant in the market and explain how they are related to short-rate models. We discuss the issue of a humped volatility structure for short-rate models and give the relevant definitions. We also present the Mercurio-Moraleda short-rate model, which allows for a parametric humped-volatility structure while exactly calibrating the initial yield curve, and briefly hint at the Moraleda-Vorst model.

We then present a method of ours for extending pre-existing time-homogeneous models to models that perfectly calibrate the initial yield curve while keeping free parameters for calibrating volatility structures. Our method preserves the possible analytical tractability of the basic model. Our extension is shown to be equivalent to HW for the Vasicek model, whereas it is original in case of the CIR model. We call CIR++ the CIR model being extended through our procedure. This model will play an important role in the final part of the book devoted to credit derivatives, in the light of the

Brigo-Alfonsi SSRD stochastic intensity and interest rate model, with the Brigo-El Bachir jump diffusion extensions (JCIR++) playing a fundamental role to attain high levels of implied volatility in CDS options. The JCIR++ model, although not studied in this chapter and delayed to the credit chapters, retains an interest of its own also for interest rate modeling, possibly also in relationship with the volatility smile problem. The reader, however, will have to adapt the model from intensity to interest rates on her own.

We then show how to extend the Dothan and EV models, as possible alternatives to the use of the popular BK model.

We explain how to price coupon-bearing bond options and swaptions with models that satisfy a specific tractability assumption, and give general comments and a few specific instructions on Monte Carlo pricing with short-rate models.

We finally analyze how the market volatility structures implied by some of the presented models change when varying the models parameters. We conclude with an example of calibration of different models to market data.

Chapter 4: Two-Factor Short-Rate Models. If the short rate is obtained as a function of all the driving diffusion components (typically a summation, leading to an additive multi-factor model), the model is said to be "multi-factor".

We start by explaining the importance of the multi-factor setting as far as more realistic correlation and volatility structures in the evolution of the interest-rate curve are concerned.

We then move to analyze two specific two-factor models.

First,we apply our above deterministic-shift method for extending pre-existing time-homogeneous models to the two-factor additive Gaussian case (G2). In doing so, we calibrate perfectly the initial yield curve while keeping five free parameters for calibrating volatility structures. As usual, our method preserves the analytical tractability of the basic model. Our extension G2++ is shown to be equivalent to the classic two-factor Hull and White model. We develop several formulas for the G2++ model and also explain how both a binomial and a trinomial tree for the two-dimensional dynamics can be obtained. We discuss the implications of the chosen dynamics as far as volatility and correlation structures are concerned, and finally present an example of calibration to market data.

The second two-factor model we consider is a deterministic-shift extension of the classic two-factor CIR (CIR2) model, which is essentially the same as extending the Longstaff and Schwartz (LS) models. Indeed, we show that CIR2 and LS are essentially the same model, as is well known. We call CIR2++ the CIR2/LS model being extended through our deterministic-shift procedure, and provide a few analytical formulas. We do not consider this model with the level of detail devoted to the G2++ model, because of the fact that its volatility structures are less flexible than the G2++'s, at least if one wishes to preserve analytical tractability. However, following some new developments coming from using this kind of model for credit derivatives, such as the Brigo-Alfonsi SSRD stochastic intensity model, we point out some further extensions and approximations that can render the CIR2++ model both flexible and tractable, and reserve their examination for further work.

Chapter 5: The Heath-Jarrow-Morton Framework. In this chapter we consider the Heath-Jarrow-Morton (HJM) framework. We introduce the general framework and point out how it can be considered the right theoretical framework for developing interest-rate theory and especially no-arbitrage. However, we also point out that the most significant models coming out concretely from such a framework are the same models we met in the short-rate approach.

We report conditions on volatilities leading to a Markovian process for the short rate. This is important for implementation of lattices, since one then obtains (linearly-growing) recombining trees, instead of exponentially-growing ones. We show that in the one-factor case, a general condition leading to Markovianity of the short rate yields the Hull-White model with all time-varying coefficients, thus confirming that, in practice, short-rate models already contained some of the most interesting and tractable cases.

We then introduce the Ritchken and Sankarasubramanian framework, which allows for Markovianity of an enlarged process, of which the short rate is a component. The related tree (Li, Ritchken and Sankarasubramanian) is presented. Finally, we present a different version of the Mercurio-Moraleda model obtained through a specification of the HJM volatility structure, pointing out its advantages for realistic volatility behavior and its analytical formula for bond options.

Part III: MARKET MODELS

Chapter 6: The LIBOR and Swap Market Models (LFM and LSM). This chapter presents one of the most popular families of interest-rate models: the market models. A fact of paramount importance is that the lognormal forward-LIBOR model (LFM) prices caps with Black's cap formula, which is the standard formula employed in the cap market. Moreover, the lognormal forward-swap model (LSM) prices swaptions with Black's swaption formula, which is the standard formula employed in the swaption market. Now, the cap and swaption markets are the two main markets in the interest-rate-derivatives world, so compatibility with the related market formulas is a very desirable property. However, even with rigorous separate compatibility with the caps and swaptions classic formulas, the LFM and LSM are not compatible with each other. Still, the separate compatibility above is so important that these models, and especially the LFM, are nowadays seen as the most promising area in interest-rate modeling.

We start the chapter with a guided tour presenting intuitively the main issues concerning the LFM and the LSM, and giving motivation for the developments to come.

We then introduce the LFM, the "natural" model for caps, modeling forward-LIBOR rates. We give several possible instantaneous-volatility structures for this model, and derive its dynamics under different measures. We explain how the model can be calibrated to the cap market, examining the impact of the different structures of instantaneous volatility on the calibration. We introduce rigorously the term structure of volatility, and again check the impact of the different parameterizations of instantaneous volatilities on its evolution in time. We point out the difference between instantaneous and terminal correlation, the latter depending also on instantaneous volatilities.

We then introduce the LSM, the "natural" model for swaptions, modeling forward-swap rates. We show that the LSM is distributionally incompatible with the LFM. We discuss possible parametric forms for instantaneous correlations in the LFM, enriching the treatment given in the first edition. We introduce several new parametric forms for instantaneous correlations, and we deal both with full rank and reduced rank matrices. We consider their impact on swaptions prices, and how, in general, Monte Carlo simulation should be used to price swaptions with the LFM instead of the LSM. Again enriching the treatment given in the first edition, we analyze the standard error of the Monte Carlo method in detail and suggest some variance reduction techniques for simulation in the LIBOR model, based on the control variate techniques. We derive several approximated analytical formulas for swaption prices in the LFM (Brace's, Rebonato's and Hull-White's). We point out that terminal correlation depends on the particular measure chosen for the joint dynamics in the LFM. We derive two analytical formulas based on "freezing the drift" for terminal correlation. These formulas clarify the relationship between instantaneous correlations and volatilities on one side and terminal correlations on the other side.

Expanding on the first edition, we introduce the problem of swaptions calibration, and illustrate the important choice concerning instantaneous correlations: should they be fixed exogenously through some historical estimation, or implied by swaptions cross-sectional data? With this new part of the book based on Massimo Morini's work we go into some detail concerning the historical instantaneous correlation matrix and some ways of smoothing it via parametric or "pivot" forms. This work is useful later on when actually calibrating the LIBOR model.

We develop a formula for transforming volatility data of semi-annual or quarterly forward rates in volatility data of annual forward rates, and test it against Monte Carlo simulation of the true quantities. This is useful for joint calibration to caps and swaptions, allowing one to consider only annual data.

We present two methods for obtaining forward LIBOR rates in the LFM over non-standard periods, i.e. over expiry/maturity pairs that are not in the family of rates modeled in the chosen LFM.

Chapter 7: Cases of Calibration of the LIBOR Market Model. In this chapter, we start from a set of market data including zero-coupon curve, caps volatilities and swaptions volatilities, and calibrate the LFM by resorting to several parameterizations of instantaneous volatilities and by several constraints on instantaneous correlations. Swaptions are evaluated through the analytical approximations derived in the previous chapter. We examine the evolution of the term structure of volatilities and the ten-year terminal correlation coming out from each calibration session, in order to assess advantages and drawbacks of every parameterization.

We finally present a particular parameterization establishing a one-to-one correspondence between LFM parameters and swaption volatilities, such that the calibration is immediate by solving a cascade of algebraic second-order equations, leading to Brigo's basic cascade calibration algorithm. No optimization is necessary in general and the calibration is instantaneous. However, if the initial swaptions data are misaligned because of illiquidity or other reasons, the calibration can lead to negative or imaginary volatilities. We show that smoothing the initial data leads again to positive real volatilities.

The first edition stopped at this point, but now we largely expanded the cascade calibration with the new work of Massimo Morini. The impact of different exogenous instantaneous correlation matrices on the swaption calibration is considered, with several numerical experiments. The interpolation of missing quotes in the original input swaption matrix seems to heavily affect the subsequent calibration of the LIBOR model. Instead of smoothing the swaption matrix, we now develop a new algorithm that makes interpolated swaptions volatilities consistent with the LIBOR model by construction, leading to Morini and Brigo's extended cascade calibration algorithm. We test this new method and see that practically all anomalies present in earlier cascade calibration experiments are surpassed. We conclude with some further remarks on joint caps/swaptions calibration and with Monte Carlo tests establishing that the swaption volatility drift freezing approximation on which the cascade calibration is based holds for the LIBOR volatilities parameterizations used in this chapter.

Chapter 8: Monte Carlo Tests for LFM Analytical Approximations. In this chapter we test Rebonato's and Hull-White's analytical formulas for swaptions prices in the LFM, presented earlier in Chapter 6, by means of a Monte Carlo simulation of the true LFM dynamics. Partial tests had already been performed at the end of Chapter 7. The new tests are done under different parametric assumptions for instantaneous volatilities and under different instantaneous correlations. We conclude that the above formulas are accurate in non-pathological situations.

We also plot the real swap-rate distribution obtained by simulation against the lognormal distribution with variance obtained by the analytical approximation. The two distributions are close in most cases, showing that the previously remarked theoretical incompatibility between LFM and

LSM (where swap rates are lognormal) does not transfer to practice in most cases.

We also test our approximated formulas for terminal correlations, and see that these too are accurate in non-pathological situations.

With respect to the first edition, based on the tests of Brigo and Capitani, we added an initial part in this chapter computing rigorously the distance between the swap rate in the LIBOR model and the lognormal family of densities, under the swap measure, resorting to Brigo and Liinev's Kullback-Leibler calculations. The distance results to be small, confirming once again the goodness of the approximation.

Part IV: THE VOLATILITY SMILE

The old section on smile modeling in the LFM has now become a whole new part of the book, consisting of four chapters.

Chapter 9: Including the Smile in the LFM. This first new smile introductory chapter introduces the smile problem with a guided tour, providing a little history and a few references. We then identify the classes of models that can be used to extend the LFM and briefly describe them; some of them are examined in detail in the following three smile chapters.

Chapter 10: Local-Volatility Models. Local-volatility models are based on asset dynamics whose absolute volatility is a deterministic transformation of time and the asset itself. Their main advantages are tractability and ease of implementation. We start by introducing the forward-LIBOR model that can be obtained by displacing a given lognormal diffusion, and also describe the constant-elasticity-of-variance model by Andersen and Andreasen. We then illustrate the class of density-mixture models proposed by Brigo and Mercurio and Brigo, Mercurio and Sartorelli, providing also an example of calibration to real market data. A seemingly paradoxical result on the correlation between the underlying and the volatility, also in relation with later uncertain parameter models, is pointed out. In this chapter mixtures resort to the eralier lognormal mixture diffusions of the first edition but also to Mercurio's Hyperbolic-Sine mixture processes. We conclude the Chapter by describing Mercurio's second general class, which combines analytical tractability with flexibility in the cap calibration.

The local-volatility models in this chapter are meant to be calibrated to the caps market, and to be only used for the pricing of LIBOR dependent derivatives. The task of a joint calibration to the cap and swaption markets and the pricing of swap-rates dependent derivatives under smile effects, is, in this book, left to stochastic-volatility models and to uncertain-parameters models, the subject of the last two smile chapters.

Chapter 11: Stochastic-Volatility Models. We then move on to describe LIBOR models with stochastic volatility. They are extensions of the LFM where the instantaneous volatility of forward rates evolves according to a diffusion process driven by a Brownian motion that is possibly instantaneously correlated with those governing the rates' evolution. When the (instantaneous) correlation between a forward rate and its volatility is zero, the existence itself of a stochastic volatility leads to smile-shaped implied volatility curves. Skew-shaped volatilities, instead, can be produced as soon as we i) introduce a non-zero (instantaneous) correlation between rate and volatility or ii) assume a displaced-diffusion dynamics or iii) assume that the rate's diffusion coefficient is a non-linear function of the rate itself. Explicit formulas for both caplets and swaptions are usually derived by calculating the characteristic function of the underlying rate under its canonical measure.

In this chapter, we will describe some of the best known extensions of the LFM allowing for stochastic volatility, namely the models of i) Andersen and Brotherton-Ratcliff, ii) Wu and Zhang, iii) Hagan, Kumar, Lesniewski and Woodward, iv) Piterbarg and v) Joshi and Rebonato.

Chapter 12: Uncertain Parameters Models. We finally consider extensions of the LFM based on parameter uncertainty. Uncertain-volatility models are an easy-to-implement alternative to stochastic-volatility models. They are based on the assumption that the asset's volatility is stochastic in the simplest possible way, modelled by a random variable rather than a diffusion process. The volatility, therefore, is not constant and one assumes several possible scenarios for its value, which is to be drawn immediately after time zero. As a consequence, option prices are mixtures of Black's option prices and implied volatilities are smile shaped with a minimum at the at-the-money level. To account for skews in implied volatilities, uncertain-volatility models are usually extended by introducing (uncertain) shift parameters.

Besides their intuitive meaning, uncertain-parameters models have a number of advantages that strongly support their use in practice. In fact, they enjoy a great deal of analytical tractability, are relatively easy to implement and are flexible enough to accommodate general implied volatility surfaces in the caps and swaptions markets. As a drawback, future implied volatilities lose the initial smile shape almost immediately. However, our empirical analysis will show that the forward implied volatilities induced by the models do not differ much from the current ones. This can further support their use in the pricing and hedging of interest rate derivatives.

In this chapter, we will describe the shifted-lognormal model with uncertain parameters, namely the extension of Gatarek's one-factor uncertain-parameters model to the general multi-factor case as considered by Errais, Mauri and Mercurio. We will derive caps and (approximated) swaptions prices in closed form. We will then consider examples of calibration to caps and swaptions data. A curious relationship between one of the simple models in this framework and the earlier lognormal-mixture local volatility dynamics, related also to underlying rates and volatility decorrelation, is pointed out as from Brigo's earlier work.

Part V: EXAMPLES OF MARKET PAYOFFS

We thought that by making your world more violent, we would make it more "realistic", more "adult". God help us if that's what it means. Maybe, for once we could try to be kind.

Grant Morrison, Animal Man 26, 1990, DC Comics.

Chapter 13: Pricing Derivatives on a Single Interest-Rate Curve. This chapter deals with pricing specific derivatives on a single interest-rate curve. Most of these are products that are found in the market and for which no standard pricing technique is available. The model choice is made on a case-by-case basis, since different products motivate different models. The differences are based on realistic behaviour, ease of implementation, analytical tractability and so on. For each product we present at least one model based on a compromise between the above features, and in some cases we present more models and compare their strong and weak points. We try to understand which model parameters affect prices with a large or small influence. The financial products we consider are: in-arrears swaps, in-arrears caps, autocaps, caps with deferred caplets, ratchet caps and floors (new for the second edition), ratchets (one-way floaters), constant-maturity swaps (introducing also the convexity-adjustment technique), average rate caps, captions and floortions, zero-coupon swaptions, Eurodollar futures, accrual swaps, trigger swaps and Bermudan-style swaptions. We add numerical examples for Bermudan swaptions. Further, in this new edition we consider target redemption notes and CMS spread options.

Chapter 14: Pricing Derivatives on Two Interest-Rate Curves. The chapter deals with pricing specific derivatives involving two interest-rate curves. Again, most of these are products that are found in the market and for which no standard pricing technique is available. As before, the model choice is made on a case-by-case basis, since different products motivate different models. The used models reduce to the LFM and the G2++ shifted two-factor Gaussian short-rate model. Under the G2++ model, we are able to model correlation between the interest rate curves of the two currencies. The financial products we consider include differential swaps, quanto caps, quanto swaptions, quanto constant-maturity swaps. A market quanto adjustment and market formulas for basic quanto derivatives are also introduced. We finally price, in a market-model setting, spread options on two-currency LIBOR rates, options on the product of two-currency LIBOR rates and trigger swaps with payments, in domestic currency, triggered by either the domestic rate or the foreign one.

Part VI: INFLATION

In this new part, we describe new derivatives, which are based on inflation rates, together with possible models to price them.

Chapter 15: Pricing of Inflation Indexed Derivatives. Inflation is defined in terms of the percentage increments of a reference index, the consumer price index, which is a representative basket of goods and services.

Floors with low strikes are the most actively traded options on inflation rates. Other extremely popular derivatives are inflation-indexed swaps, where the inflation rate is either payed on an annual basis or with a single amount at the swap maturity. All these inflation-indexed derivatives require a specific model to be valued.

Most articles on inflation modeling in the financial literature are based on the so called *foreign-currency analogy*, according to which real rates are viewed as interest rates in the real (i.e. foreign) economy, and the inflation index is interpreted as the exchange rate between the nominal (i.e. domestic) and real "currencies". In this setting, the valuation of an inflation-indexed payoff becomes equivalent to that of a cross-currency interest rate derivative.

A different approach has also been developed using the philosophy of market models. The idea is to model the evolution of forward inflation indices, so that an inflation rate can be viewed as the ratio of two consecutive "assets", and derivatives are priced accordingly.

Chapters 16, 17 and 18: Inflation-Indexed Swaps, Inflation-Indexed Caplets/Floorlets, and Calibration to Market Data. The purpose of these chapters is to define the main types of inflation-indexed swaps and caps present in the market and price them analytically and consistently with no arbitrage. To this end, we will review and use i) the Jarrow and Yildirim model, where both nominal and real rates are assumed to evolve as in a one-factor Gaussian HJM model, ii) the Mercurio application of the LFM, and iii) the market model of Kazziha, also independently developed by Belgrade, Benhamou and Koehler and by Mercurio. Examples of calibration to market data will also be presented.

Chapter 19: Introducing Stochastic Volatility. In this chapter we add stochastic volatility to the market model introduced in Chapters 16 and 17. Precisely, we describe the approach followed by Mercurio and Moreni (2006), who modelled forward CPI's with a common volatility process that evolves according to a square-root diffusion.

Modeling the stochastic volatility as in Heston (1993) has the main advantage of producing analytical formulas for options on inflation rates. In fact, we first derive an explicit expression for the characteristic function of the ratio between two consecutive forward CPI's, and then price caplets and floorlets by Carr and Madan's (1998) Fourier transform method.

Numerical examples including a calibration to market cap data are finally shown.

Chapter 20: Pricing Hybrids with an Inflation Component. In this chapter, we tackle the pricing issue of a specific hybrid payoff involving inflation features when no smile effects are taken into account. It is meant to be an important example from an increasing family of hybrid payoffs that are getting popular in the market.

Part VII: CREDIT

This new part deals with credit derivatives, counterparty risk, credit models and their analogies with interest-rate models.

Chapter 21: Introduction and Pricing under Counterparty Risk. This first chapter starts this new part of the book devoted to credit derivatives and counterparty risk. In this first chapter we introduce the financial payoffs and the families of rates we deal with in the following. We present a guided tour to give some orientation and general feeling for this credit part of the book. The guided tour also focuses on multiname credit derivatives, introducing collateralized debt obligations (CDO) and first to default (FtD) contracts as fundamental examples. The first generation pricing of these products involves copula functions, that are introduced and reviewed, including the recent family of Alfonsi and Brigo periodic copulas. The need for dynamical models of dependence is pointed out. This is the only part of the book where we mention multi-name credit derivatives. The book focuses mostly on single name credit derivatives.

Then we introduce as first credit payoffs the prototypical defaultable bonds, the Credit Default Swaps (CDS) payoffs and defaultable floaters, including a relationship between the last two. In particular, we consider some different definitions of CDS forward rates, with analogies with LIBOR vs swap rates. We explore in detail possible equivalence between CDS payoffs and rates and defaultable floaters payoffs and rates.

We then introduce CDS options payoffs, pointing out some formal analogies with the swaption payoff encountered earlier in the book. We also introduce constant maturity CDS, a product that has grown in popularity in recent times. This product presents analogies with constant maturity swaps in the default free market. Finally, we close the chapter with counterparty risk pricing in interest rate derivatives. We show how to include the event that the counterparty may default in the risk neutral valuation of the financial payoff. This is particularly important after the recent regulatory directions given by the Basel II agreement and subsequent amendments and also by the "IAS 39" (international accounting standard) system. The counterparty risk pricing formula of Brigo and Masetti for non-standard swaps and swaps under netting agreements is only hinted at.

Chapter 22: Intensity Models. In this new chapter we focus completely on intensity models, exploring in detail also the issues we have anticipated in the earlier chapter in order to be able to deal with CDS and notions of implied hazard rates and functions.

Intensity models, part of the family of reduced form models, all move from the basic idea of describing the default time as the first jump time of a

Poisson process. Default is not induced by basic market observables but has an exogenous component that is independent of all the default free market information. Monitoring the default free market (interest rates, exchange rates, etc) does not give complete information on the default process, and there is no economic rationale behind default. This family of models is particularly suited to model credit spreads and in its basic formulation is easy to calibrate to Credit Default Swap (CDS) or corporate bond data.

The basic facts from probability are essentially the theory of Poisson and Cox processes. We start from the simplest, constant intensity Poisson process and explain the interpretation of the intensity as a probability of first jumping (defaulting) per unit of time. We then move to time-inhomogeneous Poisson processes, that allow to model credit spreads without volatility. Further, we move to stochastic intensity Poisson processes, where the probability of first jumping (defaulting) is itself random and follows a stochastic process of a certain kind. This last case is referred to as "Cox process" approach, or "doubly stochastic Poisson process". This approach allows us to take into account credit spread volatility. In all three cases of constant, deterministic-time-varying and stochastic intensity we point out how the Poisson process structure allows to view survival probabilities as discount factors, the intensity as credit spread, and how this helps us in recycling the interest-rate technology for default modeling. We then analyze in detail the CDS calibration with deterministic intensity models, illustrating the notion of implied hazard function with a case study based on Parmalat CDS data. We illustrate how the only hope of inducing dependence between the default event and interest rates in a diffusion setting is through a stochastic intensity correlated with the interest rate. We explain the fundamental idea of conditioning only to the partial information of the default free market when pricing credit derivatives. This result has fundamental consequences in that it will allow us later to define the CDS market model under a measure that is equivalent to the risk neutral one. Also, our definition of forward CDS rate itself owes much to this result.

We also explain how to simulate the default time, illustrating the notion of standard error and presenting suggestions on how to keep the number of paths under control. These suggestions take into account peculiarities of default modeling that make the variance reduction more difficult than in the default free market case.

We then introduce our choice for the stochastic intensity in a diffusion setting, the Brigo-Alfonsi stochastic intensity model. We term the stochastic intensity and interest rate model SSRD: Shifted Square Root Diffusion model. It is essentially a CIR++ model for the intensity correlated with a CIR++ model for the short rate. We argue the choice is the only reasonable one in a diffusion setting for the intensity given that one wishes analytical tractability for survival probabilities (CDS calibration) and positivity of the intensity process. We show how to calibrate the SSRD model to CDS quotes and interest rate data in a separable way, and argue that the instantaneous correlation has a negligible impact on the CDS price, allowing us to maintain the separability of the calibration in practice even when correlation is not zero. We present some original numerical schemes due to Alfonsi and Brigo for the simulation of the SSRD model that preserve positivity of the discretized process and analyze the convergence of such schemes. We also introduce the Brigo-Alfonsi Gaussian mapping technique that maps the model into a two factor Gaussian model, where calculations in presence of correlation are much easier. We analyze the mapping procedure and its accuracy by means of Monte Carlo tests. We also analyze the impact of the correlation on some prototypical payoff. As an exercise we price a cancellable structure with the stochastic intensity model. We also introduce Brigo's CDS option closed form formula under deterministic interest rates and CIR++ stochastic intensity, a particular case of the SSRD model. We analyze implied CDS volatilities patterns in the full SSRD case by means of Monte Carlo simulation. Finally, we explain why the CIR++ model for the intensity cannot attain large levels (such as 50%) of implied volatilities for CDS rates, and introduce jumps in the CIR++ model, hinting at the JCIR model and at its possible calibration to both CDS and options, Brigo and El-Bachir JCIR++ model.

Chapter 23: CDS Options Market Models. In this last new chapter of the credit part we start with the payoffs and structural analogies between CDS options and callable defaultable floating rate notes (DFRN).

We then introduce the market formula for CDS options and callable DFRN, based on a rigorous change of numeraire technique as in Brigo's CDS market model, different from Schönbucher's in that it guarantees equivalence of pricing measures notwithstanding default. Numerical examples of implied volatilities from CDS option quotes are given, and are found to be rather high, in agreement with previous studies dealing with historical CDS rate volatilities (Hull and White).

We discuss possible developments towards a compete specifications of the vector dynamics of CDS forward rates under a single pricing measure, based on one-period CDS rates.

We give some hints on modeling of the volatility smile for CDS options, based on the general framework introduced earlier.

We also illustrate how to use Brigo's market model to derive an approximated formula for Constant Maturity CDS. This formula is based on a sort of convexity adjustment and bears resemblance to the formula for valuing constant maturity swaps with the LIBOR model, seen earlier in the book. The adjustment is illustrated with several numerical examples.

Part VIII: APPENDICES

Appendix A: Other Interest-Rate Models. We present a few interest-rate models that are particular in their assumptions or in the quantities they model, and that have not been treated elsewhere in the book. We do not give a detailed presentation of these models but point out their particular features, compared to the models examined earlier in the book. This was a chapter in the first edition but to simplify the layout we included it here as an appendix.

Appendix B: Pricing Equity Derivatives under Stochastic Interest Rates. The appendix treats equity-derivatives valuation under stochastic interest rates, presenting us with the challenging task of modeling stock prices and interest rates at the same time. Precisely, we consider a continuous-time economy where asset prices evolve according to a geometric Brownian motion and interest rates are either normally or lognormally distributed. Explicit formulas for European options on a given asset are provided when the instantaneous spot rate follows the Hull-White one-factor process. It is also shown how to build approximating trees for the pricing of more complex derivatives, under a more general short-rate process. This was a chapter in the first edition but to simplify the layout we included it here as an appendix.

Appendix C: a Crash Introduction to Stochastic Differential Equations and Poisson Processes.

> *There is, of course, a dearth of good mathematics teachers [...] Why subject themselves to a lifetime surrounded by a pandemonium of fresh-faced young people in uniform shouting to each other across the classroom for nine grand a year, they say, when they can do exactly the same in the trading room of any stockbrokers for ninety?*
>
> Robert Ainsley, "Bluff your way in Maths", Ravette Books, 1988

This appendix is devoted to a quick intuitive introduction on SDE's and Poisson processes. We start from deterministic differential equation and gradually introduce randomness. We introduce intuitively Brownian motion and explain how it can be used to model the "random noise" in the differential equation. We observe that Brownian motion is not differentiable, and explain that SDE's must be understood in integral form. We quickly introduce the related Ito and Stratonovich integrals, and introduce the fundamental Ito formula.

We then introduce the Euler and Milstein schemes for the time-discretization of an SDE. These schemes are essential when in need of Monte Carlo simulating the trajectories of an Ito process whose transition density is not explicitly known.

We include two important theorems: the Feynman-Kac theorem and the Girsanov theorem. The former connects PDE's to SDE's, while the latter permits to change the drift coefficient in an SDE by changing the basic probability measure. The Girsanov theorem in particular is used in the book to derive the change of numeraire toolkit.

Given its importance in default modeling, we also introduce the Poisson process, to some extent the purely jump analogous of Brownian motion.

Brownian motions and Poisson processes are among the most important random processes of probability.

Appendix D: a Useful Calculation. This appendix reports the calculation of a particular integral against a standard normal density, which is useful when dealing with Gaussian models.

Appendix E: a Second Useful Calculation. This appendix shows how to calculate analytically the price of an option on the spread between two assets, under the assumption that both assets evolve as (possibly correlated) geometric Brownian motions.

Appendix F: Approximating Diffusions with Trees. This appendix explains a general method to obtain a trinomial tree approximating the dynamics of a general diffusion process. This is then generalized to a two-dimensional diffusion process, which is approximated via a two-dimensional trinomial tree.

Appendix G: Trivia and Frequently Asked Questions (FAQ). In this appendix we answer a number of frequently asked questions concerning the book trivia and curiosities. It is a light appendix, meant as a relaxing moment in a book that at times can be rather tough.

Appendix H: Talking to the Traders. This is the ideal conclusion of the book, consisting of an interview with a quantitative trader. Several issues are discussed, also to put the book in a larger perspective. This version for the second edition has been enriched.

Abbreviations and Notation

- ATM = At the money;
- BK = Black-Karasinski model;
- bps = Basis Point ($1bps = 10^{-4} = 1E - 4 = 0.0001$);
- CC(A) = Cascade Calibration (Algorithm);
- CIR = Cox-Ingersoll-Ross model;
- CIR2++ = Shifted two-factor Cox-Ingersoll-Ross model;
- EICCA = Endogenous Interpolation Cascade Calibration Algorithm;
- EEV = Shifted (extended) exponential-Vasicek model;
- EV = Exponential-Vasicek model;
- FRN = Floating Rate Note;
- G2++ = Shifted two-factor Gaussian (Vasicek) model;
- HD = Hellinger Distance;
- HJM = Heath-Jarrow-Morton model;
- HW = Hull-White model;
- IRS = Interest Rate Swap (either payer or receiver);
- ITM = In the money;
- KLI = Kullback Leibler Information;
- LFM = Lognormal forward-Libor model (Libor market model, BGM model);
- LS = Longstaff-Schwartz short-rate model;
- LSM = Lognormal forward-swap model (swap market model);
- MC = Monte Carlo;
- OTM = Out of the money;
- PDE = Partial differential equation;
- PVBP = Present Value per Basis Point (or annuity);
- RCCA = Rectangular Cascade Calibration Algorithm; REICCA = Rectangular Endogenous-Interpolation Cascade-Calibration Algorithm;
- TSV = Term Structure of Volatilities;
- TC = Terminal Correlation;
- SDE = Stochastic differential equation;
- I_n: the $n \times n$ identity matrix;
- e_i: i-th canonical vector of $\mathbb{R}^n$, a vector with all zeroes except in the i-th entry, where a "1" is found.
- $B(t), B_t$: Money market account at time t, bank account at time t ;
- $D(t,T)$: Stochastic discount factor at time t for the maturity T;
- $P(t,T)$: Bond price at time t for the maturity T;
- $P^f(t,T)$: Foreign Bond price at time t for the maturity T;
- $r(t), r_t$: Instantaneous spot interest rate at time t;
- $B^d(t)$: Discretely rebalanced bank-account at time t;
- $R(t,T)$: Continuously compounded spot rate at time t for the maturity T;
- $L(t,T)$: Simply compounded (LIBOR) spot rate at time t for the maturity T;
- $f(t,T)$: Instantaneous forward rate at time t for the maturity T;

- $F(t;T,S)$: Simply compounded forward (LIBOR) rate at time t for the expiry–maturity pair T,S;
- $\mathrm{FP}(t;T,S)$: Forward zero-coupon-bond price at time t for maturity S as seen from expiry T, $\mathrm{FP}(t;T,S) = P(t,S)/P(t,T)$.
- $F^f(t;T,S)$: Foreign simply compounded forward (LIBOR) rate at time t for the expiry–maturity pair T,S;
- $f(t;T,S)$: Continuously compounded forward rate at time t for the expiry–maturity pair T,S;
- $T_1,T_2,\ldots,T_{i-1},T_i,\ldots$: An increasing set of maturities;
- τ_i: The year fraction between T_{i-1} and T_i;
- $F_i(t)$: $F(t;T_{i-1},T_i)$;
- $S(t;T_i,T_j), S_{i,j}(t)$: Forward swap rate at time t for a swap with first reset date T_i and payment dates $T_{i+1},\ldots,T_j$;
- $C_{i,j}(t)$: Present value of a basis point (PVBP) associated to the forward–swap rate $S_{i,j}(t)$, i.e. $\sum_{k=i+1}^{j} \tau_k P(t,T_k)$;
- Q_0: Physical/Objective/Real–World measure;
- Q: Risk-neutral measure, equivalent martingale measure, risk-adjusted measure;
- Q^U: Measure associated with the numeraire U when U is an asset;
- Q^d: Spot LIBOR measure, measure associated with the discretely rebalanced bank-account numeraire;
- Q^T: T–forward adjusted measure, i.e. measure associated with the numeraire $P(\cdot,T)$;
- Q^i: T_i–forward adjusted measure;
- $Q^{i,j}$: Swap measure between T_i,T_j, associated with the numeraire $C_{i,j}$;
- $X_t, X(t)$: Foreign exchange rate between two currencies at time T;
- W_t, Z_t: Brownian motions under the Risk Neutral measure;
- W_t^U: A Brownian motion under the measure associated with the numeraire U when U is an asset;
- W_t^T: Brownian motions under the T forward adjusted measure;
- $W_t^i, Z_t^i, W^i(t), Z^i(t)$,: Brownian motions under the T_i forward adjusted measure;
- $[x_1,\ldots,x_m]$: row vector with i-th component x_i;
- $[x_1,\ldots,x_m]'$: column vector with i-th component x_i;
- $'$: Transposition;
- 1_A, $1\{A\}$: Indicator function of the set A;
- $\#\{A\}$: Number of elements of the finite set A;
- $i \div j$: the range of integers (typically indices) going from i to j;
- $x_i \div x_j$,$x_{i \div j}$: the range of indexed quantities $x_i, x_{i+1}, \ldots, x_{j-1}, x_j$.
- E: Expectation under the risk-neutral measure;
- E^Q: Expectation under the probability measure Q;
- E^U: Expectation under the probability measure Q^U associated with the numeraire U; This may be denoted also by E^{Q^U};
- E^T: Expectation under the T-forward adjusted measure;
- E^i: Expectation under the T_i-forward adjusted measure;
- E_t, $E\{\cdot|\mathcal{F}_t\}$, $E[\cdot|\mathcal{F}_t]$, $E(\cdot|\mathcal{F}_t)$: Expectation conditional on the $\mathcal{F}_t$ σ–field;
- $\mathrm{Corr}^i(X,Y)$: correlation between X and Y under the T_i forward adjusted measure Q^i; i can be omitted if clear from the context or under the risk-neutral measure;
- $\mathrm{Var}^i(X)$: Variance of X under the T_i forward adjusted measure Q^i; i can be omitted if clear from the context or under the risk-neutral measure;
- $\mathrm{Cov}^i(X)$: covariance matrix of the random vector X under the T_i forward adjusted measure Q^i; i can be omitted if clear from the context or under the risk-neutral measure;

- $\text{Std}^i(X)$: standard deviation (acting componentwise) of the random vector X under the T_i forward adjusted measure Q^i; i can be omitted if clear from the context or under the risk-neutral measure;
- $\sim$: distributed as;
- $\mathcal{N}(\mu, V)$: Multivariate normal distribution with mean vector μ and covariance matrix V; Its density at x is at times denoted by $p_{\mathcal{N}(\mu,V)}(x)$.
- Φ: Cumulative distribution function of the standard Gaussian distribution;
- Φ_R^n: Cumulative distribution function of the n-dimensional Gaussian random vector with standard Gaussian margins and $n \times n$ correlation matrix R;
- χ_ν^2: chi-squared distribution with ν degrees of freedom;
- $\chi^2(\cdot; r, \rho)$: Cumulative distribution function of the noncentral chi-squared distribution with r degrees of freedom and noncentrality parameter ρ;
- $t_{\nu,R}^n$: n-dimensional multivariate t-distribution with ν degrees of freedom and correlation matrix R;
- $\text{Bl}(K, F, v)$: The core of Black's formula:

$$\text{Bl}(K, F, v, \omega) = F\omega\Phi(\omega d_1(K, F, v)) - K\omega\Phi(\omega d_2(K, F, v)),$$
$$d_1(K, F, v) = \frac{\ln(F/K) + v^2/2}{v},$$
$$d_2(K, F, v) = \frac{\ln(F/K) - v^2/2}{v},$$

  where ω is either -1 or 1 and is meant to be 1 when omitted. The arguments of d_1 and d_2 may be omitted if clear from the context.
- $\mathbf{CB}(t, \mathcal{T}, \tau, N, c)$: Coupon bond price at time t for a bond paying coupons $c = [c_1, \ldots, c_n]$ at times $\mathcal{T} = [T_1, \ldots, T_n]$ with year fractions $\tau = [\tau_1, \ldots, \tau_n]$ and nominal amount N; When assuming a unit nominal amount N can be omitted; When year fractions are clear from the context, τ can be omitted;
- $\mathbf{ZBC}(t, T, S, \tau_0, N, K)$: Price at time t of an European call option with maturity $T > t$ and strike–price K on a Zero–coupon bond with nominal amount N, maturing at time $S > T$; τ_0 is the year fraction between T and S and can be omitted; When assuming a unit nominal amount N can be omitted;
- $\mathbf{ZBP}(t, T, S, \tau_0, N, K)$: Same as above but for a put option;
- $\mathbf{ZBO}(t, T, S, \tau_0, N, K, \omega)$: Unified notation for the price at time t of an European option with maturity $T > t$ and strike–price K on a Zero–coupon bond with face value N maturing at time $S > T$; τ_0 is the year fraction between T and S and can be omitted; ω is $+1$ for a call option and -1 for a put option, and can be omitted;
- $\mathbf{CBC}(t, T, \mathcal{T}, \tau_0, \tau, N, c, K)$: Price at time t of an European call option with maturity $T > t$ and strike–price K on the coupon bond $\mathbf{CB}(t, \mathcal{T}, \tau, N, c)$; τ_0 is the year fraction between T and T_1 and can be omitted; When the nominal N is one it is omitted;
- $\mathbf{CBP}(t, T, \mathcal{T}, \tau_0, \tau, N, c, K)$: Same as above but for a put option;
- $\mathbf{CBO}(t, T, \mathcal{T}, \tau_0, \tau, N, c, K, \omega)$: Unified notation for the price at time t of an European option with maturity $T > t$ and strike–price K on the coupon bond $\mathbf{CB}(t, \mathcal{T}, \tau, c)$; τ_0 is the year fraction between T and T_1 and can be omitted. ω is $+1$ for a call option and -1 for a put option, and can be omitted;
- $\mathbf{Cpl}(t, T, S, \tau_0, N, y)$: Price at time t of a caplet resetting at time T and paying at time S at a fixed strike–rate y; As usual τ_0 is the year fraction between T and S and can be omitted, and N is the nominal amount and can be omitted;
- $\mathbf{Fll}(t, T, S, \tau_0, N, y)$: Price at time t of a floorlet resetting at time T and paying at time S at a fixed rate y; As usual τ_0 is the year fraction between T and S and can be omitted, and N is the nominal amount and can be omitted;

- **Cap**$(t, \mathcal{T}, \tau, N, y)$: Price at time t of a cap first resetting at time T_1 and paying at times $T_2, \ldots, T_n$ at a fixed rate y; As usual τ_i is the year fraction between T_{i-1} and T_i and can be omitted, and N is the nominal amount and can be omitted;
- **Flr**$(t, \mathcal{T}, \tau, N, y)$: Price at time t of a floor first resetting at time T_1 and paying at times $T_2, \ldots, T_n$ at a fixed rate y; As usual τ_i is the year fraction between T_{i-1} and T_i and can be omitted, and N is the nominal amount and can be omitted;
- **FRA**(t, T, S, τ, N, R): Price at time t of a forward–rate agreement with reset date T and payment date S at the fixed rate R; As usual τ is the year fraction between T and S and can be omitted, and N is the nominal amount and can be omitted;
- **PFS**$(t, \mathcal{T}, \tau, N, R)$: Price at time t of a payer forward–start interest rate swap with first reset date T_1 and payment dates $T_2, \ldots, T_n$ at the fixed rate R; As usual τ_i is the year fraction between T_{i-1} and T_i and can be omitted, and N is the nominal amount and can be omitted;
- **RFS**$(t, \mathcal{T}, \tau, N, R)$: Same as above but for a receiver swap;
- **PS**$(t, T, \mathcal{T}, \tau, N, R)$: Price of a payer swaption maturing at time T, which gives its holder the right to enter at time T an interest rate swap with first reset date T_1 and payment dates $T_2, \ldots, T_n$ (with $T_1 \geq T$) at the fixed strike–rate R; As usual τ_i is the year fraction between T_{i-1} and T_i and can be omitted, and N is the nominal amount and can be omitted;
- **RS**$(t, T, \mathcal{T}, \tau, N, R)$: Same as above but for a receiver swaption;
- **ES**$(t, T, \mathcal{T}, \tau, N, R, \omega)$: Same as above but for a general European swaption; ω is $+1$ for a payer and -1 for a receiver, and can be omitted.
- **FSCpl**$(T_j, T_{k-1}, T_k, \tau_k, \delta)$: Price at time 0 of a call option, with maturity T_k, on the LIBOR rate $F_k(T_{k-1})$, with $T_{k-1} > T_j$, where the strike price is set as a proportion δ of either the spot or forward LIBOR rate at time T_j.
- **QCpl**$(t, T, S, \tau_0, N, y, \text{curr1}, \text{curr2})$: Price at time t of a quanto caplet resetting at time T and paying at time S at a fixed strike–rate y; Rates are related to the foreign "curr2" currency, whereas the payoff is an amount in domestic "curr1" currency. As usual τ_0 is the year fraction between T and S and can be omitted, and N is the nominal amount and can be omitted; the currency names "curr1" and "curr2" can be omitted.
- **QFll**$(t, T, S, \tau_0, N, y, \text{curr1}, \text{curr2})$: As above but for a floorlet.
- **LSO**$(t, T_{i-1}, T_i, \tau_i, N, K, \omega, \psi)$: Price at time t of the spread option on two-currency LIBOR rates whose payoff is

$$N\left[\omega\big(F_i(T_{i-1}) - F_i^f(T_{i-1}) + K\big)\right]^+,$$

where N is the nominal value, K is the contract margin, τ_i is the year fraction for $(T_{i-1}, T_i]$, $\omega = 1$ for a call and $\omega = -1$ for a put, and $\psi = 0$ if the payoff is paid at time T_i, while $\psi = 1$ if the payoff is paid at time T_{i-1} ("in-arrears" case).
- **LP**$(t, T_{i-1}, T_i, \tau_i, N, K, \omega)$: Price at time t of the option on the product of the two LIBOR rates $L(T_{i-1}, T_i)$ and $L^f(T_{i-1}, T_i)$, whose payoff at time T_i, in domestic currency, is

$$\tau_i N\left[\omega\big(L(T_{i-1}, T_i)L^f(T_{i-1}, T_i) - K\big)\right]^+ = \tau_i N\left[\omega\big(F_i(T_{i-1})F_i^f(T_{i-1}) - K\big)\right]^+,$$

where N is the nominal value, K is the strike price and $\omega = 1$ for a call and $\omega = -1$ for a put.
- **TSD**$(t, T_{i-1}, T_i, \tau_i, N, K, \omega, \psi)$: Price at time t of the time-T_i payoff of a trigger swap with payment, in domestic currency, triggered by the domestic rate. The time-T_i payoff is

$$\tau_i N \left[\left(aF_i(T_{i-1}) + bF_i^f(T_{i-1}) + c \right) 1_{\{\omega F_i(T_{i-1}) \geq \omega K\}} \right] (1 + \psi \tau_i F_i(T_{i-1})),$$

where N is the nominal value, a, b, c are real constants specified by the contract, K is the trigger level, ω is either 1 or -1, $\psi = 1$ for the "in-arrears" case and $\psi = 0$ otherwise.

- **TSF**$(t, T_{i-1}, T_i, \tau_i, N, K, \omega, \psi)$: Price at time t of the time-T_i payoff of a trigger swap with payment, in domestic currency, triggered by the the foreign rate. The time-T_i payoff is

$$\tau_i N \left[\left(aF_i(T_{i-1}) + bF_i^f(T_{i-1}) + c \right) 1_{\{\omega F_i^f(T_{i-1}) \geq \omega K\}} \right] (1 + \psi \tau_i F_i(T_{i-1})),$$

  where N is the nominal value, a, b, c are real constants specified by the contract, K is the trigger level, ω is either 1 or -1, $\psi = 1$ for the "in-arrears" case and $\psi = 0$ otherwise.
- Instantaneous (absolute) volatility of a process Y is $\eta(t)$ in

$$dY_t = (\ldots)dt + \eta(t)dW_t \ .$$

- Instantaneous level-proportional (or proportional or percentage or relative or return) volatility of a process Y is $\sigma(t)$ in

$$dY_t = (\ldots)dt + \sigma(t)Y_t dW_t \ .$$

- Level-proportional (or proportional or percentage or relative) drift (or drift rate) of a process Y is $\mu(t)$ in

$$dY_t = \mu(t)Y_t dt + (\ldots)dW_t \ .$$

- DC(Y): $1 \times n$ vector diffusion coefficient of a diffusion process Y driven by the vector (correlated) Brownian motion Z with $dZ = CdW$, with C a $n \times n$ matrix, $CC' = \rho$, the $n \times n$ instantaneous correlation matrix, and W vector n-dimensional standard Brownian motion. In other terms, if

$$dY_t = (...)dt + v_t C dW_t = (...)dt + v_t dZ_t$$

  then DC(Y) $= v_t$.

INFLATION CHAPTER NOTATION:

- CPI = Consumer price index;
- ZCIIS = Zero-coupon inflation indexed swap;
- YYIIS = Year-on-year inflation indexed swap;
- IICF = Inflation-indexed caplet/floorlet;
- IICapFloor = Inflation-indexed cap/floor;
- $P_n(t, T)$: Zero-coupon bond price at time t for the maturity T in the nominal economy;
- $P_r(t, T)$: Zero-coupon bond price at time t for the maturity T in the real economy;
- $n(t)$: Instantaneous spot rate at time t in the nominal economy;
- $r(t)$: Instantaneous spot rate at time t in the real economy;
- $f_n(t, T)$: Instantaneous forward rate at time t for the maturity T in the nominal economy;
- $f_r(t, T)$: Instantaneous forward rate at time t for the maturity T in the real economy;

- $F_n(t;T,S)$: Simply compounded forward (LIBOR) rate at time t for the expiry–maturity pair T,S in the nominal economy;
- $F_r(t;T,S)$: Simply compounded forward (LIBOR) rate at time t for the expiry–maturity pair T,S in the real economy;
- $I(t)$: Value of the CPI index at time t.
- Q_n: Risk-neutral measure in the nominal economy;
- Q_r: Risk-neutral measure in the real economy;
- Q_n^T: T-forward adjusted measure in the nominal economy;
- E_n: Expectation under the risk-neutral measure in the nominal economy;
- E_r: Expectation under the risk-neutral measure in the real economy;
- E_n^T: Expectation under the T-forward adjusted measure in the nominal economy;
- $\mathcal{I}_i(t)$: Value at time t of the T_i-forward CPI index;
- $\mathbf{ZCIIS}(t,T_M,I_0,N)$: Value at time t of the inflation-indexed leg of the ZCIIS with maturity $T_M = M$ years and nominal value N, when $I(0) = I_0$;
- $\mathbf{YYIIS}(t,T_{i-1},T_i,\psi_i,N)$: Value at time t of the i-th payment (*i.e.*, the inflation rate from T_{i-1} to T_i) in the inflation-indexed leg of the YYIIS, with nominal value N and year fraction ψ_i;
- $\mathbf{IICplt}(t,T_{i-1},T_i,\psi_i,K,N,\omega)$: Value at time t of the IICF on the inflation rate from T_{i-1} to T_i, with maturity T_i, strike κ, nominal value N and year fraction ψ_i, where $K = 1+\kappa$ and $\omega = 1$ for a caplet and $\omega = -1$ for a floorlet;
- $\mathbf{IICapFloor}(0,\mathcal{T},\Psi,K,N,\omega)$: Value at time 0 of the IICapFloor paying on times $\mathcal{T} = \{T_1,\dots,T_M\}$, with strike κ, nominal value N, year fractions $\Psi = \{\psi_1,\dots,\psi_M\}$, where $K = 1+\kappa$ and $\omega = 1$ for a cap and $\omega = -1$ for a floor.

CREDIT CHAPTERS NOTATION:

- CDO: Collateralized Debt Obligation;
- CDS: Credit Default Swap.
- CMCDS: Constant-Maturity Credit Default Swap.
- DFRN: Defaultable floating rate note.
- DPVBP: Defaultable present value per basis point.
- FtD = First to Default;
- $\mathbf{PFS}^D_{a,b}(t,K)$, $\mathbf{RFS}^D_{a,b}(t,K)$: Price at time t of a payer (respectively receiver) forward–start interest rate swap taking into account counterparty (default, "D") risk. The swap has first reset date T_a and payment dates $T_{a+1},\dots,T_b$ (tenor structure) at the fixed rate K;
- $\mathbf{PS}^D_{a,b}(t;K)$, $\mathbf{RS}^D_{a,b}(t;K)$: Payer and receiver swaption prices at time t for an underlying swap with tenor $T_a,T_{a+1},\dots,T_b$ with strike K, in presence of counterparty default risk;
- $\mathbf{Cpl}^D(t,K)$, $\mathbf{Fll}^D(t,K)$: Caplet and Floorlet prices in presence of counterparty default risk;
- $\mathbf{Cap}^D_{a,b}(t,K)$, $\mathbf{Flr}^D_{a,b}(t,K)$: Cap and Floor prices under counterparty default risk;
- PR, PR2: First and second postponed CDS formulations; it is usually put in front of "CDS", so as to give "PRCDS" or "PR2CDS".
- SSRD = Shifted Square Root Diffusion;
- $\tau = \tau^C$: Default time of the reference entity "C".
- $\mathbb{Q}$, $\mathbb{E}$: Risk-neutral probability measure and expectation on the (possibly enlarged) probability space supporting τ.
- $T_a,(T_{a+1},\dots,T_{b-1}),T_b$: Initial and final dates in the protection schedule of the CDS.
- $T_{\beta(t)}$: First of the T_i's following t.
- α_i: Year fraction between T_{i-1} and T_i.

- $R_{a,b}$: Rate in the premium leg of a CDS protecting in $[T_a, T_b]$.
- REC: Recovery fraction on a unit notional.
- $\text{LGD} = 1 - \text{REC}$: CDS Protection payment against a Loss (given default of the reference entity in the protection interval).
- $\Pi_{\text{RCDS}a,b}(t)$: Discounted payoff of a running CDS as seen from the protection seller.
- $\text{CDS}_{a,b}(t, R, \text{LGD})$: Price of a running CDS to the protection seller, protecting against default of the reference entity in $[T_a, T_b]$. This is also called "price of a receiver CDS". The corresponding payer CDS price, i.e. the CDS price seen from the point of view of the protection buyer, is $-\text{CDS}_{a,b}(t, R, \text{LGD})$.
- $-\text{CDS}_{a,b}(t, R, \text{LGD})$: Price of a running CDS to the protection buyer, also called "price of a payer CDS".
- $\text{CDS}_{\text{CM}a,b,c}(0, \text{LGD})$: Price at the initial time 0 of a CMCDS as seen from the protection seller (receiver CMCDS). Protection is offered in $[T_a, T_b]$ in exchange for a periodic payment at times T_j in $T_{a+1}, \ldots, T_b$ indexed at the "c+1" long CDS rates $R_{j-1,j+c}(T_{j-1})$.
- $\Pi_{\text{DFRN}a,b}(t)$: Discounted payoff of a defaultable floating rate note (FRN) spanning $[T_a, T_b]$.
- $\text{DFRN}_{a,b}(t, X, \text{REC})$: Price of the floating rate note with spread X and recovery REC .
- $X_{a,b}$: Par spread in a prototypical floating rate note spanning $[T_a, T_b]$.
- $\Pi_{\text{CallCDS}a,b}(t; K)$: Discounted payoff of a payer CDS option to enter at T_a a CDS at strike rate K
- $\text{CallCDS}_{a,b}(t, K, \text{LGD})$: Price of payer CDS option to enter at T_a a CDS with strike rate K.
- $1_{\{\tau > T\}}$: Survival indicator, is one if default occurs after T and zero otherwise.
- $1_{\{\tau \leq T\}}$: Default indicator, is one if default occurs before or at T, and zero otherwise.
- $1_{\{\tau \in [t,t+dt)\}}$: default indicator for the infinitesimal interval $[t, t + dt)$. It is one if default is in this interval, zero otherwise. Especially when inside integrals in t, this is an informal notation for the more rigorous notation $\delta_\tau(t)dt = \delta_0(t-\tau)dt = \delta(t-\tau)dt$ (with δ_{x_0} the Dirac delta function (mass) centered in x_0; when $x_0 = 0$ it is omitted).
- $1_{\{\tau > t\}}\bar{P}(t, T)$: Defaultable Zero coupon bond at time t for maturity T.
- $\widehat{C}_{a,b}(t)$, $\widehat{\mathbb{Q}}^{a,b}$: Defaultable "Present value per basis point" (or annuity) numeraire and associated measure.
- $\mathcal{F}_t$: Default free market information up to time t.
- $\mathcal{G}_t$: Default free market information plus explicit monitoring of default up to time t.
- $\gamma(t)$: Deterministic default intensity (and then hazard rate) for the default time at time t.
- $\lambda(t)$: Possibly stochastic default intensity (and then hazard rate) for the default time at time t.
- $\Gamma(t)$: Deterministic cumulated default intensity (and then hazard function) for the default time at time t.
- $\Lambda(t)$: Possibly stochastic cumulated default intensity (and then hazard function) for the default time.
- ξ: transformation of the default time by its cumulated intensity, $\xi = \Lambda(\tau)$ or $\Gamma(\tau)$; it is exponentially distributed and independent of default free quantities. The default time can be expressed, if intensity is strictly positive, as $\tau = \Lambda^{-1}(\xi)$, otherwise we need to introduce a pseudo-inverse.

Contents

Part II. FROM SHORT RATE MODELS TO HJM

Part IV. THE VOLATILITY SMILE

Part V. EXAMPLES OF MARKET PAYOFFS

Part VI. INFLATION

Part VII. CREDIT

Part VIII. APPENDICES

Part I

BASIC DEFINITIONS AND NO ARBITRAGE

1. Definitions and Notation

"The effort to understand the universe is one of the very few things that lifts human life a little above the level of farce, and gives it some of the grace of tragedy." Steven Weinberg, Nobel Prize in Physics 1979

"What?!? Politics?!? Culture?!? We're just doing little songs!"
(Refashioned after "sono solo canzonette", Edoardo Bennato, 1980)

In this first chapter we present the main definitions that will be used throughout the book. We will introduce the basic concepts in a rigorous way while providing at the same time intuition and motivation for their introduction. However, before starting with the definitions, a remark is in order.

Remark 1.0.1. **(Interbank vs government rates, LIBOR rates).** There are different types of interest rates, and a first distinction can be made between interbank rates and government rates. Government rates are usually deduced by bonds issued by governments. By "interbank rates" we denote instead rates at which deposits are exchanged between banks, and at which swap transactions (see below) between banks occur.

Zero-coupon rates (see below) can be "stripped" either from bonds in the government sector of the market or from products in the interbank sector of the market, resulting in two different zero-coupon curves. Once this has been done, mathematical modeling of the resulting rates is analogous in the two cases. We will focus on interbank rates here, although the mathematical apparatus described in this book can be usually applied to products involving government rates as well.

The most important interbank rate usually considered as a reference for contracts is the LIBOR (London InterBank Offered Rate) rate, fixing daily in London. However, there exist analogous interbank rates fixing in other markets (e.g. the EURIBOR rate, fixing in Brussels), and when we refer to "LIBOR" we actually intend any of these interbank rates.

1.1 The Bank Account and the Short Rate

Why then didn't you put my money on deposit,
so that when I came back, I could have collected it with interest?
St. Luke XIX. 23

The concept of interest rate belongs to our every-day life and has entered our minds as something familiar we know how to deal with. When depositing a certain amount of money in a bank account, everybody expects that the amount grows (at some rate) as time goes by. The fact that lending money must be rewarded somehow, so that receiving a given amount of money tomorrow is not equivalent to receiving exactly the same amount today, is indeed common knowledge and wisdom. However, expressing such concepts in mathematical terms may be less immediate and many definitions have to be introduced to develop a consistent theoretical apparatus.

The first definition we consider is the definition of a bank account, or money-market account. A money-market account represents a (locally) riskless investments, where profit is accrued continuously at the risk-free rate prevailing in the market at every instant.

Definition 1.1.1. Bank account (Money-market account). *We define $B(t)$ to be the value of a bank account at time $t \geq 0$. We assume $B(0) = 1$ and that the bank account evolves according to the following differential equation:*

$$dB(t) = r_t B(t)dt, \quad B(0) = 1, \tag{1.1}$$

where r_t is a positive function of time. As a consequence,

$$B(t) = \exp\left(\int_0^t r_s ds\right). \tag{1.2}$$

The above definition tells us that investing a unit amount at time 0 yields at time t the value in (1.2), and r_t is the *instantaneous rate* at which the bank account accrues. This instantaneous rate is usually referred to as *instantaneous spot rate*, or briefly as *short rate*. In fact, a first order expansion in Δt gives

$$B(t + \Delta t) = B(t)(1 + r(t)\Delta t), \tag{1.3}$$

which amounts to say that, in any arbitrarily small time interval $[t, t + \Delta t)$,

$$\frac{B(t + \Delta t) - B(t)}{B(t)} = r(t)\Delta t.$$

It is then clear that the bank account grows at each time instant t at a rate $r(t)$.

The bank-account numeraire[1] B is important for relating amounts of currencies available at different times. To this end, consider the following fundamental question: What is the value at time t of one unit of currency available at time T?

Assume for simplicity that the interest rate process r, and hence B, are deterministic. We know that if we deposit A units of currency in the bank account at time 0, at time $t > 0$ we have $A \times B(t)$ units of currency. Similarly, at time $T > t$ we have $A \times B(T)$ units. If we wish to have exactly one unit of currency at time T, i.e., if we wish that

$$A\,B(T) = 1,$$

we have to initially invest the amount $A = 1/B(T)$, which is known since the process B is deterministic. Hence, the value at time t of the amount A invested at the initial time is

$$A\,B(t) = \frac{B(t)}{B(T)}.$$

We have thus seen that the value of one unit of currency payable at time T, as seen from time t, is $B(t)/B(T)$. Coming back to the initial assumption of a general (stochastic) interest-rate process, this leads to the following.

Definition 1.1.2. Stochastic discount factor. *The (stochastic) discount factor $D(t,T)$ between two time instants t and T is the amount at time t that is "equivalent" to one unit of currency payable at time T, and is given by*

$$D(t,T) = \frac{B(t)}{B(T)} = \exp\left(-\int_t^T r_s ds\right). \tag{1.4}$$

The probabilistic nature of r_t is important since it affects the nature of the basic asset of our discussion, the bank-account numeraire B. In many pricing applications, especially when applying the Black and Scholes formula in equity or foreign-exchange (FX) markets, r is assumed to be a deterministic function of time, so that both the bank account (1.2) and the discount factors (1.4) at any future time are deterministic functions of time. This is usually motivated by assuming that variability of interest rates contributes to the price of equity or FX options by a smaller order of magnitude with respect to the underlying's movements.[2]

However, when dealing with interest-rate products, the main variability that matters is clearly that of the interest rates themselves. It is therefore necessary to drop the deterministic setup and to start modeling the evolution of r in time through a stochastic process. As a consequence, the bank account (1.2) and the discount factors (1.4) will be stochastic processes, too.

---

[1] We refer to the next chapter for a formal definition of a numeraire.

[2] We will consider the possibility of removing this assumption in Chapter B.

Some particular forms (i.e., stochastic differential equations) of possible evolutions for r will be discussed later on in the book.

We now turn to other basic definitions concerning the interest-rate world.

1.2 Zero-Coupon Bonds and Spot Interest Rates

Definition 1.2.1. Zero-coupon bond. *A T-maturity zero-coupon bond (pure discount bond) is a contract that guarantees its holder the payment of one unit of currency at time T, with no intermediate payments. The contract value at time $t < T$ is denoted by $P(t,T)$. Clearly, $P(T,T) = 1$ for all T.*

If we are now at time t, a zero-coupon bond for the maturity T is a contract that establishes the present value of one unit of currency to be paid at time T (the maturity of the contract).

A natural question arising now is: What is the relationship between the discount factor $D(t,T)$ and the zero-coupon-bond price $P(t,T)$? The difference lies in the two objects being respectively an "equivalent amount of currency" and a "value of a contract".

If rates r are deterministic, then D is deterministic as well and necessarily $D(t,T) = P(t,T)$ for each pair (t,T). However, if rates are stochastic, $D(t,T)$ is a random quantity at time t depending on the future evolution of rates r between t and T. Instead, the zero-coupon-bond price $P(t,T)$, being the time t-value of a contract with payoff at time T, has to be known (deterministic) at time t. We will see later on in the book that the bond price $P(t,T)$ and the discount factor $D(t,T)$ are closely linked, in that $P(t,T)$ can be actually viewed as the *expectation* of the random variable $D(t,T)$ under a particular probability measure.

In the following, by slight abuse of notation, t and T will denote both times, as measured by a real number from an instant chosen as time origin 0, and dates expressed as days/months/years.

Definition 1.2.2. Time to maturity. *The* time to maturity $T - t$ *is the amount of time (in years) from the present time t to the maturity time $T > t$.*

The definition "$T - t$" makes sense as long as t and T are real numbers associated to two time instants. However, if t and T denote two dates expressed as day/month/year, say $D_1 = (d_1, m_1, y_1)$ and $D_2 = (d_2, m_2, y_2)$, we need to define the amount of time between the two dates in terms of the number of days between them. This choice, however, is not unique, and the market evaluates the time between t and T in different ways. Indeed, the number of days between D_1 and D_2 is calculated according to the relevant market convention, which tells you how to count these days, whether to include holidays in the counting, and so on.

Definitions 1.2.1. Year fraction, Day-count convention. *We denote by $\tau(t,T)$ the chosen time measure between t and T, which is usually referred to as* year fraction *between the dates t and T. When t and T are less than one-day distant (typically when dealing with limit quantities involving time to maturities tending to zero), $\tau(t,T)$ is to be interpreted as the time difference $T-t$ (in years). The particular choice that is made to measure the time between two dates reflects what is known as the* day-count convention.

A detailed discussion on day-count conventions is beyond the scope of the book. For a complete treatment of the subject, the interested reader is then referred to the book of Miron and Swannell (1991). However, to clarify things, we mention the following three examples of day-count conventions.

- Actual/365. With this convention a year is 365 days long and the year fraction between two dates is the actual number of days between them divided by 365. If we denote by D_2-D_1 the actual number of days between the two dates, $D_1=(d_1,m_1,y_1)$ included and $D_2=(d_2,m_2,y_2)$ excluded, we have that the year fraction in this case is
$$\frac{D_2-D_1}{365}.$$
For example, the year fraction between January 4, 2000 and July 4, 2000 (both Tuesdays) is 182/365=0.49863.
- Actual/360. A year is in this case assumed to be 360 days long. The corresponding year fraction is
$$\frac{D_2-D_1}{360}.$$
Therefore, the year fraction between January 4, 2000 and July 4, 2000 is 182/360=0.50556.
- 30/360. With this convention, months are assumed 30 days long and years are assumed 360 days long. We have that the year fraction between D_1 and D_2 is in this case given by the following formula:
$$\frac{\max(30-d_1,0)+\min(d_2,30)+360\times(y_2-y_1)+30\times(m_2-m_1-1)}{360}.$$
For example, the year fraction between January 4, 2000 and July 4, 2000 is now
$$\frac{(30-4)+4+360\times 0+30\times 5}{360}=0.5.$$

As already hinted at above, adjustments may be included in the conventions, in order to leave out holidays. If D_2 is a holiday date, it can be replaced with the first working date following it, and this changes the evaluation of the year fractions. Again we refer to Miron and Swannell (1991) for the details.

Having clarified the market practice of using different day-count conventions, we can now proceed and comment on the definition of a zero-coupon bond. It is clear that every time we need to know the present value of a future-time payment, the zero-coupon-bond price for that future time is the fundamental quantity to deal with. Zero-coupon-bond prices are the basic quantities in interest-rate theory, and all interest rates can be defined in terms of zero-coupon-bond prices, as we shall see now. Therefore, they are often used as basic auxiliary quantities from which all rates can be recovered, and in turn zero-coupon-bond prices can be defined in terms of any given family of interest rates. Notice, however, that interest rates are what is usually quoted in (interbank) financial markets, whereas zero-coupon bonds are theoretical instruments that, as such, are not directly observable in the market.

In moving from zero-coupon-bond prices to interest rates, and vice versa, we need to know two fundamental features of the rates themselves: The compounding type and the day-count convention to be applied in the rate definition. What we mean by "compounding type" will be clear from the definitions below, while the day-count convention has been discussed earlier.

Definition 1.2.3. Continuously-compounded spot interest rate. *The continuously-compounded spot interest rate prevailing at time t for the maturity T is denoted by $R(t,T)$ and is the constant rate at which an investment of $P(t,T)$ units of currency at time t accrues continuously to yield a unit amount of currency at maturity T. In formulas:*

$$R(t,T) := -\frac{\ln P(t,T)}{\tau(t,T)} \tag{1.5}$$

The continuously-compounded interest rate is therefore a constant rate that is consistent with the zero-coupon-bond prices in that

$$e^{R(t,T)\tau(t,T)} P(t,T) = 1, \tag{1.6}$$

from which we can express the bond price in terms of the continuously-compounded rate R:

$$P(t,T) = e^{-R(t,T)\tau(t,T)}. \tag{1.7}$$

The year fraction involved in continuous compounding is usually $\tau(t,T) = T - t$, the time difference expressed in years.

An alternative to continuous compounding is simple compounding, which applies when accruing occurs proportionally to the time of the investment. We indeed have the following.

Definition 1.2.4. Simply-compounded spot interest rate. *The simply-compounded spot interest rate prevailing at time t for the maturity T is denoted by $L(t,T)$ and is the constant rate at which an investment has to be made to produce an amount of one unit of currency at maturity, starting from $P(t,T)$ units of currency at time t, when accruing occurs proportionally to the investment time. In formulas:*

$$L(t,T) := \frac{1 - P(t,T)}{\tau(t,T)\ P(t,T)}. \tag{1.8}$$

The market LIBOR rates are simply-compounded rates, which motivates why we denote by L such rates. LIBOR rates are typically linked to zero-coupon-bond prices by the Actual/360 day-count convention for computing $\tau(t,T)$.

Definition (1.8) immediately leads to the simple-compounding counterpart of (1.6), i.e.,

$$P(t,T)(1 + L(t,T)\tau(t,T)) = 1, \tag{1.9}$$

so that a bond price can be expressed in terms of L as:

$$P(t,T) = \frac{1}{1 + L(t,T)\tau(t,T)}.$$

A further compounding method that is considered is annual compounding. Annual compounding is obtained as follows. If we invest today a unit of currency at the simply-compounded rate Y, in one year we will obtain the amount $A = 1(1+Y)$. Suppose that, after this year, we invest such an amount for one more year at the same rate Y, so that we will obtain $A(1+Y) = (1+Y)^2$ in two years. If we keep on reinvesting for n years, the final amount we obtain is $(1+Y)^n$. Based on this reasoning, we have the following.

Definition 1.2.5. Annually-compounded spot interest rate. *The annually-compounded spot interest rate prevailing at time t for the maturity T is denoted by $Y(t,T)$ and is the constant rate at which an investment has to be made to produce an amount of one unit of currency at maturity, starting from $P(t,T)$ units of currency at time t, when reinvesting the obtained amounts once a year. In formulas*

$$Y(t,T) := \frac{1}{[P(t,T)]^{1/\tau(t,T)}} - 1. \tag{1.10}$$

Analogously to (1.6) and (1.9), we then have

$$P(t,T)(1 + Y(t,T))^{\tau(t,T)} = 1, \tag{1.11}$$

which implies that bond prices can be expressed in terms of annually-compounded rates as

$$P(t,T) = \frac{1}{(1 + Y(t,T))^{\tau(t,T)}}. \tag{1.12}$$

A year fraction τ that can be associated to annual compounding is for example ACT/365.

A straightforward extension of the annual compounding case leads to the following definition, which is based on reinvesting k times per year.

Definition 1.2.6. k-times-per-year compounded spot interest rate. *The k-times-per-year compounded spot interest rate prevailing at time t for the maturity T is denoted by $Y^k(t,T)$ and is the constant rate (referred to a one-year period) at which an investment has to be made to produce an amount of one unit of currency at maturity, starting from $P(t,T)$ units of currency at time t, when reinvesting the obtained amounts k times a year. In formulas:*

$$Y^k(t,T) := \frac{k}{[P(t,T)]^{1/(k\tau(t,T))}} - k. \tag{1.13}$$

We then have

$$P(t,T)\left(1 + \frac{Y^k(t,T)}{k}\right)^{k\tau(t,T)} = 1, \tag{1.14}$$

so that we can write

$$P(t,T) = \frac{1}{\left(1 + \frac{Y^k(t,T)}{k}\right)^{k\tau(t,T)}}. \tag{1.15}$$

A fundamental property is that continuously-compounded rates can be obtained as the limit of k-times-per-year compounded rates for the number k of compounding times going to infinity. Indeed, we can easily show that

$$\lim_{k\to+\infty} \frac{k}{[P(t,T)]^{1/(k\tau(t,T))}} - k = -\frac{\ln(P(t,T))}{\tau(t,T)} = R(t,T),$$

which justifies the name (and definition) of the rate R. Notice also that, for each fixed Y:

$$\lim_{k\to+\infty}\left(1 + \frac{Y}{k}\right)^{k\tau(t,T)} = e^{Y\tau(t,T)}.$$

In fact, continuously-compounded rates are commonly defined through these limit relations. In this book, however, we preferred to follow a different approach.

We finally remark that all previous definitions of spot interest rates are equivalent in infinitesimal time intervals. Indeed, it can be easily proved that the short rate is obtainable as a limit of all the different rates defined above, that is, for each t,

$$\begin{aligned}
r(t) &= \lim_{T\to t^+} R(t,T) \\
&= \lim_{T\to t^+} L(t,T) \\
&= \lim_{T\to t^+} Y(t,T) \\
&= \lim_{T\to t^+} Y^k(t,T) \quad \text{for each } k.
\end{aligned}$$

1.3 Fundamental Interest-Rate Curves

"When dealing with curves, nothing ever goes straight"
Andrea Bugin, Product and Business Development, Banca IMI

A fundamental curve that can be obtained from the market data of interest rates is the zero-coupon curve at a given date t. This curve is the graph of the function mapping maturities into rates at times t. Precisely, we have the following.

Definition 1.3.1. Zero-coupon curve. *The zero-coupon curve (sometimes also referred to as "yield curve") at time t is the graph of the function*

$$T \mapsto \begin{cases} L(t,T) & t < T \leq t+1 \text{ (years)}, \\ Y(t,T) & T > t+1 \text{ (years)}. \end{cases} \tag{1.16}$$

Such a zero-coupon curve is also called the *term structure of interest rates* at time t. It is a plot at time t of simply-compounded interest rates for all maturities T up to one year and of annually compounded rates for maturities T larger than one year. An example of such a curve is shown in Figure 1.1. In this example, the curve is not monotonic, and its initially-inverted behaviour resurfaces periodically in the market. The Italian curve, for example, has been inverted for several years before becoming monotonic. The Euro curve has often shown a monotonic pattern, although in our figure we see an example of the opposite situation.

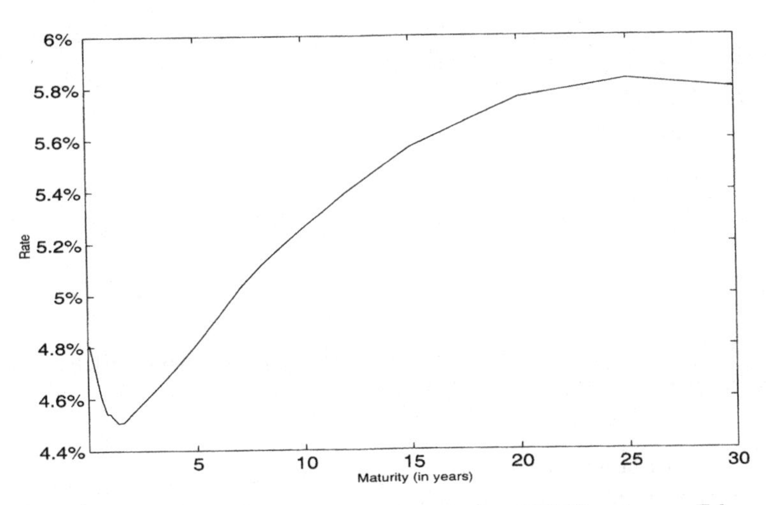

Fig. 1.1. Zero-coupon curve stripped from market EURO rates on February 13, 2001, at 5 p.m.

The term "yield curve" is often used to denote several different curves deduced from the interest-rate-market quotes, and is in fact slightly ambiguous.

When used in the book, unless differently specified, it is intended to mean "zero-coupon curve".

Finally, at times, we may consider the same plot for rates with different compounding conventions, such as for example

$$T \mapsto R(t,T), \quad T > t.$$

The term "zero-coupon curve" will be used for all such curves, no matter the compounding convention being used. In the book we will often make use of the following.

Definition 1.3.2. Zero-bond curve. *The zero-bond curve at time t is the graph of the function*

$$T \mapsto P(t,T), \quad T > t,$$

which, because of the positivity of interest rates, is a T-decreasing function starting from $P(t,t) = 1$. Such a curve is also referred to as term structure of discount factors.

An example of such a curve is shown in Figure 1.2. In a sense, at a quick

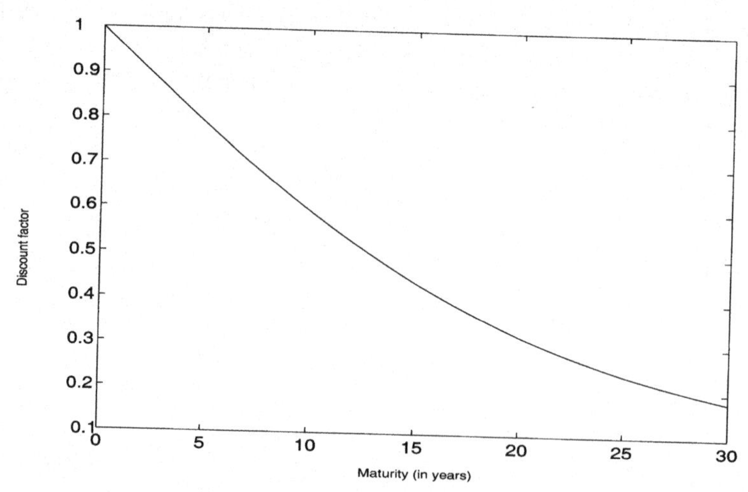

Fig. 1.2. Term structure of discount factors stripped from market EURO rates on February 13, 2001, at 5 p.m.

glance the zero-bond curve is less informative than the zero-coupon curve: The former is monotonic, while the latter can show several possible shapes. This is somehow clear from the definitions. For example, in the continuously-compounded case, the zero-coupon rates involve a logarithmic transformation of the "almost linear" zero-coupon bonds, which leads to a larger variability than that shown by the bonds themselves.

1.4 Forward Rates

We now move to the definition of forward rates. Forward rates are characterized by three time instants, namely the time t at which the rate is considered, its expiry T and its maturity S, with $t \leq T \leq S$. Forward rates are interest rates that can be locked in today for an investment in a future time period, and are set consistently with the current term structure of discount factors.

We can define a forward rate through a prototypical *forward-rate agreement* (FRA). A FRA is a contract involving three time instants: The current time t, the expiry time $T > t$, and the maturity time $S > T$. The contract gives its holder an interest-rate payment for the period between T and S. At the maturity S, a fixed payment based on a fixed rate K is exchanged against a floating payment based on the spot rate $L(T, S)$ resetting in T and with maturity S. Basically, this contract allows one to lock-in the interest rate between times T and S at a desired value K, with the rates in the contract that are simply compounded. Formally, at time S one receives $\tau(T, S)KN$ units of currency and pays the amount $\tau(T, S)L(T, S)N$, where N is the contract nominal value. The value of the contract in S is therefore

$$N\tau(T, S)(K - L(T, S)), \tag{1.17}$$

where we assume that both rates have the same day-count convention.[3] Clearly, if L is larger than K at time T, the contract value is negative, whereas in the other case it is positive. Recalling the expression (1.8) for L, we can rewrite this value as

$$N\left[\tau(T, S)K - \frac{1}{P(T, S)} + 1\right]. \tag{1.18}$$

Now consider the term $A = 1/P(T, S)$ as an amount of currency held at time S. Its value at time T is obtained by multiplying this amount A for the zero coupon price $P(T, S)$:

$$P(T, S)A = P(T, S)\frac{1}{P(T, S)} = 1,$$

so that this term is equivalent to holding one unit of currency at time T. In turn, one unit of currency at time T is worth $P(t, T)$ units of currency at time t. Therefore, the amount $1/P(T, S)$ in S is equivalent to an amount of $P(t, T)$ in t.

Then consider the other two terms in the contract value (1.18). The amount $B = \tau(T, S)K + 1$ at time S is worth

[3] It is actually market practice to apply different day-count conventions to fixed and floating rates. Here, however, we assume the same convention for simplicity, without altering the treatment of the whole subject. The generalization to different day-count conventions for the fixed and floating rates is indeed straightforward.

$$P(t,S)B = P(t,S)\tau(T,S)K + P(t,S)$$

at time t. The total value of the contract at time t is therefore

$$\mathbf{FRA}(t,T,S,\tau(T,S),N,K) = N\left[P(t,S)\tau(T,S)K - P(t,T) + P(t,S)\right]. \tag{1.19}$$

There is just one value of K that renders the contract fair at time t, i.e. such that the contract value is 0 in t. This value is of course obtained by equating to zero the FRA value. The resulting rate defines the (simply-compounded) forward rate.

Definition 1.4.1. Simply-compounded forward interest rate. *The simply-compounded forward interest rate prevailing at time t for the expiry $T > t$ and maturity $S > T$ is denoted by $F(t;T,S)$ and is defined by*

$$F(t;T,S) := \frac{1}{\tau(T,S)}\left(\frac{P(t,T)}{P(t,S)} - 1\right). \tag{1.20}$$

It is that value of the fixed rate in a prototypical FRA with expiry T and maturity S that renders the FRA a fair contract at time t.

Notice that we can rewrite the value of the above FRA (1.19) in terms of the just-defined simply compounded forward interest rate:

$$\mathbf{FRA}(t,T,S,\tau(T,S),N,K) = NP(t,S)\tau(T,S)(K - F(t;T,S)). \tag{1.21}$$

Therefore, to value a FRA, we just have to replace the LIBOR rate $L(T,S)$ in the payoff (1.17) with the corresponding forward rate $F(t;T,S)$, and then take the present value of the resulting (deterministic) quantity.

The forward rate $F(t;T,S)$ may thus be viewed as an estimate of the future spot rate $L(T,S)$, which is random at time t, based on market conditions at time t. In particular, we will see later on that $F(t;T,S)$ is the expectation of $L(T,S)$ at time t under a suitable probability measure.

When the maturity of the forward rate collapses towards its expiry, we have the notion of *instantaneous forward rate*. Indeed, let us consider the limit

$$\begin{aligned}
\lim_{S\to T^+} F(t;T,S) &= -\lim_{S\to T^+} \frac{1}{P(t,S)} \frac{P(t,S) - P(t,T)}{S-T} \\
&= -\frac{1}{P(t,T)} \frac{\partial P(t,T)}{\partial T} \\
&= -\frac{\partial \ln P(t,T)}{\partial T},
\end{aligned} \tag{1.22}$$

where we use our convention that $\tau(T,S) = S - T$ when S is extremely close to T. This leads to the following.

Definition 1.4.2. Instantaneous forward interest rate. *The instantaneous forward interest rate prevailing at time t for the maturity $T > t$ is denoted by $f(t,T)$ and is defined as*

$$f(t,T) := \lim_{S \to T^+} F(t;T,S) = -\frac{\partial \ln P(t,T)}{\partial T}, \tag{1.23}$$

so that we also have

$$P(t,T) = \exp\left(-\int_t^T f(t,u)\, du\right).$$

Clearly, for this notion to make sense, we need to assume smoothness of the zero-coupon-price function $T \mapsto P(t,T)$ for all T's.

Intuitively, the instantaneous forward interest rate $f(t,T)$ is a forward interest rate at time t whose maturity is very close to its expiry T, say $f(t,T) \approx F(t;T,T+\Delta T)$ with ΔT small.

Instantaneous forward rates are fundamental quantities in the theory of interest rates. Indeed, it turns out that one of the most general ways to express "fairness" of an interest-rate model is to relate certain quantities in the expression for the evolution of f. The "fairness" we refer to is the absence of arbitrage opportunities, as we shall define it precisely later on in the book, and the theoretical framework expressing this absence of arbitrage is the celebrated Heath, Jarrow and Morton (1992) framework, to which Chapter 5 of the present book is devoted. This framework is based on focusing on the instantaneous forward rates f as fundamental quantities to be modeled.

1.5 Interest-Rate Swaps and Forward Swap Rates

Then, a few months later Dogan got nailed by the IRS.
John Grisham, The Chamber, 1994.

We have just considered a FRA, which is a particular contract whose "fairness" can be invoked to define forward rates. A generalization of the FRA is the Interest-Rate Swap (IRS). A prototypical Payer (Forward-start) Interest-Rate Swap (PFS) is a contract that exchanges payments between two differently indexed legs, starting from a future time instant. At every instant T_i in a prespecified set of dates $T_{\alpha+1}, \ldots, T_\beta$ the fixed leg pays out the amount

$$N \tau_i K,$$

corresponding to a fixed interest rate K, a nominal value N and a year fraction τ_i between T_{i-1} and T_i, whereas the floating leg pays the amount

$$N \tau_i L(T_{i-1}, T_i),$$

corresponding to the interest rate $L(T_{i-1}, T_i)$ resetting at the previous instant T_{i-1} for the maturity given by the current payment instant T_i, with T_α a given date. Clearly, the floating-leg rate resets at dates $T_\alpha, T_{\alpha+1}, \ldots, T_{\beta-1}$ and pays at dates $T_{\alpha+1}, \ldots, T_\beta$. We set $\mathcal{T} := \{T_\alpha, \ldots, T_\beta\}$ and $\tau := \{\tau_{\alpha+1}, \ldots, \tau_\beta\}$.

In this description, we are considering that fixed-rate payments and floating-rate payments occur at the same dates and with the same year fractions. Though the generalization to different payment dates and day-count conventions is straightforward, we prefer to present a simplified version to ease the notation.[4]

When the fixed leg is paid and the floating leg is received the IRS is termed Payer IRS (PFS), whereas in the other case we have a Receiver IRS (RFS).

The *discounted* payoff at a time $t < T_\alpha$ of a PFS can be expressed as

$$\sum_{i=\alpha+1}^{\beta} D(t, T_i) N \tau_i (L(T_{i-1}, T_i) - K),$$

whereas the *discounted* payoff at a time $t < T_\alpha$ of a RFS can be expressed as

$$\sum_{i=\alpha+1}^{\beta} D(t, T_i) N \tau_i (K - L(T_{i-1}, T_i)).$$

If we view this last contract as a portfolio of FRAs, we can value each FRA through formulas (1.21) or (1.19) and then add up the resulting values. We thus obtain

$$\begin{aligned} \mathbf{RFS}(t, \mathcal{T}, \tau, N, K) &= \sum_{i=\alpha+1}^{\beta} \mathbf{FRA}(t, T_{i-1}, T_i, \tau_i, N, K) \\ &= N \sum_{i=\alpha+1}^{\beta} \tau_i P(t, T_i) \left(K - F(t; T_{i-1}, T_i)\right) \\ &= -NP(t, T_\alpha) + NP(t, T_\beta) + N \sum_{i=\alpha+1}^{\beta} \tau_i K P(t, T_i). \end{aligned} \tag{1.24}$$

The two legs of an IRS can be seen as two fundamental prototypical contracts. The fixed leg can be thought of as a coupon-bearing bond, and the floating leg can be thought of as a floating-rate note. An IRS can then be viewed as a contract for exchanging the coupon-bearing bond for the floating-rate note that are defined as follows.

---

[4] Indeed, a typical IRS in the market has a fixed leg with annual payments and a floating leg with quarterly or semiannual payments.

Definition 1.5.1. Prototypical coupon-bearing bond. *A prototypical coupon-bearing bond is a contract that ensures the payment at future times $T_{\alpha+1}, \ldots, T_\beta$ of the deterministic amounts of currency (cash-flows) $c := \{c_{\alpha+1}, \ldots, c_\beta\}$. Typically, the cash flows are defined as $c_i = N\tau_i K$ for $i < \beta$ and $c_\beta = N\tau_\beta K + N$, where K is a fixed interest rate and N is bond nominal value. The last cash flow includes the reimbursement of the notional value of the bond.*

In case $K = 0$ the bond reduces to a zero-coupon bond with maturity T_β. Since each cash flow has to be discounted back to current time t from the payment times T, the current value of the bond is

$$\mathbf{CB}(t, \mathcal{T}, c) = \sum_{i=\alpha+1}^{\beta} c_i P(t, T_i).$$

Definition 1.5.2. Prototypical floating-rate note. *A prototypical floating-rate note is a contract ensuring the payment at future times $T_{\alpha+1}, \ldots, T_\beta$ of the LIBOR rates that reset at the previous instants $T_\alpha, \ldots, T_{\beta-1}$. Moreover, the note pays a last cash flow consisting of the reimbursement of the notional value of the note at final time T_β.*

The value of the note is obtained by changing sign to the above value of the RFS with $K = 0$ (no fixed leg) and by adding to it the present value $NP(t, T_\beta)$ of the cash flow N paid at time T_β. We thus obtain

$$-\mathbf{RFS}(t, \mathcal{T}, \tau, N, 0) + NP(t, T_\beta) = NP(t, T_\alpha),$$

meaning that a prototypical floating-rate note is always equivalent to N units of currency at its first reset date T_α. In particular, if $t = T_\alpha$, the value is N, so that the value of the floating-rate note at its first reset time is always equal to its nominal value. This holds as well for $t = T_i$, for all $i = \alpha+1, \ldots, \beta-1$, in that the value of the note at all these instants is N. This is sometimes expressed by saying that "a floating-rate note always trades at par".

We have seen before that requiring a FRA to be fair leads to the definition of forward rates. Analogously, we may require the above IRS to be fair at time t, and we look for the particular rate K such that the above contract value is zero. This defines a forward swap rate.

Definition 1.5.3. *The forward swap rate $S_{\alpha,\beta}(t)$ at time t for the sets of times $\mathcal{T}$ and year fractions τ is the rate in the fixed leg of the above IRS that makes the IRS a fair contract at the present time, i.e., it is the fixed rate K for which* $\mathbf{RFS}(t, \mathcal{T}, \tau, N, K) = 0$. *We easily obtain*

$$S_{\alpha,\beta}(t) = \frac{P(t, T_\alpha) - P(t, T_\beta)}{\sum_{i=\alpha+1}^{\beta} \tau_i P(t, T_i)}. \tag{1.25}$$

Let us divide both the numerator and the denominator in (1.25) by $P(t, T_\alpha)$ and notice that the definition of F in terms of P's implies

$$\frac{P(t,T_k)}{P(t,T_\alpha)} = \prod_{j=\alpha+1}^{k} \frac{P(t,T_j)}{P(t,T_{j-1})} = \prod_{j=\alpha+1}^{k} \frac{1}{1+\tau_j F_j(t)} \quad \text{for all } k > \alpha,$$

where we have set $F_j(t) = F(t; T_{j-1}, T_j)$. Formula (1.25) can then be written in terms of forward rates as

$$S_{\alpha,\beta}(t) = \frac{1 - \prod_{j=\alpha+1}^{\beta} \frac{1}{1+\tau_j F_j(t)}}{\sum_{i=\alpha+1}^{\beta} \tau_i \prod_{j=\alpha+1}^{i} \frac{1}{1+\tau_j F_j(t)}}.$$

1.6 Interest-Rate Caps/Floors and Swaptions

We conclude this chapter by introducing the two main derivative products of the interest-rates market, namely caps and swaptions. These two products will be described more extensively later on, especially in Chapter 6, devoted to market models.

Interest-Rate Caps/Floors

A *cap* is a contract that can be viewed as a payer IRS where each exchange payment is executed only if it has positive value. The cap discounted payoff is therefore given by

$$\sum_{i=\alpha+1}^{\beta} D(t,T_i) N \tau_i (L(T_{i-1},T_i) - K)^+.$$

Analogously, a *floor* is equivalent to a receiver IRS where each exchange payment is executed only if it has positive value. The floor discounted payoff is therefore given by

$$\sum_{i=\alpha+1}^{\beta} D(t,T_i) N \tau_i (K - L(T_{i-1},T_i))^+.$$

Where do the terms "cap" and "floor" originate from? Consider the cap case. Suppose a company is LIBOR indebted and has to pay at certain times $T_{\alpha+1}, \ldots, T_\beta$ the LIBOR rates resetting at times $T_\alpha, \ldots, T_{\beta-1}$, with associated year fractions $\tau = \{\tau_{\alpha+1}, \ldots, \tau_\beta\}$, and assume that the debt notional amount is one. Set $\mathcal{T} = \{T_\alpha, \ldots, T_\beta\}$. The company is afraid that LIBOR rates will increase in the future, and wishes to protect itself by locking the payment at a maximum "cap" rate K. In order to do this, the company enters a cap with the payoff described above, pays its debt in terms of the LIBOR rate L and receives $(L-K)^+$ from the cap contract. The difference gives what is paid when considering both contracts:

$$L - (L-K)^+ = \min(L, K).$$

This implies that the company pays at most K at each payment date, since its variable (L-indexed) payments have been *capped* to the fixed rate K, which is termed the strike of the cap contract or, more briefly, *cap rate*. The cap, therefore, can be seen as a contract that can be used to prevent losses from large movements of interest rates when indebted at a variable (LIBOR) rate.

A cap contract can be decomposed additively. Indeed, its discounted payoff is a sum of terms such as

$$D(t, T_i) N \tau_i (L(T_{i-1}, T_i) - K)^+.$$

Each such term defines a contract that is termed *caplet*. The *floorlet* contracts are defined in an analogous way.

It is market practice to price a cap with the following sum of Black's formulas (at time zero)

$$\mathbf{Cap}^{\text{Black}}(0, \mathcal{T}, \tau, N, K, \sigma_{\alpha,\beta}) = N \sum_{i=\alpha+1}^{\beta} P(0, T_i) \tau_i \mathrm{Bl}(K, F(0, T_{i-1}, T_i), v_i, 1), \tag{1.26}$$

where, denoting by Φ the standard Gaussian cumulative distribution function,

$$\begin{aligned}
\mathrm{Bl}(K, F, v, \omega) &= F\omega\Phi(\omega d_1(K, F, v)) - K\omega\Phi(\omega d_2(K, F, v)),\\
d_1(K, F, v) &= \frac{\ln(F/K) + v^2/2}{v},\\
d_2(K, F, v) &= \frac{\ln(F/K) - v^2/2}{v},\\
v_i &= \sigma_{\alpha,\beta}\sqrt{T_{i-1}},
\end{aligned}$$

with the common volatility parameter $\sigma_{\alpha,\beta}$ that is retrieved from market quotes. Analogously, the corresponding floor is priced according to the formula

$$\mathbf{Flr}^{\text{Black}}(0, \mathcal{T}, \tau, N, K, \sigma_{\alpha,\beta}) = N \sum_{i=\alpha+1}^{\beta} P(0, T_i) \tau_i \mathrm{Bl}(K, F(0, T_{i-1}, T_i), v_i, -1). \tag{1.27}$$

There are several caps (floors) whose implied volatilities are quoted by the market. For example we have quotes with $\alpha = 0$, T_0 equal to three months, and all other T_i's equally three-month spaced. Another example concerns quotes with T_0 equal to three months, the next T_i's up to one year equally three-month spaced, and all other T_i's equally six-month spaced.

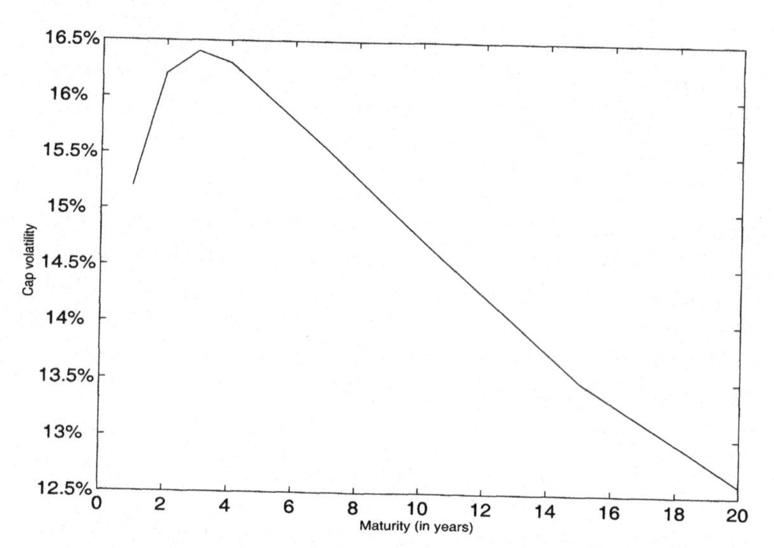

Fig. 1.3. At-the-money Euro cap volatility curve on February 13, 2001, at 5 p.m. The one-year cap has a period of three months, whereas the others (with maturities 2, 3, 4, 5, 7, 10, 15 and 20 years) of six months.

An example of market cap volatility curve is shown in Figure 1.3, where we show a plot of the $\sigma_{\alpha,\beta}$'s against their corresponding T_β's for a fixed α.

The Black formulas (1.26) and (1.27) have both an historical and a formal justification, which will be extensively explained in the guided-tour Section 6.2 and in the rest of Chapter 6, devoted to the analysis of the (interest-rate) market models.

Definition 1.6.1. *Consider a cap (floor) with payment times $T_{\alpha+1}, \ldots, T_\beta$, associated year fractions $\tau_{\alpha+1}, \ldots, \tau_\beta$ and strike K. The cap (floor) is said to be at-the-money (ATM) if and only if*

$$K = K_{ATM} := S_{\alpha,\beta}(0) = \frac{P(0, T_\alpha) - P(0, T_\beta)}{\sum_{i=\alpha+1}^{\beta} \tau_i P(0, T_i)}.$$

The cap is instead said to be in-the-money (ITM) if $K < K_{ATM}$, and out-of-the-money (OTM) if $K > K_{ATM}$, with the converse holding for a floor.

Using the equality

$$(L - K)^+ - (K - L)^+ = L - K,$$

we can note that the difference between a cap and the corresponding floor is equivalent to a forward-start swap. It is, therefore, easy to prove that a cap (floor) is ATM if and only if its price equals that of the corresponding floor (cap).

Notice also that in case the cap has only one payment date ($\alpha + 1 = \beta$), the cap collapses to a single caplet. In such a case, the at-the-money caplet strike is $K_{\text{ATM}} = F(0, T_\alpha, T_{\alpha+1})$, and the caplet is ITM if $K < F(0, T_\alpha, T_{\alpha+1})$. The reason for these terms is particularly clear in the caplet case. In fact, if the contract terminal payoff is evaluated now with the current value of the underlying replacing the corresponding terminal value, we have a positive amount in the ITM case (so that we are "in the money"), whereas we have a non-positive amount in the other case (we are "out of the money"). Just replace the terminal underlying forward rate $F(T_\alpha, T_\alpha, T_{\alpha+1})$ with its current value $F(0, T_\alpha, T_{\alpha+1})$ in the caplet terminal payoff

$$(F(T_\alpha, T_\alpha, T_{\alpha+1}) - K)^+$$

in the ITM and OTM cases to check this.

Swaptions

We finally introduce the second class of basic derivatives on interest rates. These derivatives, termed swap options or more commonly swaptions, are options on an IRS. There are two main types of swaptions, a payer version and a receiver version.

A European *payer swaption* is an option giving the right (and no obligation) to enter a payer IRS at a given future time, the swaption maturity. Usually the swaption maturity coincides with the first reset date of the underlying IRS. The underlying-IRS length ($T_\beta - T_\alpha$ in our notation) is called the *tenor* of the swaption. Sometimes the set of reset and payment dates is called the tenor structure.

We can write the discounted payoff of a payer swaption by considering the value of the underlying payer IRS at its first reset date T_α, which is also assumed to be the swaption maturity. Such a value is given by changing sign in formula (1.24), i.e.,

$$N \sum_{i=\alpha+1}^{\beta} P(T_\alpha, T_i)\tau_i(F(T_\alpha; T_{i-1}, T_i) - K).$$

The option will be exercised only if this value is positive, so that, to obtain the swaption payoff at time T_α, we have to apply the positive-part operator. The payer-swaption payoff, discounted from the maturity T_α to the current time, is thus equal to

$$N D(t, T_\alpha) \left(\sum_{i=\alpha+1}^{\beta} P(T_\alpha, T_i)\tau_i(F(T_\alpha; T_{i-1}, T_i) - K) \right)^+ .$$

Contrary to the cap case, this payoff cannot be decomposed in more elementary products, and this is a fundamental difference between the two main interest-rate derivatives. Indeed, we have seen that caps can be decomposed into the sum of the underlying caplets, each depending on a single forward rate. One can deal with each caplet separately, deriving results that can be finally put together to obtain results on the cap. The same, however, does not hold for swaptions. From an algebraic point of view, this is essentially due to the fact that the summation is *inside* the positive part operator, $(\cdots)^+$, and not outside like in the cap case. Since the positive part operator is not distributive with respect to sums, but is a piece-wise linear and convex function, we have

$$\left(\sum_{i=\alpha+1}^{\beta} P(T_\alpha, T_i)\tau_i(F(T_\alpha; T_{i-1}, T_i) - K)\right)^+$$
$$\leq \sum_{i=\alpha+1}^{\beta} P(T_\alpha, T_i)\tau_i(F(T_\alpha; T_{i-1}, T_i) - K)^+ ,$$

with no equality in general, so that the additive decomposition is not feasible. As a consequence, in order to value and manage swaptions contracts, we will need to consider the *joint* action of the rates involved in the contract payoff. From a mathematical point of view, this implies that, contrary to the cap case, *terminal correlation* between different rates could be fundamental in handling swaptions. The adjective "terminal" is somehow redundant and is used to point out that we are considering the correlation between rates instead of correlations between *infinitesimal changes* in rates, the latter being typically a dynamical property. In other words, the term "terminal" is used to stress that we are not considering *instantaneous* correlations. We will discuss the relationship between the two types of correlations at large in Section 6.6 and in other places.

Notice also that the right-hand side of the above inequality can be thought of as an alternative expression for a cap terminal payoff. *This means that a payer swaption has a value that is always smaller than the value of the corresponding cap contract.*

It is market practice to value swaptions with a Black-like formula. Precisely, the price of the above payer swaption (at time zero) is

$$\mathbf{PS}^{\text{Black}}(0, \mathcal{T}, \tau, N, K, \sigma_{\alpha,\beta}) = N\text{Bl}(K, S_{\alpha,\beta}(0), \sigma_{\alpha,\beta}\sqrt{T_\alpha}, 1) \sum_{i=\alpha+1}^{\beta} \tau_i P(0, T_i), \tag{1.28}$$

where $\sigma_{\alpha,\beta}$ is now a volatility parameter quoted in the market that is different from the corresponding $\sigma_{\alpha,\beta}$ in the caps/floors case. A similar formula is used for a *receiver swaption*, which gives the holder the right to enter at time T_α a receiver IRS, with payment dates in $\mathcal{T}$. Such a formula is

$$\mathbf{RS}^{\text{Black}}(0, \mathcal{T}, \tau, N, K, \sigma_{\alpha,\beta}) = N\text{Bl}(K, S_{\alpha,\beta}(0), \sigma_{\alpha,\beta}\sqrt{T_\alpha}, -1) \sum_{i=\alpha+1}^{\beta} \tau_i P(0, T_i). \tag{1.29}$$

An example of market swaption volatility surface is shown in Figure 1.4, where we plot volatilities $\sigma_{\alpha,\beta}$ against their corresponding maturities T_α and swap lengths (tenors) $T_\beta - T_\alpha$.

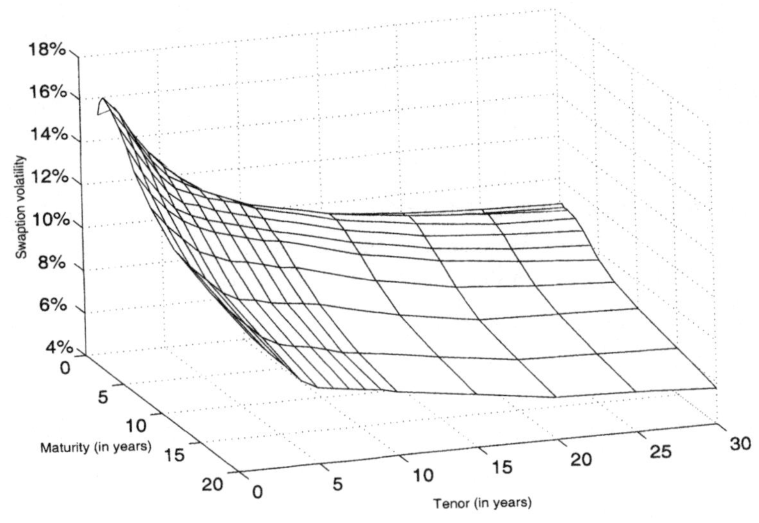

Fig. 1.4. At-the-money Euro swaption volatility surface on February 8, 2001, at 5 p.m.

Similarly to the cap/floor case, also the Black formulas (1.28) and (1.29) have a historical and a formal justification. Also in this case, a more rigorous treatment and derivation can be found in the chapter devoted to market models, and precisely in Section 6.7.

Definition 1.6.2. *Consider a payer (respectively receiver) swaption with strike K giving the holder the right to enter at time T_α a payer (receiver) IRS with payment dates $T_{\alpha+1}, \ldots, T_\beta$ and associated year fractions $\tau_{\alpha+1}, \ldots, \tau_\beta$. The swaption (either payer or receiver) is then said to be at-the-money (ATM) if and only if*

$$K = K_{ATM} := S_{\alpha,\beta}(0) = \frac{P(0,T_\alpha) - P(0,T_\beta)}{\sum_{i=\alpha+1}^{\beta} \tau_i P(0,T_i)}.$$

The payer swaption is instead said to be in-the-money (ITM) if $K < K_{ATM}$, and out of the money (OTM) if $K > K_{ATM}$. The receiver swaption is ITM if $K > K_{ATM}$, and OTM if $K < K_{ATM}$.

As in the above cap case, we can note that the difference between a payer swaption and the corresponding receiver swaption is equivalent to a forward-start swap. Therefore, it is again easy to prove that a payer swaption is ATM if and only if its price equals that of the corresponding receiver swaption.

Finally, we point out that an alternative expression for the above discounted payer-swaption payoff, expressed in terms of the relevant forward swap rate, is at time $t = 0$

$$ND(0,T_\alpha)\,(S_{\alpha,\beta}(T_\alpha) - K)^+ \sum_{i=\alpha+1}^{\beta} \tau_i P(T_\alpha, T_i).$$

This alternative expression confirms the intuitive meaning of the ITM and OTM expressions. Indeed, if you evaluate this payoff by substituting the terminal swap rate $S_{\alpha,\beta}(T_\alpha)$ with its current forward value $S_{\alpha,\beta}(0)$ you obtain a positive multiple of

$$(S_{\alpha,\beta}(0) - K)^+.$$

This amount is strictly positive (so that you are "in the money") if $S_{\alpha,\beta}(0) > K$, whereas it is worthless (you are "out of the money") in the other case. The symmetric remark applies to receiver swaptions.

2. No-Arbitrage Pricing and Numeraire Change

"They have an expression on Earth that I believe applies to this situation."
"There ain't no such thing as a free lunch"
Adam Warlock and Moondragon, in
"Silver Surfer/Warlock: Resurrection" 1, 1993, Marvel Comics.

In the context of inflationary cosmology,
it is fair to say that the universe is the ultimate free lunch.
Guth, A.H., The Inflationary Universe, Addison-Wesley, 1997.

The fundamental economic assumption in the seminal paper by Black and Scholes (1973) is the absence of arbitrage opportunities in the considered financial market. Roughly speaking, absence of arbitrage is equivalent to the impossibility to invest zero today and receive tomorrow a nonnegative amount that is positive with positive probability. In other words, two portfolios having the same payoff at a given future date must have the same price today. By constructing a suitable portfolio having the same instantaneous return as that of a riskless investment, Black and Scholes could then conclude that the portfolio instantaneous return was indeed equal to the instantaneous risk-free rate, which immediately led to their celebrated partial differential equation and, through its solution, to their option-pricing formula.

The basic Black and Scholes argument was subsequently used by Vasicek (1977) to develop a model for the evolution of the term structure of interest rates and for the pricing of interest-rate derivatives (see also next Chapter 3). However, such argument deals with infinitesimal quantities in a way that only later works have justified. Indeed, the first rigorous and mathematically sound approach for the arbitrage-free pricing of general contingent claims was developed by Harrison and Kreps (1979) and Harrison and Pliska (1981, 1983). Their general theory has then inspired the work by Heath, Jarrow and Morton (1992) who developed a general framework for interest-rates dynamics.

In this chapter, we start by reviewing the main results of Harrison and Pliska (1981, 1983). We then describe the change-of-numeraire technique as developed by Geman et al. (1995) and provide a useful toolkit explaining how the various dynamics change when changing the numeraire. As an explicit example, the pricing of a cap and a swaption is considered. We conclude the chapter by providing some results on the change-of-numeraire technique in the presence of a foreign market.

The purpose of this chapter is highlighting some fundamental results concerning the arbitrage-free pricing of contingent claims. Our treatment is not meant to be extremely rigorous and essentially aims at providing the fundamental tools for the pricing purposes of the book. The reader interested in a more formal and detailed treatment of the no-arbitrage issue under stochastic interest rates is referred, for example, to Musiela and Rutkowski (1998) or Hunt and Kennedy (2000). Finally, the technical tools we will use are basically Ito's formula and Girsanov's theorem, and are briefly outlined together with equivalent measures, Radon-Nikodym derivatives and semimartingales in Appendix C at the end of the book.

2.1 No-Arbitrage in Continuous Time

I weave a delicate strategy which rash actions could rend. Patience, please.
Adam Warlock, "The Infinity Gauntlet", 1991, Marvel Comics.

We here briefly consider the case of the continuous-time economy analyzed by Harrison and Kreps (1979) and Harrison and Pliska (1981, 1983). We do this for historical reasons, but also for introducing a terminology that will be helpful later on.

We consider a time horizon $T > 0$, a probability space $(\Omega, \mathcal{F}, Q_0)$ and a right-continuous filtration $I\!\!F = \{\mathcal{F}_t : 0 \leq t \leq T\}$. In the given economy, $K+1$ non dividend paying securities are traded continuously from time 0 until time T. Their prices are modeled by a $K+1$ dimensional adapted semimartingale $S = \{S_t : 0 \leq t \leq T\}$, whose components $S^0, S^1, \ldots, S^K$ are positive. The asset indexed by 0 is a bank account. Its price then evolves according to

$$dS_t^0 = r_t S_t^0 dt,$$

with $S_0^0 = 1$ and where r_t is the instantaneous short-term rate at time t. In the book notation (see previous chapter), $S_t^0 = B(t)$ and $1/S_t^0 = D(0,t)$ for each t.

Definitions 2.1.1. *A* trading strategy *is a ($K+1$ dimensional) process $\phi = \{\phi_t : 0 \leq t \leq T\}$, whose components $\phi^0, \phi^1, \ldots, \phi^K$ are locally bounded and predictable. The* value process *associated with a strategy ϕ is defined by*

$$V_t(\phi) = \phi_t S_t = \sum_{k=0}^{K} \phi_t^k S_t^k, \quad 0 \leq t \leq T,$$

and the gains process *associated with a strategy ϕ by*

$$G_t(\phi) = \int_0^t \phi_u dS_u = \sum_{k=0}^{K} \int_0^t \phi_u^k dS_u^k, \quad 0 \leq t \leq T.$$

The k-th component ϕ_t^k of the strategy ϕ_t at time t, for each t, is interpreted as the number of units of security k held by an investor at time t. The predictability condition on each ϕ^k means that the value ϕ_t^k is known immediately before time t. This is done to reduce the investor's freedom at any jump time. This issue, however, is not relevant if S has continuous paths (as happens for example in case S follows a diffusion process). Moreover, $V_t(\phi)$ and $G_t(\phi)$ are respectively interpreted as the market value of the portfolio ϕ_t and the cumulative gains realized by the investor until time t by adopting the strategy ϕ.

Definition 2.1.1. *A trading strategy ϕ is* self-financing *if $V(\phi) \geq 0$ and*

$$V_t(\phi) = V_0(\phi) + G_t(\phi), \quad 0 \leq t < T. \tag{2.1}$$

Intuitively, a strategy is self-financing if its value changes only due to changes in the asset prices. In other words, no additional cash inflows or outflows occur after the initial time.

A relation similar to (2.1) holds also when asset prices are all expressed in terms of the bank-account value. Indeed, Harrison and Pliska (1981) proved the following.

Proposition 2.1.1. *Let ϕ be a trading strategy. Then, ϕ is self-financing if and only if $D(0,t)V_t(\phi) = V_0(\phi) + \int_0^t \phi_u d(D(0,u)S_u)$.*

A key result in Harrison and Kreps (1979) and Harrison and Pliska (1981, 1983) is the established connection between the economic concept of absence of arbitrage and the mathematical property of existence of a probability measure, the equivalent martingale measure (or risk-neutral measure, or risk-adjusted measure), whose definition is given in the following.

Definition 2.1.2. *An* equivalent martingale measure *Q is a probability measure on the space $(\Omega, \mathcal{F})$ such that*

i) Q_0 and Q are equivalent measures, that is $Q_0(A) = 0$ if and only if $Q(A) = 0$, for every $A \in \mathcal{F}$;
ii) the Radon-Nikodym derivative dQ/dQ_0 belongs to $L^2(\Omega, \mathcal{F}, Q_0)$ (i.e. it is square integrable with respect to Q_0).
iii) the "discounted asset price" process $D(0,\cdot)S$ is an $(\mathbb{F}, Q)$-martingale, that is $E(D(0,t)S_t^k|\mathcal{F}_u) = D(0,u)S_u^k$, for all $k = 0, 1, \ldots, K$ and all $0 \leq u \leq t \leq T$, with E denoting expectation under Q.

An arbitrage opportunity is defined, in mathematical terms, as a self-financing strategy ϕ such that $V_0(\phi) = 0$ but $Q_0\{V_T(\phi) > 0\} > 0$. Harrison and Pliska (1981) then proved the fundamental result that the existence of an equivalent martingale measure implies the absence of arbitrage opportunities.

Definitions 2.1.2. *A* contingent claim *is a square-integrable and positive random variable on $(\Omega, \mathcal{F}, Q_0)$. A contingent claim H is* attainable *if there exists some self-financing ϕ such that $V_T(\phi) = H$. Such a ϕ is said to* generate *H, and $\pi_t = V_t(\phi)$ is* the price at time t *associated with H.*

The following proposition, proved by Harrison and Pliska (1981), provides the mathematical characterization of the unique no-arbitrage price associated with any attainable contingent claim.

Proposition 2.1.2. *Assume there exists an equivalent martingale measure Q and let H be an attainable contingent claim. Then, for each time t, $0 \leq t \leq T$, there exists a unique price π_t associated with H, i.e.,*

$$\pi_t = E(D(t,T)H|\mathcal{F}_t). \tag{2.2}$$

When the set of all equivalent martingale measures is nonempty, it is then possible to derive a unique no-arbitrage price associated to any attainable contingent claim. Such a price is given by the expectation of the discounted claim payoff under the measure Q equivalent to Q_0. This result generalizes that of Black and Scholes (1973) to the pricing of any claim, which, in particular, may be path-dependent. Also the underlying-asset-price dynamics are quite general, which makes formula (2.2) applicable in quite different situations.

Definition 2.1.3. *A financial market is* complete *if and only if every contingent claim is attainable.*

Harrison and Pliska (1983) proved the following fundamental result. A financial market is (arbitrage free and) complete if and only if there exists a unique equivalent martingale measure. The existence of a unique equivalent martingale measure, therefore, not only makes the markets arbitrage free, but also allows the derivation of a unique price associated with any contingent claim.

We can summarize what stated above by the following well known stylized characterization of no-arbitrage theory via martingales.

- The market is free of arbitrage if (and only if) there exists a martingale measure;
- The market is complete if and only if the martingale measure is unique;
- In an arbitrage-free market, not necessarily complete, the price of any attainable claim is uniquely given, either by the value of the associated replicating strategy, or by the risk neutral expectation of the discounted claim payoff under any of the equivalent (risk-neutral) martingale measures.

2.2 The Change-of-Numeraire Technique

Formula (2.2) gives the unique no-arbitrage price of an attainable contingent claim H in terms of the expectation of the claim payoff under the selected martingale measure Q. Geman et al. (1995) noted, however, that an equivalent martingale measure Q, as in Definition 2.1.2, is not necessarily the most natural and convenient measure for pricing the claim H. Indeed, under stochastic interest rates, for example, the presence of the stochastic discount factor $D(t,T)$ complicates considerably the calculation of the expectation. In such cases, a change of measure can be quite helpful, and this is the approach followed, for instance, by Jamshidian (1989) in the calculation of a bond-option price under the Vasicek (1977) model.

Geman et al. (1995) introduced the following.

Definition 2.2.1. *A* numeraire *is any positive non-dividend-paying asset.*

In general, a numeraire Z is identifiable with a self-financing strategy ϕ in that $Z_t = V_t(\phi)$ for each t. Intuitively, a numeraire is a reference asset that is chosen so as to normalize all other asset prices with respect to it. Choosing a numeraire Z then implies that the relative prices S^k/Z, $k = 0, 1, \ldots, K$, are considered instead of the securities prices themselves. The value of the numeraire will be often used to denote the numeraire itself.

As proven by Geman et al. (1995), Proposition 2.1.1 can be extended to any numeraire, in that self-financing strategies remain self-financing after a numeraire change. Indeed, written in differential form, the self-financing condition

$$dV_t(\phi) = \sum_{k=0}^{K} \phi_t^k dS_t^k$$

implies that

$$d\left(\frac{V_t(\phi)}{Z_t}\right) = \sum_{k=0}^{K} \phi_t^k d\left(\frac{S_t^k}{Z_t}\right).$$

Therefore, an attainable claim is also attainable under any numeraire.

In the definition of equivalent martingale measure of the previous section, it has been implicitly assumed the bank account S^0 as numeraire. However, as already pointed out, this is just one of all possible choices and it turns out, in fact, that there can be more convenient numeraires as far as the calculation of claim prices is concerned. The following proposition by Geman et al. (1995) provides a fundamental tool for the pricing of derivatives and is the natural generalization of Proposition 2.1.2 to any numeraire.

Proposition 2.2.1. *Assume there exists a numeraire N and a probability measure Q^N, equivalent to the initial Q_0, such that the price of any traded asset X (without intermediate payments) relative to N is a martingale under Q^N, i.e.,*

$$\frac{X_t}{N_t} = E^N\left\{\frac{X_T}{N_T}|\mathcal{F}_t\right\} \quad 0 \le t \le T. \tag{2.3}$$

Let U be an arbitrary numeraire. Then there exists a probability measure Q^U, equivalent to the initial Q_0, such that the price of any attainable claim Y normalized by U is a martingale under Q^U, i.e.,

$$\frac{Y_t}{U_t} = E^U\left\{\frac{Y_T}{U_T}|\mathcal{F}_t\right\} \quad 0 \le t \le T. \tag{2.4}$$

Moreover, the Radon-Nikodym derivative defining the measure Q^U is given by

$$\frac{dQ^U}{dQ^N} = \frac{U_T N_0}{U_0 N_T}. \tag{2.5}$$

The derivation of (2.5) is outlined as follows. By definition of Q^N, we know that for any tradable asset price Z,

$$E^N\left[\frac{Z_T}{N_T}\right] = E^U\left[\frac{U_0}{N_0}\frac{Z_T}{U_T}\right] \tag{2.6}$$

(both being equal to Z_0/N_0). By definition of Radon-Nikodym derivative, we know also that for all Z

$$E^N\left[\frac{Z_T}{N_T}\right] = E^U\left[\frac{Z_T}{N_T}\frac{dQ^N}{dQ^U}\right].$$

By comparing the right-hand sides of the last two equalities, from the arbitrariness of Z we obtain (2.5). The general formula (2.4) follows from immediate application of the Bayes rule for conditional expectations.

2.3 A Change of Numeraire Toolkit (Brigo & Mercurio 2001c)

'If I turn,' Ged said after some time had gone by, 'if as you say I hunt the hunter, I think the hunt will not be long. All its desire is to meet me face to face. And twice it has done so, and twice defeated me.'

'Third time is the charm', said Ogion.

Ursula Le Guin, "A Wizard of Earthsea" (1968)

In this section, we present some useful considerations and formulas on the change-of-numeraire technique developed in the previous section, following mostly the version given in the first edition but including some updates (see for example Brigo 2005c).[1]

There are basically three facts on the change of numeraire technique one should consider in practice. Before addressing the third one, namely the change of numeraire toolkit, we give a useful summary of the first two relevant change-of-numeraire facts we have encountered earlier.

---

[1] We are stressing as source our original formulation of the toolkit as from the first 2001 edition, since we have found in different later books analogous formulations of the same, even with the same identical notation. This remark extends also to other parts of the book

FACT ONE. *The price of any asset divided by a reference positive non dividend-paying asset (called numeraire) is a martingale (no drift) under the measure associated with that numeraire.* This first fact follows from Proposition 2.2.1.

As a first fundamental example, we may consider any (say non-dividend paying) asset S divided by the bank account $B(t)$. This process $S_t/B(t)$, recalling that $dB(t) = r_t B(t)dt$ and assuming for simplicity $B(0) = 1$, reads

$$\frac{S_t}{B(t)} = e^{-\int_0^t r_s ds} S_t.$$

Fact One requires this process to be a martingale under the measure Q^B associated to the bank account numeraire B, i.e. the risk neutral measure. If for a second we call $b(t, S_t)$ the risk neutral drift of S, we have[2]

$$d\left(e^{-\int_0^t r_s ds} S_t.\right) = S_t(-r_t e^{-\int_0^t r_s ds})dt + e^{-\int_0^t r_s ds} dS_t =$$

$$= -r_t S_t e^{-\int_0^t r_s ds} dt + e^{-\int_0^t r_s ds} b(t, S_t)dt + (...)dW_t =$$

$$= e^{-\int_0^t r_s ds}(-r_t S_t + b(t, S_t))dt + (...)dW_t.$$

Fact One tells us that the drift of this last equation should be zero, since the ratio has to be a martingale. By zeroing the drift we obtain

$$-r_t S_t + b(t, S_t) = 0 \quad \Rightarrow \quad b(t, S_t) = r_t S_t.$$

We see than the Fact One implies the risk neutral drift of an asset S dynamics to be the risk free interest rate r_t times the asset S_t itself, consistently with what we have seen earlier.

As a second fundamental example, we will see often in the book the forward LIBOR rate F_2 between expiry T_1 and maturity T_2:

$$F_2(t) = \frac{(P(t, T_1) - P(t, T_2))/(T_2 - T_1)}{P(t, T_2)}.$$

This is a portfolio of two zero coupon bonds divided by the zero coupon bond $P(\cdot, T_2)$. If we take the measure Q^2 associated with the numeraire $P(\cdot, T_2)$, by Fact One F_2 will be a martingale (no drift) under that measure.

Finally, a third fundamental example of Fact One concerns the forward swap rate

$$S_{\alpha,\beta}(t) = \frac{P(t, T_\alpha) - P(t, T_\beta)}{\sum_{i=\alpha+1}^{\beta}(T_i - T_{i-1})P(t, T_i)}.$$

If we take as numeraire the (positive) portfolio of zero coupon bonds

[2] No need to use the stochastic Leibnitz rule of Appendix C.2 here, since $dS_t dt = 0$

$$\sum_{i=\alpha+1}^{\beta} (T_i - T_{i-1}) P(t, T_i),$$

leading to the swap measure $Q^{\alpha,\beta}$, then our swap rate is the ratio of a tradable asset (a portfolio long one T_α zero coupon bond, short one T_β zero coupon bond) divided by our numeraire. By fact one then the forward swap rate is a martingale under $Q^{\alpha,\beta}$.

FACT TWO. *The time-t risk neutral price*

$$Price_t = E_t^{\boxed{B}}\left[\boxed{B(t)}\frac{Payoff(T)}{\boxed{B(T)}}\right]$$

is invariant by change of numeraire: If S is any other numeraire, we have

$$Price_t = E_t^{\boxed{S}}\left[\boxed{S_t}\frac{Payoff(T)}{\boxed{S_T}}\right].$$

In other terms, if we substitute the three occurrences inside the boxes of the original numeraire with a new numeraire the price does not change. This second fact is a rephrasing of Formula (2.6) seen before.

Now there is a third fact one needs to include before being able to handle most situations. This third fact is related to the purpose of this section, namely providing a useful toolkit for the derivation of the asset-price dynamics under different numeraires. The fundamental formula that will be derived here is presented in equation (2.12) below and reads, in a more stylized form,

$$\text{drift}_{\text{asset}}^{\text{Num2}} = \text{drift}_{\text{asset}}^{\text{Num1}} - \text{Vol}_{\text{asset}}\ \text{Corr}\left(\frac{\text{Vol}_{\text{Num1}}}{\text{Num1}} - \frac{\text{Vol}_{\text{Num2}}}{\text{Num2}}\right)'.$$

This formula allows us to compute the drift in the dynamics of an asset price when moving from a first numeraire (Num1) to a second one (Num2), when we know the asset drift in the original numeraire, the asset volatility (that, as all instantaneous volatilities and correlations, does not depend on the numeraire), and the instantaneous correlation in the asset price dynamics, as well as the volatilities of the two numeraires.

An equivalent informal formulation in terms of random shocks is possible. If all vector processes we are considering feature the same multi-dimensional Brownian-motion shock (with possibly correlated components, according to a given correlation matrix) under a given probability measure, then as the measure changes the random shock will no longer be a driftless (Brownian motion) shock but will acquire a drift. The formula governing this transformation will be derived later in (2.13) below, but we may reformulate it more informally now as

FACT THREE:

$$\text{BrownianShocks}^{\text{Num2}}_{\text{Corr}} = \text{BrownianShocks}^{\text{Num1}}_{\text{Corr}} - \text{Corr}\left(\frac{\text{Vol}_{\text{Num2}}}{\text{Num2}} - \frac{\text{Vol}_{\text{Num1}}}{\text{Num1}}\right)'.$$

This formula allows us to compute the correlated vector shock process when moving from a first numeraire (Num1) to a second one (Num2), when we know the correlated vector shock process under the first numeraire, and the two numeraires values and volatilities. Further, as before, we need to know the instantaneous correlation among the different components of the shock vector.

Facts one, two and three form the core of the change of numeraire paradigm and should be read carefully before proceeding further.

We now proceed more formally. We consider a numeraire S with its associated measure Q^S. We also consider an n-vector diffusion process X whose dynamics under Q^S is given by

$$dX_t = \mu_t^S(X_t)dt + \sigma_t(X_t)CdW_t^S, \quad Q^S,$$

where μ_t^S is a $n \times 1$ vector and σ_t is a $n \times n$ diagonal matrix, and where we explicitly point out the measure under which the dynamics is defined. Here W^S is a n-dimensional standard Brownian motion under Q^S, and the $n \times n$ matrix C is introduced to model correlation in the resulting noise (CdW is equivalent to an n-dimensional Brownian motion with instantaneous correlation matrix $\rho = CC'$).

We will drop the superscript in the Brownian motion when the measure is clear from the context.

Now suppose we are interested in expressing the dynamics of X under the measure associated with a new numeraire U. The new dynamics will then be

$$dX_t = \mu_t^U(X_t)dt + \sigma_t(X_t)CdW_t^U, \quad Q^U,$$

where W^U is a n-dimensional standard Brownian motion under Q^U.

We can employ Girsanov's theorem to deduce the Radon-Nikodym derivative between Q^S and Q^U from the dynamics of X under the two different measures:

$$\begin{aligned}\zeta_t := \frac{dQ^S}{dQ^U}|_{\mathcal{F}_t} = \exp\Bigg(&-\frac{1}{2}\int_0^t \left|(\sigma_s(X_s)C)^{-1}\left[\mu_s^S(X_s) - \mu_s^U(X_s)\right]\right|^2 ds \\ &+ \int_0^t \left\{(\sigma_s(X_s)C)^{-1}\left[\mu_s^S(X_s) - \mu_s^U(X_s)\right]\right\}' dW_s^U\Bigg).\end{aligned}$$

In doing so we have assumed for simplicity that C is an invertible (full-rank) matrix.

The Girsanov theorem is briefly reviewed in Appendix C in a simplified context.

The process ζ defines a measure Q^S under which X has the desired dynamics, given its dynamics under Q^U. We know that ζ is an exponential martingale, in that by setting

$$\alpha_t := \left[\mu_t^S(X_t) - \mu_t^U(X_t)\right]' ((\sigma_t(X_t)C)^{-1})'$$

we obtain the "exponential martingale" dynamics:

$$d\zeta_t = \alpha_t \zeta_t dW_t^U. \tag{2.7}$$

Now, by (2.5),

$$\zeta_T = \frac{dQ^S}{dQ^U}\bigg|_{\mathcal{F}_T} = \frac{U_0 S_T}{S_0 U_T}, \tag{2.8}$$

and, since ζ is a Q^U-martingale,

$$\zeta_t = E_t^{Q^U}(\zeta_T) = E_t^{Q^U}\left[\frac{U_0 S_T}{S_0 U_T}\right] = \frac{U_0 S_t}{S_0 U_t}. \tag{2.9}$$

It follows by differentiation that

$$d\zeta_t = \frac{U_0}{S_0} d\left(\frac{S_t}{U_t}\right) = \frac{U_0}{S_0}\sigma_t^{S/U} C dW_t^U, \tag{2.10}$$

where, since S/U is a martingale under Q^U, we have assumed the following martingale dynamics

$$d\left(\frac{S_t}{U_t}\right) = \sigma_t^{S/U} C dW_t^U, \quad Q^U,$$

where $\sigma_t^{S/U}$ is a $1 \times n$ vector. Comparing equations (2.7) and (2.10) to deduce that

$$\alpha_t \zeta_t = \frac{U_0}{S_0}\sigma_t^{S/U} C,$$

and by taking into account (2.9), we obtain the fundamental result

$$\frac{S_t}{U_t}\alpha_t = \sigma_t^{S/U} C,$$

or, by definition of α,

$$\mu_t^U(X_t) = \mu_t^S(X_t) - \frac{U_t}{S_t}\sigma_t(X_t)\rho(\sigma_t^{S/U})' \tag{2.11}$$

with $\rho = CC'$, which gives the change in the drift of a stochastic process when changing numeraire from U to S.

A useful characterization of formula (2.11) is given in the following.

Proposition 2.3.1. *Let us assume that the two numeraires S and U evolve under Q^U according to*[3]

$$dS_t = (\ldots)dt + \sigma_t^S C dW_t^U, \quad Q^U$$
$$dU_t = (\ldots)dt + \sigma_t^U C dW_t^U, \quad Q^U,$$

where both σ_t^S and σ_t^U are $1 \times n$ vectors, W^U is the usual n-dimensional driftless (under Q^U) standard Brownian motion and $CC' = \rho$. Then, the drift of the process X under the numeraire U is

$$\boxed{\mu_t^U(X_t) = \mu_t^S(X_t) - \sigma_t(X_t)\rho\left(\frac{\sigma_t^S}{S_t} - \frac{\sigma_t^U}{U_t}\right)'}. \tag{2.12}$$

Proof. Use the stochastic Leibnitz rule to compute $d(S_t/U_t)$ as

$$d\frac{S_t}{U_t} = \frac{1}{U_t}\,dS_t + S_t\,d\frac{1}{U_t} + dS_t\,d\frac{1}{U_t},$$

in combination with Ito's formula

$$d\frac{1}{U_t} = -\frac{1}{U_t^2}\,dU_t + \frac{1}{U_t^3}\,dU_t\,dU_t$$

and substitute the above dynamics for S and U to arrive easily at

$$\sigma_t^{S/U} = \frac{\sigma_t^S}{U_t} - \frac{S_t}{U_t}\frac{\sigma_t^U}{U_t},$$

which combined with (2.11) gives (2.12). □

It is sometimes helpful to consider what happens in terms of "shocks". Indeed, through straightforward passages, we can write

$$\boxed{C dW_t^S = C dW_t^U - \rho\left(\frac{\sigma_t^S}{S_t} - \frac{\sigma_t^U}{U_t}\right)' dt}. \tag{2.13}$$

Another interesting issue is the derivation of the dynamics of a tradable asset under the measure featuring the asset itself as numeraire. Let the asset be S, with diffusion coefficient as in Proposition 2.3.1, and consider the risk-neutral probability measure Q_0 associated with the money-market account numeraire B. We know that under $Q_0 = Q^B$ the process S_t/B_t is a martingale. It follows that the dynamics of S under the risk-neutral measure is

$$dS_t = r_t S_t dt + \sigma_t^S C dW_t^0.$$

[3] Or under any other equivalent measure, since the volatility coefficients in the dynamics do not depend on the particular equivalent measure that has been chosen

We now apply formula (2.13) and obtain that the desired dynamics of S under Q^S is

$$dS_t = \left[r_t S_t + \frac{\sigma_t^S \rho (\sigma_t^S)'}{S_t} \right] dt + \sigma_t^S C dW_t^S.$$

A fundamental particular case is treated in the following.

Proposition 2.3.2. *Let us assume a "level-proportional" functional form for volatilities (typical of the lognormal case), i.e.*

$$\begin{aligned}
\sigma_t^S &= v_t^S S_t, \\
\sigma_t^U &= v_t^U U_t, \\
\sigma_t(X_t) &= \operatorname{diag}(X_t) \operatorname{diag}(v_t^X),
\end{aligned}$$

where the v's are deterministic $1 \times n$-vector functions of time, and $\operatorname{diag}(X)$ *denotes a diagonal matrix whose diagonal elements are the entries of vector X. We then obtain*

$$\begin{aligned}
\mu_t^U(X_t) &= \mu_t^S(X_t) - \operatorname{diag}(X_t) \operatorname{diag}(v_t^X) \rho \left(v_t^S - v_t^U \right)' \\
&= \mu_t^S(X_t) - \operatorname{diag}(X_t) \frac{d\langle \ln X, \ln(S/U)' \rangle_t}{dt},
\end{aligned} \tag{2.14}$$

where the quadratic covariation and the logarithms, when applied to vectors, are meant to act componentwise.

In the "fully lognormal" case, where also the drift of X under Q^S is deterministically level proportional, i.e.,

$$\mu_t^S(X_t) = \operatorname{diag}(X_t)\, m_t^S,$$

with m^S a deterministic $n \times 1$ vector, it turns out that the drift under the new numeraire measure Q^U is of the same kind, i.e.,

$$\mu_t^U(X_t) = \operatorname{diag}(X_t)\, m_t^U,$$

with

$$m_t^U = m_t^S - \operatorname{diag}(v_t^X) \rho \left(v_t^S - v_t^U \right)' = m_t^S - \frac{d\langle \ln X, \ln(S/U)' \rangle_t}{dt},$$

which is often written as

$$m_t^U dt = m_t^S dt - (d \ln X_t)\, (d \ln(S_t/U_t)). \tag{2.15}$$

It is easy to see that this last formula connecting the drift rates m^U and m^S holds also in the more general case where the dynamics of the numeraires S and U are not lognormal-like. In this case one can still start from a lognormal X with deterministic drift rate m^S under Q^S, but one obtains a stochastic drift rate m^U for X under Q^U given by the above formula (where this time the covariance term is not deterministic). We will usually consider the above formula in this wider sense.

2.3.1 A helpful notation: "DC"

" Man, it's 'Diffusion Coefficient'. This has nothing to do with DC comics. Honest!"
Damiano, December 24, 2005.

The term "volatility" in the above derivations is better understood in terms of a fundamental coefficient of diffusion type stochastic differential equations: The Diffusion Coefficient (see also Appendix C.1). If we define a sort of operator extracting the vector diffusion coefficient (DC) from the dynamics of a given diffusion process, the formulation of the change of numeraire toolkit becomes clearer. Consider indeed $\mathrm{DC}_t(X)$ as the row vector v_t in

$$dX_t = (...)dt + v_t\ C\ dW_t$$

for diffusion processes X with W column vector Brownian motion[4] common to all relevant diffusion processes. This is to say that if for example $dX_1(t) = \sigma_1 X_1(t)(CdW)_1(t)$, then

$$\mathrm{DC}_t(X_1) = [\sigma_1 X_1(t),\ 0,\ 0, \ldots,\ 0] = \sigma_1 X_1(t) e_1$$

where $e_1 = [1\ 0 \ldots\ 0]$ is the first vector in the canonical basis.

We may reformulate our earlier change of numeraire results as follows. Equation (2.12) becomes

$$\mu_t^U(X_t) = \mu_t^S(X_t) - \mathrm{DC}_t(X)\rho \left(\frac{\mathrm{DC}_t(S)}{S_t} - \frac{\mathrm{DC}_t(U)}{U_t}\right)' dt \tag{2.16}$$

whereas Formula (2.13) becomes

$$CdW_t^S = CdW_t^U - \rho \left(\frac{\mathrm{DC}_t(S)}{S_t} - \frac{\mathrm{DC}_t(U)}{U_t}\right)' dt. \tag{2.17}$$

We will often omit the "t" argument in DC_t. Working with the notation "DC" can be helpful. For example, DC is linear so that

$$\mathrm{DC}(kX + hY) = k\mathrm{DC}(X) + h\mathrm{DC}(Y)$$

for any diffusions X and Y driven by CdW and deterministic constants k and h. Also, since the differential of a deterministic function of time $g(t)$ has no diffusion part, we have

$$\mathrm{DC}_t(g) = 0, \quad \mathrm{DC}_t(g_1 X + g_2) = g_1(t)\mathrm{DC}_t(X),$$

[4] Typically, later on we will consider directly the notation dZ for a correlated Brownian shock with instantaneous correlation matrix ρ. As usual, in this section we explicitly take $dZ = CdW$, $CC' = \rho$, sticking to the CdW notation. The notation $(C\ dW)_1$ is used for dZ_1

where $g_{1,2}$ are two other deterministic functions of time (or even deterministic constants).

Furthermore, through Ito's formula it is immediate to prove that for a scalar diffusion process X

$$\mathrm{DC}(\ln(X)) = \frac{\mathrm{DC}(X)}{X}, \quad \mathrm{DC}(X) = X\,\mathrm{DC}(\ln(X)).$$

This is why we may also write Equation (2.17) as

$$CdW_t^S = CdW_t^U - \rho\,(\mathrm{DC}(\ln(S)) - \mathrm{DC}(\ln(U)))'\,dt =$$

$$= CdW_t^U - \rho\,(\mathrm{DC}(\ln(S) - \ln(U)))'\,dt = CdW_t^U - \rho\,(\mathrm{DC}(\ln(S/U)))'\,dt,$$

so that

$$CdW_t^S = CdW_t^U - \rho\,\mathrm{DC}\left(\ln\left(\frac{S}{U}\right)\right)' dt, \tag{2.18}$$

and, similarly, Equation (2.16) as

$$\mu_t^U(X_t) = \mu_t^S(X_t) - \mathrm{DC}(X_t)\rho\left(\mathrm{DC}\left(\ln\left(\frac{S}{U}\right)\right)\right)'.$$

It is also helpful to notice that in general

$$\mathrm{DC}(X_1X_2\cdot\ldots\cdot X_m) = \mathrm{DC}(\ln(X_1X_2\cdot\ldots\cdot X_m))(X_1X_2\cdot\ldots\cdot X_m) =$$

$$= [\mathrm{DC}(\ln(X_1) + \ln(X_2) + \ldots + \ln(X_m))](X_1X_2\cdot\ldots\cdot X_m) =$$

$$= [\mathrm{DC}(\ln(X_1)) + \mathrm{DC}(\ln(X_2)) + \ldots + \mathrm{DC}(\ln(X_m))]X_1X_2\cdot\ldots\cdot X_m$$

$$= \left[\frac{\mathrm{DC}(X_1)}{X_1} + \frac{\mathrm{DC}(X_2)}{X_2} + \ldots + \frac{\mathrm{DC}(X_m)}{X_m}\right] X_1X_2\cdot\ldots\cdot X_m$$

$$= \mathrm{DC}(X_1)X_2\cdot\ldots\cdot X_m + \mathrm{DC}(X_2)X_1X_3\cdot\ldots\cdot X_m + \ldots + \mathrm{DC}(X_m)X_1X_2\cdot\ldots\cdot X_{m-1}$$

Remark 2.3.1. The covariation term in (2.14) is usually called "Vaillant brackets" in Rebonato (1998), who defines

$$[X, Y]_t := \frac{d\langle \ln X, \ln Y\rangle_t}{dt}.$$

However, this square-brackets operator from Rebonato (1998) is not to be confused with the one from semimartingale theory.

2.4 The Choice of a Convenient Numeraire

'[...] Yet if I run again, it will as surely find me again... and all strength is spent in the running.' Ged paced on a while, and then suddenly turned, and kneeling down before the mage he said, 'I have walked with great wizards and have lived in the Isle of the Wise, but you are my true master Ogion.' He spoke with love, and with a sombre joy.
Ursula Le Guin, "A Wizard of Earthsea" (1968)

As far as pricing derivatives is concerned, the change-of-numeraire technique is typically employed as follows. A payoff $h(X_T)$ is given, which depends on an underlying variable X (an interest rate, an exchange rate, a commodity price, etc.) at time T. Pricing such a payoff amounts to compute the risk-neutral expectation

$$E_0\{D(0,T)h(X_T)\}.$$

The risk-neutral numeraire is the money-market account

$$B(t) = D(0,t)^{-1} = \exp\left(\int_0^t r_s ds\right).$$

By using formula (2.6) for pricing under a new numeraire S, we obtain

$$E_0\{D(0,T)h(X_T)\} = S_0 E^{Q^S}\left\{\frac{h(X_T)}{S_T}\right\}. \tag{2.19}$$

Motivated by the above formula, we look for a numeraire S with the following two properties.

1. $X_t S_t$ is (the price of) a tradable asset (in such a case S is sometimes termed the *natural payoff* of X).
2. The quantity $h(X_T)/S_T$ is conveniently simple.

Why are these properties desirable?

The first condition ensures that $(X_t S_t)/S_t = X_t$ is a martingale under Q^S, so that one can assume for example a lognormal martingale dynamics for X:

$$dX_t = \sigma(t) X_t dW_t, \quad Q^S$$

with the consequent ease in computing expected values of functions of X. Indeed, in this case, the distribution of X under Q^S is known, and we have

$$\ln X_t \sim \mathcal{N}\left(\ln X_0 - \frac{1}{2}\int_0^t \sigma(s)^2 ds, \int_0^t \sigma(s)^2 ds\right).$$

The second condition ensures that the new numeraire renders the computation of the right hand side of (2.19) simpler, instead of complicating it.

Remark 2.4.1. Two standard applications of the above method are in the forward LIBOR and swap market models of Chapter 6, where the Black formulas for caps or swaptions are retrieved within a consistent no-arbitrage framework. The interested reader is therefore referred to such a chapter.

2.5 The Forward Measure

In many concrete situations, a useful numeraire is the zero-coupon bond whose maturity T coincides with that of the derivative to price. In such a case, in fact, $S_T = P(T,T) = 1$, so that pricing the derivative can be achieved by calculating an expectation of its payoff (divided by one). The measure associated with the bond maturing at time T is referred to as T-forward risk-adjusted measure, or more briefly as T-forward measure, and will be denoted by Q^T. The related expectation is denoted by E^T.

Denoting by π_t the price of the derivative at time t, formula (2.4) applied to the above numeraire yields

$$\pi_t = P(t,T)E^T\left\{H_T|\mathcal{F}_t\right\}, \tag{2.20}$$

for $0 \le t \le T$, and where H_T is the claim payoff at time T.

The reason why the measure Q^T is called forward measure is justified by the following.

Proposition 2.5.1. *Any simply-compounded forward rate spanning a time interval ending in T is a martingale under the T-forward measure, i.e.,*

$$E^T\{F(t;S,T)|\mathcal{F}_u\} = F(u;S,T),$$

for each $0 \le u \le t \le S < T$. In particular, the forward rate spanning the interval $[S,T]$ is the Q^T-expectation of the future simply-compounded spot rate at time S for the maturity T, i.e,

$$E^T\{L(S,T)|\mathcal{F}_t\} = F(t;S,T), \tag{2.21}$$

for each $0 \le t \le S < T$.

Proof. From the definition of a simply-compounded forward rate

$$F(t;S,T) = \frac{1}{\tau(S,T)}\left[\frac{P(t,S)}{P(t,T)} - 1\right],$$

with $\tau(S,T)$ the year fraction from S to T, we have that

$$F(t;S,T)P(t,T) = \frac{1}{\tau(S,T)}\left[P(t,S) - P(t,T)\right]$$

is the price at time t of a traded asset, since it is a multiple of the difference of two bonds. Therefore, by definition of the T-forward measure

$$\frac{F(t;S,T)P(t,T)}{P(t,T)} = F(t;S,T)$$

is a martingale under such a measure. The relation (2.21) then immediately follows from the equality $F(S;S,T) = L(S,T)$. □

The equality (2.21) can be extended to instantaneous rates as well. This is done in the following.

Proposition 2.5.2. *The expected value of any future instantaneous spot interest rate, under the corresponding forward measure, is equal to related instantaneous forward rate, i.e.,*

$$E^T\{r_T|\mathcal{F}_t\} = f(t,T),$$

for each $0 \leq t \leq T$.

Proof. Let us apply (2.20) with $H_T = r_T$ and remember the risk-neutral valuation formula (2.2). We then have

$$\begin{aligned} E^T\{r_T|\mathcal{F}_t\} &= \frac{1}{P(t,T)}E\left\{r_T e^{-\int_t^T r_s ds}|\mathcal{F}_t\right\} \\ &= -\frac{1}{P(t,T)}E\left\{\frac{\partial}{\partial T}e^{-\int_t^T r_s ds}|\mathcal{F}_t\right\} \\ &= -\frac{1}{P(t,T)}\frac{\partial P(t,T)}{\partial T} \\ &= f(t,T). \end{aligned}$$

□

2.6 The Fundamental Pricing Formulas

The results of the previous sections are developed in case of a finite number of basic market securities. A (theoretical) bond market, however, has a continuum of basic assets (the zero-coupon bonds), one for each possible maturity until a given time horizon. The no-arbitrage theory in such a situation is more complicated even though quite similar in spirit. A thorough analysis of this theory is, however, beyond the scope of this book. The interested reader is again referred, for example, to Musiela and Rutkowski (1998) or Hunt and Kennedy (2000). What we will do is simply to assume the existence of a risk-neutral measure Q under which the price at time t of any attainable contingent claim with payoff H_T at time $T > t$ is given by

$$\boxed{\pi_t = E\left(e^{-\int_t^T r_s ds} H_T|\mathcal{F}_t\right)}, \tag{2.22}$$

consistently with formula (2.2), where a claim is now meant to be attainable if it can be replicated by a self-financing strategy involving a finite number of basic assets at every trading time.

Remark 2.6.1. **(Attainability Assumption)** For all contingent claims that will be priced in the book we will assume attainability to hold. Therefore, when a claim is priced, we are assuming the existence of a suitable self-financing strategy that replicates the claim. However, an example of derivation of an explicit replicating strategy will be considered in Section 6.7.1.

The particular case of a European call option with maturity T, strike X and written on a unit-principal zero-coupon bond with maturity $S > T$ leads to the pricing formula

$$\mathbf{ZBC}(t,T,S,X) = E\left(e^{-\int_t^T r_s ds}(P(T,S)-X)^+|\mathcal{F}_t\right). \tag{2.23}$$

Moreover, the change-of-numeraire technique can be applied exactly as before, since it is just based on changing the underlying probability measure. For example, the forward measure Q^T is defined by the Radon-Nikodym derivative

$$\frac{dQ^T}{dQ} = \frac{P(T,T)B(0)}{P(0,T)B(T)} = \frac{e^{-\int_0^T r_s ds}}{P(0,T)} = \frac{D(0,T)}{P(0,T)},$$

so that the price at time t of the above claim is also given by (2.20), that is

$$\boxed{\pi_t = P(t,T)E^T\left(H_T|\mathcal{F}_t\right)}. \tag{2.24}$$

In the case of the above call option, we then have

$$\mathbf{ZBC}(t,T,S,X) = P(t,T)E^T\left((P(T,S)-X)^+|\mathcal{F}_t\right). \tag{2.25}$$

Of course, formula (2.24), and hence formula (2.25), turn out to be useful when the quantities defining the claim payoff have known dynamics (or distributions) under the forward measure Q^T. Typically, if $P(T,S)$ has a lognormal distribution conditional on $\mathcal{F}_t$ under the T-forward measure, the above expectation reduces to a Black-like formula with a suitable volatility input.

2.6.1 The Pricing of Caps and Floors

We now show that a cap (floor) is actually equivalent to a portfolio of European zero-coupon put (call) options. This equivalence will be repeatedly used to derive explicit formulas for cap/floor prices under the analytically tractable short-rate models we will consider in the next chapters.

We denote by $D = \{d_1, d_2, \ldots, d_n\}$ the set of the cap/floor payment dates and by $\mathcal{T} = \{t_0, t_1, \ldots, t_n\}$ the set of the corresponding times, meaning that

t_i is the difference in years between d_i and the settlement date t, and where t_0 is the first reset time. Moreover, we denote by τ_i the year fraction from d_{i-1} to d_i, $i = 1, \dots, n$ and by N the cap/floor nominal value, and we set $\tau = \{\tau_1, \dots, \tau_n\}$. The arbitrage-free price of the i-th caplet is then given by

$$\begin{aligned}
&\mathbf{Cpl}(t, t_{i-1}, t_i, \tau_i, N, X)\\
&\quad = E\left(e^{-\int_t^{t_i} r_s ds} N\tau_i (L(t_{i-1}, t_i) - X)^+ |\mathcal{F}_t\right)\\
&\quad = NE\left(e^{-\int_t^{t_{i-1}} r_s ds} P(t_{i-1}, t_i)\tau_i (L(t_{i-1}, t_i) - X)^+ |\mathcal{F}_t\right),
\end{aligned}$$

where the second equality comes from iterated conditioning, which will be explained extensively in Section 2.7.

Using the definition of the LIBOR rate $L(t_{i-1}, t_i)$, we obtain

$$\begin{aligned}
&\mathbf{Cpl}(t, t_{i-1}, t_i, \tau_i, N, X)\\
&\quad = NE\left(e^{-\int_t^{t_{i-1}} r_s ds} P(t_{i-1}, t_i)\left[\frac{1}{P(t_{i-1}, t_i)} - 1 - X\tau_i\right]^+ |\mathcal{F}_t\right)\\
&\quad = NE\left(e^{-\int_t^{t_{i-1}} r_s ds}[1 - (1 + X\tau_i)P(t_{i-1}, t_i)]^+ |\mathcal{F}_t\right),
\end{aligned}$$

so that

$$\mathbf{Cpl}(t, t_{i-1}, t_i, \tau_i, N, X) = N_i' \mathbf{ZBP}(t, t_{i-1}, t_i, X_i'), \tag{2.26}$$

where

$$\begin{aligned}
X_i' &= \frac{1}{1 + X\tau_i},\\
N_i' &= N(1 + X\tau_i).
\end{aligned}$$

Therefore, the caplet price can be written as a multiple of the price of a European put with maturity t_{i-1}, strike X_i' and written on a zero-coupon bond with maturity t_i and unit nominal amount. Alternatively, we can also write

$$\mathbf{Cpl}(t, t_{i-1}, t_i, \tau_i, N, X) = \mathbf{ZBP}(t, t_{i-1}, t_i, N_i', N). \tag{2.27}$$

The analogous formulas for the corresponding floor are then given by

$$\begin{aligned}
\mathbf{Fll}(t, t_{i-1}, t_i, \tau_i, N, X) &= N_i' \mathbf{ZBC}(t, t_{i-1}, t_i, X_i'),\\
&= \mathbf{ZBC}(t, t_{i-1}, t_i, N_i', N).
\end{aligned} \tag{2.28}$$

Finally, cap and floor prices are simply obtained by summing up the prices of the underlying caplets and floorlets, respectively. We thus obtain

$$\begin{aligned}
\mathbf{Cap}(t, \mathcal{T}, \tau, N, X) &= \sum_{i=1}^{n} N_i' \mathbf{ZBP}(t, t_{i-1}, t_i, X_i')\\
\mathbf{Flr}(t, \mathcal{T}, \tau, N, X), &= \sum_{i=1}^{n} N_i' \mathbf{ZBC}(t, t_{i-1}, t_i, X_i').
\end{aligned} \tag{2.29}$$

2.7 Pricing Claims with Deferred Payoffs

Many real-market interest-rate derivatives have (random) payoffs that depend on some interest rate whose value is set on some date prior to the derivative maturity. An easy example is given by a FRA, whose definition and pricing are described in Chapter 1. Standard interest-rate swaps themselves feature rates that usually reset a period earlier than the corresponding payment dates. In these cases, the associated time lag between reset and payment is named "natural", whereas when the time lag is different it is termed "unnatural", see Section 13.8.1 for more details.

Pricing more general claims with deferred payoffs, however, can be less straightforward. For instance, when using a tree to approximate the basic interest-rate process, the typical pricing procedure is to calculate the claim payoff on the final nodes in the tree and proceed backwards until the unique node at the initial time. In such a case, we immediately see that the payoff dependence on previously reset rates contrasts with a backwards calculation, since it would require the knowledge at time T of values that are only known at time $t < T$.

We now show a very simple way to address this type of problem.

If $t < \tau < T$ are three times and H_τ is known at time τ, then the time t-value of the payoff H_τ at time T is, by (2.22) and the tower property of conditional expectations,

$$\begin{aligned}
\pi_t &= E\left(e^{-\int_t^T r_s ds} H_\tau | \mathcal{F}_t\right) \\
&= E\left[E\left(e^{-\int_t^T r_s ds} H_\tau | \mathcal{F}_\tau\right) | \mathcal{F}_t\right] \\
&= E\left[e^{-\int_t^\tau r_s ds} H_\tau E\left(e^{-\int_\tau^T r_s ds} | \mathcal{F}_\tau\right) | \mathcal{F}_t\right] \\
&= E\left[e^{-\int_t^\tau r_s ds} H_\tau P(\tau, T) | \mathcal{F}_t\right].
\end{aligned}$$

This implies that the time-t value of the payoff H_τ "payable" at time T (first expression) is equal to the time-t value of the payoff $H_\tau P(\tau, T)$ "payable" at time τ (last expression). The above problem, therefore, can be addressed by acting as if the payoff were anticipated and multiplied by the proper discount factor. Such a result is quite intuitive and actually confirms the basic economic principle according to which receiving on a future date an amount known today is equivalent to receiving today the present value of such an amount.

2.8 Pricing Claims with Multiple Payoffs

It can be interesting to consider also the case of those interest-rate derivatives that have (random) payoffs occurring at different dates. A typical example is a cap that gives its holder a stream of option-like cashflows on some pre-defined dates until the cap maturity. In such cases, assuming there are no early-exercise features, each single cashflow can be priced by taking expectation under the associated forward measure. However, when there are path-dependent features so that we have to resort to Monte Carlo pricing, dealing with different measures is usually quite burdensome and time consuming. In this situations, it is advisable, therefore, to change measure and act as if all cash flows occurred at the same time.

We here notice that, similarly to the previous section case, there is a very simple way of doing so by means of zero-coupon bonds. Indeed, from a basic financial point of view, receiving a known amount today is equivalent to receiving the forward value of this amount at the future date corresponding to the forward maturity. Equivalently, the value at time t of a payoff x "payable" at time $T > t$, with x known at time T, is equal to the value at time t of the payoff $x/P(T,S)$ "payable" at time $S > T$. This is formally proven in the following.

Proposition 2.8.1. *If H is an $\mathcal{F}_T$-measurable random variable, we have the identity:*

$$E[D(t,T)H|\mathcal{F}_t] = E\left[\frac{D(t,S)H}{P(T,S)}|\mathcal{F}_t\right], \tag{2.30}$$

for all $t < T < S$.

Proof. From the tower property of conditional expectations, and remembering that $D(t,S) = D(t,T)D(T,S)$, we immediately have

$$\begin{aligned}
E\left[\frac{D(t,S)H}{P(T,S)}|\mathcal{F}_t\right] &= E\left(E\left[\frac{D(t,T)D(T,S)H}{P(T,S)}|\mathcal{F}_T\right]|\mathcal{F}_t\right) \\
&= E\left[\frac{D(t,T)H}{P(T,S)}E(D(T,S)|\mathcal{F}_T)|\mathcal{F}_t\right] \\
&= E\left[\frac{D(t,T)H}{P(T,S)}P(T,S)|\mathcal{F}_t\right] \\
&= E[D(t,T)H|\mathcal{F}_t].
\end{aligned}$$

□

As an example, we consider n times $T_1 < T_2 \ldots < T_n$ and an interest-rate derivative that at each time T_i pays out the quantity H_i, which is known at time T_i, with no early-exercise features. The derivative price at time $t < T_1$ is therefore

$$\begin{aligned}
\pi_t &= \sum_{i=1}^{n} E\{D(t,T_i)H_i|\mathcal{F}_t\} \\
&= \sum_{i=1}^{n} P(t,T_i)E^{T_i}\{H_i|\mathcal{F}_t\}.
\end{aligned}$$

The previous proposition, however, enables us to consider just one forward measure, precisely that relative to the longest maturity T_n. Indeed, by (2.30),

$$E\{D(t,T_i)H_i|\mathcal{F}_t\} = E\left\{\frac{D(t,T_n)H_i}{P(T_i,T_n)}|\mathcal{F}_t\right\},$$

so that we obtain

$$\begin{aligned}\pi_t &= \sum_{i=1}^{n} E\left\{\frac{D(t,T_n)H_i}{P(T_i,T_n)}|\mathcal{F}_t\right\} \\ &= P(t,T_n)E^{T_n}\left\{\sum_{i=1}^{n}\frac{H_i}{P(T_i,T_n)}|\mathcal{F}_t\right\}.\end{aligned}$$

Therefore, when resorting to Monte Carlo pricing, the last equation can be used to simulate the evolution of the underlying variables under a unique measure, namely the T_n-forward measure, often called the "terminal (forward) measure".

2.9 Foreign Markets and Numeraire Change

In this section we derive the Radon-Nikodym derivative that defines the change of measure between a foreign risk-neutral probability measure and the domestic risk-neutral probability measure. We then interpret this measure change as a change of numeraire. These results can be helpful when pricing multi-currency interest-rate derivatives, as we will show in Chapter 14.

Consider a foreign market where an asset with price X^f is traded. Denote by Q^f the corresponding (foreign) risk-adjusted martingale measure. Assume that the foreign money-market account evolves according to the process B^f. Also consider a domestic market and assume that the domestic money-market account evolves according to the process B and the exchange rate between the two corresponding currencies is modeled through the process $\mathcal{Q}$, in that 1 unit of the foreign currency is worth $\mathcal{Q}_t$ units of domestic currency at time t. Denote by $I\!F = \{\mathcal{F}_t : 0 \leq t \leq T\}$, the filtration generated by all the above processes, with $\mathcal{F}_0$ the trivial sigma-field. Assume that all expectations below are well defined.

Thinking of X^f as a derivative that pays out X^f_T at time T, from the previous sections we know that the arbitrage-free price of X^f at time t in the foreign market is (E^f denotes expectation under Q^f)

$$V^f_t = B^f_t E^f\left\{\frac{X^f_T}{B^f_T}|\mathcal{F}_t\right\},$$

which expressed in terms of the domestic currency becomes

$$V_t = \mathcal{Q}_t B_t^f E^f \left\{ \frac{X_T^f}{B_T^f} | \mathcal{F}_t \right\}.$$

From the perspective of a domestic investor, the asset X^f is perfectly equivalent to a derivative that pays out $X_T^f \mathcal{Q}_T$ at time T. In fact, since X^f is denominated in foreign currency, the actual payoff at time T for a domestic investor that buys this asset is $X_T^f \mathcal{Q}_T$. Therefore, to avoid arbitrage, the arbitrage-free price of the domestic asset $X^f \mathcal{Q}$ at time t must be equal to V_t, the foreign-currency price of X^f (i.e., V_t^f) multiplied by the exchange rate at time t. In formulas, (as usual, E denotes expectation under Q)

$$\mathcal{Q}_t B_t^f E^f \left\{ \frac{X_T^f}{B_T^f} | \mathcal{F}_t \right\} = B_t E \left\{ \frac{X_T^f \mathcal{Q}_T}{B_T} | \mathcal{F}_t \right\}. \tag{2.31}$$

Theorem 2.9.1. *The Radon-Nikodym derivative* $\frac{dQ^f}{dQ}$ *defining the change of measure from the foreign risk-neutral probability measure* Q^f *and the domestic risk-neutral probability measure* Q *is given by*

$$\frac{dQ^f}{dQ} = \frac{\mathcal{Q}_T B_T^f}{\mathcal{Q}_0 B_T}. \tag{2.32}$$

Proof. Combining equation (2.31) at time 0

$$E^f \left\{ \frac{\mathcal{Q}_0 X_T^f}{B_T^f} \right\} = E \left\{ \frac{X_T^f \mathcal{Q}_T}{B_T} \right\}$$

with the following immediate property of the measure change

$$E^f \left\{ \frac{\mathcal{Q}_0 X_T^f}{B_T^f} \right\} = E \left\{ \frac{dQ^f}{dQ} \frac{\mathcal{Q}_0 X_T^f}{B_T^f} \right\},$$

we have that (2.32) gives the right candidate for defining $\frac{dQ^f}{dQ}$. We then prove that this candidate is a positive martingale with mean one. This immediately follows from the definition of Q. In fact, we know that under Q all domestic assets divided by the domestic money-market account are martingales:

$$\begin{aligned} E \left\{ \frac{\mathcal{Q}_T B_T^f}{\mathcal{Q}_0 B_T} | \mathcal{F}_t \right\} &= \frac{1}{\mathcal{Q}_0} E \left\{ \frac{\mathcal{Q}_T B_T^f}{B_T} | \mathcal{F}_t \right\} \\ &= \frac{1}{\mathcal{Q}_0} \frac{\mathcal{Q}_t B_t^f}{B_t}, \end{aligned}$$

since $B^f \mathcal{Q}$ is a domestic asset, which proves the martingale property. As for the expectation being equal to one, just notice that the constant expected value equals the initial value of the martingale, which is trivially seen to be one. □

Corollary 2.9.1. *Changing the measure from Q^f to Q is equivalent to changing the numeraire from B^f to $\frac{B}{\mathcal{Q}}$.*

Proof. Following the discussions seen in the change-of-numeraire-toolkit section, it is enough to remember that the Radon-Nikodym derivative when changing the numeraire from B^f to any U is

$$\frac{dQ^U}{dQ^f} = \frac{U_T}{U_0 B_T^f}$$

and that

$$\frac{dQ}{dQ^f} = \frac{\mathcal{Q}_0 B_T}{\mathcal{Q}_T B_T^f}.$$

□

This corollary tells us that moving from the foreign measure Q^f to the domestic measure Q amounts to changing the numeraire from the foreign bank account to the domestic bank account translated into foreign currency through the related exchange rate.

The above result also follows from the definition of a martingale measure associated with a numeraire. In fact, under the measure associated with the numeraire $B/\mathcal{Q}$, for any (foreign) traded asset Y^f, the process

$$\left\{ \frac{Y_t^f \mathcal{Q}_t}{B_t} : 0 \leq t \leq T \right\}$$

is a martingale. Analogously, under the measure associated with the numeraire B, for any (domestic) traded asset Y, the process

$$\left\{ \frac{Y_t}{B_t} = \frac{Y_t}{\mathcal{Q}_t} \frac{\mathcal{Q}_t}{B_t} : 0 \leq t \leq T \right\}$$

is a martingale. Then, we just have to notice that Y^f is a foreign traded asset if and only if $Y^f \mathcal{Q}$ is a domestic traded asset and Y is a domestic traded asset if and only if $Y/\mathcal{Q}$ is a foreign traded asset.

Denote now by $P^f(t,T)$ the time-t discount factor for maturity T in the foreign economy. When moving from the foreign T-forward measure Q_f^T to the domestic T-forward measure Q^T, a result analogous to that of the previous corollary holds.

Theorem 2.9.2. *Changing the measure from Q_f^T to Q^T is equivalent to changing the numeraire from $P^f(\cdot,T)$ to $\frac{P(\cdot,T)}{\mathcal{Q}}$.*

Proof. We notice that, by (2.32),

$$\frac{dQ_f^T}{dQ^T} = \frac{dQ_f^T}{dQ^f}\frac{dQ^f}{dQ}\frac{dQ}{dQ^T} = \frac{1}{P^f(0,T)B_T^f}\frac{\mathcal{Q}_T B_T^f}{\mathcal{Q}_0 B_T} P(0,T)B_T,$$

and remember that, when moving from the numeraire N to U, we have

$$\frac{dQ^U}{dQ^N} = \frac{U_T N_0}{U_0 N_T}.$$

□

Moving from the foreign measure Q_f^T to the domestic measure Q^T, therefore, amounts to changing the numeraire from the foreign T-maturity zero-coupon bond to the domestic T-maturity zero-coupon bond converted into foreign currency.

Part II

FROM SHORT RATE MODELS TO HJM

3. One-factor short-rate models

"It will be short, the interim is mine.
And a man's life is no more than to say 'one' "
Hamlet, V.2

"Pilder on!" ['Piloter' on!]
Koji Kabuto, assuming control of Mazinger Z (1972)

3.1 Introduction and Guided Tour

The theory of interest-rate modeling was originally based on the assumption of specific one-dimensional dynamics for the instantaneous spot rate process r. Modeling directly such dynamics is very convenient since all fundamental quantities (rates and bonds) are readily defined, by no-arbitrage arguments, as the expectation of a functional of the process r. Indeed, the existence of a risk-neutral measure implies that the arbitrage-free price at time t of a contingent claim with payoff H_T at time T is given by

$$H_t = E_t\left\{D(t,T)\; H_T\right\} = E_t\left\{e^{-\int_t^T r(s)ds} H_T\right\}, \tag{3.1}$$

with E_t denoting the time t-conditional expectation under that measure. In particular, the zero-coupon-bond price at time t for the maturity T is characterized by a unit amount of currency available at time T, so that $H_T = 1$ and we obtain

$$P(t,T) = E_t\left\{e^{-\int_t^T r(s)ds}\right\}. \tag{3.2}$$

From this last expression it is clear that whenever we can characterize the distribution of $e^{-\int_t^T r(s)ds}$ in terms of a chosen dynamics for r, conditional on the information available at time t, we are able to compute bond prices P. As we have seen earlier in Chapter 1, from bond prices all kind of rates are available, so that indeed the whole zero-coupon curve is characterized in terms of distributional properties of r.

The pioneering approach proposed by Vasicek (1977) was based on defining the instantaneous-spot-rate dynamics under the real-world measure. His derivation of an arbitrage-free price for any interest-rate derivative followed from using the basic Black and Scholes (1973) arguments, while taking into account the non-tradable feature of interest rates.

The construction of a suitable locally-riskless portfolio, as in Black and Scholes (1973), leads to the existence of a stochastic process that only depends on the current time and instantaneous spot rate and not on the maturities of the claims constituting the portfolio. Such process, which is commonly referred to as *market price of risk*, defines a Girsanov change of measure from the real-world measure to the risk-neutral one also in case of more general dynamics than Vasicek's. Precisely, let us assume that the instantaneous spot rate evolves under the real-world measure Q_0 according to

$$dr(t) = \mu(t, r(t))dt + \sigma(t, r(t))dW^0(t),$$

where μ and σ are well-behaved functions and W^0 is a Q_0-Brownian motion. It is possible to show[1] the existence of a stochastic process λ such that if

$$dP(t,T) = \mu^T(t, r(t))dt + \sigma^T(t, r(t))dW^0(t),$$

then

$$\frac{\mu^T(t, r(t)) - r(t)P(t,T)}{\sigma^T(t, r(t))} = \lambda(t)$$

for each maturity T, with λ that may depend on r but not on T. Moreover, there exists a measure Q that is equivalent to Q_0 and is defined by the Radon-Nikodym derivative

$$\left.\frac{dQ}{dQ_0}\right|_{\mathcal{F}_t} = \exp\left(-\frac{1}{2}\int_0^t \lambda^2(s)ds - \int_0^t \lambda(s)dW^0(s)\right),$$

where $\mathcal{F}_t$ is the σ-field generated by r up to time t. As a consequence, the process r evolves under Q according to

$$dr(t) = [\mu(t, r(t)) - \lambda(t)\sigma(t, r(t))]dt + \sigma(t, r(t))dW(t),$$

where $W(t) = W^0(t) + \int_0^t \lambda(s)ds$ is a Brownian motion under Q.[2]

Let us comment briefly on this setup. The above equation for dP actually expresses the bond-price dynamics in terms of the short rate r. It expresses how the bond price P evolves over time. Now recall that r is the instantaneous-return rate of a risk-free investment, so that the difference $\mu - r$ represents a difference in returns. It tells us how much better we are doing with respect to the risk-free case, i.e. with respect to putting our money in a riskless bank account. When we divide this quantity by σ^T, we are dividing by the amount of risk we are subject to, as measured by the bond-price volatility σ^T. This is why λ is referred to as "market price of risk". An alternative term could be "excess return with respect to a risk-free investment per unit

[1] See for instance Björk (1997).

[2] The Radon-Nikodym derivative and the Girsanov change of measure are briefly reviewed in Appendix C. For a formal treatment see Musiela and Rutkowski (1998).

of risk". The crucial observation is that in order to specify completely the model, we have to provide λ. In effect, the market price of risk λ connects the real-world measure to the risk-neutral measure as the main ingredient in the mathematical object dQ/dQ_0 expressing the connection between these two "worlds". The way of moving from one world to the other is characterized by our choice of λ. However, if we are just concerned with the pricing of (interest-rate) derivatives, we can directly model the rate dynamics under the measure Q, so that λ will be implicit in our dynamics. We put ourselves in the world Q and we do not bother about the way of moving to the world Q_0. Then we would be in troubles only if we needed to move under the objective measure, but for pricing derivatives, the objective measure is not necessary, so that we can safely ignore it. Indeed, the value of the model parameters under the risk-neutral measure Q is what really matters in the pricing procedure, given also that the zero-coupon bonds are themselves derivatives under the above framework. We have then decided to present all the models we consider in this chapter under the risk-neutral measure, even when their original formulation was under the measure Q_0. We will hint at the relationship between the two measures only occasionally, and will explore the interaction of the dynamics under the two different measures in the Vasicek case as an illustration.

We start the chapter by introducing, in chronological order, some classical short-rate models: the Vasicek (1977) model, the Dothan (1978) model and the Cox, Ingersoll and Ross (1985) model followed by what we refer to as the Exponential-Vasicek model. These are all endogenous term-structure models, meaning that the current term structure of rates is an output rather than an input of the model. Their introduction is justified both for historical reasons and for easing the exposition of the more general models that follow. For example, the Vasicek model will be defined, under the risk-neutral measure Q, by the dynamics

$$dr(t) = k[\theta - r(t)]dt + \sigma dW(t), \quad r(0) = r_0 .$$

This dynamics has some peculiarities that make the model attractive. The equation is linear and can be solved explicitly, the distribution of the short rate is Gaussian, and both the expressions and the distributions of several useful quantities related to the interest-rate world are easily obtainable. Besides, the endogenous nature of the model is now clear. Since the bond price $P(t,T) = E_t\left\{e^{-\int_t^T r(s)ds}\right\}$ can be computed as a simple expression depending on k, θ, σ and $r(t)$, once the function $T \mapsto P(t,T;k,\theta,\sigma,r(t))$ is known, we know the whole interest-rate curve at time t. This means that, if $t=0$ is the initial time, the initial interest rate curve is an output of the model, depending on the parameters k, θ, σ in the dynamics (and on the initial condition r_0). However, this model features also some drawbacks. For example, rates can assume negative values with positive probability. What we are actually trying to point out with this initial hint at the Vasicek model is that the choice of a particular dynamics has several important consequences, which must be kept in mind when designing or choosing a particular short-rate model. A typical comparison is for example with the Cox Ingersoll Ross (CIR) model. Assume we take as dynamics of r the following square root process:

$$dr(t) = k[\theta - r(t)]dt + \sigma\sqrt{r(t)}dW(t), \quad r(0) = r_0 > 0 .$$

For the parameters k, θ and σ ranging in a reasonable region, this model implies positive interest rates, and the instantaneous rate is characterized by a noncentral chi-squared distribution. Moreover, this model maintains a certain degree of analytical tractability. However, the model is less tractable than the Vasicek model, especially as far as the extension to the multifactor case with correlation is concerned (see the following chapter). Therefore, the CIR dynamics has both some advantages and disadvantages with respect to the Vasicek model. In particular, when choosing a model, one should pose the following questions:

- Does the dynamics imply positive rates, i.e., $r(t) > 0$ a.s. for each t?
- What distribution does the dynamics imply for the short rate r? Is it, for instance, a fat-tailed distribution?
- Are bond prices $P(t,T) = E_t\left\{e^{-\int_t^T r(s)ds}\right\}$ (and therefore spot rates, forward rates and swap rates) explicitly computable from the dynamics?
- Are bond-option (and cap, floor, swaption) prices explicitly computable from the dynamics?
- Is the model mean reverting, in the sense that the expected value of the short rate tends to a constant value as time grows towards infinity, while its variance does not explode?
- How do the volatility structures implied by the model look like?
- Does the model allow for explicit short-rate dynamics under the forward measures?
- How suited is the model for Monte Carlo simulation?
- How suited is the model for building recombining lattices?
- Does the chosen dynamics allow for historical estimation techniques to be used for parameter estimation purposes?

These points are essential for the understanding of the theoretical and practical implications of any interest rate model. In this chapter, therefore, we will try to give an answer to the questions above for each considered short-rate model. Of course, the richness of details will vary according to the importance and practical usefulness of the model.

A classic problem with the above models is their endogenous nature. If we have the initial zero-coupon bond curve $T \mapsto P^M(0,T)$ from the market, and we wish our model to incorporate this curve, we need forcing the model parameters to produce a model curve as close as possible to the market curve. For example, again in the Vasicek case, we need to run an optimization to find the values of k, θ and σ such that the model initial curve $T \mapsto P(0,T; k, \theta, \sigma, r(0))$ is as close as possible to the market curve

$T \mapsto P^M(0,T)$. Although the values $P^M(0,T)$ are actually observed only at a finite number of maturities $P^M(0,T_i)$, three parameters are not enough to reproduce satisfactorily a given term structure. Moreover, some shapes of the zero-coupon curve $T \mapsto L^M(0,T)$ (like an inverted shape) can never be obtained with the Vasicek model, no matter the values of the parameters in the dynamics that are chosen.

The point of this digression is making clear that these models are quite hopeless: they cannot reproduce satisfactorily the initial yield curve, and so speaking of volatility structures and realism in other respects becomes partly pointless.

To improve this situation, exogenous term structure models are usually considered. Such models are built by suitably modifying the above endogenous models. The basic strategy that is used to transform an endogenous model into an exogenous model is the inclusion of "time-varying" parameters. Typically, in the Vasicek case, one does the following:

$$dr(t) = k[\theta - r(t)]dt + \sigma dW(t) \longrightarrow dr(t) = k[\vartheta(t) - r(t)]dt + \sigma dW(t) .$$

Now the function of time $\vartheta(t)$ can be defined in terms of the market curve $T \mapsto L^M(0,T)$ in such a way that the model reproduces exactly the curve itself at time 0.

In the chapter we then consider the description of some major exogenous term-structure models, i.e., models where the current term structure of rates is exogenously given. We analyze: the Hull and White (1990) extended Vasicek model, possible extensions of the Cox, Ingersoll and Ross (1985) model, the Black and Karasinski (1991) model and some humped-volatility short-rate models. Finally, we show how to extend a general time-homogeneous model so as to exactly reproduce the initial term structure of rates, with a special focus on the extensions of the Cox, Ingersoll and Ross (1985) model, the Dothan (1978) model and the Exponential-Vasicek model. The extension is based on an external shift that preserves the analytical tractability, if any, of the original model. In the context of the CIR model we present also a jump diffusion extension of the model based on Brigo and El-Bachir (2005), which can be extended further with the external shift technique. Jump diffusion processes are important in finance and it is a good idea to have at least one example from this family of models in the book.

We investigate the analytical features of each model, providing analytical formulas for zero-coupon bonds, options on zero-coupon bonds, and hence caps and floors, whenever they exist. When possible, we also explicitly write the short-rate dynamics under the forward-adjusted measure and, for few selected models, we illustrate how to build an approximating trinomial tree. When just the price of a European call is provided, the price of the corresponding put can be obtained through the put-call parity for bond options. Indeed, if the options have maturity T, strike K and are written on a zero coupon bond maturing at time τ, their prices at time t satisfy

$$\mathbf{ZBC}(t,T,\tau,K)+KP(t,T)=\mathbf{ZBP}(t,T,\tau,K)+P(t,\tau), \tag{3.3}$$

so that

$$\mathbf{ZBP}(t,T,\tau,K)=\mathbf{ZBC}(t,T,\tau,K)-P(t,\tau)+KP(t,T). \tag{3.4}$$

We then devote a section to the analytical pricing of coupon bearing bond options, and hence European swaptions, and we show how to price path-dependent derivatives through a Monte Carlo procedure.

We conclude the chapter by reporting some empirical results concerning the Black and Karasinski (1991) model and the above extensions of the Cox, Ingersoll and Ross (1985) model and the Exponential-Vasicek model. We first show and comment the cap volatility curves and swaption volatility surfaces that are implied by these models. We then consider a specific example of the models calibration to real market data and compare the resulting fitting qualities.

Throughout the chapter, we assume that the term structure of discount factors that is currently observed in the market is given by the sufficiently-smooth function $t \mapsto P^M(0,t)$. We then denote by $f^M(0,t)$ the market instantaneous forward rates at time 0 for a maturity t as associated with the bond prices $\{P^M(0,t) : t \geq 0\}$, i.e.,

$$f^M(0,t) = -\frac{\partial \ln P^M(0,t)}{\partial t}.$$

The relevant properties of the instantaneous short rate models we will analyze in this chapter are summarized in Table 3.1, where V, CIR, D, EV, HW, BK, MM, CIR++, EVV stand respectively for the Vasicek (1977) model, the Cox, Ingersoll and Ross (1985) model, the Dothan (1978) model, the Exponential Vasicek model, the Hull and White (1990) model, the Black and Karasinski (1991) model, the Mercurio and Moraleda (2000) model, the CIR++ model and the Extended Exponential Vasicek model; N and Y stand respectively for "No" and "Yes", whereas Y* means that rates are positive under suitable conditions for the deterministic function φ; $\mathcal{N}$, $L\mathcal{N}$, $NC\chi^2$, $SNC\chi^2$, $SL\mathcal{N}$ denote respectively normal, lognormal, noncentral χ^2, shifted noncentral χ^2 and shifted lognormal distributions; AB(O) stands for Analytical Bond (Option) price.

Finally, we recall that in Table 3.1 we did not include the jump-diffusion extensions such as the JCIR model we introduce later on, but only the purely diffusion models.

Model	Dynamics	$r>0$	$r\sim$	AB	AO
V	$dr_t = k[\theta - r_t]dt + \sigma dW_t$	N	$\mathcal{N}$	Y	Y
CIR	$dr_t = k[\theta - r_t]dt + \sigma\sqrt{r_t}dW_t$	Y	$NC\chi^2$	Y	Y
D	$dr_t = ar_t dt + \sigma r_t dW_t$	Y	$L\mathcal{N}$	Y	N
EV	$dr_t = r_t\left[\eta - a\ln r_t\right]dt + \sigma r_t dW_t$	Y	$L\mathcal{N}$	N	N
HW	$dr_t = k[\theta_t - r_t]dt + \sigma dW_t$	N	$\mathcal{N}$	Y	Y
BK	$dr_t = r_t\left[\eta_t - a\ln r_t\right]dt + \sigma r_t dW_t$	Y	$L\mathcal{N}$	N	N
MM	$dr_t = r_t\left[\eta_t - \left(\lambda - \frac{\gamma}{1+\gamma t}\right)\ln r_t\right]dt + \sigma r_t dW_t$	Y	$L\mathcal{N}$	N	N
CIR++	$r_t = x_t + \varphi_t,\ \ dx_t = k[\theta - x_t]dt + \sigma\sqrt{x_t}dW_t$	Y*	$SNC\chi^2$	Y	Y
EEV	$r_t = x_t + \varphi_t,\ dx_t = x_t[\eta - a\ln x_t]dt + \sigma x_t dW_t$	Y*	$SL\mathcal{N}$	N	N

Table 3.1. Summary of instantaneous short rate models.

3.2 Classical Time-Homogeneous Short-Rate Models

Has fate become so imaginatively bankrupt that I am now doomed to naught but tedious recapitulation of the past?
Adam Warlock in "Rune" 1, 1995, Malibu Comics.

The first instantaneous short rate models being proposed in the financial literature were time-homogeneous, meaning that the assumed short rate dynamics depended only on constant coefficients. The success of models like that of Vasicek (1977) and that of Cox, Ingersoll and Ross (1985) was mainly due to their possibility of pricing analytically bonds and bond options. However, as observed in the introduction, these models produce an endogenous term structure of interest rates,[3] in that the initial term structure of (e.g. continuously-compounded) rates $T \mapsto R(0,T) = -(\ln P(0,T))/T$ does not necessarily match that observed in the market, no matter how the model parameters are chosen.

[3] This is the reason why they have been also referred to as "endogenous term structure models".

Moreover, as we pointed out earlier, the small number of model parameters prevents a satisfactory calibration to market data and even the zero-coupon curve is quite likely to be badly reproduced, also because some typical shapes, like that of an inverted yield curve, may not be reproduced by the model.

In this section we present three classical time-homogeneous short-rate models, namely the Vasicek (1977), the Dothan (1978) and the Cox, Ingersoll and Ross (1985) models, and we finally introduce the Exponential-Vasicek model. As already mentioned, these models are described not only for their historical importance but also for letting us treat in a clearer way the extensions we shall illustrate in the sequel.

As to the analytical tractability of the first three models, we want to remark the following. The original derivation of the explicit formulas for bond prices was based on solving the PDE that, by no-arbitrage arguments, must be satisfied by the bond price process. In the presence of a Gaussian distribution, however, we can price bonds also by directly computing the expectation (3.2). The derivation of explicit formulas for bond options in the first and third model relied instead on a suitable change of the underlying probability measure. Indeed, once the distribution of the instantaneous short rate is known under the desired forward measure, any payoff can be priced by calculating the expectation (2.24).

Finally, in the Vasicek case we will study a few facts concerning possible uses of the dynamics under the objective measure, in order to give a feeling for the kind of considerations involved in combining these pricing and hedging short-rate models with historical data. We develop this theme only for the Vasicek model because in the rest of the book we will not use historical estimation techniques.

3.2.1 The Vasicek Model

Vasicek (1977) assumed that the instantaneous spot rate under the real-world measure evolves as an Ornstein-Uhlenbeck process with constant coefficients. For a suitable choice of the market price of risk (more on this later, see equation (3.11)), this is equivalent to assume that r follows an Ornstein-Uhlenbeck process with constant coefficients under the risk-neutral measure as well, that is

$$dr(t) = k[\theta - r(t)]dt + \sigma dW(t), \quad r(0) = r_0, \tag{3.5}$$

where r_0, k, θ and σ are positive constants.

Integrating equation (3.5), we obtain, for each $s \leq t$,

$$r(t) = r(s)e^{-k(t-s)} + \theta\left(1 - e^{-k(t-s)}\right) + \sigma\int_s^t e^{-k(t-u)}dW(u), \tag{3.6}$$

so that $r(t)$ conditional on $\mathcal{F}_s$ is normally distributed with mean and variance given respectively by

$$E\{r(t)|\mathcal{F}_s\} = r(s)e^{-k(t-s)} + \theta\left(1 - e^{-k(t-s)}\right)$$
$$\mathrm{Var}\{r(t)|\mathcal{F}_s\} = \frac{\sigma^2}{2k}\left[1 - e^{-2k(t-s)}\right]. \tag{3.7}$$

This implies that, for each time t, the rate $r(t)$ can be negative with positive probability. The possibility of negative rates is indeed a major drawback of the Vasicek model. However, the analytical tractability that is implied by a Gaussian density is hardly achieved when assuming other distributions for the process r.

As a consequence of (3.7), the short rate r is mean reverting, since the expected rate tends, for t going to infinity, to the value θ. The fact that θ can be regarded as a long term average rate could be also inferred from the dynamics (3.5) itself. Notice, indeed, that the drift of the process r is positive whenever the short rate is below θ and negative otherwise, so that r is pushed, at every time, to be closer on average to the level θ.

The price of a pure-discount bond can be derived by computing the expectation (3.2). We obtain

$$P(t,T) = A(t,T)e^{-B(t,T)r(t)}, \tag{3.8}$$

where

$$A(t,T) = \exp\left\{\left(\theta - \frac{\sigma^2}{2k^2}\right)[B(t,T) - T + t] - \frac{\sigma^2}{4k}B(t,T)^2\right\}$$
$$B(t,T) = \frac{1}{k}\left[1 - e^{-k(T-t)}\right].$$

If we fix a maturity T, the change-of-numeraire toolkit developed in Section 2.3, and formula (2.12) in particular (with $S_t = B(t)$, the bank-account numeraire, $U_t = P(t,T)$, the T-bond numeraire, and $X_t = r_t$) imply that under the T-forward measure Q^T

$$dr(t) = [k\theta - B(t,T)\sigma^2 - kr(t)]dt + \sigma dW^T(t), \tag{3.9}$$

where the Q^T-Brownian motion W^T is defined by

$$dW^T(t) = dW(t) + \sigma B(t,T)dt,$$

so that, for $s \le t \le T$,

$$r(t) = r(s)e^{-k(t-s)} + M^T(s,t) + \sigma\int_s^t e^{-k(t-u)}dW^T(u),$$

with

$$M^T(s,t) = \left(\theta - \frac{\sigma^2}{k^2}\right)\left(1 - e^{-k(t-s)}\right) + \frac{\sigma^2}{2k^2}\left[e^{-k(T-t)} - e^{-k(T+t-2s)}\right].$$

Therefore, under Q^T, the transition distribution of $r(t)$ conditional on $\mathcal{F}_s$ is still normal with mean and variance given by

$$E^T\{r(t)|\mathcal{F}_s\} = r(s)e^{-k(t-s)} + M^T(s,t)$$
$$\mathrm{Var}^T\{r(t)|\mathcal{F}_s\} = \frac{\sigma^2}{2k}\left[1 - e^{-2k(t-s)}\right].$$

The price at time t of a European option with strike X, maturity T and written on a pure discount bond maturing at time S has been derived by Jamshidian (1989). Using the known distribution of $r(T)$ under Q^T, the calculation of the expectation (3.1), where $H_T = (P(T,S)-X)^+$, yields, through the general formula (D.2),

$$\mathbf{ZBO}(t,T,S,X) = \omega\left[P(t,S)\Phi(\omega h) - XP(t,T)\Phi(\omega(h-\sigma_p))\right], \tag{3.10}$$

where $\omega = 1$ for a call and $\omega = -1$ for a put, $\Phi(\cdot)$ denotes the standard normal cumulative distribution function, and

$$\sigma_p = \sigma\sqrt{\frac{1-e^{-2k(T-t)}}{2k}}B(T,S),$$
$$h = \frac{1}{\sigma_p}\ln\frac{P(t,S)}{P(t,T)X} + \frac{\sigma_p}{2}.$$

Objective measure dynamics and historical estimation. We can consider the objective measure dynamics of the Vasicek model as a process of the form

$$dr(t) = [k\,\theta - (k + \lambda\,\sigma)r(t)]dt + \sigma dW^0(t), \quad r(0) = r_0\,, \tag{3.11}$$

where λ is a new parameter, contributing to the market price of risk. Compare this Q_0 dynamics to the Q-dynamics (3.5). Notice that for $\lambda = 0$ the two dynamics coincide, i.e. there is no difference between the risk neutral world and the objective world. More generally, the above Q_0-dynamics is expressed again as a linear Gaussian stochastic differential equation, although it depends on the new parameter λ. This is a tacit assumption on the form of the market price of risk process. Indeed, requiring that the dynamics be of the same nature under the two measures, imposes a Girsanov change of measure of the following kind to go from (3.5) to (3.11) :

$$\left.\frac{dQ}{dQ_0}\right|_{\mathcal{F}_t} = \exp\left(-\frac{1}{2}\int_0^t \lambda^2\, r(s)^2 ds + \int_0^t \lambda\, r(s)dW^0(s)\right)$$

(see Appendix C for an introduction to Girsanov's theorem).

In other terms, we are assuming that the market price of risk process $\lambda(t)$ has the functional form

$$\lambda(t) = \lambda\, r(t)$$

in the short rate. Of course, in general there is no reason why this should be the case. However, under this choice we obtain a short rate process that is tractable under both measures.[4]

It is clear why tractability under the risk-neutral measure is a desirable property: claims are priced under that measure, so that the possibility to compute expectations in a tractable way with the Q-dynamics (3.5) is important. Yet, why do we find it desirable to have a tractable dynamics under Q_0 too? In order to answer this question, suppose for a moment that we are provided with a series $r_0, r_1, r_2, \ldots, r_n$ of daily observations of a proxy of $r(t)$ (say a monthly rate, $r(t) \approx L(t, t+1m)$), and that we wish to incorporate information from this series in our model. We can estimate the model parameters on the basis of this daily series of data. However, data are collected in the real world, and their statistical properties characterize the distribution of our interest-rate process $r(t)$ under the objective measure Q_0. Therefore, what is to be estimated from historical observations is the Q_0 dynamics. The estimation technique can provide us with estimates for the objective parameters k, λ, θ and σ, or more precisely for combinations thereof.

On the other hand, prices are computed through expectations under the risk-neutral measure. When we observe prices, we observe expectations under the measure Q. Therefore, when we calibrate the model to derivative prices we need to use the Q dynamics (3.5), thus finding the parameters k, θ and σ involved in the Q-dynamics.

We could then combine the two approaches. For example, since the diffusion coefficient is the same under the two measures, we might estimate σ from historical data through a maximum-likelihood estimator, while finding k and θ through calibration to market prices. However, this procedure may be necessary when very few prices are available. Otherwise, it might be used to deduce historically a σ which can be used as initial guess when trying to find the three parameters that match the market prices of a given set of instruments.

We conclude the section by presenting the maximum-likelihood estimator for the Vasicek model. Rewrite the dynamics (3.11) as

$$dr(t) = [b - ar(t)]dt + \sigma dW^0(t), \tag{3.12}$$

with b and a suitable constants. As usual, by integration we obtain, between two any instants s and t,

$$r(t) = r(s)e^{-a(t-s)} + \frac{b}{a}(1 - e^{-a(t-s)}) + \sigma \int_s^t e^{-a(t-u)} dW^0(u). \tag{3.13}$$

As noticed earlier, conditional on $\mathcal{F}_s$, the variable $r(t)$ is normally distributed with mean $r(s)e^{-a(t-s)} + \frac{b}{a}(1 - e^{-a(t-s)})$ and variance $\frac{\sigma^2}{2a}(1 - e^{-2a(t-s)})$.

[4] Indeed, the market price of risk under the Vasicek model is usually chosen to be constant, i.e., $\lambda(t) = \lambda$. However, this is just another possible formulation.

It is natural to estimate the following functions of the parameters: $\beta := b/a$, $\alpha := e^{-a\delta}$ and $V^2 = \frac{\sigma^2}{2a}(1 - e^{-2a\delta})$, where δ denotes the time-step of the observed proxies $r_0, r_1, \ldots, r_n$ of r (typically $\delta = 1$ day). The maximum likelihood estimators for α, β and V^2 are easily derived as

$$\widehat{\alpha} = \frac{n\sum_{i=1}^n r_i r_{i-1} - \sum_{i=1}^n r_i \sum_{i=1}^n r_{i-1}}{n\sum_{i=1}^n r_{i-1}^2 - \left(\sum_{i=1}^n r_{i-1}\right)^2}, \tag{3.14}$$

$$\widehat{\beta} = \frac{\sum_{i=1}^n [r_i - \widehat{\alpha} r_{i-1}]}{n(1-\widehat{\alpha})}, \tag{3.15}$$

$$\widehat{V^2} = \frac{1}{n}\sum_{i=1}^n \left[r_i - \widehat{\alpha} r_{i-1} - \widehat{\beta}(1-\widehat{\alpha})\right]^2. \tag{3.16}$$

The estimated quantities give complete information on the δ-transition probability for the process r under Q_0, thus allowing for example simulations at one-day spaced future discrete time instants.

3.2.2 The Dothan Model

In his original paper, Dothan started from a driftless geometric Brownian motion as short-rate process under the objective probability measure Q_0:

$$dr(t) = \sigma r(t) dW^0(t), \quad r(0) = r_0,$$

where r_0 and σ are positive constants.

Subsequently, Dothan introduced a constant market price of risk, which is equivalent to directly assuming a risk-neutral dynamics of type

$$dr(t) = a r(t) dt + \sigma r(t) dW(t), \tag{3.17}$$

where a is a real constant, thus yielding a continuous-time version of the Rendleman and Bartter (1980) model.

The dynamics (3.17) are easily integrated as follows

$$r(t) = r(s) \exp\left\{\left(a - \frac{1}{2}\sigma^2\right)(t-s) + \sigma(W(t) - W(s))\right\}, \tag{3.18}$$

for $s \le t$. Hence, $r(t)$ conditional on $\mathcal{F}_s$ is lognormally distributed with mean and variance given by

$$\begin{aligned} E\{r(t)|\mathcal{F}_s\} &= r(s) e^{a(t-s)}, \\ \mathrm{Var}\{r(t)|\mathcal{F}_s\} &= r^2(s) e^{2a(t-s)}\left(e^{\sigma^2(t-s)} - 1\right). \end{aligned} \tag{3.19}$$

The lognormal distribution implies that $r(t)$ is always positive for each t, so that a main drawback of the Vasicek (1977) model is here addressed. However, as we can easily infer from (3.19), the process (3.17) is mean reverting if and only if $a < 0$ with the mean-reversion level that must be necessarily equal to zero.[5] This is a restriction on mean reversion that will be addressed by the exponential Vasicek model.

The Dothan (1978) model is the only lognormal short rate model in the literature with analytical formulas for pure discount bonds. This is the key feature that led us to consider such model and the relative extension we shall propose in a later section.

The zero-coupon bond price derived by Dothan is given by

$$P(t,T) = \frac{\bar{r}^p}{\pi^2} \int_0^\infty \sin(2\sqrt{\bar{r}} \sinh y) \int_0^\infty f(z) \sin(yz) dz dy + \frac{2}{\Gamma(2p)} \bar{r}^p K_{2p}(2\sqrt{\bar{r}}) \tag{3.20}$$

where

$$f(z) = \exp\left[\frac{-\sigma^2(4p^2 + z^2)(T-t)}{8}\right] z \left|\Gamma\left(-p + i\frac{z}{2}\right)\right|^2 \cosh\frac{\pi z}{2},$$

$$\bar{r} = \frac{2r(t)}{\sigma^2},$$

$$p = \frac{1}{2} - a,$$

and K_q denotes the modified Bessel function of the second kind of order q.

Concerning the model analytical tractability we need, however, to remark the following. Though somehow explicit, formula (3.20) is rather complex since it depends on two integrals of functions involving hyperbolic sines and cosines. A double numerical integration is needed so that the advantage of having an "explicit" formula is dramatically reduced. In particular, as far as computational issues are concerned, implementing an approximating tree for the process r may be conceptually easier and not necessarily more time consuming.

The dynamics of the process r under any T-forward measure can be derived by applying formula (2.12).

No analytical formula for an option on a zero-coupon bond is available in this model.

Finally, we need to remark a problem concerning the Dothan model, and lognormal models in general.

Explosion of the bank account for lognormal short-rate models. Assume we are at time 0 and we put one unit of currency in the bank account, for a small time Δt. We know that the expected value of our position at time Δt will be

---

[5] We should stress that mean reversion under Q does not necessarily imply mean reversion under Q_0. However, we can assume that the change of measure does not affect the asymptotic behavior of the process r.

$$E_0 B(\Delta t) = E_0 \left\{ e^{\int_0^{\Delta t} r(s)ds} \right\} \approx \dots$$

Now if Δt is small, we can approximate the integral as follows:

$$\approx E_0 \left\{ e^{\Delta t \; [r(0)+r(\Delta t)]/2} \right\} .$$

Given that the short rate $r(\Delta t)$ is lognormally distributed, we face an expectation of the type

$$E_0 \left\{ \exp(\exp(Y)) \right\}$$

where Y is normally distributed. It is easy to see that such an expectation is infinite, so that we conclude

$$E_0 \left\{ B(\Delta t) \right\} = E_0 \left\{ e^{\int_0^{\Delta t} r(s)ds} \right\} = \infty .$$

This means that in an arbitrarily small time we can make infinite money on average starting from one unit of currency. This drawback is common to all models where r is lognormally distributed. The Black Karasinski model to be introduced later on, the Dothan model, the EV model and their extensions that will be explored in the following, all share this problem. As a consequence, the price of a Eurodollar future is also infinite for all these models, too. However, this explosion problem is partially overcome when using an approximating tree, because one deals with a finite number of states, and hence with finite expectations. Since these models are always applied via trees in practice, this drawback's impact is usually less dramatic than one would expect in the first place. The problem of explosion in these models where the *continuously-compounded* instantaneous rate r is modeled through a lognormal process has been studied by Sandmann and Sondermann (1997), who observe that the problem of explosion can be avoided by modeling rates with a *strictly-positive compounding period* to be lognormal instead.

3.2.3 The Cox, Ingersoll and Ross (CIR) Model

The general equilibrium approach developed by Cox, Ingersoll and Ross (1985) led to the introduction of a "square-root" term in the diffusion coefficient of the instantaneous short-rate dynamics proposed by Vasicek (1977). The resulting model has been a benchmark for many years because of its analytical tractability and the fact that, contrary to the Vasicek (1977) model, the instantaneous short rate is always positive.

The model formulation under the risk-neutral measure Q is

$$dr(t) = k(\theta - r(t))dt + \sigma\sqrt{r(t)}\,dW(t), \quad r(0) = r_0, \tag{3.21}$$

with r_0, k, θ, σ positive constants. The condition

$$2k\theta > \sigma^2$$

has to be imposed to ensure that the origin is inaccessible to the process (3.21), so that we can grant that r remains positive.

We now consider a little digression on a tractable form for the market price of risk in this model. If we need to model the objective measure dynamics Q_0 of the model, it is a good idea to adopt the following formulation:

$$dr(t) = [k\theta - (k + \lambda\sigma)r(t)]dt + \sigma\sqrt{r(t)}\, dW^0(t), \quad r(0) = r_0. \tag{3.22}$$

Notice that in moving from Q to Q_0 the drift has been modified exactly as in the Vasicek case (3.11), and exactly for the same reason: preserving the same structure under the two measures. While in the Vasicek case the change of measure was designed so as to maintain a linear dynamics, here it has been designed so as to maintain a square-root-process structure. Since the diffusion coefficient is different, the change of measure is also different. In particular, we have

$$\left.\frac{dQ}{dQ_0}\right|_{\mathcal{F}_t} = \exp\left(-\frac{1}{2}\int_0^t \lambda^2 r(s)ds + \int_0^t \lambda\sqrt{r(s)}dW^0(s)\right).$$

In other terms, we are assuming the market price of risk process $\lambda(t)$ to be of the particular functional form

$$\lambda(t) = \lambda\sqrt{r(t)}$$

in the short rate. Of course, in general there is no reason why this should be the case. However, under this choice we obtain a short-rate process which is tractable under both measures. As for the Vasicek case, tractability under the objective measure can be helpful for historical-estimation purposes.

Let us now move back to the risk-neutral measure Q. The process r features a noncentral chi-squared distribution. Precisely, denoting by p_Y the density function of the random variable Y,

$$\begin{aligned}
p_{r(t)}(x) &= p_{\chi^2(v,\ \lambda_t)/c_t}(x) = c_t p_{\chi^2(v,\ \lambda_t)}(c_t x),\\
c_t &= \frac{4k}{\sigma^2(1-\exp(-kt))},\\
v &= 4k\theta/\sigma^2,\\
\lambda_t &= c_t r_0 \exp(-kt),
\end{aligned}$$

where the noncentral chi-squared distribution function $\chi^2(\cdot, v, \lambda)$ with v degrees of freedom and non-centrality parameter λ has density

$$p_{\chi^2(v,\ \lambda)}(z) = \sum_{i=0}^{\infty} \frac{e^{-\lambda/2}(\lambda/2)^i}{i!} p_{\Gamma(i+v/2,\ 1/2)}(z),$$

$$p_{\Gamma(i+v/2,\ 1/2)}(z) = \frac{(1/2)^{i+v/2}}{\Gamma(i+v/2)} z^{i-1+v/2} e^{-z/2} = p_{\chi^2(v+2i)}(z),$$

with $p_{\chi^2(v+2i)}(z)$ denoting the density of a (central) chi-squared distribution function with $v+2i$ degrees of freedom.[6]

The mean and the variance of $r(t)$ conditional on $\mathcal{F}_s$ are given by

$$E\{r(t)|\mathcal{F}_s\} = r(s)e^{-k(t-s)} + \theta\left(1 - e^{-k(t-s)}\right),$$
$$\mathrm{Var}\{r(t)|\mathcal{F}_s\} = r(s)\frac{\sigma^2}{k}\left(e^{-k(t-s)} - e^{-2k(t-s)}\right) + \theta\frac{\sigma^2}{2k}\left(1 - e^{-k(t-s)}\right)^2. \tag{3.23}$$

The price at time t of a zero-coupon bond with maturity T is

$$P(t,T) = A(t,T)e^{-B(t,T)r(t)}, \tag{3.24}$$

where

$$A(t,T) = \left[\frac{2h\exp\{(k+h)(T-t)/2\}}{2h + (k+h)(\exp\{(T-t)h\} - 1)}\right]^{2k\theta/\sigma^2},$$
$$B(t,T) = \frac{2(\exp\{(T-t)h\} - 1)}{2h + (k+h)(\exp\{(T-t)h\} - 1)}, \tag{3.25}$$
$$h = \sqrt{k^2 + 2\sigma^2}.$$

Under the risk-neutral measure Q, the bond price dynamics can be easily obtained via Ito's formula:

$$dP(t,T) = r(t)P(t,T)dt - B(t,T)P(t,T)\sigma\sqrt{r(t)}dW(t).$$

By inverting the bond-price formula, thus deriving r from P, we obtain

$$dP(t,T)$$
$$= \frac{1}{B(t,T)}\ln\left[\frac{A(t,T)}{P(t,T)}\right]P(t,T)dt - \sigma P(t,T)\sqrt{B(t,T)\ln\left[\frac{A(t,T)}{P(t,T)}\right]}dW(t)$$

which is better readable as

$$d\ln P(t,T) = \left(\frac{1}{B(t,T)} - \frac{1}{2}\sigma^2 B(t,T)\right)[\ln A(t,T) - \ln P(t,T)]\,dt$$
$$- \sigma\sqrt{B(t,T)\,[\ln A(t,T) - \ln P(t,T)]}dW(t).$$

We notice that the bond-price percentage volatility is not a deterministic function, but depends on the current level of the bond price.

---

[6] A useful identity concerning densities of χ^2 distributions is

$$p_{\chi^2(v,\,\lambda)}(bz) = \exp\left(\tfrac{1}{2}(1-b)(z-\lambda)\right) b^{v/2-1} p_{\chi^2(v,b\lambda)}(z).$$

The price at time t of a European call option with maturity $T > t$, strike price X, written on a zero-coupon bond maturing at $S > T$, and with the instantaneous rate at time t given by $r(t)$, is (see Cox, Ingersoll and Ross (1985))

$$\begin{aligned}
&\mathbf{ZBC}(t,T,S,X)\\
&\quad = P(t,S)\,\chi^2\left(2\bar{r}[\rho+\psi+B(T,S)];\frac{4k\theta}{\sigma^2},\frac{2\rho^2 r(t)\exp\{h(T-t)\}}{\rho+\psi+B(T,S)}\right)\\
&\quad\quad - XP(t,T)\,\chi^2\left(2\bar{r}[\rho+\psi];\frac{4k\theta}{\sigma^2},\frac{2\rho^2 r(t)\exp\{h(T-t)\}}{\rho+\psi}\right)
\end{aligned} \tag{3.26}$$

where

$$\begin{aligned}
\rho &= \rho(T-t) := \frac{2h}{\sigma^2(\exp[h(T-t)]-1)},\\
\psi &= \frac{k+h}{\sigma^2},\\
\bar{r} &= \bar{r}(S-T) := \frac{\ln(A(T,S)/X)}{B(T,S)}.
\end{aligned}$$

By applying formula (2.12) (with $S_t = B(t)$, the bank-account numeraire, $U_t = P(t,T)$, the T-bond numeraire, and $X_t = r_t$), we obtain that the short-rate dynamics under the T-forward measure Q^T is

$$dr(t) = [k\theta - (k + B(t,T)\sigma^2)r(t)]dt + \sigma\sqrt{r(t)}dW^T(t), \tag{3.27}$$

where the Q^T-Brownian motion W^T is defined by

$$dW^T(t) = dW(t) + \sigma B(t,T)\sqrt{r(t)}dt.$$

It is also possible to show that, under Q^T, the distribution of the short rate $r(t)$ conditional on $r(s)$, $s \le t \le T$, is given by

$$\begin{aligned}
p^T_{r(t)|r(s)}(x) &= p_{\chi^2(v,\delta(t,s))/q(t,s)}(x) = q(t,s)p_{\chi^2(v,\delta(t,s))}(q(t,s)x),\\
q(t,s) &= 2[\rho(t-s)+\psi+B(t,T)],\\
\delta(t,s) &= \frac{4\rho(t-s)^2 r(s)e^{h(t-s)}}{q(t,s)}.
\end{aligned} \tag{3.28}$$

This can be shown by differentiating the call-option price with respect to the strike price and by suitable decompositions.

We can also derive the forward-rate dynamics implied by the CIR short-rate dynamics. Indeed, consider the simply-compounded forward rate at time t with expiry T and maturity S, as defined by

$$F(t;T,S) = \frac{1}{\tau(T,S)}\left[\frac{P(t,T)}{P(t,S)} - 1\right],$$

where $\gamma(T,S)$ is the year fraction between T and S. By Ito's formula it is easy to check that under the forward measure Q^S the (driftless) dynamics of $F(t;T,S)$ follows

$$
\begin{aligned}
&dF(t;T,S)\\
&= \sigma \frac{A(t,T)}{A(t,S)}(B(t,S)-B(t,T))\exp\{-(B(t,T)-B(t,S))r(t)\}\sqrt{r(t)}dW^S(t).
\end{aligned}
$$

This last equation can be easily rewritten as

$$
\begin{aligned}
&dF(t;T,S)\\
&= \sigma\left(F(t;T,S)+\frac{1}{\gamma(T,S)}\right)\\
&\cdot\sqrt{(B(t,S)-B(t,T))\ln\left[(\gamma(T,S)F(t;T,S)+1)\frac{A(t,S)}{A(t,T)}\right]}dW^S(t).
\end{aligned}
$$

Notice that this is rather different from the lognormal dynamics assumed for F when pricing caps and floors with the LIBOR market model, where typically

$$dF(t;T,S) = \sigma(t)\ F(t;T,S)\ \ dW^S(t)$$

for a deterministic time function σ.

3.2.4 Affine Term-Structure Models

Affine term-structure models are interest-rate models where the continuously-compounded spot rate $R(t,T)$ is an affine function in the short rate $r(t)$, i.e.

$$R(t,T) = \alpha(t,T) + \beta(t,T)r(t),$$

where α and β are deterministic functions of time. If this happens, the model is said to possess an affine term structure. This relationship is always satisfied when the zero–coupon bond price can be written in the form

$$P(t,T) = A(t,T)e^{-B(t,T)r(t)},$$

since then clearly it suffices to set

$$\alpha(t,T) = -(\ln A(t,T))/(T-t), \quad \beta(t,T) = B(t,T)/(T-t)\ .$$

Both the Vasicek and CIR models we have seen earlier are affine models, since the bond price has an expression of the above form in both cases. The Dothan model is not an affine model.

A computation that will be helpful in the following is the instantaneous absolute volatility of instantaneous forward rates in affine models. Since in general

$$f(t,T) = -\frac{\partial \ln P(t,T)}{\partial T},$$

for affine models we have

$$f(t,T) = -\frac{\partial \ln A(t,T)}{\partial T} + \frac{\partial B(t,T)}{\partial T} r(t),$$

so that clearly

$$df(t,T) = (\cdots)dt + \frac{\partial B(t,T)}{\partial T}\sigma(t,r(t))dW(t),$$

where $\sigma(t,r(t))$ is the diffusion coefficient in the short rate dynamics. It follows that the absolute volatility of the instantaneous forward rate $f(t,T)$ at time t in a short rate model with an affine term structure is

$$\sigma_f(t,T) = \frac{\partial B(t,T)}{\partial T}\sigma(t,r(t)) \ . \tag{3.29}$$

In particular, for the Vasicek and CIR models we can obtain explicit expressions for this quantity by using the known expressions for B. We will see later on why this volatility function is important.

Given that by inspection one sees that the Vasicek and CIR models are affine models whereas Dothan is not, one may wonder whether there is a relationship between the coefficients in the risk-neutral dynamics of the short rate and affinity of the term structure in the above sense. Assume we have a general risk-neutral dynamics for the short rate,

$$dr(t) = b(t,r(t))dt + \sigma(t,r(t))dW(t) \ .$$

We may wonder whether there exist conditions on b and σ such that the resulting model displays an affine term structure. The answer is simply that the coefficients b and σ^2 need be affine functions themselves (see for example Björk (1997) or Duffie (1996)). If the coefficients b and σ^2 are of the form

$$b(t,x) = \lambda(t)x + \eta(t), \quad \sigma^2(t,x) = \gamma(t)x + \delta(t)$$

for suitable deterministic time functions $\lambda, \eta, \gamma, \delta$, then the model has an affine term structure, with α and β (or A and B) above depending on the chosen functions $\lambda, \eta, \gamma, \delta$. The functions A and B can be obtained from the coefficients $\lambda, \eta, \gamma, \delta$ by solving the following differential equations:

$$\frac{\partial}{\partial t}B(t,T) + \lambda(t)B(t,T) - \tfrac{1}{2}\gamma(t)B(t,T)^2 + 1 = 0, \quad B(T,T) = 0,$$
$$\frac{\partial}{\partial t}[\ln A(t,T)] - \eta(t)B(t,T) + \tfrac{1}{2}\delta(t)B(t,T)^2 = 0, \quad A(T,T) = 1.$$

The first equation is a Riccati differential equation that, in general, needs to be solved numerically. However, in the particular cases of Vasicek ($\lambda(t) =$

$-k, \eta(t) = k\theta, \gamma(t) = 0, \delta(t) = \sigma^2)$ or CIR $(\lambda(t) = -k, \eta(t) = k\theta, \gamma(t) = \sigma^2, \delta(t) = 0)$, we have that the equations are explicitly solvable, yielding the expressions for A and B we have written in the previous sections.

Therefore affinity in the coefficients translates into affinity of the term structure. The converse is also true, but in the time-homogeneous case. Precisely, it is possible to prove that if a model has an affine term structure and has time-homogeneous coefficients $b(t,x) = b(x)$ and $\sigma(t,x) = \sigma(x)$, then these coefficients are necessarily affine functions of x:

$$b(x) = \lambda x + \eta, \quad \sigma^2(x) = \gamma x + \delta \; ,$$

for suitable constants $\lambda, \eta, \gamma, \delta$. The relation between affine-term-structure models (ATS), affine-coefficients models (AC) and time-homogeneous models (TH) is visualized through the diagrams displayed in Figure 3.1.

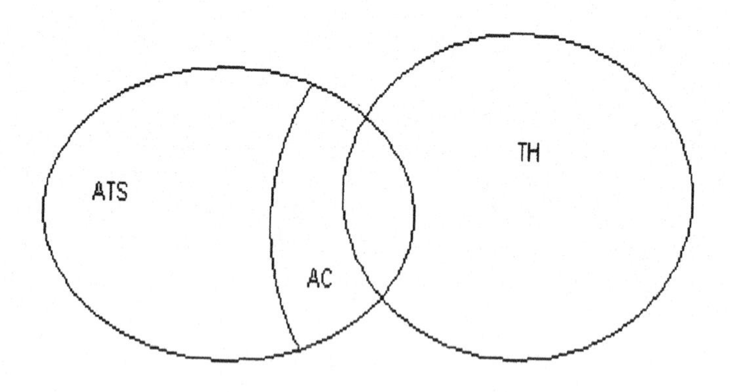

Fig. 3.1. Relation between ATS, AC and TH models.

3.2.5 The Exponential-Vasicek (EV) Model

A natural way to obtain a lognormal short-rate model that is alternative to that of Dothan (1977) is by assuming that the logarithm of r follows an Ornstein-Uhlenbeck process y under the risk-neutral measure Q. Precisely, let y be defined by

$$dy(t) = [\theta - ay(t)]dt + \sigma dW(t), \quad y(0) = y_0,$$

where θ, a and σ are positive constants and y_0 is a real number. Then if we set $r(t) = \exp(y(t))$, for each time t, we have the following dynamics for the short rate:

$$dr(t) = r(t)\left[\theta + \frac{\sigma^2}{2} - a \ln r(t)\right] dt + \sigma r(t) dW(t). \tag{3.30}$$

Since the instantaneous short rate is defined as the exponential of a process that is perfectly equivalent to that of Vasicek (1977), we shall refer to this model as to the Exponential-Vasicek model.

Remembering (3.6), the process r, for each $s \leq t$, is explicitly given by

$$r(t) = \exp\left\{\ln r(s) e^{-a(t-s)} + \frac{\theta}{a}\left(1 - e^{-a(t-s)}\right) + \sigma \int_s^t e^{-a(t-u)} dW(u)\right\},$$

so that $r(t)$ conditional on $\mathcal{F}_s$ is lognormally distributed with first and second moments given respectively by

$$E_s\{r(t)\} = \exp\left\{\ln r(s) e^{-a(t-s)} + \frac{\theta}{a}\left(1 - e^{-a(t-s)}\right) + \frac{\sigma^2}{4a}\left[1 - e^{-2a(t-s)}\right]\right\}$$

$$E_s\{r^2(t)\} = \exp\left\{2 \ln r(s) e^{-a(t-s)} + 2\frac{\theta}{a}\left(1 - e^{-a(t-s)}\right) + \frac{\sigma^2}{a}\left[1 - e^{-2a(t-s)}\right]\right\}.$$

Therefore, contrary to the Dothan (1978) process (3.17), this r is always mean reverting since

$$\lim_{t\to\infty} E\{r(t)|\mathcal{F}_s\} = \exp\left(\frac{\theta}{a} + \frac{\sigma^2}{4a}\right).$$

Notice that also the variance converges to a finite value since

$$\lim_{t\to\infty} \mathrm{Var}\{r(t)|\mathcal{F}_s\} = \exp\left(\frac{2\theta}{a} + \frac{\sigma^2}{2a}\right)\left[\exp\left(\frac{\sigma^2}{2a}\right) - 1\right]. \tag{3.31}$$

The Exponential-Vasicek model does not imply explicit formulas for either pure-discount bonds or options on them. However, when proposing in a later section an extension of this model that exactly fits the current term structure of rates, we will show how to implement fast numerical procedures that make the extension quite appealing in some practical market situations.

The EV model is not an affine term-structure model, as is clear from the criterion given in Section 3.2.4. Finally, since it implies a lognormal distribution for r, the EV model shares the explosion problem that was pointed out in the Dothan case.

3.3 The Hull-White Extended Vasicek Model

The poor fitting of the initial term structure of interest rates implied by the Vasicek model has been addressed by Hull and White in their 1990 and subsequent papers. Ho and Lee (1986) have been the first to propose an exogenous term-structure model as opposed to models that endogenously produce the current term structure of rates. However, their model was based on the assumption of a binomial tree governing the evolution of the entire term structure of rates, and even its continuous-time limit, as derived by Dybvig (1988) and Jamshidian (1988), cannot be regarded as a proper extension of the Vasicek model because of the lack of mean reversion in the short-rate dynamics.

The need for an exact fit to the currently-observed yield curve, led Hull and White to the introduction of a time-varying parameter in the Vasicek model. Notice indeed that matching the model and the market term structures of rates at the current time is equivalent to solving a system with an infinite number of equations, one for each possible maturity. Such a system can be solved in general only after introducing an infinite number of parameters, or equivalently a deterministic function of time.

By considering a further time-varying parameter, Hull and White (1990b) proposed an even more general model that is also able to fit a given term structure of volatilities. Such a model, however, may be somehow dangerous when applied to concrete market situations as we will hint at below. This is the main reason why in this section we stick to the extension where only one parameter, corresponding to the Vasicek θ, is chosen to be a deterministic function of time.

The model we analyze implies a normal distribution for the short-rate process at each time. Moreover, it is quite analytically tractable in that zero-coupon bonds and options on them can be explicitly priced. The Gaussian distribution of continuously-compounded rates then allows for the derivation of analytical formulas and the construction of efficient numerical procedures for pricing a large variety of derivative securities. On the other hand, the possibility of negative rates and the one-factor formulation make the model hardly applicable to concrete pricing problems.

However, the Hull and White extension of the Vasicek model is one of the historically most important interest-rate models, being still nowadays used for risk-management purposes. From a theoretical point of view, moreover, it allows the development of some general tools and procedures that can be easily borrowed by other short-rate models, as we will show in the following sections.

3.3.1 The Short-Rate Dynamics

Hull and White (1990) assumed that the instantaneous short-rate process evolves under the risk-neutral measure according to

$$dr(t) = [\vartheta(t) - a(t)r(t)]dt + \sigma(t)dW(t), \tag{3.32}$$

where ϑ, a and σ are deterministic functions of time.

Such a model can be fitted to the term structure of interest rates and the term structure of spot or forward-rate volatilities. However, if an exact calibration to the current yield curve is a desirable feature, the perfect fitting to a volatility term structure can be rather dangerous and must be carefully dealt with. The reason is two-fold. First, not all the volatilities that are quoted in the market are significant: some market sectors are less liquid, with the associated quotes that may be neither informative nor reliable. Second, the future volatility structures implied by (3.32) are likely to be unrealistic in that they do not conform to typical market shapes, as was remarked by Hull and White (1995b) themselves.

We therefore concentrate on the following extension of the Vasicek model being analyzed by Hull and White (1994a)

$$dr(t) = [\vartheta(t) - ar(t)]dt + \sigma dW(t), \tag{3.33}$$

where a and σ are now positive constants and ϑ is chosen so as to exactly fit the term structure of interest rates being currently observed in the market. It can be shown that, denoting by $f^M(0,T)$ the market instantaneous forward rate at time 0 for the maturity T, i.e.,

$$f^M(0,T) = -\frac{\partial \ln P^M(0,T)}{\partial T},$$

with $P^M(0,T)$ the market discount factor for the maturity T, we must have

$$\vartheta(t) = \frac{\partial f^M(0,t)}{\partial T} + af^M(0,t) + \frac{\sigma^2}{2a}(1 - e^{-2at}), \tag{3.34}$$

where $\frac{\partial f^M}{\partial T}$ denotes partial derivative of f^M with respect to its second argument.

Equation (3.33) can be easily integrated so as to yield

$$\begin{aligned} r(t) &= r(s)e^{-a(t-s)} + \int_s^t e^{-a(t-u)}\vartheta(u)du + \sigma \int_s^t e^{-a(t-u)}dW(u) \\ &= r(s)e^{-a(t-s)} + \alpha(t) - \alpha(s)e^{-a(t-s)} + \sigma \int_s^t e^{-a(t-u)}dW(u), \end{aligned} \tag{3.35}$$

where

$$\alpha(t) = f^M(0,t) + \frac{\sigma^2}{2a^2}(1 - e^{-at})^2. \tag{3.36}$$

Therefore, $r(t)$ conditional on $\mathcal{F}_s$ is normally distributed with mean and variance given respectively by

$$\begin{aligned} E\{r(t)|\mathcal{F}_s\} &= r(s)e^{-a(t-s)} + \alpha(t) - \alpha(s)e^{-a(t-s)} \\ \mathrm{Var}\{r(t)|\mathcal{F}_s\} &= \frac{\sigma^2}{2a}\left[1 - e^{-2a(t-s)}\right]. \end{aligned} \tag{3.37}$$

Notice that defining the process x by

$$dx(t) = -ax(t)dt + \sigma dW(t), \quad x(0) = 0, \tag{3.38}$$

we immediately have that, for each $s < t$,

$$x(t) = x(s)e^{-a(t-s)} + \sigma \int_s^t e^{-a(t-u)} dW(u),$$

so that we can write $r(t) = x(t) + \alpha(t)$ for each t.

As mentioned above, the theoretical possibility of r going below zero is a clear drawback of the model (3.32) in general, and of (3.33) in particular. Indeed, for model (3.33), the risk-neutral probability of negative rates at time t is explicitly given by

$$Q\{r(t) < 0\} = \Phi\left(-\frac{\alpha(t)}{\sqrt{\frac{\sigma^2}{2a}\left[1 - e^{-2at}\right]}}\right),$$

with Φ denoting the standard normal cumulative distribution function. However, such probability is almost negligible in practice. As an example, we show in Figure 3.2 the evolution over time of the two standard-deviation window, under Q, for the instantaneous short rate r, with parameters calibrated to market data as of 2 June 1999.[7]

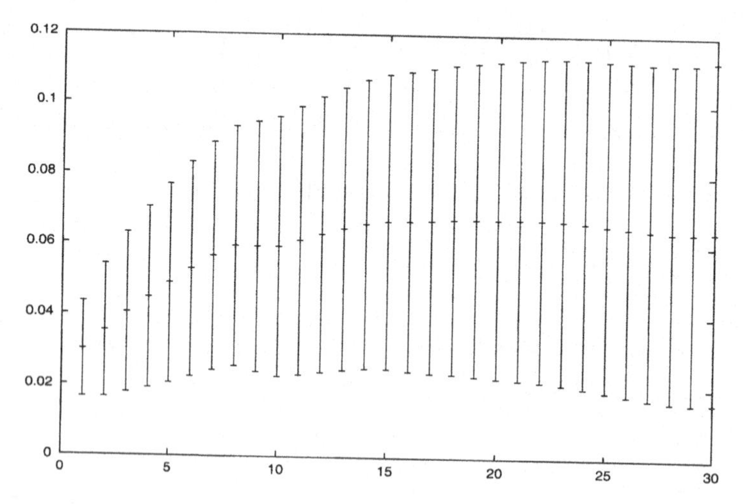

Fig. 3.2. The two standard-deviation window for the instantaneous short rate r as time goes by (market data as of 2 June 1999).

[7] We thank our colleague Francesco Rapisarda for kindly providing us with such a figure.

3.3.2 Bond and Option Pricing

The price at time t of a pure discount bond paying off 1 at time T is given by the expectation (3.2). Such expectation is relatively easy to compute under the dynamics (3.33). Notice indeed that, due to the Gaussian distribution of $r(T)$ conditional on $\mathcal{F}_t$, $t \leq T$, $\int_t^T r(u)du$ is itself normally distributed. Precisely we can show that[8]

$$\int_t^T r(u)du|\mathcal{F}_t$$
$$\sim \mathcal{N}\left(B(t,T)[r(t)-\alpha(t)] + \ln\frac{P^M(0,t)}{P^M(0,T)} + \frac{1}{2}[V(0,T)-V(0,t)], V(t,T)\right),$$

where

$$B(t,T) = \frac{1}{a}\left[1-e^{-a(T-t)}\right],$$
$$V(t,T) = \frac{\sigma^2}{a^2}\left[T-t+\frac{2}{a}e^{-a(T-t)} - \frac{1}{2a}e^{-2a(T-t)} - \frac{3}{2a}\right],$$

so that we obtain

$$P(t,T) = A(t,T)e^{-B(t,T)r(t)}, \tag{3.39}$$

where

$$A(t,T) = \frac{P^M(0,T)}{P^M(0,t)} \exp\left\{B(t,T)f^M(0,t) - \frac{\sigma^2}{4a}(1-e^{-2at})B(t,T)^2\right\}.$$

Similarly, the price $\mathbf{ZBC}(t,T,S,X)$ at time t of a European call option with strike X, maturity T and written on a pure discount bond maturing at time S is given by the expectation (2.23) or, equivalently, by (2.25). To compute the latter expectation, we need to know the distribution of the process r under the T-forward measure Q^T. Since the process x corresponds to the Vasicek's r with $\theta = 0$, we can use formula (3.9) to get

$$dx(t) = [-B(t,T)\sigma^2 - ax(t)]dt + \sigma dW^T(t),$$

where the Q^T-Brownian motion W^T is defined by $dW^T(t) = dW(t) + \sigma B(t,T)dt$, so that, for $s \leq t \leq T$,

$$x(t) = x(s)e^{-a(t-s)} - M^T(s,t) + \sigma\int_s^t e^{-a(t-u)}dW^T(u)$$

with

[8] The following formulas can be derived through the same methodology we illustrate in detail in Chapter 4 for the G2++ model

$$M^T(s,t) = \frac{\sigma^2}{a^2}\left[1 - e^{-a(t-s)}\right] - \frac{\sigma^2}{2a^2}\left[e^{-a(T-t)} - e^{-a(T+t-2s)}\right].$$

It is then easy to realize that the distribution of the short rate $r(t)$ conditional on $\mathcal{F}_s$ is, under the measure Q^T, still Gaussian with mean and variance given respectively by

$$E^T\{r(t)|\mathcal{F}_s\} = x(s)e^{-a(t-s)} - M^T(s,t) + \alpha(t),$$
$$\mathrm{Var}^T\{r(t)|\mathcal{F}_s\} = \frac{\sigma^2}{2a}\left[1 - e^{-2a(t-s)}\right].$$

As a consequence, the European call-option price is

$$\mathbf{ZBC}(t,T,S,X) = P(t,S)\Phi(h) - XP(t,T)\Phi(h - \sigma_p), \tag{3.40}$$

where

$$\sigma_p = \sigma\sqrt{\frac{1 - e^{-2a(T-t)}}{2a}}B(T,S),$$
$$h = \frac{1}{\sigma_p}\ln\frac{P(t,S)}{P(t,T)X} + \frac{\sigma_p}{2}.$$

Analogously, the price $\mathbf{ZBP}(t,T,S,X)$ at time t of a European put option with strike X, maturity T and written on a pure discount bond maturing at time S is given by

$$\mathbf{ZBP}(t,T,S,X) = XP(t,T)\Phi(-h + \sigma_p) - P(t,S)\Phi(-h). \tag{3.41}$$

Through formulas (3.40) and (3.41), we can also price caps and floors since they can be viewed as portfolios of zero-bond options. To this end, we denote by $\mathcal{D} = \{d_1, d_2, \ldots, d_n\}$ the set of the cap/floor payment dates and by $\mathcal{T} = \{t_0, t_1, \ldots, t_n\}$ the set of the corresponding times, meaning that t_i is the difference in years between d_i and the settlement date t, and where t_0 is the first reset time. Moreover, we denote by τ_i the year fraction from d_{i-1} to d_i, $i = 1, \ldots, n$. Applying formula (2.26), we then obtain that the price at time $t < t_0$ of the cap with cap rate (strike) X, nominal value N and set of times $\mathcal{T}$ is given by

$$\mathbf{Cap}(t,\mathcal{T},N,X) = N\sum_{i=1}^{n}(1 + X\tau_i)\mathbf{ZBP}\left(t, t_{i-1}, t_i, \frac{1}{1 + X\tau_i}\right),$$

or, more explicitly,

$$\mathbf{Cap}(t,\mathcal{T},N,X) = N\sum_{i=1}^{n}\left[P(t,t_{i-1})\Phi(-h_i + \sigma_p^i) - (1 + X\tau_i)P(t,t_i)\Phi(-h_i)\right], \tag{3.42}$$

where

$$\sigma_p^i = \sigma\sqrt{\frac{1-e^{-2a(t_{i-1}-t)}}{2a}}B(t_{i-1},t_i),$$
$$h_i = \frac{1}{\sigma_p^i}\ln\frac{P(t,t_i)(1+X\tau_i)}{P(t,t_{i-1})} + \frac{\sigma_p^i}{2}.$$

Analogously, the price of the corresponding floor is

$$\mathbf{Flr}(t,\mathcal{T},N,X) = N\sum_{i=1}^{n}\left[(1+X\tau_i)P(t,t_i)\Phi(h_i) - P(t,t_{i-1})\Phi(h_i-\sigma_p^i)\right]. \tag{3.43}$$

European options on coupon-bearing bonds can be explicitly priced by means of Jamshidian's (1989) decomposition. To this end, consider a European option with strike X and maturity T, written on a bond paying n coupons after the option maturity. Denote by T_i, $T_i > T$, and by c_i the payment time and value of the i-th cash flow after T. Let $\mathcal{T} := \{T_1,\ldots,T_n\}$ and $c := \{c_1,\ldots,c_n\}$. Denote by r^* the value of the spot rate at time T for which the coupon-bearing bond price equals the strike and by X_i the time-T value of a pure-discount bond maturing at T_i when the spot rate is r^*. Then the option price at time $t < T$ is

$$\mathbf{CBO}(t,T,\mathcal{T},c,X) = \sum_{i=1}^{n} c_i\mathbf{ZBO}(t,T,T_i,X_i). \tag{3.44}$$

For a formal prove of this result we refer to Section 3.11.1.

Given the analytical formula (3.44), also European swaptions can be analytically priced, since a European swaption can be viewed as an option on a coupon-bearing bond. Indeed, consider a payer swaption with strike rate X, maturity T and nominal value N, which gives the holder the right to enter at time $t_0 = T$ an interest rate swap with payment times $\mathcal{T} = \{t_1,\ldots,t_n\}$, $t_1 > T$, where he pays at the fixed rate X and receives LIBOR set "in arrears". We denote by τ_i the year fraction from t_{i-1} to t_i, $i = 1,\ldots,n$ and set $c_i := X\tau_i$ for $i = 1,\ldots,n-1$ and $c_n := 1 + X\tau_i$. Denoting by r^* the value of the spot rate at time T for which

$$\sum_{i=1}^{n} c_i A(T,t_i)e^{-B(T,t_i)r^*} = 1,$$

and setting $X_i := A(T,t_i)\exp(-B(T,t_i)r^*)$, the swaption price at time $t < T$ is then given by

$$\mathbf{PS}(t,T,\mathcal{T},N,X) = N\sum_{i=1}^{n} c_i\mathbf{ZBP}(t,T,t_i,X_i). \tag{3.45}$$

Analogously, the price of the corresponding receiver swaption is

$$\mathbf{RS}(t,T,\mathcal{T},N,X) = N\sum_{i=1}^{n} c_i \mathbf{ZBC}(t,T,t_i,X_i). \tag{3.46}$$

As a final remark, before moving to three construction techniques, we observe that the HW model is an affine term-structure model in the sense we have seen in Section 3.2.4.

3.3.3 The Construction of a Trinomial Tree

Even so, every good tree produces good fruit, but a bad tree produces bad fruit, a good tree cannot bear bad fruit, nor can a bad tree bear good fruit
St. Matthew VII.17-18

We now illustrate a procedure for the construction of a trinomial tree that approximates the evolution of the process r. This is a two-stage procedure that is basically based on those suggested by Hull and White (1993d, 1994a).

Let us fix a time horizon T and the times $0 = t_0 < t_1 < \cdots < t_N = T$, and set $\Delta t_i = t_{i+1} - t_i$, for each i. The time instants t_i need not be equally spaced. This is an essential feature when employing the tree for practical purposes.

The first stage consists in constructing a trinomial tree for the process x in (3.38) along the procedure illustrated in Appendix F. For further justifications and details we refer to such appendix.

We denote the tree nodes by (i,j) where the time index i ranges from 0 to N and the space index j ranges from some $\underline{j}_i < 0$ to some $\overline{j}_i > 0$. We denote by $x_{i,j}$ the process value on node (i,j).

Remembering formulas (3.37) and that $x(t) = r(t) - \alpha(t)$ for each t, we have

$$\begin{aligned} E\{x(t_{i+1})|x(t_i) = x_{i,j}\} &= x_{i,j}e^{-a\Delta t_i} =: M_{i,j} \\ \mathrm{Var}\{x(t_{i+1})|x(t_i) = x_{i,j}\} &= \frac{\sigma^2}{2a}\left[1 - e^{-2a\Delta t_i}\right] =: V_i^2. \end{aligned} \tag{3.47}$$

We then set $x_{i,j} = j\Delta x_i$, where

$$\Delta x_i = V_{i-1}\sqrt{3} = \sigma\sqrt{\frac{3}{2a}\left[1 - e^{-2a\Delta t_{i-1}}\right]}. \tag{3.48}$$

Assuming that at time t_i we are on node (i,j) with associated value $x_{i,j}$, the process can move to $x_{i+1,k+1}$, $x_{i+1,k}$ or $x_{i+1,k-1}$ at time t_{i+1} with probabilities p_u, p_m and p_d, respectively. The central node is therefore the k-th node at time t_{i+1}, where the level k is chosen so that $x_{i+1,k}$ is as close as possible to $M_{i,j}$, i.e.,

$$k = \mathrm{round}\left(\frac{M_{i,j}}{\Delta x_{i+1}}\right), \tag{3.49}$$

where round(x) is the closest integer to the real number x. This definition fully determines the geometry of this initial tree for x. In particular, the minimum and the maximum levels $\underline{j}_i$ and $\overline{j}_i$ at each time step i are perfectly defined.

We now derive the probability p_u, p_m and p_d such that the conditional mean and variance in (3.47) match those in the tree. We obtain

$$\begin{cases} p_u = \frac{1}{6} + \frac{\eta_{j,k}^2}{6V_i^2} + \frac{\eta_{j,k}}{2\sqrt{3}V_i}, \\ p_m = \frac{2}{3} - \frac{\eta_{j,k}^2}{3V_i^2}, \\ p_d = \frac{1}{6} + \frac{\eta_{j,k}^2}{6V_i^2} - \frac{\eta_{j,k}}{2\sqrt{3}V_i}, \end{cases} \tag{3.50}$$

where $\eta_{j,k} = M_{i,j} - x_{i+1,k}$, with the dependence on i being omitted to lighten notation.

We can easily see that both p_u and p_d are positive for every value of $\eta_{j,k}$, whereas p_m is positive if and only if $|\eta_{j,k}| \le V_i\sqrt{2}$. However, the definition of k implies that $|\eta_{j,k}| \le V_i\sqrt{3}/2$, hence the middle probability p_m is positive, too. Therefore, (3.50) are actual probabilities such that the discrete process described by the tree has conditional mean and variance that match those of the process x. An example of such a tree geometry, with varying time step, is shown in Figure 3.3.[9]

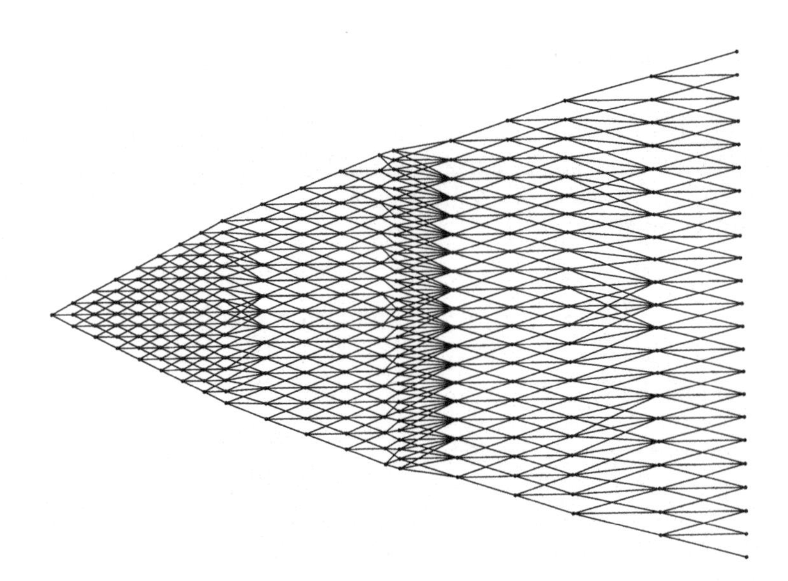

Fig. 3.3. A possible geometry for the tree approximating x.

[9] We thank our colleague Gianvittorio Mauri (aka " The Master") for kindly providing us with such a graphic "masterpiece"!

The second stage of our construction procedure consists in displacing the tree nodes to obtain the corresponding tree for r. An easy way to do so is by means of the explicit formula (3.36). This has been suggested by Pelsser (1996) and Kijima and Nagayama (1994). However, combining this exact formula with the approximate nature of the tree prevents us from retrieving the correct market discount factors at time 0. For example, the analytical displacement at time 0 is $\alpha(0) = r(0)$, so that the price of the zero-coupon bond with maturity t_1, as calculated in the tree, would be $\exp(-r(0)t_1)$, which is different in general from $P^M(0,t_1) = \exp(-R(0,t_1)t_1)$, with $R(0,t_1)$ the continuously-compounded rate at time 0 for the maturity t_1. This is the main reason why the original method proposed by Hull and White (1994a) relies on applying the displacements that perfectly reproduce the market zero-coupon curve at time 0. Notice indeed that even small errors in the pricing of discount bonds can lead to non-negligible errors in bond-option prices.

We denote by α_i the displacement at time t_i, which is common to all nodes $(i,\cdot)$. The quantity α_i is numerically calculated as follows. We denote by $Q_{i,j}$ the present value of an instrument paying 1 if node (i,j) is reached and zero otherwise (somehow discrete analogous to "Arrow-Debreu prices"). The values of α_i and $Q_{i,j}$ are calculated recursively from α_0 that is set so as to retrieve the correct discount factor for the maturity t_1, i.e., $\alpha_0 = -\ln(P^M(0,t_1))/t_1$. As soon as the value of α_i is known, the values $Q_{i+1,j}$, $j = \underline{j}_{i+1},\ldots,\bar{j}_{i+1}$, are calculated through

$$Q_{i+1,j} = \sum_h Q_{i,h}q(h,j)\exp(-(\alpha_i + h\Delta x_i)\Delta t_i),$$

where $q(h,j)$ is the probability of moving from node (i,h) to node $(i+1,j)$ and the sum is over all values of h for which such probability is non-zero. After deriving the value of $Q_{i,j}$, for each $j = \underline{j}_i,\ldots,\bar{j}_i$, the value of α_i is calculated by solving

$$P(0,t_{i+1}) = \sum_{j=\underline{j}_i}^{\bar{j}_i} Q_{i,j}\exp(-(\alpha_i + j\Delta x_i)\Delta t_i),$$

that leads to

$$\alpha_i = \frac{1}{\Delta t_i}\ln\frac{\sum_{j=\underline{j}_i}^{\bar{j}_i} Q_{i,j}\exp(-j\Delta x_i\Delta t_i)}{P(0,t_{i+1})}.$$

We finally end up with a tree where each node (i,j) has associated value $r_{i,j} = x_{i,j} + \alpha_i$. This tree geometry is displayed in Figure 3.4.

3.4 Possible Extensions of the CIR Model

Besides their extension of the Vasicek (1977) model, Hull and White (1990b) proposed an extension of the Cox, Ingersoll and Ross (1985) model based

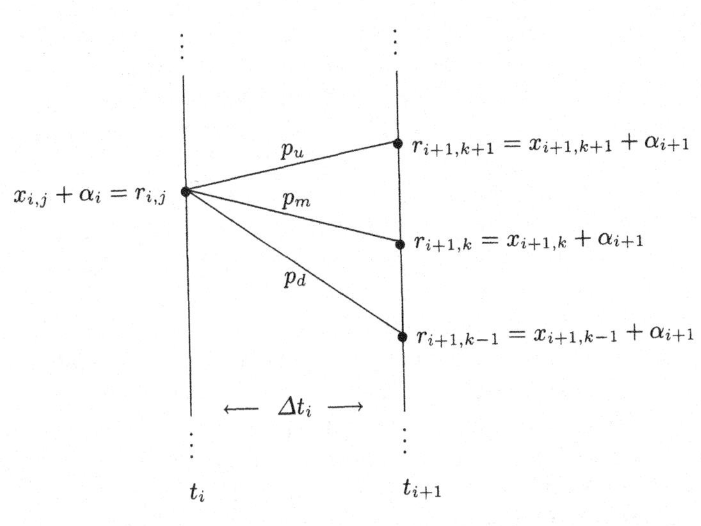

Fig. 3.4. Evolution of the process r starting from $r_{i,j}$ at time t_i and moving to $r_{i+1,k+1}$, $r_{i+1,k}$ or $r_{i+1,k-1}$ at time t_{i+1} with probabilities p_u, p_m and p_d, respectively.

on the same idea of considering time dependent coefficients. The short rate dynamics are then given by

$$dr(t) = [\vartheta(t) - a(t)r(t)]dt + \sigma(t)\sqrt{r(t)}dW(t), \tag{3.51}$$

where a, ϑ and σ are deterministic functions of time. Such extension however is not analytically tractable. Indeed, one can show that, for $t < T$, the pure-discount-bond price can be written as

$$P(t,T) = A(t,T)e^{-B(t,T)r(t)},$$

where B solves a Riccati equation and A solves a linear differential equation subject to some boundary conditions. However, the Riccati equation can be explicitly solved only for constant coefficients, so that, under the general dynamics (3.51), one has to resort to numerical procedures.[10]

Of course the same drawback holds for the simplified dynamics with constant volatility parameters

$$dr(t) = [\vartheta(t) - ar(t)]dt + \sigma\sqrt{r(t)}dW(t), \tag{3.52}$$

where a and σ are now positive constants, and only the function ϑ is assumed to be time dependent so as to exactly fit the initial term structure of interest

[10] Maghsoodi (1996) derived different formulas for the prices of bonds and bond options, but still relying on numerical integration.

rates. To our knowledge, no analytical expression for $\vartheta(t)$ in terms of the observed yield curve is available in the literature. Furthermore, there is no guarantee that a numerical approximation of $\vartheta(t)$ would keep the rate r positive, hence that the diffusion coefficient would always be well defined. These are the major reasons why this extension has been less successful than its Gaussian counterpart.

A simple version of (3.51) that turns out to be analytically tractable has been proposed by Jamshidian (1995). He assumed that, for each t, the ratio $\vartheta(t)/\sigma^2(t)$ is equal to a positive constant δ, which must be greater than 1/2 to ensure that the origin is inaccessible.

Maghsoodi (1996) and Rogers (1995) have shown that this simple version can be obtained from the classical constant coefficient CIR model by applying a deterministic time change and then multiplying the resulting process by a positive deterministic function of time. However the acquired analytical tractability is paid by having European bond-option prices that explicitly depend on the instantaneous short-term rate. This is a characteristics of all square-root models. Gaussian models feature option prices depending on the instantaneous rate only implicitly through $P(t,T)$, since $P(t,T)$ is a function of $r(t)$. Instead, in the CIR model, and in all its extensions, $r(t)$ appears explicitly, i.e, outside expressions of long-term rates such as $L(t,T)$ or $R(t,T)$ (or, equivalently, $P(t,T)$). While it is natural that long-dated option prices at time t depend on long-term rates, it may be undesirable that they depend on the instantaneous rate explicitly. This can be undesirable when pricing and hedging long-dated options.

Now we focus on a different approach to deal with negative rates than the CIR model, while we will consider a further more-effective extension of the CIR model later on, in Section 3.9.

3.5 The Black-Karasinski Model

Primary imperative of the BK1 neoroid: `Preserve the environment from possible threats`. *Primary threat for the environment:* `The human genre`. *Application of the primary imperative of the BK1 neoroid:* `Terminate the human genre`. Plan σ: The last folly of a machine, or the last folly of man, who programmed it? To Casshan, neither man nor machine, it is a folly occurring twice.

"Kyashan: Il Mito" (Casshan: Robot Hunter, 1993 remake, Tatsunoko).

The drawback of negative rates has been also addressed by Black and Karasinski (1991) in their celebrated lognormal short rate model. Black and Karasinski assumed that the instantaneous short rate process evolves as the exponential of an Ornstein-Uhlenbeck process with time dependent coefficients.[11] Since the market formulas for caps and swaptions are based on the assumption of lognormal rates, it seemed reasonable to choose the same distribution for the instantaneous short-rate process.[12] Moreover, the rather good fitting quality of the model to market data, and especially to the swaption volatility surface, has made the model quite popular among practitioners and financial engineers. However, analogously to the Exponential-Vasicek model, the Black-Karasinski (1991) model is not analytically tractable. This renders the model calibration to market data more burdensome than in the Hull and White (1990) Gaussian model, since no analytical formulas for bonds are available. Indeed, when using a tree to price an option on a zero-coupon bond, one has to construct the tree until the bond maturity, which may actually be much longer than that of the option. The same inconvenience occurs when we need to simulate rates which are not instantaneous. If we need for example to simulate the four-year rate in one year, we may simulate the short rate up to one year but then we are not done. With Hull and White's model for example, when we have simulated $r(1y)$, we can easily compute $P(1y, 5y)$ (and therefore $L(1y, 5y)$) algebraically from (3.39). The same holds for the Vasicek model, the CIR model and their deterministic shift extensions we will explore in the following. However, this desirable feature is not shared by the Black-Karasinski model, the EV model or its extension. When we have simulated $r(1y)$, we need to compute $P(1y, 5y)$ numerically for each simulated realization of r. In practice, we need to use a tree for each simulated r, and this renders the Monte Carlo approach much heavier than in the above-mentioned tractable models.

A further, and more fundamental, drawback of the model is that the expected value of the money-market account is infinite no matter which maturity is considered, as a consequence of the lognormal distribution of r. This was already remarked in the section devoted to the Dothan model and is a problem of lognormal models in general. As a consequence, the price of a Eurodollar future is infinite, too. However, this problem is partially overcome when using an approximating tree, because one deals with a finite number of states, and hence with finite expectations.

3.5.1 The Short-Rate Dynamics

Black and Karasinski (1991) assumed that the logarithm $\ln(r(t))$ of the instantaneous spot rate evolves under the risk neutral measure Q according to

$$d\ln(r(t)) = [\theta(t) - a(t)\ln(r(t))]dt + \sigma(t)dW(t), \quad r(0) = r_0, \qquad (3.53)$$

[11] The Black and Karasinski (1991) model is actually a generalization of the continuous-time formulation of the Black, Derman and Toy (1990) model.

[12] Notice, however, that a lognormal instantaneous short-rate process does not lead to lognormal simple forward rates or lognormal swap rates.

where r_0 is a positive constant, $\theta(t)$, $a(t)$ and $\sigma(t)$ are deterministic functions of time that can be chosen so as to exactly fit the initial term structure of interest rates and some market volatility curves.

As for the Hull and White (1990b) model, one can choose to set $a(t) = a$ and $\sigma(t) = \sigma$, with a and σ positive constants, leading to

$$d\ln(r(t)) = [\theta(t) - a\ln(r(t))]dt + \sigma dW(t), \quad r(0) = r_0. \tag{3.54}$$

This choice can be motivated by arguments similar to those reported in Section 3.3. By letting θ be the only time dependent function, we decide to exactly fit the current term structure of rates and to keep the other two parameters at our disposal for the calibration to option data.

As in previous models, the coefficients a and σ can be interpreted as follows: a gives a measure of the "speed" at which the logarithm of $r(t)$ tends to its long-term value; σ is the standard-deviation rate of $dr(t)/r(t)$, namely the standard deviation per time unit of the instantaneous return of $r(t)$. Note also that σ, denoting the volatility of the instantaneous spot rate, must not be confused either with the volatility of the forward rate or with the volatility of the forward swap rate that must be plugged into the Black formulas for caps/floors and swaptions, respectively.

From (3.54), by Ito's lemma, we obtain

$$dr(t) = r(t)\left[\theta(t) + \frac{\sigma^2}{2} - a\ln r(t)\right]dt + \sigma r(t)dW(t),$$

whose explicit solution satisfies, for each $s \leq t$,

$$r(t) = \exp\left\{\ln r(s)e^{-a(t-s)} + \int_s^t e^{-a(t-u)}\theta(u)du + \sigma\int_s^t e^{-a(t-u)}dW(u)\right\}.$$

Therefore, $r(t)$ conditional on $\mathcal{F}_s$ is lognormally distributed with first and second moments given respectively by

$$E_s\{r(t)\} = \exp\left\{\ln r(s)e^{-a(t-s)} + \int_s^t e^{-a(t-u)}\theta(u)du + \frac{\sigma^2}{4a}\left[1 - e^{-2a(t-s)}\right]\right\}$$

$$E_s\{r^2(t)\} = \exp\left\{2\ln r(s)e^{-a(t-s)} + 2\int_s^t e^{-a(t-u)}\theta(u)du + \frac{\sigma^2}{a}\left[1 - e^{-2a(t-s)}\right]\right\}.$$

Moreover, setting

$$\alpha(t) = \ln(r_0)e^{-at} + \int_0^t e^{-a(t-u)}\theta(u)du, \tag{3.55}$$

we have that

$$\lim_{t\to\infty} E(r(t)) = \exp\left(\lim_{t\to\infty}\alpha(t) + \frac{\sigma^2}{4a}\right).$$

The limit on the left hand side cannot be computed analytically. However, the numerical procedure below allows for the extrapolation of an asymptotic value of $\alpha(t)$.

3.5.2 The Construction of a Trinomial Tree

From the withered tree, a flower blooms
Zen saying

As we have already pointed out, the Black and Karasinski model does not yield analytical formulas either for discount bonds or for options on bonds. The pricing of these (and other more general) instruments, therefore, must be performed through numerical procedures. An efficient numerical procedure has been suggested by Hull and White (1994a) and is based on a straightforward transformation of the trinomial tree we have illustrated in Section 3.3.3. Notice, indeed, that we can write

$$r(t) = e^{\alpha(t)+x(t)}, \tag{3.56}$$

where α and x are defined as in (3.55) and (3.38), respectively. As for the Hull and White (1994a) model, we first construct a trinomial tree for x and then use (3.56) to displace the tree nodes so as to exactly retrieve the initial zero-coupon curve. For a better understanding of the construction procedure we refer to Section 3.3.3 and to Appendix F in particular.

Let us fix a time horizon T and the times $0 = t_0 < t_1 < \cdots < t_N = T$, and set $\Delta t_i = t_{i+1} - t_i$, for each i. As before, the time instants t_i need not be equally spaced. Again, we denote the tree nodes by (i, j) where the time index i ranges from 0 to N and the space index j ranges from some $\underline{j}_i < 0$ to some $\overline{j}_i > 0$.

We denote by $x_{i,j}$ the process value on node (i, j) and set $x_{i,j} = j\Delta x_i$, where Δx_i is defined as in (3.48).

Assuming that at time t_i we are on node (i, j) with associated value $x_{i,j}$, the process can move to $x_{i+1,k+1}$, $x_{i+1,k}$ or $x_{i+1,k-1}$ at time t_{i+1} with probabilities p_u, p_m and p_d, respectively. The central node is therefore the k-th node at time t_{i+1}, where k is defined as in (3.49). The probabilities p_u, p_m and p_d are defined as in (3.50). These definitions completely specify the initial tree geometry, and in particular the minimum and the maximum levels $\underline{j}_i$ and $\overline{j}_i$ at each time step i.

The final stage in our construction procedure consists in suitably shifting the tree nodes in order to obtain the proper tree for r through formula (3.56). Contrary to the Hull and White (1994a) case, the function α cannot be evaluated analytically. However, as already explained in Section 3.3.3, a numerical procedure is anyway required to exactly reproduce the initial term structure of discount factors.

We again denote by α_i the displacement at time t_i, which is common to all nodes $(i, \cdot)$. The quantity α_i is numerically calculated as in Section 3.3.3 with the only difference that now (3.56) holds. We again denote by $Q_{i,j}$ the present value of an instrument paying 1 if node (i, j) is reached and zero otherwise. The values of α_i and $Q_{i,j}$ are calculated recursively from α_0 that is set so as to retrieve the correct discount factor for the maturity t_1, i.e.,

$\alpha_0 = \ln(-\ln(P^M(0,t_1))/t_1)$. As soon as the value of α_i is known, the values $Q_{i+1,j}$, $j = \underline{j}_{i+1}, \ldots, \overline{j}_{i+1}$, are calculated through

$$Q_{i+1,j} = \sum_h Q_{i,h} q(h,j) \exp(-\exp(\alpha_i + h\Delta x_i)\Delta t_i),$$

where $q(h,j)$ is the probability of moving from node (i,h) to node $(i+1,j)$ and the sum is over all values of h for which such probability is non-zero. After deriving the value of $Q_{i,j}$, for each $j = \underline{j}_i, \ldots, \overline{j}_i$, the value of α_i is calculated by numerically solving

$$\psi(\alpha_i) := P(0,t_{i+1}) - \sum_{j=\underline{j}_i}^{\overline{j}_i} Q_{i,j} \exp(-\exp(\alpha_i + j\Delta x_i)\Delta t_i) = 0.$$

Using, for instance, the Newton-Raphson procedure, it is helpful to employ the analytical formula for the first derivative of the function ψ, that is

$$\psi'(\alpha_i) = \sum_{j=\underline{j}_i}^{\overline{j}_i} Q_{i,j} \exp(-\exp(\alpha_i + j\Delta x_i)\Delta t_i) \exp(\alpha_i + j\Delta x_i)\Delta t_i.$$

Finally, we must apply the exponential function to each node value to end up with a tree where each node (i,j) has associated value $r_{i,j} = \exp(x_{i,j} + \alpha_i)$. This tree geometry is displayed in Figure 3.5.

3.6 Volatility Structures in One-Factor Short-Rate Models

'I know what's real and what's false. In fact... I define it'
Matthew Ryder of the Linear Men, "The Kingdom" 1, 1999, DC Comics.

The aim of the present section is to clarify which volatility structures are relevant as far as the short-rate model performances are concerned. We will also point out the different volatility structures that are usually considered in the market. We will come back to this problem again in the chapters devoted to multi-factor models and to the LIBOR market model.

When approaching the interest-rate option market from a practical point of view, one immediately realizes that the *volatility* is the fundamental quantity one has to deal with. Such a quantity is so important that it is not just a sheer parameter, as theoretical researchers are tempted to view it, but it becomes an actual asset that can be bought or sold in the market. However, despite its practical importance, it may be hard to retrieve a clear definition of volatility in quantitative terms, and some confusion is likely to arise.

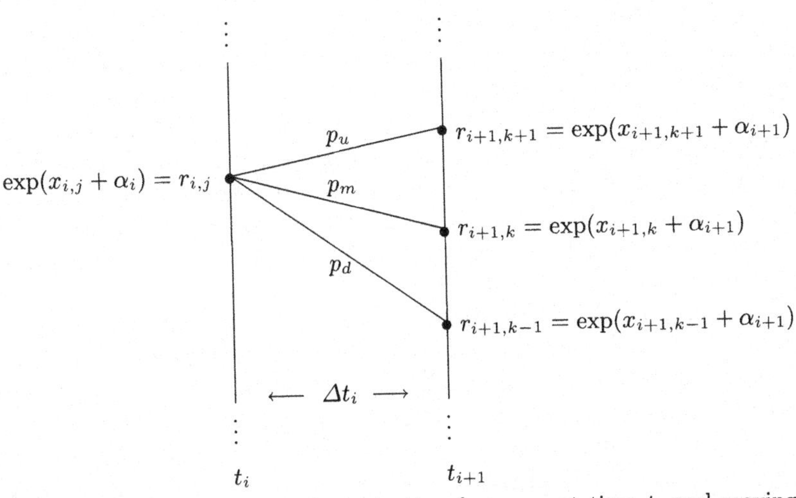

Fig. 3.5. Evolution of the process r starting from $r_{i,j}$ at time t_i and moving to $r_{i+1,k+1}$, $r_{i+1,k}$ or $r_{i+1,k-1}$ at time t_{i+1} with probabilities p_u, p_m and p_d, respectively.

A typical situation is when one hears traders pronounce sentences like: "A satisfactory interest-rate model has to allow for a humped shape in the term structure of volatilities." In this section, we will then try to clarify statements of this type.

We now describe the path we adopt in this section in introducing volatility concepts. This path does not follow a fully logical development, because we are again adopting the "market/heuristics then rigor" approach. We prefer to give the reader motivation before we introduce new concepts, even if this creates a kind of time warp in the exposition. We adopt the following plan.

- Explain how the market considers (and defines) *caplet volatilities*.
- Explain how the market builds from these the *term structure of volatilities*.
- Explain what goes wrong if we simply transliterate the caplet-volatilities definition above, consistent with the market, to short-rate models language (model-intrinsic caplet volatility).
- Explain how we can modify the definition of caplet volatility for short rate models so that things work again (model-implied caplet volatility).
- Define, consequently, the term structure of volatility for short-rate models.

Caplet Volatilities in the Market

We have already seen in Chapter 1 that it is market practice to price caps and floors by means of the Black formulas (1.26) and (1.27) and to quote, instead of the price, the volatility parameter σ that enters such formulas. The market cap volatility is then simply defined as the parameter σ that must be plugged into the Black cap formula to obtain the right market cap price. Similarly, a caplet volatility can be defined as the parameter σ to plug into the Black caplet formula to obtain the right caplet price. A fundamental difference is that the cap volatility assumes that all caplets concurring to a given cap share the same volatility, which is then set to a value matching the cap market price. On the contrary, caplet volatilities are allowed to be different also for caplets concurring to the same cap. In practice, caplet volatilities are stripped from the market cap volatilities along the procedure that will be explained in Sections 6.4 and 6.4.3 in particular, to which we refer for a more complete treatment.

Therefore, in this context, both cap and caplet volatilities are parameters obtained by inverting market prices through market established formulas. While for cap volatilities one usually stops here, the (market) caplet volatilities can instead be defined in an alternative way, which sheds further light on their actual meaning. Indeed, let us consider a caplet resetting at time T and paying at time $T+\tau$ the LIBOR rate $\tau L(T, T+\tau) = \tau F(T; T, T+\tau)$, where τ (measured in years) is typically three or six months. Denote the T-expiry caplet (percentage) volatility by

$$\begin{aligned} v_{T-\text{caplet}}^2 &:= \frac{1}{T}\int_0^T (d\ln F(t;T,T+\tau))(d\ln F(t;T,T+\tau)) \\ &= \frac{1}{T}\int_0^T \sigma(t;T,T+\tau)^2\,dt. \end{aligned}$$

The quantity $\sigma(t; T, T+\tau)$ is the (percentage) instantaneous volatility at time t of the simply-compounded forward rate $F(t; T, T+\tau)$ that underlies the T-expiry caplet. The instantaneous percentage volatilities $\sigma(t; T, T+\tau)$ are modeled deterministically in Black's market model for the cap market, so that the caplet volatility $v_{T-\text{caplet}}$, obtained by integrating deterministic functions, is also deterministic.

The Term Structure of (Caplet) Volatilities

The term structure of volatilities at time 0 is then to be intended as the graph of the map

$$T \mapsto v_{T-\text{caplet}},$$

and *this* is the graph that is observed to be humped most of times in the market.

An important question arises now: In what sense are caplet volatilities to be understood for short-rate models? There are two possible ways to define caplet (and cap) volatilities for short-rate models. We may call them the "model-intrinsic caplet volatility" and the "model-implied caplet volatility".

Why not Simply Transliterating the Definition to Short-Rate Models?

Let us begin with the "model-intrinsic caplet volatility". Before giving the formal definition, let us illustrate this idea by resorting to a specific model. Consider for example the CIR model. In a model like CIR the integrals in the definition of $v_{T-\text{caplet}}$ above are stochastic, since the corresponding function $\sigma(t;T,T+\tau)$ is not deterministic. One can indeed convince oneself of this by recalling the dynamics of F in the CIR model as from Section 3.2.3,

$$dF(t;T,S) = \sigma\left(F(t;T,S) + \frac{1}{\gamma(T,S)}\right) \cdot \sqrt{(B(t,S)-B(t,T))\ln\left[(\gamma(T,S)F(t;T,S)+1)\frac{A(t,S)}{A(t,T)}\right]}\,dW^S(t).$$

Here we have a stochastic percentage instantaneous volatility that is given by

$$\sigma(t;T,S) = \sigma\left(1 + \frac{1}{\gamma(T,S)F(t;T,S)}\right) \cdot \sqrt{(B(t,S)-B(t,T))\ln\left[(\gamma(T,S)F(t;T,S)+1)\frac{A(t,S)}{A(t,T)}\right]}.$$

Since the quantities $v_{T-\text{caplet}}$'s are (square roots of) integrals of the instantaneous variances $\sigma(t;T,T+\tau)^2$, in the CIR model the $v_{T-\text{caplet}}$'s are stochastic. More generally, we may define, for a short rate model,

Definition 3.6.1. Model intrinsic T-caplet volatility. *The* model intrinsic T-caplet volatility *at time* 0 *is defined as the random variable*

$$\sqrt{\frac{1}{T}\int_0^T \sigma(t;T,T+\tau)^2\,dt}\ .$$

As the CIR example pointed out, however, this definition presents us with a problem: if we define caplet volatilities like this, by simply mimicking the definition we have seen above for the market, we obtain random caplet volatilities. But caplet volatilities are not random in the market. How can we modify the definition of caplet volatility for short-rate models so as to go back to a deterministic setup?

Modify the Definition so as to have Deterministic Caplet Volatilities

The necessity of dealing with deterministic quantities leads to the following procedure. Compute the model price of the at-the-money T-caplet at time

0, thus obtaining a function $\mathbf{Cpl}(0,T,T+\tau,F(0;T,T+\tau))$ of the model parameters, and then invert the T-expiry Black market formula for caplets to find the percentage Black volatility $v_{T-\text{caplet}}$ that once plugged into such a formula gives the model price. More precisely, one solves the following equation for $v^{\text{MODEL}}_{T-\text{caplet}}$:

$$P(0,T+\tau)\ \tau F(0;T,T+\tau)\left(2\Phi\left(\frac{v^{\text{MODEL}}_{T-\text{caplet}}\sqrt{T}}{2}\right)-1\right) \tag{3.57}$$
$$= \mathbf{Cpl}(0,T,T+\tau,F(0;T,T+\tau)).$$

The left-hand side is the market Black's formula for a T-expiry $T+\tau$-maturity at-the-money caplet, whereas the right-hand side is the corresponding model formula. We are now ready to introduce the following.

Definition 3.6.2. Model-implied T-caplet volatility. *The* model-implied T-caplet volatility *at time 0 is the (deterministic) solution $v^{\text{MODEL}}_{T-\text{caplet}}$ of the above equation (3.57).*

Implied cap volatilities can be defined in an analogous way. Precisely, let us consider a set of reset times $\{T_\alpha,\dots,T_{\beta-1}\}$ with the final payment time T_β, and the set of the associated year fractions $\{\tau_\alpha,\dots,\tau_\beta\}$. Then, setting $\mathcal{T}_i=\{T_\alpha,\dots,T_i\}$ and $\bar{\tau}_i=\{\tau_{\alpha+1},\dots,\tau_i\}$, we have the following.

Definition 3.6.3. Model-implied $\mathcal{T}_i$-cap volatility. *The* model-implied $\mathcal{T}_i$-cap volatility *at time 0 is the (deterministic) solution $v^{\text{MODEL}}_{\mathcal{T}_i-\text{cap}}$ of the equation:*

$$\sum_{j=\alpha+1}^{i} P(0,T_j)\tau_j \text{Bl}(S_{\alpha,\beta}(0),F(0,T_{j-1},T_j),v^{\text{MODEL}}_{\mathcal{T}_i-\text{cap}}\sqrt{T_{j-1}})$$
$$= \mathbf{Cap}(0,\mathcal{T}_i,\bar{\tau}_i,S_{\alpha,\beta}(0)),$$

where the forward swap rate $S_{\alpha,\beta}(0)$ is defined as in (1.25).

Term Structure of Volatilities for a Short-Rate Model

We can now define easily two types of term structure of volatilities associated with a short-rate model.

Definition 3.6.4. Term structure of caplet volatilities implied by a short-rate model. *The* term structure of caplet volatilities implied by a short-rate model *is the graph of the model-implied T-caplet volatility against T, i.e. the graph of the function $T\mapsto v^{\text{MODEL}}_{T-\text{caplet}}$.*

Definition 3.6.5. Term structure of cap volatilities implied by a short-rate model. *The* term structure of cap volatilities implied by a short-rate model *is the graph of the model-implied $\mathcal{T}_i$-cap volatility against T_i, i.e. the graph of the function $T_i\mapsto v^{\text{MODEL}}_{\mathcal{T}_i-\text{cap}}$.*

In the remainder of the book, when dealing with short-rate models, $v^{\text{MODEL}}_{T-\text{caplet}}$ will always denote the model-*implied* T-caplet volatility, unless differently specified, and we will often omit the "MODEL" superscript. Notice that an analogous graph for the model-intrinsic T-caplet volatility would consist of a bunch of curves, one for each trajectory ω of $F(\cdot;T,T+\tau)(\omega)$.

A similar argument applies to the model-*implied* T-cap volatility. Notice also that for a cap with a single payment in T_i (coinciding with a T_{i-1}-resetting caplet), we have $\mathcal{T}_i = \{T_i\}$, i.e. $v_{T_{i-1}-\text{caplet}} = v_{\{T_i\}-\text{cap}}$. A little attention can avoid being disoriented by this "i" versus "$i-1$" notation.

Now that we have motivated and introduced the relevant definitions, we continue the discussion. Going back to the "humped shape observed in the market", what is actually observed to have a humped shape is the curve of the cap volatilities for different maturities at time 0. From cap volatilities one can strip caplet volatilities, and also this second curve shows a humped shape when observed in the market.[13]

In practice, it is commonly seen that this term structure is able to feature large humps if the related absolute instantaneous volatilities of instantaneous forward rates,

$$T \mapsto \sqrt{\frac{\text{Var}(df(t,T))}{dt}} =: \sigma_f(t,T),$$

allow for a hump themselves. In other terms, there is a link between potentially large humps of the term structure of volatilities (as observed in the market) and possible humps in the volatility of instantaneous forward rates. We used the term "large humps" because *small* humps in the model caplet and cap curves $T \mapsto v_{T-\text{caplet}}$ and $\mathcal{T}_i \mapsto v_{\mathcal{T}_i-\text{cap}}$ are possible even when the model instantaneous-forward volatility curve $T \mapsto \sigma_f(t,T)$ is monotonically decreasing. In short, it usually happens that:

1. No humps in $T \mapsto \sigma_f(t,T) \Rightarrow$ only small humps for $T \mapsto v^{\text{MODEL}}_{T-\text{caplet}}$ are possible;
2. Humps in $T \mapsto \sigma_f(t,T) \Rightarrow$ large humps for $T \mapsto v^{\text{MODEL}}_{T-\text{caplet}}$ are possible.

A typical example is the CIR++ model (3.76) we will introduce later on. After computing the absolute volatility of instantaneous forward rates $\sigma_f(t,T)$ through formula (3.29), a little analysis shows that under this model $T \mapsto \sigma_f(t,T)$ is indeed monotonically decreasing, while the model calibration to caps data usually leads to implied cap volatilities displaying a slightly humped shape (so that also the initial caplet volatilities display a small hump). See also the related section in this chapter. Summarizing, *small* humps in the caplet curve are possible even with monotonically decreasing instantaneous-forward volatilities. However, for these models *large* humps in caplets volatilities are often ruled out by monotonicity of $T \mapsto \sigma_f(t,T)$.

[13] What we are saying here is a little redundant with the material appearing in other chapters. This is done to maintain the single chapters self-contained to a certain degree.

These considerations on the amplitude of the cap-volatility hump mainly apply when the zero-coupon curve is increasing or slightly inverted, like that in Figure 1.1. However, in case of decreasing yield curves, it turns out that large humps can be produced even by those models for which the map $T \mapsto \sigma_f(t,T)$ is monotone. For instance, in case of the Hull and White model (3.33), we can prove that

$$v^{\mathrm{HW}}_{T-\mathrm{caplet}} \approx \sigma \frac{1-e^{-a\tau}}{a}\left[1+\frac{1}{\tau F(0,T,T+\tau)}\right], \tag{3.58}$$

for a and T positive and small enough. Therefore, in case of a (initially) decreasing forward rate curve $T \mapsto F(0,T,T+\tau)$, the implied caplet curve is initially upward sloping, and then necessarily humped since $v^{\mathrm{HW}}_{T-\mathrm{caplet}}$ goes to zero for T going to infinity.

We noticed above that confusion often arises when speaking of "allowing a humped shape for forward-rates volatilities". In other words, it is not clear what kind of volatilities one is considering: instantaneous-forward absolute volatilities, caplet volatilities, cap volatilities, etc. Some further specification is then needed, in general. As we have just seen, in fact, some models allow for humps in all of these structures. Some others, instead, only allow for small humps in the caplets or caps volatilities.

The framework just introduced can be used for calibration purposes. For instance, when the CIR++ model is calibrated to the cap market, the parameters k,θ,σ in the CIR dynamics are set to values such that the market cap volatilities $v^{\mathrm{MKT}}_{T_i-\mathrm{cap}}$ are as close as possible to the model volatilities $v^{\mathrm{CIR}}_{T_i-\mathrm{cap}}(k,\theta,\sigma)$ for all relevant i's. Alternatively, if one has already derived caplet volatilities from cap quotes along the lines of the method suggested in Section 6.4.3, the calibration can be made on caplet volatilities directly. The parameters k,θ,σ in the CIR dynamics are then set to values such that the market caplet volatilities $v^{\mathrm{MKT}}_{T-\mathrm{caplet}}$ are as close as possible to the model volatilities $v^{\mathrm{CIR}}_{T-\mathrm{caplet}}(k,\theta,\sigma)$ for all (relevant) T's.

We finally stress that the considered short-rate model need not be analytically tractable. Indeed, we just have to notice that, in non analytically tractable models, we need to resort to numerical procedures in order to compute the caplet prices $\mathbf{Cpl}(0,T,T+\tau,F(0;T,T+\tau))$ and retrieve $v^{\mathrm{MODEL}}_{T-\mathrm{caplet}}$ through inversion of the Black caplet formula.

3.7 Humped-Volatility Short-Rate Models

As we have seen in Section 3.6, allowing for a humped shape in the volatility structure of instantaneous forward rates results in possibly large humps in the caplet or cap volatility curves, as is often observed in the market.

Motivated by such considerations, as well as by a number of empirical works, Mercurio and Moraleda (2000) proposed an extension of the Hull and

White model (3.33) that in general is outside the general Hull and White class (3.32) and that allows for humped shapes in the volatility structure of instantaneous forward rates. They selected a suitable volatility function in the Heath, Jarrow and Morton (1992) framework and derived closed form formulas for European bond option prices. Precisely, they assumed the following form for the absolute volatility of the instantaneous forward rate process $f(t,T)$

$$\sqrt{\frac{\mathrm{Var}(df(t,T))}{dt}} = \sigma[1+\gamma(T-t)]e^{-\lambda(T-t)}, \tag{3.59}$$

where σ, λ and γ are positive constants. Their model, which will be better described in Chapter 5, has however two major drawbacks. First, the implied instantaneous spot rate process is not Markov and no explicit formula for zero-coupon bond prices can be derived. Second, the process distribution is Gaussian so that negative rates are actually possible.

The first drawback has been addressed by Moraleda and Vorst (1997) who considered the following dynamics for the instantaneous spot rate

$$\begin{cases} dr(t) = [\theta(t) - \beta(t)r(t)]\,dt + \sigma dW(t) \\ \beta(t) = \lambda - \frac{\gamma}{1+\gamma t} \end{cases} \tag{3.60}$$

which is a particular case of the full Hull and White model (3.32) which coincides with (3.33) for $\gamma = 0$ (and setting $\lambda = a$), and where $\theta(t)$ is a deterministic function of time. Equivalently, for each $s \le t$,

$$\begin{aligned} r(t) &= r(s)e^{-\int_s^t \beta(u)du} + \int_s^t e^{-\int_u^t \beta(v)dv}\vartheta(u)du + \sigma \int_s^t e^{-\int_u^t \beta(v)dv} dW(u) \\ &= r(s)\mathcal{B}(s,t) + \int_s^t \mathcal{B}(u,t)\vartheta(u)du + \sigma \int_s^t \mathcal{B}(u,t)dW(u) \end{aligned}$$

where

$$\mathcal{B}(s,t) := e^{-\int_s^t \beta(u)du} = \frac{1+\gamma t}{1+\gamma s}e^{-\lambda(t-s)}. \tag{3.61}$$

The absolute volatility of the instantaneous-forward-rate process $f(t,T)$ implied by (3.60) is

$$\sqrt{\frac{\mathrm{Var}(df(t,T))}{dt}} = \sigma\frac{1+\gamma T}{1+\gamma t}e^{-\lambda(T-t)} \tag{3.62}$$

and coincides, at first order in $T-t$, with (3.59). The model (3.60) is analytically tractable and leads to analytical formulas for zero coupon bonds as well. Indeed, Moraleda and Vorst proved that the price at time t of a pure discount bond with maturity T is given by

$$P(t,T) = \frac{P(0,T)}{P(0,t)} \exp\left\{-\frac{1}{2}\Lambda^2(t,T)\phi(t) + \Lambda(t,T)[f(0,t) - r(t)]\right\},$$

where

$$\Lambda(t,T) = \frac{1}{\lambda^2(1+\gamma t)}\left[\gamma\lambda t + \gamma + \lambda - (\gamma\lambda T + \gamma + \lambda)e^{-\lambda(T-t)}\right],$$

$$\phi(t) = \frac{\sigma^2(1+\gamma t)}{\gamma^2}\left[2\lambda(1+\gamma t)\mathrm{Ei}\left(\frac{2\lambda(1+\gamma t)}{\gamma}\right)e^{-\frac{2\lambda(1+\gamma t)}{\gamma}} - \gamma\right],$$
$$- \frac{\sigma^2(1+\gamma t)^2}{\gamma^2}\left[2\lambda\mathrm{Ei}\left(\frac{2\lambda}{\gamma}\right)e^{-\frac{2\lambda(1+\gamma t)}{\gamma}} - \gamma e^{-2\lambda t}\right],$$

with Ei denoting the exponential integral function

$$\mathrm{Ei}(x) = \int_{-\infty}^{x} \frac{e^t}{t}dt.$$

It is clear that this model is affine in the sense explained in Section 3.2.4, as is immediate also by the drift and diffusion coefficients in its short-rate dynamics.

Moreover, the time-t price of a European call option with maturity T, strike price X and written on a pure discount bond with maturity S is given by

$$\mathbf{ZBC}(t,T,S,X) = P(t,S)\Phi(h) - XP(t,T)\Phi(h-\sigma_p),$$

where

$$h = \frac{\ln\frac{P(t,S)}{XP(t,T)} + \frac{1}{2}\sigma_p^2}{\sigma_p},$$

$$\sigma_p^2 = \frac{\sigma^2 A^2}{\lambda^4\gamma^2}\left[B(T) - B(t)\right],$$

$$A = (\gamma\lambda S + \gamma + \lambda)e^{-\lambda S} - (\gamma\lambda T + \gamma + \lambda)e^{-\lambda T},$$

$$B(\tau) = 2\lambda e^{-\frac{2\lambda}{\gamma}}\mathrm{Ei}\left(\frac{2\lambda(1+\gamma\tau)}{\gamma}\right) - \frac{\gamma e^{2\lambda\tau}}{1+\gamma\tau}.$$

However, the short-rate process is still normally distributed, and since (3.62) is not just a function of $T-t$, the humped shape of the forward-rate volatility is lost as soon as t increases.

The second drawback of the Mercurio and Moraleda (2000) model has been addressed by Mercurio and Moraleda (2001) by assuming the following dynamics for the instantaneous spot rate

$$\begin{cases} dx(t) = [\theta(t) - \beta(t)x(t)]\,dt + \sigma dW(t) \\ \beta(t) = \lambda - \frac{\gamma}{1+\gamma t} \\ r(t) = e^{x(t)} \end{cases} \tag{3.63}$$

where x is an underlying Gaussian process. This model can be viewed as a characterization of (3.53), where the function σ is set to be a positive constant and $a(t) = \beta(t)$ for each t.[14] Indeed, for each $s \leq t$,

$$r(t) = \exp\left\{\ln r(s)\mathcal{B}(s,t) + \int_s^t \mathcal{B}(u,t)\theta(u)du + \sigma \int_s^t \mathcal{B}(u,t)dW(u)\right\},$$

with $\mathcal{B}(s,t)$ defined as in (3.61). Notice that, for $\gamma = 0$, we retrieve the model (3.54).

Though sharing the same drawbacks of the model (3.54), the model (3.63) implies, however, a much better fitting of the cap-volatility curve, which indeed justifies the introduction of the extra parameter γ for practical purposes. For related empirical results we refer to the original paper by Mercurio and Moraleda (2001) and to a later section in this chapter. Such results are based on the implementation of a trinomial tree, which is constructed in a similar fashion to that of Section 3.5 by taking into account that the value of a changes at each time step.

3.8 A General Deterministic-Shift Extension

In this section we illustrate a simple method to extend any time-homogeneous short-rate model, so as to exactly reproduce any observed term structure of interest rates while preserving the possible analytical tractability of the original model. This method has been extensively described by Brigo and Mercurio (1998, 2001a). A similar approach has inspired the independent works by Scott (1995), Dybvig (1997), and Avellaneda and Newman (1998).

In the case of the Vasicek (1977) model, the extension is perfectly equivalent to that of Hull and White (1990b), see Section 3.3. In the case of the Cox-Ingersoll-Ross (1985) model, instead, the extension is more analytically tractable than those of Section 3.4 and avoids problems concerning the use of numerical solutions. In fact, we can exactly fit any observed term structure of interest rates and derive analytical formulas both for pure discount bonds and for European bond options. The unique drawback is that in principle we can guarantee the positivity of rates only through restrictions on the parameters, which might worsen the quality of the calibration to caps/floors or swaption prices.

The CIR model is the most relevant case to which this procedure can be applied. Indeed, the extension yields the unique short-rate model featuring the following three properties: i) Exact fit of any observed term structure; ii) Analytical formulas for bond prices, bond-option prices, swaptions and caps prices; iii) The distribution of the instantaneous spot rate has tails that are fatter than in the Gaussian case and, through restriction on the parameters, it is always possible to guarantee positive rates without worsening the

[14] The specification of the function a is motivated by its interpretation in the above Gaussian case.

volatility calibration in most situations. Moreover, one further property of the extended model is that the term structure is affine in the short rate. The above uniqueness is the reason why we devote more space to the CIR case.

The extension procedure is also applied to the Dothan (1978) model, thus yielding a shifted lognormal short-rate model that fits any given yield curve and for which there exist analytical formulas for zero-coupon bonds.

Though conceived for analytically tractable models, the method we describe in this section can be employed to extend more general time-homogeneous models. As a clarification, we consider the example of an original short-rate process that evolves as an Exponential-Vasicek process. The only requirement that is needed in general is a numerical procedure for pricing interest-rate derivatives under the original model.

3.8.1 The Basic Assumptions

We consider a time-homogeneous stochastic process x^α, whose dynamics under a given measure Q^x is expressed by

$$dx_t^\alpha = \mu(x_t^\alpha;\alpha)dt + \sigma(x_t^\alpha;\alpha)dW_t^x \,, \qquad (3.64)$$

where W^x is a standard Brownian motion, x_0^α is a given real number, $\alpha = \{\alpha_1, \ldots, \alpha_n\} \in I\!R^n$, $n \geq 1$, is a vector of parameters, and μ and σ are sufficiently well behaved real functions. We set $\mathcal{F}_t^x$ to be the sigma-field generated by x^α up to time t.

We assume that the process x^α describes the evolution of the instantaneous spot interest rate under the risk-adjusted martingale measure, and refer to this model as to the "reference model". We denote by $P^x(t,T)$ the price at time t of a zero-coupon bond maturing at T and with unit face value, so that

$$P^x(t,T) = E^x\left\{\exp\left[-\int_t^T x_s^\alpha ds\right] |\mathcal{F}_t^x\right\},$$

where E^x denotes the expectation under the risk-adjusted measure Q^x.

We also assume there exists an explicit real function Π^x, defined on a suitable subset of $I\!R^{n+3}$, such that

$$P^x(t,T) = \Pi^x(t,T,x_t^\alpha;\alpha). \qquad (3.65)$$

The best known examples of spot-rate models satisfying the assumptions above are the Vasicek (1977) model (3.5), the Dothan (1978) model (3.17) and the Cox-Ingersoll-Ross (1985) model (3.21).

We now illustrate a simple approach for extending the time-homogeneous spot-rate model (3.64), in such a way that the extended version preserves the analytical tractability of the initial model while exactly fitting the observed term structure of interest rates. Precisely, we define the instantaneous short rate under the risk neutral measure Q by

$$r_t = x_t + \varphi(t;\alpha)\ , \quad t \geq 0, \tag{3.66}$$

where x is a stochastic process that has under Q the same dynamics as x^α under Q^x, and φ is a deterministic function, depending on the parameter vector (α, x_0), that is integrable on closed intervals. Notice that x_0 is one more parameter at our disposal. We are free to select its value as long as

$$\varphi(0;\alpha) = r_0 - x_0\ .$$

The function φ can be chosen so as to fit exactly the initial term structure of interest rates. We set $\mathcal{F}_t$ to be the sigma-field generated by x up to time t.

We notice that the process r depends on the parameters $\alpha_1, \ldots, \alpha_n, x_0$ both through the process x and through the function φ. As a common practice, we can determine $\alpha_1, \ldots, \alpha_n, x_0$ by calibrating the model to the current term structure of volatilities, fitting for example cap and floor prices or a few swaption prices.

Notice that, if φ is differentiable, the stochastic differential equation for the short-rate process (3.66) is,

$$dr_t = \left[\frac{d\varphi(t;\alpha)}{dt} + \mu(r_t - \varphi(t;\alpha);\alpha)\right] dt + \sigma(r_t - \varphi(t;\alpha);\alpha)dW_t\ .$$

As we have seen in Section 3.2.4, in case of time-homogeneous coefficients an affine term structure in the short-rate is equivalent to affine drift and squared diffusion coefficients. It follows immediately that if the reference model has an affine term structure so does the extended model. We may then anticipate that the shifted Vasicek (equivalent to HW) and CIR (CIR++) models will be affine term-structure models.

3.8.2 Fitting the Initial Term Structure of Interest Rates

Definition (3.66) immediately leads to the following.

Theorem 3.8.1. *The price at time t of a zero-coupon bond maturing at T is*

$$P(t,T) = \exp\left[-\int_t^T \varphi(s;\alpha)ds\right] \Pi^x(t,T,r_t - \varphi(t;\alpha);\alpha). \tag{3.67}$$

Proof. Denoting by E the expectation under the measure Q, we simply have to notice that

$$\begin{aligned} P(t,T) &= E\left\{\exp\left[-\int_t^T (x_s + \varphi(s;\alpha))ds\right] |\mathcal{F}_t\right\} \\ &= \exp\left[-\int_t^T \varphi(s;\alpha)ds\right] E\left\{\exp\left[-\int_t^T x_s ds\right] |\mathcal{F}_t\right\} \\ &= \exp\left[-\int_t^T \varphi(s;\alpha)ds\right] \Pi^x(t,T,x_t;\alpha), \end{aligned}$$

where in the last step we use the equivalence of the dynamics of x under Q and x^α under Q^x. □

If we denote by $f^x(0,t;\alpha)$ the instantaneous forward rate at time 0 for a maturity t as associated with the bond price $\{P^x(0,t) : t \geq 0\}$, i.e.,

$$f^x(0,t;\alpha) = -\frac{\partial \ln P^x(0,t)}{\partial t} = -\frac{\partial \ln \Pi^x(0,t,x_0;\alpha)}{\partial t},$$

we then have the following.

Corollary 3.8.1. *The model (3.66) fits the currently observed term structure of discount factors if and only if*

$$\varphi(t;\alpha) = \varphi^*(t;\alpha) := f^M(0,t) - f^x(0,t;\alpha), \tag{3.68}$$

i.e., if and only if

$$\exp\left[-\int_t^T \varphi(s;\alpha)ds\right] = \Phi^*(t,T,x_0;\alpha) := \frac{P^M(0,T)}{\Pi^x(0,T,x_0;\alpha)} \frac{\Pi^x(0,t,x_0;\alpha)}{P^M(0,t)}. \tag{3.69}$$

Moreover, the corresponding zero-coupon-bond prices at time t are given by $P(t,T) = \Pi(t,T,r_t;\alpha)$, where

$$\Pi(t,T,r_t;\alpha) = \Phi^*(t,T,x_0;\alpha)\, \Pi^x(t,T,r_t - \varphi^*(t;\alpha);\alpha) \tag{3.70}$$

Proof. From the equality

$$P^M(0,t) = \exp\left[-\int_0^t \varphi(s;\alpha)ds\right] \Pi^x(0,t,x_0;\alpha),$$

we obtain (3.68) by taking the natural logarithm of both members and then differentiating. From the same equality, we also obtain (3.69) by noting that

$$\begin{aligned}\exp\left[-\int_t^T \varphi(s;\alpha)ds\right] &= \exp\left[-\int_0^T \varphi(s;\alpha)ds\right] \exp\left[\int_0^t \varphi(s;\alpha)ds\right] \\ &= \frac{P^M(0,T)}{\Pi^x(0,T,x_0;\alpha)} \frac{\Pi^x(0,t,x_0;\alpha)}{P^M(0,t)},\end{aligned}$$

which, combined with (3.67), gives (3.70). □

Notice that by choosing $\varphi(t;\alpha)$ as in (3.68), the model (3.66) exactly fits the observed term structure of interest rates, no matter which values of α and x_0 are chosen.

3.8.3 Explicit Formulas for European Options

The extension (3.66) is even more interesting when the reference model (3.64) allows for analytical formulas for zero-coupon-bond options as well. It is easily seen that the extended model preserves the analytical tractability for option prices by means of analytical correction factors that are defined in terms of φ.

Noting that under the model (3.64), the price at time t of a European call option with maturity T, strike K and written on a zero-coupon bond maturing at time τ is

$$V^x(t,T,\tau,K) = E^x\left\{\exp\left[-\int_t^T x_s^\alpha ds\right](P^x(T,\tau)-K)^+|\mathcal{F}_t^x\right\},$$

we assume there exists an explicit real function Ψ^x defined on a suitable subset of $\mathbb{R}^{n+5}$, such that

$$V^x(t,T,\tau,K) = \Psi^x(t,T,\tau,K,x_t^\alpha;\alpha). \tag{3.71}$$

The best known examples of models (3.64) for which this holds are again the Vasicek (1977) model (3.5) and the Cox-Ingersoll-Ross (1985) model (3.21).

Straightforward algebra leads to the following.

Theorem 3.8.2. *Under the model (3.66), the price at time t of a European call option with maturity T, strike K and written on a zero-coupon bond maturing at time τ is*

$$\begin{aligned}\mathbf{ZBC}(t,T,\tau,K) &= \exp\left[-\int_t^T \varphi(s;\alpha)ds\right]\\ &\cdot\Psi^x\left(t,T,\tau,K\exp\left[\int_T^\tau \varphi(s;\alpha)ds\right],r_t-\varphi(t;\alpha);\alpha\right).\end{aligned} \tag{3.72}$$

Proof. We simply have to notice that

$$\begin{aligned}\mathbf{ZBC}(t,T,\tau,K) &= E\left\{\exp\left[-\int_t^T (x_s+\varphi(s;\alpha))ds\right](P(T,\tau)-K)^+|\mathcal{F}_t\right\}\\ &= \exp\left[-\int_t^T \varphi(s;\alpha)ds\right]E\left\{\exp\left[-\int_t^T x_s ds\right]\right.\\ &\left.\cdot\left(\exp\left[-\int_T^\tau \varphi(s;\alpha)ds\right]\Pi^x(T,\tau,x_T;\alpha)-K\right)^+|\mathcal{F}_t\right\}\end{aligned}$$

$$= \exp\left[-\int_t^T \varphi(s;\alpha)ds\right] E\left\{\exp\left[-\int_t^T x_s^\alpha ds\right]\right.$$

$$\left.\cdot\left(\Pi^x(T,\tau,x_T;\alpha) - K\exp\left[\int_T^\tau \varphi(s;\alpha)ds\right]\right)^+ |\mathcal{F}_t^x\right\}$$

$$= \exp\left[-\int_t^T \varphi(s;\alpha)ds\right] \Psi^x\left(t,T,\tau,K\exp\left[\int_T^\tau \varphi(s;\alpha)ds\right],x_t;\alpha\right),$$

where in the last step we use the equivalence of the dynamics of x under Q and x^α under Q^x. □

The price of a European put option can be obtained through the put-call parity for bond options, and formula (3.4) in particular. As immediate consequence, caps and floors can be priced analytically as well.

The previous formula for a European option holds for any specification of the function φ. In particular, when exactly fitting the initial term structure of interest rates, the equality (3.68) must be used to yield $\mathbf{ZBC}(t,T,\tau,K) = \Psi(t,T,\tau,K,r_t;\alpha)$, where

$$\Psi(t,T,\tau,K,r_t;\alpha) = \Phi^*(t,\tau,x_0;\alpha)\Psi^x(t,T,\tau,K\Phi^*(\tau,T,x_0;\alpha),r_t - \varphi^*(t;\alpha);\alpha). \tag{3.73}$$

To this end, we notice that, if prices are to be calculated at time 0, we need not explicitly compute $\varphi^*(t;\alpha)$ since the relevant quantities are the discount factors at time zero.

Moreover, if Jamshidian (1989)'s decomposition for valuing coupon-bearing bond options, and hence swaptions, can be applied to the model (3.64), the same decomposition is still feasible under (3.66) through straightforward modifications, so that also in the extended model we can price analytically coupon-bearing bond options and swaptions. See also the later section in this chapter being devoted to such issue.

3.8.4 The Vasicek Case

The first application we consider is based on the Vasicek (1977) model. In this case, the basic time-homogeneous model x^α evolves according to (3.5), where the parameter vector is $\alpha = (k,\theta,\sigma)$. Extending this model through (3.66) amounts to have the following short-rate dynamics:

$$dr_t = \left[k\theta + k\varphi(t;\alpha) + \frac{d\varphi(t;\alpha)}{dt} - kr_t\right]dt + \sigma dW_t\,. \tag{3.74}$$

Moreover, under the specification (3.68), $\varphi(t;\alpha) = \varphi^{VAS}(t;\alpha)$, where

$$\varphi^{VAS}(t;\alpha) = f^M(0,t) + (e^{-kt}-1)\frac{k^2\theta - \sigma^2/2}{k^2} - \frac{\sigma^2}{2k^2}e^{-kt}(1-e^{-kt}) - x_0e^{-kt}\,.$$

The price at time t of a zero-coupon bond maturing at time T is

$$P(t,T) = \frac{P^M(0,T)A(0,t)\exp\{-B(0,t)x_0\}}{P^M(0,t)A(0,T)\exp\{-B(0,T)x_0\}} A(t,T)\exp\{-B(t,T)[r_t - \varphi^{VAS}(t;\alpha)]\},$$

and the price at time t of a European call option with strike K, maturity T and written on a zero-coupon bond maturing at time τ is

$$\mathbf{ZBC}(t,T,\tau,K) = \frac{P^M(0,\tau)A(0,t)\exp\{-B(0,t)x_0\}}{P^M(0,t)A(0,\tau)\exp\{-B(0,\tau)x_0\}} \cdot \Psi^{VAS}\left(t,T,\tau,K\frac{P^M(0,T)A(0,\tau)\exp\{-B(0,\tau)x_0\}}{P^M(0,\tau)A(0,T)\exp\{-B(0,T)x_0\}}, r_t - \varphi^{VAS}(t;\alpha);\alpha\right),$$

where

$$\Psi^{VAS}(t,T,\tau,X,x;\alpha) = A(t,\tau)\exp\{-B(t,\tau)x\}\Phi(h) - XA(t,T)\exp\{-B(t,T)x\}\Phi(h-\bar{\sigma})$$

and

$$h = \frac{1}{\bar{\sigma}}\ln\frac{A(t,\tau)\exp\{-B(t,\tau)x\}}{XA(t,T)\exp\{-B(t,T)x\}} + \frac{\bar{\sigma}}{2},$$

$$\bar{\sigma} = \sigma B(T,\tau)\sqrt{\frac{1-e^{-2k(T-t)}}{2k}}$$

$$A(t,T) = \exp\left[\frac{(B(t,T)-T+t)(k^2\theta - \sigma^2/2)}{k^2} - \frac{\sigma^2 B(t,T)^2}{4k}\right],$$

$$B(t,T) = \frac{1-e^{-k(T-t)}}{k}.$$

We now notice that by defining

$$\vartheta(t) = \theta + \varphi(t;\alpha) + \frac{1}{k}\frac{d\varphi(t;\alpha)}{dt},$$

model (3.74) can be written as

$$dr_t = k(\vartheta(t) - r_t)dt + \sigma\, dW_t \tag{3.75}$$

which coincides with the Hull and White (1994a) extended Vasicek model (3.33).

Vice versa, from (3.75), one can obtain the extension (3.74) by setting

$$\varphi(t;\alpha) = e^{-kt}\varphi(0;\alpha) + k\int_0^t e^{-k(t-s)}\vartheta(s)ds - \theta(1-e^{-kt}).$$

Therefore, this extension of the Vasicek (1977) model is perfectly equivalent to that of Hull and White (1994a), as we can verify also by developing more explicitly our bond and option-pricing formulas. This equivalence is basically due to the linearity of the reference-model equation (3.5). In fact, the extra parameter θ turns out to be redundant since it is absorbed by the time-dependent function ϑ that is completely determined through the fitting of the current term structure of interest rates.

We notice that the function φ^{VAS}, for $\theta = 0$, is related to the function α in (3.36). The only difference is that, in (3.36), r_0 is completely absorbed by the function α, whereas here r_0 is partly absorbed by the reference-model initial condition x_0 and partly by $\varphi^{VAS}(0;\alpha)$.

We finally notice that in the Vasicek case keeping a general x_0 adds no further flexibility to the extended model, because of linearity of the short-rate equation. Therefore, here we can safely set $x_0 = r_0$ and $\varphi^{VAS}(0;\alpha) = 0$ without affecting the model fitting quality.

In the following sections we will develop further examples of models that can be obtained via the deterministic shift extension (3.66).

3.9 The CIR++ Model

The most relevant application of the results of the previous section is the extension of the Cox-Ingersoll-Ross (1985) model, referred to as CIR++. In this case, the process x^α is defined as in (3.21), where the parameter vector is $\alpha = (k, \theta, \sigma)$. The short-rate dynamics is then given by

$$\begin{aligned} dx(t) &= k(\theta - x(t))dt + \sigma\sqrt{x(t)}dW(t), \quad x(0) = x_0, \\ r(t) &= x(t) + \varphi(t), \end{aligned} \tag{3.76}$$

where x_0, k, θ and σ are positive constants such that $2k\theta > \sigma^2$, thus ensuring that the origin is inaccessible to x, and hence that the process x remains positive.

We calculate the analytical formulas implied by such extension, by simply retrieving the explicit expressions for Π^x and Ψ^x as given in (3.24) and (3.26). Then, assuming exact fitting of the initial term structure of discount factors, we have that $\varphi(t) = \varphi^{CIR}(t;\alpha)$ where

$$\begin{aligned} \varphi^{CIR}(t;\alpha) &= f^M(0,t) - f^{CIR}(0,t;\alpha), \\ f^{CIR}(0,t;\alpha) &= \frac{2k\theta(\exp\{th\}-1)}{2h+(k+h)(\exp\{th\}-1)} + x_0\frac{4h^2\exp\{th\}}{[2h+(k+h)(\exp\{th\}-1)]^2} \end{aligned} \tag{3.77}$$

with $h = \sqrt{k^2 + 2\sigma^2}$. Moreover, the price at time t of a zero-coupon bond maturing at time T is

$$P(t,T) = \bar{A}(t,T)e^{-B(t,T)r(t)},$$

where

$$\bar{A}(t,T) = \frac{P^M(0,T)A(0,t)\exp\{-B(0,t)x_0\}}{P^M(0,t)A(0,T)\exp\{-B(0,T)x_0\}} A(t,T)e^{B(t,T)\varphi^{CIR}(t;\alpha)},$$

and $A(t,T)$ and $B(t,T)$ are defined as in (3.25).

The spot interest rate at time t for the maturity T is therefore

$$R(t,T) = \frac{\ln \frac{P^M(0,t)A(0,T)\exp\{-B(0,T)x_0\}}{A(t,T)P^M(0,T)A(0,t)\exp\{-B(0,t)x_0\}}}{T-t} - \frac{B(t,T)\varphi^{CIR}(t;\alpha) - B(t,T)r(t)}{T-t}$$

which is still affine in $r(t)$.

The price at time t of a European call option with maturity $T > t$ and strike price K on a zero-coupon bond maturing at $\tau > T$ is

$$\mathbf{ZBC}(t,T,\tau,K) = \frac{P^M(0,\tau)A(0,t)\exp\{-B(0,t)x_0\}}{P^M(0,t)A(0,\tau)\exp\{-B(0,\tau)x_0\}} \cdot \Psi^{CIR}\left(t,T,\tau,K\frac{P^M(0,T)A(0,\tau)\exp\{-B(0,\tau)x_0\}}{P^M(0,\tau)A(0,T)\exp\{-B(0,T)x_0\}}, r(t) - \varphi^{CIR}(t;\alpha);\alpha\right),$$

where $\Psi^{CIR}(t,T,\tau,X,x;\alpha)$ is the CIR option price as defined in (3.26) with $r(t) = x$. By further simplifying this formula, we obtain

$$\begin{aligned}\mathbf{ZBC}(t,T,\tau,K) = {} & P(t,\tau)\chi^2\left(2\hat{r}[\rho+\psi+B(T,\tau)]; \frac{4k\theta}{\sigma^2}, \frac{2\rho^2[r(t)-\varphi^{CIR}(t;\alpha)]\exp\{h(T-t)\}}{\rho+\psi+B(T,\tau)}\right) \\ & - KP(t,T)\chi^2\left(2\hat{r}[\rho+\psi]; \frac{4k\theta}{\sigma^2}, \frac{2\rho^2[r(t)-\varphi^{CIR}(t;\alpha)]\exp\{h(T-t)\}}{\rho+\psi}\right),\end{aligned} \tag{3.78}$$

with

$$\hat{r} = \frac{1}{B(T,\tau)}\left[\ln\frac{A(T,\tau)}{K} - \ln\frac{P^M(0,T)A(0,\tau)\exp\{-B(0,\tau)x_0\}}{P^M(0,\tau)A(0,T)\exp\{-B(0,T)x_0\}}\right].$$

The analogous put-option price is obtained through the put-call parity (3.4).

Through formula (3.78), we can also price caps and floors since they can be viewed as portfolios of zero bond options.

Let us start by a single caplet value at time t. The caplet resets at time T, pays at time $T+\tau$ and has strike X and nominal amount N.

$$\mathbf{Cpl}(t,T,T+\tau,N,X) = N(1+X\tau)\mathbf{ZBP}\left(t,T,T+\tau,\frac{1}{1+X\tau}\right). \quad (3.79)$$

More generally, as far as a cap is concerned, we denote by $\mathcal{T} = \{t_0, t_1, \ldots, t_n\}$ the set of the cap payment times augmented with the first reset date t_0, and by τ_i the year fraction from t_{i-1} to t_i, $i = 1, \ldots, n$. Applying formulas (2.29), we then obtain that the price at time $t < t_0$ of the cap with cap rate (strike) X, nominal value N and set of times $\mathcal{T}$ is given by

$$\mathbf{Cap}(t,\mathcal{T},N,X) = N\sum_{i=1}^{n}(1+X\tau_i)\mathbf{ZBP}\left(t,t_{i-1},t_i,\frac{1}{1+X\tau_i}\right), \quad (3.80)$$

whereas the price of the corresponding floor is

$$\mathbf{Flr}(t,\mathcal{T},N,X) = N\sum_{i=1}^{n}(1+X\tau_i)\mathbf{ZBC}\left(t,t_{i-1},t_i,\frac{1}{1+X\tau_i}\right). \quad (3.81)$$

European swaptions can be explicitly priced by means of Jamshidian's (1989) decomposition. Indeed, consider a payer swaption with strike rate X, maturity T and nominal value N, which gives the holder the right to enter at time $t_0 = T$ an interest-rate swap with payment times $\mathcal{T} = \{t_1, \ldots, t_n\}$, $t_1 > T$, where he pays at the fixed rate X and receives LIBOR set "in arrears". We denote by τ_i the year fraction from t_{i-1} to t_i, $i = 1, \ldots, n$ and set $c_i := X\tau_i$ for $i = 1, \ldots, n-1$ and $c_n := 1 + X\tau_i$. Denoting by r^* the value of the spot rate at time T for which

$$\sum_{i=1}^{n} c_i \bar{A}(T,t_i)e^{-B(T,t_i)r^*} = 1,$$

and setting $X_i := \bar{A}(T,t_i)\exp(-B(T,t_i)r^*)$, the swaption price at time $t < T$ is then given by

$$\mathbf{PS}(t,T,\mathcal{T},N,X) = N\sum_{i=1}^{n} c_i \mathbf{ZBP}(t,T,t_i,X_i). \quad (3.82)$$

Analogously, the price of the corresponding receiver swaption is

$$\mathbf{RS}(t,T,\mathcal{T},N,X) = N\sum_{i=1}^{n} c_i \mathbf{ZBC}(t,T,t_i,X_i). \quad (3.83)$$

Remark 3.9.1. The T-forward-measure dynamics and distribution of the short rate are easily obtained through (3.66) from (3.27) and (3.28) where r is replaced with x.

3.9.1 The Construction of a Trinomial Tree

Legolas Greenleaf long under tree
In joy thou hast lived. Beware of the Sea!
J.R.R. Tolkien, "The Lord of The Rings"

A binomial tree for the CIR model has been suggested by Nelson and Ramaswamy (1990). Here, instead, we propose an alternative trinomial tree that is constructed along the procedure illustrated in Appendix F.

For convergence purposes, we define the process y as

$$y(t) = \sqrt{x(t)}, \tag{3.84}$$

where the process x is defined in (3.76). By Ito's lemma,

$$dy(t) = \left[\left(\frac{k\theta}{2} - \frac{1}{8}\sigma^2\right)\frac{1}{y(t)} - \frac{k}{2}y(t)\right]dt + \frac{\sigma}{2}dW(t) \tag{3.85}$$

We first construct a trinomial tree for y, we then use (3.84) and we finally displace the tree nodes so as to exactly retrieve the initial zero-coupon curve.

Let us fix a time horizon T and the times $0 = t_0 < t_1 < \cdots < t_N = T$, and set $\Delta t_i = t_{i+1} - t_i$, for each i. As usual, the time instants t_i need not be equally spaced. Again, we denote the tree nodes by (i, j) where the time index i ranges from 0 to N and the space index j ranges from some $\underline{j}_i$ to some $\overline{j}_i$.

We denote by $y_{i,j}$ the process value on node (i, j) and set $y_{i,j} = j\Delta y_i$, where $\Delta y_i := V_{i-1}\sqrt{3}$ and $V_i := \sigma\sqrt{\Delta t_i}/2$. We set

$$M_{i,j} := y_{i,j} + \left[\left(\frac{k\theta}{2} - \frac{1}{8}\sigma^2\right)\frac{1}{y_{i,j}} - \frac{k}{2}y_{i,j}\right]\Delta t_i.$$

Assuming that at time t_i we are on node (i, j), with associated value $y_{i,j}$, the process can move to $y_{i+1,k+1}$, $y_{i+1,k}$ or $y_{i+1,k-1}$ at time t_{i+1} with probabilities p_u, p_m and p_d, respectively. The central node is therefore the k-th node at time t_{i+1}, where k is defined by

$$k = \text{round}\left(\frac{M_{i,j}}{\Delta y_{i+1}}\right).$$

Setting $\eta_{j,k} = M_{i,j} - y_{i+1,k}$,[15] we finally obtain

$$\begin{cases} p_u = \frac{1}{6} + \frac{\eta_{j,k}^2}{6V_i^2} + \frac{\eta_{j,k}}{2\sqrt{3}V_i}, \\ p_m = \frac{2}{3} - \frac{\eta_{j,k}^2}{3V_i^2}, \\ p_d = \frac{1}{6} + \frac{\eta_{j,k}^2}{6V_i^2} - \frac{\eta_{j,k}}{2\sqrt{3}V_i}. \end{cases}$$

[15] We again omit to express the dependence on the index i to lighten notation.

However, the tree thus defined has the drawback that some nodes may lie below the zero level. Since the tree must approximate a positive process, we truncate the tree below some predefined level $\epsilon > 0$, which can be chosen arbitrarily close to zero, and then suitably define the tree geometry and probabilities around this level.

All the definitions above completely specify the initial-tree geometry, and in particular the minimum and the maximum levels $\underline{j}_i$ and $\overline{j}_i$ at each time step i. The tree for the process x is then built by remembering that, by (3.84), $x(t) = y^2(t)$.

The final stage in our construction procedure consists in suitably shifting the tree nodes in order to obtain the proper tree for r. To this end, we can either apply formula (3.76) with the analytical formula for φ given in (3.77), or employ a similar procedure to that illustrated in Section 3.3 to retrieve the correct discount factors at the initial time.

3.9.2 Early Exercise Pricing via Dynamic Programming

When discussing intensity models for credit derivatives, in Section 22.7.10 we will give some references to pricing Bermudan options with the CIR++ model. The algorithm can be easily recast for interest rate options. The method represents a possible alternative to the trinomial tree approach for pricing options under early exercise features. The interested reader is referred also to Ben Ameur, Brigo and Errais (2005).

3.9.3 The Positivity of Rates and Fitting Quality

The use of CIR++ as a pricing model concerns mostly non-standard interest-rate derivatives. The analytical formulas given in this section are used to determine the model parameters α such that the model prices are as close as possible to the selected subset of market prices. An important issue for the CIR++ model is whether calibration to market prices is feasible while imposing positive rates. We know that the CIR++ rates are always positive if

$$\varphi^{CIR}(t;\alpha) > 0 \quad \text{for all } t \geq 0.$$

In turn, this condition is satisfied if

$$f^{CIR}(0,t;\alpha) < f^M(0,t) \quad \text{for all } t \geq 0. \tag{3.86}$$

Studying the behaviour of the function $t \mapsto f^{CIR}(0,t;\alpha)$ is helpful in addressing the positivity issue. In particular, we are interested in its supremum

$$f^*(\alpha) := \sup_{t\geq 0} f^{CIR}(0,t;\alpha)\ .$$

There are three possible cases of interest:

i) $x_0 \leq \theta h/k$: In this case $t \mapsto f^{CIR}(0,t;\alpha)$ is monotonically increasing and the supremum of all its values is

$$f_1^*(\alpha) = \lim_{t\to\infty} f^{CIR}(0,t;\alpha) = \frac{2k\theta}{k+h};$$

ii) $\theta h/k < x_0 < \theta$: In this case $t \mapsto f^{CIR}(0,t;\alpha)$ takes its maximum value in

$$t^* = \frac{1}{h}\ln\frac{(x_0 h + k\theta)(h-k)}{(x_0 h - k\theta)(h+k)} > 0$$

and such a value is given by

$$f_2^*(\alpha) = x_0 + \frac{(x_0-\theta)^2 k^2}{2\sigma^2 x_0};$$

iii) $x_0 \geq \theta$: In this case $t \mapsto f^{CIR}(0,t;\alpha)$ is monotonically decreasing for $t > 0$ and the supremum of all its values is

$$f_3^*(\alpha) = f^{CIR}(0,0;\alpha) = x_0.$$

We can try and enforce positivity of φ^{CIR} analytically in a number of ways. To ensure (3.86) one can for example impose that the market curve $t \mapsto f^M(0,t)$ remains above the corresponding CIR curve by requiring

$$f^*(\alpha) \leq \inf_{t\geq 0} f^M(0,t)\ .$$

This condition ensures (3.86), but appears to be too restrictive. To fix ideas, assume that, calibrating the model parameters to market cap and floor prices, we constrain the parameters to satisfy

$$x_0 > \theta\ .$$

Then we are in case iii) above. All we need for ensuring the positivity of φ^{CIR} in this case is

$$x_0 < \inf_{t\geq 0} f^M(0,t)\ .$$

This amounts to say that the initial condition of the time-homogeneous part of the model has to be placed below the whole market forward curve. On the other hand, since $\theta < x_0$ and θ is the mean-reversion level of the time-homogeneous part, this means that the time-homogeneous part will tend to decrease, and indeed we have seen that in case iii) $t \mapsto f^{CIR}(0,t;\alpha)$ is monotonically decreasing. If, on the contrary, the market forward curve is increasing (as is happening in the most liquid markets nowadays) the "reconciling role" of φ^{CIR} between the time-homogeneous part and the market forward curve will be stronger. Part of the flexibility of the time-homogeneous part of the model is then lost in this strong "reconciliation", thus subtracting freedom which the model can otherwise use to improve calibration to caps and floors prices.

In general, it turns out in real applications that the above requirements are too strong. Constraining the parameters to satisfy

$$\theta < \; x_0 < \inf_{t \geq 0} f^M(0,t)$$

leads to a caps/floors calibration whose quality is much lower than we have with less restrictive constraints. For practical purposes it is in fact enough to impose weaker restrictions. Such restrictions, though not guaranteeing positivity of φ^{CIR} analytically, work well in all the market situation we tested. First consider the case of a monotonically increasing market curve $t \mapsto f^M(0,t)$. We can choose a similarly increasing $t \mapsto f^{CIR}(0,t;\alpha)$ (case i)) starting from below the market curve, $x_0 < r_0$, and impose that its asymptotic limit

$$\frac{2k\theta}{k+h} \approx \theta$$

be below the corresponding market-curve limit. If we do this, calibration results are satisfactory and we obtain usually positive rates. Of course we have no analytical certainty. The f^{CIR} curve might increase quicker than the market one, f^M, for small t's, cross it, and then increase more slowly so as to return below f^M for large $t's$. In such a case φ^{CIR} would be negative in the in–between interval. However, this situation is very unlikely and never occurred in the real-market situations we tested.

Next, consider a case with a decreasing market curve $t \mapsto f^M(0,t)$. In this case we take again $x_0 < r_0$ and we impose the same condition as before on the terminal limit for $t \to \infty$.

Similar considerations apply in the case of an upwardly humped market curve $t \mapsto f^M(0,t)$, which can be reproduced qualitatively by the time-homogeneous-model curve $t \mapsto f^{CIR}(0,t;\alpha)$. In this case one makes sure that the initial point, the analytical maximum and the asymptotic value of the time-homogeneous-CIR curve remain below the corresponding points of the market curve.

Finally, the only critical situation is the case of an inverted yield curve, as was observed for example in the Italian market in the past years. The forward curve $t \mapsto f^{CIR}(0,t;\alpha)$ of the CIR model cannot mimic such a shape. Therefore, either we constrain it to stay below the inverted market curve by choosing a decreasing CIR curve (case iii)) starting below the market curve, i.e. $x_0 < r_0$, or we make the CIR curve start from a very small x_0 and increase, though not too steeply. The discrepancy between f^{CIR} and f^M becomes very large for large t's in the first case and for small t's in the second one. This feature lowers the quality of the caps/floors fitting in the case of an inverted yield curve if one wishes to maintain positive rates. However, highly inverted curves are not so common in liquid markets, so that this problem can be generally avoided.

As far as the quality of fitting for the caps volatility curve is concerned, we notice the following. If the initial point in the market caps volatility curve is much smaller than the second one, so as to produce a large hump, the CIR++ model has difficulties in fitting the volatility structure. We observed through numerical simulations that, ceteris paribus, lowering the initial point of the caps volatility structure implied by the model roughly amounts to lowering the parameter x_0. This is consistent with the following formula

$$\sigma\sqrt{x_0}\,\frac{2h\,\exp(Th)}{[2h+(k+h)(\exp\{Th\}-1)]^2}$$

for the volatility of instantaneous forward rates $f^{CIR}(t,T;\alpha)$ at the initial time $t=0$ for the maturity T. Therefore, a desirable fitting of the caps volatility curve can require low values of x_0, in agreement with one of the conditions needed to preserve positive rates.

3.9.4 Monte Carlo Simulation

We give some hints on Monte Carlo simulation for the CIR++ model in Section (3.11.2) below. However, most of the ad-hoc schemes for the CIR and CIR++ models are given in the Credit Chapters, in Sections 22.7.3 and 22.7.4, where we present the schemes of Brigo and Alfonsi (2003, 2005) and of Alfonsi (2005). Although these sections are mostly devoted to CIR++ intensity models, they can be easily re-cast in terms of CIR++ short rate models. The reader interested in implementing a Monte Carlo scheme for the CIR++ model is advised to read first Section 3.11.2 and then Sections 22.7.3 and 22.7.4.

3.9.5 Jump Diffusion CIR and CIR++ models (JCIR, JCIR++)

It is possible to add a jump component to the CIR and CIR++ models, while retaining analytical tractability for bond prices, part of the feasibility of Jamshidian's decomposition for coupon bond bearing options and swaptions, and other attractive features. For a discussion see also Brigo and El-Bachir (2005). This extension assumes

$$dr_t = k(\theta - r_t)dt + \sigma\sqrt{r_t}dW_t + dJ_t,$$

where J is a pure jump process with jumps arrival rate $\alpha > 0$ and jump sizes distribution π on $\mathbb{R}^+$.

Notice that we restrict the jumps to be positive, preserving the attractive feature of positive interest rates implied by the basic CIR dynamics. Further, assume that π is an exponential distribution with mean $\gamma > 0$, and that

$$J_t = \sum_{i=1}^{M_t} Y_i$$

where M is a time-homogeneous Poisson process with intensity α, the Ys being exponentially distributed with parameter γ. The larger α, the more frequent the jumps, and the larger γ, the larger the sizes of the occurring jumps. We denote the resulting jump process by $J^{\alpha,\gamma}$, to point out the parameters influencing its dynamics. We write

$$dr_t = k(\theta - r_t)dt + \sigma\sqrt{r_t}dW_t + dJ_t^{\alpha,\gamma}, \tag{3.87}$$

so that we can see all the parameters in the dynamics.

This model is an affine model, in that the bond price formula maintains the familiar log-affine shape. Many more details on this model, including the shifted (++) version, are available in Section 22.8 of the credit chapters, whereas an introduction on Poisson processes is given in the Appendices in Section C.6. Again in Section 22.8 detailed formulas and algorithms are given and further references are suggested. We also show how this model may attain high implied volatilities (for swaptions or caps, in the present context) when the basic CIR++ model fails to do so. The reader may have to translate back the notation from stochastic intensity modeling to short rate modeling, but this is rather straightforward.

3.10 Deterministic-Shift Extension of Lognormal Models

The third example we consider is the extension of the Dothan (1978) model. This extension yields a "quasi" lognormal short-rate model that fits any given yield curve and for which there exist analytical formulas for zero-coupon bonds, i.e.,

$$\begin{aligned} r(t) &= x(t) + \varphi(t), \\ x(t) &= x_0 \exp\left\{\left(a - \frac{1}{2}\sigma^2\right)t + \sigma W(t)\right\}, \end{aligned} \tag{3.88}$$

where x_0, a and σ are real constants and φ is a deterministic function of time.

The introduction of a deterministic shift in the Dothan (1978) short-rate dynamics (3.17) implies that the expected long-term rate is not just zero or infinity, depending on the sign of the parameter a, but can have any value, given the asymptotic behaviour of the function φ.

The analytical formulas for bond prices in the extended Dothan model are simply obtained by combining (3.70) with (3.20). Equivalently to (3.20), however, the resulting bond price formula is rather involved, requiring double numerical integrations, so that implementing an approximating tree for the process x may be easier and not necessarily more time consuming.

The construction of a binomial tree for pricing interest-rate derivatives under this extended Dothan model is rather straightforward. In fact, we just have to build a tree for x, the time-homogeneous part of the process r, and then shift the tree nodes at each time period by the corresponding value of φ.

Since (3.17) is a geometric Brownian motion, the well consolidated Cox-Ross-Rubinstein (1979) procedure can be used here, thus rendering the tree construction extremely simple while simultaneously ensuring well known results of convergence in law. To this end, we denote by N the number of time-steps in the tree and with T a fixed maturity. We then define the following coefficients

$$u = e^{\sigma\sqrt{\frac{T}{N}}}, \tag{3.89}$$

$$d = e^{-\sigma\sqrt{\frac{T}{N}}}, \tag{3.90}$$

$$p = \frac{e^{a\frac{T}{N}} - d}{u - d}, \tag{3.91}$$

and build inductively the tree for x starting from x_0. Denoting by $\hat{x}$ the value of x at a certain node of the tree, the value of x in the subsequent period can either go up to $\hat{x}u$ with probability p or go down to $\hat{x}d$ with probability $1-p$. Notice that the probabilities p and $1-p$ are always well defined for a sufficiently large N, both tending to $\frac{1}{2}$ for N going to infinity.

The tree for the short-rate process r is finally constructed by displacing the previous nodes through the function φ. When exactly fitting the current term-structure of interest rates, the displacement of the tree nodes can be done, for instance, through a procedure that is similar to that illustrated in Section 3.3.3.

The last example we consider is the extension of the Exponential-Vasicek model (3.30). To this end, we remark that the extension procedure of the previous section, though particularly meaningful when the original model is time-homogeneous and analytically tractable, is quite general and can be in principle applied to any endogenous term-structure model.

The short-rate process under such extension is explicitly given by

$$\begin{aligned} r(t) &= x(t) + \varphi(t) \\ x(t) &= \exp\left\{\ln x_0 e^{-at} + \frac{\theta}{a}\left(1 - e^{-at}\right) + \sigma \int_0^t e^{-a(t-u)} dW(u)\right\}, \end{aligned} \tag{3.92}$$

where x_0, θ, a and σ are real constants and φ is a deterministic function of time.

The Exponential-Vasicek model does not imply explicit formulas for pure discount bonds. However, we can easily construct an approximating tree for the reference process x, whose dynamics are defined by (3.30). In fact, defining the process z as

$$dz(t) = -az(t)dt + \sigma dW(t), \quad z(0) = 0,$$

by Ito's lemma, we obtain

$$x(t) = \exp\left(z(t) + \left(\ln x_0 - \frac{\theta}{a}\right)e^{-at} + \frac{\theta}{a}\right). \tag{3.93}$$

We can then use the procedure illustrated in Section 3.3.3 to construct a trinomial model for z, apply (3.93) and finally displace the tree node so as to exactly retrieve the initial term structure of discount factors.

This Extended Exponential-Vasicek (EEV) model leads to a fairly good calibration to cap prices in that the resulting model prices lie within the band formed by the market bid and ask prices. Indeed, this model clearly outperforms the classical lognormal short-rate model (3.54). An example of the quality of the model fitting to at-the-money cap prices is shown later in this chapter.

A final word has to be spent on a further advantage of the above extensions of classical lognormal model. In fact, the assumption of a shifted lognormal distribution can be quite helpful when fitting volatility smiles. Notice, indeed, that the Black formula for cap prices is based on a lognormal distribution for the forward rates, and that a practical way to obtain implied-volatility smiles is by shifting the support of this distribution by a quantity to be suitably determined. However, the major drawback of these extensions is that the positivity of the short-rate process cannot be guaranteed any more since its distribution is obtained by shifting a lognormal distribution by a possibly negative quantity.

3.11 Some Further Remarks on Derivatives Pricing

In this section we first show how to price analytically a European option on a coupon-bearing bond under an analytically-tractable short-rate model. The method we describe is a natural generalization of the Jamshidian (1989) decomposition, originally developed for the Vasicek (1977) model. We then show how to price path-dependent derivatives through a Monte Carlo simulation approach. We finally explain how to price early-exercise products by means of a (trinomial) tree.

3.11.1 Pricing European Options on a Coupon-Bearing Bond

We assume that the short-rate model is analytically tractable and denote the analytical price at time t of the zero coupon bond maturing at time T by $\Pi(t, T, r(t))$, where the dependence on the short rate at time t is explicitly written.

We consider a coupon-bearing bond paying the cash flows $\mathcal{C} = [c_1, \ldots, c_n]$ at maturities $\mathcal{T} = [T_1, \ldots, T_n]$. Let $T \leq T_1$. The price of our coupon-bearing bond in T is given by

$$\mathbf{CB}(T, \mathcal{T}, \mathcal{C}) = \sum_{i=1}^{n} c_i P(T, T_i) = \sum_{i=1}^{n} c_i \Pi(T, T_i, r(T)).$$

We want to price at time t a European put option on the coupon-bearing bond with strike price K and maturity T. The option payoff is

$$[K - \mathbf{CB}(T, \mathcal{T}, \mathcal{C})]^+ = \left[K - \sum_{i=1}^{n} c_i \Pi(T, T_i, r(T))\right]^+.$$

Jamshidian (1989) devised a simple method to convert this positive part of a sum into the sum of positive parts. His trick was based on finding the solution r^* of the following equation

$$\sum_{i=1}^{n} c_i \Pi(T, T_i, r^*) = K,$$

and rewriting the payoff as

$$\left[\sum_{i=1}^{n} c_i \big(\Pi(T, T_i, r^*) - \Pi(T, T_i, r(T))\big)\right]^+.$$

To achieve the desired decomposition we need a condition that was automatically verified in the case of the Vasicek (1977) model treated by Jamshidian. We therefore assume that the short-rate model satisfies the following assumption:

$$\frac{\partial \Pi(t, s, r)}{\partial r} < 0 \quad \text{for all } 0 < t < s.$$

It is easy to see, for instance, that both the Hull and White (3.33) and the CIR++ (3.76) models satisfy this assumption. Under such condition, the payoff can be rewritten as

$$\sum_{i=1}^{n} c_i \big[\Pi(T, T_i, r^*) - \Pi(T, T_i, r(T))\big]^+,$$

so that pricing our coupon-bond option becomes equivalent to value a portfolio of put options on zero-coupon bonds. If we take the risk-neutral expectation of the discounted payoff, we obtain the price at time t of the coupon-bearing-bond option with maturity T and strike K:

$$\mathbf{CBP}(t, T, \mathcal{T}, \mathcal{C}, K) = \sum_{i=1}^{n} c_i \mathbf{ZBP}(t, T, T_i, \Pi(T, T_i, r^*)).$$

An analogous relationship holds for a European call so that we can write

$$\mathbf{CBO}(t, T, \mathcal{T}, \mathcal{C}, K, \omega) = \sum_{i=1}^{n} c_i \mathbf{ZBO}(t, T, T_i, \Pi(T, T_i, r^*), \omega),$$

where $\omega = 1$ for a call and $\omega = -1$ for a put.

3.11.2 The Monte Carlo Simulation

This subsection comments on the use of a short-rate model for Monte Carlo pricing of path-dependent interest-rate derivatives. We may start with a few remarks on the Monte Carlo method. We will reconsider this type of remarks in Section 13.14.

The category of products that is usually considered for Monte Carlo pricing is the family of "path-dependent" payoffs. These products are to be exercised only at a final time, and their final payoffs involve the history of the underlying variable up to the final time, and not only the final value of the underlying variable. The Monte Carlo method works through forward propagation in time of the key variables, by simulating their transition density between dates where the key-variables history matters to the final payoff. Monte Carlo is thus ideally suited to "travel forward in time". Instead, the Monte Carlo method has problems with early exercise. Since with Monte Carlo we propagate trajectories forward in time, we have no means to know whether at a certain point in time it is optimal to continue or to exercise. Therefore, standard Monte Carlo cannot be used for products involving early exercise, although in Section 13.14 we present a method that proposes a remedy to this situation.

Now assume we want to price, at the current time $t = 0$, a path-dependent payoff with European exercise features. The payoff is a function of the values $r(t_1), \ldots, r(t_m)$ of the instantaneous interest rate at preassigned time instants $0 < t_1 < t_2 < \ldots < t_m = T$, where T is the final maturity. Let us denote the given discounted payoff (payments of different additive components occur at different times and are discounted accordingly) by

$$\sum_{j=1}^{m} \exp\left[-\int_0^{t_j} r(s)\, ds\right] H(r(t_1), \ldots, r(t_j)). \tag{3.94}$$

Usually the payoff depends on the instantaneous rates $r(t_i)$ through the simple spot rates $L(t_i, t_i + \tau, r(t_i)) := [1/P(t_i, t_i + \tau) - 1]/\tau$ (typically six-month rates, i.e. $\tau = 0.5$). The availability of analytical formulas for such L's simplifies considerably the Monte Carlo pricing. Indeed, without analytical formulas one should compute the L's from bond prices obtained through trees or through numerical approximations of the solution of the bond price PDE. This should be done for every path, so as to increase dramatically the computational burden.

Pricing the generic additive term in the payoff (3.94) by a Monte Carlo method involves simulation of p paths of the short-rate process r and computing the arithmetic mean of the p values assumed by the discounted payoff along each path.[16]

[16] If no variance reduction technique is employed, p typically ranges from $p = 100,000$ to $p = 1,000,000$.

These paths are simulated after defining $q+1$ sampling times $0 = s_0 < s_1 < s_2 < \ldots < s_q = T$, which have of course to include the times t_j's. We set $\Delta s_i := s_{i+1} - s_i$, $i = 0, \ldots, q-1$.

When pricing under the risk-neutral measure one typically approximates a discounting term like

$$\exp\left[-\int_{s_k}^{s_j} r(s)ds\right]$$

with

$$\exp\left(-\sum_{i=k}^{j-1} r(s_i)\Delta s_i\right).$$

In order for the approximation to be accurate, the Δs_i's have to be small, which can be quite burdensome from the computational point of view. If the short-rate model is analytically tractable, it is therefore advisable to work under the T-forward adjusted measure, since (see also Chapter 2)

$$E\left\{\exp\left[-\int_0^{t_j} r(s)ds\right] H(r(t_1), \ldots, r(t_j))\right\} = P(0,T)E^T\left\{\frac{H(r(t_1), \ldots, r(t_j))}{P(t_j,T)}\right\}. \tag{3.95}$$

Notice that the term $P(t_j,T)$ is determined analytically from the simulated $r(t_j)$, so that no further simulation is required to compute it.

In order to obtain the simulated paths, two approaches are possible:

1. Sample at each time step the exact transition density from $r(s_i)$ to $r(s_{i+1})$, $i = 0, \ldots, q-1$, under the T-forward-adjusted measure;
2. Discretize the SDE for r, under the T-forward-adjusted measure, via the Euler or the Milstein scheme (see Klöden and Platen (1995)).

In the latter case, one has simply to sample at each time s_i the distribution of $W^T_{s_i+\Delta s_i} - W^T_{s_i}$, which is normal with mean 0 and variance Δs_i. Since the increments of the Brownian motion are independent, these normal samples are drawn from variables which are independent in different time intervals, so that one can generate a priori $p \times q$ independent realizations of such Gaussian random variables.

Considering as an example the CIR++ model, we can then either sample the exact noncentral-chi-squared transition density from s_i to s_{i+1}, as given by combining (3.66) with (3.28) and where $s = s_i$ and $t = s_{i+1}$, or employ the following Milstein scheme for the reference process x

$$\begin{aligned} x(s_i + \Delta s_i) &= x(s_i) + [k\theta - (k + B(s_i,T)\sigma^2)x(s_i)]\ \Delta s_i \\ &\quad - \frac{\sigma^2}{4}\left((W^T_{s_i+\Delta s_i} - W^T_{s_i})^2 - \Delta s_i\right) + \sigma\sqrt{x(s_i)}\left(W^T_{s_i+\Delta s_i} - W^T_{s_i}\right). \end{aligned}$$

The price of our derivative is finally obtained by calculating the payoff value

$$h_i := \sum_{j=1}^{m} \frac{H(r(t_1), \ldots, r(t_j))}{P(t_j, T)}$$

associated with each simulated path, averaging all payoff values and then discounting to the current time, thus yielding

$$P(0,T)\frac{\sum_{i=1}^{p} h_i}{p}.$$

We will have more to say on the simulation of CIR++ processes in Sections 22.7.3 and 22.7.4, where we introduce the schemes of Brigo and Alfonsi (2003, 2005) and of Alfonsi (2005).

3.11.3 Pricing Early-Exercise Derivatives with a Tree

Soldati
Bosco di Courton luglio 1918
Si sta come
d'autunno
sugli alberi
le foglie

[Soldiers
We remain
like leaves
on the trees
in autumn]

Giuseppe Ungaretti, translation by William Fense Weaver.

All the short-rate models we have described in this chapter can be approximated with lattices. For some of the models we introduced, we have detailed the construction of a trinomial tree based on the discretization of the rate dynamics under the risk-neutral measure. In this subsection, we show how to price a general (non-path-dependent) derivative with such a numerical procedure. This is particularly useful when pricing early-exercise products, as we will also see by analyzing the particular case of a Bermudan-style swaption. For a general and informal introduction to the use of trees as opposed to Monte Carlo and other methods see also Section 13.14. We now present a few general remarks on pricing with trees, and then move to the details.

The pricing of early-exercise products can be carried out through binomial/trinomial trees, when the fundamental underlying variable is low-dimensional (say one or two-dimensional). This is the case of the short-rate models we are considering here and in the next chapter. In these cases, the tree is the ideal instrument, given its "backward-in-time" nature. The value of the payoff in each final node is known, and we can move backward in time, thus updating the value of continuation through discounting. At each node of the tree we compare the "backwardly-cumulated" value of continuation with the payoff evaluated at that node ("immediate-exercise value"), thus deciding whether exercise is to be considered or not at that point. Once this exercise decision has been taken, the backward induction restarts and we continue to propagate backwards the updated value. Upon reaching the initial node of the tree (at time 0) we have the approximated price of our early-exercise product. It is thus clear that trees are ideally suited to "travel backward in time".

Trees may have problems with path-dependent products, for which we have seen Monte Carlo to be ideally suited. Indeed, in this case when we try and propagate backwards the contract value from the final nodes we are immediately in trouble, since to value the payoff at any final node we need to know the past history of the underlying variable. But this past history is not determined yet, since we move backward in time. Although there are ad-hoc procedures to render trees able to price particular path-dependent products in the Black and Scholes setup (barrier and lookback options), in general there is no efficient and universally accepted method for using a tree with path-dependent payoffs.

Now assume we have selected a short-rate model and a related tree. In order to fix ideas, assume we have selected a trinomial tree, constructed according to our general procedure of Appendix F. We thus have a finite set of times $0 = t_0 < t_1 < \cdots < t_n = T$ and, at each time t_i, a finite number of states. The time-horizon T is the longest maturity that is relevant for pricing a given derivative. The j-th node at time t_i is denoted by (i, j), with associated short-rate value $r_{i,j}$, and $i = 0, \ldots, n$ with j ranging, for each fixed i, from some $\underline{j}_i$ to some $\overline{j}_i$.

The tree branching is shown in Figure 3.6: assuming that, at time t_{i-1}, we are on node $(i-1, j)$, we can move to the three nodes $(i, k+1)$, (i, k) or $(i, k-1)$ at time t_i, with probabilities p_u, p_m and p_d, respectively, where the indices i, j and k in the probabilities are omitted for brevity.

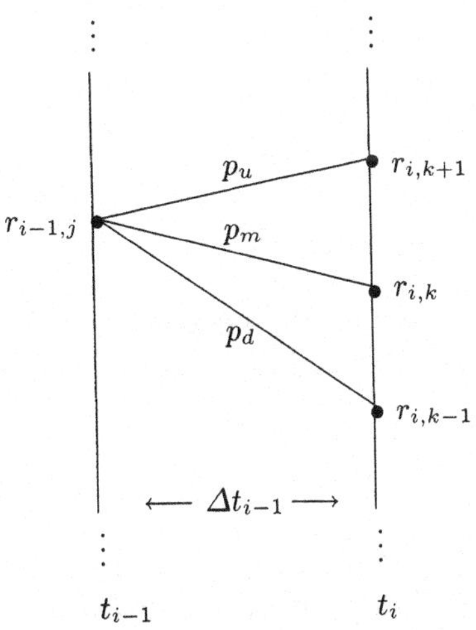

Fig. 3.6. Evolution of the process r starting from $r_{i-1,j}$ at time t_{i-1} and moving to $r_{i,k+1}$, $r_{i,k}$ or $r_{i,k-1}$ at time t_i with probabilities p_u, p_m and p_d, respectively.

A first simple payoff. We start by considering the simple case of a derivative whose payoff at time T is given by a function $H(T, r_T)$. Denoting by $h(t, r_t)$ the arbitrage-free price of the claim at time t, when the short rate is r_t, we have clearly $h(T, r_T) = H(T, r_T)$ and

$$
\begin{aligned}
h(0, r_0) &= E\left\{e^{-\int_0^T r_t dt} h(T, r_T)\right\} \\
&= E\left\{e^{-\int_0^{t_{n-1}} r_t dt} E\left[e^{-\int_{t_{n-1}}^T r_t dt} h(T, r_T) | \mathcal{F}_{t_{n-1}}\right]\right\} \\
&= E\left\{e^{-\int_0^{t_{n-1}} r_t dt} h(t_{n-1}, r_{t_{n-1}})\right\} \\
&= E\left\{e^{-\int_0^{t_1} r_t dt} h(t_1, r_{t_1})\right\},
\end{aligned}
$$

where we have applied the tower property of conditional expectations. The generic value $h(t_i, r_{t_i})$ is calculated by iteratively taking expectations of the discounted values at later times, i.e.

$$
h(t_{i-1}, r_{t_{i-1}}) = E\left[e^{-\int_{t_{i-1}}^{t_i} r_t dt} h(t_i, r_{t_i}) | \mathcal{F}_{t_{i-1}}\right], \tag{3.96}
$$

so that the derivative value at any time can be calculated iteratively, starting from its time-T payoff. This is the property that is employed for the price calculation with the tree. Precisely, the following procedure based on backward induction is used.

We denote by $h_{i,j}$ the derivative value on node (i, j) and set at the final nodes

$$
h_{n,j} := h(T, r_{n,j}) = H(T, r_{n,j}). \tag{3.97}
$$

At the final time $t = t_n = T$, the derivative values on the tree nodes are known through this payoff condition (3.97). We now mode backwards to the time-t_{n-1} nodes and apply the general rule (3.96) with $i = n$.

Starting from the lowest level $j = \underline{j}_{n-1}$ up to $j = \overline{j}_{n-1}$, we use the approximation

$$
h(t_{n-1}, r_{t_{n-1}}) \approx e^{-r_{t_{n-1}}(T - t_{n-1})} E\left[h(T, r_T) | \mathcal{F}_{t_{n-1}}\right]
$$

to calculate the derivative value on the generic node $(n-1, j)$ as the discounted expectation of the corresponding values on nodes $(n, k+1)$, (n, k) and $(n, k-1)$:

$$
h_{n-1,j} = e^{-r_{n-1,j}(T - t_{n-1})}[p_u h_{n,k+1} + p_m h_{n,k} + p_d h_{n,k-1}].
$$

We then move to time-t_{n-2} nodes, apply again (3.96) with $i = n - 1$, and calculate discounted expectations as in the previous step.

Proceeding like this, the generic step backwards between time t_{i+1} and t_i is described by

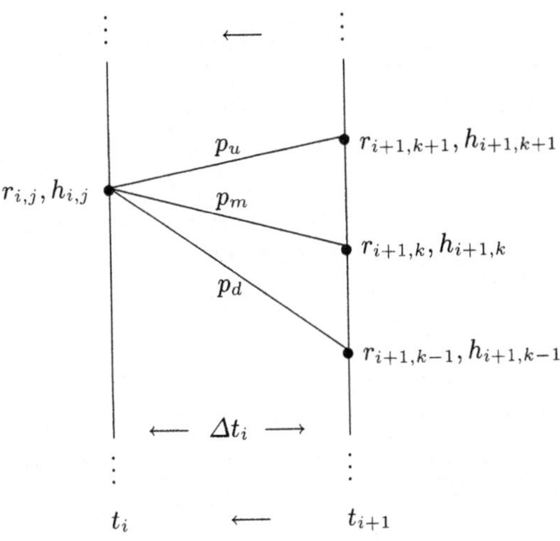

Fig. 3.7. Backward-induction step for the payoff evaluation on the tree moving back from time t_{i+1} to time t_i.

$$h_{i,j} = e^{-r_{i,j}(t_{i+1}-t_i)}[p_u h_{i+1,k+1} + p_m h_{i+1,k} + p_d h_{i+1,k-1}],$$

and is illustrated in Figure 3.7. We remind that the indices i, j and k in the probabilities have been omitted for brevity.

We keep on moving backwards along the tree until we reach the initial node $(0,0)$, whose associated value $h_{0,0}$ gives the desired approximation for the derivative price $h(0, r_0)$.

A first simple payoff with early exercise. Now take the same payoff as before but assume that the holder of the contract has the right to exercise it at any instant t before final time T, receiving upon exercise at time t the amount $H(t, r_t)$ (typical examples are American options). Since we plan to use a tree with time instants t_i's, with no loss of generality we assume that early exercise can occur "only" at the instants t_i's.

We need to modify the generic backward-induction step in the tree as follows: the generic step backwards between time t_{i+1} and t_i is now described by

$$h_{i,j} = \max\left(e^{-r_{i,j}(t_{i+1}-t_i)}[p_u h_{i+1,k+1} + p_m h_{i+1,k} + p_d h_{i+1,k-1}], H(t_i, r_{i,j})\right).$$

Indeed, what we are doing is the following. We roll back the "backwardly-cumulated" value $h_{i+1,\cdot}$ at time t_{i+1} down to time t_i, and then compare this backwardly-cumulated value

$$e^{-r_{i,j}(t_{i+1}-t_i)}[p_u h_{i+1,k+1} + p_m h_{i+1,k} + p_d h_{i+1,k-1}]$$

with the "immediate-exercise" value we would obtain by exercising the contract, i.e. with

$$H(t_i, r_{i,j}).$$

We then take the best of the two choices, i.e. the one that maximizes the value. Some contracts such as Bermudan-style swaptions only allow for exercise at a selected set of times that are typically one-year spaced, thus resulting in a much reduced subset of exercise times than the tree's t_i's. Then, when moving backwards along the tree, the comparison between the "backwardly-cumulated" value and the "immediate-exercise" value occurs only at times where exercise is allowed. In the remaining instants, the backward-induction step is the same as in the previous "no-early-exercise" case.

A second payoff. A second slightly more complicated example of payoff at time T is the following:

$$g(T, P(T, T_1), P(T, T_2), \ldots, P(T, T_m)), \tag{3.98}$$

where $P(T, T_1), P(T, T_2), \ldots, P(T, T_m)$ are the bond prices at time T for increasing maturities $T_1, T_2, \ldots, T_m$, with $T_1 > T$. This is the typical case one encounters for payoffs depending, for instance, on LIBOR or swap rates, since they can be written in terms of zero-coupon bond prices.

If we initially selected an analytically-tractable short-rate model, for example the Hull-White (1994a) model (3.33) or the CIR++ model (3.76), explicit formulas for bond prices are available as functions of time and short-rate value: $P(t, S) = \Pi(t, S; r_t)$ with Π an explicit function. Therefore, in such a case, the payoff (3.98) can be replaced with some $\hat{g}(T, r_T)$ and thus be priced exactly as the first payoff above, with the tree being constructed until time T.

Instead, if we are dealing with a model like Black-Karasinski's (3.54), for instance, no analytical formula for bonds is available. In such a case, we are then compelled to construct the tree until the last relevant maturity, i.e. T_m.

Each bond value at time T is obtained, through backward induction, by assigning value 1 to all the tree nodes at the corresponding maturity. Notice that we need to propagate the whole vector of bond prices backwards through time, together with the short-rate and the backwardly-cumulated value. Therefore, in each tree node at a given time t_i we need to store all bond prices that have already come to life at that time, i.e. all $P(t_i, T_l)$ with $T_l > t_i$, and keep on propagating all of them backwards. Moreover, each time we reach a new maturity $t_i = T_l$, we need to add a component to the vector, and set to one the value of that new component. This component represents the bond $P(\cdot, T_l)$, that has come to life now and whose (terminal) current value is clearly 1. We thus see that, as we move backwards, the dimension of the bond vector to be stored at each node may increase. For an example see the scheme for Bermudan swaptions below.

Since in each node we have the whole zero-bond curve at the relevant instants, now from time T down to time 0 the calculation procedure is exactly equivalent to the previous ones. And again, in case early exercise is introduced, in the tree nodes at each time where early exercise is allowed we have to take the maximum between the backwardly-cumulated value and the immediate-exercise value.

Remark 3.11.1. **(Pricing early-exercise derivatives with a tree in multi-factor short-rate models).** The backward-induction paradigm illustrated here can be adapted to the two-factor models we will discuss in the next chapter, provided the tree for r is replaced with the two-dimensional tree for the two factors x and y and that the backward induction is modified accordingly.

3.11.4 A Fundamental Case of Early Exercise: Bermudan-Style Swaptions.

An important case is given by Bermudan swaptions, which we discuss now. Bermudan swaptions will be defined in larger detail in Section 13.15. We briefly introduce them here in order to illustrate how the tree works in an important case. Consider an interest-rate swap first resetting in T_α and paying at $T_{\alpha+1}, \ldots, T_\beta$, with fixed rate K, and year fractions $\tau_{\alpha+1}, \ldots, \tau_\beta$. Assume one has the right to enter the IRS at any of the following reset times:

$$T_h, T_{h+1}, \ldots, T_k, \quad \text{with} \quad T_\alpha \leq T_h < T_k < T_\beta.$$

The contract thus described is a Bermudan swaption of payer (receiver) type if the IRS is a payer (receiver) IRS.

Here we have a clear case of early exercise. Suppose we hold the contract. We can either wait for the final maturity T_k and then exercise our right to enter the "final" IRS with first reset date T_k and last payment in T_β, or exercise at any earlier time T_l (with $T_h \leq T_l < T_k$) and then enter the IRS with first reset date T_l and last payment in T_β.

In order to value such a contract, we have to discretize time.

For each l, select a set of times t^l forming a partition of $[T_l, T_{l+1}]$ as follows:

$$T_l = t_1^l < t_2^l < \cdots < t_{d(l)-1}^l < t_{d(l)}^l = T_{l+1}.$$

We then price the Bermudan swaption through the following method.

Build the usual trinomial tree for the short rate r with time instants

$$t_1^1, \ldots, t_{d(1)}^1 = t_1^2, \ldots, t_{d(2)}^2 = t_1^3, \ldots, t_{d(\beta)}^\beta.$$

Denote the short-rate value on the generic j-node at time t_i^l by $r_{i,j}^l$. Denote by $P_{i,j}^l(T_s)$ the bond price $P(t_i^l, T_s)$ for the maturity $T_s > t_i^l$ in the generic tree node hosting $r_{i,j}^l$. Lest one be lost in indices, we recall that, in $P_{i,j}^l(T_s)$:

- l specifies inside which $[T_l, T_{l+1}]$ discretization is occurring;
- i specifies at which time instant t_i inside $[T_l, T_{l+1}]$ the current discretization step is occurring;
- j specifies in which (vertical) "spatial" node in the "time-t_i column" of the tree we are acting;
- T_s specifies which bond we are propagating.

The pricing scheme runs as follows.

1. Set $l + 1 = \beta$.
2. (Positioning in a new $[T_l, T_{l+1}]$ and adding a new zero-coupon bond). Set $P^l_{d(l),j}(T_{l+1}) = 1$ for all j. Set $i + 1 = d(l)$.
3. (Backward induction inside $[T_l, T_{l+1}]$).
   While going backwards from time t^l_{i+1} to t^l_i in the tree, propagate backwards the vector of bond prices as follows:

$$P^l_{i,j}(T_s) = e^{-r^l_{i,j}(t^l_{i+1}-t^l_i)}[p_u P^l_{i+1,k+1}(T_s) + p_m P^l_{i+1,k}(T_s) + p_d P^l_{i+1,k-1}(T_s)],$$

   for all $s = l + 1, \dots, \beta$. Store in the j-node of the current time t^l_i the values of the above bond prices. This step is completely analogous to rolling back a general payoff h, as we have seen in earlier sections and as is illustrated in Figure 3.7. The notation is slightly more complicated due to the presence of the l partition index.
4. if $i > 1$ then decrease i by one and go back to the preceding point.
5. (Initial time T_l of the current partition reached).
   Since now $i = 1$, we have reached $t^l_1 = T_l$. If $l > k$, decrease l by one and go back to point 2, otherwise move on to point 6.
6. (Last exercise-time reached).
   Since now $l = k$, by going backwards we have reached the last point in time (first in our backward direction) where the swaption can be exercised.
7. (Checking the exercise opportunity in each node of the current time).
   Compute, for each level j in the current "column" of the tree, the time-T_l value of the underlying IRS with first reset date T_l and last payment in T_β, based on the backwardly-propagated bond prices

$$P^l_{1,j}(T_{l+1}), P^l_{1,j}(T_{l+2}), \dots, P^l_{1,j}(T_\beta),$$

   i.e.

$$\mathrm{IRS}^l_j = 1 - P^l_{1,j}(T_\beta) - \sum_{s=l+1}^{\beta} \tau_s K P^l_{1,j}(T_s).$$

   If $l = k$ then define the backwardly-Cumulated value from Continuation (CC) of the Bermudan swaption as this IRS value in each node j of the current time level in the tree,

$$\mathrm{CC}^{k-1}_{d(k-1),j} := \mathrm{IRS}^k_j.$$

Else, if $l < k$, check the exercise opportunity as follows: if the underlying IRS is larger than the backwardly-cumulated value from continuation, then set the CC value equal to the IRS value. In our notation, for each node j in the current column of the tree:
If $\mathrm{IRS}^l_j > \mathrm{CC}^l_{1,j}$ then $\mathrm{CC}^l_{1,j} \leftarrow \mathrm{IRS}^l_j$.
Store such CC values in the corresponding nodes of the tree. Decrease l by one and move to the next step.

8. (Positioning in a new $[T_l, T_{l+1}]$).
Set $P^l_{d(l),j}(T_{l+1}) = 1$ for all nodes j in the current time level. Set $i+1 = d(l)$.
9. (Backward induction inside $[T_l, T_{l+1}]$).
Calculate backwards from time t^l_{i+1} to t^l_i: i) the vector of bond prices, and ii) the backwardly-cumulated value from continuation (CC) of the Bermudan swaption.
The P's propagation is as in point 3 above, and the CC update is analogous:

$$\mathrm{CC}^l_{i,j} = e^{-r^l_{i,j}(t^l_{i+1}-t^l_i)}[p_u \mathrm{CC}^l_{i+1,k+1} + p_m \mathrm{CC}^l_{i+1,k} + p_d \mathrm{CC}^l_{i+1,k-1}]. \quad (3.99)$$

Store in each node j of the current time t^l_i the values of the above bond prices and the value from continuation $\mathrm{CC}^l_{i,j}$. Again, this step is completely analogous to rolling back a general payoff h, as we have seen in earlier sections and as is illustrated in Figure 3.7.
10. if $i > 1$ then decrease i by one and go back to the preceding point 9, otherwise move on to the next point.
11. (New exercise time T_l reached).
Since now $i = 1$, we have reached $t^l_1 = T_l$. If $l > h$, we are still in the exercise region: move back to point 7 to check the related exercise opportunity and go on from there. Otherwise, if $l = h$, check the last (in a backward sense) exercise opportunity (indeed the first one specified by the contract) as follows. In each node j of the current column of the tree, if the value of the underlying IRS, namely

$$\mathrm{IRS}^h_j = 1 - P^h_{1,j}(T_\beta) - \sum_{s=h+1}^{\beta} \tau_s K P^h_{1,j}(T_s),$$

is larger than the backwardly-cumulated value from continuation of the Bermudan swaption, set the CC value equal to the IRS value. In our notation:
If $\mathrm{IRS}^h_j > \mathrm{CC}^h_{1,j}$ then $\mathrm{CC}^h_{1,j} \leftarrow \mathrm{IRS}^h_j$.
Then move on to the next point.
12. (No more exercise times left).

We have reached the first allowed exercise time T_h, and now there are no exercise opportunities left as we keep on moving backward in time. We can now start rolling backwards the current backwardly cumulated value from continuation from current time $t_1^h = T_h$ until time 0 along the tree as in (3.99) above. Now it is no longer necessary to propagate backwards the P's, but only CC.

13. When we reach time 0, the value of CC at the corresponding initial node of the tree is the Bermudan-option price.

We conclude this example with a few remarks. The method we just sketched allows for different discretizations in different partitions $[T_l, T_{l+1}]$. As l changes, we are free to consider either a more refined or a coarser discretization in the tree construction. This is to say that we may decide to refine the discretization in zones of the tree that are more relevant to the payoff evaluation. Typically, one takes a finer discretization in the regions where early exercise is to be checked, and a coarser (but not too coarse) discretization after the last possible exercise date (Mauri (2001)).

An example of tree with different discretization steps is shown in Figure 3.3. Here it is clearly seen that, in different regions, the discretization is different. Notice in particular that in the initial part of the tree the discretization is particularly refined. Notice also that at the end of the tree the discretization step is larger.

Finally, as far as the above scheme is concerned, points 1 to 6 can be avoided if the short-rate model being used allows for analytical formulas for zero-coupon-bond prices. Indeed, in such a case, knowledge of the short rate amounts to knowledge of the bond price itself through the related analytical formulas, and therefore we do not need backward propagation to obtain the zero-coupon bonds at the last exercise date, since we can directly start from the short rate at that date. Also, in principle, there would be no need to store the vector of bond prices at each node, since they can be recovered analytically in terms of the short-rate value at each node. Nonetheless, there are times where it can be convenient to keep on propagating the bond-price vectors rather than resorting to the analytical formulas, for a series of reasons involving, for instance, the tradeoff between speed of convergence and computational complexity (Mauri (2001)).

3.12 Implied Cap Volatility Curves

When using an interest-rate model for pricing related derivatives, it can be useful to understand how the implied volatility structures vary due to changes in the model parameters values. To this end, we study the cap volatility curves that are implied by some of the short-rate models we have previously described. Precisely, we concentrate our analysis on the Black-Karasinski (3.54), the CIR++ (3.76) and the Extended Exponential-Vasicek (3.92) models.

The implied cap volatility curves are obtained by first pricing market caps with the chosen model and then inverting the Black formula (1.26) to retrieve the implied volatility associated with each cap maturity.

The model prices are calculated with the trinomial trees we have proposed, for each model, in the related sections. The correct market conventions and payment dates are taken into account, with the variable time steps being roughly equal to 0.02 years. The initial values for each model parameters are displayed in Table 3.2.[17] We carry out our analysis by moving the model

BK	CIR++	EEV
a= 0.0771118	x_0= 0.0059379	x_0= 1.009423
σ= 0.2286	k= 0.394529	a= 0.4512079
	θ = 0.271373	θ= -1.47198
	σ = 0.0545128	σ= 0.6926841

Table 3.2. The initial set of parameters for the BK, CIR++ and EEV models

parameters around their starting values. We then look for a qualitative relationship between shifts in the parameters' values and changes in shape of the implied cap volatility curve. Our conclusions, however, should not be regarded as irrefutable statements, but simply as helpful remarks for a better understanding of the above models behaviour.

In our examples, we consider caps with integer maturities ranging from 1 to 15 years. The 2 to 15 year caps have semiannual frequency, whereas the frequency of the 1 year cap is quarterly.

3.12.1 The Black and Karasinski Model

The cap volatility curves that are implied by the Black and Karasinski model (3.54) are displayed in Figure 3.8. The graphs on the left are plotted by keeping the σ parameter fixed and varying a, whereas those on the right are plotted by keeping the a parameter fixed and varying σ. These graphs suggest that cap volatilities increase as σ increases and decrease as a increases. In addition, a has a bigger influence on longer maturities. Such features are indeed consistent with the theoretical meaning of the model parameters.

If we plot, for each given maturity, cap volatilities against σ, we almost obtain straight lines passing by the origin, as is shown in Figure 3.9. This suggests that the functional dependence of the implied cap volatilities on the model parameters is roughly of type $f(a,\sigma) = \sigma g(a)$.

[17] These parameters values derive from the model calibration to the actual Euro ATM caps volatility curve on January 17, 2000.

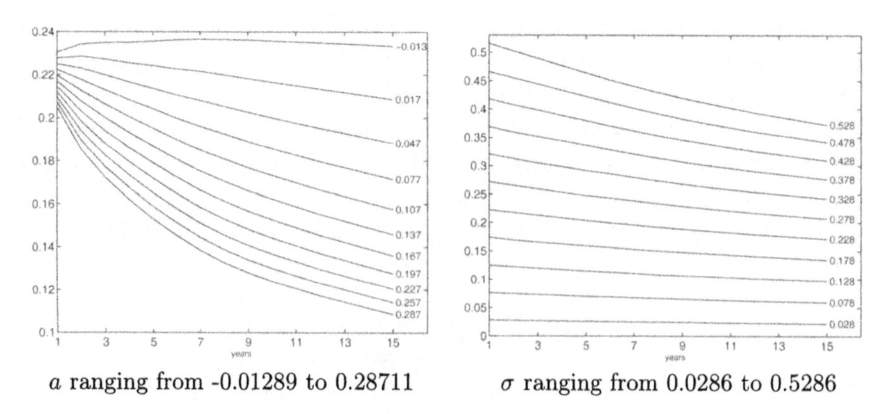

a ranging from -0.01289 to 0.28711 σ ranging from 0.0286 to 0.5286

Fig. 3.8. Cap volatility curves implied by the Black and Karasinski (1991) model.

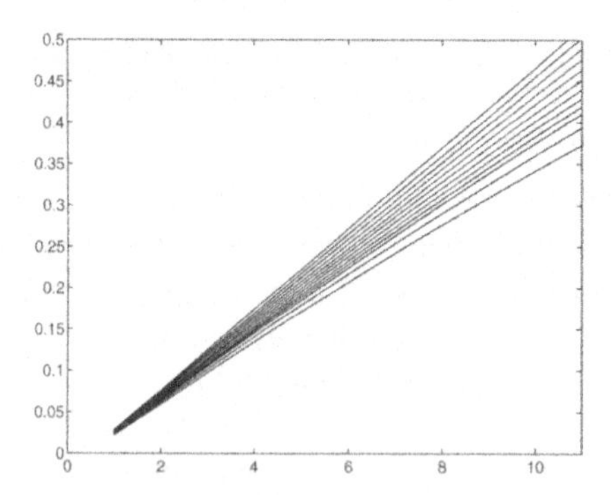

Fig. 3.9. Volatility-vs-σ plots for each maturity in the Black Karasinski (1991) model.

3.12.2 The CIR++ Model

Examples of the cap volatility curves that are implied by the CIR++ model (3.76) are displayed in Figure 3.10, where we set $\bar{\theta} := k\theta$.[18] The volatility curves are often humped and tend to be decreasing from the five to six year maturities on. More precisely, we can state the following.

Increasing x_0 makes the volatilities increase. The effect is much more pronounced for short-maturity volatilities. Indeed, x_0 does not affect the asymptotic variance of the spot rate r since, for a positive k,

$$\lim_{t \to \infty} \operatorname{Var}\{r(t)\} = \theta \frac{\sigma^2}{2k}.$$

It is reasonable, therefore, that long-maturity volatilities are less sensitive to changes in x_0 and that increasing x_0 has the effect to decrease the initial slope of the volatility curve.

[18] Notice that, when studying the implied cap volatility curve, varying θ is equivalent to varying $\bar{\theta}$ as soon as k is fixed and both θ and k are positive.

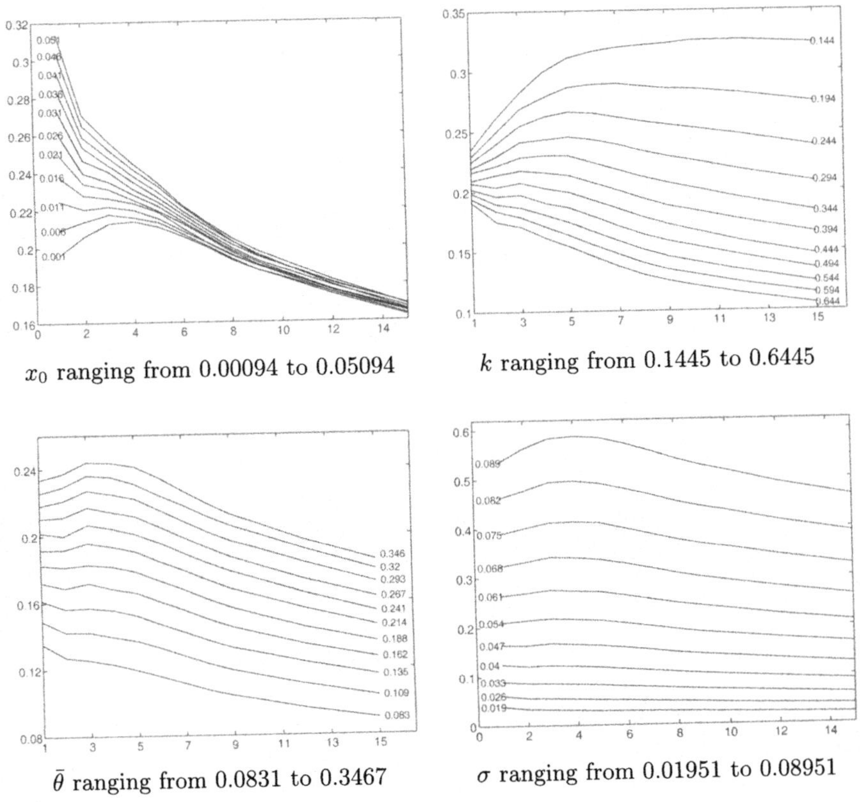

x_0 ranging from 0.00094 to 0.05094
k ranging from 0.1445 to 0.6445
$\bar{\theta}$ ranging from 0.0831 to 0.3467
σ ranging from 0.01951 to 0.08951

Fig. 3.10. Cap volatility curves implied by the CIR++ model.

Increasing the mean reverting parameter k makes the volatility level decrease. The effect is less pronounced for short maturity caps. Accordingly, the volatility curve is initially increasing when k is small and decreasing when k is large.

Increasing the parameter θ makes the volatility level increase. The absolute change is fairly similar for all maturities so that the curve shape hardly changes with θ. Moreover, plotting, for each fixed maturity, implied volatilities against θ gives almost linear graphs. We may then guess that, for each fixed maturity, the implied volatility has roughly a functional dependence on the model parameters of the following type:

$$f(x_0, k, \theta, \sigma) = g(x_0, k, \sigma) + \theta h(x_0, k, \sigma).$$

The parameter σ affects the implied volatility curve as expected. Volatilities increase as σ grows and the volatility curve moves with almost parallel shifts. In our examples, the four and five year volatilities grow a bit faster, so that, as σ increases, a humped shape appears. Plotting, for each fixed maturity, implied volatilities against σ we obtain the increasing convex curves that are shown in Figure 3.11. Indeed, there is an empirical confirmation that the functional dependence of the implied volatility on the parameter σ is of type $f(\sigma) = k_1\sigma^2 + k_2$.

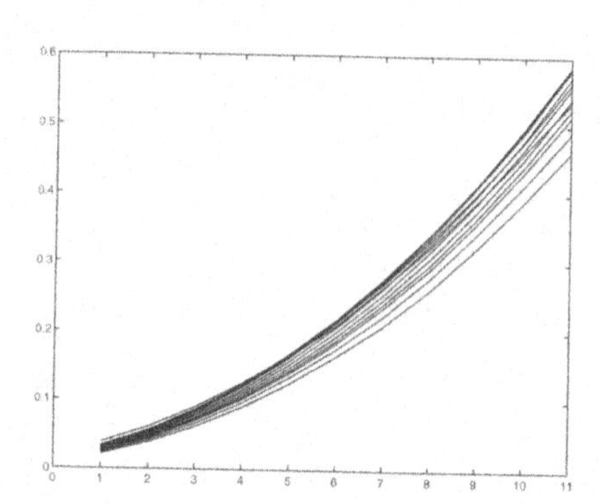

Fig. 3.11. Volatility-vs-σ plot for each maturity in the CIR++ model

3.12.3 The Extended Exponential-Vasicek Model

Examples of the cap volatility curves that are implied by the Extended Exponential-Vasicek model (3.92) are plotted in Figure 3.12, where we set $y_0 := \ln x_0$. Similarly to the CIR++ case, also in the examples for the EEV model we often find implied volatility curves that are humped and tend to be decreasing from the six year maturity on. More precisely, we have the following.

Increasing x_0 makes the volatilities increase and the absolute change is more relevant for short-maturity volatilities. Indeed, as in the CIR++ case, x_0 does not affect the asymptotic variance of the spot rate since, for a positive a (see also (3.31)),

$$\lim_{t\to\infty} \mathrm{Var}\{r(t)\} = \exp\left(\frac{2\theta}{a} + \frac{\sigma^2}{2a}\right)\left[\exp\left(\frac{\sigma^2}{2a}\right) - 1\right].$$

Again, it is reasonable that long-maturity volatilities are less sensitive to changes in x_0 and that increasing x_0 has the effect to decrease the initial slope of the volatility curve.

The drift parameters a and θ have a similar impact on the implied volatility curve, making mostly the five-year volatility change, with the one-year volatility that is the least affected. This naturally leads to a humped shape as soon as these parameters grow.

Finally σ behaves as expected. The implied volatilities increase as σ increases, and changes in σ lead to almost parallel shifts in the volatility curve. In the given examples, the five-year volatility is subject to larger moves,

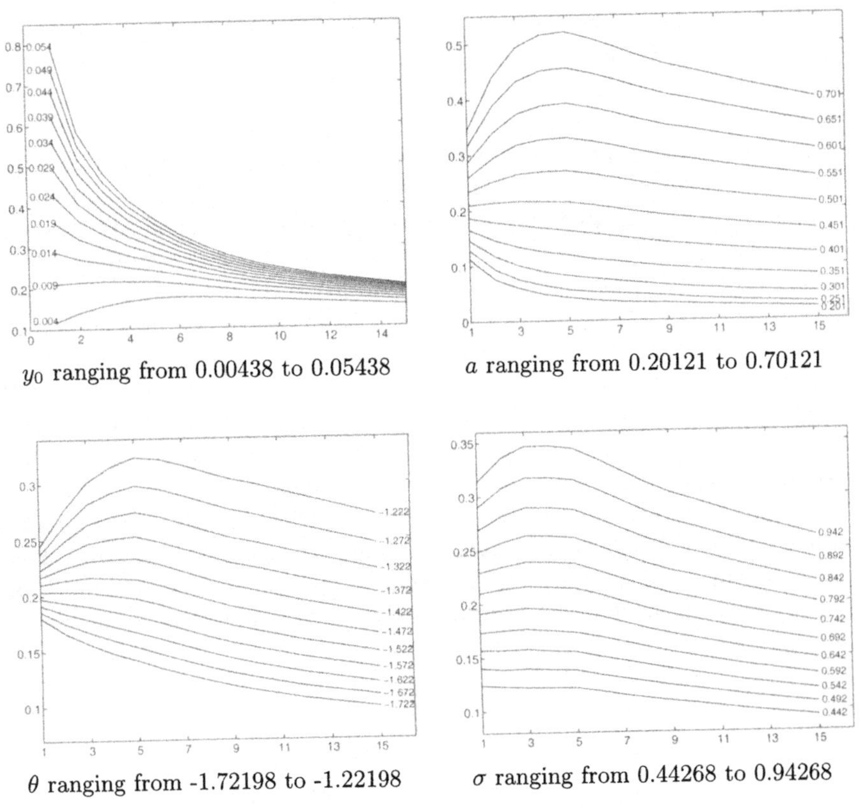

y_0 ranging from 0.00438 to 0.05438 — a ranging from 0.20121 to 0.70121

θ ranging from -1.72198 to -1.22198 — σ ranging from 0.44268 to 0.94268

Fig. 3.12. Cap volatility curves implied by the EEV model.

which leads to a hump as soon as σ increases. Plotting, for each fixed maturity, implied volatilities against σ produces almost linear graphs. However, the hypothesis that implied volatilities depend linearly on the parameter σ does not have a fully satisfactory empirical confirmation.

3.13 Implied Swaption Volatility Surfaces

Similarly to the implied cap volatility curve, it can be useful to understand how the implied swaption volatility structures vary due to changes in the model-parameters values. To this end, we study the swaptions volatility curves and surfaces that are implied by the Black and Karasinski (3.54) and the Extended Exponential-Vasicek (3.92) models.[19]

[19] We just concentrate on these two models for practical purposes. Indeed, among the short rate models developed in this chapter, the BK and EEV models are likely to imply the best fitting to the swaption volatility surfaces in many concrete market situations.

The implied swaption volatility curves are obtained by first pricing market swaptions with the model and then inverting the Black formula (1.28) to retrieve the implied volatility associated with the selected pair of maturity and tenor.

As before, the model prices are calculated with the corresponding trinomial trees. The correct market conventions and payment dates are taken into account, with the variable time steps being again roughly equal to 0.02 years.

We consider ATM European swaptions whose maturities go from one to twenty years and whose tenors (i.e., the durations of the underlying swaps) go from one to twenty years as well. For our examples, we first choose the initial set of parameters shown in Table 3.3 and shock each parameter ceteris paribus.[20] We then plot the implied volatility curves that are obtained by fixing respectively the one, five and ten-year tenors. We also show the implied volatility surfaces that correspond to the highest and lowest values being chosen for each parameter.

As for the cap-volatilities case, the purpose of this section is to comment on how the model parameters affect the shape of the implied volatility structures. Again, our conclusions are far from being "universal statements" and simply aim to provide some intuitions on the practical effects of the parameters changes.

BK	EEV
a= 0.1318301	x_0= 1.014098
σ=0.2342142	a = 0.1073888
	θ =-0.457712
	σ = 0.692684

Table 3.3. The initial set of parameters for the BK and EEV models

3.13.1 The Black and Karasinski Model

Figure 3.13 shows the swaption volatility curves that are obtained by fixing the tenor and by separately varying the two parameters a and σ. As to these examples, we can infer the following.

Implied volatilities decrease as a increases. We can see that, for small tenors, a has a small effect on short maturity volatilities, whereas for large tenors, changing a seems to affect short and long maturities with the same extent.

Volatilities also grow as σ grows. Similarly to the cap-volatilities case, if we fix tenor and maturity and plot volatilities against σ, we obtain an almost

[20] These initial parameters values derive from the models calibration to the actual Euro ATM swaption volatility surface on January 17, 2000.

linear graph. However, if the maturity is large enough, this linearity is lost, revealing a more complex relation between volatilities and σ. Examples of

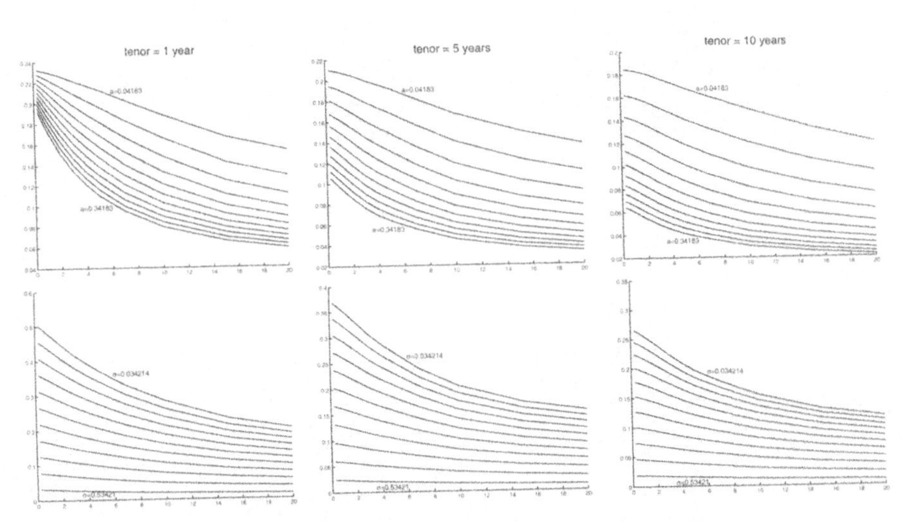

Fig. 3.13. Fixed-tenor volatility curves implied by the BK model. In the first row a varies ceteris paribus from 0.04183 to 0.34183 with step 0.03, whereas in the second row σ varies ceteris paribus from 0.034214 to 0.534214 with step 0.05.

the swaption volatility surfaces that are implied by the Black and Karasinski (1991) model are shown in Figure 3.14 for the displayed parameter values.

3.13.2 The Extended Exponential-Vasicek Model

Examples of the swaption volatility curves that, for fixed tenors, are implied by the extended exponential-Vasicek model (3.92) are plotted in Figure 3.15, where again $y_0 := \ln x_0$. The swaption volatility surfaces in this case are more complex than those implied by the Black and Karasinski (1991) model. Indeed, our examples reveal that humped shapes are possible along the maturity dimension. More specifically, we have the following.

The dependence of the volatilities on the y_0 parameter is similar to that in the caps-volatilities case. Only short-maturity volatilities seem to be affected, increasing as y_0 increases. Moreover, the tenor does not seem to affect the dependence on y_0, and the volatilities decrease as the tenor increases.

When a increases, the volatility level decreases. Moreover, increasing a leads to a humped shape, and this holds for each tenor. The influence of the parameter a on the swaption volatilities is smaller when both maturity and tenor are small.

The influence of the parameter θ on the implied volatility curves is similar to that of a. The volatilities increase as θ increases. For small tenors, the

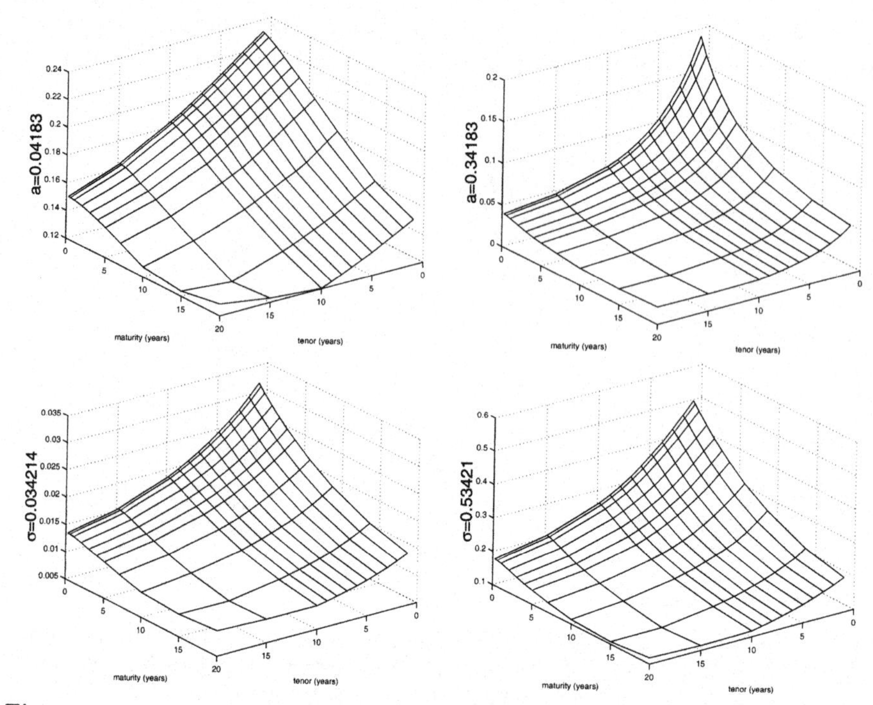

Fig. 3.14. Swaption volatility surfaces implied by the Black and Karasinski (1991) model.

long-maturity volatilities change more than the short-maturity ones. Finally, a hump appears as soon as θ increases.

As expected, increasing σ causes volatilities to increase. Indeed, for each fixed tenor, changing σ produces an almost parallel shift in the implied volatility curve.

Examples of the swaption volatility surfaces implied by the EEV model are displayed in Figure 3.17.

3.14 An Example of Calibration to Real-Market Data

We conclude the chapter with an example of calibration to real-market volatility data, which illustrates the fitting capability of the main one-factor (exogenous-term-structure) short-rate models we have reviewed in the previous sections. To this end, we use the at-the-money Euro cap-volatility quotes on February 13, 2001, at 5 p.m., which are reported in Table 3.4.. The zero-coupon curve, on the same day and time, is that shown in Figure 1.1 of Chapter 1.

We focus on the Hull and White model (3.33), the Black and Karasinski model (3.54), the Mercurio and Moraleda model (3.63), the CIR++ model

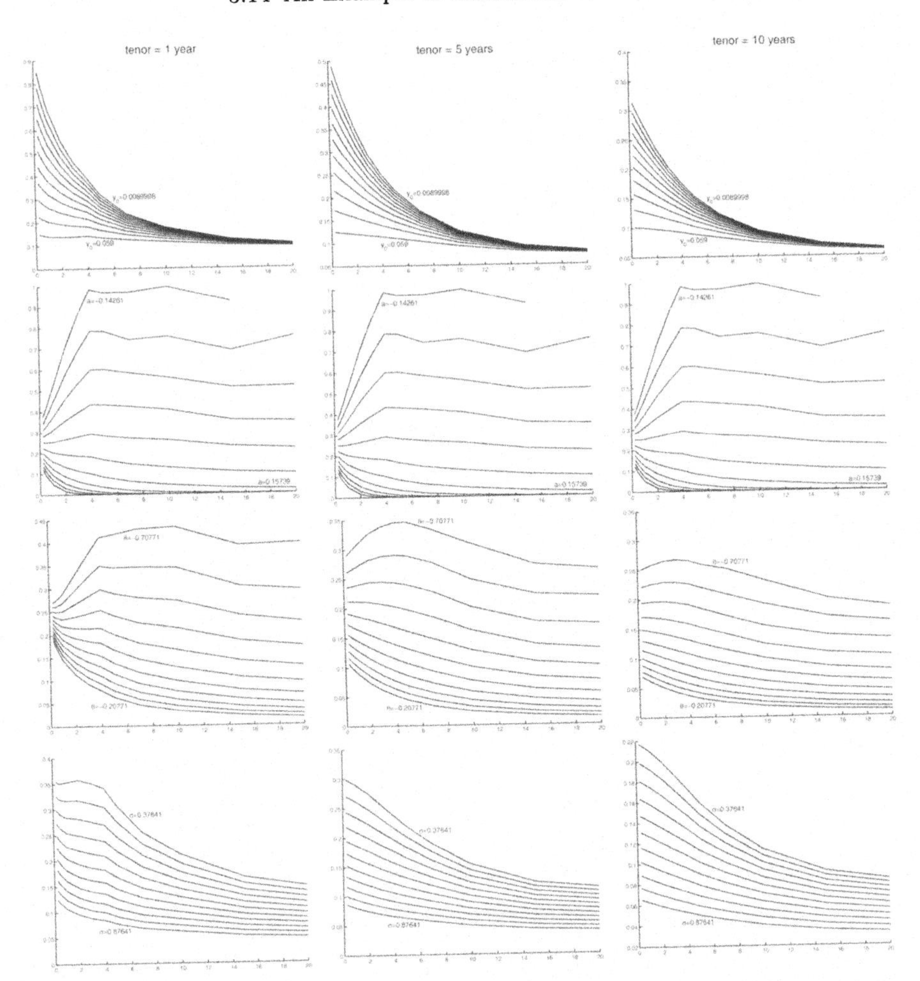

Fig. 3.15. Fixed-tenor swaption volatility curves implied by the EEV model. In the first row, y_0 varies ceteris paribus from 0.0089998 to 0.0599998 with step 0.005; In the second row, a varies ceteris paribus from -0.14261 to 0.15739 with step 0.03; In the third row, θ varies ceteris paribus from -0.70771 to -0.20771 with step 0.05; In the fourth row σ varies ceteris paribus from 0.37641 to 0.87641 with step 0.05.

(3.76) and the Extended exponential-Vasicek model (3.92). Minimizing the sum of the square percentage differences between model and market cap prices,[21] we obtain the results that are shown in Figure 3.16, where the models implied cap volatility curves are compared to the market one. Recall that a model cap implied volatility is the volatility parameter to be plugged into Black's formula (1.26) to match the observed model price.

[21] We use analytical formulas for the HW model, whereas, for the other models, we build a trinomial tree with a variable (small) time step in the first year, and an average of 50 time steps per year afterwards.

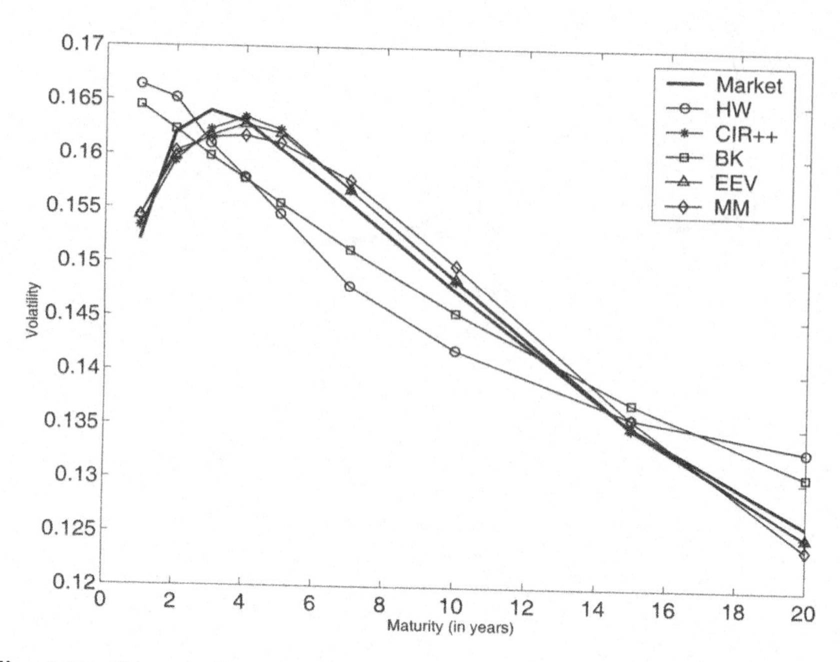

Fig. 3.16. Cap volatility curves implied by some short-rate models calibrated to the at-the-money Euro cap volatility curve on February 13, 2001, at 5 p.m.

Maturity	Volatility
1	0.152
2	0.162
3	0.164
4	0.163
5	0.1605
7	0.1555
10	0.1475
15	0.135
20	0.126

Table 3.4. At-the-money Euro cap-volatility quotes on February 13, 2001, at 5 p.m.

Now examine carefully Figure 3.16, and notice that both the HW model and the BK model imply decreasing cap volatility curves as best fitting to the humped market curve. In addition, the calibrated value of the mean-reversion parameter a in the HW model is negative, actually leading to a mean-diverging model where the volatility of instantaneous forward rates diverges too. This is a common situation when calibrating the HW model to cap volatilities, and has often been interpreted as the result of a possible predominance of the increasing part of the market volatility curve.

We also see that the introduction of an extra parameter (in a suitable time-dependent function), as in the MM model, helps recover the typical shape of the market cap-volatility curve. We can then see that the "award for the best fitting quality" is shared by the CIR++ model and the EEV model, whose implied curves are almost overlapped. This could be expected since both models have the highest number of parameters, amounting to four.

Remark 3.14.1. A model fitting quality can be deeply affected by the specific market conditions one is trying to reproduce. We must indeed remember that in some particular conditions even the HW and BK models can lead to an implied cap volatility curve that follows the market hump. This happens, for instance, in case of a decreasing forward-rate curve. However, the "usual" situation illustrated in Figure 3.16 represents what we have been commonly witnessing in the Euro market in the last years.

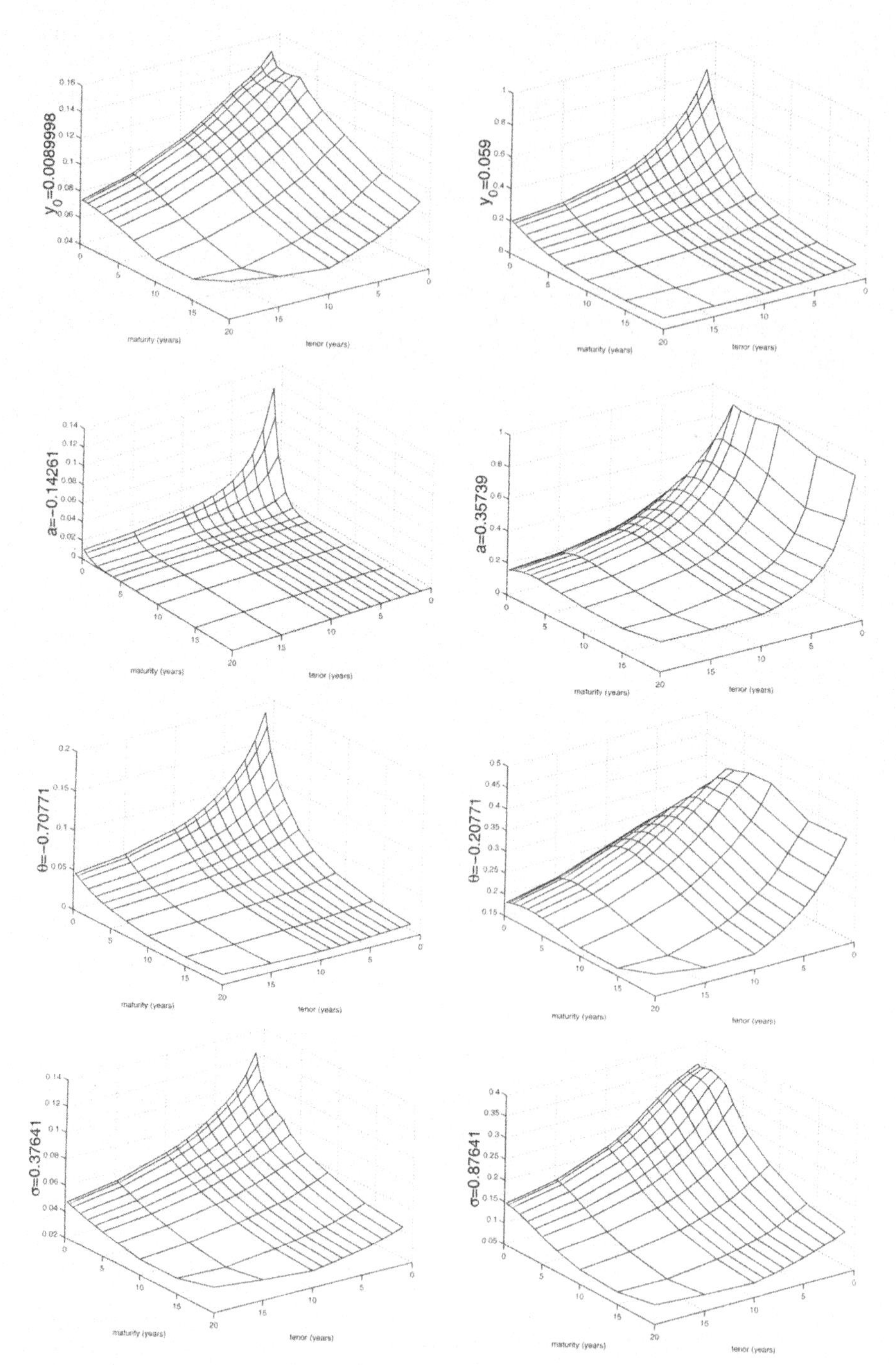

Fig. 3.17. Swaption volatility surfaces implied by the EEV model.

4. Two-Factor Short-Rate Models

"Entering secondary dimension". Adam Warlock, in:
"Silver Surfer/Warlock: Resurrection" 1, 1993, Marvel Comics.

4.1 Introduction and Motivation

However, motivation never harmed anyone (least of all, pure mathematicians), [...]
L.C.G. Rogers and D. Williams, in Chapter III.4 of "Diffusions, Markov Processes and Martingales", Vol. 1, 1994, Wiley and Sons

In the present chapter we introduce two major two-factor short-rate models. Before starting with the actual models, we would like to motivate two-factor models by pointing out the weaknesses of the one-factor models of the previous chapter. This is the purpose of this introductory section.

We have seen previously that the short rate r_t may constitute the fundamental coordinate with which the whole yield curve can be characterized. Indeed, knowledge of the short rate and of its distributional properties leads to knowledge of bond prices as from the usual relationship

$$P(t,T) = E_t \left[\exp\left(- \int_t^T r_s \, ds \right) \right] .$$

From all bond prices $P(t,T)$ at a given time t one can reconstruct the whole zero-coupon interest-rate curve at the same time t, so that indeed the evolution of the whole curve is characterized by the evolution of the single quantity r. However, choosing a poor model for the evolution of r will result in a poor model for the evolution of the yield curve. In order to clarify this rather important point, let us consider for a moment the Vasicek model (3.5).

Recall from the previous chapter that the Vasicek model assumes the evolution of the short-rate process r to be given by the linear-Gaussian SDE

$$dr_t = k(\theta - r_t)dt + \sigma dW_t .$$

Recall also the bond price formula $P(t,T) = A(t,T)\exp(-B(t,T)r_t)$, from which all rates can be computed in terms of r. In particular, continuously-compounded spot rates are given by the following affine transformation of the fundamental quantity r

$$R(t,T) = -\frac{\ln A(t,T)}{T-t} + \frac{B(t,T)}{T-t} r_t =: a(t,T) + b(t,T) r_t \,.$$

Consider now a payoff depending on the joint distribution of two such rates at time t. For example, we may set $T_1 = t + 1$ years and $T_2 = t + 10$ years. The payoff would then depend on the joint distribution of the one-year and ten-year continuously-compounded spot interest rates at "terminal" time t. In particular, since the joint distribution is involved, the correlation between the two rates plays a crucial role. With the Vasicek model such terminal correlation is easily computed as

$$\mathrm{Corr}(R(t,T_1), R(t,T_2)) = \mathrm{Corr}(a(t,T_1) + b(t,T_1) r_t, a(t,T_2) + b(t,T_2) r_t) = 1,$$

so that at every time instant rates for all maturities in the curve are perfectly correlated. For example, the thirty-year interest rate at a given instant is perfectly correlated with the three-month rate at the same instant. This means that a shock to the interest rate curve at time t is transmitted equally through all maturities, and the curve, when its initial point (the short rate r_t) is shocked, moves almost rigidly in the same direction. Clearly, it is hard to accept this perfect-correlation feature of the model. Truly, interest rates are known to exhibit some decorrelation (i.e. non-perfect correlation), so that a more satisfactory model of curve evolution has to be found.

One-factor models such as HW, BK, CIR++, EEV may still prove useful when the product to be priced does not depend on the correlations of different rates but depends at every instant on a single rate of the whole interest-rate curve (say for example the six-month rate). Otherwise, the approximation can still be acceptable, especially for "risk-management-like" purposes, when the rates that jointly influence the payoff at every instant are close (say for example the six-month and one-year rates). Indeed, the real correlation between such near rates is likely to be rather high anyway, so that the perfect correlation induced by the one-factor model will not be unacceptable in principle.

But in general, whenever the correlation plays a more relevant role, or when a higher precision is needed anyway, we need to move to a model allowing for more realistic correlation patterns. This can be achieved with multifactor models, and in particular with two-factor models. Indeed, suppose for a moment that we replace the Gaussian Vasicek model with its hypothetical two-factor version (G2):

$$\begin{aligned} r_t &= x_t + y_t, \\ dx_t &= k_x(\theta_x - x_t)dt + \sigma_x dW_1(t), \\ dy_t &= k_y(\theta_y - y_t)dt + \sigma_y dW_2(t), \end{aligned} \tag{4.1}$$

with instantaneously-correlated sources of randomness, $dW_1 dW_2 = \rho\, dt$. Again, we will see later on in the chapter that also for this kind of models the bond price is an affine function, this time of the two factors x and y,

$$P(t,T) = A(t,T)\exp(-B^x(t,T)x_t - B^y(t,T)y_t),$$

where quantities with the superscripts "x" or "y" denote the analogous quantities for the one-factor model where the short rate is given by x or y, respectively. Taking this for granted at the moment, we can see easily that now

$$\begin{aligned}&\text{Corr}(R(t,T_1),R(t,T_2))\\&\quad= \text{Corr}(b^x(t,T_1)x_t + b^y(t,T_1)y_t, b^x(t,T_2)x_t + b^y(t,T_2)y_t),\end{aligned}$$

and this quantity is not identically equal to one, but depends crucially on the correlation between the two factors x and y, which in turn depends, among other quantities, on the instantaneous correlation ρ in their joint dynamics. How much flexibility is gained in the correlation structure and whether this is sufficient for practical purposes will be debated in the chapter. It is however clear that the choice of a multi-factor model is a step forth in that correlation between different rates of the curve at a given instant is not necessarily equal to one.

Another question that arises naturally is: How many factors should one use for practical purposes? Indeed, what we have suggested with two factors can be extended to three or more factors. The choice of the number of factors then involves a compromise between numerically-efficient implementation and capability of the model to represent realistic correlation patterns (and covariance structures in general) and to fit satisfactorily enough market data in most concrete situations.

Usually, historical analysis of the whole yield curve, based on principal component analysis or factor analysis, suggests that under the objective measure two components can explain 85% to 90% of variations in the yield curve, as illustrated for example by Jamshidian and Zhu (1997) (in their Table 1), who consider JPY, USD and DEM data. They show that one principal component explains from 68% to 76% of the total variation, whereas three principal components can explain from 93% to 94%. A related analysis is carried out in Chapter 3 of Rebonato (1998) (in his Table 3.2) for the UK market, where results seem to be more optimistic: One component explains 92% of the total variance, whereas two components already explain 99.1% of the total variance. In some works an interpretation is given to the components in terms of average level, slope and curvature of the zero-coupon curve, see for example again Jamshidian and Zhu (1997).

What we learn from these analyses is that, in the objective world, a two- or three-dimensional process is needed to provide a realistic evolution of the whole zero-coupon curve. Since the instantaneous-covariance structure of the same process when moving from the objective probability measure to the risk-neutral probability measure does not change, we may guess that also in the risk-neutral world a two- or three-dimensional process may be needed in order to obtain satisfactory results. This is a further motivation for introducing a two- or three-factor model for the short rate. In this book, we have decided to focus on two-factor models for their better tractability and implementability, especially as far as recombining lattices are concerned. In particular, we will consider additive models of the form

$$r_t = x_t + y_t + \varphi(t), \tag{4.2}$$

where φ is a deterministic shift which is added in order to fit exactly the initial zero-coupon curve, as in the one-factor case of Section 3.8. This formulation encompasses the classical Hull and White two-factor model as a deterministically-shifted two-factor Vasicek (G2++), and an extension of the Longstaff and Schwartz (LS) model that is capable of fitting the initial term structure of rates (CIR2++), where the basic LS model is obtained as a two-factor additive CIR model. These are the two-factor models we will consider, and we will focus especially on the two-factor additive Gaussian model G2++. The main advantage of the G2++ model (4.2) with x and y as in (4.1) over the shifted Longstaff and Schwartz CIR2++ given by (4.2) with x and y as in

$$\begin{aligned} dx_t &= k_x(\theta_x - x_t)dt + \sigma_x\sqrt{x_t}dW_1(t), \\ dy_t &= k_y(\theta_y - y_t)dt + \sigma_y\sqrt{y_t}dW_2(t), \end{aligned} \tag{4.3}$$

is that in the latter we are forced to take $dW_1 dW_2 = 0\,dt$ in order to maintain analytical tractability, whereas in the former we do not need to do so. The reason why we are forced to take $\rho = 0$ in the CIR2++ case lies in the fact that square-root non-central chi-square processes do not work as well as linear-Gaussian processes when adding nonzero instantaneous correlations. Requiring $dW_1 dW_2 = \rho\,dt$ with $\rho \neq 0$ in the above CIR2++ model would indeed destroy analytical tractability: It would no longer be possible to compute analytically bond prices and rates starting from the short-rate factors. Moreover, the distribution of r would become more involved than that implied by a simple sum of independent non-central chi-square random variables. Why is the possibility that the parameter ρ be different than zero so important as to render G2++ preferable to CIR2++? As we said before, the presence of the parameter ρ renders the correlation structure of the two-factor model more flexible. Moreover, $\rho < 0$ allows for a humped volatility curve of the instantaneous forward rates. Indeed, if we consider at a given time instant t the graph of the T function

$$T \mapsto \sqrt{\mathrm{Var}[d\,f(t,T)]/dt}$$

where the instantaneous forward rate $f(t,T)$ comes from the G2++ model, it can be seen that for $\rho = 0$ this function is decreasing and upwardly concave. This function can assume a humped shape for suitable values of k_x and k_y only when $\rho < 0$. Since such a humped shape is a desirable feature of the model which is in agreement with market behaviour, it is important to allow for nonzero instantaneous correlation in the G2++ model. The situation is somewhat analogous in the CIR2++ case: Choosing $\rho = 0$ does not allow for humped shapes in the curve

$$T \mapsto \sqrt{\mathrm{Var}[d\ f(t,T)]/dt},$$

which consequently results monotonically decreasing and upwardly concave, exactly as in the G2++ case with $\rho = 0$, as we will see later on in the chapter. In turn, the advantage of CIR2++ over G2++ is that, as in the one-factor case where HW is compared to CIR++, it can maintain positive rates through reasonable restrictions on the parameters. Moreover, the distribution of the short rate is the distribution of the sum of two independent noncentral chi-square variables, and as such it has fatter tails than the Gaussian distribution in G2++. This is considered a desirable property, especially because in such a way (continuously-compounded) spot rates for any maturity are affine transformations of such non-central chi-squared variables and are closer to the lognormal distribution than the Gaussian distribution for the same rates implied by the G2++ model. Therefore, both from a point of view of positivity and distribution of rates, the CIR2++ model would be preferable to the G2++ model. However, the humped shape for the instantaneous forward rates volatility curve is very important for the model to be able to fit market data in a satisfactory way. Furthermore, the G2++ model is more analytically tractable and easier to implement. These overall considerations then imply that the G2++ model is more suitable for practical applications, even though we should not neglect the advantages that a model like CIR2++ may have. This is particularly true after the introduction of the Gaussian mapping technique by Brigo and Alfonsi (2003, 2005), that allows to map a CIR2++ model *with nonzero correlation between the shocks* into a suitable G2++ model. The nonzero correlation might allow for a humped volatility shape. The mapped G2++ model can then be used to compute prices while maintaining the link with the original correlated CIR2++ model. Indeed, in the credit chapters, the Gaussian dependence mapping is illustrated in detail and tested on some first examples concerning credit derivatives. The interested reader is advised to read Section 22.7.5 and to re-cast the relevant results for correlated CIR++ processes in terms of the notation adopted in the present chapter. This can be a way to have the CIR2++ model benefits while maintaining the G2++ model tractability.

In general, when analyzing an interest rate model from a practical point of view, one should try to answer questions like the following. Is a two-factor model like G2++ flexible enough to be calibrated to a large set of swaptions, or even to caps and swaptions at the same time? How many swaptions can be calibrated in a sufficiently satisfactory way? What is the evolution of the term structure of volatilities as implied by the calibrated model? Is this realistic? How can one implement trees for models such as G2++? Is Monte

Carlo simulation feasible? Can the model be profitably used for quanto-like products and for products depending on more than an interest rate curve when taking into account correlations between different interest-rate curves and also with exchange rates?

In this chapter, we will focus mainly on the G2++ model and we will try to deal with some of the above questions. Other questions will be then addressed in the second part of the book. We will then deal with the CIR2++ model, but for this model we will not consider the same level of detail we devoted to the G2++ model. We will try and keep the two sections on the two different models as self-contained as possible, in order for a reader interested only in one of the models to be able to concentrate on the interested section while skipping the other one. This will cause of course a little overlap in the presentation of the two models.

4.2 The Two-Additive-Factor Gaussian Model G2++

In this section we consider an interest-rate model where the instantaneous-short-rate process is given by the sum of two correlated Gaussian factors plus a deterministic function that is properly chosen so as to exactly fit the current term structure of discount factors. The model is quite analytically tractable in that explicit formulas for discount bonds, European options on pure discount bonds, hence caps and floors, can be readily derived.

Gaussian models like this G2++ model are very useful in practice, despite their unpleasant feature of the theoretical possibility of negative rates. Indeed, their analytical tractability considerably ease the task of pricing exotic products. The Gaussian distribution allows the derivation of explicit formulas for a number of non-plain-vanilla instruments and, combined with the analytical expression for zero-coupon bonds, leads to efficient and fairly fast numerical procedures for pricing any possible payoff. Also, finite spot and forward rates at a given time for any maturity and accrual conventions can be given an explicit analytical expression in terms of the short-rate factors at the relevant instant. This allows for easy propagation of the whole zero-coupon curve in terms of the two factors. Another consequence of the presence of two factors is that the actual variability of market rates is described in a better way: Among other improvements, a non-perfect correlation between rates of different maturities is introduced. This results in a more precise calibration to correlation-based products like European swaptions. These major advantages are the main reason why we devote so much attention to a two-factor Gaussian model. Such a model can also be helpful when pricing out-of-the-money exotic instruments after calibration to at-the-money plain-vanilla products. In fact, the smile effect that is present in the market can be better captured by a distribution with a significant mass around zero. This is the case, for instance, of a receiver out-of-the-money Bermudan swaption with the model parameters being calibrated to the market prices of the corresponding European at-the-money swaptions.

The Gaussian model of this section is naturally related to the Hull-White (1994c) two-factor model in that we can actually prove the equivalence between these two approaches. However, the formulation with two additive factors leads to less complicated formulas and is easier to implement in practice, even though we may lose some insight and intuition on the nature and the interpretation of the two factors.

This section is structured as follows. In the first subsection, we introduce the short-rate dynamics and explain the resulting distributional features. In the second subsection, we derive an analytical formula for the price of a zero-coupon bond. In the third subsection we derive the dynamics of forward rates and analyze their volatility and correlation structures. In the fourth subsection, we derive the short-rate dynamics and distribution under a general forward measure. As a consequence, we price European options on zero-coupon bonds, caps and floors and finally swaptions. The knowledge of the short-rate distribution under any forward measure will be quite helpful when pricing most of the specific interest-rate derivatives we analyze in the second part of the book. In the fifth subsection we study the analogy of this model with the two-factor model being proposed by Hull and White (1994c). In the sixth subsection we show how to construct a two-dimensional binomial tree that approximates the short-rate dynamics. The longest proofs are written in separate appendices.

4.2.1 The Short-Rate Dynamics

We assume that the dynamics of the instantaneous-short-rate process under the risk-adjusted measure Q is given by

$$r(t) = x(t) + y(t) + \varphi(t), \quad r(0) = r_0, \tag{4.4}$$

where the processes $\{x(t) : t \geq 0\}$ and $\{y(t) : t \geq 0\}$ satisfy

$$\begin{aligned} dx(t) &= -ax(t)dt + \sigma dW_1(t), \quad x(0) = 0, \\ dy(t) &= -by(t)dt + \eta dW_2(t), \quad y(0) = 0, \end{aligned} \tag{4.5}$$

where (W_1, W_2) is a two-dimensional Brownian motion with instantaneous correlation ρ as from

$$dW_1(t)dW_2(t) = \rho dt,$$

where r_0, a, b, σ, η are positive constants, and where $-1 \leq \rho \leq 1$. The function φ is deterministic and well defined in the time interval $[0, T^*]$, with T^* a given time horizon, typically 10, 30 or 50 (years). In particular, $\varphi(0) = r_0$. We denote by $\mathcal{F}_t$ the sigma-field generated by the pair (x, y) up to time t.

Simple integration of equations (4.5) implies that for each $s < t$

$$r(t) = x(s)e^{-a(t-s)} + y(s)e^{-b(t-s)}$$
$$+\sigma \int_s^t e^{-a(t-u)} dW_1(u) + \eta \int_s^t e^{-b(t-u)} dW_2(u) + \varphi(t),$$

meaning that $r(t)$ conditional on $\mathcal{F}_s$ is normally distributed with mean and variance given respectively by

$$\begin{aligned} E\{r(t)|\mathcal{F}_s\} &= x(s)e^{-a(t-s)} + y(s)e^{-b(t-s)} + \varphi(t), \\ \mathrm{Var}\{r(t)|\mathcal{F}_s\} &= \frac{\sigma^2}{2a}\left[1 - e^{-2a(t-s)}\right] + \frac{\eta^2}{2b}\left[1 - e^{-2b(t-s)}\right] \\ &\quad + 2\rho\frac{\sigma\eta}{a+b}\left[1 - e^{-(a+b)(t-s)}\right]. \end{aligned} \tag{4.6}$$

In particular

$$r(t) = \sigma \int_0^t e^{-a(t-u)} dW_1(u) + \eta \int_0^t e^{-b(t-u)} dW_2(u) + \varphi(t). \tag{4.7}$$

The dynamics of the processes x and y can be also expressed in terms of two independent Brownian motions $\widetilde{W}_1$ and $\widetilde{W}_2$ as follows:[1]

$$\begin{aligned} dx(t) &= -ax(t)dt + \sigma d\widetilde{W}_1(t), \\ dy(t) &= -by(t)dt + \eta\rho d\widetilde{W}_1(t) + \eta\sqrt{1-\rho^2} d\widetilde{W}_2(t), \end{aligned} \tag{4.8}$$

where

$$\begin{aligned} dW_1(t) &= d\widetilde{W}_1, \\ dW_2(t) &= \rho d\widetilde{W}_1(t) + \sqrt{1-\rho^2} d\widetilde{W}_2(t), \end{aligned}$$

so that we can also write

$$r(t) = x(s)e^{-a(t-s)} + y(s)e^{-b(t-s)} + \sigma \int_s^t e^{-a(t-u)} d\widetilde{W}_1(u)$$
$$+ \eta\rho \int_s^t e^{-b(t-u)} d\widetilde{W}_1(u) + \eta\sqrt{1-\rho^2} \int_s^t e^{-b(t-u)} d\widetilde{W}_2(u) + \varphi(t).$$

4.2.2 The Pricing of a Zero-Coupon Bond

We denote by $P(t,T)$ the price at time t of a zero-coupon bond maturing at T and with unit face value, so that

$$P(t,T) = E\left\{e^{-\int_t^T r_s ds} | \mathcal{F}_t\right\},$$

where E denotes the expectation under the risk-adjusted measure Q. In order to explicitly compute this expectation, we need the following.

[1] This is equivalent to performing a Cholesky decomposition on the variance-covariance matrix of the pair $(W_1(t), W_2(t))$.

Lemma 4.2.1. *For each t, T the random variable*

$$I(t,T) := \int_t^T [x(u) + y(u)]du$$

conditional to the sigma-field $\mathcal{F}_t$ is normally distributed with mean $M(t,T)$ and variance $V(t,T)$, respectively given by

$$M(t,T) = \frac{1 - e^{-a(T-t)}}{a} x(t) + \frac{1 - e^{-b(T-t)}}{b} y(t) \tag{4.9}$$

and

$$\begin{aligned} V(t,T) &= \frac{\sigma^2}{a^2}\left[T - t + \frac{2}{a}e^{-a(T-t)} - \frac{1}{2a}e^{-2a(T-t)} - \frac{3}{2a}\right] \\ &+ \frac{\eta^2}{b^2}\left[T - t + \frac{2}{b}e^{-b(T-t)} - \frac{1}{2b}e^{-2b(T-t)} - \frac{3}{2b}\right] \\ &+ 2\rho\frac{\sigma\eta}{ab}\left[T - t + \frac{e^{-a(T-t)} - 1}{a} + \frac{e^{-b(T-t)} - 1}{b} - \frac{e^{-(a+b)(T-t)} - 1}{a+b}\right]. \end{aligned} \tag{4.10}$$

Proof. See Appendix A in this chapter. □

Theorem 4.2.1. *The price at time t of a zero-coupon bond maturing at time T and with unit face value is*

$$\begin{aligned} P(t,T) = \exp\Bigg\{ &- \int_t^T \varphi(u)du - \frac{1 - e^{-a(T-t)}}{a} x(t) \\ &- \frac{1 - e^{-b(T-t)}}{b} y(t) + \frac{1}{2} V(t,T) \Bigg\}. \end{aligned} \tag{4.11}$$

Proof. Being φ a deterministic function, the theorem follows from straightforward application of Lemma 4.2.1 and the fact that if Z is a normal random variable with mean m_Z and variance σ_Z^2, then $E\{\exp(Z)\} = \exp(m_Z + \frac{1}{2}\sigma_Z^2)$. □

Let us now assume that the term structure of discount factors that is currently observed in the market is given by the sufficiently smooth function $T \mapsto P^M(0,T)$.

If we denote by $f^M(0,T)$ the instantaneous forward rate at time 0 for a maturity T implied by the term structure $T \mapsto P^M(0,T)$, i.e.,

$$f^M(0,T) = -\frac{\partial \ln P^M(0,T)}{\partial T},$$

we then have the following.

Corollary 4.2.1. *The model (4.4) fits the currently-observed term structure of discount factors if and only if, for each T,*

$$\varphi(T) = f^M(0,T) + \frac{\sigma^2}{2a^2}\left(1-e^{-aT}\right)^2 + \frac{\eta^2}{2b^2}\left(1-e^{-bT}\right)^2 + \rho\frac{\sigma\eta}{ab}\left(1-e^{-aT}\right)\left(1-e^{-bT}\right), \quad (4.12)$$

i.e., if and only if

$$\exp\left\{-\int_t^T \varphi(u)du\right\} = \frac{P^M(0,T)}{P^M(0,t)}\exp\left\{-\frac{1}{2}[V(0,T)-V(0,t)]\right\}, \quad (4.13)$$

so that the corresponding zero-coupon-bond prices at time t are given by

$$\begin{aligned} P(t,T) &= \frac{P^M(0,T)}{P^M(0,t)}\exp\{\mathcal{A}(t,T)\} \\ \mathcal{A}(t,T) &:= \frac{1}{2}[V(t,T)-V(0,T)+V(0,t)] \\ &\quad - \frac{1-e^{-a(T-t)}}{a}x(t) - \frac{1-e^{-b(T-t)}}{b}y(t). \end{aligned} \quad (4.14)$$

Proof. The model (4.4) fits the currently-observed term structure of discount factors if and only if for each maturity $T \le T^*$ the discount factor $P(0,T)$ produced by the model (4.4) coincides with the one observed in the market, i.e., if and only if

$$P^M(0,T) = \exp\left\{-\int_0^T \varphi(u)du + \frac{1}{2}V(0,T)\right\}.$$

Now let us take logs of both sides and differentiate with respect to T, so as to obtain (4.12) by noting that (see also Appendix A in this chapter)

$$\begin{aligned} V(t,T) =& \frac{\sigma^2}{a^2}\int_t^T \left[1-e^{-a(T-u)}\right]^2 du + \frac{\eta^2}{b^2}\int_t^T \left[1-e^{-b(T-u)}\right]^2 du \\ &+ 2\rho\frac{\sigma\eta}{ab}\int_t^T \left[1-e^{-a(T-u)}\right]\left[1-e^{-b(T-u)}\right] du. \end{aligned}$$

Equality (4.13) follows from noting that, under the specification (4.12),

$$\begin{aligned} \exp\left\{-\int_t^T \varphi(u)du\right\} &= \exp\left\{-\int_0^T \varphi(u)du\right\}\exp\left\{\int_0^t \varphi(u)du\right\} \\ &= \frac{P^M(0,T)\exp\left\{-\frac{1}{2}V(0,T)\right\}}{P^M(0,t)\exp\left\{-\frac{1}{2}V(0,t)\right\}}. \end{aligned}$$

Equality (4.14) immediately follows from (4.11) and (4.13). □

Remark 4.2.1. **(Is it really necessary to derive the market instantaneous forward curve?)** Notice that, at a first sight, one may have the impression that in order to implement the G2++ model we need to derive the whole φ curve, and therefore the market instantaneous forward curve $T \mapsto f^M(0,T)$. Now, this curve involves differentiating the market discount curve $T \mapsto P^M(0,T)$, which is usually obtained from a finite set of maturities via interpolation. Interpolation and differentiation may induce a certain degree of approximation, since the particular interpolation technique being used has a certain impact on (first) derivatives.

However, it turns out that one does not really need the whole φ curve. Indeed, what matters is the integral of φ between two given instants. This integral has been computed in (4.13). From this expression, we see that the only curve needed is the market discount curve, which need not be differentiated, and only at times corresponding to the maturities of the bond prices and rates desired, thus limiting also the need for interpolation.

Remark 4.2.2. **(Short-rate distribution and probability of negative rates).** By fitting the currently-observed term structure of discount factors, i.e. by applying (4.12), we obtain that the expected instantaneous short rate at time t, $\mu_r(t)$, is

$$\begin{aligned}\mu_r(t) := E\{r(t)\} = f^M(0,t) + \frac{\sigma^2}{2a^2}\left(1-e^{-at}\right)^2 + \frac{\eta^2}{2b^2}\left(1-e^{-bt}\right)^2 \\ + \rho\frac{\sigma\eta}{ab}\left(1-e^{-at}\right)\left(1-e^{-bt}\right),\end{aligned}$$

while the variance $\sigma_r^2(t)$ of the instantaneous short rate at time t, see (4.6), is

$$\sigma_r^2(t) = \mathrm{Var}\{r(t)\} = \frac{\sigma^2}{2a}\left(1-e^{-2at}\right)+\frac{\eta^2}{2b}\left(1-e^{-2bt}\right)+2\frac{\rho\sigma\eta}{a+b}\left(1-e^{-(a+b)t}\right).$$

This implies that the risk-neutral probability of negative rates at time t is

$$Q\{r(t)<0\} = \Phi\left(-\frac{\mu_r(t)}{\sigma_r(t)}\right),$$

which is often negligible in many concrete situations, with Φ denoting the standard normal cumulative distribution function.

Furthermore, we have that the limit distribution of the process r is Gaussian with mean $\mu_r(\infty)$ and variance $\sigma_r^2(\infty)$ given by

$$\begin{aligned}\mu_r(\infty) &:= \lim_{t\to\infty} E\{r(t)\} = f^M(0,\infty) + \frac{\sigma^2}{2a^2} + \frac{\eta^2}{2b^2} + \rho\frac{\sigma\eta}{ab}, \\ \sigma_r^2(\infty) &:= \lim_{t\to\infty} \mathrm{Var}\{r(t)\} = \frac{\sigma^2}{2a} + \frac{\eta^2}{2b} + 2\rho\frac{\sigma\eta}{a+b},\end{aligned}$$

where

$$f^M(0,\infty) = \lim_{t\to\infty} f^M(0,t).$$

4.2.3 Volatility and Correlation Structures in Two-Factor Models

We now derive the dynamics of forward rates under the risk-neutral measure to obtain an equivalent formulation of the two-additive-factor Gaussian model in the Heath-Jarrow-Morton (1992) framework. In particular, we explicitly derive the volatility structure of forward rates. This also allows us to understand which market-volatility structures can be fitted by the model.

Let us define $A(t,T)$ and $B(z,t,T)$ by

$$\begin{aligned} A(t,T) =& \frac{P^M(0,T)}{P^M(0,t)} \exp\left\{\frac{1}{2}[V(t,T) - V(0,T) + V(0,t)]\right\}, \\ B(z,t,T) =& \frac{1-e^{-z(T-t)}}{z}, \end{aligned}$$

so that we can write

$$P(t,T) = A(t,T)\exp\{-B(a,t,T)x(t) - B(b,t,T)y(t)\}. \tag{4.15}$$

The (continuously-compounded) instantaneous forward rate at time t for the maturity T is then given by

$$\begin{aligned} f(t,T) =& -\frac{\partial}{\partial T}\ln P(t,T) \\ =& -\frac{\partial}{\partial T}\ln A(t,T) + \frac{\partial B}{\partial T}(a,t,T)x(t) + \frac{\partial B}{\partial T}(b,t,T)y(t), \end{aligned}$$

whose differential form can be written as

$$df(t,T) = \ldots dt + \frac{\partial B}{\partial T}(a,t,T)\sigma dW_1(t) + \frac{\partial B}{\partial T}(b,t,T)\eta dW_2(t).$$

Therefore

$$\begin{aligned} &\frac{\mathrm{Var}(df(t,T))}{dt} \\ &= \left(\frac{\partial B}{\partial T}(a,t,T)\sigma\right)^2 + \left(\frac{\partial B}{\partial T}(b,t,T)\eta\right)^2 + 2\rho\sigma\eta\frac{\partial B}{\partial T}(a,t,T)\frac{\partial B}{\partial T}(b,t,T) \\ &= \sigma^2 e^{-2a(T-t)} + \eta^2 e^{-2b(T-t)} + 2\rho\sigma\eta e^{-(a+b)(T-t)}, \end{aligned}$$

which implies that the absolute volatility of the instantaneous forward rate $f(t,T)$ is

$$\sigma_f(t,T) = \sqrt{\sigma^2 e^{-2a(T-t)} + \eta^2 e^{-2b(T-t)} + 2\rho\sigma\eta e^{-(a+b)(T-t)}}. \tag{4.16}$$

From (4.16), we immediately see that the desirable feature, as far as calibration to the market is concerned, of a humped volatility structure similar to what is commonly observed in the market for the caplets volatility, may be only reproduced for negative values of ρ. Notice indeed that if ρ is positive, the terms $\sigma^2 e^{-2a(T-t)}$, $\eta^2 e^{-2b(T-t)}$ and $2\rho\sigma\eta e^{-(a+b)(T-t)}$ are all decreasing functions of the time to maturity $T-t$ and no hump is possible. This does not mean, in turn, that every combination of the parameter values with a negative ρ leads to a volatility hump. A simple study of $\sigma_f(t,T)$ as a function of $T-t$, however, shows that there exist suitable choices of the parameter values that produce the desired shape.

Notice that the instantaneous-forwards humped-volatility shape has been considered, also in relation with market calibration, in Mercurio and Moraleda (2000a, 2000b) and in references given therein. See also Sections 3.6 and 3.7 of the present book.

Lest confusion may arise in the reader, we detail the humped-shape issue further, as we did in Section 3.6 for the one-factor case. We will allow for a little redundancy in the treatment, so as to preserve partially self-contained chapters.

We will go through the following points.

- Recall how cap and caplet volatilities are defined in the market model for caps.
- Recall how the term structure of volatility in the market is obtained by these, and recall its usually humped shape.
- Transliterate the market definition of caplet volatility above to short-rate models (model-intrinsic caplet volatility) and show what goes wrong.
- Modify the definition of caplet volatility for short-rate models so that things work again (model-implied caplet volatility and model-implied cap volatility), and then explain how the term structure of volatilities is consequently defined for a short-rate model.

Recall the Market-Like Definition of Caplet Volatilities

As we have seen earlier, cap or single caplet volatilities can be retrieved from market prices by inverting the related market formulas (see also Sections 6.4 and 6.4.3 in particular). Alternatively, caplet volatilities can be defined as suitable integrals (averages) as

$$\begin{aligned} v^2_{T-\text{caplet}} &:= \frac{1}{T}\int_0^T (d\ln F(t;T,T+\tau))(d\ln F(t;T,T+\tau)) \\ &= \frac{1}{T}\int_0^T \sigma(t;T,T+\tau)^2\,dt, \end{aligned}$$

where τ is typically six months, and where $\sigma(t;T,T+\tau)$ is the (percentage) instantaneous volatility at time t of the simply-compounded forward rate $F(t;T,T+\tau)$ underlying the T-expiry $(T+\tau)$-maturity caplet. The instantaneous percentage volatilities $\sigma(t;T,T+\tau)$ are deterministic in Black's model for the cap market, so that the caplet volatility $v_{T-\text{caplet}}$ is also deterministic.

What is observed to have a humped shape in the market is the curve of the caplet volatilities for different maturities at time 0, i.e., $T \mapsto v_{T-\text{caplet}}$.

Recall the Term Structure of Volatilities in the Market

The above-mentioned caplet curve $T \mapsto v_{T-\text{caplet}}$ is called the term structure of (caplet) volatilities at time 0.

Transliterating the Market Definition to Short-Rate Models: What Goes Wrong

Recall also that, in a model like G2++, the integrals in the above definition of $v_{T-\text{caplet}}$ would be stochastic since $\sigma(t; T, T+\tau)$ is not deterministic. This is an important point. One can convince oneself of this by deriving the expression for

$$d \ln F(t; T, T + \tau) = d \ln \left(\frac{P(t, T)}{P(t, T + \tau)} - 1 \right)$$

in the G2++ model through Ito's formula when expressing $P(t, T)$ and $P(t, T + \tau)$ according to (4.15).[2] It is easy to see that in both cases the (log) volatility (diffusion coefficient) will not be deterministic, but rather a stochastic quantity depending on $x(t)$ and $y(t)$. As usual, the logarithm is considered because the diffusion coefficient for the logarithm of a certain process amounts to the process percentage instantaneous volatility.

Since the quantity $v_{T-\text{caplet}}$ is an integral of the instantaneous variance $\sigma(t; T, T + \tau)^2$, in the G2++ model the $v_{T-\text{caplet}}$ would be stochastic if computed as

$$\sqrt{\frac{1}{T} \int_0^T \sigma(t; T, T + \tau)^2 \, dt}.$$

This is what we defined as the model-intrinsic T-caplet volatility in Section 3.6. We also pointed out that this is not the way caplet volatilities can be defined for models such as G2++.

Modify the Definition of Caplet Volatility for Short-Rate Models so that Things Work Again

Indeed, for models different from Black's market model for caps, such as the G2++ model considered here, the $v_{T-\text{caplet}}$'s are defined so as to be again deterministic and are usually understood as implied volatilities. One prices an at-the-money T-expiry caplet $\mathbf{Cpl}(0, T, T + \tau, F(0; T, T + \tau))$ with the model, and then inverts the T-expiry Black's market formula and finds the percentage Black volatility $v_{T-\text{caplet}}$ that, plugged into such a formula, yields the model price. More precisely, one solves the following equation for $v_{T-\text{caplet}}^{\text{G2++}}$:

[2] Below we consider an analogous calculation for the corresponding continuously-compounded forward rate $f(t; T, T + \tau)$.

$$P(0,T+\tau)F(0;T,T+\tau)\left(2\Phi\left(\frac{v^{\mathrm{G2++}}_{T-\mathrm{caplet}}\sqrt{T}}{2}\right)-1\right)$$
$$=\mathbf{Cpl}(0,T,T+\tau,F(0;T,T+\tau)).$$

The left-hand side is the market Black's formula for a T-expiry $T+\tau$-maturity at-the-money caplet, whereas the right-hand side is the corresponding G2++ model formula (4.27) given later on. When this is done for all expiries T one can plot the term structure of (caplet) volatilities implied by the G2++ model,

$$T \mapsto v^{\mathrm{G2++}}_{T-\mathrm{caplet}}.$$

Implied cap volatilities can be defined similarly by considering a set of reset times $\{T_\alpha,\ldots,T_{\beta-1}\}$ with the final payment time T_β, and the set of the associated year fractions $\{\tau_\alpha,\ldots,\tau_\beta\}$. Then, setting $\mathcal{T}_i=\{T_\alpha,\ldots,T_i\}$ and $\bar{\tau}_i=\{\tau_{\alpha+1},\ldots,\tau_i\}$, the *model implied $\mathcal{T}_i$-cap volatility* at time 0 is the (deterministic) solution $v^{\mathrm{G2++}}_{\mathcal{T}_i-\mathrm{cap}}$ of the equation:

$$\sum_{j=\alpha+1}^{i} P(0,T_j)\tau_j \mathrm{Bl}(S_{\alpha,\beta}(0),F(0,T_{j-1},T_j),v^{\mathrm{G2++}}_{\mathcal{T}_i-\mathrm{cap}}\sqrt{T_{j-1}})$$
$$=\mathbf{Cap}(0,\mathcal{T}_i,\bar{\tau}_i,S_{\alpha,\beta}(0)),$$

where the forward swap rate $S_{\alpha,\beta}(0)$ is defined in (1.25).

When this is done for all i's, one can plot the term structure of cap volatilities implied by the G2++ model,

$$\mathcal{T}_i \mapsto v^{\mathrm{G2++}}_{\mathcal{T}_i-\mathrm{cap}}.$$

When the G2++ model is calibrated to the cap market, the parameters a,σ,b,η,ρ in the G2++ dynamics are set to values such that the model volatilities $v^{\mathrm{G2++}}_{\mathcal{T}_i-\mathrm{cap}}(a,\sigma,b,\eta,\rho)$ are as close as possible to the market cap volatilities $v^{\mathrm{MKT}}_{\mathcal{T}_i-\mathrm{cap}}$.

Alternatively, if one has already obtained caplet volatilities by stripping them from cap volatilities along the lines of Section 6.4.3, one can calibrate the model directly to caplet volatilities. The parameters a,σ,b,η,ρ in the G2++ dynamics are then set to values such that the model volatilities $v^{\mathrm{G2++}}_{T-\mathrm{caplet}}(a,\sigma,b,\eta,\rho)$ are as close as possible to the market caplet volatilities $v^{\mathrm{MKT}}_{T-\mathrm{caplet}}$.

We have also observed in Section 3.6 that, for this term structure to be able to feature large humps, the model (absolute) instantaneous volatilities of instantaneous forward rates,

$$T \mapsto \sigma_f(t,T),$$

usually need to allow for a hump themselves. We have also seen that in order to obtain *small* humps in such a term structure, it is not necessary to allow for a corresponding hump in the model absolute volatility of instantaneous forward rates. In short, one usually observes the following.

1. No humps in $T \mapsto \sigma_f(t,T) \Rightarrow$ only small humps for $T \mapsto v_{T-\text{caplet}}^{\text{MODEL}}$ are possible;
2. Humps in $T \mapsto \sigma_f(t,T) \Rightarrow$ large humps for $T \mapsto v_{T-\text{caplet}}^{\text{MODEL}}$ are possible.

We have remarked that, in the typical example of the CIR++ model, a little analysis of the related analytical formulas shows how $T \mapsto \sigma_f(t,T)$ is monotonically decreasing, thus usually implying only small humps in the model caplet (and cap) volatilities.

Let us repeat once again that confusion often arises when speaking of "allowing a humped shape for forward-rates volatilities". Indeed, one has to specify what kind of volatilities are considered: instantaneous-forward absolute volatilities, caplet volatilities, cap volatilities. As we have seen, some models allow for humps in all of these structures; some other models only allow for small humps in the caplets or caps volatilities.

After the instantaneous volatility of forward rates, we can consider an analogous calculation for the instantaneous covariance per unit time between the two forward rates $f(t,T_1)$ and $f(t,T_2)$, obtaining

$$
\begin{aligned}
&\frac{\mathrm{Cov}(df(t,T_1), df(t,T_2))}{dt} \\
&= \sigma^2 \frac{\partial B}{\partial T}(a,t,T_1)\frac{\partial B}{\partial T}(a,t,T_2) + \eta^2 \frac{\partial B}{\partial T}(b,t,T_1)\frac{\partial B}{\partial T}(b,t,T_2) \\
&\quad + \rho\sigma\eta \left[\frac{\partial B}{\partial T}(a,t,T_1)\frac{\partial B}{\partial T}(b,t,T_2) + \frac{\partial B}{\partial T}(a,t,T_2)\frac{\partial B}{\partial T}(b,t,T_1)\right] \\
&= \sigma^2 e^{-a(T_1+T_2-2t)} + \eta^2 e^{-b(T_1+T_2-2t)} \\
&\quad + \rho\sigma\eta \left[e^{-aT_1-bT_2+(a+b)t} + e^{-aT_2-bT_1+(a+b)t}\right],
\end{aligned}
$$

so that the instantaneous correlation between the two forward rates $f(t,T_1)$ and $f(t,T_2)$ is

$$
\begin{aligned}
\mathrm{Corr}(df(t,T_1), df(t,T_2)) &= \frac{\sigma^2 e^{-a(T_1+T_2-2t)} + \eta^2 e^{-b(T_1+T_2-2t)}}{\sigma_f(t,T_1)\sigma_f(t,T_2)} \\
&\quad + \frac{\rho\sigma\eta\left[e^{-aT_1-bT_2+(a+b)t} + e^{-aT_2-bT_1+(a+b)t}\right]}{\sigma_f(t,T_1)\sigma_f(t,T_2)}.
\end{aligned}
$$

As expected, the absolute value of such a correlation is smaller than one for general parameter values for which the model is non-degenerate.[3] The previous analyses and remarks also apply to forward rates spanning finite time intervals. We have the following results.

The (continuously-compounded) forward rate at time t between times T_1 and T_2 is

---

[3] The model is said to be degenerate if the two underlying factors are driven by the same noise.

$$f(t,T_1,T_2) = \frac{\ln P(t,T_1) - \ln P(t,T_2)}{T_2 - T_1},$$

whose differential form can be written as

$$df(t,T_1,T_2) = \ldots dt + \frac{B(a,t,T_2) - B(a,t,T_1)}{T_2 - T_1}\sigma dW_1(t) + \frac{B(b,t,T_2) - B(b,t,T_1)}{T_2 - T_1}\eta dW_2(t).$$

Therefore the absolute volatility of the forward rate $f(t,T_1,T_2)$ is

$$\sigma_f(t,T_1,T_2) = \sqrt{\sigma^2\beta(a,t,T_1,T_2)^2 + \eta^2\beta(b,t,T_1,T_2)^2 + 2\rho\sigma\eta\beta(a,t,T_1,T_2)\beta(b,t,T_1,T_2)},$$

where

$$\beta(z,t,T_1,T_2) = \frac{B(z,t,T_2) - B(z,t,T_1)}{T_2 - T_1}.$$

Analogously, the instantaneous covariance per unit time between the two forward rates $f(t,T_1,T_2)$ and $f(t,T_3,T_4)$ is

$$\begin{aligned}
&\frac{\mathrm{Cov}(df(t,T_1,T_2), df(t,T_3,T_4))}{dt} \\
&= \sigma^2 \frac{B(a,t,T_2) - B(a,t,T_1)}{T_2 - T_1}\frac{B(a,t,T_4) - B(a,t,T_3)}{T_4 - T_3} \\
&\quad + \eta^2 \frac{B(b,t,T_2) - B(b,t,T_1)}{T_2 - T_1}\frac{B(b,t,T_4) - B(b,t,T_3)}{T_4 - T_3} \\
&\quad + \rho\sigma\eta \left[\frac{B(a,t,T_2) - B(a,t,T_1)}{T_2 - T_1}\frac{B(b,t,T_4) - B(b,t,T_3)}{T_4 - T_3}\right. \\
&\quad \left. + \frac{B(a,t,T_4) - B(a,t,T_3)}{T_4 - T_3}\frac{B(b,t,T_2) - B(b,t,T_1)}{T_2 - T_1}\right].
\end{aligned}$$

4.2.4 The Pricing of a European Option on a Zero-Coupon Bond

The price at time t of a European call option with maturity T and strike K, written on a zero-coupon bond with unit face value and maturity τ is

$$\mathbf{ZBC}(t,T,S,K) = E\left\{ e^{-\int_t^T r(s)ds}(P(T,S) - K)^+ \middle| \mathcal{F}_t \right\}.$$

In order to explicitly compute this expectation we need to change probability measure as indicated by Jamshidian (1989) and more generally by Geman et al. (1995). Precisely, for any fixed maturity T, we denote by Q^T the probability measure defined by the Radon-Nikodym derivative (see Appendix C at the end of the book on the Radon-Nikodym derivative)

$$
\begin{aligned}
\frac{dQ^T}{dQ} &= \frac{B(0)P(T,T)}{B(T)P(0,T)} = \frac{\exp\left\{-\int_0^T r(u)du\right\}}{P(0,T)} \\
&= \frac{\exp\left\{-\int_0^T \varphi(u)du - \int_0^T [x(u)+y(u)]du\right\}}{P(0,T)} \\
&= \exp\left\{-\frac{1}{2}V(0,T) - \int_0^T [x(u)+y(u)]du\right\},
\end{aligned}
\tag{4.17}
$$

where B here is the bank-account numeraire. The measure Q^T is the well known T-forward (risk-adjusted) measure. The following lemma yields the dynamics of the processes x and y under Q^T.

Lemma 4.2.2. *The processes x and y under the forward measure Q^T evolve according to*

$$
\begin{aligned}
dx(t) &= \left[-ax(t) - \frac{\sigma^2}{a}(1-e^{-a(T-t)}) - \rho\frac{\sigma\eta}{b}(1-e^{-b(T-t)})\right] dt + \sigma dW_1^T(t), \\
dy(t) &= \left[-by(t) - \frac{\eta^2}{b}(1-e^{-b(T-t)}) - \rho\frac{\sigma\eta}{a}(1-e^{-a(T-t)})\right] dt + \eta dW_2^T(t),
\end{aligned}
\tag{4.18}
$$

where W_1^T and W_2^T are two correlated Brownian motions under Q^T with $dW_1^T(t)dW_2^T(t) = \rho\, dt$.

Moreover, the explicit solutions of equations (4.18) are, for $s \le t \le T$,

$$
\begin{aligned}
x(t) &= x(s)e^{-a(t-s)} - M_x^T(s,t) + \sigma \int_s^t e^{-a(t-u)} dW_1^T(u) \\
y(t) &= y(s)e^{-b(t-s)} - M_y^T(s,t) + \eta \int_s^t e^{-b(t-u)} dW_2^T(u),
\end{aligned}
\tag{4.19}
$$

where

$$
\begin{aligned}
M_x^T(s,t) &= \left(\frac{\sigma^2}{a^2} + \rho\frac{\sigma\eta}{ab}\right)\left[1 - e^{-a(t-s)}\right] - \frac{\sigma^2}{2a^2}\left[e^{-a(T-t)} - e^{-a(T+t-2s)}\right] \\
&\quad - \frac{\rho\sigma\eta}{b(a+b)}\left[e^{-b(T-t)} - e^{-bT-at+(a+b)s}\right], \\
M_y^T(s,t) &= \left(\frac{\eta^2}{b^2} + \rho\frac{\sigma\eta}{ab}\right)\left[1 - e^{-b(t-s)}\right] - \frac{\eta^2}{2b^2}\left[e^{-b(T-t)} - e^{-b(T+t-2s)}\right] \\
&\quad - \frac{\rho\sigma\eta}{a(a+b)}\left[e^{-a(T-t)} - e^{-aT-bt+(a+b)s}\right],
\end{aligned}
$$

so that, under Q^T, the distribution of $r(t)$ conditional on $\mathcal{F}_s$ is normal with mean and variance given respectively by

$$
\begin{aligned}
E^{Q^T}\{r(t)|\mathcal{F}_s\} &= x(s)e^{-a(t-s)} - M_x^T(s,t) + y(s)e^{-b(t-s)} - M_y^T(s,t) + \varphi(t), \\
Var^{Q^T}\{r(t)|\mathcal{F}_s\} &= \frac{\sigma^2}{2a}\left[1 - e^{-2a(t-s)}\right] + \frac{\eta^2}{2b}\left[1 - e^{-2b(t-s)}\right] \\
&\quad + 2\rho\frac{\sigma\eta}{a+b}\left[1 - e^{-(a+b)(t-s)}\right].
\end{aligned}
\tag{4.20}
$$

Proof. See Appendix B in this chapter for a detailed proof, or else apply directly formula (2.12) with $U = P(\cdot, T)$, $S = B$, and $X = [x\ y]'$. □

Formulas (4.20) are very useful when pricing path-dependent derivatives through Monte Carlo generation of scenarios. To this end, we refer to the second part of this book.

We can now state the following.

Theorem 4.2.2. *The price at time t of a European call option with maturity T and strike K, written on a zero-coupon bond with unit face value and maturity S is given by*

$$
\begin{aligned}
\mathbf{ZBC}(t,T,S,K) &= P(t,S)\Phi\left(\frac{\ln\frac{P(t,S)}{KP(t,T)}}{\Sigma(t,T,S)} + \frac{1}{2}\Sigma(t,T,S)\right) \\
&\quad - P(t,T)K\Phi\left(\frac{\ln\frac{P(t,S)}{KP(t,T)}}{\Sigma(t,T,S)} - \frac{1}{2}\Sigma(t,T,S)\right),
\end{aligned}
\tag{4.21}
$$

where

$$
\begin{aligned}
\Sigma(t,T,S)^2 &= \frac{\sigma^2}{2a^3}\left[1 - e^{-a(S-T)}\right]^2\left[1 - e^{-2a(T-t)}\right] \\
&\quad + \frac{\eta^2}{2b^3}\left[1 - e^{-b(S-T)}\right]^2\left[1 - e^{-2b(T-t)}\right] \\
&\quad + 2\rho\frac{\sigma\eta}{ab(a+b)}\left[1 - e^{-a(S-T)}\right]\left[1 - e^{-b(S-T)}\right]\left[1 - e^{-(a+b)(T-t)}\right].
\end{aligned}
$$

Analogously, the price at time t of a European put option with maturity T and strike K, written on a zero-coupon bond with unit face value and maturity S is given by

$$
\begin{aligned}
\mathbf{ZBP}(t,T,S,K) &= -P(t,S)\Phi\left(\frac{\ln\frac{KP(t,T)}{P(t,S)}}{\Sigma(t,T,S)} - \frac{1}{2}\Sigma(t,T,S)\right) \\
&\quad + P(t,T)K\Phi\left(\frac{\ln\frac{KP(t,T)}{P(t,S)}}{\Sigma(t,T,S)} + \frac{1}{2}\Sigma(t,T,S)\right).
\end{aligned}
\tag{4.22}
$$

Proof. See Appendix C in this chapter. □

We then have the following obvious generalization to the case where the underlying bond has an arbitrary face value.

Corollary 4.2.2. *The price at time t of a European call option with maturity T and strike K, written on a zero-coupon bond with face value N and maturity S is given by*

$$\begin{aligned}\mathbf{ZBC}(t,T,S,N,K) &= NP(t,S)\Phi\left(\frac{\ln\frac{NP(t,S)}{KP(t,T)}}{\Sigma(t,T,S)}+\frac{1}{2}\Sigma(t,T,S)\right)\\ &\quad - P(t,T)K\Phi\left(\frac{\ln\frac{NP(t,S)}{KP(t,T)}}{\Sigma(t,T,S)}-\frac{1}{2}\Sigma(t,T,S)\right).\end{aligned} \tag{4.23}$$

Analogously, the price at time t of the corresponding put option is

$$\begin{aligned}\mathbf{ZBP}(t,T,S,N,K) &= -NP(t,S)\Phi\left(\frac{\ln\frac{KP(t,T)}{NP(t,S)}}{\Sigma(t,T,S)}-\frac{1}{2}\Sigma(t,T,S)\right)\\ &\quad + P(t,T)K\Phi\left(\frac{\ln\frac{KP(t,T)}{NP(t,S)}}{\Sigma(t,T,S)}+\frac{1}{2}\Sigma(t,T,S)\right).\end{aligned} \tag{4.24}$$

The pricing of caplets and floorlets. Given the current time t and the future times T_1 and T_2, an "in-arrears" caplet pays off at time T_2

$$[L(T_1,T_2)-X]^+\,\alpha(T_1,T_2)N, \tag{4.25}$$

where N is the nominal value, X is the caplet rate (strike), $\alpha(T_1,T_2)$ is the year fraction between times T_1 and T_2 and $L(T_1,T_2)$ is the LIBOR rate at time T_1 for the maturity T_2, i.e.,

$$L(T_1,T_2)=\frac{1}{\alpha(T_1,T_2)}\left[\frac{1}{P(T_1,T_2)}-1\right]. \tag{4.26}$$

The no-arbitrage value at time t of the payoff (4.25) is, by (2.26) and (2.27),

$$\begin{aligned}\mathbf{Cpl}(t,T_1,T_2,N,X) &= N'\mathbf{ZBP}(t,T_1,T_2,X')\\ &= \mathbf{ZBP}(t,T_1,T_2,N',N),\end{aligned} \tag{4.27}$$

where

$$\begin{aligned}X' &= \frac{1}{1+X\alpha(T_1,T_2)},\\ N' &= N(1+X\alpha(T_1,T_2)).\end{aligned}$$

Explicitly,

$$\mathbf{Cpl}(t,T_1,T_2,N,X) = -N'P(t,T_2)\Phi\left(\frac{\ln\frac{NP(t,T_1)}{N'P(t,T_2)}}{\Sigma(t,T_1,T_2)} - \frac{1}{2}\Sigma(t,T_1,T_2)\right) + P(t,T_1)N\Phi\left(\frac{\ln\frac{NP(t,T_1)}{N'P(t,T_2)}}{\Sigma(t,T_1,T_2)} + \frac{1}{2}\Sigma(t,T_1,T_2)\right). \tag{4.28}$$

Analogously, the no-arbitrage value at time t of the floorlet that pays off

$$[X - L(T_1,T_2)]^+ \alpha(T_1,T_2)N$$

at time T_2 is

$$\begin{aligned}\mathbf{Fll}(t,T_1,T_2,N,X) &= N'\mathbf{ZBC}(t,T_1,T_2,X')\\ &= \mathbf{ZBC}(t,T_1,T_2,N',N).\end{aligned}$$

Explicitly,

$$\mathbf{Fll}(t,T_1,T_2,N,X) = N'P(t,T_2)\Phi\left(\frac{\ln\frac{N'P(t,T_2)}{NP(t,T_1)}}{\Sigma(t,T_1,T_2)} + \frac{1}{2}\Sigma(t,T_1,T_2)\right) - P(t,T_1)N\Phi\left(\frac{\ln\frac{N'P(t,T_2)}{NP(t,T_1)}}{\Sigma(t,T_1,T_2)} - \frac{1}{2}\Sigma(t,T_1,T_2)\right).$$

The pricing of caps and floors. We denote by $\mathcal{T} = \{T_0, T_1, T_2, \ldots, T_n\}$ the set of the cap/floor payment dates, augmented with the first reset date T_0, and by $\tau = \{\tau_1, \ldots, \tau_n\}$ the set of the corresponding year fractions, meaning that τ_i is the year fraction between T_{i-1} and T_i.

Since the price of a cap (floor) is the sum of the prices of the underlying caplets (floorlets), the price at time t of a cap with cap rate (strike) X, nominal value N, set of times $\mathcal{T}$ and year fractions τ is then given by

$$\begin{aligned}&\mathbf{Cap}(t,\mathcal{T},\tau,N,X)\\ &= \sum_{i=1}^{n}\Bigg[-N(1+X\tau_i)P(t,T_i)\Phi\left(\frac{\ln\frac{P(t,T_{i-1})}{(1+X\tau_i)P(t,T_i)}}{\Sigma(t,T_{i-1},T_i)} - \frac{1}{2}\Sigma(t,T_{i-1},T_i)\right)\\ &\quad + P(t,T_{i-1})N\Phi\left(\frac{\ln\frac{P(t,T_{i-1})}{(1+X\tau_i)P(t,T_i)}}{\Sigma(t,T_{i-1},T_i)} + \frac{1}{2}\Sigma(t,T_{i-1},T_i)\right)\Bigg],\end{aligned} \tag{4.29}$$

and the price of the corresponding floor is

$$
\begin{aligned}
&\mathbf{Flr}(t,\mathcal{T},N,X) \\
&= \sum_{i=1}^{n}\left[N(1+X\tau_i)P(t,T_i)\Phi\left(\frac{\ln\frac{(1+X\tau_i)P(t,T_i)}{P(t,T_{i-1})}}{\Sigma(t,T_{i-1},T_i)}+\frac{1}{2}\Sigma(t,T_{i-1},T_i)\right)\right. \\
&\qquad \left.-P(t,T_{i-1})N\Phi\left(\frac{\ln\frac{(1+X\tau_i)P(t,T_i)}{P(t,T_{i-1})}}{\Sigma(t,T_{i-1},T_i)}-\frac{1}{2}\Sigma(t,T_{i-1},T_i)\right)\right].
\end{aligned}
\tag{4.30}
$$

The pricing of European swaptions.

God does not care about our mathematical difficulties.
He integrates empirically.
Albert Einstein (1879-1955)

Consider a European swaption with strike rate X, maturity T and nominal value N, which gives the holder the right to enter at time $t_0 = T$ an interest-rate swap with payment times $\mathcal{T} = \{t_1,\ldots,t_n\}$, $t_1 > T$, where he pays (receives) at the fixed rate X and receives (pays) LIBOR set "in arrears". We denote by τ_i the year fraction from t_{i-1} to t_i, $i = 1,\ldots,n$ and set $c_i := X\tau_i$ for $i = 1,\ldots,n-1$ and $c_n := 1 + X\tau_n$. We then have the following theorem.

Theorem 4.2.3. *The arbitrage-free price at time $t = 0$ of the above European swaption is given by numerically computing the following one-dimensional integral:*

$$
\mathbf{ES}(0,T,\mathcal{T},N,X,\omega) =
$$

$$
N\omega P(0,T)\int_{-\infty}^{+\infty}\frac{e^{-\frac{1}{2}\left(\frac{x-\mu_x}{\sigma_x}\right)^2}}{\sigma_x\sqrt{2\pi}}\left[\Phi(-\omega h_1(x))-\sum_{i=1}^{n}\lambda_i(x)e^{\kappa_i(x)}\Phi(-\omega h_2(x))\right]dx,
\tag{4.31}
$$

where $\omega = 1$ ($\omega = -1$) for a payer (receiver) swaption,

$$
\begin{aligned}
h_1(x) &:= \frac{\bar{y}-\mu_y}{\sigma_y\sqrt{1-\rho_{xy}^2}}-\frac{\rho_{xy}(x-\mu_x)}{\sigma_x\sqrt{1-\rho_{xy}^2}} \\
h_2(x) &:= h_1(x)+B(b,T,t_i)\sigma_y\sqrt{1-\rho_{xy}^2} \\
\lambda_i(x) &:= c_i A(T,t_i)e^{-B(a,T,t_i)x} \\
\kappa_i(x) &:= -B(b,T,t_i)\left[\mu_y-\frac{1}{2}(1-\rho_{xy}^2)\sigma_y^2 B(b,T,t_i)+\rho_{xy}\sigma_y\frac{x-\mu_x}{\sigma_x}\right],
\end{aligned}
$$

$\bar{y} = \bar{y}(x)$ is the unique solution of the following equation

$$
\sum_{i=1}^{n} c_i A(T,t_i)e^{-B(a,T,t_i)x-B(b,T,t_i)\bar{y}} = 1,
$$

and

$$\begin{aligned}
\mu_x &:= -M_x^T(0,T),\\
\mu_y &:= -M_y^T(0,T),\\
\sigma_x &:= \sigma\sqrt{\frac{1-e^{-2aT}}{2a}},\\
\sigma_y &:= \eta\sqrt{\frac{1-e^{-2bT}}{2b}},\\
\rho_{xy} &:= \frac{\rho\sigma\eta}{(a+b)\sigma_x\sigma_y}\left[1-e^{-(a+b)T}\right].
\end{aligned}$$

Proof. See Appendix D in this chapter. □

4.2.5 The Analogy with the Hull-White Two-Factor Model

The Hull-White (1994c) two-factor model assumes that the instantaneous short rate evolves in the risk-adjusted measure according to

$$dr(t) = [\theta(t) + u(t) - \bar{a}r(t)]dt + \sigma_1 dZ_1(t), \quad r(0) = r_0, \tag{4.32}$$

where the stochastic mean-reversion level satisfies

$$du(t) = -\bar{b}u(t)dt + \sigma_2 dZ_2(t), \quad u(0) = 0,$$

with (Z_1, Z_2) a two-dimensional Brownian motion with $dZ_1(t)dZ_2(t) = \bar{\rho}dt$, r_0, $\bar{a}$, $\bar{b}$, σ_1 and σ_2 positive constants, and $-1 \le \bar{\rho} \le 1$. The function θ is deterministic and properly chosen so as to exactly fit the current term structure of interest rates.

Simple integration leads to

$$\begin{aligned}
r(t) &= r(s)e^{-\bar{a}(t-s)} + \int_s^t \theta(v)e^{-\bar{a}(t-v)}dv + \int_s^t u(v)e^{-\bar{a}(t-v)}dv\\
&\quad + \sigma_1\int_s^t e^{-\bar{a}(t-v)}dZ_1(v),\\
u(t) &= u(s)e^{-\bar{b}(t-s)} + \sigma_2\int_s^t e^{-\bar{b}(t-v)}dZ_2(v).
\end{aligned}$$

Assuming $\bar{a} \neq \bar{b}$, we have

$$\begin{aligned}
&\int_s^t u(v)e^{-\bar{a}(t-v)}dv\\
&= \int_s^t u(s)e^{-\bar{b}(v-s)-\bar{a}(t-v)}dv + \sigma_2\int_s^t e^{-\bar{a}(t-v)}\int_s^v e^{-\bar{b}(v-x)}dZ_2(x)dv\\
&= u(s)\frac{e^{-\bar{b}(t-s)} - e^{-\bar{a}(t-s)}}{\bar{a}-\bar{b}} + \sigma_2 e^{-\bar{a}t}\int_s^t e^{(\bar{a}-\bar{b})v}\int_s^v e^{\bar{b}x}dZ_2(x)dv.
\end{aligned}$$

By integration by parts we then have

$$\begin{aligned}
&\int_s^t e^{(\bar{a}-\bar{b})v} \int_s^v e^{\bar{b}x} dZ_2(x) dv \\
&= \frac{1}{\bar{a}-\bar{b}} \int_s^t \left(\int_s^v e^{\bar{b}x} dZ_2(x) \right) d_v\left(e^{(\bar{a}-\bar{b})v}\right) \\
&= \frac{1}{\bar{a}-\bar{b}} \left[e^{(\bar{a}-\bar{b})t} \int_s^t e^{\bar{b}x} dZ_2(x) - \int_s^t e^{(\bar{a}-\bar{b})v} d_v\left(\int_s^v e^{\bar{b}x} dZ_2(x) \right) \right] \\
&= \frac{1}{\bar{a}-\bar{b}} \int_s^t \left[e^{(\bar{a}-\bar{b})t} - e^{(\bar{a}-\bar{b})v} \right] d_v\left(\int_s^v e^{\bar{b}x} dZ_2(x) \right) \\
&= \frac{1}{\bar{a}-\bar{b}} \int_s^t \left[e^{\bar{a}t-\bar{b}(t-v)} - e^{\bar{a}v} \right] dZ_2(v),
\end{aligned}$$

so that we finally obtain

$$\begin{aligned}
r(t) = r(s)e^{-\bar{a}(t-s)} &+ \int_s^t \theta(v)e^{-\bar{a}(t-v)}dv + \sigma_1 \int_s^t e^{-\bar{a}(t-v)}dZ_1(v) \\
&+ u(s)\frac{e^{-\bar{b}(t-s)} - e^{-\bar{a}(t-s)}}{\bar{a}-\bar{b}} + \frac{\sigma_2}{\bar{a}-\bar{b}} \int_s^t \left[e^{-\bar{b}(t-v)} - e^{-\bar{a}(t-v)} \right] dZ_2(v),
\end{aligned}$$

and in particular,

$$\begin{aligned}
r(t) = r_0 e^{-\bar{a}t} &+ \int_0^t \theta(v)e^{-\bar{a}(t-v)}dv + \sigma_1 \int_0^t e^{-\bar{a}(t-v)}dZ_1(v) \\
&+ \frac{\sigma_2}{\bar{a}-\bar{b}} \int_0^t \left[e^{-\bar{b}(t-v)} - e^{-\bar{a}(t-v)} \right] dZ_2(v).
\end{aligned}$$

Now if we assume $\bar{a} > \bar{b}$ (the case "$\bar{a} < \bar{b}$" is analogous) and define

$$\begin{aligned}
\sigma_3 &= \sqrt{\sigma_1^2 + \frac{\sigma_2^2}{(\bar{a}-\bar{b})^2} + 2\bar{\rho}\frac{\sigma_1\sigma_2}{\bar{b}-\bar{a}}} \\
dZ_3(t) &= \frac{\sigma_1 dZ_1(t) - \frac{\sigma_2}{\bar{a}-\bar{b}} dZ_2(t)}{\sigma_3} \\
\sigma_4 &= \frac{\sigma_2}{\bar{a}-\bar{b}},
\end{aligned}$$

we can write

$$
\begin{aligned}
r(t) &= r_0 e^{-\bar{a}t} + \int_0^t \theta(v) e^{-\bar{a}(t-v)} dv + \int_0^t e^{-\bar{a}(t-v)} \left[\sigma_1 dZ_1(v) + \frac{\sigma_2}{\bar{b}-\bar{a}} dZ_2(v) \right] \\
&\quad + \frac{\sigma_2}{\bar{a}-\bar{b}} \int_0^t e^{-\bar{b}(t-v)} dZ_2(v) \\
&= r_0 e^{-\bar{a}t} + \int_0^t \theta(v) e^{-\bar{a}(t-v)} dv \\
&\quad + \sigma_3 \int_0^t e^{-\bar{a}(t-v)} dZ_3(v) + \sigma_4 \int_0^t e^{-\bar{b}(t-v)} dZ_2(v).
\end{aligned}
$$

At this stage, the analogy with the G2++ model (4.4) becomes clear. Precisely, by setting

$$
\begin{aligned}
a &= \bar{a} \\
b &= \bar{b} \\
\sigma &= \sigma_3 \\
\eta &= \sigma_4 \\
\rho &= \frac{\sigma_1 \bar{\rho} - \sigma_4}{\sigma_3} \\
\varphi(t) &= r_0 e^{-\bar{a}t} + \int_0^t \theta(v) e^{-\bar{a}(t-v)} dv,
\end{aligned}
$$

we exactly recover the expression (4.7) for the short rate in the G2++ model. Conversely, given the G2++ model (4.4), we can recover the classical two-factor Hull-White model (4.32), by setting

$$
\begin{aligned}
\bar{a} &= a \\
\bar{b} &= b \\
\sigma_1 &= \sqrt{\sigma^2 + \eta^2 + 2\rho\sigma\eta} \\
\sigma_2 &= \eta(a-b) \\
\bar{\rho} &= \frac{\sigma\rho + \eta}{\sqrt{\sigma^2 + \eta^2 + 2\rho\sigma\eta}} \\
\theta(t) &= \frac{d\varphi(t)}{dt} + a\varphi(t).
\end{aligned}
$$

A different way to prove this analogy is by defining the new stochastic process

$$
\chi(t) = r(t) + \delta u(t),
$$

where $\delta = 1/(\bar{b} - \bar{a})$. In fact,

$$
\begin{aligned}
d\chi(t) &= [\theta(t) + u(t) - \bar{a}r(t)]dt + \sigma_1 dZ_1(t) - \delta\bar{b}u(t)dt + \delta\sigma_2 dZ_2(t) \\
&= [\theta(t) + u(t) - \bar{a}\chi(t) + \bar{a}\delta u(t) - \bar{b}\delta u(t)]dt + \sigma_1 dZ_1(t) + \delta\sigma_2 dZ_2(t) \\
&= [\theta(t) - \bar{a}\chi(t)]dt + \sigma_3 dZ_3(t),
\end{aligned}
$$

with σ_3 and dZ_3 defined previously.

Moreover, if we define

$$\psi(t) = \frac{u(t)}{\bar{a} - \bar{b}} = -\delta u(t),$$

then

$$\begin{aligned} d\psi(t) &= -\frac{\bar{b}}{\bar{a} - \bar{b}} u(t)dt + \frac{\sigma_2}{\bar{a} - \bar{b}} dZ_2(t) \\ &= -\bar{b}\psi(t)dt + \sigma_4 dZ_2(t), \end{aligned}$$

with σ_4 defined previously. Therefore, we again obtain that $r(t)$ can be written as

$$r(t) = \tilde{\chi}(t) + \psi(t) + \varphi(t),$$

where

$$\begin{aligned} d\tilde{\chi}(t) &= -\bar{a}\tilde{\chi}(t)dt + \sigma_3 dZ_3(t), \\ d\psi(t) &= -\bar{b}\psi(t)dt + \sigma_4 dZ_2(t), \\ \varphi(t) &= r_0 e^{-\bar{a}t} + \int_0^t \theta(v) e^{-\bar{a}(t-v)} dv. \end{aligned}$$

4.2.6 The Construction of an Approximating Binomial Tree

The purpose of this section is the construction of an approximating tree for the G2++ process (4.4). Such a tree is a fundamental tool when pricing exotic interest rate derivatives.

A two-dimensional tree, trinomial in both dimensions, can be constructed according to the procedure suggested by Hull-White (1994c). We just have to follow the general method illustrated in Appendix F at the end of the book and apply it to the dynamics underlying (4.4).

Alternatively, we can build a simpler tree, which is binomial in both dimensions. The construction of such a tree is, for sake of completeness, outlined in this section.

We start by constructing two binomial trees approximating respectively the dynamics of the processes x and y given in (4.5).

We first remember that, for any t and $\Delta t > 0$, we have

$$\begin{cases} E\{x(t+\Delta t)|\mathcal{F}_t\} = x(t)e^{-a\Delta t}, \\ \text{Var}\{x(t+\Delta t)|\mathcal{F}_t\} = \frac{\sigma^2}{2a}(1 - e^{-2a\Delta t}), \end{cases}$$

$$\begin{cases} E\{y(t+\Delta t)|\mathcal{F}_t\} = y(t)e^{-b\Delta t}, \\ \text{Var}\{y(t+\Delta t)|\mathcal{F}_t\} = \frac{\eta^2}{2b}(1 - e^{-2b\Delta t}), \end{cases}$$

and

$$
\begin{aligned}
&\mathrm{Cov}\{x(t+\Delta t), y(t+\Delta t)|\mathcal{F}_t\} \\
&= E\left\{[x(t+\Delta t) - E\{x(t+\Delta t)|\mathcal{F}_t\}]\,[y(t+\Delta t) - E\{y(t+\Delta t)|\mathcal{F}_t\}]\,|\mathcal{F}_t\right\} \\
&= \sigma\eta E\left\{\int_t^{t+\Delta t} e^{-a(t+\Delta t-u)}dW_1(u) \int_t^{t+\Delta t} e^{-b(t+\Delta t-u)}dW_2(u)|\mathcal{F}_t\right\} \\
&= \sigma\eta\rho \int_t^{t+\Delta t} e^{-(a+b)(t+\Delta t-u)}du \\
&= \frac{\sigma\eta\rho}{a+b}\left[1 - e^{-(a+b)\Delta t}\right].
\end{aligned}
$$

By expanding up to first order in Δt, we have

$$
\begin{cases}
E\{x(t+\Delta t)|\mathcal{F}_t\} = x(t)(1 - a\Delta t), \\
\mathrm{Var}\{x(t+\Delta t)|\mathcal{F}_t\} = \sigma^2\Delta t,
\end{cases}
\qquad
\begin{cases}
E\{y(t+\Delta t)|\mathcal{F}_t\} = y(t)(1 - b\Delta t), \\
\mathrm{Var}\{y(t+\Delta t)|\mathcal{F}_t\} = \eta^2\Delta t,
\end{cases}
$$

and

$$
\mathrm{Cov}\{x(t+\Delta t), y(t+\Delta t)|\mathcal{F}_t\} = \sigma\eta\rho\Delta t.
$$

The Binomial Trees for x and y. The binomial trees approximating the processes x and y are reproduced in Figure 4.1. Precisely, we assume that if at time t we have a value $x(t)$ (resp. $y(t)$), then at time $t + \Delta t$, the process x (resp. y) can either move up to $x(t) + \Delta x$ (resp. $y(t) + \Delta y$) with probability p (resp. q) or down to $x(t) - \Delta x$ (resp. $y(t) - \Delta y$) with probability $1 - p$ (resp. $1 - q$). The quantities Δx, Δy, p and q are to be properly chosen in order to match (at first order in Δt) the conditional mean and variance of the (continuous-time) processes x and y. Precisely, we have to solve

$$
\begin{cases}
p(x(t) + \Delta x) + (1 - p)(x(t) - \Delta x) = x(t)(1 - a\Delta t), \\
p(x(t) + \Delta x)^2 + (1 - p)(x(t) - \Delta x)^2 - [x(t)(1 - a\Delta t)]^2 = \sigma^2\Delta t,
\end{cases}
$$

and, equivalently,

$$
\begin{cases}
q(y(t) + \Delta y) + (1 - q)(y(t) - \Delta y) = y(t)(1 - b\Delta t), \\
q(y(t) + \Delta y)^2 + (1 - q)(y(t) - \Delta y)^2 - [y(t)(1 - b\Delta t)]^2 = \eta^2\Delta t.
\end{cases}
$$

Neglecting all terms with higher order than $\sqrt{\Delta t}$, we obtain

$$
\begin{cases}
\Delta x = \sigma\sqrt{\Delta t}, \\
p = \dfrac{1}{2} - \dfrac{x(t)a\Delta t}{2\Delta x} = \dfrac{1}{2} - \dfrac{x(t)a}{2\sigma}\sqrt{\Delta t},
\end{cases}
$$

$$
\begin{cases}
\Delta y = \eta\sqrt{\Delta t}, \\
q = \dfrac{1}{2} - \dfrac{y(t)b\Delta t}{2\Delta y} = \dfrac{1}{2} - \dfrac{y(t)b}{2\eta}\sqrt{\Delta t},
\end{cases}
$$

so that both p and q only depend on the values of $x(t)$ and $y(t)$ respectively and not explicitly on time t. It is also easy to see that

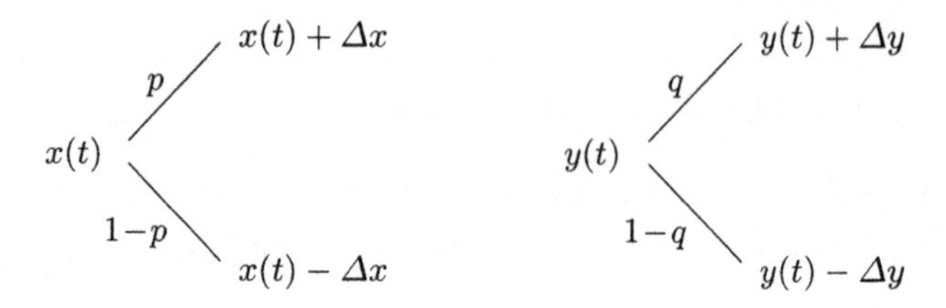

Fig. 4.1. Evolution of the processes x and y starting respectively from $x(t)$ and $y(t)$ at time t and moving upwards to $x(t) + \Delta x$ and to $y(t) + \Delta y$ at time $t + \Delta t$ with probabilities p and q, downwards to $x(t) - \Delta x$ and to $y(t) - \Delta y$ at time $t + \Delta t$ with probabilities $1 - p$ and $1 - q$.

$$0 \le p \le 1 \text{ if and only if } |x(t)| \le \frac{\sigma}{a\sqrt{\Delta t}},$$
$$0 \le q \le 1 \text{ if and only if } |y(t)| \le \frac{\eta}{b\sqrt{\Delta t}}.$$

The Approximating Tree for r. If the two factors x and y are both approximated through the previous binomial lattices, the process r can be approximated through a quadrinomial tree as represented in Figure 4.2. Precisely, we assume that if at time t we start from a pair $(x(t), y(t))$, then at time $t + \Delta t$, the pair (x, y) can move to

- $(x(t) + \Delta x, y(t) + \Delta y)$ with probability π_1;
- $(x(t) + \Delta x, y(t) - \Delta y)$ with probability π_2;
- $(x(t) - \Delta x, y(t) + \Delta y)$ with probability π_3;
- $(x(t) - \Delta x, y(t) - \Delta y)$ with probability π_4;

where $0 \le \pi_1, \pi_2, \pi_3, \pi_4 \le 1$ and $\pi_1 + \pi_2 + \pi_3 + \pi_4 = 1$. The probabilities π_1, π_2, π_3 and π_4 are to be chosen in order to match the marginal distributions of the binomial trees for x and y and the conditional covariance (at first order in Δt) between the (continuous-time) processes x and y. Matching the marginal distributions and imposing that the probabilities sum up to one, we get

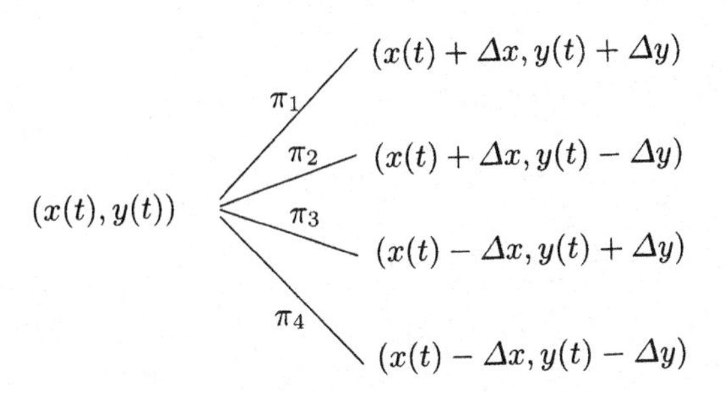

Fig. 4.2. Evolution of the pair process (x, y) starting from $(x(t), y(t))$ and moving to i) $(x(t) + \Delta x, y(t) + \Delta y)$ with probability π_1; ii) $(x(t) + \Delta x, y(t) - \Delta y)$ with probability π_2; iii) $(x(t) - \Delta x, y(t) + \Delta y)$ with probability π_3; iv) $(x(t) - \Delta x, y(t) - \Delta y)$ with probability π_4.

$$\begin{cases} \pi_1 + \pi_2 + \pi_3 + \pi_4 = 1 \\ \pi_1 + \pi_2 = p \\ \pi_3 + \pi_4 = 1 - p \\ \pi_1 + \pi_3 = q \\ \pi_2 + \pi_4 = 1 - q \end{cases}$$

which implies that

$$\begin{cases} \pi_1 = \pi_4 - 1 + q + p \\ \pi_2 = -\pi_4 + 1 - q \\ \pi_3 = -\pi_4 + 1 - p \\ \pi_4 = \pi_4 \end{cases} \tag{4.33}$$

Matching the conditional covariance, we have the additional constraint

$$\begin{aligned} &(\Delta x + ax(t)\Delta t)(\Delta y + by(t)\Delta t)\pi_1 + (\Delta x + ax(t)\Delta t)(-\Delta y + by(t)\Delta t)\pi_2 \\ &\quad + (ax(t)\Delta t - \Delta x)(\Delta y + by(t)\Delta t)\pi_3 + (ax(t)\Delta t - \Delta x)(-\Delta y + by(t)\Delta t)\pi_4 \\ &= \rho\sigma\eta\Delta t \end{aligned} \tag{4.34}$$

which leads to

$$\begin{cases} \pi_1 = \frac{1+\rho}{4} - \frac{b\sigma y(t) + a\eta x(t)}{4\sigma\eta}\sqrt{\Delta t} \\ \pi_2 = \frac{1-\rho}{4} + \frac{b\sigma y(t) - a\eta x(t)}{4\sigma\eta}\sqrt{\Delta t} \\ \pi_3 = \frac{1-\rho}{4} - \frac{b\sigma y(t) - a\eta x(t)}{4\sigma\eta}\sqrt{\Delta t} \\ \pi_4 = \frac{1+\rho}{4} + \frac{b\sigma y(t) + a\eta x(t)}{4\sigma\eta}\sqrt{\Delta t} \end{cases}$$

It is clear, however, that, for fixed a, σ, b, η, ρ and Δt, the conditions

$$\begin{cases} 0 \leq \frac{1+\rho}{4} - \frac{b\sigma y(t) + a\eta x(t)}{4\sigma\eta}\sqrt{\Delta t} \leq 1 \\ 0 \leq \frac{1-\rho}{4} + \frac{b\sigma y(t) - a\eta x(t)}{4\sigma\eta}\sqrt{\Delta t} \leq 1 \\ 0 \leq \frac{1-\rho}{4} - \frac{b\sigma y(t) - a\eta x(t)}{4\sigma\eta}\sqrt{\Delta t} \leq 1 \\ 0 \leq \frac{1+\rho}{4} + \frac{b\sigma y(t) + a\eta x(t)}{4\sigma\eta}\sqrt{\Delta t} \leq 1 \end{cases} \tag{4.35}$$

are not necessarily satisfied for every choice of $(x(t), y(t))$. This is exactly the same problem that Hull and White (1994c) encountered in the construction of their two-dimensional trinomial tree. As also suggested by them, a possible way out is to start from (4.33) and solve (4.34) by suitably changing the value of ρ in such a way that the (4.35)'s are fulfilled. This should not affect in a relevant way the pricing of a general claim if we choose Δt to be sufficiently small. Notice, in fact, that the limit values of the probabilities for Δt going to zero always fall in-between zero and one (since $|\rho| \leq 1$), so that, for sufficiently small Δt, the nodes in the quadrinomial tree where we impose a different (and wrong) correlation give actually a negligible contribution.

4.2.7 Examples of Calibration to Real-Market Data

We now propose two examples of calibration to real-market volatility data, which illustrate the fitting capability of our two-factor Gaussian model. We first consider a calibration to cap volatilities and then a calibration to swaption volatilities.

In our first example, we use the same at-the-money Euro cap-volatility data as in Section 3.14, see Table 3.4 or Table 4.1 below. The zero-coupon curve, on the same day and time, is again that shown in Figure 1.1 of Chapter 1.

Given its high number of parameters (five) and the possibility of humped structures in the volatility of instantaneous forward rates, one can expect the fitting quality of the model to be high in general. Indeed, according to our experience, the G2++ model can reproduce market cap-volatility data very accurately. This is confirmed by our calibration result, which is shown in Table 4.1. An important point is that this accurate calibration did not require the introduction of some "all-fitting" time-varying parameters for the volatility part. This guarantees a more regular evolution in time of the market volatility structures associated with the calibrated dynamics. Indeed, analogously to what we have observed when commenting the one-factor dynamics (3.32) in Section 3.3.1, too many time-varying parameters can lead to over-fitting. Moreover, here too the future volatility structures implied by a G2++ model with more time-varying parameters are likely to be unrealistic. This is not to say that the basic G2++ model always produces realistic future structures, but simply that such structures are easier to control in terms of the model parameters.

The calibration is performed by minimizing the sum of the squares of the percentage differences between model and market cap prices, and leads to the following parameters: $a = 0.543009105$, $b = 0.075716774$, $\sigma = 0.005837408$, $\eta = 0.011657837$ and $\rho = -0.991401219$.[4]

Contrary to its one-factor analogue (3.33), we here obtain positive best-fitting values for the mean-reversion parameters a and b. Indeed, the increasing part of the market hump can be retrieved with the help of a highly negative ρ.

However, as in this case, it often happens that the ρ value is quite close to minus one, which implies that the G2++ model tends to degenerate into a one-factor (non-Markov) short-rate process. This is also intuitive. In fact, as already explained in Chapter 1, caps prices do not depend on the correlation of forward rates, so that even a one-factor model that implies perfectly-correlated rates can fit caps data well in many situations. Notice, moreover, that the degenerate process for the short rate is still non-Markovian (if $a \neq b$), which explains what really makes the G2++ model outperform its one-factor version (3.33).

Maturity	Market volatility	G2++ implied volatility
1	0.1520	0.1520
2	0.1620	0.1622
3	0.1640	0.1631
4	0.1630	0.1631
5	0.1605	0.1614
7	0.1555	0.1554
10	0.1475	0.1472
15	0.1350	0.1349
20	0.1260	0.1261

Table 4.1. At-the-money Euro cap-volatility quotes on February 13, 2001, at 5 p.m., and corresponding volatilities implied by the G2++ model.

[4] The minimization is performed with a simulated-annealing method followed by a local-search algorithm to refine the last solution found.

We now move to our second example of calibration. We again consider data as of February 13, 2001, at 5 p.m., with the related swaption-volatility quotes being shown in Table 4.2 below. Swaption maturities are one, two, three, four, five, seven and ten years, and the tenors of the underlying swaps go from one to ten years. For a detailed explanation of how ATM-swaptions data are organized in such a table, see Section 6.17 later on. Indeed, it is with market models that swaptions enter the picture at full power, and this is the reason why we present this explanation in the related Chapter 6.

	1y	2y	3y	4y	5y	6y	7y	8y	9y	10y
1y	0.1640	0.1550	0.1430	0.1310	0.1240	0.1190	0.1160	0.1120	0.1100	0.1070
2y	0.1600	0.1500	0.1390	0.1290	0.1220	0.1190	0.1160	0.1130	0.1100	0.1080
3y	0.1570	0.1450	0.1340	0.1240	0.1190	0.1150	0.1130	0.1100	0.1080	0.1060
4y	0.1480	0.1360	0.1260	0.1190	0.1140	0.1120	0.1090	0.1070	0.1050	0.1030
5y	0.1400	0.1280	0.1210	0.1140	0.1100	0.1070	0.1050	0.1030	0.1020	0.1000
7y	0.1300	0.1190	0.1130	0.1050	0.1010	0.0990	0.0970	0.0960	0.0950	0.0930
10y	0.1160	0.1070	0.1000	0.0930	0.0900	0.0890	0.0870	0.0860	0.0850	0.0840

Table 4.2. At-the-money Euro swaption-volatility quotes on February 13, 2001, at 5 p.m.

Minimization of the sum of the squares of the percentage differences between model and market swaption prices produces the following calibrated parameters: $a = 0.773511777$, $b = 0.082013014$, $\sigma = 0.022284644$, $\eta = 0.010382461$ and $\rho = -0.701985206$.

Notice that the value of ρ is now far from minus one. This could be expected since swaption prices contain information on the correlation between forward rates, and indeed a possible way to incorporate such information into the model consists in assigning a non-trivial value to the ρ coefficient.

The calibration results are summarized in Tables 4.3 and 4.4. In the first, we show the fitted swaption volatilities as implied by the G2++ model, whereas in the second we report the percentage differences

$$\frac{\text{G2++ implied volatility - market volatility}}{\text{market volatility}}.$$

We can see that the calibration result is rather satisfactory. Indeed, apart from few exceptions in the first two columns, the percentage errors are rather low given that we have tried to fit seventy prices with only five parameters.

A few concluding remarks are in order.

It may be a good idea to calibrate the G2++ model only to the most significant swaptions data, leaving out the illiquid entries (see again Section 6.17 for more details). Or, when in need to price a particular product that is influenced only by a certain set of swap rates, it may be reasonable to calibrate the model only to the relevant swaptions. For instance, to price a Bermudan swaption, one typically fits the model around the (diagonal) volatilities of the underlying European swaptions. It is however necessary to check how the

	1y	2y	3y	4y	5y	6y	7y	8y	9y	10y
1y	0.1870	0.1529	0.1395	0.1327	0.1276	0.1231	0.1190	0.1154	0.1120	0.1085
2y	0.1603	0.1427	0.1348	0.1295	0.1251	0.1210	0.1174	0.1139	0.1103	0.1068
3y	0.1509	0.1376	0.1307	0.1258	0.1216	0.1180	0.1146	0.1109	0.1073	0.1041
4y	0.1422	0.1311	0.1252	0.1210	0.1175	0.1142	0.1106	0.1069	0.1037	0.1005
5y	0.1335	0.1247	0.1199	0.1165	0.1134	0.1098	0.1062	0.1030	0.0998	0.0968
7y	0.1218	0.1161	0.1124	0.1087	0.1050	0.1018	0.0986	0.0954	0.0928	0.0902
10y	0.1090	0.1029	0.0997	0.0965	0.0934	0.0909	0.0884	0.0858	0.0833	0.0809

Table 4.3. G2++ calibrated swaptions volatilities.

	1y	2y	3y	4y	5y	6y	7y	8y	9y	10y
1y	14.01%	-1.35%	-2.43%	1.27%	2.86%	3.44%	2.59%	3.03%	1.81%	1.40%
2y	0.17%	-4.88%	-3.03%	0.39%	2.51%	1.67%	1.19%	0.84%	0.31%	-1.12%
3y	-3.91%	-5.10%	-2.49%	1.44%	2.21%	2.64%	1.42%	0.86%	-0.64%	-1.83%
4y	-3.92%	-3.59%	-0.61%	1.65%	3.08%	1.97%	1.44%	-0.07%	-1.26%	-2.41%
5y	-4.61%	-2.54%	-0.91%	2.23%	3.09%	2.65%	1.16%	-0.01%	-2.13%	-3.25%
7y	-6.29%	-2.47%	-0.50%	3.51%	3.94%	2.82%	1.65%	-0.58%	-2.33%	-3.01%
10y	-6.08%	-3.81%	-0.34%	3.80%	3.79%	2.17%	1.58%	-0.21%	-2.00%	-3.69%

Table 4.4. Swaptions calibration results: percentage differences.

resulting calibrated model prices swaptions that had not been included in the calibration set. This is important in order to verify that the behaviour of the model out of the calibration range is not wild or too weird.

We also point out that both for the cap and swaptions calibrations we have resorted to the related analytical formulas of the G2++ model. In particular, for the swaptions we have used formula (4.31), resorting to a numerical integration against a Gaussian distribution, whose support can be reduced to a convenient number of standard deviations. The related optimization requires a reasonable time when running on a PC. A global optimization involving the seventy swaptions requires a few minutes, whereas a local optimization takes about one minute. Clearly, such times reduce when considering only a few swaption instead of the whole table.

Finally, one can try a joint calibration to caps and swaptions data. Results are usually not completely satisfactory. This may be due both to misalignments between the two markets and to the low number of parameters in the model. In order to allow for a full-power joint calibration we will need to resort to the LIBOR market model. The related cases will be presented in Chapter 7.

Appendix A: Proof of Lemma 4.2.1

Stochastic integration by parts implies that

$$\int_t^T x(u)du = Tx(T) - tx(t) - \int_t^T u dx(u) = \int_t^T (T-u)dx(u) + (T-t)x(t). \tag{4.36}$$

By definition of x, the integral in the right-hand side can be written as

$$\int_t^T (T-u)dx(u) = -a\int_t^T (T-u)x(u)du + \sigma\int_t^T (T-u)dW_1(u)$$

by substituting the expression for $dx(u)$, and

$$\int_t^T (T-u)x(u)du = x(t)\int_t^T (T-u)e^{-a(u-t)}du$$
$$+\sigma\int_t^T (T-u)\int_t^u e^{-a(u-s)}dW_1(s)du.$$

by substituting the expression for $x(u)$. Calculating separately the last two integrals (multiplied by $-a$), we have

$$-ax(t)\int_t^T (T-u)e^{-a(u-t)}du = -x(t)(T-t) - \frac{e^{-a(T-t)}-1}{a}x(t)$$

and, again by integration by parts,

$$\begin{aligned}
&-a\sigma\int_t^T (T-u)\int_t^u e^{-a(u-s)}dW_1(s)du \\
&= -a\sigma\int_t^T \left(\int_t^u e^{as}dW_1(s)\right)d_u\left(\int_t^u (T-v)e^{-av}dv\right) \\
&= -a\sigma\left[\left(\int_t^T e^{au}dW_1(u)\right)\left(\int_t^T (T-v)e^{-av}dv\right)\right. \\
&\qquad \left.-\int_t^T\left(\int_t^u (T-v)e^{-av}dv\right)e^{au}dW_1(u)\right] \\
&= -a\sigma\int_t^T\left(\int_u^T (T-v)e^{-av}dv\right)e^{au}dW_1(u) \\
&= -\sigma\int_t^T\left[(T-u)+\frac{e^{-a(T-u)}-1}{a}\right]dW_1(u),
\end{aligned}$$

where in the last step we have used the fact that

$$\int_u^T (T-v)e^{-av}dv = \frac{(T-u)e^{-au}}{a} + \frac{e^{-aT}-e^{-au}}{a^2}.$$

Adding up the previous terms, we obtain

$$\int_t^T x(u)du = \frac{1-e^{-a(T-t)}}{a}x(t) + \frac{\sigma}{a}\int_t^T\left[1-e^{-a(T-u)}\right]dW_1(u).$$

The analogous expression for y is obtained by replacing a, σ and W_1 respectively with b, η and W_2 , i.e.,

$$\int_t^T y(u)du = \frac{1-e^{-b(T-t)}}{b}y(t) + \frac{\eta}{b}\int_t^T \left[1-e^{-b(T-u)}\right] dW_2(u),$$

so that (4.9) is immediately verified. As to the calculation of the conditional variance, we have

$$\begin{aligned}
\mathrm{Var}\{I(t,T)|\mathcal{F}_t\} &= \mathrm{Var}\left\{\frac{\sigma}{a}\int_t^T \left[1-e^{-a(T-u)}\right] dW_1(u) \right.\\
&\qquad \left. +\frac{\eta}{b}\int_t^T \left[1-e^{-b(T-u)}\right] dW_2(u) \right| \mathcal{F}_t\Bigg\}\\
&= \frac{\sigma^2}{a^2}\int_t^T \left[1-e^{-a(T-u)}\right]^2 du + \frac{\eta^2}{b^2}\int_t^T \left[1-e^{-b(T-u)}\right]^2 du\\
&\quad + 2\rho\frac{\sigma\eta}{ab}\int_t^T \left[1-e^{-a(T-u)}\right]\left[1-e^{-b(T-u)}\right] du.
\end{aligned}$$

Simple integration then leads to (4.10).

Appendix B: proof of Lemma 4.2.2

The Radon-Nikodym derivative (4.17) can be written in terms of the two independent Brownian motions $\widetilde{W}_1$ and $\widetilde{W}_2$ as follows

$$\begin{aligned}
\frac{dQ^T}{dQ} = \exp\Bigg\{&-\frac{1}{2}V(0,T) - \int_0^T \left[\frac{\sigma}{a}(1-e^{-a(T-u)}) + \rho\frac{\eta}{b}(1-e^{-b(T-u)})\right] d\widetilde{W}_1(u)\\
&-\frac{\eta}{b}\sqrt{1-\rho^2}\int_0^T (1-e^{-b(T-u)})d\widetilde{W}_2(u)\Bigg\}.
\end{aligned}$$

Since

$$\begin{aligned}
\int_0^T &\left[\frac{\sigma}{a}(1-e^{-a(T-u)}) + \rho\frac{\eta}{b}(1-e^{-b(T-u)})\right]^2 du\\
&+\frac{\eta^2}{b^2}(1-\rho^2)\int_0^T (1-e^{-b(T-u)})^2 du = V(0,T),
\end{aligned}$$

the Girsanov theorem implies that the two processes $\widetilde{W}_1^T$ and $\widetilde{W}_2^T$ defined by

$$\begin{aligned}
d\widetilde{W}_1^T(t) &= d\widetilde{W}_1(t) + \left[\frac{\sigma}{a}(1-e^{-a(T-t)}) + \rho\frac{\eta}{b}(1-e^{-b(T-t)})\right] dt\\
d\widetilde{W}_2^T(t) &= d\widetilde{W}_2(t) + \frac{\eta}{b}\sqrt{1-\rho^2}(1-e^{-b(T-t)})dt
\end{aligned} \tag{4.37}$$

are two independent Brownian motions under the measure Q^T. Then defining W_1^T and W_2^T by

$$dW_1^T(t) = d\widetilde{W}_1^T(t),$$
$$dW_2^T(t) = \rho d\widetilde{W}_1^T(t) + \sqrt{1-\rho^2} d\widetilde{W}_2^T(t),$$

and combining (4.8) with (4.37) we obtain (4.18).

Formulas (4.19) follow from integration of (4.18) where the following equalities are being used:

$$\int_s^t \left[\frac{\sigma^2}{a}(1-e^{-a(T-u)}) + \rho\frac{\sigma\eta}{b}(1-e^{-b(T-u)})\right] e^{-a(t-u)}du = M_x^T(s,t)$$
$$\int_s^t \left[\frac{\eta^2}{b}(1-e^{-b(T-u)}) + \rho\frac{\sigma\eta}{a}(1-e^{-a(T-u)})\right] e^{-b(t-u)}du = M_y^T(s,t).$$

Formulas (4.20) are straightforward.

Appendix C: Proof of Theorem 4.2.2

The change-of-numeraire technique, which basically consists in changing the probability measure from Q to Q^T, implies that

$$\mathbf{ZBC}(t,T,\tau,K) = P(t,T)E^{Q^T}\{(P(T,\tau)-K)^+|\mathcal{F}_t\}$$

(see formula 2.24). Since

$$P(T,\tau) = \frac{P^M(0,\tau)}{P^M(0,T)}\exp\left\{\frac{1}{2}[V(T,\tau)-V(0,\tau)+V(0,T)] - \frac{1-e^{-a(\tau-T)}}{a}x(T) - \frac{1-e^{-b(\tau-T)}}{b}y(T)\right\},$$

under Q^T the logarithm of $P(T,\tau)$ conditional on $\mathcal{F}_t$ is normally distributed with mean

$$M_p = \ln\frac{P^M(0,\tau)}{P^M(0,T)} + \frac{1}{2}[V(T,\tau)-V(0,\tau)+V(0,T)] - \frac{1-e^{-a(\tau-T)}}{a}E^{Q^T}\{x(T)|\mathcal{F}_t\} - \frac{1-e^{-b(\tau-T)}}{b}E^{Q^T}\{y(T)|\mathcal{F}_t\}$$

and variance

$$(V_p)^2 = \frac{\sigma^2}{2a^3}(1-e^{-a(\tau-T)})^2(1-e^{-2a(T-t)}) + \frac{\eta^2}{2b^3}(1-e^{-b(\tau-T)})^2(1-e^{-2b(T-t)}) + 2\rho\frac{\sigma\eta}{ab(a+b)}(1-e^{-a(\tau-T)})(1-e^{-b(\tau-T)})(1-e^{-(a+b)(T-t)}).$$

Since (see also formula (D.1) in Appendix D at the end of the book)

$$\int_{-\infty}^{+\infty} \frac{1}{\sqrt{2\pi}V_p}(e^z - K)^+ e^{-\frac{1}{2}\frac{(z-M_p)^2}{V_p^2}} dz$$
$$= e^{M_p+\frac{1}{2}V_p^2}\Phi\left(\frac{M_p - \ln K + V_p^2}{V_p}\right) - K\Phi\left(\frac{M_p - \ln K}{V_p}\right),$$

we have that

$$\mathbf{ZBC}(t,T,\tau,K)$$
$$= P(t,T)\left[e^{M_p+\frac{1}{2}V_p^2}\Phi\left(\frac{M_p - \ln K + V_p^2}{V_p}\right) - K\Phi\left(\frac{M_p - \ln K}{V_p}\right)\right].$$

Noting that $\Sigma(t,T,\tau)^2 = V_p^2$, to retrieve (4.21) we have to use the equality

$$M_p = \ln\frac{P(t,\tau)}{P(t,T)} - \frac{1}{2}\Sigma(t,T,\tau)^2,$$

that can be proved by noting that $\frac{P(t,\tau)}{P(t,T)}$ is a martingale under Q^T, hence

$$\frac{P(t,\tau)}{P(t,T)} = E^{Q^T}\{P(T,\tau)|\mathcal{F}_t\} = e^{M_p+\frac{1}{2}V_p^2}.$$

Formula (4.22) immediately follows from the put-call parity:

$$\mathbf{ZBC}(t,T,\tau,K) + KP(t,T) = \mathbf{ZBP}(t,T,\tau,K) + P(t,\tau).$$

Appendix D: Proof of Theorem 4.2.3

By the general formula (2.24), the arbitrage-free price of the above European swaption is

$$\mathbf{ES}(0,T,\mathcal{T},N,X,\omega)$$
$$= NP(0,T)E^T\left\{\left[\omega\left(1-\sum_{i=1}^n c_i P(T,t_i)\right)\right]^+\right\}$$
$$= NP(0,T)\int_{\mathbb{R}^2}\left[\omega\left(1-\sum_{i=1}^n c_i A(T,t_i)e^{-B(a,T,t_i)x-B(b,T,t_i)y}\right)\right]^+ f(x,y)dydx,$$

where f is the density of the random vector $(x(T), y(T))$, i.e.,

$$f(x,y) := \frac{\exp\left\{-\frac{1}{2(1-\rho_{xy}^2)}\left[\left(\frac{x-\mu_x}{\sigma_x}\right)^2 - 2\rho_{xy}\frac{(x-\mu_x)(y-\mu_y)}{\sigma_x\sigma_y} + \left(\frac{y-\mu_y}{\sigma_y}\right)^2\right]\right\}}{2\pi\sigma_x\sigma_y\sqrt{1-\rho_{xy}^2}}.$$

Freezing x in the integrand and integrating over y, from $-\infty$ to $+\infty$, we obtain

$$\int_{\bar{y}(x)}^{+\infty\cdot\omega}\left(1-\sum_{i=1}^{n}\lambda_i e^{-B(b,T,t_i)y}\right)\gamma e^{E+F(y-\mu_y)-G(y-\mu_y)^2}dy,$$

where

$$\gamma := \frac{1}{2\pi\sigma_x\sigma_y\sqrt{1-\rho_{xy}^2}},$$

$$E := -\frac{1}{2(1-\rho_{xy}^2)}\left(\frac{x-\mu_x}{\sigma_x}\right)^2$$

$$F := \frac{\rho_{xy}}{1-\rho_{xy}^2}\frac{x-\mu_x}{\sigma_x\sigma_y}$$

$$G := \frac{1}{2(1-\rho_{xy}^2)\sigma_y^2},$$

and the dependence on x has been omitted for simplicity. Using the general formula

$$\int_a^b e^{-Ax^2+Bx}dx = \frac{\sqrt{\pi}}{\sqrt{A}}e^{\frac{B^2}{4A}}\left[\Phi\left(b\sqrt{2A}-\frac{B}{\sqrt{2A}}\right)-\Phi\left(a\sqrt{2A}-\frac{B}{\sqrt{2A}}\right)\right],$$

with $A > 0$, a, b and B real constants, the integral becomes

$$\gamma\frac{\sqrt{\pi}}{\sqrt{G}}e^{E+\frac{F^2}{4G}}\left[\Phi\left(+\infty\cdot\omega\sqrt{2G}-\frac{F}{\sqrt{2G}}\right)-\Phi\left((\bar{y}-\mu_y)\sqrt{2G}-\frac{F}{\sqrt{2G}}\right)\right]$$
$$-\gamma\frac{\sqrt{\pi}}{\sqrt{G}}e^{E}\sum_{i=1}^{n}\lambda_i e^{-B(b,T,t_i)\mu_y+\frac{[F-B(b,T,t_i)]^2}{4G}}$$
$$\cdot\left[\Phi\left(+\infty\cdot\omega\sqrt{2G}-\frac{F-B(b,T,t_i)}{\sqrt{2G}}\right)-\Phi\left((\bar{y}-\mu_y)\sqrt{2G}-\frac{F-B(b,T,t_i)}{\sqrt{2G}}\right)\right].$$

A little algebra then leads to

$$\gamma\frac{\sqrt{\pi}}{\sqrt{G}}e^{E+\frac{F^2}{4G}}\left\{\frac{\omega+1}{2}-\Phi\left((\bar{y}-\mu_y)\sqrt{2G}-\frac{F}{\sqrt{2G}}\right)\right.$$
$$-\sum_{i=1}^{n}\lambda_i e^{-B(b,T,t_i)\mu_y+\frac{B(b,T,t_i)[B(b,T,t_i)-2F]}{4G}}$$
$$\left.\cdot\left[\frac{\omega+1}{2}-\Phi\left((\bar{y}-\mu_y)\sqrt{2G}-\frac{F-B(b,T,t_i)}{\sqrt{2G}}\right)\right]\right\},$$

Since $(\omega+1)/2-\Phi(z)=\omega\Phi(-\omega z)$, for each real constant z, we obtain (4.31) by noting that

$$\gamma\frac{\sqrt{\pi}}{\sqrt{G}} = \frac{1}{\sigma_x\sqrt{2\pi}},$$

$$E + \frac{F^2}{4G} = -\frac{1}{2}\left(\frac{x-\mu_x}{\sigma_x}\right)^2,$$

$$\frac{F}{\sqrt{2G}} = \frac{\rho_{xy}}{\sqrt{1-\rho_{xy}^2}}\frac{x-\mu_x}{\sigma_x},$$

$$\sqrt{2G} = \frac{1}{\sigma_y\sqrt{1-\rho_{xy}^2}}.$$

4.3 The Two-Additive-Factor Extended CIR/LS Model CIR2++

In this section, we propose an alternative two-factor short-rate model, which is based on adding a deterministic shift to a sum of two independent square-root processes. This model, which we will refer to as CIR2++, can be viewed as the natural two-factor generalization of the CIR++ model of Section 3.9. As for the CIR++ model, also in this two-factor formulation the deterministic shift is added so as to exactly retrieve the term structure of zero-coupon rates at the initial time.

We start by studying the non-shifted (time-homogeneous) two-factor CIR model without requiring an exact fitting of the initial term structure, and observe that it is equivalent to the Longstaff and Schwartz's (1992b) model (LS). We then derive the model formulas by using the one-factor case as a guide. The interested reader is therefore advised to acquire at least a basic knowledge of the CIR and CIR++ models as illustrated in the previous chapter. Indeed, the model presented in this section is largely built upon its one-factor version, using in particular its explicit formulas.

As stated earlier, a good feature of the CIR2++ model is that it can maintain positive rates through reasonable restrictions on the parameters. We do not explicitly consider such restrictions here, and just remember that a flavor of the type of restrictions involved in maintaining positive rates can be found in the one-factor CIR++ case in Section 3.9.3.

Distributionally, the short rate is a sum of two independent non-central chi-square variables, and as such it has fat tails, which is a desirable distributional property improving on the G2++ case. However, the volatility structures allowed here are poorer than in the Gaussian case. Consider once again the curve

$$T \mapsto \sqrt{\mathrm{Var}[d\ f(t,T)]/dt}.$$

It can be easily checked that in the CIR2++ model this curve can only be decreasing, as hinted at below, after formula (4.44). As a consequence, no hump can be allowed in the absolute volatilities of instantaneous forward rates. It follows that the caplet-volatility term structure implied by the model may not reproduce large humps, see the discussion in Section 4.2.3 and especially the discussion towards the end of Section 4.1. Hence, if the market features a largely-humped term structure of cap or caplet volatilities, the CIR2++ model may yield an unsatisfactory calibration to such market data.. A possibility to circumvent this difficulty is introducing instantaneous correlation between the Brownian shocks W_1 and W_2 of the factors x and y. In doing so we spoil the analytical tractability and do not know any longer the joint distribution of the factors (x, y). However, in the credit chapters, the Gaussian dependence mapping by Brigo and Alfonsi (2003, 2005) can be used to restore, to some extent, analytical tractability by mapping the CIR2++ model into a suitable G2++ model. This allows to use the G2++ formulas under the CIR2++ model. The interested reader is advised to read Section 22.7.5 and to re-cast the relevant results for correlated CIR++ processes in terms of the notation adopted in the present chapter.

4.3.1 The Basic Two-Factor CIR2 Model

In the two-factor CIR model the instantaneous interest rate is obtained by adding two independent processes like x in (3.21) under the risk-neutral measure:

$$\begin{aligned} dx(t) &= k_1(\theta_1 - x(t))dt + \sigma_1\sqrt{x(t)}\, dW_1(t), \\ dy(t) &= k_2(\theta_2 - y(t))dt + \sigma_2\sqrt{y(t)}\, dW_2(t), \end{aligned}$$

where W_1 and W_2 are independent Brownian motions under the risk-neutral measure, and $k_1, \theta_1, \sigma_1, k_2, \theta_2$ and σ_2 are positive constants such that $2k_1\theta_1 > \sigma_1^2$ and $2k_2\theta_2 > \sigma_2^2$.

The short rate is then defined as

$$\xi_t^\alpha := x(t) + y(t), \quad \alpha = (\alpha_1, \alpha_2),$$

with $\alpha_1 = (k_1, \theta_1, \sigma_1)$ and $\alpha_2 = (k_2, \theta_2, \sigma_2)$.

Due to independence of the factors, it is immediate to derive the following formulas (based on the notation for the one-factor case).

The price at time t of a zero-coupon bond with maturity T is explicitly given by

$$P^\xi(t, T; x(t), y(t), \alpha) = P^1(t, T; x(t), \alpha_1)\; P^1(t, T; y(t), \alpha_2), \tag{4.38}$$

where P^1 denotes the bond-price formula for the one-factor CIR model as a function of the one-factor instantaneous short rate and of the parameters, which has been given in (3.24). For example, $P^1(t, T; x(t), \alpha_1)$ is formula (3.24) with $r(t)$ replaced by $x(t)$ and k, θ, σ replaced by $(k_1, \theta_1, \sigma_1) = \alpha_1$.

The continuously-compounded spot rate at time t for the maturity T is given by

$$R^{\xi}(t,T;x(t),y(t),\alpha) = R^1(t,T;x(t),\alpha_1) + R^1(t,T;y(t),\alpha_2), \tag{4.39}$$

where R^1 denotes the spot rate for the one-factor CIR model, whose formula is obtained immediately from that of P^1. From the structure of the R's we see that we have an affine term structure in two dimensions.

Under the risk-neutral measure the bond price dynamics can be easily obtained via Itô's formula:

$$\begin{aligned} dP^{\xi}(t,T;\alpha) = P^{\xi}(t,T;\alpha)[\xi_t^{\alpha}\, dt - B(t,T;\alpha_1)\sigma_1\sqrt{x(t)}\, dW_1(t) \\ -B(t,T;\alpha_2)\sigma_2\sqrt{y(t)}\, dW_2(t)], \end{aligned}$$

where the deterministic function B is defined as in (3.25).

4.3.2 Relationship with the Longstaff and Schwartz Model (LS)

Longstaff and Schwartz (1992b) consider an interest-rate model where the short rate ξ_t, under the risk-neutral measure, is obtained as a linear combination of two basic processes X and Y as follows:

$$\begin{aligned} dX_t &= a(b - X_t)dt + \sqrt{X_t}\, dW_1(t), \\ dY_t &= c(e - Y_t)dt + \sqrt{Y_t}\, dW_2(t), \\ \xi_t &= \mu_x X_t + \mu_y Y_t, \end{aligned}$$

where all parameters have positive values with $\mu_x \neq \mu_y$, and again W_1 and W_2 are independent Brownian motions under the risk-neutral measure.

Longstaff and Schwartz give an equilibrium derivation of their model and show how it can be expressed as a stochastic-volatility model via a change of variable. Here our purpose is to show that the model is essentially a two-factor CIR model. Set

$$x(t) = \mu_x X_t, \quad y(t) = \mu_y Y_t,$$

so that

$$\xi_t = x(t) + y(t).$$

It is immediate to check that

$$\begin{aligned} dx(t) &= a(\mu_x\, b - x(t))dt + \sqrt{\mu_x}\sqrt{x(t)}\, dW_1(t) \\ &=: k_x(\theta_x - x(t))dt + \sigma_x\sqrt{x(t)}\, dW_1(t), \\ dy(t) &= c(\mu_y\, e - y(t))dt + \sqrt{\mu_y}\sqrt{y(t)}\, dW_2(t) \\ &=: k_y(\theta_y - y(t))dt + \sigma_y\sqrt{y(t)}\, dW_2(t). \end{aligned}$$

Since both x and y describe a one-factor CIR model, we see that the LS model can be interpreted as a two-factor CIR model. The parameters are linked by the relationships

$$
\begin{aligned}
k_x &= a, \\
\theta_x &= \mu_x b, \\
\sigma_x &= \sqrt{\mu_x}, \\
k_y &= c, \\
\theta_y &= \mu_y e, \\
\sigma_y &= \sqrt{\mu_y}\,.
\end{aligned}
$$

Therefore, when extending a two-factor CIR model in order to exactly fit the observed term structure, we are implicitly extending also the Longstaff and Schwartz (1992b) model. The only requirement is that in the basic CIR model $\sigma_x \neq \sigma_y$, so that in the LS formulation the condition $\mu_x \neq \mu_y$ is satisfied.

4.3.3 Forward-Measure Dynamics and Option Pricing for CIR2

Since the two factors $x(t)$ and $y(t)$ are independent, the T-forward measure dynamics for each factor is exactly the same as in (3.27)

$$
\begin{aligned}
dx(t) &= [k_1\theta_1 - (k_1 + B(t,T;\alpha_1)\sigma_1^2)x(t)]dt + \sigma_1\sqrt{x(t)}dW_1^T(t), \\
dy(t) &= [k_2\theta_2 - (k_2 + B(t,T;\alpha_2)\sigma_2^2)y(t)]dt + \sigma_2\sqrt{y(t)}dW_2^T(t),
\end{aligned}
$$

where W_1^T and W_2^T are independent standard Brownian motions under the T-forward adjusted measure. Pricing a zero-coupon-bond option can be carried out as follows. From the general formula for a call-option price at time t with maturity $T > t$ for an underlying zero-coupon bond with maturity $S > T$ and notional amount N and with strike price K we obtain

$$
\begin{aligned}
&C^\xi(t,T,S,N,K;x(t),y(t),\alpha) \\
&= E_t\left\{\exp\left[-\int_t^T \xi_u^\alpha du\right][NP^\xi(T,S;x(T),y(T),\alpha) - K]^+\right\} \\
&= P^\xi(t,T;x(t),y(t),\alpha)\, E_t^T\{[NP^\xi(T,S;x(T),y(T),\alpha) - K]^+\}\,.
\end{aligned}
$$

We recall from (3.28) that we know the densities of each x_T and y_T conditional on x_t and y_t respectively, under the T-forward measure. Therefore, we obtain

$$
\begin{aligned}
&C^\xi(t,T,S,N,K;x(t),y(t),\alpha) \\
&= P^\xi(t,T;x(t),y(t),\alpha)\int_0^{+\infty}\int_0^{+\infty}[NP^1(T,S;x_1,\alpha_1)\,P^1(T,S;x_2,\alpha_2) - K]^+ \\
&\quad \cdot p^T_{x(T)|x(t)}(x_1)p^T_{y(T)|y(t)}(x_2)dx_1dx_2,
\end{aligned}
\tag{4.40}
$$

which is an integral against the product of two non-central chi-square densities. Such a formula is the one being proposed by Longstaff and Schwartz (1992b). We need to remark that Chen and Scott (1992) develop a specific method for computing this integral through a suitable change of variable, reducing the calculation to that of a one-dimensional integral.

4.3.4 The CIR2++ Model and Option Pricing

In perfect analogy with the general method developed in Section 3.8, used in Section 3.9 for the one-factor case, let us define the instantaneous spot rate, under the risk-neutral measure, by

$$r_t = \varphi(t;\alpha) + \xi_t^\alpha = \varphi(t;\alpha) + x(t) + y(t) , \tag{4.41}$$

with $x(0) = x_0$, $y(0) = y_0$, and where $\varphi(t;\alpha)$ is a deterministic function of time, depending on the augmented parameter vector

$$\alpha := (x_0, y_0, k_1, \theta_1, \sigma_1, k_2, \theta_2, \sigma_2),$$

which is chosen so as to exactly retrieve the initial zero-coupon curve. In particular, $\varphi(0;\alpha) = r_0 - x_0 - y_0$.

It is easy to see that in order to fit exactly the zero-coupon curve observed in the market, it suffices to set

$$\varphi(t;\alpha) = f^M(0,t) - f^1(0,t;x_0,\alpha_1) - f^1(0,t;y_0,\alpha_2), \tag{4.42}$$

where f^1 is the one-factor instantaneous forward rate f^{CIR} as given in formula (3.77) and f^M is the market instantaneous forward rate.

In what follows, it is helpful to define the quantity

$$\begin{aligned}
\Phi^\xi(u,v;\alpha) &:= \exp\left[-\int_u^v \varphi(s;\alpha)ds\right] \\
&= \frac{P^M(0,v)}{P^M(0,u)}\frac{P^\xi(0,u;\alpha)}{P^\xi(0,v;\alpha)} \\
&= \exp\left\{\left[R^\xi(0,v;\alpha) - R^M(0,v)\right]v - \left[R^\xi(0,u;\alpha) - R^M(0,u)\right]u\right\},
\end{aligned} \tag{4.43}$$

which is known based on the initial market term structures of discount factors $T \to P^M(0,T)$ or spot rates $T \to R^M(0,T)$, on the initial instantaneous short rate r_0 and on the analytical expressions (4.38) or (4.39).

Zero-Coupon Bonds

As extensively explained in Section 3.8, adding a deterministic shift to the process ξ^α that admits explicit bond-price formulas, preserves the original analytical tractability in that also the new process implies explicit bond-price formulas.

Indeed, it is straightforward to verify that the price at time t of a zero-coupon bond maturing at time T and with unit face value is given by

$$P(t,T;x(t),y(t),\alpha) = \Phi^\xi(t,T;\alpha)P^\xi(t,T;x(t),y(t),\alpha) .$$

The derivation of this formula is perfectly analogous to that of (3.70) in the general one-factor case.

Instantaneous Forward Rates

It is immediate to check that, from the definition of the instantaneous forward rate

$$f(t,T) = -\frac{\partial}{\partial T} \ln P(t,T),$$

we have

$$\begin{aligned}\mathrm{Var}[df(t,T;x(t),y(t),\alpha)] &= \left(\frac{\partial B}{\partial T}(t,T;\alpha_1)\sigma_1\sqrt{x(t)}\right)^2 dt \\ &+ \left(\frac{\partial B}{\partial T}(t,T;\alpha_2)\sigma_2\sqrt{y(t)}\right)^2 dt,\end{aligned} \tag{4.44}$$

where again the deterministic function B is that figuring in the bond-price formulas for the one-factor CIR model, being defined as in (3.25).

It is a little laborious, though not difficult, to check analytically, by substituting the B's expressions in (3.25), that this is a monotonically-decreasing function which allows no humped shape for the curve

$$T \mapsto \mathrm{Var}[df(t,T;x(t),y(t),\alpha)],$$

with consequences discussed in Section 4.1.

Zero-Coupon-Bond Options and Caps and Floors

The price at time t of an European call option with maturity $T > t$ and strike price K on a zero-coupon bond with face value N maturing at $S > T$ is given by the following straightforward modification of the CIR2 formula:

$$\begin{aligned}&\mathbf{ZBC}(t,T,S,N,K;x(t),y(t),\alpha) \\ &\quad = N\Phi^{\xi}(t,S;\alpha)C^{\xi}(t,T,S,N,K/\Phi^{\xi}(T,S);x(t),y(t),\alpha)\end{aligned}$$

where C^{ξ} is given by (4.40). This formula is analogous to (3.73), as derived in the general one-factor case, and can be proven exactly in the same manner.

The analogous put option price is obtained from the put-call parity (3.3) and from the call-option-price formula:

$$\begin{aligned}\mathbf{ZBP}(t,T,S,N,K;x(t),y(t),\alpha) = \mathbf{ZBC}&(t,T,S,N,K;x(t),y(t),\alpha) \\ &-NP(t,S;x(t),y(t),\alpha) \\ &+ KP(t,T;x(t),y(t),\alpha).\end{aligned}$$

Since caps and floors can be viewed as portfolios of options on zero-coupon bonds, the general formulas (2.29), see also (3.80) and (3.81), remain valid in the two-factor case provided one considers them as expressed in terms of the two factors $x(t)$ and $y(t)$.

European Swaptions and Path-Dependent Payoffs

In the two-factor case Jamshidian's decomposition for coupon-bearing-bond options and European swaptions (see for instance Section 3.11.1) is not applicable. Therefore, such products need to be priced via alternative methods, such as numerical integration of the payoff or Monte Carlo simulation of scenarios. Alternatively, trees might be used as in the G2++ model. Since the two factors x and y are independent, conceptually the tree implementation follows from the implementation of two different one-dimensional CIR-like trees, whose construction has been explained in Section 3.9.1. Once the model has been calibrated to a set of market data, pricing path-dependent payoffs can be achieved through Monte Carlo simulation in complete analogy with the one-factor case. The only difference is that now we need to simulate a two-dimensional process x, y. However, since the two components are independent processes, the generation of paths can take place exactly in the same way as in the one-factor case, with the chosen method applied independently to each factor. Then the pricing of exotic interest-rate derivatives can be carried out exactly as in the one-factor case, by simulating the two factors separately, each as explained in Section 3.11.2. All other considerations of that section remain valid once the single factor process used there is replaced by the two factors considered here.

5. The Heath-Jarrow-Morton (HJM) Framework

I decided I'd spent too much time philosophizing.
It is, unfortunately, one of my character flaws.
J'onn J'onnz in "Martian Manhunter annual" 2, 1999, DC Comics.

Modeling the interest-rate evolution through the instantaneous short rate has some advantages, mostly the large liberty one has in choosing the related dynamics. For example, for one-factor short-rate models one is free to choose the drift and instantaneous volatility coefficient in the related diffusion dynamics as one deems fit, with no general restrictions. We have seen several examples of possible choices in Chapter 3. However, short-rate models have also some clear drawbacks. For example, an exact calibration to the initial curve of discount factors and a clear understanding of the covariance structure of forward rates are both difficult to achieve, especially for models that are not analytically tractable.

The first historically important alternative to short-rate models has been proposed by Ho and Lee (1986), who modeled the evolution of the entire yield curve in a binomial-tree setting. Their basic intuition was then translated in continuous time by Heath, Jarrow and Morton (HJM) (1992) who developed a quite general framework for the modeling of interest-rate dynamics. Precisely, by choosing the instantaneous forward rates as fundamental quantities to model, they derived an arbitrage-free framework for the stochastic evolution of the entire yield curve, where the forward-rates dynamics are fully specified through their instantaneous volatility structures. This is a major difference with arbitrage free one-factor short-rate dynamics, where the volatility of the short rate alone does not suffice to characterize the relevant interest-rate model. But in order to clarify the matter, let us consider the Merton (1973) toy short-rate model.

Assume we take the following equation for the short rate under the risk-neutral measure:

$$dr_t = \theta dt + \sigma dW_t, \quad r_0.$$

If you wish, this is a very particular toy version of the Hull-White model (3.33) seen in Chapter 3 with constant coefficient θ. Now, for this (affine) model, one can easily compute the bond price,

$$P(t,T) = \exp\left[\frac{\sigma^2}{6}(T-t)^3 - \frac{\theta}{2}(T-t)^2 - (T-t)r_t\right],$$

and the instantaneous forward rate

$$f(t,T) = -\frac{\partial \ln P(t,T)}{\partial T} = -\frac{\sigma^2}{2}(T-t)^2 + \theta(T-t) + r_t.$$

Differentiate this and substitute the short-rate dynamics to obtain

$$df(t,T) = (\sigma^2(T-t) - \theta)dt + \theta dt + \sigma dW_t,$$

or

$$df(t,T) = \sigma^2(T-t)dt + \sigma dW_t.$$

Look at this last equation. *The drift $\sigma^2(T-t)$ in the f's dynamics is determined as a suitable transformation of the diffusion coefficient σ in the same dynamics.* This is no mere coincidence due to the simplicity of our toy model, but a general fact that Heath, Jarrow and Morton (1992) expressed in full generality.

Clearly, if one wishes to directly model this instantaneous forward rate, there is no liberty in selecting the drift of its process, as it is completely determined by the chosen volatility coefficient. This is essentially due to the fact that we are modeling a derived quantity f and not the fundamental quantity r. Indeed, f is expressed in terms of the more fundamental r by

$$f(t,T) = -\frac{\partial \ln E_t\left[\exp\left(-\int_t^T r(s)ds\right)\right]}{\partial T},$$

and, as you see, an expectation has already acted in the definition of f, adding structure and taking away freedom, so to say.

More generally, under the HJM framework, one assumes that, for each T, the forward rate $f(t,T)$ evolves according to

$$df(t,T) = \alpha(t,T)dt + \sigma(t,T)dW(t),$$

where W is a (possibly multi-dimensional) Brownian motion. As we have just seen in the above example, contrary to the short-rate modeling case, where one is free to specify the drift of the considered diffusion, here the function α is completely determined by the choice of the (vector) diffusion coefficient σ.

The importance of the HJM theory lies in the fact that virtually any (exogenous term-structure) interest-rate model can be derived within such a framework.[1] However, only a restricted class of volatilities is known to imply a Markovian short-rate process. This means that, in general, burdensome procedures, like those based on non-recombining lattices, are needed to price interest-rate derivatives. Substantially, the problem remains of defining

[1] Even the celebrated LIBOR market model was developed starting from instantaneous-forward-rate dynamics in Brace, Gatarek and Musiela (1997), although it is possible to obtain it also through the change-of-numeraire approach, as we will see in the next chapter.

a suitable volatility function for practical purposes. This is the reason why in this book we do not devote too much attention to the HJM theory, preferring to deal with explicitly formulated models.

In this chapter, we briefly review the HJM framework and explicitly write the HJM no-arbitrage condition. We then describe some analogies with instantaneous short-rate models. We show, in particular, that a one-factor HJM model with deterministic volatility is equivalent to the Hull and White (1990b) short-rate model (3.32). We then introduce the Ritchken and Sankarasubramanian (1995) (RS) framework and briefly illustrate the Li, Ritchken and Sankarasubramanian (1995a, 1995b) algorithm for pricing derivatives. We finally mention the Mercurio and Moraleda (2000) humped-volatility model as a specific example of a one-factor Gaussian model within the HJM framework.

5.1 The HJM Forward-Rate Dynamics

Heath, Jarrow and Morton (1992) assumed that, for a fixed a maturity T, the instantaneous forward rate $f(t,T)$ evolves, under a given measure, according the following diffusion process:

$$\begin{aligned} df(t,T) &= \alpha(t,T)dt + \sigma(t,T)dW(t), \\ f(0,T) &= f^M(0,T), \end{aligned} \tag{5.1}$$

with $T \mapsto f^M(0,T)$ the market instantaneous-forward curve at time $t = 0$, and where $W = (W_1, \ldots, W_N)$ is an N-dimensional Brownian motion, $\sigma(t,T) = (\sigma_1(t,T), \ldots, \sigma_N(t,T))$ is a vector of adapted processes and $\alpha(t,T)$ is itself an adapted process. The product $\sigma(t,T)dW(t)$ is intended to be the scalar product between the two vectors $\sigma(t,T)$ and $dW(t)$.

The advantage of modeling forward rates as in (5.1) is that the current term structure of rates is, by construction, an input of the selected model. Remember, in fact, that the following relations between zero-bond prices and forward rates hold (see (1.23)):

$$\begin{aligned} f(t,T) &= -\frac{\partial \ln P(t,T)}{\partial T}, \\ P(t,T) &= e^{-\int_t^T f(t,u)du}. \end{aligned}$$

The dynamics in (5.1) is not necessarily arbitrage-free. Following an approach similar to that of Harrison and Pliska (1981) (see Chapter 2) Heath, Jarrow and Morton proved that, in order for a unique equivalent martingale measure to exist, the function α cannot be arbitrarily chosen, but it must equal a quantity depending on the vector volatility σ and on the drift rates in the dynamics of N selected zero-coupon bond prices. In particular, if the dynamics (5.1) are under the risk-neutral measure, then we must have

$$\alpha(t,T) = \sigma(t,T) \int_t^T \sigma(t,s)ds = \sum_{i=1}^N \sigma_i(t,T) \int_t^T \sigma_i(t,s)ds, \qquad (5.2)$$

so that the integrated dynamics of $f(t,T)$ under the risk-neutral measure are

$$\begin{aligned} f(t,T) &= f(0,T) + \int_0^t \sigma(u,T) \int_u^T \sigma(u,s)ds\, du + \int_0^t \sigma(s,T)dW(s) \\ &= f(0,T) + \sum_{i=1}^N \int_0^t \sigma_i(u,T) \int_u^T \sigma_i(u,s)ds\, du + \sum_{i=1}^N \int_0^t \sigma_i(s,T)dW_i(s) \end{aligned}$$

and are fully specified once the vector volatility function σ is provided. Given this dynamics of the instantaneous forward rate $f(t,T)$, application of Ito's lemma gives the following dynamics of the zero-coupon bond price $P(t,T)$:

$$dP(t,T) = P(t,T) \left[r(t)dt - \left(\int_t^T \sigma(t,s)ds \right) dW(t) \right],$$

where $r(t)$ is the instantaneous short term interest rate at time t, that is

$$\begin{aligned} r(t) = f(t,t) &= f(0,t) + \int_0^t \sigma(u,t) \int_u^t \sigma(u,s)ds\, du + \int_0^t \sigma(s,t)dW(s) \\ &= f(0,t) + \sum_{i=1}^N \int_0^t \sigma_i(u,t) \int_u^t \sigma_i(u,s)ds\, du + \sum_{i=1}^N \int_0^t \sigma_i(s,t)dW_i(s). \end{aligned} \qquad (5.3)$$

5.2 Markovianity of the Short-Rate Process

The short-rate process (5.3) is not a Markov process in general. Notice in fact that the time t appears in the stochastic integral both as extreme of integration and inside the integrand function. However, there are suitable specifications of σ for which r is indeed a Markov process. As proven by Carverhill (1994), this happens, for example, if we can write, for each $i = 1, \ldots, N$,

$$\sigma_i(t,T) = \xi_i(t)\psi_i(T), \qquad (5.4)$$

with ξ_i and ψ_i strictly positive and deterministic functions of time. Under such a separable specification, the short-rate process becomes

$$\begin{aligned} r(t) &= f(0,t) + \sum_{i=1}^N \int_0^t \xi_i(u)\psi_i(t) \int_u^t \xi_i(u)\psi_i(s)ds\, du + \sum_{i=1}^N \int_0^t \xi_i(s)\psi_i(t)dW_i(s) \\ &= f(0,t) + \sum_{i=1}^N \psi_i(t) \int_0^t \xi_i^2(u) \int_u^t \psi_i(s)ds\, du + \sum_{i=1}^N \psi_i(t) \int_0^t \xi_i(s)dW_i(s). \end{aligned}$$

Notice that in the one-factor case ($N = 1$), if we define the (strictly-positive) deterministic function A by

$$A(t) := f(0,t) + \psi_1(t) \int_0^t \xi_1^2(u) \int_u^t \psi_1(s) ds \, du,$$

and assume its differentiability, we can write

$$\begin{aligned} dr(t) &= A'(t)dt + \psi_1'(t) \int_0^t \xi_1(s) dW_1(s) + \psi_1(t)\xi_1(t)dW_1(t) \\ &= \left[A'(t) + \psi_1'(t) \frac{r(t) - A(t)}{\psi_1(t)} \right] dt + \psi_1(t)\xi_1(t)dW_1(t) \\ &= [a(t) + b(t)r(t)] \, dt + c(t)dW_1(t), \end{aligned}$$

with obvious definition of the coefficients a, b and c.

We therefore end up with the general short-rate dynamics proposed by Hull and White (1990b), see (3.32), thus establishing an equivalence between the HJM one-factor model for which (5.4) holds and the general formulation of the Gaussian one-factor short-rate model of Hull and White (1990b). In particular, we can easily derive the HJM forward-rate dynamics that is equivalent to the short-rate dynamics (3.33). To this end, let us set

$$\sigma_1(t,T) = \sigma e^{-a(T-t)},$$

where a and σ are now real constants, so that

$$\begin{aligned} \xi_1(t) &= \sigma e^{at}, \\ \psi_1(T) &= e^{-aT}, \\ A(t) &= f(0,t) + \frac{\sigma^2}{2a^2} \left(1 - e^{-at}\right)^2. \end{aligned}$$

The resulting short-rate dynamics is then given by

$$\begin{aligned} dr(t) = &\left[\frac{\partial f}{\partial T}(0,t) + \frac{\sigma^2}{a} \left(e^{-at} - e^{-2at}\right) \right. \\ &\left. -a\left(r(t) - f(0,t) - \frac{\sigma^2}{2a^2} \left(1 - e^{-at}\right)^2 \right) \right] dt + \sigma dW_1(t) \\ = &\left[\frac{\partial f}{\partial T}(0,t) + af(0,t) + \frac{\sigma^2}{2a} \left(1 - e^{-2at}\right) - ar(t) \right] dt + \sigma dW_1(t), \end{aligned}$$

which is equivalent to (3.33) when combined with (3.34).

5.3 The Ritchken and Sankarasubramanian Framework

It is now clear that an arbitrary specification of the forward-rate volatility will likely lead to a non-Markovian instantaneous short-rate process. In such a case, we would soon encounter major computational problems when discretizing the dynamics (5.3) for the pricing of a general derivative. In fact, the approximating lattice will not be recombining, and the number of nodes in the tree will grow exponentially with the number of steps. This will make the numerical procedure quite difficult to handle, especially as far as execution time (combined with a pricing accuracy) is concerned.

These pricing problems can be addressed by noting that, even though the short-rate process is not Markovian, there may yet exist a higher-dimensional Markov process having the short rate as one of its components. Exploiting such intuition, Ritchken and Sankarasubramanian (1995) have identified necessary and sufficient conditions on the volatility structure of forward rates for capturing the path dependence of r through a single sufficient statistic. Precisely, they proved the following.

Proposition 5.3.1. *Consider a one-factor HJM model. If the volatility function $\sigma(t,T)$ is differentiable with respect to T, a necessary and sufficient condition for the price of any (interest-rate) derivative to be completely determined by a two-state Markov process $\chi(\cdot) = (r(\cdot), \phi(\cdot))$ is that the following condition holds:*

$$\sigma(t,T) = \sigma_{RS}(t,T) := \eta(t)e^{-\int_t^T \kappa(x)dx}, \tag{5.5}$$

where η is an adapted process and κ is a deterministic (integrable) function. In such a case, the second component of the process χ is defined by

$$\phi(t) = \int_0^t \sigma_{RS}^2(s,t)ds.$$

Accordingly, zero-coupon-bond prices are explicitly given by

$$P(t,T) = \frac{P(0,T)}{P(0,t)} \exp\left\{-\frac{1}{2}\Lambda^2(t,T)\phi(t) + \Lambda(t,T)[f(0,t) - r(t)]\right\},$$

where

$$\Lambda(t,T) = \int_t^T e^{-\int_t^u \kappa(x)dx} du.$$

Differentiation of equation (5.3) shows that, under the RS class of volatilities (5.5), the process χ, and hence the instantaneous short-rate r, evolve according to

$$d\chi(t) = \begin{pmatrix} dr(t) \\ d\phi(t) \end{pmatrix} = \begin{pmatrix} \mu(r,t)dt + \eta(t)dW(t) \\ \left[\eta^2(t) - 2\kappa(t)\phi(t)\right] dt \end{pmatrix} \tag{5.6}$$

with

$$\mu(r,t) = \kappa(t)[f(0,t) - r(t)] + \phi(t) + \frac{\partial}{\partial t} f(0,t).$$

We can now see that η is nothing but the instantaneous short-rate volatility process.

The yield curve dynamics described by (5.6) can be, therefore, discretized in a Markovian (recombining) lattice in terms of the two variables r and ϕ. This was suggested by Li, Ritchken and Sankarasubramanian (1995a, 1995b) (LRS), who developed an efficient lattice to approximate the processes (5.6). Their tree construction procedure is briefly outlined in the following, under the particular case where

$$\eta(t) = \hat{\sigma}(r(t))$$
$$\hat{\sigma}(x) := v x^{\rho}$$

with v and ρ positive constants, $\rho \in [0,1]$, so that

$$\sigma_{RS}(t,T) = v[r(t)]^{\rho} e^{-\int_t^T \kappa(x)dx}.$$

Li, Ritchken and Sankarasubramanian considered the following transformation, which yields a process with constant volatility

$$Y(t) = \int \frac{1}{\hat{\sigma}(x)} dx \bigg|_{x=r(t)}$$

where the right-hand side denotes a primitive of $1/\hat{\sigma}(x)$ calculated in $x = r(t)$ and where the constant in the primitive is set to zero. This is exactly the transformation of r needed to have a unit diffusion coefficient when applying Ito's formula to compute its differential, as we shall see in a moment. Notice that, by substituting the expression for $\hat{\sigma}$ and by integrating, we have

$$Y(t) = \bar{Y}(r(t)) := \begin{cases} \frac{1}{v} \ln(r(t)) & \text{if } \rho = 1 \\ \frac{1}{v(1-\rho)} [r(t)]^{1-\rho} & \text{if } 0 \le \rho < 1 \end{cases} \tag{5.7}$$

with the function Y defined in a suitable domain $\mathcal{D}_\rho$ depending on the value of ρ. Denoting by $x = \varphi(y)$ the inverse function of $y = \bar{Y}(x)$ on $\mathcal{D}_\rho$, we have

$$\varphi(y) = \begin{cases} e^{vy} & \text{if } \rho = 1 \\ (v(1-\rho)y)^{1/(1-\rho)} & \text{if } 0 \le \rho < 1 \end{cases} \tag{5.8}$$

Application of Ito's lemma and straightforward algebra show that

$$dY(t) = m(Y,\phi,t)dt + dW(t)$$
$$d\phi(t) = [\hat{\sigma}(r(t)) - 2\kappa(t)\phi(t)]\, dt$$

where indeed the diffusion coefficient in the Y dynamics is one, and where

$$m(Y,\phi,t) = \frac{\kappa(t)[f(0,t) - \varphi(Y(t))] + \phi(t) + \frac{\partial f}{\partial t}(0,t)}{v[\varphi(Y(t))]^{\rho}} - \frac{v\rho}{2[\varphi(Y(t))]^{1-\rho}} \tag{5.9}$$

for any $0 \le \rho \le 1$.

If, for instance, $\rho = 0.5$, we have that, for $r(t) > 0$, $Y(t) = 2\sqrt{r(t)}/v$, $\varphi(y) = \frac{v^2 y^2}{4}$, and $m(Y, \phi, t)$ in (5.9) becomes

$$m(Y, \phi, t) = \frac{\kappa(t)[f(0,t) - \frac{1}{4}v^2(Y(t))^2] + \phi(t) + \frac{\partial f}{\partial t}(0,t)}{\frac{1}{2}v^2 Y(t)} - \frac{1}{2Y(t)},$$

whereas, if $\rho = 1$, we have that, for $r(t) > 0$, $Y(t) = \ln[r(t)]/v$, $\varphi(y) = e^{vy}$ and

$$m(Y, \phi, t) = \frac{\kappa(t)[f(0,t) - e^{vY(t)}] + \phi(t) + \frac{\partial f}{\partial t}(0,t)}{v e^{vY(t)}} - \frac{1}{2}v.$$

Building a lattice for Y is eased by the presence of a unit diffusion coefficient, which is the reason for adopting such a transformation in the first place. Also, since ϕ's dynamics (5.6) has no diffusion part, the related lattice component need not branch. The approximating lattice is then constructed as follows. Divide the given time horizon into intervals of equal length Δt, and suppose that, at the beginning t of some time interval, the state variables are y and ϕ, and that, in the next time period, they move either to (y^+, ϕ^*) or to (y^-, ϕ^*) where

$$\begin{aligned} y^+ &= y + (J(y, \phi) + 1)\sqrt{\Delta t}, \\ y^- &= y + (J(y, \phi) - 1)\sqrt{\Delta t}, \\ \phi^* &= \phi + [\hat{\sigma}^2(\varphi(y)) - 2\kappa(t)\phi]\Delta t \end{aligned}$$

Setting $Z(y, \phi) = \text{int}\left[m(y, \phi, t)\sqrt{\Delta t}\right]$, where $\text{int}[x]$ denotes the largest integer smaller or equal than the real x, the function J is defined by

$$J(y, \phi) = \begin{cases} |Z(y, \phi)| & \text{if } Z(y, \phi) \text{ is even} \\ Z(y, \phi) + 1 & \text{otherwise.} \end{cases}$$

The branching probabilities, p for an up-move and $1 - p$ for a down-move, are then derived by solving

$$p(y^+ - y) + (1 - p)(y^- - y) = m(y, \phi, t)\Delta t,$$

thus obtaining

$$p = \frac{m(y, \phi, t)\Delta t + y - y^-}{y^+ - y^-}.$$

The above choices ensure that

$$y^+ \geq y + m(y, \phi, t)\Delta t \geq y^-,$$

and hence that the probabilities of moving from one node to another in the lattice lie always in $[0, 1]$.[2]

Once the LRS tree has been built, derivatives prices can then be calculated in a quite standard way.

[2] We refer to Li, Ritchken and Sankarasubramanian (1995a) for a detailed description of all the calculations above.

5.4 The Mercurio and Moraleda Model

We conclude the chapter by briefly reviewing the Mercurio and Moraleda (2000) model, which explicitly assumes a humped volatility structure in the instantaneous-forward-rate dynamics. Such assumption is motivated by the fact that the volatility structure of forward rates, as implied by market quotes, is commonly humped. We also refer to Section 3.6 for a detailed explanation of the opportunity and practical relevance of such an assumption.

Mercurio and Moraleda (2000) proposed a one-factor Gaussian model in the HJM framework by considering the following form for the volatility of instantaneous forward rates:

$$\sigma(t,T) = \sigma[\gamma(T-t)+1]e^{-\frac{\lambda}{2}(T-t)}, \tag{5.10}$$

where σ, γ and λ are non-negative constants. Under such specification, (5.1) becomes

$$\begin{aligned} df(t,T) &= \bar{\alpha}(T-t)dt + \sigma[\gamma(T-t)+1]e^{-\frac{\lambda}{2}(T-t)}dW(t), \\ \bar{\alpha}(\tau) &= -\frac{2\sigma^2}{\lambda}[\gamma\tau+1]\,e^{-\lambda\tau}\left[\gamma\tau + \left(\frac{2\gamma}{\lambda}+1\right)\left(1-e^{\frac{\lambda}{2}\tau}\right)\right], \end{aligned}$$

which implies that instantaneous (forward and spot) rates are normally distributed.

The choice of the volatility function (5.10) is motivated by the following features.

(a) It provides a humped volatility structure for strictly positive σ, γ and λ, and $2\gamma > \lambda$;
(b) It depends only on the "time to maturity" $T-t$ rather than on time t and maturity T separately;
(c) It leads to analytical formulas for European options on discount bonds;
(d) It generalizes the volatility specification of the Hull and White model (3.33), in that for $\gamma = 0$ and $\lambda = 2a$ we get the volatility of forward rates as implied by (3.33).[3]

It is obvious that (b) and (d) hold. Basic calculus shows that (a) is also true. In fact, for strictly positive σ, γ and λ, the function $f(x) = \sigma[\gamma x + 1]\exp\{-\frac{\lambda}{2}x\}$ has the following features: (i) it is strictly positive for $x \geq 0$; (ii) it is increasing and concave in the interval $[0, (2\gamma-\lambda)/(\gamma\lambda)]$; (iii) it has a maximum in $(2\gamma-\lambda)/(\gamma\lambda)$ whose value is $(2\sigma\gamma/\lambda)\exp\{(-2\gamma+\lambda)/(2\gamma)\}$; (iv) it is decreasing and concave in the interval $[(2\gamma-\lambda)/(\gamma\lambda), (4\gamma-\lambda)/(\gamma\lambda)]$;

[3] When modeling humped volatility structures, many other specifications can of course be considered. For example the term between square brackets in (5.10) can be generalized to be any polynomial in $(T-t)$. It is disputable however whether there exists a simpler characterization than (5.10) and for which (a), (b), (c) and (d) hold.

(v) it is decreasing and convex from $(4\gamma - \lambda)/(\gamma\lambda)$ onwards; (vi) it tends asymptotically to zero.

As for property (c), Mercurio and Moraleda (2000) used the results of Merton (1973) to prove that the time t-price of a European call option with maturity T and strike price X on a pure discount bond with maturity S is given by

$$\mathbf{ZBC}(t, T, S, X) = P(t, S)\Phi(d_1(t)) - XP(t, T)\Phi(d_2(t)),$$

where

$$\begin{aligned} d_1(t) &:= \frac{\ln(P(t,S)/(XP(t,T)) + \frac{1}{2}v_t^2}{v_t}, \\ d_2(t) &:= d_1(t) - v_t, \end{aligned}$$

and

$$\begin{aligned} v_t^2 = &\frac{4\sigma^2}{\lambda^7}(A^2\lambda^2 + 2AB\lambda + 2B^2)\left(e^{\lambda T} - e^{\lambda t}\right) - \frac{8\sigma^2 B}{\lambda^6}(A\lambda + B)\left(Te^{\lambda T} - te^{\lambda t}\right) \\ &+ \frac{4\sigma^2 B^2}{\lambda^5}\left(T^2 e^{\lambda T} - t^2 e^{\lambda t}\right), \end{aligned}$$

where

$$\begin{aligned} A &:= (\lambda + 2\gamma)\left(e^{-\frac{\lambda}{2}S} - e^{-\frac{\lambda}{2}T}\right) + \gamma\lambda\left(Se^{-\frac{\lambda}{2}S} - Te^{-\frac{\lambda}{2}T}\right), \\ B &:= \gamma\lambda\left(e^{-\frac{\lambda}{2}S} - e^{-\frac{\lambda}{2}T}\right). \end{aligned}$$

However, no analytical formula for pure discount bonds can be derived. Notice, in fact, that the instantaneous-short-rate process is not Markovian since (5.10) does not belong to the RS class (5.5).

Mercurio and Moraleda also tested empirically their model. They considered a time series of market cap prices and compared the fitting quality of their model with that implied by the Hull and White model (3.33). Using the Schwarz-Information-Criterion test, they concluded that, in most situations, their humped-volatility model is indeed preferable. We refer to Mercurio and Moraleda (2000) for a detailed description of the calibration results.

Part III

MARKET MODELS

6. The LIBOR and Swap Market Models (LFM and LSM)

There is so much they don't tell. Is that why I became a cop?
To learn what they don't tell?
John Jones (J'onn J'onnz) in "Martian Manhunter: American Secrets" 1, DC Comics, 1992

6.1 Introduction

In this chapter we consider one of the most popular and promising families of interest-rate models: The market models.

Why are such models so popular? The main reason lies in the agreement between such models and well-established market formulas for two basic derivative products. Indeed, the lognormal forward-LIBOR model (LFM) prices caps with Black's cap formula, which is the standard formula employed in the cap market. Moreover, the lognormal forward-swap model (LSM) prices swaptions with Black's swaption formula, which again is the standard formula employed in the swaption market. Since the caps and swaptions markets are the two main markets in the interest-rate-options world, it is important for a model to be compatible with such market formulas.

Before market models were introduced, there was no interest-rate dynamics compatible with either Black's formula for caps or Black's formula for swaptions. These formulas were actually based on mimicking the Black and Scholes model for stock options under some simplifying and inexact assumptions on the interest-rates distributions. The introduction of market models provided a new derivation of Black's formulas based on rigorous interest-rate dynamics.

However, even with full rigor given separately to the caps and swaptions classic formulas, we point out the now classic problem of this setup: The LFM and the LSM are not compatible. Roughly speaking, if forward LIBOR rates are lognormal each under its measure, as assumed by the LFM, forward swap rates cannot be lognormal at the same time under their measure, as assumed by the LSM. Although forward swap rates obtained from lognormal forward LIBOR rates are not far from being lognormal themselves under the relevant measure (as shown in some empirical works and also in Chapter 8 of the present book), the problem still stands and reduces one's enthusiasm for the theoretical setup of market models, if not for the practical one.

In this chapter, we will derive the LFM dynamics under different measures by resorting to the change-of-numeraire technique. We will show how caps are priced in agreement with Black's cap formula, and explain how swaptions can be priced through a Monte Carlo method in general. Analytical approximations leading to swaption-pricing formulas are also presented, as well as closed-form formulas for terminal correlations based on similar approximations.

We will suggest parametric forms for the instantaneous covariance structure (volatilities and correlations) in the LFM. Part of the parameters in this structure can be obtained directly from market-quoted cap volatilities, whereas other parameters can be obtained by calibrating the model to swaption prices. The calibration to swaption prices can be made computationally efficient through the analytical approximations mentioned above.

We will derive results and approximations connecting semi-annual caplet volatilities to volatilities of swaptions whose underlying swap is one-year long. We will also show how one can obtain forward rates over non-standard periods from standard forward rates either through drift interpolation or via a bridging technique.

We will then introduce the LSM and show how swaptions are priced in agreement with Black's swaptions formula, although Black's formula for caps does not hold under this model.

We will finally consider the "smile problem" for the cap market, and introduce some possible extensions of the basic LFM that are analytically tractable and allow for a volatility smile.

6.2 Market Models: a Guided Tour

"Fischer had many remarkable qualities that became apparent almost as soon as you met him. If you knew about the Black-Scholes breakthrough that allowed the somehow miraculous determination of the fair price of an option independent of what you thought about the stock, and appreciated what a giant leap forward that was in the world of financial economics, then you expected him to be deep and brilliant. But what struck you even more forcefully was how meticulous was his devotion to clarity and simplicity in presentation and speaking. [...]". Emanuel Derman, December 3, 1996.

Induction Speech for Fischer Black (1938-1995).

Before market models were introduced, short-rate models used to be the main choice for pricing and hedging interest-rate derivatives. Short-rate models are still chosen for many applications and are based on modeling the instantaneous spot interest rate ("short rate") via a (possibly multi-dimensional) diffusion process. This diffusion process characterizes the evolution of the complete yield curve in time. We have seen examples of such models in Chapters 3 and 4.

To fix ideas, let us consider the time-0 price of a T_2-maturity caplet resetting at time T_1 $(0 < T_1 < T_2)$ with strike X and a notional amount of 1. Caplets and caps have been defined in Chapter 1 and will be described more generally in Section 6.4. Let τ denote the year fraction between T_1 and T_2. Such a contract pays out the amount

$$\tau(L(T_1, T_2) - X)^+$$

at time T_2, where in general $L(u, s)$ is the LIBOR rate at time u for maturity s.

Again to fix ideas, let us choose a specific short-rate model and assume we are using the shifted two-factor Vasicek model G2++ given in (4.4). The parameters of this two-factor Gaussian additive short-rate model are here denoted by $\theta = (a, \sigma, b, \eta, \rho)$. Then the short rate r_t is obtained as the sum of two linear diffusion processes x_t and y_t, plus a deterministic shift φ that is used for fitting the initially observed yield curve at time 0:

$$r_t = x_t + y_t + \varphi(t; \theta)\,.$$

Such model allows for an analytical formula for forward LIBOR rates F,

$$\begin{aligned} F(t; T_1, T_2) &= F(t; T_1, T_2; x_t, y_t, \theta), \\ L(T_1, T_2) &= F(T_1; T_1, T_2; x_{T_1}, y_{T_1}, \theta)\,. \end{aligned}$$

At this point one can try and price a caplet. To this end, one can compute the risk-neutral expectation of the payoff discounted with respect to the bank account numeraire $\exp\left(\int_0^{T_2} r_s ds\right)$ so that one has

$$E\left[\exp\left(-\int_0^{T_2} r_s ds\right) \tau(F(T_1; T_1, T_2, x_{T_1}, y_{T_1}, \theta) - X)^+\right].$$

This too turns out to be feasible, and leads to a function

$$U_C(0, T_1, T_2, X, \theta).$$

On the other hand, the market has been pricing caplets (actually caps) with Black's formula for years. One possible derivation of Black's formula for caplets is based on the following approximation. When pricing the discounted payoff

$$E\left[\exp\left(-\int_0^{T_2} r_s ds\right) \tau(L(T_1, T_2) - X)^+\right] = \cdots$$

one first assumes the discount factor $\exp\left(-\int_0^{T_2} r_s ds\right)$ to be deterministic and identifies it with the corresponding bond price $P(0,T_2)$. Then one factors out the discount factor to obtain:

$$\cdots \approx P(0,T_2)\tau E\left[(L(T_1,T_2)-X)^+\right] = P(0,T_2)\tau E\left[(F(T_1;T_1,T_2)-X)^+\right] .$$

Now, inconsistently with the previous approximation, one goes back to assuming rates to be stochastic, and models the forward LIBOR rate $F(t;T_1,T_2)$ as in the classical Black and Scholes option pricing setup, i.e as a (driftless) geometric Brownian motion:

$$dF(t;T_1,T_2) = vF(t;T_1,T_2)dW_t \ , \tag{6.1}$$

where v is the instantaneous volatility, assumed here to be constant for simplicity, and W is a standard Brownian motion under the risk-neutral measure Q.

Then the expectation

$$E\left[(F(T_1;T_1,T_2)-X)^+\right]$$

can be viewed simply as a T_1-maturity call-option price with strike X and whose underlying asset has volatility v, in a market with zero risk-free rate.

We therefore obtain:

$$\begin{aligned}
\mathbf{Cpl}(0,T_1,T_2,X) &:= P(0,T_2)\tau E(F(T_1;T_1,T_2)-X)^+ \\
&= P(0,T_2)\tau[F(0;T_1,T_2)\Phi(d_1(X,F(0;T_1,T_2),v\sqrt{T_1})) \\
&\qquad -X\Phi(d_2(X,F(0;T_1,T_2),v\sqrt{T_1}))],
\end{aligned}$$

$$d_1(X,F,u) = \frac{\ln(F/X)+u^2/2}{u}, \quad d_2(X,F,u) = \frac{\ln(F/X)-u^2/2}{u},$$

where Φ is the standard Gaussian cumulative distribution function.

From the way we just introduced it, this formula seems to be partly based on inconsistencies. However, within the change-of-numeraire setup, the formula can be given full mathematical rigor as follows. Denote by Q^2 the measure associated with the T_2-bond-price numeraire $P(\cdot,T_2)$ (T_2-forward measure) and by E^2 the corresponding expectation. Then, by the change-of-numeraire approach, we can switch from the bank-account numeraire $B(t) = B_0 \exp\left(\int_0^t r_s ds\right)$ associated with the risk-neutral measure $Q = Q^B$ to the bond-price numeraire $P(t,T_2)$ and obtain (set $F_2 = F(\cdot,T_1,T_2)$ for brevity, so that $L(T_1,T_2) = F(T_1;T_1,T_2) = F_2(T_1)$)

$$E\left[\exp\left(-\int_0^{T_2} r_s ds\right)\tau(L(T_1,T_2)-X)^+\right] =$$

$$= E\left[\exp\left(-\int_0^{T_2} r_s ds\right)\tau(F_2(T_1)-X)^+\right] =$$

$$= E^{\boxed{B}}\left[\frac{\boxed{B(0)}}{\boxed{B(T_2)}}\tau(F_2(T_1)-X)^+\right] = E^{\boxed{2}}\left[\frac{\boxed{P(0,T_2)}}{\boxed{P(T_2,T_2)}}\tau(F_2(T_1)-X)^+\right] = \dots$$

where we have used Fact Two on the change of numeraire technique (Section 2.3), obtaining an invariant quantity by changing numeraire in the three boxes. Take out $P(0,T_2)$ and recall that $P(T_2,T_2)=1$. We have

$$E\left[\exp\left(-\int_0^{T_2} r_s ds\right)\tau(L(T_1,T_2)-X)^+\right] = P(0,T_2)\tau E^2\left[(L(T_1,T_2)-X)^+\right].$$

What has just been done, rather than assuming deterministic discount factors, is a change of measure. We have "factored out" the stochastic discount factor and replaced it with the related bond price, but in order to do so we had to change the probability measure under which the expectation is taken. Now the last expectation is no longer taken under the risk-neutral measure but rather under the T_2-forward measure. Since by definition $F(t;T_1,T_2)$ can be written as the price of a tradable asset divided by $P(t,T_2)$, it needs follow a martingale under the measure associated with the numeraire $P(t,T_2)$, i.e. under Q^2 (Fact One on the change of numeraire, again Section 2.3). As we have hinted at in Appendix C, martingale means "driftless" when dealing with diffusion processes. Therefore, the dynamics of $F(t;T_1,T_2)$ under Q^2 is driftless, so that the dynamics

$$dF(t;T_1,T_2) = vF(t;T_1,T_2)dW_t \tag{6.2}$$

is correct under the measure Q^2, where W is a standard Brownian motion under Q^2. Notice that the driftless (lognormal) dynamics above is precisely the dynamics we need in order to recover *exactly* Black's formula, without approximation. *We can say that the choice of the numeraire $P(\cdot,T_2)$ is based on this fact: It makes the dynamics (6.2) of F driftless under the related Q^2 measure, thus replacing rigorously the earlier arbitrary assumption on the F dynamics (6.1) under the risk-neutral measure Q.* Following this rigorous approach we indeed obtain Black's formula, since the process F has the same distribution as in the approximated case above, and hence the expected value has the same value as before. Nonetheless, it can be instructive to derive Black's formula in detail at least once in the book. We do so now. The reader that is not interested in the derivation may skip the derivation and continue with the guided tour.

Begin(detailed derivation of Black's formula for caplets)

I wear black on the outside / 'cause black is how I feel on the inside
The Smiths, 1987

"Blackness coming... this Earth next..."
Superboy from Hyper-time, "Hyper-tension" , 1999, DC Comics

We change slightly the notation to derive a formula for time-varying volatilities. As seen above, by Fact One on the change of numeraire (again Section 2.3) F_2 is a martingale (no drift) under Q^2. Take a geometric Brownian motion

$$dF(t; T_1, T_2) = \sigma_2(t)\ F(t; T_1, T_2) dW(t),$$

where σ_2 is the instantaneous volatility, and W is a standard Brownian motion under the measure Q^2.

Let us solve this equation and compute the caplet price term $E^{Q^2}\left[(F_2(T_1) - X)^+\right]$. By Ito's formula:

$$\begin{aligned} d\ln(F_2(t)) &= \ln'(F_2)dF_2 + \tfrac{1}{2}\ln''(F_2)\ dF_2\ dF_2 \\ &= \frac{1}{F_2}dF_2 + \tfrac{1}{2}\left(-\frac{1}{(F_2)^2}\right) dF_2\ dF_2 = \\ = \frac{1}{F_2}\sigma_2 F_2 dW &- \tfrac{1}{2}\frac{1}{(F_2)^2}(\sigma_2 F_2 dW)(\sigma_2 F_2 dW) = \\ &= \sigma_2 dW - \tfrac{1}{2}\frac{1}{(F_2)^2}\sigma_2^2 F_2^2 dW dW = \\ &= \sigma_2(t)dW(t) - \tfrac{1}{2}\sigma_2^2(t)dt \end{aligned}$$

(where $'$ and $''$ denote here the first and second derivative and where we used $dW\ dW = dt$). So we have

$$d\ln(F_2(t)) = \sigma_2(t)dW(t) - \tfrac{1}{2}\sigma_2^2(t)dt.$$

Integrate both sides:

$$\begin{aligned} \int_0^T d\ln(F_2(t)) &= \int_0^T \sigma_2(t)dW(t) - \tfrac{1}{2}\int_0^T \sigma_2^2(t)dt \\ \ln(F_2(T)) - \ln(F_2(0)) &= \int_0^T \sigma_2(t)dW(t) - \tfrac{1}{2}\int_0^T \sigma_2^2(t)dt \\ \ln\frac{F_2(T)}{F_2(0)} &= \int_0^T \sigma_2(t)dW(t) - \tfrac{1}{2}\int_0^T \sigma_2^2(t)dt \\ \frac{F_2(T)}{F_2(0)} &= \exp\left(\int_0^T \sigma_2(t)dW(t) - \tfrac{1}{2}\int_0^T \sigma_2^2(t)dt\right) \\ F_2(T) = F_2(0)\exp&\left(\int_0^T \sigma_2(t)dW(t) - \tfrac{1}{2}\int_0^T \sigma_2^2(t)dt\right). \end{aligned}$$

Compute the distribution of the random variable in the exponent.

It is Gaussian, since it is a stochastic integral of a deterministic function times a Brownian motion (roughly, sum of independent Gaussians is Gaussian).

Compute the expectation:

$$E\left[\int_0^T \sigma_2(t)dW(t) - \tfrac{1}{2}\int_0^T \sigma_2^2(t)dt\right] = 0 - \tfrac{1}{2}\int_0^T \sigma_2^2(t)dt$$

and the variance

$$\mathrm{Var}\left[\int_0^T \sigma_2(t)dW(t) - \tfrac{1}{2}\int_0^T \sigma_2^2(t)dt\right] = \mathrm{Var}\left[\int_0^T \sigma_2(t)dW(t)\right]$$
$$= E\left[\left(\int_0^T \sigma_2(t)dW(t)\right)^2\right] - 0^2 = \int_0^T \sigma_2(t)^2 dt$$

where we have used Ito's isometry in the last step. We thus have

$$I(T) := \int_0^T \sigma_2(t)dW(t) - \tfrac{1}{2}\int_0^T \sigma_2^2(t)dt \sim$$
$$\sim m + V\mathcal{N}(0,1), \quad m = -\tfrac{1}{2}\int_0^T \sigma_2(t)^2 dt,$$
$$V^2 = \int_0^T \sigma_2(t)^2 dt.$$

Recall that we have

$$F_2(T) = F_2(0)\exp(I(T)) = F_2(0)e^{m+V\mathcal{N}(0,1)}.$$

Compute now the option price term

$$E^{Q^2}[(F_2(T_1) - X)^+] = E^{Q^2}[(F_2(0)e^{m+V\mathcal{N}(0,1)} - X)^+]$$
$$= \int_{-\infty}^{+\infty} (F_2(0)e^{m+Vy} - X)^+ p_{\mathcal{N}(0,1)}(y)dy = \ldots$$

Note that $F_2(0)\exp(m+Vy) - X > 0$ if and only if

$$y > \frac{-\ln\left(\frac{F_2(0)}{X}\right) - m}{V} =: \bar{y}$$

so that

$$\ldots = \int_{\bar{y}}^{+\infty} (F_2(0)\exp(m+Vy) - X)p_{\mathcal{N}(0,1)}(y)dy =$$
$$= F_2(0)\int_{\bar{y}}^{+\infty} e^{m+Vy}p_{\mathcal{N}(0,1)}(y)dy - X\int_{\bar{y}}^{+\infty} p_{\mathcal{N}(0,1)}(y)dy$$

$$= F_2(0)\frac{1}{\sqrt{2\pi}}\int_{\bar{y}}^{+\infty} e^{-\frac{1}{2}y^2+Vy+m}dy - X(1-\Phi(\bar{y}))$$
$$= F_2(0)\frac{1}{\sqrt{2\pi}}\int_{\bar{y}}^{+\infty} e^{-\frac{1}{2}(y-V)^2+m-\frac{1}{2}V^2}dy - X(1-\Phi(\bar{y})) =$$

$$= F_2(0)e^{m-\frac{1}{2}V^2}\frac{1}{\sqrt{2\pi}}\int_{\bar{y}}^{+\infty} e^{-\frac{1}{2}(y-V)^2}dy - X(1-\Phi(\bar{y})) =$$

$$= F_2(0)e^{m-\frac{1}{2}V^2}\frac{1}{\sqrt{2\pi}}\int_{\bar{y}-V}^{+\infty} e^{-\frac{1}{2}z^2}dz - X(1-\Phi(\bar{y})) =$$

$$= F_2(0)e^{m-\frac{1}{2}V^2}(1-\Phi(\bar{y}-V)) - X(1-\Phi(\bar{y})) =$$

$$= F_2(0)e^{m-\frac{1}{2}V^2}\Phi(-\bar{y}+V) - X\Phi(-\bar{y}) =$$

$$= F_2(0)\Phi(d_1) - X\Phi(d_2), \quad d_{1,2} = \frac{\ln\frac{F_2(0)}{X} \pm \frac{1}{2}\int_0^{T_1}\sigma_2^2(t)dt}{\sqrt{\int_0^{T_1}\sigma_2^2(t)dt}}.$$

When including the initial discount factor $P(0, T_2)$ and the year fraction τ this is exactly the classic market Black's formula for the $T_1 - T_2$ caplet.

End(detailed derivation of Black's formula for caplets)

The example just introduced is a simple case of what is known as "lognormal forward-LIBOR model". It is known also as Brace-Gatarek-Musiela (1997) model, from the name of the authors of one of the first papers where it was introduced rigorously. This model was also introduced by Miltersen, Sandmann and Sondermann (1997). Jamshidian (1997) also contributed significantly to its development. At times in the literature and in conversations, especially in Europe, the LFM is referred to as "BGM" model, from the initials of the three above authors. In other cases, colleagues in the U.S. called it simply an "HJM model", related perhaps to the fact that the BGM derivation was based on the HJM framework rather than on the change-of-numeraire technique. However, a common terminology is now emerging and the model is generally known as "LIBOR Market Model". We will stick to the "Lognormal Forward-LIBOR Model", since this is more informative on the properties of the model: Modeling forward LIBOR rates through a lognormal distribution (under the relevant measures).

Let us now go back to our short-rate model formula U_C and ask ourselves whether this formula can be compatible with the above reported Black's market formula. It is well known that the two formulas are not compatible. Indeed, by the two-dimensional version of Ito's formula we may derive the Q^2-dynamics of the forward LIBOR rate between T_1 and T_2 under the short-rate model,

$$dF(t; T_1, T_2; x_t, y_t, \theta) = \frac{\partial F}{\partial(t, x, y)}\, d[t\ x_t\ y_t]' + \frac{1}{2}\, d[x_t\ y_t]\, \frac{\partial^2 F}{\partial^2(x, y)}\, d[x_t\ y_t]',\ Q^2, \tag{6.3}$$

where the Jacobian vector and the Hessian matrix have been denoted by their partial derivative notation. This dynamics clearly depends on the linear-Gaussian dynamics of x and y under the T_2-forward measure. The thus obtained dynamics is easily seen to be incompatible with the lognormal dynamics leading to Black's formula. More specifically, for no choice of the parameters θ does the distribution of the forward rate F in (6.3) produced by the short-rate model coincide with the distribution of the "Black"-like forward rate F following (6.2). In general, no known short-rate model can lead to Black's formula for caplets (and more generally for caps).

What is then done with short-rate models is the following. After setting the deterministic shift φ so as to obtain a perfect fit of the initial term structure, one looks for the parameters θ that produce caplet (actually cap) prices U_C that are closest to a number of observed market cap prices. The model is thus calibrated to (part of) the cap market and should reproduce well the observed prices. Still, the prices U_C are complicated nonlinear functions of the parameters θ, and this renders the parameters themselves difficult to interpret. On the contrary, the parameter v in the above "market model" for F has a clear interpretation as a lognormal percentage (instantaneous) volatility, and traders feel confident in handling such kind of parameters.

When dealing with several caplets involving different forward rates, different structures of instantaneous volatilities can be employed. One can select a different v for each forward rate by assuming each forward rate to have a constant instantaneous volatility. Alternatively, one can select piecewise-constant instantaneous volatilities for each forward rate. Moreover, different forward rates can be modeled as each having different random sources W that are instantaneously correlated. Modeling correlation is necessary for pricing payoffs depending on more than a single rate at a given time, such as swaptions. Possible volatility and correlation structures are discussed in Sections 6.3.1 and 6.9. The implications of such structures as far as caplets and caps are concerned are discussed in Section 6.4, and their consequences on the term structure of volatilities as a whole are discussed in Section 6.5.

As hinted at above, the model we briefly introduced is the market model for "half" of the interest-rate-derivatives world, i.e. the cap market. But what happens when dealing with basic products from the other "half" of this world, such as swaptions? Swaptions are options on interest-rate swaps. Interest rate swaps and swaptions have been defined in Chapter 1 and will be again described in Section 6.7. Swaptions are priced by a Black-like market formula that is, in many respects, similar to the cap formula. This market formula can be given full rigor as in the case of the caps formula. However, doing so involves choosing a numeraire under which the relevant forward *swap* rate (rather than a particular forward LIBOR rate) is driftless and lognormal. This numeraire is indeed different from any of the bond-price numeraires used in the derivation of Black's formula for caps according to the LFM. The obtained model is known as "lognormal (forward) swap model" (LSM) and is also referred to as the Jamshidian (1997) market model or "swap market model".

One may wonder whether the two models are distributionally compatible or not, similarly to our previous comparison of the LFM with the shifted two-factor Vasicek model G2++. As before, the two models are incompatible. If we adopt the LFM for caps we cannot recover the market formula given by the LSM for swaptions. The two models (LFM and LSM) collide. We will point out this incompatibility in Section 6.8.

There are some works investigating the "size" of the discrepancy between these two models, and we will address this issue in Chapter 8 by comparing swap-rates distributions under the two models. Results seem to suggest that the difference is not large in most cases. However, the problem remains of choosing either of the two models for the whole market.

When the choice is made, the half market consistent with the model is calibrated almost automatically, see for example the cap calibration with the LFM in Section 6.4. But one has still the problem of calibrating the chosen model to the remaining half, e.g. the swaption market in case the LFM is adopted for both markets.

Indeed, Brace, Dun and Barton (1998) suggest to adopt the LFM as central model for the two markets, mainly for its mathematical tractability. We will stick to their suggestion, also because of the fact that forward rates are somehow more natural and more representative coordinates of the yield-curve than swap rates. Indeed, it is more natural to express forward swap rates in terms of a suitable preselected family of LIBOR forward rates, rather than doing the converse.

We are now left with the problem of finding a way to compute swaption prices with the LFM. In order to understand the difficulties of this task, let us consider a very simple swaption. Assume we are at time 0. The underlying interest-rate swap (IRS) starts at T_1 and pays at T_2 and T_3. All times are equally spaced by a year fraction denoted by τ, and we take a unit notional amount. The (payer) swaption payoff can be written as

$$[P(T_1,T_2)\tau(F_2(T_1)-K)+P(T_1,T_3)\tau(F_3(T_2)-K)]^+ ,$$

where in general we set $F_k(t) = F(t;T_{k-1},T_k)$. Recall that the discount factors $P(T_1,T_2)$ and $P(T_1,T_3)$ can be expressed in terms of $F_2(T_1)$ and $F_3(T_2)$, so that this payoff actually depends only on the rates $F_2(T_1)$ and $F_3(T_2)$. The key point, however, is the following. The payoff is not additively "separable" with respect to the different rates. As a consequence, when you take expectation of such a payoff, the *joint* distribution of the two rates F_2 and F_3 is involved in the calculation, so that the correlation between the two rates F_2 and F_3 has an impact on the value of the contract. This does not happen with caps. Indeed, let us go back to caps for a moment and consider a cap consisting of the T_2- and T_3-caplets. The cap payoff as seen from time 0 would be

$$\exp\left(-\int_0^{T_2} r_u du\right)\tau(F_2(T_1)-K)^+ + \exp\left(-\int_0^{T_3} r_u du\right)\tau(F_3(T_2)-K)^+ .$$

This time, the payoff is additively separated with respect to different rates. Indeed, we can compute

$$
\begin{aligned}
& E\left[\exp\left(-\int_0^{T_2} r_u du\right)\tau(F_2(T_1)-K)^+\right. \\
& \qquad \left. +\exp\left(-\int_0^{T_3} r_u du\right)\tau(F_3(T_2)-K)^+\right] \\
& = E\left[\exp\left(-\int_0^{T_2} r_u du\right)\tau(F_2(T_1)-K)^+\right] \\
& \qquad +E\left[\exp\left(-\int_0^{T_3} r_u du\right)\tau(F_3(T_2)-K)^+\right] \\
& = P(0,T_2)\tau E^2\left[(F_2(T_1)-K)^+\right] + P(0,T_3)\tau E^3\left[(F_3(T_2)-K)^+\right].
\end{aligned}
$$

In this last expression we have two expectations, each one involving a *single* rate. The joint distribution of the two rates F_2 and F_3 is not involved, therefore, in the calculation of this last expression, since it is enough to know the marginal distributions of F_2 and F_3 separately. Accordingly, the terminal correlation between rates F_2 and F_3 does not affect this payoff.

As a consequence of this simple example, it is clear that adequately modeling correlation can be important in defining a model that can be effectively calibrated to swaption prices. When the number of swaptions to which the model has to be calibrated is large, correlation becomes definitely relevant. If a short-rate model has to be chosen, it is better to choose a multi-factor model. Multi-dimensional instantaneous sources of randomness guarantee correlation patterns among terminal rates F that are clearly more general than in the one-factor case. However, producing a realistic correlation pattern with, for instance, a two-factor short-rate model is not always possible.

As far as the LFM is concerned, as we hinted at above, the solution is usually to assign a different Brownian motion to each forward rate and to assume such Brownian motions to be instantaneously correlated. Manipulating instantaneous correlation leads to manipulation of correlation of simple rates (terminal correlation), although terminal correlation is also influenced by the way in which average volatility is distributed among instantaneous volatilities, see Section 6.6 on this point.

Choosing an instantaneous-correlation structure flexible enough to express a large number of swaption prices and, at the same time, parsimonious enough to be tractable is a delicate task. The integrated covariance matrix need not have full rank. Usually ranks two or three are sufficient, see Brace Dun and Barton (1998) on this point. We will suggest some parametric forms for the covariance structure of instantaneous forward rates in Section 6.9.

Brace (1996) proposed an approximated formula to evaluate analytically swaptions in the LFM. The formula is based on a rank-one approximation of the integrated covariance matrix plus a drift approximation and works well in specific contexts and in non-pathological situations. More generally, a rank-r approximation is considered in Brace (1997). See again Brace, Dun and Barton (1998) for numerical experiments and results.

The rank-one approximated formula is reviewed in Section 6.13, while the rank-r approximation is reviewed in Section 6.14. There are also analytical swaption-pricing formulas that are simpler and still accurate enough for most practical purposes. Such formulas are based on expressing forward swap rates as linear combinations of forward LIBOR rates, to then take variance on both sides and integrate while freezing some coefficients. This "freezing the drift and collapsing all measures" approximation is reviewed in Section 6.15 and has also been tested by Brace, Dun and Barton (1998). Given its importance, we performed numerical tests of our own. These are reported in Section 8.3 of Chapter 8, and confirm that the formula works well.

More generally, to evaluate swaptions and other payoffs with the LFM one has usually to resort to Monte Carlo simulation. Once the numeraire is chosen, one simulates all forward rates involved in the payoff by discretizing their joint dynamics with a numerical scheme for stochastic differential equations (SDEs).

Notice that each forward rate is driftless only under its associated measure. Indeed, for example, while F_2 is driftless under Q^2 (recall (6.2)), it is not driftless under Q^1. The dynamics of F_2 under Q^1 is derived below and leads to a process that we need to discretize in order to obtain simulations, whereas in the driftless case (6.2) the transition distribution of F is known to be lognormal and no numerical scheme is needed. This point is discussed in detail again in Section 6.10. As an introductory illustration, consider again the above swaption payoff, now discounted at time 0, and take its risk-neutral expectation:

$$E\left\{\exp\left(-\int_0^{T_1} r_u du\right)[P(T_1,T_2)\tau(F_2(T_1)-K)\right.$$
$$\left.+P(T_1,T_3)\tau(F_3(T_2)-K)]^+\right\} = \cdots$$

Take $P(\cdot,T_1)$ as numeraire, which corresponds to choose the T_1-forward-adjusted measure Q^1. One then obtains:

$$\cdots = P(0,T_1)E^1\left[P(T_1,T_2)\tau(F_2(T_1)-K)+P(T_1,T_3)\tau(F_3(T_2)-K)\right]^+ .$$

It is known that F_1 is driftless under the chosen measure Q^1, while F_2 and F_3 are not. For example, take for simplicity the one-factor case, so that

$$dF_2(t) = v_2F_2(t)dW_t, \tag{6.4}$$

where W is now a standard Brownian motion under Q^2. Then if Z is a standard Brownian motion under Q^1, it can be shown that the change-of-numeraire technique leads to

$$dF_2(t) = \frac{v_2^2 F_2^2(t)\tau}{1 + \tau F_2(t)} dt + v_2 F_2(t) dZ_t. \tag{6.5}$$

Now it is clear that, as we stated above, the no-arbitrage dynamics (6.5) of F_2 under Q^1 is not driftless, and moreover its transition distribution (which is necessary to perform exact simulations) is not known, contrary to the Q^2 driftless case (6.4).

In order to price the swaption, it is the Q^1 dynamics of F_2 that matters, so that we need to discretize the related equation (6.5) in order to be able to simulate it. Numerical schemes for SDEs are available, like the Euler or Milstein scheme, see also Appendix C. Alternative schemes that guarantee the (weak) no-arbitrage condition to be maintained in discrete time and not just in the continuous-time limit have been proposed by Glasserman and Zhao (2000).

Whichever scheme is chosen, Monte Carlo pricing is to be performed by simulating the relevant forward LIBOR rates. While path-dependent derivatives can be priced by this approach in general, as far as early-exercise (e.g. American-style or Bermudan-style) products are concerned, the situation is delicate, since the joint dynamics of the LFM usually does not lead to a recombining lattice for the short rate, so that it is not immediately clear how to evaluate a Bermudan- or American-style product with a tree in the LFM.

Usually, ad-hoc techniques are needed, such as Carr and Yang's (1997) who provide a method for simulating Bermudan-style derivatives with the LFM via a Markov chain approximation, or Andersen's (1999) who approximates the early exercise boundary as a function of intrinsic value and "still-alive" nested European swaptions. However, a general method for combining backward induction with Monte Carlo simulation has been proposed by Longstaff and Schwartz (2001), and this method is rather promising, especially because of its generality. We briefly review all these methods in Chapter 13.

Several other problems remain, like for example the possibility of including a volatility smile in the model. This will be addressed in Part IV of the book, which represent a large extension of the material presented in the first edition.

6.3 The Lognormal Forward-LIBOR Model (LFM)

Let $t = 0$ be the current time. Consider a set $\mathcal{E} = \{T_0, \ldots, T_M\}$ from which expiry-maturity pairs of dates (T_{i-1}, T_i) for a family of spanning forward rates are taken. We shall denote by $\{\tau_0, \ldots, \tau_M\}$ the corresponding year fractions, meaning that τ_i is the year fraction associated with the expiry-maturity pair (T_{i-1}, T_i) for $i > 0$, and τ_0 is the year fraction from settlement to T_0. Times T_i will be usually expressed in years from the current time. We set $T_{-1} := 0$.

Consider the generic forward rate $F_k(t) = F(t; T_{k-1}, T_k)$, $k = 1, \ldots, M$, which is "alive" up to time T_{k-1}, where it coincides with the simply-compounded spot rate $F_k(T_{k-1}) = L(T_{k-1}, T_k)$. In general $L(S,T)$ is the simply compounded spot rate prevailing at time S for the maturity T.

Consider now the probability measure Q^k associated with the numeraire $P(\cdot, T_k)$, i.e. to the price of the bond whose maturity coincides with the maturity of the forward rate. Q^k is often called the *forward (adjusted) measure for the maturity* T_k. Under simple compounding, it follows immediately by definition that

$$F_k(t)P(t,T_k) = [P(t,T_{k-1}) - P(t,T_k)]/\tau_k.$$

Therefore, $F_k(t)P(t,T_k)$ is the price of a tradable asset (difference between two discount bonds with notional amounts $1/\tau_k$). As such, when its price is expressed with respect to the numeraire $P(\cdot, T_k)$, it has to be a martingale under the measure Q^k associated with that numeraire (by definition of measure associated with a numeraire). But the price $F_k(t)P(t,T_k)$ of our tradable asset divided by this numeraire is simply $F_k(t)$ itself. Therefore, F_k follows a martingale under Q^k. It follows that if F_k is modeled according to a diffusion process, it needs to be driftless under Q^k.

We assume the following driftless dynamics for F_k under Q^k:

$$dF_k(t) = \underline{\sigma}_k(t)F_k(t)dZ^k(t), \quad t \le T_{k-1}, \tag{6.6}$$

where $Z^k(t)$ is an M-dimensional column-vector Brownian motion (under Q^k) with instantaneous covariance $\rho = (\rho_{i,j})_{i,j=1,\dots,M}$,

$$dZ^k(t)\, dZ^k(t)' = \rho\, dt\,,$$

and where $\underline{\sigma}_k(t)$ is the horizontal M-vector volatility coefficient for the forward rate $F_k(t)$.

Unless differently stated, from now on we will assume that

$$\underline{\sigma}_j(t) = [0\ 0\ \dots\ \sigma_j(t)\ \dots\ 0\ 0]\,,$$

with the only non-zero entry $\sigma_j(t)$ occurring at the j-th position in the vector $\underline{\sigma}_j(t)$.

We hope the reader is not disoriented by our notation. The point of switching from vector to scalar volatility is that in some computations the former will be more convenient, whereas in other cases the latter will be useful. Just notice that the above equation for F_k can be rewritten under Q^k as

$$dF_k(t) = \sigma_k(t)F_k(t)dZ_k^k(t), \quad t \le T_{k-1},$$

where $Z_k^k(t)$ is the k-th component of the vector Brownian motion Z^k and is thus a standard Brownian motion. Lower indices indicate the component we are considering in a vector, whereas upper indices show under which measure we are working. Upper indices are usually omitted when the context is clear, so that we write

$$dF_k(t) = \sigma_k(t)F_k(t)dZ_k(t), \quad t \leq T_{k-1}. \tag{6.7}$$

With this scalar notation, $\sigma_k(t)$ now bears the usual interpretation of instantaneous volatility at time t for the forward LIBOR rate F_k.

Notice that in case σ is bounded (as in all of our volatility parameterizations below), this last equation has a unique strong solution, since it describes a geometric Brownian motion. This can be checked immediately also by writing, through Ito's formula,

$$d\ln F_k(t) = -\frac{\sigma_k(t)^2}{2}dt + \sigma_k(t)dZ_k(t), \quad t \leq T_{k-1},$$

and by observing that the coefficients of this last equation are bounded, so that there exists trivially a unique strong solution. Indeed, we have immediately the unique solution

$$\ln F_k(T) = \ln F_k(0) - \int_0^T \frac{\sigma_k(t)^2}{2}dt + \int_0^T \sigma_k(t)dZ_k(t).$$

We will often consider piecewise-constant instantaneous volatilities:

$$\sigma_k(t) = \sigma_{k,\beta(t)}\,, \; t > T_{-1} = 0,$$

where in general $\beta(t) = m$ if $T_{m-2} < t \leq T_{m-1}$, $m \geq 1$, so that

$$t \in (T_{\beta(t)-2}, T_{\beta(t)-1}].$$

In the particular case $t = T_{-1} = 0$ we set $\sigma_k(0) = \sigma_{k,1}$.

Notice that $\beta(t)$ is the index of the first forward rate that has not expired by t: $F_{\beta(t)}$. All preceding forward rates $F_{\beta(t)-1}, F_{\beta(t)-2}..$ have expired by t.

At times we will use the notation $Z_t = Z(t)$.

As is clear from our assumption on the vector "noise" dZ above, the single "noises" in the dynamics of different forward rates are assumed to be instantaneously correlated according to

$$dZ_i(t)\,dZ_j(t) = d\langle Z_i, Z_j\rangle_t = \rho_{i,j}dt.$$

Remark 6.3.1. **(Achieving decorrelation)**. Historical one-factor short-rate models, such as for instance Hull and White's (1990b) or Black and Karasinski's (1991), imply forward-rate dynamics that are perfectly instantaneously correlated. This means that for such models we would have $\rho_{i,j} = 1$ for all i, j. As a consequence, we can say that forward rates in these models are usually too correlated.

When trying to improve these models, one of the objectives is to lower the correlation of the forward rates implied by the model. Some authors refer to this objective as to achieving *decorrelation*. As we shall see later on in Section 6.6, decorrelation can be achieved not only by "lowering" instantaneous correlations, but also by carefully redistributing integrated variances of forward rates over time.

6.3.1 Some Specifications of the Instantaneous Volatility of Forward Rates

A few general remarks are now in order. First, as we said before, we will often assume that the forward rate $F_k(t)$ has a piecewise-constant instantaneous volatility. In particular, the instantaneous volatility of $F_k(t)$ is constant in each "expiry-maturity" time interval (associated with any other forward rate) $T_{m-2} < t \leq T_{m-1}$ where it is "alive".

Under this assumption, it is possible to organize instantaneous volatilities in the following matrix, where "Instant. Vols" and "Fwd" are abbreviations for Instantaneous Volatility and Forward respectively:

TABLE 1

Instant. Vols	Time: $t \in (0, T_0]$	$(T_0, T_1]$	$(T_1, T_2]$	...	$(T_{M-2}, T_{M-1}]$
Fwd Rate: $F_1(t)$	$\sigma_{1,1}$	Dead	Dead	...	Dead
$F_2(t)$	$\sigma_{2,1}$	$\sigma_{2,2}$	Dead	...	Dead
$\vdots$	...	...	...	...	...
$F_M(t)$	$\sigma_{M,1}$	$\sigma_{M,2}$	$\sigma_{M,3}$	...	$\sigma_{M,M}$

A first assumption can be made on the entries of TABLE 1 so as to reduce the number of volatility parameters. We can assume that volatilities depend only on the time-to-maturity $T_k - T_{\beta(t)-1}$ of a forward rate rather than on time t and maturity T_k separately. In such a case, by assuming

$$\sigma_k(t) = \sigma_{k,\beta(t)} =: \eta_{k-(\beta(t)-1)}, \tag{6.8}$$

the above matrix looks like:

TABLE 2

Instant. Vols	Time: $t \in (0, T_0]$	$(T_0, T_1]$	$(T_1, T_2]$	...	$(T_{M-2}, T_{M-1}]$
Fwd Rate: $F_1(t)$	η_1	Dead	Dead	...	Dead
$F_2(t)$	η_2	η_1	Dead	...	Dead
$\vdots$	...	...	...	...	...
$F_M(t)$	η_M	η_{M-1}	η_{M-2}	...	η_1

A second alternative assumption can be made on the entries of TABLE 1 so as to reduce the number of parameters. We can assume

$$\sigma_k(t) = \sigma_{k,\beta(t)} := s_k \tag{6.9}$$

for all t, which amounts to assuming that a forward rate $F_k(t)$ has constant instantaneous volatility s_k regardless of t.

This leads to the following table:

TABLE 3

Instant. Vols	Time: $t \in (0, T_0]$	$(T_0, T_1]$	$(T_1, T_2]$	...	$(T_{M-2}, T_{M-1}]$
Fwd Rate: $F_1(t)$	s_1	Dead	Dead	...	Dead
$F_2(t)$	s_2	s_2	Dead	...	Dead
$\vdots$	...	...	...	...	...
$F_M(t)$	s_M	s_M	s_M	...	s_M

A third alternative assumption can be made on the entries of TABLE 1 so as to reduce the number of parameters with respect to the general case. We can assume

$$\sigma_k(t) = \sigma_{k,\beta(t)} := \Phi_k \Psi_{\beta(t)} \tag{6.10}$$

for all t. This leads to the following table:

TABLE 4

Instant. Vols	Time: $t \in (0, T_0]$	$(T_0, T_1]$	$(T_1, T_2]$	...	$(T_{M-2}, T_{M-1}]$
Fwd Rate: $F_1(t)$	$\Phi_1 \Psi_1$	Dead	Dead	...	Dead
$F_2(t)$	$\Phi_2 \Psi_1$	$\Phi_2 \Psi_2$	Dead	...	Dead
$\vdots$	...	...	...	...	...
$F_M(t)$	$\Phi_M \Psi_1$	$\Phi_M \Psi_2$	$\Phi_M \Psi_3$	...	$\Phi_M \Psi_M$

This last formulation includes formula (6.9) as a special case, for example by taking all Ψ's equal to one and the Φ's equal to the s's. It is the most general formulation leading to a rank-one integrated covariance matrix in case one takes all instantaneous correlations ρ equal to one (one-factor model), see for example Brace, Dun and Barton (1998) for a proof.

A fourth (final) alternative assumption can be made on the entries of TABLE 1 so as to reduce the number of parameters. We can assume

$$\sigma_k(t) = \sigma_{k,\beta(t)} := \Phi_k \psi_{k-(\beta(t)-1)} \tag{6.11}$$

for all t. This leads to the following table:

TABLE 5

Instant. Vols	Time: $t \in (0, T_0]$	$(T_0, T_1]$	$(T_1, T_2]$	...	$(T_{M-2}, T_{M-1}]$
Fwd Rate: $F_1(t)$	$\Phi_1 \psi_1$	Dead	Dead	...	Dead
$F_2(t)$	$\Phi_2 \psi_2$	$\Phi_2 \psi_1$	Dead	...	Dead
$\vdots$	...	...	...	...	...
$F_M(t)$	$\Phi_M \psi_M$	$\Phi_M \psi_{M-1}$	$\Phi_M \psi_{M-2}$	...	$\Phi_M \psi_1$

This formulation includes formula (6.8) as a special case, for example by taking all Φ's equal to one and the ψ's equal to the η's. It is the product of a structure dependent only on the time to maturity (the ψ's) by a structure dependent only on the maturity (the Φ's). It does not lead to a rank-one integrated covariance matrix in general, not even in the case of perfect instantaneous correlations ρ all equal to one. We will clarify the peculiarities of this last formulation later on. At the moment we just anticipate that this form has the potential for maintaining the qualitative shape of the term structure of volatilities in time through its ψ part, as we will explain in Section 6.5, while being rich enough to allow a satisfactory calibration to market data.

The five tables conclude the piecewise-constant models for instantaneous volatilities we planned to present. Now we introduce instead some parametric forms for the same quantities. A first possibility is the following:

FORMULATION 6

$$\boxed{\sigma_i(t) = \psi(T_{i-1} - t; a, b, c, d) := [a(T_{i-1} - t) + d]e^{-b(T_{i-1}-t)} + c} \quad (6.12)$$

This form allows a humped shape in the graph of the instantaneous volatility of the generic forward rate F_i as a function of the time to maturity,

$$T_{i-1} - t \mapsto \sigma_i(t)\,.$$

We will see later on that this parametric form possesses some desirable and undesirable qualitative features, and that its flexibility is not sufficient for practical purposes when jointly calibrating to the caps and swaptions markets. Dependence on the time to maturity, rather than time and maturity separately, renders this form a parametric analogue of the piecewise-constant case (6.8) of TABLE 2, the function ψ being an analogue of the η's.

The above formulation can be perfected into a richer parametric form. Indeed consider

FORMULATION 7

$$\boxed{\sigma_i(t) = \Phi_i\, \psi(T_{i-1} - t; a, b, c, d) := \Phi_i\, \left([a(T_{i-1} - t) + d]e^{-b(T_{i-1}-t)} + c\right)} \quad (6.13)$$

This parametric form reduces to the previous one when all the Φ's are set to one. This form can be seen to have a parametric core ψ that is locally altered for each maturity T_i by the Φ's. These local modifications, if small, do not destroy the essential dependence on the time to maturity, so as to maintain the desirable qualitative properties we will describe in Section 6.5; at the same time, such modifications add flexibility to the parametric form, flexibility that can be used in order to improve the joint calibration of the model to the caps and swaptions markets, as we will observe later on. When seen as a local modification of a structure depending only on the time to maturity, this parametric form is the analogue of the piecewise-constant case (6.11) of TABLE 5.

We will discuss the benefits of this parametric assumption in Section 6.5, when describing the associated term structure of volatilities.

We will come back to matters concerning the assumptions on instantaneous volatilities later on.

6.3.2 Forward-Rate Dynamics under Different Numeraires

"He [J'onn] has a fine tactical mind. He has been with the justice league since the beginning and understands group dynamics better than anyone I've ever met." Batman to Superman, "JLA: New World Order".

For reasons that will be clear afterwards, we are interested in finding the dynamics of $F_k(t)$ under a measure Q^i different from Q^k, for $t \leq \min(T_i, T_{k-1})$ (both the chosen numeraire and the forward rate being modeled have to be alive in t). This can be done by writing down the general rule under which asset price dynamics changes when one changes the numeraire (Chapter 2), and then by applying this tool to the specific case of forward-rates dynamics.

Proposition 6.3.1. (Forward-measure dynamics in the LFM). *Under the lognormal assumption, we obtain that the dynamics of F_k under the forward-adjusted measure Q^i in the three cases $i < k$, $i = k$ and $i > k$ are, respectively,*

$$
\begin{aligned}
i < k,\ t \leq T_i: \quad dF_k(t) &= \sigma_k(t) F_k(t) \sum_{j=i+1}^{k} \frac{\rho_{k,j}\, \tau_j\, \sigma_j(t)\, F_j(t)}{1 + \tau_j F_j(t)}\, dt \\
&\quad + \sigma_k(t) F_k(t)\, dZ_k(t), \qquad (6.14)
\end{aligned}
$$

$$
i = k,\ t \leq T_{k-1}: \quad dF_k(t) = \sigma_k(t) F_k(t)\, dZ_k(t),
$$

$$
\begin{aligned}
i > k,\ t \leq T_{k-1}: \quad dF_k(t) &= -\sigma_k(t) F_k(t) \sum_{j=k+1}^{i} \frac{\rho_{k,j}\, \tau_j\, \sigma_j(t)\, F_j(t)}{1 + \tau_j F_j(t)}\, dt \\
&\quad + \sigma_k(t) F_k(t)\, dZ_k(t),
\end{aligned}
$$

where, as explained before, $Z = Z^i$ is a Brownian motion under Q^i. All of the above equations admit a unique strong solution if the coefficients $\sigma(\cdot)$ are bounded.

Proof. Consider the forward rate $F_k(t) = F(t, T_{k-1}, T_k)$ and suppose we wish to derive its dynamics first under the T_i-forward measure Q^i with $i < k$. We know that the dynamics under the T_k-forward measure Q^k has null drift.

From this dynamics, we propose to recover the dynamics under Q^i. Let us apply (2.15) to $X = F_k = F(\cdot, T_{k-1}, T_k)$, where we set $S = P(\cdot, T_k)$ and $U = P(\cdot, T_i)$, and the dynamics of X under Q^k is given by (6.6).

We obtain the percentage drift as

$$m_t^i = -\frac{d\langle \ln X, \ln(P(\cdot, T_k)/P(\cdot, T_i))\rangle_t}{dt}.$$

Now notice that

$$\ln(P(t,T_k)/P(t,T_i)) = \ln\left(1/\left[\prod_{j=i+1}^{k}(1+\tau_j F_j(t))\right]\right) = -\sum_{j=i+1}^{k}\ln(1+\tau_j F_j(t))$$

from which

$$m_t^i = -\frac{d\langle \ln F_k, \ln(P(\cdot,T_k)/P(\cdot,T_i))\rangle_t}{dt} = \sum_{j=i+1}^{k}\frac{d\langle \ln F_k, \ln(1+\tau_j F_j(t))\rangle_t}{dt}$$
$$= \sum_{j=i+1}^{k}\frac{\tau_j}{1+\tau_j F_j(t)}\frac{d\langle \ln F_k, F_j\rangle_t}{dt} = \sum_{j=i+1}^{k}\frac{\rho_{j,k}\tau_j\sigma_k(t)\sigma_j(t)F_j(t)}{1+\tau_j F_j(t)}.$$

Secondly, consider the case $i > k$, where the numeraire is a bond whose maturity is longer than the maturity of the forward rate being modeled. In such a case, the derivation is similar, but now

$$\ln(P(t,T_k)/P(t,T_i)) = \ln\left(\left[\prod_{j=k+1}^{i}(1+\tau_j F_j(t))\right]\right) = \sum_{j=k+1}^{i}\ln(1+\tau_j F_j(t))$$

from which

$$m_t^i = -\sum_{j=k+1}^{i}\frac{\rho_{j,k}\,\tau_j\,\sigma_k(t)\,\sigma_j(t)\,F_j(t)}{1+\tau_j\,F_j(t)}.$$

As for existence and uniqueness of the solution, we have treated earlier the trivial case $i = k$. In the case $i < k$, compute through Ito's formula

$$d\ln F_k(t) = \sigma_k(t)\sum_{j=i+1}^{k}\frac{\rho_{k,j}\tau_j\sigma_j(t)F_j(t)}{1+\tau_j F_j(t)}dt - \frac{\sigma_k(t)^2}{2}dt + \sigma_k(t)dZ_k(t).$$

Now notice that the diffusion coefficient of this equation is both deterministic and bounded. Moreover, since

$$0 < \frac{\tau_j F_j(t)}{1 + \tau_j F_j(t)} < 1,$$

also the drift is bounded, besides being smooth in the F's (that are positive). This ensures existence and uniqueness of a strong solution for the above SDE. The case $i > k$ is analogous. □

Remark 6.3.2. **(Alternative derivation of the forward measure dynamics.)** The proof can be also carried out in terms of the DC operator introduced in Section 2.3.1 and working with the shocks, as in Fact Three of Section 2.3. Let us apply (2.18) with $S = P(\cdot, T_k)$ and $U = P(\cdot, T_i)$, again for the case $i < k$:

$$dZ_t^k = dZ_t^i - \rho \mathrm{DC}(\ln(P(\cdot, T_k)/P(\cdot, T_i)))' \, dt. \tag{6.15}$$

Compute

$$\ln \frac{P(t, T_k)}{P(t, T_i)} = \ln \left(\frac{P(t, T_k)}{P(t, T_{k-1})} \frac{P(t, T_{k-1})}{P(t, T_{k-2})} \cdots \frac{P(t, T_{i+1})}{P(t, T_i)} \right) =$$

$$= \ln \left(\frac{1}{1 + \tau_k F_k(t)} \cdot \frac{1}{1 + \tau_{k-1} F_{k-1}(t)} \cdots \frac{1}{1 + \tau_{i+1} F_{i+1}(t)} \right) =$$

$$= \ln \left(1 / \left[\prod_{j=i+1}^{k} (1 + \tau_j F_j(t)) \right] \right) = - \sum_{j=i+1}^{k} \ln (1 + \tau_j F_j(t))$$

so that from linearity of DC (see the linear and logarithmic DC properties given in Section 2.3.1)

$$\mathrm{DC} \left(\ln \frac{P(t, T_k)}{P(t, T_i)} \right) = - \sum_{j=i+1}^{k} \mathrm{DC} \ln (1 + \tau_j F_j(t))$$

$$= - \sum_{j=i+1}^{k} \frac{\mathrm{DC}(1 + \tau_j F_j(t))}{1 + \tau_j F_j(t)} = - \sum_{j=i+1}^{k} \tau_j \frac{\mathrm{DC}(F_j(t))}{1 + \tau_j F_j(t)} =$$

$$= - \sum_{j=i+1}^{k} \tau_j \frac{\sigma_j(t) F_j(t) e_j}{1 + \tau_j F_j(t)}$$

where e_j is a zero row vector except in the j-th position, where we have 1 (vector diffusion coefficient for dF_j is $\sigma_j F_j e_j$). Substituting in (6.15) we may now write

$$dZ_t^k = dZ_t^i + \rho \sum_{j=i+1}^{k} \tau_j \frac{\sigma_j(t) F_j(t) e_j'}{1 + \tau_j F_j(t)} \, dt.$$

Pre-multiply both sides by e_k. We obtain (omitting t)

$$dZ_k^k = dZ_k^i + [\rho_{k,1} \; \rho_{k,2} \ldots \rho_{k,n}] \sum_{j=i+1}^{k} \tau_j \frac{\sigma_j(t) F_j(t) e_j'}{1 + \tau_j F_j(t)} \, dt$$

$$= dZ_k^i + \sum_{j=i+1}^{k} \tau_j \frac{\sigma_j(t) F_j(t) \rho_{k,j}}{1 + \tau_j F_j(t)} dt.$$

Substitute this in our usual basic equation $dF_k = \sigma_k F_k dZ_k^k$ to obtain

$$dF_k = \sigma_k F_k \left(dZ_k^i + \sum_{j=i+1}^{k} \tau_j \frac{\sigma_j(t) F_j(t) \rho_{k,j}}{1 + \tau_j F_j(t)} dt \right)$$

that is finally the equation showing the dynamics of a forward rate with maturity k under the forward measure with maturity i when $i < k$. Again, the case $i > k$ is analogous.

Remark 6.3.3. **(Relationship between Brownian motions under different numeraires)**. In particular, by applying formula (2.13) (the change-of-numeraire toolkit) to the above situation for changing from CdW^k to CdW^{k+1} (i.e. with $S = P(\cdot, T_{k+1})$ and $U = P(\cdot, T_k)$) we obtain, after straightforward calculations similar to the ones in the above proof:

$$dZ^{k+1} = dZ^k + \frac{\tau_{k+1} F_{k+1}(t)}{1 + \tau_{k+1} F_{k+1}(t)} \rho \, \sigma_{k+1}(t) e'_{k+1} \, dt. \tag{6.16}$$

This relationship connects Brownian motions under two adjacent forward measures. It will be useful in the sequel when in need to recover non-standard expiry-maturity forward rates from our original family $F_1, \ldots, F_M$. Also, Brace, Musiela and Schlögl (1998), for example, point out the importance of this relationship for simulations.

The above dynamics constitute the **lognormal forward-LIBOR model (LFM)**. The name is due to the fact that for $i = k$ the distribution of F_k is lognormal. The above dynamics (except in the case $i = k$) do not feature known transition densities, in that knowledge of σ's and ρ's above does not allow one to write an analytical expression for the density of forward rates in a future instant given the forward rates in a past instant. As a consequence, no analytical formula or simple numeric integration can be used in order to price contingent claims depending on the joint dynamics. Moreover, when generating paths for the forward rates, they cannot be generated *one-shot*, but rather have to be obtained from a time-discretization of equation (6.14). In other words, if we need for example one million realizations of

$$\varphi(t) := [F(t; T_0, T_1) \;\; F(t; T_1, T_2) \;\; \ldots \;\; F(t; T_{i-1}, T_i)]$$

for a certain $t \leq T_0$ under Q^0, we cannot generate them directly, because the distribution of $\varphi(t)$ is not known. One then has to discretize equations (6.14) between 0 and t with a sufficiently (but not too) small time step Δt, and generate the distributionally-known Gaussian shocks $Z_{t+\Delta t} - Z_t$.

We are not yet done with numeraire changes, since some products (typically Eurodollar futures) require the *risk-neutral* dynamics of forward rates. Recall that the risk-neutral measure Q is the measure associated with the bank-account numeraire $B(t) = 1/D(0,t)$, where $D(0,t)$ is the stochastic discount factor for maturity t.

The problem with deriving the risk-neutral dynamics with the LFM is that the numeraire $B(t)$ is not natural for forward rates with preassigned tenor and maturities, and thus leads to a "residual" in the change of measure, which is somehow awkward and mixes discrete tenor and continuous tenor quantities. Since the LFM has as a cornerstone feature the idea of modeling discrete tenor quantities, the reappearance of continuous-tenor terms à la HJM in the dynamics is an undesirable ingredient.

There is a rather obvious way to avoid this undesirable mixture of continuous and discrete features, and we will come to that below. For now, let us derive the risk-neutral dynamics of the LFM.

Recall from Chapter 1 that typically, in terms of the instantaneous spot rate r,

$$D(0,T) = \exp\left(-\int_0^T r_t dt\right), \quad B(T) = \exp\left(\int_0^T r_t dt\right),$$

$$P(0,T) = E\left[\exp\left(-\int_0^T r_t dt\right)\right].$$

By remembering that B has dynamics

$$dB(t) = r_t B(t) dt$$

with zero volatility, one derives the risk-neutral dynamics of F_k by applying our change-of-numeraire toolkit to move from the numeraire $P(\cdot, T_k)$ to the numeraire $B(t)$. Indeed, by using formula (2.12) with $U = B$, $S = P(\cdot, T_k)$ and $X = F_k$, one obtains after straightforward calculations:

$$dF_k(t) = -\underline{\sigma}_{P(\cdot,T_k)} \rho \underline{\sigma}_k(t)' F_k(t) dt + \underline{\sigma}_k(t) F_k(t) d\widetilde{Z}(t)$$

where $\underline{\sigma}_{P(\cdot,T_k)}$ is the vector percentage volatility of the bond price (diffusion coefficient of the log-price). Since the bond price can be expressed in terms of spanning forward rates as

$$P(t,T_k) = P(t, T_{\beta(t)-1}) \prod_{j=\beta(t)}^{k} \frac{1}{1+\tau_j F_j(t)}$$

we can compute the above drift as (as usual "DC" is an abbreviation for vector diffusion coefficient)

$$\begin{aligned}\widetilde{\mu}_k(t) &:= -\underline{\sigma}_{P(t,T_k)}\bar{\rho}\underline{\sigma}_k(t)' \\ &= -\mathrm{DC}(\ln P(t,T_k))\rho\underline{\sigma}_k(t)' \\ &= -\mathrm{DC}\left[\ln P(t,T_{\beta(t)-1}) + \sum_{j=\beta(t)}^{k} \ln\left(\frac{1}{1+\tau_j F_j(t)}\right)\right]\rho\underline{\sigma}_k(t)'\,.\end{aligned}$$

Now notice that since in general

$$\ln P(t,T) = -\int_t^T f(t,u)\,du\,,$$

we have easily that

$$\mathrm{DC}[\ln P(t,T)] = -\int_t^T \underline{\sigma}_f(t,u)\,du\,,$$

where $\underline{\sigma}_f(t,u)$ is the absolute instantaneous vector volatility of the instantaneous forward rate $f(t,u)$. We therefore have the following.

Proposition 6.3.2. (Risk-neutral dynamics in the LFM). *The risk-neutral dynamics of forward LIBOR rates in the LFM is:*

$$dF_k(t) = \widetilde{\mu}_k(t)F_k(t)\,dt + \sigma_k(t)F_k(t)\,d\widetilde{Z}_k(t)\,, \tag{6.17}$$

where

$$\begin{aligned}\widetilde{\mu}_k(t) &= \sum_{j=\beta(t)}^{k} \frac{\tau_j\,\rho_{j,k}\,\sigma_j(t)\,\sigma_k(t)\,F_j(t)}{1+\tau_j F_j(t)} + \underline{\sigma}_k(t)\,\rho\int_t^{T_{\beta(t)-1}} \underline{\sigma}_f(t,u)'du \\ &= \sum_{j=\beta(t)}^{k} \frac{\tau_j\,\rho_{j,k}\,\sigma_j(t)\,\sigma_k(t)\,F_j(t)}{1+\tau_j F_j(t)} + \sum_{j=\beta(t)}^{k} \rho_{k,j}\,\sigma_k(t)\int_t^{T_{\beta(t)-1}} (\underline{\sigma}_f)_j(t,u)du.\end{aligned}$$

As we had anticipated, the drift in (6.17) looks awkward, because of the second summation. The second summation originates from the continuous-tenor part coming from the evolution of $P(t,T_{\beta(t)-1})$, which cannot be deduced from forward rates in our family. We are thus forced to model the instantaneous forward rate volatility in order to close the equation, but this is something *external* to the discrete-tenor setting we have adopted.

We can partly improve this situation by considering a discretely rebalanced bank-account numeraire as an alternative to the continuously rebalanced bank account $B(t)$, whose value, at any time t, changes according to $dB(t) = r_t B(t)dt$. We then introduce a bank account that is rebalanced only on the times in our discrete-tenor structure. To this end, consider the numeraire asset

$$B_d(t) = \frac{P(t,T_{\beta(t)-1})}{\prod_{j=0}^{\beta(t)-1} P(T_{j-1},T_j)} = \prod_{j=0}^{\beta(t)-1}(1+\tau_j F_j(T_{j-1}))\,P(t,T_{\beta(t)-1})\,.$$

The interpretation of $B_d(t)$ is that of the value at time t of a portfolio defined as follows. The portfolio starts with one unit of currency at the initial time $t = 0$, exactly as in the continuous-bank-account case ($B(0)=1$), but this unit amount is now invested in a quantity X_0 of T_0 zero-coupon bonds. Such X_0 is readily found by noticing that, since we invested one unit of currency, the present value of the bonds needs to be one, so that $X_0 P(0, T_0) = 1$, and hence $X_0 = 1/P(0, T_0)$. At time T_0, we cash the bonds payoff X_0 and invest it in a quantity $X_1 = X_0/P(T_0, T_1) = 1/(P(0, T_0)P(T_0, T_1))$ of T_1 zero-coupon bonds. We continue this procedure until we reach the last tenor date $T_{\beta(t)-2}$ preceding the current time t, where we invest

$$X_{\beta(t)-1} = 1/ \prod_{j=0}^{\beta(t)-1} P(T_{j-1}, T_j)$$

in $T_{\beta(t)-1}$ zero-coupon bonds. The present value at the current time t of this investment is $X_{\beta(t)-1} P(t, T_{\beta(t)-1})$, i.e. our $B_d(t)$ above. Thus, $B_d(t)$ is obtained by starting from one unit of currency and reinvesting at each tenor date in zero-coupon bonds for the next tenor. This gives a discrete-tenor counterpart of the continuous bank account B, and the subscript "d" in B_d stands for "discrete".

Now choose B_d as numeraire and apply the change-of-numeraire technique starting from the dynamics (6.6) under Q^k, to obtain the dynamics under B_d. If you go through the (change-of-numeraire toolkit) calculations, you will notice that now the "awkward" contributions of the term $P(t, T_{\beta(t)-1})$ cancel out, since this term appears both in $P(t, T_k)$ and in $B_d(t)$. The measure Q^d associated with B_d is called *spot LIBOR measure*. We then have the following.

Proposition 6.3.3. (Spot-LIBOR-measure dynamics in the LFM). *The spot LIBOR measure dynamics of forward LIBOR rates in the LFM is:*

$$dF_k(t) = \sigma_k(t) F_k(t) \sum_{j=\beta(t)}^{k} \frac{\tau_j \rho_{j,k} \sigma_j(t) F_j(t)}{1 + \tau_j F_j(t)} dt + \sigma_k(t) F_k(t) dZ_k^d(t). \quad (6.18)$$

Notice, indeed, that, under this last numeraire, no continuous-tenor term appears in the drift.

A comment on the type of benefits obtained when resorting to the spot LIBOR measure instead of any forward LIBOR measure is in order. Assume we are in need to value a payoff involving rates $F_1, \ldots, F_{10}$ from time 0 to time T_9. Consider two possible measures under which we can do pricing.

First Q^{10}. Under this measure, consider each rate F_j in each interval with the number of terms in the drift summation of each rate dynamics shown between square brackets:

$$\begin{array}{ll} 0 - T_0: & F_1[9], F_2[8], F_3[7], \ldots, F_9[1], F_{10}[0] \\ T_0 - T_1: & \quad\quad\; F_2[8], F_3[7], \ldots, F_9[1], F_{10}[0] \\ T_1 - T_2: & \quad\quad\quad\quad\quad\; F_3[7], \ldots, F_9[1], F_{10}[0] \end{array}$$

etc. Notice that if we discretize the exact continuous time dynamics with a numerical scheme, as we will do for example in Section 6.10, some rates will be more biased than others. Indeed, F_{10} will be much less biased than F_3, since while F_3 has seven drift terms that need to be discretized, F_{10} has none. Thus the bias is indeed distributed differently on different rates.

Instead, with the spot LIBOR measure

$$\begin{array}{rl} 0-T_0: & F_1[1], F_2[2], F_3[3], \ldots, F_9[9], F_{10}[10] \\ T_0-T_1: & F_2[1], F_3[2], \ldots, F_9[8], F_{10}[9] \\ T_1-T_2: & F_3[1], \ldots, F_9[7], F_{10}[8] \end{array}$$

etc. Now the possible bias coming from the discretized drift, if any, is more distributed among different rates.

A final remark concerning the dynamics under forward or spot measures is the following.

Remark 6.3.4. Both the spot-measure dynamics and the risk-neutral dynamics admit no known transition densities, so that the related equations need to be discretized in order to perform simulations.

We finally notice that a full no-arbitrage derivation of the LFM dynamics above, based on the change-of-numeraire technique, can be found in Musiela and Rutkowski (1997).

6.4 Calibration of the LFM to Caps and Floors Prices

The calibration to caps and floors prices for the LFM is almost automatic, since one can simply input in the model volatilities σ given by the market in form of Black-like implied volatilities for cap prices.

Indeed, we have already hinted in Section 6.2 that the LFM cap pricing formula coincides with that used in the market for pricing caps, namely the Black formula (1.26), which can be then re-derived under a fully consistent theoretical apparatus.

We now shortly review this alternative derivation of Black's cap formula, in order also to clarify some points on the calibration.

The discounted payoff at time 0 of a cap with first reset date T_α and payment dates $T_{\alpha+1}, \ldots, T_\beta$ is given by

$$\sum_{i=\alpha+1}^{\beta} \tau_i D(0, T_i)(F(T_{i-1}, T_{i-1}, T_i) - K)^+,$$

where we assume a unit nominal amount.

We remind that the caps whose implied volatilities are quoted by the market typically have either T_0 equal to three months, $\alpha = 0$ and all other

T's equally three-months spaced, or T_0 equal to six months, $\alpha = 0$ and all other T's equally six-months spaced.

The pricing of the cap can be obtained by considering the risk-neutral expectation E of its discounted payoff:

$$E\left\{ \sum_{i=\alpha+1}^{\beta} \tau_i D(0, T_i)(F_i(T_{i-1}) - K)^+ \right\}$$
$$= \sum_{i=\alpha+1}^{\beta} \tau_i P(0, T_i) E^i \left[(F_i(T_{i-1}) - K)^+ \right]$$

where we used the forward-adjusted measure Q^i. The above payoff is therefore reduced to a sum of payoffs. The single additive term

$$(F_i(T_{i-1}) - K)^+$$

in the above payoff is associated with a contract called T_{i-1}-*caplet*, and we will often identify the contract with its payoff, thus naming "caplet" the payoff itself. Caps and caplets were introduced earlier in Chapter 1. Recall that a T_{i-1}-caplet is a contract paying at time T_i the difference between the T_i-maturity spot rate reset at time T_{i-i} and a strike rate K, if this difference is positive, and zero otherwise. The price at the initial time 0 of the T_{i-1}-caplet is then given by

$$P(0, T_i) \; E^i (F_i(T_{i-1}) - K)^+ \; .$$

Remark 6.4.1. **(Correlations have no impact on caps).** As we have remarked earlier in the "guided tour" to market models, notice carefully that the joint dynamics of forward rates is not involved in this payoff. As a consequence, the correlation between different rates does not afflict the payoff, since marginal distributions of the single $F's$ are enough to compute the expectations appearing in the payoff. Indeed, there are no expectations involving two or more forward rates at the same time, so that correlations are not relevant.

Recall also that the T_{i-1}-caplet is said to be "at-the-money" when its strike price $K = K_i$ equals the current value $F_i(0)$ of the underlying forward rate. The caplet is said to be "in-the-money" when $F_i(0) > K_i$, reflecting the fact that if rates were deterministic, we would get money (amounting to $F_i(0) - K_i$) from the caplet contract at maturity. The caplet is said to be "out-of-the-money" when $F_i(0) < K_i$.

It is advisable to keep in mind also the notion of "at-the-money" (ATM) cap. As we just recalled, a cap is a collection of caplets with a common strike K. Each caplet, however, would be ATM for a different strike K_i. We cannot therefore define the ATM cap strike K in terms of the single ATM caplets strikes K_i, since they are different in general. We can however select a single rate that takes into account all the forward rates of the underlying caplets. This is chosen to be the forward swap rate $K = S_{\alpha,\beta}(0)$, which we will reintroduce in Section 6.7.

Going back to a general strike K, in order to compute

$$E^i[(F_i(T_{i-1}) - K)^+],$$

remember that under Q^i the process F_i follows the martingale

$$dF_i(t) = \sigma_i(t)F_i(t)dZ_i(t), \quad Q^i, \quad t \leq T_{i-1}.$$

Given the lognormal distribution for F_i, the above expectation is easily computed as a Black and Scholes price for a stock call option whose underlying "stock" is F_i, struck at K, with maturity T_{i-1}, with zero constant "risk-free rate" and instantaneous percentage volatility $\sigma_i(t)$. We thus obtain the following.

Proposition 6.4.1. (Equivalence between LFM and Black's caplet prices). *The price of the T_{i-1}-caplet implied by the LFM coincides with that given by the corresponding Black caplet formula, i.e.*

$$\begin{aligned}\mathbf{Cpl}^{LFM}(0, T_{i-1}, T_i, K) &= \mathbf{Cpl}^{Black}(0, T_{i-1}, T_i, K, v_i) \\ &= P(0, T_i)\tau_i \, \mathrm{Bl}(K, F_i(0), v_i) \, ,\end{aligned}$$

$$\begin{aligned}\mathrm{Bl}(K, F_i(0), v_i) &= E^i(F_i(T_{i-1}) - K)^+ \\ &= F_i(0)\Phi(d_1(K, F_i(0), v_i)) - K\Phi(d_2(K, F_i(0), v_i)),\end{aligned}$$

$$\begin{aligned}d_1(K, F, v) &= \frac{\ln(F/K) + v^2/2}{v}, \\ d_2(K, F, v) &= \frac{\ln(F/K) - v^2/2}{v},\end{aligned}$$

where

$$v_i^2 = T_{i-1} \, v_{T_{i-1}-caplet}^2, \tag{6.19}$$

$$v_{T_{i-1}-caplet}^2 := \frac{1}{T_{i-1}} \int_0^{T_{i-1}} \sigma_i(t)^2 dt \, .$$

Notice that in our notation v_i^2 will always denote the integrated instantaneous variance not divided by the time amount, contrary to $v_{T_{i-1}-\mathrm{caplet}}^2$, which will be always standardized with respect to time (average instantaneous variance over time). The quantity $v_{T_{i-1}-\mathrm{caplet}}$ is termed *T_{i-1}-caplet volatility* and has thus been (implicitly) defined as the square root of the average percentage variance of the forward rate $F_i(t)$ for $t \in [0, T_{i-1})$.

We now review the implications of the different volatility structures introduced in Section 6.3 as far as calibration is concerned.

6.4.1 Piecewise-Constant Instantaneous-Volatility Structures

These are the structures described in TABLEs 1–5. In general, we have:

$$v^2_{T_{i-1}-\text{caplet}} = \frac{1}{T_{i-1}} \int_0^{T_{i-1}} \sigma^2_{i,\beta(t)} dt = \frac{1}{T_{i-1}} \sum_{j=1}^{i} \tau_{j-2,j-1}\, \sigma^2_{i,j}\,, \tag{6.20}$$

where we denote by $\tau_{i,j} = T_j - T_i$ the time between T_i and T_j in years.

Formulation of TABLE 2. If we assume that the piecewise-constant instantaneous volatilities $\sigma_{i,\beta(t)}$ depend only on the time to maturity, according to the formulation of TABLE 2, where $\sigma_{i,\beta(t)} = \eta_{i-\beta(t)+1}$, we then obtain:

$$v_i^2 = \sum_{j=1}^{i} \tau_{j-2,j-1}\, \eta^2_{i-j+1}. \tag{6.21}$$

In this case the parameters η can be used to exactly fit the (squares of the) market caplet volatilities (multiplied by time) v_i^2. Although calibration to swaptions will be dealt with later on, we already hint at the fact that, with this formulation, the only parameters left to tackle swaptions calibration are the instantaneous correlations of forward rates.

Formulation of TABLE 3. If we instead assume that the piecewise-constant instantaneous volatilities $\sigma_{i,\beta(t)}$ depend only on the maturity T_i of the considered forward rate, according to the formulation leading to TABLE 3 where $\sigma_{i,\beta(t)} = s_i$, we obtain:

$$v_i^2 = T_{i-1} s_i^2, \quad v_{T_{i-1}-\text{caplet}} = s_i. \tag{6.22}$$

Again, the parameters s can be used to exactly fit the (squares of the) market caplet volatilities (multiplied by time) v_i^2, and the only parameters left to tackle swaptions calibration are again the instantaneous correlations of forward rates.

Formulation of TABLE 4. On the other hand, if we assume that the piecewise-constant instantaneous volatilities $\sigma_{i,\beta(t)}$ follow the separable structure of the formulation leading to TABLE 4, where $\sigma_{i,\beta(t)} = \Phi_i \Psi_{\beta(t)}$, we obtain:

$$v_i^2 = \Phi_i^2 \sum_{j=1}^{i} \tau_{j-2,j-1}\, \Psi_j^2\,. \tag{6.23}$$

If the squares of the caplet volatilities (multiplied by time) v_i^2 are read from the market, $(v_i^{\text{MKT}})^2$, the parameters Φ can be given in terms of the parameters Ψ as

$$\Phi_i^2 = \frac{(v_i^{\mathrm{MKT}})^2}{\sum_{j=1}^{i} \tau_{j-2,j-1} \Psi_j^2}.$$

Therefore the caplet prices are incorporated in the model by determining the Φ's in term of the Ψ's. The parameters Ψ, together with the instantaneous correlation of forward rates, can be then used in the calibration to swaption prices.

Formulation of TABLE 5. Finally, if we assume that the piecewise-constant instantaneous volatilities $\sigma_{i,\beta(t)}$ follow the separable structure of the formulation leading to TABLE 5, where $\sigma_{i,\beta(t)} = \Phi_i \psi_{i-(\beta(t)-1)}$, we obtain:

$$v_i^2 = \Phi_i^2 \sum_{j=1}^{i} \tau_{j-2,j-1} \psi_{i-j+1}^2 . \tag{6.24}$$

If the squares of the caplet volatilities (multiplied by time) v_i^2 are read from the market, the parameters Φ can now be given in terms of the parameters ψ as

$$\Phi_i^2 = \frac{(v_i^{\mathrm{MKT}})^2}{\sum_{j=1}^{i} \tau_{j-2,j-1} \psi_{i-j+1}^2}. \tag{6.25}$$

Therefore, the caplet prices are incorporated in the model by determining the Φ's in terms of the ψ's. As in the previous case, the parameters ψ, together with the instantaneous correlation of forward rates, can then be used in the calibration to swaption prices.

6.4.2 Parametric Volatility Structures

As far as the parametric structures of Formulations 6 and 7 are concerned, we observe the following.

Formulation 6. If we assume Formulation 6 for instantaneous volatilities, as from (6.12), we obtain the following expression for the squares of caplet volatilities (multiplied by time):

$$v_i^2 = \int_0^{T_{i-1}} \left([a(T_{i-1} - t) + d] e^{-b(T_{i-1}-t)} + c \right)^2 dt =: I^2(T_{i-1}; a, b, c, d) . \tag{6.26}$$

In this case the parameters a, b, c, d can be used to fit the (squares of the) market caplet volatilities (multiplied by time) v_i^2. Therefore, with this formulation the only parameters left to tackle swaptions calibration are the instantaneous correlations of forward rates.

The function I can be computed also through a software for formal manipulations, with a command line such as

```
int(((a*(T-t)+d)*exp(-b*(T-t))+c)
        *((a*(T-t)+d)*exp(-b*(T-t))+c),t=0..T);
```

Formulation 7. If we assume Formulation 7 for instantaneous volatilities, as from (6.13), we obtain the following expression for the squares of caplet volatilities (multiplied by time):

$$v_i^2 = \Phi_i^2 \int_0^{T_{i-1}} \left([a(T_{i-1} - t) + d]e^{-b(T_{i-1}-t)} + c \right)^2 dt = \Phi_i^2 \, I^2(T_{i-1}; a, b, c, d) \, . \tag{6.27}$$

Now the Φ's parameters can be used to calibrate automatically caplet volatilities. Indeed, if the v_i are inferred from market data, we may set

$$\Phi_i^2 = \frac{(v_i^{\mathrm{MKT}})^2}{I^2(T_{i-1}; a, b, c, d)} \, . \tag{6.28}$$

Thus caplet volatilities are incorporated by expressing the parameters Φ as functions of the parameters a, b, c, d, which are still free. In this way, for swaptions calibration we can rely upon the parameters a, b, c, d and upon the instantaneous correlations between forward rates.

6.4.3 Cap Quotes in the Market

As already pointed out, the market typically quotes volatilities for caps with first reset date either in three months (T_0 equal to three months, $\alpha = 0$ and all other T's equally three-months spaced) or in six months (T_0 equal to six months, $\alpha = 0$ and all other T's equally six-months spaced), and progressively increasing maturities. However, in this chapter we will essentially consider examples of caps with a six-month reset period. We set $\mathcal{T}_j = [T_0, \ldots, T_j]$ for all j.

An equation is considered between the market price $\mathbf{Cap}^{\mathrm{MKT}}(0, \mathcal{T}_j, K)$ of the cap with $\alpha = 0$ and $\beta = j$ and the sum of the first j caplets prices:

$$\mathbf{Cap}^{\mathrm{MKT}}(0, \mathcal{T}_j, K) = \sum_{i=1}^{j} \tau_i P(0, T_i) \, \mathrm{Bl}(K, F_i(0), \sqrt{T_{i-1}} \, v_{\mathcal{T}_j-\mathrm{cap}})$$

where a same average-volatility value $v_{\mathcal{T}_j-\mathrm{cap}}$ has been put in all caplets up to j. The quantities $v_{\mathcal{T}_j-\mathrm{cap}}$ are called sometimes *forward volatilities.* The market solves the above equation in $v_{\mathcal{T}_j-\mathrm{cap}}$ and quotes $v_{\mathcal{T}_j-\mathrm{cap}}$, annualized and in percentages.

Remark 6.4.2. **(One kind of inconsistency in cap volatilities)**. Notice carefully that *the same* average volatility $v_{\mathcal{T}_j-\mathrm{cap}}$ is assumed for all caplets concurring to the $\mathcal{T}_j$-maturity cap. However, when the same caplets concur to a different cap, say a T_{j+1}-maturity cap, their average volatility is changed. This appears to be somehow inconsistent. In the cap volatility system, the same caplet is linked to different volatilities when concurring to different caps.

Clearly, to recover correctly cap prices with our forward-rate dynamics we need to have

$$\sum_{i=1}^{j} \tau_i P(0,T_i)\,\mathrm{Bl}(K, F_i(0), \sqrt{T_{i-1}}\, v_{T_j-\mathrm{cap}})$$
$$= \sum_{i=1}^{j} \tau_i P(0,T_i)\,\mathrm{Bl}(K, F_i(0), \sqrt{T_{i-1}}\, v_{T_{i-1}-\mathrm{caplet}}).$$

The quantities $v_{T_{i-1}-\mathrm{caplet}}$ are called sometimes *forward forward volatilities.* Notice that *different* average volatilities $v_{T_{i-1}-\mathrm{caplet}}$ are assumed for different caplets concurring to the T_j-maturity cap.

A stripping algorithm can be used for recovering the v_{caplet}'s from the market quoted v_{cap}'s based on the last equality applied to $j = 1, 2, 3, \ldots$. This can be done off-line, before any model-calibration/pricing procedure is tackled. In turn, from the v_{caplet}'s we can go back to the σ's via (6.20), to the η's via (6.21), or to the s's via (6.22). Notice that with the general formulation (6.20) deriving from TABLE 1, or with the last reduced formulations (6.23) and (6.24) of TABLE 4 and TABLE 5, we cannot recover the whole table, since we have more unknown than equations. However, having more parameters can be helpful at later stages, when calibration to swaptions is considered, especially when choosing few instantaneous correlation parameters. On the contrary, formulations (6.8) and (6.9), by reducing the number of unknowns, allow the complete determination of TABLE 2 and TABLE 3, respectively, based on cap prices.

Therefore, modulo some trivial algebra and numerical non-linear equation solving, the calibration to cap prices with these formulations, leading to (6.21) and (6.22), is automatic since the instantaneous volatilities parameters η's and s's can be deduced directly from the market via an algebraic (numerical) procedure.

Finally, for the purpose of jointly calibrating caps and swaptions, we notice that the formulation leading to (6.23) would be particularly handy when using Brace's analytical approximation, reviewed in Sections 6.13 and 6.14, although the formulation leading to (6.24), or its parametric analogue leading to (6.27), turn out to better combine both flexibility for a good initial fitting and "controllability" for a qualitatively acceptable evolution of the term structure of volatilities, as we are about to see.

6.5 The Term Structure of Volatility

Consider again the set of times $\mathcal{E} = \{T_0, \ldots, T_M\}$ representing adjacent expiry-maturity pairs for a family of spanning forward rates. The term structure of volatility at time T_j is a graph of expiry times T_{h-1} against average volatilities $V(T_j, T_{h-1})$ of the forward rates $F_h(t)$ up to that expiry time itself, i.e. for $t \in (T_j, T_{h-1})$. In other terms, at time $t = T_j$, the volatility term structure is the graph of points

$$\{(T_{j+1}, V(T_j, T_{j+1})), (T_{j+2}, V(T_j, T_{j+2})), \ldots, (T_{M-1}, V(T_j, T_{M-1}))\}$$

where

$$V^2(T_j, T_{h-1}) = \frac{1}{\tau_{j,h-1}} \int_{T_j}^{T_{h-1}} \frac{dF_h(t)\, dF_h(t)}{F_h(t) F_h(t)} = \frac{1}{\tau_{j,h-1}} \int_{T_j}^{T_{h-1}} \sigma_h^2(t) dt,$$

for $h > j+1$ (we remind that $\tau_{i,j} = T_j - T_i$). The term structure of volatilities at time 0 is

$$\begin{aligned} &\{(T_0, V(0, T_0)), \ldots, (T_{M-1}, V(0, T_{M-1}))\} \\ &= \{(T_0, v_{T_0-\text{caplet}}), \ldots, (T_{M-1}, v_{T_{M-1}-\text{caplet}})\} \end{aligned}$$

These are simply (pairs of expiries with corresponding) forward forward volatilities at time 0, and a typical example (with annualized versions of the caplet volatilities, as from Section 6.20) from the Euro market is displayed in Figure 6.1. Notice the humped structure, starting from about 15% for the

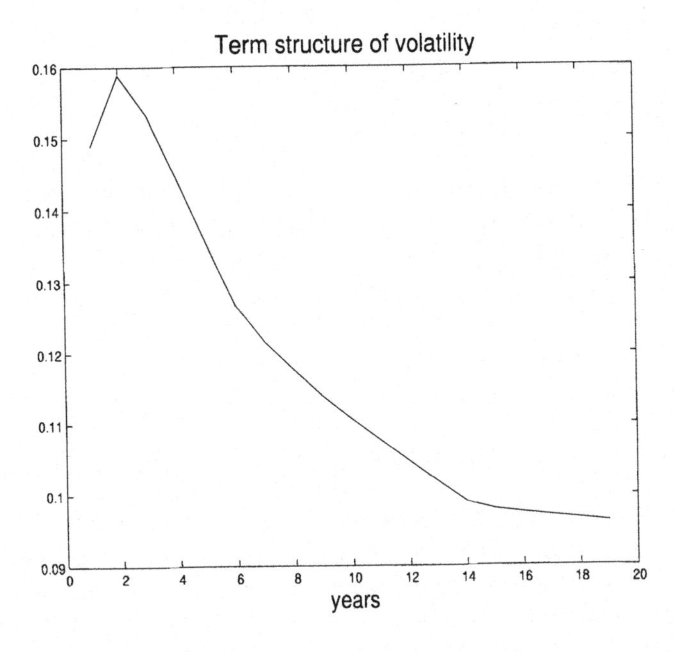

Fig. 6.1. Example of term structure of volatility, $T \mapsto v_{T-\text{caplet}}$, from the Euro market.

one-year caplet, up to about 16% for the two-year caplet, and afterwards always decreasing, down to less than 10% for the nineteen-year caplet.

Different assumptions on the behaviour of instantaneous volatilities imply different evolutions for the term structure of volatilities. We now examine the impact of different formulations of instantaneous volatilities on the evolution of the term structure.

6.5.1 Piecewise-Constant Instantaneous Volatility Structures

We first examine the two cases corresponding to TABLE 2 and TABLE 3 respectively.

Formulation of TABLE 2. Formulation (6.8) of TABLE 2 gives

$$V^2(T_j, T_{h-1}) = \frac{1}{\tau_{j,h-1}} \sum_{k=j+1}^{h-1} \tau_{k-1,k} \eta_{h-k}^2 .$$

Assume for simplicity that $\tau_{i-1,i} = \tau > 0$, so that $\tau_{j,h-1} = (h-j-1)\tau$. It follows easily that, in such a case,

$$V^2(T_j, T_{h-1}) = \frac{1}{h-j-1} \sum_{k=j+1}^{h-1} \eta_{h-k}^2, \tag{6.29}$$

which immediately implies that

$$V(T_j, T_{h-1}) = V(T_{j+1}, T_h).$$

Therefore, when time moves from T_j to T_{j+1}, the volatility term structure remains the same, except that now it is shorter, since the number of possible expiries has decreased by one. We move in fact from

$$\{(T_{j+1}, V(T_j, T_{j+1})), (T_{j+2}, V(T_j, T_{j+2})), (T_{j+3}, V(T_j, T_{j+3})), \ldots \\ \ldots, (T_{M-1}, V(T_j, T_{M-1}))\}$$

to

$$\{(T_{j+2}, V(T_j, T_{j+1})), (T_{j+3}, V(T_j, T_{j+2})), \ldots, (T_{M-1}, V(T_j, T_{M-2}))\}.$$

Summarizing, with instantaneous volatilities of forward rates like in TABLE 2, as time passes from an expiry date to the next, the volatility term structure remains the same, except that the "tail" of the graph is cut away. In particular, the volatility term structure does not change over time. It has been often observed that *qualitatively* this is a desirable property, since the actual shape of the market term structure (typically humped) does not change too much over time (see also Rebonato (1998)). We present an example of such a desirable evolution in the first part of Figure 6.2.

Still, there remains the fundamental question of whether this formulation allows for a humped structure to be fitted at the initial time. What we have just seen, in fact, is that, under this volatility parameterization, *if* the initial term structure is humped then it remains humped at future times. But does this volatility specification allow for a humped-volatility structure in the first place?

Consider formula (6.29) for $j = -1$, i.e. the volatility term structure at time 0. It is easy to see that the map

$$T_{h-1} \mapsto \sqrt{T_{h-1}} V(0, T_{h-1}) = \sqrt{\tau \sum_{k=0}^{h-1} \eta_{h-k}^2}$$

is increasing. Therefore, if the market data imply an initial volatility term structure such that $\sqrt{T} V(0,T)$ is humped as a function of T, the instantaneous-volatility parameterization of TABLE 2 cannot be used since it implies, instead, an increasing behaviour.

However, one is usually interested in $T \mapsto V(0,T)$ to be humped, rather than $T \mapsto \sqrt{T} V(0,T)$. This gives us further hope. In fact, we want a decreasing structure following the hump, namely

$$V(0, T_{h-1}) \geq V(0, T_h)$$

for T_h (typically) larger than three (or four-five) years. On the other hand, we have just seen that the parameterization obtained through the η's imposes that

$$\sqrt{T_{h-1}}\, V(0, T_{h-1}) \leq \sqrt{T_h}\, V(0, T_h)\,.$$

Putting these two constraints together yields:

$$1 \leq \frac{V(0, T_{h-1})}{V(0, T_h)} \leq \sqrt{\frac{h}{h-1}}.$$

We observe that, for large h, the last term is close to one, meaning that the term structure gets almost flat at the end. Indeed, the above constraints imply $V(0, T_h) \approx V(0, T_{h-1})$ for large h. In case the term structure is steeply decreasing also for large maturities, the parameterization of TABLE 2 is thus to be avoided.

Formulation of TABLE 3. Consider now formulation (6.9) for instantaneous volatilities, which leads to TABLE 3. In such a case

$$V^2(T_j, T_{h-1}) = s_h^2.$$

This is the simplest case of volatility term structure. When moving from expiry time T_j to T_{j+1}, the term structure changes from

$$\{(T_{j+1}, s_{j+2}), (T_{j+2}, s_{j+3}), (T_{j+3}, s_{j+4}), \ldots, (T_{M-1}, s_M)\}$$

to

$$\{(T_{j+2}, s_{j+3}), (T_{j+3}, s_{j+4}), \ldots, (T_{M-1}, s_M)\}.$$

With instantaneous volatilities of forward rates like in TABLE 3, as time passes from an expiry date to the next, the volatility term structure is obtained from the previous one by "cutting off" the head instead of the tail. This is less desirable than in the previous formulation, since now the qualitative behaviour of the term structure can be altered over time. Typically, if the term structure features a hump around two years, this hump will be absent in the term structure occurring in three years. See the second part of Figure 6.2 for an example.

Formulations of TABLES 4 and 5. Now consider the separable case of TABLE 4, as from (6.10). In such a case

$$V^2(T_j, T_{h-1}) = \frac{\Phi_h^2}{\tau_{j,h-1}} \sum_{k=j+2}^{h} \tau_{k-2,k-1}\Psi_k^2.$$

Here the qualitative behaviour depends on the particular specification of both the Φ's and the Ψ's, and there is no a priori clear qualitative pattern for the evolution of the term structure of volatilities as a whole. Moreover, it is hard to control the future term structure of volatilities by acting on the parameters with this formulation.

This task results to be easier under the last formulation (6.11), which yields

$$V^2(T_j, T_{h-1}) = \frac{\Phi_h^2}{\tau_{j,h-1}} \sum_{k=j+1}^{h-1} \tau_{k-1,k}\psi_{h-k}^2.$$

By assuming again all adjacent τ's to be equal, this reduces to

$$V^2(T_j, T_{h-1}) = \frac{\Phi_h^2}{h-j-1} \sum_{k=j+1}^{h-1} \psi_{h-k}^2.$$

If the weights Φ are all equal, this formulation is equivalent to that of TABLE 2. Therefore, in such a case the term structure of volatility remains unchanged as time passes and in particular it maintains the hump over time. The presence of the Φ terms, when they are not all equal, can change this situation, but if we make sure that the Φ's remain sufficiently close to each other, the qualitative behaviour will not be affected and the hump in the evolution of the term structure in time will be preserved. An example of such a desirable behaviour is given in Figure 6.3 for the similar parametric formulation 7 below.

The abundance of parameters and the "controllability" of the future term structure make this parameterization particularly appealing among the piecewise-constant ones.

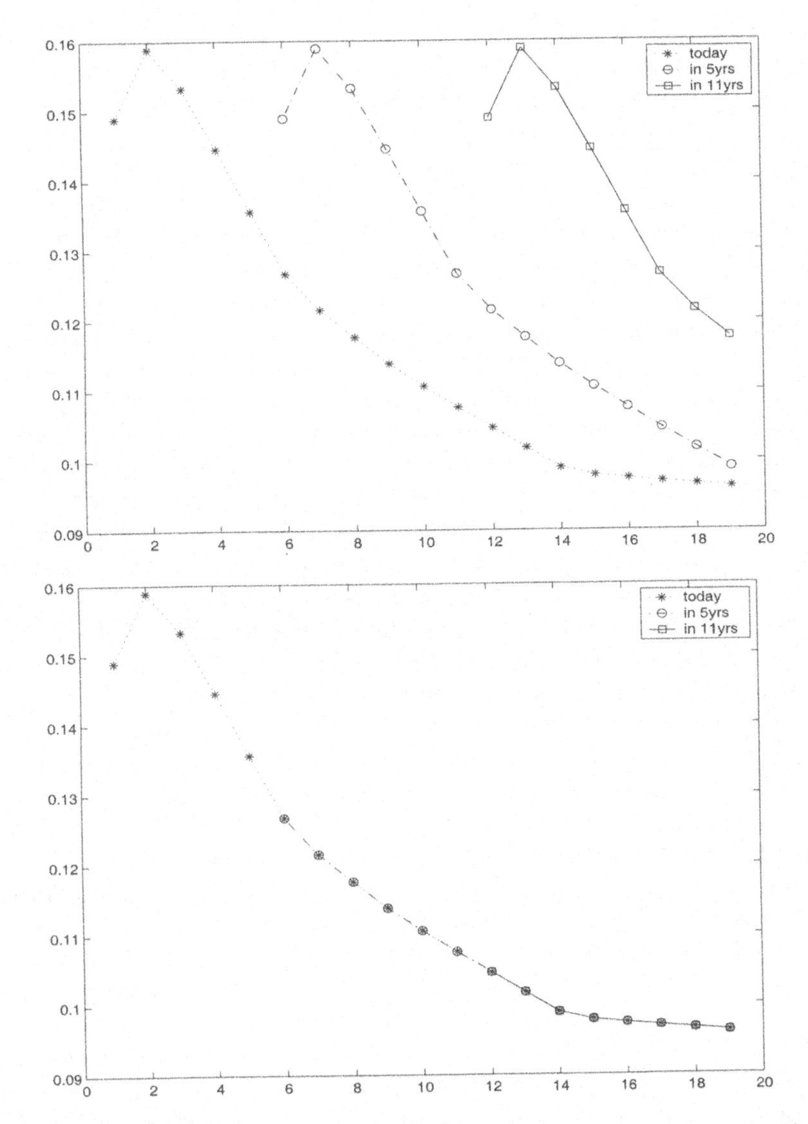

Fig. 6.2. Examples of evolution of the term structure of volatilities under Formulations 2 and 3.

6.5.2 Parametric Volatility Structures

We now consider the evolution of the term structure of volatilities in time as implied by Formulations 6 and 7.

Evolution under Formulation 6. We have already noticed that Formulation 6 is the analogue of the piecewise-constant formulation of TABLE 2. As such, the same qualitative behaviour can be expected. The term structure remains the same as time passes, and in particular it maintains its humped shape if initially humped. Nevertheless, the problem of a non-decreasing structure is still present. If the initial term structure is decreasing for large maturities, it cannot be calibrated with this formulation, as observed for the case of TABLE 2.

Evolution under Formulation 7. We have also already noticed that Formulation 7 is the analogue of the piecewise-constant formulation of TABLE 5. Again, the same qualitative behaviour can be expected. The term structure remains the same as time passes, and in particular it can maintain its humped shape if initially humped and if all Φ's are not too different from one another. This form has been suggested, among others, by Rebonato (1999d).

An example of the evolution of the term structure of volatilities, as implied by this formulation, is given in Figure 6.3. We start with the same initial term structure of volatilities as in the previous Figure 6.2, which is given here by setting $a = 0.19085664, b = 0.97462314, c = 0.08089168, d = 0.01344948$, and with the Φ vector plotted in the first graph of this figure. Clearly, this short example is just a sample of the kind of tests and games one has to play with a formulation before feeling confident about it. In particular, we give it as an example of the kind of questions that are posed by traders and practitioners, in general, when a model is presented to them.

A final remark is that, compared to other models, the LFM allows for an immediate calculation of the future term structures of volatilities, which is a rather important advantage. This calculation requires no simulation and leads to a deterministic evolution. For other models this is not the case. Usually the instantaneous percentage volatility of forward rates (i.e. the diffusion coefficient of $d\ln F_k(t)$ or of $dF_k(t)/F_k(t)$) is not deterministic, so that the integrated instantaneous variance leading to future volatilities is a random variable. Displaying scenarios of future term structures of volatility implies, therefore, simulation of trajectories for the underlying processes.

Indeed, if we were using, for example, a one-factor short-rate model, say Vasicek's $dr_t = k(\theta - r_t)dt + \sigma dW_t$, we would have $\ln F_k(t) = \phi_k(r_t)$ for some function ϕ_k. Then by Ito's formula, the instantaneous percentage volatility would be given by the diffusion coefficient of $d\ln F_k(t)$, which is equal to $\phi_k'(r_t)\sigma$ (where $'$ denotes differentiation). Now take $T_1 = 2y$, $T_2 = 3y$ and compute, for instance,

$$V^2(T_1, T_2) = \int_{T_1}^{T_2} d\ln F_3(t) d\ln F_3(t) = \int_{T_1}^{T_2} \phi_3'(r_t)^2 \sigma^2 dt\,.$$

This is a random variable depending on the trajectory of r between T_1 and T_2. If $T_1 > 0$, there is no escape from this situation, and we need to cope with a random future term structure of volatilities. However, if we are considering the term structure of volatilities at time 0, there is a way out. We can in

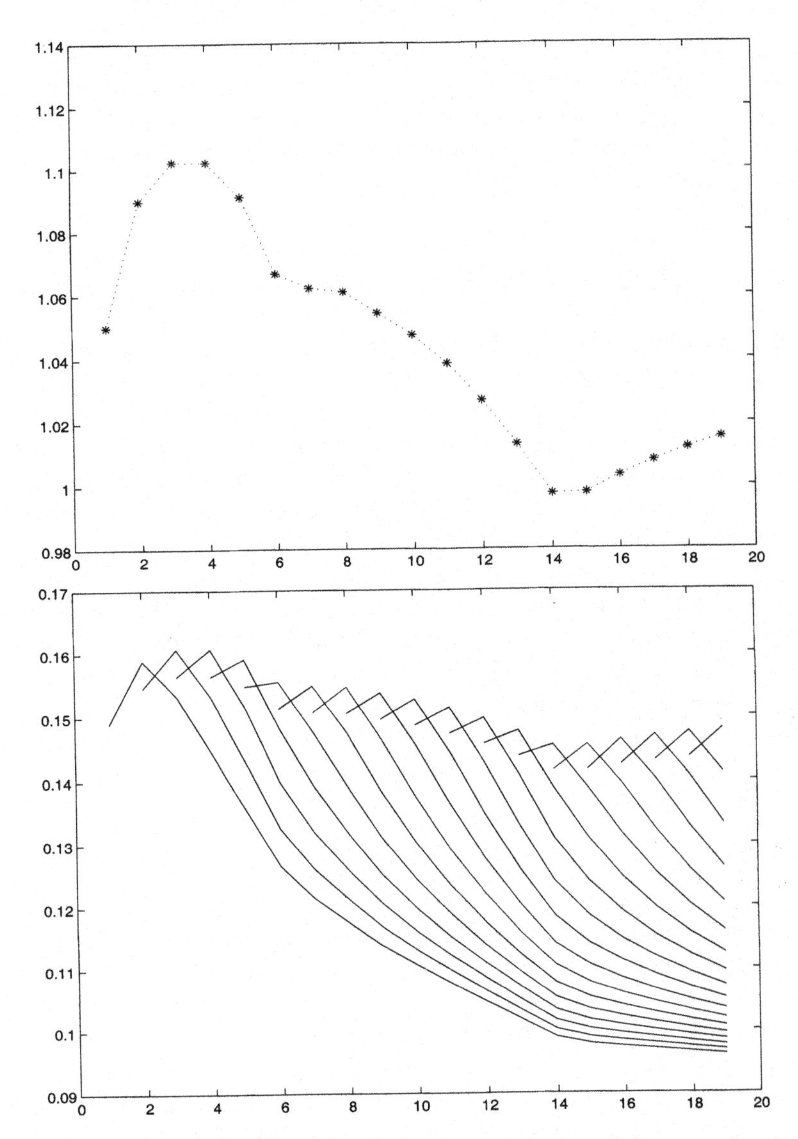

Fig. 6.3. An example of Φ (first graph) and the evolution of the term structure of volatilities (second graph) under Formulation 7.

fact obtain a deterministic volatility by pricing at-the-money caplets with the model and then by inverting Black's formula for caplets to derive an implied Black-like caplet volatility for the Vasicek model. With the LFM this is not necessary, and there is no difference between the model "intrinsic" caplet volatility and the model "implied" caplet volatility, as in the Vasicek case, for example (for a discussion of such matters in the short-rate world see Sections 3.6 and 4.2.3).

To sum up, the LFM has the advantage over other models of allowing one to display (deterministic) future term structures of volatilities once the model has been calibrated. With other models such structures would be stochastic, except for the structure at time 0 when its definition can be modified in "implied" terms.

6.6 Instantaneous Correlation and Terminal Correlation

In the following we shall consider payoffs whose evaluation will be based on expected values of quantities involving several rates at the same time, typically swaptions. As a consequence, prices will depend on the terminal correlation between different forward rates. The point of this section is showing that the terminal correlation between different forward rates depends not only on the instantaneous correlations of forward-rate dynamics, but also on the instantaneous volatility specification.

Indeed, we have assumed earlier each forward rate F_k to be driven by a different Brownian motion Z_k, and we assumed also different Brownian motions to be instantaneously correlated. Now it is important to understand how this instantaneous correlation in the forward-rate dynamics translates into a terminal correlation of simple rates.

Recall that the *instantaneous* correlation is a quantity summarizing the degree of "dependence" between *changes* of different forward rates. Roughly speaking, for example,

$$\rho_{2,3} = \frac{dF_2(t)\,dF_3(t)}{\mathrm{Std}(dF_2(t))\,\mathrm{Std}(dF_3(t))},$$

where "Std" denotes the standard deviation conditional on the information available at time t at which the change occurs. From this formula it is clear that indeed the instantaneous correlation ρ is related to changes dF in forward rates. Instead, the *terminal* correlation is a quantity summarizing the degree of "dependence" between two different forward rates at a given "terminal" time-instant. Typically, the T_1 terminal correlation between F_2 and F_3 is the correlation between $F_2(T_1)$ and $F_3(T_1)$. Is this terminal correlation completely determined by the instantaneous correlations $\rho_{2,3}$ between Z_2 and Z_3?

The answer is negative in general. The terminal correlation between different forward rates certainly depends on the introduced instantaneous correlations, but not only on them. We will show, in fact, that such a terminal correlation depends also on the way the "total" average volatility of each forward rate (caplet volatility) is "disintegrated" or "distributed" in instantaneous volatility. Precisely, the correlation between $F_2(T)$ and $F_3(T)$ depends also on the particular functions of time $\sigma_{2,\beta(t)}$ and $\sigma_{3,\beta(t)}$ that are used to recover the average volatilities v_2 and v_3 in $[0, T_1]$ through integration. Keeping the same instantaneous correlations and decomposing v_2 and v_3 in two different ways "a" ($\sigma^a_{2,\beta(t)}, \sigma^a_{3,\beta(t)}$) and "$b$" ($\sigma^b_{2,\beta(t)}, \sigma^b_{3,\beta(t)}$), leads to different correlations between $F_2(T_1)$ and $F_3(T_1)$ in the two cases. This elementary and fundamental feature of terminal correlation was noticed and pointed out in Rebonato (1998, 1999d).

We notice also that, while the instantaneous correlation does not depend on the particular probability measure (or numeraire asset) under which we are working, the terminal correlation does. Indeed, the Girsanov theorem establishes that the instantaneous covariance structure is the same for all the equivalent measures under which a process can be expressed, so that the particular measure under which we work makes no difference. This is not the case for terminal correlation. However, we will see further on in Section 6.16 that approximated terminal correlations based on "freeze part of the drift"-like approximations in the dynamics will not depend on the particular probability measure being chosen.

To clarify dependence of the terminal correlation on the volatility decomposition we now consider two examples, the first of which, though not addressing completely this issue, has the advantage of being based on exact calculations.

As first example, consider a payoff depending on the two forward rates $F_2(t), F_3(t)$ for $t \leq T_1$. Take $T_0 > 0$, $T_1 = 2T_0$, and take as underlying measure Q^2, under which F_2 is a martingale. We thus have:

$$dF_2(t) = \sigma_{2,\beta(t)} F_2(t) dZ_2(t)$$

$$dF_3(t) = \frac{\tau_3\, \sigma^2_{3,\beta(t)}\, F_3^2(t)}{1 + \tau_3 F_3(t)} dt + \sigma_{3,\beta(t)} F_3(t) dZ_3(t), \quad t \leq T_1 \; .$$

Assume the payoff depends on $\ln F_3(T_1) - \ln F_3(0)$ and on $F_2(T_1) - F_2(0)$ jointly, and consider the product of these two variables as an academic example of a quantity whose expectation depends on the T_1-terminal correlation between F_2 and F_3. Ito's isometry plus some straightforward algebra give

$$E^2\left[(F_2(T_1) - F_2(0))(\ln F_3(T_1) - \ln F_3(0))\right] = F_2(0)\rho_{2,3} \int_0^{T_1} \sigma_{2,\beta(t)}\sigma_{3,\beta(t)} dt$$
$$= F_2(0)\rho_{2,3}(\sigma_{2,1}\sigma_{3,1}T_0 + \sigma_{2,2}\sigma_{3,2}\tau_{0,1}). \tag{6.30}$$

Now notice that any caplet price involving the above rates would simply depend on the average volatilities

$$v_2^2 = T_0\, \sigma^2_{2,1} + \tau_{0,1}\, \sigma^2_{2,2},$$

and

$$v_3^2 = T_0\, \sigma^2_{3,1} + \tau_{0,1}\, \sigma^2_{3,2} + \tau_{1,2}\, \sigma^2_{3,3}$$

no matter how the total variances v_2^2 and v_3^2 are distributed between F_2 and F_3, i.e. no matter the particular values of $\sigma_{2,1}, \sigma_{2,2}, \sigma_{3,1}, \sigma_{3,2}$ are chosen, as long as v_2 and v_3 are preserved.

On the contrary, the correlation-dependent term (6.30) does depend on this choice. Consider for example $\tau_{i-1,i} = T_i - T_{i-1} = 0.5$ for all i, $F_2(0) = F_3(0) = 0.05$, $\rho_{2,3} = 0.75$, $\sigma_{3,3} = 0.1$, and finally take the two cases

$$a)\ \sigma_{2,1} = 0.5,\ \ \sigma_{3,1} = 0.1,\ \ \sigma_{2,2} = 0.1,\ \ \sigma_{3,2} = 0.5$$

and

$$b)\ \sigma_{2,1} = 0.1,\ \ \sigma_{3,1} = 0.1,\ \ \sigma_{2,2} = 0.5,\ \ \sigma_{3,2} = 0.5.$$

In cases a) and b) v_2 and v_3 have the same values, whereas the correlation-dependent quantity (6.30) gives $1.875E-3$ in case a) and $2.34375E-4$ in case b). This example shows that the value of a payoff depending on the terminal correlation between different rates may depend not only on the instantaneous correlations ρ of the forward-rate dynamics, but also on the way the instantaneous volatilities σ are modeled from average volatilities v coming from cap prices.

A more extreme and direct example is as follows. As stated earlier, we will derive in Section 6.16 an approximated formula for computing the terminal correlation of forward rates. The formula, in this particular example, would read

$$\mathrm{Corr}(F_2(T_1), F_3(T_1)) \approx \rho_{2,3} \frac{\int_0^{T_1} \sigma_2(t)\sigma_3(t)dt}{\sqrt{\int_0^{T_1} \sigma_2^2(t)dt}\ \sqrt{\int_0^{T_1} \sigma_3^2(t)dt}}\,.$$

Take $T_0 = 1y$, $T_1 = 2y$, $T_2 = 3y$.

Given the piecewise-constant volatility formulation, and the fact that T_1 is exactly the expiry time for the forward rate F_2, we can write

$$\mathrm{Corr}(F_2(T_1), F_3(T_1)) \approx \rho_{2,3} \frac{\sigma_{2,1}\sigma_{3,1} + \sigma_{2,2}\sigma_{3,2}}{v_2\sqrt{\sigma_{3,1}^2 + \sigma_{3,2}^2}}\,.$$

Let us fix the instantaneous correlation $\rho_{2,3} = 1$ and, as usual, require that our piecewise-constant functions $\sigma_2(t)$ and $\sigma_3(t)$ be compatible with T_1- and T_2-caplet volatilities. We therefore require

$$v_2^2 = \sigma_{2,1}^2 + \sigma_{2,2}^2,$$

and

$$v_3^2 = \sigma_{3,1}^2 + \sigma_{3,2}^2 + \sigma_{3,3}^2\,.$$

There are two extreme cases of interest. The first is obtained as

$$\sigma_{2,1} = v_2,\ \ \sigma_{2,2} = 0\,;$$

$$\sigma_{3,1} = v_3, \sigma_{3,2} = 0,\ \ \sigma_{3,3} = 0\,.$$

In this case, the above formula yields easily

$$\mathrm{Corr}(F_2(T_1), F_3(T_1)) \approx \rho_{2,3} = 1\,.$$

The second extreme case is obtained as

$$\sigma_{2,1} = 0, \ \sigma_{2,2} = v_2 ;$$

$$\sigma_{3,1} = v_3, \sigma_{3,2} = 0, \ \sigma_{3,3} = 0 .$$

In this second case, the above formula yields immediately

$$\text{Corr}(F_2(T_1), F_3(T_1)) \approx 0 \, \rho_{2,3} = 0 .$$

So we have seen two cases with the same instantaneous correlation and caplet volatilities, where a different rebalancing of instantaneous volatilities leads respectively to terminal correlations of 0 and 1.

Whichever choice of the instantaneous volatilities structure is considered, say for example (6.8), (6.9), or (6.10), the point of this section is making clear that this choice influences also the terminal correlation between different rates.

6.7 Swaptions and the Lognormal Forward-Swap Model (LSM)

Assume a unit notional amount. Recall that a (prototypical) interest-rate swap (IRS) is a contract that exchanges payments between two differently indexed legs. At every instant T_j in a prespecified set of dates $T_{\alpha+1}, ..., T_\beta$ the fixed leg pays out an amount corresponding to a fixed interest rate K,

$$\tau_j K ,$$

where τ_j is the year fraction from T_{j-1} to T_j, whereas the floating leg pays an amount corresponding to the interest rate $F_j(T_{j-1})$ set at the previous instant T_{j-1} for the maturity given by the current payment instant T_j.

Clearly, the floating-leg rate is reset at dates $T_\alpha, T_{\alpha+1}, \ldots, T_{\beta-1}$ and paid at dates $T_{\alpha+1}, \ldots, T_\beta$.

The payoff at time T_α for the payer party of such an IRS can be expressed as

$$\sum_{i=\alpha+1}^{\beta} D(T_\alpha, T_i) \tau_i (F_i(T_{i-1}) - K).$$

Accordingly, the *discounted* payoff at a time $t < T_\alpha$ is

$$\sum_{i=\alpha+1}^{\beta} D(t, T_i) \tau_i (F_i(T_{i-1}) - K). \tag{6.31}$$

The value of such a contract is easily computed as

$$\begin{aligned}
\mathbf{PFS}(t, [T_\alpha, \dots, T_\beta], K) &= E_t \left\{ \sum_{i=\alpha+1}^{\beta} D(t, T_i)\tau_i (F_i(T_{i-1}) - K) \right\} \\
&= \sum_{i=\alpha+1}^{\beta} P(t, T_i)\tau_i E_t^i (F_i(T_{i-1}) - K) \\
&= \sum_{i=\alpha+1}^{\beta} P(t, T_i)\tau_i (F_i(t) - K) \\
&= \sum_{i=\alpha+1}^{\beta} [P(t, T_{i-1}) - (1 + \tau_i K) P(t, T_i)] \, .
\end{aligned} \tag{6.32}$$

Notice that the same value is obtained by defining as T_α-payoff

$$\sum_{i=\alpha+1}^{\beta} P(T_\alpha, T_i)\tau_i (F_i(T_\alpha) - K),$$

leading to the t-discounted payoff

$$D(t, T_\alpha) \sum_{i=\alpha+1}^{\beta} P(T_\alpha, T_i)\tau_i (F_i(T_\alpha) - K),$$

and by taking the risk-neutral expectation.

From the last formula we can also notice that, as is well known, neither volatility nor correlation of rates affect the pricing of this financial product.

The *forward swap rate* corresponding to the above IRS is the particular value of the fixed-leg rate K that makes the contract fair, i.e. that makes its present value equal to zero. The forward swap rate associated with the above IRS is therefore obtained by equating to zero the last expression in (6.32) and by solving it in K. We obtain

$$\begin{aligned}
S_{\alpha,\beta}(t) &= \frac{P(t, T_\alpha) - P(t, T_\beta)}{\sum_{i=\alpha+1}^{\beta} \tau_i P(t, T_i)} = \frac{1 - \mathrm{FP}(t; T_\alpha, T_\beta)}{\sum_{i=\alpha+1}^{\beta} \tau_i \mathrm{FP}(t; T_\alpha, T_i)} \\
&=: \exp\left(\psi(F_{\alpha+1}(t), F_{\alpha+2}(t), \dots, F_\beta(t))\right)
\end{aligned}$$

$$\mathrm{FP}(t; T_\alpha, T_i) = \frac{P(t, T_i)}{P(t, T_\alpha)} = \prod_{j=\alpha+1}^{i} \mathrm{FP}_j(t), \quad \mathrm{FP}_j(t) = \frac{1}{1 + \tau_j F_j(t)},$$

where FP denotes the "forward discount factor". The expression in terms of an exponential of a function of the underlying forward rates, which is implicitly defined by the last equality, will be useful in the following and is written also to point out that a forward swap rate is actually a (nonlinear) function of the underlying forward LIBOR rates. Indeed, we can rewrite the above formula directly as

$$S_{\alpha,\beta}(t) = \frac{1 - \prod_{j=\alpha+1}^{\beta} \frac{1}{1+\tau_j F_j(t)}}{\sum_{i=\alpha+1}^{\beta} \tau_i \prod_{j=\alpha+1}^{i} \frac{1}{1+\tau_j F_j(t)}} \tag{6.33}$$

We may also derive an alternative expression for the forward swap rate by equating to zero the expression (6.32) given in terms of F's, and obtain

$$\begin{aligned} S_{\alpha,\beta}(t) &= \sum_{i=\alpha+1}^{\beta} w_i(t)\, F_i(t) \\ w_i(t) &= \frac{\tau_i \mathrm{FP}(t, T_\alpha, T_i)}{\sum_{k=\alpha+1}^{\beta} \tau_k \mathrm{FP}(t, T_\alpha, T_k)} = \frac{\tau_i P(t, T_i)}{\sum_{k=\alpha+1}^{\beta} \tau_k P(t, T_k)}. \end{aligned} \tag{6.34}$$

This last expression for S is important because it can lead to useful approximations as follows. It looks like a weighted average, so that forward swap rates can be interpreted as weighted averages of spanning forward rates. However, notice carefully that the weights w's depend on the F's, so that we do not have properly a weighted average. Based on empirical studies showing the variability of the w's to be small compared to the variability of the F's, one can approximate the w's by their (deterministic) initial values $w(0)$ and obtain

$$S_{\alpha,\beta}(t) \approx \sum_{i=\alpha+1}^{\beta} w_i(0) F_i(t)\,.$$

This can be helpful for example in estimating the absolute volatility of swap rates from the absolute volatility of forward rates.

Finally, notice that the IRS discounted payoff (6.31) for a K different from the swap rate can be expressed, at time $t = 0$, also in terms of swap rates as

$$D(0, T_\alpha)\, (S_{\alpha,\beta}(T_\alpha) - K) \sum_{i=\alpha+1}^{\beta} \tau_i P(T_\alpha, T_i), \tag{6.35}$$

We can now move to define a swaption. A swaption is a contract that gives its holder the right (but not the obligation) to enter at a future time $T_\alpha > 0$ an IRS, whose first reset time usually coincides with T_α, with payments occurring at dates $T_{\alpha+1}, T_{\alpha+2}, \ldots, T_\beta$. The fixed rate K of the underlying swap is usually called the swaption strike. As explained in Chapter 1, for the payer swaption this right will be exercised only when the swap rate at the exercise time T_α is larger than the IRS fixed rate K (the resulting IRS has positive value). Consequently, if we assume unit notional amount, the (payer) swaption payoff can be written as

$$D(0,T_\alpha)\,(S_{\alpha,\beta}(T_\alpha)-K)^+\sum_{i=\alpha+1}^{\beta}\tau_i P(T_\alpha,T_i). \tag{6.36}$$

Recall that a swaption is said to be "at-the-money" when its strike price K equals the current value $S_{\alpha,\beta}(0)$ of the underlying forward swap rate. The (payer) swaption is said to be "in-the-money" when $S_{\alpha,\beta}(0) > K$, reflecting the fact that if rates were deterministic, we would get money (a positive multiple of $S_{\alpha,\beta}(0) - K$) from the (payer) swaption contract. The (payer) swaption is said to be "out-of-the-money" when $S_{\alpha,\beta}(0) < K$. Clearly, moneyness for a receiver swaption is defined in the opposite way. A numeraire under which the above forward swap rate $S_{\alpha,\beta}$ follows a martingale, is

$$C_{\alpha,\beta}(t)=\sum_{i=\alpha+1}^{\beta}\tau_i P(t,T_i).$$

Indeed, the product $C_{\alpha,\beta}(t)S_{\alpha,\beta}(t) = P(t,T_\alpha) - P(t,T_\beta)$ gives the price of a tradable asset that, expressed in $C_{\alpha,\beta}$ units, coincides with our forward swap rate. Therefore, when choosing the swap's "present value for basis point" $C_{\alpha,\beta}$ as numeraire, the forward swap rate $S_{\alpha,\beta}(t)$ evolves according to a martingale under the measure $Q^{\alpha,\beta}$ associated with the numeraire $C_{\alpha,\beta}$. The measure $Q^{\alpha,\beta}$ is called *forward-swap measure* or simply swap measure.

By assuming a lognormal dynamics, we obtain

$$dS_{\alpha,\beta}(t)=\sigma^{(\alpha,\beta)}(t)S_{\alpha,\beta}(t)\,dW_t^{\alpha,\beta}, \tag{6.37}$$

where the instantaneous percentage volatility $\sigma^{(\alpha,\beta)}(t)$ is deterministic and $W^{\alpha,\beta}$ is a standard Brownian motion under $Q^{\alpha,\beta}$.

We denote by $v^2_{\alpha,\beta}(T)$ the average percentage variance of the forward swap rate in the interval $[0,T]$ times the interval length:

$$v^2_{\alpha,\beta}(T)=\int_0^T(\sigma^{(\alpha,\beta)}(t))^2\,dt\,.$$

This model for the evolution of forward swap rates is known as **lognormal forward-swap model (LSM)**, since each swap rate $S_{\alpha,\beta}$ has a lognormal distribution under its swap measure $Q^{\alpha,\beta}$.

When pricing a swaption, this model is particularly convenient, since it yields the well-known Black formula for swaptions , as we state in the following.

Proposition 6.7.1. (Equivalence between LSM and Black's swaption prices). *The price of the above payer swaption, as implied by the LSM, coincides with that given by Black's formula for swaptions, i.e.*

$$\mathbf{PS}^{LSM}(0, T_\alpha, [T_\alpha, \ldots, T_\beta], K) = \mathbf{PS}^{Black}(0, T_\alpha, [T_\alpha, \ldots, T_\beta], K)$$
$$= C_{\alpha,\beta}(0) \, \mathrm{Bl}(K, S_{\alpha,\beta}(0), v_{\alpha,\beta}(T_\alpha)),$$

where $\mathrm{Bl}(\cdot,\cdot,\cdot)$ *was defined in Section (6.4).*

Proof. The swaption price is the risk-neutral expectation of the discounted payoff (6.36),

$$E\left(D(0, T_\alpha)\,(S_{\alpha,\beta}(T_\alpha) - K)^+ \, C_{\alpha,\beta}(T_\alpha)\right) \tag{6.38}$$
$$= E^{\boxed{B}}\left(\frac{\boxed{B(0)}}{\boxed{B(T_\alpha)}}\,(S_{\alpha,\beta}(T_\alpha) - K)^+ \, C_{\alpha,\beta}(T_\alpha)\right)$$
$$= E^{\boxed{\alpha,\beta}}\left\{\frac{\boxed{C_{\alpha,\beta}(0)}}{\boxed{C_{\alpha,\beta}(T_\alpha)}}(S_{\alpha,\beta}(T_\alpha) - K)^+ C_{\alpha,\beta}(T_\alpha)\right\}$$
$$= C_{\alpha,\beta}(0)\, E^{\alpha,\beta}\left\{(S_{\alpha,\beta}(T_\alpha) - K)^+\right\},$$

as follows immediately from the definition of measure associated with a numeraire (formula (2.6) with $Z(T) = (S_{\alpha,\beta}(T_\alpha) - K)^+ C_{\alpha,\beta}(T_\alpha)$, $U = B$, $N = C_{\alpha,\beta}$). Boxes refer to Fact Two on the change of numeraire (quantities are invariant when changing numeraire in the three boxes). Now notice that, given the lognormal distribution of S, computing the last expectation in the above formula with the dynamics (6.37) leads to Black's formula for swaptions. Indeed, the above expectation is the classical Black and Scholes price for a call option whose underlying "asset" is $S_{\alpha,\beta}$, struck at K, with maturity T_α, with 0 constant "risk-free rate" and instantaneous percentage volatility $\sigma^{(\alpha,\beta)}(t)$. □

6.7.1 Swaptions Hedging

The knowledge of the self-financing strategy replicating a given (attainable) derivative is extremely useful when in need of managing the risk of short/long positions in the claim itself. For instance, when selling a stock option, the resulting exposure can be hedged by (continuously) trading in the underlying stock and in bonds, if some simplified assumptions are introduced as in Black and Scholes (1973). Another classical example concerns options on zero-coupon bonds, which can be dynamically hedged with two bonds: the underlying one and the zero-coupon bond expiring at the option maturity, see Musiela and Rutkowski (1998).

In general, it is rather difficult to come up with a suitable hedging strategy associated to a given claim. However, in case of European swaptions, and hence caps and floors (which can be viewed as one-period swaptions), a replicating strategy can be easily derived.

Exploiting the similarities between Black's formula for swaptions and the classical Black-Scholes option-pricing formula, Jamshidian (1996) constructed a self-financing strategy exactly replicating the swaption payoff at maturity. His derivation is outlined in the following.

Black's formula for a (payer) swaption at a time $t \geq 0$ is

$$C_{\alpha,\beta}(t)\,\mathrm{Bl}(K, S_{\alpha,\beta}(t), v_{\alpha,\beta}(t, T_\alpha)),$$

where we set

$$v^2_{\alpha,\beta}(t,T) := \int_t^T (\sigma^{(\alpha,\beta)}(u))^2\,du\,.$$

Simple algebra shows that such a formula can be rewritten as:

$$\alpha_+(t)[P(t,T_\alpha) - P(t,T_\beta)] - K\alpha_-(t)\,C_{\alpha,\beta}(t)\,,$$

where

$$\alpha_+(t) = \Phi\left(\frac{\ln(S_{\alpha,\beta}(t)/K) + v^2_{\alpha,\beta}(t,T_\alpha)/2}{v_{\alpha,\beta}(t,T_\alpha)}\right),$$

$$\alpha_-(t) = \Phi\left(\frac{\ln(S_{\alpha,\beta}(t)/K) - v^2_{\alpha,\beta}(t,T_\alpha)/2}{v_{\alpha,\beta}(t,T_\alpha)}\right).$$

Jamshidian then considered the portfolio that at any time t is made of a long position in $\alpha_+(t)$ bonds $P(t,T_\alpha)$, a short position in $\alpha_+(t)$ bonds $P(t,T_\beta)$ and short positions in $K\alpha_-(t)\tau_i$ bonds $P(t,T_i)$, $i = \alpha+1, \ldots, \beta$. This portfolio has a value $V(t)$ that coincides, by construction, with the Black swaption price at every time t, and at time T_α in particular. Moreover, it is self-financing in the sense of Chapter 2, since

$$dV(t) = \alpha_+(t)\,dP(t,T_\alpha) - \alpha_+(t)\,dP(t,T_\beta) - \sum_{i=\alpha+1}^{\beta} K\alpha_-(t)\tau_i\,dP(t,T_i)\,.$$

This has been proved by Jamshidian (1996) under the assumption of a deterministic $v_{\alpha,\beta}(t,T_\alpha)$, which is trivially true in our case given the deterministic nature of the instantaneous percentage volatility $\sigma^{(\alpha,\beta)}(t)$.

The just stated result shows that we can exactly replicate a European-swaption payoff by trading a portfolio of zero-coupon bonds, which must be continuously rebalanced so as to hold the proper amounts of bonds at each time until maturity. However, a due care is needed. Indeed, the self-financing nature of the above replicating strategy deeply relies on the assumption of a deterministic swap-rate volatility, which basically implies the knowledge today of all future implied volatility structures. As a result, the replicating portfolio is just made of bonds, a clear sign that we are only hedging the risk due to fluctuations of interest rates.

In practice, a trader will always try and hedge her exposure to volatility risk, which can even be the largest portion of the risk involved in a deal. However, there is no universal recipe for this and hedging strategies are usually constructed so as to be insensitive to local variations of the risk parameters (interest rates, volatilities, ...). With this respect, the trader's experience and sensibility is invaluable and cannot be replaced with the sheer output of any quantitative model.

6.7.2 Cash-Settled Swaptions

The swaptions that are actually traded in the Euro market are usually defined with a different payoff at maturity T_α, namely

$$(S_{\alpha,\beta}(T_\alpha) - K)^+ \sum_{i=\alpha+1}^{\beta} \tau_i \frac{1}{(1 + S_{\alpha,\beta}(T_\alpha))^{\tau_{\alpha,i}}} .$$

Such swaptions are said to be "cash settled". Here $\tau_{\alpha,i}$ denotes the year fraction between T_α and T_i. At maturity, the relevant swap rate in the payoff comes from the average of swap rates quoted by a number of major banks operating in the Euro market.

In this formulation, instead of the proper discount factors contributing to the usual numeraire

$$\sum_{i=\alpha+1}^{\beta} \tau_i P(t, T_i),$$

a flat yield curve at the swap-rate level at maturity is used to discount the swap payments at maturity. In the Euro market, this is done in order to drastically simplify the determination of the cash settlement. This settlement simplification is not used in the US market, where the classic swaption payoff is kept. The "flat-curve" payoff amounts to choosing the numeraire

$$G_{\alpha,\beta}(t) := \sum_{i=\alpha+1}^{\beta} \tau_i \frac{1}{(1 + S_{\alpha,\beta}(t))^{\tau(t,T_i)}},$$

where we denote by $\tau(t, T_i)$ the year fraction between t and T_i. The correct price of the related cash-settled swaption would be, by risk-neutral valuation:

$$\begin{aligned} &E\left[D(0, T_\alpha)(S_{\alpha,\beta}(T_\alpha) - K)^+ G_{\alpha,\beta}(T_\alpha)\right] \\ &= G_{\alpha,\beta}(0) E^{G_{\alpha,\beta}}\left[(S_{\alpha,\beta}(T_\alpha) - K)^+\right]. \end{aligned}$$

However, we cannot impose a suitable martingale-dynamics distribution of $S_{\alpha,\beta}$ under the numeraire $G_{\alpha,\beta}$, since $S_{\alpha,\beta}G_{\alpha,\beta}$ is not a tradable asset. In order to derive an analytical formula one can use instead the LSM numeraire in the expectation:

$$\begin{aligned} &E\left[D(0, T_\alpha)(S_{\alpha,\beta}(T_\alpha) - K)^+ G_{\alpha,\beta}(T_\alpha)\right] \\ &\approx G_{\alpha,\beta}(0)\, E^{\alpha,\beta}\left[(S_{\alpha,\beta}(T_\alpha) - K)^+\right] \\ &= G_{\alpha,\beta}(0)\, \mathrm{Bl}(K, S_{\alpha,\beta}(0), v_{\alpha,\beta}(T_\alpha)). \end{aligned}$$

The expectation can now be computed as the classical Black-like formula, as we have recalled after (6.38) above.

This last formula could be used to connect swaptions volatilities to swaptions prices. In such a case, when reading volatilities from the Euro market, we should recover the related swaptions prices through this last formula, instead of the LSM Black formula given in Proposition 6.7.1. Nonetheless, we consider classic swaption payoffs and keep the original LSM as swaptions pricing model. We are therefore assuming that prices associated with the two different numeraires are close enough, as seems to happen in most situations. We will therefore ignore from now on the "cash-settled" feature of swaptions in the Euro market and treat them as swaptions in the US market, as suggested by experienced practitioners working in the market (Castagna (2001)).

6.8 Incompatibility between the LFM and the LSM

Recall the LSM dynamics introduced in Section 6.7. When pricing a swaption, this model is particularly convenient, since it yields the well-known Black formula for swaptions. However, for different products involving swap rates there are no analytical formulae in general. Since we selected forward LIBOR rates, rather than forward swap rates, for a basic description of the yield curve, we will rarely consider the dynamics of forward swap rates. The only computation we shall consider here concerns the dynamics of the forward swap rates under the numeraire $P(\cdot, T_\alpha)$, which is a possible numeraire for the forward-rate dynamics given earlier. We can thus express forward-rates and forward-swap-rates dynamics under the unique measure $Q^{P(\cdot,T_\alpha)}$. This unique-measure setup will be useful in the following.

By applying the change-of-numeraire toolkit (formula (2.15)), we obtain the percentage drift $m^\alpha(t)$ for $S_{\alpha,\beta}(t)$ under $Q^{P(\cdot,T_\alpha)}$ as follows:

$$m^\alpha dt = 0\,dt \ - d\ln(S_{\alpha,\beta}(t))\, d\ln\left(\frac{C_{\alpha,\beta}(t)}{P(t,T_\alpha)}\right).$$

The covariation term can be computed as follows. Notice that

$$\ln\left(\frac{C_{\alpha,\beta}(t)}{P(t,T_\alpha)}\right) = \ln\left(\sum_{i=\alpha+1}^{\beta} \tau_i \mathrm{FP}(t;T_\alpha,T_i)\right) =: \chi(F_{\alpha+1}(t), F_{\alpha+2}(t), \ldots, F_\beta(t)),$$

and recall the function ψ defined earlier by

$$S_{\alpha,\beta}(t) = \exp\left(\psi(F_{\alpha+1}(t), F_{\alpha+2}(t), \ldots, F_\beta(t))\right).$$

It follows that

$$m^\alpha dt = \sum_{i,j=\alpha+1}^{\beta} \frac{\partial\psi}{\partial F_i}\frac{\partial\chi}{\partial F_j}\, dF_i(t)dF_j(t).$$

After straightforward but lengthy computations we obtain the following.

Proposition 6.8.1. (Forward-swap-rate dynamics under the numeraire $P(\cdot, T_\alpha)$). *The dynamics of the forward swap rate $S_{\alpha,\beta}$ under the numeraire $P(\cdot, T_\alpha)$ is given by*

$$dS_{\alpha,\beta}(t) = m^\alpha(t) S_{\alpha,\beta}(t)dt + \sigma^S(t) S_{\alpha,\beta}(t) dW_t,$$

$$m^\alpha(t) = \frac{\sum_{h,k=\alpha+1}^{\beta} \mu_{h,k}(t)\tau_h\tau_k FP_h(t)FP_k(t)\rho_{h,k}\sigma_h(t)\sigma_k(t)F_h(t)F_k(t)}{1 - FP(t;T_\alpha,T_\beta)}$$

$$\mu_{h,k}(t) = \frac{\left[FP(t;T_\alpha,T_\beta)\sum_{i=\alpha+1}^{h-1}\tau_i FP(t;T_\alpha,T_i) + \sum_{i=h}^{\beta}\tau_i FP(t;T_\alpha,T_i)\right]}{\left(\sum_{i=\alpha+1}^{\beta}\tau_i FP(t;T_\alpha,T_i)\right)^2} \cdot \sum_{i=k}^{\beta}\tau_i FP(t;T_\alpha,T_i) \tag{6.39}$$

where W is a $Q^{P(\cdot,T_\alpha)}$ standard Brownian motion and where we set $FP_k(t) = FP(t;T_\alpha,T_k)$ for all k for brevity.

Symmetrically, it is possible to work out the dynamics of forward LIBOR rates F under the LSM numeraire $C_{\alpha,\beta}$. Again by applying the change-of-numeraire technique, we have the following.

Proposition 6.8.2. (Forward-rate dynamics under $Q^{\alpha,\beta}$). *The forward-rate dynamics under the forward-swap measure $Q^{\alpha,\beta}$ is given by*

$$dF_k(t) = \sigma_k(t)F_k(t)\left(\mu_k^{\alpha,\beta}(t)dt + dZ_k(t)\right), \tag{6.40}$$

$$\mu_k^{\alpha,\beta}(t) = \sum_{j=\alpha+1}^{\beta}(2\,1_{(j\le k)} - 1)\tau_j\frac{P(t,T_j)}{C_{\alpha,\beta}(t)}\sum_{i=\min(k+1,j+1)}^{\max(k,j)}\frac{\tau_i\rho_{k,i}\sigma_i(t)F_i(t)}{1+\tau_i F_i(t)},$$

where the Z's are now Brownian motions under $Q^{\alpha,\beta}$.

Notice that (6.40) is a closed set of SDEs when k ranges from $\alpha+1$ to β, since the terms

$$\frac{P(t,T_j)}{C_{\alpha,\beta}(t)}$$

can be easily expressed as suitable functions of the spanning forward rates

$$F_{\alpha+1}(t), \ldots, F_\beta(t).$$

We are now able to appreciate the theoretical incompatibility of the two market models, the LFM and the LSM. Lest this appreciation be lost because of an excess of details and formulas, we now recap the situation. When computing the swaption price as

$$E\left(D(0,T_\alpha)\left(S_{\alpha,\beta}(T_\alpha)-K\right)^+ C_{\alpha,\beta}(T_\alpha)\right) = C_{\alpha,\beta}(0)E^{\alpha,\beta}[(S_{\alpha,\beta}(T_\alpha)-K)^+]$$

we have two possibilities. We can either do this with the LFM where the swap rate (6.33) is expressed in terms of the LFM dynamics (6.40), or with the LSM (6.37). Once we have selected the swap measure as our basic measure for pricing, corresponding to choosing $C_{\alpha,\beta}$ as numeraire, we can compare the situations induced by the two different models. Indeed, for a distributional comparison to make sense, we need to work with both models under the same measure, the swap measure in this case.

Pricing the swaption through the above expectation with the LSM means we are basing our computations directly on swap-rate dynamics such as (6.37). In this case the swap-rate distribution is exactly lognormal and the expectation reduces to the well-known Black formula.

Instead, pricing the swaption through the above expectation with the LFM means we are basing our computations on the dynamics of forward LIBOR rates (6.40). These forward LIBOR rates define a swap rate through the algebraic relationship (6.33), and there is no reason for the distribution of the swap rate thus obtained to be lognormal. This therefore explains the incompatibility issue from a theoretical point of view.

However, one may wonder whether this incompatibility holds also from a practical point of view. The answer seems to be negative in general. Indeed, Brace, Dun and Barton (1998) argue that the distribution of swap rates, as implied by the LFM, is not far from being lognormal *in practice.*

We will test ourselves lognormality of the swap rates $S_{\alpha,\beta}(T_\alpha)$, computed as functions of the forward rates, in several situations in Section 8.3, where we will take the forward measure Q^α (instead of the swap measure $Q^{\alpha,\beta}$) as reference measure. In Section 8.3, we will also present several plots of the LFM Q^α-density of the swap rate versus a related lognormal density, and we will show that the two agree well most of times.

We can finally resume the difference between the LFM and the LSM in the following remark.

Remark 6.8.1. **(Distributional Incompatibility of LFM and LSM).** While the swap rate coming from the LSM dynamics (6.37) is lognormally distributed, the swap rate coming from the LFM dynamics (6.33, 6.40) is not lognormal. This results in the two models being theoretically incompatible. However, this incompatibility is mostly theoretical, since in practice we have that the LFM distribution for $S_{\alpha,\beta}$ is almost lognormal.

6.9 The Structure of Instantaneous Correlations

Here we borrow from Brigo (2002). In the previous sections we discussed several possible volatility structures for the LFM and considered the possible consequences of each formulation as far as some fundamental market structures are concerned. We have also seen that both the instantaneous-volatility formulation and the chosen instantaneous correlation can contribute to terminal correlations, which are not determined by instantaneous correlations alone. Before going into details on the possible parameterization of correlations, we recall some typical qualities one would like an instantaneous correlation matrix ρ associated with a LIBOR market model to have.

The main properties are

- We typically expect positive correlations, $\rho_{i,j} \geq 0$ for all i, j.
- When moving away from a "1" diagonal entry of the matrix along a column or a row, we should find a monotonically decreasing pattern. Clearly, joint movements of far away rates are less correlated than movements of rates with close maturities. Formally, the map

$$i \mapsto \rho_{i,j}$$

  has to be decreasing for $i \geq j$.
- When we move along the yield curve, the larger the tenor, the more correlated changes in adjacent forward rates become. This corresponds to require sub-diagonals of the correlation matrix to be increasing, or formally the map

$$i \mapsto \rho_{i+p,i}$$

  has to be increasing for a fixed p. We will come back to this requirement shortly.

Now we focus on instantaneous correlations and present a possible parametric formulation for them.

We start with a given $M \times M$ full-rank correlation matrix, where $M = \beta - \alpha$ is the number of forward rates. In general the full instantaneous correlation matrix is characterized by $M\ (M-1)/2$ entries, given symmetry and the ones in the diagonal. This number of entries can be too high for practical purposes. Therefore, a parsimonious parametric form has to be found for ρ, based on a reduced number of parameters.

As a first possibility, we may decide to maintain a *full-rank* correlation matrix involving a number of parameters that is smaller than $M(M-1)/2$. Such an approach has been followed by Schoenmakers and Coffey (2000), in connection to swaptions calibration in the LIBOR market model. We summarize some of their findings in Section 6.9.1.

Alternatively, *reduced-rank* correlations can be obtained, either through Rebonato's angles parameterization, or through eigenvalues zeroing. We review both possibilities in Section 6.9.2. An interesting iterative method extending the latter approach to achieve a higher accuracy is given in Morini and Webber (2003).

Finally, we mention Zhang and Wu (2001) for their work on optimal low-rank approximations of correlation matrices based on Lagrange multipliers techniques.

6.9.1 Some convenient full rank parameterizations

Schoenmakers and Coffey (2000) propose the following full-rank parametric form for the correlation matrix ρ. They consider a finite sequence of positive real numbers

$$1 = c_1 < c_2 < \ldots < c_M, \quad \frac{c_1}{c_2} < \frac{c_2}{c_3} < \ldots < \frac{c_{M-1}}{c_M},$$

and they set

$$\rho^F(c)_{i,j} := c_i/c_j, \quad i \leq j, \quad i,j = 1,\ldots,M.$$

Notice that the correlation between changes in adjacent rates is $\rho^F_{i+1,i} = c_i/c_{i+1}$. The above requirements on c's translate into the requirement that the sub-diagonal of the resulting correlation matrix $\rho^F(c)$ be increasing when moving from NW ("North-West") to SE ("South-East"). This bears the interpretation that when we move along the yield curve, the larger the tenor, the more correlated changes in adjacent forward rates become. This corresponds to the experienced fact that the forward curve tends to flatten and to move in a more "correlated" way for large maturities than for small ones. Correlations in the sub-diagonal (i.e. referring to changes in adjacent rates) thus tend to grow as the maturity increases (as the index in the sub-diagonal increases). This holds also for lower levels below the diagonal: In general the map

$$i \mapsto \rho_{i+p,i}$$

is increasing for all p. All changes between equally spaced forward rates become more correlated as their expiries increase, not only changes in adjacent forward rates.

The number of parameters needed in this formulation is M, versus the $M(M-1)/2$ number of entries in the general correlation matrix. It is easy to prove that $\rho^F(c)$ is always a viable correlation matrix if defined as above (symmetric, positive semidefinite and with ones in the diagonal).

Schoenmakers and Coffey (2000) observe also that this parameterization can be always characterized in terms of a finite sequence of non-negative numbers $\Delta_2, \ldots, \Delta_M$:

$$c_i = \exp\left[\sum_{j=2}^{i} j\Delta_j + \sum_{j=i+1}^{M} (i-1)\Delta_j\right].$$

Some particular cases in this class of parameterizations that Schoenmakers and Coffey (2000) consider to be promising can be formulated through suitable changes of variables as follows. The first is the case where all Δ's are zero except the last two: by a change of variable one has

Stable, full rank, two-parameters, "increasing along sub-diagonals" parameterization for instantaneous correlation:

$$\rho_{i,j} = \exp\left[-\frac{|i-j|}{M-1}\left(-\ln\rho_\infty + \eta\frac{M-1-i-j}{M-2}\right)\right]. \tag{6.41}$$

Stability here is meant to point out that relatively small movements in the c-parameters connected to this form (and thus in the correlations themselves) cause relatively small changes in ρ_∞ and η. Notice that $\rho_\infty = \rho_{1,M}$ is the correlation between the farthest forward rates in the family considered, whereas η is related to the first non-zero Δ, i.e. $\eta = \Delta_{M-1}(M-1)(M-2)/2$.

A three-parameters form is obtained with the Δ_i's following a straight line (two parameters) for $i = 2, 3, \ldots, M-1$ and set to a third parameter for $i = M$. One finds

$$\rho_{i,j} = \exp\left[-|i-j|\left(\beta - \frac{\alpha_2}{6M-18}\left(i^2+j^2+ij-6i-6j-3M^2+15M-7\right)\right.\right.$$
$$\left.\left.+\frac{\alpha_1}{6M-18}\left(i^2+j^2+ij-3Mi-3Mj+3i+3j+3M^2-6M+2\right)\right)\right]. \tag{6.42}$$

where the parameters should be constrained to be non-negative, if one wants to be sure all the typical desirable properties are indeed present. In order to get parameter stability, Schoenmakers and Coffey introduce a change of variables, thus obtaining a laborious expression generalizing (6.41). The calibration experiments of Schoenmakers and Coffey (2000) pointed out, however, that the parameter associated with the final point Δ_{M-1} of our straight line in the Δ's is practically always close to zero. Setting thus $\Delta_{M-1} = 0$ and maintaining the other characteristics of the last parameterization leads to the following

Improved, stable, full rank, two-parameters, "increasing along sub-diagonals" parameterization for instantaneous correlations:

$$\rho_{i,j} = \exp\left[-\frac{|i-j|}{M-1}\left(-\ln\rho_\infty\right.\right. \tag{6.43}$$
$$\left.\left.+\eta\frac{i^2+j^2+ij-3Mi-3Mj+3i+3j+2M^2-M-4}{(M-2)(M-3)}\right)\right].$$

As before, $\rho_\infty = \rho_{1,M}$, whereas η is related to the steepness of the straight line in the Δ's.

Finally, consider the

Classical, two-parameters, exponentially decreasing parameterization

$$\rho_{i,j} = \rho_\infty + (1-\rho_\infty)\exp[-\beta|i-j|], \quad \beta \geq 0. \tag{6.44}$$

where now ρ_∞ is only asymptotically representing the correlation between the farthest rates in the family.

Schoenmakers and Coffey (2000) point out that Rebonato's (1999c,d) full-rank parameterization, consisting in the following perturbation of the classical structure:

Rebonato's three parameters full rank parameterization

$$\boxed{\rho_{i,j} = \rho_\infty + (1 - \rho_\infty) \exp[-|i - j|(\beta - \alpha (\max(i, j) - 1))],} \tag{6.45}$$

has still the desirable property of being increasing along sub-diagonals. However, this form does not fit the above general framework based on the c parameters, and moreover the domain of positivity for the resulting matrix is not specified "off-line" in terms of the parameters $\alpha, \beta, \rho_\infty$. One has to check at every step of a hypothetical calibration/optimization that the resulting matrix is positive semidefinite. On the contrary, the above formulations based on the c parameters are automatically positive semidefinite and thus do not require an iterative check in a calibration session. Since an unconstrained optimization is preferable to a constrained one, the above parameterizations of Schoenmakers and Coffey (2000) can be preferred from this point of view. Nonetheless, often Rebonato's formulation is preferred because the parameters are easier to calibrate while getting sensible outputs (positive real correlations), and hardly violates the positive (semi)definitiveness of the resulting correlation matrix in practical situations. We ourselves used this last formulation in several situations. See also the discussion in Section 6.19.2 below.

6.9.2 Reduced-rank formulations: Rebonato's angles and eigenvalues zeroing

We know that, being ρ a positive definite symmetric matrix, it can be written as

$$\rho = PHP',$$

where P is a real orthogonal matrix, $P'P = PP' = I_M$, and H is a diagonal matrix of the positive eigenvalues of ρ. *The columns of P are the eigenvectors of ρ, associated to the eigenvalues located in the corresponding position in the diagonal matrix H.* Let Λ be the diagonal matrix whose entries are the square roots of the corresponding entries of H, so that if we set $A := P\Lambda$ we have both

$$AA' = \rho, \quad A'A = H \ .$$

We can try and mimic the decomposition $\rho = AA'$ by means of a suitable n-rank $M \times n$ matrix B such that BB' is an n-rank correlation matrix, with typically $n << M$.

The advantage of doing so is that we may take as new noise a standard n-dimensional Brownian motion W and replace the original M-dimensional random shocks $dZ(t)$ by $B\ dW(t)$. In other terms, we move from a noise correlation structure

$$dZ\ dZ' = \rho\ dt$$

to

$$B\ dW(B\ dW)' = B\ dW dW'\ B' = BB'dt.$$

Therefore, with noise given by $B\ dW$ our new instantaneous noise-correlation matrix is BB' whose rank is $n << M$, and the dimension of our random shocks has decreased to n. We set

$$\rho^B = BB'.$$

If we decide to adopt indeed a reduced-rank approach, we are left with the problem of choosing a suitable parametric form for the B matrix, such that BB' is a possible correlation matrix.

Rebonato (1999d) suggests the following general form for the i-th row of the above B matrix:

$$\begin{aligned} b_{i,1} &= \cos\theta_{i,1} \\ b_{i,k} &= \cos\theta_{i,k}\sin\theta_{i,1}\cdots\sin\theta_{i,k-1}, \quad 1 < k < n, \\ b_{i,n} &= \sin\theta_{i,1}\cdots\sin\theta_{i,n-1}, \end{aligned} \tag{6.46}$$

for $i = 1, 2, \ldots, M$. Notice that with this parameterization ρ^B is clearly positive semidefinite and its diagonal terms are ones. It follows that ρ^B is a possible correlation matrix. The number of parameters in this case is $M \times (n-1)$.

The first attempt is a simple two-factor structure, $n = 2$, consisting of M parameters. This is obtained as

$$b_{i,1} = \cos\theta_{i,1}, \quad b_{i,2} = \sin\theta_{i,1}\ . \tag{6.47}$$

Dropping the second subscript for θ, we have

$$\rho^B_{i,j} = b_{i,1}b_{j,1} + b_{i,2}b_{j,2} = \cos(\theta_i - \theta_j). \tag{6.48}$$

This structure consists of M parameters $\theta_1, \ldots, \theta_M$.

A three-factor structure is given by

$$b_{i,1} = \cos\theta_{i,1}, \quad b_{i,2} = \cos\theta_{i,2}\sin\theta_{i,1}, \quad b_{i,3} = \sin\theta_{i,1}\sin\theta_{i,2}\ , \tag{6.49}$$

so that it follows easily

$$\begin{aligned} \rho^B_{i,j} &= b_{i,1}b_{j,1} + b_{i,2}b_{j,2} + b_{i,3}b_{j,3} \\ &= \cos\theta_{i,1}\cos\theta_{j,1} + \sin\theta_{i,1}\sin\theta_{j,1}\cos(\theta_{i,2} - \theta_{j,2})\ . \end{aligned}$$

Clearly, if $\theta_{k,2}$ is almost constant in k we are back to the two-factor case.

Notice the following interesting feature of the angle parameterization. Assume we use this parameterization while keeping full-rank $n = M$. The number of angle parameters is then $M(M-1)$, and is twice the $M(M-1)/2$ significant entries in the correlation matrix. The angle parameterization in this case increases the dimension, and features too many parameters. The excess in parameters is understood with a toy example in dimension two. Take $n = M = 2$. Then we have only one correlation parameter $\rho_{1,2} = \rho_{2,1} = c$, but two angles:

$$\begin{bmatrix} 1 & c \\ c & 1 \end{bmatrix} = \begin{bmatrix} 1 & \cos(\theta_2 - \theta_1) \\ \cos(\theta_2 - \theta_1) & 1 \end{bmatrix}.$$

As we see, what matters here is the difference between the angles (1 parameter) and not the particular angles themselves (2 parameters).

Incidentally, we see that without further ideas the reduction of the number of parameters in the correlation matrix cannot be solved simply by slightly reducing the rank through the angle formulation. If we wish the number $M(n-1)$ of parameters of the angle parameterization to be much smaller than the significant number of $M(M-1)/2$ entries in the full rank matrix, we need ask not only $n < M$, as the rank reduction implies, but rather

$$n << (M+1)/2.$$

However, there is a way to restore the correct dimension of the angle parameterization. We briefly address this issue in Section 6.9.3.

Now let us assume we have selected $n = 2$ as above. If M is large (typically 20), we can still have troubles with a too large number of parameters θ. We can then select a subparameterization for the θ's of the type

$$\theta_k = \vartheta(k) ,$$

where $\vartheta(\cdot)$ is a function depending on a small (say four or five) number of parameters. Such a function could be for example a linear-exponential combination. However, in our experience keeping the θ's free from a subparameterization can be necessary if we are calibrating the LIBOR market model to a large number of swaptions, when using instantaneous correlations as swaptions-fitting parameters.

In general, when we calibrate the LIBOR market model to swaptions using instantaneous correlations ρ as fitting parameters, we are free to select a priori a parametric form for the correlation matrix. We may for example take the n rank matrix $\rho^B = BB'$ defined above in terms of angles θ. Here the reduced rank is built into the parameterization and we are sure that, once the model has been calibrated, an n-dimensional independent shocks structure $dW(t)$ will be sufficient to perform simulations with M dimensional correlated shocks $B\ dW(t)$.

However, in some parameterizations of the instantaneous covariance structure of the LIBOR market model we will see later on, see also Brigo and

Mercurio (2002a), the correlation matrix is given exogenously (say through historical estimation), instead of being calibrated to the swaption market. When the correlation matrix ρ is given as an exogenous input to the calibration, instead of being kept as a fitting parameter, it will have full rank M in general. The problem in this situation is that when decomposing ρ as BB', the resulting B will be a $M \times M$ matrix, so that the independent random shocks structure $dW(t)$ in the LIBOR market model will have full dimension M, exactly as the correlated noise structure. When in need of performing Monte Carlo simulation to value exotics, this full-dimensionality is an undesirable feature.

This problem can be alleviated if we can find a way to obtain a reduced rank correlation matrix $\rho^{(n)}$ that is, in some sense, the best approximation of the exogenously-given full-rank correlation matrix ρ. Then the swaption calibration proceeds by keeping $\rho^{(n)}$ as exogenous correlation matrix.

We now list some possible approaches to this problem,

$$M\text{-rank-}\rho \quad \longrightarrow n\text{-rank-}\rho^{(n)}.$$

Summing up, we are given in input a full rank $M \times M$ correlation matrix ρ. We aim at approximating this matrix through an at most n-rank correlation matrix $\rho^{(n)}$, $n < M$.

The approach of zeroing the smallest eigenvalues and rescaling. As before, we can try and mimic the decomposition $\rho = AA'$ by means of a suitable n-rank $M \times n$ matrix B such that BB' is an n-rank correlation matrix, with typically $n << M$. But this time, instead of taking B as an angle-parameterized matrix, we define B as follows.

Consider the diagonal matrix $\bar{\Lambda}^{(n)}$ defined as the matrix Λ with the $M-n$ smallest diagonal terms set to zero.

Define then $\bar{B}^{(n)} := P\bar{\Lambda}^{(n)}$, and the related candidate correlation matrix

$$\bar{\rho}^{(n)} := \bar{B}^{(n)}(\bar{B}^{(n)})'.$$

Notice that we can also equivalently define $\Lambda^{(n)}$ as the $n \times n$ (instead of $M \times M$) diagonal matrix obtained from Λ by taking away (instead of zeroing) the $M - n$ smallest diagonal elements and shrinking the matrix correspondingly. Analogously, we can define the $M \times n$ matrix $P^{(n)}$ as the matrix P from which we take away the columns corresponding to the diagonal elements we took away from Λ. More precisely, if we call

$$\Lambda_{i_1,i_1}\ ,\ \Lambda_{i_2,i_2}\ ,\ \ldots,\ \Lambda_{i_{M-n},i_{M-n}}$$

the $M-n$ smallest diagonal elements of Λ, then $P^{(n)}$ is obtained from P by taking away the columns $i_1, i_2, \ldots, i_{M-n}$. The result does not change, in that if we define the $M \times n$ matrix $B^{(n)} = P^{(n)}\Lambda^{(n)}$ we have

$$\bar{\rho}^{(n)} = \bar{B}^{(n)}(\bar{B}^{(n)})' = B^{(n)}(B^{(n)})'.$$

We keep the $B^{(n)}$ formulation. Now the problem is that, in general, while $\bar{\rho}^{(n)}$ is positive semidefinite, it does not feature ones in the diagonal. Throwing away some eigenvalues from Λ has altered the diagonal. The solution is to interpret $\bar{\rho}^{(n)}$ as a *covariance* matrix, and to derive the correlation matrix associated with it. We can do this immediately by defining

$$\rho_{i,j}^{(n)} := \frac{\bar{\rho}_{i,j}^{(n)}}{\sqrt{\bar{\rho}_{i,i}^{(n)}\ \bar{\rho}_{j,j}^{(n)}}}.$$

Now $\rho_{i,j}^{(n)}$ is an n-rank approximation of the original matrix ρ. But how good is the approximation, and are there more precise methods to approximate a full rank correlation matrix with a n-rank matrix?

The approach of optimizing on a low rank parametric form. Rebonato and Jäckel (1999) suggest the following alternative to the above procedure and then compare the two results.

We can start from an n-rank matrix $\rho(\theta) = B(\theta)B'(\theta)$ defined in terms of the angles as before: the i-th row of B is again

$$\begin{aligned} b_{i,1}(\theta) &= \cos\theta_{i,1} \\ b_{i,k}(\theta) &= \cos\theta_{i,k}\sin\theta_{i,1}\cdots\sin\theta_{i,k-1}, \quad 1 < k < n, \\ b_{i,n}(\theta) &= \sin\theta_{i,1}\cdots\sin\theta_{i,n-1}, \end{aligned}$$

where now the dependence on θ has been pointed out explicitly.

When a target full-rank correlation matrix $\hat{\rho}$ is given as input, we can try and find the parameters θ that minimize a norm of the difference between the target matrix ρ and our parameterized matrix $\rho(\theta)$. In other terms, we try and minimize with respect to θ the quantity

$$\sum_{i,j=1}^{M} (|\rho_{i,j} - \rho_{i,j}(\theta)|^2).$$

An important feature of this formulation is that the θ's are completely free: The resulting matrix $\rho(\theta)$ is *always* symmetric, positive semidefinite and with ones in the diagonal, so that we need to pose no constraints in the optimization.

Rebonato and Jäckel (1999) argue that the differences between this method, giving the optimal solution, and the "eigenvalues zeroing" method above is typically small, and show some examples. We are going to confirm these results with a few numerical experiments of our own in the following.

Assume we have a 10×10 full-rank correlation matrix coming from the full-rank classical parametric form:

$$\hat{\rho}_{i,j} = 0.5 + (1 - 0.5)\exp[-0.05|i - j|],$$

and that we first try to fit this matrix with a rank-2 correlation structure.

The input matrix $\hat{\rho}$ to be fitted is

1	0.9756	0.9524	0.9304	0.9094	0.8894	0.8704	0.8523	0.8352	0.8188
0.9756	1	0.9756	0.9524	0.9304	0.9094	0.8894	0.8704	0.8523	0.8352
0.9524	0.9756	1	0.9756	0.9524	0.9304	0.9094	0.8894	0.8704	0.8523
0.9304	0.9524	0.9756	1	0.9756	0.9524	0.9304	0.9094	0.8894	0.8704
0.9094	0.9304	0.9524	0.9756	1	0.9756	0.9524	0.9304	0.9094	0.8894
0.8894	0.9094	0.9304	0.9524	0.9756	1	0.9756	0.9524	0.9304	0.9094
0.8704	0.8894	0.9094	0.9304	0.9524	0.9756	1	0.9756	0.9524	0.9304
0.8523	0.8704	0.8894	0.9094	0.9304	0.9524	0.9756	1	0.9756	0.9524
0.8352	0.8523	0.8704	0.8894	0.9094	0.9304	0.9524	0.9756	1	0.9756
0.8188	0.8352	0.8523	0.8704	0.8894	0.9094	0.9304	0.9524	0.9756	1

In this case we show also the orthogonal (eigenvectors) matrix P and the diagonal (eigenvalues) matrix H such that $\rho = PHP'$. The matrix P is

-0.0714	0.1398	0.2072	0.2657	0.3224	0.3652	-0.4045	-0.4261	0.4293	0.3081
0.2039	-0.3624	-0.4416	-0.424	-0.3099	-0.1325	-0.0844	-0.2703	0.4003	0.3135
-0.3164	0.4468	0.3137	-0.0024	-0.3193	-0.4466	0.3067	-0.0081	0.3247	0.3175
0.3982	-0.3612	0.0715	0.4252	0.3131	-0.1406	0.4401	0.2577	0.2113	0.3202
-0.4412	0.1379	-0.3981	-0.2623	0.3162	0.3603	0.2038	0.423	0.0733	0.3216
0.4412	0.1379	0.3981	-0.2623	-0.3162	0.3603	-0.2038	0.423	-0.0733	0.3216
-0.3982	-0.3612	-0.0715	0.4252	-0.3131	-0.1406	-0.4401	0.2577	-0.2113	0.3202
0.3164	0.4468	-0.3137	-0.0024	0.3193	-0.4466	-0.3067	-0.0081	-0.3247	0.3175
-0.2039	-0.3624	0.4416	-0.424	0.3099	-0.1325	0.0844	-0.2703	-0.4003	0.3135
0.0714	0.1398	-0.2072	0.2657	-0.3224	0.3652	0.4045	-0.4261	-0.4293	0.3081

whereas H is the 10×10 diagonal matrix whose diagonal elements ($\hat{\rho}$'s eigenvalues) are, from NW to SE:

0.0128 0.0138 0.0157 0.0191 0.0249 0.0359 0.0594 0.1268 0.4207 9.2709

In this case the smallest eight eigenvalues are the first eight. We are thus left with the matrix $P^{(2)}$ given by the last two columns of P, and $\Lambda^{(2)}$ given by

$\sqrt{0.4207}$	0
0	$\sqrt{9.2709}$

By defining the 10×2 matrix $B^{(2)} := P^{(2)} \Lambda^{(2)}$ and by following the above procedure we obtain, through scaling of $\bar{\rho}^{(2)} = B^{(2)}(B^{(2)})'$ the final rank-2 correlation matrix $\rho^{(2)}$:

1	0.9997	0.9973	0.9889	0.9713	0.9437	0.9097	0.8761	0.8503	0.838
0.9997	1	0.9987	0.9921	0.9765	0.9511	0.919	0.887	0.8622	0.8503
0.9973	0.9987	1	0.9972	0.9863	0.9656	0.938	0.9094	0.887	0.8761
0.9889	0.9921	0.9972	1	0.9959	0.9824	0.9613	0.938	0.919	0.9097
0.9713	0.9765	0.9863	0.9959	1	0.9953	0.9824	0.9656	0.9511	0.9437
0.9437	0.9511	0.9656	0.9824	0.9953	1	0.9959	0.9863	0.9765	0.9713
0.9097	0.919	0.938	0.9613	0.9824	0.9959	1	0.9972	0.9921	0.9889
0.8761	0.887	0.9094	0.938	0.9656	0.9863	0.9972	1	0.9987	0.9973
0.8503	0.8622	0.887	0.919	0.9511	0.9765	0.9921	0.9987	1	0.9997
0.838	0.8503	0.8761	0.9097	0.9437	0.9713	0.9889	0.9973	0.9997	1

This is the rank-2 matrix resulting from the "eigenvalues zeroing" quick procedure. Notice that we can express this rank-2 matrix by means of its angles parameters. By inverting the relevant transformation, we see that this matrix is associated with the angles $\theta^{(2)}$ given by

$$\theta^{(2)} = [1.2886 \;\; 1.3081 \;\; 1.3586 \;\; 1.4333 \;\; 1.5233 \;\; 1.6183 \;\; 1.7083 \;\; 1.7830 \;\; 1.8335 \;\; 1.8530]'.$$

As a more refined alternative, we now consider the above rank-2 parameterization in the θ's and minimize the difference

$$\sum_{i,j=1}^{M} (|\hat{\rho}_{i,j} - \rho_{i,j}(\theta)|^2)$$

through a numerical optimization.

The numerical optimization takes a few seconds in an interpreted language and yields

$$\theta^{*(2)} = [1.2367 \;\; 1.2812 \;\; 1.3319 \;\; 1.3961 \;\; 1.4947 \;\; 1.6469 \;\; 1.7455 \;\; 1.8097 \;\; 1.8604 \;\; 1.9049].$$

Compare these angles $\theta^{*(2)}$ to $\theta^{(2)}$ to see how close the quickly-computed zeroed-eigenvalues matrix $\rho^{(2)} = \rho(\theta^{(2)})$ is to the optimal rank-2 correlation matrix $\rho(\theta^{*(2)})$.

The resulting optimal rank-2 matrix $\rho(\theta^{*(2)})$ is given by

1	0.999	0.9955	0.9873	0.9669	0.917	0.8733	0.8403	0.8117	0.7849
0.999	1	0.9987	0.9934	0.9773	0.9339	0.8941	0.8636	0.8369	0.8117
0.9955	0.9987	1	0.9979	0.9868	0.9508	0.9157	0.888	0.8636	0.8403
0.9873	0.9934	0.9979	1	0.9951	0.9687	0.9396	0.9157	0.8941	0.8733
0.9669	0.9773	0.9868	0.9951	1	0.9885	0.9687	0.9508	0.9339	0.917
0.917	0.9339	0.9508	0.9687	0.9885	1	0.9951	0.9868	0.9773	0.9669
0.8733	0.8941	0.9157	0.9396	0.9687	0.9951	1	0.9979	0.9934	0.9873
0.8403	0.8636	0.888	0.9157	0.9508	0.9868	0.9979	1	0.9987	0.9955
0.8117	0.8369	0.8636	0.8941	0.9339	0.9773	0.9934	0.9987	1	0.999
0.7849	0.8117	0.8403	0.8733	0.917	0.9669	0.9873	0.9955	0.999	1

To compare the difference between the original matrix $\hat{\rho}$, the zeroed-eigenvalues rank-2 matrix $\rho^{(2)}$ and finally the optimal rank-2 matrix $\rho(\theta^{*(2)})$, we plot the second columns of the three matrices in the upper part of figure 6.4. From the upper figure we see that both rank-2 formulations are in troubles because of the sigmoid-like shape they are forced to assume. On the contrary, the original correlations move away from the diagonal in an almost straight pattern.

The optimal rank-2 matrix is the best we can do with a rank-2 approximation, and we see that "on average" it recovers the true correlation columns, in that it stays partly above and partly below the true correlation (almost straight) plot. Indeed, being forced to assume sigmoid-like columns, the rank-2 optimal approximation oscillates from above to below said column. The rougher zeroed-eigenvalues rank-2 reduction does not even do that, remaining always above the true correlation column. Differences are not huge but are however not negligible. Moreover, if decorrelation is more pronounced, the approximation can worsen.

Rebonato (1998) explains that this is due to the low rank (two) of the parameterized correlation matrix and that there is no way to circumvent this when retaining only two factors. Only when the number of factors approaches the number of forward rates the sigmoid-like shape can be changed in a shape closer to a straight line. Rebonato (1998) devotes a whole chapter to the statistical approach to yield curve models, where matters such as the sigmoid shape are discussed at length.

We now check the analogous approximations if we resort to rank 4 approximation. The resulting correlation matrices are as follows. The zeroed-eigenvalues matrix $\rho^{(4)}$ is given by

1	0.9951	0.9708	0.9379	0.9148	0.8979	0.8821	0.867	0.8448	0.8237
0.9951	1	0.9897	0.966	0.9435	0.9206	0.8985	0.8819	0.8627	0.8448
0.9708	0.9897	1	0.9922	0.9742	0.9463	0.9174	0.8986	0.8819	0.867
0.9379	0.966	0.9922	1	0.9925	0.9685	0.9388	0.9174	0.8985	0.8821
0.9148	0.9435	0.9742	0.9925	1	0.9904	0.9685	0.9463	0.9206	0.8979
0.8979	0.9206	0.9463	0.9685	0.9904	1	0.9925	0.9742	0.9435	0.9148
0.8821	0.8985	0.9174	0.9388	0.9685	0.9925	1	0.9922	0.966	0.9379
0.867	0.8819	0.8986	0.9174	0.9463	0.9742	0.9922	1	0.9897	0.9708
0.8448	0.8627	0.8819	0.8985	0.9206	0.9435	0.966	0.9897	1	0.9951
0.8237	0.8448	0.867	0.8821	0.8979	0.9148	0.9379	0.9708	0.9951	1

whereas the optimal four-rank matrix $\rho(\theta^{*(4)})$ based on the optimal angles is this time

1	0.9957	0.9633	0.9267	0.8999	0.8867	0.8752	0.8598	0.8346	0.8136
0.9957	1	0.9839	0.9559	0.9295	0.9078	0.8893	0.8715	0.8524	0.8346
0.9633	0.9839	1	0.9904	0.9673	0.9317	0.9012	0.8796	0.8715	0.8598
0.9267	0.9559	0.9904	1	0.9911	0.9603	0.9276	0.9012	0.8893	0.8752
0.8999	0.9295	0.9673	0.9911	1	0.9865	0.9603	0.9317	0.9078	0.8867
0.8867	0.9078	0.9317	0.9603	0.9865	1	0.9911	0.9673	0.9295	0.8999
0.8752	0.8893	0.9012	0.9276	0.9603	0.9911	1	0.9904	0.9559	0.9267
0.8598	0.8715	0.8796	0.9012	0.9317	0.9673	0.9904	1	0.9839	0.9633
0.8346	0.8524	0.8715	0.8893	0.9078	0.9295	0.9559	0.9839	1	0.9957
0.8136	0.8346	0.8598	0.8752	0.8867	0.8999	0.9267	0.9633	0.9957	1

The two matrices are associated respectively with the angles

	$\theta^{(4)}$			$\theta^{*(4)}$	
1.6695	1.7239	1.2837	1.6844	1.7328	1.2775
1.5914	1.6672	1.3068	1.6088	1.6828	1.2965
1.496	1.5737	1.358	1.4688	1.581	1.3444
1.4634	1.4784	1.4319	1.4435	1.4708	1.4267
1.5211	1.4194	1.5226	1.5051	1.3957	1.5203
1.6205	1.4194	1.6189	1.6365	1.3957	1.6213
1.6782	1.4784	1.7097	1.6981	1.4708	1.7149
1.6456	1.5737	1.7836	1.6728	1.581	1.7972
1.5502	1.6672	1.8348	1.5328	1.6828	1.8451
1.4721	1.7239	1.8579	1.4571	1.7328	1.864

We can plot again the second columns of the three matrices in the lower part of figure 6.4.

As we see both from the figures and the tables, the approximation has improved.

When moving to rank 7 matrices, the approximation becomes very good and few differences are noticeable in the three matrices.

We now examine a more extreme case where the rates decorrelate very quickly and steeply from 1 to 0. Consider the 10×10 full-rank correlation matrix $\hat{\rho}$ coming from the full-rank classical parametric form:

$$\hat{\rho}_{i,j} = \exp[-|i - j|].$$

The matrix $\hat{\rho}$ is this time

1	0.3679	0.1353	0.0498	0.0183	0.0067	0.0025	0.0009	0.0003	0.0001
0.3679	1	0.3679	0.1353	0.0498	0.0183	0.0067	0.0025	0.0009	0.0003
0.1353	0.3679	1	0.3679	0.1353	0.0498	0.0183	0.0067	0.0025	0.0009
0.0498	0.1353	0.3679	1	0.3679	0.1353	0.0498	0.0183	0.0067	0.0025
0.0183	0.0498	0.1353	0.3679	1	0.3679	0.1353	0.0498	0.0183	0.0067
0.0067	0.0183	0.0498	0.1353	0.3679	1	0.3679	0.1353	0.0498	0.0183
0.0025	0.0067	0.0183	0.0498	0.1353	0.3679	1	0.3679	0.1353	0.0498
0.0009	0.0025	0.0067	0.0183	0.0498	0.1353	0.3679	1	0.3679	0.1353
0.0003	0.0009	0.0025	0.0067	0.0183	0.0498	0.1353	0.3679	1	0.3679
0.0001	0.0003	0.0009	0.0025	0.0067	0.0183	0.0498	0.1353	0.3679	1

If we focus on a rank-4 approximation, the zeroed-eigenvalues procedure yields a matrix $\rho^{(4)}$ given by

1	0.9474	0.5343	-0.0116	-0.1967	-0.0427	0.1425	0.1378	-0.042	-0.1511
0.9474	1	0.775	0.2884	0.0164	-0.03	0.0316	0.0538	0	-0.042
0.5343	0.775	1	0.8137	0.4993	0.0979	-0.1229	-0.1035	0.0538	0.1378
-0.0116	0.2884	0.8137	1	0.8583	0.3725	-0.0336	-0.1229	0.0316	0.1425
-0.1967	0.0164	0.4993	0.8583	1	0.7658	0.3725	0.0979	-0.03	-0.0427
-0.0427	-0.03	0.0979	0.3725	0.7658	1	0.8583	0.4993	0.0164	-0.1967
0.1425	0.0316	-0.1229	-0.0336	0.3725	0.8583	1	0.8137	0.2884	-0.0116
0.1378	0.0538	-0.1035	-0.1229	0.0979	0.4993	0.8137	1	0.775	0.5343
-0.042	0	0.0538	0.0316	-0.03	0.0164	0.2884	0.775	1	0.9474
-0.1511	-0.042	0.1378	0.1425	-0.0427	-0.1967	-0.0116	0.5343	0.9474	1

The optimization on an angle-parameterized rank-4 matrix yields the following output matrix $\rho(\theta^{*(4)})$:

1	0.9399	0.4826	-0.0863	-0.2715	-0.0437	0.1861	0.1808	-0.077	-0.2189
0.9399	1	0.7515	0.234	-0.0587	-0.0572	0.0496	0.0843	-0.0135	-0.077
0.4826	0.7515	1	0.7935	0.4329	0.015	-0.1745	-0.1195	0.0843	0.1808
-0.0863	0.234	0.7935	1	0.8432	0.3222	-0.0872	-0.1745	0.0496	0.1861
-0.2715	-0.0587	0.4329	0.8432	1	0.7421	0.3222	0.015	-0.0572	-0.0437
-0.0437	-0.0572	0.015	0.3222	0.7421	1	0.8432	0.4329	-0.0587	-0.2715
0.1861	0.0496	-0.1745	-0.0872	0.3222	0.8432	1	0.7935	0.234	-0.0863
0.1808	0.0843	-0.1195	-0.1745	0.015	0.4329	0.7935	1	0.7515	0.4826
-0.077	-0.0135	0.0843	0.0496	-0.0572	-0.0587	0.234	0.7515	1	0.9399
-0.2189	-0.077	0.1808	0.1861	-0.0437	-0.2715	-0.0863	0.4826	0.9399	1

If we resort to a rank-7 approximation, the zeroed-eigenvalues approach yields the following matrix $\rho^{(7)}$:

1	0.5481	0.0465	0.0944	0.0507	-0.0493	0.034	0.0169	-0.0441	0.0284
0.5481	1	0.6737	0.0647	0.0312	0.112	-0.0477	-0.0162	0.0691	-0.0441
0.0465	0.6737	1	0.579	0.1227	0.0353	0.0562	0.0012	-0.0162	0.0169
0.0944	0.0647	0.579	1	0.5822	0.0674	0.0806	0.0562	-0.0477	0.034
0.0507	0.0312	0.1227	0.5822	1	0.6472	0.0674	0.0353	0.112	-0.0493
-0.0493	0.112	0.0353	0.0674	0.6472	1	0.5822	0.1227	0.0312	0.0507
0.034	-0.0477	0.0562	0.0806	0.0674	0.5822	1	0.579	0.0647	0.0944
0.0169	-0.0162	0.0012	0.0562	0.0353	0.1227	0.579	1	0.6737	0.0465
-0.0441	0.0691	-0.0162	-0.0477	0.112	0.0312	0.0647	0.6737	1	0.5481
0.0284	-0.0441	0.0169	0.034	-0.0493	0.0507	0.0944	0.0465	0.5481	1

The optimization on an angle-parameterized rank-7 matrix yields the following output matrix $\rho(\theta^{*(7)})$:

1	0.5592	-0.0177	0.1085	0.0602	-0.0795	0.0589	0.018	-0.0734	0.0667
0.5592	1	0.5992	0.0202	0.0277	0.1123	-0.0652	-0.008	0.0797	-0.0734
-0.0177	0.5992	1	0.5464	0.0618	0.0401	0.0561	-0.012	-0.008	0.018
0.1085	0.0202	0.5464	1	0.5556	0.018	0.0834	0.0561	-0.0652	0.0589
0.0602	0.0277	0.0618	0.5556	1	0.5819	0.018	0.0401	0.1123	-0.0795
-0.0795	0.1123	0.0401	0.018	0.5819	1	0.5556	0.0618	0.0277	0.0602
0.0589	-0.0652	0.0561	0.0834	0.018	0.5556	1	0.5464	0.0202	0.1085
0.018	-0.008	-0.012	0.0561	0.0401	0.0618	0.5464	1	0.5992	-0.0177
-0.0734	0.0797	-0.008	-0.0652	0.1123	0.0277	0.0202	0.5992	1	0.5592
0.0667	-0.0734	0.018	0.0589	-0.0795	0.0602	0.1085	-0.0177	0.5592	1

We plot the fifth columns of the original matrix and of the rank-4 approximations in the upper part of figure 6.5. It is clear that, due to the steep decorrelation pattern when moving off the diagonal, both 4-rank approximations are worse than with our earlier example. To obtain a better approximation we have to resort to 7-rank matrices, and the related plots are on the lower part of figure 6.5. Keep in mind, however, that this is an extreme case and that in general rank-4 approximations are satisfactory in most situations.

6.9.3 Reducing the angles

There are some problems left with the general angles parameterization. We pointed out earlier that the θ parameters are redundant. Let us consider the full rank case $n = M$ for a start. In this case we have $M(M-1)$ angle parameters θ, whereas we just need half these parameters to characterize a correlation matrix. Rapisarda found an interesting solution to this problem, based on a geometric interpretation of the angles parameterization, as explained in Brigo, Mercurio and Rapisarda (2002). Rapisarda bases his interpretation on the idea of subsequent Jacobi rotations. Roughly speaking, it turns out that the correlation matrix can be obtained by mutual projections of versors obtained by subsequent rotations of a starting versor. The redundancy in the angles parameterization comes essentially from not fixing this initial versor and not constraining properly the rotation angles.

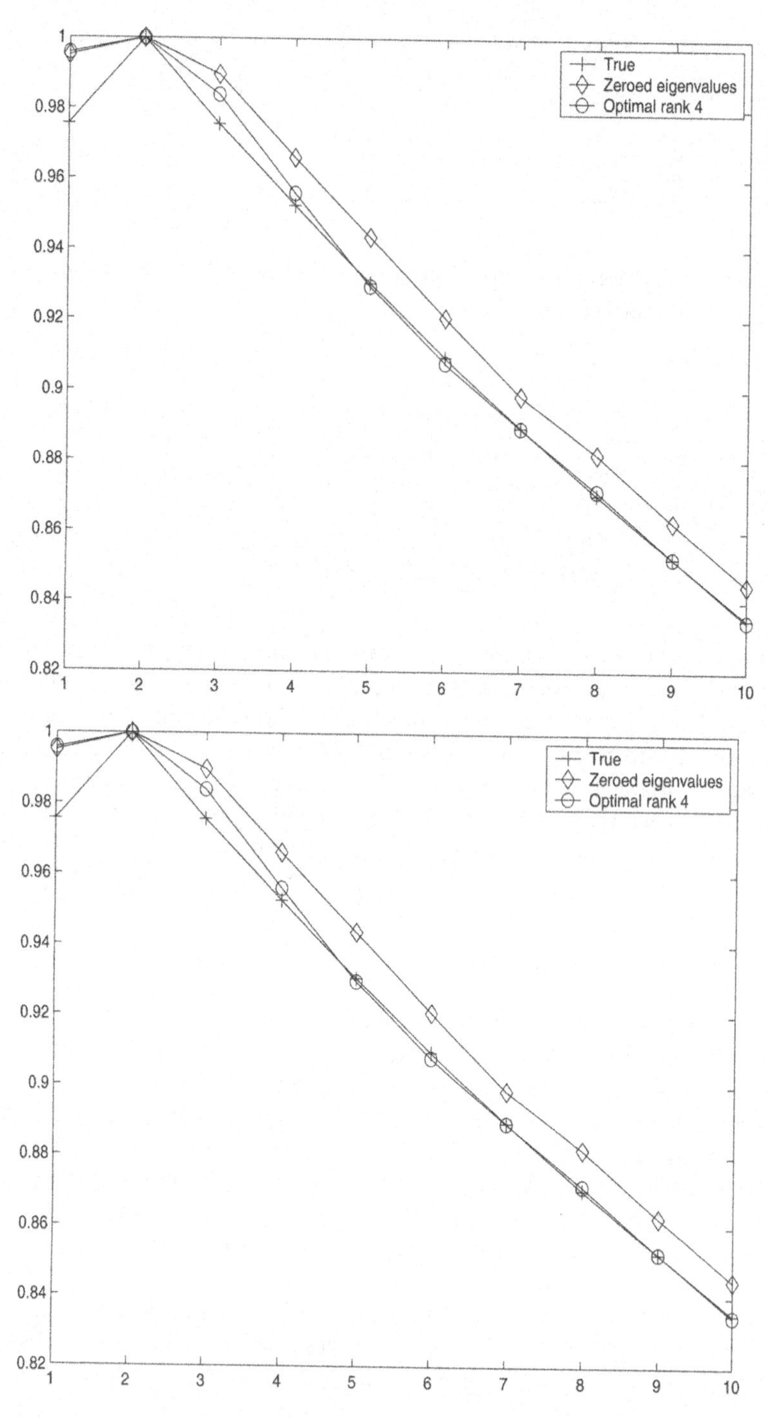

Fig. 6.4. Above: Second columns of $\hat{\rho}$, $\rho^{(2)}$, and $\rho(\theta^{*(2)})$. Below: Second columns of $\hat{\rho}$, $\rho^{(4)}$, and $\rho(\theta^{*(4)})$

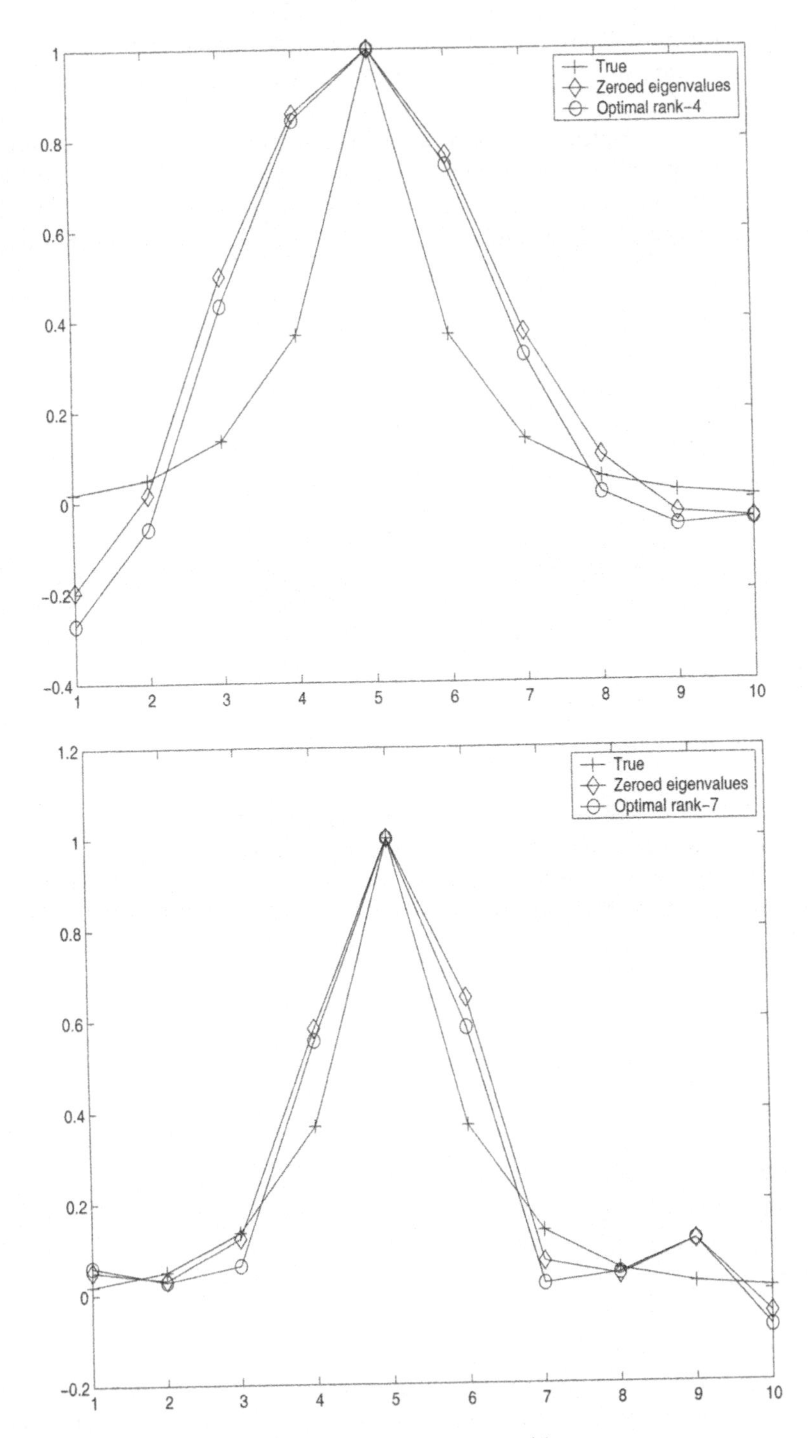

Fig. 6.5. Above: Fifth columns of $\hat{\rho}$, $\rho^{(4)}$, and $\rho(\theta^{*(4)})$. Below: Fifth columns of $\hat{\rho}$, $\rho^{(7)}$, and $\rho(\theta^{*(7)})$

If we establish this initial versor to be for example $[1, 0, \ldots, 0]'$, then redundancy vanishes, and if further we take $\theta_{i,j} = 0$ for each $j \geq i$ when $i < n$, we obtain a "canonical" version of the angle parameterization that is as rich as the initial one in terms of correlation matrices. Incidentally, notice that now the number of nonzero angles is $M(M-1)/2$ in the full-rank case ($n = M$), exactly the number of different entries in a general full-rank correlation matrix. For example, in case $M = n = 5$, we have the angles matrix

$$\begin{array}{|cccc|} 0 & 0 & 0 & 0 \\ \theta_{2,1} & 0 & 0 & 0 \\ \theta_{3,1} & \theta_{3,2} & 0 & 0 \\ \theta_{4,1} & \theta_{4,2} & \theta_{4,3} & 0 \\ \theta_{5,1} & \theta_{5,2} & \theta_{5,3} & \theta_{5,4} \end{array}$$

to which corresponds a triangular matrix $B(\theta)$:

$$\begin{array}{|ccccc|} 1 & 0 & 0 & 0 & 0 \\ b_{2,1} & b_{2,2} & 0 & 0 & 0 \\ b_{3,1} & b_{3,2} & b_{3,3} & 0 & 0 \\ b_{4,1} & b_{4,2} & b_{4,3} & b_{4,4} & 0 \\ b_{5,1} & b_{5,2} & b_{5,3} & b_{5,4} & b_{5,5} \end{array}$$

which is the Cholesky decomposition of the correlation matrix $\rho(\theta)$.

More generally, when the rank is smaller than the matrix size, $n < M$, we have a similar triangular matrix on the top n rows, followed below by a rectangular matrix. For example, in case $n = 5$ and $M = 10$, we have

$$\begin{array}{|cccc|} 0 & 0 & 0 & 0 \\ \theta_{2,1} & 0 & 0 & 0 \\ \theta_{3,1} & \theta_{3,2} & 0 & 0 \\ \theta_{4,1} & \theta_{4,2} & \theta_{4,3} & 0 \\ \theta_{5,1} & \theta_{5,2} & \theta_{5,3} & \theta_{5,4} \\ \theta_{6,1} & \theta_{6,2} & \theta_{6,3} & \theta_{6,4} \\ \theta_{7,1} & \theta_{7,2} & \theta_{7,3} & \theta_{7,4} \\ \theta_{8,1} & \theta_{8,2} & \theta_{8,3} & \theta_{8,4} \\ \theta_{9,1} & \theta_{9,2} & \theta_{9,3} & \theta_{9,4} \\ \theta_{10,1} & \theta_{10,2} & \theta_{10,3} & \theta_{10,4} \end{array}$$

and a similar structure holds for the $B(\theta)$ matrix.

The above standardization of the θ's (and the induced standardization on the B's) helps by reducing the number of parameters in the optimization and also in other respects. It is further possible to constrain the angles in suitable intervals in order to avoid further redundancy. These problems are discussed in Brigo, Mercurio and Rapisarda (2002).

As an example, we consider again the matrix

$$\hat{\rho}_{i,j} = 0.5 + (1 - 0.5)\exp[-0.05|i - j|]$$

given earlier, with $M = 10$. If we take $n = 4$, the resulting canonical "triangularized" angles, replacing the partly redundant $\theta^{*(4)}$ given earlier, are

$\theta^{**(4)}$		
0	0	0
-0.0907	0	0
-0.2709	0.062	0
0.3845	2.8322	2.8972
0.4508	2.5452	2.8267
0.4803	2.1913	2.7029
0.5049	1.9868	2.483
0.5365	1.9199	2.1985
0.5841	2.0295	1.8625
0.621	2.0721	1.7017

The related matrices $\rho(\theta^{*(4)})$ and $\rho(\theta^{**(4)})$, although being very close, are not exactly equal, since the final result depends on approximations and details concerning the optimization procedure. Indeed, $\rho(\theta^{**(4)})$ reads

1	0.9959	0.9635	0.927	0.9001	0.8869	0.8752	0.8595	0.8342	0.8133
0.9959	1	0.9837	0.9555	0.929	0.9075	0.8893	0.8718	0.8529	0.8353
0.9635	0.9837	1	0.9904	0.9674	0.9319	0.9013	0.8795	0.8714	0.8594
0.927	0.9555	0.9904	1	0.991	0.9602	0.9274	0.9012	0.8894	0.8752
0.9001	0.929	0.9674	0.991	1	0.9864	0.9602	0.9318	0.9079	0.8867
0.8869	0.9075	0.9319	0.9602	0.9864	1	0.991	0.9673	0.9294	0.8998
0.8752	0.8893	0.9013	0.9274	0.9602	0.991	1	0.9904	0.9558	0.9268
0.8595	0.8718	0.8795	0.9012	0.9318	0.9673	0.9904	1	0.9839	0.9634
0.8342	0.8529	0.8714	0.8894	0.9079	0.9294	0.9558	0.9839	1	0.9958
0.8133	0.8353	0.8594	0.8752	0.8867	0.8998	0.9268	0.9634	0.9958	1

However, the two matrices can be considered to be equal for all practical purposes.

Since by taking away the first (zeroed) row of this canonical θ's matrix one is left with a lower triangular matrix on top, we refer to this canonical θ parameterization as to the "triangular angles parameterization".

For more details and a geometric insight on this procedure see Brigo, Mercurio and Rapisarda (2002).

We will address further issues on instantaneous correlation, especially as concerns market data calibration, later on in Section 6.18.

6.10 Monte Carlo Pricing of Swaptions with the LFM

In this section we try and address Monte Carlo evaluation with the LFM. We use the fundamental swaption payoff as a guide for our exposition, but the treatment is rather general and can be used to value different payoffs.

Consider again the swaption price

$$E\left(D(0,T_\alpha)\,(S_{\alpha,\beta}-K)^+\sum_{i=\alpha+1}^{\beta}\tau_i P(T_\alpha,T_i)\right)$$
$$= P(0,T_\alpha)E^\alpha\left[(S_{\alpha,\beta}-K)^+\sum_{i=\alpha+1}^{\beta}\tau_i P(T_\alpha,T_i)\right],$$

where, this time, we take the LFM numeraire $P(\cdot,T_\alpha)$ rather than the LSM numeraire $C_{\alpha,\beta}$. The above expectation is "closed" within the LFM because, based on (6.33), the swap rate can be expressed in terms of spanning forward rates at time T_α.

Now, while keeping in mind (6.33), notice carefully that this last expectation depends on the *joint* distribution of spanning forward rates

$$F_{\alpha+1}(T_\alpha), F_{\alpha+2}(T_\alpha), \ldots, F_\beta(T_\alpha).$$

Therefore, as already mentioned, correlations between different forward rates *do matter* in pricing swaptions.

We now consider the Monte Carlo pricing of swaptions. Recalling the dynamics of forward rates under Q^α:

$$dF_k(t) = \sigma_k(t)F_k(t)\sum_{j=\alpha+1}^{k}\frac{\rho_{k,j}\tau_j\sigma_j(t)F_j(t)}{1+\tau_jF_j(t)}dt + \sigma_k(t)F_k(t)dZ_k(t), \quad (6.50)$$

for $k = \alpha+1,\ldots,\beta$, we need to generate, according to such dynamics, m realizations of

$$F_{\alpha+1}(T_\alpha), F_{\alpha+2}(T_\alpha), \ldots, F_\beta(T_\alpha).$$

Subsequently, we can evaluate the payoff

$$(S_{\alpha,\beta}(T_\alpha)-K)^+\sum_{i=\alpha+1}^{\beta}\tau_i P(T_\alpha,T_i) \qquad (6.51)$$

in each realization, and average. This leads to the Monte Carlo price of the swaption.

We now explain how to generate m realizations of the $M=\beta-\alpha$ forward rates

$$F_{\alpha+1}(T_\alpha), F_{\alpha+2}(T_\alpha), \ldots, F_\beta(T_\alpha),$$

which are consistent with the dynamics (6.50). Since the dynamics (6.50) does not lead to a distributionally known process, we need to discretize it with a sufficiently (but not too) small time step Δt, in order to reduce the random inputs to the distributionally known (independent Gaussian) shocks $Z_{t+\Delta t} - Z_t$.

In doing so, "taking logs" can be helpful. By Ito's formula,

$$d \ln F_k(t) = \sigma_k(t) \sum_{j=\alpha+1}^{k} \frac{\rho_{k,j}\tau_j\sigma_j(t)F_j(t)}{1+\tau_j F_j(t)} dt - \frac{\sigma_k(t)^2}{2} dt + \sigma_k(t) dZ_k(t). \quad (6.52)$$

This last equation has the advantage that the diffusion coefficient is deterministic. As a consequence, the naive Euler scheme coincides with the more sophisticated Milstein scheme, see for example Klöden and Platen (1995) or Appendix C, so that the discretization

$$\begin{aligned} \ln F_k^{\Delta t}(t+\Delta t) = \ln F_k^{\Delta t}(t) + \sigma_k(t) \sum_{j=\alpha+1}^{k} \frac{\rho_{k,j}\,\tau_j\,\sigma_j(t)\,F_j^{\Delta t}(t)}{1+\tau_j F_j^{\Delta t}(t)} \Delta t \\ - \frac{\sigma_k(t)^2}{2}\Delta t + \sigma_k(t)(Z_k(t+\Delta t) - Z_k(t)), \end{aligned} \quad (6.53)$$

leads to an approximation of the true process such that there exists a δ_0 with

$$E^\alpha\{|\ln F_k^{\Delta t}(T_\alpha) - \ln F_k(T_\alpha)|\} \le c(T_\alpha)\Delta t \quad \text{for all} \quad \Delta t \le \delta_0$$

where $c(T_\alpha)$ is a positive constant (strong convergence of order 1, from the exponent of Δt on the right-hand side). Recall that $Z_{t+\Delta t} - Z_t \sim \sqrt{\Delta t}\,\mathcal{N}(0,\rho)$, and that $Z_{t_3} - Z_{t_2}$ is independent of $Z_{t_2} - Z_{t_1}$ for all $t_1 < t_2 < t_3$, so that the joint shocks can be taken as independent draws from a multivariate normal distribution $\mathcal{N}(0,\rho)$.

Remark 6.10.1. **(A refined variance for simulating the shocks).** Notice that in integrating equation (6.52) between t and $t+\Delta t$, the resulting Brownian-motion part, in vector notation, is

$$\Delta\zeta_t := \int_t^{t+\Delta t} \underline{\sigma}(s) dZ(s) \sim \mathcal{N}(0, \mathrm{COV}_t)$$

(here the product of vectors acts component by component), where the matrix COV_t is given by

$$(\mathrm{COV}_t)_{h,k} = \int_t^{t+\Delta t} \rho_{h,k}\sigma_h(s)\sigma_k(s)\, ds.$$

Therefore, in principle we have no need to approximate this term by

$$\underline{\sigma}(t)(Z(t+\Delta t) - Z(t)) \sim \mathcal{N}(0,\, \Delta t\, \underline{\sigma}(t)\, \rho\, \underline{\sigma}(t)')$$

as is done in the classical general scheme (6.53). Indeed, we may consider a more refined scheme coming from (6.53) where the following substitution occurs:

$$\underline{\sigma}(t)(Z(t+\Delta t) - Z(t)) \longrightarrow \Delta\zeta_t.$$

The new shocks vector $\Delta\zeta_t$ can be simulated easily through its Gaussian distribution given above. This is the technique we have employed ourselves when implementing numerically the LFM.

Remark 6.10.2. **(The Glasserman and Zhao (2000) no-arbitrage discretization scheme).** Discretizing the continuous-time exact dynamics does not lead to discrete-time interest-rate processes that are compatible with discrete-time no-arbitrage. In other terms, the discretized process leads to bond prices that are not martingales when expressed with respect to the relevant numeraires. Alternatively, one can introduce a discretization scheme that maintains the martingale property required by no-arbitrage in discrete time. This matter is addressed in Glasserman and Zhao (2000). We do not address it here, since we assume that the violation of the no-arbitrage condition due to the time-discretization of no-arbitrage continuous-time processes is negligible enough when choosing sufficiently small discretization steps (as is known to happen in most practical situations).

6.11 Monte Carlo Standard Error

Before briefly introducing the fundamental *variance reduction technique* for Monte Carlo simulation (the control variate technique) we introduce the notion of standard error associated with a Monte Carlo simulation. We introduce such general issues on Monte Carlo simulation at this point of the book because the LIBOR model is probably the interest-rate model where Monte Carlo simulation is most necessary.

The idea is as follows. Assume we need to value a payoff $\Pi(T)$ depending on the realization of different forward LIBOR rates

$$F(t) = [F_{\alpha+1}(t), \ldots, F_\beta(t)]'$$

in a time interval $t \in [0, T]$, where typically $T \leq T_\alpha$.

We have seen a particular case of $\Pi(T) = \Pi(T_\alpha)$ in (6.51). The simulation scheme (6.53) for the rates entering the payoff provides us with the F's needed to form scenarios on $\Pi(T)$. Let us denote by a superscript the scenario (or path) under which a quantity is considered, and let n_p denote the number of simulated paths.

The Monte Carlo price of our payoff is computed, based on the simulated paths, as

$$E[D(0,T)\Pi(T)] = P(0,T)E^T(\Pi(T)) = P(0,T)\sum_{j=1}^{n_p} \Pi^j(T)/n_p,$$

where the forward rates F^j entering $\Pi^j(T)$ have been simulated under the T-forward measure. We omit the T-argument in $\Pi(T), E^T$ and Std^T to contain notation: from now on in this section and in the next one all distributions, expectations and statistics are under the T-forward measure. However, the reasoning is general and holds under any other measure. Incidentally, what we say here can be used with Monte Carlo simulation involving short-rate models: one just has to replace the underlying F's with the additive terms (x and y for the G2++ model (4.4) for example) contributing to the definition of the short-rate r.

We wish to have an estimate of the error me have when estimating the true expectation $E(\Pi)$ by its Monte Carlo estimate $\sum_{j=1}^{n_p} \Pi^j / n_p$. To do so, the classic reasoning is as follows.

Let us view $(\Pi^j)_j$ as a sequence of independent identically distributed (iid) random variables, distributed as Π. By the central limit theorem, we know that under suitable assumptions one has

$$\frac{\sum_{j=1}^{n_p}(\Pi^j - E(\Pi))}{\sqrt{n_p}\,\mathrm{Std}(\Pi)} \to \mathcal{N}(0,1),$$

in law, as $n_p \to \infty$, from which we have that we may write, approximately and for large n_p:

$$\frac{\sum_{j=1}^{n_p} \Pi^j}{n_p} - E(\Pi) \sim \frac{\mathrm{Std}(\Pi)}{\sqrt{n_p}}\,\mathcal{N}(0,1).$$

It follows that

$$\begin{aligned} Q^T\left\{\left|\frac{\sum_{j=1}^{n_p} \Pi^j}{n_p} - E(\Pi)\right| < \epsilon\right\} &= Q^T\left\{|\mathcal{N}(0,1)| < \epsilon\,\frac{\sqrt{n_p}}{\mathrm{Std}(\Pi)}\right\} \\ &= 2\Phi\left(\epsilon\,\frac{\sqrt{n_p}}{\mathrm{Std}(\Pi)}\right) - 1, \end{aligned}$$

where as usual Φ denotes the cumulative distribution function of the standard Gaussian random variable.

The above equation gives the probability that our Monte Carlo estimate $\sum_{j=1}^{n_p} \Pi^j / n_p$ is not farther than ϵ from the true expectation $E(\Pi)$ we wish to estimate. Typically, one sets a desired value for this probability, say 0.98, and derives ϵ by solving

$$2\Phi\left(\epsilon\,\frac{\sqrt{n_p}}{\mathrm{Std}(\Pi)}\right) - 1 = 0.98.$$

For example, since we know from the Φ tables that

$$2\Phi(z) - 1 = 0.98 \iff \Phi(z) = 0.99 \iff z \approx 2.33,$$

we have that

$$\epsilon = 2.33 \frac{\mathrm{Std}(\Pi)}{\sqrt{n_p}}.$$

This means that the true value of $E(\Pi)$ is inside the "window"

$$\left[\frac{\sum_{j=1}^{n_p} \Pi^j}{n_p} - 2.33 \frac{\mathrm{Std}(\Pi)}{\sqrt{n_p}}, \quad \frac{\sum_{j=1}^{n_p} \Pi^j}{n_p} + 2.33 \frac{\mathrm{Std}(\Pi)}{\sqrt{n_p}}\right]$$

with a 98% probability. This window is sometimes called a 98% confidence interval for the mean $E(\Pi)$. Other typical confidence levels are given in Table 6.1. We can see that, ceteris paribus, as n_p increases, the window shrinks

$2\Phi(z) - 1$	$z \approx$
99%	2.58
98%	2.33
95.45%	2
95%	1.96
90%	1.65
68.27%	1

Table 6.1. Confidence levels

as $1/\sqrt{n_p}$, which is worse than the $1/n_p$ rate one would typically wish. This means that if we need to reduce the window size to one tenth, we have to increase the number of scenarios by a factor 100. This implies that sometimes, to reach a chosen accuracy (a small enough window), we need to take a huge number of scenarios n_p. When this is too time-consuming, there are "variance-reduction" techniques that may be used to reduce the above window size. We will hint at these techniques in the next section.

A more fundamental problem with the above window is that the true standard deviation $\mathrm{Std}(\Pi)$ of the payoff is usually unknown. This is typically replaced by the known sample standard deviation obtained by the simulated paths,

$$(\widehat{\mathrm{Std}}(\Pi; n_p))^2 := \sum_{j=1}^{n_p} (\Pi^j)^2/n_p - (\sum_{j=1}^{n_p} \Pi^j/n_p)^2$$

and the actual 98% Monte Carlo window we compute is

$$\left[\frac{\sum_{j=1}^{n_p} \Pi^j}{n_p} - 2.33 \frac{\widehat{\mathrm{Std}}(\Pi; n_p)}{\sqrt{n_p}}, \quad \frac{\sum_{j=1}^{n_p} \Pi^j}{n_p} + 2.33 \frac{\widehat{\mathrm{Std}}(\Pi; n_p)}{\sqrt{n_p}}\right]. \tag{6.54}$$

To obtain a 95% (narrower) window it is enough to replace 2.33 by 1.96, and to obtain a (still narrower) 90% window it is enough to replace 2.33 by 1.65. All other sizes may be derived by the Φ tables.

6.12 Monte Carlo Variance Reduction: Control Variate Estimator

We know that in some cases, to obtain a 98% window (6.54) whose (half-) width $2.33\ \widehat{\mathrm{Std}}(\Pi; n_p)/\sqrt{n_p}$ is small enough, we are forced to take a huge number of paths n_p. This can be a problem for computational time. A way to reduce the impact of this problem is, for a given n_p that we deem to be large enough, to find alternatives that reduce the variance $(\widehat{\mathrm{Std}}(\Pi; n_p))^2$, thus narrowing the above window without increasing n_p.

One of the most effective methods to do this is the control variate technique.

We begin by selecting an alternative payoff Π^{an} which we know how to evaluate analytically, in that

$$E(\Pi^{\mathrm{an}}) = \pi^{\mathrm{an}}$$

is known. When we simulate our original payoff Π we now simulate also the analytical payoff Π^{an} as a function of the same scenarios for the underlying variables F. We define a new control-variate estimator for $E\Pi$ as

$$\widehat{\Pi}_c(\gamma; n_p) := \frac{\sum_{j=1}^{n_p} \Pi^j}{n_p} + \gamma\left(\frac{\sum_{j=1}^{n_p} \Pi^{\mathrm{an},j}}{n_p} - \pi^{\mathrm{an}}\right),$$

with γ a constant to be determined. When viewing Π^j as iid copies of Π and $\Pi^{\mathrm{an},j}$ as iid copies of Π^{an}, the above estimator remains unbiased, since we are subtracting the true known mean π^{an} from the correction term in γ. So, once we have found that the estimator has not been biased by our correction, we may wonder whether our correction can be used to lower the variance.

Consider the random variable

$$\Pi_c(\gamma) := \Pi + \gamma(\Pi^{\mathrm{an}} - \pi^{\mathrm{an}})$$

whose expectation is the $E(\Pi)$ we are estimating, and compute its variance:

$$\mathrm{Var}(\Pi_c(\gamma)) = \mathrm{Var}(\Pi) + \gamma^2\mathrm{Var}(\Pi^{\mathrm{an}}) + 2\gamma\mathrm{Corr}(\Pi, \Pi^{\mathrm{an}})\mathrm{Std}(\Pi)\mathrm{Std}(\Pi^{\mathrm{an}}),$$

We may minimize this function of γ by differentiating and setting the first derivative to zero. We obtain easily that the variance is minimized by the following value of γ:

$$\gamma^* := -\mathrm{Corr}(\Pi, \Pi^{\mathrm{an}})\mathrm{Std}(\Pi)\ /\mathrm{Std}(\Pi^{\mathrm{an}}).$$

By plugging $\gamma = \gamma^*$ into the above expression, we obtain easily

$$\mathrm{Var}(\Pi_c(\gamma^*)) = \mathrm{Var}(\Pi)(1 - \mathrm{Corr}(\Pi, \Pi^{\mathrm{an}})^2),$$

from which we see that $\Pi_c(\gamma^*)$ has a smaller variance than our original Π, the smaller this variance the larger (in absolute value) the correlation between Π and Π^{an}. Accordingly, when moving to simulated quantities, we set

$$\widehat{\mathrm{Std}}(\Pi_c(\gamma^*); n_p) = \widehat{\mathrm{Std}}(\Pi; n_p)(1 - \widehat{\mathrm{Corr}}(\Pi, \Pi^{\mathrm{an}}; n_p)^2)^{1/2}, \tag{6.55}$$

where $\widehat{\mathrm{Corr}}(\Pi, \Pi^{\mathrm{an}}; n_p)$ is the sample correlation

$$\widehat{\mathrm{Corr}}(\Pi, \Pi^{\mathrm{an}}; n_p) = \frac{\widehat{\mathrm{Cov}}(\Pi, \Pi^{\mathrm{an}}; n_p)}{\widehat{\mathrm{Std}}(\Pi; n_p)\, \widehat{\mathrm{Std}}(\Pi^{\mathrm{an}}; n_p)}$$

and the sample covariance is

$$\widehat{\mathrm{Cov}}(\Pi, \Pi^{\mathrm{an}}; n_p) = \sum_{j=1}^{n_p} \Pi^j \Pi^{\mathrm{an},j}/n_p - (\sum_{j=1}^{n_p} \Pi^j)(\sum_{j=1}^{n_p} \Pi^{\mathrm{an},j})/(n_p^2)$$

and

$$(\widehat{\mathrm{Std}}(\Pi^{\mathrm{an}}; n_p))^2 := \sum_{j=1}^{n_p} (\Pi^{\mathrm{an},j})^2/n_p - (\sum_{j=1}^{n_p} \Pi^{\mathrm{an},j}/n_p)^2.$$

One may include the correction factor $n_p/(n_p - 1)$ to correct for the bias of the variance estimator, although the correction is irrelevant for large n_p.

We see from (6.55) that for the variance reduction to be relevant, we need to choose the analytical payoff Π^{an} to be as (positively or negatively) correlated as possible with the original payoff Π we need to evaluate. Notice that in the limit case of correlation equal to one the variance shrinks to zero.

The window for our control-variate Monte Carlo estimate $\widehat{\Pi}_c(\gamma; n_p)$ of $E(\Pi)$ is now:

$$\left[\widehat{\Pi}_c(\gamma; n_p) - 2.33\,\frac{\widehat{\mathrm{Std}}(\Pi_c(\gamma^*); n_p)}{\sqrt{n_p}},\quad \widehat{\Pi}_c(\gamma; n_p) + 2.33\,\frac{\widehat{\mathrm{Std}}(\Pi_c(\gamma^*); n_p)}{\sqrt{n_p}}\right], \tag{6.56}$$

where $\widehat{\mathrm{Std}}(\Pi_c(\gamma^*); n_p)$ is computed as in (6.55). This window is narrower than (6.54) by a factor $(1 - \widehat{\mathrm{Corr}}(\Pi, \Pi^{\mathrm{an}}; n_p)^2)^{1/2}$.

What we introduced so far is rather general and can be applied to risk-neutral Monte Carlo pricing in any market. We may wonder about a good possible Π^{an} in case the Π we need value is an interest-rate derivative. As a general rule, the first attempt is with the underlying variables themselves. In case of the LIBOR model, we may select as Π^{an} the simplest payoff depending on the underlying rates

$$F(t) = [F_{\alpha+1}(t), \ldots, F_\beta(t)]'.$$

This is given by the FRA contract we introduced in Section 1.4. We consider the sum of at-the-money FRA payoffs, each on a single forward rate included in our family.

In other terms, if we are simulating under the T_j forward measure a payoff paying at T_α, with , the payoff we consider is

$$\Pi^{\mathrm{an}}(T_\alpha) = \sum_{i=\alpha+1}^{\beta} \tau_i P(T_\alpha, T_i)(F_i(T_\alpha) - F_i(0))/P(T_\alpha, T_j)$$

whose expected value under the Q^j measure is easily seen to be 0 by remembering that quantities featuring $P(\cdot, T_j)$ as denominator are martingales. Thus in our case $\pi^{\mathrm{an}} = 0$ and we may use the related control-variate estimator. Somehow surprisingly, this simple correction has allowed us to reduce the number of paths of up to a factor 10 in several cases, including for example Monte Carlo evaluation of ratchet caps, a payoff we will describe later in the book.

6.13 Rank-One Analytical Swaption Prices

In this section we present approximations that lead to analytical formulae for swaption prices. This can be helpful for calibration and hedging purposes. The presentation is based on Brace (1996). Consider the same swaption as in the previous section, whose underlying is $S_{\alpha,\beta}$. In order to obtain an analytical formula for swaption prices, some simplifications are needed. Indeed, the swaption price can be also written

$$E\left[D(0,T_\alpha)\left(\sum_{i=\alpha+1}^{\beta} P(T_\alpha,T_i)\tau_i(F_i(T_\alpha)-K)\right)^+\right]$$

$$= \sum_{i=\alpha+1}^{\beta} \tau_i P(0,T_i)\, E^i\left[(F_i(T_\alpha)-K)\, 1_A\right],$$

where

$$A := \left\{\left(\sum_{i=\alpha+1}^{\beta} P(T_\alpha,T_i)\tau_i(F_i(T_\alpha)-K)\right) > 0\right\} = \{S_{\alpha,\beta}(T_\alpha) > K\} \;,$$

and where 1_A denotes the indicator function of the set A. The problem consists of finding approximations leading to analytical formulas for

$$E^i\left[(F_i(T_\alpha)-K)\, 1_A\right].$$

We can approach the problem as follows. Consider the vector-dynamics for forward rates $F_{\alpha+1}, \ldots, F_\beta$. Denote by

$$Z_t = [Z_{\alpha+1}(t), \ldots, Z_\beta(t)]'$$

the vector-collected Brownian motions of the forward rates, which are still "alive" at time $t < T_\alpha$, and by F the corresponding vector of forward rates

$$F(t) = [F_{\alpha+1}(t), \dots, F_\beta(t)]'.$$

We know that the dynamics under Q^γ, with $\gamma \geq \alpha$, is

$$dF_k(t) = -\sigma_k(t)F_k(t) \sum_{j=k+1}^{\gamma} \frac{\rho_{k,j}\tau_j\sigma_j(t)F_j(t)}{1+\tau_jF_j(t)} dt + \sigma_k(t)F_k(t)dZ_k(t),$$
$$\text{for } k = \alpha+1, \dots, \gamma-1,$$
$$dF_k(t) = \sigma_k(t)F_k(t)dZ_k(t), \quad \text{for } k = \gamma,$$
$$dF_k(t) = \sigma_k(t)F_k(t) \sum_{j=\gamma+1}^{k} \frac{\rho_{k,j}\tau_j\sigma_j(t)F_j(t)}{1+\tau_jF_j(t)} dt + \sigma_k(t)F_k(t)dZ_k(t),$$
$$\text{for } k = \gamma+1, \dots, \beta.$$

A first approximation consists in replacing the drift with deterministic coefficients:

$$-\sum_{j=k+1}^{\gamma} \frac{\rho_{k,j}\tau_j\sigma_j(t)F_j(t)}{1+\tau_jF_j(t)} \approx -\sum_{j=k+1}^{\gamma} \frac{\rho_{k,j}\tau_j\sigma_j(t)F_j(0)}{1+\tau_jF_j(0)} =: \mu_{\gamma,k}(t), \ k < \gamma,$$
$$0 =: \mu_{\gamma,\gamma}(t), \ k = \gamma,$$
$$\sum_{j=\gamma+1}^{k} \frac{\rho_{k,j}\ \tau_j\sigma_j(t)F_j(t)}{1+\tau_jF_j(t)} \approx \sum_{j=\gamma+1}^{k} \frac{\rho_{k,j}\tau_j\sigma_j(t)\ F_j(0)}{1+\tau_jF_j(0)} =: \mu_{\gamma,k}(t), \ k > \gamma.$$

The forward-rate dynamics is now

$$dF_k(t) = \sigma_k(t)\mu_{\gamma,k}(t)F_k(t)dt + \sigma_k(t)F_k(t)dZ_k(t)$$

under Q^γ. This equation describes a geometric Brownian motion and can be easily integrated. It turns out, by doing so, that

$$F(T_\alpha) = F(0)\exp\left(\int_0^{T_\alpha} \sigma(t)\mu_{\gamma,\cdot}(t)dt - \frac{1}{2}\int_0^{T_\alpha} \sigma(t)^2 dt\right)\exp(X^\gamma) \quad (6.57)$$

where operators are supposed to act componentwise and

$$X^\gamma \sim N(0, V), \quad V_{i,j} = \int_0^{T_\alpha} \sigma_i(t)\sigma_j(t)\rho_{i,j}dt.$$

It is easy to check that

$$X^\gamma = X^\alpha - \int_0^{T_\alpha} \sigma(t)[\mu_{\gamma,\cdot}(t) - \mu_{\alpha,\cdot}(t)]dt.$$

Now a second approximation can be introduced. We approximate the (generally $(\beta - \alpha)$-rank) matrix V by the rank-one matrix obtained through V's dominant eigenvalue. To do so, *we need to assume* $\rho > 0$. Indeed, in such a case, all components of V are strictly positive. It follows that V is an irreducible matrix, and as such (by the Perron-Frobenius theorem) it admits a unique dominant eigenvalue whose associate dominant eigenvector has positive components. Define

$$\Gamma := \sqrt{\lambda_1(V)}\, e_1(V),$$

where $\lambda_1(V)$ denotes the largest eigenvalue of V and $e_1(V)$ the corresponding eigenvector. Positivity of Γ, ensured by positivity of ρ, will be needed in the following.

More generally, denote by $\lambda_k(V)$ the k-th largest eigenvalue and by $e_k(V)$ the corresponding eigenvector.

We approximate V by the rank-one matrix

$$V^1 := \Gamma\Gamma',$$

and replace V with V^1 in the previous dynamics. Therefore, we still consider (6.57) but now we take

$$X^\gamma \sim N(0, V^1),$$

which, distributionally, amounts to setting

$$X_i^\gamma = \Gamma_i U^\gamma,$$

with U^γ a scalar standard Gaussian random variable under Q^γ. The advantage is that now the whole vector $F(t)$ has all its randomness condensed in the scalar random variable U^γ.

By backward substitution we see that

$$\begin{aligned}
\int_0^{T_\alpha} \sigma_k(t)\mu_{\gamma,k}(t)dt &= -\Gamma_k \sum_{j=k+1}^{\gamma} \frac{\tau_j \Gamma_j F_j(0)}{1+\tau_j F_j(0)} =: \Gamma_k\, q_{\gamma,k},\ k < \gamma\,, \\
\int_0^{T_\alpha} \sigma_\gamma(t)\mu_{\gamma,\gamma}(t)dt &= -\Gamma_\gamma\, 0 =: \Gamma_\gamma\, q_{\gamma,\gamma},\ k = \gamma\,, \qquad (6.58) \\
\int_0^{T_\alpha} \sigma_k(t)\mu_{\gamma,k}(t)dt &= \Gamma_k \sum_{j=\gamma+1}^{k} \frac{\tau_j \Gamma_j F_j(0)}{1+\tau_j F_j(0)} =: \Gamma_k\, q_{\gamma,k},\ k > \gamma\,.
\end{aligned}$$

The corresponding forward-rate vector is then

$$F(T_\alpha) = F(0)\exp(\Gamma q_{\gamma,\cdot} - \tfrac{1}{2}\Gamma^2)\exp(X^\gamma)\,. \qquad (6.59)$$

Notice that

$$X^\gamma = X^g - \Gamma\,(q_{\gamma,\cdot} - q_{g,\cdot}),\ \ \gamma \ge g,\ \ g = \alpha, \ldots, \beta\,,$$

which can be rewritten in terms of U's as

$$X^\gamma = \Gamma\,(U^g + q_{g,\cdot} - q_{\gamma,\cdot})\,.$$

Since in our treatment below the maturity T_α will play a central role, we specialize our notation for the case $g = \alpha$. We set

$$p = q_{\alpha,\cdot}\,, \quad U = U^\alpha,$$

so that

$$X^\gamma = \Gamma\,(U + p - q_{\gamma,\cdot})\,, \quad \gamma = \alpha, \ldots, \beta.$$

Notice also that the q's can be easily expressed in terms of p's as

$$q_{\gamma,j} = p_j - p_\gamma\,.$$

We are now able to express the event A in a convenient way. Denote by $F_j(T_\alpha; U^j)$ the forward rate obtained according to the above approximation when $\gamma = j$:

$$\begin{aligned} F_j(T_\alpha; U^j) &:= F_j(0)\exp(-\tfrac{1}{2}\Gamma_j^2)\exp(\Gamma_j U^j) \\ &= F_j(0)\exp(-\tfrac{1}{2}\Gamma_j^2)\exp(\Gamma_j(U + p_j)) =: F_j(T_\alpha; U)\,, \end{aligned} \tag{6.60}$$

from which we deduce that

$$U^j = U + p_j\,.$$

Then rewrite A as

$$\begin{aligned} &\sum_{j=\alpha+1}^{\beta}\left(\prod_{k=\alpha+1}^{j}\frac{1}{1+\tau_k F_k(T_\alpha;U)}\right)\tau_j(F_j(T_\alpha;U) - K) \\ &\quad =: \sum_{j=\alpha+1}^{\beta} P(T_\alpha, T_j; U)\tau_j(F_j(T_\alpha;U) - K) > 0\,, \end{aligned} \tag{6.61}$$

or, by expressing the forward rates in terms of bond prices,

$$1 - P(T_\alpha, T_\beta; U) - \sum_{j=\alpha+1}^{\beta} \tau_j K P(T_\alpha, T_j; U) > 0\,.$$

It is possible to show that the partial derivative with respect to U of the left hand side of this inequality is always positive, *provided that* $\Gamma > 0$ (and this is why we need irreducibility of V, guaranteed by $\rho > 0$). Indeed, it is enough to show that (the logarithm of) each term in the above summation has positive partial derivative with respect to U. This amounts to showing that

$$\frac{\partial}{\partial U} \ln P(T_\alpha, T_j; U) < 0, \quad j = \alpha + 1, \ldots, \beta.$$

This can be rewritten as

$$\frac{\partial}{\partial U} \left[\sum_{k=\alpha+1}^{j} \ln(1 + \tau_k F_k(T_\alpha; U)) \right] > 0\,,$$

and, again, it is enough to prove that each term in the summation has partial derivative with respect to U with the right sign. In doing so we can take away logarithms. Eventually, a sufficient condition for the partial derivative of the left-hand side of (6.61) to be positive is

$$\frac{\partial}{\partial U} F_k(T_\alpha; U) > 0, \quad k = \alpha + 1, \ldots, \beta\,.$$

A quick investigation shows this to be the case, provided that $\Gamma > 0$. As a consequence, the left-hand side of (6.61) is an increasing function of U, and therefore the equation

$$\sum_{j=\alpha+1}^{\beta} \left(\prod_{k=\alpha+1}^{j} \frac{1}{1 + \tau_k F_k(T_\alpha; U_*)} \right) \tau_j (F_j(T_\alpha; U_*) - K) = 0$$

has a unique solution U_*. Moreover, from monotonicity, inequality (6.61) is equivalent to

$$U > U_*\,,$$

or

$$U^j > U_* + p_j\,.$$

We can finally compute the above expectation with the following approximation

$$E^i\left[(F_i(T_\alpha) - K)\, 1_A\right] = E^i\left[(F_i(T_\alpha) - K)\ 1_{(U^i > U_* + p_i)}\right],$$

where $U^i \sim \mathcal{N}(0, 1)$. More explicitly,

$$\begin{aligned}
&E^i\left[(F_i(T_\alpha) - K)\, 1_A\right] \\
&= E^i\left[(F_i(T_\alpha) - K)\, 1_{(U^i > U_* + p_i)}\right] \\
&= E^i\left[\left(F_i(0) \exp(-\tfrac{1}{2}\Gamma_i^2 + \Gamma_i U^i) - K\right)\ 1_{(U^i > U_* + p_i)}\right] \\
&= E^i\left[\left(F_i(0) \exp(-\tfrac{1}{2}\Gamma_i^2 + \Gamma_i U^i) - K\right)\ 1_{(F_i(0)\exp(-\frac{1}{2}\Gamma_i^2 + \Gamma_i U^i) > F_i^*)}\right] \\
&= E^i\left[\left(F_i(0) \exp(-\Gamma_i^2 + \Gamma_i U^i) - F_i^* + F_i^* - K\right)\ 1_{(F_i(0)\exp(-\frac{1}{2}\Gamma_i^2 + \Gamma_i U^i) > F_i^*)}\right] \\
&= E^i\left[F_i(0) \exp(-\Gamma_i^2 + \Gamma_i U^i) - F_i^*\right]^+ \\
&\qquad + (F_i^* - K)\ E^i\left[1_{(F_i(0)\exp(-\frac{1}{2}\Gamma_i^2 + \Gamma_i U^i) > F_i^*)}\right] \\
&= F_i(0)\Phi(d_1(F_i^*, F_i(0), \Gamma_i)) - K\Phi(d_2(F_i^*, F_i(0), \Gamma_i))
\end{aligned}$$

where the last expression is obtained by adding a call price to a cash-or-nothing-call price, and

$$F_i^* := F_i(0)\exp(-\tfrac{1}{2}\Gamma_i^2 + \Gamma_i(U_* + p_i)).$$

We then have the following.

Proposition 6.13.1. (Brace's rank-one formula). *The approximated price of the above swaption is given by*

$$\sum_{i=\alpha+1}^{\beta} \tau_i P(0,T_i)\left[F_i(0)\Phi(\Gamma_i - U_* - p_i) - K\Phi(-U_* - p_i)\right] . \tag{6.62}$$

This formula is analytical and just involves a root-searching procedure to find U_*.

The formulation we presented here works for any possible choice of the instantaneous volatilities $(\sigma_k(t))_k$ and correlation ρ structures. However, in applying this apparatus, if one starts from a parametric form for volatilities and correlations, one might as well choose a parameterization that renders the computation of the matrix Γ as simple as possible. Another desirable feature is the approximation between the real integrated covariance matrix V and its Γ-based approximation V^1 to be as good as possible.

Now suppose to choose a say rank-two instantaneous-correlation matrix ρ^B and then apply the formula given here. The problem with this setup is that, since the integrated covariance matrix V will be approximated via a rank-one matrix, we may as well start from a rank-one instantaneous covariance matrix. This is obtained by taking a one-factor model, i.e. by setting all ρ's components equal to one. If one does this, the approximation needed to obtain a rank one matrix V^1 from V will be minimal.

But we can do even better. We can check whether there is a volatility formulation among (6.8), (6.9), (6.10) and (6.11) leading to an integrated covariance matrix V that has automatically rank one. It turns out that this is possible. Indeed, when taking a one-factor model ($\rho_{i,j} = 1$ for all i, j), the integrated covariance matrix has rank one if (and only if) the volatility structure is separable as in (6.10), (and in particular as in (6.9)), see Brace, Dun and Barton (1998) for the "only if" part. Indeed, under formulation (6.10) we have

$$V_{i,j} = \Phi_i \Phi_j \sum_{k=0}^{\alpha} \tau_{j-1,j} \Psi_{j+1}^2 ,$$

which gives a rank-one matrix, so that it suffices to set

$$\Gamma = \sqrt{\sum_{k=0}^{\alpha} \tau_{j-1,j} \Psi_{j+1}^2} \;\; [\Phi_{\alpha+1}, \ldots, \Phi_\beta]' ,$$

and no eigenvalue/eigenvector analysis or calculations are needed.

A simpler formulation is obtained with the volatility structure (6.9) (and $\rho_{i,j} = 1$ for all i, j as before). Indeed, in such a case, $V_{i,j} = s_i s_j \tau_{0,\alpha}$, so that we can directly take $\Gamma_i = s_i \sqrt{\tau_{0,\alpha}}$.

Finally notice that even in the one-factor case ($\rho_{i,j} = 1$ for all i, j), formulations (6.8) and more generally (6.11) do not lead to a rank-one integrated covariance matrix V in general, so that if we choose (6.8) or (6.11) we actually need to include an eigenvector/eigenvalue calculation in order to apply Brace's formula (6.62). The integrated covariance matrix with formulation (6.11) is given by

$$V_{i,j} = \Phi_i \Phi_j \sum_{k=0}^{\alpha} \tau_{j-1,j} \psi_{j-k} \psi_{i-k} ,$$

and Γ has to be derived through the dominant eigenvector of this matrix. This last approach has the disadvantage of requiring an explicit eigenvalues/eigenvectors calculation. However, as stated in Section 6.5, it has the advantage of giving a possibly good qualitative behaviour for the term structure of volatilities over time, and is also more controllable than the previous one.

6.14 Rank-r Analytical Swaption Prices

The rank-one analytical approximation of Brace (1996) has been extended to the rank-r case by Brace (1997) himself as follows. The derivation is the same as in Section 6.13 up to the definition of Γ. This time we select Γ to be the following matrix

$$\Gamma := \left[\sqrt{\lambda_1(V)}\, e_1(V) \ \ \sqrt{\lambda_2(V)}\, e_2(V) \ \ \ldots \ \ \sqrt{\lambda_r(V)}\, e_r(V) \right],$$

leading to the following rank-r approximation of V,

$$V^r := \Gamma \Gamma'.$$

Again, we substitute backwards in the dynamics. Therefore, we still consider (6.57) but now we take

$$X^\gamma \sim N(0, V^r),$$

which amounts to setting

$$X_i^\gamma = \Gamma_{(i)} U^\gamma ,$$

where $\Gamma_{(i)}$ denotes the i-th row of the Γ matrix, and with U^γ a r-th dimensional standard Gaussian random vector under Q^γ. Now the whole $(\beta - \alpha)$-dimensional vector $F(t)$ has all its randomness condensed in the r-dimensional random vector U^γ, with typically $r << \beta - \alpha$.

Formulas (6.58) still hold when replacing Γ_k with $\Gamma_{(k)}$ and Γ_j with $\Gamma'_{(j)}$. Now the q's are r-vectors, and so are the p_j's. Under the same adjustments, (6.60) still holds, and the "exercise set" A can still be expressed as (6.61). However, this time the situation is more delicate since U is a vector. We can deal with this more general case as follows.

Brace (1997) observed that close numerical examination of the U-surface

$$\sum_{j=\alpha+1}^{\beta}\left(\prod_{k=\alpha+1}^{j}\frac{1}{1+\tau_k F_k(T_\alpha;U)}\right)\tau_j(F_j(T_\alpha;U)-K) \tag{6.63}$$
$$=: \sum_{j=\alpha+1}^{\beta} P(T_\alpha,T_j;U)\tau_j(F_j(T_\alpha;U)-K)=0$$

inside the hypercube $[-5,5]^r$ (where almost all of the probability density of U is concentrated) reveals a slightly curved surface close to be a hyperplane, which is almost perpendicular to the U_1 axis. We therefore assume the solution of (6.63) to be a hyperplane:

$$U_1 = s_1 + \sum_{k=2}^{r} s_k U_k\,. \tag{6.64}$$

Under this assumption, if the partial derivative with respect to U_1 of the left-hand side $f(U)$ of (6.63) is positive, we can deduce that

$$A=\left\{[U_1,U_2,\dots,U_r]:\ U_1 > s_1+\sum_{k=2}^{r}s_kU_k\right\}. \tag{6.65}$$

Indeed, in such a case,

$$f(U_1,U_2,\dots,U_k) > f\left(s_1+\sum_{k=2}^{r}s_kU_k,U_2,\dots,U_k\right)=0\,,$$

which was the inequality to be solved.

The partial derivative with respect to U_1 of the left-hand side of (6.63) can be shown to be positive in exactly the same way as in the rank-one case. This is guaranteed by positivity of the first column of Γ, which, in turn, is guaranteed by positivity of ρ (and therefore of V) through the Perron-Frobenius theorem.

At this point we need to determine $s_1, s_2, \dots, s_r$. To do so, we need to perform $2r-1$ root searches similar to the ones of the rank-one case. We proceed as follows.

- Set $U=[\alpha_1,0,\dots,0]$ and solve numerically (6.63) thus finding α_1. As a solution, $U=[\alpha_1,0,\dots,0]$ satisfies (6.64) and therefore

$$s_1=\alpha_1.$$

- Set $U = [\alpha_2^-, -1/2, 0, \dots, 0]$ and solve numerically (6.63), thus finding α_2^-. Set $U = [\alpha_2^+, 1/2, 0, \dots, 0]$ and solve numerically (6.63), thus finding α_2^+. Being solutions, both $[\alpha_2^-, -1/2, 0, \dots, 0]$ and $[\alpha_2^+, 1/2, 0, \dots, 0]$ satisfy (6.64), from which it follows easily
$$s_2 = \alpha_2^+ - \alpha_2^- .$$
- For all other k, set $U = [\alpha_k^-, 0, \dots, 0, -1/2, 0, \dots, 0]$, with 1/2 as k-th component of the vector U, and solve numerically (6.63), thus finding α_k^-. Set $U = [\alpha_k^+, 0, \dots, 0, 1/2, 0, \dots, 0]$ and solve numerically (6.63), thus finding α_k^+. Being solutions, these two vectors U both satisfy (6.64), from which it follows easily
$$s_k = \alpha_k^+ - \alpha_k^- .$$

All the above equations admit a unique solution, since the partial derivative with respect to U_1 of the left hand side of our equation (6.63) is positive.

Once the s's have been determined, we can proceed and compute the swaption price. To do so, notice that we can express A in terms of U_i by combining (6.65) with $U^i = U + p_i$, thus obtaining

$$A = \left\{ U^i : U_1^i - \sum_{k=2}^r s_k U_k^i > s_1 + (p_i)_1 - \sum_{k=2}^r s_k (p_i)_k \right\} .$$

To ease notation, set

$$w := [1, -s_2, \dots, -s_r], \quad s_i^* := s_1 + (p_i)_1 - \sum_{k=2}^r s_k (p_i)_k ,$$

so that
$$A = \{U^i : wU^i > s_i^*\}.$$

This time we compute expectations as follows:

$$\begin{aligned}
&E^i \left[(F_i(T_\alpha) - K)\, 1_A \right] \\
&= E^i \left[(F_i(T_\alpha; U^i) - K)\, 1_{(wU^i > s_i^*)} \right], \quad U^i \sim \mathcal{N}(0, I_r) \\
&= E^i \left[\left(F_i(0) \exp(\Gamma_{(i)} U^i - \tfrac{1}{2} |\Gamma_{(i)}|^2) - K \right)\, 1_{(wU^i > s_i^*)} \right] \\
&= F_i(0) E^i \left[\exp(\Gamma_{(i)} U^i - \tfrac{1}{2} |\Gamma_{(i)}|^2) 1_{(wU^i > s_i^*)} \right] - K E^i 1_{(wU^i > s_i^*)} \\
&= F_i(0) E^i \left[1_{(w[U^i + \Gamma'_{(i)}] > s_i^*)} \right] - K \Phi\left(- \frac{s_i^*}{|w|} \right) \\
&= F_i(0) \Phi\left(- \frac{s_i^* - w\Gamma'_{(i)}}{|w|} \right) - K \Phi\left(- \frac{s_i^*}{|w|} \right)
\end{aligned}$$

where use has been made of the general property

$$E^i\left\{\exp(b'\,U^i - \tfrac{1}{2}b'b)g(U^i)\right\} = E^i\{g(U^i + b)\}, \quad b = [b_1, \ldots, b_r]'.$$

We then have the following.

Proposition 6.14.1. (Brace's rank-r formula). *The swaption price can be approximated as*

$$\sum_{i=\alpha+1}^{\beta} \tau_i P(0,T_i)\left[F_i(0)\Phi\left(-\frac{s_i^* - w\Gamma'_{(i)}}{|w|}\right) - K\Phi\left(-\frac{s_i^*}{|w|}\right)\right]. \tag{6.66}$$

This formula is analytical and involves $2r-1$ root-searching procedures to find

$$\alpha_1, \alpha_2^-, \alpha_2^+, \ldots, \alpha_r^-, \alpha_r^+,$$

needed to determine $s_1, \ldots, s_r$ and, therefore, w and s^* appearing in the formula.

Brace (1997) argued that the rank-two approximation ($r = 2$) is enough for obtaining sufficient accuracy, based on numerical examination of the U-surface.

Let us therefore take $r = 2$ and let us see which volatility and correlation parametric structures are handy when one tries to apply formula (6.66) with $r = 2$.

Exactly as in the rank-one case, the formulation we presented here works for any possible choice of the instantaneous volatilities $(\sigma_k(t))_k$ and correlation-ρ structures. As before, in applying this apparatus, one may choose a parameterization that renders the Γ matrix computation as simple as possible and the approximation of the real integrated covariance matrix V, with its Γ-based approximation V^2, as good as possible. In particular, since the integrated covariance matrix V will be approximated via a rank-two matrix, we may as well start from a rank-two instantaneous covariance matrix. This is obtained, for example, by taking a two-factor model, i.e. by taking $\rho = \rho^B$ as in (6.48),

$$\rho_{i,j}^B = b_{i,1}b_{j,1} + b_{i,2}b_{j,2} = \cos(\theta_i - \theta_j),$$

and the separable instantaneous-volatility structure (6.10). Notice carefully that the above formula holds under the assumption $\rho_{i,j} > 0$ for all i, j, as required in order to apply the Perron-Frobenius theorem. This is guaranteed by

$$|\theta_i - \theta_j| < \pi/2,$$

which in turn is guaranteed by taking all θ's in an interval whose length is less than $\pi/2$, say $(-\pi/4,\ \pi/4)$. With this choice, the resulting V has automatically rank 2 and we can set

$$\Gamma = \sqrt{\sum_{k=0}^{\alpha} \tau_{j-1,j}\Psi_{j+1}^2}\ \ \mathrm{Diag}(\Phi_{\alpha+1}, \ldots, \Phi_\beta)B$$

where $\mathrm{Diag}(x_a, \ldots, x_b)$ denotes a diagonal matrix whose diagonal entries are $x_a, x_{a+1}, \ldots, x_b$, and where $B = (b_{i,j})$ is the matrix defined in (6.47),

$$b_{i,1} = \cos\theta_i, \quad b_{i,2} = \sin\theta_i \,.$$

Finally notice again that the alternative formulations (6.8) and (6.11) do not lead to a rank-2 integrated covariance matrix V in general, so that if we choose (6.11) we actually need to include an eigenvector/eigenvalue calculation in order to apply Brace's formula (6.66). The integrated covariance matrix with formulation (6.11) is now given by

$$V_{i,j} = \Phi_i \Phi_j \cos(\theta_i - \theta_j) \sum_{k=0}^{\alpha} \tau_{j-1,j} \psi_{j-k} \psi_{i-k} \,,$$

and Γ has to be derived through the first two dominant eigenvectors of this matrix.

As observed before, this last approach has the disadvantage of requiring an explicit eigenvalues/eigenvectors calculation. But once again, as stated in Section 6.5, it has the advantage of giving a possibly good qualitative behaviour for the term structure of volatilities over time.

6.15 A Simpler LFM Formula for Swaptions Volatilities

It takes one moment to compute this variance! Well, maybe two.
Andrea Bugin, Product and Business Development, Banca IMI

There is a further approximation method to compute swaption prices with the LFM without resorting to Monte Carlo simulation. This method is rather simple and its quality has been tested, for example, by Brace, Dun, and Barton (1999). We have tested the method ourselves and will present the related numerical results in Chapter 8.

Recall from Section 6.7 the forward-swap-rate dynamics underlying the LSM (leading to Black's formula for swaptions):

$$dS_{\alpha,\beta}(t) = \sigma^{(\alpha,\beta)}(t) S_{\alpha,\beta}(t) dW_t^{\alpha,\beta}, \quad Q^{\alpha,\beta} \,.$$

A crucial role in the LSM is played by the (squared) Black swaption volatility (multiplied by T_α)

$$(v_{\alpha,\beta}(T_\alpha))^2 := \int_0^{T_\alpha} \sigma^2_{\alpha,\beta}(t)dt = \int_0^{T_\alpha} (d\ln S_{\alpha,\beta}(t))(d\ln S_{\alpha,\beta}(t))$$

entering Black's formula for swaptions. We plan to compute, under a number of approximations, an analogous quantity $v_{\alpha,\beta}^{\mathrm{LFM}}$ in the LFM.

Recall from formula (6.34) that forward swap rates can be thought of as algebraic transformations of forward rates, according to

$$S_{\alpha,\beta}(t) = \sum_{i=\alpha+1}^{\beta} w_i(t) F_i(t),$$

where

$$w_i(t) = w_i(F_{\alpha+1}(t), F_{\alpha+2}(t), \ldots, F_\beta(t)) = \frac{\tau_i \mathrm{FP}(t, T_\alpha, T_i)}{\sum_{k=\alpha+1}^{\beta} \tau_k \mathrm{FP}(t, T_\alpha, T_k)}$$

$$= \frac{\tau_i \prod_{j=\alpha+1}^{i} \frac{1}{1+\tau_j F_j(t)}}{\sum_{k=\alpha+1}^{\beta} \tau_k \prod_{j=\alpha+1}^{k} \frac{1}{1+\tau_j F_j(t)}}.$$

A first approximation is derived as follows. Start by freezing the w's at time 0, so as to obtain

$$S_{\alpha,\beta}(t) \approx \sum_{i=\alpha+1}^{\beta} w_i(0) F_i(t).$$

This approximation is justified by the fact that the variability of the w's is much smaller than the variability of the F's. This can be tested both historically and through simulation of the F's (and therefore of the w's) via a Monte Carlo method.

Then differentiate both sides

$$dS_{\alpha,\beta}(t) \approx \sum_{i=\alpha+1}^{\beta} w_i(0) dF_i(t) = (\ldots)\, dt + \sum_{i=\alpha+1}^{\beta} w_i(0) \sigma_i(t) F_i(t)\, dZ_i(t),$$

under any of the forward-adjusted measures, and compute the quadratic variation

$$dS_{\alpha,\beta}(t)\, dS_{\alpha,\beta}(t) \approx \sum_{i,j=\alpha+1}^{\beta} w_i(0) w_j(0) F_i(t) F_j(t) \rho_{i,j} \sigma_i(t) \sigma_j(t)\, dt.$$

The percentage quadratic variation is

$$\left(\frac{dS_{\alpha,\beta}(t)}{S_{\alpha,\beta}(t)}\right)\left(\frac{dS_{\alpha,\beta}(t)}{S_{\alpha,\beta}(t)}\right) = (d\ln S_{\alpha,\beta}(t))(d\ln S_{\alpha,\beta}(t))$$

$$\approx \frac{\sum_{i,j=\alpha+1}^{\beta} w_i(0) w_j(0) F_i(t) F_j(t) \rho_{i,j} \sigma_i(t) \sigma_j(t)}{S_{\alpha,\beta}(t)^2}\, dt.$$

Introduce now a further approximation by freezing all F's in the above formula (as was done earlier for the w's) to their time-zero value:

$$(d\ln S_{\alpha,\beta}(t))(d\ln S_{\alpha,\beta}(t)) \approx \sum_{i,j=\alpha+1}^{\beta} \frac{w_i(0) w_j(0) F_i(0) F_j(0) \rho_{i,j}}{S_{\alpha,\beta}(0)^2} \sigma_i(t) \sigma_j(t)\, dt.$$

Using this last formula, finally compute an approximation $(v^{\mathrm{LFM}}_{\alpha,\beta})^2$ of the integrated percentage variance of S as

$$\int_0^{T_\alpha} (d\ln S_{\alpha,\beta}(t))(d\ln S_{\alpha,\beta}(t))$$
$$\approx \sum_{i,j=\alpha+1}^{\beta} \frac{w_i(0)w_j(0)F_i(0)F_j(0)\rho_{i,j}}{S_{\alpha,\beta}(0)^2} \int_0^{T_\alpha} \sigma_i(t)\sigma_j(t)\, dt =: (v^{\mathrm{LFM}}_{\alpha,\beta})^2 \,,$$

so that we have the following.

Proposition 6.15.1. (Rebonato's formula). *The LFM Black-like (squared) swaption volatility (multiplied by T_α) can be approximated by*

$$\boxed{(v^{LFM}_{\alpha,\beta})^2 = \sum_{i,j=\alpha+1}^{\beta} \frac{w_i(0)w_j(0)F_i(0)F_j(0)\rho_{i,j}}{S_{\alpha,\beta}(0)^2} \int_0^{T_\alpha} \sigma_i(t)\sigma_j(t)\, dt} \,. \tag{6.67}$$

We refer to formula (6.67) as to "Rebonato's formula" for brevity, since a sketchy version of it can be found in Rebonato's (1998) book, and since Rebonato was one of the first to explicitly point out the interpretation of swap rates as linear combinations of forward rates.

The quantity $v^{\mathrm{LFM}}_{\alpha,\beta}$ can be used as a proxy for the Black volatility $v_{\alpha,\beta}(T_\alpha)$ of the swap rate $S_{\alpha,\beta}$. Putting this quantity in Black's formula for swaptions allows one to compute approximated swaptions prices with the LFM. Formula (6.67) is obtained under a number of assumptions, and at first one would imagine its quality to be rather poor. However, it turns out that the approximation is not at all bad, as also pointed out by Brace, Dun and Barton (1998). In Chapter 8, we will present numerical investigations of our own and confirm that the approximation is indeed satisfactory in general.

A slightly more sophisticated version of the above procedure has been proposed, for example, by Hull and White (1999). Hull and White differentiate $S_{\alpha,\beta}(t)$ without freezing the w's, thus obtaining

$$dS_{\alpha,\beta}(t) = \sum_{i=\alpha+1}^{\beta} (w_i(t)\, dF_i(t) + F_i(t)\, dw_i(t)) + (\ldots)\, dt$$
$$= \sum_{i,h=\alpha+1}^{\beta} \left(w_h(t)\delta_{\{h,i\}} + F_i(t)\frac{\partial w_i(t)}{\partial F_h} \right) dF_h(t) + (\ldots)\, dt \,,$$

where $\delta_{\{i,i\}} = 1$ and $\delta_{\{i,h\}} = 0$ for $h \neq i$. By straightforward calculations one obtains

$$\frac{\partial w_i(t)}{\partial F_h} = \frac{w_i(t)\tau_h}{1+\tau_h F_h(t)} \left[\frac{\sum_{k=h}^{\beta} \tau_k \prod_{j=\alpha+1}^{k} \frac{1}{1+\tau_j F_j(t)}}{\sum_{k=\alpha+1}^{\beta} \tau_k \prod_{j=\alpha+1}^{k} \frac{1}{1+\tau_j F_j(t)}} - 1_{\{i \geq h\}} \right],$$

so that by setting

$$\bar{w}_h(t) := w_h(t) + \sum_{i=\alpha+1}^{\beta} F_i(t) \frac{\partial w_i(t)}{\partial F_h} ,$$

we have

$$dS_{\alpha,\beta}(t) = \sum_{h=\alpha+1}^{\beta} \bar{w}_h(t)\, dF_h(t) + (\ldots)\, dt ,$$

where the $\bar{w}(t)$'s are completely determined in terms of the forward rates $F(t)$'s. Now we may freeze all F's (and therefore $\bar{w}$'s) at time 0 and obtain

$$dS_{\alpha,\beta}(t) \approx \sum_{h=\alpha+1}^{\beta} \bar{w}_h(0)\, dF_h(t) + (\ldots)\, dt .$$

Now, in order to derive an approximated Black swaption volatility $\bar{v}_{\alpha,\beta}^{\text{LFM}}$, based on the weights $\bar{w}$ rather than on the weights w, we may reason as in the previous derivation so as to obtain the following.

Proposition 6.15.2. (Hull and White's formula). *The LFM Black-like (squared) swaption volatility (multiplied by T_α) can be better approximated by*

$$\boxed{(\bar{v}_{\alpha,\beta}^{LFM})^2 := \sum_{i,j=\alpha+1}^{\beta} \frac{\bar{w}_i(0)\bar{w}_j(0)F_i(0)F_j(0)\rho_{i,j}}{S_{\alpha,\beta}(0)^2} \int_0^{T_\alpha} \sigma_i(t)\sigma_j(t)\, dt} . \quad (6.68)$$

We have investigated, among other features, the difference between the outputs of formulas (6.67) and (6.68) in a number of situation, and will describe the results in Section 8.3. We anticipate that the difference between the two formulas is practically negligible in most situations.

6.16 A Formula for Terminal Correlations of Forward Rates

We have seen earlier in Section 6.6 that the terminal correlation between forward rates depends not only on the instantaneous correlations of the forward-rate dynamics, but also on the way the instantaneous volatilities are modeled from average volatilities coming from caps, and possibly swaptions, prices.

In general, if one is interested in terminal correlations of forward rates at a future time instant, as implied by the LFM, the computation has to be based on a Monte Carlo simulation technique. Indeed, assume we are interested in computing the terminal correlation between the forward rates $F_i = F(\cdot; T_{i-1}, T_i)$ and $F_j = F(\cdot; T_{j-1}, T_j)$ at time T_α, $\alpha \leq i-1 < j$, say under the measure Q^γ, $\gamma \geq \alpha$. Then we need to compute

$$
\begin{aligned}
&\mathrm{Corr}^\gamma(F_i(T_\alpha), F_j(T_\alpha)) \qquad (6.69)\\
&= \frac{E^\gamma\left[(F_i(T_\alpha) - E^\gamma F_i(T_\alpha))(F_j(T_\alpha) - E^\gamma F_j(T_\alpha))\right]}{\sqrt{E^\gamma[(F_i(T_\alpha) - E^\gamma F_i(T_\alpha))^2]}\ \sqrt{E^\gamma[(F_j(T_\alpha) - E^\gamma F_j(T_\alpha))^2]}}\,.
\end{aligned}
$$

Recall the dynamics of F_i and F_j under Q^γ:

$$
\begin{aligned}
dF_k(t) &= -\sigma_k(t)F_k(t)\sum_{j=k+1}^{\gamma}\frac{\rho_{k,j}\tau_j\sigma_j(t)F_j(t)}{1+\tau_jF_j(t)}\,dt + \sigma_k(t)F_k(t)\,dZ_k(t),\\
&\qquad k = \alpha+1,\ldots,\gamma-1\,,\\
dF_k(t) &= \sigma_k(t)F_k(t)\,dZ_k(t), \quad k=\gamma\,,\\
dF_k(t) &= \sigma_k(t)F_k(t)\sum_{j=\gamma+1}^{k}\frac{\rho_{k,j}\tau_j\sigma_j(t)F_j(t)}{1+\tau_jF_j(t)}\,dt + \sigma_k(t)F_k(t)\,dZ_k(t),\\
&\qquad k = \gamma+1,\ldots,\beta\,,
\end{aligned}
$$

where Z is a Brownian motion under Q^γ. The expected values appearing in the expression (6.69) for the terminal correlation can be obtained by simulating the above dynamics for $k=i$ and $k=j$ respectively, thus simulating F_i and F_j up to time T_α. The simulation can be based on a discretized Milstein dynamics analogous to (6.53), which was introduced to Monte Carlo price swaptions.

However, at times, traders may need to quickly check reliability of the model's terminal correlations, so that there could be no time to run a Monte Carlo simulation. Fortunately, there does exist an approximated formula, in the spirit of the approximated formulas for Black's swaption volatilities, which allows us to compute terminal correlations algebraically from the LFM parameters ρ and $\sigma(\cdot)$. We now derive such an approximated formula.

The first approximation we introduce is a partial freezing of the drift in the dynamics, analogous to what was done for Brace's rank-one formula:

$$
\begin{aligned}
-\sum_{j=k+1}^{\gamma}\frac{\rho_{k,j}\tau_j\sigma_j(t)F_j(t)}{1+\tau_jF_j(t)} \approx -\sum_{j=k+1}^{\gamma}\frac{\rho_{k,j}\tau_j\sigma_j(t)F_j(0)}{1+\tau_jF_j(0)} &=: \mu_{\gamma,k}(t)\,, k<\gamma\,,\\
0 &=: \mu_{\gamma,\gamma}(t)\,, k=\gamma\,,\\
\sum_{j=\gamma+1}^{k}\frac{\rho_{k,j}\tau_j\sigma_j(t)F_j(t)}{1+\tau_jF_j(t)} \approx \sum_{j=\gamma+1}^{k}\frac{\rho_{k,j}\tau_j\sigma_j(t)\ F_j(0)}{1+\tau_jF_j(0)} &=: \mu_{\gamma,k}(t)\,, k>\gamma\,.
\end{aligned}
$$

Under this approximation, the forward-rate dynamics, under Q^γ, is

$$
dF_k(t) = \bar\mu_{\gamma,k}(t)F_k(t)\,dt + \sigma_k(t)F_k(t)\,dZ_k(t), \quad \bar\mu_{\gamma,k}(t) := \sigma_k(t)\mu_{\gamma,k}(t)\,,
$$

which describes a geometric Brownian motion and can thus be easily integrated. If considered for $k=i$ and $k=j$, this equation leads to jointly normally distributed variables $\ln F_i(T_\alpha)$ and $\ln F_j(T_\alpha)$ under the measure Q^γ.

This allows for an exact evaluation of the expected value in the numerator of (6.69). Indeed, we obtain easily

$$F_k(T_\alpha) = F_k(0) \exp\left[\int_0^{T_\alpha} \left(\bar{\mu}_{\gamma,k}(t) - \frac{\sigma_k^2(t)}{2}\right) dt + \int_0^{T_\alpha} \sigma_k(t)\, dZ_k(t)\right],$$

for $k \in \{i, j\}$, and

$$F_i(T_\alpha)F_j(T_\alpha) = F_i(0)F_j(0) \exp\left[\int_0^{T_\alpha} \left(\bar{\mu}_{\gamma,i}(t) + \bar{\mu}_{\gamma,j}(t) - \frac{\sigma_i^2(t) + \sigma_j^2(t)}{2}\right) dt \right.$$
$$\left. + \int_0^{T_\alpha} \sigma_i(t)\, dZ_i(t) + \int_0^{T_\alpha} \sigma_j(t)\, dZ_j(t)\right].$$

Recalling that, by a trivial application of Ito's isometry, the two-dimensional random vector

$$\left[\int_0^{T_\alpha} \sigma_i(t)\, dZ_i(t)\,,\ \int_0^{T_\alpha} \sigma_j(t)\, dZ_j(t)\right]'$$

is jointly normally distributed with mean $[0, 0]'$ and covariance matrix

$$\begin{bmatrix} \int_0^{T_\alpha} \sigma_i^2(t)dt & \rho_{i,j} \int_0^{T_\alpha} \sigma_i(t)\sigma_j(t)dt \\ \rho_{i,j} \int_0^{T_\alpha} \sigma_i(t)\sigma_j(t)dt & \int_0^{T_\alpha} \sigma_j^2(t)dt \end{bmatrix},$$

we see that the correlation approximated according to this distribution is easily obtained. We present the related formula in the following

Proposition 6.16.1. (Analytical terminal-correlation formula). *The terminal correlation between the forward rates F_i and F_j at time T_α, $\alpha \leq i - 1 < j$, under the measure Q^γ, $\gamma \geq \alpha$, can be approximated as follows:*

$$\boxed{\mathrm{Corr}^\gamma(F_i(T_\alpha), F_j(T_\alpha)) \approx \frac{\exp\left(\int_0^{T_\alpha} \sigma_i(t)\sigma_j(t)\rho_{i,j}\, dt\right) - 1}{\sqrt{\exp\left(\int_0^{T_\alpha} \sigma_i^2(t)\, dt\right) - 1}\ \sqrt{\exp\left(\int_0^{T_\alpha} \sigma_j^2(t)\, dt\right) - 1}}}. \tag{6.70}$$

Notice that a first order expansion of the exponentials appearing in formula (6.70) yields

Rebonato's terminal-correlation formula:

$$\boxed{\mathrm{Corr}^{\mathrm{REB}}(F_i(T_\alpha), F_j(T_\alpha)) = \rho_{i,j} \frac{\int_0^{T_\alpha} \sigma_i(t)\sigma_j(t)\, dt}{\sqrt{\int_0^{T_\alpha} \sigma_i^2(t)\, dt}\ \sqrt{\int_0^{T_\alpha} \sigma_j^2(t)\, dt}}}. \tag{6.71}$$

This formula shows, through the Schwartz inequality, that

$$\mathrm{Corr}^{\mathrm{REB}}(F_i(T_\alpha), F_j(T_\alpha)) \leq \rho_{i,j} \quad \text{if} \quad \rho_{i,j} \geq 0\,, \tag{6.72}$$

$$\mathrm{Corr}^{\mathrm{REB}}(F_i(T_\alpha), F_j(T_\alpha)) \geq \rho_{i,j} \quad \text{if} \quad \rho_{i,j} < 0\,. \tag{6.73}$$

In words, we can say that *terminal correlations are, in absolute value, always smaller than or equal to instantaneous correlations.* In agreement with this general observation, recall that through a clever repartition of integrated volatilities in instantaneous volatilities $\sigma_i(t)$ and $\sigma_j(t)$ we can make the terminal correlation $\mathrm{Corr}^{\mathrm{REB}}(F_i(T_\alpha), F_j(T_\alpha))$ arbitrarily close to zero, even when the instantaneous correlation $\rho_{i,j}$ is one (see Section 6.6).

The approximate formulas above are particularly appealing. They do not depend on the measure Q^γ under which the correlation has been computed. Moreover, they can be computed algebraically based on the volatility and correlation parameters σ and ρ of the LFM. Therefore, once the LFM parameters determining σ and ρ have been obtained through calibration to the cap and swaption markets, we can immediately check the terminal-correlation structure at any future date. There remains the problem of actually checking that the above formulas lead to acceptable approximation errors. This can be done by rigorously evaluating the correlation Corr^γ through a Monte Carlo method, as we will do in Chapter 8.

Finally, it is important to consider the following

Remark 6.16.1. **(Terminal Correlation and Swaptions Volatilities).** By comparing the right hand side of Formula (6.71) with the right hand side of Formula (6.67), we see that swaption volatilities in the LIBOR market model are directly linked with terminal correlations, rather than with instantaneous ones. This will be further discussed in relationship with swaptions calibration later on.

6.17 Calibration to Swaptions Prices

Errors using inadequate data are much less than those using no data at all.
Charles Babbage (1791-1871)

We have seen several ways to compute swaption prices with the LFM. However, as already mentioned, computing market (plain-vanilla) swaption prices is not the purpose of an interest-rate model. In fact, it is common practice in the market to compute such prices through a Black-like formula. At most, the LFM can be used to determine the price of illiquid swaptions or of standard swaptions for which the Black volatility is not quoted or is judged to be not completely reliable. But in general, as far as standard (plain-vanilla) swaptions are concerned, the market is happy with Black's formula. Black's formula, indeed, is a "metric" by which traders translate prices into implied volatilities, and it is both pointless and hopeless to expect traders to give up such a formula.

Since traders already know standard-swaptions prices from the market, they wish a chosen model to incorporate as many such prices as possible. In case we are adopting the LFM, we need to find the instantaneous-volatility and correlation parameters in the LFM dynamics that reflect the swaptions prices observed in the market.

First, let us quickly see how the market organizes swaption prices in a table. To simplify ideas, assume we are interested only in swaptions with maturity and underlying-swap length (tenor) given by multiples of one year. Traders typically consider a matrix of at-the-money Black's swaption volatilities, where each row is indexed by the swaption maturity T_α, whereas each column is indexed in terms of the underlying swap length, $T_\beta - T_\alpha$. The $x \times y$-swaption is then the swaption in the table whose maturity is x years and whose underlying swap is y years long. Thus a 2×10 swaption is a swaption maturing in two years and giving then the right to enter a ten-year swap. Here, we consider maturities of $1, 2, 3, 4, 5, 7, 10$ years and underlying-swap lengths of $1, 2, 3, 4, 5, 6, 7, 8, 9, 10$ years.

A typical example of table of swaption volatilities is shown below.

An example of Black's implied volatilities of at-the-money swaptions, May 16, 2000.

	1y	2y	3y	4y	5y	6y	7y	8y	9y	10y
1y	16.4	15.8	14.6	13.8	13.3	12.9	12.6	12.3	12.0	11.7
2y	17.7	15.6	14.1	13.1	12.7	12.4	12.2	11.9	11.7	11.4
3y	17.6	15.5	13.9	12.7	12.3	12.1	11.9	11.7	11.5	11.3
4y	16.9	14.6	12.9	11.9	11.6	11.4	11.3	11.1	11.0	10.8
5y	15.8	13.9	12.4	11.5	11.1	10.9	10.8	10.7	10.5	10.4
7y	14.5	12.9	11.6	10.8	10.4	10.3	10.1	9.9	9.8	9.6
10y	13.5	11.5	10.4	9.8	9.4	9.3	9.1	8.8	8.6	8.4

This is a submatrix of the complete table. Its entries are the implied volatilities obtained by inverting the related at-the-money swaption prices through Black's formula for swaptions. A remark at this point is in order.

Remark 6.17.1. **(Possible misalignments in the swaption matrix).** Usually, one should not completely rely on the swaption matrix provided by a single broker, and, in any case, one should not take it for granted. The problem is that the matrix is not necessarily uniformly updated. The most liquid swaptions are updated regularly, whereas some entries of the matrix refer to older market situations. This "temporal misalignment" in the swaptions matrix can cause troubles, since, when we try a calibration, the model parameters might reflect this misalignment by assuming "weird" values (we will see in Chapter 7 examples leading to imaginary and complex forward-rate volatilities). If one trusts the model, this can indeed be used to detect such misalignments. The model can then be calibrated to the liquid swaptions, and used to price the remaining swaptions, looking at the values that most differ from the corresponding market ones. This may give an indication of which swaptions can cause troubles.

What is one to do with the above table when presented with the problem of calibrating the LFM? The problem is incorporating as much information as possible from such a table into the LFM parameters. To focus ideas, consider the LFM with (volatility) Formulation 7 (given by formula (6.13)) and rank-two correlations expressed by the angles θ. We have seen earlier several ways to compute swaption prices with the LFM: Monte Carlo simulation, Brace's rank-one and rank-two formulas, Rebonato's formula and Hull and White's formula. Whichever method is chosen, for given values of a, b, c, d and θ we can price all the swaptions in the table, thus obtaining a table of LFM prices for swaptions corresponding to the chosen values of the LFM parameters.

Indeed, each price of an at-the-money European (payer) swaption with underlying swap $S_{\alpha,\beta}$ with the LFM is a function of the LFM parameters a, b, c, d and θ's (the Φ's being determined through caplet volatilities via (6.28)). We can then try and change the parameters a, b, c, d and θ and consider the related Φ's (6.28) in such a way that the LFM table approaches as much as possible, in some sense, the market swaptions table. We can, for instance, minimize the sum of the squares of the differences of the corresponding swaption prices in the two tables. Such a sum will be a function of a, b, c, d and θ, and when we find the parameters that minimize it, we can say we have calibrated the LFM to the swaption market.

We will approach this problem by using Rebonato's formula to price swaptions under the LFM. As we will see in Section 8.3, we have tested this formula in a number of situations and we have found it to be sufficiently accurate. The formula gives directly the swaption volatility as a simple function of the parameters. Neither simulations (as in the Monte Carlo method) nor root searches (as in Brace's formulas) are needed, so that this method is ideally suited to calibrate a large table with a contained computational effort.

We need also to point out that some attention has moved on a pre-selected instantaneous-correlation matrix. Typically, one estimates the instantaneous correlation ρ historically from time series of zero rates at a given set of maturities, and then approximates it by a lower rank matrix, thus finding, for example, the θ's. With this approach, the parameters that are left for the swaptions calibration are the free parameters in the volatility structure. Take Formulation 7 as an example. The Φ's are determined by the cap market as functions of a, b, c and d, which are parameters to be used in the calibration to swaptions prices. However, if the number of swaptions is large, four parameters are not sufficient for practical purposes. One then needs to consider richer parametric forms for the instantaneous volatility, such as for example the formulation of TABLE 5, although in this case care must be taken for the resulting market structures to be regular enough.

6.18 Instantaneous Correlations: Inputs (Historical Estimation) or Outputs (Fitting Parameters)?

> *"How did you get these correlations? Did you roll the dice?"*
> [pulls the lever of an imaginary slot machine]
> *Former Head of the Interest-Rate-Derivatives Desk of Banca IMI*

Now that we have properly introduced the issue of swaptions calibration, we resume our discussion on instantaneous correlations, started in Section 6.9. In that earlier section we have seen some possible parametric forms for instantaneous correlation. A question is in order, at this point. Will instantaneous correlations be inputs or outputs in our models calibration to market quotes of (caps and) swaptions we will consider later on? Indeed, whichever volatility and correlation parameterizations are chosen, the following step consists of calibrating the model to swaption prices. A set of European swaptions, whose prices are quoted in the market, is thus considered. These swaptions are priced with one of the methods we will see later on, and the model prices will depend on the correlation parameters appearing in ρ. Should we infer ρ itself from swaption market quotes or should we estimate ρ exogenously and impose it, leaving the calibration only to volatility parameters? Are the parameters in ρ inputs or outputs to the calibration?

Inputs? Indeed, we might consider a time series of past interest-rate curves data, which are observed under the real world probability measure. This would allow us, through interpolation, to obtain a corresponding time series for the particular forward LIBOR rates being modelled in our LIBOR model. These series would be observed under the objective or real-world measure. Thanks to the Girsanov theorem this is not a problem, since instantaneous correlations, considered as instantaneous covariations between driving Brownian motions in forward rate dynamics, do not depend on the probability measure under which we are specifying the joint forward rate dynamics. Only the drifts depend on the measure. Then, by using some historical estimation technique, we can obtain an historical estimate of the instantaneous correlation matrix. This historical matrix ρ, or a stylized version of it, can be considered as a given ρ for our LIBOR model, and the remaining free parameters σ are to be used to calibrate market derivatives data at a given instant. In this case (caps and) swaptions calibration will consist in finding the σ's such that the model (caps and) swaptions prices match, as close as possible, the corresponding market prices. In this "matching" procedure (often an optimization) ρ is fixed from the start to the found historical estimate and we play on the volatility parameters σ to achieve our matching.

Outputs? This second possibility considers instantaneous correlations as fitting parameters. The model swaptions prices are functions of ρ^B, and possibly of some remaining instantaneous volatility parameters, that are forced to match as much as possible the corresponding market swaptions prices, so that the parameters values implied by the market, $\rho^B = \rho^B_{\text{MKT}}$, are found. In the two-factor angles case for example, one obtains the values of $\theta_1, \ldots, \theta_M$ (and of the volatility parameters not determined by the calibration to caps) that are implied by the market.

Which of the two methods is preferable? We will consider again this question later on. In the next section, we try and address the issue of determining a decent historical ρ in case we are to decide later for the "inputs" approach.

6.19 The exogenous correlation matrix

The exogenous nature of the correlation matrix is an interesting opportunity to introduce in the calibration correlation structures bearing resemblance to real market patterns. Therefore we identify market patterns through historical estimation. However, plugging a historical estimation directly into a calibration routine is not always desirable. We address some of these issues in the present section, following Brigo and Morini (2002).

Rebonato and Jäckel (1999) recall that historical estimations are usually characterized by problems such as outliers, non-synchronous data and, in particular with reference to interest rates, discontinuities in correlation surfaces due to the use of discount factors extracted from different financial instruments. Consequently, Rebonato and Jäckel (1999) propose to use possibly parameterized correlation matrices approximating the results of the estimation, but designed to remain smooth, regular and enjoying good properties by construction. Such good properties are those mentioned in the beginning of Section 6.9. The idea is then to fit a suitable parametric form onto the historically estimated matrix. In fact there exist in the literature parsimonious (low-parametric) forms for correlation matrices, and such parameterizations are designed to enforce the desirable properties. Some of these forms have been presented earlier in Section 6.9.

The importance of Rebonato and Jäckel's (1999) suggestion depends on the context, in particular on the relevance assigned to the details of the historical estimations. European swaption prices are relatively insensitive to correlation details, so that simple, standardized correlations typically produce prices very little different from those obtained using more complex and realistic instantaneous correlation matrices, as pointed out for instance by Jäckel and Rebonato (2000). On the other hand a more regular correlation structure can be helpful in calibration to obtain improved diagonstics: more regular volatilities, a more stable evolution of the volatility term structure, and better terminal correlations. In the special case of the swaptions cascade calibration we will see later on, swaption prices are recovered exactly and are not altered by correlations, while the regularity of the σ's may depend on the smoothness and regularity of the instantaneous correlations ρ.

Consequently, later on for swaptions cascade calibration we will use historically estimated exogenous correlation matrices, but we will also try smooth parametric forms fitted upon them. Alternatively, we will also avoid fitting but present some methods to obtain a good and smooth parametric matrix resembling, in some sense, the original historically estimated correlation matrix.

6.19.1 Historical Estimation

Since the procedures here considered are mainly related to cross-section calibration, we are not going into the details of an econometric analysis. However, we outline here the main features of the historical estimates used in the following.

Notice that, in estimating correlations for the LFM, one has to take into account the nature of forward rates in the model. They are characterized by a fixed maturity, contrary to market quotations, where a fixed time-to-maturity is usually considered as time passes. In other words, we observe arrays of discount factors of the following kind

$$P(t, t+Z), P(t+1, t+1+Z), \dots, P(t+n, t+n+Z),$$

where Z is time-to-maturity, ranging in a standard set of times, whereas what we need for LFM forward rates are sequences such as

$$P(t, T), P(t+1, T), \dots, P(t+n, T),$$

for the maturities T included in the chosen tenor structure. Accordingly, an interpolation (log-linear) between discount factors has been carried out, and only one year of data has been used, since the first forward rate in the family expires in one year from the starting date.

From the above daily quotations of notional zero-coupon bonds, whose maturities range from one to twenty years from today, we extracted daily log-returns of the annual forward rates involved in the model. Starting from the following usual Gaussian approximation

$$\left[\ln\left(\frac{F_1(t+\Delta t)}{F_1(t)}\right), \dots, \ln\left(\frac{F_{19}(t+\Delta t)}{F_{19}(t)}\right)\right] \sim \mathcal{N}(\mu, V),$$

where $\Delta t = 1$ day, our estimations of the parameters are based on sample mean and covariance for Gaussian variables, and are given by

$$\hat{\mu}_i = \frac{1}{m}\sum_{k=0}^{m-1} \ln\left(\frac{F_i(t_{k+1})}{F_i(t_k)}\right),$$

$$\hat{V}_{i,j} = \frac{1}{m}\sum_{k=0}^{m-1}\left[\left(\ln\left(\frac{F_i(t_{k+1})}{F_i(t_k)}\right) - \hat{\mu}_i\right)\left(\ln\left(\frac{F_j(t_{k+1})}{F_j(t_k)}\right) - \hat{\mu}_j\right)\right],$$

where m is the number of observed log-returns for each rate, so that our estimation of the general correlation element $\rho_{i,j}$ is

$$\hat{\rho}_{i,j} = \frac{\hat{V}_{i,j}}{\sqrt{\hat{V}_{i,i}}\sqrt{\hat{V}_{j,j}}}.$$

The first correlation matrix obtained, given in Table 6.2, comes from zero coupon bond data spanning the year from February 1, 2001 to February 1, 2002, being the latter the day the data on swaption market volatilities we will use in Section 7.6 refer to.

Examining the estimated correlation matrix in Table 6.2, what is clearly visible is a pronounced and approximately monotonic decorrelation along the columns, when moving away from the diagonal. Particularly noteworthy are the initial steepness of the decorrelation and its proportions. Conversely, the second tendency considered typical of correlation amongst rates, that is the upward trend along the sub-diagonals, is definitely not remarkable. That might be due to the smaller extent of such a phenomenon, more likely to be hidden by noise or differences in liquidity amongst longer rates. Not very different features are visible also in the previous similar estimate showed in Brace, Gatarek and Musiela (1997).

We did some tests on the stability of the estimates, finding out that the values remain rather stable when changing the sample size or its time positioning. We carried out a basic principal component analysis, revealing that as many as 7 factors are required to explain 90% of the overall variability. Indeed, eigenvalues of the historical correlation matrix given in Table 6.2 are reported, in decreasing order, in Table 6.3. The third column shows the percentage variance explained by each eigenvalue, while the fourth one explains the cumulative variance explained by all eigenvalues up to the considered index.

7 is a larger number of factors than in most other analogous surveys, see Rebonato (1998) or Jamshidian and Zhu (1997).

Many reasons can be put forward to explain such a difference: the use of correlations instead of covariances, the sample size, or, of course, an actually different market situation. In addition, the use of interpolation might have been relevant, even though our empirical tests do not underpin such hypothesis.

However we consider 19 rates, a number of rates higher than usual in defining the term structure. If we move to consider only 8 forward rates, 4 factors become enough to reach and exceed 91%. The number of initial forward rates included in the estimation appears relevant, whether it is due to added noise or to a real increase in dimensionality.

Now we have a realistic correlation benchmark, consistent with market tendencies, to be used directly in calibration or as a blueprint for determining the parameters in the above synthetic forms. In the latter case, parameters can be fixed so as to minimize some metric for the distance from the historical

1.00	0.82	0.69	0.65	0.58	0.47	0.29	0.23	0.43	0.47	0.33	0.43	0.29	0.23	0.26	0.21	0.23	0.29	0.25
0.82	1.00	0.80	0.73	0.68	0.55	0.45	0.40	0.53	0.57	0.42	0.45	0.48	0.34	0.35	0.32	0.32	0.31	0.32
0.69	0.80	1.00	0.76	0.72	0.63	0.47	0.56	0.67	0.61	0.48	0.52	0.48	0.54	0.46	0.42	0.45	0.42	0.35
0.65	0.73	0.76	1.00	0.78	0.67	0.58	0.56	0.68	0.70	0.56	0.59	0.58	0.50	0.50	0.48	0.49	0.44	0.35
0.58	0.68	0.72	0.78	1.00	0.84	0.66	0.67	0.71	0.73	0.70	0.67	0.64	0.59	0.58	0.65	0.65	0.53	0.42
0.47	0.55	0.63	0.67	0.84	1.00	0.77	0.68	0.73	0.69	0.77	0.69	0.66	0.63	0.61	0.68	0.70	0.57	0.45
0.29	0.45	0.47	0.58	0.66	0.77	1.00	0.72	0.71	0.65	0.65	0.62	0.71	0.62	0.63	0.66	0.64	0.52	0.38
0.23	0.40	0.56	0.56	0.67	0.68	0.72	1.00	0.73	0.66	0.64	0.56	0.61	0.72	0.59	0.64	0.64	0.49	0.46
0.43	0.53	0.67	0.68	0.71	0.73	0.71	0.73	1.00	0.75	0.59	0.66	0.69	0.69	0.69	0.63	0.64	0.52	0.40
0.47	0.57	0.61	0.70	0.73	0.69	0.65	0.66	0.75	1.00	0.63	0.68	0.70	0.63	0.64	0.65	0.62	0.52	0.40
0.33	0.42	0.48	0.56	0.70	0.77	0.65	0.64	0.59	0.63	1.00	0.83	0.72	0.64	0.58	0.68	0.73	0.57	0.45
0.43	0.45	0.52	0.59	0.67	0.69	0.62	0.56	0.66	0.68	0.83	1.00	0.82	0.69	0.67	0.70	0.69	0.65	0.43
0.29	0.48	0.48	0.58	0.64	0.66	0.71	0.61	0.69	0.70	0.72	0.82	1.00	0.79	0.78	0.79	0.72	0.59	0.42
0.23	0.34	0.54	0.50	0.59	0.63	0.62	0.72	0.69	0.63	0.64	0.69	0.79	1.00	0.82	0.83	0.79	0.60	0.45
0.26	0.35	0.46	0.50	0.58	0.61	0.63	0.59	0.69	0.64	0.58	0.67	0.78	0.82	1.00	0.90	0.80	0.50	0.22
0.21	0.32	0.42	0.48	0.65	0.68	0.66	0.64	0.63	0.65	0.68	0.70	0.79	0.83	0.90	1.00	0.94	0.71	0.46
0.23	0.32	0.45	0.49	0.65	0.70	0.64	0.64	0.64	0.62	0.73	0.69	0.72	0.79	0.80	0.94	1.00	0.82	0.66
0.29	0.31	0.42	0.44	0.53	0.57	0.52	0.49	0.52	0.52	0.57	0.65	0.59	0.60	0.50	0.71	0.82	1.00	0.84
0.25	0.32	0.35	0.35	0.42	0.45	0.38	0.46	0.40	0.40	0.45	0.43	0.42	0.45	0.22	0.46	0.66	0.84	1.00

Table 6.2. Historically estimated instantaneous correlation matrix for forward LIBOR rates, February 1, 2002.

1	11.6992	61.575%	61.575%
2	2.1478	11.304%	72.879%
3	1.1803	6.212%	79.091%
4	0.7166	3.772%	82.863%
5	0.6413	3.375%	86.238%
6	0.4273	2.249%	88.487%
7	0.386	2.032%	90.519%
8	0.3389	1.784%	92.303%
9	0.2805	1.476%	93.779%
10	0.2542	1.338%	95.117%
11	0.1995	1.050%	96.167%
12	0.1692	0.891%	97.057%
13	0.1611	0.848%	97.905%
14	0.1503	0.791%	98.696%
15	0.0877	0.462%	99.158%
16	0.0601	0.316%	99.474%
17	0.0515	0.271%	99.745%
18	0.0333	0.175%	99.921%
19	0.0151	0.079%	100.000%

Table 6.3. Eigenvalues of the historical correlation matrix given in Table 6.2 and percentage variance explained by them

estimation, as proposed for instance by Schoenmakers and Coffey (2002). Besides, we suggest another simple method to determine the parameters taking into account the major features of the historical matrix. It is described in the following section.

6.19.2 Pivot matrices

Here we concentrate on the parsimonious parameterizations seen in Section 6.9.1, that is on forms with two or three parameters. The classic methodology to find their parameters taking into account the estimation results, as proposed for instance by Schoenmakers and Coffey (2002), is fitting a parametric form to historical estimates by minimizing some loss function. But there is at least another procedure worth mentioning, resulting from the possibility to invert the functional structure of the parametric forms. In this way, we can express the parameters as functions of specific elements of the target matrix, so that such elements will be exactly reproduced in the resulting matrix. We can dub such elements "pivot points" of the target matrix, hence the resulting matrices will be referred to as pivot matrices.

The pivot points we select for three-parameters structures are the entries $\rho_{1,2}$, $\rho_{1,M}$ and $\rho_{M-1,M}$. Such elements are the vertices of the lower triangle of a correlation matrix, and indeed embed basic information about the last two typical properties of correlations amongst rates we have mentioned in the beginning of Section 6.9. We show below the relationships we obtain by computing the pivot versions of the 3 parameter matrices seen in Section 6.9, adding also the numerical values when the correlation in Table 6.2 is the reference matrix.

Starting with Rebonato's form (6.45), we find the following expressions for the three parameters ρ_∞, α and β. First

$$\left(\frac{\rho_{1,M}-\rho_\infty}{1-\rho_\infty}\right)=\left(\frac{\rho_{M-1,M}-\rho_\infty}{1-\rho_\infty}\right)^{(M-1)},$$

from which one can, simply albeit numerically, extract the value for ρ_∞, entering the following relationships for α and β

$$\alpha=\frac{\ln\left(\frac{\rho_{1,2}-\rho_\infty}{\rho_{M-1,M}-\rho_\infty}\right)}{2-M},\qquad \beta=\alpha-\ln\left(\frac{\rho_{1,2}-\rho_\infty}{1-\rho_\infty}\right).$$

Considering the numerical values for the three pivot correlations in the historical matrix, the results are

$$\rho_\infty=0.23551,\quad \alpha=0.00126,\quad \beta=0.26388.$$

Let us now move on to form (6.42) by Schoenmakers and Coffey, called in short S&C3 in what follows. The first expression is as simple as

$$\beta=-\ln\left(\rho_{M-1,M}\right).$$

while the remaining two, computationally slightly longer, finally read

$$\alpha_1=\frac{6\ln\rho_{1,M}}{(M-1)(M-2)}-\frac{2\ln\rho_{M-1,M}}{(M-2)}-\frac{4\ln\rho_{1,2}}{(M-2)},$$
$$\alpha_2=-\frac{6\ln\rho_{1,M}}{(M-1)(M-2)}+\frac{4\ln\rho_{M-1,M}}{(M-2)}+\frac{2\ln\rho_{1,2}}{(M-2)},$$

leading to

$$\alpha_1=0.03923,\quad \alpha_2=-0.03743,\quad \beta=0.17897.$$

Also the pivot version of (6.43), called S&C2 in the following, will turn out to be useful. So, let us briefly expose the relationships obtained, using as pivot points $\rho_{1,M}$ and $\rho_{1,2}$. We obtain

$$\rho_\infty=\rho_{1,M},\quad \eta=\frac{(-\ln\rho_{1,2})(M-1)+\ln\rho_\infty}{2},$$

and consequently

$$\rho_\infty=0.24545,\quad \eta=1.04617.$$

We compared the two three-parameters pivot forms with respect to the goodness of fit (to the historical matrix). We found that S&C3 pivot is superior when we take as loss function the simple average squared difference (denoted by MSE), whilst Rebonato pivot form is better if considering the average squared relative difference expressed in percentage terms with respect to the corresponding entry of the estimated matrix (denoted by MSE%). This is shown in the following table.

	MSE	MSE%	$\sqrt{\text{MSE}}$	$\sqrt{\text{MSE\%}}$
Reb. 3 pivot	0.030121	0.09542	0.173554	0.30890
S&C3 pivot	0.024127	0.10277	0.155327	0.32058

The latter error quantity (MSE%) appears to be particularly meaningful in our context, in that there are no entries so small as to lead to an over-estimation of slight errors, while, due to the differences in magnitude of the correlation entries, the relative importance of the discrepancies can be more informative than their absolute value.

Some more reasons for considering Rebonato pivot form preferable in this context arise from the graphical observation of the behaviour of these matrices. As visible in the first figure below, showing the plot of the first columns, such matrix seems a better approximation of the estimated tendency, whereas S&C3 pivot tends to keep higher than the historical matrix. Moreover, in matching the estimated values selected, the parameter α_2 in S&C3 has turned out to be negative. This has led to a non-monotonic trend for sub-diagonals, see in fact the humped shape for the first sub-diagonal, plotted in Figure 6.7.A similar problem is hinted at also by Schoenmakers and Coffey (2000), who report that, in their constrained tests, α_2 tends to assume always the minimum value allowed, namely zero, and therefore they propose the form (6.43). We performed some more tests building different pivot versions of this last matrix. The results seem to suggest that this very faint increasing tendency along sub-diagonals, joined with the level of decorrelation along the columns seen in the historical estimate, represent a configuration very hard to replicate with this parameterization maintaining at the same time its typical qualitative properties. Indeed, building a pivot S&C2 keeping out information upon the sub-diagonal behaviour, one gets a matrix spontaneously featuring a strong increase along such sub-diagonals. On the other hand, including information on this estimated behaviour, a far larger decorrelation is implied than in the historically estimated matrix. More elements and details on such tests are given in Morini (2002). This simple example shows that the pivot methodology can prove useful to pick out some peculiarities of different parameterizations.

No such problem has emerged for Rebonato's form, that seems to allow for an easier separation of the tendency along sub-diagonal from the one along the columns. Moreover, notice that Rebonato pivot form, with our data, turns out to be positive definite, so that its main theoretical limitation does not represent a problem in practice.

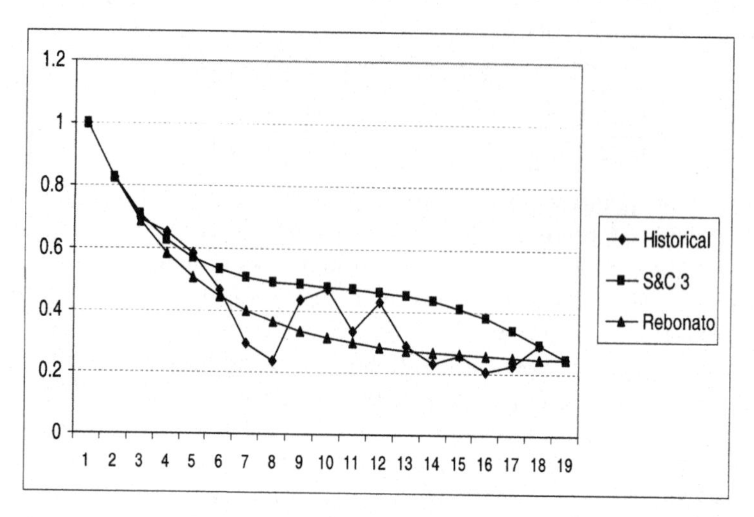

Fig. 6.6. First columns of the historical matrix and of some fitted "pivot" matrices

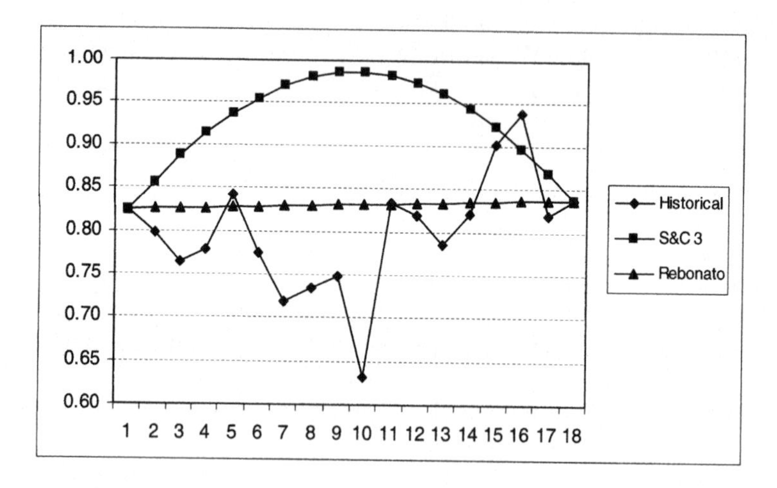

Fig. 6.7. Corresponding sub-diagonals

If one is only looking for a parametric matrix as close as possible to the target one, the classic minimization procedure can be seen as the natural choice. Otherwise, there are several points that make the "pivot" proposal somewhat appealing:

1. It does not need any optimization routine, and can be instantaneously carried out by simple formulas.
2. If the pivot points are chosen appropriately, the resulting matrix will replicate effectively the major characteristics of the historically estimated matrix.
3. It reduces the negative effects of irregularities and outliers in the historical estimations.
4. Since parameters are expressed in terms of key correlation entries, they have a clear intuitive meaning.

The last point can be of crucial importance in practice. In fact it allows to easily modify the matrix, which is often needed on the market. This allows operators to incorporate personal views, recent market shifts, or to carry out scenario analysis for risk management and hedging purposes. A trader or a risk manager can be interested in assessing how the model outputs change in relation to changes in correlations, for instance between short and long term rates. In this context, this task can be trivially performed by simply modifying the relevant pivot points. The matrix will vary as desired, while maintaining all regularity properties of the selected parametric form. This is a valuable feature, since when correlation entries are modified outside a well defined structure it is hard to make sure that the resulting matrix is a proper correlation matrix (e.g. positive semidefinite). Indeed, under uncontrolled or naive manipulations of the entries of a viable correlation matrix one may obtain irregular or even meaningless matrices.

Now we have still to check the divergence between pivot matrices and matrices optimally fitted to the entire target matrix through minimization of a loss function expressing the difference between the two matrices. We have already seen that the error committed is reasonable, in particular considering that *only 3 points, out of 171, have been used.* Let us now see how it improves making use of the totality of information. We will compare the pivot version of Rebonato's parameterization with two optimal specifications of the same form obtained by minimizing the aforementioned loss functions. In the following table we present for each optimal form the square root of the corresponding error, besides the value obtained, for the same error quantity, when considering the pivot form.

	$\sqrt{\text{MSE}}$	$\sqrt{\text{MSE\%}}$
Fitted vs Historical	0.108434	0.25949
Pivot vs Historical	0.173554	0.30890

Apparently, *the addition of 168 data to the 3 pivot points* brought about quite a narrow improvement, especially as regards the percentage error, particularly meaningful in our context. This is confirmed by the graphical comparison of the columns, as visible, for instance, in Figure 6.8, where we plot first columns. Overall, considering also the other columns, no parameterization stands clearly out.

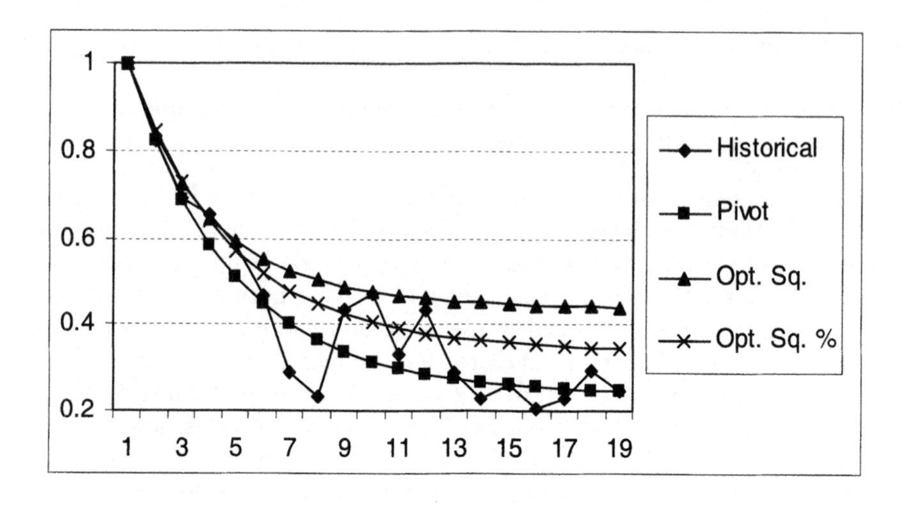

Fig. 6.8. First columns of correlation matrices

The improvement seen in the error cannot be negligible, if one is only looking for a replication of the target matrix as accurate as possible. On the contrary, this is not really noteworthy when what is needed is mainly a fast and intuitive way to build up a correlation matrix, if possible synthetically consistent with major market tendencies, but above all regular and easy to modify. This is just the case in the cascade calibration market tests we will consider further on, so that the level of precision given by the pivot forms is more than enough for our purposes.

6.20 Connecting Caplet and $S \times 1$-Swaption Volatilities

We now present a result that may facilitate the joint calibration of the LFM to the caps and swaptions markets. Indeed, an important problem is that in the cap market forward rates are mostly semi-annual, whereas those entering the forward-swap-rate expressions are typically annual rates. Therefore, when considering both markets at the same time, we may have to reconcile volatilities of semi-annual forward rates and volatilities of annual forward rates. In this section we address this problem. Further investigations will be presented once other calibration issues are clear, in Section 7.9.

Consider three instants $0 < S < T < U$, all six-months spaced. To fix ideas, assume we are dealing with an $S \times 1$ swaption and with S and T-

expiry six-month caplets. For instance, we might have $S = 5$ years, $T = 5.5$ years and $U = 6$ years. We aim at deriving a relationship between the Black swaption volatility and the two Black caplet volatilities.

Consider the three following forward rates at a generic instant $t < S$:

$$F_1(t) := F(t; S, T), \quad F_2(t) := F(t; T, U), \quad F(t) := F(t; S, U).$$

The first two, F_1 and F_2, are semi-annual forward rates, whereas the third one is the annual forward rate in which the two previous forward rates are "nested". We assume year fractions of 0.5 for F_1 and F_2 and of 1 for F.

The algebraic relationship between F, F_1 and F_2 is easily derived by expressing all forward rates in terms of zero-coupon-bond prices. Start from

$$F_1(t) = \frac{1}{0.5}\left[\frac{P(t,S)}{P(t,T)} - 1\right], \quad F_2(t) = \frac{1}{0.5}\left[\frac{P(t,T)}{P(t,U)} - 1\right]$$

and

$$F(t) = \frac{1}{1}\left[\frac{P(t,S)}{P(t,U)} - 1\right].$$

Now observe that

$$F(t) = \frac{P(t,S)}{P(t,T)}\frac{P(t,T)}{P(t,U)} - 1,$$

and then substitute the expression for the two inner fractions of discount factors from the above expressions of F_1 and F_2. One easily obtains

$$\boxed{F(t) = \frac{F_1(t) + F_2(t)}{2} + \frac{F_1(t)F_2(t)}{4}} \tag{6.74}$$

so that if F_1 and F_2 are lognormal, F cannot be exactly lognormal at the same time.

Assume now the following dynamics

$$\begin{aligned} dF_1(t) &= (\ldots)\, dt + \sigma_1(t) F_1(t)\, dZ_1(t), \\ dF_2(t) &= (\ldots)\, dt + \sigma_2(t) F_2(t)\, dZ_2(t), \\ dZ_1\, dZ_2 &= \rho\, dt \end{aligned}$$

for the two semi-annual rates. The quantity ρ is the "infra-correlation" between the "inner rates" F_1 and F_2. We obtain easily by differentiation

$$\begin{aligned} dF(t) = (\ldots)\, dt + \sigma_1(t)\left(\frac{F_1(t)}{2} + \frac{F_1(t)F_2(t)}{4}\right) dZ_1(t) \\ + \sigma_2(t)\left(\frac{F_2(t)}{2} + \frac{F_1(t)F_2(t)}{4}\right) dZ_2(t). \end{aligned}$$

By taking variance on both sides, conditional on the information available at time t, and by calling $\sigma(t)$ the percentage volatility of F we have

$$\sigma^2(t)F^2(t) = \sigma_1(t)^2 \left(\frac{F_1(t)}{2} + \frac{F_1(t)F_2(t)}{4}\right)^2 + \sigma_2(t)^2 \left(\frac{F_2(t)}{2} + \frac{F_1(t)F_2(t)}{4}\right)^2 + 2\rho\sigma_1(t)\sigma_2(t) \left(\frac{F_1(t)}{2} + \frac{F_1(t)F_2(t)}{4}\right) \left(\frac{F_2(t)}{2} + \frac{F_1(t)F_2(t)}{4}\right).$$

Set

$$\boxed{u_1(t) := \frac{1}{F(t)} \left(\frac{F_1(t)}{2} + \frac{F_1(t)F_2(t)}{4}\right)}$$

$$\boxed{u_2(t) := \frac{1}{F(t)} \left(\frac{F_2(t)}{2} + \frac{F_1(t)F_2(t)}{4}\right)}$$

so that

$$\sigma^2(t) = u_1^2(t)\sigma_1(t)^2 + u_2^2(t)\sigma_2(t)^2 + 2\rho\sigma_1(t)\sigma_2(t)u_1(t)u_2(t).$$

Let us introduce a first (deterministic) approximation by freezing all F's (and therefore u's) at their time-zero value:

$$\boxed{\sigma^2_{\text{appr}}(t) = u_1^2(0)\sigma_1(t)^2 + u_2^2(0)\sigma_2(t)^2 + 2\rho\sigma_1(t)\sigma_2(t)u_1(0)u_2(0)}\,.$$

Now recall that F is the particular (one-period) swap rate underlying the $S \times 1$ swaption, whose (squared) Black's swaption volatility is therefore

$$v^2_{\text{Black}} \approx \frac{1}{S}\int_0^S \sigma^2_{\text{appr}}(t)\,dt = \frac{1}{S}\left[u_1^2(0)\int_0^S \sigma_1(t)^2\,dt + u_2^2(0)\int_0^S \sigma_2(t)^2\,dt + 2\rho u_1(0)u_2(0)\int_0^S \sigma_1(t)\sigma_2(t)\,dt\right]. \tag{6.75}$$

The problem is evaluating the last three integrals. Consider the first one:

$$\frac{1}{S}\int_0^S \sigma_1(t)^2\,dt = v^2_{S-\text{caplet}},$$

since S is exactly the expiry of the semi-annual rate F_1. Therefore the first integral can be inputed directly as a market caplet volatility. Not so for the second and third integrals. They both require some parametric assumption on the instantaneous volatility structure of rates in order to be computed. The simplest solution is to assume that forward rates have constant volatilities. In such a case, we compute immediately the second integral as

$$\frac{1}{S}\int_0^S \sigma_2(t)^2\,dt = \frac{1}{S}\int_0^S v^2_{T-\text{caplet}}\,dt = v^2_{T-\text{caplet}},$$

while the third one is

$$\frac{1}{S}\int_0^S \sigma_1(t)\sigma_2(t)\,dt = \frac{1}{S}\int_0^S v_{T-\text{caplet}} v_{S-\text{caplet}}\,dt = v_{T-\text{caplet}} v_{S-\text{caplet}}\,.$$

Under this assumption of constant volatility affecting the second and third integral we obtain

$$\boxed{v^2_{\text{Black}} \approx u_1^2(0)v^2_{S-\text{caplet}} + u_2^2(0)v^2_{T-\text{caplet}} + 2\rho u_1(0)u_2(0)v_{S-\text{caplet}} v_{T-\text{caplet}}}\,. \tag{6.76}$$

Other assumptions on the volatility term structure are possible, but the simple one above is one of the few rendering the approximated formula self-sufficient on the basis of direct market quantities. When dealing with swaptions, De Jong, Driessen and Pelsser (1999) noticed that flat volatilities in forward rates tend to overprice swaptions, so that we can argue that the approximated formula above can give volatilities that are slightly larger than the actual ones. This is partly confirmed analytically as follows. As far as the second integral approximation is concerned, we may note that when S is large and the instantaneous volatility $\sigma_2(t)$ does not differ largely in $[S,T]$ from its average value in $[0,S]$, we can approximate the integral as

$$\frac{1}{S}\int_0^S \sigma_2(t)^2\,dt \approx \frac{1}{T}\int_0^T \sigma_2(t)^2\,dt = v^2_{T-\text{caplet}}.$$

As far as the third integral is concerned, by the Schwartz inequality we have:

$$\int_0^S \sigma_1(t)\sigma_2(t)\,dt \le \sqrt{\int_0^S \sigma_1(t)^2\,dt}\,\sqrt{\int_0^S \sigma_2(t)^2\,dt} \approx S\, v_{S-\text{caplet}}\, v_{T-\text{caplet}}\,,$$

which, in case of positive correlation, shows that (6.76) overestimates volatility with respect to (6.75).

The above formula (6.76) can also be used to back out an implied correlation between adjacent semi-annual rates starting from the $S \times 1$-Black swaption volatility and from the two Black caplet volatilities for expiries S and T. Indeed, one reads from the swaption market v_{Black} and from the cap market one calculates $v_{S-\text{caplet}}$ and $v_{T-\text{caplet}}$. Then one inverts formula (6.76) and obtains ρ. One has to keep in mind, however, that this correlation depends on the constant-volatility assumption above.

A fundamental use for formula (6.76) is easing the joint calibration of the LFM to the caps and swaptions markets. Indeed, the problem (ignored by us in earlier sections and chapters) is that in the cap market forward rates are semi-annual, whereas the forward rates concurring to a swap rate are annual forward rates. Therefore, when calibrating the LFM, we have volatilities of semi-annual forward rates, such as F_1 and F_2, from the cap market but we need to put volatilities of annual forward rates, such as F, in the swaption calibration apparatus. What can be helpful here is formula (6.75). Whichever structure is assumed for the instantaneous volatilities of semi-annual rates, this formula allows us to compute the corresponding integrated volatility for annual forward rates. Hence, we can treat caplets as if they were on annual forward rates, and the joint calibration can now be based on the same family of (annual) forward rates. Notice that the ρ's between adjacent semi-annual forward rates are further parameters that can be used in the calibration, even though they should not be allowed to take small or negative values, since instantaneous correlations between adjacent forward rates are usually close to one.

We close this section by presenting some numerical tests of our formulas above.

We adopt the piecewise-constant instantaneous-volatility formulation (6.8) that through the η's preserves the humped shape of the term structure of volatilities through time.

The η's have been deduced by a typical Euro-market caplet-volatility table and are reported below in Table 6.4. In the first column we have the expiry of the relevant semi-annual forward rate, in the second column the related caplet volatility, in the third column the time interval where the instantaneous volatility of the forward rate, whose expiry is the terminal point of the interval, is equal to the value of η reported in the corresponding position of the last column. So, for example, in the third row we read that the caplet resetting at 1.5 years (and paying at 2 years) has a caplet volatility of 19.78%. The related forward rate is $F(\cdot; 1.5, 2)$ and has piecewise-constant instantaneous volatility given by 0.225627879 for $t \in [0,\ 0.5)$, with subsequent values in $[0.5,\ 1)$ and $[1,\ 1.5)$ obtained by moving up along the column (amounting respectively to 0.208017523 and 0.1523).

We plan to test formula (6.75) (approximation 1, shortly "a1") versus formula (6.76) ("a2"), and finally compare them to the real implied volatility obtained by inverting the Monte Carlo price of the related annual caplet through Black's formula. In doing this, we shall always assume for each pair of adjacent rows the first forward rate to be $F_1(0) = 0.04$ and the second to be $F_2(0) = 0.05$, while always taking as caplet strike $K = 0.0455 = F(0)$.

The "true" Monte Carlo annual caplet volatility is obtained as follows. Given the instantaneous volatilities η, one simulates the discretized Milstein dynamics of F_1 and F_2 up to the expiry of F_1, under the "canonical" forward measure for F_2. Then one calculates the annual forward rate F along each path based on the simulated F_1 and F_2 through Formula (6.74).

At this point one can evaluate the annual caplet, whose underlying rate is F, through a Monte Carlo pricing, by averaging the discounted payoff in F along all paths. By including 1.65 times the standard deviation of the simulated payoff divided by the square root of the number of paths we can obtain a 90% window for the true annual caplet price, and we can check

Expiries	Caplet volatilities	Time intervals	η's
0.5	0.1523	[0, 0.5]	0.1523
1	0.1823	[0.5, 1]	0.208017523
1.5	0.1978	[1, 1.5]	0.225627879
2	0.1985	[1.5, 2]	0.200585343
2.5	0.1999	[2, 2.5]	0.205404601
3	0.1928		0.152417158
3.5	0.1868		0.145700515
4	0.1807		0.130231486
4.5	0.1754		0.12516597
5	0.1715		0.131286176
5.5	0.1677	[5, 5.5]	0.123424835
6	0.1638		0.112290204
6.5	0.16		0.104089961
7	0.1576		0.122182814
7.5	0.1552		0.116520213
8	0.1528		0.110724162
8.5	0.1504		0.104772515
9	0.148		0.098637113
9.5	0.1456		0.092281309
10	0.145		0.133087039
10.5	0.1444	[10, 10.5]	0.131827766
11	0.1439		0.132966387
11.5	0.1433		0.129398029
12	0.1427		0.128126851
12.5	0.1422		0.129622683
13	0.1416		0.125672431
13.5	0.141		0.124388263
14	0.1405		0.126253713
14.5	0.1399		0.121906891
15	0.1403		0.151441111
15.5	0.1406	[15, 15.5]	0.149319992
16	0.141		0.152881784
16.5	0.1414		0.153651163
17	0.1418		0.154419817
17.5	0.1421		0.151947985
18	0.1425	[17.5, 18]	0.15585458
18.5	0.1429	[18, 18.5]	0.156621103
19	0.1433	[18.5, 19]	0.157386944
19.5	0.1436	[19, 19.5]	0.154569143
20	0.1436	[19.5, 20]	0.1436

Table 6.4. Testing "a1" and "a2" approximations: inputs (expiries and time intervals are in years, caplets are semi-annual).

whether the approximated analytical formulas (6.75) and (6.76) are inside this window ("in") or not ("out"). Our results are reported in Table 6.5.

In this table, we can see what happens in three different cases of infra-instantaneous correlations ρ between F_1 and F_2. The upper table concerns the case $\rho = 1$, which should be close to the appropriate value of correlation between adjacent rates. In the middle table, we see the case with $\rho = 0.5$, where adjacent rates are decorrelated to a large extent, whereas in the final table, we see the case of complete decorrelation $\rho = 0$, which sounds rather unrealistic for close rates.

In each case we reported the annual caplet volatilities under both approximations "a1" and "a2", the "true" Monte Carlo (MC) implied volatility and the related 95% window. We have written "in" or "out" to signal when

$\rho = 1$						95%	Window	Errors (%)			Vols		
mat.	a1		a2		MC	inf	sup	a1-a2	a1-MC	a2-MC	F_1 a1	F_2 a2	F_2 a1
1y	20.34	in	19.30	out	20.40	20.23	20.58	-5.14	-0.30	-5.42	18.23	19.78	21.70
2y	20.67	in	20.15	out	20.78	20.59	20.97	-2.51	-0.55	-3.05	19.85	19.99	21.01
5y	17.15	in	17.13	in	17.19	17.02	17.36	-0.17	-0.20	-0.37	17.15	16.77	16.92
10y	14.56	in	14.63	out	14.46	14.31	14.61	0.44	0.68	1.12	14.5	14.44	14.40
19y	14.47	in	14.50	in	14.39	14.21	14.56	0.26	0.54	0.81	14.33	14.36	14.34
$\rho = .5$						95%	Window	Errors (%)			Vols		
mat.	a1		a2		MC	inf	sup	a1-a2	a1-MC	a2-MC	F_1 a1	F_2 a2	F_2 a1
1y	17.74	in	16.78	out	17.64	17.49	17.79	-5.43	0.57	-4.89	18.23	19.78	21.70
2y	17.97	in	17.48	out	17.89	17.72	18.05	-2.70	0.46	-2.25	19.85	19.99	21.01
5y	14.90	in	14.85	in	14.84	14.70	14.98	-0.29	0.40	0.11	17.15	16.77	16.92
10y	12.65	out	12.69	in	12.79	12.66	12.91	0.35	-1.10	-0.76	14.5	14.44	14.40
19y	12.56	out	12.59	out	12.73	12.59	12.87	0.21	-1.34	-1.14	14.33	14.36	14.34
$\rho = 0$						95%	Window	Errors (%)			Vols		
mat.	a1		a2		MC	inf	sup	a1-a2	a1-MC	a2-MC	F_1 a1	F_2 a2	F_2 a1
1y	14.68	out	13.80	out	14.89	14.77	15.02	-5.99	-1.47	-7.37	18.23	19.78	21.70
2y	14.79	in	14.34	out	14.91	14.78	15.04	-3.06	-0.80	-3.83	19.85	19.99	21.01
5y	12.23	out	12.17	out	12.41	12.29	12.52	-0.52	-1.44	-1.95	17.15	16.77	16.92
10y	10.38	out	10.40	out	10.57	10.46	10.67	0.17	-1.73	-1.56	14.5	14.44	14.40
19y	10.31	out	10.32	out	10.62	10.51	10.73	0.10	-2.94	-2.84	14.33	14.36	14.34

Table 6.5. Testing "a1" and "a2" approximations: results.

an approximation is inside or outside the window. The related percentage differences are also reported.

Finally, if we call T_{row} the expiry in the first column of the table, in the last three columns we have reported the following.

- "F_1 a1" is the average volatility of the first forward rate, F_1, up to its expiry T_{row}, and is therefore the F_1 caplet volatility for that row:

$$\sqrt{\frac{1}{T_{\text{row}}} \int_0^{T_{\text{row}}} \sigma(t; T_{\text{row}}, T_{\text{row}} + 6m)^2 \, dt} = v_{T_{\text{row}}-\text{caplet}}$$

- "F_2 a1" is the average volatility of the second forward rate, F_2, up to the expiry T_{row} of F_1, which, under "a1", is computed exactly:

$$\sqrt{\frac{1}{T_{\text{row}}} \int_0^{T_{\text{row}}} \sigma(t; T_{\text{row}} + 6m, T_{\text{row}} + 1y)^2 \, dt}$$

- "F_2 a2" is the average volatility of the second forward rate, F_2, up to the expiry T_{row}, which, under "a2", is approximated by the $T_{\text{row}} + 6m$-semi-annual caplet volatility for F_2:

$$\sqrt{\frac{1}{T_{\text{row}}} \int_0^{T_{\text{row}}} \sigma(t; T_{\text{row}} + 6m, T_{\text{row}} + 1y)^2 \, dt} \approx v_{T_{\text{row}}+6m-\text{caplet}} \, .$$

In this way we can see the difference between the approximations "a1" and "a2" for the second integral. This is important because "a2" is obtained directly as a caplet volatility, and requires no explicit knowledge of instantaneous volatilities.

Results seem to show that the differences between "a1" and "a2" are small for large expiries, say $T_{\text{row}} \geq 5y$. This is intuitive: "a2" is based on assuming the averages of $\sigma_2^2(\cdot)$ in $[0, T_{\text{row}}]$ and $[0, T_{\text{row}} + 6m]$, respectively, to be close, which is more likely to happen for large T_{row}.

Moreover, "a1" results to be acceptable compared to the true volatility in most situations, and especially when $\rho = 1$. Instead, in such a comparison, "a2" is in trouble for short maturities, exactly as in its comparison with "a1".

6.21 Forward and Spot Rates over Non-Standard Periods

Assume that, in order to price a financial product, we need to know spot LIBOR rates $L(S) := L(S, S + \delta)$ at certain dates $S = s_1, \ldots, S = s_n$, where δ is the common time-to-maturity of the considered rates (typically half a year or one year). This set of rates can be obtained through a suitable family of forward rates defined over expiry-maturity pairs that are non-standard, i.e. that are not contained in our starting set $\{T_0, \ldots, T_M\}$. This can help when pricing products such as trigger or accrual swaps, depending on the daily or weekly evolution of the spot LIBOR rate. Specifically, suppose we are interested in propagating the forward rates

$$F(\cdot; s_1, U_1),\ F(\cdot; s_2, U_2), \ldots,\ F(\cdot; s_n, U_n),$$

with $s_1 < s_2 < \ldots < s_n$ and $U_1 < U_2 < \ldots < U_n$, from time 0 to time s_n. Typically, as we said above, $U_i = s_i + \delta$. Notice that this propagation provides us also with the spot LIBOR rates

$$L(s_1, U_1),\ L(s_2, U_2), \ldots,\ L(s_n, U_n).$$

How can we obtain the dynamics of the above forward rates from the dynamics of the original family $F_1, \ldots, F_M$? Let us examine two possible methods.

6.21.1 Drift Interpolation

Marry, sir, here's my drift Polonius in Hamlet, II.1

To fix ideas, take a generic maturity U and assume it to be included in $[T_k, T_{k+1}]$, and assume the related expiry S to be not too far back from U (i.e. $S - U$ should not be too much larger than the average τ_k).

The key idea here is to use formula (6.16) as a guide to write

$$dZ^U \approx dZ^k + \frac{(U - T_k)F_{k+1}(t)}{1 + \tau_{k+1}F_{k+1}(t)} \rho \underline{\sigma}_{k+1}(t)'\, dt\,. \tag{6.77}$$

Notice that for $U = T_{k+1}$ we obtain formula (6.16) by assuming $\tau_{k+1} = T_{k+1} - T_k$. Also, for $U = T_k$ we obtain an identity, as should be. We need also to define instantaneous correlations between these shocks. We set

$$dZ_t^{U_a} dZ_t^{U_b} = \rho_{k,j}\, dt \quad \text{if} \quad k - 1 < U_a \le k,\ j - 1 < U_b \le j\,,$$

so as to extend the original instantaneous correlations to non-standard maturities in the simplest possible way, that is by taking the same instantaneous correlations of the original rates where the considered non-standard rates are "nested". Now use the last approximated shock above in the martingale dynamics

$$dF(t; S, U) = \underline{\sigma}(t; S, U)F(t; S, U)\, dZ^U(t)$$

to deduce the following.

Proposition 6.21.1. (Interpolated-drift dynamics for non-standard rates in the LFM). *The dynamics of the non-standard forward rate $F(t; S, U)$, derived through the above drift interpolation, is given by*

$$\begin{aligned} dF(t; S, U) = {} & \frac{(U - T_k)F_{k+1}(t)}{1 + \tau_{k+1}F_{k+1}(t)} F(t; S, U)\underline{\sigma}(t; S, U)\rho\underline{\sigma}_{k+1}'(t)\, dt \\ & + \underline{\sigma}(t; S, U)F(t; S, U)\, dZ^k(t)\,, \quad T_k < U < T_{k+1}\,. \end{aligned} \tag{6.78}$$

The (deterministic) volatility $\underline{\sigma}(t; S, U)$ is readily obtained if we have calibrated a certain functional form of instantaneous volatilities of the LFM, such as for example Formulations 6 or 7. As for volatility parameters depending on maturities, such as the Φ's of Formulation 7 or the volatilities themselves in the piecewise-constant formulations, the parameters for $\underline{\sigma}(t; S, U)$ can be obtained by suitably interpolating the corresponding parameters of $\underline{\sigma}_k(t)$ and $\underline{\sigma}_{k+1}(t)$, i.e. by interpolating the instantaneous volatilities of the closest adjacent rates in the original family. We may as well set directly $\underline{\sigma}(t; S, U) \approx \underline{\sigma}_{k+1}(t)$.

For Monte Carlo simulation, the obtained dynamics can be discretized first by taking logarithms and then by applying the Milstein scheme to the obtained dynamics:

$$\begin{aligned}\ln F(t+\Delta t; S, U) = \ln F(t; S, U) &+ \frac{(U - T_k)F_{k+1}(t)}{1 + \tau_{k+1}F_{k+1}(t)} \underline{\sigma}(t; S, U)\rho\underline{\sigma}_{k+1}'(t)\Delta t \\ &- \tfrac{1}{2}|\underline{\sigma}(t; S, U)|^2 \Delta t + \underline{\sigma}(t; S, U)(Z^k(t+\Delta t) - Z^k(t)) \,.\end{aligned}$$

Now, to obtain the required family of forward rates, we just set $U = U_1$, $U = U_2$ up to $U = U_n$ and use the above scheme from time 0 to time s_n, making sure that we choose a time step Δt such that all the dates $s_1, s_2, \ldots, s_n$ are encountered in the discretization. Notice that here too the analogous of Remark 6.10.1 applies for the shocks' simulation.

There is still a possible problem. Assume we need the daily evolution of the one-year spot LIBOR rate up to 5 years for valuing an accrual swap with daily accruing and yearly payment period. Let us focus on the second payment, and assume that the swap leg we are to price pays out at the second year the amount given by the one-year LIBOR rate resetting after one year, times the number of days the spot rate has remained in-between two given levels in the period from one to two years, divided by 365. More formally, the discounted payoff reads

$$D(0,2)L(1,2) \, \frac{\sum_{t=1y}^{2y-1d} 1_{\{B_1 < L(t+1y, t+2y) < B_2\}}}{365},$$

B_1 and B_2 being the levels, $d = 1/365y$ being a shorthand notation for "day" and the step in the summation being one day.

To evaluate this payment of the swap leg at time 0, we would need to set $s_1 = 1y$ up to $s_{365} = 2y - 1d$, and $U_1 = 2y$ up to $U_{365} = 3y - 1d$. However, it is clearly undesirable to simulate jointly a 365 dimensional vector. It is true that, as time passes, the vector length diminishes by one each day and the "alive" forward rates reduce in number. Yet, the task can still be too demanding. What can be done is simulating say weekly or three-monthly forward rates, and then at each simulated instant one can interpolate directly the forward rates themselves. A naive way to do this, again with reference to the considered example, is as follows.

Suppose we simulate monthly rates by choosing $s_1 = 1y, \ldots, s_n = 1y + 12m$ and the corresponding U's one-year shifted, thus having $\delta = 1y$. Suppose further that we are at day $1y + 35d = 1y + 1m + 5d$ and we need the one-year spot LIBOR rate $L(1y + 35d, 2y + 35d) = F(1y + 35d; 1y + 35d, 2y + 35d)$ to be put in the second-payment payoff. Our numerical Monte Carlo scheme provides us with $F(1y{+}35d; 1y{+}1m, 2y{+}1m) := F(1y{+}1m; 1y{+}1m, 2y{+}1m)$ (by extending forward rates after their "death" in the trivial "frozen" way) and $F(1y + 35d; 1y + 2m, 2y + 2m)$. It is now easy to interpolate between the points $(1y + 1m, F(1y + 35d; 1y + 1m, 2y + 1m))$ and $(1y + 2m, F(1y + 35d; 1y + 2m, 2y + 2m))$ to obtain $F(1y + 35d; 1y + 35d, 2y + 35d)$. There are several classical interpolation methods that can be used.

Of course, the above "drift interpolation" technique, possibly partially combined with the above "directly interpolate rates" reductions, is not the only possibility. We propose below a technique based on a "Brownian bridge"-like method.

6.21.2 The Bridging Technique

We propose a different solution to the problem of simulating spot-rate values $L(t) := L(t, t + \delta)$ in-between the two spot-rate values $L(T_{\beta(t)-2})$ and $L(T_{\beta(t)-1})$ obtained from the original family of spanning forward rates as $F_{\beta(t)-1}(T_{\beta(t)-2})$ and $F_{\beta(t)}(T_{\beta(t)-1})$, respectively. We here assume that the T_i's are equally δ-spaced.

To fix ideas, assume the current time t to be between T_{i-1} and T_i,

$$t \in [T_{i-1}, T_i] = [T_{\beta(t)-2}, T_{\beta(t)-1}],$$

and that we need the LIBOR rate $L(t)$ at times $s_1 = T_{i-1}, s_2, \ldots, s_{l-1}, s_l = T_i$. This second method is based on ideas mutuated from the notion of Brownian bridge. It works according to the following steps:

a) Through the discretized forward-rate dynamics analogous to (6.53) we have generated m realizations of $L(T_{i-1}) = F_i(T_{i-1})$ and m corresponding realizations of $L(T_i) = F_{i+1}(T_i)$. We denote the j-th realization by a superscript j, so that, for example, $L^j(T_{i-1})$ denotes the realization of $L(T_{i-1})$ obtained under the j-th scenario.

b) Assume a "geometric-Brownian-motion"-like dynamics for the spot rate $L(t)$ in-between the two already-generated values $L(T_{i-1})$ and $L(T_i)$,

$$dL(t) = \mu_i L(t)\, dt + v_i L(t)\, dZ(t - T_{i-1}), \; t \in [T_{i-1}, T_i], \qquad (6.79)$$

where Z is a standard Brownian motion under the same measure used for generating the forward-rate dynamics.

c) Consider

$$\ln \frac{L(T_i)}{L(T_{i-1})} = (\mu_i - v_i^2/2)(T_i - T_{i-1}) + v_i Z(T_i - T_{i-1}) \qquad (6.80)$$

$$\sim \mathcal{N}\left((\mu_i - v_i^2/2)(T_i - T_{i-1}), v_i^2(T_i - T_{i-1})\right).$$

Since we know the maximum-likelihood estimators for mean and variance of Gaussian random variables, we can use such estimators to estimate μ_i and v_i as implied from the m values generated for $L(T_{i-1})$ and $L(T_i)$. The maximum-likelihood estimators consist simply on the sample mean and variance,

$$(T_i - T_{i-1})(\hat{\mu}_i - \hat{v}_i^2/2) = \frac{1}{m}\sum_{j=1}^{m} \ln \frac{L^j(T_i)}{L^j(T_{i-1})},$$

$$(T_i - T_{i-1})\hat{v}_i^2 = \left(\frac{1}{m}\sum_{j=1}^{m} \ln^2 \frac{L^j(T_i)}{L^j(T_{i-1})} - (T_i - T_{i-1})^2(\hat{\mu}_i - \hat{v}_i^2/2)^2\right)$$

which can be easily inverted to obtain $\hat{\mu}_i$ and $\hat{v}_i$. The superscript j denotes the scenario under which the considered variables have been generated.

d) Consider again (6.80), replace μ_i and v_i with their estimates $\hat{\mu}_i$ and $\hat{v}_i$, and solve for $Z(T_i - T_{i-1})$. We find

$$Z(T_i - T_{i-1}) = \ln \frac{L(T_i)}{L(T_{i-1})} - (\hat{\mu}_i - \hat{v}_i^2/2)(T_i - T_{i-1}).$$

The final value of Z is therefore known in advance. In particular, under the j-th scenario, we have the final value of Z given by

$$Z^j(T_i - T_{i-1}) = \ln \frac{L^j(T_i)}{L^j(T_{i-1})} - (\hat{\mu}_i - \hat{v}_i^2/2)(T_i - T_{i-1}).$$

e) For each scenario j, we have now the initial and final values $L^j(T_{i-1})$ and $L^j(T_i)$, and the final value of the Brownian motion Z in the L-dynamics (6.79). We now need to connect the initial and final values by finding a path

$$L(s_1) = L^j(T_{i-1}),\ L(s_2),\ L(s_3), \ldots, L(s_{l-2}),\ L(s_{l-1}),\ L(s_l) = L^j(T_i)$$

describing all the spot-rate values needed under the j-th scenario.
Now assume we are in scenario j. All we need to do is simulating the shocks dZ consistently with the known final value $Z^j(T_i - T_{i-1})$. In order to do so, we replace $Z(t - T_{i-1})$ with a process with the same initial and final known values and whose increments are as close as possible to the increments of a Brownian motion. The process we consider is

$$\zeta(t - T_{i-1}) := V(t - T_{i-1}) - \frac{t - T_{i-1}}{T_i - T_{i-1}} \left(V(T_i - T_{i-1}) - Z^j(T_i - T_{i-1})\right), \tag{6.81}$$

where V is a new standard Brownian motion, independent of the previous one. By construction, $\zeta(0) = Z(0) = 0$, $\zeta(T_i - T_{i-1}) = Z^j(T_i - T_{i-1})$.
Clearly,

$$d\zeta(t - T_{i-1}) = dV(t - T_{i-1}) - \frac{V(T_i - T_{i-1}) - Z^j(T_i - T_{i-1})}{T_i - T_{i-1}}\,dt,$$

so that this is a Brownian motion with constant drift. Of course

$$d\zeta(t - T_{i-1})\, d\zeta(t - T_{i-1}) = dt$$

as for a standard Brownian motion.

f) At this point, we simulate the intermediate values of the process (6.79) under the j-th scenario through a single path from the following modified version of its exact discrete dynamics:

$$L^j(s_{h+1}) = L^j(s_h)\exp\left[(\hat{\mu}_i - \hat{v}_i^2/2)(s_{h+1} - s_h) + \hat{v}_i(\zeta(s_{h+1}) - \zeta(s_h))\right],$$

for $h = 1, \dots, l-1$. This is perfectly feasible at this point, since all the quantities above are known. In particular, one path for ζ through $s_1, \dots, s_k$ can be obtained by simulating a priori one path for the independent standard Brownian motion V and subsequently using (6.81).

The above "bridging" technique, coupled with the considered forward-rate dynamics, is a possible tool for evaluating derivatives whose payoff depends on the daily or weekly evolution of spot rates. The only limitation lies in the geometric Brownian motion assumption, which features no mean reversion for the spot rate. However, notice that the final value of the considered rates is consistent with forward-rate dynamics by construction, since the drift $\hat{\mu}$ and volatility $\hat{v}$ of the in-between geometric Brownian motion reflect the correct initial and final values for the spot-rate process as implied from the correct forward-rates dynamics. Moreover, the fact that usually the geometric Brownian motion is reset at each payment date T_i to the "true" spot rate –as implied by forward-rate dynamics– renders the absence of mean reversion quite bearable, since mean reversion is usually intended to act on longer periods than the typically six-month long $[T_{i-1}, T_i]$. Finally, one can combine this method with "directly interpolating rates" so as to propagate a much smaller number of forward rates, similarly to what was done for the previous method. However, both the feasibility and the quality of this method need to be tested with simulations and numerical investigations before drawing any conclusion about its performances.

7. Cases of Calibration of the LIBOR Market Model

"How did you get that to work first time?"
"I didn't. I just tried a thousand different combinations at super-speed"
Green Lantern/Kyle and Flash/Wally, JLA: New World Order.

On two occasions I have been asked [by members of Parliament], 'Pray, Mr. Babbage, if you put into the machine wrong figures, will the right answers come out?' I am not able rightly to apprehend the kind of confusion of ideas that could provoke such a question.
Charles Babbage, Passages from the Life of a Philosopher, 1864, London.

In this chapter we present some numerical examples concerning the goodness of fit of the LFM to both the caps and swaptions markets, based on market data. We study several cases based on different instantaneous-volatility parameterizations. We will also point out a particular parameterization allowing for a closed-form-formulas calibration to swaption volatilities and establishing a one to one correspondence between swaption volatilities and LFM covariance parameters.

The whole chapter deals with the lognormal LIBOR market model, with no volatility smile modeling. The purpose of this first treatment is showing that even the calibration of the at-the-money swaption market quotes poses difficult problems and forces one to make hard choices, and it can be quite educational for someone tackling the smile calibration problem. Further, since even as we write the smile in the swaptions market is not quoted massively, it may be enough to deal with at the money data. Inclusion of a volatility smile model, and attempts to link the caplet smile to the swaption smile, would in general destroy the precious one-to-one correspondence above, or would at least render the cascade calibration impossible, with subsequent computational and numerical problems that would render the calibration to a large set of market data unfeasible.

However, even in the limited context of at-the-money data, we would like to make ourselves clear by pointing out that the examples presented here are a first attempt at explaining the relevant choices as far as the LFM parameterization is concerned. We do not pretend to be exhaustive in these examples, and we do not employ statistical testing or econometric techniques in our analysis. We will base our considerations only on cross-sectional calibration to the market-quoted volatilities, although we will check diagnostics implications of the obtained calibrations as far as the future time-evolution of key structures of the market are concerned. Therefore, to decide upon the quality of a calibration, we will look at the future starting from the present, rather than using the past to elicit information on the present. Within such cross-sectional approach we are not being totally systematic. We are aware there are several other issues concerning further possible parameterizations, modeling correlations implicitly as inner products of vector instantaneous volatilities, and so on. Yet, most of the available literature on interest-rate models does not deal with the questions and examples we raise here on the market model. We thought about presenting the examples below in order to let the reader appreciate what are the current problems with the LFM, especially as far as practitioners and traders are concerned. Indeed, the kind of problems we used to read about in interest-rate modeling before starting our work in a bank was rather different from the problems presented here. We hope the reader will have a clear grasp of the main issues at stake nowadays, and this is all we are trying to obtain with the examples below.

The chapter is structured as follows: Section 7.1 explains the kind of inputs we adopt for the first calibration tests, Section 7.2 deals with a joint caps-swaptions calibration when instantaneous correlations ρ are used as fitting parameters and under some specific piecewise constant parameterization for volatilities σ, whereas Section 7.3 does the same under a parametric form for σ's.

Section 7.4 deals with the choice of calibrating exactly all swaptions under an exogenously given instantaneous correlation matrix ρ and when assuming the general piecewise constant (GPC) parameterization for σ's as from TABLE 1 of Chapter 6. The devised method leads to an immediate algorithm easy to implement even in a spreadsheet and requiring no optimization or advanced numerical tools.

Section 7.5 presents a first break to collect ideas and leads to further developments on the cascade method in Section 7.6.

The important section 7.7 presents a method for avoiding interpolation of missing input swaption volatilities that are inconsistent with the assumed LIBOR market model dynamics, and investigates on the robustness induced by this choice on the calibration procedure.

Section 7.9 presents a possible way to calibrate jointly caps and swaptions when taking the semi-annual tenor of caps into account, thus establishing some relationships between semi-annual and annual forward rates and volatilities. Section 7.10 concludes the chapter.

7.1 Inputs for the First Cases

We will try and calibrate the following data: "annualized" initial curve of forward rates, "annualized" caplet volatilities, and swaptions volatilities. Let us examine these data in more detail.

We actually take as input the vector

$$F0 = [F(0;0,0.5), F(0;0.5,1), \ldots, F(0;19.5,20)]$$

of initial semi-annual forward rates as of May 16, 2000, and the semi-annual at-the-money caplet volatilities (y stands for year/years)

$$v0 = \left[v_{1y-\text{caplet}}, v_{1.5y-\text{caplet}}, \ldots, v_{19.5y-\text{caplet}}\right],$$

with the first semi-annual caplet resetting in one year and paying at 1.5 (years), the last semi-annual caplet resetting in 19.5 (years) and paying at 20 (years), all other reset dates being six-months spaced.

These volatilities have been provided by our interest-rate traders, based on a stripping algorithm combined with personal adjustments applied to cap volatilities. All basic data are as of May 16, 2000. We transform semi-annual data in annual data through the methods highlighted in Section 6.20, and work with the annual forward rates

$$[F(\cdot;1y,2y), F(\cdot;2y,3y), \ldots, F(\cdot;19y,20y)]$$

and their associated annual caplet volatilities. In the transformation formula, infra-correlations are set to one. Notice that infra-correlations might be kept as further parameters to ease the calibration, see again Section 6.20 for a discussion.

In our data, the initial spot rate is $F(0;0,1y) = 0.0469$, and the other initial forward rates and caplet volatilities are shown in Table 7.1

Finally, the values of swaptions volatilities are the same as in the table given in Section 6.17. We will change swaptions volatilities when calibrating only to swaptions, in Section 7.4.

7.2 Joint Calibration with Piecewise-Constant Volatilities as in TABLE 5

In order to satisfactorily calibrate the above data with the LFM, we first try the volatility structure of TABLE 5, with a local algorithm of minimization for finding the best-fitting parameters $\psi_1, \ldots, \psi_{19}$ and $\theta_1, \ldots, \theta_{19}$, starting from the initial guesses $\psi_i = 1$ and $\theta_i = \pi/2$. We thus adopt formula (6.11) for instantaneous volatilities and obtain the Φ's directly as functions of the parameters ψ by using the (annualized) caplet volatilities and formula (6.25).

initial $F0$	v_{caplet}
0.050114	0.180253
0.055973	0.191478
0.058387	0.186154
0.060027	0.177294
0.061315	0.167887
0.062779	0.158123
0.062747	0.152688
0.062926	0.148709
0.062286	0.144703
0.063009	0.141259
0.063554	0.137982
0.064257	0.134708
0.064784	0.131428
0.065312	0.128148
0.063976	0.1271
0.062997	0.126822
0.06184	0.126539
0.060682	0.126257
0.05936	0.12597

Table 7.1. Initial (annualized) forward rates and caplet volatilities.

We compute swaptions prices as functions of the ψ's and θ's by using Rebonato's formula (6.67). Since we are using piecewise-constant instantaneous volatilities, the formula reduces to a summation of products of volatility parameters. We also impose the constraints

$$-\pi/2 < \theta_i - \theta_{i-1} < \pi/2$$

to the correlation angles, which implies that $\rho_{i,i-1} > 0$. We thus require that adjacent rates have positive correlations. As we shall see, this requirement is obviously too weak to guarantee the instantaneous correlation matrix coming from the calibration to be "reasonable".

We obtain the parameters that are shown in Table 7.2, where the Φ's have been computed through (6.25).

The fitting quality is as follows. The caplets are fitted exactly, whereas we calibrated the whole swaptions volatility matrix except for the first column of $S \times 1$-swaptions. This is left aside because of possible misalignments with the annualized caplet volatilities, since we are basically quoting twice the same volatilities. A more complete approach can be obtained by keeping semi-annual volatilities and by introducing infra-correlations as new parameters as indicated in Section 6.20. The matrix of percentage errors in the swaptions calibration,

$$100 \cdot \frac{\text{Market swaption volatility - LFM swaption volatility}}{\text{Market swaption volatility}}$$

is reported below.

Index	ψ	Φ	θ
1	2.5114	0.0718	1.7864
2	1.5530	0.0917	2.0767
3	1.2238	0.1009	1.5122
4	1.0413	0.1055	1.6088
5	0.9597	0.1074	2.3713
6	1.1523	0.1052	1.6031
7	1.2030	0.1043	1.1241
8	0.9516	0.1055	1.8323
9	1.3539	0.1031	2.3955
10	1.1912	0.1021	2.5439
11	0	0.1046	1.6118
12	3.3778	0.0844	1.3172
13	0	0.0857	1.2225
14	1.2223	0.0847	1.0995
15	0	0.0869	1.2602
16	0	0.0896	1.0905
17	0	0.0921	0.8006
18	0.1156	0.0946	0.8739
19	0.5753	0.0965	1.7096

Table 7.2. Calibration results under the volatility formulation of TABLE 5: parameter values.

	2y	3y	4y	5y	6y	7y	8y	9y	10y
1y	-0.71	0.90	1.67	4.93	3.00	3.25	2.81	0.83	0.11
2y	-2.43	-3.48	-1.54	-0.70	0.70	0.01	-0.22	-0.45	0.49
3y	-3.84	1.28	-2.44	-0.69	-1.18	0.21	1.51	1.57	-0.01
4y	1.87	-2.52	-2.65	-3.34	-2.17	-0.44	-0.11	-0.63	-0.38
5y	1.80	4.15	-1.40	-1.89	-1.74	-0.79	-0.34	-0.07	1.28
7y	-0.33	2.27	1.47	-0.97	-0.77	-0.65	-0.57	-0.15	0.19
10y	-0.02	0.61	0.45	-0.31	0.02	-0.03	0.01	0.23	-0.30

Errors are actually small, and from this point of view, such calibration seems to be satisfactory, considering that we are trying to fit 19 caplets and 63 swaption volatilities! However, since the LFM allows for a quick check of future term structures of volatilities and terminal correlations, let us have a look at these quantities.

The first observation is that the calibrated θ's above imply quite erratic instantaneous correlations. Consider the resulting instantaneous-correlation matrix as given in Table 7.3. As you can see, correlations fluctuate occasionally between positive and negative values. This is too a weird behaviour to be trusted. Terminal correlations computed through formula (6.71) are in this case, after ten years,

	10y	11y	12y	13y	14y	15y	16y	17y	18y	19
10y	1.00	0.56	0.27	0.19	0.09	0.21	0.08	-0.10	-0.06	0.37
11y	0.56	1.00	0.61	0.75	0.67	0.68	0.64	0.44	0.42	0.50
12y	0.27	0.61	1.00	0.42	0.71	0.53	0.48	0.43	0.40	0.42
13y	0.19	0.75	0.42	1.00	0.36	0.71	0.50	0.41	0.43	0.34
14y	0.09	0.67	0.71	0.36	1.00	0.32	0.67	0.43	0.40	0.36
15y	0.21	0.68	0.53	0.71	0.32	1.00	0.28	0.59	0.39	0.33
16y	0.08	0.64	0.48	0.50	0.67	0.28	1.00	0.22	0.62	0.30
17y	-0.10	0.44	0.43	0.41	0.43	0.59	0.22	1.00	0.17	0.36
18y	-0.06	0.42	0.40	0.43	0.40	0.39	0.62	0.17	1.00	0.07
19y	0.37	0.50	0.42	0.34	0.36	0.33	0.30	0.36	0.07	1.00

and they still look erratic.

	1y	2y	3y	4y	5y	6y	7y	8y	9y	10y
1y	1.000	0.958	0.963	0.984	0.834	0.983	0.789	0.999	0.820	0.727
2y	0.958	1.000	0.845	0.893	0.957	0.890	0.580	0.970	0.950	0.893
3y	0.963	0.845	1.000	0.995	0.653	0.996	0.926	0.949	0.635	0.513
4y	0.984	0.893	0.995	1.000	0.723	1.000	0.885	0.975	0.706	0.594
5y	0.834	0.957	0.653	0.723	1.000	0.719	0.318	0.858	1.000	0.985
6y	0.983	0.890	0.996	1.000	0.719	1.000	0.888	0.974	0.702	0.589
7y	0.789	0.580	0.926	0.885	0.318	0.888	1.000	0.760	0.295	0.150
8y	0.999	0.970	0.949	0.975	0.858	0.974	0.760	1.000	0.846	0.757
9y	0.820	0.950	0.635	0.706	1.000	0.702	0.295	0.846	1.000	0.989
10y	0.727	0.893	0.513	0.594	0.985	0.589	0.150	0.757	0.989	1.000
11y	0.985	0.894	0.995	1.000	0.725	1.000	0.883	0.976	0.708	0.596
12y	0.892	0.725	0.981	0.958	0.494	0.959	0.981	0.870	0.473	0.337
13y	0.845	0.657	0.958	0.926	0.410	0.928	0.995	0.820	0.387	0.247
14y	0.773	0.559	0.916	0.873	0.295	0.876	1.000	0.743	0.271	0.126
15y	0.865	0.685	0.968	0.940	0.444	0.942	0.991	0.841	0.422	0.283
16y	0.768	0.552	0.912	0.869	0.286	0.872	0.999	0.737	0.263	0.117
17y	0.552	0.291	0.757	0.691	0.000	0.695	0.948	0.513	-0.024	-0.172
18y	0.612	0.360	0.803	0.742	0.073	0.746	0.969	0.575	0.049	-0.099
19y	0.997	0.933	0.981	0.995	0.789	0.994	0.833	0.993	0.774	0.672

Table 7.3. Calibration results under the volatility formulation of TABLE 5: instantaneous-correlation matrix .

Finally, let us have a look at the time evolution of caplet volatilities. We know that the model reproduces exactly the initial caplet volatility structure observed in the market. However, as time passes, the above ψ and Φ parameters imply the evolution shown in Figure 7.1. This evolution shows that the

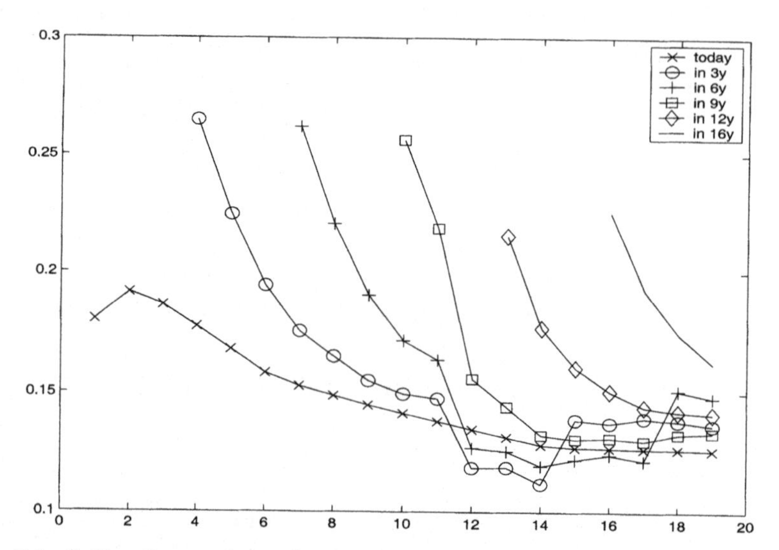

Fig. 7.1. Calibration results under the volatility formulation of TABLE 5: evolution of the term structure of caplet volatilities.

structure loses its humped shape after a short time. Moreover, it becomes somehow "noisy".

What is one to learn from such an example? The idea is that the fitting quality is not the only criterion by which a calibration session has to be judged. The so called "uncertainty principle of modeling", "the more a model fits the less it explains", has to be carefully considered here. A trader has then to decide whether he is willing to sacrifice the fitting quality for a better evolution in time of the key structures.

We have tried several other calibrations with the separable piecewise-constant parameterization of TABLE 5. We tried to impose more stringent constraints on the angles θ's, and we even fixed them both to typical and atypical values, leaving the calibration to the volatility parameters. We also let all instantaneous correlations go to one, so as to have a one-factor LFM to be calibrated only through its instantaneous volatility parameters.

We reached the following conclusions. In order to have a good calibration to swaptions data we need to allow for at least partially oscillating patterns in the correlation matrix. If we force a given "smooth/monotonic" correlation matrix into the calibration and rely upon volatilities, the results are the same as in the case of a one-factor LFM where correlations are all set to one. This kind of results suggests some considerations. Since by fixing rather different instantaneous correlations the calibration does not change that much, probably instantaneous correlations do not have a strong link with European swaptions prices. Therefore swaptions volatilities do not always contain clear and precise information on instantaneous correlations of forward rates. This was clearly stated also in Rebonato (1999d). But what is then the link between correlations and swaptions? Compare the right hand sides of formulas (6.67) and (6.71). It is immediate to notice that there is a common term, so that swaptions clearly contain information on *terminal* correlation rather than on the instantaneous one. Since the former depends also on forward rates volatilities, we cannot extract clear information on the ρ's from swaptions prices, since the effects of the σ's and of the ρ's are mixing. Thus in order to better calibrate swaptions we may have to allow for more general *terminal* correlation structures, i.e. for more general volatility or *instantaneous* correlation structures. We will choose the former solution in Section 7.4, where we will fix exogenously instantaneous correlations.

7.3 Joint Calibration with Parameterized Volatilities as in Formulation 7

We now adopt Formulation 7 for instantaneous volatilities. Instantaneous volatilities are now given by formula (6.13), and depend on the parameters a, b, c, d and Φ's. We again try a local algorithm of minimization for finding the fitted parameters a, b, c, d and $\theta_1, \ldots, \theta_{19}$ starting from the initial guesses

$a = 0.0285$, $b = 0.20004$, $c = 0.1100$, $d = 0.0570$, and initial θ components ranging from $\theta_1 = 0$ to $\theta_{19} = 2\pi$ and equally spaced. The Φ's are obtained as functions of a, b, c, d, $\Phi = \Phi(a, b, c, d)$, through caplet volatilities according to formula (6.28), and this is the caplet calibration part. As for swaptions, we compute swaptions prices as functions of a, b, c, d and the θ's by using Rebonato's formula (6.67). To this end, the lengthy computation of terms such as

$$\int_0^T \psi(T_{i-1} - t; a, b, c, d)\, \psi(T_{j-1} - t; a, b, c, d)\, dt$$

has to be carried out. This can be done easily with a software for formal manipulations, with a command line like

```
int(((a*(S-t)+d)*exp(-b*(S-t))+c)*((a*(T-t)+d)
                    *exp(-b*(T-t))+c),t=t1..t2);
```

We also impose the constraints

$$-\pi/3 < \theta_i - \theta_{i-1} < \pi/3, \quad 0 < \theta_i < \pi,$$

to the correlation angles. Putting $\pi/2$ in the first constraint would ensure that $\rho_{i,i-1} > 0$, but here we impose a stronger constraint, although not as strong as in the $\pi/4$ case. Finally, the local minimization is constrained by the requirement

$$1 - 0.1 \leq \Phi_i(a, b, c, d) \leq 1 + 0.1\,,$$

for all i. This constraint ensures that all Φ's will be close to one, so that the qualitative behaviour of the term structure should be preserved in time. Moreover, with this parameterization, we can expect a smooth shape for the term structure of volatilities at all instants, since with linear/exponential functions we avoid the typical erratic behaviour of piecewise-constant formulations.

For the calibration, we use only volatilities in the swaptions matrix corresponding to the 2y, 5y and 10y columns, in order to speed up the constrained optimization. The local optimization routine produced the following parameters:

$$a = 0.29342753, \quad b = 1.25080230, \quad c = 0.13145869, \quad d = 0.$$

$$\begin{aligned}
\theta_{1\div 6} &= [1.754112 \;\; 0.577818 \;\; 1.685018 \;\; 0.581761 \;\; 1.538243 \;\; 2.436329],\\
\theta_{7\div 12} &= [0.880112 \;\; 1.896454 \;\; 0.486056 \;\; 1.280206 \;\; 2.440311 \;\; 0.944809],\\
\theta_{13\div 19} &= [1.340539 \;\; 2.911335 \;\; 1.996228 \;\; 0.700425 \;\; 0 \;\; 0.815189 \;\; 2.383766].
\end{aligned}$$

Notice that $d = 0$ has reached the lowest value allowed by the positivity constraint, meaning that possibly the optimization would have improved with a negative d. The instantaneous correlations resulting from this calibration are again oscillating and non-monotonic. The first ten rows and columns of the instantaneous-correlation matrix are, for example,

	1y	2y	3y	4y	5y	6y	7y	8y	9y	10y
1y	1.000	0.384	0.998	0.388	0.977	0.776	0.642	0.990	0.298	0.890
2y	0.384	1.000	0.447	1.000	0.573	-0.284	0.955	0.249	0.996	0.763
3y	0.998	0.447	1.000	0.451	0.989	0.731	0.693	0.978	0.363	0.919
4y	0.388	1.000	0.451	1.000	0.576	-0.280	0.956	0.253	0.995	0.766
5y	0.977	0.573	0.989	0.576	1.000	0.623	0.791	0.937	0.496	0.967
6y	0.776	-0.284	0.731	-0.280	0.623	1.000	0.015	0.858	-0.370	0.403
7y	0.642	0.955	0.693	0.956	0.791	0.015	1.000	0.526	0.923	0.921
8y	0.990	0.249	0.978	0.253	0.937	0.858	0.526	1.000	0.160	0.816
9y	0.298	0.996	0.363	0.995	0.496	-0.370	0.923	0.160	1.000	0.701
10y	0.890	0.763	0.919	0.766	0.967	0.403	0.921	0.816	0.701	1.000

We find some repeated oscillations between positive and negative values that are not desirable. Terminal correlations share part of this negative behaviour. However, the evolution in time of the term structure of caplet volatilities is now reasonable, as shown in Figure 7.2.

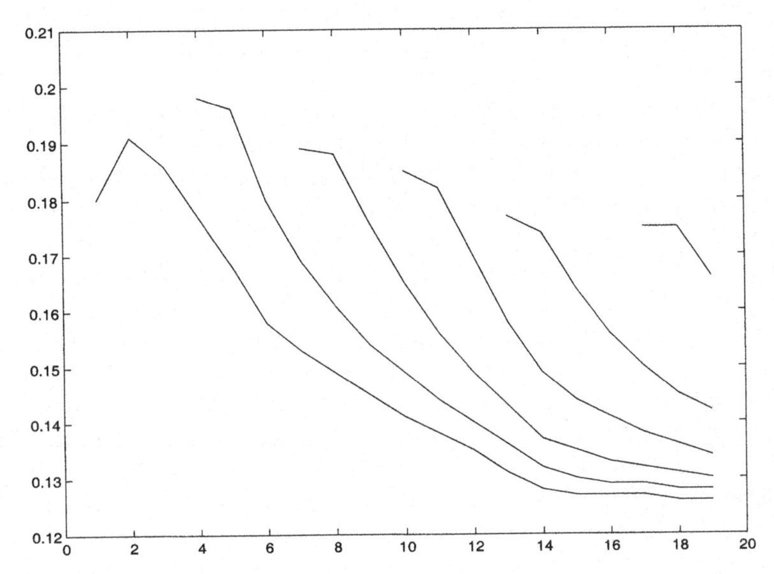

Fig. 7.2. Calibration results under Formulation 7: evolution of the term structure of caplet volatilities.

There remains to check the fitting quality. Caplets are fitted exactly, whereas the swaption volatilities are fitted with the following percentage differences:

	2y	3y	4y	5y	6y	7y	8y	9y	10y
1y	2.28%	-3.74%	-3.19%	-4.68%	2.46%	1.50%	0.72%	1.33%	-1.42%
2y	-1.23%	-7.67%	-9.97%	2.10%	0.49%	1.33%	1.56%	-0.44%	1.88%
3y	2.23%	-6.20%	-1.30%	-1.32%	-1.43%	1.86%	-0.19%	2.42%	1.17%
4y	-2.59%	9.02%	1.70%	0.79%	3.22%	1.19%	4.85%	3.75%	1.21%
5y	-3.26%	-0.28%	-8.16%	-0.81%	-3.56%	-0.23%	-0.08%	-2.63%	2.62%
7y	0.10%	-2.59%	-10.85%	-2.00%	-3.67%	-6.84%	2.15%	1.19%	0.00%
10y	0.29%	-3.44%	-11.83%	-1.31%	-4.69%	-2.60%	4.07%	1.11%	0.00%

Recall that here we have fitted only the 2y, 5y and 10y columns. In such columns differences, in absolute value, reach at most 4.68%, and are usually much smaller. The remaining data are also reproduced with small errors, even if they have not been included in the calibration, with a number of exceptions. For example, the 10×4-swaption features a difference of 11.83%. However, the last columns show relatively small errors, so that the 7y, 8y and 9y columns seem to be rather aligned with the 5y and 10y columns . On the contrary, the 3y and 4y columns seem to be rather misaligned with the 2y and 5y columns, since they show larger errors.

We have performed many more experiments with this choice of volatility, and some are reported later on. We have tried rank-three correlations structures, less or more stringent constraints on the angles and on the Φ's, and so on. A variety of results have been obtained. In general, we can say that the fitting to the whole swaption matrix can be improved, but at the cost of an erratic behaviour of both correlations and of the evolution of the term structure of volatilities in time. In particular, the three-factor choice does not seem to help that much.

The above example is sufficient to let one appreciate both the potential and the disadvantages of this parameterization with respect to the piecewise-constant case. In general, this parameterization allows for an easier control of the evolution of the term structure of volatilities, but produces more erratic correlation structures, since most of the "noise" in the swaptions data now ends up in the angles, because of the fact that we have only four volatility parameters a, b, c, d that can be used to calibrate swaption volatilities. A different possible use of the model, however, is to limit the calibration to act only on swaption prices, by ignoring the cap market, or by keeping it for testing the caps/swaptions misalignment a posteriori. With this approach the Φ's become again free parameters to be used in the swaption calibration, and are no longer functions of a, b, c, d imposed by the caplet volatilities. In this case we obtained a good fitting to market data, not so good instantaneous correlations, interesting terminal correlations and relatively satisfactory evolution of the term structure of volatilities in time.

7.4 Exact Swaptions "Cascade" Calibration with Volatilities as in TABLE 1

We now examine the case of the instantaneous-volatility structure summarized in TABLE 1, which is the one with the largest number of parameters. One would expect this structure to lead to a complex calibration routine, requiring optimization in a space of huge dimension. Instead, we have devised a method such that, by assuming exogenously-given instantaneous correlations ρ, *the calibration can be carried out through closed-form formulas having as inputs the exogenous correlations and the swaption volatilities.* Our method is illustrated in what follows.

Recall that the instantaneous-volatility structure considered here is defined by

$$\sigma_k(t) = \sigma_{k,\beta(t)},$$

where in general $T_{\beta(t)-2} < t \le T_{\beta(t)-1}$, and TABLE 1 reads as

Instant. Vols	Time: $t \in (0, T_0]$	$(T_0, T_1]$	$(T_1, T_2]$	...	$(T_{M-2}, T_{M-1}]$
Fwd Rate: $F_1(t)$	$\sigma_{1,1}$	Dead	Dead	...	Dead
$F_2(t)$	$\sigma_{2,1}$	$\sigma_{2,2}$	Dead	...	Dead
⋮	...	...	...	...	...
$F_M(t)$	$\sigma_{M,1}$	$\sigma_{M,2}$	$\sigma_{M,3}$	...	$\sigma_{M,M}$

Dealing with swaptions, we denote as usual by $V_{\alpha,\beta}$ the Black volatility for the swaption whose underlying swap rate is $S_{\alpha,\beta}$ (T_α is the swap starting date and $T_{\alpha+1}, \ldots, T_\beta$ are the swap payment dates). Recall also the approximated formula (6.67) for valuing swaptions volatilities $v_{\alpha,\beta}$ in the LFM. Apply this formula, dividing by time T_α, to the volatility formulation of TABLE 1 to obtain

$$(V_{\alpha,\beta})^2 \approx \sum_{i,j=\alpha+1}^{\beta} \frac{w_i(0)w_j(0)F_i(0)F_j(0)\rho_{i,j}}{T_\alpha S_{\alpha,\beta}(0)^2} \sum_{h=0}^{\alpha} \tau_{h-1,h}\sigma_{i,h+1}\sigma_{j,h+1} \quad (7.1)$$

with $\tau_{h-1,h} = T_h - T_{h-1}$, and $T_{-1} = 0$.

We remind that the weights w are specific of the swaption being considered, i.e. they depend on α and β. As usual we will omit such dependence to shorten notation, but each time we change swaption the corresponding w's change.

In order to effectively illustrate the calibration results in this case, without getting lost in notation and details, we work out an example with just six swaptions. We will then show how to generalize our procedure to an arbitrary number of contracts.

Suppose we start from the swaptions volatilities in the upper half of the swaption matrix:

Length Maturity	1y	2y	3y
$T_0 = 1$y	$V_{0,1}$	$V_{0,2}$	$V_{0,3}$
$T_1 = 2$y	$V_{1,2}$	$V_{1,3}$	-
$T_2 = 3$y	$V_{2,3}$	-	-

Let us move along this table, starting from the $(1,1)$ entry $V_{0,1}$. Using the approximating formula (7.1), we compute, after straightforward simplifications,

$$(V_{0,1})^2 \approx \sigma_{1,1}^2$$

This formula is immediately invertible and yields the volatility parameter $\sigma_{1,1}$ in the forward-rate dynamics as a function of the swaption volatility $V_{0,1}$, which we read from our matrix.

We then move on to the right, to entry $(1,2)$, containing $V_{0,2}$. The same formula gives, this time,

$$S_{0,2}(0)^2(V_{0,2})^2 \approx w_1(0)^2 F_1(0)^2 \sigma_{1,1}^2 + w_2(0)^2 F_2(0)^2 \sigma_{2,1}^2 \\ +2\rho_{1,2} w_1(0) F_1(0) w_2(0) F_2(0) \sigma_{1,1} \sigma_{2,1} \,.$$

Everything in this formula is known, except $\sigma_{2,1}$. We then solve the "elementary-school" algebraic second-order equation in $\sigma_{2,1}$, and, assuming existence and uniqueness of a positive solution, we analytically recover $\sigma_{2,1}$ in terms of the previously found $\sigma_{1,1}$ and of the known swaptions data. More generally, we assume for the moment that all the next algebraic second-order equations admit a unique positive solution.

We keep on moving to the right, to entry $(1,3)$, containing $V_{0,3}$. Formula (7.1) gives, this time,

$$S_{0,3}(0)^2(V_{0,3})^2 \approx w_1(0)^2 F_1(0)^2 \sigma_{1,1}^2 + w_2(0)^2 F_2(0)^2 \sigma_{2,1}^2 \\ + w_3(0)^2 F_3(0)^2 \sigma_{3,1}^2 + 2\rho_{1,2} w_1(0) F_1(0) w_2(0) F_2(0) \sigma_{1,1} \sigma_{2,1} \\ + 2\rho_{1,3} w_1(0) F_1(0) w_3(0) F_3(0) \sigma_{1,1} \sigma_{3,1} + 2\rho_{2,3} w_2(0) F_2(0) w_3(0) F_3(0) \sigma_{2,1} \sigma_{3,1} .$$

Similarly to the previous formula, here everything is known except for $\sigma_{3,1}$. We then solve the algebraic second-order equation in $\sigma_{3,1}$, and recover analytically $\sigma_{3,1}$ in terms of the previously found $\sigma_{1,1}$, $\sigma_{2,1}$ and of the known swaptions data.

We now move on to the second row of the swaptions matrix, entry (2,1), containing $V_{1,2}$. Our formula gives:

$$T_1 \, V_{1,2}^2 \approx T_0 \, \sigma_{2,1}^2 + (T_1 - T_0) \, \sigma_{2,2}^2 \,.$$

This time, everything is known except $\sigma_{2,2}$. Once again, we solve explicitly this equation for $\sigma_{2,2}$, being $\sigma_{2,1}$ known from previous passages.

We move on to the right, entry (2,2), containing $V_{1,3}$. Formula (7.1) gives:

$$T_1 S_{1,3}(0)^2 V_{1,3}^2 \approx w_2(0)^2 F_2(0)^2 (\tau_{-1,0} \sigma_{2,1}^2 + \tau_{0,1} \sigma_{2,2}^2) \\ +w_3(0)^2 F_3(0)^2 (\tau_{-1,0} \sigma_{3,1}^2 + \tau_{0,1} \sigma_{3,2}^2) \\ +2\rho_{2,3} w_2(0) F_2(0) w_3(0) F_3(0) (\tau_{-1,0} \sigma_{2,1} \sigma_{3,1} + \tau_{0,1} \sigma_{2,2} \sigma_{3,2}).$$

Here everything is known except $\sigma_{3,2}$. Once again, we solve explicitly this equation for $\sigma_{3,2}$, being $\sigma_{2,1}, \sigma_{2,2}$ and $\sigma_{3,1}$ known from previous passages.

Finally, we move to the only entry (3,1) of the third row, containing $V_{2,3}$. The usual formula gives:

$$T_2 \, V_{2,3}^2 \approx \tau_{-1,0} \sigma_{3,1}^2 + \tau_{0,1} \sigma_{3,2}^2 + \tau_{1,2} \sigma_{3,3}^2 \,.$$

The only unknown entry at this point is $\sigma_{3,3}$, which can be easily found by explicitly solving this last equation.

We have thus been able to find all instantaneous volatilities

Instant. Vols	Time: $t \in (0, T_0]$	$(T_0, T_1]$	$(T_1, T_2]$
Fwd Rate: $F_1(t)$	$\sigma_{1,1}$	Dead	Dead
$F_2(t)$	$\sigma_{2,1}$	$\sigma_{2,2}$	Dead
$F_3(t)$	$\sigma_{3,1}$	$\sigma_{3,2}$	$\sigma_{3,3}$

in terms of swaptions volatilities. The following table summarizes the dependence of the swaptions volatilities V on the instantaneous forward volatilities σ.

Length / Maturity	1y	2y	3y
$T_0 = 1y$	$V_{0,1}$ $\sigma_{1,1}$	$V_{0,2}$ $\sigma_{1,1}, \sigma_{2,1}$	$V_{0,3}$ $\sigma_{1,1}, \sigma_{2,1}, \sigma_{3,1}$
$T_1 = 2y$	$V_{1,2}$ $\sigma_{2,1}$ $\sigma_{2,2}$	$V_{1,3}$ $\sigma_{2,1}, \sigma_{3,1}$ $\sigma_{2,2}, \sigma_{3,2}$	-
$T_2 = 3y$	$V_{2,3}$ $\sigma_{3,1}$ $\sigma_{3,2}$ $\sigma_{3,3}$	-	-

In this table, we have put in each entry the related swaption volatility and the instantaneous volatilities upon which it depends. In reading the table left to right and top down, you realize that, each time, only one new σ appears, and this makes the relationship between the V's and the σ's invertible (analytically).

We now give the general method for calibrating our volatility formulation of TABLE 1 to the upper-triangular part of the swaption matrix when an arbitrary number s of rows of the matrix is given. We thus generalize the just seen case, where $s = 3$, to a generic positive integer s. At times we will refer to the resulting scheme as to the "Cascade Calibration Algorithm" (CCA). This name is due to the fact that the method is essentially given by a cascade of second order equations.

Let us rewrite formula (7.1) as follows:

$$
\begin{aligned}
T_\alpha S^2_{\alpha,\beta}(0) V^2_{\alpha,\beta} = & \sum_{i,j=\alpha+1}^{\beta-1} w_i(0) w_j(0) F_i(0) F_j(0) \rho_{i,j} \sum_{h=0}^{\alpha} \tau_{h-1,h} \sigma_{i,h+1} \sigma_{j,h+1} \\
& + 2 \sum_{j=\alpha+1}^{\beta-1} w_\beta(0) w_j(0) F_\beta(0) F_j(0) \rho_{\beta,j} \sum_{h=0}^{\alpha-1} \tau_{h-1,h} \sigma_{\beta,h+1} \sigma_{j,h+1} \\
& + 2 \sum_{j=\alpha+1}^{\beta-1} w_\beta(0) w_j(0) F_\beta(0) F_j(0) \rho_{\beta,j} \tau_{\alpha-1,\alpha} \sigma_{j,\alpha+1} \boxed{\sigma_{\beta,\alpha+1}} \\
& + w_\beta(0)^2 F_\beta(0)^2 \sum_{h=0}^{\alpha-1} \tau_{h-1,h} \sigma^2_{\beta,h+1} \\
& + w_\beta(0)^2 F_\beta(0)^2 \tau_{\alpha-1,\alpha} \boxed{\sigma^2_{\beta,\alpha+1}} . \qquad (7.2)
\end{aligned}
$$

In turn, by suitable definition of the coefficients A, B and C, this equation can be rewritten as:

$$
A_{\alpha,\beta} \sigma^2_{\beta,\alpha+1} + B_{\alpha,\beta} \sigma_{\beta,\alpha+1} + C_{\alpha,\beta} = 0 \,, \qquad (7.3)
$$

and thus it can be solved analytically. It is important to realize (as from the above $s = 3$ example) that when one solves this equation, all quantities are indeed known with the exception of $\sigma_{\beta,\alpha+1}$ *if* the swaption matrix is visited from left to right and top down.

Our procedure can be written in algorithmic form as follows.

Algorithm 7.4.1. ***Cascade Calibration Algorithm (CCA)*** *(Brigo and Mercurio (2001c, 2002a)).*

1. *Select the number s of rows in the swaption matrix that are of interest for the calibration;*
2. *Set $\alpha = 0$;*
3. *Set $\beta = \alpha + 1$;*
4. *Solve equation (7.3) in $\sigma_{\beta,\alpha+1}$. Since both $A_{\alpha,\beta}$ and $B_{\alpha,\beta}$ are strictly positive, if we assume positive instantaneous correlations, (7.3) has at most one positive solution, namely*

$$
\sigma_{\beta,\alpha+1} = \frac{-B_{\alpha,\beta} + \sqrt{B^2_{\alpha,\beta} - 4 A_{\alpha,\beta} C_{\alpha,\beta}}}{2 A_{\alpha,\beta}},
$$

   if and only if $C_{\alpha,\beta} < 0$.
5. *Increase β by one. If β is smaller than or equal to s, go back to point 4, otherwise increase α by one.*
6. *If $\alpha < s$ go back to 3, otherwise stop.*

Of course, this recipe works if and only if the condition $C_{\alpha,\beta} < 0$ is verified every time we must solve an equation like (7.3). Our practical experience is that such condition is always met for non-pathological swaptions data. However, for sake of completeness, we will illustrate later on some examples leading to $C_{\alpha,\beta} > 0$.

We may wonder about what happens when in need to recover the whole matrix and not only the upper-triangular part. By continuing our example with $s = 3$ above, we realize that the dependence of the whole set of swaptions volatilities V on the instantaneous forward volatilities σ is as follows:

Length / Maturity	1y	2y	3y
$T_0 = 1$y	$V_{0,1}$	$V_{0,2}$	$V_{0,3}$
	$\sigma_{1,1}$	$\sigma_{1,1}, \sigma_{2,1}$	$\sigma_{1,1}, \sigma_{2,1}, \sigma_{3,1}$
$T_1 = 2$y	$V_{1,2}$	$V_{1,3}$	$V_{1,4}$
	$\sigma_{2,1}$	$\sigma_{2,1}, \sigma_{3,1}$	$\sigma_{2,1}, \sigma_{3,1}, \sigma_{4,1}$
	$\sigma_{2,2}$	$\sigma_{2,2}, \sigma_{3,2}$	$\sigma_{2,2}, \sigma_{3,2}, \sigma_{4,2}$
$T_2 = 3$y	$V_{2,3}$	$V_{2,4}$	$V_{2,5}$
	$\sigma_{3,1}$	$\sigma_{3,1}, \sigma_{4,1}$	$\sigma_{3,1}, \sigma_{4,1}, \sigma_{5,1}$
	$\sigma_{3,2}$	$\sigma_{3,2}, \sigma_{4,2}$	$\sigma_{3,2}, \sigma_{4,2}, \sigma_{5,2}$
	$\sigma_{3,3}$	$\sigma_{3,3}, \sigma_{4,3}$	$\sigma_{3,3}, \sigma_{4,3}, \sigma_{5,3}$

To analytically determine the σ's from the V's, we proceed initially as before, thus ending up with the upper-triangular part of the matrix:

Length / Maturity	1y	2y	3y
$T_0 = 1$y	$V_{0,1}$	$V_{0,2}$	$V_{0,3}$
	$\sigma_{1,1}$	$\sigma_{1,1}, \sigma_{2,1}$	$\sigma_{1,1}, \sigma_{2,1}, \sigma_{3,1}$
$T_1 = 2$y	$V_{1,2}$	$V_{1,3}$	-
	$\sigma_{2,1}$	$\sigma_{2,1}, \sigma_{3,1}$	
	$\sigma_{2,2}$	$\sigma_{2,2}, \sigma_{3,2}$	
$T_2 = 3$y	$V_{2,3}$	-	-
	$\sigma_{3,1}$		
	$\sigma_{3,2}$		
	$\sigma_{3,3}$		

We then move off the diagonal of one level, starting from the upper entry (2,3), containing $V_{1,4}$.

Length Maturity	1y	2y	3y
$T_0 = 1\text{y}$	$V_{0,1}$ $\sigma_{1,1}$	$V_{0,2}$ $\sigma_{1,1}, \sigma_{2,1}$	$V_{0,3}$ $\sigma_{1,1}, \sigma_{2,1}, \sigma_{3,1}$
$T_1 = 2\text{y}$	$V_{1,2}$ $\sigma_{2,1}$ $\sigma_{2,2}$	$V_{1,3}$ $\sigma_{2,1}, \sigma_{3,1}$ $\sigma_{2,2}, \sigma_{3,2}$	$V_{1,4}$ $\sigma_{2,1}, \sigma_{3,1}, \sigma_{4,1}$ $\sigma_{2,2}, \sigma_{3,2}, \sigma_{4,2}$
$T_2 = 3\text{y}$	$V_{2,3}$ $\sigma_{3,1}$ $\sigma_{3,2}$ $\sigma_{3,3}$	-	-

This presents us with *two* new unknowns, rather than one, and precisely $\sigma_{4,1}$ and $\sigma_{4,2}$. We can now consider again the analogue of equation (7.3), but we have two unknowns. An easy way out is to assume a relationship between them, and one of the easiest possibilities is to assume the two unknowns to be equal, $\sigma_{4,1} = \sigma_{4,2}$. By doing so, we end up again with an analytically-solvable second-order equation, we solve it (assuming existence and uniqueness of a positive solution), and we move down along the subdiagonal to entry (3,2) containing $V_{2,4}$:

Length Maturity	1y	2y	3y
$T_0 = 1\text{y}$	$V_{0,1}$ $\sigma_{1,1}$	$V_{0,2}$ $\sigma_{1,1}, \sigma_{2,1}$	$V_{0,3}$ $\sigma_{1,1}, \sigma_{2,1}, \sigma_{3,1}$
$T_1 = 2\text{y}$	$V_{1,2}$ $\sigma_{2,1}$ $\sigma_{2,2}$	$V_{1,3}$ $\sigma_{2,1}, \sigma_{3,1}$ $\sigma_{2,2}, \sigma_{3,2}$	$V_{1,4}$ $\sigma_{2,1}, \sigma_{3,1}, \sigma_{4,1}$ $\sigma_{2,2}, \sigma_{3,2}, \sigma_{4,2}$
$T_2 = 3\text{y}$	$V_{2,3}$ $\sigma_{3,1}$ $\sigma_{3,2}$ $\sigma_{3,3}$	$V_{2,4}$ $\sigma_{3,1}, \sigma_{4,1}$ $\sigma_{3,2}, \sigma_{4,2}$ $\sigma_{3,3}, \sigma_{4,3}$	-

This time, we just have one new unknown, $\sigma_{4,3}$, so that we can solve the related second-order equation and we are done.

We then move one level further below the diagonal, i.e. to entry (3,3) containing $V_{2,5}$. Looking at the full table,

Length / Maturity	1y	2y	3y
$T_0 = 1$y	$V_{0,1}$	$V_{0,2}$	$V_{0,3}$
	$\sigma_{1,1}$	$\sigma_{1,1}, \sigma_{2,1}$	$\sigma_{1,1}, \sigma_{2,1}, \sigma_{3,1}$
$T_1 = 2$y	$V_{1,2}$	$V_{1,3}$	$V_{1,4}$
	$\sigma_{2,1}$	$\sigma_{2,1}, \sigma_{3,1}$	$\sigma_{2,1}, \sigma_{3,1}, \sigma_{4,1}$
	$\sigma_{2,2}$	$\sigma_{2,2}, \sigma_{3,2}$	$\sigma_{2,2}, \sigma_{3,2}, \sigma_{4,2}$
$T_2 = 3$y	$V_{2,3}$	$V_{2,4}$	$V_{2,5}$
	$\sigma_{3,1}$	$\sigma_{3,1}, \sigma_{4,1}$	$\sigma_{3,1}, \sigma_{4,1}, \sigma_{5,1}$
	$\sigma_{3,2}$	$\sigma_{3,2}, \sigma_{4,2}$	$\sigma_{3,2}, \sigma_{4,2}, \sigma_{5,2}$
	$\sigma_{3,3}$	$\sigma_{3,3}, \sigma_{4,3}$	$\sigma_{3,3}, \sigma_{4,3}, \sigma_{5,3}$

we see that we now have three new unknowns, namely $\sigma_{5,1}$, $\sigma_{5,2}$ and $\sigma_{5,3}$. Again, if we assume them to be equal, we can solve the related second-order equation and fill the table, so that the calibration is now completed.

Following Brigo and Morini (2002) and Morini (2002), we can see the analytical details of this extension of the triangular CCA, given in Remark 7.4.1, to a general rectangular calibration to the entire swaption matrix. The entries considered in the triangular CCA are those such that, denoting by $(i+j)$ the sum of the column and row indices and by s the matrix dimension, the following condition holds

$$(i+j) \leq (s+1),$$

so that we refer to swaption volatilities $V_{\alpha,\beta}$ for which $\beta \leq s$.

A relevant feature of the latter calibration is that results are independent of s, which means that the output of the calibration to a sub-matrix of the swaption matrix V will be a subset of the output of a calibration to V. By reasoning the other way around, this implies also that any swaption matrix V such as Table 7.11 below can be seen in principle as a sub-matrix of a larger one, say $\bar{V}$, including V itself in its upper triangular part, so that all entries of V, including those in its lower triangular part, will be recovered by applying the triangular CCA algorithm to the upper part of the larger matrix $\bar{V}$. In other words, this "nested consistency" means that, if we had available so many market quotations that matrix V of interest could always be embedded in a sufficiently large market $\bar{V}$, the "upper part" algorithm given in Remark 7.4.1 would not need to be extended. Of course this is usually not the case, since there is no larger market $\bar{V}$ to be exploited, so we need an extended algorithm to calibrate to the entire rectangular swaption matrix V.

Suppose that we visit the rectangular swaption matrix from left to right and top down. Suppose further that we apply the above algorithm each time we have a single unknown, and that all multiple unknowns are determined as soon as they occur by inverting the usual formula, but with the assumption that they are equal to each other. In this case, it is easy to see that the multiple-unknowns situation will occur only when reaching swaptions volatilities in the last column of the matrix (with the exception, of course, of the first-row element of such column, still belonging to the upper part of the swaption matrix). Further details are given in Morini (2002).

Following these observations we can generalize the triangular CCA given in Remark 7.4.1 finding the following algorithm for the complete matrix.

Algorithm 7.4.2. *(**Rectangular Cascade Calibration Algorithm (RCCA)**). Consider the algorithm given in Remark 7.4.1 and modify it as follows. At point 5 the condition is no longer $\beta \leq s$, but $(\beta - \alpha) \leq s$. Furthermore, in case $\beta = s + \alpha$, i.e. when one reaches one entry on the last column (with the exception of the first one), the new point 4 requires to assume all the unknowns to be equal to the standard unknown $\sigma_{\beta,\alpha+1}$:*

$$\sigma_{\beta,\alpha+1} = \sigma_{\beta,\alpha} = \ldots = \sigma_{\beta,1} \quad \textit{for} \quad \beta = s + \alpha.$$

Hence the new equation to solve is

$$A^*_{\alpha,\beta}\sigma^2_{\beta,\alpha+1} + B^*_{\alpha,\beta}\sigma_{\beta,\alpha+1} + C^*_{\alpha,\beta} = 0, \tag{7.4}$$

where

$$A^*_{\alpha,\beta} = w_\beta(0)^2 F_\beta(0)^2 (T_\alpha - T_{\alpha-1}) + w_\beta(0)^2 F_\beta(0)^2 \sum_{h=0}^{\alpha-1} (T_h - T_{h-1}),$$

$$\begin{aligned} B^*_{\alpha,\beta} = 2 &\sum_{j=\alpha+1}^{\beta-1} w_\beta(0) w_j(0) F_\beta(0) F_j(0) \rho_{\beta,j} \; (T_\alpha - T_{\alpha-1}) \; \sigma_{j,\alpha+1} \\ &+2 \sum_{j=\alpha+1}^{\beta-1} w_\beta(0) w_j(0) F_\beta(0) F_j(0) \rho_{\beta,j} \sum_{h=0}^{\alpha-1} (T_h - T_{h-1}) \; \sigma_{j,h+1}, \end{aligned}$$

$$\begin{aligned} C^*_{\alpha,\beta} = &\sum_{i,j=\alpha+1}^{\beta-1} w_i(0) w_j(0) F_i(0) F_j(0) \rho_{i,j} \sum_{h=0}^{\alpha} (T_h - T_{h-1}) \; \sigma_{i,h+1} \; \sigma_{j,h+1} \\ &-T_\alpha \; S_{\alpha,\beta}(0)^2 \; V^2_{\alpha,\beta}. \end{aligned}$$

The rest of the algorithm keeps unchanged.

7.4.1 Some Numerical Results

Here, we plan to present some numerical results. We start from the simplest case of a swaption matrix with $s = 3$, and with swaptions volatilities only in the upper half of the table. Then we move to the full matrix, and, finally, we consider the case with $s = 10$.

Under the case with $s = 3$, we consider the three subcases: a) market swaption-volatility matrix as of alternative set, given in Table 7.4 below; b) the same volatility matrix with modified upper corners; c) the same volatility matrix with modified upper and lower corners.

We assume the usual typical nice rank-two correlation structure given exogenously, corresponding to the angles

$$\theta_{1\div 9} = [\ 0.0147\ 0.0643\ 0.1032\ 0.1502\ 0.1969\ 0.2239\ 0.2771\ 0.2950\ 0.3630\ ],$$
$$\theta_{10\div 19} = [\ 0.3810\ 0.4217\ 0.4836\ 0.5204\ 0.5418\ 0.5791\ 0.6496\ 0.6679\ 0.7126\ 0.7659\ ].$$

We have the following inputs.

Input: the three swaptions volatility matrices a), b) and c)

0.180	0.167	0.154
0.181	0.162	
0.178		

0.300	0.167	0.100
0.181	0.162	
0.178		

0.300	0.167	0.100
0.181	0.162	
0.100		

Calibrating to these upper-triangular swaption matrices yields the following instantaneous-volatility parameters in the corresponding TABLE 1:

Output: instantaneous volatility parameters for a), b) and c)

0.1800	-	-
0.1540	0.2050	-
0.1270	0.1570	0.2340

0.3000	-	-
0.0287	0.2540	-
-0.0424	0.2030	0.2280

0.3000	-	-
0.0287	0.2540	-
-0.0424	0.2030	0+0.1136i

This first example helps us realize what can go wrong with our automatic and algebraic technique. While for the first swaption matrix a) we see that the parameters σ are all positive real numbers, we immediately notice that, by tilting the first row of the swaption matrix as in case b), we get a negative entry in the corresponding σ table. Even worse, by tilting also the first column, we obtain both negative and imaginary instantaneous volatilities σ, as is clear from case c). Therefore caution is in order when running a calibration. In fact, too steep matrices can yield both negative and imaginary instantaneous volatilities.

We find further confirmation of this fact by calibrating to the whole 3×3 swaptions matrices analogous to a), b) and c):

Input: the three swaptions volatility matrices d), e) and f)

0.180	0.167	0.154
0.181	0.162	0.145
0.178	0.155	0.137

0.300	0.167	0.100
0.181	0.162	0.145
0.178	0.155	0.137

0.300	0.167	0.100
0.181	0.162	0.145
0.100	0.155	0.280

We obtain the following tables of instantaneous-volatility parameters σ:

	1y	2y	3y	4y	5y	6y	7y	8y	9y	10y
1y	0.180	0.167	0.154	0.145	0.138	0.134	0.130	0.126	0.124	0.122
2y	0.181	0.162	0.145	0.135	0.127	0.123	0.120	0.117	0.115	0.113
3y	0.178	0.155	0.137	0.125	0.117	0.114	0.111	0.108	0.106	0.104
4y	0.167	0.143	0.126	0.115	0.108	0.105	0.103	0.100	0.098	0.096
5y	0.154	0.132	0.118	0.109	0.104	0.104	0.099	0.096	0.094	0.092
6y	0.147	0.127	0.113	0.104	0.098	0.098	0.094	0.092	0.090	0.089
7y	0.140	0.121	0.107	0.098	0.092	0.091	0.089	0.087	0.086	0.085
8y	0.137	0.117	0.103	0.095	0.089	0.088	0.086	0.084	0.083	0.082
9y	0.133	0.114	0.100	0.091	0.086	0.085	0.083	0.082	0.081	0.080
10y	0.130	0.110	0.096	0.088	0.083	0.082	0.080	0.079	0.078	0.077

Table 7.4. Alternative Swaptions Volatilities

Output: instantaneous-volatility parameters for d), e) and f)

0.1800	-	-
0.1548	0.2039	-
0.1285	0.1559	0.2329
0.1105	0.1105	0.1660
0.1012	0.1012	0.1012

0.300	-	-
0.041	0.253	-
- 0.032	0.205	0.228
0.138	0.138	0.171
0.106	0.106	0.106

0.300	-	-
0.041	0.253	-
- 0.032	0.205	0 + 0.1147i
0.138	0.138	0.4053 - 0.1182i
0.5629 + 0.0001i	0.5629 + 0.0001i	0.5629 + 0.0001i

where in the equations for σ's involving more than one unknown we have assumed all unknowns to be equal. Again, we obtain first negative and then imaginary volatilities in cases e) and f).

We now move from our 3×3 toy input matrices to the full 10×10 swaption matrix, obtained from an alternative modified input 7×10 market matrix of 16 May, 2000, where the 6y, 8y, 9y lines have been added by linear interpolation. This matrix is given in Table 7.4.

A plot of the implied surface is shown in Figure 7.3.

Calibrating to such a matrix yields the GPC instantaneous volatilities, collected as in TABLE 1 of Chapter 6, that are shown in Table 7.5.

This "real-market" calibration shows several negative signs in instantaneous volatilities. Recall that these undesirable negative entries might be due to "temporal misalignments" caused by illiquidity in the swaption matrix. As observed before, these misalignments can cause troubles, since after a calibration, the model parameters might reflect these misalignments. To avoid this inconvenience, we smooth the above market swaption matrix by means of the following parametric form:

$$\mathrm{vol}(S,T) = \gamma(S) + \left(\frac{\exp(f\ln(T))}{e\,S} + D(S)\right)\exp(-\beta\exp(p\ln(T))),$$

where

0.1800	-	-	-	-	-	-	-	-	-
0.1548	0.2039	-	-	-	-	-	-	-	-
0.1285	0.1559	0.2329	-	-	-	-	-	-	-
0.1178	0.1042	0.1656	0.2437	-	-	-	-	-	-
0.1091	0.0988	0.0973	0.1606	0.2483	-	-	-	-	-
0.1131	0.0734	0.0781	0.1009	0.1618	0.2627	-	-	-	-
0.1040	0.0984	0.0502	0.0737	0.1128	0.1633	0.2633	-	-	-
0.0940	0.1052	0.0938	0.0319	0.0864	0.0969	0.1684	0.2731	-	-
0.1065	0.0790	0.0857	0.0822	0.0684	0.0536	0.0921	0.1763	0.2848	-
0.1013	0.0916	0.0579	0.1030	0.1514	-0.0316	0.0389	0.0845	0.1634	0.2777
0.0916	0.0916	0.0787	0.0431	0.0299	0.2088	-0.0383	0.0746	0.0948	0.1854
0.0827	0.0827	0.0827	0.0709	0.0488	0.0624	0.1561	-0.0103	0.0731	0.0911
0.0744	0.0744	0.0744	0.0744	0.0801	0.0576	0.0941	0.1231	-0.0159	0.0610
0.0704	0.0704	0.0704	0.0704	0.0704	0.1009	0.0507	0.0817	0.1203	-0.0210
0.0725	0.0725	0.0725	0.0725	0.0725	0.0725	0.1002	0.0432	0.0619	0.1179
0.0753	0.0753	0.0753	0.0753	0.0753	0.0753	0.0753	0.0736	0.0551	0.0329
0.0719	0.0719	0.0719	0.0719	0.0719	0.0719	0.0719	0.0719	0.0708	0.0702
0.0690	0.0690	0.0690	0.0690	0.0690	0.0690	0.0690	0.0690	0.0690	0.0680
0.0663	0.0663	0.0663	0.0663	0.0663	0.0663	0.0663	0.0663	0.0663	0.0663

Table 7.5. Fitted GPC instantaneous volatilities σ, collected as in TABLE 1.

a	b	c	d	e	f	del	bet
0.000359	1.432288	2.5269	-1.93552	5.751286	0.065589	0.02871	-5.41842
g	h	m	p	q	r	s	t
-0.02129	17.64259	2.043768	-0.06907	-0.09817	-0.87881	2.017844	0.600784

Table 7.6. Parameters corresponding to the smoothed swaption matrix

$$\gamma(S) = c + (\exp(h \ln(S))a + d) \exp(-b \exp(m \ln(S))),$$
$$D(S) = (\exp(g \ln(S))q + r) \exp(-s \exp(t \ln(S))) + \delta,$$

and S and T are respectively the maturity and the tenor of the related swaption. So, for example, vol$(2,3)$ is the volatility of the swaption whose underlying swap starts in two years and lasts three years (entry $(2,3)$ of the swaption matrix).

We do not claim that this form has any appealing characteristic or that it always yields the precision needed by a trader, but we use it to point out the effect of smoothing.

The smoothing yields the parameter values given in Table 7.6.

These parameters lead to the smoothed matrix, and the absolute difference between the market and the smoothed matrices is given in Table 7.7.

If we run our algebraic calibration with the smoothed swaptions data as input, the instantaneous volatility values are all real and positive, as we can see in Table 7.8.

We thus conclude that irregularity and illiquidity in the input swaption matrix can cause negative or even imaginary values in the calibrated instan-

	1y	2y	3y	4y	5y	6y	7y	8y	9y	10y
1y	-0.46	0.49	0.33	0.16	-0.01	0.01	-0.06	-0.18	-0.14	-0.14
2y	-0.39	0.53	0.18	0.03	-0.17	-0.11	-0.05	-0.05	0.01	0.03
3y	0.03	0.64	0.22	-0.13	-0.32	-0.16	-0.10	-0.10	-0.05	-0.03
4y	0.01	0.43	0.05	-0.23	-0.35	-0.21	-0.06	-0.08	-0.04	-0.03
5y	-0.36	0.12	-0.02	-0.15	-0.10	0.31	0.14	0.11	0.14	0.13
6y	-0.31	0.19	-0.02	-0.18	-0.21	0.13	0.09	0.10	0.16	0.20
7y	-0.27	0.25	-0.01	-0.21	-0.32	-0.05	0.05	0.09	0.19	0.27
8y	-0.13	0.27	-0.04	-0.22	-0.32	-0.06	0.02	0.09	0.18	0.25
9y	0.00	0.30	-0.05	-0.24	-0.32	-0.07	0.00	0.10	0.18	0.25
10y	0.15	0.32	-0.07	-0.25	-0.31	-0.08	-0.02	0.09	0.17	0.23

Table 7.7. 100× absolute difference between the market and the smoothed swaption matrices

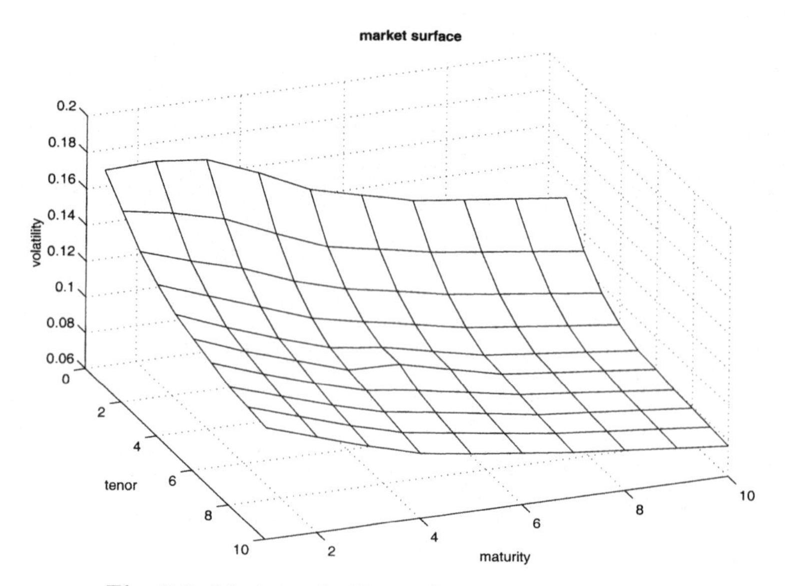

Fig. 7.3. Market volatility surface as of May 16, 2000

taneous volatilities. However, by smoothing the input data before calibration, usually this undesirable features can be avoided.

In the following we show the terminal correlations and the evolution of the term structure of volatility for the smoothed case. But first let us have a look at ten-year terminal correlations for the non-smoothed case in Table 7.9.

Compare this matrix with the corresponding ten-year terminal correlation matrix obtained in the smoothed case, given in Table 7.10.

We see that the non-smoothed case is much worse: It shows terminal correlations that deviate significantly from a monotonic behaviour, especially in the first column, roughly corresponding to the portion of instantaneous volatilities that go negative in the calibration. The non-smoothed case shows

18.46	-	-	-	-	-	-	-	-	-
14.09	22.03	-	-	-	-	-	-	-	-
12.84	13.11	24.71	-	-	-	-	-	-	-
12.14	11.17	13.00	25.94	-	-	-	-	-	-
11.64	10.11	10.59	12.54	27.10	-	-	-	-	-
11.19	9.51	9.44	9.87	12.73	28.06	-	-	-	-
10.94	8.88	8.47	8.53	9.82	13.01	28.58	-	-	-
10.59	8.61	7.82	7.57	8.58	10.06	12.92	29.62	-	-
10.37	8.25	7.53	6.81	7.52	8.61	9.74	13.51	30.20	-
10.26	7.73	7.21	6.43	7.14	7.65	8.31	10.45	13.56	30.35
8.89	8.89	7.08	6.31	6.39	7.23	7.38	8.73	10.40	13.41
8.07	8.07	8.07	6.23	6.30	6.82	6.79	7.96	8.63	10.10
7.35	7.35	7.35	7.35	6.27	6.43	6.29	7.38	7.96	8.44
7.01	7.01	7.01	7.01	7.01	6.39	5.85	6.89	6.70	7.46
6.53	6.53	6.53	6.53	6.53	6.53	6.29	5.96	6.92	6.68
6.23	6.23	6.23	6.23	6.23	6.23	6.23	6.97	5.58	6.57
6.06	6.06	6.06	6.06	6.06	6.06	6.06	6.06	6.57	5.77
5.76	5.76	5.76	5.76	5.76	5.76	5.76	5.76	5.76	6.35
5.62	5.62	5.62	5.62	5.62	5.62	5.62	5.62	5.62	5.62

Table 7.8. 100× LFM TABLE 1 GPC volatilities σ following calibration of smoothed swaptions

	10y	11y	12y	13y	14y	15y	16y	17y	18y	19y
10y	1.000	0.677	0.695	0.640	0.544	0.817	0.666	0.762	0.753	0.740
11y	0.677	1.000	0.614	0.617	0.665	0.768	0.696	0.760	0.752	0.740
12y	0.695	0.614	1.000	0.758	0.716	0.938	0.848	0.870	0.862	0.850
13y	0.640	0.617	0.758	1.000	0.740	0.866	0.914	0.894	0.885	0.875
14y	0.544	0.665	0.716	0.740	1.000	0.771	0.919	0.885	0.879	0.868
15y	0.817	0.768	0.938	0.866	0.771	1.000	0.923	0.965	0.960	0.953
16y	0.666	0.696	0.848	0.914	0.919	0.923	1.000	0.983	0.980	0.975
17y	0.762	0.760	0.870	0.894	0.885	0.965	0.983	1.000	0.999	0.995
18y	0.753	0.752	0.862	0.885	0.879	0.960	0.980	0.999	1.000	0.999
19y	0.740	0.740	0.850	0.875	0.868	0.953	0.975	0.995	0.999	1.000

Table 7.9. Ten-year terminal correlations on non-smoothed swaption data

	10y	11y	12y	13y	14y	15y	16y	17y	18y	19y
10y	1.000	0.939	0.898	0.872	0.851	0.838	0.823	0.809	0.817	0.787
11y	0.939	1.000	0.992	0.980	0.969	0.962	0.947	0.941	0.936	0.915
12y	0.898	0.992	1.000	0.996	0.990	0.986	0.975	0.972	0.966	0.950
13y	0.872	0.980	0.996	1.000	0.997	0.995	0.986	0.984	0.979	0.966
14y	0.851	0.969	0.990	0.997	1.000	0.997	0.992	0.989	0.984	0.973
15y	0.838	0.962	0.986	0.995	0.997	1.000	0.994	0.995	0.990	0.982
16y	0.823	0.947	0.975	0.986	0.992	0.994	1.000	0.997	0.997	0.992
17y	0.809	0.941	0.972	0.984	0.989	0.995	0.997	1.000	0.998	0.995
18y	0.817	0.936	0.966	0.979	0.984	0.990	0.997	0.998	1.000	0.998
19y	0.787	0.915	0.950	0.966	0.973	0.982	0.992	0.995	0.998	1.000

Table 7.10. Ten-year terminal correlations on smoothed swaptions data

also a slightly erratic evolution of the term structure of volatilities compared to the smoothed case.

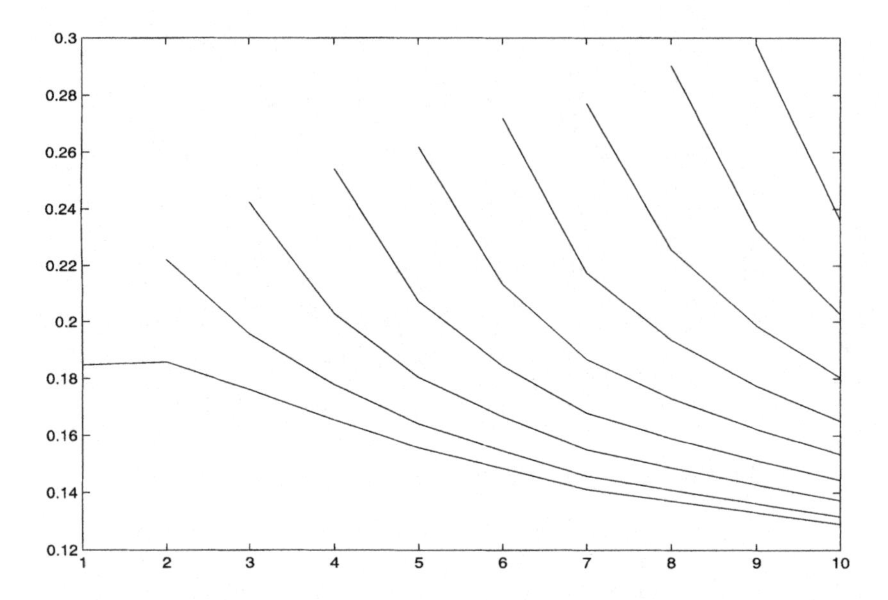

Fig. 7.4. Term structure evolution corresponding to the smoothed volatility swaption data

As we can see in Figure 7.4, with the smoothed swaption prices in input, the evolution of the term structure of volatility resulting from the calibration is interesting, and the related terminal correlations given in Table 7.10 are decreasing and non-negative.

Figure 7.4, although being interesting and displaying a smooth evolution of the term structure of volatilities following the calibration, presents the worrying feature of caplet-like swaption volatilities increasing up to a level of 30%, when the maximum initial value is below 19%. To see whether this undesirable feature is typical of the decomposition of swaptions volatilities into LIBOR σ's implicit in the cascade calibration, or rather it is due to peculiar characteristics of this swaption matrix (featuring a first column much higher than those who follow), we need to perform more tests. This is done in Sections 7.6 and 7.7, where we also tackle the issue of discarding the numerical problems encountered, starting from the considerations in the next section.

7.5 A Pause for Thought

Far from being conclusive or even systematic, so far our examples have served the purpose of pointing out many of the actual problems involved in the LFM calibration, with some discussion on possible solutions.

In general, one would like a calibration of the LFM to have the following features:

1. A small calibration error, i.e. small percentage differences

$$100 \cdot \frac{\text{Market swaption volatility - LFM swaption volatility}}{\text{Market swaption volatility}}$$

2. Regular instantaneous correlations. One would like to avoid large oscillations along the columns of the correlation matrix. Indeed, one would appreciate a monotonically decreasing pattern when moving away from a diagonal term of the matrix along the related row or column.
3. Regular terminal correlations.
4. Smooth and qualitatively stable evolution of the term structure of caplet volatilities over time.
5. In addition, one would like to accomplish these points, which are related to each other, with a reasonable computational effort.

7.5.1 First summary

None of the structures we proposed so far can perfectly meet the above requirements at the same time, although the formulation of TABLE 1 seems to go in the right direction. This shows once again that the core of the basic no-smile LFM is a clever choice of the instantaneous covariance as a function of time. The rest are just mathematical details. One may try and combine many of the ideas presented here to come up with a different approach that might work. Research in this issue is still quite open. Before moving to further considerations, we mention that a synthesis of the results presented in this chapter so far, concerning the LFM calibration to market data, had already appeared in Brigo, Capitani and Mercurio (2000) and, in a perfected version, in Brigo, Mercurio and Morini (2002, 2005).

Based on the points highlighted above, we just dare say that the GPC formulation of TABLE 1 in Chapter 6, coupled with our "automatic closed-form algebraic" exact calibration, seems the most promising. In fact, if the industry approximation (6.67) is used for pricing, market prices can be matched exactly, with an immediate inversion procedure and without calibration errors. At the same time a decent instantaneous-correlation matrix can be imposed exogenously. This seems to be of great help to obtain also decent terminal correlations.

Moreover, our method maps, in a one-to-one correspondence, swaptions volatilities into pieces of instantaneous volatilities of forward rates. This can help also in computing sensitivities with respect to swaptions volatilities, since one knows on which σ's one needs to act in order to influence a single swaption volatility. By a kind of chain rule we can translate sensitivity with respect to the σ's in sensitivity with respect to the swaption volatilities used in the calibration.

On the other hand, in applying the method to the single set of data seen in the above sections we have encountered numerical problems, so that the results in terms of LIBOR volatilities are not satisfactory. However, we have used one single swaption matrix as quoted in the market from one single broker and a correlation matrix coming from a different calibration. As observed in Remark 6.17.1, we must be somehow careful. We have actually found that with smoothed data in input, the outputs improve drastically, a reason for suspecting that the model is detecting a misalignment in the market, instead of concluding that the model is not suitable for the joint calibration. The "truth" might also lie in-between. Before concluding for sure that limitations concern only the model formulations and not a possible inconsistency present in the data we used, further extensive investigation and tests are needed, a task we face in the following sections.

7.5.2 An automatic fast analytical calibration of LFM to swaptions. Motivations and plan

When considering the promising features of the cascade calibration, we are led naturally to further investigate this method and to try and render the cascade calibration empirically efficient. But, following Brigo and Morini (2003), the reasons for tackling such a further investigation may be even more general and related to the basic purposes for which market models have been designed.

The LFM owes its popularity among practitioners to the compatibility with Black's market formula for caps. A feature that, as seen in previous sections, allows an immediate calibration to cap prices. Further, the calibration is exact and unambiguous, in the sense that given some homogeneity assumptions on the volatility time-dependence, there is a unique solution reproducing cap quotes. On the other hand, calibration to the swaption market, the other fundamental reference market for interest rate derivatives, is still an open problem. In fact, being the LFM distributional assumptions not compatible with the Black formula as used in the swaption market (as underlined in Section 6.8) it is not possible to solve the LFM calibration problem to swaptions by the simple inversion of a formula *à la* Black & Scholes.

Therefore most proposals, as those we presented in Sections 7.2 and 7.3, have focused on minimizing the overall pricing error on a panel of current swaption prices. This implies persistence of calibration errors, however minimized they may be, and possibly time-consuming optimization procedures. More importantly, considering the high number of parameters typical of this model, one could suspect the problem to be largely undetermined and the solutions found to be of limited meaning.

The use of analytical approximations, such as the industry formula (6.67), has allowed us to avoid simulation in each calibration iteration and has led to rather efficient calibration methods. However, these methods are still based on minimization of a loss function expressing the distance between market and model prices. From the point of view of calibration, this makes such methods analogous in spirit to the methods typical of term structure models of earlier generations, modeling purely theoretical and non-observable state variables. But such methods are very far from the "unique solution", "immediate and exact" calibration typical of the Black formula, the market standard which market models would like to incorporate and replace.

Consequently the use of this kind of methodologies could in this sense turn out to be a limitation in the application of the LFM, a somewhat inconsistent choice and a violation of the spirit of market model. Indeed the LFM, as a market model, has been designed to allow for a perfect fitting to the reference products used in calibration, a goal that could hardly be achieved by means of earlier models, based on variables and assumptions radically incompatible with Black's standard market formula.

Indeed, short-rate models are at times preferred because they are synthetic in terms of variables and parameters, and can easily incorporate explicitly features that are considered to be economically important, such as for example mean reversion. Given their synthetic nature and the economic explanatory features, when using these models one is willing to tolerate non-negligible calibration errors. The same errors would be undesirable if the dominant goal were pricing exotic products as consistently as possible with reference market prices. This is the case when using market models such as the LFM. Therefore, if market models are applied losing the basic property for which they have been designed, that is immediate exact calibration, the advantages they allow can be less apparent. And while it is true that it would have been naive to expect an immediate way to calibrate the LFM to swaptions, given the fact that the correct models to do this would be the other market models (swap market models), we may still wonder whether there exists a way to "close the loop", i.e. to obtain an immediate calibration of the LFM to swaptions, thus having a unique central market model respecting the philosophy for which market models have been designed.

Indeed, considering that the LFM is increasingly used as a "central" model for pricing also swaption-related products, it becomes of great interest the possibility of keeping such a "exact immediate calibration" feature also when calibrating the model to swaptions. And this is where the Cascade Calibration comes in. The cascade calibration, based on inverting an industry formula proven to be reliable in pricing swaptions with the LFM, seems to move exactly towards this goal. In fact, through closed form formulas, it gives an analytical, practically instantaneous and unique solution to the calibration problem, leading to the exact recovery of market prices by means of the industry formula (6.67) coupled with Black formula.

However the fact of being analytical, and not requiring any optimization routine, is not only a major advantage of this procedure, making it faster and exact. There is also the other side of the coin: no constraints can be set on the output. Indeed, we have seen that the solution found in the example above did not satisfy the obvious non-negativity constraints on $\sigma_{k,\beta(t)}$ and $\sigma^2_{k,\beta(t)}$, $\forall k, t$. This happened even for parameters depending only on the basic calibration to the upper part of the swaption matrix, which does not involve any additional simplifying assumption.

Such a result seems to suggest a non-negligible inconsistency between the model underlying the cascade and the way prices are really determined in the market (the *data generating process* in an evocative terminology, or more simply the assumptions of the market makers). The smoothing shown above is not a solution, since it alters market quotations and introduces not negligible calibration errors, as noticed in Brigo and Mercurio (2002a), but points out that inconsistencies or misalignments in input data could be responsible for the encountered troubles. So in the following we will concentrate on detecting the reasons leading to these troubles, in order to rule them out.

Our work will be split into four different parts. First, in Section 7.6 it will be shown that it is possible to obtain satisfactory results with cascade calibration as in the CC Algorithm 7.4.1, but depending on restrictive choices on the input correlations. Thus we will move to a deeper analysis of the reasons for this weakness, leading in Section 7.7 to a modified algorithm proving to be empirically much more satisfactory and robust.

Finally, two other important points are addressed. One is a fundamental question: is approximation (6.67) reliable even in this special situation of GPC volatilities with various exogenous correlations at various ranks, cases not addressed in Brigo and Mercurio (2001c) (reported also here in the test chapter 8)? We will answer this question, by comparison with montecarlo simulation results, in Section 7.8. Then a possible broadening in the application of this method is considered, that is embedding cap data for a joint calibration. We hint at possible choices, and related dangers, in Section 7.9, before drawing some conclusions and making some proposals for future research.

7.6 Further Numerical Studies on the Cascade Calibration Algorithm

The examples and considerations given in this section are fundamental steps towards the development of a more efficient approach to the cascade calibration; in particular, the investigations carried out in Section 7.7 have appeared earlier in Morini (2002) and in Brigo and Morini (2002).

The input swaption matrix is given in Table 7.11, referring to February 1, 2002.

Notice that the rows associated with the swaptions maturities of $6, 8$ and 9 years have been written in italics. In fact they do not refer to market

	1	2	3	4	5	6	7	8	9	10
1	**17.90**	**16.50**	**15.30**	**14.40**	**13.70**	**13.20**	**12.80**	**12.50**	**12.30**	**12.00**
2	**15.40**	**14.20**	**13.60**	**13.00**	**12.60**	**12.20**	**12.00**	**11.70**	**11.50**	11.30
3	**14.30**	**13.30**	**12.70**	**12.20**	**11.90**	**11.70**	**11.50**	**11.30**	11.10	10.90
4	**13.60**	**12.70**	**12.10**	**11.70**	**11.40**	**11.30**	**11.10**	10.90	10.80	10.70
5	**12.90**	**12.10**	**11.70**	**11.30**	**11.10**	**10.90**	10.80	10.60	10.50	10.40
6	***12.50***	***11.80***	***11.40***	***10.95***	***10.75***	*10.60*	*10.50*	*10.40*	*10.35*	*10.25*
7	**12.10**	**11.50**	**11.10**	**10.60**	10.40	10.30	10.20	10.20	10.20	10.10
8	***11.80***	***11.20***	***10.83***	*10.40*	*10.23*	*10.17*	*10.10*	*10.10*	*10.07*	*10.00*
9	***11.50***	***10.90***	*10.57*	*10.20*	*10.07*	*10.03*	*10.00*	*10.00*	*9.93*	*9.90*
10	**11.20**	10.60	10.30	10.00	9.90	9.90	9.90	9.90	9.80	9.80

Table 7.11. Black implied volatilities of ATM swaptions on February 1, 2002.

$F(0;0,1)$:	1	0.036712	11	0.058399
$F(0;1,2)$:	2	0.04632	12	0.058458
...:	3	0.050171	13	0.058569
	4	0.05222	14	0.058339
	5	0.054595	15	0.057951
	6	0.056231	16	0.057833
	7	0.057006	17	0.057555
	8	0.057699	18	0.057297
	9	0.05691	19	0.056872
	10	0.057746	20	0.056738

Table 7.12. Initial annualized forward LIBOR rate vector $F(0)$ from market data of February 1, 2002.

quotations, since the corresponding maturities are not quoted on the Euro market, and have been obtained, like in Section 7.4.1, by a simple linear interpolation between the adjacent market values on the same columns. This aspect is specifically tackled in Section 7.6.3.

Also remark that the upper triangular part, with reference to the skew diagonal included, is bold-faced. This is to underline once again that the basic CCA algorithm applies only to these values, and these will be our main reference in evaluating results in the first part of our investigation. In fact in this first analysis we want to keep out possible numerical problems due to the simple "equal multiple unknowns" assumption, made in Section 7.4, that is needed to extend our algorithm to the whole swaption matrix. Indeed, we have seen earlier that problems are already present when considering only the upper part of the input swaption matrix. When in Section 7.7 we introduce a more robust cascade calibration, we will consider the entire swaption matrix, since numerical problems become very rare even extending the basic CCA with the simplest "equal multiple unknowns" assumption.

The annualized forward LIBOR rates from the corresponding zero curve on the same date are given in Table 7.12. We will consider the historical correlation matrix and its parametric pivot forms given in Section 6.19.

7.6.1 Cascade Calibration under Various Correlations and Ranks

The focus in this section is on assessing if it possible to find acceptable cascade calibration outputs without any modifications of market swaption data, i.e. by acting only on the exogenously imposed instantaneous correlation matrix ρ. Although, as we have seen earlier, one can build extremely stylized toy input swaption matrices leading to negative or imaginary σ's for any choice of correlations, we wonder whether with realistic input market swaptions data exogenous correlations can be relevant in obtaining a robust calibration.

The first exogenous correlation matrix we apply, in view of the results of Section 6.19.2, is Rebonato 3 parameters pivot. We will consider different ranks, obtained by fitting the angles form seen in Section 6.9.2, at the desired rank, onto the chosen full rank correlation, thus obtaining $\rho(\theta^{*(\text{ Rank})})$. We can recall the principal component analysis we hinted at in Section 6.19.1 and decide to start with rank 7. The calibrated volatilities $\sigma's$ are:

0.179									
0.153	0.155								
0.144	0.129	0.154							
0.144	0.134	0.105	0.156						
0.140	0.122	0.112	0.112	0.154					
0.143	0.134	0.103	0.101	0.106	0.153				
0.143	0.127	0.143	0.088	0.097	0.086	0.144			
0.146	0.153	0.128	0.078	0.070	0.098	0.093	0.145		
0.157	0.109	0.155	0.160	0.067	0.007	0.101	0.081	0.107	
0.136	0.152	0.126	0.123	0.121	0.108	-0.040	0.120	0.077	0.067

There is a negative volatility, $\sigma_{10,7}$. What can we do to avoid this problem? Let us start by changing the rank of the correlation matrix. A calibration with full rank, equal to 19, gives not only the same negative volatility, but also one with non-null imaginary part, $\sigma_{10,10}$. The same bad results persist and even worsen a little when imposing the original historically estimated instantaneous correlation matrix as exogenous ρ. In this case no Rebonato-3-parameters or other smoothing techniques occur, before or after the possible rank reduction. We do not present the results with the original non-smoothed historical exogenous ρ, since they are rather similar to the results we obtained with the Rebonato 3-parameters pivot and its reduced rank versions.

When reducing the rank, down to rank 5 we get the same negative volatility, though reduced in absolute value. At rank 4, 3 and 2 the negative entry disappears and the output is acceptable, as shown below for rank 3:

0.179									
0.152	0.156								
0.130	0.130	0.166							
0.119	0.131	0.122	0.167						
0.114	0.117	0.120	0.125	0.163					
0.120	0.120	0.102	0.111	0.122	0.166				
0.130	0.113	0.125	0.094	0.113	0.107	0.156			
0.142	0.137	0.111	0.075	0.079	0.109	0.108	0.157		
0.156	0.099	0.135	0.128	0.056	0.025	0.114	0.111	0.145	
0.136	0.141	0.108	0.094	0.082	0.078	0.008	0.142	0.119	0.139

What might cause a similar behaviour? Recall that lowering the rank of a correlation matrix amounts to impose an oscillating tendency to the columns, that for very low ranks is represented by a sigmoid-like shape, as visible in Figure 7.5

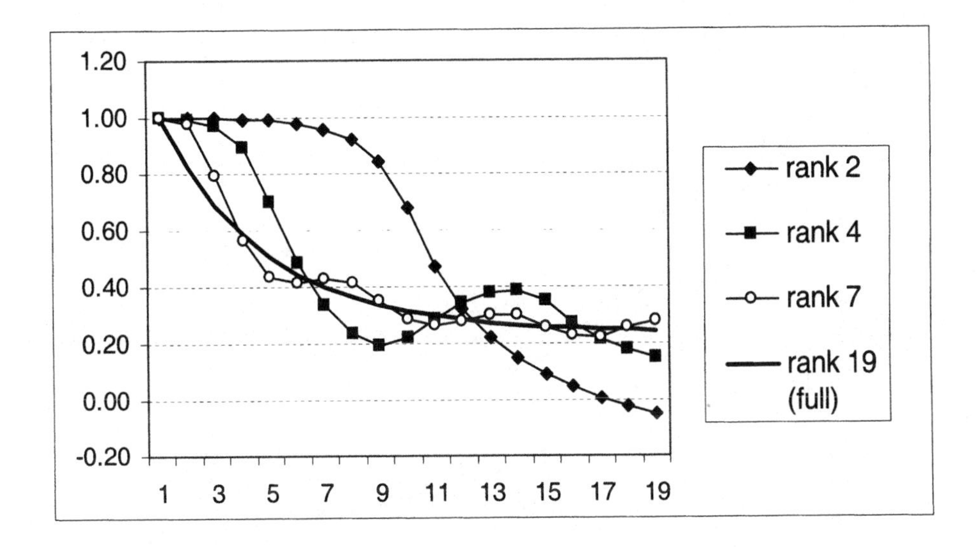

Fig. 7.5. First columns of correlation matrices of different ranks fitted on the "Rebonato 3 parameters pivot" correlation matrix

An interpretation of such results might be based on recalling that usually PCA's results suggest that only a reduced number of factors is relevant in explaining term-structure dynamics, or on the relative efficiency of low dimension swaption pricing. A less far-reaching, but more apparent interpretation simply suggests that some features of the lower rank correlation matrices are better suited to these swaptions data.

The most relevant feature discriminating between lower and higher rank matrices seems to be the slope of the columns in their initial part, that is the initial steepness of the decorrelation among rates. Remark that, for columns different from the first one, "initial" means starting by the unitary element on the principal diagonal. We might elicit that correlation matrices characterized by less steep initial decorrelation allow for acceptable results, whereas those featuring more marked steepness force some σ's to assume weird configurations.

To verify such an interpretation we made a number of further tests, especially making use of simple, synthetic correlation matrices, whose essential features can be easily modified and controlled. We varied the parameters ρ_∞ and β of the classic exponential structure given in (6.44), so as to obtain different configurations in terms of correlation patterns, represented in Figure 7.6. The parameters are modified, each time starting from the previous set, as follows:

a) $\rho_\infty = 0.5$. $\beta = 0.05$; b) Reduce ρ_∞ to 0; c) Set β to 0.2; d) Set ρ_∞ up to 0.5; e) Set β to 0.4; f) Take $\beta = 0.2$ and $\rho_\infty = 0.4$; g) $\rho_\infty = 0$, $\beta = 0.1$.

A detailed description of the implications of the parameters values and a longer discussion on these tests can be found in Morini (2002).

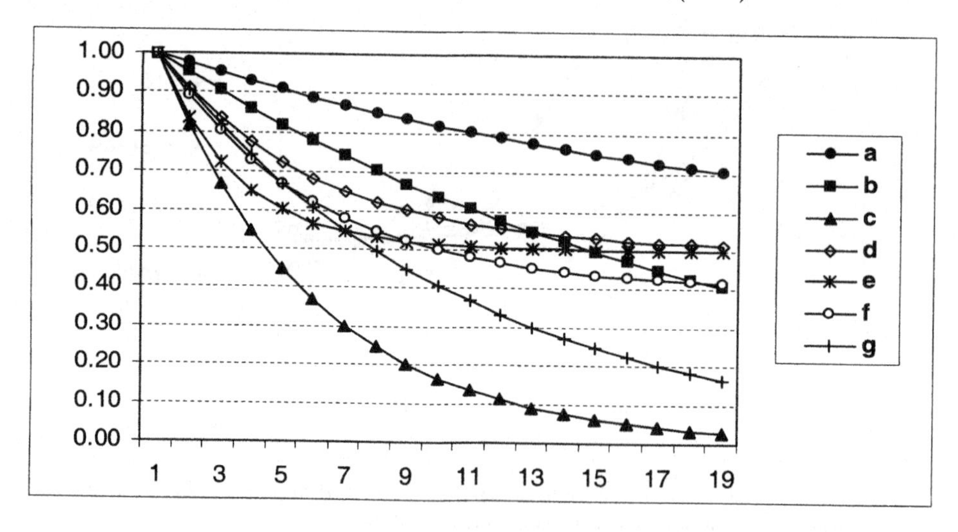

Fig. 7.6. First columns of classic exponential structure for several values of the parameters

We start with case **a**, giving volatilities all real and positive, at any rank and considering the entire matrix. Keeping as a reference rank 5 (first problematic level when increasing the rank of Rebonato three-parameters form) and the upper part of the swaption matrix, we have the following outcomes: with the correlations corresponding to **b**, **d** and **g**, we keep avoiding negative or complex volatilities, whereas **c**, **e** and **f** give a negative volatility in the same position as in our previous tests. Neither the final level of correlation nor the inclination of the terminal part of the columns appear decisive features to distinguish problematic from non-problematic correlations. Instead, what shows up is that we find non-meaningful results for those correlations featuring columns initially steeper, while the four configurations characterized by less initial steepness lead to real and positive volatilities.

Based on these results, S&C2 pivot looks interesting, since characterized by a more pronounced increase along sub-diagonals which is here coupled with, in general, initially less steep decorrelation along columns. As one might by now expect, actually this correlation gives us volatilities all real and positive, for rank from full 19 down to 2, in the last case even calibrating to the entire matrix, as shown in the next table.

0.179									
0.152	0.156								
0.130	0.130	0.166							
0.119	0.131	0.122	0.167						
0.112	0.115	0.120	0.126	0.164					
0.112	0.115	0.100	0.113	0.126	0.171				
0.113	0.103	0.119	0.098	0.120	0.119	0.163			
0.122	0.124	0.108	0.082	0.091	0.121	0.119	0.160		
0.138	0.093	0.130	0.129	0.073	0.047	0.123	0.113	0.149	
0.121	0.129	0.106	0.098	0.092	0.090	0.023	0.144	0.118	0.147
0.120	0.120	0.101	0.093	0.134	0.063	0.060	0.045	0.142	0.108
0.107	0.107	0.107	0.142	0.036	0.135	0.078	0.063	0.051	0.143
0.112	0.112	0.112	0.112	0.084	0.084	0.074	0.108	0.062	0.052
0.103	0.103	0.103	0.103	0.103	0.123	0.116	0.043	0.105	0.061
0.097	0.097	0.097	0.097	0.097	0.097	0.169	0.088	0.068	0.108
0.093	0.093	0.093	0.093	0.093	0.093	0.093	0.153	0.117	0.089
0.094	0.094	0.094	0.094	0.094	0.094	0.094	0.094	0.090	0.155
0.097	0.097	0.097	0.097	0.097	0.097	0.097	0.097	0.097	0.016
0.099	0.099	0.099	0.099	0.099	0.099	0.099	0.099	0.099	0.099

Although parameters found here are consistent with their probabilistic meaning, avoiding the numerical problems encountered before, it is important to check if they are also financially meaningful, namely reasonably regular and realistic. Such an automatic calibration, in a model extremely rich of volatility and correlation parameters such as the LFM, might tempt the user to treat such quantities as pure free-fitting parameters, irrespective of considerations on their configuration, internal consistency, and structural implications. In order to keep away from such a danger, in line with remarks given in Rebonato (1998) and earlier in this book, we require calibration to allow for instantaneous and terminal correlations, and the evolution of the term structure of volatilities, displaying those qualitative requirements mentioned in the beginning of Section 7.5.

7.6.2 Cascade Calibration Diagnostics: Terminal Correlation and Evolution of Volatilities

Let us start by considering the evolution of the term structure of volatilities (TSV). First, we see below how it appears in case of a calibration with Rebonato three-parameters pivot correlation matrix at rank 2.

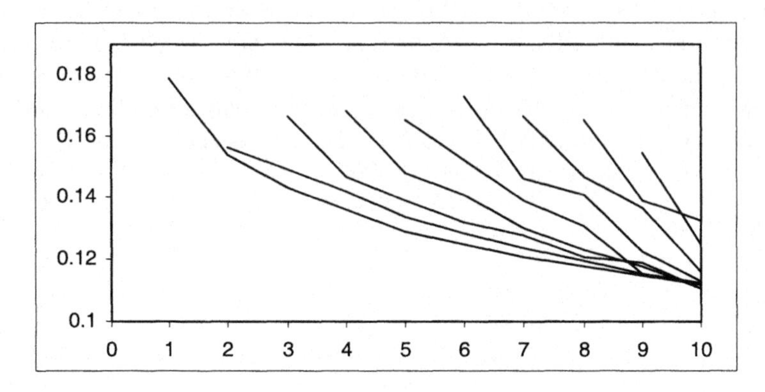

Considering the generality of the GPC volatility parameterization a la TABLE 1, such evolution appears surprisingly regular, smooth and stable over time, as well as being rather realistic. In fact, although it features no hump, such a characteristic is consistent with the swaptions data we are calibrating to. These are the properties we are usually looking for.

In order to assess the relevance of the chosen correlation parameterization, we show in the figure below the evolution we obtain with the S&C2 choice, keeping the same rank.

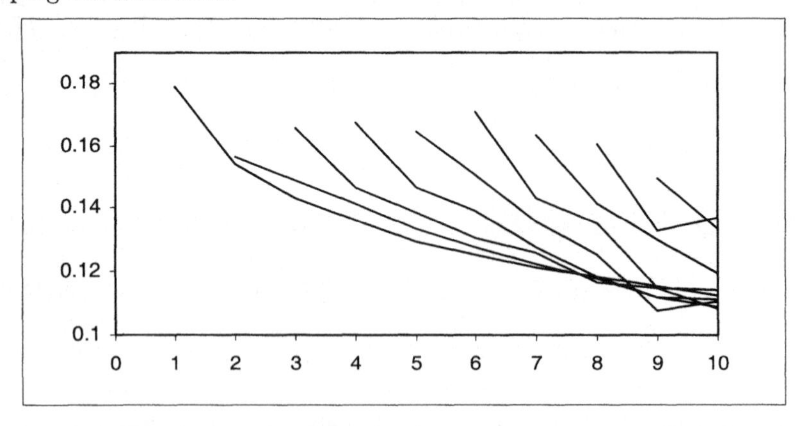

This less flexible form has brought about some worsening, but the general features we appreciated earlier are still present. Let us see what happens when increasing the rank. We tried different ranks of the two correlation matrices above, and we plot now below the evolution of the TSV under S&C2 pivot at rank 10.

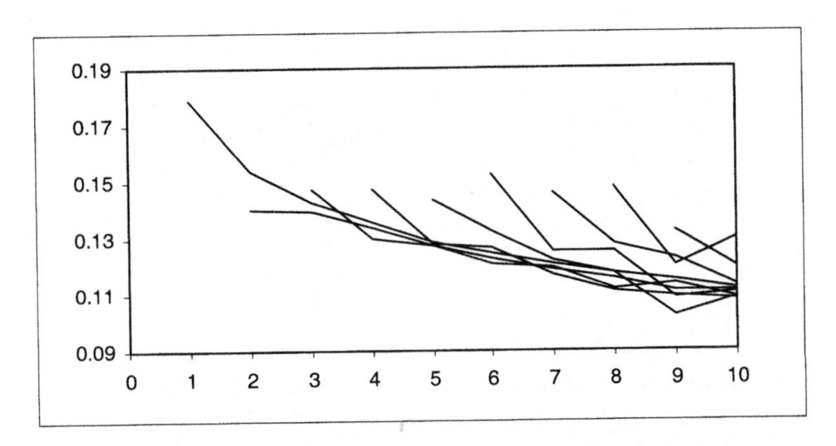

We see that the evolution is still acceptably realistic. However, we see also that when exogenous correlation matrices are chosen with high rank, and are thus more structured and less smooth, we can find less regular and less stable evolutions of the term structure of volatilities. This is confirmed also by other tests we performed. However, we point out that, as one might expect, by using particularly smooth and stylized correlations it is possible to attain a regular evolution even at full rank.

Now we move on to examining terminal correlations (TC's). Following previous tests we present the TC's after ten years. In the following tests, we stick to the S&C2 pivot instantaneous correlation matrix, and fit upon it reduced rank angles correlation matrices. Starting with rank 2, we obtain the following TC matrix:

	10	11	12	13	14	15	16	17
10	1.000	0.928	0.895	0.920	0.855	0.846	0.928	0.924
11	0.928	1.000	0.863	0.909	0.933	0.881	0.901	0.923
12	0.895	0.863	1.000	0.916	0.908	0.910	0.878	0.939
13	0.920	0.909	0.916	1.000	0.944	0.931	0.956	0.926
14	0.855	0.933	0.908	0.944	1.000	0.954	0.923	0.928
15	0.846	0.881	0.910	0.931	0.954	1.000	0.937	0.958
16	0.928	0.901	0.878	0.956	0.923	0.937	1.000	0.957
17	0.924	0.923	0.939	0.926	0.928	0.958	0.957	1.000

This matrix presents the peculiar feature of entries very high and close to each other. However, these features are already present in the corresponding portion of the instantaneous correlation matrix, due both to the peculiarities of S&C2 matrix and to the severe rank reduction. Thus the terminal correlations are replicating the features we imposed, as desired, only slightly perturbed by the dynamics of instantaneous volatility functions.

To see financially more meaningful TC's we must increase the rank, for instance up to 10, as shown below.

	10	11	12	13	14	15	16	17
10	1.000	0.887	0.806	0.792	0.708	0.690	0.757	0.734
11	0.887	1.000	0.822	0.837	0.825	0.746	0.749	0.753
12	0.806	0.822	1.000	0.877	0.841	0.820	0.758	0.801
13	0.792	0.837	0.877	1.000	0.919	0.877	0.881	0.806
14	0.708	0.825	0.841	0.919	1.000	0.932	0.878	0.840
15	0.690	0.746	0.820	0.877	0.932	1.000	0.915	0.914
16	0.757	0.749	0.758	0.881	0.878	0.915	1.000	0.934
17	0.734	0.753	0.801	0.806	0.840	0.914	0.934	1.000

This matrix looks in general satisfactory, and the same applies to the matrices we get when going up to full rank 19. The typical forward rates correlation patterns expected are clearly visible, following our qualitative desiderata. Moreover, such patterns look still very similar to those in instantaneous correlations, although again time-varying volatilities, while reducing correlations, have moderately perturbed the typical properties. Similar results have been obtained also when using different forms for instantaneous correlations.

Remark 7.6.1. **(A trade off between TSV and TC?)** Notice that such results suggest that, with unconstrained automatic cascade calibration with exogenous correlation, there might be a trade-off between regularity of the evolution of the TSV and satisfactory TC's, since they vary in opposite directions when rank is changed. Obviously, for different data and different parameterizations one should check again if such a trade-off is still relevant. In such a case, a possible criterion to make a choice is considering whether the financial instruments one is applying the model to depend essentially on instantaneous volatilities or are strongly influenced by the configuration of correlations.

The ease in analytical calibration and the more realistic TSV obtained when rank is low could be related to a practice of quoting swaption prices underlying the use of low dimensional models, with simple correlation structures. But the investigation of such a possible interpretation is not our aim here. We are interested in obtaining a calibration procedure as robust as possible with respect to changes in the input structure.

The above results show that through the CCA it is possible to obtain parameters not only perfectly acceptable as model volatilities, but also implying diagnostics structures which are regular and financially realistic. This excludes cascade calibration to be inapplicable for radical misspecification.

However, these satisfactory results are in general limited to the cascade calibration of the upper part of the swaption matrix, and depend on particular features of parametric structures, not easy to be detected and pondered in detail under the continuous flow of market data. Therefore these partial results can still be considered not sufficient for an automatic, trouble-free, everyday market application. In fact nonsensical results persist in circumstances that, even though traditionally not very common in modeling, such as high-rank historically estimated matrices, are reasonable and relevant. So they represent a sign of scarce robustness of the algorithm, with respect to changes both in the market data and in modeling assumptions on the correlation structure.

Detecting and then removing some of the reasons for this weakness is the focus of the next part of the work, starting with an analysis of the way input data are treated in the original algorithm.

7.6.3 The interpolation for the swaption matrix and its impact on the CCA

We can start to consider this point with an interesting remark, as from Morini (2002).

Remark 7.6.2. **(Negative/imaginary entries of calibrated σ's occur in correspondence of linearly interpolated swaptions volatilities).** In all previous real market cascade calibration tests, including those in Section 7.4.1, negative or imaginary values occur only for volatilities depending also on the artificial swaption volatilities obtained by a local linear interpolation along the columns of the swaption matrix. On the contrary, volatilities obtained before such artificial interpolated values enter the algorithm are all real and positive.

Following this remark, we are naturally led to suspect the interpolation used for missing market quotations to bear some responsibility for the problems found. The first step might be checking whether the linear interpolation is really the most suited to replicate the typical patterns in the swaption market. We tried out some fitting forms for the columns of the swaption matrix. Let us see the results for fitting a log-linear (or "power") functional form in the maturity. The best-fitting function, for the first column in our data, is

$$Y = 0.1785\,(X)^{-0.201}, \quad \text{or} \ln(Y) = \ln(0.1785) - 0.201\ln(X),$$

where Y denotes the swaption volatility and X the maturity. In Figure 7.7 the results, based on all values fo the column, are shown.

The power fitting form appears clearly closer to the real market pattern than the linear one, as further confirmed by standard diagnostics concerning the optimization output. Also looking at other columns and other quotations, referring to some months later, the picture does not change.

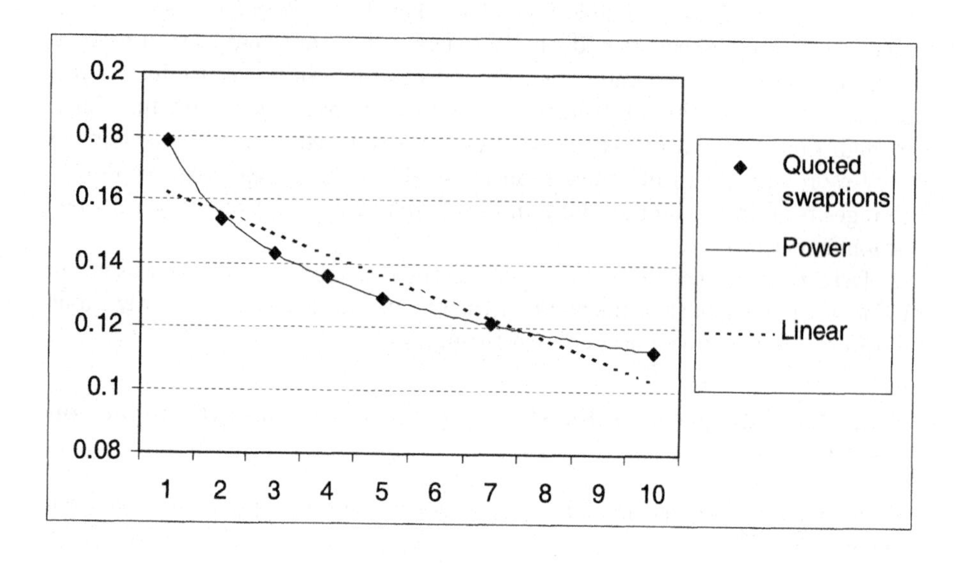

Fig. 7.7. First columns of the swaptions data with fitted linear and log-linear parametric forms

However, we must recall what is reported in Rebonato and Joshi (2001) about typical swaption configurations. Two are the common shape patterns that can be found in the Euro swaption market: a humped one, called *normal* and typical of periods of stability, and a monotonically decreasing one, called *excited* since associated with periods immediately following large movement in the yield curve and in the swaption matrix. Our data appear to belong to the second pattern, so in case of humped patterns a similar form would be likely to prove inadequate.

It is natural to wonder whether, using such a more realistic interpolation for missing maturities, it is possible to change the output of the cascade calibration. In the tests we performed, negative entries previously found with linear local interpolation do not disappear with the new one, but they reduce in absolute value. Thus, seemingly, acting on the methodology used for filling missing entries and replacing input entries can improve the output.

This conjecture is underpinned also by results of tests on using this realistic functional form also to replace market values suspected to be misaligned. In our case, for instance, numerical problems appear in correspondence with interpolated quotations and with market swaption volatilities of 7 year maturity, the maturity isolated between missing quotations. Considering that experienced market operators confirm it to be rather illiquid (Dominici (2002)), and hence prone to possible misalignments, both 7 year-maturity quotations and missing values are replaced through the power form above. Although based on similar considerations, such a more focused and specific manipulation has a much less drastic influence on the input market data than the general smoothing we saw in Section 7.4.1. In a triangular calibration, only 4 (instead of 55) market values have been substituted, and the errors induced are definitely lower:

Errors	-0.00028	-0.00119	-0.00079	0.00049
% Errors	-0.23272%	-1.03388%	-0.70780%	0.45776%

Keeping our first choice correlation matrix (Rebonato three-parameters pivot), the negative value found in previous calibrations now disappears at any rank considered, including full 19 rank. This confirms that inconsistencies in the input swaption matrix can have negative effects on calibration results, showing how analytical, unconstrained calibration can be of use to detect the possible presence of such inconsistencies in input data. Indeed, looking at the patterns in the columns of the input swaption matrix we see that linearly interpolated missing entries introduce inconsistencies within such patterns, and the subsequent cascade algorithm detects such inconsistencies by producing some negative/imaginary σ's. Here we have seen that the use of interpolation based on a functional form more consistent with market patterns can be helpful.

However, the need for replacing specific market values still brings about some loss of information, and can become arbitrary. A remark that applies also to the choice of the method used to interpolate on market values, even though clearly better than simple local linear interpolation. Two problems that can be dealt with at once, as shown in the next section.

7.7 Empirically efficient Cascade Calibration

The developments on improving the output of the CCA presented in this Section, and related tests, are based on Brigo and Morini (2003).

Remark 7.6.2 and results with more realistic interpolation techniques indicate that the relevance of artificial inconsistent input swaption volatilities in provoking numerical problems might have been overlooked. Since the cascade of second order equations in the calibration requires a complete swaption matrix, featuring values for each and every maturity and length in the range, quotations for the few nontraded maturities on the Euro market had been replaced by simple local linear interpolation. Such modification of input data could appear harmless and negligible, and is not unusual in the market or in the literature (see for instance Rebonato and Joshi (2001)). However our tests have shown a clear empirical link between the use of linearly interpolated values in the algorithm and the appearance of negative or imaginary volatilities. An influence confirmed by the improved results obtained by making use of fitting forms clearly more consistent with market patterns.

Nevertheless, no exogenous interpolation we attempted has proven enough to be trouble-free enough to avoid numerical problems in a complete calibration and with a broad range of model dimensions. Generally satisfactory results have been ensured only through modifications of market data, making clear the importance of input consistency but spoiling exact calibration, and leaving room for mistrust of model specification. In order to shed light on this point, namely whether problems are due to artificial input data or to model misspecification, the direct way is making cascade calibration independent of artificial data at all. Once this is done, one can see how the cascade calibration performs on pure market data. If calibration assumptions are, at least in general, consistent with market assumptions and practice in computing prices, when used only with pure market data the algorithm should become robust. Meaningless results should disappear in normal situations and become unlikely exceptions, without any need to manipulate correlations or altering market data.

Is it possible to modify cascade calibration so as to obtain an algorithm relying only on market data, without giving up its benefits?

7.7.1 CCA with Endogenous Interpolation and Based Only on Pure Market Data

A cascade calibration relying only on directly available market data with no exogenous data interpolation can be constructed simply by following the same procedure used to elaborate algorithm 7.4.2. One assumes σ parameters to be related in a pre-specified way when they surface as multiple unknowns, with multiplicity due to the lack of market data needed to make a specific discernment. This assumption allows one to compute all σ's consistently with the directly available data. Following this method, we worked out a general modification of the CCA, allowing us to invert industry formula 6.67 even in presence of "holes" in the swaption matrix. This algorithm permits to calibrate exactly to market data without the need to resort to artificial data, unlike the previous version of the algorithm. At the same time, the new algorithm keeps all valuable features with respect to the previous one, and in particular remains analytical, fast and exact.

The new algorithm we present amounts to carrying out an *endogenous interpolation*. In fact all volatilities σ, including those in principle depending on missing swaption volatilities quotes, are computed based only on available quotes by means of the "equal multiple unknowns" assumption. This procedure allows us to use the obtained LIBOR model σ's to compute missing swaptions volatilities quotes and fill the holes in the input swaption matrix.

This "filling" is an interpolation based on internal quantities of the model obtained from the calibration to the directly available data only, thus allowing artificial values to be coherent with the underlying model by construction. This method allows us to have a consistent calibration based on *all* available market swaption quotes, and *only* on them.

Let us see the details of this algorithm in case the assumption is the simplest and more natural, namely constancy of the volatility of a forward rate when no data are available to infer possible changes. Showing immediately the rectangular version for a complete calibration to the whole quoted swaption matrix, the algorithm reads:

Algorithm 7.7.1. ***(Rectangular Cascade Calibration with Endogenous Interpolation Algorithm (RCCAEI) (Morini (2003), Brigo and Morini (2003)))****. With reference to equations (7.3) and (7.4), we must follow the steps:*

1. *Select the number s of rows in the swaption matrix that are of interest for the calibration, including those non quoted. The sth row must be available from the market. Let us define the set*

$$K := \{k \in \mathbb{Z} : 0 \le k < s-1, \text{ and } V_{k,y} \text{ are missing for all } \ y : k < y \le k+s\},$$

namely the list of missing maturities.

2. *Set $\alpha = 0$;*
3. *a. If $\alpha \in K$, define*

$$m = m^{\alpha} = \min\{i \in \mathbb{Z} : \alpha < i < s, \ \ i \notin K\},$$

namely the first market quoted maturity higher than α. Then set

$$\sigma_{j,m+1} = \sigma_{j,m} = \ldots = \sigma_{j,\alpha+1} =: \sigma_j \ \text{ for all } j: \quad m+1 \le j < s+\alpha, \tag{7.5}$$

that is we assume that all the involved forward rates have constant volatility in the period. Set $\gamma = \alpha$ (first missing maturity involved) and then $\alpha = m$.

b. If $\alpha \notin K$, set $\gamma = \alpha$.

Set $\beta = \alpha + 1$.

4. *a. If $\gamma \in K$, solve in σ_β equation (7.3) after adjusting to take into account constraint (7.5).*

b. If $\gamma \notin K$, solve in $\sigma_{\beta,\alpha+1}$ equation (7.3).

5. *Increase β by 1. If $\beta < s+\gamma$ go back to point 4. If $\beta = s+\gamma$, set*

$$\sigma_{\beta,\alpha+1} = \sigma_{\beta,\alpha} = \ldots = \sigma_{\beta,1}$$

and solve in $\sigma_{\beta,\alpha+1}$ equation (7.4).

If $\beta < s+\alpha$, repeat point 5. Otherwise increase α by 1.

6. *If $\alpha < s$, go back to point 3, otherwise stop.*

When algorithm 7.7.1 is applied to a swaption matrix such as in Table 7.11, we have

$$K = \{5, 7, 8\},$$

namely the maturities at 6, 8 and 9 years after today. Notice that in this case there are four σ's which are not determined, precisely $\sigma_{6,6}$, $\sigma_{8,8}$, $\sigma_{9,8}$ and $\sigma_{9,9}$. In fact, no market quoted swaption volatility depends on them, as one can easily verify considering that the swaption volatilities contributing (together with $V_{\alpha,\beta}$) to determine the value of $\sigma_{\beta,\alpha+1}$ are those marked with "$\times$" in the table below. The table refers to a 5×5 swaption matrix, taking for example $\alpha = 3$, $\beta = 7$.

	1	2	3	4	5		
1				$\times$	$\times$	$\times$	$\times$
2			$\times$	$\times$	$\times$	$\times$	
3		$\times$	$\times$	$\times$	$\times$		
4	$\times$	$\times$	$\times$	$v_{3,7}$			
5							

The marked area is a parallelogram whose bottom/right vertex is given by $V_{\alpha,\beta}$ itself and extending as far as the first column on the left and the first row upwards. If any of the volatilities in the parallelogram are missing, their information content in determining the σ's could be replaced, through suitable hypothesis, by the information content of $V_{\alpha,\beta}$. But nothing is known of volatilities out of all parallelograms built on quoted swaptions, so there is no way to determine consistently σ's depending only on these "outside-all-parallelograms" missing volatilities. In particular, with the full input swaption matrix with missing $K = \{5, 7, 8\}$, we have no way of computing $\sigma_{6,6}$, $\sigma_{8,8}$, $\sigma_{9,8}$ and $\sigma_{9,9}$without additional assumptions.

With other kinds of calibration this fact might be overlooked, but with this non-parametric analytical cascade calibration no data are available to recover the precise information required. Anyway, for the same reason, all other volatility buckets are independent of the missing V's, so that this does not affect the workings of the algorithm beyond the inability of recovering the "isolated" σ's.

Thus it is not possible to determine the above 4 parameters $\sigma_{6,6}$, $\sigma_{8,8}$, $\sigma_{9,8}$ and $\sigma_{9,9}$ consistently with the algorithm, and furthermore these four σ's have no impact on the recovery of available swaption market quotes. This implies immediately that when these parameters are needed anyway (for example for diagnostics purposes) they must be recovered outside the proposed precise calibration algorithm. One simple possibility is to use one of the possible homogeneity assumptions on piecewise constant σ's seen in Section 6.3.1, say $\sigma_{k,\beta(t)} =: \eta_{k-(\beta(t)-1)}$, corresponding to TABLE 2 in said section. We may apply this assumption to the missing σ's and thus get $\sigma_{6,6} := \sigma_{5,5}$, $\sigma_{8,8} := \sigma_{7,7}$, $\sigma_{9,8} := \sigma_{8,7}$ and $\sigma_{9,9} := \sigma_{8,8}$.

Why this particular homogeneity assumption? Since these missing σ's are needed to display usual diagnostics structures and in particular to plot complete TSV's, we have chosen one assumption that seems to have a limited effect on the shape of TSV, namely that of TABLE 2, with reference to the preceding rates. However, any different choice is possible with no effect on any other values.

The obvious interest is now to check how algorithm 7.7.1 performs in practice. Having discarded the influence of artificial data, it is now possible to see how cascade calibration really works on market data. We will consider the two sets used earlier for the cascade calibration empirical tests, namely market data of 16 May 2000 and 1 February 2002, in order to compare with previous results. We will also use two more recent data sets, of 10 December 2002 and of 10 October 2003, to have some evidence on the stability over time of properties found.

Let us start with swaptions and forward rates from Table 7.11 and 7.12 with historically estimated correlations at full rank 19, namely no rank reduction and no parametric smoothing, calibrating to the entire swaption matrix. This situation corresponds to the worst possible situation according to the results discussed in Section 7.6, and in fact such calibration with CCA from Algorithm 7.4.1 and linearly exogenously interpolated values gave negative and imaginary σ's even when restricting to the upper part of the matrix, and many more when considering a "rectangular" calibration to all available swaption quotes.

Now, with only market data and endogenous interpolation, the output is as follows:

0.179									
0.167	0.140								
0.153	0.138	0.138							
0.142	0.148	0.130	0.122						
0.135	0.131	0.134	0.135	0.109					
0.142	0.135	0.106	0.118	0.112	0.109				
0.155	0.126	0.145	0.098	0.130	0.087	0.087			
0.150	0.141	0.118	0.099	0.103	0.142	0.142	0.087		
0.130	0.092	0.136	0.153	0.095	0.122	0.122	0.142	0.087	
0.109	0.127	0.116	0.116	0.130	0.088	0.088	0.112	0.112	0.112
0.123	0.123	0.115	0.112	0.166	0.115	0.115	0.118	0.118	0.118
0.111	0.111	0.111	0.165	0.056	0.147	0.147	0.081	0.081	0.081
0.118	0.118	0.118	0.118	0.107	0.102	0.102	0.083	0.083	0.083
0.117	0.117	0.117	0.117	0.117	0.145	0.145	0.097	0.097	0.097
0.127	0.127	0.127	0.127	0.127	0.127	0.127	0.106	0.106	0.106
0.104	0.104	0.104	0.104	0.104	0.104	0.104	0.135	0.135	0.135
0.114	0.114	0.114	0.114	0.114	0.114	0.114	0.114	0.114	0.114
0.120	0.120	0.120	0.120	0.120	0.120	0.120	0.120	0.120	0.120
0.166	0.166	0.166	0.166	0.166	0.166	0.166	0.166	0.166	0.166

All meaningless results have disappeared, we have only real and positive σ's still allowing a perfect recovery of all market prices, as can be easily checked by Black formula coupled with industry approximation (6.67). And this applies not only to the triangular calibration which was our reference in previous tests, but also in a calibration extended to the entire swaption matrix by means of the simple assumption of constant volatility under multiple unknowns.

The same full rank test gives satisfactory results also when moving to the first parametric form seen in Section 7.6, namely Rebonato 3 parameters pivot. For S&C2 pivot exogenous ρ (i.e. the stylized instantaneous correlation matrix working well in previous tests on the upper part of the swaption matrix), we found no numerical problems even when extending the calibration to the entire swaption matrix and at any correlation rank.

Considering the first set used for the Cascade Calibration, namely the highly problematic set from Table 7.4 with its rank 2 typical correlations, the σ volatility matrix we obtain with the new algorithm is given below

0.180									
0.155	0.204								
0.129	0.156	0.233							
0.118	0.104	0.166	0.244						
0.109	0.099	0.097	0.161	0.248					
0.113	0.073	0.078	0.101	0.162	0.248				
0.104	0.098	0.050	0.074	0.113	0.219	0.219			
0.094	0.105	0.094	0.032	0.086	0.137	0.137	0.219		
0.106	0.079	0.086	0.082	0.068	0.075	0.075	0.137	0.219	
0.101	0.092	0.058	0.103	0.151	0.013	0.013	0.194	0.194	0.194
0.092	0.092	0.079	0.043	0.030	0.061	0.061	0.135	0.135	0.135
0.083	0.083	0.083	0.071	0.049	0.113	0.113	0.058	0.058	0.058
0.074	0.074	0.074	0.074	0.080	0.078	0.078	0.045	0.045	0.045
0.070	0.070	0.070	0.070	0.070	0.073	0.073	0.045	0.045	0.045
0.077	0.077	0.077	0.077	0.077	0.077	0.077	0.077	0.077	0.077
0.075	0.075	0.075	0.075	0.075	0.075	0.075	0.049	0.049	0.049
0.071	0.071	0.071	0.071	0.071	0.071	0.071	0.071	0.071	0.071
0.069	0.069	0.069	0.069	0.069	0.069	0.069	0.069	0.069	0.069
0.066	0.066	0.066	0.066	0.066	0.066	0.066	0.066	0.066	0.066

Compare this with the old cascade calibration output, given in Table 7.5.

Moving to more recent data, below we show the results for swaptions and zero curve market data on October 2003, with exogenous full rank historically estimated correlation matrix computed on the same day.

0.291									
0.232	0.242								
0.199	0.183	0.222							
0.169	0.173	0.177	0.199						
0.167	0.120	0.169	0.180	0.180					
0.156	0.136	0.146	0.158	0.175	0.180				
0.132	0.147	0.097	0.155	0.129	0.155	0.155			
0.133	0.137	0.113	0.063	0.148	0.165	0.165	0.155		
0.128	0.122	0.103	0.107	0.097	0.141	0.141	0.165	0.155	
0.108	0.133	0.100	0.101	0.112	0.093	0.093	0.155	0.155	0.155
0.125	0.125	0.099	0.098	0.104	0.094	0.094	0.123	0.123	0.123
0.122	0.122	0.122	0.051	0.097	0.103	0.103	0.110	0.110	0.110
0.129	0.129	0.129	0.129	0.045	0.070	0.070	0.090	0.090	0.090
0.130	0.130	0.130	0.130	0.130	0.060	0.060	0.067	0.067	0.067
0.116	0.116	0.116	0.116	0.116	0.116	0.116	0.068	0.068	0.068
0.110	0.110	0.110	0.110	0.110	0.110	0.110	0.085	0.085	0.085
0.093	0.093	0.093	0.093	0.093	0.093	0.093	0.093	0.093	0.093
0.108	0.108	0.108	0.108	0.108	0.108	0.108	0.108	0.108	0.108
0.126	0.126	0.126	0.126	0.126	0.126	0.126	0.126	0.126	0.126

In applying the RCCAEI to this most recent data set we never came across any numerical problem giving rise to inconsistent results. Overall, all data sets mentioned above have been examined, each with synchronously estimated historical correlations, and results are summarized in the following.

No anomalous results or numerical problems have been found in any test outputs, at any rank considered, when calibrating only to the upper part of the swaption matrix. This upper part represents the market quotations retrieved through the cascade calibration principle with no need for additional "equal multiple unknowns" hypothesis, besides the homogeneity assumptions required for endogenously interpolated missing quotations. This upper part calibration was the basic reference case in previous tests, and with the endogenous interpolation none of the problems seen in earlier tests have appeared again.

Although full rank tests suggest high rank not to be a problem any longer, notice that these results do not include limited rank reduction tests, corresponding to discarding only a few factors. This is due to the inefficiency of the angles rank-reduction technique introduced in Section 6.9.2 when the required output rank is high. As pointed out in Morini and Webber (2003), to which we refer for details, in such a case the angles parameterization optimization can become computationally intensive, and very time consuming in interpreted languages, while the accuracy advantage it can provide with respect to faster methods tightens. In the same work, and always in the context of market interest rate correlations, an alternative algorithm is presented, called *eigenvalue zeroing by iteration*, which is intuitively appealing and much faster than angles optimization. This rank reduction algorithm requires a low number of iterations for producing a level of accuracy similar to that of the angles method. This methodology renders cascade calibration easy to implement and fast at all possible ranks for the exogenous instantaneous correlation matrix ρ. We did so, finding confirmation of the good results above at all ranks.

Nevertheless, the promising results lead us to consider, in every earlier test, also the output of a complete "rectangular" calibration. These further "rectangular" tests involved a relevant use of the simplifying assumption of constant volatilities in presence of multiple σ unknowns. While the complete RCCA was almost always problematic with exogenous interpolation of the input swaption matrix (be it linear or power), now we try and extend the previous considerations for the upper input swaption matrix to the whole matrix in the context of endogenous interpolation. In doing so, moving from the upper triangular matrix to the whole rectangular matrix does not spoil the good results we obtain with the upper triangular case alone. We show this by testing the method on rectangular input swaption matrices.

With the available data sets we experiment all lower ranks matrices obtained from the historically estimated correlation matrix through the eigenvalue zeroing by iteration rank reduction method. We find only regular results and no numerical problems. For lower required ranks we considered also the optimal angles rank reduction method introduced in Section 6.9.2, finding very similar results and the same generally satisfactory behaviour, with one exception. In this case numerical problems appeared, for the 2002 data sets, in a test at rank 4. Although a rank 4 exogenous correlation matrix is far from being realistic, due to the severe rank reduction, it is richer and not as simple and smooth as a rank two matrix. The problems we find in this single case is represented by two negative volatilities, both very close to zero. This problem occurs only when extending calibration to the lower part of the matrix and when both the "equal multiple σ unknowns" and the "homogeneity for isolated σ values" assumptions are enforced.

Therefore this only single case with a minor problem could be likely avoided by more realistic and flexible hypothesis on ρ's or on the σ's, without altering the cascade algorithm substantially. We do not investigate this case in detail, since the problem is easy to avoid, for instance by changing rank or moving to the S&C2 parametric form, or more simply by using the different rank reduction technique mentioned earlier. With any of these choices the small negative entry we found is replaced by a small positive one.

Such results suggest a couple of remarks. First, while the rough features of the caplet volatility structures are qualitatively preserved in general under different exogenous instantaneous correlation choices, the fine details of volatility parameters, and their regularity, have a precise dependence on the details and regularity of the chosen exogenous correlations structure. This will be clear also in next section when analysing the implied evolutions of the term structure of volatility. This strong sensitivity of the σ details on the exogenous ρ choice could be seen as a flaw of the cascade calibration, especially considering that usually *instantaneous* correlations are deemed not have a strong influence on swaption prices. On the other hand, such a dependency is an unavoidable feature of any exact analytical calibration method. And what looks like a flaw can be instead an opportunity, providing us with one of the few methods giving a precise indication on the influence of the chosen instantaneous correlations on calibration .

Secondly the particularly good behaviour of the eigenvalue zeroing by iteration method, coupled with the Cascade Calibration is interesting, since the use of angles optimization for rank reduction represents a possible bottleneck in computational efficiency in obtaining the rank reduced exogenous ρ to be imposed before the calibration. If the eigenvalue zeroing by iteration algorithm is used, the rank reduction (and all other) calibration steps are instantaneous for all practical purposes, keeping out any traditional optimization routines. Alternatively, simple non-iterative eigenvalue zeroing, although being less precise with respect to the full rank matrix, is even faster.

7.7.2 Financial Diagnostics of the RCCAEI test results

Even if numerical problems giving rise to σ parameters of no statistical meaning (negative/imaginary) appear now unlikely exceptions and easy to avoid, one has to check if the found parameters imply financial quantities meaningful enough to be applied safely to valuation in the market. This analysis can also help to understand if the satisfactory results found are simply related to a numerical ease of computation or also to a truly increased capability to pick out implied market structures.

Let us start by considering, as in previous sections, the evolution of the term structure of caplet-like volatility. First we show results for the highly problematic set from Table 7.4 with its typical rank-2 correlations.

Compare this output with Figure 7.4. Although some peculiarities related to this set are still visible, the most worrying features have remarkably reduced. So bad features have improved with the new algorithm. But let us check that earlier good results have not worsened with our new endogenous interpolation method. Below is the plot of TSV for test with the Rebonato 3 parameters pivot matrix at rank two on February 1, 2002, to be compared with the analogous plot seen in Section 7.6.2

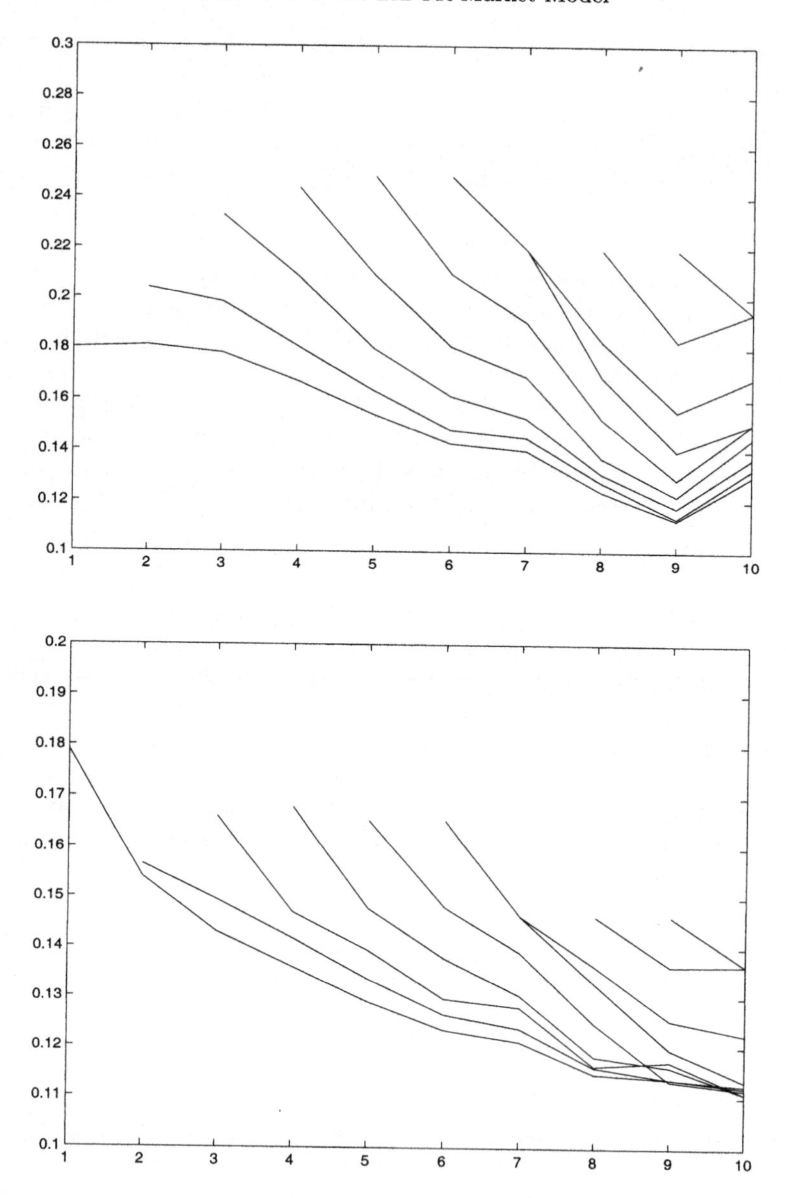

The plot appears very similar and at least as good as the one we obtained earlier. It is somewhat smoother and more regular, although the overall level is slightly less constant over time. These two features are more apparent when we increase the rank, while still keeping a smooth parametric form. See for instance how the TSV appears when considering the S&C2 matrix from the earlier tests, now at full rank:

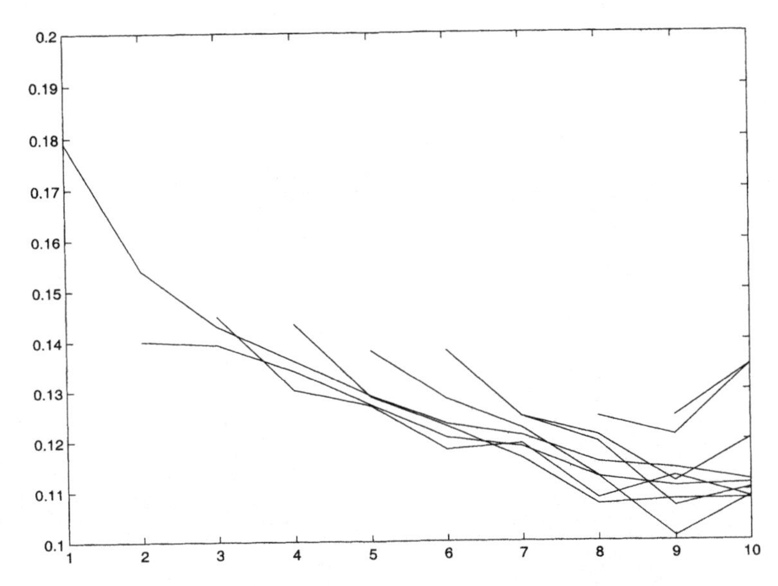

Even though previous considerations on the effect of smoothness and the exogenous-correlation-matrix rank are basically confirmed, here with endogenous interpolation we find a structure featuring characteristics of regularity and realism also at full rank. Notice that some irregularities concentrate in those parts of the structure corresponding to missing quotes, and also the few missing quotes have been determined independently of the algorithm only to plot a complete term structure of volatility. A similar phenomenon can be seen also in the previous parametric calibration tests, but only with non-parametric calibration (such as the cascade calibration) this can be directly related to the lack of some data on the market.

Due to the increased robustness of the algorithm, also the historically estimated instantaneous correlation matrices with no smoothing become an important possible choice. Let us see which kind of TSV these exogenous historical ρ matrices lead to. With a very low rank choice, say 2, one can have a uniform and flat evolution, as shown below for market data on 10 December 2002.

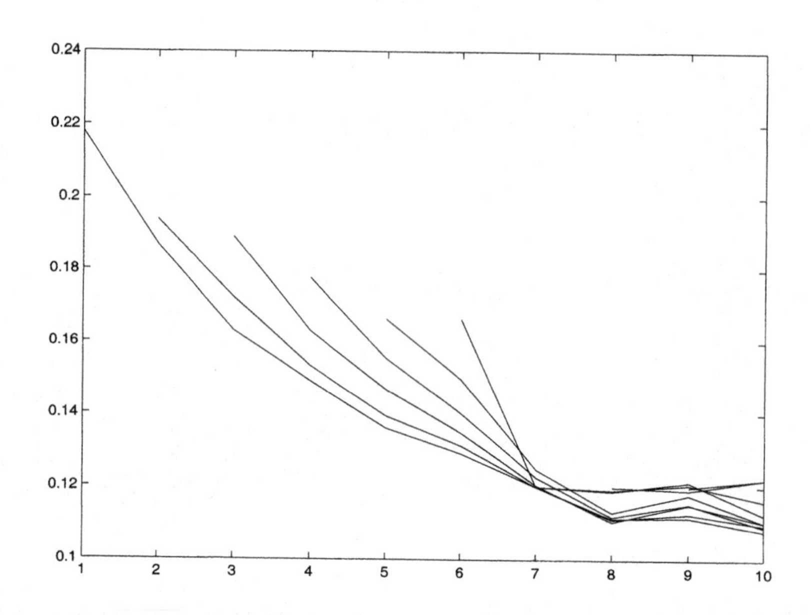

With maximum rank the exogenous instantaneous correlation matrix is exactly the historically estimated one, that is correlation has not been smoothed in any way. One can expect the resulting TSV evolution to be highly irregular. Let us see results both for data referring to 1 February 2002 and for October 2003.

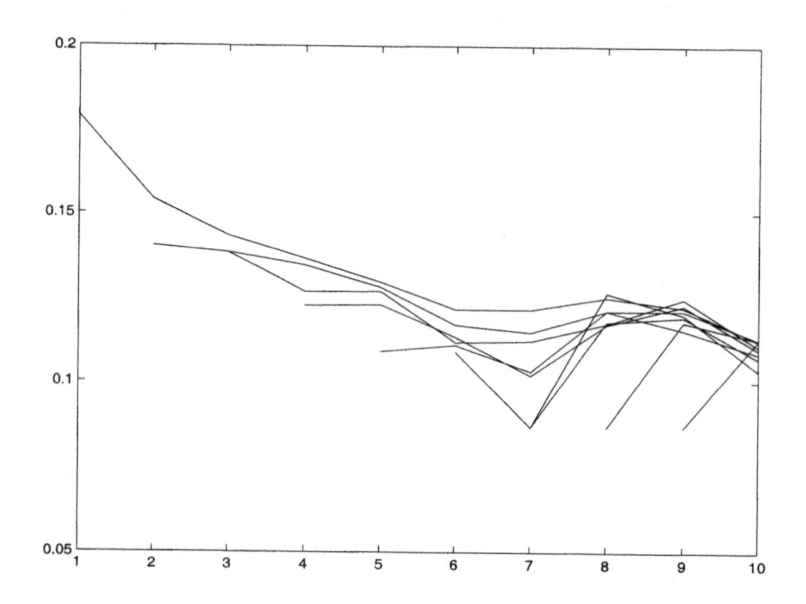

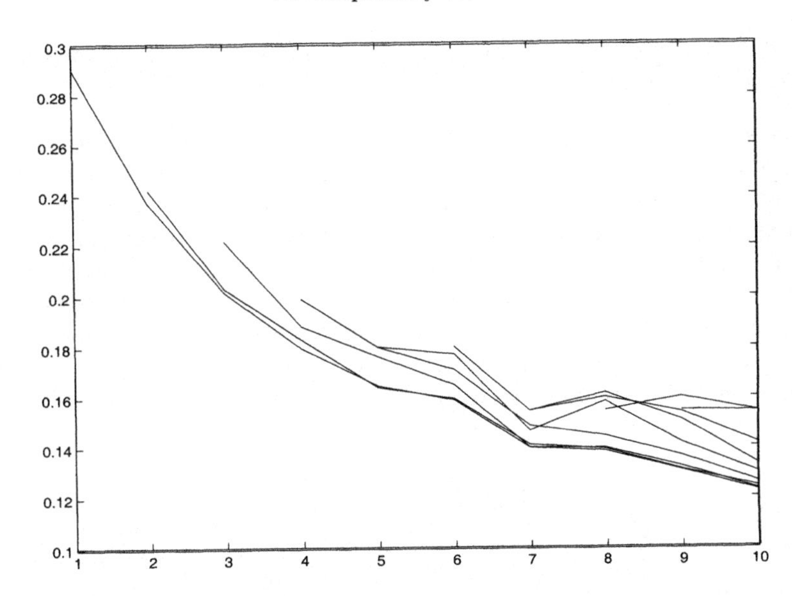

Apart from local irregularities in the first one still concomitant with missing values, the patterns are still qualitatively stable and regular. In terms of level stability something is lost when increasing the rank, confirming previous observations. Yet, regularity and smoothness, typically hard to obtain with unconstrained and non parametric procedures, are generally preserved.

Moving now to terminal correlations, it appears that the conclusions based on our previous results can be extended to the present case. The patterns and the properties are similar to those of the exogenously imposed instantaneous historically estimated correlation matrix, partially perturbed by the obtained σ patterns, as on can expect. In the table below we see the terminal correlation referring to the October 2003 full rank calibration.

	1	2	3	4	5	6	7	8	9	10
1	1.000	0.946	0.871	0.774	0.648	0.638	0.681	0.670	0.603	0.557
2	0.946	1.000	0.922	0.824	0.694	0.679	0.718	0.701	0.634	0.592
3	0.871	0.922	1.000	0.882	0.746	0.736	0.761	0.734	0.659	0.603
4	0.774	0.824	0.882	1.000	0.905	0.863	0.840	0.757	0.644	0.548
5	0.648	0.694	0.746	0.905	1.000	0.944	0.879	0.751	0.605	0.486
6	0.638	0.679	0.736	0.863	0.944	1.000	0.958	0.845	0.711	0.589
7	0.681	0.718	0.761	0.840	0.879	0.958	1.000	0.958	0.868	0.775
8	0.670	0.701	0.734	0.757	0.751	0.845	0.958	1.000	0.967	0.907
9	0.603	0.634	0.659	0.644	0.605	0.711	0.868	0.967	1.000	0.974
10	0.557	0.592	0.603	0.548	0.486	0.589	0.775	0.907	0.974	1.000

7.7.3 Endogenous Cascade Interpolation for missing swaptions volatilities quotes

Another interesting feature of this RCCAEI is that, as a by-product, we can compute the values for the missing swaptions volatilities implicit in the parameters obtained through the LIBOR model cascade calibration. This can be seen as an endogenous interpolation technique that is consistent with the (cascade calibrated) LIBOR model by construction. Observing how they look like can give us an important insight on the coherence between cascade mechanism and market practice. The results appear generally uniform across our tests. As an example we report below the output for the October 2003 exogenous full rank historical correlation.

	1	2	3	4	5	6	7	8	9	10
1	0.291	0.252	0.228	0.209	0.197	0.185	0.173	0.165	0.159	0.153
2	0.237	0.209	0.193	0.178	0.167	0.158	0.152	0.147	0.143	0.140
3	0.202	0.185	0.172	0.161	0.150	0.144	0.139	0.135	0.132	0.130
4	0.180	0.168	0.157	0.147	0.138	0.133	0.129	0.126	0.123	0.122
5	0.165	0.155	0.143	0.135	0.129	0.125	0.122	0.119	0.117	0.116
6	*0.159*	**0.143**	**0.135**	**0.129**	**0.123**	**0.120**	**0.117**	**0.114**	**0.112**	**0.111**
7	0.140	0.134	0.127	0.120	0.116	0.113	0.110	0.108	0.107	0.106
8	*0.139*	*0.130*	**0.122**	**0.117**	**0.114**	**0.110**	**0.107**	**0.105**	**0.104**	**0.102**
9	*0.131*	**0.123**	**0.117**	**0.113**	**0.109**	**0.105**	**0.103**	**0.102**	**0.100**	**0.099**
10	0.123	0.116	0.111	0.106	0.102	0.100	0.099	0.097	0.096	0.096

The bold-faced lines represent the missing quoted maturities whose values are obtained as outputs of the cascade calibration. Among these the values based on the 4 $\sigma'{s}$ determined outside the algorithm have been written in italics. The regularities typical of market quotations, such as the decreasing tendencies along maturity and along length, are respected, and are not in contrast with the assumption of fundamental consistency between the specification of the model and market practice. Notice also that in our tests such market patterns are not as easily recovered for the 4 values in italics determined outside the cascade calibration. This seems to confirm that cascade calibration endogenous interpolation grasps non trivial market regularities.

This aspect should be investigated further, considering also its relationships with no-arbitrage requirements. Indeed, knowing that behind the interpolation of missing market quotes there is an arbitrage free dynamical model whose output patterns and diagnostics make sense can be of some comfort. If this consistency appears to hold, the cascade calibration can be used, in various contexts, also as a reasonable and precise methodology to interpolate among available quotations to recover missing swaption volatilities.

7.7.4 A first partial check on the calibrated σ parameters stability

Lastly we address another relevant question in assessing the reliability of the implementation of a theoretical model: the stability of the calibrated parameters. At times this point is overlooked in modern finance, since daily recalibration is becoming a widely accepted standard. Nevertheless it is reasonable to expect parameters calibrated to a cross section of simultaneous quotes not only to be regular and financially meaningful, but also to show some stability over time. Indeed, if parametric structures really extrapolate the relevant information on the market, they should not experience sudden major changes unless, of course, outstanding market shifts take place.

Again this is of particular relevance for the cascade calibration, since its precise dependence from the details of input data can lead to suspect calibrated parameters to overreact to market movements. We do not carry out a rigorous historical analysis of calibrated parameters, which we suggest as an interesting subject for future research, but simply compare the outputs for our 3 sets, all about 10 month spaced. Calibration outputs appear very similar in structure, and this can be appreciated also by looking at the rough and synthetic indication given in tables below, with reference to an exogenous full rank historical correlation matrix cascade calibration. The tables report the average and the maximum variation over all calibrated $\sigma's$ in the range from one to another of the considered trading dates.

	Feb-02	Dec-02	Oct-03
Feb-02	0.000	0.018	0.029
Dec-02	0.018	0.000	0.023
Oct-03	0.029	0.023	0.000

Average $|\Delta\sigma|$

	Feb-02	Dec-02	Oct-03
Feb-02	0.000	0.066	0.113
Dec-02	0.066	0.000	0.109
Oct-03	0.113	0.109	0.000

Max $|\Delta\sigma|$

Compared to the range of volatility values that can be observed in results reported in Section 7.7.1, these changes appear limited and reasonable. Although the 10 month interim considered can be indeed a long time on financial markets, the differences do not appear to be too small either, since the shifts occurred in the swaption market have been limited as well in the considered period.

	Feb-02	Dec-02	Oct-03
Feb-02	0.000	0.009	0.029
Dec-02	0.009	0.000	0.020
Oct-03	0.029	0.020	0.000

Average $|\Delta V|$

	Feb-02	Dec-02	Oct-03
Feb-02	0.000	0.039	0.112
Dec-02	0.039	0.000	0.073
Oct-03	0.112	0.073	0.000

Max $|\Delta v|$

In conclusion, in spite of the lack of smoothing and constraints, cascade calibrated parameters appear to show an acceptable behaviour in terms of regularity and financial significance. However one has to remind that we considered a limited number of data sets and market situations. Therefore the tests presented here are more examples of the sort of controls and methods one can regularly enact on the behaviour of a calibration methodology, rather than definitive conclusions.

7.8 Reliability: Monte Carlo tests

We have already pointed out that the reliability of the exact swaptions cascade calibration depends also on the accuracy of the underlying approximated formula (6.67). This formula has already been tested, for instance by Brigo and Mercurio (2001c) and the results are reported later in this book in Chapter 8. The formula has been tested also in Jäckel and Rebonato (2000). What we do here is extending similar tests, based on Monte Carlo simulation of the true LFM dynamics, to different or more general conditions, either peculiar to the cascade calibration, such as general piecewise constant (GPC, TABLE 1) volatility structure, or particularly easy to implement in this context, such as various correlations with various ranks.

First of all, we need the discretized LIBOR model dynamics, seen in (6.53). Around the obtained Monte Carlo price, we build a two-side 98% window, based on the standard error of the method, as we have seen in Section 6.11. Then, by inverting Black's formula, we obtain the corresponding values for volatilities, which are compared with the result of formula (6.67). In particular, we check if the approximation is in-between the extremes of the Monte Carlo window, denoted by "inf" and "sup". In most cases, input volatilities are the output of a cascade calibration, so that formula (6.67) yields exactly the market original value. In such cases, this is what is compared with the value implied by the true Libor Market Model Monte Carlo dynamics specified by the same covariance parameters.

All following tests refer to simulations with 4 time-steps per year and 200000 paths, "doubled" by the use of the variance-reduction technique known as antithetic variates. We selected a sample of test results, particularly meaningful since covering a differentiated range of conditions.

a) Let us start by checking how the approximation works under GPC parameterization with low rank and "normal" conditions as for swaption characteristics, so that we consider a 5×6 swaption with Rebonato pivot matrix at rank 2. We obtain:

MC volatility	MC inf	MC sup	Approximation
0.108612	0.108112	0.109112	0.109000

b) Now we try out an increase in swaption maturity and underlying swap length. Indeed we consider a 10×10 swaption, with S&C2 matrix at rank 2, obtaining:

MC volatility	MC inf	MC sup	Approximation
0.097970	0.097531	0.098409	0.098000

c) We see here a little higher rank, with a 5×6 swaption and S&C 2 correlation matrix now at rank 5.

MC volatility	MC inf	MC sup	Approximation
0.108824	0.108323	0.109325	0.109000

In the three cases we considered, with market calibrated volatilities, the volatility approximation (6.67) appears definitely good. This confirms its accuracy also in case of GPC volatility, with low or intermediate rank, and even for long maturities and lengths.

d) We move now on to consider artificially modified instantaneous volatilities, so as to test the approximation in case of anomalous values. First, we upwardly shift all calibrated volatilities ("Cal. vol's" in the following tables) by multiplying them by 1.2. Then we experiment stronger increase, by adding 0.2 to all volatilities, following what we will do later on in Chapter 8. In this case, it appears useful to show also what is returned by a simple Monte Carlo without variance-reduction techniques, denoted by "MC vol (no a.v.)". The results, maintaining the remaining conditions as in test a), are as follows:

	MC volatility	MC inf	MC sup	Approx.
Cal. vol's $\times 1.2$	0.130261	0.129644	0.130878	0.130800
Cal. vol's $+0.2$	0.300693	0.298988	0.302399	0.303820
	MC vol (no a.v.)	MC inf	MC sup	Approx.
Cal. vol's $+0.2$	0.301934	0.299010	0.304861	0.303820

e) Let us perform the same set of tests, but this time on a 6×7 swaption, with S&C 2 correlation matrix and rank increased up to 10. We obtain:

	MC volatility	MC inf	MC sup	Approx.
Cal. vol's	0.104608	0.104131	0.105085	0.105000
Cal. vol's $\times 1.2$	0.125469	0.124881	0.126057	0.126000
Cal. vol's $+0.2$	0.286702	0.285133	0.288272	0.289777
	MC vol (no a.v.)	MC inf	MC sup	Approx.
Cal. vol's $+0.2$	0.287380	0.284671	0.290091	0.289777

Both in point d) and e) we see that even with moderately increased volatilities, the approximated formula still works well in this context of GPC volatilities, different correlations and low or intermediate rank. In case of very high volatilities, the approximation is outside the antithetic variates Monte Carlo window, but inside the non-reduced variance simple Monte Carlo. Therefore, we conclude that the approximation keeps even in this case an acceptable behaviour, but looses accuracy.

f) Then we performed some tests in more extreme conditions. See below the results obtained for a 10×10 swaption, with historically estimated correlation for 10 December 2002 at full rank 19, and EICCA inputs.

	MC volatility	MC inf	MC sup	Approx.
Cal. vol's	0.094937	0.094512	0.095363	0.095000
Cal. vol's $\times 1.2$	0.113853	0.113333	0.114372	0.114000
Cal. vol's $+0.2$	0.273425	0.272064	0.274788	0.277094
	MC vol (no a.v.)	MC inf	MC sup	Approx.
Cal. vol's $+0.2$	0.273317	0.270865	0.275774	0.277094

In case of maximum rank coupled with a "far" and "long" swaption, the approximation appears good for normal or slightly increased volatilities, but gets less reliable for very high volatilities, being outside both MC windows. Further tests of ours confirm this observations and seem to suggest that the factors that most influenced such a worsening are the elevated maturity and length, rather than the high rank.

g) In this last set of tests, we investigate the case of higher input forward rates. Thus we uniformly upwardly shifted the initial forward-rates vector by adding 0.02 to all original rates. Initially we apply that to the same covariance structure as in test c), obtaining:

MC volatility	MC inf	MC sup	Approximation
0.108923	0.108437	0.109410	0.109136

h) And then we consider the extreme situation of point f), with the following results:

MC volatility	MC inf	MC sup	Approximation
0.097716	0.097301	0.098130	0.098048

In the typical context of cascade calibration, the approximation (6.67) appears reliable even with higher forward rates, irrespective of the level of the rank and the features of the swaption considered.

To sum up, our tests seem to allow us to extend the generally positive judgement on the reliability of industry approximation (6.67) also to configurations typical of this work such as GPC parameterization and different correlation structures with various ranks, ranging from 2 to 19. This holds true even for high maturity and length swaptions, moderately increased volatilities and upwardly shifted initial forward rates. On the other hand, we can also extend to such cases the more negative valuation expressed for instance in Chapter 8 on the accuracy of the formula in case of market situations characterized by very high volatility that we might consider pathological.

Similar remarks apply therefore also to the automatic swaptions cascade calibration, which is based on this formula joined with GPC volatilities and exogenous correlations.

	Swaption volatilities	Semi-annual	rates	Caplet	volatilities
1	0,1790	0,0436	0,0480	0,1805	0,1720
2	0,1540	0,0483	0,0508	0,1911	0,1745
3	0,1430	0,0508	0,0523	0,1641	0,1575
4	0,1360	0,0532	0,0545	0,1546	0,1517
5	0,1290	0,0550	0,0560	0,1516	0,1480
6	0,1250	0,0559	0,0566	0,1445	0,1409
7	0,1210	0,0566	0,0572	0,1374	0,1352
8	0,1180	0,0560	0,0562	0,1329	0,1307
9	0,1150	0,0568	0,0571	0,1285	0,1262
10	0,1120	0,0575	0,0577	0,1240	0,1231

Table 7.13. Volatilities and forward rates on February 1, 2002

7.9 Cascade Calibration and the cap market

We have seen earlier in Sections 7.2 and 7.3 some attempts at a stylized joint calibration of the LIBOR market model to caps and swaptions. In general, even though some sound objections have been raised to the convenience of carrying out joint calibrations to cap and swaptions, one might find it interesting to consider how information coming from cap data can be embedded in the cascade calibration. Or at least it would be interesting to investigate how the two sets of information can be given mutual consistency. From our experience, this is a typical question traders ask when this methodology is presented.

In the following, we provide some hints and discussions on the matter, following Morini (2002). Firstly, making use the annualization of semi-annual caps data presented in Section 6.20, we can introduce a simple, rather brute-force opportunity to incorporate cap data into the CCA. In fact, based on the formulas given in Section 6.20, we can simply replace the first column of the input swaption matrix, containing volatilities for unitary length swaptions, with the corresponding array of annualized caplet volatilities.

In this way, we actually lose some information on the swaption market, now overlapping with the artificially annualized cap data. In turn, we achieve a perfect fit to annualized caplet volatilities. To implement such an approach with our data, included in Table 7.13, and referring to February 1, 2002, we used formula (6.76), in that it depends only on market quantities. We took all ρ's equal to one, as we did earlier in Section 6.20. We obtained the following values.

ANNUALIZED CAPLET VOLATILITIES									
0.178	0.185	0.163	0.155	0.152	0.145	0.138	0.134	0.129	0.125

Except for the first one, these values are all higher than the swaption volatilities they are to replace. It is of interest to see the effect of this substitution in calibration. An appealing feature of cascade calibration in this context is that now in the first column we have no missing quotations, and this enables to calibrate through Endogenous Interplation CC determining all parameters within the algorithm, avoiding the external assumptions required with all-swaption data.

We present results with two different exogenous instantaneous correlations, rank 2 Rebonato 3-parameter form and the non-smoothed historically estimated correlations at full rank. The σ's we obtain now are still real and positive, but many entries are markedly different from the corresponding ones in the previous corresponding tests. This leads to an implied evolution of the term structure of volatility which is smooth but shows a peculiar increasing tendency, as shown below first for rank 2 and then for full rank test.

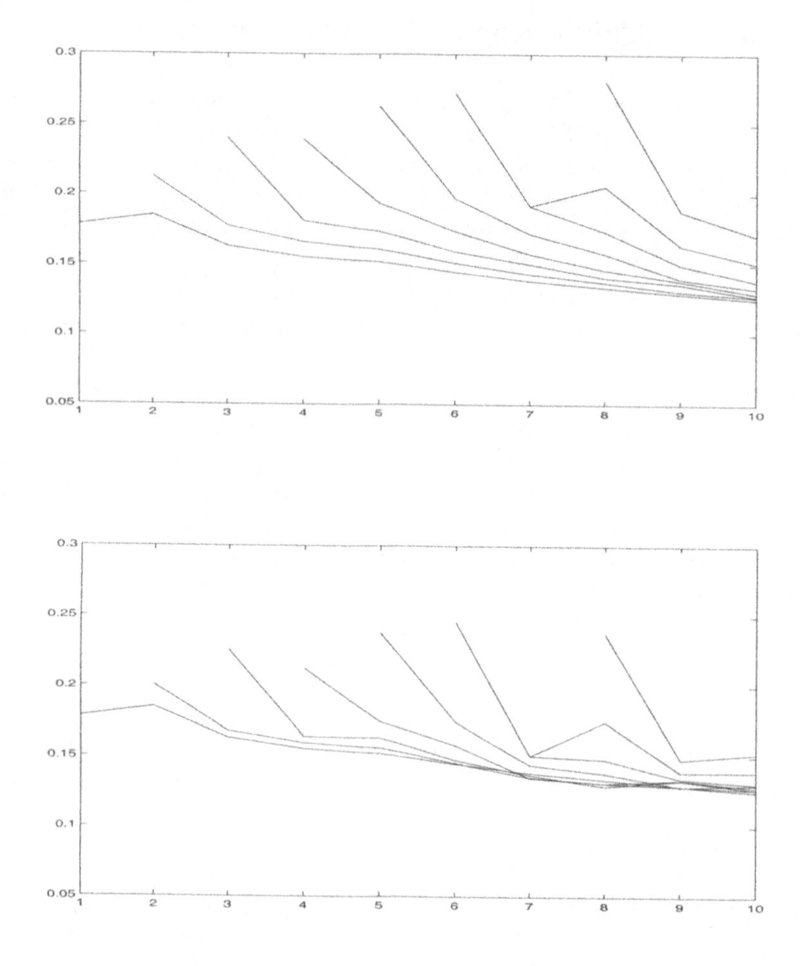

Although we do not obtain nonsensical instantaneous volatilities, a similar evolution of the TSV can cause some perplexity. This behavior might be due to a possible inconsistency between the caps and swaptions data we mixed together. However, we should consider that a similar result is found also in Section 7.4.1, where the original all-swaption matrix simply featured a humped shape in the first column, like our matrix does after the substitution. Moreover, in Section 6.20 we pointed out that formula (6.76) tends to overprice swaptions, a bias that might have been further increased by fixing infra-correlations to their maximum value. Consequently, better-founded conclusions might be induced by further investigations, including different, more realistic assumptions on instantaneous volatilities, and different values for the ρ's. For instance, ρ's could be based on the analysis of historical or implied typical infra-correlation patterns.

The latter point is partially addressed in the following. In fact, we use the above setting to calculate the infra-correlations implied in our data, since this can be seen as a different opportunity to link a swaptions cascade calibration with the information coming from the cap market. Indeed, the calibration considered earlier provides us with an annual tenor LIBOR Market Model exactly and automatically calibrated to swaptions. Now this can be coupled with a semi-annual LFM exactly and automatically calibrated to semi-annual tenor caps, being the relationships between the two models represented by a specification of formula (6.75), and hence by the corresponding values for the infra-correlation parameters ρ. For example, if the specification chosen is formula (6.76), then by extracting v^2_{Black} from the swaption market and $v_{S-caplet}$ and $v_{T-caplet}$ from the cap market we can invert the formula easily obtaining the implied value for semi-annual ρ's. This is the scheme we implemented, starting from the data in Table 7.13, and obtaining the ρ's reported below, each one associated with a different unitary swaption:

INFRA-CORRELATIONS									
1.022	0.388	0.543	0.536	0.444	0.493	0.533	0.560	0.586	0.598

If one deeply trusts the model and the formulas presented above, such results might be considered a possible confirmation of the underlying assumptions, in particular the existence of a well-defined relationship between cap and swaption volatilities. In fact, the values we obtained are indeed viable correlations, except for the first one. And also the anomaly regarding that value is not particularly relevant. From such a viewpoint, this might even be considered a possible arbitrage opportunity to exploit.

However, a more cautious and critical approach is likely to give rise to a different kind of remarks. First, besides the fact that the first value is outside the viable range for correlations, the other values appear too low to represent real correlations between adjacent rates. Although the variables are different, one may compare them to the results of our earlier historically estimated correlation matrix. A possible reason for these values is the aforementioned bias due to the chosen volatility parameterization. Again, more realistic hypothesis can lead to different results. But the really relevant reasons calling for a cautious interpretation of such results are of a different nature. Indeed, relations and discrepancies between caps and swaptions tend to be influenced by causes concerning the market fundamentals. This point is closely related to a more general problem, namely whether or not there exists a basic congruence between the cap and swaption markets, that a model can successfully detect and incorporate. Rebonato (2001) seems to warn against excessive enthusiasm in considering such a possibility. Rebonato recalls that problems such as illiquidities, agency problems and value-at-risk based limit structures strongly reduce the effectiveness of those market operators, called *quasi-arbitrageurs*, that are supposed to maintain the internal consistency between the two markets. Thus swaption and caplet volatility surfaces may turn out to be non-congruent.

Accordingly, simple artificial values such as the infra-correlations above, expressing synthetically the relationships between those surfaces in the context of a modeling framework, are likely to be actually influenced by many different external factors that are hard to detect and measure. Maybe some more light might be shed on this matter by comparing implied and historically estimated infra-correlations, and analyzing their stability over time.

To conclude, even more caution is required in considering whether to apply similar outputs to a trading strategy. We recall the words by Rebonato, quoted in Alexander (2002), suggesting that apparent arbitrages arising from mispricing of either caps or swaptions by a general model would be too risky to trade upon. And a further warning comes from recalling that the losses involving several banks at the end of the 90's can be attributed to the fact that the used models relied on a fixed relation between caplet and swaption volatilities. As recalled by Pelsser (2000), such an assumption was finally belied by the true market behaviour.

7.10 Cascade Calibration: Conclusions

Vega Vega Vega è qua!" [Vega Vega Vega is here]

"Vega", Atlas-Ufo-Robot (Ufo robot Grendizer) Italian song, 1978

In this chapter we have considered several issues concerning the calibration of the LIBOR market model to caps and/or swaptions data. In the first part, whose conclusions have been summarized earlier in Section 7.5, we noticed that our first examples where we tried to imply directly instantaneous correlations from swaptions volatilities, led to a number of problems. We then introduced the Cascade calibration, pointing out both the achieved improvements and the new problems.

In the second part we remarked that some fundamental features make Cascade Calibration particularly appealing, giving motivations for analyzing this method in more detail. In particular we focused on detecting how to avoid nonsensical results, such as negative or complex volatility parameters, which were typically occurring in the first part of the chapter.

Accordingly, we performed calibration tests with exogenous historically estimated correlation matrices and synthetic parameterization versions of such matrices. We pointed out empirical links between correlation features and the occurrence of anomalous outputs, identifying a number of structures giving rise to acceptable results. We showed that regular terminal correlations and a satisfactory evolution of the term structure of volatilities are possible, although there can be a trade-off between regularity of the evolution of the TSV and realism of TC's, depending on the rank of the exogenous correlation matrix.

Then we addressed the related matter of the interpolation used for missing market quotations, pointing out its relevance on the quality of the results, as confirmed by improvements when using an alternative log-linear (power) form, appearing much closer to real market patterns than the usual linear interpolation. These turned out to be mainly steps towards the introduction of a different and more satisfactory general approach, presented in Section 7.7 following Brigo and Morini (2003). In fact the above observations led us to a new cascade calibration algorithm, called (Rectangular) Exogenous Interpolation CC algorithm (RCCAEI), relying only on available market data. This allowed us to discard earlier requirements for exogenously interpolated artificial inputs. The endogenous interpolation version not only maintained all positive features of the earlier cascade calibration, but has also proven to be empirically much more efficient. In fact, in terms of regularity and significance of the obtained σ parameters, it shows remarkable characteristics of robustness with respect to changes both in the market and in the modeling assumptions on the exogenous instantaneous correlation structure.

Thus, in the context of the LIBOR market model, the RCCAEI method is possibly the most promising swaption calibration method we introduced. We can summarize the RCCAEI features as follows.

1. The calibration can be carried out through closed form formulas. Because of that:
   a. The solution to the calibration problem is unique.
   b. The algorithm is, for practical purposes, instantaneous.
2. If the industry formula (6.67) is used for pricing swaptions in combination with Black's formula, market swaption prices are recovered exactly;
3. The method establishes a one-to-one correspondence between model volatility parameters σ and market swaption volatilities.
4. The instantaneous correlation matrix ρ is an exogenous input. This allows instantaneous correlations to be regular and realistic, and our tests have shown that these features are not spoiled in moving to terminal correlation.
5. The output of the calibration is uncostrained and thus there are no a-priori guarantees on the obtained σ parameters to belong to acceptable

and reasonable ranges. However, the method appears to be empirically robust. This means that our tests on different sets of market data have in general given resulting σ parameters that were automatically satisfying the required constraints. The σ's also appear to be stable over time, consistent with market patterns if determined through endogenous interpolation, and they give rise to acceptably regular evolutions of the term structures of volatility.

This confirms the correctness of our original intuition about the possibility to invert the industry formula for obtaining a set of parameters that are meaningful under both a probabilistic and a financial point of view, thus finding instantaneously a unique and exact solution to the problem of LFM calibration to swaptions.

Another issue that needed to be assessed, before considering cascade calibration as a reliable tool for market applications, was the accuracy of the industry approximation (6.67) (for swaption volatilities in the LIBOR market model) this method is based on, under the covariance structures typical of this context. We tested it via Monte Carlo simulation, concluding the approximation to be sufficiently accurate even under these different conditions, except in case of anomalously high volatility conditions. Consequently this can be extended also to the CCA in all of its variations. We are perfectly aware that a scarcely flexible algorithm like the CCA cannot suit all situations, and all needs of an applied mathematician working as a quant. Such an exact methodology with no smoothing or range constraints might appear, at a first glance, to be nothing but a special case of an overfitting approach. But after checking the meaningfulness and the regularity of the results that can be obtained when, and only when, real market data are used in a fully consistent way, it turns out to be more likely the closest available approximation of the concept of implied volatility in applying the LIBOR Market Model to swaptions.

Obviously any cascade calibration method cannot be considered to be as robust as a method based on a constrained optimization. In fact the CCA provides us with a unique solution and we cannot force it to satisfy acceptability or regularity requirements. However, this is an essential feature of any inversion procedure returning implied values of parameters, such as the Black & Scholes classic model itself. And, provided that acceptable and regular results keep on being generally obtained under normal market conditions, as it was in our tests, the possible unlikely appearance of anomalous outputs can act as a useful warning of inconsistency in input data, or between data and our assumptions on instantaneous covariance structures. A warning that more flexible methodologies based on optimization rather than on exact inversion might fail to communicate.

Finally, we remark that further investigations, not considered here, can be carried out to assess or improve the effectiveness of this calibration. For instance, alternative sub-parameterizations of volatilities could replace our constancy assumption on multiple unknowns. One might prefer hypothesis reflecting typical patterns detected in the market, or devised for having some desired properties of volatilities.

The possible extension of the method to a joint calibration setting, embedding also cap market data, has been here only sketched, showing some possibilities and related dangers, but could be developed further if required in real market operations. Moreover, although we considered sets from different points in times and noticed that parameters appear relatively stable under a financial point view, the stability over time of the calibrated parameters could be checked by a rigorous historical analysis of the fitted volatilities, as suggested above and by Brigo and Mercurio (2002a). In the same paper, also a possible alternative use of cascade calibration is proposed: an analytical calibration as the one here presented can be used (once negative and complex parameters are set to zero, if they had appeared) as an initial guess for a constrained optimization, easing computation. Here we did not investigate such an opportunity, since by relying only on market data we found directly satisfactory results without resorting to traditional optimizations. Among further alternative uses of the cascade calibration, the endogenous interpolation cascade calibration based only on directly quoted market data is also a methodology to compute missing quotes consistently with directly available market quotes and the no-arbitrage LIBOR model paradigm. If further investigation confirms the reliability shown here, such endogenously interpolated values can be used also for applications different from cascade calibration.

Other developments that might reveal particular interest for market application are a further extension of tests and analysis on the effects of different situations and structures, and the application of this method to hedging and sensitivity analysis. In fact such features as the absence of calibration errors and direct one-to-one correspondence induced between model parameters and market swaption volatilities make this approach particularly fit for computing meaningful and precise swaptions vega bucketing.

Yuke yuke DUKE FLEED ⟨*Go! Go DUKE FLEED!*⟩
tobe tobe GRENDIZER ⟨*Fly! Fly GRENDIZER!*⟩
daichi to umi to aozora to ⟨*Mother Earth, the sea, the blue sky and all friends:*⟩
tomo to chikatta kono heiwa ⟨*you promised them to defend peace*⟩
mamori mo kataku tachiagare ⟨*Bravely rise.*⟩
chikyū wa konna ni chisai keredo ⟨*Earth is so small, yet*⟩
seigi to ai to de kagayaku hoshi da ⟨*it is a star shining with justice and love*⟩
mamore mamore mamore ⟨*protect protect protect her*⟩
ningen no hoshi minna no chikyū ⟨*star of humanity, Earth of all people*⟩.

"Tobe Grendizer" ⟨Fly Grendizer!⟩, Ufo Robot Grendizer opening theme, 1975.
(Duke Fleed protected Earth from the Vega invasion in 1975-1977)

Damiano is grateful to dottoressa Cristina Dalle Vedove from Noventa di Piave for fundamental help with the Romaji transcription of the original song and with the Italian translation.

8. Monte Carlo Tests for LFM Analytical Approximations

Hey, Houston, we've got a problem here.
Jack Swigert, April 13 1970, Apollo 13 mission to the moon

"Are you ready to think faster than the ultimate computer, Mr. West?"
Batman to Flash, DC One Million (1999), DC Comics.

In this chapter we test the analytical approximations leading to closed-form formulas for both swaption volatilities and terminal correlations under the Libor market model (LFM), by resorting to Monte Carlo simulation of the LFM dynamics. We aim at establishing whether the approximations based on drift freezing and approximating lognormal distributions are accurate. We adopt two different contexts.

First Part: Tests via Distance between Probability Distributions

As a first attempt, following Brigo and Liinev (2002), we attack the problem by resorting to a rigorous notion of distance on the space of probability distributions. More precisely, we are concerned with the distributional difference of forward swap rates between the lognormal forward–Libor model (LFM) or "Libor market model", and the lognormal forward-swap model (LSM) or "swap market model", the two modern models for interest-rate derivatives we encountered earlier in the book. To measure this distributional difference, we resort to a "metric" in the space of distributions, the well known Kullback-Leibler information (KLI). We explain how the KLI can be used to measure the distance of a given distribution from the lognormal (exponential) family of densities, and then apply this to our models' comparison. The volatility of (i.e. standard deviation associated with) the projection of the LFM swap-rate distribution onto the lognormal family is compared to the industry synthetic swap volatility approximation obtained via "freezing the drift" techniques in the LFM (Formula (6.67)). Finally, for some instantaneous covariance parameterizations of the LFM we analyze how the above distance changes according to the parameter values and to the parameterizations themselves, in an attempt to characterize the situations where LFM and LSM are really distributionally close, as is assumed by the market. This first part of tests based on KLI is carried out in Section 8.1.

Second Part: Classical and More Exhaustive Tests

In the second part, starting from Section 8.2 and until the end of the chapter, we simply test numerically the approximated swaption formula against an implied swaption volatility backed out from the Monte Carlo price of the swaption in the Libor market model (LFM). We consider similar tests for terminal correlations.

More in detail, we first explain what kind of rates we are dealing with, and then move to the volatility part. Section 8.3 gives a plan of the tests on the swaption-volatility approximations and the subsequent section presents results in detail. In particular, we plot, in several cases, the real swap-rate probability density as implied by the LFM dynamics versus a lognormal density characterized by our analytically approximated volatility. We thus measure indirectly the discrepancy between the LFM swap-rate distribution and the lognormal-distribution assumption for the swap rate, as implied instead by the swap market model LSM. The direct measure of this discrepancy has been investigated in the first part of the chapter.

Subsequently, we consider our analytical approximation for terminal correlations, and present our related testing plans in Section 8.5. Again, detailed results follow in the subsequent section. A general section of conclusions on this second part closes the chapter.

8.1 First Part. Tests Based on the Kullback Leibler Information (KLI)

> [...] *This introduces an asymmetry in the evolution of the Universe which resembles to a large extent the asymmetry that appears in the growth of entropy and it might be related to the question of the arrow of time.*
>
> D. S. Goldwirth and T. Piran (1991) Class. Quantum Grav. 8

From now on, until Section 8.2, we consider the first approach via KLI (related to entropy), based on Brigo and Liinev (2002), to which we refer for further details. To proceed, we need to introduce first the KLI itself.

8.1.1 Distance between distributions: The Kullback Leibler information

In this section we introduce briefly the Kullback-Leibler information and we explain its importance for our problem, see also Brigo and Hanzon (1998). Suppose we are given the space H of all the densities of probability measures on the real line equipped with its Borel field, which are absolutely continuous w.r.t. the Lebesgue measure. Then define

$$D(p_1,p_2) := E_{p_1}\{\log p_1 - \log p_2\} \geq 0, \quad p_1,p_2 \in H, \tag{8.1}$$

where in general

$$E_p\{\phi\} = \int \phi(x)p(x)dx, \quad p \in H.$$

The above quantity is the well-known Kullback-Leibler information (KLI). Its non-negativity follows from the Jensen inequality. It gives a measure of how much the density p_2 is displaced w.r.t. the density p_1. We remark the important fact that D is not a distance: in order to be a metric, it should be symmetric and satisfy the triangular inequality, which is not the case.

However, the KLI features many properties of a distance in a generalized geometric setting (see for instance Amari (1985)). For example, it is well-known that the KLI is infinitesimally equivalent to the Fisher information metric around every point of a finite–dimensional manifold of densities such as $EM(c)$ defined below. For this reason, we will refer to the KLI as to a "distance" even if it is not a metric.

Consider a finite dimensional manifold of exponential probability densities such as

$$\begin{aligned} EM(c) &= \{p(\cdot,\theta) : \theta \in \Theta \subset \mathbb{R}^m\}, \quad \Theta \text{ open in } \mathbb{R}^m, \\ p(\cdot,\theta) &= \exp[\theta_1 c_1(\cdot) + \dots + \theta_m c_m(\cdot) - \psi(\theta)], \end{aligned} \tag{8.2}$$

expressed w.r.t the expectation parameters η defined by

$$\eta_i(\theta) = E_{p(\cdot,\theta)}\{c_i\} = \partial_{\theta_i}\psi(\theta), \quad i = 1,..,m \tag{8.3}$$

(see for example Amari (1985), Brigo (1999) or Brigo, Hanzon and Le Gland (1999) for more details on the geometry of exponential families).

We define $p(x;\eta(\theta)) := p(x,\theta)$ (the semicolon/colon notation identifies the parameterization).

Now suppose we are given a density $p \in H$, and we want to approximate it by a density of the finite dimensional manifold $EM(c)$. It seems then reasonable to find a density $p(\cdot,\theta)$ in $EM(c)$ which minimizes the Kullback Leibler information $D(p,.)$. Compute

$$\begin{aligned} \min_\theta D(p,p(\cdot,\theta)) &= \min_\theta\{E_p[\log p - \log p(\cdot,\theta)]\} \\ &= E_p \log p - \max_\theta\{\theta_1 E_p c_1 + \dots + \theta_m E_p c_m - \psi(\theta)\} \\ &= E_p \log p - \max_\theta V(\theta), \\ V(\theta) &:= \theta_1 E_p c_1 + \dots + \theta_m E_p c_m - \psi(\theta). \end{aligned}$$

It follows immediately that a necessary condition for the minimum to be attained at θ^* is

$\partial_{\theta_i} V(\theta^*) = 0, \quad i = 1,...,m$

which yields

$$E_p c_i - \partial_{\theta_i}\psi(\theta^*) = E_p c_i - E_{p(\cdot,\theta^*)} c_i = 0, \qquad i = 1,..,m$$

i.e. $E_p c_i = \eta_i(\theta^*)$, $i = 1,..,m$. This last result indicates that according to the Kullback Leibler information, the best approximation of p in the manifold $EM(c)$ is given by the density of $EM(c)$ which shares the same c_i expectations (c_i-moments) as the given density p. This means that in order to approximate p we only need its c_i moments, $i = 1, 2, .., m$.

The above discussion provides also a way to compute the distance of the density p from the exponential family $EM(c)$ as the distance between p and its projection $p(\cdot,\theta^*)$ onto $EM(c)$ in the KL sense. We have

$$\begin{aligned} D(p, EM(c)) &= E_p \log p \; - (\theta_1^* E_p c_1 + ... + \theta_m^* E_p c_m - \psi(\theta^*)) \\ &= E_p \log p \; - (\theta_1^* \eta_1(\theta^*) + ... + \theta_m^* \eta_m(\theta^*) - \psi(\theta^*)). \end{aligned} \tag{8.4}$$

One can look at the problem from the opposite point of view. Suppose we decide to approximate the density p by taking into account only its m c_i–moments. It can be proved (see Kagan, Linnik, and Rao (1973), Theorem 13.2.1) that the maximum entropy distribution which shares the c–moments with the given p belongs to the family $EM(c)$.

Summarizing: If we decide to approximate by using c–moments, then entropy analysis supplies arguments to use the family $EM(c)$; and if we decide to use the approximating family $EM(c)$, Kullback–Leibler says that the "closest" approximating density in $EM(c)$ shares the c–moments with the given density.

This moments-matching characterization of the projected density for exponential families is the main reason why we resort to the KLI as a "distance" between distributions. Alternatively, we might use the Hellinger distance (HD), which is defined, for two densities $p_1, p_2 \in H$ as

$$H(p_1, p_2) := 2 - 2\int \sqrt{p_1(x)p_2(x)}dx, \tag{8.5}$$

from which we see that the HD takes values in $[0, 2]$ and is a real metric. It is well-known, however, that the KLI is infinitesimally equivalent to the Hellinger distance around every point of a finite–dimensional manifold of densities such as $EM(c)$ defined above. For this reason one refers to the KLI as to a "distance" even if it is not a metric. Indeed, consider the two densities $p(\cdot,\theta)$ and $p(\cdot,\theta + d\theta)$ of $EM(c)$. By expanding in Taylor series, we obtain easily

$$K(p(\cdot,\theta), p(\cdot,\theta + d\theta)) = -\sum_{i=1}^{m} E_{p(\cdot,\theta)}\{\frac{\partial \log p(\cdot,\theta)}{\partial \theta_i}\} d\theta_i$$

$$- \sum_{i,j=1}^{m} E_{p(\cdot,\theta)}\{\frac{\partial^2 \log p(\cdot,\theta)}{\partial\theta_i \partial\theta_j}\} d\theta_i \, d\theta_j + O(|d\theta|^3)$$

which is the same expression we obtain by expanding $H(p(\cdot,\theta),p(\cdot,\theta+d\theta))$. Given this first-order relationship, we expect that the Hellinger distance would lead us to the same results as the KLI, since the KLI distances we will find are rather small.

8.1.2 Distance of the LFM swap rate from the lognormal family of distributions

Since under the swap measure the LSM dynamics for the swap rate follows a (driftless) geometric Brownian motion, we consider here the SDE describing a general geometric Brownian motion

$$dS_t = \mu(t)S_t\,dt + \sigma(t)\,S_t\,dW_t\,,\quad S_0 = s_0$$

whose solution is

$$S_t = s_0 \exp\left[\int_0^t (\mu(u) - \tfrac{1}{2}\sigma^2(u))du + \int_0^t \sigma(u)dW_u\right],$$

so that

$$\log S_t \sim \mathcal{N}\left(\log s_0 + \int_0^t (\mu(u) - \tfrac{1}{2}\sigma^2(u))du, \int_0^t \sigma^2(u)du\right). \tag{8.6}$$

The probability density p_{S_t} of S_t, at any time t, is therefore given by

$$p_{S_t}(x) = p(x,\theta(t)) = \exp\left\{\theta_1(t)\ln\frac{x}{s_0} + \theta_2(t)\ln^2\frac{x}{s_0} - \psi(\theta_1(t),\theta_2(t))\right\},$$

$$\theta_1(t) = \frac{\int_0^t \mu(u)\,du}{\int_0^t \sigma^2(u)du} - \frac{3}{2},\quad \theta_2(t) = -\frac{1}{2\int_0^t \sigma^2(u)du},$$

$$\psi(\theta_1(t),\theta_2(t)) = -\frac{(\theta_1(t)+1)^2}{4\theta_2(t)} + \tfrac{1}{2}\ln\left(\frac{-\pi\,s_0^2}{\theta_2(t)}\right),$$

where $x > 0$, and is clearly in the exponential class, with $c_1(x) = \ln(x/s_0)$, $c_2 = c_1^2$. We will denote by $\mathcal{L}$ the related exponential family $EM(c)$. As concerns the expectation parameters for this family, they are readily computed as follows:

$$\eta_1 = E_\theta \ln(x/s_0) = \partial_{\theta_1}\psi(\theta_1,\theta_2) = -\frac{\theta_1+1}{2\theta_2}$$

$$\eta_2 = E_\theta \ln^2(x/s_0) = \partial_{\theta_2}\psi(\theta_1,\theta_2) = \left(\frac{\theta_1+1}{2\theta_2}\right)^2 - \frac{1}{2\theta_2}.$$

As for the Gaussian family, in this particular family the θ parameters can be computed back from the η parameters by inverting the above formulae:

$$\theta_1 = \frac{\eta_1}{\eta_2 - \eta_1^2} - 1 , \tag{8.7}$$

$$\theta_2 = -\frac{1}{2(\eta_2 - \eta_1^2)} ,$$

$$\psi(\theta_1, \theta_2) = \frac{1}{2}\left[\frac{\eta_1^2}{\eta_2 - \eta_1^2} + \ln(2\pi(\eta_2 - \eta_1^2)s_0^2)\right] .$$

We can now compute the distance of a density p from the lognormal family $\mathcal{L}$ by applying formula (8.4):

$$D(p, \mathcal{L}) = E_p \ln p \; - (\theta_1^* \eta_1(\theta^*) + \theta_2^* \eta_2(\theta^*) - \psi(\theta^*)),$$

where, as previously seen, minimizing the distance implies finding the parameters θ^* such that

$$\eta_1(\theta^*) = E_p \ln(x/s_0), \quad \eta_2(\theta^*) = E_p \ln^2(x/s_0) .$$

By substituting (8.7), omitting the argument θ^* and simplifying, we obtain

$$D(p, \mathcal{L}) = E_p \ln p \; + \tfrac{1}{2} + \eta_1 + \tfrac{1}{2}\ln(2\pi(\eta_2 - \eta_1^2)s_0^2) .$$

Actually, the s_0 term is kind of redundant when computing the distance. We can thus resort to the simpler moments

$$\bar{\eta}_1(\bar{\theta}^*) = E_p \ln(x), \quad \bar{\eta}_2(\bar{\theta}^*) = E_p \ln^2(x) ,$$

and compute the distance as

$$D(p, \mathcal{L}) = E_p \ln p \; + \tfrac{1}{2} + \bar{\eta}_1 + \tfrac{1}{2}\ln(2\pi(\bar{\eta}_2 - \bar{\eta}_1{}^2)) . \tag{8.8}$$

As noticed before, the LSM swap–rate density under the swap measure belongs to the family $\mathcal{L}$: Such density is given by $p(\cdot, \theta(T_\alpha))$ above when taking $s_0 = S_{\alpha,\beta}(0)$, $\mu(t) = 0$ and $\sigma(t) = \sigma_{\alpha,\beta}(t)$.

Now consider instead the LFM swap rate under the swap measure, obtained once again through (6.33) and (6.40). This second swap rate will not be lognormally distributed. Let $p_{\alpha,\beta}$ denote the probability density of the LFM swap rate $S_{\alpha,\beta}(T_\alpha)$ under the swap measure $Q^{\alpha,\beta}$.

We plan to compute numerically the distance of the LFM swap density $p_{\alpha,\beta}$ from the lognormal exponential family $\mathcal{L}$ where the LSM swap density lies. But we also plan to verify the volatility–approximation provided by the quantity $v_{\alpha,\beta}^{\text{LFM}}$ of formula (6.67) as follows: The density $p(\cdot, \theta^*) = p(\cdot; \eta(\theta^*))$ represents the lognormal density which is closest (in the Kullback-Leibler sense) to the swap-rate density $p_{\alpha,\beta}$ implied by the LFM under the swap measure. Incidentally, we can compute the terminal volatility implied by this lognormal density as

$$v_{\alpha,\beta}^{\text{KLI}} = \sqrt{\eta_2(\theta^*) - \eta_1(\theta^*)^2} = \sqrt{\bar{\eta}_2 - \bar{\eta}_1^2} .$$

This is the best approximation of the volatility of the swap rate based on the lognormal approximation. It can be interesting to compare such best approximation to the much handier approximation $v^{\text{LFM}}_{\alpha,\beta}$ considered earlier in (6.67).

Recall that $v^{\text{LFM}}_{\alpha,\beta}$ is obtained by "ignore the drifts" and "freeze stochastic coefficients" arguments, whereas $v^{\text{KLI}}_{\alpha,\beta}$ is obtained by minimizing the distance from the lognormal densities. Should the two results be close, this would represent a further confirmation of the validity of the industry formula $v^{\text{LFM}}_{\alpha,\beta}$.

Now we proceed by applying formula (8.8) according to the following scheme:

1. Simulate p realizations of the forward rates

$$F_{\alpha+1}(T_\alpha), F_{\alpha+2}(T_\alpha), \dots, F_\beta(T_\alpha)$$

   under the swap measure $Q^{\alpha,\beta}$ through the discretized dynamics (6.53) with a sufficiently small time step;
2. Compute p realizations of the swap rate $S_{\alpha,\beta}(T_\alpha)$ of the LFM under the swap measure $Q^{\alpha,\beta}$ through (6.33) applied to each realization of the forward rates F vector obtained in the previous point;
3. Based on the simulated $S_{\alpha,\beta}(T_\alpha)$, compute firstly

$$\bar{\eta}_1 = E^{\alpha,\beta} \ln(S_{\alpha,\beta}(T_\alpha)), \quad \bar{\eta}_2 = E^{\alpha,\beta} \ln^2(S_{\alpha,\beta}(T_\alpha)) \, , v^{\text{KLI}}_{\alpha,\beta} = \sqrt{\bar{\eta}_2 - \bar{\eta}_1{}^2} \, ,$$

   and secondly an approximation of

$$E_p \ln p := \int (\ln p_{\alpha,\beta}(x)) p_{\alpha,\beta}(x) dx \, .$$

   This quantity is the opposite of entropy and can be estimated from the simulated $S_{\alpha,\beta}(T_\alpha)$'s through an entropy estimator. For example, Vasicek's (1976) estimator reads, in our case,

$$H_V(q,p) = -\frac{1}{p} \sum_{i=1}^{p} \ln \left[\frac{p}{2q} (S_{\alpha,\beta}(T_\alpha)_{[i+q]} - S_{\alpha,\beta}(T_\alpha)_{[i-q]}) \right] ,$$

   where $S_{\alpha,\beta}(T_\alpha)_{[j]}$ is the j-th order statistics from our sample. We set $S_{\alpha,\beta}(T_\alpha)_{[j]} = S_{\alpha,\beta}(T_\alpha)_{[1]}$ for $j < 1$ and $S_{\alpha,\beta}(T_\alpha)_{[j]} = S_{\alpha,\beta}(T_\alpha)_{[p]}$ for $j > p$.
   This estimator converges to the desired integral in probability as

$$p \to \infty, \quad q \to \infty, \quad q/p \to 0.$$

   However, in our numerical simulations we considered a different type of entropy estimator. We used the plug-in estimate of entropy based on a cross-validation density estimate, proposed by Ivanov and Rozhkova (1981). For an overview of entropy estimators see for example Dudewicz and van der Meulen (1987). The estimator we used can be briefly summarized as follows. Let $S^1_{\alpha,\beta}(T_\alpha), \ldots, S^p_{\alpha,\beta}(T_\alpha)$ be i.i.d. sample of swap rates with unknown probability density function $p_{\alpha,\beta}(x)$, and consider

$$p^{p,i}_{\alpha,\beta}(S^i_{\alpha,\beta}(T_\alpha)) = \frac{1}{pa_p} \sum_{j \neq i} K\left(\frac{S^i_{\alpha,\beta}(T_\alpha) - S^j_{\alpha,\beta}(T_\alpha)}{a_p}\right),$$

where $\{a_p\}$ satisfies the condition that $a_p \to 0$, $pa_p \to \infty$, and K is a kernel function. Note that we used Gaussian kernel in our computations (this choice was based on smoothing experiments with several kernels, and a Gaussian kernel was more precise in capturing the shape of the underlying data histogram). The estimator of Ivanov and Rozhkova (1981) can be written in the following form:

$$H_{IR}(p) = -\frac{1}{p} \sum_{i=1}^{p} \left\{ \ln p^{p,i}_{\alpha,\beta}(S^i_{\alpha,\beta}(T_\alpha)) \right\} I_{[S^i_{\alpha,\beta}(T_\alpha) \in A_p]}, \qquad (8.9)$$

where with the set A_p one typically excludes the small and the tail values of $p^{p,i}_{\alpha,\beta}(S^i_{\alpha,\beta}(T_\alpha))$. Ivanov and Rozhkova (1981) showed that under certain conditions on K, $p_{\alpha,\beta}$, a_p and A_p, $H_{IR}(p)$ converges with probability 1 to the desired integral as $p \to \infty$. We also tried some alternative estimators, such as (4) and (6) in Miller (2003). This did not change our numerical results significantly.

4. With the quantities obtained from the previous point apply formula (8.8) and obtain $D(p_{\alpha,\beta}, \mathcal{L})$.
5. Compute $v^{\text{LFM}}_{\alpha,\beta}$ through formula (6.67) and compare this to $v^{\text{KLI}}_{\alpha,\beta}$ obtained above.

It is interesting to plot $D(p_{\alpha,\beta}, \mathcal{L})$ and the difference $|v^{\text{KLI}}_{\alpha,\beta} - v^{\text{LFM}}_{\alpha,\beta}|$, as α and β change, and also analyze these quantities in the different formulations of instantaneous volatilities and correlations. For each formulation, which are the parameters to which the distance is more sensitive?

8.1.3 Monte Carlo tests for measuring KLI

In this section we numerically test how the KLI "distance" between a simulated swap rate density and the family of lognormal densities changes for different parameterizations. We will consider a family of forward rates whose expiry/maturity pairs are $T_0 = 1y$, $T_1 = 2y$ up to $T_{18} = 19y$, and our initial input is the forward rate vector $F = [F_{0-9}, F_{10-19}]$:

$$F_{0-9}(0) = [4.69 \;\; 5.01 \;\; 5.60 \;\; 5.84 \;\; 6.00 \;\; 6.13 \;\; 6.28 \;\; 6.27 \;\; 6.29 \;\; 6.23]/100$$

$$F_{10-19}(0) = [6.30 \;\; 6.36 \;\; 6.43 \;\; 6.48 \;\; 6.53 \;\; 6.40 \;\; 6.30 \;\; 6.18 \;\; 6.07 \;\; 5.94]/100,$$

$F_0(0)$ being the initial one-year spot rate. These values are consistent with the volatility and correlation values below, in that all such initial inputs will reflect a possible calibration of the LFM to caps and swaptions. The corresponding swap rates we consider are $S_{5,10}(0) = 0.06238$, $S_{10,15}(0) = 0.06411$ $S_{15,20}(0) = 0.06191$, $S_{5,15}(0) = 0.06312$, $S_{10,20}(0) = 0.06318$, $S_{5,20}(0) = 0.06283$. We adopt the linear-exponential (LE) formulation (6.13) (i.e. the "Formulation 7" of Chapter 6) for instantaneous volatilities, which we report here:

FORMULATION 7

$$\boxed{\sigma_i(t) = \Phi_i\, \psi(T_{i-1} - t; a, b, c, d) := \Phi_i \left([a(T_{i-1} - t) + d]e^{-b(T_{i-1}-t)} + c \right)} \tag{8.10}$$

In some cases we will let it collapse to the THPC (time-homogeneous piecewise constant) formulation, given by

THPC

$$\sigma_i(t) = \Phi_i \tag{8.11}$$

For instantaneous correlations we resort to the angle form (6.48), which we report below:

$$\rho_{i,j} = \cos(\theta_i - \theta_j) \tag{8.12}$$

in the parameters $\theta = [\theta_1 \ldots \theta_{19}]$. The values of the parameters a, b, c, d and the values of the θ's in the general case of the full LE formulation have been built so as to reflect possible joint calibrations of the LFM to caplets and swaptions. Such values are reported in points (2.a–c) below. For a discussion on which forms are to be preferred from the point of view of realistic behaviour of future volatilty structures (typically the forms of cases (2.a–c) below) see the discussion on the term structure of volatilties given in Chapter 6. A short map of our sets of testing parameters is given in the following.

(1.a): Constant instantaneous (THPC) volatilities, typical rank-two correlations

LE formulation with $a = 0,\ b = 0,\ c = 1,\ d = 0$. This is actually the THPC formulation, since the ψ-part of the LE formulation is collapsed to one. We set

$$\begin{aligned}
\Phi_{1-9} &= [0.1490\, 0.1589\, 0.1533\, 0.1445\, 0.1356\, 0.1267\, 0.1215\, 0.1176\, 0.1138],\\
\Phi_{10-17} &= [0.1106\, 0.1076\, 0.1046\, 0.1017\, 0.0989\, 0.0978\, 0.0974\, 0.0969\, 0.0965],\\
\Phi_{19} &= 0.0961.
\end{aligned}$$

The correlation angles are taken as

$$\theta_{1-8} = [0.0147\ 0.0643\ 0.1032\ 0.1502\ 0.1969\ 0.2239\ 0.2771\ 0.2950],$$
$$\theta_{9-16} = [0.3630\ 0.3810\ 0.4217\ 0.4836\ 0.5204\ 0.5418\ 0.5791\ 0.6496]$$
$$\theta_{17-19} = [0.6679\ 0.7126\ 0.7659].$$

This set of angles implies positive and decreasing instantaneous correlations when moving away from the "1" diagonal entries along columns.

(1.b): Constant instantaneous (THPC) volatilities, perfect correlation

LE formulation with a, b, c, d and Φ's as in (1.a) and $\theta = [\ 0\ 0\ \dots\ 0\ 0\ ]$, implying that all instantaneous correlations are set to one.

(1.c): Constant inst. (THPC) volatilities, some negative rank-two correlations

LE formulation with a, b, c, d and Φ's as in (1.a) and $\theta = [\ \theta_{1-9},\ \theta_{10-19}\ ]$, where

$$\theta_{1-9} = [\ 0\ 0.0000\ 0.0013\ 0.0044\ 0.0096\ 0.0178\ 0.0299\ 0.0474\ 0.0728\ ],$$
$$\theta_{10-18} = [\ 0.1100\ 0.1659\ 0.2534\ 0.3989\ 0.6565\ 1.1025\ 1.6605\ 2.0703\ 2.2825\ ]$$
$$\theta_{19} = 2.2260.$$

These parameters θ imply some negative correlations while maintaining a decreasing correlation pattern when moving away from the diagonal in the resulting correlation matrix.

(2.a): Humped and expiry-adjusted (LE) instantaneous volatilities depending only on time to expiry, typical rank-two correlations

LE formulation with $a = 0.1908$, $b = 0.9746$, $c = 0.0808$, $d = 0.0134$ and θ as in (1). This is the "most normal" situation, in that it reflects a joint calibration to caplets and swaptions volatilities. The parameters Φ's are, in our case:

$$\Phi_{1-9} = [\,1.0500\ 1.0900\ 1.1025\ 1.1025\ 1.0913\ 1.0669\ 1.0624\ 1.0611\ 1.0544\,],$$
$$\Phi_{10-19} = [\,1.0475\ 1.0386\ 1.0270\ 1.0132\ 0.9975\ 0.9979\ 1.0033\ 1.0079\ 1.0119\ 1.0152\,].$$

(2.b): Humped and maturity-adjusted (LE) instantaneous volatilities depending only on time to expiry, perfect correlation

LE formulation with a, b, c, d and Φ's as in (2.a) and $\theta = [\ 0\ 0\ \dots\ 0\ 0\ ]$.

(2.c): Humped and maturity-adjusted (LE) instantaneous volatilities depending only on time to expiry, some negative rank-two correlations

LE formulation with a, b, c, d and Φ's as in (2.a) and θ as in (1.c).

After the test summary, we now present the results obtained in evaluating the KLI for swap rates through the Monte Carlo method with antithetic shocks and with 2×100000 paths.

In the tables, we denote by *KLI* the estimated value of (8.8) and by *absdiff* the absolute differences $|v^{\text{KLI}}_{\alpha,\beta} - v^{\text{LFM}}_{\alpha,\beta}|$. First we would like to have a feeling for what it means to have a KLI distance of 0.006 between two distributions. We may resort to the KLI distance of two lognormals, which is easily computed analytically. Indeed, if we call θ_1, θ_2 the parameters of the first lognormal density (with corresponding expectation parameters η_1 and η_2) and $\hat{\theta}_1, \hat{\theta}_2$ the parameters of the second lognormal density (with corresponding expectation parameters $\hat{\eta}_1$ and $\hat{\eta}_2$), it is easy to compute the KLI distance as $(\hat{\theta}_1 - \theta_1)\hat{\eta}_1 + (\hat{\theta}_2 - \theta_2)\hat{\eta}_2 + \psi(\theta) - \psi(\hat{\theta})$. If we take for example $\eta_1 = \hat{\eta}_1 = 0.06$ (same mean) and then $\eta_2 = 0.04, \hat{\eta}_2 = 0.0404$, corresponding to $\theta_1 = 0.6484$, $\theta_2 = -27.4725$, $\hat{\theta}_1 = 0.6304$, $\hat{\theta}_2 = -27.1739$, we find a KLI distance of 0.00606, comparable in size to our distances above. Compute the standard deviations (volatilities) of the two distributions according to $\sqrt{\eta_2 - \eta_1^2}$. One has $\sqrt{\eta_2 - \eta_1^2} = 0.1908$, $\sqrt{\hat{\eta}_2 - \hat{\eta}_1^2} = 0.1918$. Recall that the two lognormal densities have the same mean of 0.06. Therefore, in a lognormal world with the mean fixed at 0.06, a KLI distance of 0.006 would amount to an absolute difference in volatility of about 0.001 for volatilities ranging around 0.19. This amounts to a percentage difference in standard deviations of 0.55%. This gives a feeling for the size of the distributional discrepancy our distances imply in the *worst* case we obtain from our simulations.

Before presenting our final results, a remark is in order on the accuracy of the KLI "distances" obtained by simulation. Typically, the standard error ranges about $3E-5$ with 200000 paths for the first distance 0.0001857 in Table (8.1), and for all other cases varies roughly in the same proportion with respect to the distance, so that for example the distance of 0.0068614 of Table (8.3) has a standard error of about 1E-3, and so on. Every time the standard error makes the distance pattern uncertain we add a question mark in the summary tables.

Fortunately, this happens only in five of the eighteen cases we analyzed, namely (i) in case (1.a) with $\uparrow \beta$, (ii) in case (1.c) with $\uparrow (\alpha, \beta)$, (iii) in case (2.a) with $\uparrow (\alpha, \beta)$, (iv) in case (2.a) with $\uparrow \beta$ and (v) in case (2.c) with $\uparrow (\alpha, \beta)$.

Now let us consider our test results, starting from Table 8.1 for case (1.a).

Swap	*KLI*	$v^{\text{KLI}}_{\alpha,\beta}/\sqrt{T_\alpha}$	$v^{\text{LFM}}_{\alpha,\beta}/\sqrt{T_\alpha}$	*absdiff*
$S_{5,10}(5)$	0.0001857	0.12376	0.12360	0.00016
$S_{10,15}(10)$	0.0000357	0.10516	0.10512	0.00004
$S_{15,20}(15)$	0.0000638	0.09705	0.09682	0.00023
$S_{5,10}(5)$	0.0001857	0.12376	0.12360	0.00016
$S_{5,15}(5)$	0.0002088	0.11466	0.11509	0.00043
$S_{5,20}(5)$	0.0003604	0.10951	0.10985	0.00034
$S_{5,20}(5)$	0.0003604	0.10951	0.10985	0.00034
$S_{10,20}(10)$	0.0001105	0.10150	0.10116	0.00034
$S_{15,20}(15)$	0.0000638	0.09705	0.09682	0.00023

Table 8.1. Results for case (1.a)

First consider distances from the lognormal family for $S_{5,10}(5), S_{10,15}(10)$, and $S_{15,20}(15)$ respectively (increasing maturity T_α, constant tenor $T_\beta - T_\alpha$, denoted "$\uparrow (\alpha, \beta)$"). The distance first decreases and then increases, displaying a "V" shape when plotted for example against α. This is interesting and might be due to the shape of the instantaneous volatility functions $\sigma(t)$ in this formulation. If not for particular shapes of volatilities, one would expect instead the distance to increase with the maturity, since the more the "non-lognormal dynamics" goes on, the more it is likely that one moves away from the lognormal distribution. Then consider $S_{5,10}(5), S_{5,15}(5)$, and $S_{5,20}(5)$ (increasing tenor, constant maturity, denoted "$\uparrow \beta$"). The distance increases this time, as expected, although the first two distances differ less than the standard error, so that we may not exclude a "V" shape a priori. Anyway, an increasing pattern is to be expected: if the tenor increases we are adding more forward rates to form the swap rate, and intuitively we move farther away from the lognormal family. Consider also $S_{5,20}(5), S_{10,20}(10)$, and $S_{15,20}(15)$ (increasing maturity, decreasing tenor, denoted "$\uparrow \alpha$"). The distance decreases, as is partly expected by the fact that at the final time we are adding less forward rates, even though the dynamics is propagated for longer times. Finally consider $S_{10,15}(10)$ and $S_{10,20}(10)$ (again increasing tenor, constant maturity). This time the distance increases, as expected.

At this point, as from (1.b), we set all correlations to one and recompute all distances. Table 8.2 shows our results in this case. The only pattern that has changed qualitatively concerns increasing maturity and decreasing tenor, denoted "$\uparrow \alpha$", which is now V-shaped instead of decreasing as before. Enlarging instantaneous correlation between far rates has caused the distance to increase with maturity in the final step.

Lowering now correlations again, including some negative entries in the correlation matrix, according to (1.c), gives us the results of Table 8.3. We see that negative correlations give the same qualitative results as positive correlations in (1.a), with the exception of the first pattern "$\uparrow (\alpha, \beta)$" which is now decreasing or humped (uncertainty coming from the standard error size). At this point, with negative correlation, this pattern is more counter-

Swap	*KLI*	$v^{\text{KLI}}_{\alpha,\beta}/\sqrt{T_\alpha}$	$v^{\text{LFM}}_{\alpha,\beta}/\sqrt{T_\alpha}$	*absdiff*
$S_{5,10}(5)$	0.0002073	0.12405	0.12380	0.00025
$S_{10,15}(10)$	0.0000178	0.10504	0.10530	0.00026
$S_{15,20}(15)$	0.0001339	0.09741	0.09701	0.00040
$S_{5,10}(5)$	0.0002073	0.12405	0.12380	0.00025
$S_{5,15}(5)$	0.0002814	0.11552	0.11583	0.00031
$S_{5,20}(5)$	0.0003917	0.11124	0.11143	0.00019
$S_{5,20}(5)$	0.0003917	0.11124	0.11143	0.00019
$S_{10,20}(10)$	0.0000370	0.10220	0.10186	0.00034
$S_{15,20}(15)$	0.0001339	0.09741	0.09701	0.00040

Table 8.2. Results for case (1.b)

intuitive. The only reason for a decreasing distance of the swap rate distribution from the lognormal family when propagating a "non lognormal dynamics" for longer times and with the same tenor is given by lower $\sigma(\cdot)$ functions. Indeed, with the volatility formulation of cases (1.a)–(1.c) the term structure of future volatilities decreases considerably in time, so that farther forward rates have much lower volatilities than forward rates involved in "earlier" swap rates. For example, the swap rate $S_{15,20}$ involves F_{19}, whose volatility in (1.a)–(1.c) is set to 0.0961, whereas earlier swap rates may involve F_2, whose volatility is much higher and set to 0.1589. So even if the dynamics propagates for longer times, it does so with lower randomness, and this effect dominates the other one.

Swap	*KLI*	$v^{\text{KLI}}_{\alpha,\beta}/\sqrt{T_\alpha}$	$v^{\text{LFM}}_{\alpha,\beta}/\sqrt{T_\alpha}$	*absdiff*
$S_{5,10}(5)$	0.0001589	0.12402	0.12377	0.00025
$S_{10,15}(10)$	0.0001126	0.10348	0.10343	0.00005
$S_{15,20}(15)$	0.0000609	0.08767	0.08735	0.00032
$S_{5,10}(5)$	0.0001589	0.12402	0.12377	0.00025
$S_{5,15}(5)$	0.0002497	0.11370	0.11407	0.00037
$S_{5,20}(5)$	0.0068614	0.08629	0.08720	0.00091
$S_{5,20}(5)$	0.0068614	0.08629	0.08720	0.00091
$S_{10,20}(10)$	0.0023644	0.07192	0.07161	0.00031
$S_{15,20}(15)$	0.0000609	0.08767	0.08735	0.00032

Table 8.3. Results for case (1.c)

As a summary of patterns for the cases with constant instantaneous volatilities we display in Table 8.4 the behavior of the distance for different correlation configurations (1.a)–(1.c).

Now let us move to commenting our results for cases (2.a)–(2.c), given in Tables 8.5, 8.6, and 8.7.

These results are qualitatively analogous to the results of the corresponding cases (1.a)–(1.c), with one strong exception and two weaker exceptions.

Action	Positive correlations	Perfect correlations	Some negative correl.
$\uparrow (\alpha, \beta)$	V shaped	V shaped	decreas. (humped?)
$\uparrow \beta$	increas. (V-shaped?)	increasing	increasing
$\uparrow \alpha$	decreasing	V shaped	decreasing

Table 8.4. Distance patterns against (α, β), β and α respectively, for cases (1.a), (1.b), (1.c)

Swap	*KLI*	$v^{\mathrm{KLI}}_{\alpha,\beta}/\sqrt{T_\alpha}$	$v^{\mathrm{LFM}}_{\alpha,\beta}/\sqrt{T_\alpha}$	*absdiff*
$S_{5,10}(5)$	0.00016634	0.11033	0.11017	0.00016
$S_{10,15}(10)$	0.00017273	0.09541	0.09534	0.00007
$S_{15,20}(15)$	0.00009037	0.08976	0.08969	0.00007
$S_{5,10}(5)$	0.00016634	0.11033	0.11017	0.00016
$S_{5,15}(5)$	0.00041173	0.09778	0.09803	0.00025
$S_{5,20}(5)$	0.00047629	0.09306	0.09320	0.00014
$S_{5,20}(5)$	0.00047629	0.09306	0.09320	0.00014
$S_{10,20}(10)$	0.00023379	0.08921	0.08895	0.00026
$S_{15,20}(15)$	0.00009037	0.08976	0.08969	0.00007

Table 8.5. Results for case (2.a)

Swap	*KLI*	$v^{\mathrm{KLI}}_{\alpha,\beta}/\sqrt{T_\alpha}$	$v^{\mathrm{LFM}}_{\alpha,\beta}/\sqrt{T_\alpha}$	*absdiff*
$S_{5,10}(5)$	0.00021961	0.11052	0.11035	0.00017
$S_{10,15}(10)$	0.00007206	0.09554	0.09552	0.00002
$S_{15,20}(15)$	0.00015586	0.09003	0.08987	0.00016
$S_{5,10}(5)$	0.00021961	0.11052	0.11035	0.00017
$S_{5,15}(5)$	0.00044884	0.09848	0.09867	0.00019
$S_{5,20}(5)$	0.00057553	0.09453	0.09457	0.00004
$S_{5,20}(5)$	0.00057553	0.09453	0.09457	0.00004
$S_{10,20}(10)$	0.00011407	0.08982	0.08957	0.00025
$S_{15,20}(15)$	0.00015586	0.09003	0.08987	0.00016

Table 8.6. Results for case (2.b)

The strong exception concerns the pattern "$\uparrow (\alpha, \beta)$" for the typical rank two correlations (case (2.a)), where we have an opposite humped pattern with respect to the earlier V-shaped case. This is due to the different volatility structure, that is now homogeneous with respect to time-to-maturity. Consider, however, that in this case we have uncertainty in the pattern due to the standard error, and that in fact the pattern in (2.a) could be decreasing.

The first weaker exception concerns the pattern "$\uparrow (\alpha, \beta)$" for the case with some negative correlations (case (2.c)), where we have a humped pattern instead of a decreasing one, although both patterns are uncertain, with the possibility of the patterns coinciding in a decreasing or humped configuration. The second possible weaker exception is for $\uparrow \beta$ with positive correlation (case (2.a)), where the two patterns coincide unless the standard error changes them in two opposite configurations.

Swap	*KLI*	$v^{\mathrm{KLI}}_{\alpha,\beta}/\sqrt{T_\alpha}$	$v^{\mathrm{LFM}}_{\alpha,\beta}/\sqrt{T_\alpha}$	*absdiff*
$S_{5,10}(5)$	0.00017029	0.11050	0.11032	0.00018
$S_{10,15}(10)$	0.00017182	0.09392	0.09388	0.00004
$S_{15,20}(15)$	0.00008607	0.08101	0.08078	0.00023
$S_{5,10}(5)$	0.00017029	0.11050	0.11032	0.00018
$S_{5,15}(5)$	0.00046520	0.09670	0.09721	0.00051
$S_{5,20}(5)$	0.00635040	0.07363	0.07409	0.00046
$S_{5,20}(5)$	0.00635040	0.07363	0.07409	0.00046
$S_{10,20}(10)$	0.00265217	0.06370	0.06344	0.00026
$S_{15,20}(15)$	0.00008607	0.08101	0.08078	0.00023

Table 8.7. Results for case (2.c)

As a summary we display in Table 8.8 the behavior of the distance for different correlation configurations in cases (2).

Action	Positive correlation	Perfect correlation	Some negative correl.
$\uparrow (\alpha, \beta)$	humped (decreasing?)	V shaped	humped (decreasing?)
$\uparrow \beta$	increasing (humped?)	increasing	increasing
$\uparrow \alpha$	decreasing	V shaped	decreasing

Table 8.8. Distance patterns against (α, β), β and α respectively, for cases (2.a), (2.b), (2.c)

In closing, we notice that *absdiff* is always small, meaning that the industry approximation $v^{\mathrm{LFM}}_{\alpha,\beta}$ is good since it is always close to $v^{\mathrm{KLI}}_{\alpha,\beta}$, i.e. to the best one can do with a lognormal family.

8.1.4 Conclusions on the KLI-based approach

Our KLI analysis confirms that swap rates associated with the LIBOR market model are close to being log normal. This has been checked via a distributional distance obtained through Monte Carlo simulation. Our analysis also confirms the goodness of the standard market approximation for swaption volatilities in the LIBOR market model, based on freezing the drift in the forward rate dynamics. In the following we proceed with more classical tests and abandon the geometric framework by Brigo and Liinev (2002).

8.2 Second Part: Classical Tests

Now we abandon the rigorous geometric framework of Brigo and Liinev (2002) based on distances on the spaces of distributions. We move to the more classical tests appeared in the first edition of this same book (Brigo and Mercurio (2001c)).

The Specification of Rates

In our tests, both on instantaneous volatilities and on terminal correlations, we will consider a family of forward rates whose expiry/maturity pairs are adjacent elements in the array

$$0y \;\; 1y \;\; 2y \;\; \dots \;\; 19y \;\; 20y.$$

We thus take as resetting times (expires) of the rates in our family, $T_0 = 1y$, $T_1 = 2y$ up to $T_{18} = 19y$, meaning that we take as initial input the forward rates

$$F_0(0),\; F_1(0),\; F_2(0),\; \dots,\; F_{19}(0)\,,$$

$F_0(0) = L(0, 1y)$ being the initial spot rate.

The chosen values for these initial rates will always be

$$F_{0-9}(0) = [4.69 \;\; 5.01 \;\; 5.60 \;\; 5.84 \;\; 6.00 \;\; 6.13 \;\; 6.28 \;\; 6.27 \;\; 6.29 \;\; 6.23]$$
$$F_{10-19}(0) = [6.30 \;\; 6.36 \;\; 6.43 \;\; 6.48 \;\; 6.53 \;\; 6.40 \;\; 6.30 \;\; 6.18 \;\; 6.07 \;\; 5.94],$$

where all rates are expressed as percentages. These values are consistent with the volatility and correlation values to be illustrated later on, in that all such initial inputs reflect a possible calibration of the LFM to caps and swaptions.

At times, we will stress the initial forward rates by uniformly shifting upwards all their values by 2%, thus taking as initial rates

$$F_{0-9}(0) = [6.69 \;\; 7.01 \;\; 7.60 \;\; 7.84 \;\; 8.00 \;\; 8.13 \;\; 8.28 \;\; 8.27 \;\; 8.29 \;\; 8.23]$$
$$F_{10-19}(0) = [8.30 \;\; 8.36 \;\; 8.43 \;\; 8.48 \;\; 8.53 \;\; 8.40 \;\; 8.30 \;\; 8.18 \;\; 8.07 \;\; 7.94].$$

8.3 The "Testing Plan" for Volatilities

If the art of the detective began and ended in reasoning from an arm-chair, my brother would be the greatest criminal agent that ever lived.
Sherlock Holmes

We plan to test Rebonato's and Hull and White's formulas, (6.67) and (6.68), against a Monte Carlo evaluation as follows.

We adopt Formulation 7 for instantaneous volatilities, although in some cases we will let it collapse to Formulation 6, or to the formulations of TABLEs 2 or 3 respectively.

We recall the linear-exponential Formulation 7:

$$\sigma_i(t) = \Phi_i \left([a(T_{i-1} - t) + d] \, e^{-b(T_{i-1}-t)} + c \right),$$

where, as usual, T_{i-1} is the expiry of the relevant forward rate. We will often consider restrictions on the possible parameters values.

Recall the two-factor parametric form in θ for instantaneous correlations,

$$\rho^B_{i,j} = \cos(\theta_i - \theta_j), \qquad \theta = [\, \theta_1 \, \dots \, \theta_{19} \,].$$

The values of the parameters a, b, c, d and the values of the θ's, in the general case of Formulation 7, have been built so as to reflect possible joint calibrations of the LFM to caps and swaptions. Such values are reported in point (3.a) below.

A short map of our sets of testing parameters is given in the following.

- (1.a) *Constant instantaneous volatilities, typical rank-two correlations.* Formulation 7 with $a = 0, \; b = 0, \; c = 1, \; d = 0$. This amounts actually to Formulation 3, since the ψ part of Formulation 7 is collapsed to one and the Φ's act as the constant instantaneous volatilities s of TABLE 3, which are set to the caplet volatilities taken in input. The correlation angles are $\theta = [\theta_{1-8} \; \theta_{9-16} \; \theta_{17-19}]$, where

$$\theta_{1-8} = [0.0147 \; 0.0643 \; 0.1032 \; 0.1502 \; 0.1969 \; 0.2239 \; 0.2771 \; 0.2950],$$
$$\theta_{9-16} = [0.3630 \; 0.3810 \; 0.4217 \; 0.4836 \; 0.5204 \; 0.5418 \; 0.5791 \; 0.6496]$$
$$\theta_{17-19} = [0.6679 \; 0.7126 \; 0.7659].$$

  This set of angles implies positive and decreasing instantaneous correlations as from Figure 8.1. In the same figure, the values of the Φ's are also shown.
- (1.b) *Constant instantaneous volatilities, perfect correlation.* Formulation 7 with a, b, c and d as in (1.a) and $\theta = [0 \; 0 \; \dots \; 0 \; 0]$, implying that all instantaneous correlations are set to one.
- (1.c) *Constant instantaneous volatilities, some negative rank-two correlations.* Formulation 7 with a, b, c and d as in (1.a) and $\theta = [\theta_{1-9} \; \theta_{10-17} \; \theta_{18,19}]$, where

$$\theta_{1-9} = [0 \; 0.0000 \; 0.0013 \; 0.0044 \; 0.0096 \; 0.0178 \; 0.0299 \; 0.0474 \; 0.0728],$$
$$\theta_{10-17} = [0.1100 \; 0.1659 \; 0.2534 \; 0.3989 \; 0.6565 \; 1.1025 \; 1.6605 \; 2.0703]$$
$$\theta_{18,19} = [2.2825 \; 2.2260].$$

  This set of angles implies instantaneous correlations as from Figure 8.6.
- (2.a) *Humped instantaneous volatilities depending only on time to maturity, perfect correlation.* Formulation 6 with $a = 0.1908, \; b = 0.9746, \; c = 0.0808, \; d = 0.0134$ and $\theta = [0 \; 0 \; \dots \; 0 \; 0]$.

- (2.b) *Humped instantaneous volatilities depending only on time to maturity, some negative rank-two correlations.*
  Formulation 6 with a, b, c and d as in (2.a) and θ as in (1.c).
- (3.a) *Humped and maturity-adjusted instantaneous volatilities depending only on time to maturity, typical rank-two correlations.*
  Formulation 7 with $a = 0.1908,\ b = 0.9746,\ c = 0.0808,\ d = 0.0134$ as in (2.a) and θ as in (1.a). This is the "most normal" situation, in that it reflects a qualitatively acceptable evolution of the term structure of volatilities in time and allows for a satisfactory joint calibration to caplets and swaptions volatilities.
- (3.b) *Humped and maturity-adjusted instantaneous volatilities depending only on time to maturity, perfect correlation.*
  Formulation 7 with a, b, c and d as in (3.a) and $\theta = [0\ 0\ \ldots\ 0\ 0]$.
- (3.c) *Humped and maturity-adjusted instantaneous volatilities depending only on time to maturity, some negative rank-two correlations.*
  Formulation 7 with a, b, c and d as in (3.a) and θ as in (1.c).

In each of the above situations, we will display some numerical results and also some plots of probability densities.

The results are based on a comparison of Rebonato's and Hull and White's formulas with the volatilities that plugged into Black's formula lead to the Monte Carlo prices of the corresponding at-the-money swaptions. Indeed, we price a chosen swaption with underlying swap $S_{\alpha,\beta}$ through the Monte Carlo method, and then invert Black's formula by solving the following equation in $v^{\mathrm{MC}}_{\alpha,\beta}$

$$C_{\alpha,\beta}(0)\,\mathrm{Bl}(S_{\alpha,\beta}(0), S_{\alpha,\beta}(0), v^{\mathrm{MC}}_{\alpha,\beta}) = \mathrm{MCprice}_{\alpha,\beta}\,,$$

thus deriving the Black volatility implied by the LFM Monte Carlo price.

Furthermore, we compute the standard error of the method, corresponding to a two-side 98% window around the mean:

$$\mathrm{MCerr}_{\alpha,\beta} = 2.33\,\frac{\mathrm{Std}\{(P(0,T_\alpha)\,C^j_{\alpha,\beta}(T_\alpha)\,(S^j_{\alpha,\beta}(T_\alpha) - K)^+\}}{\sqrt{\mathrm{npath}}}\,,$$

where j denotes the scenario, "npath" is the number of scenarios in the Monte Carlo method, and the standard deviation is taken over the simulated scenarios. We use this quantity to compute the "inf" and "sup" volatilities $v^{\mathrm{MCinf}}_{\alpha,\beta}$, $v^{\mathrm{MCsup}}_{\alpha,\beta}$ obtained from solving the equations

$$C_{\alpha,\beta}(0)\,\mathrm{Bl}(S_{\alpha,\beta}(0), S_{\alpha,\beta}(0), v^{\mathrm{MCinf}}_{\alpha,\beta}) = \mathrm{MCprice}_{\alpha,\beta} - \mathrm{MCerr}_{\alpha,\beta},$$
$$C_{\alpha,\beta}(0)\,\mathrm{Bl}(S_{\alpha,\beta}(0), S_{\alpha,\beta}(0), v^{\mathrm{MCsup}}_{\alpha,\beta}) = \mathrm{MCprice}_{\alpha,\beta} + \mathrm{MCerr}_{\alpha,\beta},$$

thus deriving the Black implied volatilities corresponding to the extremes of the 98% price window.

Since the differences between Rebonato's and Hull and White's formulas will be shown to be typically negligible, we will just consider the following percentage differences based on Rebonato's volatility:

$$100\,(v^{\text{MC}}_{\alpha,\beta} - v^{\text{LFM}}_{\alpha,\beta})/v^{\text{MC}}_{\alpha,\beta}\,,$$
$$100\,(v^{\text{MCinf}}_{\alpha,\beta} - v^{\text{LFM}}_{\alpha,\beta})/v^{\text{MCinf}}_{\alpha,\beta}\,,$$
$$100\,(v^{\text{MCsup}}_{\alpha,\beta} - v^{\text{LFM}}_{\alpha,\beta})/v^{\text{MCsup}}_{\alpha,\beta}\,.$$

As far as distributions are concerned, we reason as follows. The Rebonato formula is derived under the assumption that the swap rate is driftless and lognormal under any of the forward measures Q^k, with integrated percentage variance $(v^{\text{LFM}}_{\alpha,\beta})^2$. In particular, the swap rate is assumed to be driftless and lognormal under Q^α, with mean $S_{\alpha,\beta}(0)$ and log-variance $(v^{\text{LFM}}_{\alpha,\beta})^2$. Indeed, as far as the swaption corresponding to $S_{\alpha,\beta}$ is concerned, this is equivalent to assuming $S_{\alpha,\beta}(t) = \overline{S}_{\alpha,\beta}(t)$, the classic driftless geometric Brownian motion:

$$d\overline{S}_{\alpha,\beta}(t) = v(t)\overline{S}_{\alpha,\beta}(t)dW^\alpha\,, \tag{8.13}$$

so that we can write

$$Y = \overline{S}_{\alpha,\beta}(T_\alpha) = \overline{S}_{\alpha,\beta}(0)\exp\left\{-\frac{1}{2}(v^{\text{LFM}}_{\alpha,\beta})^2 + v^{\text{LFM}}_{\alpha,\beta}\overline{W}\right\},$$

where v is any function recovering the correct terminal variance, $\int_0^T v(t)^2 dt = (v^{\text{LFM}}_{\alpha,\beta})^2$, and $\overline{W} \sim \mathcal{N}(0,1)$ under Q^α.

To test the quality of this approximation, we compare the sampled distribution under Q^α of the (Monte Carlo) simulated $S_{\alpha,\beta}(T_\alpha)$ with a lognormal distribution corresponding to an initial condition $S_{\alpha,\beta}(0)$ and an integrated percentage variance $(v^{\text{LFM}}_{\alpha,\beta})^2$, consistently with the approximated dynamics (8.13).

The density of this lognormal distribution is given by

$$p_Y(y) = \frac{1}{y v^{\text{LFM}}_{\alpha,\beta}}\frac{1}{\sqrt{2\pi}}\exp\left\{-\frac{1}{2(v^{\text{LFM}}_{\alpha,\beta})^2}\left[\ln\frac{y}{S_{\alpha,\beta}(0)} + \frac{1}{2}(v^{\text{LFM}}_{\alpha,\beta})^2\right]^2\right\}.$$

However, we know that the true S-dynamics is driftless under $Q^{\alpha,\beta}$ but not under Q^α. Therefore, there will be a bias in assuming

$$E_0^\alpha\{S_{\alpha,\beta}(T_\alpha)\} \approx E_0^\alpha\{\overline{S}_{\alpha,\beta}(T_\alpha)\} = S_{\alpha,\beta}(0)$$

since what is true is, instead,

$$E_0^{\alpha,\beta}\{S_{\alpha,\beta}(T_\alpha)\} = S_{\alpha,\beta}(0)\,.$$

To measure this bias, we consider a new lognormal distribution where we replace $S_{\alpha,\beta}(0)$ in the previous one with $\mu_\alpha(0) := E_0^\alpha\{S_{\alpha,\beta}(T_\alpha)\}$. Moreover, we calculate the third central moment and the kurtosis of the swap-rate logarithm, and compare them with those of a standard normal distribution, amounting respectively to 0 and 3.

Further, we plot the density obtained through the Monte Carlo method against the analytical one in both the biased and unbiased cases, so that we can isolate the part of the error in the distribution due to the bias in the drift. We will see that, in almost all cases, the correction in the mean accounts for the whole difference, in that the "true" (Monte Carlo) distribution of S under Q^α is practically indistinguishable from a lognormal distribution with mean $\mu_\alpha(0)$ and log-variance $(v_{\alpha,\beta}^{\text{LFM}})^2$.

8.4 Test Results for Volatilities

I'm running tests, but I don't begin to understand it
Querl Dox (Brainiac 5), Legion Lost 6, 2000, DC Comics

In this section, we present the results obtained when deriving implied volatilities from the swaptions prices calculated through a Monte Carlo method, with four time steps per year and 200000 paths, against Rebonato's and Hull and White's analytical formulas. This will be done for several formulations of instantaneous volatilities and correlations, and for different "expiry/maturity" pairs of forward rates.

In our numerical tests, we will follow the map of cases given in the previous section. For each point of the testing plan, we will also consider the "stressed" results obtained when upwardly shifting the Φ's values by 0.2. Furthermore, going back to unstressed Φ's, we will consider the "stressed" results obtained when upwardly shifting the initial forward-rate vector $[F(0,0,1),\ldots,F(0,19,20)]$ by 2%.

With reference to the previous section, before showing our numerical results and plots of probability densities, we explain the terms that will appear in the tables.

We denote the first reset date T_α by "res" and take, in all cases, the last payment date to be $T_\beta = 20y$. The volatilities $v_{\alpha,\beta}^{\text{MC}}$, $v_{\alpha,\beta}^{\text{MCinf}}$, $v_{\alpha,\beta}^{\text{MCsup}}$, $v_{\alpha,\beta}^{\text{LFM}}$, $\overline{v}_{\alpha,\beta}^{\text{LFM}}$, divided by $\sqrt{T_\alpha}$, are denoted, respectively, by "MCVol", "MCinf", "MCsup", "RebVol", "HWVol".

We denote by "differr" and "errperc", respectively, the absolute and percentage differences (errors) between Monte Carlo implied volatilities and Rebonato's volatilities. We also denote by "errinf" and "errsup" the percentage difference between the lower ("inf") and upper ("sup") extremes of the 98% Monte Carlo window and Rebonato's volatility.

We then report the initial swap rate $S_{\alpha,\beta}(0)$, "sw0", the mean $\mu_\alpha(0)$ of the "Monte Carlo" swap rate (generated under the measure Q^α), "mean", and finally the third central moment and the kurtosis of the simulated swap-rate logarithm, "third" and "kurt" (again under the measure Q^α).

8.4.1 Case (1): Constant Instantaneous Volatilities

The first case we analyze is that of constant instantaneous volatilities. We have seen that with such a formulation, as time passes from an expiry date to the next, the volatility term structure is obtained from the previous one by "cutting off" the head, so that its qualitative behaviour can be altered over time. Typically, if the term structure features a hump around two years, this hump will be absent in the term structure occurring in three years.

We can impose structures of this kind via the formulation of TABLE 3. We actually resort to Formulation 7 and take $a = b = d = 0$ and $c = 1$, so that the caplet fitting forces the Φ's to equal the corresponding caplet volatilities, as obtained from the market, $\Phi_i = v_{T_{i-1}-\text{caplet}}$.

(1.a): Typical rank-two correlations

The instantaneous-correlation matrix implied by this choice of parameters comes from a possible joint caps/swaptions calibration and is reported in Figure 8.1. The graph of the parameters Φ's is shown in Figure 8.2.

Numerical results are given in the following:

res	MCVol	MCinf	MCsup	RebVol	HWVol	differr	errperc
9	0.1019	0.1014	0.1023	0.1015	0.1017	0.0003	0.3116

errinf	errsup	sw0	mean	third	kurt
-0.1389	0.7581	0.0635	0.0655	0.0000	3.0076

The volatility approximation appears to be excellent. The biased- and unbiased-density plots are shown in Figure 8.3.

(1.b): Perfect correlations

The set of parameters in this case is as in (1.a) but with perfect instantaneous correlation (matrix of ones, obtained putting $\theta = [0 \ldots 0]$). We obtain numerical results similar to those in the previous case.

res	MCVol	MCinf	MCsup	RebVol	HWVol	differr	errperc
9	0.1026	0.1021	0.1030	0.1021	0.1022	0.0004	0.4381

errinf	errsup	sw0	mean	third	kurt
-0.0123	0.8844	0.0635	0.0655	0.0000	2.9986

Again, the volatility approximations seem to be quite satisfactory. As a "stress test", in this case we also upwardly shifted the parameters Φ by 20%, and, as expected, our results worsened. More specifically, since under this choice of a, b, c and d the Φ's are given by the caplet volatilities, our shift amounts to

$$\Phi_i = v_{T_{i-1}-\text{caplet}} \quad \rightarrow \quad \Phi_i = v_{T_{i-1}-\text{caplet}} + 0.2.$$

This leads to rather large volatilities. In fact, just add 20% to all the original caplet volatilities in input, shown in Figure 8.2, to see that the new ones are more than doubled. Since the "freezing" approximation works better when variability is small, and increasing the Φ's amounts to increasing volatility, it is natural to expect the approximation to worsen, as is happening here.

Phi	0.1490	0.1590	0.1530	0.1450	0.1360	0.1270	0.1210	0.1180	0.1140	0.1110	0.1080	0.1050	0.1020	0.0989	0.0978	0.0974	0.0969	0.0965	0.0961
thetas	0.0147	0.0643	0.1030	0.1500	0.1970	0.2240	0.2770	0.2950	0.3630	0.3810	0.4220	0.4840	0.5200	0.5420	0.5790	0.6500	0.6680	0.7130	0.7660
corr	1y	2y	3y	4y	5y	6y	7y	8y	9y	10y	11y	12y	13y	14y	15y	16y	17y	18y	19y
1y	1.0000	0.9988	0.9961	0.9908	0.9834	0.9782	0.9658	0.9610	0.9399	0.9337	0.9183	0.8920	0.8748	0.8642	0.8449	0.8051	0.7941	0.7662	0.7309
2y	0.9988	1.0000	0.9992	0.9963	0.9912	0.9873	0.9774	0.9735	0.9557	0.9503	0.9368	0.9134	0.8978	0.8881	0.8704	0.8335	0.8233	0.7971	0.7638
3y	0.9961	0.9992	1.0000	0.9989	0.9956	0.9927	0.9849	0.9817	0.9664	0.9617	0.9497	0.9285	0.9142	0.9053	0.8889	0.8544	0.8447	0.8200	0.7883
4y	0.9908	0.9963	0.9989	1.0000	0.9989	0.9973	0.9920	0.9895	0.9774	0.9735	0.9634	0.9449	0.9323	0.9243	0.9095	0.8779	0.8690	0.8460	0.8164
5y	0.9834	0.9912	0.9956	0.9989	1.0000	0.9996	0.9968	0.9952	0.9862	0.9831	0.9749	0.9592	0.9481	0.9411	0.9279	0.8993	0.8911	0.8700	0.8425
6y	0.9782	0.9873	0.9927	0.9973	0.9996	1.0000	0.9986	0.9975	0.9903	0.9877	0.9805	0.9665	0.9564	0.9499	0.9376	0.9108	0.9030	0.8830	0.8567
7y	0.9658	0.9774	0.9849	0.9920	0.9968	0.9986	1.0000	0.9998	0.9963	0.9946	0.9896	0.9787	0.9705	0.9652	0.9548	0.9314	0.9246	0.9067	0.8829
8y	0.9610	0.9735	0.9817	0.9895	0.9952	0.9975	0.9998	1.0000	0.9977	0.9963	0.9920	0.9823	0.9747	0.9697	0.9599	0.9378	0.9313	0.9141	0.8912
9y	0.9399	0.9557	0.9664	0.9774	0.9862	0.9903	0.9963	0.9977	1.0000	0.9998	0.9983	0.9927	0.9876	0.9841	0.9768	0.9592	0.9539	0.9395	0.9200
10y	0.9337	0.9503	0.9617	0.9735	0.9831	0.9877	0.9946	0.9963	0.9998	1.0000	0.9992	0.9947	0.9903	0.9871	0.9804	0.9641	0.9591	0.9455	0.9268
11y	0.9183	0.9368	0.9497	0.9634	0.9749	0.9805	0.9896	0.9920	0.9983	0.9992	1.0000	0.9981	0.9951	0.9928	0.9876	0.9741	0.9698	0.9580	0.9413
12y	0.8920	0.9134	0.9285	0.9449	0.9592	0.9665	0.9787	0.9823	0.9927	0.9947	0.9981	1.0000	0.9993	0.9983	0.9954	0.9863	0.9831	0.9739	0.9604
13y	0.8748	0.8978	0.9142	0.9323	0.9481	0.9564	0.9705	0.9747	0.9876	0.9903	0.9951	0.9993	1.0000	0.9998	0.9983	0.9917	0.9891	0.9816	0.9700
14y	0.8642	0.8881	0.9053	0.9243	0.9411	0.9499	0.9652	0.9697	0.9841	0.9871	0.9928	0.9983	0.9998	1.0000	0.9993	0.9942	0.9921	0.9855	0.9750
15y	0.8449	0.8704	0.8889	0.9095	0.9279	0.9376	0.9548	0.9599	0.9768	0.9804	0.9876	0.9954	0.9983	0.9993	1.0000	0.9975	0.9961	0.9911	0.9826
16y	0.8051	0.8335	0.8544	0.8779	0.8993	0.9108	0.9314	0.9378	0.9592	0.9641	0.9741	0.9863	0.9917	0.9942	0.9975	1.0000	0.9998	0.9980	0.9932
17y	0.7941	0.8233	0.8447	0.8690	0.8911	0.9030	0.9246	0.9313	0.9539	0.9591	0.9698	0.9831	0.9891	0.9921	0.9961	0.9998	1.0000	0.9990	0.9952
18y	0.7662	0.7971	0.8200	0.8460	0.8700	0.8830	0.9067	0.9141	0.9395	0.9455	0.9580	0.9739	0.9816	0.9855	0.9911	0.9980	0.9990	1.0000	0.9986
19y	0.7309	0.7638	0.7883	0.8164	0.8425	0.8567	0.8829	0.8912	0.9200	0.9268	0.9413	0.9604	0.9700	0.9750	0.9826	0.9932	0.9952	0.9986	1.0000

Fig. 8.1. Case (1.a): parameters Φ, θ and matrix of instantaneous correlations.

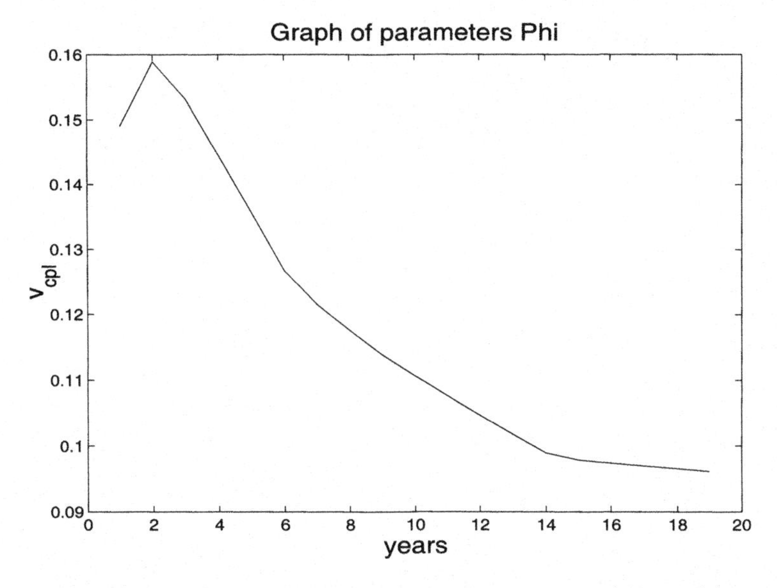

Fig. 8.2. Case (1.a): graph of caplet volatilities against their expiries.

res	MCVol	MCinf	MCsup	RebVol	HWVol	differr	errperc
2	0.3106	0.3094	0.3119	0.3184	0.3168	-0.0077	-2.4907
9	0.2972	0.2958	0.2987	0.3021	0.3025	-0.0049	-1.6355

errinf	errsup	sw0	mean	third	kurt
- 2.8959	-2.0886	0.0623	0.0718	0.0010	2.9995
- 2.1440	-1.1315	0.0635	0.0894	0.0011	2.8877

Nevertheless, we still believe the approximation to be acceptable, although not as good as in the "unstressed" case.

The biased- and unbiased-density plots, in both the "unstressed" and "stressed" cases, are shown in Figure 8.4 and 8.5, respectively. The last graph shows that a kurtosis far from 3 yields more marked differences even in the unbiased-density plot.

(1.c): Some negative rank-two correlations

Another situation we study is that of a partially negative rank-two correlation matrix, obtained with a new set of θ's. This matrix is reported below, in Figure 8.6.

We now present the results obtained with this set of parameters and, in the second row, with the Φ's upwardly shifted by 20%:

res	MCVol	MCinf	MCsup	RebVol	HWVol	differr	errperc
15	0.0939	0.0934	0.0944	0.0937	0.0938	0.0002	0.1952
15	0.2862	0.2839	0.2885	0.2870	0.2872	-0.0009	-0.2984

errinf	errsup	sw0	mean	third	kurt
-0.3303	0.7153	0.0619	0.0630	0.0000	2.9933
-1.1097	0.5021	0.0619	0.0817	0.0028	3.0334

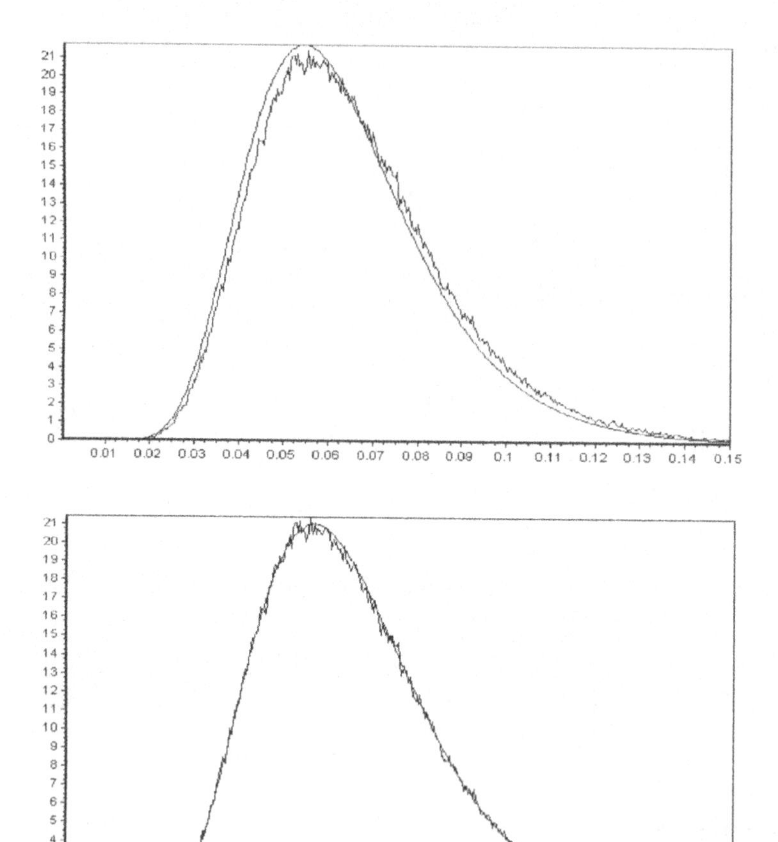

Fig. 8.3. Case (1.a): biased- and unbiased-density plots; res=9y.

Again we obtain small percentage differences for unstressed Φ's, while stressing the Φ's yields relatively small percentage differences for volatilities.

The biased-and unbiased-density plots, in both the "unstressed" and "stressed" cases, are shown in Figures 8.7 and 8.8, respectively.

From Figure 8.8, we see that large differences in both density plots are still observed. Notice also that while in (1.b) we had a log-kurtosis faraway from 3, in this case the kurtosis is close to the correct value, and yet densities appear to be quite different. This might be due to differences in higher order moments. Notice also the large bias in the mean. However, keep in mind that volatilities here were at pathological values, given that we upwardly shifted typical market percentage caplet volatilities by $0.2 = 20\%$.

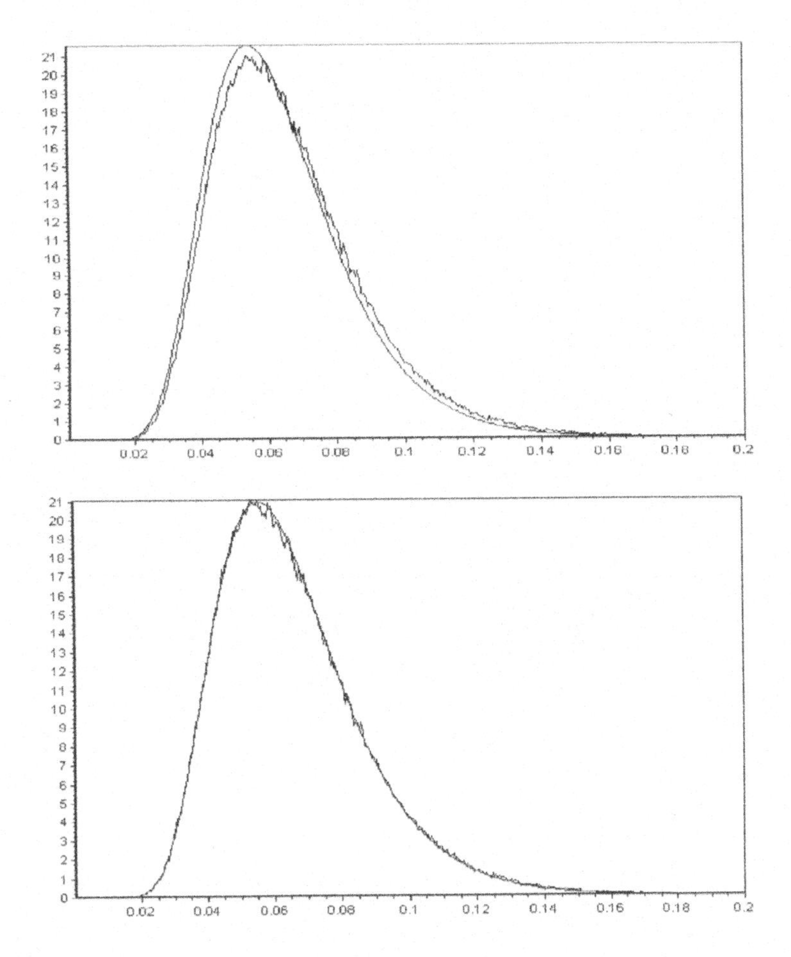

Fig. 8.4. Case (1.b): biased- and unbiased-density plots; $\theta = 0$, res=9y.

8.4.2 Case (2): Volatilities as Functions of Time to Maturity

The second case we analyze is that of instantaneous volatilities depending only on time to maturity.

We obtain this structure when starting from a parametric form allowing for a humped shape in the graph of the instantaneous volatility of the generic forward rate F_i as a function of time to expiry, i.e.

$$T_{i-1} - t \mapsto \sigma_i(t).$$

This is obtained by setting all Φ's equal to one in the previous Formulation 7, so that:

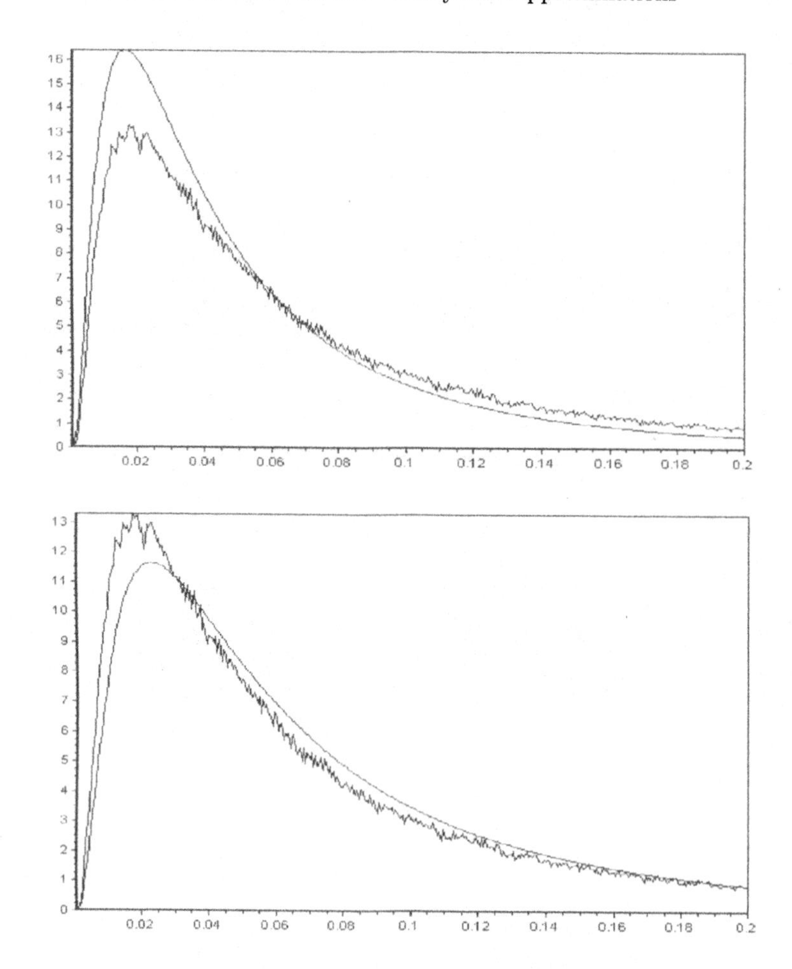

Fig. 8.5. Case (1.b): biased- and unbiased-density plots; $\theta = 0$ and Φ increased of 20%, res=9y.

$$\sigma_i(t) = \psi(T_{i-1} - t; a, b, c, d) = [a(T_{i-1} - t) + d]\, e^{-b(T_{i-1}-t)} + c.$$

The term structure of volatilities and the function $t \mapsto \psi(t; a, b, c, d)$ for $a = 0.19085664$, $b = 0.97462314$, $c = 0.08089167$, $d = 0.013449479$ are plotted in Figure 8.9. The values of a, b, c, d have been taken from the next case (3).

(2.a): Perfect correlations

We test this instantaneous-volatility structure starting with a perfect instantaneous-correlation matrix, by setting all θ's to be equal to zero. As usual, this implies that all ρ's are equal to one. We obtain the following numerical results:

thetas	0.0000	0.0000	0.0013	0.0044	0.0096	0.0178	0.0299	0.0474	0.0728	0.1100	0.1660	0.2530	0.3990	0.6570	1.1000	1.6600	2.0700	2.2800	2.2300
corr	1y	2y	3y	4y	5y	6y	7y	8y	9y	10y	11y	12y	13y	14y	15y	16y	17y	18y	19y
1y	1.0000	1.0000	1.0000	1.0000	1.0000	0.9998	0.9996	0.9989	0.9974	0.9940	0.9863	0.9681	0.9215	0.7921	0.4514	- 0.0896	- 0.4790	- 0.6531	- 0.6093
2y	1.0000	1.0000	1.0000	1.0000	1.0000	0.9998	0.9996	0.9989	0.9974	0.9940	0.9863	0.9681	0.9215	0.7921	0.4514	- 0.0896	- 0.4790	- 0.6531	- 0.6093
3y	1.0000	1.0000	1.0000	1.0000	1.0000	0.9999	0.9996	0.9989	0.9974	0.9941	0.9865	0.9684	0.9220	0.7929	0.4525	- 0.0883	- 0.4778	- 0.6521	- 0.6083
4y	1.0000	1.0000	1.0000	1.0000	1.0000	0.9999	0.9997	0.9991	0.9977	0.9944	0.9870	0.9692	0.9232	0.7948	0.4553	- 0.0852	- 0.4751	- 0.6498	- 0.6058
5y	1.0000	1.0000	1.0000	1.0000	1.0000	1.0000	0.9998	0.9993	0.9980	0.9950	0.9878	0.9704	0.9252	0.7980	0.4599	- 0.0800	- 0.4705	- 0.6458	- 0.6017
6y	0.9998	0.9998	0.9999	0.9999	1.0000	1.0000	0.9999	0.9996	0.9985	0.9958	0.9891	0.9724	0.9283	0.8029	0.4672	- 0.0718	- 0.4633	- 0.6395	- 0.5951
7y	0.9996	0.9996	0.9996	0.9997	0.9998	0.9999	1.0000	0.9998	0.9991	0.9968	0.9908	0.9751	0.9327	0.8100	0.4778	- 0.0598	- 0.4525	- 0.6302	- 0.5853
8y	0.9989	0.9989	0.9989	0.9991	0.9993	0.9996	0.9998	1.0000	0.9997	0.9980	0.9930	0.9789	0.9389	0.8202	0.4931	- 0.0423	- 0.4369	- 0.6165	- 0.5711
9y	0.9974	0.9974	0.9974	0.9977	0.9980	0.9985	0.9991	0.9997	1.0000	0.9993	0.9957	0.9837	0.9473	0.8344	0.5151	- 0.0169	- 0.4139	- 0.5963	- 0.5500
10y	0.9940	0.9940	0.9941	0.9944	0.9950	0.9958	0.9968	0.9980	0.9993	1.0000	0.9984	0.9897	0.9586	0.8543	0.5466	0.0203	- 0.3797	- 0.5660	- 0.5186
11y	0.9863	0.9863	0.9865	0.9870	0.9878	0.9891	0.9908	0.9930	0.9957	0.9984	1.0000	0.9962	0.9730	0.8821	0.5925	0.0761	- 0.3275	- 0.5191	- 0.4700
12y	0.9681	0.9681	0.9684	0.9692	0.9704	0.9724	0.9751	0.9789	0.9837	0.9897	0.9962	1.0000	0.9894	0.9198	0.6607	0.1630	- 0.2436	- 0.4424	- 0.3911
13y	0.9215	0.9215	0.9220	0.9232	0.9252	0.9283	0.9327	0.9389	0.9473	0.9586	0.9730	0.9894	1.0000	0.9670	0.7625	0.3043	- 0.1004	- 0.3077	- 0.2535
14y	0.7921	0.7921	0.7929	0.7948	0.7980	0.8029	0.8100	0.8202	0.8344	0.8543	0.8821	0.9198	0.9670	1.0000	0.9022	0.5369	0.1564	- 0.0552	0.0013
15y	0.4514	0.4514	0.4525	0.4553	0.4599	0.4672	0.4778	0.4931	0.5151	0.5466	0.5925	0.6607	0.7625	0.9022	1.0000	0.8483	0.5671	0.3809	0.4325
16y	- 0.0896	- 0.0896	- 0.0883	- 0.0852	- 0.0800	- 0.0718	- 0.0598	- 0.0423	- 0.0169	0.0203	0.0761	0.1630	0.3043	0.5369	0.8483	1.0000	0.9172	0.8127	0.8443
17y	- 0.4790	- 0.4790	- 0.4778	- 0.4751	- 0.4705	- 0.4633	- 0.4525	- 0.4369	- 0.4139	- 0.3797	- 0.3275	- 0.2436	- 0.1004	0.1564	0.5671	0.9172	1.0000	0.9776	0.9879
18y	- 0.6531	- 0.6531	- 0.6521	- 0.6498	- 0.6458	- 0.6395	- 0.6302	- 0.6165	- 0.5963	- 0.5660	- 0.5191	- 0.4424	- 0.3077	- 0.0552	0.3809	0.8127	0.9776	1.0000	0.9984
19y	- 0.6093	- 0.6093	- 0.6083	- 0.6058	- 0.6017	- 0.5951	- 0.5853	- 0.5711	- 0.5500	- 0.5186	- 0.4700	- 0.3911	- 0.2535	0.0013	0.4325	0.8443	0.9879	0.9984	1.0000

Fig. 8.6. Case (1.c): parameters θ and matrix of instantaneous correlation.

res	MCVol	MCinf	MCsup	RebVol	HWVol	differr	errperc
2	0.0930	0.0927	0.0934	0.0939	0.0935	-0.0009	-0.9584
9	0.0883	0.0879	0.0887	0.0884	0.0885	-0.0001	-0.1112
17	0.0956	0.0951	0.0962	0.0954	0.0954	0.0002	0.2404

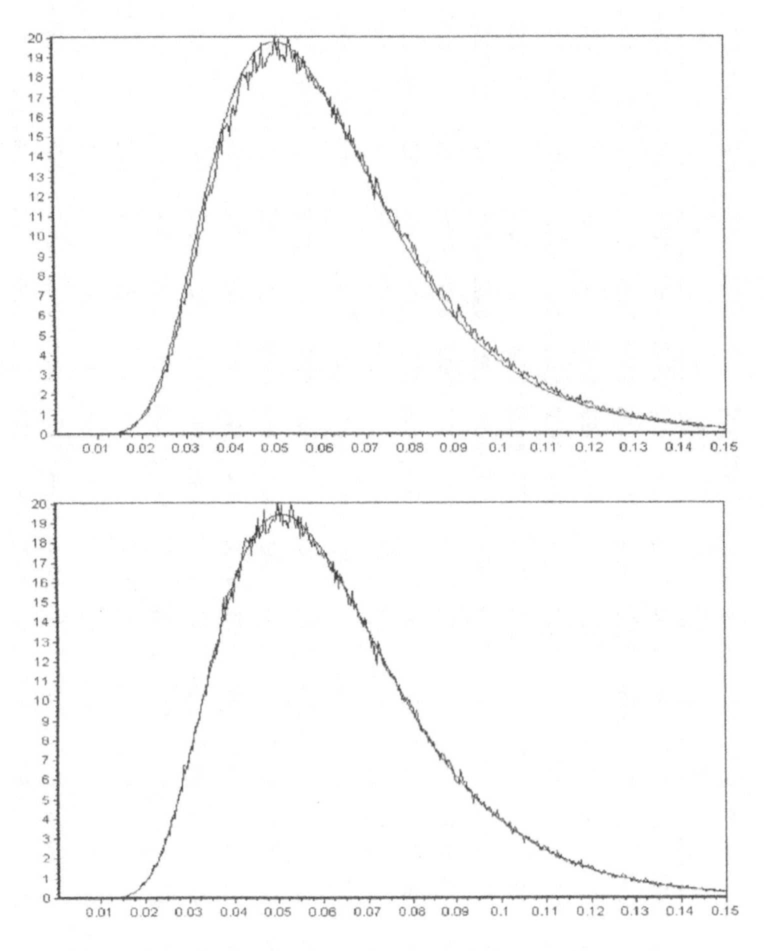

Fig. 8.7. Case (1.c): biased- and unbiased-density plots; res=15y.

errinf	errsup	sw0	mean	third	kurt
-1.3715	-0.5486	0.0623	0.0630	0.0000	3.0289
-0.5594	0.3330	0.0635	0.0650	0.0000	3.0098
-0.3273	0.8018	0.0607	0.0613	0.0000	3.0024

The approximation appears to be satisfactory in this case too. Subsequently, as in the previous cases (1), we upwardly shift all Φ's by 20%. However, we need to keep in mind that this shift has a different impact, due to the different formulation for instantaneous volatilities we are using. We obtain:

res	MCVol	MCinf	MCsup	RebVol	HWVol	differr	errperc
9	0.1064	0.1059	0.1069	0.1061	0.1062	0.0003	0.2649

errinf	errsup	sw0	mean	third	kurt
-0.1899	0.7156	0.0635	0.0657	0.0001	3.0106

The approximation is still good, contrary to the corresponding case (1.b) with shifted Φ's. We explain the reason for this in the next case (2.b).

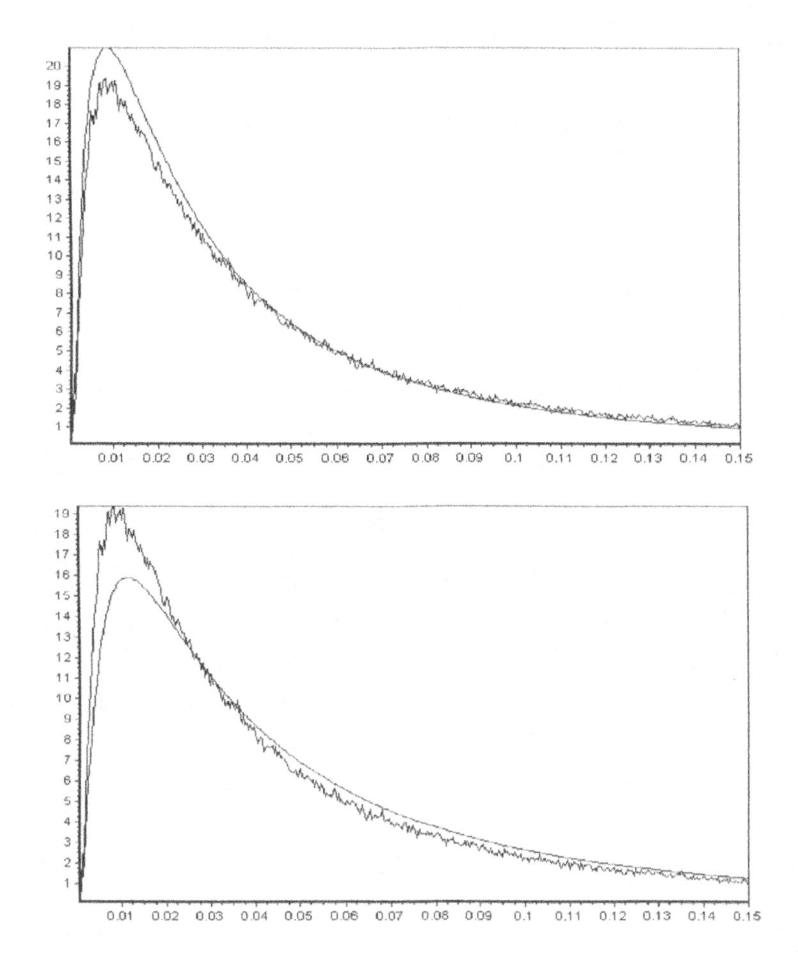

Fig. 8.8. Case (1.c): biased- and unbiased-density plots; Φ increased of 20%; res=15y.

Finally, after restoring the original Φ's, we increase of 2% the initial forward rates $[F(0,0,1),\ldots,F(0,19,20)]$, with the following numerical results:

res	MCVol	MCinf	MCsup	RebVol	HWVol	differr	errperc
9	0.0887	0.0883	0.0890	0.0888	0.0889	-0.0001	-0.1473

errinf	errsup	sw0	mean	third	kurt
-0.5763	0.2781	0.0835	0.0860	0.0000	3.0138

Here the percentage error is close to that of the original unstressed case (2.a).

By looking also at the density plots in Figures 8.10, 8.11, 8.12 and 8.13, we confirm the impression that unlike case (1.b) (volatility-structure constant in time, perfect instantaneous correlation), where the error clearly increases as the Φ's increase, here we do not observe large differences or a pronounced

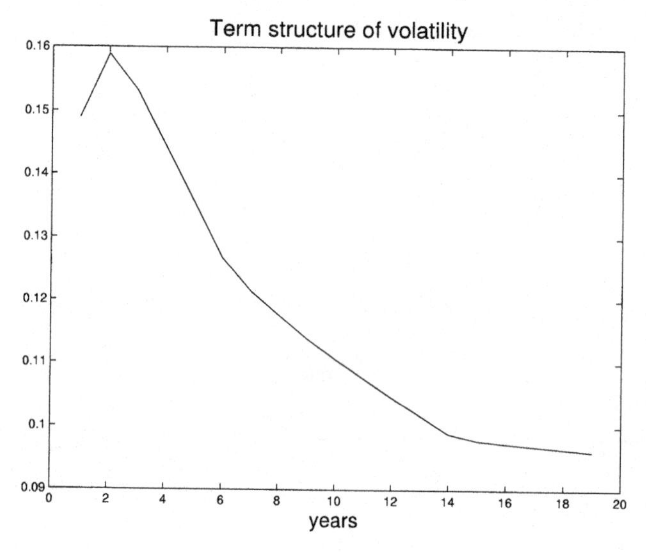

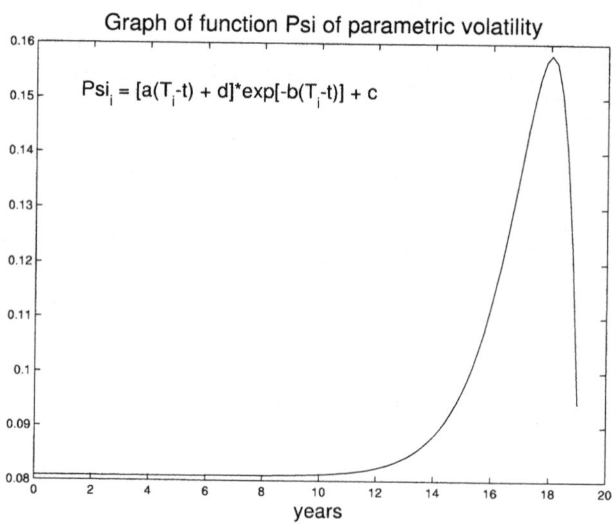

Fig. 8.9. Case (2): term structure of volatility and graph of the function $t \mapsto \psi(t; a, b, c, d)$.

worsening. This may indeed be due to the fact that now the shift in volatilities is *relative*, and that we have only a 20% increase in the Φ values.

(2.b): Negative rank-two correlations

We now repeat the same test with some negative correlations, and precisely with the same instantaneous correlations as in (1.c).

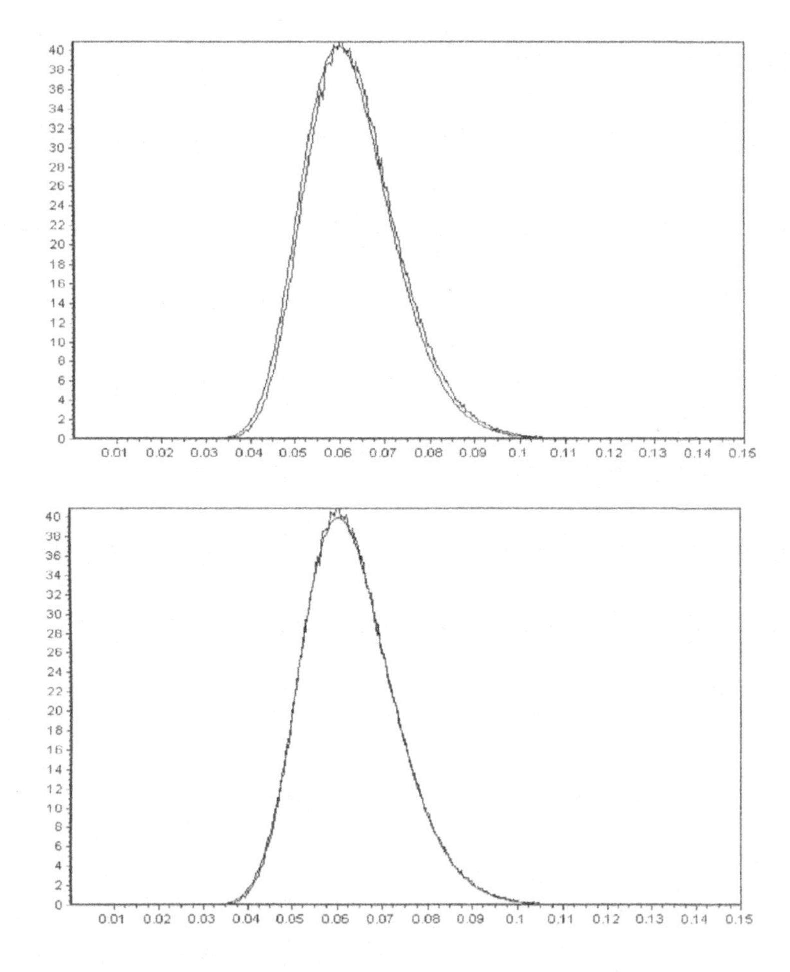

Fig. 8.10. Case (2.a): biased- and unbiased-density plots; $\theta = 0$; res=2y.

The following table and the density graphs in Figures 8.14, 8.15 and 8.16, show the results obtained respectively with the original parameters (first row) and the two usual shifts, first on the Φ's (second row) and then on $F(0)$'s (third row).

res	MCVol	MCinf	MCsup	RebVol	HWVol	differr	errperc
15	0.0891	0.0886	0.0895	0.0892	0.0893	-0.0001	-0.1644
15	0.1070	0.1064	0.1076	0.1070	0.1071	-0.0000	-0.0382
15	0.1160	0.1150	0.1160	0.1160	0.1160	-0.0006	-0.5020

errinf	errsup	sw0	mean	third	kurt
-0.6872	0.3531	0.0619	0.0629	0.0000	3.0096
-0.5869	0.5047	0.0619	0.0634	0.0000	3.0134
-0.9160	-0.0903	0.0623	0.0635	0.0001	3.0400

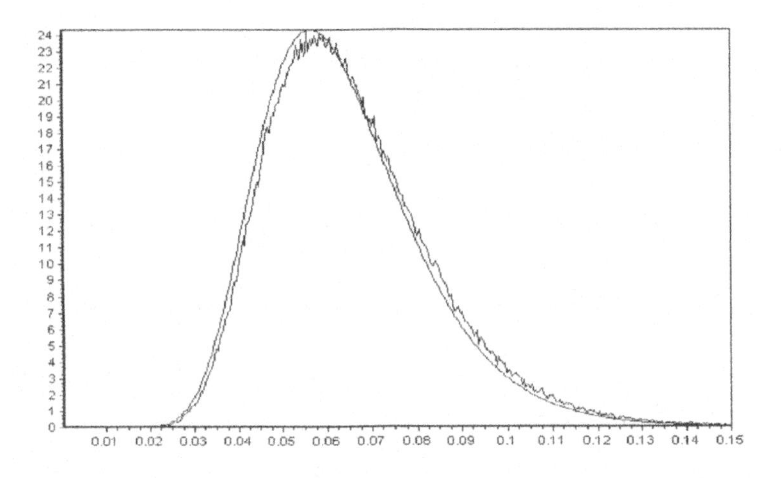

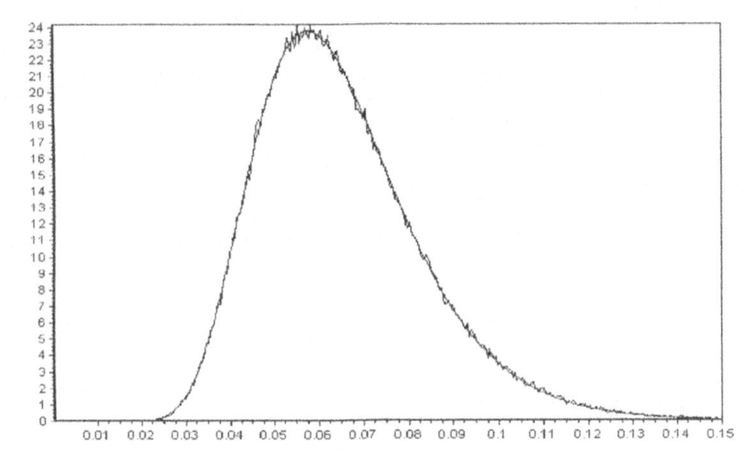

Fig. 8.11. Case (2.a): biased- and unbiased-density plots; $\theta = 0$; res=9y.

Again, the approximation appears to be satisfactory in all cases. In particular, shifting the Φ's does not seem to affect sensibly the density plots, contrary to the corresponding case (1.c) of the constant-volatility formulation. As in the previous case (2.a), we recall that now the shift in volatilities is relative. For instance, if we had a caplet volatility of say 16%, now we have a caplet volatility of $1.2 * 16\% = 19.2\%$, and not of $16 + 20 = 36\%$ as before.

We may try and increase volatilities with the current formulation so as to reach again levels such as 36%. This can be done by leaving the Φ's unaffected and by instead increasing both c and d by $0.2 = 20\%$. By doing so, we obtain the following results:

res	MCVol	MCinf	MCsup	RebVol	HWVol	differr	errperc
15	0.2883	0.2859	0.2906	0.2893	0.2895	-0.0010	-0.3475

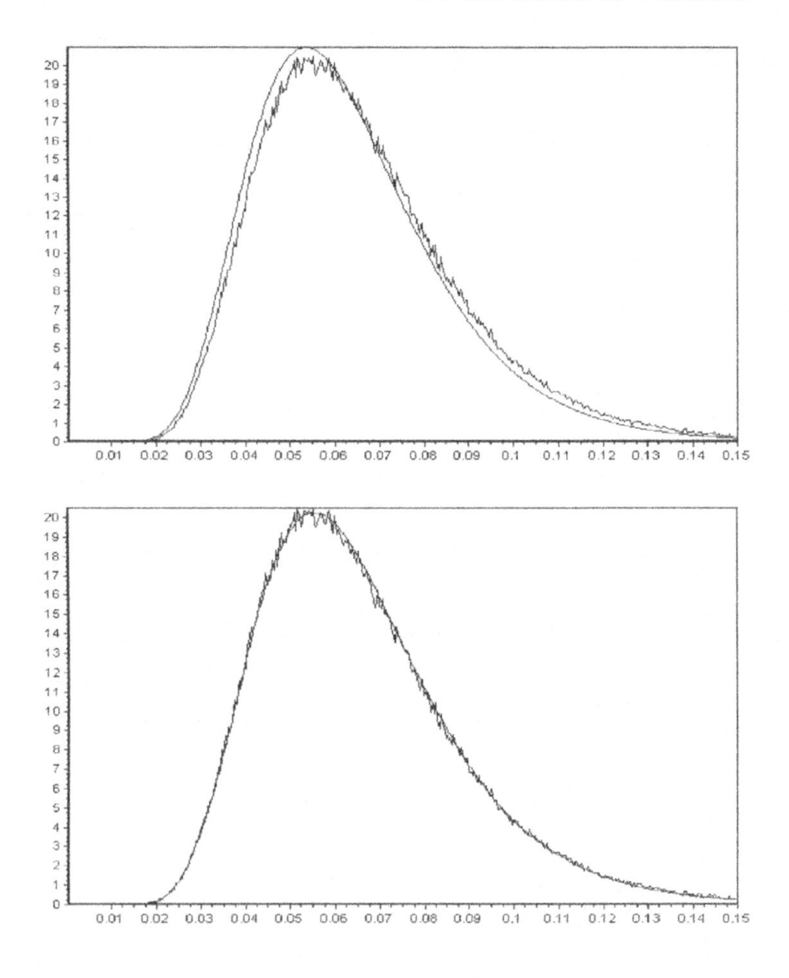

Fig. 8.12. Case (2.a): biased- and unbiased-density plots; $\theta = 0$ and Φ increased of 20%; res=9y.

errinf	errsup	sw0	mean	third	kurt
-1.1630	0.4572	0.0619	0.0835	0.0033	3.0817

We also show the related biased- and unbiased-density plots in Figure 8.17. Notice that the plots are completely analogous to those of case (1.c), as one could have expected, since the uniform shift of the initial volatility structure amounts to 0.2 in both cases. This seems to suggest that with huge volatility values, our distributional approximations can become rough, no matter the particular volatility formulation chosen.

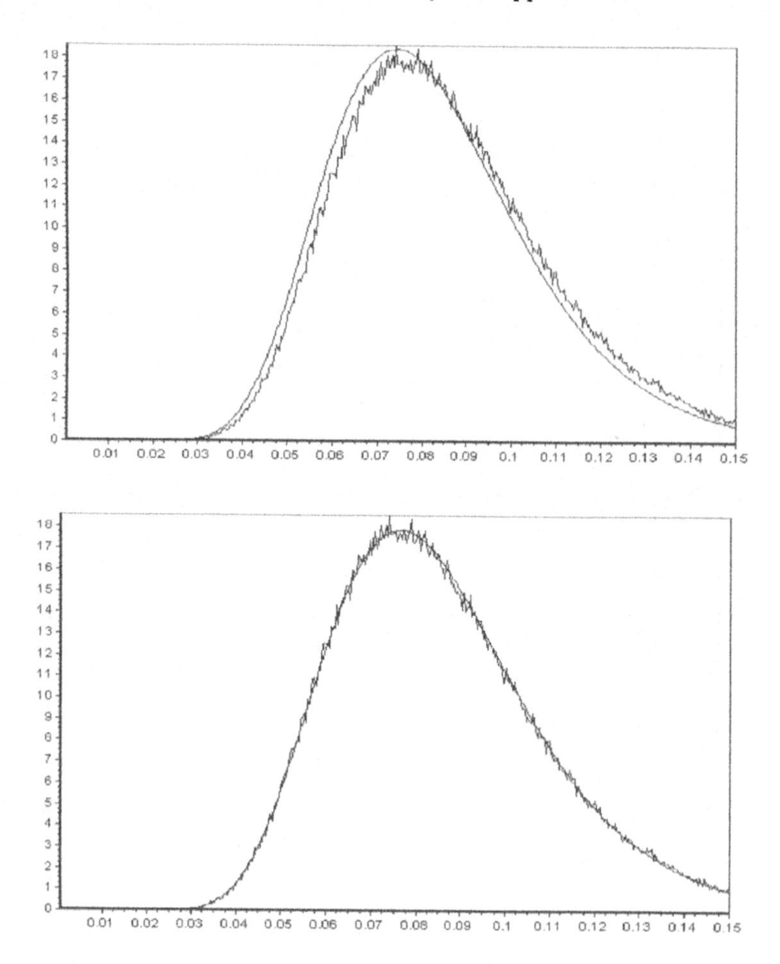

Fig. 8.13. Case (2.a): biased- and unbiased-density plots; $\theta = 0$ and $F(0)$ increased of 2%; res=9y.

8.4.3 Case (3): Humped and Maturity-Adjusted Instantaneous Volatilities Depending only on Time to Maturity

The last case we consider is the most general, since it uses the full Formulation 7 for instantaneous volatility:

$$\sigma_i(t) = \Phi_i \left([a(T_{i-1} - t) + d]e^{-b(T_{i-1}-t)} + c \right),$$

which reduces to the previous case (2) when all the Φ's are set to one. Recall that this form can be seen as having a parametric core ψ, which is locally altered for each expiry T_{i-1} by the Φ's. As we already noticed in Section 6.5,

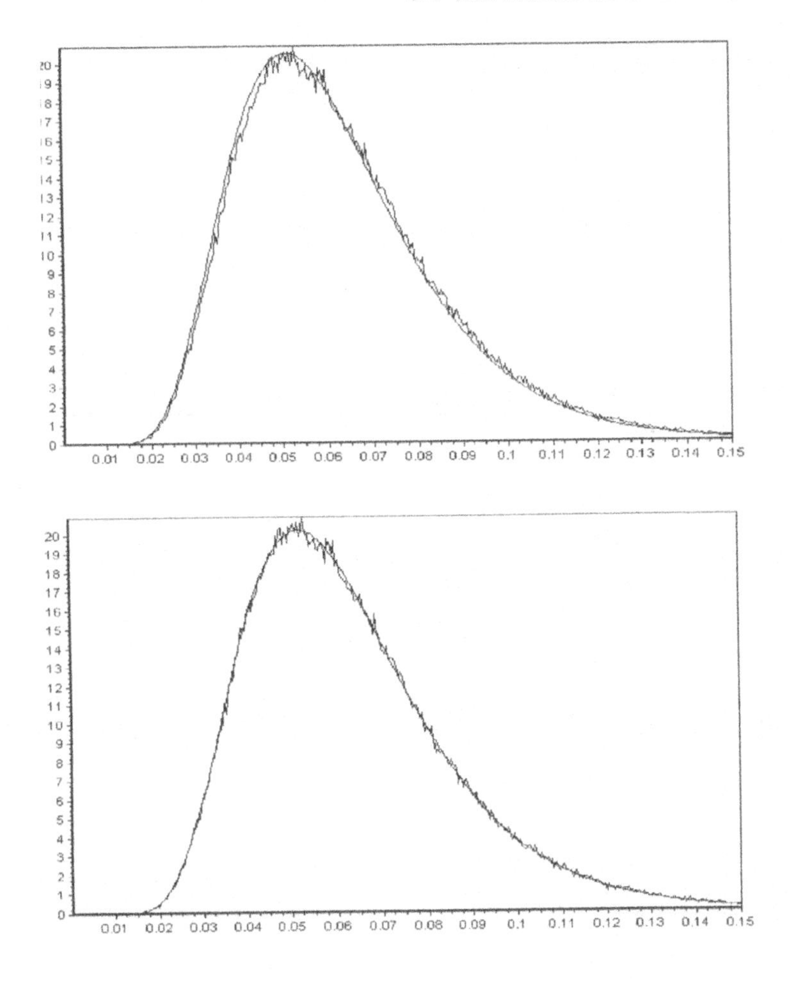

Fig. 8.14. Case (2.b): biased- and unbiased-density plots; res=15y.

these local modifications, if small, do not destroy the essential dependence on time to maturity, so as to maintain the desirable behaviour of the term structure of volatilities. In fact, the term structure remains qualitatively the same as time passes, and in particular it can maintain its humped shape if initially humped and if all Φ's are not too different.

We consider the following values of a, b, c and d: $a = 0.19085664$, $b = 0.97462314$, $c = 0.08089167$, $d = 0.013449479$.

Recall that, if the caplet volatilities are given as market input, we may express the Φ's as functions of a, b, c, d through formula (6.28). The computed parameters $\Phi = [\Phi_{1-9}, \Phi_{10-19}]$, matching the caplet volatilities, are in our case:

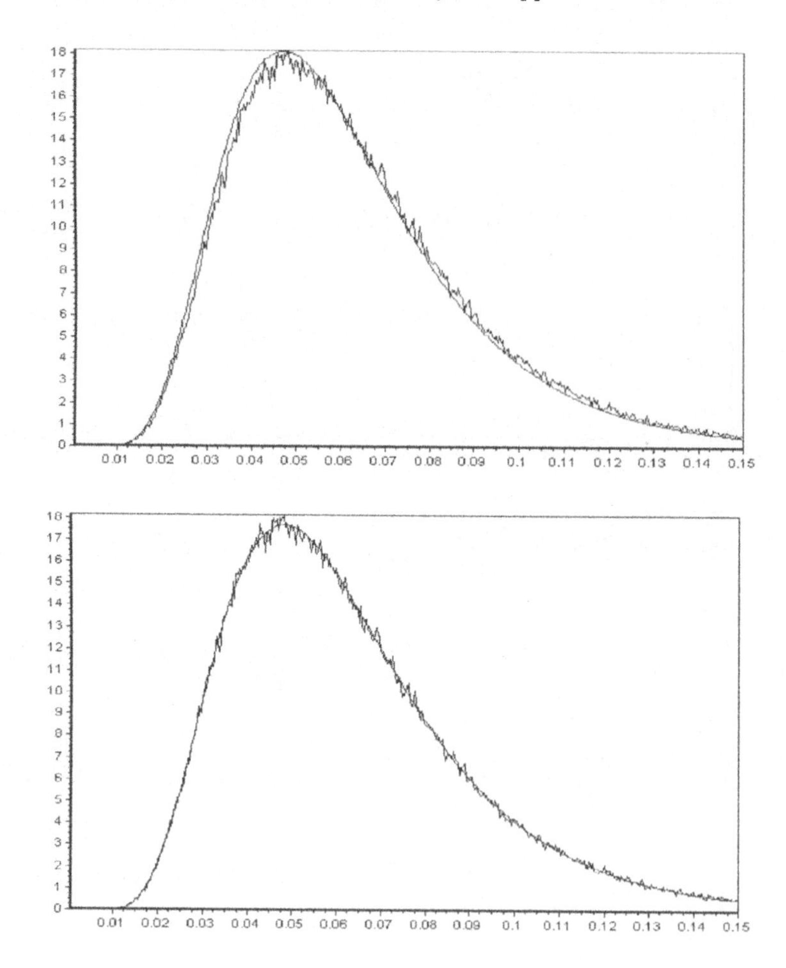

Fig. 8.15. Case (2.b): biased- and unbiased-density plots; Φ increased of 20%; res=15y.

$$\begin{aligned}
\Phi_{1-8} &= [\,1.0500\;1.0900\;1.1025\;1.1025\;1.0913\;1.0669\;1.0624\;1.0611\,]\\
\Phi_{9-17} &= [\,1.0544\;1.0475\;1.0386\;1.0270\;1.0132\;0.9975\;0.9979\;1.0033\;1.0079\,]\\
\Phi_{18-19} &= [1.0119\;1.0152\,].
\end{aligned}$$

A plot of the parameters Φ and of the core function $t \mapsto \psi(t; a, b, c, d)$ is shown below in Figure 8.18.

(3.a): Typical rank-two correlations

Here we analyze the case where the instantaneous-correlation matrix is the same as in (1.a). We obtain the following results:

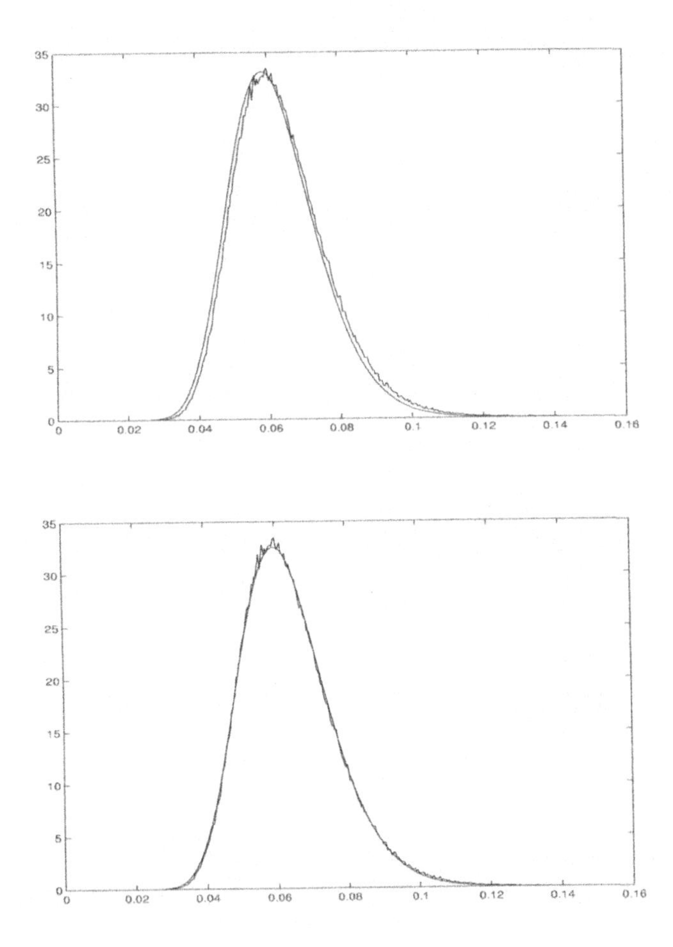

Fig. 8.16. Case (2.b): biased- and unbiased-density plots; $F(0)$ increased of 2%; res=15y.

res	MCVol	MCinf	MCsup	RebVol	HWVol	differr	errperc
2	0.0969	0.0966	0.0973	0.0977	0.0973	-0.0008	-0.7825
9	0.0895	0.0891	0.0899	0.0897	0.0898	-0.0002	-0.2270
17	0.0968	0.0962	0.0973	0.0965	0.0965	0.0002	0.2401

errinf	errsup	sw0	mean	third	kurt
-1.1977	-0.3707	0.0623	0.0631	0.0001	3.0538
-0.6771	0.2191	0.0635	0.0650	0.0000	3.0186
-0.3296	0.8034	0.0607	0.0613	0.0000	3.0010

The approximation appears to be satisfactory. Also with this parametric form of case (3), we consider the usual upward shifts by 20% and 2% respectively for the Φ's and the $F(0)$'s. Notice again that here, too, increasing the Φ's amounts to a relative increase of 20% of the volatility structure.

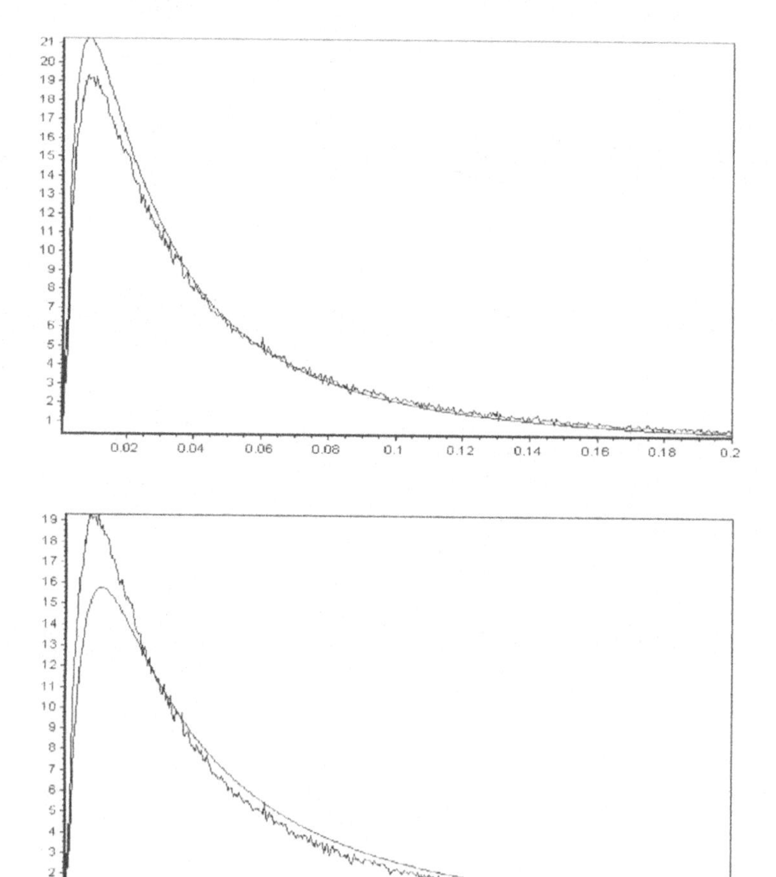

Fig. 8.17. Case (2.b): biased- and unbiased-density plots; c, d increased of 20%; res=15y.

When we increase the computed Φ's of 20%, we find the following numerical results:

res	MCVol	MCinf	MCsup	RebVol	HWVol	differr	errperc
2	0.1150	0.1150	0.1160	0.1160	0.1160	-0.0012	-0.9980
9	0.1073	0.1068	0.1078	0.1073	0.1074	0.0000	0.0099
17	0.1160	0.1153	0.1167	0.1156	0.1156	0.0004	0.3559

errinf	errsup	sw0	mean	third	kurt
-1.4200	-0.5840	0.0623	0.0634	0.0001	3.0500
-0.4483	0.4639	0.0635	0.0658	0.0001	3.0209
-0.2543	0.9589	0.0607	0.0617	0.0000	3.0124

Results appear to be satisfactory in all cases. We then restore the original Φ's and increase the $F(0)$'s of 2%, obtaining:

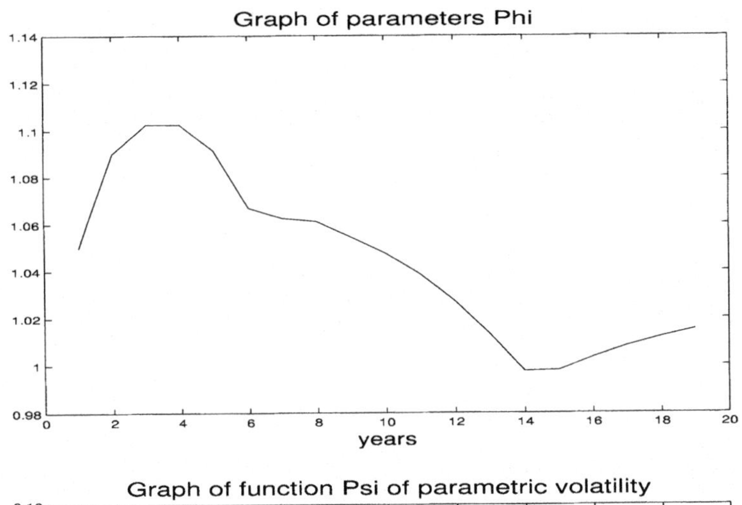

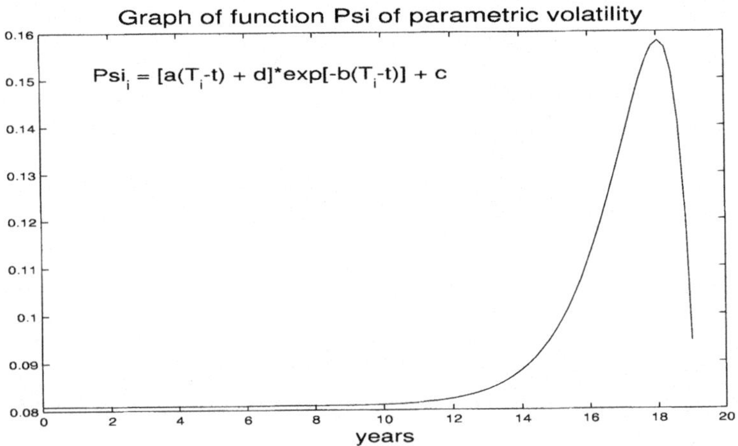

Fig. 8.18. Case (3): parameters Φ and function $t \mapsto \psi(t; a, b, c, d)$.

res	MCVol	MCinf	MCsup	RebVol	HWVol	differr	errperc
2	0.0987	0.0983	0.0991	0.0995	0.0991	-0.0009	-0.8840
9	0.0898	0.0894	0.0902	0.0901	0.0902	-0.0003	-0.3174
17	0.0967	0.0961	0.0972	0.0965	0.0965	0.0002	0.1565

errinf	errsup	sw0	mean	third	kurt
-1.2900	-0.4850	0.0822	0.0836	0.0001	3.0600
-0.7489	0.1104	0.0835	0.0861	0.0000	3.0230
-0.4029	0.7099	0.0807	0.0818	0.0000	2.9838

As we can see from these results, the approximation error remains roughly at the same level in all stressed and unstressed subcases, and is larger for the swaption with first date of reset equal to 2 years (first payment at 3 years) and last payment in 19 years. The error appears to be lower for swaptions whose underlying swap is one year long.

The biased- and unbiased-density plots, in the above "unstressed" and "stressed" cases, are shown in Figures 8.19, 8.20, 8.21 and 8.22.

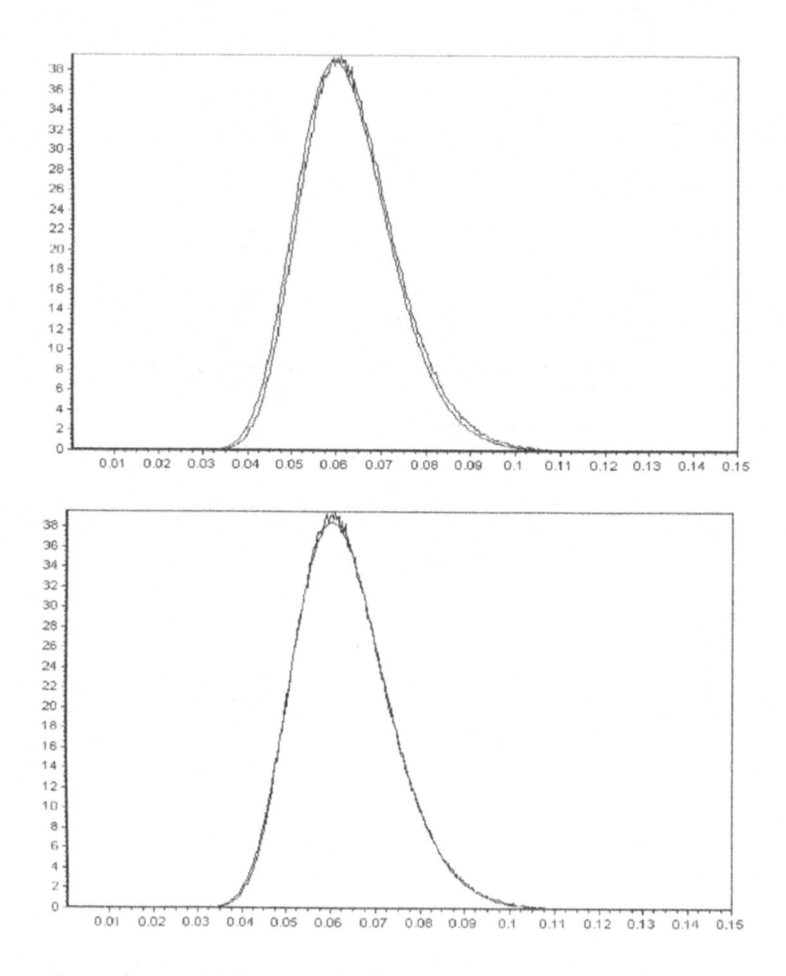

Fig. 8.19. Case (3.a): biased- and unbiased-density plots; res=2y.

(3.b): Perfect correlations

Now we select the perfect instantaneous-correlation matrix, setting θ's to zero, while maintaining the original a, b, c, d and Φ's. We thus have all ρ's set to one. We obtain the following results:

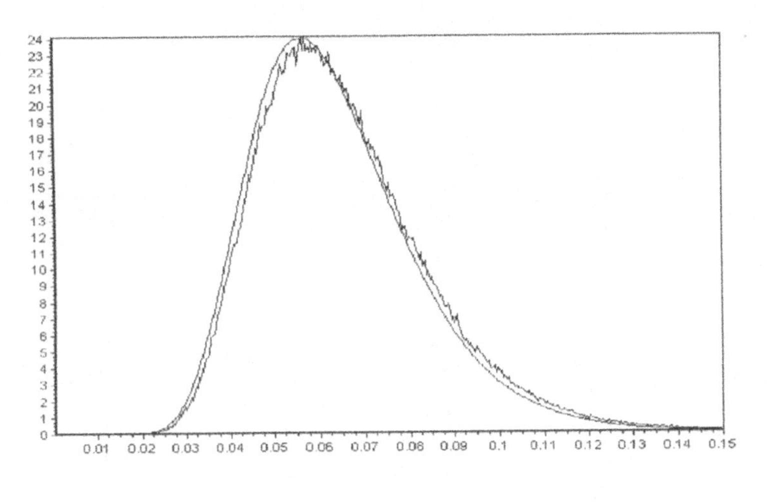

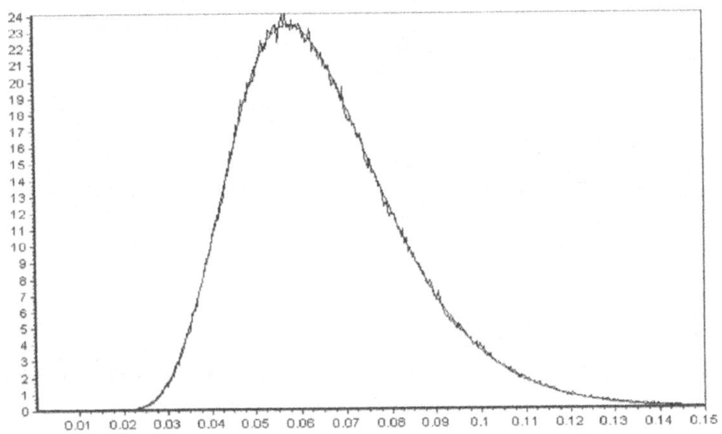

Fig. 8.20. Case (3.a): biased- and unbiased-density plots; res=9y.

res	MCVol	MCinf	MCsup	RebVol	HWVol	differr	errperc
2	0.0984	0.0980	0.0988	0.0993	0.0989	-0.0010	-1.0025
9	0.0902	0.0898	0.0906	0.0902	0.0903	0.0000	0.0382
17	0.0964	0.0958	0.0969	0.0965	0.0965	-0.0002	-0.1613

errinf	errsup	sw0	mean	third	kurt
-1.4180	-0.5904	0.0623	0.0631	0.0001	3.0390
-0.4096	0.4821	0.0635	0.0650	0.0000	3.0011
-0.7323	0.4033	0.0607	0.0613	0.0000	2.9869

The approximation seems to be working in this case too.

The next step is the usual increase of 20% for the vector Φ. We observe a small worsening in the percentage difference between Monte Carlo's and Rebonato's volatilities. This difference is lower for the swaption resetting at 9 years, as we can see from the following:

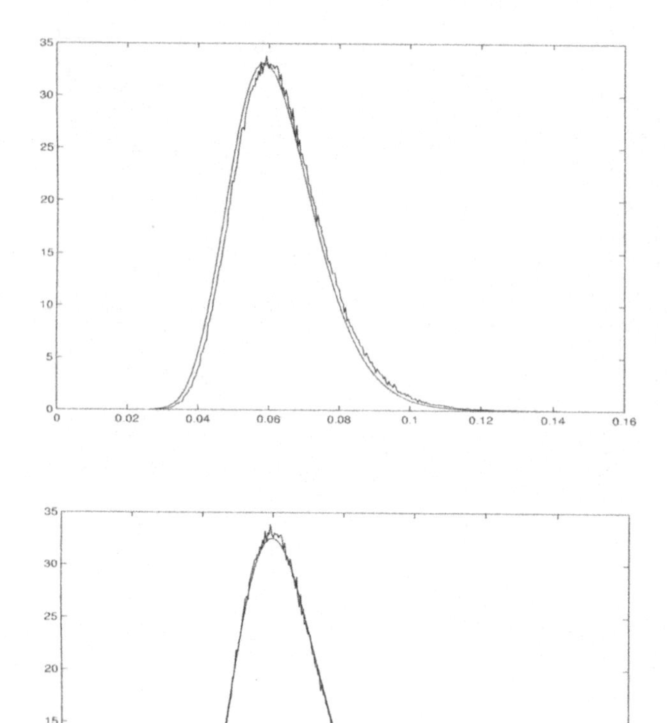

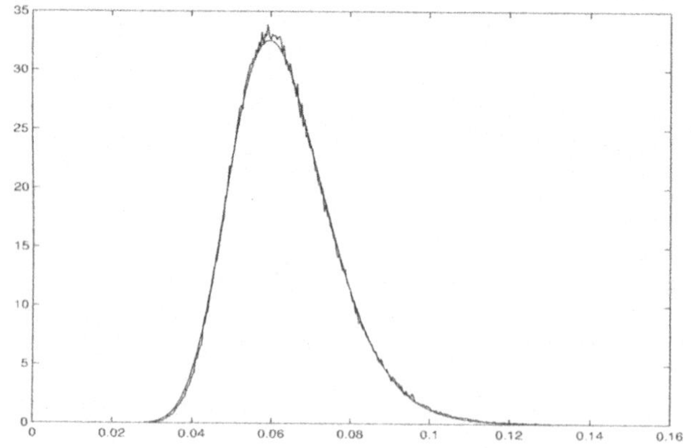

Fig. 8.21. Case (3.a): biased- and unbiased-density plots; Φ increased of 20% ; res=2y.

res	MCVol	MCinf	MCsup	RebVol	HWVol	differr	errperc	errinf	errsup	sw0	mean	third	kurt
2	0.1168	0.1163	0.1173	0.1181	0.1176	-0.0013	-1.1196	-1.5374	-0.7052	0.0623	0.0635	0.0001	3.0508
9	0.1078	0.1073	0.1083	0.1079	0.1080	-0.0001	-0.0624	-0.5195	0.3906	0.0635	0.0658	0.0001	3.0083
17	0.1160	0.1153	0.1167	0.1156	0.1156	0.0004	0.3353	-0.2735	0.9370	0.0607	0.0617	0.0000	2.9968

Next, as usual, we restore the Φ's and increase the $F(0)$'s of 2%. In this case, we notice that the most pronounced (and yet small in absolute terms) worsening with respect to the unstressed situation is obtained for a 9 year reset date.

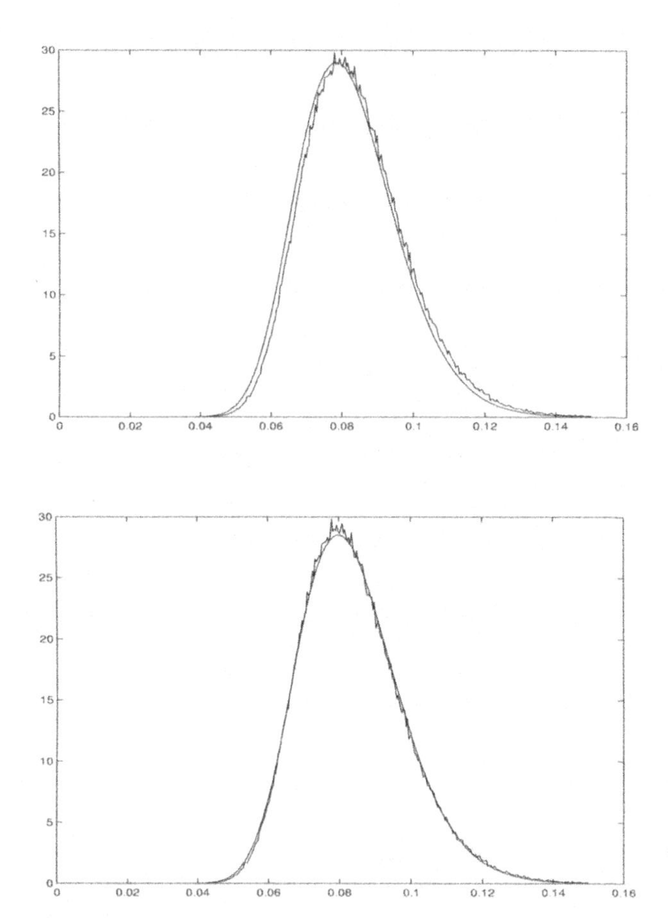

Fig. 8.22. Case (3.a): biased- and unbiased-density plots; $F(0)$ increased of 2%; res=2y.

res	MCVol	MCinf	MCsup	RebVol	HWVol	differr	errperc	errinf	errsup	sw0	mean	third	kurt
2	0.1150	0.1150	0.1160	0.1160	0.1160	-0.0011	-0.9590	-1.3700	-0.5460	0.0623	0.0634	0.0001	3.0300
9	0.0904	0.0900	0.0908	0.0906	0.0907	-0.0002	-0.2465	-0.6778	0.1811	0.0835	0.0861	0.0000	3.0183
17	0.0965	0.0959	0.0970	0.0965	0.0965	-0.0000	-0.0481	-0.6102	0.5078	0.0807	0.0818	0.0000	3.0030

The approximation appears to be working in both stressed cases too. See also the related density plots in Figures 8.23, 8.24 and 8.25.

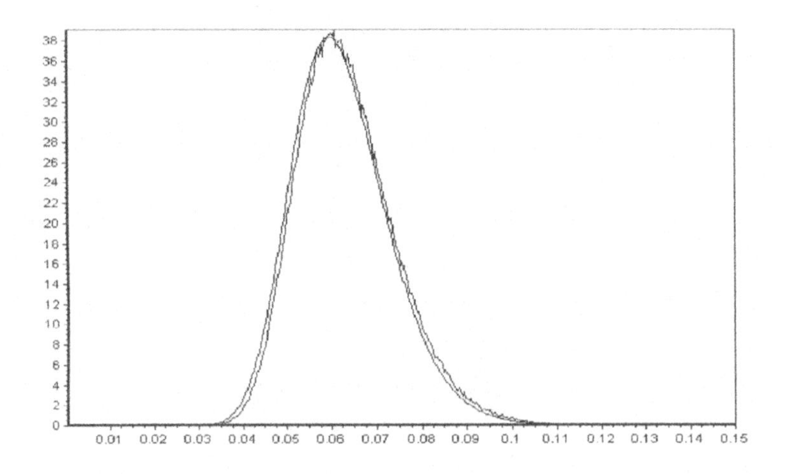

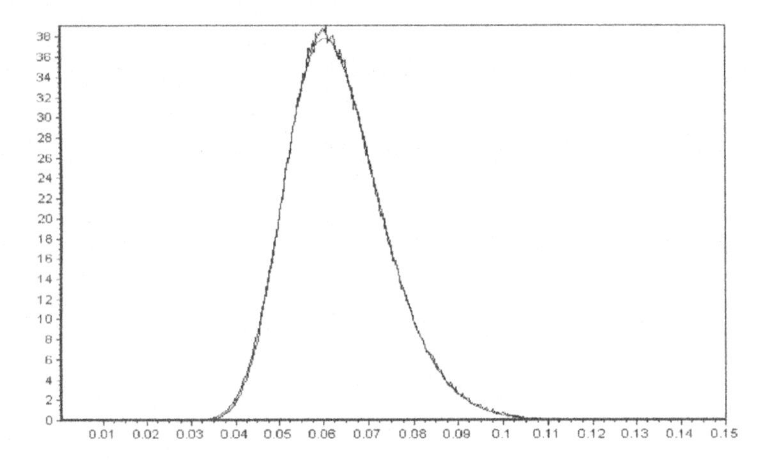

Fig. 8.23. Case (3.b): biased- and unbiased-density plots; $\theta = 0$; res=2y.

(3.c): Some negative rank-two correlations

Concerning our chosen Formulation 7 of case (3), we finally present the results obtained when the rank-two instantaneous correlation matrix is partially negative, with the same values as in cases (1.c) and (2.b). Again, we first consider the unstressed situation (first row), then we increase Φ's by 20% (second row) and finally, after resetting the Φ's, we increase the $F(0)$'s by 2% (third row). We obtain:

res	MCVol	MCinf	MCsup	RebVol	HWVol	differr	errperc
15	0.0905	0.0900	0.0909	0.0899	0.0899	0.0006	0.6758
15	0.1075	0.1069	0.1081	0.1077	0.1078	-0.0002	-0.2134
15	0.0901	0.0896	0.0905	0.0899	0.0900	0.0002	0.2084

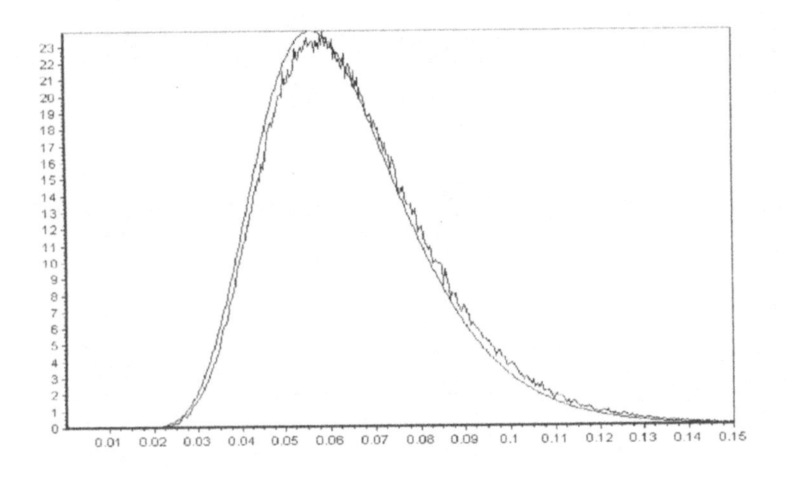

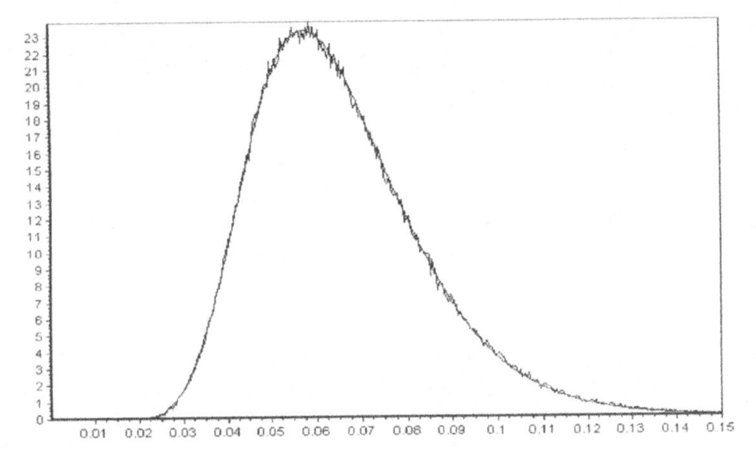

Fig. 8.24. Case (3.b): biased- and unbiased-density plots; $\theta = 0$; res=9y.

errinf	errsup	sw0	mean	third	kurt
0.1584	1.1879	0.0619	0.0629	0.0000	2.9957
-0.7640	0.3315	0.0619	0.0634	0.0000	3.0123
-0.3002	0.7120	0.0819	0.0836	0.0000	3.0183

The approximation appears to be working in these last cases too. See also the related density plots in Figures 8.26, 8.27 and 8.28.

8.5 The "Testing Plan" for Terminal Correlations

"I appreciate your confidence in me, Plastic Man. Of course I have a plan."
Batman in JLA 26, 1998, DC Comics.

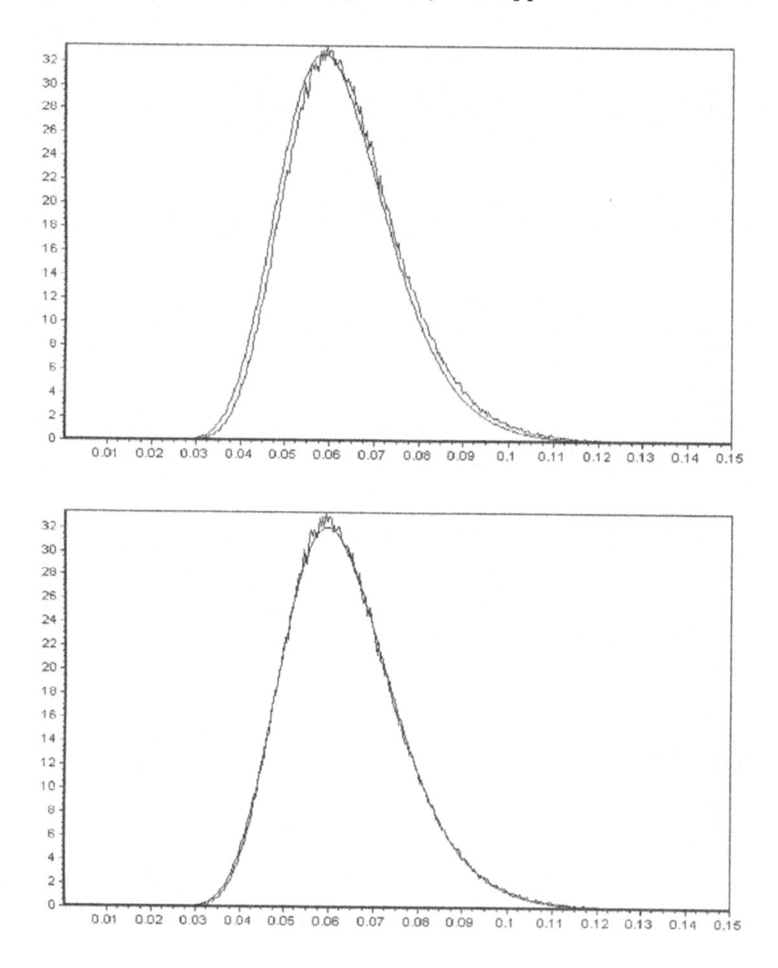

Fig. 8.25. Case (3.b): biased- and unbiased-density plots; Φ increased of 20%; res=2y.

We plan to test both our analytical and Rebonato's terminal correlation formulas (6.70) and (6.71) as follows.

Once the initial reset time T_α has been fixed, we will deal with calculations for the terminal correlation between two any forward rates $F_i(T_\alpha) = F(T_\alpha; T_{i-1}, T_i)$ and $F_j(T_\alpha) = F(T_\alpha; T_{j-1}, T_j)$ at time T_α, $\alpha \le i-1 < j \le \beta$, under the T_α-forward measure Q^α. We will consider the three cases $\alpha = 1,\ \alpha = 9,\ \alpha = 15$, where the last forward rate has always maturity equal to $T_\beta = 20$ years. We adopt again Formulation 7 for instantaneous volatilities, although in some cases we will let it collapse to the formulation of TABLE 3, here too. Again, we adopt the two-factor parametric form in θ for instantaneous correlations, $\rho^B_{i,j} = \cos(\theta_i - \theta_j)$.

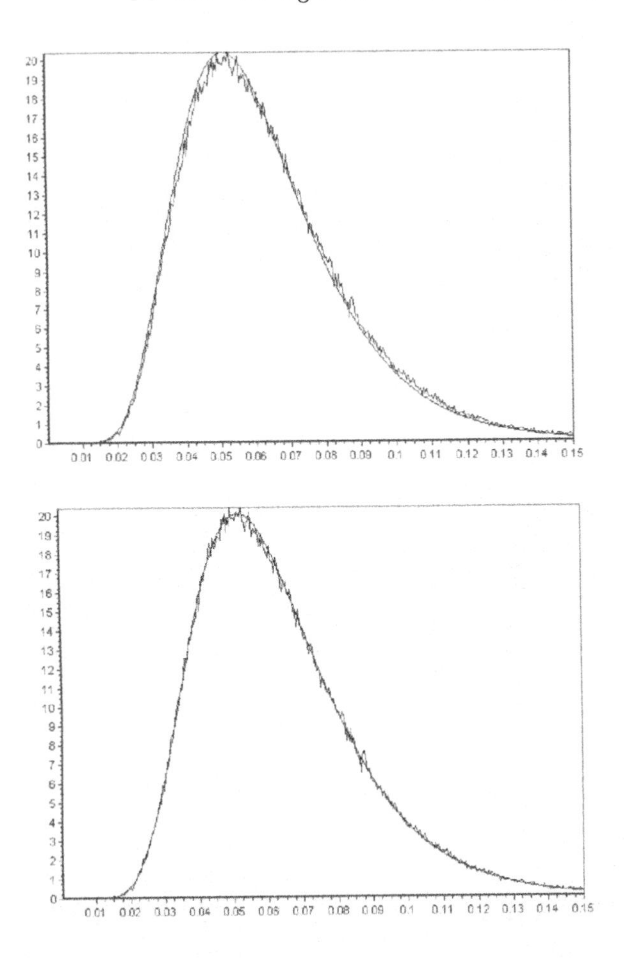

Fig. 8.26. Case (3.c): biased- and unbiased-density plots; res=15y.

In the following, we present a short map of our sets of testing parameters, most of which are retained from our previous volatility testing plan.

- (i) *Humped and maturity-adjusted instantaneous volatilities depending only on time to maturity, typical rank-two correlations.*
  These are the same volatility and correlation parameters as in case (3.a) of the volatility testing plan, and we report them below. We use Formulation 7 with $a = 0.1908$, $b = 0.9746$, $c = 0.0808$, $d = 0.0134$ and $\theta = [\theta_{1-8}\ \theta_{9-17}\ \theta_{18,19}]$, where

$$\theta_{1-8} = [0.0147\ 0.0643\ 0.1032\ 0.1502\ 0.1969\ 0.2239\ 0.2771\ 0.2950],$$
$$\theta_{9-17} = [0.3630\ 0.3810\ 0.4217\ 0.4836\ 0.5204\ 0.5418\ 0.5791\ 0.6496\ 0.6679],$$
$$\theta_{18,19} = [0.7126\ 0.7659].$$

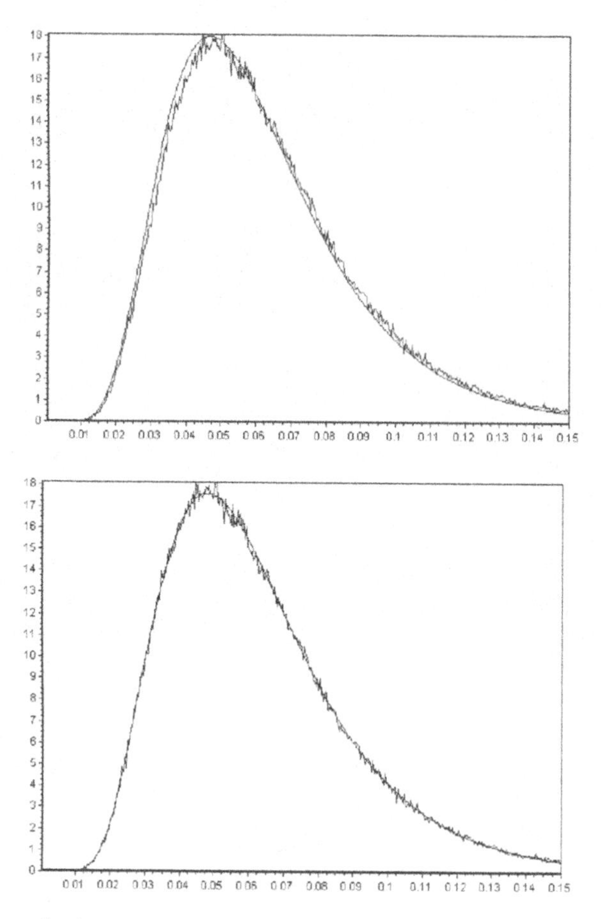

Fig. 8.27. Case (3.c): biased- and unbiased-density plots; Φ increased of 20%; res=15y.

This set of angles implies positive and decreasing instantaneous correlations as in Figure 8.1. The values of the Φ's in the volatility structure are shown in the same figure.

- (ii) *Constant instantaneous volatilities, typical rank-two correlations.* These are the same volatility and correlation parameters as in case (1.a) of the volatility testing plan: Formulation 7 with $a = 0,\ b = 0,\ c = 1,\ d = 0$. This amounts to using the formulation of TABLE 3, since the ψ part of Formulation 7 is forced to one and the Φ's now act as the constant volatilities s in TABLE 3. The correlation angles are taken as in the previous case (i).

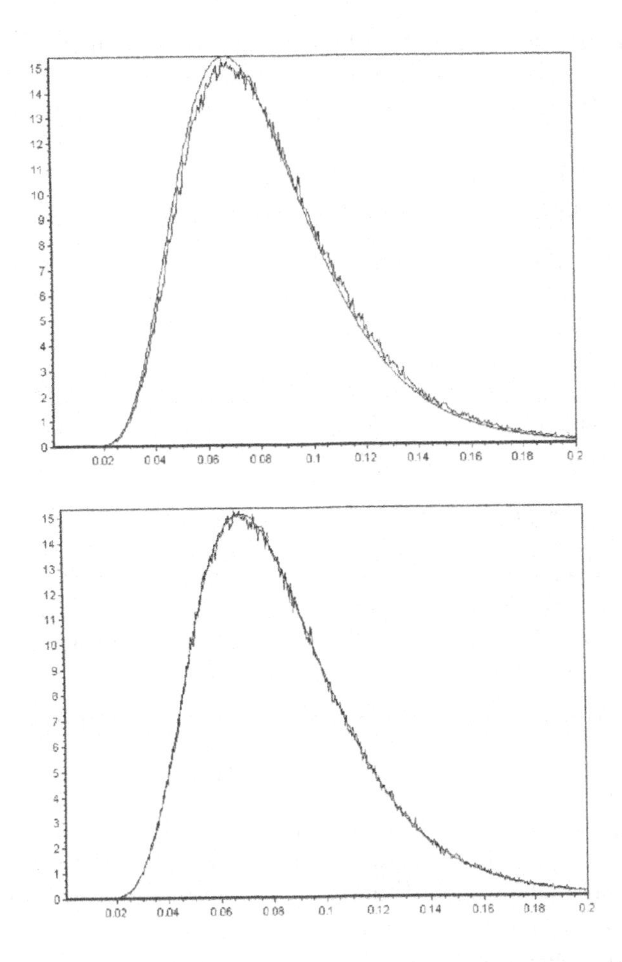

Fig. 8.28. Case (3.c): biased- and unbiased-density plots; $F(0)$ increased of 2%; res=15y.

- (iii) *Humped and maturity-adjusted instantaneous volatilities depending only on time to maturity, some negative rank-two correlations.*
  These are the same volatility and correlation parameters as in case (3.c) of the volatility testing plan: Formulation 7 with $a = 0.1908,\ b = 0.9746,\ c = 0.0808,\ d = 0.0134$ as in case (i) and $\theta = [\theta_{1-9}\ \theta_{10-17}\ \theta_{18,19}]$, where

$$\begin{aligned}\theta_{1-9} &= [0\ 0.0000\ 0.0013\ 0.0044\ 0.0096\ 0.0178\ 0.0299\ 0.0474\ 0.0728],\\ \theta_{10-17} &= [0.1100\ 0.1659\ 0.2534\ 0.3989\ 0.6565\ 1.1025\ 1.6605\ 2.0703],\\ \theta_{18,19} &= [2.2825\ 2.2260].\end{aligned}$$

- (iv) *Constant instantaneous volatilities, some negative rank-two correlations.*

Formulation 7 with a, b, c and d as in case (ii) and θ as in case (iii) (analogous to case (1.c) of the volatility testing plan).

- (v) *Constant instantaneous volatilities, perfect correlation, upwardly shifted Φ's.*
  Formulation 7 with a, b, c and d as in (ii), $\theta = [0\ 0\ \ldots\ 0\ 0]$ implying $\rho^B_{i,j} = 1$ for all i, j, and Φ's upwardly shifted by 0.2.

In each of the above situations we will display some numerical results, consisting of a comparison of both our analytical and Rebonato's formulas with the corresponding terminal correlations of forward rates (as implied by the LFM dynamics) whose computation is based on a Monte Carlo simulation with four time steps per year and 400000 simulated paths. More specifically, we will compare the three formulas (6.69), (6.70), and (6.71).

We thus compute the three correlation matrices:

$$\mathrm{Corr}^{\alpha}_{\mathrm{MC}}(F_i(T_\alpha), F_j(T_\alpha)),\ \mathrm{Corr}^{\mathrm{AN}}(F_i(T_\alpha), F_j(T_\alpha)),\ \mathrm{Corr}^{\mathrm{REB}}(F_i(T_\alpha), F_j(T_\alpha)),$$

which we denote, respectively, by "MCcorr", "ancorr", "Rebcorr".

To estimate the Monte Carlo error in the "real" correlation matrix we reason as follows. The i, j entry of the (Monte Carlo) estimated correlation matrix is the mean over scenarios s of the terms

$$\left(\frac{(F_i^s(T_\alpha) - \mathrm{Mean}\{F_i(T_\alpha)\})(F_j^s(T_\alpha) - \mathrm{Mean}\{F_j(T_\alpha)\})}{\mathrm{Std}\,\{F_i(T_\alpha)\}\mathrm{Std}\{F_j(T_\alpha)\}} \right)_s,$$

where s denotes the scenario index. Actually, this term already contains statistics built upon the scenarios, such as "Mean" and "Std", which denote the mean and standard deviation of the given variables over the different scenarios. We ignore the error in such "inner" statistics leading to the Monte Carlo estimation of the correlation, and consider as an approximation of the standard error of the method, corresponding to a two-side 98% window around the mean, the quantity:

$$\mathrm{MCerr} = \frac{2.33}{\sqrt{\mathrm{npath}}}\,\mathrm{Std}\left\{ \frac{(F_i^s(T_\alpha) - \mathrm{Mean}\{F_i(T_\alpha)\})(F_j^s(T_\alpha) - \mathrm{Mean}\{F_j(T_\alpha)\})}{\mathrm{Std}\{F_i(T_\alpha)\}\,\mathrm{Std}\{F_j(T_\alpha)\}} \right\}$$

where "npath" denotes again the number of paths in the Monte Carlo method. We are well aware that this definition of Monte Carlo error is far from being ideal, but it is easily computable and gives us a quick estimate of the true error.

We use the quantity "MCerr" to compute the "inf" and "sup" correlation matrices, "corrinf" and "corrsup", corresponding to the extremes of the 98% terminal correlation window, simply by adding and subtracting "MCerr" to the estimated correlation matrix. We will see later on, in several cases, that we can have entries in "corrsup" larger than one. Of course, each of these entries should be replaced by 1 to obtain a tighter upper bound for the correlation matrix. However, we will report, for completeness, the values generated by our definition.

Again, we consider the following percentage differences (errors):

$$100\,(\mathrm{Corr}^{\alpha}_{\mathrm{MC}} - \mathrm{Corr}^{\mathrm{AN}})/\mathrm{Corr}^{\alpha}_{\mathrm{MC}},$$
$$100\,(\mathrm{Corr}^{\alpha}_{\mathrm{MC}} - \mathrm{Corr}^{\mathrm{REB}})/\mathrm{Corr}^{\alpha}_{\mathrm{MC}},$$
$$100\,(\mathrm{Corr}^{\mathrm{REB}} - \mathrm{Corr}^{\mathrm{AN}})/\mathrm{Corr}^{\mathrm{AN}},$$

where each operator acts componentwise. We denote these matrices, respectively, by "percan", "percreb" and "anreb".

As we said before, for each point in the above testing plan we consider three possible values for the initial expiry T_α, the time where the terminal correlation is computed: $T_\alpha = 1y$, $9y$ and $15y$, and the last forward rate is always taken with maturity $T_\beta = 20$.

We display only a few of the obtained results for each choice of α, by taking into consideration the main cases, although we comment all possible situations.

In every case we analyze, we will see that our analytical approximation, formula (6.70), represents a better approximation of the true value (6.69) than Rebonato's formula (6.71), although the difference will be usually rather small. Moreover, we find that the percentage (non-zero) differences between the Monte Carlo method and Rebonato's formula are always negative.

The smallest differences occur in proximity of the diagonal of the correlation matrix, i.e. for (almost) adjacent forward rates, and there the analytical formulas provide an excellent approximation.

The largest differences are observed in case (v): constant instantaneous volatilities, perfect correlation, upwardly shifted Φ's. This was somehow expected, since shifting the volatilities parameters Φ by a large amount, such as 0.2, increases randomness. The increased randomness implies that the freezing procedure, implicit in our analytical formulas, has much heavier consequences on the approximations than in the previous cases (i)-(iv), exactly as we saw for volatility approximations.

8.6 Test Results for Terminal Correlations

8.6.1 Case (i): Humped and Maturity-Adjusted Instantaneous Volatilities Depending only on Time to Maturity, Typical Rank-Two Correlations

We consider Formulation 7 with the volatility and correlation parameters indicated in the previous section for this case (i). By recalling that we may express the parameters Φ's as functions of a, b, c, d through formula (6.28), the computed $\Phi = [\Phi_{1-9}\ \Phi_{10-18}\ \Phi_{19}]$ matching the caplet volatilities are:

$$\Phi_{1-9} = [1.0500\,1.0900\,1.1025\,1.1025\,1.0913\,1.0669\,1.0624\,1.0611\,1.0544]$$
$$\Phi_{10-18} = [1.0475\,1.0386\,1.0270\,1.0132\,0.9975\,0.9979\,1.0033\,1.0079\,1.0119],$$
$$\Phi_{19} = 1.0152,$$

exactly as in the case (3) of the volatility test plan.

We have now three subcases according to the different time T_α we choose.

(i.a): $T_\alpha = 1$

In this case, the correlation matrices are 18×18 matrices. We only show the correlation matrix obtained with the Monte Carlo method in Tables 8.9 and 8.10, and the first four columns of the percentage difference matrices, "percan", "percreb", which are the most significant ones, in Table 8.11.

MCcorr	2y	3y	4y	5y	6y	7y	8y	9y	10y
2y	1.0000	0.9989	0.9957	0.9900	0.9855	0.9754	0.9712	0.9533	0.9478
3y	0.9989	1.0000	0.9987	0.9948	0.9912	0.9830	0.9795	0.9642	0.9594
4y	0.9957	0.9987	1.0000	0.9987	0.9967	0.9911	0.9885	0.9763	0.9724
5y	0.9900	0.9948	0.9987	1.0000	0.9995	0.9966	0.9949	0.9859	0.9827
6y	0.9855	0.9912	0.9967	0.9995	1.0000	0.9986	0.9974	0.9902	0.9876
7y	0.9754	0.9830	0.9911	0.9966	0.9986	1.0000	0.9998	0.9963	0.9946
8y	0.9712	0.9795	0.9885	0.9949	0.9974	0.9998	1.0000	0.9977	0.9963
9y	0.9533	0.9642	0.9763	0.9859	0.9902	0.9963	0.9977	1.0000	0.9998
10y	0.9478	0.9594	0.9724	0.9827	0.9876	0.9946	0.9963	0.9998	1.0000
11y	0.9343	0.9474	0.9622	0.9744	0.9804	0.9895	0.9920	0.9983	0.9992
12y	0.9108	0.9261	0.9437	0.9587	0.9663	0.9787	0.9822	0.9927	0.9947
13y	0.8952	0.9118	0.9310	0.9477	0.9562	0.9705	0.9747	0.9876	0.9903
14y	0.8855	0.9029	0.9230	0.9406	0.9497	0.9651	0.9697	0.9840	0.9871
15y	0.8677	0.8865	0.9082	0.9274	0.9374	0.9547	0.9599	0.9767	0.9804
16y	0.8309	0.8520	0.8766	0.8988	0.9106	0.9313	0.9377	0.9592	0.9641
17y	0.8206	0.8423	0.8677	0.8906	0.9028	0.9245	0.9312	0.9538	0.9591
18y	0.7945	0.8176	0.8447	0.8694	0.8827	0.9065	0.9140	0.9395	0.9455
19y	0.7612	0.7860	0.8151	0.8419	0.8564	0.8827	0.8911	0.9199	0.9267

Table 8.9. Case (i.a): Monte Carlo correlation matrix, first nine columns.

As we can see from these results, the percentage errors are rather low, although it is already clear that our analytical formula works slightly better than Rebonato's.

(i.b): $T_\alpha = 9$

The accuracy of our analytical formula for terminal correlations is again pointed out by the results shown in Table 8.12.

(i.c): $T_\alpha = 15$

In this case, we have 4×4 correlation matrices, so that we can show all the obtained matrices in Table 8.13.

MCcorr	11y	12y	13y	14y	15y	16y	17y	18y	19y
2y	0.9343	0.9108	0.8952	0.8855	0.8677	0.8309	0.8206	0.7945	0.7612
3y	0.9474	0.9261	0.9118	0.9029	0.8865	0.8520	0.8423	0.8176	0.7860
4y	0.9622	0.9437	0.9310	0.9230	0.9082	0.8766	0.8677	0.8447	0.8151
5y	0.9744	0.9587	0.9477	0.9406	0.9274	0.8988	0.8906	0.8694	0.8419
6y	0.9804	0.9663	0.9562	0.9497	0.9374	0.9106	0.9028	0.8827	0.8564
7y	0.9895	0.9787	0.9705	0.9651	0.9547	0.9313	0.9245	0.9065	0.8827
8y	0.9920	0.9822	0.9747	0.9697	0.9599	0.9377	0.9312	0.9140	0.8911
9y	0.9983	0.9927	0.9876	0.9840	0.9767	0.9592	0.9538	0.9395	0.9199
10y	0.9992	0.9947	0.9903	0.9871	0.9804	0.9641	0.9591	0.9455	0.9267
11y	1.0000	0.9981	0.9951	0.9928	0.9876	0.9741	0.9698	0.9579	0.9413
12y	0.9981	1.0000	0.9993	0.9983	0.9954	0.9862	0.9830	0.9739	0.9604
13y	0.9951	0.9993	1.0000	0.9998	0.9983	0.9917	0.9891	0.9816	0.9700
14y	0.9928	0.9983	0.9998	1.0000	0.9993	0.9942	0.9921	0.9854	0.9750
15y	0.9876	0.9954	0.9983	0.9993	1.0000	0.9975	0.9960	0.9911	0.9826
16y	0.9741	0.9862	0.9917	0.9942	0.9975	1.0000	0.9998	0.9980	0.9932
17y	0.9698	0.9830	0.9891	0.9921	0.9960	0.9998	1.0000	0.9990	0.9952
18y	0.9579	0.9739	0.9816	0.9854	0.9911	0.9980	0.9990	1.0000	0.9986
19y	0.9413	0.9604	0.9700	0.9750	0.9826	0.9932	0.9952	0.9986	1.0000

Table 8.10. Case (i.a): Monte Carlo correlation matrix, second nine columns.

percan	2y	3y	4y	5y
2y	0.0000	0.0025	0.0028	0.0040
3y	0.0025	-0.0000	0.0007	0.0035
4y	0.0028	0.0007	-0.0000	0.0010
5y	0.0040	0.0035	0.0010	0.0000
6y	0.0057	0.0065	0.0029	0.0005
7y	0.0093	0.0101	0.0054	0.0017
8y	0.0106	0.0115	0.0064	0.0023
9y	0.0161	0.0160	0.0099	0.0049
10y	0.0176	0.0171	0.0108	0.0056
11y	0.0216	0.0203	0.0134	0.0078
12y	0.0288	0.0262	0.0184	0.0120
13y	0.0334	0.0300	0.0217	0.0148
14y	0.0362	0.0321	0.0236	0.0166
15y	0.0415	0.0365	0.0274	0.0200
16y	0.0527	0.0461	0.0357	0.0272
17y	0.0555	0.0483	0.0376	0.0291
18y	0.0632	0.0547	0.0433	0.0341
19y	0.0729	0.0628	0.0503	0.0403

percreb	2y	3y	4y	5y
2y	0.0000	-0.0155	-0.0601	-0.1021
3y	-0.0155	-0.0000	-0.0130	-0.0336
4y	-0.0601	-0.0130	-0.0000	-0.0046
5y	-0.1021	-0.0336	-0.0046	0.0000
6y	-0.1320	-0.0503	-0.0117	-0.0016
7y	-0.1478	-0.0595	-0.0162	-0.0034
8y	-0.1547	-0.0636	-0.0183	-0.0045
9y	-0.1659	-0.0709	-0.0224	-0.0066
10y	-0.1711	-0.0745	-0.0245	-0.0078
11y	-0.1793	-0.0802	-0.0279	-0.0097
12y	-0.1914	-0.0886	-0.0330	-0.0127
13y	-0.2010	-0.0955	-0.0371	-0.0152
14y	-0.2088	-0.1010	-0.0405	-0.0172
15y	-0.2151	-0.1057	-0.0435	-0.0192
16y	-0.2270	-0.1144	-0.0494	-0.0233
17y	-0.2300	-0.1170	-0.0513	-0.0247
18y	-0.2389	-0.1238	-0.0561	-0.0282
19y	-0.2510	-0.1332	-0.0628	-0.0333

Table 8.11. Case (i.a): percentage differences, first fourth columns.

MCcorr	10y	11y	12y	13y	14y	15y	16y	17y	18y	19y
10y	1.0000	0.9956	0.9828	0.9707	0.9625	0.9534	0.9360	0.9304	0.9167	0.8982
11y	0.9956	1.0000	0.9955	0.9882	0.9827	0.9758	0.9614	0.9567	0.9447	0.9280
12y	0.9828	0.9955	1.0000	0.9983	0.9958	0.9920	0.9823	0.9789	0.9696	0.9560
13y	0.9707	0.9882	0.9983	1.0000	0.9995	0.9976	0.9907	0.9881	0.9804	0.9688
14y	0.9625	0.9827	0.9958	0.9995	1.0000	0.9992	0.9940	0.9918	0.9852	0.9747
15y	0.9534	0.9758	0.9920	0.9976	0.9992	1.0000	0.9975	0.9960	0.9910	0.9825
16y	0.9360	0.9614	0.9823	0.9907	0.9940	0.9975	1.0000	0.9998	0.9980	0.9932
17y	0.9304	0.9567	0.9789	0.9881	0.9918	0.9960	0.9998	1.0000	0.9990	0.9952
18y	0.9167	0.9447	0.9696	0.9804	0.9852	0.9910	0.9980	0.9990	1.0000	0.9986
19y	0.8982	0.9280	0.9560	0.9688	0.9747	0.9825	0.9932	0.9952	0.9986	1.0000

percan	10y	11y	12y	13y	14y	15y	16y	17y	18y	19y
10y	0.0000	-0.0069	-0.0189	-0.0310	-0.0441	-0.0509	-0.0343	-0.0423	-0.0255	0.0107
11y	-0.0069	-0.0000	-0.0015	-0.0051	-0.0110	-0.0122	0.0072	0.0052	0.0256	0.0641
12y	-0.0189	-0.0015	-0.0000	-0.0010	-0.0037	-0.0036	0.0123	0.0136	0.0329	0.0683
13y	-0.0310	-0.0051	-0.0010	-0.0000	-0.0007	-0.0001	0.0133	0.0164	0.0349	0.0680
14y	-0.0441	-0.0110	-0.0037	-0.0007	0.0000	0.0009	0.0125	0.0164	0.0340	0.0655
15y	-0.0509	-0.0122	-0.0036	-0.0001	0.0009	0.0000	0.0062	0.0093	0.0232	0.0495
16y	-0.0343	0.0072	0.0123	0.0133	0.0125	0.0062	-0.0000	0.0002	0.0054	0.0206
17y	-0.0423	0.0052	0.0136	0.0164	0.0164	0.0093	0.0002	-0.0000	0.0030	0.0154
18y	-0.0255	0.0256	0.0329	0.0349	0.0340	0.0232	0.0054	0.0030	-0.0000	0.0047
19y	0.0107	0.0641	0.0683	0.0680	0.0655	0.0495	0.0206	0.0154	0.0047	-0.0000

percreb	10y	11y	12y	13y	14y	15y	16y	17y	18y	19y
10y	0.0000	-0.0497	-0.1539	-0.2444	-0.3107	-0.3571	-0.4053	-0.4324	-0.4665	-0.5010
11y	-0.0497	-0.0000	-0.0285	-0.0725	-0.1098	-0.1374	-0.1667	-0.1838	-0.2042	-0.2235
12y	-0.1539	-0.0285	-0.0000	-0.0101	-0.0265	-0.0400	-0.0548	-0.0640	-0.0743	-0.0829
13y	-0.2444	-0.0725	-0.0101	-0.0000	-0.0038	-0.0096	-0.0171	-0.0220	-0.0271	-0.0303
14y	-0.3107	-0.1098	-0.0265	-0.0038	0.0000	-0.0015	-0.0054	-0.0082	-0.0108	-0.0114
15y	-0.3571	-0.1374	-0.0400	-0.0096	-0.0015	0.0000	-0.0013	-0.0027	-0.0039	-0.0035
16y	-0.4053	-0.1667	-0.0548	-0.0171	-0.0054	-0.0013	-0.0000	-0.0003	-0.0007	-0.0001
17y	-0.4324	-0.1838	-0.0640	-0.0220	-0.0082	-0.0027	-0.0003	-0.0000	-0.0000	0.0007
18y	-0.4665	-0.2042	-0.0743	-0.0271	-0.0108	-0.0039	-0.0007	-0.0000	-0.0000	0.0004
19y	-0.5010	-0.2235	-0.0829	-0.0303	-0.0114	-0.0035	-0.0001	0.0007	0.0004	-0.0000

Table 8.12. Case (i.b): Monte Carlo correlation matrix and percentage differences.

In particular, look at "corrinf" and "corrsup", and notice that their entries are quite close to the corresponding entries of "MCcorr", thus suggesting that, in case (i), the Monte Carlo error is small and that we should not bother about it.

As for the remaining aspects, we can see results analogous to the previous cases.

8.6.2 Case (ii): Constant Instantaneous Volatilities, Typical Rank-Two Correlations.

Although this would more properly be the formulation in TABLE 3, as usual we equivalently resort to Formulation 7 and take $a = b = d = 0$ and $c = 1$ so that the caplet fitting forces the Φ's to equal the caplet volatilities, which are obtained from the market: $\Phi_i = v_{T_{i-1}-\text{caplet}}$. Such Φ values were plotted in Figure 8.2. The θ's are the same as in the previous case (i).

MCcorr	16y	17y	18y	19y
16y	1.0000	0.9964	0.9871	0.9758
17y	0.9964	1.0000	0.9968	0.9895
18y	0.9871	0.9968	1.0000	0.9977
19y	0.9758	0.9895	0.9977	1.0000

ancorr	16y	17y	18y	19y
16y	1.0000	0.9965	0.9873	0.9762
17y	0.9965	1.0000	0.9968	0.9896
18y	0.9873	0.9968	1.0000	0.9977
19y	0.9762	0.9896	0.9977	1.0000

anreb	16y	17y	18y	19y
16y	0.0000	-0.0319	-0.1000	-0.1720
17y	-0.0319	0.0000	-0.0208	-0.0634
18y	-0.1000	-0.0208	0.0000	-0.0124
19y	-0.1720	-0.0634	-0.0124	0.0000

Rebcorr	16y	17y	18y	19y
16y	1.0000	0.9968	0.9883	0.9779
17y	0.9968	1.0000	0.9970	0.9902
18y	0.9883	0.9970	1.0000	0.9979
19y	0.9779	0.9902	0.9979	1.0000

percan	16y	17y	18y	19y
16y	-0.0000	-0.0101	-0.0287	-0.0422
17y	-0.0101	0.0000	-0.0035	-0.0059
18y	-0.0287	-0.0035	0.0000	0.0005
19y	-0.0422	-0.0059	0.0005	-0.0000

corrinf	16y	17y	18y	19y
16y	0.9926	0.9889	0.9796	0.9683
17y	0.9889	0.9926	0.9894	0.9821
18y	0.9796	0.9894	0.9926	0.9903
19y	0.9683	0.9821	0.9903	0.9926

percreb	16y	17y	18y	19y
16y	-0.0000	-0.0421	-0.1287	-0.2143
17y	-0.0421	0.0000	-0.0243	-0.0693
18y	-0.1287	-0.0243	0.0000	-0.0118
19y	-0.2143	-0.0693	-0.0118	-0.0000

corrsup	16y	17y	18y	19y
16y	1.0074	1.0038	0.9945	0.9832
17y	1.0038	1.0074	1.0042	0.9970
18y	0.9945	1.0042	1.0074	1.0052
19y	0.9832	0.9970	1.0052	1.0074

Table 8.13. Case (i.c): Monte Carlo correlation matrix, correlation matrices from analytical formulas and percentage differences.

(ii.a): $T_\alpha = 1$

We show the Monte Carlo terminal correlation matrix in Tables 8.14 and 8.15, the first four columns of the analytical-approximations matrices, "ancorr", "Rebcorr", in Table 8.16 and the related percentage-difference matrices, "percan", "percreb", in Table 8.17.

It may be curious to notice that although, in this case, the percentage differences are of the same order of magnitude as in the previous case, the entries of "MCcorr" are centered between the corresponding entries of "ancorr" (slightly smaller) and "Rebcorr" (slightly larger). In this case too, both analytical formulas seem to provide excellent approximations.

(ii.b): $T_\alpha = 9$

The results in this case are similar to previous ones obtained with $T_\alpha = 1$, so that we show only the significant part of the percentage-differences matrix in Table 8.18.

(ii.c): $T_\alpha = 15$

In this long-maturity case, the percentage differences for a constant instantaneous-volatility structure (with typical rank-two correlations) are lower than the corresponding differences obtained from a humped and maturity-adjusted instantaneous-volatilities structure as in case (i.c). Also in this case, both approximations provide us with excellent results, as we can see from Table 8.19.

MCcorr	2y	3y	4y	5y	6y	7y	8y	9y	10y
2y	1.0000	0.9992	0.9962	0.9910	0.9869	0.9770	0.9729	0.9550	0.9495
3y	0.9992	1.0000	0.9989	0.9955	0.9925	0.9846	0.9812	0.9659	0.9611
4y	0.9962	0.9989	1.0000	0.9989	0.9972	0.9918	0.9893	0.9771	0.9731
5y	0.9910	0.9955	0.9989	1.0000	0.9996	0.9967	0.9951	0.9861	0.9829
6y	0.9869	0.9925	0.9972	0.9996	1.0000	0.9986	0.9974	0.9903	0.9876
7y	0.9770	0.9846	0.9918	0.9967	0.9986	1.0000	0.9998	0.9963	0.9946
8y	0.9729	0.9812	0.9893	0.9951	0.9974	0.9998	1.0000	0.9977	0.9963
9y	0.9550	0.9659	0.9771	0.9861	0.9903	0.9963	0.9977	1.0000	0.9998
10y	0.9495	0.9611	0.9731	0.9829	0.9876	0.9946	0.9963	0.9998	1.0000
11y	0.9359	0.9490	0.9630	0.9746	0.9804	0.9895	0.9920	0.9983	0.9992
12y	0.9124	0.9278	0.9444	0.9589	0.9663	0.9787	0.9822	0.9927	0.9947
13y	0.8968	0.9134	0.9317	0.9478	0.9562	0.9704	0.9747	0.9876	0.9903
14y	0.8870	0.9045	0.9237	0.9407	0.9497	0.9650	0.9696	0.9840	0.9871
15y	0.8693	0.8880	0.9089	0.9275	0.9374	0.9546	0.9599	0.9767	0.9804
16y	0.8325	0.8535	0.8773	0.8989	0.9106	0.9313	0.9377	0.9592	0.9641
17y	0.8222	0.8439	0.8684	0.8907	0.9028	0.9245	0.9312	0.9538	0.9591
18y	0.7961	0.8192	0.8454	0.8696	0.8828	0.9066	0.9140	0.9395	0.9455
19y	0.7628	0.7875	0.8159	0.8421	0.8565	0.8828	0.8912	0.9199	0.9269

Table 8.14. Case (ii.a): Monte Carlo correlation matrix, first nine columns.

MCcorr	11y	12y	13y	14y	15y	16y	17y	18y	19y
2y	0.9359	0.9124	0.8968	0.8870	0.8693	0.8325	0.8222	0.7961	0.7628
3y	0.9490	0.9278	0.9134	0.9045	0.8880	0.8535	0.8439	0.8192	0.7875
4y	0.9630	0.9444	0.9317	0.9237	0.9089	0.8773	0.8684	0.8454	0.8159
5y	0.9746	0.9589	0.9478	0.9407	0.9275	0.8989	0.8907	0.8696	0.8421
6y	0.9804	0.9663	0.9562	0.9497	0.9374	0.9106	0.9028	0.8828	0.8565
7y	0.9895	0.9787	0.9704	0.9650	0.9546	0.9313	0.9245	0.9066	0.8828
8y	0.9920	0.9822	0.9747	0.9696	0.9599	0.9377	0.9312	0.9140	0.8912
9y	0.9983	0.9927	0.9876	0.9840	0.9767	0.9592	0.9538	0.9395	0.9199
10y	0.9992	0.9947	0.9903	0.9871	0.9804	0.9641	0.9591	0.9455	0.9269
11y	1.0000	0.9981	0.9951	0.9928	0.9876	0.9741	0.9698	0.9580	0.9414
12y	0.9981	1.0000	0.9993	0.9983	0.9954	0.9863	0.9831	0.9739	0.9605
13y	0.9951	0.9993	1.0000	0.9998	0.9983	0.9917	0.9891	0.9816	0.9700
14y	0.9928	0.9983	0.9998	1.0000	0.9993	0.9942	0.9921	0.9855	0.9750
15y	0.9876	0.9954	0.9983	0.9993	1.0000	0.9975	0.9961	0.9911	0.9826
16y	0.9741	0.9863	0.9917	0.9942	0.9975	1.0000	0.9998	0.9980	0.9932
17y	0.9698	0.9831	0.9891	0.9921	0.9961	0.9998	1.0000	0.9990	0.9952
18y	0.9580	0.9739	0.9816	0.9855	0.9911	0.9980	0.9990	1.0000	0.9986
19y	0.9414	0.9605	0.9700	0.9750	0.9826	0.9932	0.9952	0.9986	1.0000

Table 8.15. Case (ii.a): Monte Carlo correlation matrix, second nine columns.

8.6.3 Case (iii): Humped and Maturity-Adjusted Instantaneous Volatilities Depending only on Time to Maturity, Some Negative Rank-Two Correlations.

Here instantaneous volatilities follow again Formulation 7 with $a = 0.1908$, $b = 0.9746$, $c = 0.0808$, $d = 0.0134$, as in case (i), and $\theta = [\theta_{1-9}\ \theta_{10-18}\ \theta_{19}]$, where

$$\begin{aligned}
\theta_{1-9} &= [0\ 0.0000\ 0.0013\ 0.0044\ 0.0096\ 0.0178\ 0.0299\ 0.0474\ 0.0728],\\
\theta_{10-18} &= [0.1100\ 0.1659\ 0.2534\ 0.3989\ 0.6565\ 1.1025\ 1.6605\ 2.0703\ 2.2825],\\
\theta_{19} &= 2.2260.
\end{aligned}$$

ancorr	2y	3y	4y	5y
2y	1.0000	0.9992	0.9962	0.9910
3y	0.9992	1.0000	0.9989	0.9955
4y	0.9962	0.9989	1.0000	0.9989
5y	0.9910	0.9955	0.9989	1.0000
6y	0.9869	0.9925	0.9972	0.9996
7y	0.9769	0.9845	0.9918	0.9967
8y	0.9728	0.9812	0.9893	0.9951
9y	0.9548	0.9658	0.9770	0.9860
10y	0.9493	0.9609	0.9730	0.9828
11y	0.9357	0.9488	0.9628	0.9745
12y	0.9120	0.9274	0.9442	0.9587
13y	0.8963	0.9130	0.9314	0.9475
14y	0.8865	0.9040	0.9233	0.9404
15y	0.8687	0.8875	0.9084	0.9271
16y	0.8317	0.8528	0.8767	0.8984
17y	0.8213	0.8431	0.8677	0.8901
18y	0.7951	0.8182	0.8446	0.8689
19y	0.7617	0.7864	0.8149	0.8413

Rebcorr	2y	3y	4y	5y
2y	1.0000	0.9992	0.9963	0.9912
3y	0.9992	1.0000	0.9989	0.9956
4y	0.9963	0.9989	1.0000	0.9989
5y	0.9912	0.9956	0.9989	1.0000
6y	0.9873	0.9927	0.9973	0.9996
7y	0.9774	0.9849	0.9920	0.9968
8y	0.9735	0.9817	0.9895	0.9952
9y	0.9557	0.9664	0.9774	0.9862
10y	0.9503	0.9617	0.9735	0.9831
11y	0.9368	0.9497	0.9634	0.9749
12y	0.9134	0.9285	0.9449	0.9592
13y	0.8978	0.9142	0.9323	0.9481
14y	0.8881	0.9053	0.9243	0.9411
15y	0.8704	0.8889	0.9095	0.9279
16y	0.8335	0.8544	0.8779	0.8993
17y	0.8233	0.8447	0.8690	0.8911
18y	0.7971	0.8200	0.8460	0.8700
19y	0.7638	0.7883	0.8164	0.8425

Table 8.16. Case (ii.a): correlation matrices from analytical formulas.

The Φ's obtained from caplet volatilities are as in case (i), while the instantaneous-correlation matrix implied by the chosen θ is again the matrix shown earlier in Figure 8.6.

We will see that, especially in the cases (iii.a) and (iii.b) the existence of some negative instantaneous correlations can induce larger percentage errors.

(iii.a): $T_\alpha = 1$

We start by showing in Table 8.20 the last columns of the Monte Carlo terminal correlations matrix "MCcorr". Subsequently, we show in Table 8.21 the corresponding matrix "percan" of percentage differences between "MCcorr" and the analytical approximations "ancorr".

We also include in Table 8.22 part of the matrix "anreb" of percentage differences between the analytical approximation "ancorr" and Rebonato's approximation "Rebcorr".

percan	2y	3y	4y	5y
2y	0.0000	0.0002	0.0010	0.0027
3y	0.0002	0.0000	0.0003	0.0015
4y	0.0010	0.0003	0.0000	0.0004
5y	0.0027	0.0015	0.0004	0.0000
6y	0.0042	0.0026	0.0011	0.0002
7y	0.0083	0.0059	0.0033	0.0014
8y	0.0101	0.0074	0.0045	0.0022
9y	0.0186	0.0149	0.0105	0.0068
10y	0.0215	0.0174	0.0126	0.0084
11y	0.0289	0.0240	0.0183	0.0131
12y	0.0426	0.0364	0.0291	0.0223
13y	0.0523	0.0453	0.0369	0.0291
14y	0.0586	0.0510	0.0420	0.0336
15y	0.0706	0.0621	0.0519	0.0423
16y	0.0969	0.0863	0.0736	0.0617
17y	0.1046	0.0934	0.0800	0.0674
18y	0.1249	0.1121	0.0969	0.0826
19y	0.1520	0.1371	0.1194	0.1028

percreb	2y	3y	4y	5y
2y	0.0000	-0.0015	-0.0084	-0.0204
3y	-0.0015	0.0000	-0.0028	-0.0109
4y	-0.0084	-0.0028	0.0000	-0.0027
5y	-0.0204	-0.0109	-0.0027	0.0000
6y	-0.0347	-0.0222	-0.0094	-0.0022
7y	-0.0488	-0.0336	-0.0171	-0.0062
8y	-0.0578	-0.0412	-0.0227	-0.0098
9y	-0.0727	-0.0537	-0.0318	-0.0159
10y	-0.0811	-0.0610	-0.0376	-0.0200
11y	-0.0916	-0.0701	-0.0446	-0.0251
12y	-0.1037	-0.0805	-0.0527	-0.0308
13y	-0.1128	-0.0885	-0.0591	-0.0356
14y	-0.1202	-0.0952	-0.0646	-0.0399
15y	-0.1242	-0.0987	-0.0671	-0.0415
16y	-0.1274	-0.1011	-0.0684	-0.0417
17y	-0.1284	-0.1020	-0.0689	-0.0419
18y	-0.1289	-0.1022	-0.0686	-0.0411
19y	-0.1279	-0.1012	-0.0672	-0.0392

Table 8.17. Case (ii.a): percentage differences, first fourth columns.

By first looking at the "anreb" matrix we see that, once again, our analytical formula and Rebonato's formula agree well. But how do they relate to the "true" Monte Carlo terminal-correlation matrix? The matrix "percan" shows us that in most cases the approximation is still excellent, with a few seeming exceptions. Take, for example, the percentage difference relative to the terminal correlations between the $14y$-expiry and the $19y$-expiry forward rates. Such percentage difference is 155%, and, in "percan", there are also differences of 6%, 2% etc.

However, notice that these large percentage errors are due mostly to the fact that the found correlations are very close to zero. Clearly, if the analytical formula presents us with 0.0013 as an approximation of a Monte Carlo value of -0.0023, we do not worry too much, although the *percentage* difference between this two values is large. Moreover, in our simulations, the analytical values are usually well inside the Monte Carlo window. For exam-

percan	10y	11y	12y	13y
10y	-0.0000	0.0012	0.0079	0.0170
11y	0.0012	0.0000	0.0031	0.0092
12y	0.0079	0.0031	0.0000	0.0016
13y	0.0170	0.0092	0.0016	-0.0000
14y	0.0256	0.0157	0.0049	0.0009
15y	0.0425	0.0292	0.0131	0.0054
16y	0.0886	0.0684	0.0411	0.0263
17y	0.1045	0.0823	0.0519	0.0348
18y	0.1520	0.1243	0.0852	0.0627
19y	0.2243	0.1892	0.1388	0.1092

percreb	10y	11y	12y	13y
10y	-0.0000	-0.0055	-0.0282	-0.0512
11y	-0.0055	0.0000	-0.0088	-0.0230
12y	-0.0282	-0.0088	0.0000	-0.0036
13y	-0.0512	-0.0230	-0.0036	-0.0000
14y	-0.0707	-0.0368	-0.0107	-0.0020
15y	-0.0920	-0.0522	-0.0189	-0.0060
16y	-0.1285	-0.0795	-0.0350	-0.0158
17y	-0.1391	-0.0876	-0.0403	-0.0194
18y	-0.1596	-0.1033	-0.0503	-0.0261
19y	-0.1795	-0.1181	-0.0596	-0.0319

Table 8.18. Case (ii.b): percentage differences, the most significant columns.

MCcorr	16y	17y	18y	19y	percan	16y	17y	18y	19y
16y	1.0000	0.9998	0.9979	0.9928	16y	0.0000	-0.0002	0.0013	0.0095
17y	0.9998	1.0000	0.9989	0.9949	17y	-0.0002	0.0000	0.0012	0.0084
18y	0.9979	0.9989	1.0000	0.9985	18y	0.0013	0.0012	0.0000	0.0031
19y	0.9928	0.9949	0.9985	1.0000	19y	0.0095	0.0084	0.0031	0.0000

anreb	16y	17y	18y	19y	percreb	16y	17y	18y	19y
16y	0.0000	-0.0013	-0.0146	-0.0492	16y	0.0000	-0.0015	-0.0133	-0.0396
17y	-0.0013	0.0000	-0.0072	-0.0346	17y	-0.0015	0.0000	-0.0060	-0.0261
18y	-0.0146	-0.0072	0.0000	-0.0102	18y	-0.0133	-0.0060	0.0000	-0.0071
19y	-0.0492	-0.0346	-0.0102	0.0000	19y	-0.0396	-0.0261	-0.0071	0.0000

Table 8.19. Case (ii.c): Monte Carlo correlation matrix and percentage differences.

ple, the Monte Carlo value -0.0023 corresponds to a Monte Carlo window $[-0.0077\ 0.0030]$.

In order to avoid this kind of misleading percentage errors, we also show in Table 8.23 the matrix of absolute differences, "absan", between the Monte Carlo terminal-correlation matrix and our analytical-approximation matrix. These absolute differences appear to be small, and once again confirm that both our analytical approximation and Rebonato's work indeed well in many situations.

(iii.b): $T_\alpha = 9$

In this case, we find a situation similar to the preceding one. Here, besides showing in Table 8.24 the Monte Carlo terminal correlations matrix, we display directly the absolute differences, "absan", in Table 8.25, ignoring the percentage differences. Once again all approximations work well.

MCcorr	11y	12y	13y	14y	15y	16y	17y	18y	19y
2y	0.9839	0.9657	0.9192	0.7901	0.4499	-0.0891	-0.4745	-0.6457	-0.6027
3y	0.9842	0.9662	0.9199	0.7912	0.4514	-0.0880	-0.4742	-0.6460	-0.6028
4y	0.9859	0.9681	0.9223	0.7942	0.4549	-0.0851	-0.4727	-0.6454	-0.6020
5y	0.9875	0.9701	0.9250	0.7980	0.4600	-0.0801	-0.4690	-0.6426	-0.5989
6y	0.9890	0.9724	0.9284	0.8032	0.4674	-0.0720	-0.4622	-0.6370	-0.5930
7y	0.9908	0.9752	0.9329	0.8105	0.4781	-0.0601	-0.4517	-0.6280	-0.5835
8y	0.9930	0.9789	0.9390	0.8206	0.4934	-0.0427	-0.4362	-0.6145	-0.5695
9y	0.9957	0.9838	0.9475	0.8348	0.5152	-0.0175	-0.4135	-0.5947	-0.5488
10y	0.9984	0.9898	0.9587	0.8547	0.5466	0.0194	-0.3798	-0.5648	-0.5177
11y	1.0000	0.9962	0.9731	0.8823	0.5924	0.0749	-0.3280	-0.5185	-0.4697
12y	0.9962	1.0000	0.9895	0.9200	0.6603	0.1612	-0.2450	-0.4427	-0.3917
13y	0.9731	0.9895	1.0000	0.9670	0.7620	0.3019	-0.1030	-0.3094	-0.2555
14y	0.8823	0.9200	0.9670	1.0000	0.9017	0.5342	0.1525	-0.0586	-0.0023
15y	0.5924	0.6603	0.7620	0.9017	1.0000	0.8469	0.5638	0.3771	0.4288
16y	0.0749	0.1612	0.3019	0.5342	0.8469	1.0000	0.9165	0.8114	0.8431
17y	-0.3280	-0.2450	-0.1030	0.1525	0.5638	0.9165	1.0000	0.9774	0.9878
18y	-0.5185	-0.4427	-0.3094	-0.0586	0.3771	0.8114	0.9774	1.0000	0.9984
19y	-0.4697	-0.3917	-0.2555	-0.0023	0.4288	0.8431	0.9878	0.9984	1.0000

Table 8.20. Case (iii.a): Monte Carlo correlation matrix, last nine columns.

percan	11y	12y	13y	14y	15y	16y	17y	18y	19y
2y	0.0138	0.0270	0.0605	0.1451	0.2957	0.4690	0.2978	0.2175	0.2392
3y	0.0168	0.0298	0.0631	0.1473	0.2971	0.4761	0.2999	0.2192	0.2407
4y	0.0131	0.0259	0.0588	0.1419	0.2881	0.5107	0.3015	0.2186	0.2408
5y	0.0097	0.0222	0.0545	0.1361	0.2771	0.5733	0.3072	0.2211	0.2442
6y	0.0079	0.0199	0.0513	0.1308	0.2647	0.6832	0.3188	0.2278	0.2521
7y	0.0066	0.0179	0.0479	0.1245	0.2493	0.8919	0.3367	0.2387	0.2645
8y	0.0050	0.0152	0.0433	0.1158	0.2286	1.3977	0.3642	0.2551	0.2833
9y	0.0031	0.0117	0.0369	0.1038	0.2008	3.9210	0.4085	0.2808	0.3127
10y	0.0011	0.0074	0.0285	0.0875	0.1644	-4.2307	0.4845	0.3230	0.3613
11y	0.0000	0.0027	0.0179	0.0660	0.1185	-1.3629	0.6310	0.3980	0.4489
12y	0.0027	0.0000	0.0066	0.0396	0.0641	-0.8143	0.9885	0.5510	0.6333
13y	0.0179	0.0066	0.0000	0.0127	0.0098	-0.5617	2.8466	0.9744	1.1906
14y	0.0660	0.0396	0.0127	0.0000	-0.0175	-0.3574	-2.2216	6.2191	155.30
15y	0.1185	0.0641	0.0098	-0.0175	0.0000	-0.1189	-0.4488	-0.8075	-0.6877
16y	-1.3629	-0.8143	-0.5617	-0.3574	-0.1189	0.0000	-0.0510	-0.1003	-0.0873
17y	0.6310	0.9885	2.8466	-2.2216	-0.4488	-0.0510	0.0000	-0.0048	-0.0031
18y	0.3980	0.5510	0.9744	6.2191	-0.8075	-0.1003	-0.0048	0.0000	-0.0001
19y	0.4489	0.6333	1.1906	155.305	-0.6877	-0.0873	-0.0031	-0.0001	0.0000

Table 8.21. Case (iii.a): percentage differences between "MCcorr" and our analytical approximations "ancorr".

(iii.c): $T_\alpha = 15$

Also in this long-maturity subcase, we confirm our previous findings. Moreover, we do not need to resort to absolute differences, since all terminal correlations are far away from zero. From Table 8.26, we can see that percentage differences between the Monte Carlo matrix and our analytical-approximation matrix are at most 0.1348%. Notice that instead Rebonato's approximation leads to percentage differences up to 1%. In this case, our analytical approximation seems to be doing again sensibly better than Rebonato's, although Rebonato's is still good enough for practical purposes.

anreb	11y	12y	13y	14y	15y	16y	17y	18y	19y
2y	-0.1669	-0.1831	-0.2186	-0.3085	-0.5337	-0.8942	-1.1571	-1.2771	-1.2500
3y	-0.0793	-0.0921	-0.1211	-0.1955	-0.3842	-0.6867	-0.9074	-1.0082	-0.9857
4y	-0.0299	-0.0399	-0.0631	-0.1240	-0.2812	-0.5339	-0.7185	-0.8030	-0.7843
5y	-0.0120	-0.0200	-0.0395	-0.0919	-0.2292	-0.4507	-0.6130	-0.6875	-0.6712
6y	-0.0057	-0.0125	-0.0295	-0.0762	-0.2008	-0.4030	-0.5518	-0.6203	-0.6054
7y	-0.0039	-0.0098	-0.0254	-0.0693	-0.1880	-0.3823	-0.5260	-0.5925	-0.5780
8y	-0.0028	-0.0080	-0.0224	-0.0641	-0.1790	-0.3692	-0.5112	-0.5773	-0.5628
9y	-0.0016	-0.0060	-0.0189	-0.0580	-0.1687	-0.3552	-0.4960	-0.5622	-0.5477
10y	-0.0006	-0.0037	-0.0147	-0.0504	-0.1561	-0.3388	-0.4793	-0.5462	-0.5313
11y	0.0000	-0.0014	-0.0094	-0.0404	-0.1388	-0.3162	-0.4564	-0.5244	-0.5091
12y	-0.0014	0.0000	-0.0036	-0.0271	-0.1141	-0.2830	-0.4225	-0.4920	-0.4760
13y	-0.0094	-0.0036	0.0000	-0.0110	-0.0787	-0.2319	-0.3686	-0.4398	-0.4230
14y	-0.0404	-0.0271	-0.0110	0.0000	-0.0319	-0.1519	-0.2781	-0.3493	-0.3317
15y	-0.1388	-0.1141	-0.0787	-0.0319	0.0000	-0.0498	-0.1427	-0.2049	-0.1885
16y	-0.3162	-0.2830	-0.2319	-0.1519	-0.0498	0.0000	-0.0274	-0.0623	-0.0520
17y	-0.4564	-0.4225	-0.3686	-0.2781	-0.1427	-0.0274	0.0000	-0.0075	-0.0041
18y	-0.5244	-0.4920	-0.4398	-0.3493	-0.2049	-0.0623	-0.0075	0.0000	-0.0005
19y	-0.5091	-0.4760	-0.4230	-0.3317	-0.1885	-0.0520	-0.0041	-0.0005	0.0000

Table 8.22. Case (iii.a): percentage differences between our analytical approximation "ancorr" and Rebonato's approximation "Rebcorr".

absan	11y	12y	13y	14y	15y	16y	17y	18y	19y
2y	0.0001	0.0003	0.0006	0.0011	0.0013	-0.0004	-0.0014	-0.0014	-0.0014
3y	0.0002	0.0003	0.0006	0.0012	0.0013	-0.0004	-0.0014	-0.0014	-0.0015
4y	0.0001	0.0003	0.0005	0.0011	0.0013	-0.0004	-0.0014	-0.0014	-0.0014
5y	0.0001	0.0002	0.0005	0.0011	0.0013	-0.0005	-0.0014	-0.0014	-0.0015
6y	0.0001	0.0002	0.0005	0.0011	0.0012	-0.0005	-0.0015	-0.0015	-0.0015
7y	0.0001	0.0002	0.0004	0.0010	0.0012	-0.0005	-0.0015	-0.0015	-0.0015
8y	0.0000	0.0001	0.0004	0.0010	0.0011	-0.0006	-0.0016	-0.0016	-0.0016
9y	0.0000	0.0001	0.0003	0.0009	0.0010	-0.0007	-0.0017	-0.0017	-0.0017
10y	0.0000	0.0001	0.0003	0.0007	0.0009	-0.0008	-0.0018	-0.0018	-0.0019
11y	0.0000	0.0000	0.0002	0.0006	0.0007	-0.0010	-0.0021	-0.0021	-0.0021
12y	0.0000	0.0000	0.0001	0.0004	0.0004	-0.0013	-0.0024	-0.0024	-0.0025
13y	0.0002	0.0001	0.0000	0.0001	0.0001	-0.0017	-0.0029	-0.0030	-0.0030
14y	0.0006	0.0004	0.0001	0.0000	-0.0002	-0.0019	-0.0034	-0.0036	-0.0036
15y	0.0007	0.0004	0.0001	-0.0002	0.0000	-0.0010	-0.0025	-0.0030	-0.0029
16y	-0.0010	-0.0013	-0.0017	-0.0019	-0.0010	0.0000	-0.0005	-0.0008	-0.0007
17y	-0.0021	-0.0024	-0.0029	-0.0034	-0.0025	-0.0005	0.0000	0.0000	0.0000
18y	-0.0021	-0.0024	-0.0030	-0.0036	-0.0030	-0.0008	0.0000	0.0000	0.0000
19y	-0.0021	-0.0025	-0.0030	-0.0036	-0.0029	-0.0007	0.0000	0.0000	0.0000

Table 8.23. Case (iii.a): absolute differences between the Monte Carlo terminal-correlation matrix and our analytical-approximation matrix.

MCcorr	10y	11y	12y	13y	14y	15y	16y	17y	18y	19y
10y	1.0000	0.9949	0.9778	0.9389	0.8309	0.5272	0.0209	-0.3517	-0.5203	-0.4774
11y	0.9949	1.0000	0.9935	0.9658	0.8714	0.5816	0.0746	-0.3105	-0.4886	-0.4431
12y	0.9778	0.9935	1.0000	0.9882	0.9165	0.6548	0.1593	-0.2355	-0.4237	-0.3753
13y	0.9389	0.9658	0.9882	1.0000	0.9662	0.7587	0.2977	-0.1006	-0.2998	-0.2480
14y	0.8309	0.8714	0.9165	0.9662	1.0000	0.9001	0.5277	0.1465	-0.0606	-0.0057
15y	0.5272	0.5816	0.6548	0.7587	0.9001	1.0000	0.8425	0.5534	0.3655	0.4171
16y	0.0209	0.0746	0.1593	0.2977	0.5277	0.8425	1.0000	0.9132	0.8048	0.8374
17y	-0.3517	-0.3105	-0.2355	-0.1006	0.1465	0.5534	0.9132	1.0000	0.9766	0.9873
18y	-0.5203	-0.4886	-0.4237	-0.2998	-0.0606	0.3655	0.8048	0.9766	1.0000	0.9983
19y	-0.4774	-0.4431	-0.3753	-0.2480	-0.0057	0.4171	0.8374	0.9873	0.9983	1.0000

Table 8.24. Case (iii.b): Monte Carlo terminal-correlation matrix.

absan	10y	11y	12y	13y	14y	15y	16y	17y	18y	19y
10y	0.0000	0.0000	-0.0001	0.0003	0.0016	0.0041	0.0019	-0.0014	-0.0020	-0.0018
11y	0.0000	0.0000	0.0000	0.0003	0.0016	0.0041	0.0018	-0.0017	-0.0024	-0.0023
12y	-0.0001	0.0000	0.0000	0.0002	0.0012	0.0035	0.0013	-0.0024	-0.0032	-0.0030
13y	0.0003	0.0003	0.0002	0.0000	0.0005	0.0023	0.0001	-0.0037	-0.0046	-0.0045
14y	0.0016	0.0016	0.0012	0.0005	0.0000	0.0007	-0.0017	-0.0059	-0.0072	-0.0070
15y	0.0041	0.0041	0.0035	0.0023	0.0007	0.0000	-0.0019	-0.0064	-0.0084	-0.0080
16y	0.0019	0.0018	0.0013	0.0001	-0.0017	-0.0019	0.0000	-0.0017	-0.0033	-0.0030
17y	-0.0014	-0.0017	-0.0024	-0.0037	-0.0059	-0.0064	-0.0017	0.0000	-0.0003	-0.0002
18y	-0.0020	-0.0024	-0.0032	-0.0046	-0.0072	-0.0084	-0.0033	-0.0003	0.0000	0.0000
19y	-0.0018	-0.0023	-0.0030	-0.0045	-0.0070	-0.0080	-0.0030	-0.0002	0.0000	0.0000

Table 8.25. Case (iii.b): absolute differences between the Monte Carlo terminal-correlation matrix and our analytical-approximation matrix.

MCcorr	16y	17y	18y	19y
16y	1.0000	0.9098	0.7962	0.8231
17y	0.9098	1.0000	0.9744	0.9818
18y	0.7962	0.9744	1.0000	0.9975
19y	0.8231	0.9818	0.9975	1.0000

percan	16y	17y	18y	19y
16y	0.0000	0.0457	0.1348	0.0587
17y	0.0457	0.0000	0.0170	-0.0092
18y	0.1348	0.0170	0.0000	-0.0013
19y	0.0587	-0.0092	-0.0013	0.0000

anreb	16y	17y	18y	19y
16y	0.0000	-0.5534	-1.2219	-1.0526
17y	-0.5534	0.0000	-0.1434	-0.1043
18y	-1.2219	-0.1434	0.0000	-0.0133
19y	-1.0526	-0.1043	-0.0133	0.0000

percreb	16y	17y	18y	19y
16y	0.0000	-0.5074	-1.0855	-0.9932
17y	-0.5074	0.0000	-0.1263	-0.1135
18y	-1.0855	-0.1263	0.0000	-0.0146
19y	-0.9932	-0.1135	-0.0146	0.0000

Table 8.26. Case (iii.c): Monte Carlo correlation matrix and percentage differences.

8.6.4 Case (iv): Constant Instantaneous Volatilities, Some Negative Rank-Two Correlations.

Instantaneous volatilities follow again Formulation 7 with a, b, c, d as in case (ii) and θ as in case (iii).

As we observed in the previous case (iii), the presence of negative correlations can lead to some isolated large percentage differences, independently of the values of the instantaneous volatilities. Here, comments are completely analogous to those in the corresponding subcases of case (iii).

(iv.a): $T_\alpha = 1$

Our tests produced a "MCcorr" close to the one obtained in subcase (iii.a), and similar results for "ancorr" and "Rebcorr", so that we show only the matrices "percan" and "absan" in Tables 8.27 and 8.28, respectively. These are shown to point out that the large differences are in the same positions as in case (iii.a) (and mostly concern terminal correlations between the 14y-expiry and 19y-expiry forward rates). Again, all approximations work well, the large percentage differences being due to proximity to zero.

(iv.b): $T_\alpha = 9$

Analogous results and comments as in (iii.b).

percan	11y	12y	13y	14y	15y	16y	17y	18y	19y
2y	0.0098	0.0224	0.0556	0.1447	0.3414	0.0001	0.2246	0.1755	0.1914
3y	0.0093	0.0219	0.0548	0.1436	0.3390	0.0034	0.2250	0.1754	0.1915
4y	0.0087	0.0210	0.0537	0.1416	0.3344	0.0121	0.2275	0.1768	0.1931
5y	0.0080	0.0200	0.0521	0.1388	0.3274	0.0281	0.2324	0.1798	0.1965
6y	0.0071	0.0187	0.0498	0.1348	0.3169	0.0594	0.2407	0.1851	0.2025
7y	0.0060	0.0169	0.0468	0.1290	0.3021	0.1219	0.2537	0.1932	0.2117
8y	0.0046	0.0145	0.0425	0.1209	0.2813	0.2841	0.2745	0.2061	0.2263
9y	0.0029	0.0112	0.0366	0.1093	0.2527	1.1446	0.3095	0.2273	0.2503
10y	0.0011	0.0071	0.0286	0.0934	0.2145	-1.5707	0.3711	0.2630	0.2911
11y	0.0000	0.0027	0.0183	0.0719	0.1648	-0.6609	0.4929	0.3281	0.3666
12y	0.0027	0.0000	0.0069	0.0445	0.1036	-0.4870	0.7965	0.4645	0.5295
13y	0.0183	0.0069	0.0000	0.0152	0.0378	-0.3983	2.4041	0.8513	1.0331
14y	0.0719	0.0445	0.0152	0.0000	-0.0062	-0.2905	-1.9780	5.7456	163.88
15y	0.1648	0.1036	0.0378	-0.0062	0.0000	-0.1074	-0.4316	-0.7992	-0.6760
16y	-0.6609	-0.4870	-0.3983	-0.2905	-0.1074	0.0000	-0.0542	-0.1118	-0.0961
17y	0.4929	0.7965	2.4041	-1.9780	-0.4316	-0.0542	0.0000	-0.0067	-0.0042
18y	0.3281	0.4645	0.8513	5.7456	-0.7992	-0.1118	-0.0067	0.0000	-0.0003
19y	0.3666	0.5295	1.0331	163.88	-0.6760	-0.0961	-0.0042	-0.0003	0.0000

Table 8.27. Case (iv.a): percentage differences between "MCcorr" and our analytical approximations "ancorr".

absan	11y	12y	13y	14y	15y	16y	17y	18y	19y
2y	0.0001	0.0002	0.0005	0.0011	0.0015	0.0000	-0.0011	-0.0011	-0.0012
3y	0.0001	0.0002	0.0005	0.0011	0.0015	0.0000	-0.0011	-0.0011	-0.0012
4y	0.0001	0.0002	0.0005	0.0011	0.0015	0.0000	-0.0011	-0.0011	-0.0012
5y	0.0001	0.0002	0.0005	0.0011	0.0015	0.0000	-0.0011	-0.0012	-0.0012
6y	0.0001	0.0002	0.0005	0.0011	0.0015	0.0000	-0.0011	-0.0012	-0.0012
7y	0.0001	0.0002	0.0004	0.0010	0.0014	-0.0001	-0.0011	-0.0012	-0.0012
8y	0.0000	0.0001	0.0004	0.0010	0.0014	-0.0001	-0.0012	-0.0013	-0.0013
9y	0.0000	0.0001	0.0003	0.0009	0.0013	-0.0002	-0.0013	-0.0013	-0.0014
10y	0.0000	0.0001	0.0003	0.0008	0.0012	-0.0003	-0.0014	-0.0015	-0.0015
11y	0.0000	0.0000	0.0002	0.0006	0.0010	-0.0005	-0.0016	-0.0017	-0.0017
12y	0.0000	0.0000	0.0001	0.0004	0.0007	-0.0008	-0.0019	-0.0020	-0.0021
13y	0.0002	0.0001	0.0000	0.0001	0.0003	-0.0012	-0.0025	-0.0026	-0.0026
14y	0.0006	0.0004	0.0001	0.0000	-0.0001	-0.0016	-0.0030	-0.0033	-0.0033
15y	0.0010	0.0007	0.0003	-0.0001	0.0000	-0.0009	-0.0024	-0.0030	-0.0029
16y	-0.0005	-0.0008	-0.0012	-0.0016	-0.0009	0.0000	-0.0005	-0.0009	-0.0008
17y	-0.0016	-0.0019	-0.0025	-0.0030	-0.0024	-0.0005	0.0000	-0.0001	0.0000
18y	-0.0017	-0.0020	-0.0026	-0.0033	-0.0030	-0.0009	-0.0001	0.0000	0.0000
19y	-0.0017	-0.0021	-0.0026	-0.0033	-0.0029	-0.0008	0.0000	0.0000	0.0000

Table 8.28. Case (iv.a): absolute differences between the Monte Carlo terminal-correlation matrix and our analytical-approximation matrix.

(iv.c): $T_\alpha = 15$

Analogous results and comments as in (iii.c).

8.6.5 Case (v): Constant Instantaneous Volatilities, Perfect Correlations, Upwardly Shifted Φ's

We consider again Formulation 7 with a, b, c, d as in (ii) and $\theta = [0 \ \dots \ 0]$, leading to $\rho^B_{i,j} = 1$ for all i, j, but now we upwardly shift the Φ's by $0.2 = 20\%$. Since with this choice of a, b, c and d the Φ's coincide with the caplet volatilities, this amounts to upwardly shifting the caplet volatilities by 0.2%:

$$\Phi_i = v_{T_{i-1}-\text{caplet}} \rightarrow \Phi_i = v_{T_{i-1}-\text{caplet}} + 0.2.$$

This leads to quite large volatilities, since it amounts to more than doubling all the original caplet volatilities taken in input (whose graph is shown in Figure 8.2 and which roughly range from 10% to 16%).

As one can expect, in this case, the approximation results worsen. This is due to the fact that the "freezing" approximation works better when variability is small, and increasing the Φ's amounts to increasing variability. We will also notice that the errors in the Monte Carlo method will be huge, so as to render the comparison difficult to interpret.[1]

When all instantaneous correlations are equal to one and instantaneous volatilities are constant as in this case, Rebonato's formula yields immediately a terminal correlation matrix where all entries are equal to one. This does not happen with our analytical formula, which, once again, does better.

(v.a): $T_\alpha = 1$

In this subcase, our results show small percentage errors, lower than 0.1% in the case of Rebonato's formula, and even below 0.013% when using our analytical formula. Being all matrices close to each other, we show only a part of the terminal correlation matrix computed with the Monte Carlo method in Table 8.29.

MCcorr	2y	3y	4y	5y	6y	7y	8y	9y	10y
2y	1.0000	1.0000	1.0000	0.9999	0.9998	0.9997	0.9996	0.9996	0.9995
3y	1.0000	1.0000	1.0000	0.9999	0.9998	0.9998	0.9997	0.9997	0.9996
4y	1.0000	1.0000	1.0000	1.0000	0.9999	0.9999	0.9998	0.9998	0.9997
5y	0.9999	0.9999	1.0000	1.0000	1.0000	1.0000	0.9999	0.9999	0.9999
6y	0.9998	0.9998	0.9999	1.0000	1.0000	1.0000	1.0000	1.0000	0.9999
7y	0.9997	0.9998	0.9999	1.0000	1.0000	1.0000	1.0000	1.0000	1.0000
8y	0.9996	0.9997	0.9998	0.9999	1.0000	1.0000	1.0000	1.0000	1.0000
9y	0.9996	0.9997	0.9998	0.9999	1.0000	1.0000	1.0000	1.0000	1.0000
10y	0.9995	0.9996	0.9997	0.9999	0.9999	1.0000	1.0000	1.0000	1.0000
11y	0.9994	0.9995	0.9997	0.9998	0.9999	1.0000	1.0000	1.0000	1.0000
12y	0.9993	0.9995	0.9996	0.9998	0.9999	0.9999	1.0000	1.0000	1.0000
13y	0.9993	0.9994	0.9996	0.9997	0.9998	0.9999	0.9999	1.0000	1.0000
14y	0.9992	0.9993	0.9995	0.9997	0.9998	0.9999	0.9999	0.9999	1.0000
15y	0.9992	0.9993	0.9995	0.9997	0.9998	0.9998	0.9999	0.9999	1.0000
16y	0.9991	0.9993	0.9995	0.9996	0.9998	0.9998	0.9999	0.9999	0.9999
17y	0.9991	0.9993	0.9994	0.9996	0.9997	0.9998	0.9999	0.9999	0.9999
18y	0.9991	0.9992	0.9994	0.9996	0.9997	0.9998	0.9998	0.9999	0.9999
19y	0.9990	0.9992	0.9994	0.9996	0.9997	0.9998	0.9998	0.9999	0.9999

Table 8.29. Case (v.a): Monte Carlo correlation matrix, first nine columns.

In this case both approximated formulas still work well.

[1] We have also tried to increase the number of Monte Carlo paths and decrease the time step, but the results did not change substantially.

(v.b): $T_\alpha = 9$

In this subcase, we find the worst results of all our correlation tests. This can be due to the fact that terminal correlations are computed at a much later time, so that the freezing procedure has an heavier impact than in the short-time subcase (v.a).

We find that our analytical formula and Rebonato's work about in the same way, so that we show only percentage differences for our analytical matrix. Yet, although both formulas agree, the values of the terminal correlations involving the farthest forward rates and obtained by the Monte Carlo method are very different from the corresponding correlations computed with our analytical method. Differences reach up to 43%, as we can see from Table 8.30.

MCcorr	10y	11y	12y	13y	14y	15y	16y	17y	18y	19y
10y	1.000	0.998	0.989	0.974	0.951	0.917	0.872	0.818	0.758	0.697
11y	0.998	1.000	0.997	0.987	0.970	0.942	0.903	0.854	0.799	0.742
12y	0.989	0.997	1.000	0.997	0.986	0.965	0.932	0.890	0.841	0.788
13y	0.974	0.987	0.997	1.000	0.996	0.983	0.958	0.924	0.881	0.833
14y	0.951	0.970	0.986	0.996	1.000	0.995	0.979	0.953	0.918	0.877
15y	0.917	0.942	0.965	0.983	0.995	1.000	0.994	0.978	0.952	0.919
16y	0.872	0.903	0.932	0.958	0.979	0.994	1.000	0.994	0.979	0.955
17y	0.818	0.854	0.890	0.924	0.953	0.978	0.994	1.000	0.995	0.981
18y	0.758	0.799	0.841	0.881	0.918	0.952	0.979	0.995	1.000	0.995
19y	0.697	0.742	0.788	0.833	0.877	0.919	0.955	0.981	0.995	1.000

percan	10y	11y	12y	13y	14y	15y	16y	17y	18y	19y
10y	0.000	-0.240	-1.057	-2.612	-5.070	-8.995	-14.661	-22.211	-31.762	-43.302
11y	-0.240	0.000	-0.287	-1.251	-3.042	-6.123	-10.752	-17.042	-25.067	-34.775
12y	-1.057	-0.287	0.000	-0.335	-1.433	-3.660	-7.263	-12.337	-18.920	-26.934
13y	-2.612	-1.251	-0.335	0.000	-0.377	-1.739	-4.334	-8.247	-13.489	-19.964
14y	-5.070	-3.042	-1.433	-0.377	0.000	-0.488	-2.097	-4.911	-8.921	-14.019
15y	-8.995	-6.123	-3.660	-1.739	-0.488	0.000	-0.550	-2.229	-5.011	-8.781
16y	-14.661	-10.752	-7.263	-4.334	-2.097	-0.550	0.000	-0.552	-2.165	-4.716
17y	-22.211	-17.042	-12.337	-8.247	-4.911	-2.229	-0.552	0.000	-0.519	-1.974
18y	-31.762	-25.067	-18.920	-13.489	-8.921	-5.011	-2.165	-0.519	0.000	-0.459
19y	-43.302	-34.775	-26.934	-19.964	-14.019	-8.781	-4.716	-1.974	-0.459	0.000

Table 8.30. Case (v.b): Monte Carlo correlation matrix and percentage differences.

We also report in Table 8.31 the two matrices "corrinf" and "corrsup" corresponding to the extremes of the Monte Carlo 98% terminal-correlation window. The two matrices are quite different, showing that the Monte Carlo estimate is still rather uncertain. Caution is therefore in order when comparing the Monte Carlo terminal-correlation matrix to our analytical approximations, since the Monte Carlo window appears to be large.

(v.c): $T_\alpha = 15$

The situation is analogous to the previous subcase. Here, we can report all matrices in Table 8.32, since their size is rather reduced.

corrinf	10y	11y	12y	13y	14y	15y	16y	17y	18y	19y
10y	0.3561	0.3537	0.3455	0.3304	0.3075	0.2731	0.2278	0.1739	0.1146	0.0535
11y	0.3537	0.3561	0.3532	0.3436	0.3264	0.2981	0.2587	0.2102	0.1554	0.0978
12y	0.3455	0.3532	0.3561	0.3527	0.3419	0.3206	0.2882	0.2461	0.1968	0.1437
13y	0.3304	0.3436	0.3527	0.3561	0.3523	0.3389	0.3145	0.2798	0.2372	0.1896
14y	0.3075	0.3264	0.3419	0.3523	0.3561	0.3512	0.3355	0.3093	0.2742	0.2331
15y	0.2731	0.2981	0.3206	0.3389	0.3512	0.3561	0.3506	0.3343	0.3084	0.2754
16y	0.2278	0.2587	0.2882	0.3145	0.3355	0.3506	0.3561	0.3506	0.3349	0.3110
17y	0.1739	0.2102	0.2461	0.2798	0.3093	0.3343	0.3506	0.3561	0.3509	0.3367
18y	0.1146	0.1554	0.1968	0.2372	0.2742	0.3084	0.3349	0.3509	0.3561	0.3515
19y	0.0535	0.0978	0.1437	0.1896	0.2331	0.2754	0.3110	0.3367	0.3515	0.3561

corrsup	10y	11y	12y	13y	14y	15y	16y	17y	18y	19y
10y	1.6439	1.6415	1.6334	1.6182	1.5953	1.5610	1.5156	1.4617	1.4024	1.3413
11y	1.6415	1.6439	1.6410	1.6315	1.6142	1.5860	1.5466	1.4980	1.4432	1.3856
12y	1.6334	1.6410	1.6439	1.6406	1.6297	1.6085	1.5761	1.5339	1.4847	1.4316
13y	1.6182	1.6315	1.6406	1.6439	1.6401	1.6268	1.6023	1.5677	1.5250	1.4774
14y	1.5953	1.6142	1.6297	1.6401	1.6439	1.6391	1.6234	1.5971	1.5620	1.5209
15y	1.5610	1.5860	1.6085	1.6268	1.6391	1.6439	1.6385	1.6221	1.5962	1.5632
16y	1.5156	1.5466	1.5761	1.6023	1.6234	1.6385	1.6439	1.6384	1.6227	1.5989
17y	1.4617	1.4980	1.5339	1.5677	1.5971	1.6221	1.6384	1.6439	1.6388	1.6246
18y	1.4024	1.4432	1.4847	1.5250	1.5620	1.5962	1.6227	1.6388	1.6439	1.6393
19y	1.3413	1.3856	1.4316	1.4774	1.5209	1.5632	1.5989	1.6246	1.6393	1.6439

Table 8.31. Case (v.b): the two matrices corresponding to the extremes of the Monte Carlo 98% terminal-correlation window.

MCcorr	16y	17y	18y	19y
16y	1.0000	0.9851	0.9370	0.8605
17y	0.9851	1.0000	0.9828	0.9337
18y	0.9370	0.9828	1.0000	0.9835
19y	0.8605	0.9337	0.9835	1.0000

percan	16y	17y	18y	19y
16y	0.0000	-1.5084	-6.7230	-16.2117
17y	-1.5084	0.0000	-1.7487	-7.1032
18y	-6.7230	-1.7487	0.0000	-1.6813
19y	-16.2117	-7.1032	-1.6813	0.0000

anreb	16y	17y	18y	19y
16y	0.0000	-0.0001	-0.0004	-0.0008
17y	-0.0001	0.0000	-0.0001	-0.0004
18y	-0.0004	-0.0001	0.0000	-0.0001
19y	-0.0008	-0.0004	-0.0001	0.0000

percreb	16y	17y	18y	19y
16y	0.0000	-1.5085	-6.7234	-16.2126
17y	-1.5085	0.0000	-1.7488	-7.1036
18y	-6.7234	-1.7488	0.0000	-1.6814
19y	-16.2126	-7.1036	-1.6814	0.0000

corrinf	16y	17y	18y	19y
16y	0.7043	0.6895	0.6413	0.5648
17y	0.6895	0.7043	0.6871	0.6380
18y	0.6413	0.6871	0.7043	0.6878
19y	0.5648	0.6380	0.6878	0.7043

corrsup	16y	17y	18y	19y
16y	1.2957	1.2808	1.2327	1.1562
17y	1.2808	1.2957	1.2785	1.2293
18y	1.2327	1.2785	1.2957	1.2791
19y	1.1562	1.2293	1.2791	1.2957

Table 8.32. Case (v.c): Monte Carlo correlation matrix, percentage differences and extreme matrices of the Monte Carlo 98% terminal-correlation window.

8.7 Test Results: Stylized Conclusions

In all our tests, we have found uncertain or even negative results in the cases where shifting the Φ's amounted to shift the caplet volatilities by the corresponding absolute amount of 20%. In such cases, our Monte Carlo windows for terminal correlations become too large, and the volatility approximations seem to worsen with respect to the unstressed cases. Yet, the volatility approximation does not worsen excessively, as one can see from the shifted cases (1.b) and (1.c).

The density plots, instead, appear to be rather affected by the shift in the Φ's, thus confirming that large volatilities render both the "freezing the drift" and the "collapsing all measures" approximations less reliable, bringing the swap-rate distribution far away from the lognormal family of distributions.

However, besides these pathological cases, all suggested analytical approximations seem to work well in "normal" situations.

Part IV

THE VOLATILITY SMILE

9. Including the Smile in the LFM

What's the use of worrying?
It never was worth while, so
Pack up your troubles in your old kit-bag,
And smile, smile, smile.
George Asaf, "Pack up your Troubles" (song), 1915.

9.1 A Mini-tour on the Smile Problem

We have seen in previous chapters that Black's formula for caplets is the standard in the cap market. This formula is consistent with the LFM, in that it comes as the expected value of the discounted caplet payoff under the related forward measure when the forward-rate dynamics is given by the LFM.

To fix ideas, let us consider again the time-0 price of a T_2-maturity caplet resetting at time T_1 ($0 < T_1 < T_2$) with strike K and a notional amount of 1. Caplets and caps have been described more generally in Section 6.4. Let τ denote the year fraction between T_1 and T_2. Such a contract pays out the amount

$$\tau(F(T_1; T_1, T_2) - K)^+,$$

at time T_2, so that its value at time 0 is

$$P(0, T_2)\tau E_0^2[(F(T_1; T_1, T_2) - K)^+].$$

The dynamics for F in the above expectation under the T_2-forward measure is the lognormal LFM dynamics

$$dF(t; T_1, T_2) = \sigma_2(t)F(t; T_1, T_2)\, dW_t\,. \tag{9.1}$$

Lognormality of the T_1-marginal distribution of this dynamics implies that the above expectation results in Black's formula

$$\mathbf{Cpl}^{\text{Black}}(0, T_1, T_2, K) = P(0, T_2)\tau \text{Bl}(K, F_2(0), v_2(T_1))\,,$$

$$v_2(T_1)^2 = \int_0^{T_1} \sigma_2^2(t)dt\,.$$

It is clear that in this derivation, the average volatility of the forward rate in $[0, T_1]$, i.e. $v_2(T_1)/\sqrt{T_1}$, does not depend on the strike K of the option. Indeed, in this formulation, volatility is a characteristic of the forward rate underlying the contract, and has nothing to do with the nature of the contract itself. In particular, it has nothing to do with the strike K of the contract.

Now take two different strikes K_1 and K_2. Suppose that the market provides us with the prices of the two related caplets $\mathbf{Cpl}^{\text{MKT}}(0, T_1, T_2, K_1)$ and $\mathbf{Cpl}^{\text{MKT}}(0, T_1, T_2, K_2)$. Both caplets have the same underlying forward rates and the same maturity.

Life would be simple if the market followed Black's formula in a consistent way. But is this the case? Does there exist a *single* volatility parameter $v_2(T_1)$ such that both

$$\mathbf{Cpl}^{\text{MKT}}(0, T_1, T_2, K_1) = P(0, T_2)\tau \text{Bl}(K_1, F_2(0), v_2(T_1))$$

and

$$\mathbf{Cpl}^{\text{MKT}}(0, T_1, T_2, K_2) = P(0, T_2)\tau \text{Bl}(K_2, F_2(0), v_2(T_1))$$

hold? The answer is a resounding "no". In general, market caplet prices do not behave like this. What one sees when looking at the market is that two *different* volatilities $v_2(T_1, K_1)$ and $v_2(T_1, K_2)$ are required to match the observed market prices if one is to use Black's formula:

$$\begin{aligned}
\mathbf{Cpl}^{\text{MKT}}(0, T_1, T_2, K_1) &= P(0, T_2)\tau \text{Bl}(K_1, F_2(0), v_2^{\text{MKT}}(T_1, K_1)), \\
\mathbf{Cpl}^{\text{MKT}}(0, T_1, T_2, K_2) &= P(0, T_2)\tau \text{Bl}(K_2, F_2(0), v_2^{\text{MKT}}(T_1, K_2)).
\end{aligned}$$

In other terms, each caplet market price requires its own Black volatility $v_2^{\text{MKT}}(T_1, K)$ depending on the caplet strike K.

The market therefore uses Black's formula simply as a metric to express caplet prices as volatilities. The curve $K \mapsto v_2^{\text{MKT}}(T_1, K)/\sqrt{T_1}$ is the so called volatility smile of the T_1-expiry caplet. If Black's formula were consistent along different strikes, this curve would be flat, since volatility should not depend on the strike K. Instead, this curve is commonly seen to exhibit "smiley" or "skewed" shapes. The term skew is generally used for those structures where, for a fixed maturity, low-strikes implied volatilities are higher than high-strikes implied volatilities.[1] The term smile is used instead to denote those structures where, again for a fixed maturity, the volatility has a minimum value around the current value of underlying forward rate.

Another way of experimenting the inconsistency of Black's formula in practice is through the implied distribution of forward rates. Indeed, suppose we have market caplet prices with expiry T_1 and for a set of strikes $K = K_i$. By interpolation, we can obtain the price for every other possible K, i.e. we can build a function $K \mapsto \mathbf{Cpl}^{\text{MKT}}(0, T_1, T_2, K)$. Now, if this strike-$K$ price really corresponds to an expectation, we have

[1] A skew can be either monotonically decreasing or initially decreasing and then increasing, with a negative slope at the ATM level.

$$\mathbf{Cpl}^{\text{MKT}}(0,T_1,T_2,K) = P(0,T_2)\tau E_0^2(F(T_1;T_1,T_2)-K)^+ \quad (9.2)$$
$$= P(0,T_2)\tau \int (x-K)^+ p_2(x)\,dx, \quad (9.3)$$

where p_2 is the probability density function of $F_2(T_1)$ under the T_2-forward measure.

The density p_2 of the forward rate at time T_1 that is compatible with the given interpolated prices can be obtained by differentiating the above integral twice with respect to K, see also Breeden and Litzenberger (1978),

$$\frac{\partial^2 \mathbf{Cpl}^{\text{MKT}}(0,T_1,T_2,K)}{\partial K^2} = P(0,T_2)\tau p_2(K). \quad (9.4)$$

If Black's formula were consistent along strikes, this density would be the lognormal density, coming for example from a dynamics such as (9.1). As we have seen above, this is not the case in real markets, for any interpolation method we may choose.

Having established that the dynamics (9.1) can not properly accommodate market implied volatilities, the natural question is whether there exist alternative models that are suitable for this purpose. To state it differently, what kind of dynamics, alternative to (9.1), does the density p_2 come from?

This question has been addressed, in the equity or foreign-exchange markets, by Dupire (1994, 1997). For a possible solution in the interest-rate case, we refer to Section 10.13. The resulting dynamics, however, are not uniquely determined, since only a few strikes $K = K_i$ are quoted by the market. Notice, in fact, that the density p_2 in (9.4) depends on the chosen interpolation method through the second derivative of the interpolated caplet prices.

A different answer to the above issues can be given the other way around, namely by directly starting from an alternative dynamics. For example, assume that

$$dF(t;T_1,T_2) = \nu(t, F(t;T_1,T_2))\,dW_t \quad (9.5)$$

under the T_2-forward measure, where ν can be either a deterministic or a stochastic function of $F(t;T_1,T_2)$.

A deterministic $\nu(t,\cdot)$ leads to a so called "local-volatility model". We can set, for instance, $\nu(t,F) = \sigma_2(t)F^\gamma$, where $0 \le \gamma \le 1$ and σ_2 is deterministic. The latter case, instead, corresponds to a "stochastic-volatility model", where, for example, $\nu(t,F) = \xi(t)F$, with ξ following a second stochastic differential equation.

The alternative dynamics (9.5) leads to a volatility smile to be fitted to the market one. The model's smile is explicitly generated as follows.

1. Set K to a starting value;
2. Compute the model caplet price

$$\Pi(K) = P(0,T_2)\tau E_0^2(F(T_1;T_1,T_2)-K)^+$$

with F obtained through the alternative dynamics (9.5).

3. Invert Black's formula for this strike, i.e. solve

$$\Pi(K) = P(0,T_2)\tau \mathrm{Bl}(K, F_2(0), v(K)\sqrt{T_1})$$

in $v(K)$, thus obtaining the (average) model implied volatility $v(K)$.
4. Change K and restart from point 2.

The fact that the alternative dynamics is not lognormal implies that we obtain a curve $K \mapsto v(K)$ that is not flat. Clearly, one needs to choose $\nu(t,\cdot)$ flexible enough for this curve to be able to resemble or even match the corresponding volatility curves coming from the market. Indeed, the model implied volatilities $v(K_i)$ corresponding to the observed strikes have to be made as close as possible to the corresponding market volatilities $v_2^{\mathrm{MKT}}(T_1, K_i)/\sqrt{T_1}$, by acting on the coefficient $\nu(\cdot, F)$ in the alternative dynamics. We will address this problem in the following sections.

We finally point out that one has to deal, in general, with an implied-volatility surface, since we have a caplet-volatility curve for each considered expiry. The calibration issues, however, are essentially unchanged, apart from the obviously larger computational effort required when trying to fit a bigger set of data.

9.2 Modeling the Smile

Similarly to what happens in the equity or foreign-exchange markets, also in the interest-rate market non-flat structures are normally observed when plotting implied volatilities against strikes and maturities. For example, you may plot the curve $K \mapsto v_2^{\mathrm{MKT}}(T_1, K)/\sqrt{T_1}$ given in the earlier section and you may find it to have a smile-like shape in several cases.

As we have just seen in the mini-tour, modeling the dynamics of forward LIBOR rates as in the LFM leads to implied caplet (and hence cap) volatilities that are constant for each fixed caplet (cap) maturity. The LFM, therefore, can be used to exactly retrieve ATM cap prices, but it fails to reproduce non-flat volatility surfaces when a whole range of strikes is considered.

The above considerations suggest the need for an alternative model that is capable of suitably fitting the larger set of prices that is usually available to a trader. Many researchers have tried to address the problem of a good, possibly exact, fitting of market option data. We now briefly review the major approaches proposed in the existing literature.[2]

Local-volatility models (LVMs). LVMs have been introduced as straightforward analytical extensions of a geometric Brownian motion that allow for

[2] An excellent reference for a thorough description and comparison of different LIBOR market models, allowing for smiles and skews, is the Diploma Thesis of Meister (2004).

skews in the implied volatility. The main examples are the constant-elasticity-of-variance (CEV) processes of Cox (1975) and Cox and Ross (1976), with the related application to the LIBOR market model developed by Andersen and Andreasen (2000), and the displaced diffusion of Rubinstein (1983).

More flexible LVMs, allowing for smile-shaped implied volatilities, have been proposed by Brigo, Mercurio and Sartorelli (2003) and Brigo and Mercurio (2003). These and other LVMs will be described in detail in Section 10.

Stochastic-volatility models (SVMs). SVMs have been designed both to capture the stochastic behavior of volatility and to accommodate market smiles and skews. The main examples are those of Hull and White (1987) and Heston (1993), with the related application to the LIBOR market model developed by Wu and Zhang (2002), see Section 11.2.

Other extensions of the LFM allowing for stochastic volatility are those of Andersen and Brotherton-Ratcliffe (2001), Hagan, Kumar, Lesniewski and Woodward (2002), Piterbarg (2003) and Joshi and Rebonato (2003). These SVMs will be reviewed in Section 11.

Jump-diffusion models (JDMs). JDMs have been introduced to model discontinuities in the underlying stochastic process, namely the possibility of finite changes in the value of the related financial variable over infinitesimal time intervals. Discontinuous dynamics seem ideally suited for the interest rate market, where short-term rates can suddenly jump due to central banks' interventions.

The first example of JDM in the financial literature is due to Merton (1976). Jump-diffusion LIBOR models have been developed by Glasserman and Merener (2001) and Glasserman and Kou (2003).

In options markets, JDMs are usually employed with the purpose of calibrating pronounced smiles or skews for short maturities. In the interest rate market, however, their use is rather limited because of both their implementation difficulties and the lack of short-term quotes, which makes LVMs or SVMs good enough in many situations.

Market models of implied volatility (MMsIV). MMsIV have been developed based on the stylized fact that implied volatility is directly quoted in many options markets. The first examples are in Schönbucher (1999) and Ledoit and Santa Clara (1998). An application to the LIBOR market model is due to Brace, Goldys, Klebaner and Womersley (2001).

The problem to face in a MMIV is the derivation of the implied-volatility's drift under a given pricing measure, since implied volatility is not the price of a tradable asset. This issue is typically addressed by imposing classical no-arbitrage conditions on the corresponding option prices. The resulting drift, however, may be difficult to calculate in practice and the stochastic form for the volatility of forward rates may be hard to infer from market quotes.

Levy-driven models (LDMs). LDMs have been designed to allow for stochastic evolutions governed by general Levy processes, as for instance time-changed Brownian motions. The first examples of LDM in the financial literature are due to Madan and Seneta (1990) and Eberlein and Keller (1995). Applications to the LIBOR and swap market models are due, respectively, to Eberlein and Özkan (2005) and to Liinev and Eberlein (2004).

Pure-jump and infinite-activity LDMs are found to perform quite well, especially in the equity markets, both from the historical and the cross-section point of views. However, these models may be not so straightforward to calibrate, and numerical procedures for pricing exotic derivatives are usually rather difficult to implement.

Uncertain-parameters models (UPMs). UPMs have been introduced as the most straightforward extension to a LFM that is analytically tractable and able to accommodate a large variety of implied-volatility curve and surfaces. They are extremely simple SVMs where the underlying-asset volatility is assumed to be a random variable, whose value is drawn at an infinitesimal time.

UPMs have been proposed by Mercurio (2002), Alexander, Brintalos and Nogueira (2003), Gatarek (2003) and Brigo, Mercurio and Rapisarda (2004). The LFM extensions, based on parameter uncertainty, due to Gatarek (2003) and Errais, Mauri and Mercurio (2004) will be reviewed in Section 12.

In the following, we will illustrate some major LVMs, SVMs and UPMs. The models we will describe have been selected because of either their tractability and ease of implementation or their popularity and appeal. In fact, JDMs, MMsIV and LDMs are still difficult to be successfully applied in the interest rate market.[3] For this reason, we decided not to review any of them in this book, even though we do not exclude that their popularity and possibility of application may increase in the future.

[3] In the equity and foreign-exchange markets, jump diffusions and Levy-driven models are instead rather popular and used in many practical applications.

10. Local-Volatility Models

In the equity or foreign-exchange markets, under the assumption that a whole surface of option prices (in strike and maturity) is available for the underlying asset, Dupire (1994, 1997) has derived a candidate LVM that is compatible with the given implied-volatility surface. Balland and Hughston (2000) and Brigo and Mercurio (2003), see Section 10.13, have addressed a similar issue in the interest-rate case, where a single caplet maturity is available for each market forward rate.

The problem of finding a distribution (dynamics) that consistently prices all quoted options in a given market is, however, largely undetermined in general, since there are infinitely many curves connecting (smoothly) finitely many points. A possible solution is given by assuming a particular parametric distribution depending on several, time-dependent parameters. But the question remains of finding forward-rate dynamics consistent with the chosen parametric density.

A possible answer to this issue is given by Brigo and Mercurio (2000b, 2001b, 2002b) who find local-volatility dynamics leading to a parametric distribution that may be flexible enough for practical purposes. Their approach is reviewed in Section 10.3.

LVMs have been used in the market because of their tractability and ease of implementation. Their fitting quality, however, can vary considerably from one model to another. In fact, some LVMs lead only to skewed implied volatilities or to smile-shaped ones, whereas some others are more flexible and can accommodate rather general structures.

When selecting a LVM, therefore, one must be aware of its possible limitations in terms of calibration to options data. Moreover, LVMs may suffer from parameter instability and from the fact that implied volatilities in the future can move in an opposite direction to that typically observed in the market, see for instance Hagan, Kumar, Lesniewski and Woodward (2002). This is certainly a drawback, and also for this reason SVMs can be eventually preferred. But the clear advantages of LVMs, especially in terms of tractability, still make them useful in certain practical applications.

The LVMs in this section are meant to be calibrated to the caps market, and to be only used for the pricing of LIBOR dependent derivatives. No swaption pricing is explicitly considered, even though one can try and approximate swaption prices with the usual "freezing the drift" and "collapsing all measures" approach. In fact, the task of a joint calibration to the caps and swaptions markets and the pricing of swap-rates dependent derivatives, under smile effects, is, in this book, left to the SVMs and UPMs we will describe in Sections 11 and 12, respectively.

In the following, we first introduce the forward-LIBOR model that can be obtained by displacing a given lognormal diffusion, and also describe the CEV model used by Andersen and Andreasen (2000) to model the evolution of the forward-rate process. We then study some dynamics in the class of density-mixture models proposed by Brigo and Mercurio (2000b, 2001b, 2002b) and Brigo, Mercurio and Sartorelli (2003), providing also an example of calibration to real market data. We conclude the section by describing a second general class of LVMs introduced by Brigo and Mercurio (2003), with the main purpose of combining analytical tractability with flexibility in the calibration to caps data.

10.1 The Shifted-Lognormal Model

A very simple way of constructing forward-rate dynamics that implies non-flat volatility structures is by shifting the generic lognormal dynamics analogous to (9.1). Indeed, let us assume that the forward rate F_j evolves, under its associated T_j-forward measure, according to

$$\begin{aligned} F_j(t) &= X_j(t) + \alpha, \\ dX_j(t) &= \beta(t) X_j(t)\, dW_t, \end{aligned} \tag{10.1}$$

where α is a real constant, β is a deterministic function of time and W is a standard Brownian motion. We immediately have that

$$dF_j(t) = \beta(t)(F_j(t) - \alpha)\, dW_t, \tag{10.2}$$

so that, for $t < T \leq T_{j-1}$, the forward rate F_j can be explicitly written as

$$F_j(T) = \alpha + (F_j(t) - \alpha) e^{-\frac{1}{2}\int_t^T \beta^2(u)\, du + \int_t^T \beta(u)\, dW_u}. \tag{10.3}$$

The distribution of $F_j(T)$, conditional on $F_j(t)$, $t < T \leq T_{j-1}$, is then a shifted lognormal distribution with density

$$p_{F_j(T)|F_j(t)}(x) = \frac{1}{(x-\alpha)U(t,T)\sqrt{2\pi}} \exp\left\{ -\frac{1}{2} \left(\frac{\ln \frac{x-\alpha}{F_j(t)-\alpha} + \frac{1}{2}U^2(t,T)}{U(t,T)} \right)^2 \right\}, \tag{10.4}$$

for $x > \alpha$, where

$$U(t,T) := \sqrt{\int_t^T \beta^2(u) du}. \tag{10.5}$$

The resulting model for F_j preserves the analytically tractability of the geometric Brownian motion X. Notice indeed that

$$P(t,T_j)E^j\{[F_j(T_{j-1})-K]^+|\mathcal{F}_t\} = P(t,T_j)E^j\{[X_j(T_{j-1})-(K-\alpha)]^+|\mathcal{F}_t\},$$

so that, for $\alpha < K$, the caplet price $\mathbf{Cpl}(t,T_{j-1},T_j,\tau_j,N,K)$ associated with (10.2) is simply given by

$$\mathbf{Cpl}(t,T_{j-1},T_j,\tau_j,N,K) = \tau_j N P(t,T_j)\text{Bl}(K-\alpha, F_j(t)-\alpha, U(t,T_{j-1})). \tag{10.6}$$

The implied Black volatility $\hat{\sigma} = \hat{\sigma}(K,\alpha)$ corresponding to a given strike K and to a chosen α is obtained by backing out the volatility parameter $\hat{\sigma}$ in Black's formula that matches the model price:

$$\begin{aligned}&\tau_j N P(t,T_j)\text{Bl}(K, F_j(t), \hat{\sigma}(K,\alpha)\sqrt{T_{j-1}-t})\\ &= \tau_j N P(t,T_j)\text{Bl}(K-\alpha, F_j(t)-\alpha, U(t,T_{j-1})).\end{aligned}$$

We can now understand why the simple affine transformation (10.1) can be useful in practice. The resulting forward-rate process, in fact, besides having explicit dynamics and known marginal density, immediately leads to closed-form formulas for caplet prices that allow for skews in the caplet implied volatility. An example of the skewed volatility structure $K \mapsto \hat{\sigma}(K,\alpha)$ that is implied by such a model is shown in Figure 10.1.[1]

Introducing a non-zero parameter α has two effects on the implied caplet volatility structure, which for $\alpha = 0$ is flat at the constant level $U(0,T_{j-1})$. First, it leads to a strictly decreasing ($\alpha < 0$) or increasing ($\alpha > 0$) curve. Second, it moves the curve upwards ($\alpha < 0$) or downwards ($\alpha > 0$). More generally, ceteris paribus, increasing α shifts the volatility curve $K \mapsto \hat{\sigma}(K,\alpha)$ down, whereas decreasing α shifts the curve up. The formal proof of these properties is straightforward. Notice, for example, that at time $t = 0$ the implied at-the-money ($K = F_j(0)$) caplet volatility $\hat{\sigma}$ satisfies

$$\text{Bl}(F_j(0), F_j(0), \sqrt{T_{j-1}}\hat{\sigma}(F_j(0),\alpha)) = \text{Bl}(F_j(0)-\alpha, F_j(0)-\alpha, U(0,T_{j-1})),$$

which reads

$$(F_j(0)-\alpha)\left[2\Phi\left(\frac{1}{2}U(0,T_{j-1})\right)-1\right] = F_j(0)\left[2\Phi\left(\frac{1}{2}\sqrt{T_{j-1}}\hat{\sigma}(F_j(0),\alpha)\right)-1\right].$$

When increasing α the left hand side of this equation decreases, thus decreasing the $\hat{\sigma}$ in the right-hand side that is needed to match the decreased left-hand side. Moreover, when differentiating (10.6) with respect to α we obtain a quantity that is always negative.

[1] Such a figure shows a decreasing caplet-volatility curve. In real markets, however, different structures can be encountered too (smile-shaped, skewed to the right,...).

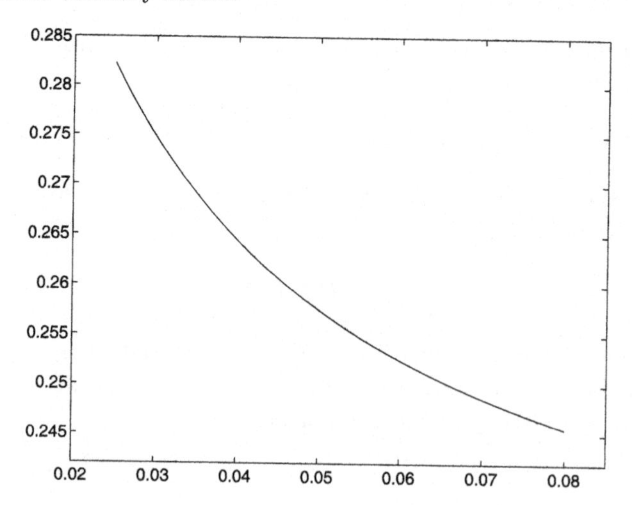

Fig. 10.1. Caplet volatility structure $\hat{\sigma}(K,\alpha)$ plotted against K implied, at time $t = 0$, by the forward-rate dynamics (10.2), where we set $T_{j-1} = 1$, $T_j = 1.5$, $\alpha = -0.015$, $\beta(t) = 0.2$ for all t and $F_j(0) = 0.055$.

Shifting a lognormal diffusion can then help in recovering skewed volatility structures. However, such structures are often too rigid, and highly negative slopes are impossible to recover. Moreover, the best fitting of market data is often achieved for decreasing implied volatility curves, which correspond to negative values of the α parameter, and hence to a support of the forward-rate density containing negative values. Even though the probability of negative rates may be negligible in practice, many people regard this drawback as an undesirable feature.

The next models we illustrate may offer the properties and flexibility required for a satisfactory fitting of market data.

10.2 The Constant Elasticity of Variance Model

Another classical model leading to skews in the implied caplet-volatility structure is the CEV model of Cox (1975) and Cox and Ross (1976). Andersen and Andreasen (2000) applied the CEV dynamics as a model of the evolution of forward LIBOR rates.

Andersen and Andreasen start with a general forward-LIBOR dynamics of the following type:

$$dF_j(t) = \phi(F_j(t))\sigma_j(t)\, dZ_j^j(t),$$

where ϕ is a general function. Andersen and Andreasen suggest as a particularly tractable case in this family the CEV model, where

$$\phi(F_j(t)) = [F_j(t)]^\gamma,$$

with $0 < \gamma < 1$. Notice that the "border" cases $\gamma = 0$ and $\gamma = 1$ would lead respectively to a normal and a lognormal dynamics.

The model then reads

$$dF_j(t) = \sigma_j(t)[F_j(t)]^\gamma \, dW_t, \quad F_j = 0 \text{ absorbing boundary when } 0 < \gamma < 1/2, \tag{10.7}$$

where we set $W = Z_j^j$, a one-dimensional Brownian motion under the T_j forward measure.

For $0 < \gamma < 1/2$ equation (10.7) does not have a unique solution unless we specify a boundary condition at $F_j = 0$. This is why we take $F_j = 0$ as an absorbing boundary for the above SDE when $0 < \gamma < 1/2$.[2]

Time dependence of σ_j can be dealt with through a deterministic time change. Indeed, by first setting

$$v(\tau, T) = \int_\tau^T \sigma_j(s)^2 ds$$

and then

$$\widetilde{W}(v(0,t)) := \int_0^t \sigma_j(s) dW(s),$$

we obtain a Brownian motion $\widetilde{W}$ with time parameter v. We substitute this time change in equation (10.7) by setting $f_j(v(t)) := F_j(t)$ and obtain

$$df_j(v) = f_j(v)^\gamma d\widetilde{W}(v), \quad f_j = 0 \text{ absorbing boundary when } 0 < \gamma < 1/2. \tag{10.8}$$

This is a process that can be easily transformed into a Bessel process via a change of variable. Straightforward manipulations lead then to the transition density function of f. By also remembering our time change, we can finally go back to the transition density for the continuous part of our original forward-rate dynamics. The continuous part of the density function of $F_j(T)$ conditional on $F_j(t)$, $t < T \le T_{j-1}$, is then given by

$$\begin{aligned} p_{F_j(T)|F_j(t)}(x) &= 2(1-\gamma)k^{1/(2-2\gamma)}(uw^{1-4\gamma})^{1/(4-4\gamma)} e^{-u-w} I_{1/(2-2\gamma)}(2\sqrt{uw}), \\ k &= \frac{1}{2v(t,T)(1-\gamma)^2}, \\ u &= k[F_j(t)]^{2(1-\gamma)}, \\ w &= kx^{2(1-\gamma)}, \end{aligned} \tag{10.9}$$

with I_q denoting the modified Bessel function of the first kind of order q. Moreover, denoting by $g(y,z) = \frac{e^{-z} z^{y-1}}{\Gamma(y)}$ the gamma density function and by

[2] Andersen and Andreasen (2000) also extend their treatment to the case $\gamma > 1$, while noticing that this can lead to explosion when leaving the T_j-forward measure (under which the process has null drift).

$G(y,x) = \int_x^{+\infty} g(y,z)dz$ the complementary gamma distribution, the probability that $F_j(T) = 0$ conditional on $F_j(t)$ is $G\left(\frac{1}{2(1-\gamma)}, u\right)$.

A major advantage of the model (10.7) is its analytical tractability, allowing for the above transition density function. This transition density can be useful, for example, in Monte Carlo simulations. From knowledge of the density follows also the possibility to price simple claims. In particular, the following explicit formula for a caplet price can be derived:

$$\mathbf{Cpl}(t, T_{j-1}, T_j, \tau_j, N, K) = \tau_j N P(t, T_j) \Bigg[F_j(t) \cdot \sum_{n=0}^{+\infty} g(n+1, u)\, G\left(c_n, kK^{2(1-\gamma)}\right) - K \sum_{n=0}^{+\infty} g\,(c_n, u)\, G\left(n+1, kK^{2(1-\gamma)}\right) \Bigg], \tag{10.10}$$

where k and u are defined as in (10.9) and

$$c_n := n + 1 + \frac{1}{2(1-\gamma)}.$$

This price can be expressed also in terms of the non-central chi-squared distribution function we have encountered in the CIR model. Recall that we denote by $\chi^2(x; r, \rho)$ the cumulative distribution function for a non-central chi-squared distribution with r degrees of freedom and non-centrality parameter ρ, computed at point x. Then the above price can be rewritten as

$$\mathbf{Cpl}(t, T_{j-1}, T_j, \tau_j, N, K) = \tau_j N P(t, T_j) \cdot \left[F_j(t) \left(1 - \chi^2\left(2K^{1-\gamma}; \frac{1}{1-\gamma} + 2, 2u\right)\right) - K\chi^2\left(2u; \frac{1}{1-\gamma}, 2kK^{1-\gamma}\right)\right]. \tag{10.11}$$

As hinted at above, the caplet price (10.10) leads to skews in the implied volatility structure. An example of the structure that can be implied is shown in Figure 10.2. As previously done in the case of a geometric Brownian motion, an extension of the above model can be proposed based on displacing the CEV process (10.7) and defining accordingly the forward-rate dynamics. The introduction of the extra parameter determining the density shifting may improve the calibration to market data.

Finally, there is the possibly annoying feature of absorption in $F = 0$. While this does not necessarily constitute a problem for caplet pricing, it can be an undesirable feature from an empirical point of view. Also, it is not clear whether there could be some problems when pricing more exotic structures. As a remedy to this absorption problem, Andersen and Andreasen (2000) propose a "Limited" CEV (LCEV) process, where instead of $\phi(F) = F^\gamma$ they set

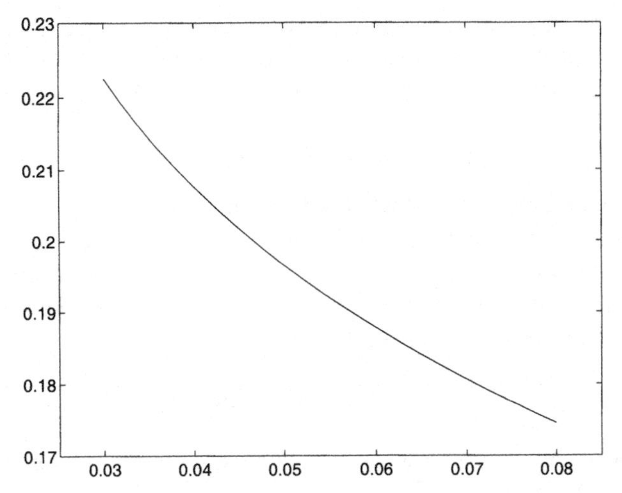

Fig. 10.2. Caplet volatility structure implied by (10.10) at time $t = 0$, where we set $T_{j-1} = 1$, $T_j = 1.5$, $\sigma_j(t) = 0.045$ for all t, $\gamma = 0.5$ and $F_j(0) = 0.055$.

$$\phi(F) = F \ \min(\epsilon^{\gamma-1}, F^{\gamma-1}) \ ,$$

where ϵ is a small positive real number. This function collapses the CEV diffusion coefficient F^γ to a (lognormal) level-proportional diffusion coefficient $F\epsilon^{\gamma-1}$ when F is small enough to make little difference (smaller than ϵ itself). Andersen and Andreasen (2000) compare the LCEV and CEV models as far as cap prices are concerned and conclude that the differences are small and tend to vanish when $\epsilon \to 0$. They also investigate, to some extent, the speed of convergence. A Crank-Nicholson scheme is used to compute cap prices within the LCEV model. As for the CEV model itself, Andersen and Andreasen allow for $\gamma > 1$ also in the LCEV case, with the difference that then ϵ has to be taken very large.

As far as the calibration of the CEV model to swaptions is concerned, approximated swaption prices based on "freezing the drift" and "collapsing all measures" are also derived (analogous to Rebonato's formula in the LFM). See Andersen and Andreasen (2000) for the details.

10.3 A Class of Analytically-Tractable Models

A natural way to obtain more general implied volatility structures is by assuming a forward-rate distribution that depends on several, time-dependent parameters. To this end, Brigo and Mercurio (2001b, 2002b) proposed a class of analytically tractable diffusions that are flexible enough to recover a large variety of market volatility surfaces.[3] We here briefly review their approach. Fundamental particular cases will then be considered in the next sections.

Let the dynamics of the forward rate F_j under the forward measure Q^j be expressed by

$$dF_j(t) = \sigma(t, F_j(t))F_j(t)\, dW_t, \tag{10.12}$$

where σ is a well-behaved deterministic function.

The function σ, which is usually termed *local volatility* in the financial literature, must be chosen so as to grant a unique strong solution to the SDE (10.12). In particular, we assume that $\sigma(\cdot,\cdot)$ satisfies, for a suitable positive constant L, the linear-growth condition

$$\sigma^2(t, y)y^2 \le L(1 + y^2) \;\text{ uniformly in } t, \tag{10.13}$$

which basically ensures existence of a strong solution (uniqueness will be addressed on a case-by-case basis).

Let us then consider N diffusion processes with dynamics given by

$$dG_i(t) = v_i(t, G_i(t))\, dW_t, \;\; i = 1, \ldots, N, \;\; G_i(0) = F_j(0)\;, \tag{10.14}$$

with common initial value $F_j(0)$, and where $v_i(t, y)$'s are real functions satisfying regularity conditions to ensure existence and uniqueness of the solution to the SDE (10.14). In particular we assume that, for suitable positive constants L_i's, the following linear-growth conditions hold:

$$v_i^2(t, y) \le L_i(1 + y^2) \;\text{ uniformly in } t, \;\; i = 1, \ldots, N. \tag{10.15}$$

For each t, we denote by $p_t^i(\cdot)$ the density function of $G_i(t)$, i.e., $p_t^i(y) = d(Q^j\{G_i(t) \le y\})/dy$, where, in particular, p_0^i is the δ-Dirac function centered in $G_i(0)$.

The problem addressed by Brigo and Mercurio (2001b, 2002b) is the derivation of the local volatility $\sigma(t, S_t)$ such that the Q^j-density of $F_j(t)$ satisfies, for each time t,

$$p_t(y) := \frac{d}{dy}\, Q^j\{F_j(t) \le y\} = \sum_{i=1}^{N} \lambda_i \frac{d}{dy}\, Q^j\{G_i(t) \le y\} = \sum_{i=1}^{N} \lambda_i p_t^i(y), \tag{10.16}$$

where the λ_i's are strictly positive constants such that $\sum_{i=1}^{N} \lambda_i = 1$. Indeed, $p_t(\cdot)$ is a proper Q^j-density function since, by definition,

$$\int_0^{+\infty} y p_t(y)dy = \sum_{i=1}^{N} \lambda_i \int_0^{+\infty} y p_t^i(y)dy = \sum_{i=1}^{N} \lambda_i G_i(0) = F_j(0).$$

[3] In fact, their underlying asset is quite general and not necessarily a forward LIBOR rate.

Remark 10.3.1. Notice that in the last calculation we were able to recover the proper Q^j-expectation thanks to our assumption that all processes (10.14) share the same null drift. However, the role of the processes G_i is merely instrumental, and there is no need to assume their drift to be of that form if not for simplifying calculations. In particular, what matters in obtaining the right expectation as in the last formula above is the marginal distribution p_i.

The above problem is essentially the reverse to that of finding the marginal density function of the solution of an SDE when the coefficients are known. In particular, $\sigma(t, F_j(t))$ can be found by solving the Fokker-Planck equation

$$\frac{\partial}{\partial t} p_t(y) = \frac{1}{2} \frac{\partial^2}{\partial y^2} \left(\sigma^2(t, y) y^2 p_t(y) \right), \tag{10.17}$$

given that each density $p_t^i(y)$ satisfies itself the Fokker-Planck equation

$$\frac{\partial}{\partial t} p_t^i(y) = \frac{1}{2} \frac{\partial^2}{\partial y^2} \left(v_i^2(t, y) p_t^i(y) \right). \tag{10.18}$$

Applying the definition (10.16) and the linearity of the derivative operator, (10.17) can be written as

$$\sum_{i=1}^N \lambda_i \frac{\partial}{\partial t} p_t^i(y) = \sum_{i=1}^N \lambda_i \left[-\frac{\partial}{\partial y} \left(\mu y p_t^i(y) \right) \right] + \sum_{i=1}^N \lambda_i \left[\frac{1}{2} \frac{\partial^2}{\partial y^2} \left(\sigma^2(t, y) y^2 p_t^i(y) \right) \right],$$

which by substituting from (10.18) becomes

$$\sum_{i=1}^N \lambda_i \left[\frac{1}{2} \frac{\partial^2}{\partial y^2} \left(v_i^2(t, y) p_t^i(y) \right) \right] = \sum_{i=1}^N \lambda_i \left[\frac{1}{2} \frac{\partial^2}{\partial y^2} \left(\sigma^2(t, y) y^2 p_t^i(y) \right) \right].$$

Using again linearity of the second order derivative operator, we obtain

$$\frac{\partial^2}{\partial y^2} \left[\sum_{i=1}^N \lambda_i v_i^2(t, y) p_t^i(y) \right] = \frac{\partial^2}{\partial y^2} \left[\sigma^2(t, y) y^2 \sum_{i=1}^N \lambda_i p_t^i(y) \right].$$

If we look at this last equation as to a second-order differential equation for $\sigma(t, \cdot)$, we find easily its general solution

$$\sigma^2(t, y) y^2 \sum_{i=1}^N \lambda_i p_t^i(y) = \sum_{i=1}^N \lambda_i v_i^2(t, y) p_t^i(y) + A_t y + B_t, \tag{10.19}$$

with A and B suitable real functions of time. The regularity conditions (10.15) and (10.13) imply that the LHS of the equation has zero limit for $y \to \infty$. As a consequence, the RHS must have a zero limit as well. This holds if and only if $A_t = B_t = 0$, for each t. We therefore obtain that the expression for $\sigma(t, y)$

that is consistent with the marginal density (10.16) and with the regularity constraint (10.13) is, for $(t, y) > (0, 0)$,

$$\sigma(t, y) = \sqrt{\frac{\sum_{i=1}^{N} \lambda_i v_i^2(t, y) p_t^i(y)}{\sum_{i=1}^{N} \lambda_i y^2 p_t^i(y)}}. \tag{10.20}$$

Indeed, notice that by setting

$$\Lambda_i(t, y) := \frac{\lambda_i p_t^i(y)}{\sum_{i=1}^{N} \lambda_i p_t^i(y)} \tag{10.21}$$

for each $i = 1, \ldots, N$ and $(t, y) > (0, 0)$, we can write

$$\sigma^2(t, y) = \sum_{i=1}^{N} \Lambda_i(t, y) \frac{v_i^2(t, y)}{y^2}, \tag{10.22}$$

so that the square of the volatility σ can be written as a (stochastic) convex combination of the squared volatilities of the basic processes (10.14). In fact, for each (t, y), $\Lambda_i(t, y) \geq 0$ for each i and $\sum_{i=1}^{N} \Lambda_i(t, y) = 1$. Moreover, by (10.15) and setting $L := \max_{i=1,\ldots,N} L_i$, the condition (10.13) is fulfilled since

$$\sigma^2(t, y) y^2 = \sum_{i=1}^{N} \Lambda_i(t, y) v_i^2(t, y) \leq \sum_{i=1}^{N} \Lambda_i(t, y) L_i (1 + y^2) \leq L(1 + y^2).$$

The function σ may be then extended to the semi-axes $\{(t, 0) : t > 0\}$ and $\{(0, y) : y > 0\}$ according to the specific choice of the basic densities $p_t^i(\cdot)$.

Formula (10.20) leads to the following SDE for the forward rate under measure Q^j, see Brigo and Mercurio (2001b, 2002b):

$$dF_j(t) = \sqrt{\frac{\sum_{i=1}^{N} \lambda_i v_i^2(t, F_j(t)) p_t^i(F_j(t))}{\sum_{i=1}^{N} \lambda_i F_j(t)^2 p_t^i(F_j(t))}} F_j(t)\, dW_t. \tag{10.23}$$

This SDE, however, must be regarded as defining some candidate dynamics that leads to the marginal density (10.16). Indeed, if σ is bounded, then the SDE is well defined, but the conditions imposed so far are not sufficient to grant the uniqueness of the strong solution, so that a verification must be done on a case-by-case basis.

Let us now assume that the SDE (10.23) has a unique strong solution.[4] Remembering the definition (10.16), it is straightforward to derive the model caplet prices in terms of the caplet prices associated to the basic models (10.14). Indeed, let us consider a caplet with strike K associated to the given forward rate. Then, the caplet price at the initial time $t = 0$ is given by

[4] We will see later on a fundamental case where this assumption holds.

$$\begin{aligned}\mathbf{Cpl}(0, T_{j-1}, T_j, \tau_j, K) &= \tau_j P(0, T_j) E^j \left\{ [F_j(T) - K)]^+ \right\} \\ &= \tau_j P(0, T_j) \sum_{i=1}^{N} \lambda_i \int_0^{+\infty} [y - K]^+ p_{T_j}^i(y) dy \\ &= \sum_{i=1}^{N} \lambda_i \mathbf{Cpl}^i(0, T_{j-1}, T_j, \tau_j, K),\end{aligned} \tag{10.24}$$

where $\mathbf{Cpl}^i(0, T_{j-1}, T_j, \tau_j, K)$ denotes the caplet price, with unit notional amount, associated with (10.14).

The assumption that the forward rate marginal density be given by the mixture of known basic densities can now be easily justified. When proposing alternative dynamics, it is usually quite problematic to come up with analytical formulas for caplets. Here, instead, such a problem can be avoided since the beginning if we use analytically-tractable densities p^i.[5] Moreover, the absence of bounds on the parameter N implies that a virtually unlimited number of parameters can be introduced in the dynamics so as to be used for a better calibration to market data. In practice, however, one has to find the correct tradeoff between model flexibility and number of parameters so as to avoid both poor calibration and over-parametrization issues (over-fitting, parameter instability, ...) .

A last remark concerns the classical economic interpretation of a mixture of densities: We can view F_j as a process whose density at time t coincides with the basic density p_t^i with probability λ_i.

10.4 A Lognormal-Mixture (LM) Model

Le nostre misture riproducono grazie all'arte l'essenza che si vuole invocare, moltiplicano il potere di ciascun elemento.
(Our compounds, thanks to art, reproduce the essence that one wants to evoke, they multiply the power of each element.)
Ipazia in "Baudolino", Umberto Eco, 2000.

Brigo and Mercurio (2000a, 2000b) proposed an alternative forward-LIBOR model based on a diffusion process that is consistent with a given mixture of lognormal densities. Their model can be obtained as a particular case of the general dynamics (10.23), where the densities p_t^i's are all lognormal. Precisely, Brigo and Mercurio assumed that, for each i,

$$v_i(t, y) = \sigma_i(t) y, \tag{10.25}$$

where all σ_i's are deterministic and continuous functions of time that are bounded from above and below by (strictly) positive constants.

[5] Note that, due to the linearity of the derivative operator, the same convex combination applies to all Greeks.

The marginal density of $G_i(t)$, for each time t, is then lognormal and given by

$$p_t^i(y) = \frac{1}{yV_i(t)\sqrt{2\pi}} \exp\left\{-\frac{1}{2V_i^2(t)}\left[\ln\frac{y}{F_j(0)} + \frac{1}{2}V_i^2(t)\right]^2\right\},$$
$$V_i(t) := \sqrt{\int_0^t \sigma_i^2(u)du}. \tag{10.26}$$

Proposition 10.4.1. *Let us assume that each σ_i is continuous and bounded from above and below by (strictly) positive constants, and that there exists an $\varepsilon > 0$ such that $\sigma_i(t) = \sigma_0 > 0$, for each t in $[0, \varepsilon]$ and $i = 1, \ldots, N$. Then, if we set*

$$\nu(t,y) := \sqrt{\frac{\sum_{i=1}^N \lambda_i \sigma_i^2(t) \frac{1}{V_i(t)} \exp\left\{-\frac{1}{2V_i^2(t)}\left[\ln\frac{y}{F_j(0)} + \frac{1}{2}V_i^2(t)\right]^2\right\}}{\sum_{i=1}^N \lambda_i \frac{1}{V_i(t)} \exp\left\{-\frac{1}{2V_i^2(t)}\left[\ln\frac{y}{F_j(0)} + \frac{1}{2}V_i^2(t)\right]^2\right\}}}, \tag{10.27}$$

for $(t, y) > (0, 0)$ and $\nu(t, y) = \sigma_0$ for $(t, y) = (0, F_j(0))$, the SDE

$$dF_j(t) = \nu(t, F_j(t))F_j(t)\, dW_t \tag{10.28}$$

has a unique strong solution whose marginal density is given by the mixture (10.16) of lognormals (10.26).

The above proposition provides us with the analytical expression for the diffusion coefficient in the SDE (10.12) such that the resulting equation has a unique strong solution whose marginal density is given by (10.16). Applying (10.22), the square of the local volatility $\nu(t, y)$ can be viewed as a weighted average of the squared basic volatilities $\sigma_1^2(t), \ldots, \sigma_N^2(t)$, where the weights are all functions of the lognormal marginal densities (10.26). That is, for each $i = 1, \ldots, N$ and $(t, y) > (0, 0)$, we can write

$$\nu^2(t,y) = \sum_{i=1}^N \Lambda_i(t,y)\sigma_i^2(t),$$
$$\Lambda_i(t,y) := \frac{\lambda_i p_t^i(y)}{\sum_{i=1}^N \lambda_i p_t^i(y)}.$$

As a consequence, for each $t > 0$ and $y > 0$, the function ν is bounded from below and above by (strictly) positive constants. In fact

$$\sigma_* \leq \nu(t,y) \leq \sigma^* \quad \text{for each } t, y > 0, \tag{10.29}$$

where

$$\sigma_* := \inf_{t \geq 0} \left\{ \min_{i=1,\ldots,N} \sigma_i(t) \right\} > 0,$$

$$\sigma^* := \sup_{t \geq 0} \left\{ \max_{i=1,\ldots,N} \sigma_i(t) \right\} < +\infty.$$

Remark 10.4.1. The function $\nu(t, y)$ can be extended by continuity to the semi-axes $\{(0, y) : y > 0\}$ and $\{(t, 0) : t \geq 0\}$ by setting $\nu(0, y) = \sigma_0$ and $\nu(t, 0) = \nu^*(t)$, where $\nu^*(t) := \sigma_{i^*}(t)$ and $i^* = i^*(t)$ is such that $V_{i^*}(t) = \max_{i=1,\ldots,N} V_i(t)$. In particular, $\nu(0, 0) = \sigma_0$. Indeed, for every $\bar{y} > 0$ and every $\bar{t} \geq 0$,

$$\lim_{t \to 0} \nu(t, \bar{y}) = \sigma_0,$$
$$\lim_{y \to 0} \nu(\bar{t}, y) = \nu^*(t).$$

Proof of Proposition 10.4.1 (Brigo and Mercurio, 2001b). Expression (10.27), for$(t, y) > (0, 0)$, is trivially obtained from (10.20) by using (10.25) and (10.26). If we write $F_j(t) = \exp(Z_t)$, where

$$dZ_t = -\frac{1}{2}\sigma^2\left(t, e^{Z_t}\right) dt + \sigma\left(t, e^{Z_t}\right) dW_t, \tag{10.30}$$

the SDE (10.12) then admits a unique strong solution since the SDE (10.30) admits a unique strong solution. In fact, its coefficients are bounded and hence satisfy the usual linear-growth condition. Moreover, setting $u(t, z) := \sigma(t, e^z)$, we have

$$\frac{\partial u^2}{\partial z}(t, z) = (\ln(F_j(0)) - z) \frac{\sum_{i=1}^N \sum_{j=1}^N \lambda_i \lambda_j A_i A_j \frac{\sigma_i^2(t)}{V_i(t) V_j(t)} \left(\frac{1}{V_i^2(t)} - \frac{1}{V_j^2(t)} \right)}{\sum_{i=1}^N \sum_{j=1}^N \lambda_i \lambda_j A_i A_j \frac{1}{V_i(t) V_j(t)}},$$

where

$$A_i := \exp\left[-\frac{1}{2V_i^2(t)} \left(z - \ln(F_j(0)) + \frac{1}{2} V_i^2(t) \right)^2 \right],$$

so that $\frac{\partial u^2}{\partial z}(t, z)$ is well defined and continuous for each $(t, z) \in (0, M] \times (-\infty, +\infty)$, $M > 0$, due to the continuity of each σ_i and V_i, and

$$\lim_{t \to 0} \frac{\partial u^2}{\partial z}(t, z) = 0,$$

since u is constant for $t \in [0, \epsilon]$. Therefore, $\frac{\partial u^2}{\partial z}(t, z)$ is bounded on each compact set $[0, M] \times [-M, M]$, and so is $\frac{\partial u}{\partial z}(t, z) = \frac{1}{2u(t,z)} \frac{\partial u^2}{\partial z}(t, z)$ since σ is bounded from below. Hence, the function u is locally Lipschitz and, as a consequence, the SDE (10.30) admits a unique strong solution, see Rogers and Williams (1987). □

At time $t = 0$, the pricing of caplets under our forward-rate dynamics (10.28) is quite straightforward. Indeed,

$$P(0,T_j)E^j\{[F_j(T_{j-1}) - K]^+\} = P(0,T_j)\int_0^{+\infty}(y-K)^+ p_{T_{j-1}}(y)dy$$
$$= P(0,T_j)\sum_{i=1}^N \lambda_i \int_0^{+\infty}(y-K)^+ p^i_{T_{j-1}}(y)dy$$

so that, the caplet price $\mathbf{Cpl}(0,T_{j-1},T_j,\tau_j,\bar{N},K)$ associated with our dynamics (10.28) is simply given by

$$\mathbf{Cpl}(0,T_{j-1},T_j,\tau_j,\bar{N},K) = \tau_j \bar{N} P(0,T_j)\sum_{i=1}^N \lambda_i \mathrm{Bl}(K,F_j(0),V_i(T_{j-1})). \tag{10.31}$$

The caplet price (10.31) leads to smiles in the implied volatility structure. An example of the shape that can be reproduced is shown in Figure 10.3.[6] Observe that the implied volatility curve has a minimum exactly at a strike equal to the initial forward rate $F_j(0)$. This property, which is formally proven in Brigo and Mercurio (2000a), makes the model suitable for recovering smile-shaped volatility surfaces. In fact, also skewed shapes can be retrieved, but with zero slope at the ATM level.

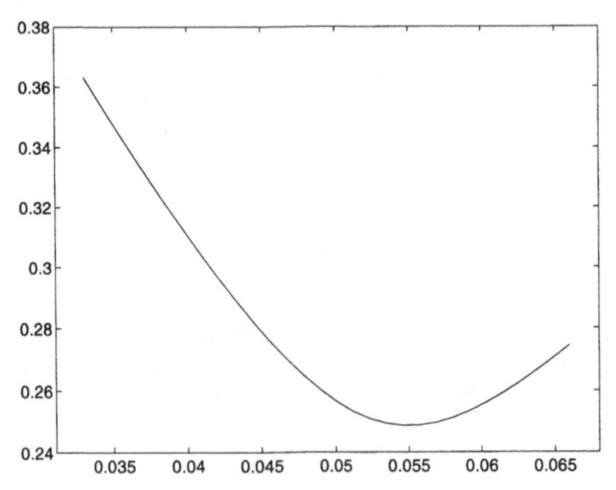

Fig. 10.3. Caplet volatility structure implied by the option prices (10.31), where we set, $T_{j-1} = 1$, $N = 3$, $(V_1(1), V_2(1), V_3(1)) = (0.6, 0.1, 0.2)$, $(\lambda_1, \lambda_2, \lambda_3) = (0.2, 0.3, 0.5)$ and $F_j(0) = 0.055$.

[6] In such a figure, we consider directly the values of the V_i's. Notice that one can easily find some σ_i's satisfying our technical assumptions that are consistent with the chosen V_i's.

Given the above analytical tractability, we can easily derive an explicit approximation for the caplet implied volatility as a function of the caplet strike price. More precisely, define the moneyness m as the logarithm of the ratio between the forward rate and the strike, i.e.,

$$m := \ln \frac{F_j(0)}{K}.$$

The implied volatility $\hat{\sigma}(m)$ for the moneyness m is implicitly defined by equating the Black caplet price in $\hat{\sigma}(m)$ to the price implied by our model according to

$$\begin{aligned} &\left[\Phi\left(\frac{m+\frac{1}{2}\hat{\sigma}(m)^2 T_{j-1}}{\hat{\sigma}(m)\sqrt{T_{j-1}}}\right) - e^{-m}\Phi\left(\frac{m-\frac{1}{2}\hat{\sigma}(m)^2 T_{j-1}}{\hat{\sigma}(m)\sqrt{T_{j-1}}}\right)\right] \\ &\quad = \sum_{i=1}^{N} \lambda_i \left[\Phi\left(\frac{m+\frac{1}{2}V_i^2(T_{j-1})}{V_i(T_{j-1})}\right) - e^{-m}\Phi\left(\frac{m-\frac{1}{2}V_i^2(T_{j-1})}{V_i(T_{j-1})}\right)\right]. \end{aligned} \tag{10.32}$$

A repeated application of Dini's implicit function theorem and a Taylor's expansion around $m = 0$, lead to

$$\begin{aligned} \hat{\sigma}(m) = \hat{\sigma}(0) + \frac{1}{2\hat{\sigma}(0)T_{j-1}} \sum_{i=1}^{N} \lambda_i \left[\frac{\hat{\sigma}(0)\sqrt{T_{j-1}}}{V_i(T_{j-1})} e^{\frac{1}{8}\left(\hat{\sigma}(0)^2 T_{j-1} - V_i^2(T_{j-1})\right)} - 1\right] m^2 \\ + o(m^3) \end{aligned} \tag{10.33}$$

where the ATM implied volatility, $\hat{\sigma}(0)$, is explicitly given by

$$\hat{\sigma}(0) = \frac{2}{\sqrt{T_{j-1}}} \Phi^{-1}\left(\sum_{i=1}^{N} \lambda_i \Phi\left(\frac{1}{2} V_i(T_{j-1})\right)\right). \tag{10.34}$$

10.5 Forward Rates Dynamics under Different Measures

So far we have just considered the dynamics of a single forward rate F_j under its canonical measure Q^j. Indeed, as far as caps-calibration issues are concerned this is all that matters.

However, in order to price exotic derivatives, one typically needs to propagate the whole term structure of rates under a common reference measure.[7] To this end, one resorts to the same procedure of Section 6.3 and finds the drift corrections induced by the considered measure change. This applies to

[7] The forward measure associated to the derivative's final maturity is often a convenient choice.

all LVMs in this section, under the implicit assumption that the related measure changes are well defined. As an example, we now explicitly deal with the case of the LM model of Section 10.4.

Let $t = 0$ be the current time. As in the general LFM of Section 6.3, consider a set $\{T_0, \ldots, T_M\}$ from which expiry-maturity pairs of dates (T_{i-1}, T_i) for a family of spanning forward rates are taken. We shall again denote by $\{\tau_1, \ldots, \tau_M\}$ the corresponding year fractions, meaning that τ_i is the year fraction associated with the expiry-maturity pair (T_{i-1}, T_i) for $i > 0$. Times T_i will be usually expressed in years from the current time.

Proposition 10.4.1 applies to every forward rate, provided one considers different coefficients for different rates. Precisely, assume $\sigma_{i,j}$'s are deterministic and continuous functions of time that are bounded from above and below by (strictly) positive constants, and that there exists an $\varepsilon > 0$ such that $\sigma_{i,j}(t) = \sigma_j^0 > 0$, for each t in $[0, \varepsilon]$ and $i = 1, \ldots, N$. Define

$$V_{i,j}(t) := \sqrt{\int_0^t \sigma_{i,j}^2(u)du}$$

$$\nu_j(t,y) := \sqrt{\frac{\sum_{i=1}^N \lambda_{i,j}\, \sigma_{i,j}^2(t) \frac{1}{V_{i,j}(t)} \exp\left\{-\frac{1}{2V_{i,j}^2(t)}\left[\ln\frac{y}{F_j(0)} + \frac{1}{2}V_{i,j}^2(t)\right]^2\right\}}{\sum_{i=1}^N \lambda_{i,j} \frac{1}{V_{i,j}(t)} \exp\left\{-\frac{1}{2V_{i,j}^2(t)}\left[\ln\frac{y}{F_j(0)} + \frac{1}{2}V_{i,j}^2(t)\right]^2\right\}}},$$

with $\lambda_{i,j} > 0$, for each i, j, and $\sum_{i=1}^N \lambda_{i,j} = 1$ for each j. We then have the following.

Proposition 10.5.1. *The dynamics of $F_j = F(\cdot; T_{j-1}, T_j)$ under the forward measure Q^i in the three cases $i < j$, $i = j$ and $i > j$ are, respectively,*

$$\begin{aligned}
i < j,\ t \le T_i: \quad & dF_j(t) = \nu_j(t, F_j(t))F_j(t) \sum_{k=i+1}^{j} \frac{\rho_{j,k}\,\tau_k\,\nu_k(t, F_k(t))\,F_k(t)}{1 + \tau_k F_k(t)}\,dt \\
& \qquad + \nu_j(t, F_j(t))F_j(t)\,dW_j^i(t), \\
i = j,\ t \le T_{j-1}: \quad & dF_j(t) = \nu_j(t, F_j(t))F_j(t)\,dW_j^i(t), \\
i > j,\ t \le T_{j-1}: \quad & dF_j(t) = -\nu_j(t, F_j(t))F_j(t) \sum_{k=j+1}^{i} \frac{\rho_{j,k}\,\tau_k\,\nu_k(t, F_k(t))\,F_k(t)}{1 + \tau_k F_k(t)}\,dt \\
& \qquad + \nu_j(t, F_j(t))F_j(t)\,dW_j^i(t),
\end{aligned}$$

where $W^i = (W_1^i, \ldots, W_M^i)$ is an M-dimensional Brownian motion under Q^i, with instantaneous correlation matrix $(\rho_{j,k})$, meaning that $dW_j^i(t)\,dW_k^i(t) = \rho_{j,k}\,dt$.

Moreover, all of the above equations admit a unique strong solution.

Proof. The proof is a direct consequence of Proposition 6.3.1 and of the fact that all volatility coefficients ν_j's are bounded. □

10.5.1 Decorrelation Between Underlying and Volatility

A final curious aspect on this LVM concerns the correlation between the underlying forward rate level and the instantaneous (squared local) volatility. While it is true that for this model we have, as for any local volatility model,

$$\text{“Corr”}(dF_j(t),\ d\nu_j^2(t, F_j(t))) = 1, \tag{10.35}$$

(quadratic covariation would be a better notion here), something quite funny happens with what we might call “terminal correlations” between the underlying asset and its “time-averaged” or local (squared) volatility.

Remark 10.5.1. **(Terminal correlation between underlying forward rate asset and (possibly averaged) squared volatility).** Set

$$v_j^2(T) := \frac{1}{T}\int_0^T \nu_j^2(t, F_j(t))dt\ ,$$

the “time-averaged squared volatility” of the process for F_j. Then

$$\text{Corr}(\nu_j^2(T, F_j(T)), F_j(T)) = 0, \quad \text{and} \quad \text{Corr}(v_j^2(T), F_j(T)) = 0 \quad \text{for all } T\ . \tag{10.36}$$

This seemingly counter-intuitive result of (10.35) and (10.36) holding at the same time is at first surprising. We have two quantities that are perfectly instantaneously correlated but totally decorrelated at any finite time horizon. This is actually linked to our LVM above to be the projection on the local volatility family of an uncertain volatility model for which the above instantaneous correlation is zero. Indeed, in Subsection 12.1.1 we will explain this after introducing the relevant uncertain parameter model. See also Brigo Mercurio and Rapisarda (2004) and Brigo (2002b) for more details.

10.6 Shifting the LM Dynamics

I say we attack again. I can get through. I know I can
Lar Gand, “Panic in the Sky”, 1993, DC Comics

Brigo and Mercurio (2000b) proposed a simple way to generalize the dynamics (10.28). With the main target consisting of retrieving a larger variety of volatility structures, the basic LM model was combined with the displaced-diffusion technique by assuming that the forward-rate process is given by

$$F_j(t) = \alpha + \bar{F}_j(t), \tag{10.37}$$

where α is a real constant and $\bar{F}_j$ evolves according to the basic LM dynamics (10.28). It is easy to prove that this is actually the most general affine transformation for which the forward-rate process is still a martingale under its canonical measure.

The analytical expression for the marginal density of such a process is given by the shifted mixture of lognormals

$$p_t(y) = \sum_{i=1}^{N} \lambda_i \frac{1}{(y-\alpha)V_i(t)\sqrt{2\pi}} \exp\left\{ -\frac{1}{2V_i^2(t)} \left[\ln \frac{y-\alpha}{F_j(0)-\alpha} + \tfrac{1}{2}V_i^2(t) \right]^2 \right\},$$

with $y > \alpha$.

By Ito's formula, we obtain that the forward-rate process evolves according to

$$dF_j(t) = \nu(t, F_j(t) - \alpha)\,(F_j(t) - \alpha)dW_t. \tag{10.38}$$

This model for the forward-rate process preserves the analytical tractability of the original process $\bar{F}_j$. Indeed,

$$P(0,T_j)E^j\left\{[F_j(T_{j-1}) - K]^+\right\} = P(0,T_j)E^j\left\{\left[\bar{F}(T_{j-1}) - (K-\alpha)\right]^+\right\},$$

so that, for $\alpha < K$, the caplet price $\mathbf{Cpl}(0, T_{j-1}, T_j, \tau_j, \bar{N}, K)$ associated with (10.37) is simply given by

$$\mathbf{Cpl}(t, T_{j-1}, T_j, \tau_j, \bar{N}, K) = \tau_j \bar{N} P(t, T_j) \sum_{i=1}^{N} \lambda_i \mathrm{Bl}(K-\alpha, F_j(0)-\alpha, V_i(T_{j-1})). \tag{10.39}$$

Moreover, the caplet implied volatility (as a function of m) can be approximated as follows:

$$\hat{\sigma}(m) = \hat{\sigma}(0) + \hat{\sigma}'(0)m + \frac{1}{2}\hat{\sigma}''(0)m^2 + o(m^2)$$

$$\hat{\sigma}'(0) = \alpha\sqrt{\frac{2\pi}{T_{j-1}}}\frac{e^{\frac{1}{8}\hat{\sigma}(0)^2 T_{j-1}}}{F_j(0)}\left(-\sum_{i=1}^{N}\lambda_i\Phi\left(\frac{1}{2}V_i(T_{j-1})\right)+\frac{1}{2}\right)$$

$$\hat{\sigma}''(0) = \frac{F_j(0)}{F_j(0)-\alpha}\sum_{i=1}^{N}\lambda_i\frac{e^{\frac{1}{8}(\hat{\sigma}(0)^2 T_{j-1} - V_i(T_{j-1})^2)}}{V_i(T_{j-1})\sqrt{T_{j-1}}} - \frac{4-\hat{\sigma}(0)^2\hat{\sigma}'(0)^2 T_{j-1}^2}{4\hat{\sigma}(0)T_{j-1}},$$

where the ATM implied volatility, $\hat{\sigma}(0)$, is explicitly given by

$$\hat{\sigma}(0) = \frac{2}{\sqrt{T_{j-1}}}\Phi^{-1}\left(\frac{F_j(0)-\alpha}{F_j(0)}\sum_{i=1}^{N}\lambda_i\Phi\left(\frac{1}{2}V_i(T_{j-1})\right)+\frac{\alpha}{2F_j(0)}\right).$$

For $\alpha = 0$ the process F_j obviously coincides with $\bar{F}_j$ while preserving the correct zero drift. The introduction of the new parameter α has the effect that, decreasing α, the variance of the forward rate at each time increases while maintaining the correct expectation. Indeed:

$$E^j(F_j(t)) = F_j(0),$$

$$\mathrm{Var}^j(F_j(t)) = (F_j(0) - \alpha)^2 \left(\sum_{i=1}^{N} \lambda_i e^{V_i^2(t)} - 1 \right).$$

As for the model (10.1), the parameter α affects the shape of the implied volatility curve in two ways. First, it concurs to determine the level of such curve in that changing α leads to an almost parallel shift of the curve itself. Second, it moves the strike with minimum volatility. Precisely, if $\alpha > 0$ (< 0) the minimum is attained for strikes lower (higher) than the ATM's. When varying all parameters, the parameter α can be used to add asymmetry around the ATM volatility without shifting the curve.

We now consider two further examples in the class of Section 10.3, which have been proposed by Brigo, Mercurio and Sartorelli (2003). The related forward-rate processes, though slightly more involved than (10.28), have the major advantage of being more flexible as far as the implied caplet volatility curves are concerned.

10.7 A Lognormal-Mixture with Different Means (LMDM)

In the first example we consider, the densities p_t^i's are still lognormal, but their means are now assumed to be different. Precisely, we assume that the instrumental processes G_i evolve, under Q^j, according to

$$dG_i(t) = \mu_i(t)G_i(t)dt + \sigma_i(t)G_i(t)\,dW_t, \quad i = 1, \dots, N, \quad G_i(0) = F_j(0)\,,$$

where σ_i's satisfy the conditions of Section 10.4, and μ_i's are deterministic functions of time. The density of G_i at time t is thus given by

$$p_t^i(y) = \frac{1}{yV_i(t)\sqrt{2\pi}} \exp\left\{ -\frac{1}{2V_i^2(t)} \left[\ln \frac{y}{F_j(0)} - M_i(t) + \tfrac{1}{2}V_i^2(t) \right]^2 \right\},$$
$$M_i(t) := \int_0^t \mu_i(u)du, \tag{10.41}$$

with V_i defined as before. The functions μ_i's can not be given arbitrarily, but must be chosen so that

$$\sum_{i=1}^{N} \lambda_i e^{M_i(t)} = 1, \quad \forall t > 0. \tag{10.42}$$

This is because $p_t(y) = \sum_{i=1}^{N} \lambda_i p_t^i(y)$ must have a constant mean equal to $F_j(0)$.

As in Section 10.3, we look for a diffusion coefficient $\psi(\cdot,\cdot)$ such that the SDE

$$dF_j(t) = \psi(t, F_j(t))F_j(t)\,dW_t \tag{10.43}$$

has a solution with marginal density $p_t(y) = \sum_{i=1}^N \lambda_i p_t^i(y)$. Using, as before, the Fokker-Planck equations for processes F_j and G_i's, we then find that

$$\begin{aligned}\psi(t,y)^2 &:= \nu(t,y)^2 + \frac{2\sum_{i=1}^N \lambda_i \mu_i(t) \int_y^{+\infty} x p_t^i(x)dx}{y^2 \sum_{i=1}^N \lambda_i p_t^i(y)} \\ &= \nu(t,y)^2 + \frac{2F_j(0)\sum_{i=1}^N \lambda_i \mu_i(t)\, e^{M_i(t)} \Phi\left(\frac{\ln\frac{F_j(0)}{y} + M_i(t) + \frac{1}{2}V_i^2(t)}{V_i(t)}\right)}{y^2 \sum_{i=1}^N \lambda_i p_t^i(y)},\end{aligned} \tag{10.44}$$

with ν defined as in (10.27), namely

$$\nu(t,y)^2 = \frac{\sum_{i=1}^N \lambda_i \sigma_i(t)^2 p_t^i(y)}{\sum_{i=1}^M \lambda_i p_t^i(y)},$$

where the new p_t^i's are to be used.[8] The coefficient ψ is not necessarily well defined, since the second term in the RHS of (10.44) can become negative for some choices of the basic parameters. However, Brigo, Mercurio and Sartorelli (2003) have derived conditions under which the strict positivity of $\psi(t,y)^2$ is granted. Such conditions are reported in the following.

Lemma 10.7.1. *Assume that:*

i) there exists $n \in \{1, 2, \ldots, N\}$ such that, for each $t \in [0, T_{j-1}]$, $\mu_i(t) \geq 0$ for each $i = 1, \ldots, N$, $i \neq n$, and $\mu_n(t) \leq 0$;
ii) the condition

$$\frac{V_i^2(t)}{2} - \frac{2V_i^2(t)}{\sigma_i^2(t)}\mu_i(t) > \frac{V_n^2(t)}{2} - \frac{2V_n^2(t)}{\sigma_n^2(t)}\mu_n(t) \tag{10.45}$$

is satisfied for each $t \in (0, T_{j-1}]$ and for each $i \neq n$.

Then the function ψ^2 in (10.44) is strictly positive on $(0, T_{j-1}] \times (0, +\infty)$.

Brigo, Mercurio and Sartorelli (2003) also proved the following result concerning the existence and uniqueness of the solution to the SDE (10.43).

Proposition 10.7.1. *Let us assume that each σ_i is continuous and bounded from below by a positive constant, and that there exists an $\varepsilon > 0$ such that $\sigma_i(t) = \sigma_0 > 0$, for each t in $[0, \varepsilon]$ and $i = 1, \ldots, N$. Let us further assume*

[8] We notice that the integrals in the numerator of the second term in the RHS of (10.44) are quantities proportional the Black-Scholes prices of asset or nothing options for the instrumental processes G_i.

that each μ_i is continuous, that (10.42) is satisfied, and that $\mu_i(t) = \mu > 0$, for each t in $[0, \varepsilon]$ and $i = 1, \ldots, N$. Then, under the assumptions of Lemma 10.7.1, the SDE (10.43) has a unique strong solution whose marginal density is given by the mixture of lognormal densities (10.41).

The pricing of caplets, under dynamics (10.43), is again quite straightforward. Indeed, the caplet price $\mathbf{Cpl}(0, T_{j-1}, T_j, \tau_j, K)$ is simply given by

$$\begin{aligned}&\mathbf{Cpl}(0, T_{j-1}, T_j, \tau_j, K)\\&= \tau_j P(0, T_j) \sum_{i=1}^{N} \lambda_i e^{M_i(T_{j-1})} \mathrm{Bl}\left(K\, e^{-M_i(T_{j-1})}, F_j(0), V_i(T_{j-1})\right).\end{aligned} \quad (10.46)$$

Also this price leads to smiles in the implied volatility structure. However, the non-zero drifts in the G_i-dynamics allows us to reproduce steeper and more skewed curves than in the zero-drifts case, with minimums that can be shifted far away from the ATM level.

10.8 The Case of Hyperbolic-Sine Processes

The second case we consider lies in the class of dynamics (10.23). We in fact assume that the basic processes G_i evolve, under Q^j, according to a hyperbolic-sine process, i.e.[9]

$$G_i(t) = \beta_i(t) \sinh\left[\int_0^t \alpha_i(u) dW_u - L_i\right], \quad i = 1, \ldots, N, \quad G_i(0) = F_j(0), \quad (10.47)$$

where α_i's are positive and deterministic functions of time, L_i's are negative constants, and β_i's are chosen so as to render the G_i's martingales, namely

$$\beta_i(t) = \frac{F_j(0)\, e^{-\frac{1}{2}A_i^2(t)}}{\sinh(-\alpha_i L_i)},$$

where we set $A_i(t) := \sqrt{\int_0^t \alpha_i^2(u) du}$.

Each G_i is thus defined as an increasing function of a time-changed Brownian motion, and evolves according to

$$dG_i(t) = \alpha_i(t) \sqrt{\beta_i^2(t) + G_i^2(t)}\, dW_t, \quad i = 1, \ldots, N.$$

Looking at this SDE's diffusion coefficient we immediately notice that it is roughly deterministic for small values of $G_i(t)$, whereas it is roughly proportional to $G_i(t)$ for large values of $G_i(t)$. Therefore, in the former case, the dynamics are approximately of Gaussian type, whereas in the latter they are

[9] We remind that $\sinh(x) = \frac{e^x - e^{-x}}{2}$, and that $\sinh^{-1}(x) = \ln(x + \sqrt{1 + x^2})$.

approximately of lognormal type. For further details on such a process we refer to Carr, Tari and Zariphopoulou (1999).[10]

The hyperbolic-sine process (10.47) shares all the analytical tractability of the classical geometric Brownian motion. This is intuitive, since (10.47) is basically the difference of two geometric Brownian motions (with perfectly negatively correlated logarithms).

The cumulative distribution function of process G_i at each time t is easily derived as follows:

$$\begin{aligned} Q^j\{G_i(t) \le y\} &= Q^j\left\{\int_0^t \alpha_i(u)dW_u \le L_i + \sinh^{-1}\left(\frac{y}{\beta_i(t)}\right)\right\} \\ &= \Phi\left(\frac{L_i}{A_i(t)} + \frac{1}{A_i(t)}\sinh^{-1}\left(\frac{y}{\beta_i(t)}\right)\right), \end{aligned}$$

so that the time-t marginal density of G_i is

$$p_t^i(y) = \frac{\exp\left\{-\frac{1}{2A_i^2(t)}\left[L_i + \sinh^{-1}\left(\frac{y}{\beta_i(t)}\right)\right]^2\right\}}{A_i(t)\sqrt{2\pi}\sqrt{\beta_i^2(t) + y^2}}. \tag{10.48}$$

Moreover, through a straightforward integration, we obtain the associated caplet price as

$$\begin{aligned} \mathbf{Cpl}(0, T_{j-1}, T_j, \tau_j, K) &= \tau_j P(0, T_j)\left[\frac{F_j(0)}{2\sinh(-L_i)}\left(e^{-L_i}\right.\right. \\ &\cdot \Phi\left(\bar{y}_i(T_{j-1}) + A_i(T_{j-1})\right) - e^{L_i}\Phi\left(\bar{y}_i(T_{j-1}) - A_i(T_{j-1})\right)\Big) - K\Phi\big(\bar{y}_i(T_{j-1})\big)\Big], \end{aligned} \tag{10.49}$$

where we set, for a general T,

$$\bar{y}_i(T) := -\frac{L_i}{A_i(T)} - \frac{1}{A_i(T)}\sinh^{-1}\left(\frac{K}{\beta_i(T)}\right).$$

The pricing function (10.49) leads to steeply decreasing patterns in the implied volatility curve. Therefore, we can hope that a mixture of densities (10.48) leads to steeper implied volatility skews than in the LM model. Indeed, this turns out to be the case.

The results in Section 10.3, and equation (10.23) in particular, immediately yield the following SDE for the forward rate under measure Q^j:

[10] Carr, Tari and Zariphopoulou (1999) actually considered a process where negative values are absorbed into zero. Their process is slightly more complicated, but does not lose in tractability.

$$dF_j(t) = \chi(t, F_j(t))\, dW_t$$

$$\chi(t,y) := \sqrt{\frac{\sum_{i=1}^N \lambda_i \frac{\alpha_i^2(t)\sqrt{\beta_i(t)^2+y^2}}{A_i(t)} \exp\left\{-\frac{1}{2A_i^2(t)}\left[L_i + \sinh^{-1}\left(\frac{y}{\beta_i(t)}\right)\right]^2\right\}}{\sum_{i=1}^N \frac{\lambda_i}{A_i(t)\sqrt{\beta_i(t)^2+y^2}} \exp\left\{-\frac{1}{2A_i^2(t)}\left[L_i + \sinh^{-1}\left(\frac{y}{\beta_i(t)}\right)\right]^2\right\}}}, \tag{10.50}$$

for $(t, y) > (0, 0)$. As in the previous lognormal-mixtures cases, this equation must be handled with due care since the function χ is in general discontinuous in $(0, F_j(0))$. However, Brigo, Mercurio and Sartorelli (2003) have proven the following result stating the existence and uniqueness of a solution of such SDE, under mild assumptions on the model coefficients.

Proposition 10.8.1. *Let us assume that each α_i is continuous and bounded from below by a positive constant, that there exists an $\varepsilon > 0$ such that $\alpha_i(t) = \alpha_0 > 0$, for each t in $[0, \varepsilon]$ and $i = 1, \dots, N$, and that all L_i's are equal. Then, setting $\chi(0, S(0)) = \alpha_0$, we have that for $t \in [0, T_{j-1}]$,*

$$C \le \chi^2(t, y) \le D(1 + y^2). \tag{10.51}$$

Moreover, the SDE (10.50) admits a unique strong solution.

We refer to (10.50) as to hyperbolic-sine density mixture (HSDM) model. The general treatment of Section 10.3 implies that the caplet price associated to such model is

$$\mathbf{Cpl}(0, T_{j-1}, T_j, \tau_j, K) = \tau_j P(0, T_j) \sum_{i=1}^N \lambda_i \left[\frac{F_j(0)}{2\sinh(-L_i)}\left(e^{-L_i}\right.\right.$$
$$\left.\left.\cdot \Phi\left(\bar{y}_i(T_{j-1}) + A_i(T_{j-1})\right) - e^{L_i}\Phi\left(\bar{y}_i(T_{j-1}) - A_i(T_{j-1})\right)\right) - K\Phi(\bar{y}_i(T_{j-1}))\right]. \tag{10.52}$$

As anticipated, this caplet price leads to steep skews in the implied volatility curve. An example of the shape that can be reproduced is shown in Figure 10.4.

10.9 Testing the Above Mixture-Models on Market Data

In this section we compare the performance of the above mixture models by testing them on real market data. To this end, we consider an example of calibration to the implied caplet-volatility surface, in the Euro market, as of December 16th 2002. This surface, which is shown in figure 10.5, yields a valuable test for a model flexibility, since its caplet-volatility curves are highly skewed and even V-shaped for a number of maturities.

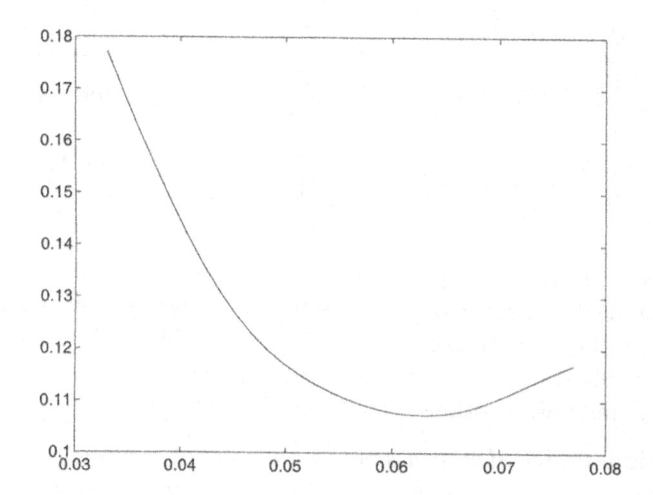

Fig. 10.4. Caplet volatility curve implied by price (10.52), where we set, $T_{j-1} = 1$, $T_j = 1.5$, $\tau_j = 0.5$ $N = 2$, $(A_1(1), A_2(1)) = (0.01, 0.04)$, $(L_1, L_2) = (-0.056, -0.408)$, $(\lambda_1, \lambda_2) = (0.1, 0.9)$ and $F_j(0) = 0.055$.

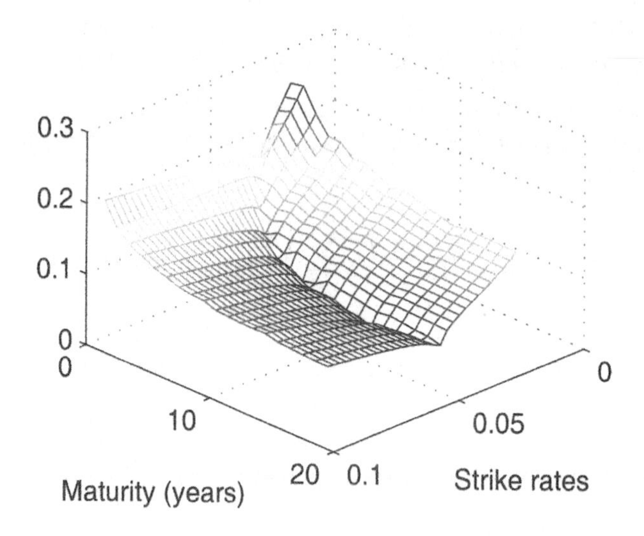

Fig. 10.5. Euro caplet (implied) volatility surface as of December 16th 2002.

We have $M = 19$ caplet maturities $T_1 < T_2 < \ldots < T_M$ with $T_j = j + 1$ years and 19 forward rates $F_j(t)$ with expiry date $T_j - 0.5$ and maturity T_j. For each maturity T_j, we calibrate the three models to the related implied volatility curve by minimizing the sum of the square relative errors between theoretical and market caplet prices. We thus run several calibrations to single-maturity implied-volatility curves rather than a unique calibration to a global volatility surface. Notice, in fact, that when we change the caplet maturity, we change the underlying forward rate as well.

For the LM model we have implemented a three-density version, while the other two models have been tested in their two-density-mixture formulation. Therefore, the LM and HSDM models have five free parameters, while the free parameters in the LMDM model are only four.[11]

The absolute errors of calibration for the LM, LMDM and HSDM models are shown in Figures 10.6, 10.7, 10.8, respectively. In Figures 10.9, 10.10 and 10.11, instead, we compare the fitting behavior of the three models on specific maturity dates, also to better highlight their respective performances.

What we can infer from this test, and similar other ones we have made, can be summarized as follows. When dealing with smile-shaped implied volatilities, the LM model is usually good enough to achieve a satisfactory calibration. In the presence of clear asymmetries, however, one has to allow for different means in the basic lognormal densities, especially to reproduce a non-zero slope at the ATM level. To better accommodate skew-shaped curves, like those of Figure 10.5, one should preferably resort to the HSDM model, because of its higher degree of flexibility and capability of recovering general skews. One must bear in mind, nonetheless, that steep skews for low strikes and short maturities may still be difficult to accommodate, as the considered case of calibration also shows.

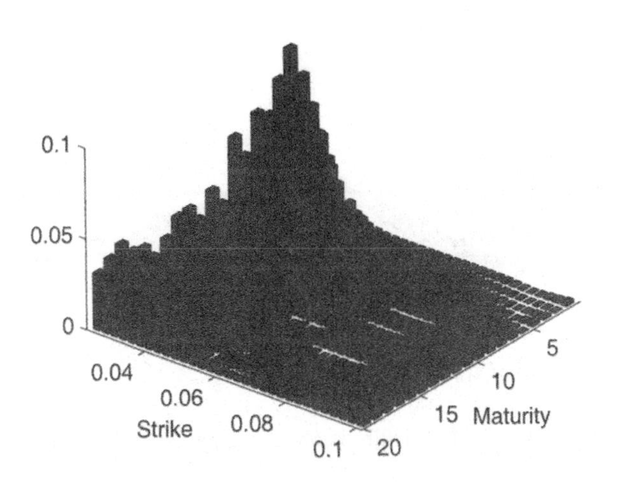

Fig. 10.6. Absolute calibration errors for the LM model.

[11] We note that the time-dependent parameters in each model enter the corresponding option pricing formulas only through their integrated value from zero to the expiry time (six months before the maturity time). This implies that the calibration is actually performed on such integrated value.

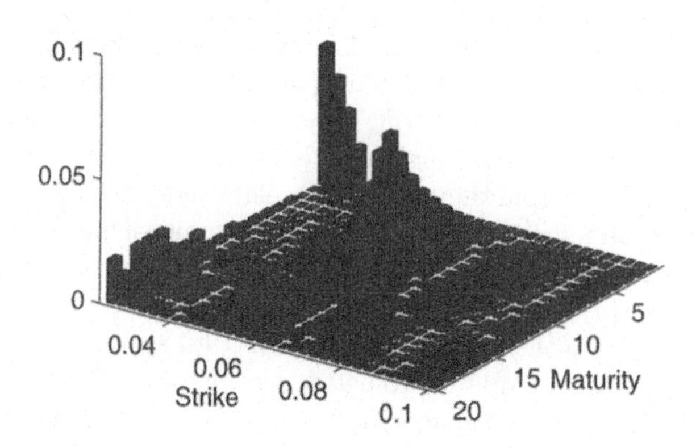

Fig. 10.7. Absolute calibration errors for the LMDM model.

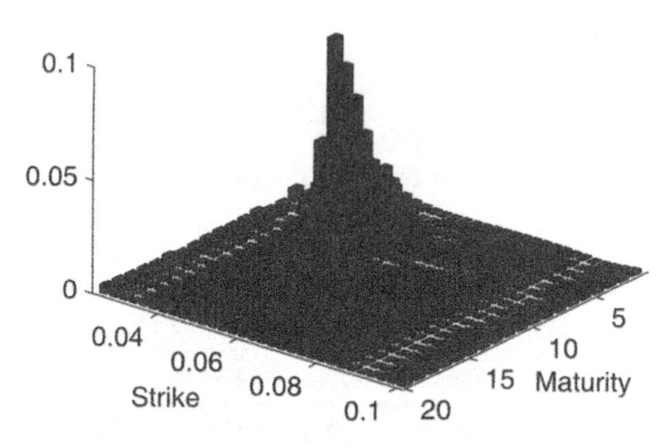

Fig. 10.8. Absolute calibration errors for the HSDM model.

10.10 A Second General Class

Analytical tractability is often a key property for financial models. In general, the calibration to market options data can be extremely cumbersome and time consuming when resorting to numerical methods. This can be a serious drawback for a trader who has to price new contracts or, even worse, re-evaluate an entire book. Having closed-form formulas for (plain-vanilla)

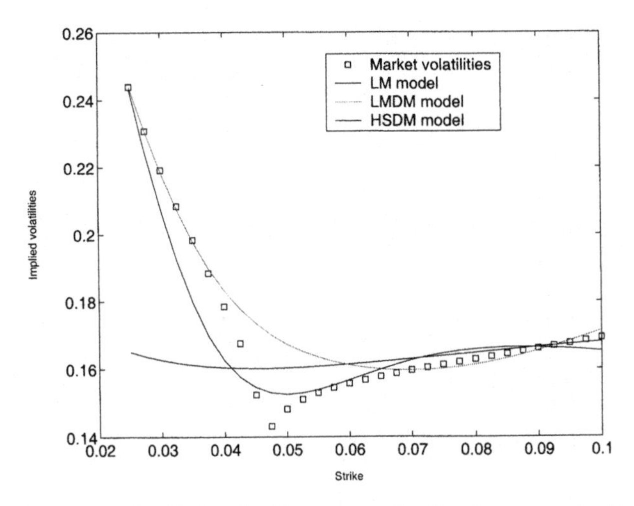

Fig. 10.9. Implied volatility curves for the 4 year maturity.

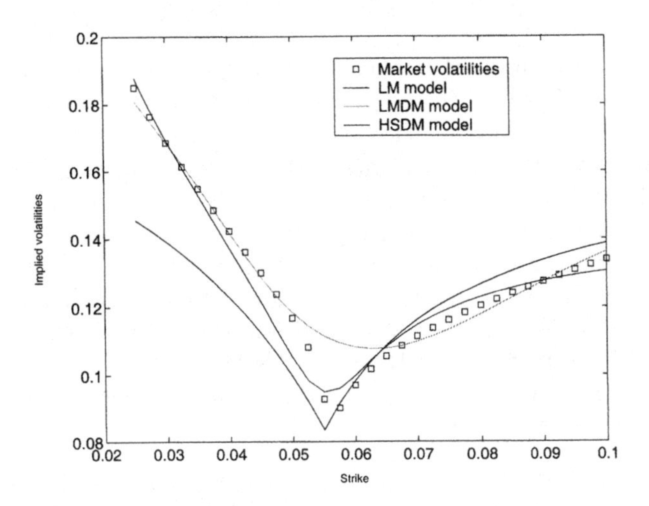

Fig. 10.10. Implied volatility curves for the 10 year maturity.

options, therefore, may translate into the actual possibility of applying a given model in practice.

Explicit formulas for options, moreover, must be combined with a good fitting to market data. In this respect, models (10.1) and (10.7) are likely to be outperformed by (10.28). Traders love small calibration errors and naturally prefer models that best reproduce their market conditions. However, the goodness of a calibration is also measured, in a less direct way, by checking the evolution of the relevant implied-volatility structures in the future, see also Chapter 7. In fact, unusual or unrealistic patterns for future volatilities may have a strong (negative) impact on the pricing, and especially hedging, of interest rates derivatives.

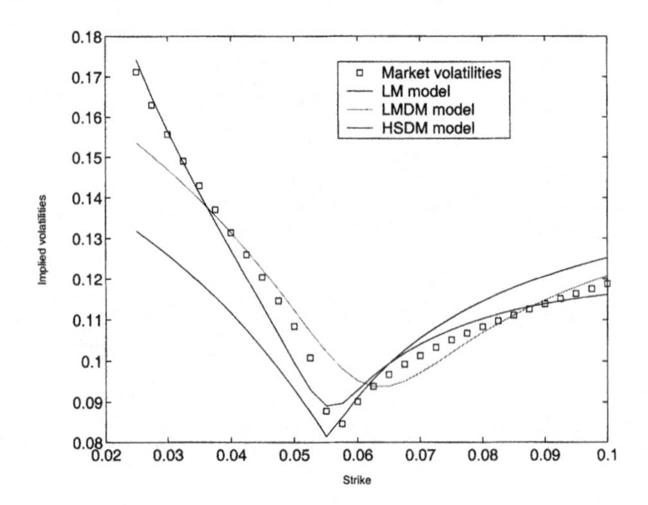

Fig. 10.11. Implied volatility curves for the 18 year maturity.

The future evolution of implied volatilities can be rapidly checked only in the presence of explicit future caplet prices, which is directly connected to the possibility of expressing analytically the transition density of the forward rate process. In this respect, models (10.1) and (10.7), with their formulas (10.4) and (10.9), are certainly preferable to (10.28), for which we only know the marginal density (10.16) under (10.26).

There is, therefore, the need for a LIBOR model that is capable of combining, at the same time, a good fitting of caps data with the highest possible degree of tractability. This is why we define the following class of models, which has been introduced by Brigo and Mercurio (2003).

Assume now that F_j can be expressed by the following (time-dependent) transformation of the Brownian motion W:

$$F_j(t) = h(t, W_t) \text{ for each } 0 \leq t \leq T_{j-1}, \tag{10.53}$$

where the function h satisfies:[12]

A1) h belongs to $C^{1,2}(\mathcal{D}_j)$, with $\mathcal{D}_j := [0, T_{j-1}] \times I\!R$;
A2) $h(t, w) > 0$ for each $(t, w) \in \mathcal{D}_j$;
A3) for each $t > 0$, the function $h_t : I\!R \to I\!R^+$, $w \mapsto h_t(w) := h(t, w)$ has zero limit at minus infinity, $\lim_{w \to -\infty} h_t(w) = 0$, and is (strictly) increasing, i.e. $dh_t(w)/dw > 0$ (equivalently, $\partial h(t, w)/\partial w > 0$), so that, for each $t > 0$, the function h_t is invertible and the inverse function h_t^{-1} is differentiable;
A4) $E^j\{h(T_{j-1}, W_{T_{j-1}})\}$ exists finite and $E^j\{h(T_{j-1}, W_{T_{j-1}})|\mathcal{F}_t\} = h(t, W_t)$, for each $0 \leq t \leq T_{j-1}$, so that F_j is indeed a martingale.

---

[12] We actually consider different functions $h = h^j$ for different j's. We drop the superscript j to lighten the notation.

A simple example of function h that fulfills these requirements is the exponential $h(t,w) = a\exp(-b^2t/2+bw)$, with $a,b>0$, which leads to a geometric Brownian motion, namely to the classical LFM of Section 6.3.

The purpose of this section is to design a general framework which preserves the analytical tractability of a geometric Brownian motion, but includes models that can produce realistic implied-volatility smiles. In the following, we show the advantages of our assumptions.

The related SDE. The SDE followed by the forward rate F_j is immediately derived by applying Ito's lemma:

$$\begin{aligned} dF_j(t) &= \left[\frac{\partial h}{\partial t}(t,W_t) + \tfrac{1}{2}\frac{\partial^2 h}{\partial w^2}(t,W_t)\right]dt + \frac{\partial h}{\partial w}(t,W_t)\,dW_t \\ &= \frac{\partial h}{\partial w}\left(t, h_t^{-1}(F_j(t))\right)dW_t, \\ &= \sigma(t,F_j(t))F_j(t)\,dW_t, \end{aligned} \tag{10.54}$$

with the obvious definition of the local-volatility function $\sigma(\cdot,\cdot)$, and where the drift term is zero due to the last assumption on h and the Feynman-Kač theorem.[13] The process F_j is therefore a (one-dimensional) diffusion.

Marginal density. Denote by p_t the marginal density of $F_j(t)$, $t \le T_{j-1}$. We have:

$$Q^j\{F_j(t) \le x\} = Q^j\{h_t(W_t) \le x\} = Q^j\{W_t \le h_t^{-1}(x)\} = \Phi\left(\frac{h_t^{-1}(x)}{\sqrt{t}}\right). \tag{10.55}$$

Differentiating with respect to x yields the marginal density

$$p_t(x) = \frac{d}{dx}Q^j\{F_j(t)\le x\} = \frac{1}{\sqrt{2\pi t}}\,e^{-\frac{1}{2t}[h_t^{-1}(x)]^2}\frac{d}{dx}h_t^{-1}(x). \tag{10.56}$$

Transition density. Given two any instants $t < T \le T_{j-1}$, from the definition (10.53) we get that the forward rate $F_j(T)$ conditional on $F_j(t)$ can be written as

$$F_j(T) = h\left(T, h_t^{-1}(F_j(t)) + W_T - W_t\right). \tag{10.57}$$

Equation (10.57) allows us to derive the transition density of the process F_j. In fact, denoting by $p(t,y;T,x)$ the density of $F_j(T)$ conditional on $F_j(t) = y$ we have

$$\begin{aligned} Q^j\left\{F_j(T) \le x | F_j(t) = y\right\} &= Q^j\left\{h\left(T, h_t^{-1}(y) + W_T - W_t\right) \le x | F_j(t) = y\right\} \\ &= Q^j\left\{W_T - W_t \le h_T^{-1}(x) - h_t^{-1}(y) | F_j(t) = y\right\} \\ &= \Phi\left(\frac{h_T^{-1}(x) - h_t^{-1}(y)}{\sqrt{T-t}}\right), \end{aligned}$$

[13] This is obviously consistent with the martingale assumption on F_j.

so that $p(t,y;T,x)$ is immediately obtained through differentiation with respect to x, namely

$$\begin{aligned}p(t,y;T,x) &= \frac{d}{dx}Q^j\left\{F_j(T)\le x|F_j(t)=y\right\}\\ &= \frac{1}{\sqrt{2\pi(T-t)}}e^{-\frac{1}{2(T-t)}\left[h_T^{-1}(x)-h_t^{-1}(y)\right]^2}\frac{d}{dx}h_T^{-1}(x).\end{aligned} \tag{10.58}$$

Useful characterizations. We first notice that assumptions A3) and A4) imply the following.

Lemma 10.10.1. *For each $t < T_{j-1}$, the function h_t can be written in terms of $h_{T_{j-1}}^{-1}$ as*

$$h_t(w) = \int_0^{+\infty}\Phi\left(\frac{w-h_{T_{j-1}}^{-1}(z)}{\sqrt{T_{j-1}-t}}\right)dz. \tag{10.59}$$

Proof. From assumption A4), and by integration by parts, we get

$$\begin{aligned}h_t(w) &= \int_{-\infty}^{+\infty} h_{T_{j-1}}(x)\frac{1}{\sqrt{2\pi(T_{j-1}-t)}}e^{-\frac{(x-w)^2}{2(T_{j-1}-t)}}dx\\ &= \left[-h_{T_{j-1}}(x)\Phi\left(\frac{w-x}{\sqrt{T_{j-1}-t}}\right)\right]_{-\infty}^{+\infty} + \int_{-\infty}^{+\infty}\frac{d}{dx}h_{T_{j-1}}(x)\Phi\left(\frac{w-x}{\sqrt{T_{j-1}-t}}\right)dx.\end{aligned}$$

The first term in the right-hand side of the last equality is zero since

$$\lim_{x\to-\infty} h_{T_{j-1}}(x)\Phi\left(\frac{w-x}{\sqrt{T_{j-1}-t}}\right) = 0$$

by assumption A3), whereas

$$\lim_{x\to+\infty} h_{T_{j-1}}(x)\Phi\left(\frac{w-x}{\sqrt{T_{j-1}-t}}\right) = 0$$

by assumption A4) and the fact that, for sufficiently large x,

$$\Phi\left(\frac{w-x}{\sqrt{T_{j-1}-t}}\right) \le \frac{\sqrt{T_{j-1}-t}}{x-w}\frac{1}{\sqrt{2\pi}}e^{-\frac{(x-w)^2}{2(T_{j-1}-t)}}.$$

Equality (10.59) is then obtained through the change of variable $z = h_{T_{j-1}}(x)$. □

Assuming we can swap integrals and derivatives, straightforward application of the derivation rule for a parameter-dependent integral leads to the following.

Corollary 10.10.1. *For each $t < T_{j-1}$, the derivative of function h_t can be written as*

$$\frac{d}{dw} h_t(w) = \int_0^{+\infty} \frac{\exp\left\{ -\frac{1}{2(T_{j-1}-t)} \left[w - h_{T_{j-1}}^{-1}(z) \right]^2 \right\}}{\sqrt{2\pi(T_{j-1}-t)}} dz. \tag{10.60}$$

Caplet pricing. The price at time t of the caplet setting in T_{j-1}, paying in T_j and with strike K is given by

$$\begin{aligned}
&\mathbf{Cpl}(t, T_{j-1}, T_j, \tau_j, K) = \tau_j P(t, T_j) E^j\{[F_j(T_{j-1}) - K]^+ | \mathcal{F}_t\} \\
&= \tau_j P(t, T_j) \int_{-\infty}^{+\infty} \frac{\left[h_{T_{j-1}}\left(h_t^{-1}(F_j(t)) + w \right) - K \right]^+}{\sqrt{2\pi(T_{j-1}-t)}} e^{-\frac{w^2}{2(T_{j-1}-t)}} dw \\
&= \tau_j P(t, T_j) \left[\int_{h_{T_{j-1}}^{-1}(K) - h_t^{-1}(F_j(t))}^{+\infty} \frac{h_{T_{j-1}}\left(h_t^{-1}(F_j(t)) + w \right)}{\sqrt{2\pi(T_{j-1}-t)}} e^{-\frac{w^2}{2(T_{j-1}-t)}} dw \right. \\
&\left. - K \int_{h_{T_{j-1}}^{-1}(K) - h_t^{-1}(F_j(t))}^{+\infty} \frac{1}{\sqrt{2\pi(T_{j-1}-t)}} e^{-\frac{w^2}{2(T_{j-1}-t)}} dw \right] \\
&= \tau_j P(t, T_j) \int_{h_{T_{j-1}}^{-1}(K) - h_t^{-1}(F_j(t))}^{+\infty} \frac{h_{T_{j-1}}\left(h_t^{-1}(F_j(t)) + w \right)}{\sqrt{2\pi(T_{j-1}-t)}} e^{-\frac{w^2}{2(T_{j-1}-t)}} dw \\
&- K \tau_j P(t, T_j) \Phi\left(\frac{h_t^{-1}(F_j(t)) - h_{T_{j-1}}^{-1}(K)}{\sqrt{T_{j-1}-t}} \right),
\end{aligned} \tag{10.61}$$

where, in general, the final integral must be calculated numerically. In the following, however, we will consider an example where this integral can be calculated in an explicit fashion.

10.11 A Particular Case: a Mixture of GBM's

As a particular case of the general dynamics (10.53), Brigo and Mercurio (2003) considered a linear combination of N driftless geometric Brownian motions that are perfectly instantaneously correlated:

$$\begin{aligned}
F_j(t) &= h(t, W_t) \quad \text{for each } t \geq 0, \\
h(t, w) &= h_t(w) = \sum_{i=1}^{N} \psi_i \, e^{-\frac{1}{2}\beta_i^2 t + \beta_i w},
\end{aligned} \tag{10.62}$$

where $F_j(0)$, β_i's and ψ_i's are positive constants.

It is straightforward to show that this function fulfills our assumptions A1) to A4) and that the derivative of its inverse h_t^{-1} is

$$\frac{d}{dx}h_t^{-1}(x) = \frac{1}{\frac{d}{dw}h_t(h_t^{-1}(x))} = \frac{1}{\sum_{i=1}^N \psi_i\beta_i \, e^{-\frac{1}{2}\beta_i^2 t + \beta_i h_t^{-1}(x)}}.$$

The initial condition imposes that

$$\sum_{i=1}^N \psi_i = F_j(0),$$

so that setting, for each i, $\lambda_i := \psi_i/F_j(0)$, we can write F_j as a mixture of N (driftless) geometric Brownian motions starting at $F_j(0)$:

$$\begin{aligned} F_j(t) &= \sum_{i=1}^N \lambda_i Y_i(t) \\ dY_i(t) &= Y_i(t)\beta_i \, dW_t, \quad Y_i(0) = F_j(0). \end{aligned} \tag{10.63}$$

Straightforward application of Ito's lemma implies that

$$\begin{aligned} dF_j(t) &= \sum_{i=1}^N \lambda_i Y_i(t)\beta_i \, dW_t \\ &= \sum_{i=1}^N \psi_i\beta_i e^{-\frac{1}{2}\beta_i^2 t + \beta_i h_t^{-1}(F_j(t))} \, dW_t \\ &= \bar{\sigma}(t, F_j(t))F_j(t) \, dW_t, \end{aligned} \tag{10.64}$$

which is obviously consistent with (10.54), and where the local-volatility function $\bar{\sigma}$ is defined by the last equality. Notice we can also write

$$\begin{aligned} dF_j(t) &= F_j(t)\sum_{i=1}^N \Lambda_i(t, F_j(t))\beta_i \, dW_t \\ \Lambda_i(t, z) &:= \frac{\psi_i \, e^{-\frac{1}{2}\beta_i^2 t + \beta_i h_t^{-1}(z)}}{\sum_{k=1}^N \psi_k \, e^{-\frac{1}{2}\beta_k^2 t + \beta_k h_t^{-1}(z)}}. \end{aligned}$$

The local volatility $\bar{\sigma}(\cdot, \cdot)$ can thus be viewed as a stochastic weighted average of the basic volatilities β_i's, since the Λ_i's are positive and sum up to one.

There is a clear analogy between this mixture of GBM's and the LM model of Section 10.4. When mixing (lognormal) densities, the square of the resulting local volatility is a (stochastic) weighted average of the squared basic volatilities, whereas when mixing (lognormal) processes, a similar property applies to the basic volatilities themselves.

Marginal and transition densities. From the general formulas (10.56) and (10.58), we get that the marginal and transition density functions of F_j are given respectively by

$$p_t(x) = \frac{e^{-\frac{1}{2t}\left(h_t^{-1}(x)\right)^2}}{\sqrt{2\pi t}\sum_{i=1}^N \psi_i\beta_i e^{-\frac{1}{2}\beta_i^2 t+\beta_i h_t^{-1}(x)}}, \tag{10.65}$$

and

$$p(t,y;T,x) = \frac{e^{-\frac{1}{2(T-t)}\left[h_T^{-1}(x)-h_t^{-1}(y)\right]^2}}{\sqrt{2\pi(T-t)}\sum_{i=1}^N \psi_i\beta_i\, e^{-\frac{1}{2}\beta_i^2 T+\beta_i h_T^{-1}(x)}}. \tag{10.66}$$

Caplet pricing. We calculate the integral in the last member of (10.61) and obtain

$$\begin{aligned}
&\int_{h_{T_{j-1}}^{-1}(K)-h_t^{-1}(F_j(t))}^{+\infty} \frac{h_{T_{j-1}}\left(h_t^{-1}(F_j(t))+w\right)}{\sqrt{2\pi(T_{j-1}-t)}}\, e^{-\frac{w^2}{2(T_{j-1}-t)}}\,dw\\
&= \int_{h_{T_{j-1}}^{-1}(K)-h_t^{-1}(F_j(t))}^{+\infty} \sum_{i=1}^N \psi_i\, e^{-\frac{1}{2}\beta_i^2 T_{j-1}+\beta_i h_t^{-1}(F_j(t))+\beta_i w}\,\frac{e^{-\frac{w^2}{2(T_{j-1}-t)}}}{\sqrt{2\pi(T_{j-1}-t)}}\,dw\\
&= \sum_{i=1}^N \psi_i\, e^{-\frac{1}{2}\beta_i^2 T_{j-1}+\beta_i h_t^{-1}(F_j(t))} \int_{h_{T_{j-1}}^{-1}(K)-h_t^{-1}(F_j(t))}^{+\infty} \frac{e^{\beta_i w-\frac{w^2}{2(T_{j-1}-t)}}}{\sqrt{2\pi(T_{j-1}-t)}}\,dw\\
&= \sum_{i=1}^N \psi_i\, e^{-\frac{1}{2}\beta_i^2 t+\beta_i h_t^{-1}(F_j(t))} \int_{h_{T_{j-1}}^{-1}(K)-h_t^{-1}(F_j(t))}^{+\infty} \frac{e^{-\frac{[w-\beta_i(T_{j-1}-t)]^2}{2(T_{j-1}-t)}}}{\sqrt{2\pi(T_{j-1}-t)}}\,dw
\end{aligned}$$

which immediately leads to the caplet price

$$\begin{aligned}
&\mathbf{Cpl}(t,T_{j-1},T_j,\tau_j,K) = \tau_j P(t,T_j)\\
&\cdot\sum_{i=1}^N \psi_i\, e^{-\frac{1}{2}\beta_i^2 t+\beta_i h_t^{-1}(F_j(t))}\Phi\left(\frac{\beta_i(T_{j-1}-t)-h_{T_{j-1}}^{-1}(K)+h_t^{-1}(F_j(t))}{\sqrt{T_{j-1}-t}}\right)\\
&-K\tau_j P(t,T_j)\Phi\left(\frac{h_t^{-1}(F_j(t))-h_{T_{j-1}}^{-1}(K)}{\sqrt{T_{j-1}-t}}\right).
\end{aligned} \tag{10.67}$$

In particular, the caplet price at time $t=0$ reduces to

$$\begin{aligned}
&\mathbf{Cpl}(0,T_{j-1},T_j,\tau_j,K)\\
&= \tau_j P(0,T_j)\left[\sum_{i=1}^N \psi_i\,\Phi\left(\frac{\beta_i T_{j-1}-h_{T_{j-1}}^{-1}(K)}{\sqrt{T_{j-1}}}\right) - K\Phi\left(-\frac{h_{T_{j-1}}^{-1}(K)}{\sqrt{T_{j-1}}}\right)\right].
\end{aligned} \tag{10.68}$$

The implied volatility curves obtained from the caplet prices (10.68) typically show weird patterns (e.g. are increasing and concave in the strike),[14]

[14] This behavior is also typical of the shifted-lognormal model (10.1) when the shift parameter α is positive. This analogy is not surprising. Notice in fact that the

which renders our GBM mixture model hardly suitable for calibration to market data. To overcome this drawback, we can resort to a slightly more general model, which is described in the following.

10.12 An Extension of the GBM Mixture Model Allowing for Implied Volatility Skews

Dealing with (strictly) positive combinators ψ turns out to be too restrictive as far as the shape of the implied volatility surface is concerned. We now relax this assumption and allow for negative combinators, without possibly losing the analytical tractability of the initial model.

One of the key features in the definition (10.62) is that, for each fixed t, the function $h(t,\cdot)$ is increasing and invertible, property which is easily lost if some ψ_i is negative. In order to preserve the desired behavior of h, we assume that each volatility parameter β_i has the same sign of the corresponding combinator ψ_i. Summarizing:

$$\text{for each } i = 1, \ldots, N: \ \psi_i, \ \beta_i \in \mathbb{R}, \ \text{but sign}(\psi_i \beta_i) = 1. \tag{10.69}$$

Trivially, under (10.69) the function h_t is still differentiable, increasing and invertible. This basically means that formulas (10.65), (10.66) and (10.67) still hold true, with no modification. However, we have lost something: the positivity of function h, and hence of the price process F_j. This is a rather undesirable feature, which we must accommodate somehow.

We proceed as follows. We fix $\delta > 0$ and set $\bar{w} := h_{T_{j-1}}^{-1}(0) + \delta$, which is well and uniquely defined (we omit the dependence on j for brevity), and $\varepsilon := h_{T_{j-1}}(\bar{w})$. We also set

$$\begin{aligned} \alpha &:= \frac{dh_{T_{j-1}}}{dw}(\bar{w}), \\ \beta &:= \frac{d^2 h_{T_{j-1}}}{dw^2}(\bar{w}), \end{aligned} \tag{10.70}$$

and assume that $\beta < \alpha^2/\varepsilon$. If we define:

$$f(w) := a\, e^{bw + cw^2},$$

where

$$\begin{aligned} a &:= \varepsilon\, e^{-b\bar{w} - c\bar{w}^2}, \\ b &:= \frac{\alpha}{\varepsilon} - \bar{w}\left(\frac{\beta}{\varepsilon} - \frac{\alpha^2}{\varepsilon^2}\right), \\ c &:= \frac{\beta}{2\varepsilon} - \frac{\alpha^2}{2\varepsilon^2} < 0, \end{aligned} \tag{10.71}$$

---

shifted-lognormal model is a particular case of GBM mixture model (take $N = 2$, $\beta_1 = 0$ and $\beta_2 > 0$).

we have that f satisfies

$$f(w) > 0 \text{ for each } w \in \mathbb{R}, \quad f(w) \in C^2(\mathbb{R}),$$
$$\frac{df}{dw}(w) > 0 \text{ for each } w < \bar{w}, \quad \lim_{w \to -\infty} f(w) = 0,$$
$$f(\bar{w}) = \varepsilon, \quad \frac{df}{dw}(\bar{w}) = \alpha, \quad \frac{d^2 f}{dw^2}(\bar{w}) = \beta.$$

We can then define a new function $h_{T_{j-1}}$, which we shall denote by $\bar{h}_{T_{j-1}}$, such that

$$\bar{h}_{T_{j-1}}(w) := \begin{cases} h_{T_{j-1}}(w) & \text{if } w \geq \bar{w}, \\ f(w) & \text{if } w < \bar{w}. \end{cases} \tag{10.72}$$

We can immediately notice that this function $\bar{h}_{T_{j-1}}$ possesses all the advantages of function $h_{T_{j-1}}$ in (10.62), including positivity. Moreover, provided we choose strikes $K \geq \varepsilon$, prices of T_j-maturity caplets under $\bar{h}_{T_{j-1}}$ coincide with the corresponding option prices (10.67) under $h_{T_{j-1}}$. For strikes lower than ε we can still derive a closed form formula for a caplet price. However, if we take δ in such a way that ε is the smallest quoted strike, the caplet price (10.67) is all that really matters.

We have thus solved the positivity issue at expiry T_{j-1}. The price we paid is that $\bar{h}_{T_{j-1}}(W_{T_{j-1}})$ does not have the same expected value as $h_{T_{j-1}}(W_{T_{j-1}})$, which essentially means we are violating no-arbitrage. However, we can replace the coefficients ψ_i's, implicit in (10.70), (10.71) and (10.72), with new (and smaller) $\bar{\psi}_i$'s such that:[15]

$$E[\bar{h}_{T_{j-1}}(W_{T_{j-1}})] = F_j(0),$$

and we are basically done. Two examples of volatility curves implied by the option price (10.67), under (10.69), are shown in Figure 10.12.

One is then tempted to extend the above procedure to every time instant $t < T_{j-1}$, leading to (positive) functions $\bar{h}_t(\cdot)$. However, if we did so we would miss a crucial feature: the forward rate process must be a martingale under Q^j. We therefore proceed as follows.

To meet the no-arbitrage requirement, we define

$$F_j(t) = \bar{h}_t(W_t),$$
$$\bar{h}_t(w) = E^j[\bar{h}_{T_{j-1}}(W_{T_{j-1}})|W_t = w].$$

This conditional expectation can be calculated by writing $W_{T_{j-1}} = W_t + (W_{T_{j-1}} - W_t)$ and remembering that $W_{T_{j-1}} - W_t$ is independent of W_t. By completing the squares in the exponents and setting $\Delta t := T_{j-1} - t$, we obtain:

[15] We can indeed prove the existence of such $\bar{\psi}_i$'s.

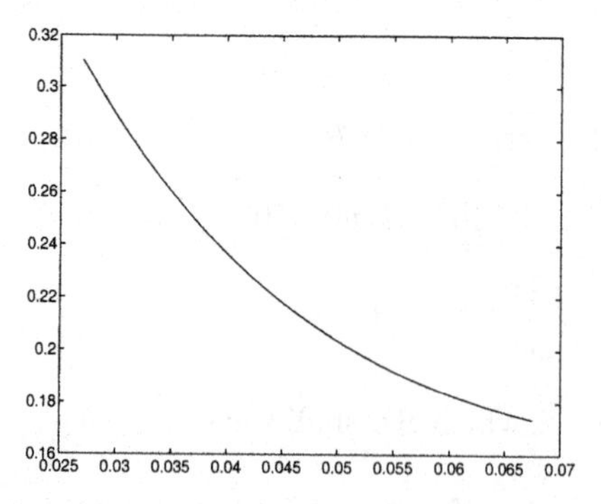

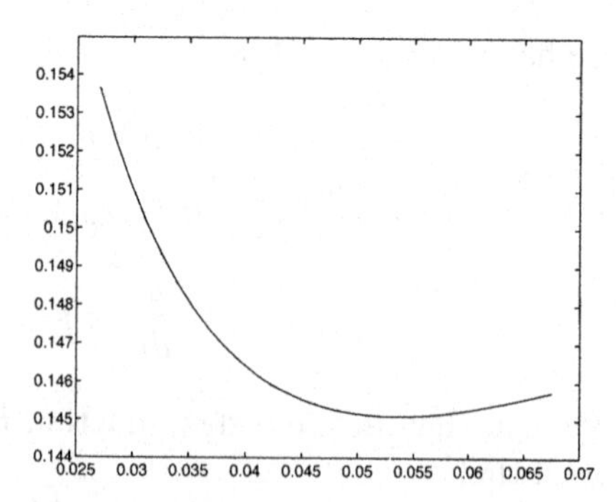

Fig. 10.12. Implied volatility curves for the caplet price (10.67), where $F_j(0) = 0.045$, $T_{j-1} = 1$, $(\psi_1, \psi_2, \psi_3) = \{0.02475, -0.01125, 0.0315\}$ and $\bar{w} = -2$, so that $(\bar{\psi}_1, \bar{\psi}_2, \bar{\psi}_3) = \{0.02474376, -0.01124716, 0.031492056\}$ and $\varepsilon = 0.02368$. Left: $(\beta_1, \beta_2, \beta_3) = \{0.2, -0.4, 0.0155\}$. Right: $(\beta_1, \beta_2, \beta_3) = \{0.2, -0.1, 0.0155\}$.

$$
\begin{aligned}
E^j[\bar{h}_{T_{j-1}}(W_{T_{j-1}})|W_t = w] &= \int_{-\infty}^{+\infty} \bar{h}_{T_{j-1}}(w+x) \frac{1}{\sqrt{2\pi \Delta t}}\, e^{-\frac{x^2}{2\Delta t}} dx \\
&= \int_{-\infty}^{\bar{w}-w} a \frac{e^{b(w+x)+c(w+x)^2-\frac{x^2}{2\Delta t}}}{\sqrt{2\pi\Delta t}} dx \\
&\quad + \int_{\bar{w}-w}^{+\infty} \frac{1}{\sqrt{2\pi\Delta t}} \sum_{i=1}^{N} \bar{\psi}_i\, e^{-\frac{1}{2}\beta_i^2 T_{j-1} + \beta_i (w+x) - \frac{x^2}{2\Delta t}} dx \\
&= a\, e^{bw+cw^2} \int_{-\infty}^{\bar{w}-w} \frac{e^{bx+2cwx+cx^2-\frac{x^2}{2\Delta t}}}{\sqrt{2\pi\Delta t}} dx \\
&\quad + \sum_{i=1}^{N} \bar{\psi}_i\, e^{-\frac{1}{2}\beta_i^2 T_{j-1} + \beta_i w} \int_{\bar{w}-w}^{+\infty} \frac{1}{\sqrt{2\pi\Delta t}}\, e^{\beta_i x - \frac{x^2}{2\Delta t}} dx \\
&= a\, e^{bw+cw^2+\frac{\Delta t(b+2cw)^2}{2(1-2c\Delta t)}} \int_{-\infty}^{\bar{w}-w} \frac{1}{\sqrt{2\pi\Delta t}}\, e^{-\frac{1-2c\Delta t}{2\Delta t}\left[x - \frac{(b+2cw)\Delta t}{1-2c\Delta t}\right]^2} dx \\
&\quad + \sum_{i=1}^{N} \bar{\psi}_i\, e^{-\frac{1}{2}\beta_i^2 t + \beta_i w} \int_{\bar{w}-w}^{+\infty} \frac{1}{\sqrt{2\pi\Delta t}}\, e^{-\frac{1}{2\Delta t}(x-\beta_i\Delta t)^2} dx \\
&= \frac{a\, e^{\frac{2bw+2cw^2+b^2\Delta t}{2(1-2c\Delta t)}}}{\sqrt{1-2c\Delta t}} \Phi\left(\frac{\bar{w}(1-2c\Delta t) - w - b\Delta t}{\sqrt{(1-2c\Delta t)\Delta t}} \right) \\
&\quad + \sum_{i=1}^{N} \bar{\psi}_i\, e^{-\frac{1}{2}\beta_i^2 t + \beta_i w} \Phi\left(-\frac{\bar{w} - w - \beta_i \Delta t}{\sqrt{\Delta t}} \right)
\end{aligned}
$$

We thus have:

$$\bar{h}_t(w) = \frac{a\, e^{\frac{2b\bar{w}+2c\bar{w}^2+b^2(T_{j-1}-t)}{2(1-2c(T_{j-1}-t))}}}{\sqrt{1-2c(T_{j-1}-t)}} \Phi\left(\frac{\bar{w}(1-2c(T_{j-1}-t)) - w - b(T_{j-1}-t)}{\sqrt{(1-2c(T_{j-1}-t))(T_{j-1}-t)}}\right)$$
$$+ \sum_{i=1}^{N} \bar{\psi}_i\, e^{-\frac{1}{2}\beta_i^2 t + \beta_i w} \Phi\left(\frac{w - \bar{w} + \beta_i(T_{j-1}-t)}{\sqrt{T_{j-1}-t}}\right) \tag{10.73}$$

The function $w \mapsto \bar{h}_t(w)$ possesses all the advantages of the original function h_t:

- $\bar{h}_t > 0$ for each $(t, w) \in \mathcal{D}_j$ since $\bar{h}_{T_{j-1}}$ is positive and the sign is preserved when taking expectation;
- for each $t > 0$, the function $\bar{h}_t$ is differentiable in w and $d\bar{h}_t(w)/dw > 0$; in fact,

$$\frac{d\bar{h}_t(w)}{dw} = \int_{-\infty}^{+\infty} \frac{d}{dw}\bar{h}_{T_{j-1}}(w+x) \frac{1}{\sqrt{2\pi(T_{j-1}-t)}} e^{-\frac{x^2}{2(T_{j-1}-t)}}\, dx > 0,$$

  since $d\bar{h}_{T_{j-1}}(z)/dz > 0$ for each z. Accordingly, $\bar{h}_t$ is (strictly) increasing;
- for each $t > 0$, the function h_t is invertible;
- for each $0 < s < t$, the transition density $p(s, y; t, x)$ can be obtained through numerical inversion of $\bar{h}_t$.

Remark 10.12.1. The above construction procedure is quite general and can be applied to any initial function $h_{T_{j-1}}$ that is twice differentiable and increasing.

10.13 A General Dynamics à la Dupire (1994)

We now consider a second LIBOR model belonging to the general class (10.53). This model has the advantage of exactly retrieving the caplet volatility smile for the associated forward rate. With respect to the previous GBM-mixture model, the relevant formulas are less explicit, requiring numerical integrations. However, the calibration is automatic and calculations are still fast and efficient.[16]

Assume now that caplet prices with reset and maturity dates in T_{j-1} and T_j, respectively, are available (in the market) for a continuum of strikes. Precisely, denoting by $C_j(K)$ the caplet price with strike K, the following no-arbitrage conditions are assumed to hold:

B1) $C_j \in C^2((0, +\infty))$;

[16] This model has been proposed by Brigo and Mercurio (2003). Similarly, Balland and Hughston (2000) proved the existence of a LIBOR forward model that depends on some given Markovian factors and exactly calibrates the market caplet volatility curves.

B2) $\lim_{x\to 0^+} C_j(x) = \tau_j P(0,T_j) F_j(0)$ and $\lim_{x\to +\infty} C_j(x) = 0$;
B3) $\lim_{x\to 0^+} \frac{dC_j}{dx}(x) = -\tau_j P(0,T_j)$ and $\lim_{x\to +\infty} x\frac{dC_j}{dx}(x) = 0$;
B4) $\frac{d^2 C_j}{dx^2}(x) > 0$ for each $x > 0$, implying $-\tau_j P(0,T_j) < \frac{dC_j}{dx}(x) < 0$ for each $x > 0$.

Proposition 10.13.1. *The function $h_{T_{j-1}}$ that is consistent with the given caplet prices is implicitly defined by*

$$h^{-1}_{T_{j-1}}(x) = -\sqrt{T_{j-1}}\,\Phi^{-1}\left(-\frac{\frac{dC_j}{dx}(x)}{\tau_j P(0,T_j)}\right), \quad x > 0, \qquad (10.74)$$

which is well defined due to the assumptions on C_j.

Proof. Following Breeden and Litzenberger (1978), see also (9.4), and applying (10.55), we get that

$$\begin{aligned}\frac{\partial}{\partial K}\mathbf{Cpl}(0,T_{j-1},T_j,\tau_j,K) &= \tau_j P(0,T_j)\left[Q^j\{F_j(T_{j-1}) \le K\} - 1\right]\\ &= \tau_j P(0,T_j)\left[\Phi\left(\frac{h^{-1}_{T_{j-1}}(K)}{\sqrt{T_{j-1}}}\right) - 1\right]\\ &= -\tau_j P(0,T_j)\Phi\left(-\frac{h^{-1}_{T_{j-1}}(K)}{\sqrt{T_{j-1}}}\right).\end{aligned}$$

Imposing an exact match of the market caplet prices, we must have

$$\frac{dC_j}{dK}(K) = -\tau_j P(0,T_j)\Phi\left(-\frac{h^{-1}_{T_{j-1}}(K)}{\sqrt{T_{j-1}}}\right),$$

which immediately leads to (10.74) through inversion of the standard normal cumulative distribution function, given the bounds on the derivative $dC_j(x)/dx$. □

Corollary 10.13.1. *The function $h_{T_{j-1}}$ can be explicitly written as*

$$h_{T_{j-1}}(w) = \left(\frac{dC_j}{dx}\right)^{-1}\left(-\tau_j P(0,T_j)\Phi\left(-\frac{w}{\sqrt{T_{j-1}}}\right)\right), \quad w \in \mathbb{R}, \qquad (10.75)$$

with $\left(\frac{dC_j}{dx}\right)^{-1}$ denoting the inverse function of the first derivative of C_j.

Proof. We take $x = h_{T_{j-1}}(w)$ in (10.74) obtaining

$$-\frac{w}{\sqrt{T_{j-1}}} = \Phi^{-1}\left(-\frac{\frac{dC_j}{dx}(h_{T_{j-1}}(w))}{\tau_j P(0,T_j)}\right),$$

and apply function Φ to both sides. □

Corollary 10.13.2. *The function $h_{T_{j-1}}$ is strictly positive, differentiable, increasing and with zero limit at minus infinity,*

$$\lim_{w \to -\infty} h_{T_{j-1}}(w) = 0. \tag{10.76}$$

Moreover, the Q^j-expectation of $h_{T_{j-1}}(W_{T_{j-1}})$ is finite and equal to $F_j(0)$.

Proof. Since (10.74) holds for $x > 0$ (the domain of $h^{-1}_{T_{j-1}}$ is the range of $h_{T_{j-1}}$), $h_{T_{j-1}}$ is (strictly) positive. Moreover,

$$\frac{d}{dw} h_{T_{j-1}}(w) = \frac{1}{\frac{d^2 C_j}{dx^2}(h_{T_{j-1}}(w))} \frac{\tau_j P(0,T_j)}{\sqrt{2\pi T_{j-1}}} e^{-\frac{w^2}{2T_{j-1}}},$$

which is (strictly) positive due to B4). The limit in (10.76) then follows from (10.75) and assumptions B3) and B4). Finally, remembering (10.55),

$$\begin{aligned}
&E^j[h_{T_{j-1}}(W_{T_{j-1}})] \\
&= - \lim_{x \to +\infty} x\, Q^j\{F_j(T_{j-1}) \geq x\} + \int_0^{+\infty} Q^j\{F_j(T_{j-1}) \geq x\} dx \\
&= - \lim_{x \to +\infty} x\, \Phi\left(-\frac{h^{-1}_{T_{j-1}}(x)}{\sqrt{T_{j-1}}}\right) + \int_0^{+\infty} \Phi\left(-\frac{h^{-1}_{T_{j-1}}(x)}{\sqrt{T_{j-1}}}\right) dx \\
&= \frac{1}{\tau_j P(0,T_j)} \lim_{x \to +\infty} x \frac{dC_j}{dx}(x) - \frac{1}{\tau_j P(0,T_j)} \int_0^{+\infty} \frac{dC_j}{dx}(x) \\
&= 0 - \frac{1}{\tau_j P(0,T_j)} \left[\lim_{x \to +\infty} C_j(x) - \lim_{x \to 0+} C_j(x)\right] \\
&= F_j(0),
\end{aligned}$$

where we have used (10.74) and assumptions B2) and B3). □

From a computational point of view, the value of function $h_{T_{j-1}}$ in any point w can be obtained either by means of the explicit definition (10.75) or by numerically solving the equation

$$w - h^{-1}_{T_{j-1}}(x) = 0$$

in the variable x.[17]

Proposition 10.13.2. *The value of the forward rate F_j at any time $t < T_{j-1}$ that is consistent with $h_{T_{j-1}}$ is*

$$F_j(t) = h_t(W_t), \tag{10.77}$$

[17] Several algorithms are available for solving nonlinear equations. In this specific case, the solution search is extremely efficient since the nonlinear function $h^{-1}_{T_{j-1}}$ is monotone and with an analytical gradient.

where

$$h_t(w) = E^j[h_{T_{j-1}}(W_{T_{j-1}})|W_t = w]$$
$$= \int_0^{+\infty} \Phi\left(\frac{w + \sqrt{T_{j-1}}\,\Phi^{-1}\left(-\frac{1}{\tau_j P(0,T_j)}\frac{dC_j}{dz}(z)\right)}{\sqrt{T_{j-1}-t}}\right) dz. \tag{10.78}$$

Proof. Since the expectation of $h_{T_{j-1}}(W_{T_{j-1}})$ under Q^j is finite, and equal to $F_j(0)$, the first equation in (10.78) defines the unique $(Q^j, \mathcal{F}_t)$-martingale with value $h_{T_{j-1}}(W_{T_{j-1}})$ at time T_{j-1} due to the Markov property of Brownian motion. The second equation in (10.78) then follows from (10.59) and (10.74).[18] □

From (10.78) we immediately see that h_t is positive for each $t < T_{j-1}$. Assuming twice differentiability for h_t and the possibility of swapping integrals and derivatives, the forward rate dynamics is then given by the following.

Corollary 10.13.3. *The (driftless) dynamics of F_j that is consistent with the forward rate value (10.78) is*

$$dF_j(t) = \frac{\partial h_t}{\partial w}\left(h_t^{-1}(F_j(t))\right) dW_t$$
$$= \int_0^{+\infty} \frac{\exp\left\{-\frac{\left[h_t^{-1}(F_j(t))+\sqrt{T_{j-1}}\,\Phi^{-1}\left(-\frac{1}{\tau_j P(0,T_j)}\frac{dC_j}{dz}(z)\right)\right]^2}{2(T_{j-1}-t)}\right\}}{\sqrt{2\pi(T_{j-1}-t)}}\, dz\, dW_t. \tag{10.79}$$

Proof. The result follows from the differentiability assumption on h_t and from (10.60) with (10.74). □

In order to check the evolution of the implied volatility curves produced by the forward rate process in the future, we need to analytically price caplets at any future time $0 < t < T_{j-1}$. From the general pricing formula (10.61), we obtain the following.

Proposition 10.13.3. *The price at time $t < T_{j-1}$ of the caplet setting in T_{j-1}, paying in T_j and with strike K is given by*

$$\mathbf{Cpl}(t, T_{j-1}, T_j, \tau_j, K)$$
$$= \tau_j P(t,T_j)\left[F_j(t) - K + \int_0^K \Phi\left(\frac{h_{T_{j-1}}^{-1}(z) - h_t^{-1}(F_j(t))}{\sqrt{T_{j-1}-t}}\right) dz\right]$$
$$= \tau_j P(t,T_j)$$
$$\cdot \left[F_j(t) - \int_0^K \Phi\left(\frac{\sqrt{T_{j-1}}\,\Phi^{-1}\left(-\frac{1}{\tau_j P(0,T_j)}\frac{dC_j}{dz}(z)\right) + h_t^{-1}(F_j(t))}{\sqrt{T_{j-1}-t}}\right) dz\right] \tag{10.80}$$

[18] We can apply (10.59) since $\lim_{w\to-\infty} h_{T_{j-1}}(w) = 0$ and $E^j\{h_{T_{j-1}}(W_{T_{j-1}})\}$ exists finite.

and can easily be calculated with numerical methods.

Proof. The final integral in (10.61) can be rewritten as follows:

$$
\begin{aligned}
&F_j(t) - \int_{-\infty}^{h_{T_{j-1}}^{-1}(K)-h_t^{-1}(F_j(t))} \frac{h_{T_{j-1}}\left(h_t^{-1}(F_j(t)) + w\right)}{\sqrt{2\pi(T_{j-1}-t)}} e^{-\frac{w^2}{2(T_{j-1}-t)}}\, dw \\
&= F_j(t) - K\Phi\left(\frac{h_{T_{j-1}}^{-1}(K) - h_t^{-1}(F_j(t))}{\sqrt{T_{j-1}-t}}\right) \\
&\quad + \int_{-\infty}^{h_{T_{j-1}}^{-1}(K)-h_t^{-1}(F_j(t))} \frac{d}{dw} h_{T_{j-1}}\left(h_t^{-1}(F_j(t)) + w\right) \Phi\left(\frac{w}{\sqrt{T_{j-1}-t}}\right) dw \\
&= F_j(t) - K\Phi\left(\frac{h_{T_{j-1}}^{-1}(K) - h_t^{-1}(F_j(t))}{\sqrt{T_{j-1}-t}}\right) \\
&\quad + \int_0^K \Phi\left(\frac{h_{T_{j-1}}^{-1}(z) - h_t^{-1}(F_j(t))}{\sqrt{T_{j-1}-t}}\right) dz,
\end{aligned}
$$

where we have used the definition of $F_j(t)$ (first equality), the integration by parts (second equality) and the change of variable $z = h_{T_{j-1}}(h_t^{-1}(F_j(t)) + w)$ (third equality).

Putting pieces together, we finally obtain the first equality in (10.80), whereas the second follows from (10.74). □

Remark 10.13.1. Propositions 10.13.1 and 10.13.2 can be equivalently restated for a general pricing function C_j, which does not necessarily match all market quotes. In fact, we can assume a flexible parametric distribution for $F_j(T_{j-1}) = h_{T_{j-1}}(W_{T_{j-1}})$, thus avoiding classic interpolation issues on option prices. For instance, we can consider a mixture of lognormal densities as in (10.41), where we set $t = T_{j-1}$ and take constant coefficients,

$$
\begin{aligned}
&p_{T_{j-1}}(x) \\
&= \sum_{i=1}^N \lambda_i \frac{1}{x\sigma_i\sqrt{2\pi T_{j-1}}} \exp\left\{-\frac{1}{2\sigma_i^2 T_{j-1}}\left[\ln\frac{x}{F_j(0)} - \mu_i T_{j-1} + \tfrac{1}{2}\sigma_i^2 T_{j-1}\right]^2\right\},
\end{aligned}
$$

with $\sum_{i=1}^N \lambda_i\, e^{\mu_i T_{j-1}} = 1$. In this case, we have

$$
C_j(K) = \tau_j P(0,T_j) \sum_{i=1}^N \lambda_i e^{\mu_i T_{j-1}} \mathrm{Bl}\left(K\, e^{-\mu_i T_{j-1}}, F_j(0), \sigma_i\sqrt{T_{j-1}}\right),
$$

from which (10.74) and (10.77) can be derived.

We can then compare the resulting forward rate dynamics (10.79) with (10.43) under (10.44). It is evident that the latter has a more explicit formulation involving no numerical procedure (besides the standard ones for the computation of transcendent functions). However, the former has several advantages: it allows for i) less severe restrictions on the model parameters to ensure that the local volatility is well defined, ii) constant coefficients, and hence for a more parsimonious parametrization, iii) a rapid calculation of future caplet prices, and hence for a quick check of the future evolution of the caplet volatility structure.

11. Stochastic-Volatility Models

LIBOR models with stochastic volatility are extensions of the LFM where the instantaneous volatility of relevant rates evolves according to a diffusion process driven by a Brownian motion that is possibly instantaneously correlated with those governing the rates' evolution. Formally, the general forward rate F_j is assumed to evolve under its canonical measure Q^j according to

$$\begin{aligned} dF_j(t) &= a_j(t)\varphi(F_j(t))[V(t)]^\gamma \, dZ_j(t), \\ dV(t) &= a_V \, dt + b_V \, dW(t), \end{aligned} \tag{11.1}$$

where a_j and φ are deterministic functions, $\gamma \in \{1/2, 1\}$, a_V and b_V are adapted processes, and Z_j and W are possibly correlated Brownian motions.

Besides an empirical evidence of stochastic volatility for interest rates, SVMs are also introduced to accommodate market smiles and skews while predicting a realistic evolution for the implied volatilities in the future.[1]

The best known example of SVM in the financial literature is that of Heston (1993), reformulated in a LIBOR model setting by Wu and Zhang (2002), see Section 11.2. The possibility of pricing options analytically, while assuming a non-zero correlation between the underlying asset and its volatility, is the main reason for its popularity and widespread use.

Model implied volatilities. When the correlation between Z_j and W is zero, the existence itself of a stochastic volatility leads to smile-shaped implied volatility curves. In fact, assuming that φ is the identity function, Renault and Touzi (1996) proved that, under mild assumptions on the model coefficients, a SVM like (11.1) automatically generates an implied volatility function that has a local minimum at the ATM value. Skew-shaped volatilities, instead, can be produced as soon as we relax some of the previous assumptions. In the existing financial literature, three are the solutions commonly proposed to obtain non-zero slopes at the ATM level:

- Introducing a non-zero (instantaneous) correlation between rates and volatility.
- Assuming a displaced-diffusion dynamics, which corresponds to setting $\varphi(x) = x + \beta$, where β is a real constant.

[1] The latter feature is typically granted by considering time-homogeneous equations in (11.1).

- Assuming a non-linear function φ as, for instance, $\varphi(x) = x^\alpha$ with $\alpha \in (0,1)$.

The second and third possibilities should not come as a surprise. We have seen in fact, at the beginning of this chapter, that both the shifted-lognormal model (10.2) and the CEV model (10.7) lead to skewed implied volatilities. One can then expect that a similar behavior also holds when adding stochastic volatility.

We have already seen in previous sections that caplet-volatility curves in real markets are typically skew-shaped. Any reasonable LIBOR model with stochastic volatility, therefore, should be characterized by at least one of the three features above. The SVMs we illustrate in this section all fulfil this requirement.

Analytical tractability. In order to construct an analytically-tractable model, it is often convenient to assume that the Brownian motions Z_j and W are uncorrelated, so that one can apply the Hull and White (1987) methodology for SVMs. One first prices a caplet for a given path of $V(t)$ and then averages such a price with respect to the density of a suitable functional of the volatility path. For example, if a_j is constant, φ is the identity function and $\gamma = 1/2$, then a caplet's price with expiry T is equal to the integral of corresponding Black's prices multiplied by the density of the root-mean-square volatility $\left(\int_0^T V(t)\,dt\right)^{1/2}$. With this procedure, one obtains analytical prices, in terms of a one-dimensional integral, only if the first equation in (11.1) can be solved explicitly and the density of the volatility-path functional is known. However, even for extremely simple volatility dynamics, such a density can be hard to calculate analytically. One then typically resorts to numerical methods as in the Joshi and Rebonato (2003) model, see Section 11.5.

An alternative and more efficient procedure aiming at explicit formulas is based on the knowledge of the characteristic function of F_j under Q^j and the calculation of a caplet price by inversion of a Fourier Transform (FT). The characteristic function can be derived by solving an associated PDE, which is found by applying the Feynman-Kac theorem. This method, pioneered by Heston (1993), is rather general and can also be applied under a non-zero correlation between rates and volatility, see the Wu and Zhang (2002) model in Section 11.2.

In the following, we will describe some of the best known extensions of the LFM allowing for stochastic volatility, namely the SVMs of Andersen and Brotherton-Ratcliff (2001), Wu and Zhang (2002), Hagan, Kumar, Lesniewski and Woodward (2002), Piterbarg (2003) and Joshi and Rebonato (2003). A summary of the main properties of these models is reported in Table 11.4.

The notation of Section 10.5 will be in force. In particular, we denote by $\mathcal{T} = \{T_0, \ldots, T_M\}$ the set of times defining a family of spanning forward rates F_j, $j = 1, \ldots, M$, and by $\tau_0, \ldots, \tau_M$ the corresponding year fractions.

11.1 The Andersen and Brotherton-Ratcliffe (2001) Model

Andersen and Brotherton-Ratcliffe model forward rates $F_j(t) = F(t; T_{j-1}, T_j)$, $j = 1, \ldots, M$, under the spot LIBOR measure Q^d, as follows

$$
\begin{aligned}
dF_j(t) &= \sigma_j(t)\sqrt{V(t)}\varphi(F_j(t))\left[\sqrt{V(t)} \sum_{k=\beta(t)}^{j} \frac{\tau_k \rho_{j,k}\sigma_k(t)\varphi(F_k(t))}{1+\tau_k F_k(t)}\, dt + dZ_j(t)\right] \\
dV(t) &= \kappa(\theta - V(t))\, dt + \epsilon\psi(V(t))\, dW(t),
\end{aligned}
\tag{11.2}
$$

where $Z = (Z_1, \ldots, Z_M)$ and W are Q^d-Brownian motions with

$$
\begin{aligned}
dZ_j dZ_k &= \rho_{j,k}\, dt, \quad j, k = 1, \ldots, M, \\
dZ_j dW &= 0, \quad j = 1, \ldots, M,
\end{aligned}
$$

and where the σ_j's are deterministic scalar functions, κ, θ and ϵ are positive constants and $\psi : \mathbb{R}^+ \to \mathbb{R}^+$ is a regular function.

The volatility process is mean reverting, with mean-reversion level θ and speed κ, and is always positive provided that ψ satisfies suitable conditions, such as, for instance, $\psi(0) = 0$. In most of their empirical work, Andersen and Brotherton-Ratcliffe choose to set $\psi(x) = x^{3/4}$, so that the resulting V is non-explosive and the origin is inaccessible. To ensure maximum flexibility, however, their theoretical treatment allows ψ to be a general function.

The assumption of zero correlation between the two Brownian motions Z and W ensures that the dynamics of V remains the same under a different numeraire, since no correction term is to be added to the drift of V when changing measure. For instance, when we move to the forward measure Q^j, we obtain (by the change-of-numeraire toolkit in Section 2.3)

$$
\begin{aligned}
dF_j(t) &= \sigma_j(t)\sqrt{V(t)}\varphi(F_j(t))\, dZ_j^j(t) \\
dV(t) &= \kappa(\theta - V(t))\, dt + \epsilon\psi(V(t))\, dW^j(t)
\end{aligned}
\tag{11.3}
$$

where $Z^j = (Z_1^j, \ldots, Z_M^j)$ and W^j are uncorrelated Q^j-Brownian motions.

If one lets φ and ψ be arbitrary functions, the dynamics (11.3) do not allow, in general, the derivation of exact formulas for options on LIBOR rates. Andersen and Brotherton-Ratcliffe, however, develop an efficient approximation of a caplet's price, based on an asymptotic expansion, as follows.

The pricing of caplets. The price at time t of a caplet whose payoff is set at T_{j-1} and paid at T_j can be written as (assuming unit notional)

$$
\begin{aligned}
\mathbf{Cpl}(t, T_{j-1}, T_j, \tau_j, K) &= P(t, T_j)\tau_j E_t^j \left[(F_j(T_{j-1}) - K)^+\right] \\
&= P(t, T_j)\tau_j G_j(t, F_j(t), V(t)),
\end{aligned}
\tag{11.4}
$$

for some function $G_j(t, F, V)$.

Applying Feynman-Kac's formula one finds that G_j must satisfy the PDE:

$$\frac{\partial G_j}{\partial t} + \kappa(\theta - V)\frac{\partial G_j}{\partial V} + \frac{1}{2}\epsilon^2\psi(V)^2\frac{\partial^2 G_j}{\partial V^2} + \frac{1}{2}\varphi(F)^2 V\sigma_j(t)^2\frac{\partial^2 G_j}{\partial F^2} = 0 \quad (11.5)$$

with boundary condition

$$G_j(T_{j-1}, F, V) = (F - K)^+.$$

Andersen and Brotherton-Ratcliffe first assume that V is constant and equal to one.[2] They then approximate G_j around the known solution for $\varphi(x) = x$ (the corresponding Black price), obtaining the following formula through an asymptotic expansion of the implied volatility in powers of $\delta_j = T_{j-1} - t$:

$$\begin{aligned} G_j(t, F, 1) &= g_j\left(t, F; (T_{j-1} - t)^{-1}\int_t^{T_{j-1}} \sigma_j(s)^2 ds\right) \\ g_j(t, F; x) &= F\Phi(d_+^j) - K\Phi(d_-^j) \\ d_\pm^j &= \frac{\ln(F/K) \pm \frac{1}{2}\Omega^j(t, F, x)^2}{\Omega^j(t, F, x)} \\ \Omega^j(t, F, x) &= \sum_{i\geq 0} x^{i+1/2}\delta_j^{i+1/2}\Omega_i(F), \end{aligned} \quad (11.6)$$

where the functions Ω_i, which do not depend on δ_j, are obtained by substituting (11.6) into (11.5) and matching terms of the same order in δ_j. The first two terms, Ω_0 and Ω_1, can be found by solving ordinary differential equations of the Bernoulli type and imposing a finite limit for $F \to K$. One gets:

$$\begin{aligned} \Omega_0(F) &= \frac{\ln(F/K)}{\int_K^F \varphi(s)^{-1}ds}, \\ \Omega_1(F) &= -\frac{\Omega_0(F)}{\left(\int_K^F \varphi(s)^{-1}ds\right)^2}\ln\left[\Omega_0(F)\left(\frac{FK}{\varphi(F)\varphi(K)}\right)^{\frac{1}{2}}\right]. \end{aligned} \quad (11.7)$$

Andersen and Brotherton-Ratcliffe claim that the approximation

$$\Omega^j(t, F, x) \approx \Omega_0(F)x^{1/2}\delta_j^{1/2} + \Omega_1(F)x^{3/2}\delta_j^{3/2}$$

is often accurate also for long maturities. They also stress that, for many choices of φ, the accuracy of their asymptotic expansion can be improved by expanding G_j around the known solution for either a shifted-lognormal diffusion or a Gaussian diffusion rather than the geometric Brownian motion chosen previously.

---

[2] The case $V \neq 1$ but still constant is easily obtained by scaling the σ_j's.

Andersen and Brotherton-Ratcliffe finally introduce stochastic volatility as in (11.3). They note that, by the Hull and White (1987) decomposition result, the function G_j can be written as

$$G_j(t,F,V) = E_t^j\left[g_j(t,F;(T_{j-1}-t)^{-1}U_j)\right],$$
$$U_j := \int_t^{T_{j-1}} \sigma_j(s)^2 V(s)\,ds,$$

and then consider the fourth-order Taylor expansion:

$$G_j(t,F,V) \approx \sum_{n=0}^{4} \frac{1}{n!}\frac{\partial^n g_j}{\partial x^n}(t,F;\bar{x}_j(t,V))E_t^j\left[\left(\frac{1}{\delta_j}U_j - \bar{x}_j(t,V)\right)^n\right], \quad (11.8)$$

where $\partial^0 g_j/\partial x^0 = g_j$ and

$$\bar{x}_j(t,V(t)) := \delta_j^{-1}E_t^j[U_j] = \delta_j^{-1}\int_t^{T_{j-1}} \sigma_j(s)^2[\theta + (V(t)-\theta)\mathrm{e}^{-\kappa(s-t)}]\,ds.$$

The central moments in (11.8) are calculated by finding an analytical approximation for the Laplace transform $E_t^j[\exp(-sU_j)]$ of the density of U_j and then taking partial derivatives at $s=0$.

Setting $Y := \ln(F/K)$, Andersen and Brotherton-Ratcliffe eventually obtain the following approximation for the function G_j in terms of a double expansion in powers of ϵ and δ_j:

$$G_j(t,F,V) = g_j(t,F,x_j(t,V)), \quad (11.9)$$

where $x_j(t,V)$ is defined by

$$x_j(t,V) = \bar{x}_j(t,V) + \epsilon^2(\alpha_0 + \alpha_1 Y^2) + \epsilon^4(\beta_0 + \beta_1 Y^2 + \beta_2 Y^4 \mathrm{e}^{-\Lambda\epsilon^2 Y^2}) + \mathcal{O}(\epsilon^6), \quad (11.10)$$

and where α_i's and β_i's are coefficients depending on the Ω_i's, on ψ and on δ_j,[3] and Λ is an arbitrary positive number introduced to avoid explosion of the fourth-order term for away-from-the-money options.[4]

Thanks to the previous approximation procedure, an estimate for the implied volatility in the Andersen and Brotherton-Ratcliffe SVM is straightforward and given by:

$$V_{impl}(F,V,\delta_j,K) = \Omega_0(F)\sqrt{x_j(t,V)} + \Omega_1(F)x_j(t,V)^{3/2}\delta_j + \mathcal{O}(\delta_j^2).$$

[3] The expressions for coefficients α_i's and β_i's are rather involved, and are not reported here. The interested reader is then referred to Andersen and Brotherton-Ratcliffe (2001).

[4] Andersen and Brotherton-Ratcliffe suggest a value for Λ between 1 and 10.

The pricing of swaptions. The pricing of swaptions in the SVM (11.2) can be tackled in quite a similar fashion to that of caplets. In fact, consider the swap rate

$$S_{\alpha,\beta}(t) = \frac{P(t,T_\alpha) - P(t,T_\beta)}{C_{\alpha,\beta}(t)}$$

where

$$C_{\alpha,\beta}(t) = \sum_{j=\alpha+1}^{\beta} \tau_j P(t,T_j)$$

and $T_0 < t < T_\alpha < T_\beta \le T_M$.

Since the swap rate is a function of forward rates $F_{\alpha+1},\dots,F_\beta$, see also equation (6.33), one has that, under the forward swap measure $Q^{\alpha,\beta}$ (with numeraire $C_{\alpha,\beta}$), and given $t \le u < T_\alpha$,

$$\begin{aligned} dS_{\alpha,\beta}(u) &= \sum_{j=\alpha+1}^{\beta} \frac{\partial S_{\alpha,\beta}(u)}{\partial F_j(u)} \varphi(F_j(u))\sigma_j(u)\sqrt{V(u)}\, dZ_j^{\alpha,\beta}(u) \\ &\approx \varphi(S_{\alpha,\beta}(u)) \sum_{j=\alpha+1}^{\beta} w_j(t)\sigma_j(u)\sqrt{V(u)}\, dZ_j^{\alpha,\beta}(u), \end{aligned} \tag{11.11}$$

where the approximation comes from the usual weight-freezing technique, $Z_j^{\alpha,\beta}$'s are $Q^{\alpha,\beta}$-Brownian motions and

$$w_j(t) := \frac{\partial S_{\alpha,\beta}(t)}{\partial F_j(t)} \frac{\varphi(F_j(t))}{\varphi(S_{\alpha,\beta}(t))}.$$

Thanks to this approximation,[5] and given that one can write

$$\sum_{j=\alpha+1}^{\beta} w_j(t)\,\sigma_j(u) dZ_j^{\alpha,\beta}(u) = \sigma_{\alpha,\beta}(t;u)\, d\tilde{Z}(u)$$

where $\tilde{Z}$ is a standard (one-dimensional) Brownian motion (uncorrelated to the Brownian motion of V) and $\sigma_{\alpha,\beta}(t;u)$ is a deterministic function, swap rates under their associated swap measures and forward rates under their canonical forward measures follow similar SDEs. Therefore, the pricing procedure developed for caplets can be equivalently applied for the pricing of swaptions.

Precisely, as a payer-swaption price on the swap rate $S_{\alpha,\beta}$ is given, at time t, by

$$C_{\alpha,\beta}(t) E_t^{\alpha,\beta}[(S_{\alpha,\beta}(T_\alpha) - K)^+],$$

one can value this expectation by means of the previous formula (11.9), provided that one replaces F with $S_{\alpha,\beta}$, σ_j with $\sigma_{\alpha,\beta}$ and T_{j-1} with T_α.

[5] Whenever φ is a power function, this approximation seems to be quite accurate, as also stressed by Andersen and Andreasen (2000).

11.2 The Wu and Zhang (2002) Model

Wu and Zhang (2002) model the (squared) volatility of forward rates through a square-root process like (3.21), as in Heston (1993). Contrary to Andersen and Brotherton-Ratcliffe (2001), they allow for a non-zero correlation between rates and volatility. This has the clear advantage that skew effects can be captured by keeping a linear diffusion term in the forward-rates dynamics as in the LFM. However, the price to pay is that V evolves as a square-root process only under a given reference measure and the volatility dynamics changes as soon as we change numeraire, unless some ad-hoc approximation is introduced as we will see in the following.

Wu and Zhang choose the risk-neutral measure Q as reference measure and assume that forward rates F_j, $j = 1, \dots, M$, evolve, under Q, as follows

$$\begin{aligned} dF_j(t) &= \sqrt{V(t)} F_j(t) \underline{\sigma}_j(t) \cdot \left[-\sqrt{V(t)} \underline{\Gamma}_j(t)\, dt + d\underline{Z}(t) \right] \\ dV(t) &= \kappa(\theta - V(t))\, dt + \epsilon \sqrt{V(t)}\, dW(t) \end{aligned} \tag{11.12}$$

where $\underline{Z}$ is a vector of independent Q-Brownian motions, $\underline{\sigma}_j$'s are deterministic vector functions, "$\cdot$" denotes scalar product, κ, θ and ϵ are positive constants, W is another Q-Brownian motion such that

$$\frac{(\underline{\sigma}_j(t) \cdot d\underline{Z}(t))\, dW(t)}{|\underline{\sigma}_j(t)|} = \rho_j(t)\, dt,$$

with ρ_j's time-dependent correlations and $|\underline{\sigma}_j(t)|$ the norm of $\underline{\sigma}_j(t)$, and where $\sqrt{V(t)} \underline{\Gamma}_j(t)$ is the percentage volatility of the T_j-maturity bond, with $\underline{\Gamma}_j$'s deterministic vector functions such that

$$\underline{\Gamma}_{j-1}(t) - \underline{\Gamma}_j(t) = \frac{\tau_j F_j(t)}{1 + \tau_j F_j(t)} \underline{\sigma}_j(t).$$

In this model, functions $\underline{\sigma}_j$'s are responsible for the level of the caplets smiles, parameter ϵ controls the smiles' curvature and correlations ρ_j account for the slopes at the ATM level.

Allowing for non-zero correlations ρ's, when one moves to the forward measure Q^j, the stochastic-volatility process $V(t)$ gains an extra drift term. Precisely, the dynamics of F_j under Q^j (applying the change-of-numeraire toolkit in Section 2.3) is given by

$$\begin{aligned} dF_j(t) &= \sqrt{V(t)} F_j(t) \underline{\sigma}_j(t) \cdot d\underline{Z}^j(t) \\ dV(t) &= [\kappa\theta - (\kappa + \epsilon \xi_j(t)) V(t)]\, dt + \epsilon \sqrt{V(t)}\, dW^j(t), \end{aligned} \tag{11.13}$$

where $\underline{Z}^j$ is a vector of independent Q^j-Brownian motions and W^j is a Q^j-Brownian motion such that $\underline{Z}^j$ and W^j have the same correlation structure as $\underline{Z}$ and W, and

$$\xi_j(t) = \sum_{k=1}^{j} \frac{\tau_k F_k(t)}{1 + \tau_k F_k(t)} |\underline{\sigma}_k(t)| \rho_k(t).$$

The pricing of caplets. The presence of forward rates in the drift of V makes such a model untractable when allowing for non-zero correlations. Wu and Zhang, therefore, resort to the classical freezing technique and approximate the ξ_i's by freezing forward rates at time 0:

$$\xi_j(t) \approx \sum_{k=1}^{j} \frac{\tau_k F_k(0)}{1 + \tau_k F_k(0)} |\underline{\sigma}_k(t)| \rho_k(t),$$

thus obtaining a "square-root type" equation for V that involves no other process:

$$\begin{aligned} dV(t) &= \kappa[\theta - \tilde{\xi}_j(t) V(t)]\, dt + \epsilon \sqrt{V(t)}\, dW^j(t), \\ \tilde{\xi}_j(t) &:= 1 + \frac{\epsilon}{\kappa} \xi_j(t). \end{aligned}$$

Similarly to Heston (1993), Wu and Zhang write the price at time 0 of the j-th caplet as (assuming unit notional)

$$\begin{aligned} \mathbf{Cpl}(0, T_{j-1}, T_j, \tau_j, K) &= P(0, T_j) \tau_j \\ &\cdot \left[F_j(0) E^j \left(e^{X(T_{j-1})} 1_{\{F_j(T_{j-1}) > K\}} \right) - K E^j (1_{\{F_j(T_{j-1}) > K\}}) \right], \end{aligned} \tag{11.14}$$

where we set $X(t) := \ln(F_j(t)/F_j(0))$.

Denoting by ϕ the moment generating function of $X(T_{j-1})$,

$$\phi(X(t), V(t), t; z) = E_t^j \left(e^{zX(T_{j-1})} \right), \quad z \in \mathbb{C},$$

the expectations in (11.14) can be written in terms of ϕ as

$$\begin{aligned} E^j[1_{\{F_j(T_{j-1}) > K\}}] &= \frac{1}{2} + \frac{1}{\pi} \int_0^\infty \frac{\mathrm{Im}[e^{-iu \ln(K/F_j(0))}] \tilde{\phi}(iu)}{u} du, \\ E^j \left[e^{X(T_{j-1})} 1_{\{F_j(T_{j-1}) > K\}} \right] &= \frac{1}{2} + \frac{1}{\pi} \int_0^\infty \frac{\mathrm{Im}[e^{-iu \ln(K/F_j(0))}] \tilde{\phi}(iu + 1)}{u} du, \end{aligned} \tag{11.15}$$

with $\tilde{\phi}(z) := \phi(0, V(0), 0; z)$ and $\mathrm{Im}(w)$ denoting the imaginary part of the complex number w.

Explicit formulas for caplets, therefore, can be derived as soon as one finds an analytical expression for ϕ.

Thanks to Feynman-Kac's formula, see Appendix C, we have that ϕ satisfies the PDE

$$\frac{\partial\phi}{\partial t} + \kappa[\theta - \tilde{\xi}_j(t)V]\frac{\partial\phi}{\partial V} - \frac{1}{2}|\underline{\sigma}_j(t)|^2 V\left(\frac{\partial\phi}{\partial x} - \frac{\partial^2\phi}{\partial x^2}\right) + \frac{1}{2}\epsilon^2 V\frac{\partial^2\phi}{\partial V^2} + \epsilon\rho_j(t)V|\underline{\sigma}_j(t)|\frac{\partial^2\phi}{\partial V\partial x} = 0, \tag{11.16}$$

with terminal condition

$$\phi(x, V, T_{j-1}; z) = e^{zx}.$$

This PDE can be solved explicitly, see also Heston (1993), by considering a solution of the form

$$\phi(x, V, t; z) = \bar{\phi}(x, V, \delta; z) := \exp[A(\delta, z) + B(\delta, z)V + zx],$$

where $\delta = T_{j-1} - t$, and assuming that the coefficients of (11.16) are piecewise constant on a grid $0 = \delta_0 < \delta_1 < \cdots < \delta_n = T_{j-1}$. One gets $A(0, z) = B(0, z) = 0$ and, for $\delta_i \le \delta < \delta_{i+1}$, $i = 0, \ldots, n-1$,

$$A(\delta, z) = A(\delta_i, z) + \frac{\kappa\theta}{\epsilon^2}\left\{(a_i + d_i)(\delta - \delta_i) - 2\ln\left[\frac{1 - g_i e^{d_i(\delta-\delta_i)}}{1 - g_i}\right]\right\},$$

$$B(\delta, z) = B(\delta_i, z) + \frac{(a_i + d_i - \epsilon^2 B(\delta_i, z))(1 - e^{d_i(\delta-\delta_i)})}{\epsilon^2(1 - g_i e^{d_i(\delta-\delta_i)})},$$

where

$$a_i = \kappa\tilde{\xi}_{j,i} - \rho_{j,i}\epsilon\sigma_{j,i}z,$$

$$d_i = \sqrt{a_i^2 - \sigma_{j,i}^2\epsilon^2(z^2 - z)},$$

$$g_i = \frac{a_i + d_i - \epsilon^2 B(\delta_i, z)}{a_i - d_i - \epsilon^2 B(\delta_i, z)},$$

and where $\tilde{\xi}_{j,i}$, $\rho_{j,i}$ and $\sigma_{j,i}$ are the (constant) values that the respective functions $\tilde{\xi}_j$, ρ_j and σ_j take on the interval $(T_{j-1} - \delta_{i+1}, T_{j-1} - \delta_i)$.[6]

The pricing of swaptions. Using again the freezing technique, Wu and Zhang show that the swap rate $S_{\alpha,\beta}$ evolves under the related swap measure $Q^{\alpha,\beta}$ (approximately) as

$$dS_{\alpha,\beta}(t) = S_{\alpha,\beta}(t)\sqrt{V(t)}\sum_{j=\alpha+1}^{\beta} w_j(0)\underline{\sigma}_j(t)\cdot d\underline{Z}^{\alpha,\beta}(t)$$
$$dV(t) = \kappa[\theta - \tilde{\xi}_{\alpha,\beta}(t)V(t)]\,dt + \epsilon\sqrt{V(t)}\,dW^{\alpha,\beta}(t) \tag{11.17}$$

where $\underline{Z}^{\alpha,\beta}$ is a vector of $Q^{\alpha,\beta}$-Brownian motions and $W^{\alpha,\beta}$ is a $Q^{\alpha,\beta}$-Brownian motion such that $\underline{Z}^{\alpha,\beta}$ and $W^{\alpha,\beta}$ have the same correlation structure as $\underline{Z}$ and W, and

[6] We mention that Wu and Zhang derive also an alternative caplet formula based on the Fourier-Transform method proposed by Carr and Madan (1998).

$$w_j(0) = \frac{\partial S_{\alpha,\beta}(0)}{\partial F_j(0)} \frac{F_j(0)}{S_{\alpha,\beta}(0)},$$

$$\tilde{\xi}_{\alpha,\beta}(t) = 1 + \frac{\epsilon}{\kappa} \sum_{j=\alpha+1}^{\beta} \left[\tau_j P(0,T_j)\xi_j(t)\right] / C_{\alpha,\beta}(0)$$

Since equation (11.17) is similar to (11.13), the calculation of a payer-swaption price at time 0,

$$C_{\alpha,\beta}(0) E^{\alpha,\beta}[(S_{\alpha,\beta}(T_\alpha) - K)^+],$$

is equivalent to that of a caplet. Precisely, this last expectation can be valued as the term between square brackets in (11.14) by applying the following substitutions:

$$E^j \to E^{\alpha,\beta}, \quad F_j \to S_{\alpha,\beta}, \quad T_{j-1} \to T_\alpha, \quad \underline{\sigma}_j(t) \to \underline{\sigma}^{\alpha,\beta}(t) := \sum_{j=\alpha+1}^{\beta} w_j(0)\underline{\sigma}_j(t),$$

$$\tilde{\xi}_j(t) \to \tilde{\xi}_{\alpha,\beta}(t), \quad \rho_j(t) \to \rho^{\alpha,\beta}(t) := \frac{\sum_{j=\alpha+1}^{\beta} w_j(0)|\underline{\sigma}_j(t)|\rho_j(t)}{|\sum_{j=\alpha+1}^{\beta} w_j(0)\underline{\sigma}_j(t)|}. \tag{11.18}$$

11.3 The Piterbarg (2003) Model

Piterbarg proposes an alternative forward LIBOR model with stochastic volatility, deriving efficient approximations for swaption prices that allow calibration of the model to all European swaptions across all maturities, tenors and strikes. In this model, forward rates are assumed to follow a shifted-lognormal diffusion with a stochastic (squared) volatility that: i) follows a square-root process, ii) is mean-reverting towards its initial value, and iii) is uncorrelated with the Brownian motions governing the rates' dynamics.

Piterbarg assumes that forward rates evolves, under a generic measure $\bar{Q}$, according to

$$\begin{aligned} dF_j(t) &= \sqrt{V(t)}[b_j(t)F_j(t) + (1 - b_j(t))F_j(0)]\underline{\sigma}_j(t) \cdot \left[\sqrt{V(t)}\underline{\mu}^{\bar{Q}}(t)\, dt + d\underline{Z}(t)\right] \\ dV(t) &= \kappa(V(0) - V(t))\, dt + \epsilon\sqrt{V(t)}\, dW(t), \end{aligned} \tag{11.19}$$

where $\underline{Z}$ is a vector of independent $\bar{Q}$-Brownian motions, $\underline{\sigma}_j$'s are deterministic vector functions, κ and ϵ are positive constants, b_j's are deterministic functions, W is another $\bar{Q}$-Brownian motion with $d\underline{Z}(t)\, dW(t) = 0$, and $\underline{\mu}^{\bar{Q}}$ is a suitable adapted vector process.

In (11.19), similarly to the SVM of Wu and Zhang (2002), functions $\underline{\sigma}_j$'s are responsible for the level of the caplets smiles and parameter ϵ accounts for the smiles' curvature. Contrary to (11.12), slopes at ATM strikes are here controlled by functions b_j's and not by correlations between volatility and rates, which are in fact set to zero.

Allowing for time-dependent coefficients complicates the task of finding analytical formulas for options on swap rates. The technical contribution of Piterbarg's paper is the derivation of fast and accurate European option prices under general time-dependent parameters, and not only piecewise constant as in the Wu and Zhang (2002) model.

Consider the swap rate

$$S_{\alpha,\beta}(t) = \frac{P(t,T_\alpha) - P(t,T_\beta)}{\sum_{j=\alpha+1}^{\beta} \tau_j P(t,T_j)},$$

which, under the forward swap measure $Q^{\alpha,\beta}$, follows

$$dS_{\alpha,\beta}(t) = \sum_{j=\alpha+1}^{\beta} \frac{\partial S_{\alpha,\beta}(t)}{\partial F_j(t)} [F_j(0) + b_j(t)(F_j(t) - F_j(0))] \sqrt{V(t)} \underline{\sigma}_j(t) \cdot d\underline{Z}^{\alpha,\beta}(t). \tag{11.20}$$

The first step in Piterbarg's derivation is approximating this SDE with

$$dS_{\alpha,\beta}(t) = [S_{\alpha,\beta}(0) + b_{\alpha,\beta}(t)(S_{\alpha,\beta}(t) - S_{\alpha,\beta}(0))] \sqrt{V(t)} \underline{\sigma}_{\alpha,\beta}(t) \cdot d\underline{Z}^{\alpha,\beta}(t). \tag{11.21}$$

By imposing equivalence under constant paths $S_{\alpha,\beta}(t) = S_{\alpha,\beta}(0)$ and $F_j(t) = F_j(0)$, he first obtains

$$\underline{\sigma}_{\alpha,\beta}(t) = \sum_{j=\alpha+1}^{\beta} \frac{F_j(0)}{S_{\alpha,\beta}(0)} \frac{\partial S_{\alpha,\beta}(0)}{\partial F_j(0)} \underline{\sigma}_j(t), \tag{11.22}$$

and then, differentiating the right-hand-sides of (11.20) and (11.21) and discarding second-order terms,

$$b_{\alpha,\beta}(t) \underline{\sigma}_{\alpha,\beta}(t) = b_j(t) \underline{\sigma}_j(t) \quad \forall j = \alpha+1, \dots, \beta. \tag{11.23}$$

This system of equations, however, can not be solved unless the product $b_j(t)\underline{\sigma}_j(t)$ is constant over j, which is not the case in general. To address this issue, Piterbarg reformulates the problem in the least-squares sense and finds

$$\begin{aligned} b_{\alpha,\beta}(t) &= \arg\min_g \sum_{j=\alpha+1}^{\beta} |g \underline{\sigma}_{\alpha,\beta}(t) - b_j(t)\underline{\sigma}_j(t)|^2 \\ &= \frac{\sum_{j=\alpha+1}^{\beta} b_j(t)\underline{\sigma}_j(t) \cdot \underline{\sigma}_{\alpha,\beta}(t)}{(\beta-\alpha)|\underline{\sigma}_{\alpha,\beta}(t)|^2}. \end{aligned} \tag{11.24}$$

The second step in Piterbarg's approach is finding *effective* (constant) parameters λ and b such that the SDE (11.21) can be approximated by

$$dS_{\alpha,\beta}(t) = \lambda[S_{\alpha,\beta}(0) + b(S_{\alpha,\beta}(t) - S_{\alpha,\beta}(0))]\sqrt{V(t)}\,dZ(t), \qquad (11.25)$$

where Z is a (one-dimensional) standard Brownian motion. The approximation is based on effective parameters such that option prices implied by the time-homogeneous model (11.25) are as close as possible, for all strikes, to those implied by (11.21).[7]

Piterbarg finds that the *effective skew* b can be written as a weighted average of the "skew" function $b_{\alpha,\beta}(t)$:

$$\begin{aligned} b = \bar{b} &:= \int_0^{T_\alpha} b_{\alpha,\beta}(t)\omega(t)\,dt \\ \omega(t) &= \frac{v^2(t)|\underline{\sigma}_{\alpha,\beta}(t)|^2}{\int_0^{T_\alpha} v^2(t)|\underline{\sigma}_{\alpha,\beta}(t)|^2 dt} \\ v^2(t) &= V(0)^2 \int_0^t |\underline{\sigma}_{\alpha,\beta}(s)|^2\,ds + V(0)\epsilon^2 \mathrm{e}^{-\kappa t} \int_0^t |\underline{\sigma}_{\alpha,\beta}(s)|^2 \frac{\mathrm{e}^{\kappa s} - \mathrm{e}^{-\kappa s}}{2\kappa}\,ds \end{aligned} \qquad (11.26)$$

To calculate the *effective volatility*, we first notice that the expectation defining the ATM payer-swaption price, on the swap rate $S_{\alpha,\beta}$, is given by

$$\begin{aligned} &E^{\alpha,\beta}[S_{\alpha,\beta}(T_\alpha) - S_{\alpha,\beta}(0)]^+ \\ &= E^{\alpha,\beta}\{E^{\alpha,\beta}[(S_{\alpha,\beta}(T_\alpha) - S_{\alpha,\beta}(0))^+ | V(t),\, 0 \le t \le T_\alpha]\} \\ &= E^{\alpha,\beta}\left[g\left(\int_0^{T_\alpha} |\underline{\sigma}_{\alpha,\beta}(s)|^2 V(s)\,ds\right)\right] \end{aligned}$$

where

$$g(x) = \frac{S_{\alpha,\beta}(0)}{\bar{b}}(2\Phi(\bar{b}\sqrt{x}/2) - 1).$$

The *effective volatility* $\lambda = \bar{\lambda}$ is then obtained by matching the ATM prices implied by (11.25) and (11.21):

$$E^{\alpha,\beta}\left[g\left(\int_0^{T_\alpha} |\underline{\sigma}_{\alpha,\beta}(s)|^2 V(s)\,ds\right)\right] = E^{\alpha,\beta}\left[g\left(\bar{\lambda}^2 \int_0^{T_\alpha} V(s)\,ds\right)\right]. \qquad (11.27)$$

Since the Laplace transform

$$m(u) = E^{\alpha,\beta}\left[e^{-u\mathcal{Z}(T_\alpha)}\right], \quad \mathcal{Z}(T_\alpha) := \int_0^{T_\alpha} |\underline{\sigma}_{\alpha,\beta}(s)|^2 V(s)\,ds$$

can be easily calculated numerically, Piterbarg approximates $g(x)$ with $a_0 + a_1 e^{-a_2 x}$, and finds a_0, a_1, a_2 by matching the second-order Taylor expansions at point

[7] Piterbarg minimizes the integral of the difference between option prices over all strikes.

$$x = \bar{x} := E^{\alpha,\beta}[\mathcal{Z}(T_\alpha)] = V(0)\int_0^{T_\alpha} |\underline{\sigma}_{\alpha,\beta}(s)|^2\, ds.$$

He gets

$$a_2 = -\frac{g''(\bar{x})}{g'(\bar{x})}, \quad a_1 = -\frac{g'(\bar{x})}{a_2} e^{a_2 \bar{x}}, \quad a_0 = g(\bar{x}) + \frac{1}{a_2} g'(\bar{x}).$$

The problem (11.27) is thus approximated with

$$E^{\alpha,\beta}\left[\exp\left(-a_2 \int_0^{T_\alpha} |\underline{\sigma}_{\alpha,\beta}(s)|^2 V(s)\, ds\right)\right] = E^{\alpha,\beta}\left[\exp\left(-a_2 \bar{\lambda}^2 \int_0^{T_\alpha} V(s)\, ds\right)\right], \tag{11.28}$$

so that, defining

$$m_0(u) := E^{\alpha,\beta}\left[e^{-u\int_0^{T_\alpha} V(s)\, ds}\right],$$

the parameter $\bar{\lambda}$ can be found by numerically solving

$$m_0\left(-\frac{g''(\bar{x})}{g'(\bar{x})}\bar{\lambda}^2\right) = m\left(-\frac{g''(\bar{x})}{g'(\bar{x})}\right). \tag{11.29}$$

Remark 11.3.1. **(Caplet pricing).** The dynamics of the generic forward rate F_k, under the associated measure Q^k, is given by (11.19) with $\underline{\mu}^{\bar{Q}} = \underline{\mu}^{Q^k} = 0$, and is similar to (11.21). The valuation of caplets can thus be tackled by applying the previous procedure. In fact, an explicit approximation for an (ATM) option on the forward rate F_k can be obtained by setting $\alpha = k - 1$ and $\beta = k$.

Calibration to the swaption smile. The purpose of model calibration is to obtain model parameters $b_j(t)$ and $\underline{\sigma}_j(t)$ from the "market" ones $\{\lambda^*_{\alpha,\beta}, b^*_{\alpha,\beta}, \epsilon^* : \alpha, \beta\}$.

To this end, one first finds, for each market pair (α, β), the local parameters $\lambda^*_{\alpha,\beta}$, $b^*_{\alpha,\beta}$ and the global parameter ϵ^* by calibrating model (11.25) to the prices of European swaptions with underlying swap rate $S_{\alpha,\beta}$. One then solves the following two minimization problems

$$\begin{aligned} &\min \sum_{\alpha,\beta} (\bar{b}_{\alpha,\beta} - b^*_{\alpha,\beta})^2, \\ &\min \sum_{\alpha,\beta} (\bar{\lambda}_{\alpha,\beta} - \lambda^*_{\alpha,\beta})^2, \end{aligned} \tag{11.30}$$

over the model parameters $b_j(t)$ and $\underline{\sigma}_j(t)$, where $\bar{b}_{\alpha,\beta}$ and $\bar{\lambda}_{\alpha,\beta}$ are, respectively, the effective skew and volatility for the swap rate $S_{\alpha,\beta}$, which depend on the model parameters through (11.26) and (11.29).

It is now clear the advantage of defining effective parameters. In fact, calibration is performed by directly fitting $\bar{b}$ to b^* and $\bar{\lambda}$ to λ^*, without resorting to any approximation for option prices.

The optimization problem (11.30) can be made more efficient by splitting it into two smaller subproblems. The first is fitting the term structure of swaption skews, the second is fitting the term structure of swaption volatilities. This is possible since the local volatility in (11.25) does not depend on b when $S_{\alpha,\beta}(t) = S_{\alpha,\beta}(0)$, so that the skew parameter b is almost independent of the volatility parameter λ.

The calibration procedure, therefore, can be carried out in three steps:

1. The model skew parameters $b_j(t)$ are all set to a common value, for instance the average of $b^*_{\alpha,\beta}$ over all market swaptions. The model volatilities $\underline{\sigma}_j(t)$ are then found by fitting $\bar{\lambda}_{\alpha,\beta}$ to $\lambda^*_{\alpha,\beta}$.
2. Using the $\underline{\sigma}_j$'s obtained in the previous step, the model skews $b_j(t)$ are found by fitting $\bar{b}_{\alpha,\beta}$ to $b^*_{\alpha,\beta}$.
3. The model volatilities $\underline{\sigma}_j(t)$ are re-calibrated to $\lambda^*_{\alpha,\beta}$, using the $b_j(t)$ derived in the previous step.

The last step is optional, and Piterbarg claims that one cycle is already enough to achieve a very good fit to swaptions data.

11.4 The Hagan, Kumar, Lesniewski and Woodward (2002) Model

Hagan, Kumar, Lesniewski and Woodward propose a SVM for the evolution of the forward price of an asset under the asset's canonical measure. The forward-asset dynamics are of CEV type, see (10.7), with a stochastic volatility that follows a driftless geometric Brownian motion, possibly instantaneously correlated with the forward price itself.

This model, which is commonly known as the SABR model (acronym for stochastic, alpha, beta and rho, three of the four model parameters) is widely used in practice because of its simplicity and tractability, and in particular the possibility of deriving analytical approximations for implied volatilities. Another reason for the model's popularity is the intuitive meaning of its parameters, which play specific roles in the generation of smiles and skews.

In the SABR model, the forward rate F_j is assumed to evolve under the associated measure Q^j according to

$$\begin{aligned} dF_j(t) &= V(t)F_j(t)^\beta \, dZ_j^j(t), \\ dV(t) &= \epsilon V(t) \, dW^j(t), \\ V(0) &= \alpha, \end{aligned} \tag{11.31}$$

where Z_j^j and W^j are Q^j-standard Brownian motions with

$$dZ_j^j(t) \, dW^j(t) = \rho \, dt,$$

and where $\beta \in (0, 1]$, ϵ and α are positive constants and $\rho \in [-1, 1]$.[8]

As shown by Jourdain (2004), the forward-rate process (11.31) is always a martingale when $\beta < 1$. In the "lognormal case" $\beta = 1$, instead, F_j is a martingale if and only if $\rho \leq 0$, which puts some constraint on the admissible values for the model parameters.

Using singular perturbation techniques, Hagan, Kumar, Lesniewski and Woodward obtain a closed-form approximation for the price at time $t = 0$ of a European option on the asset. In particular, the price of a T_j-maturity caplet is

$$\begin{aligned}\mathbf{Cpl}(0, T_{j-1}, T_j, \tau_j, K) &= \tau_j P(0, T_j)[F_j(0)\Phi(d_+) - K\Phi(d_-)] \\ d_\pm &= \frac{\ln(F_j(0)/K) \pm \frac{1}{2}\sigma^{\text{imp}}(K, F_j(0))^2 T_{j-1}}{\sigma^{\text{imp}}(K, F_j(0))\sqrt{T_{j-1}}}\end{aligned} \tag{11.32}$$

where the implied volatility $\sigma^{\text{imp}}(K, F)$ is given by the following analytical approximation:

$$\begin{aligned}\sigma^{\text{imp}}(K, F) =& \frac{\alpha}{(FK)^{\frac{1-\beta}{2}}\left[1 + \frac{(1-\beta)^2}{24}\ln^2\left(\frac{F}{K}\right) + \frac{(1-\beta)^4}{1920}\ln^4\left(\frac{F}{K}\right) + \cdots\right]}\frac{z}{x(z)} \\ &\cdot\left\{1 + \left[\frac{(1-\beta)^2\alpha^2}{24(FK)^{1-\beta}} + \frac{\rho\beta\epsilon\alpha}{4(FK)^{\frac{1-\beta}{2}}} + \epsilon^2\frac{2-3\rho^2}{24}\right]T_{j-1} + \cdots\right\},\end{aligned} \tag{11.33}$$

with

$$z := \frac{\epsilon}{\alpha}(FK)^{\frac{1-\beta}{2}}\ln\left(\frac{F}{K}\right)$$

and

$$x(z) := \ln\left\{\frac{\sqrt{1 - 2\rho z + z^2} + z - \rho}{1 - \rho}\right\}.$$

The dots in (11.33) stand for higher-order terms that are usually negligible, especially in case of short maturities.

The ATM (caplet) implied volatility is readily obtained from (11.33) by setting $K = F = F_j(0)$:

$$\begin{aligned}\sigma^{ATM} &= \sigma^{\text{imp}}(F_j(0), F_j(0)) \\ &= \frac{\alpha}{F_j(0)^{1-\beta}}\left\{1 + \left[\frac{(1-\beta)^2\alpha^2}{24F_j(0)^{2-2\beta}} + \frac{\rho\beta\epsilon\alpha}{4F_j(0)^{1-\beta}} + \epsilon^2\frac{2-3\rho^2}{24}\right]T_{j-1} + \cdots\right\}.\end{aligned} \tag{11.34}$$

The ATM volatility, as the forward rate $F_j(0)$ varies, traces a curve that is called *backbone*. The leading term in σ^{ATM} is $\alpha/F_j(0)^{1-\beta}$, meaning that

[8] Hagan, Kumar, Lesniewski and Woodward also consider the case of time-dependent parameters.

α and β concur in determining both the level and slope of ATM implied volatilities (the other parameters have less relevant impacts).

With the aim of better understanding the role played by each parameter, Hagan, Kumar, Lesniewski and Woodward propose a different approximation based on an expansion of (11.33) in powers of $\ln(K/F_j(0))$:

$$\sigma^{\text{imp}}(K, F_j(0)) = \frac{\alpha}{F_j(0)^{1-\beta}} \left\{ 1 - \frac{1}{2}(1 - \beta - \rho\lambda) \ln \frac{K}{F_j(0)} \right.$$
$$\left. + \frac{1}{12}[(1-\beta)^2 + (2 - 3\rho^2)\lambda^2] \ln^2 \frac{K}{F_j(0)} + \cdots \right\}, \quad (11.35)$$

where

$$\lambda := \epsilon F_j(0)^{(1-\beta)}/\alpha.$$

The first term in (11.35) is exactly the leading term in σ^{ATM}, which well approximates the ATM implied volatility. The linear term in $\ln(K/F_j(0))$ is approximatively the slope of the implied volatility with respect to the strike and results from two contributions: the first is the "beta skew", proportional to $-\frac{1}{2}(1-\beta)$, the second the "vanna skew", proportional to $\frac{1}{2}\rho\lambda$ and determined by the correlation between the forward rate and its volatility. When such a correlation is negative, the implied volatility is downward sloping.

Also the term proportional to $\ln^2(K/F_j(0))$, which is responsible for convexity, has two contributions: the first is proportional to the square of the beta skew, the second is a "volga" (volatility-gamma) term, proportional to λ^2, namely to the volatility of volatility.

We have just seen that the dynamics (11.31) leads to skews in the implied volatilities both through a β smaller than one ("non-lognormal" case) and through a non-zero correlation. In practice, however, it can be difficult to disentangle the contributions of the two parameters, since market implied volatilities can be fitted equally well by different choices of $\beta \in (0, 1]$. Hagan, Kumar, Lesniewski and Woodward, therefore, suggest to determine β either by historical calibration or by a-priori choice based on personal taste. In the former case, taking logs in both members of (11.34), we have

$$\ln \sigma^{ATM} = \ln \alpha - (1 - \beta) \ln F_j(0) + \ln\{1 + \cdots\},$$

so that β can be found with a linear regression applied to a historical plot of $(\ln F_j(0), \ln \sigma^{ATM})$, ignoring terms involving expiry times.

Remark 11.4.1. Hagan, Kumar, Lesniewski and Woodward postulate the evolution of a single forward asset, and show that their model accommodates quite well implied volatilities for a single maturity. Their SVM, however, is not an extension of the LFM. In a LIBOR market model, in fact, not only has one to specify the joint evolution of forward rates under a common measure, but also to clarify the relations among the volatility dynamics of each forward rate.

The pricing of swaptions. The SABR model can be equivalently used for modeling a swap-rate evolution and, consequently, for the (analytical) pricing of swaptions. In fact, one has simply to replace the forward rate $F_j(t)$ with the swap rate $S_{a,b}(t)$, obtaining under the swap measure $Q^{a,b}$:

$$\begin{aligned} dS_{a,b}(t) &= V(t) S_{a,b}(t)^{\beta}\, dZ^{a,b}(t), \\ dV(t) &= \epsilon V(t)\, dW^{a,b}(t), \\ V(0) &= \alpha. \end{aligned}$$

In practice, despite the shortcomings stressed in Remark 11.4.1, this model is widely used by financial institutions to quote implied volatility smiles and skews for swaptions. Precisely, the implied volatility $\sigma^{\mathrm{imp}}(K, S_{a,b}(0))$ of the swaption with maturity T_a, payments in $T_{a+1}, \ldots, T_b$ and strike K is obtained in terms of model parameters as follows:

$$\begin{aligned} &\sigma^{\mathrm{imp}}(K, S_{a,b}(0)) \\ &= \frac{\alpha}{(S_{a,b}(0)K)^{\frac{1-\beta}{2}}\left[1 + \frac{(1-\beta)^2}{24}\ln^2\left(\frac{S_{a,b}(0)}{K}\right) + \frac{(1-\beta)^4}{1920}\ln^4\left(\frac{S_{a,b}(0)}{K}\right)\right]} \frac{z}{x(z)} \\ &\quad \cdot \left\{1 + \left[\frac{(1-\beta)^2\alpha^2}{24(S_{a,b}(0)K)^{1-\beta}} + \frac{\rho\beta\epsilon\alpha}{4(S_{a,b}(0)K)^{\frac{1-\beta}{2}}} + \epsilon^2\frac{2-3\rho^2}{24}\right] T_a\right\}, \end{aligned} \tag{11.36}$$

where

$$z := \frac{\epsilon}{\alpha}(S_{a,b}(0)K)^{\frac{1-\beta}{2}} \ln\left(\frac{S_{a,b}(0)}{K}\right)$$

and

$$x(z) := \ln\left\{\frac{\sqrt{1 - 2\rho z + z^2} + z - \rho}{1 - \rho}\right\}.$$

This formula comes from the approximation (11.33) for the SABR implied volatility. It is market practice, however, to consider (11.36) as exact and to use it as a functional form mapping strikes into implied volatilities.

Based on this consideration, we now illustrate an example of calibration to EUR market data as of September 28th, 2005. Swaption smile quotes are reported in Table 11.3. Discount factors and ATM volatilities are shown in Tables 11.1 and 11.2, respectively.

Swaption volatilities are quoted (in the EUR market), for different strikes K, as a difference $\Delta\sigma^{\mathrm{M}}_{a,b}$ with respect to the ATM level

$$\Delta\sigma^{\mathrm{M}}_{a,b}(\Delta K) := \sigma^{\mathrm{M}}_{a,b}(K^{\mathrm{ATM}} + \Delta K) - \sigma^{\mathrm{ATM}}_{a,b}$$

for values of ΔK usually equal to $\pm 200, \pm 100, \pm 50, \pm 25$ basis points.[9]

[9] Smile quotes are not necessarily provided for all the swaption tenors and expiries, for which ATM volatilities are available. Interpolation schemes are then to be employed to complete the missing quotes.

Denoting by $\mathcal{S}$ the set of market pairs (a, b), we assume that each swap rate $S_{a,b}$, $(a, b) \in \mathcal{S}$, is associated with different parameters α, ϵ and ρ. The parameter β is instead assumed to be equal across different maturities and tenors.[10]

Our calibration is performed by minimizing the square percentage difference between model volatilities and the corresponding market ones. We set $\beta = 0.5$ and, for each expiry-tenor pair $(a, b) \in \mathcal{S}$, we calibrate the associated parameters $\alpha = \alpha_{a,b}$, $\epsilon = \epsilon_{a,b}$ and $\rho = \rho_{a,b}$ to the corresponding swaption volatility smile (including the ATM values). The resulting calibration errors, as absolute differences between market and model volatilities, are shown in Table 11.4.

T	$P(0,T)$	T	$P(0,T)$	T	$P(0,T)$
29-Sep-05	0.999942	30-Sep-10	0.870275	30-Sep-25	0.476246
03-Oct-05	0.999710	30-Sep-11	0.841442	30-Sep-26	0.456786
07-Oct-05	0.999470	28-Sep-12	0.812133	30-Sep-27	0.438279
31-Oct-05	0.998059	30-Sep-13	0.782237	29-Sep-28	0.420569
30-Nov-05	0.996281	30-Sep-14	0.752476	28-Sep-29	0.403782
21-Mar-06	0.989593	30-Sep-15	0.723171	30-Sep-30	0.387987
15-Jun-06	0.984243	30-Sep-16	0.694473	30-Sep-31	0.372828
21-Sep-06	0.977920	29-Sep-17	0.666446	30-Sep-32	0.358414
20-Dec-06	0.971923	28-Sep-18	0.639278	30-Sep-33	0.344706
20-Mar-07	0.965760	30-Sep-19	0.612743	29-Sep-34	0.331698
21-Jun-07	0.959287	30-Sep-20	0.587727	28-Sep-35	0.319318
20-Sep-07	0.952856	30-Sep-21	0.563523	29-Sep-45	0.217295
19-Dec-07	0.946414	30-Sep-22	0.540313	30-Sep-55	0.149179
30-Sep-08	0.925790	29-Sep-23	0.518115		
30-Sep-09	0.898422	30-Sep-24	0.496713		

Table 11.1. Market discount factors.

	Tenor								
Expiry	1y	2y	3y	4y	5y	7y	10y	20y	30y
1y	21.50%	21.90%	21.50%	21.00%	20.40%	19.10%	17.60%	15.30%	14.60%
2y	21.60%	21.30%	20.70%	20.00%	19.50%	18.50%	17.30%	15.20%	14.70%
3y	20.90%	20.50%	19.80%	19.20%	18.60%	17.80%	16.80%	15.10%	14.60%
4y	20.00%	19.60%	18.90%	18.40%	17.80%	17.20%	16.40%	15.00%	14.40%
5y	19.10%	18.70%	18.10%	17.60%	17.20%	16.60%	16.00%	14.80%	14.30%
7y	17.80%	17.00%	16.60%	16.20%	15.90%	15.60%	15.20%	14.30%	14.00%
10y	15.90%	15.20%	14.80%	14.70%	14.60%	14.40%	14.40%	13.60%	13.10%
15y	14.10%	13.70%	13.50%	13.40%	13.30%	13.20%	13.20%	12.40%	12.00%
20y	13.20%	13.10%	13.10%	13.10%	13.10%	13.10%	13.10%	12.10%	11.90%
25y	12.90%	12.90%	13.00%	13.10%	13.10%	13.20%	13.10%	12.20%	12.20%
30y	13.00%	13.00%	13.00%	12.90%	12.90%	12.90%	12.90%	12.30%	12.30%

Table 11.2. Market ATM volatilities.

MODELS OVERVIEW

[10] We stress that here we are not assuming a swap model in a strict sense, but simply that swaption volatilities are given in terms of the functional form (11.36).

Expiry	Tenor	Strike -200	-100	-50	-25	25	50	100	200
5y	2y	8.71%	2.57%	0.93%	0.38%	-0.25%	-0.37%	-0.34%	0.29%
10y	2y	5.73%	1.79%	0.64%	0.27%	-0.19%	-0.28%	-0.25%	0.27%
20y	2y	5.32%	1.84%	0.74%	0.32%	-0.25%	-0.42%	-0.55%	-0.35%
30y	2y	5.14%	1.76%	0.71%	0.32%	-0.26%	-0.44%	-0.64%	-0.60%
5y	5y	8.64%	2.85%	1.10%	0.47%	-0.34%	-0.55%	-0.69%	-0.24%
10y	5y	6.34%	2.22%	0.89%	0.39%	-0.31%	-0.51%	-0.69%	-0.45%
20y	5y	5.62%	1.99%	0.81%	0.36%	-0.29%	-0.49%	-0.70%	-0.59%
30y	5y	5.52%	1.93%	0.79%	0.35%	-0.29%	-0.49%	-0.72%	-0.70%
5y	10y	7.80%	2.63%	1.02%	0.44%	-0.33%	-0.53%	-0.63%	-0.17%
10y	10y	6.39%	2.25%	0.91%	0.40%	-0.31%	-0.52%	-0.71%	-0.47%
20y	10y	5.86%	2.07%	0.85%	0.37%	-0.30%	-0.51%	-0.73%	-0.62%
30y	10y	5.44%	1.92%	0.79%	0.35%	-0.29%	-0.52%	-0.79%	-0.85%
5y	20y	7.43%	2.56%	1.00%	0.43%	-0.32%	-0.51%	-0.60%	-0.10%
10y	20y	6.59%	2.34%	0.94%	0.41%	-0.32%	-0.54%	-0.72%	-0.43%
20y	20y	6.11%	2.19%	0.90%	0.40%	-0.32%	-0.55%	-0.77%	-0.61%
30y	20y	5.46%	1.92%	0.79%	0.35%	-0.29%	-0.50%	-0.72%	-0.69%
5y	30y	7.45%	2.58%	1.01%	0.44%	-0.33%	-0.52%	-0.61%	-0.13%
10y	30y	6.73%	2.38%	0.96%	0.42%	-0.33%	-0.53%	-0.68%	-0.35%
20y	30y	6.20%	2.22%	0.91%	0.40%	-0.32%	-0.54%	-0.74%	-0.55%
30y	30y	5.39%	1.90%	0.78%	0.35%	-0.28%	-0.50%	-0.72%	-0.68%

Table 11.3. Market volatility smiles for the selected expiry-tenor pairs.

Expiry	Tenor	Strike -200	-100	-50	-25	0	25	50	100	200
5y	2y	2.0	1.1	1.9	0.8	0.3	1.5	1.5	1.3	2.0
10y	2y	1.5	1.4	0.0	1.1	1.0	0.7	1.8	1.1	1.5
20y	2y	2.1	1.4	1.7	0.5	0.3	0.6	2.0	1.0	1.6
30y	2y	2.5	2.1	1.5	1.3	0.4	1.7	1.6	1.7	2.0
5y	5y	1.7	1.3	0.8	0.3	0.4	0.5	0.7	1.7	1.5
10y	5y	1.9	1.3	1.1	0.7	0.6	0.9	1.4	1.2	1.4
20y	5y	1.7	1.2	1.0	0.7	0.5	1.0	0.6	1.4	1.3
30y	5y	2.6	2.1	1.7	0.7	0.3	1.3	1.2	1.7	1.8
5y	10y	2.1	1.2	0.9	1.0	1.0	1.2	1.5	1.4	1.7
10y	10y	1.5	0.7	1.1	0.7	0.5	0.7	1.1	1.2	1.2
20y	10y	1.9	1.1	1.7	0.3	0.5	0.8	1.1	1.4	1.4
30y	10y	2.0	1.7	1.1	0.3	0.6	0.2	1.6	1.7	1.6
5y	20y	2.6	1.8	1.1	0.7	0.7	0.8	1.4	1.9	2.0
10y	20y	1.9	1.2	0.8	0.4	0.8	0.4	1.0	1.7	1.5
20y	20y	2.4	1.2	1.4	0.9	0.6	0.4	1.6	1.8	1.7
30y	20y	3.3	2.4	2.2	1.0	0.5	1.6	2.1	1.7	2.1
5y	30y	2.3	1.5	0.8	1.1	0.8	1.5	1.1	1.1	1.5
10y	30y	1.5	0.5	1.1	0.8	0.7	1.0	1.1	0.8	1.1
20y	30y	2.7	1.7	1.7	0.6	0.5	0.8	1.4	1.6	1.7
30y	30y	3.9	3.0	2.3	1.4	0.4	1.1	2.8	2.5	2.8

Table 11.4. Absolute differences in basis points between market and model volatilities for the selected expiry-tenor pairs.

$$dX(t) = \cdots dt + \text{Vol}(X)\, dZ(t)$$
$$dV(t) = \cdots dt + \text{Vol}(V)\, dW(t)$$

11.5 The Joshi and Rebonato (2003) Model

Joshi and Rebonato propose a shifted-lognormal LIBOR model with a volatility parametrization based on a functional form with stochastic coefficients. Compared with the other extensions of the LFM we describe in this chapter,

Model	Vol(X)	Vol(V)	V is mean reverting	X and V are correlated	Explicit implied volatility
AB-R	$\varphi(X(t))\sigma(t)\sqrt{V(t)}$	$\epsilon\psi(V(t))$	YES	NO	YES
WZ	$X(t)\sigma(t)\sqrt{V(t)}$	$\epsilon\sqrt{V(t)}$	YES	YES	NO
P	$[X(t)+a(t)]\sigma(t)\sqrt{V(t)}$	$\epsilon\sqrt{V(t)}$	YES	NO	YES
SABR	$X(t)^{\beta}V(t)$	$\epsilon V(t)$	NO	YES	YES

Table 11.5. Summary of stochastic-volatility LIBOR models, where X denotes a general forward or swap rate.

their model has the drawback of not being analytically tractable. In fact, even caplet prices need to be calculated numerically via simulation, which renders the calibration to market data rather cumbersome and time consuming unless some clever trick is implemented in the Monte Carlo simulations.

However, the Joshi and Rebonato model has the advantage of being inspired by the joint historical evolution of swap rates and associated implied volatilities. In fact, Joshi and Rebonato note that, even though implied volatilities are manifestly stochastic, they may result from deterministic transformations of the corresponding swap rates. Precisely, they stress that, given a swap rate S and the associated ATM implied volatility σ^{imp}, for a suitable constant β the quantity

$$Y(t) := \sigma^{\mathrm{imp}}(t)[S(t)]^{\beta}$$

displays a very low variability, at least in the 1x1 swaption case.

From a historical point of view, therefore, ATM implied volatilities seem to be approximately of CEV type. Since also the ATM implied volatility in a CEV model is (approximately) proportional to a power of the underlying asset, see the first-order term in (11.34), Joshi and Rebonato suggest to use a CEV model for each forward rate $F_j(t)$, $j = 1, \ldots, M$, under Q^j,

$$dF_j(t) = \eta_j(t)[F_j(t)]^{\gamma_j}\, dZ_j^j(t), \tag{11.37}$$

where γ_j's are positive constants, $\eta_j(t)$'s are deterministic functions and $Z_1^j, \ldots, Z_M^j$ are Brownian motions with $dZ_j^j dZ_k^j = \rho_{j,k}\, dt$.

Given the high degree of similarity between CEV models and displaced diffusions, see Marris (1999) and Muck (2003),[11] Joshi and Rebonato replace the dynamics (11.37) with the more tractable shifted-lognormal process

[11] Given the CEV process

$$dF_j(t) = \sigma_j(t)[F_j(t) + \alpha_j]\, dZ_j^j(t)$$

where α_j's are positive constants and $\sigma_j(t)$'s are deterministic functions.

To achieve a better fit to market data, they then introduce stochastic volatility by assuming that

$$\sigma_j(t) = [a(t) + b(t)(T_{j-1} - t)]e^{-c(t)(T_{j-1}-t)} + d(t),$$

with the parameters a, b, c and d that are common to all forward rates and follow

$$dx(t) = \lambda_x(r_x - x(t))\, dt + \sigma_x\, dW_x(t) \quad x \in \{a, b, \ln(c), \ln(d)\} \qquad (11.38)$$

where λ_x, r_x and σ_x are positive constants. The Brownian motions W_x are assumed to be independent among each other and also independent of the M-dimensional Brownian motion governing the forward rates dynamics. Equations (11.38), therefore, are the same under any forward measure or under the spot LIBOR measure.

As mentioned above, option prices in the Joshi and Rebonato model can only be calculated numerically. Precisely, given the independence between forward rates and their volatility, and following Hull and White (1987), caplet prices can be written as integrals of adjusted Black's prices:

$$\begin{aligned}&\mathbf{Cpl}(0, T_{j-1}, T_j, \tau_j, K)\\&= \tau_j P(0, T_j) \int_0^{+\infty} \mathrm{Bl}(K + \alpha_j, F_j(0) + \alpha_j, y\sqrt{T_{j-1}}) f_j(y)\, dy,\end{aligned} \qquad (11.39)$$

where $f_j(y)$ is the density of the root-mean-square volatility

$$\sqrt{\frac{1}{T_{j-1}} \int_0^{T_{j-1}} \sigma_j^2(t)\, dt},$$

which needs to be simulated by Monte Carlo.

Calibrating with Monte Carlo simulations can be rather time consuming. However, it is worth mentioning that the same volatility path can be used for different time horizons and that many strikes can be priced simultaneously by means of the same volatility distribution.

$$dF(t) = \eta F(t)^\gamma\, dZ(t),$$

one can construct an "equivalent" displaced diffusion

$$dF(t) = \sigma[F(t) + \alpha]\, dZ(t),$$

by matching values and first derivatives in $t = 0$ of their squared diffusion coefficients. One gets:

$$\alpha = F(0)\frac{1-\gamma}{\gamma}, \quad \sigma = \eta\gamma F(0)^{\gamma-1}.$$

As to swaptions, assuming that the shift parameters α_j are all equal to a real constant $\bar{\alpha}$, the classical "freezing" technique leads to the following (approximated) implied volatility for the swap rate $S_{\alpha,\beta}$, for each given path of forward volatilities,

$$\sqrt{\frac{\sum_{k,h=\alpha+1}^{\beta} (F_k(0)+\bar{\alpha})(F_h(0)+\bar{\alpha})\frac{\partial S_{\alpha,\beta}}{\partial F_k}\frac{\partial S_{\alpha,\beta}}{\partial F_h}\Big|_{t=0}\rho_{k,h}\int_0^{T_\alpha}\sigma_k(t)\sigma_h(t)\,dt}{T_\alpha[S_{\alpha,\beta}(0)+\bar{\alpha}]^2}}.$$

The corresponding swaption price can then be calculated with a formula similar to (11.39), where one must replace the forward-rate volatility with this approximated swaption volatility.

12. Uncertain-Parameter Models

I suspect my path is as uncertain as yours.
Bishop Aringarosa in "The Da Vinci Code", Dan Brown, 2003.

In order to accommodate market option prices while preserving analytical tractability, uncertain-volatility models (UVMs) have been recently proposed in the financial literature as an easy-to-implement alternative to SVMs. UVMs are based on the assumption that the asset's volatility is stochastic in the simplest possible way, modelled by a random variable rather than a diffusion process. Precisely, the dynamics of a general forward rate F under the associated forward measure is assumed to be given by

$$dF_t = \sigma F_t \, dZ_t, \tag{12.1}$$

where σ is a discrete random variable independent of the Brownian motion Z, which can take values $\sigma_1, \sigma_2, \ldots, \sigma_N$, with probabilities $\lambda_1, \lambda_2, \ldots, \lambda_N$, respectively. A UVM, therefore, can be viewed as a Black (1976) model where the volatility σ is not constant and one assumes several possible scenarios for its value, which is to be drawn immediately after time zero.

The independence of σ and W implies that the density of F is a mixture of lognormal densities, that caplet prices are mixtures of Black's caplet prices and that caplet volatilities are smile-shaped with a minimum in F_0.[1]

To account for skews in implied volatilities, the UVM (12.1) can be extended by introducing a shift parameter α. The dynamics of the forward rate F thus becomes

$$dF_t = \sigma(F_t + \alpha) \, dZ_t, \tag{12.2}$$

where the pair (α, σ) is assumed to be a discrete random vector taking values $(\alpha_1, \sigma_1), (\alpha_2, \sigma_2), \ldots, (\alpha_N, \sigma_N)$, with probabilities $\lambda_1, \lambda_2, \ldots, \lambda_N$, respectively. As before, the random value of (α, σ) is drawn immediately after time zero.

The formulation (12.2) was suggested, in a one-factor case, by Gatarek (2003), and independently proposed by Brigo, Mercurio and Rapisarda (2004)

---

[1] From a distributional point of view, there is a clear analogy between (12.1) and the LVM (10.28) since the two processes have the same marginal density. The two models, however, differ for a number of reasons: transitions densities, exotic-option prices, forward or future implied volatilities, etc.

for modeling a general asset. The intuition behind the model is as follows. Forward rates dynamics are given by displaced geometric Brownian motions where the model parameters are not known at the initial time, and one assumes different scenarios for them. The forward-rates volatilities and displacements are random variables whose values will be known immediately after time zero.

Given the initial uncertainty on the volatility and the shift parameters, both (12.1) and (12.2) belong to the class of UPMs.

Besides their intuitive meaning, UPMs have a number of advantages that strongly support their use in practice. In fact, UPMs enjoy a great deal of analytical tractability, are relatively easy to implement and are flexible enough to accommodate general implied volatility surfaces in the caps and swaptions markets. This will be clarified below, where practical examples with market data will also be provided. Nonetheless, the initial lack of knowledge on the true parameters values, which are suddenly revealed an instant later one prices a given claim, may be rather disturbing, especially if one pretends meaningful dynamics also from an historical point of view. However, models should be judged also from their implications and not only their assumptions. In this respect, no other alternative to the LFM is as simple, as tractable and as flexible as the UPM (12.2).

A drawback of UPMs is that future implied volatilities lose the initial smile shape almost immediately, i.e. as soon as the random value of the model parameters is drawn, since from that moment on forward rates evolves according to displaced geometric Brownian motions under their respective measures.[2] However, our empirical analysis will show that the forward implied volatilities induced by the model do not differ much from the current ones, which can further support its use in the pricing and hedging of interest rate derivatives.

In the following sections, we will describe the extension of Gatarek's one-factor UPM to the general multi-factor case as considered by Errais, Mauri and Mercurio (2004). We will derive caps and (approximated) swaptions prices in closed form. We will then illustrate how the model can accommodate market caps data and how the instantaneous correlation parameters can be used for a calibration to swaptions prices. We will also analyze important model's implications, inferring the swaptions smile implied by our joint calibration, and plotting the evolution of forward volatilities implied by the model. Finally, we will consider a more realistic and thorough example of simultaneous calibration to caps and swaptions data.

[2] When the shift parameters are zero, as in a UVM, implied volatilities become flat in the future.

12.1 The Shifted-Lognormal Model with Uncertain Parameters (SLMUP)

In the shifted-lognormal model with uncertain parameters, each forward rate F_j evolves under the corresponding forward measure Q^j as in (12.2). Precisely, the instantaneous volatility of F_j is assumed to be given by

$$\sigma_j^I(t)\left[F_j(t)+\alpha_j^I\right], \tag{12.3}$$

where I is a random variable that takes values in the set $\{1,2,\dots,N\}$, with N a natural number, $Q^j(I=i)=\lambda_i$, $\lambda_i>0$ and $\sum_{i=1}^N \lambda_i=1$, and where α_j's are real constants and σ_j's are deterministic functions of time. The value of the random index I is drawn at time $t=0^+$, namely at an infinitesimal time after zero.

Following Gatarek (2003) and Errais, Mauri and Mercurio (2004), we choose to model the joint evolution of all rates F_j, $j=1,\dots,M$, under the spot LIBOR measure Q^d, which has the discretely rebalanced bank-account

$$B_d(t)=\frac{P(t,T_{\beta(t)-1})}{\prod_{j=0}^{\beta(t)-1}P(T_{j-1},T_j)}$$

as associated numeraire. Given the volatility coefficients (12.3), and remembering (6.18), we thus have that under Q^d each F_j follows

$$\begin{aligned} dF_j(t)=&\sigma_j^I(t)(F_j(t)+\alpha_j^I)\sum_{k=\beta(t)}^{j}\frac{\tau_k\rho_{k,j}\sigma_k^I(t)(F_k(t)+\alpha_k^I)}{1+\tau_kF_k(t)}\,dt \\ &+\sigma_j^I(t)(F_j(t)+\alpha_j^I)\,dZ_j^d(t), \end{aligned} \tag{12.4}$$

where Z^d is an M-dimensional Brownian motion with $dZ_i^d(t)\,dZ_k^d(t)=\rho_{i,k}\,dt$, I is independent of Z^d and takes values in the set $\{1,2,\dots,N\}$ with $Q^d(I=i)=\lambda_i$, and $\beta(t)=\min\{i: t\le T_{i-1}\}$, namely $\beta(t)=m$ if $T_{m-2}<t\le T_{m-1}$.

Dynamics (12.4) are obtained by applying the change-of-numeraire toolkit in Section 2.3 under the assumption that the volatility of F_j is given by (12.3), so that the dynamics of F_j under Q^j is given by

$$dF_j(t)=\sigma_j^I(t)\left(F_j(t)+\alpha_j^I\right)dZ_j^j(t), \tag{12.5}$$

where Z_j^j is a standard Brownian motion under Q^j.

The random variable I is also independent of Z_j^j and, for this reason, its distribution remains the same when moving from the spot LIBOR measure to a forward one, or vice versa.

12.1.1 Relationship with the Lognormal-Mixture LVM

We close this section with a curious remark on the link between this model with $\alpha_j^I = 0$ and the similar LVM given in Section 10.5 (the W Brownian motion processes there are the same as the processes Z we are using here). In this subsection we then assume

$$\alpha_j^I = 0.$$

See also the discussion at the end of Subsection 10.5.1.

Proposition 12.1.1. *The local-volatility lognormal-mixture diffusion dynamics in Proposition 10.5.1 (case $i = j$) is the local-volatility version of the uncertain-volatility mixture dynamics (12.5) with $\alpha_j^I = 0$. The two models are linked by the relationship*

$$\nu_j^2(t,x) = E\{\sigma_j^I(t)^2 | F_j(t) = x\}.$$

The proof is immediate by resorting to Bayes' formula, see for example Brigo, Mercurio and Rapisarda (2004) or Brigo (2002b).

Remark 12.1.1. **(Casting some light on the "zero terminal correlation" result of Subsection 10.5.1).** The terminal correlation computed at time 0 between the underlying rate $F_j(T)$ and its averaged squared volatility $(1/T)\int_0^T \sigma_j^I(t)^2 dt$ is easily seen to be zero at any horizon T, due to independence of I and Z's. As was pointed out in Subsection 10.5.1, the same property is shared by the local-volatility version, which maintains the decorrelation pattern between volatility and underlying asset that is so natural for its uncertain-volatility originator. Now, the result of Subsection 10.5.1 looks less surprising in the light of this result for the uncertain-volatility version.

12.2 Calibration to Caplets

In this section we show that caplet prices can be calculated in closed form and explain how to perform the calibration to the corresponding market volatilities.

The price at time $t = 0$ of a caplet whose payoff is set at T_{j-1} and paid at T_j can be calculated, assuming unit notional, by applying an iterated conditioning:

$$\begin{aligned}
\mathbf{Cpl}(0, T_{j-1}, T_j, \tau_j, K) &= P(0,T_j)\tau_j E^j\left[(F_j(T_{j-1}) - K)^+\right] \\
&= \tau_j P(0,T_j) E^j\left[E^j\left[(F_j(T_{j-1}) - K)^+ |I\right]\right] \\
&= \tau_j P(0,T_j) E^j\left[E^j\left[\left(F_j(T_{j-1}) + \alpha_j^I - (K + \alpha_j^I)\right)^+ |I\right]\right] \qquad (12.6) \\
&= \tau_j P(0,T_j) \sum_{i=1}^N \lambda_i E^j\left[\left(F_j(T_{j-1}) + \alpha_j^I - (K + \alpha_j^I)\right)^+ |I = i\right].
\end{aligned}$$

Since F_j evolves according to (12.5), conditional on I, $F_j(T_{j-1}) + \alpha_j^I$ is a lognormal random variable. The last expectation, therefore, is nothing but the adjusted Black's caplet price that comes from a shifted geometric Brownian motion, see also (10.6),

$$\mathbf{Cpl}(0, T_{j-1}, T_j, \tau_j, K) = \tau_j P(0, T_j) \sum_{i=1}^{N} \lambda_i \mathrm{Bl}\left(K + \alpha_j^i, F_j(0) + \alpha_j^i, V_j^i(T_{j-1})\right)$$

where

$$V_j^i(T_{j-1}) = \sqrt{\int_0^{T_{j-1}} \sigma_j^i(s)^2 \, ds}.$$

To describe the calibration problem one has to deal with, we remember that, in the market, caplets are priced through Black's formula:

$$\mathbf{Cpl}^{\mathrm{Mkt}}(T_{j-1}, T_j, \tau_j, K) = \tau_j P(0, T_j) \mathrm{Bl}\left(K, F_j(0), V^{\mathrm{impl}}(T_{j-1}, K)\sqrt{T_{j-1}}\right),$$

where $V^{\mathrm{impl}}(T_{j-1}, K)$ is the market implied volatility of a caplet setting at T_{j-1} and paying at T_{j-1} with strike K.

Setting $\Phi_j := \left(V_j^1(T_{j-1}), \ldots, V_j^N(T_{k-1}), \alpha_j^1, \ldots, \alpha_j^N\right)$, the calibration to market caplets data is performed by determining Φ_j as follows:

$$\Phi_j = \arg\min \left[\sum_l g\left(\mathbf{Cpl}^{\mathrm{Mkt}}(T_{j-1}, T_j, \tau_j, K_l), \mathbf{Cpl}(0, T_{j-1}, T_j, \tau_j, K_l)\right)\right],$$

where the sum is taken over all available strikes for the maturity T_j, and g is some "distance" function. In our practical examples, we will set $g(x, y) = (x - y)^2$. The parameters Φ_j are found iteratively starting from $j = 1$ up to $j = M$, for some a priori given values of the probabilities λ_i.[3]

Since the optimization produces as output the integral of instantaneous volatilities and not their point value, we have to introduce assumptions on the volatility functions and infer their values from the calibrated integrals V_j^i. For instance, if we assume the σ_j^i's to be constant, we come up with the following formula:

$$\sigma_j^i = \frac{V_j^i(T_{j-1})}{\sqrt{T_{j-1}}}$$

Another possibility is to assume a functional form like (6.12), namely

$$\sigma_j^i(t) := \left[a(T_{j-1} - t) + d\right] \exp\left(-b(T_{j-1} - t)\right) + c.$$

In this case, to find the parameters a, b, c and d, we need to solve

[3] Mercurio and Morini (2005) identify a particular SLMUP that allows for an automatic calibration to ATM prices.

$$V_j^i(T_{j-1}) = \sqrt{\int_0^{T_{j-1}} \left[\left(a(T_{j-1} - t) + d \right) \exp\left(-b(T_{j-1} - t) \right) + c \right]^2 dt}.$$

For simplicity, we will adopt the first (non parametric) assumption. An example of calibration to real market data will be considered in Section 12.6.

12.3 Swaption Pricing

We now derive the approximation of swaption prices that was first proposed by Gatarek (2003) in a one-factor case, and then extended by Errais, Mauri and Mercurio (2004) to account for a general number of factors. The approximation is, once again, based on a "freezing" technique.

We remember, see (6.34), that the time-t forward swap rate

$$S_{\alpha,\beta}(t) = \frac{P(t, T_\alpha) - P(t, T_\beta)}{\sum_{k=\alpha+1}^{\beta} \tau_k P(t, T_k)}$$

can be written as a smooth function of the forward rates $F_{\alpha+1}, \dots, F_\beta$ as

$$S_{\alpha,\beta}(t) = \frac{\sum_{k=\alpha+1}^{\beta} \tau_k P(t, T_k) F_k(t)}{\sum_{k=\alpha+1}^{\beta} \tau_k P(t, T_k)} = \sum_{k=\alpha+1}^{\beta} \omega_k(t) F_k(t) \tag{12.7}$$

where

$$\omega_k(t) = \frac{\tau_k P(t, T_k)}{\sum_{h=\alpha+1}^{\beta} \tau_h P(t, T_h)}.$$

Since $S_{\alpha,\beta}$ is a martingale under the associated forward swap measure $Q^{\alpha,\beta}$, Ito's lemma implies the following dynamics for the forward swap rate under $Q^{\alpha,\beta}$

$$\begin{aligned} dS_{\alpha,\beta}(t) &= \sum_{k=\alpha+1}^{\beta} \frac{\partial S_{\alpha,\beta}(t)}{\partial F_k(t)} \sigma_k^I(t) \left(F_k(t) + \alpha_k^I \right) dZ_k^{\alpha,\beta}(t) \\ &= \sum_{k=\alpha+1}^{\beta} \gamma_k(t)\, dZ_k^{\alpha,\beta}(t), \end{aligned}$$

where

$$\gamma_k(t) := \frac{\partial S_{\alpha,\beta}(t)}{\partial F_k(t)} \sigma_k^I(t) \left(F_k(t) + \alpha_k^I \right),$$

and $Z^{\alpha,\beta}$ is an M-dimensional Brownian motion with $dZ_i^{\alpha,\beta}(t)\, dZ_k^{\alpha,\beta}(t) = \rho_{i,k}\, dt$.

The terms γ_k can be approximated by neglecting the dependence of $\omega_k(t)$ on $F_k(t)$. We thus obtain:

$$
\begin{aligned}
\gamma_k(t) &\approx \frac{\tau_k P(t,T_k)\,\sigma_k^I(t)\left(F_k(t)+\alpha_k^I\right)}{\sum_{h=\alpha+1}^{\beta}\tau_h P(t,T_h)} \\
&= \frac{\tau_k P(t,T_k)\,\sigma_k^I(t)\left(F_k(t)+\alpha_k^I\right)}{\sum_{h=\alpha+1}^{\beta}\tau_h P(t,T_h)\left(F_h(t)+\alpha_h^I\right)}\,\frac{\sum_{h=\alpha+1}^{\beta}\tau_h P(t,T_h)\left(F_h(t)+\alpha_h^I\right)}{\sum_{h=\alpha+1}^{\beta}\tau_h P(t,T_h)} \\
&= \frac{\tau_k P(t,T_k)\,\sigma_k^I(t)\left(F_k(t)+\alpha_k^I\right)}{\sum_{h=\alpha+1}^{\beta}\tau_h P(t,T_h)\left(F_h(t)+\alpha_h^I\right)}\left[S_{\alpha,\beta}(t)+\frac{\sum_{h=\alpha+1}^{\beta}\tau_h P(t,T_h)\,\alpha_h^I}{\sum_{h=\alpha+1}^{\beta}\tau_h P(t,T_h)}\right].
\end{aligned}
$$

The last expression can be further approximated by freezing the forward rates and the discount factors at their time-zero values:

$$
\begin{aligned}
\gamma_k(t) &\approx \frac{\tau_k P(0,T_k)\sigma_k^I(t)\left(F_k(0)+\alpha_k^I\right)}{\sum_{h=\alpha+1}^{\beta}\tau_h P(0,T_h)\left(F_h(0)+\alpha_h^I\right)}\left[S_{\alpha,\beta}(t)+\frac{\sum_{h=\alpha+1}^{\beta}\tau_h P(0,T_h)\alpha_h^I}{\sum_{h=\alpha+1}^{\beta}\tau_h P(0,T_h)}\right] \\
&= \gamma_k^I(t)\left[S_{\alpha,\beta}(t)+\eta_{\alpha,\beta}^I\right],
\end{aligned}
$$

where we define

$$
\begin{aligned}
\gamma_k^I &:= \frac{\tau_k P(0,T_k)\sigma_k^I(t)\left(F_k(0)+\alpha_k^I\right)}{\sum_{h=\alpha+1}^{\beta}\tau_h P(0,T_h)\left(F_h(0)+\alpha_h^I\right)}, \\
\eta_{\alpha,\beta}^I &:= \frac{\sum_{h=\alpha+1}^{\beta}\tau_h P(0,T_h)\alpha_h^I}{\sum_{h=\alpha+1}^{\beta}\tau_h P(0,T_h)}.
\end{aligned}
$$

Therefore, the dynamics of $S_{\alpha,\beta}(t)$ under $Q^{\alpha,\beta}$ approximately reads as :

$$
\begin{aligned}
dS_{\alpha,\beta}(t) &= \sum_{k=\alpha+1}^{\beta}\gamma_k^I(t)\left[S_{\alpha,\beta}(t)+\eta_{\alpha,\beta}^I\right]dZ_k^{\alpha,\beta}(t) \\
&= \left[S_{\alpha,\beta}(t)+\eta_{\alpha,\beta}^I\right]\sum_{k=\alpha+1}^{\beta}\gamma_k^I(t)\,dZ_k^{\alpha,\beta}(t) \\
&= \gamma_{\alpha,\beta}^I(t)\left[S_{\alpha,\beta}(t)+\eta_{\alpha,\beta}^I\right]dW_I^{\alpha,\beta}(t),
\end{aligned}
$$

where we set

$$
\gamma_{\alpha,\beta}^I(t) := \sqrt{\sum_{k,h=\alpha+1}^{\beta}\gamma_k^I(t)\,\gamma_h^I(t)\,\rho_{k,h}}
$$

and

$$
dW_I^{\alpha,\beta}(t) := \frac{\sum_{k=\alpha+1}^{\beta}\gamma_k^I(t)\,dZ_k^{\alpha,\beta}(t)}{\gamma_{\alpha,\beta}^I(t)}.
$$

It can easily be verified that, conditional on I, $W_I^{\alpha,\beta}$ is a $Q^{\alpha,\beta}$-Brownian motion, so that the (approximated) dynamics of $S_{\alpha,\beta}$ under its canonical measure is equivalent to that of the general forward rate F_j under the associated Q^j. Precisely, defining

$$X_{\alpha,\beta}^{I}(t) := S_{\alpha,\beta}(t) + \eta_{\alpha,\beta}^{I}$$

we have that, conditional on I, $X_{\alpha,\beta}^{I}$ is a geometric Brownian motion whose dynamics, under $Q^{\alpha,\beta}$, is given by

$$dX_{\alpha,\beta}^{I}(t) = \gamma_{\alpha,\beta}^{I}(t)\, X_{\alpha,\beta}^{I}(t)\, dW_{I}^{\alpha,\beta}(t).$$

We can thus price swaptions by following the same procedure as in the previous caplet case.

Consider a European payer swaption with maturity T_α and strike K, whose underlying swap pays on times $T_{\alpha+1}, \ldots, T_\beta$. Assuming unit notional, the swaption price at time zero can be calculated as follows:

$$\begin{aligned}
\mathbf{PS}(0;\alpha,\beta,K) &= \sum_{h=\alpha+1}^{\beta} \tau_h P(0,T_h)\, E^{\alpha,\beta}\left[\left(S_{\alpha,\beta}(T_\alpha) - K\right)^+\right] \\
&= \sum_{h=\alpha+1}^{\beta} \tau_h P(0,T_h)\, E^{\alpha,\beta}\left[\left(X_{\alpha,\beta}^{I}(T_\alpha) - (K + \eta_{\alpha,\beta}^{I})\right)^+\right] \\
&= \sum_{h=\alpha+1}^{\beta} \tau_h P(0,T_h) \sum_{i=1}^{N} \lambda_i E^{\alpha,\beta}\left[\left(X_{\alpha,\beta}^{I}(T_\alpha) - (K + \eta_{\alpha,\beta}^{I})\right)^+ \mid I = i\right].
\end{aligned}$$

We thus obtain:

$$\mathbf{PS}(0;\alpha,\beta,K) = \sum_{h=\alpha+1}^{\beta} \tau_h P(0,T_h) \sum_{i=1}^{N} \lambda_i \mathrm{Bl}\left[K + \eta_{\alpha,\beta}^{i}, S_{\alpha,\beta}(t) + \eta_{\alpha,\beta}^{i}, \Gamma_{\alpha,\beta}^{i}\right], \tag{12.8}$$

where

$$\Gamma_{\alpha,\beta}^{i} := \sqrt{\int_0^{T_\alpha} \left[\gamma_{\alpha,\beta}^{i}(s)\right]^2 ds} = \sqrt{\sum_{k,h=\alpha+1}^{\beta} \rho_{k,h} \int_0^{T_\alpha} \gamma_k^i(s)\, \gamma_h^i(s)\, ds}.$$

Similarly to the caplet case, therefore, the swaption price is simply a mixture of adjusted Black's swaption prices.

12.4 Monte-Carlo Swaption Pricing

To verify the accuracy of the above swaption-price approximation, we compare the analytical formula (12.8) with the price obtained by Monte Carlo simulation. We now illustrate how this simulation is performed.

Using the spot LIBOR measure Q^d as reference measure, we can write the swaption price, at time zero, as:

$$\mathbf{PS}(0;\alpha,\beta,K) = E^d\left[\frac{(S_{\alpha,\beta}(T_\alpha)-K)^+\sum_{h=\alpha+1}^{\beta}\tau_h P(T_\alpha,T_h)}{B_d(T_\alpha)}\right].$$

This expectation can be calculated numerically by simulating the values, at time T_α, of the forward rates spanning the swap rate interval,

$$F_{\alpha+1}(T_\alpha), F_{\alpha+2}(T_\alpha), \ldots, F_\beta(T_\alpha),$$

which evolve under Q^d according to (12.4) and can be generated by means of consecutive Euler approximations.

A more efficient simulation is obtained by considering the processes $\bar{F}_k^{\alpha,I}(t) := F_k(t) + \alpha_k^I$ whose logarithms evolve, by Ito's lemma, according to

$$d\ln \bar{F}_k^{\alpha,I}(t) = \sigma_k^I(t)\sum_{j=\beta(t)}^{k}\frac{\tau_j\rho_{j,k}\sigma_j^I(t)\bar{F}_j^{\alpha,I}(t)}{1+\tau_j\left(\bar{F}_j^{\alpha,I}(t)-\alpha_j^I\right)}dt - \frac{\sigma_k^I(t)^2}{2}dt + \sigma_k^I(t)\,dZ_k^\alpha(t).$$

These processes have the advantage of having, conditional on I, deterministic diffusion coefficients, so that the Euler scheme for them coincides with the more sophisticated Milstein scheme.

Our simulation is performed by first drawing randomly a value i of I and then generating paths of the processes $\bar{F}_k^{\alpha,i}$ according to

$$\begin{aligned}\ln \bar{F}_k^{\alpha,i}(t+\Delta t) - \ln \bar{F}_k^{\alpha,i}(t) &= \sigma_k^i(t)\sum_{j=\beta(t)}^{k}\frac{\tau_j\rho_{j,k}\sigma_j^i(t)\bar{F}_k^{\alpha,i}(t)}{1+\tau_j\left(\bar{F}_j^{\alpha,i}(t)-\alpha_j^i\right)}\Delta t \\ &\quad - \frac{\sigma_k^i(t)^2}{2}\Delta t + \sigma_k^i(t)\left(Z_k^\alpha(t+\Delta t) - Z_k^\alpha(t)\right),\end{aligned}$$

through simulation of the jointly-normal increments $Z_k^\alpha(t+\Delta t) - Z_k^\alpha(t)$.

Once the forward rates are simulated for the value i of the random index, we can evaluate the normalized swaption payoff

$$\frac{(S_{\alpha,\beta}(T_\alpha;i)-K)^+\sum_{h=\alpha+1}^{\beta}\tau_h P(T_\alpha,T_h;i)}{B_d(T_\alpha;i)}$$

along each trajectory and finally average the resulting values over all simulated paths. This leads to the Monte Carlo price of the swaption:

$$\mathbf{PS}^{\mathrm{MC}}(0;\alpha,\beta,K) = \frac{1}{n}\sum_{j=1}^{n}\frac{(S_{\alpha,\beta}(T_\alpha;i,w_j)-K)^+\sum_{h=\alpha+1}^{\beta}\tau_h P(T_\alpha,T_h;i,w_j)}{B_d(T_\alpha;i,w_j)},$$

where w_j denotes the j-th path and n is the total number of simulations.

To test the accuracy of our approximation, we will compare the analytical price (12.8) with the 99% Monte Carlo confidence interval:

$$\left[\mathbf{PS}^{\mathrm{MC}}(0;\alpha,\beta,K) - 2.576\frac{\sigma_{PS}^{\mathrm{MC}}}{\sqrt{n}},\ \mathbf{PS}^{\mathrm{MC}}(0;\alpha,\beta,K) + 2.576\frac{\sigma_{PS}^{\mathrm{MC}}}{\sqrt{n}}\right],$$

where $\sigma_{PS}^{\mathrm{MC}}$ is the standard deviation of the Monte Carlo price.

The accuracy test can also be performed on implied volatilities rather than on prices. The implied volatility associated to (12.8) is defined as the parameter $\sigma_{\alpha,\beta}$ to plug into the market formula

$$\mathbf{PS}^{\mathrm{Mkt}}(0;\alpha,\beta,K) = \mathrm{Bl}\big(K, S_{\alpha,\beta}(0), \sigma_{\alpha,\beta}\sqrt{T_\alpha}\big) \sum_{h=\alpha+1}^{\beta} \tau_h P(0,T_h)$$

to match the model's price. Such a volatility, which is denoted by $\sigma_{\alpha,\beta}^{\mathrm{SLMUP}}$, is thus implicitly defined by

$$\mathrm{Bl}\big(K, S_{\alpha,\beta}(0), \sigma_{\alpha,\beta}^{\mathrm{SLMUP}}\sqrt{T_\alpha}\big) = \sum_{i=1}^{N} \lambda_i \mathrm{Bl}\left[K + \eta_{\alpha,\beta}^i, S_{\alpha,\beta}(0) + \eta_{\alpha,\beta}^i, \Gamma_{\alpha,\beta}^i\right].$$

A similar definition applies to the Monte Carlo implied volatility $\sigma_{\alpha,\beta}^{\mathrm{MC}}$, which is implicitly defined by:

$$\mathrm{Bl}\big(K, S_{\alpha,\beta}(0), \sigma_{\alpha,\beta}^{\mathrm{MC}}\sqrt{T_\alpha}\big) \sum_{h=\alpha+1}^{\beta} \tau_h P(0,T_h) = \mathbf{PS}^{\mathrm{MC}}(0;\alpha,\beta,K)$$

Numerical results and illustrations of these tests are presented in Section 12.7.

12.5 Calibration to Swaptions

The first goal of model (12.4) is to capture the skew that is commonly seen in the cap market. After reaching this objective, we can move further and try to calibrate the model to ATM swaption prices using the degrees of freedom left, namely the instantaneous correlations between the forward rates in our family. More precisely, after finding σ_k^I's and α_k^I's through a calibration to caplet prices, as explained in Section 12.6, we look for the correlation matrix best fitting a selected set of swaptions prices. Clearly, a joint calibration to both interest rate markets would require an overall optimization procedure, where the parameters σ_k^I's and α_k^I's contribute to accommodate the swaptions implied volatilities, see Section 12.9 below. Here, instead, we resort to this two-stage procedure for simplicity, also because some model advantages are already evident in the simplified approach.

A typical problem faced when we perform a joint calibration to caps and swaptions is the difference between fixed- and floating-leg tenors, as we have already seen in Chapter 7. To this end, we apply the procedure of Section 6.20

to find a correspondence between semiannual parameters and their one-year counterparts.

Precisely, consider three six-month spaced maturities S, T and N, where $0 < S < T < N$. Denote by $\sigma^I_{S,1}$, $\sigma^I_{S,.5}$ and $\sigma^I_{T,.5}$, respectively, the volatilities of the forward rates $F(\cdot; S, N)$, $F(\cdot; S, T)$ and $F(\cdot; T, N)$.

Assuming, for the moment, that the shift parameters α are all zero, we obtain, see also (6.76):

$$\begin{aligned}\sigma^I_{S,1}(t)^2 \approx\, & u_1(0)^2 \sigma^I_{S,.5}(t)^2 + u_2(0)^2 \sigma^I_{T,.5}(t)^2 \\ & + 2\rho_{T,N} u_1(0)\, u_2(0)\, \sigma^I_{S,.5}(t)\, \sigma^I_{T,.5}(t),\end{aligned}$$

where

$$\begin{aligned}u_1(0) &= \frac{1}{F_{S,N}(0)}\left[\frac{F_{S,T}(0)}{2} + \frac{F_{S,T}(0)F_{T,N}(0)}{4}\right],\\ u_2(0) &= \frac{1}{F_{S,N}(0)}\left[\frac{F_{T,N}(0)}{2} + \frac{F_{S,T}(0)F_{T,N}(0)}{4}\right].\end{aligned}$$

To assign a value to the α's of the annual forward rate $F_{S,N}$, we simply set $\alpha^I_{S,1} = \alpha^I_{S,.5}$ where $\alpha^I_{S,.5}$ is the α parameter for the semi-annual forward rate $F_{S,T}$. This is also motivated by the fact that the calibrated values of α^I_j are usually rather similar (with the same order of magnitude).

After converting semi-annual parameters into annual ones, we then parameterize the instantaneous-correlation matrix to proceed to the calibration to swaptions prices.

We choose to test two different parameterizations. The first is a two-parameter correlation structure:

$$\rho_{i,j} = \bar{\rho} + (1 - \bar{\rho}) \sin\left(\frac{\pi}{2} \exp\left(-a\frac{|T_j - T_i|}{\bar{T}}\right)\right) \tag{12.9}$$

where $\bar{\rho}$ is the long term correlation coefficient, a is the (positive) "speed" of decorrelation, and $\bar{T}$ is the maximum distance between any two maturities. The functional form (12.9) is decreasing in the difference $|T_j - T_i|$ and can lead to sigmoid shapes in the correlations.

The correlation matrix generated by (12.9) is not necessarily positive semi-definite. However, for $\bar{\rho} > 0$ and a approaching zero, we can consider the following first order Taylor expansion:

$$\begin{aligned}\rho_{i,j} &\simeq \bar{\rho} + (1 - \bar{\rho}) \sin\left(\frac{\pi}{2}\left(1 - a\frac{|T_j - T_i|}{\bar{T}}\right)\right)\\ &\simeq \bar{\rho} + (1 - \bar{\rho}) \cos\left(\frac{\pi}{2} a \frac{|T_j - T_i|}{\bar{T}}\right),\end{aligned}$$

so that we approximately have

$$\rho_{i,j} \simeq \bar{\rho} + (1-\bar{\rho})\cos(\theta_j - \theta_i), \quad \theta_l = \frac{\pi}{2} a \frac{T_l}{\bar{T}}, \text{ for } l \in \{i,j\},$$

so that $(\rho_{i,j})$ is a positive semi-definite matrix. Moreover, for $\bar{\rho} > 0$ and a approaching ∞, we obtain:

$$\begin{aligned} \rho_{i,j} &\simeq \bar{\rho} + (1-\bar{\rho}) \left(\frac{\pi}{2} \exp\left(-a\frac{|T_j - T_i|}{\bar{T}}\right) - \frac{1}{6}\left(\frac{\pi}{2}\exp\left(-a\frac{|T_j - T_i|}{\bar{T}}\right)\right)^3 \right) \\ &\simeq \bar{\rho}, \end{aligned}$$

and hence $(\rho_{i,j})$ is again positive semi-definite.

The second parametrization we test is Rebonato's, see also (6.44):

$$\rho_{i,j} = \bar{\rho} + (1-\bar{\rho})\exp(-a|T_j - T_i|). \tag{12.10}$$

The above Taylor expansion shows that, for a large enough, the two correlation structures are quite similar. However, they diverge for a small a, resulting in a different fitting quality to the same set of swaptions prices, as we will see in a later section.

The calibration to swaptions prices is performed as follows. We recall, from the previous section, the model and market formulas for a payer swaption price, and we define Ψ as the pair of optimization parameters, $\Psi := (\bar{\rho}, a)$. The calibration's purpose is finding Ψ such that

$$\Psi = \arg\min \left(\sum_{\alpha,\beta} g\left(\mathbf{PS}^{\mathrm{Mkt}}(0;\alpha,\beta,K), \mathbf{PS}^{\mathrm{SLMUP}}(0;\alpha,\beta,K)\right) \right),$$

where α and β range over our selected ATM swaption maturities and tenors. Some numerical results are presented in Section 12.6.

12.6 Calibration to Market Data

In this section, we illustrate some examples of calibration to caps and swaptions data, which are carried out along the lines we have previously mentioned. To simplify our analysis, we assume the volatility parameters $\sigma_j^i(t)$ to be constant over time, but we allow them to be different for different scenarios i or forward rates j.[4]

We start by testing model (12.4) on the caps market by considering different numbers of scenarios. Our numerical experiments, based on market data, show that three scenarios are usually enough for a satisfactory calibration, even though the fitting quality worsen for high strikes and low maturities.

[4] Further empirical work can be carried out by trying different parameterizations for the time-dependent functions $\sigma_j^i(t)$.

We then move to consider swaption volatilities. As previously mentioned, our purpose here is not to perform a true joint calibration, which will instead be tackled in Section 12.9, but rather to infer reasonable forward rates correlations to test the model in terms of the swaption volatilities it implies.

Calibration to caplet quotes. We calibrate the model to the cap volatilities data quoted on August 11, 2004 in the Euro market, using caplets with six-month tenors. The caplets strikes range from 2.5% to 7%, and are 25 basis points spaced. The caplets maturities spectrum ranges from 2 to 15 years (expiries from 1.5 to 14.5 years). The volatility data used is plotted in Figure 12.1, while a sample of the same data is displayed in Table 12.1.

We consider two, three and finally four different scenarios ($N = 2, 3, 4$), and come to the conclusion that three scenarios are usually the best choice. In fact, a two-scenario model does not seem to accommodate the caplets skew accurately. Moreover, the marginal improvement in the fitting quality implied by four scenarios tends to be negligible. Therefore, in the remainder of our numerical tests, we decide to stick to a three-scenario model with a-priori given probabilities λ.[5] We take $\lambda_1 = .6$, $\lambda_2 = .3$ and $\lambda_3 = .1$, thus assuming a high probability scenario, an average one, and a low probability scenario.

The calibrated values of σ_k^I and α_k^I in the three different scenarios, and for expiries up to 14.5 years, are displayed in Tables 12.2 and 12.3. The resulting caplet volatilities are plotted in Figure 12.2. Figures 12.3 to 12.7, instead, show a graphical comparison between the market and model implied caplet volatilities for a few selected expiries, namely 2, 3, 5, 10 and 12 years (maturities are six months later).

Calibration to swaptions quotes. We consider the ATM swaption volatilities quoted on August 11, 2004 in the Euro market. In the swaption's price formula (12.8), the λ parameters have been assigned a priori, whereas the α and σ parameters for annual rates are obtained, as explained in the previous section, from the semi-annual ones coming from the initial calibration to caplets. The only free parameters are the correlations ρ, which are parameterized first through Rebonato's function (12.10), and then according to form (12.9).

Table 12.4 displays the data used for the calibration. Tables 12.5 and 12.6 present the calibration results with Rebonato's formula, whereas Tables 12.7 and 12.8 present the results obtained with the correlation function (12.9).

---

[5] Substantially equivalent results, in terms of fitting, can be achieved through different choices for the scenarios probabilities. This is why we fix their value initially.

	Maturity						
Strike(%)	2.5y	3y	3.5y	4y	4.5y	5y	5.5y
2.5	26.32	25.68	24.77	23.99	23.53	23.18	23.04
2.75	25.04	24.43	23.6	22.85	22.44	22.13	22.03
3	23.97	23.33	22.55	21.82	21.46	21.17	21.11
3.25	23.12	22.4	21.65	20.9	20.59	20.3	20.28
3.5	22.62	21.64	20.89	20.11	19.82	19.52	19.53
3.75	22.66	21.09	20.31	19.43	19.16	18.84	18.86
4	22.81	21.04	20.04	18.91	18.62	18.26	18.28
4.25	23.075	21.17	20.11	18.795	18.42	17.87	17.86
4.5	23.34	21.3	20.18	18.68	18.22	17.48	17.44
4.75	23.7	21.54	20.36	18.795	18.3	17.5	17.4
5	24.06	21.78	20.54	18.91	18.38	17.52	17.36
5.25	24.49	22.07	20.78	19.09	18.52	17.61	17.42
5.5	24.94	22.4	21.04	19.3	18.68	17.74	17.51
5.75	25.43	22.76	21.34	19.54	18.88	17.89	17.63
6	25.94	23.14	21.65	19.8	19.1	18.06	17.77
6.25	26.48	23.55	21.99	20.08	19.33	18.26	17.93
6.5	27.04	23.97	22.35	20.38	19.59	18.47	18.11
6.75	27.62	24.42	22.73	20.7	19.86	18.7	18.3
7	28.23	24.89	23.13	21.03	20.15	18.95	18.51

Table 12.1. Sample of market caplet volatilities (in %) by strike and maturity.

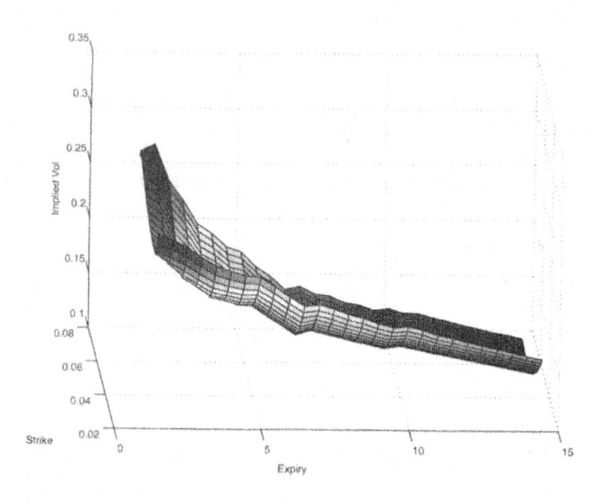

Fig. 12.1. Market caplet volatilities as of 11 August 2004 by strike and expiry.

12.7 Testing the Approximation for Swaptions Prices

In this section we test the quality of the approximation (12.8) and the related implied volatility, by comparing their values with those coming from a Monte Carlo simulation.

In our test, we consider payer swaptions with five-year tenors. The model parameters used in the test are those coming from the previous calibrations

Expiry	Scenario 1	Scenario 2	Scenario 3
1.5	0.16304434	0.06422773	0.06332962
2	0.18921474	0.05634155	0.05316215
2.5	0.16791712	0.05989887	0.05520289
3	0.17362959	0.06458859	0.06557753
3.5	0.15172758	0.03617556	0.03240995
4	0.15829939	0.04999707	0.03960524
4.5	0.14123047	0.04344321	0.03870800
5	0.14422573	0.05366341	0.04914180
5.5	0.12441415	0.05057469	0.04322650
6	0.12480429	0.05173816	0.04849220
6.5	0.10744960	0.03105830	0.02892788
7	0.12148858	0.05141806	0.04410974
7.5	0.10697440	0.04382249	0.04019843
8	0.11604050	0.04996482	0.03892146
8.5	0.10106829	0.04246569	0.04085141
9	0.10789990	0.04728225	0.04185310
9.5	0.09942587	0.04068057	0.03929327
10	0.10961703	0.04656102	0.04349751
10.5	0.09826417	0.04100323	0.04260043
11	0.10773474	0.04301775	0.04531406
11.5	0.09486490	0.03734212	0.04308718
12	0.09552005	0.03850786	0.04121609
12.5	0.09505042	0.03621755	0.03896017
13	0.09450712	0.03610641	0.04549258
13.5	0.08950882	0.03703596	0.03723106
14	0.09231341	0.03448961	0.03534253
14.5	0.08597086	0.03469210	0.03425724

Table 12.2. The calibrated values of σ_k^I for different expiries (in years) and scenarios.

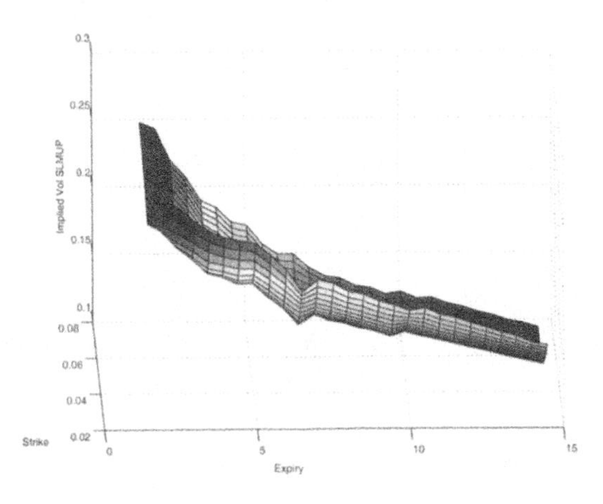

Fig. 12.2. Caplet volatilities implied by model (12.4), calibrated with three scenarios to the overall market skews.

Expiry	Scenario 1	Scenario 2	Scenario3
1.5	0.02801044	0.03841978	0.03326450
2	0.02091647	0.03346086	0.03674928
2.5	0.02596937	0.02426207	0.03236464
3	0.02202359	0.01782716	0.02032976
3.5	0.02776751	0.05817754	0.06401326
4	0.02418964	0.02997589	0.04818130
4.5	0.03074806	0.03732896	0.04390922
5	0.02946808	0.02161852	0.02895950
5.5	0.03719965	0.01976091	0.03170472
6	0.03330999	0.01331947	0.01965747
6.5	0.04104563	0.04570878	0.04912145
7	0.03306023	0.01305767	0.01541584
7.5	0.04193459	0.01881899	0.02348780
8	0.03346430	0.01242679	0.01402056
8.5	0.04336819	0.01611235	0.02000735
9	0.03598035	0.00976178	0.01192281
9.5	0.04088379	0.01448904	0.01933940
10	0.03408434	0.00878642	0.01169428
10.5	0.04182204	0.01214344	0.01838369
11	0.03248840	0.00903381	0.01301825
11.5	0.04134638	0.01417237	0.02011819
12	0.03909778	0.01134578	0.01687170
12.5	0.03775666	0.01239276	0.01803944
13	0.03635414	0.00716251	0.01400981
13.5	0.03920179	0.00779229	0.01037969
14	0.03476033	0.00719563	0.02012456
14.5	0.03879874	0.00730602	0.01047056

Table 12.3. The calibrated values of α_k^I for different expiries (in years) and scenarios.

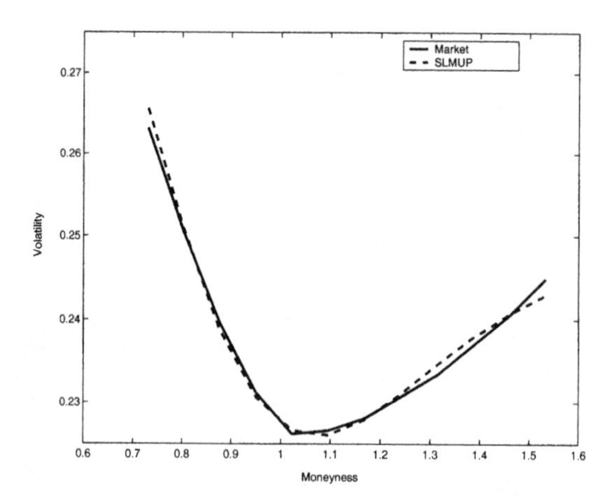

Fig. 12.3. Calibration of model (12.4) to the caplets setting in two-years.

to caps and swaptions, with instantaneous correlations parameterized by Rebonato's function (12.10).

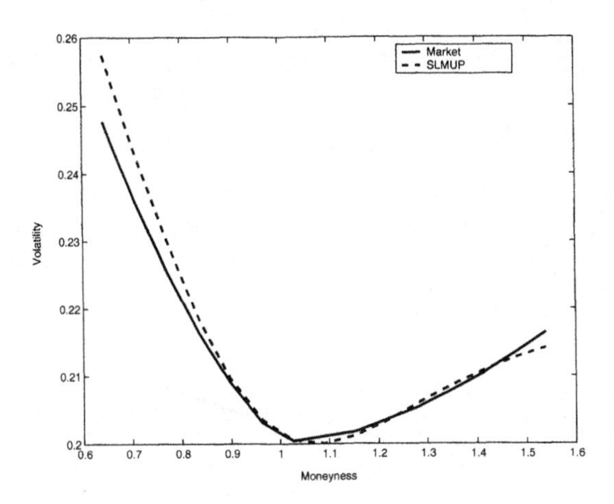

Fig. 12.4. Calibration of model (12.4) to the caplets setting in three-years.

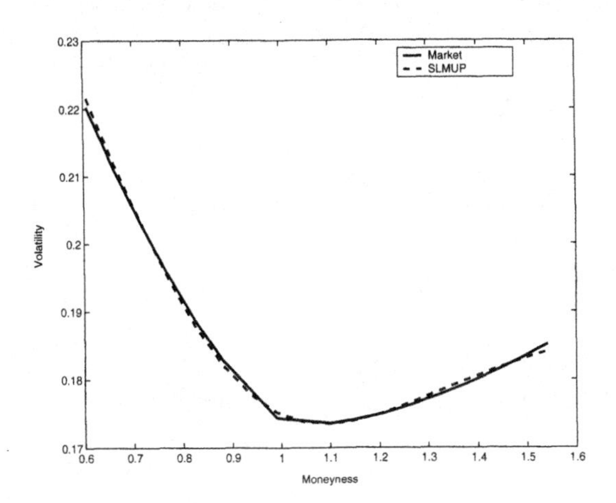

Fig. 12.5. Calibration of model (12.4) to the caplets setting in five-years.

	Maturity			
Tenor	2y	3y	4y	5y
2y	0.201	0.179	0.163	0.15
3y	0.187	0.168	0.153	0.142
4y	0.174	0.157	0.144	0.134
5y	0.163	0.147	0.135	0.127

Table 12.4. Market ATM swaption volatilities as of 11 August 2004.

Tables 12.9, 12.10, 12.11 and 12.12 present the results of our analysis. In each table, we report the approximation price, the 99% confidence interval of the Monte Carlo price and the implied volatilities associated to the

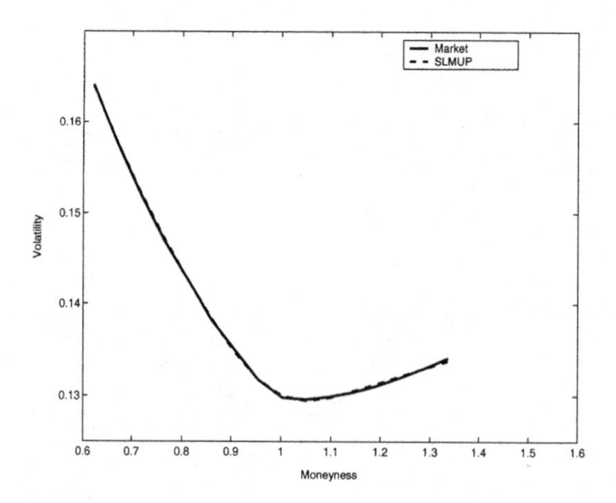

Fig. 12.6. Calibration of model (12.4) to the caplets setting in ten-years.

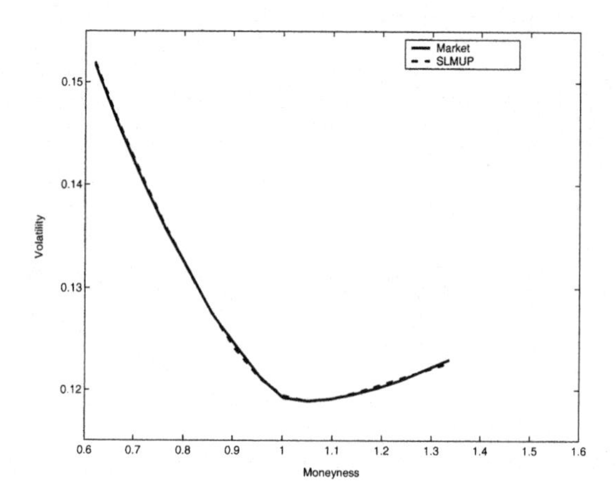

Fig. 12.7. Calibration of model (12.4) to the caplets setting in twelve-years.

	Maturity			
Tenor	2y	3y	4y	5y
2y	0.20268	0.18589	0.16783	0.15550
3y	0.18438	0.17016	0.15458	0.14416
4y	0.17016	0.15697	0.14401	0.13497
5y	0.15774	0.14632	0.13519	0.12671

Table 12.5. Model's swaption volatilities under Rebonato's correlation function (12.10).

approximation and the Monte Carlo prices.,[6] These tables show that the ap-

[6] We simulated ten million paths with a three-month time step. We also applied major variance-reduction techniques.

$\lvert T_j - T_i \rvert$	$\rho_{i,j}$
0	1.00000
1	0.78098
2	0.61347
3	0.48535
4	0.38737
5	0.31244
6	0.25512
7	0.21129
8	0.17777
9	0.15213
10	0.13252
11	0.11752
12	0.10605

Table 12.6. Correlations under Rebonato's parametrization.

	Maturity			
Tenor	2y	3y	4y	5y
2y	0.20040	0.18379	0.16591	0.15373
3y	0.18204	0.16799	0.15251	0.14224
4y	0.16895	0.15596	0.14298	0.13411
5y	0.15836	0.14699	0.13583	0.12726

Table 12.7. Model's swaption volatilities under the correlation function (12.9).

$\lvert T_j - T_i \rvert$	$\rho_{i,j}$
0	1.00000
1	0.74093
2	0.59764
3	0.55401
4	0.54125
5	0.53754
6	0.53646
7	0.53614
8	0.53605
9	0.53602
10	0.53601
11	0.53601
12	0.53601

Table 12.8. Correlations under formulation (12.9).

proximation is quite accurate for every considered strike, both in terms of prices and implied volatilities.

12.8 Further Model Implications

Besides the possibility of accommodating market caps and swaptions prices, model (12.4) has two further appealing features.

Strikes	Approx	Confidence Interval		$\sigma_{\alpha,\beta}^{\text{SLMUP}}$	$\sigma_{\alpha,\beta}^{\text{MC}}$
1	13.82129	13.81992	13.82305	29.69013	30.93328
1.5	11.7133	11.71163	11.7148	26.16912	26.09091
2	9.616925	9.614254	9.617442	23.74041	23.53314
2.5	7.558583	7.553763	7.556938	21.73473	21.53948
3	5.585923	5.578241	5.581351	19.80649	19.64416
3.5	3.76625	3.75582	3.758774	17.80746	17.6729
4	2.237119	2.224806	2.227436	16.19233	16.07505
4.5	1.205713	1.194609	1.196827	15.71271	15.61379
5	0.652548	0.645185	0.647076	16.18942	16.11517
5.5	0.361724	0.35747	0.359052	16.8489	16.79595
6	0.19664	0.19451	0.195787	17.29857	17.26596
6.5	0.103012	0.102198	0.103189	17.54427	17.53361
7	0.052057	0.05184	0.052581	17.6619	17.67022

Table 12.9. Approximation price, Monte Carlo price-window and related volatilities for a 2x5 swaption.

Strikes	Approx	Confidence Interval		$\sigma_{\alpha,\beta}^{\text{SLMUP}}$	$\sigma_{\alpha,\beta}^{\text{MC}}$
1	14.41191	14.40977	14.41306	28.06964	26.61654
1.5	12.39908	12.39643	12.39975	24.68383	24.24306
2	10.402	10.39805	10.40138	22.36748	22.09171
2.5	8.444132	8.438116	8.44144	20.50211	20.30066
3	6.560787	6.552229	6.555507	18.79316	18.63539
3.5	4.796158	4.785429	4.788599	17.09941	16.97469
4	3.22747	3.215358	3.218309	15.58011	15.47835
4.5	1.998144	1.986538	1.989155	14.68867	14.60644
5	1.196614	1.187471	1.189798	14.59899	14.53471
5.5	0.731471	0.725521	0.727613	14.9988	14.95317
6	0.455184	0.451807	0.453648	15.46998	15.44153
6.5	0.280946	0.279435	0.281013	15.81599	15.80503
7	0.1699	0.169553	0.170871	16.02736	16.03384

Table 12.10. Approximation price, Monte Carlo price-window and related volatilities for a 3x5 swaption.

The first is that it is possible to define a swaptions' smile or skew based on the cap ones and on ATM swaption volatilities. Swaptions are mainly quoted ATM and volatilities for away-from-the-money strikes may be difficult to obtain. However, practitioners who trade OTM and ITM swaptions agree on pricing them with a different volatility (usually higher) than the ATM one. Therefore, in the absence of OTM and ITM quotes, model (12.4) can provide a guide to infer a possible swaption's smile.

Figure 12.8 shows the implied volatilities curves induced by the model for the 2x5, 3x5, 4x5, 5x5 swaptions, with the same parameters used in the previous section to test the model's swaption price formula. These plots should only be regarded as examples of possible implications of the model. In fact, more "realistic" curves could be obtained through a more accurate calibration to ATM swaptions prices.

Strikes	Approx	Confidence Interval		$\sigma_{\alpha,\beta}^{SLMUP}$	$\sigma_{\alpha,\beta}^{MC}$
1	14.63908	14.63835	14.64162	26.82	27.69253
1.5	12.72157	12.72022	12.72351	23.53826	23.61617
2	10.82197	10.81942	10.82273	21.30599	21.22553
2.5	8.960791	8.956455	8.959773	19.53616	19.43209
3	7.16581	7.15942	7.162715	17.95743	17.85965
3.5	5.468099	5.459839	5.463064	16.41676	16.33159
4	3.913081	3.903392	3.906463	14.93485	14.86101
4.5	2.603121	2.592899	2.595692	13.8419	13.77778
5	1.658358	1.649555	1.652049	13.4545	13.40327
5.5	1.067966	1.061611	1.063902	13.6628	13.62509
6	0.707301	0.703289	0.705397	14.0972	14.07241
6.5	0.471931	0.469642	0.471546	14.48584	14.47222
7	0.311619	0.310611	0.312297	14.75378	14.75166

Table 12.11. Approximation price, Monte Carlo price-window and related volatilities for a 4x5 swaption.

Strikes	Approx	Confidence Interval		$\sigma_{\alpha,\beta}^{SLMUP}$	$\sigma_{\alpha,\beta}^{MC}$
1	14.59393	14.5914	14.59454	25.50006	24.68521
1.5	12.76953	12.7665	12.76966	22.35811	22.04726
2	10.96356	10.95958	10.96276	20.22934	20.04543
2.5	9.193978	9.188817	9.192016	18.55779	18.43177
3	7.483998	7.477476	7.480671	17.09149	16.99571
3.5	5.85859	5.850897	5.854049	15.68794	15.61252
4	4.350503	4.341947	4.344986	14.32488	14.26323
4.5	3.03345	3.024703	3.027518	13.2165	13.16572
5	2.017548	2.009981	2.012516	12.65815	12.61917
5.5	1.337118	1.331628	1.333973	12.66687	12.63976
6	0.908976	0.905317	0.907527	12.98943	12.97167
6.5	0.628781	0.626678	0.628735	13.35872	13.35002
7	0.434486	0.433625	0.435504	13.64837	13.64914

Table 12.12. Approximation price, Monte Carlo price-window and related volatilities for a 5x5 swaption.

The second important implication of the model is that it produces self-similar forward caplets volatilities. Having realistic forward volatility skews is important as far as pricing exotic options is concerned since it allows traders to implement a better hedging strategy for their derivatives books.

A forward implied volatility is defined as the volatility parameter to plug into the Black-Scholes formula for a forward-starting option to match the model's price.[7] A forward-start caplet (FSC) with forward start date T_j and maturity T_k is a call option on the LIBOR rate $F_k(T_{k-1})$, with $T_{k-1} > T_j$, where the strike price is set as a proportion δ of the spot LIBOR rate at time T_j. The FSC payoff at time T_k is (we assume a unit nominal amount):

$$\tau_k \left[F_k\left(T_{k-1}\right) - \delta F_{j+1}\left(T_j\right)\right]^+ . \tag{12.11}$$

[7] While in equity and foreign exchange markets forward-starting options are commonly traded, these derivatives are less present in the fixed income market.

This payoff, however, involves two different forward rates, and as such can not lead to a consistent definition of forward implied volatility. To this end, we replace (12.11) with the following:

$$\tau_k \left[F_k\left(T_{k-1}\right) - \delta F_k\left(T_j\right)\right]^+ . \tag{12.12}$$

The difference between the two formulations is that in the former the strike is defined by the spot LIBOR rate at time T_j for maturity T_{j+1}, whereas in the latter, it is defined by the time-T_j forward LIBOR rate between T_{k-1} and T_k.

Even though the payoff (12.12) is somehow more intuitive, it can be hard to trade in practice since forward rates are not directly quoted by the market but only stripped from zero-coupon bonds. However, having to define a forward volatility, and not to price a true market derivative, (12.12) is a more convenient choice. In fact, this payoff is based on the difference between values of the same "asset" computed at two different times, involving as such only one dynamics and not two as in the case of the former payoff.

Following Rubinstein (1991), the price of a FSC at time zero in a LFM is given by

$$\begin{aligned}
&\mathbf{FSCpl}\left(T_j, T_{k-1}, T_k, \tau_k, \delta\right) \\
&= \tau_k P\left(0, T_k\right) \mathrm{Bl}\left(\delta F_k\left(0\right), F_k\left(0\right), V^{\mathrm{impl}}\left(T_j, T_{k-1}\right)\sqrt{T_{k-1} - T_j}\right)
\end{aligned}$$

where $V^{\mathrm{impl}}\left(T_j, T_{k-1}\right)$ is the time-T_j forward implied volatility for the expiry T_{k-1}.

Similarly, the price of a FSC with model (12.4) is given by

$$\begin{aligned}
&\mathbf{FSCpl}\left(T_j, T_{k-1}, T_k, \tau_k, \delta\right) \\
&= \tau_k P\left(0, T_k\right) \sum_{i=1}^{N} \lambda_i E^k \left[\left(\left(F_k\left(T_{k-1}\right) + \alpha_k^i\right) - \left(\delta F_k\left(T_j\right) + \alpha_k^i\right)\right)^+ | I = i\right],
\end{aligned}$$

where the expectation is equal to

$$\begin{aligned}
&= E^k\left[E^k\left[\left(\left(F_k\left(T_{k-1}\right) + \alpha_k^i\right) - \left(\delta F_k\left(T_j\right) + \alpha_k^i\right)^+\right) | \{F_k(T_j), I = i\}\right] | I = i\right] \\
&= E^k\left[\mathrm{Bl}\left(\delta F_k\left(T_j\right) + \alpha_k^i, F_k\left(T_j\right) + \alpha_k^i, V_{T_j}^i\left(T_{k-1}\right)\right)\right] \\
&= \int_{-\infty}^{+\infty} \left[\mathrm{Bl}\left(\delta F_k\left(T_j\right) + \alpha_k^i, F_k\left(T_j\right) + \alpha_k^i, V_{T_j}^i\left(T_{k-1}\right)\right)\right] d\widetilde{\Phi}\left(F_k\left(T_j\right)\right),
\end{aligned}$$

where we set $V_{T_j}^i\left(T_{k-1}\right) = \sqrt{\int_{T_j}^{T_{k-1}} \sigma_k^i\left(s\right)^2 ds}$ and $\widetilde{\Phi}\left(F_k\left(T_j\right)\right)$ is the (lognormal) cumulative distribution function of $F_k\left(T_j\right)$ under Q^k.

The time-T_j forward implied volatility for the expiry T_{k-1} is, by definition, the parameter $V^{\mathrm{impl}}\left(T_j, T_{k-1}\right)$ that satisfies:

$$\mathbf{FSCpl}^{\mathrm{LFM}}\left(T_j, T_{k-1}, T_k, \tau_k, \delta\right) = \mathbf{FSCpl}^{\mathrm{SLMUP}}\left(T_j, T_{k-1}, T_k, \tau_k, \delta\right).$$

To test the forward implied volatilities generated by the model in a quick and simpler way, we assume the shift-parameters α_k^I's to be zero in each scenario. In this case the formula for the model's price simplifies to

$$\mathbf{FSCpl}(T_j, T_{k-1}, T_k, \tau_k, \delta) = \tau_k P(0, T_k) \sum_{i=1}^{N} \lambda_i \mathrm{Bl}\Big(\delta F_k(0), F_k(0), V_{T_j}^i(T_{k-1})\Big).$$

We priced FSCs starting in one year and setting after 2, 3, 5 and 10 years. Figure 12.9 shows the results of our test.

From our analysis, model (12.4) seems to imply forward volatilities that are very similar to the current ones. Note also that a slight difference is due to the fact that we priced the FSCs by neglecting the α_k^I's (the forward-volatility graphs do not display a clear skew), as opposed to the spot caplet-volatility curves.

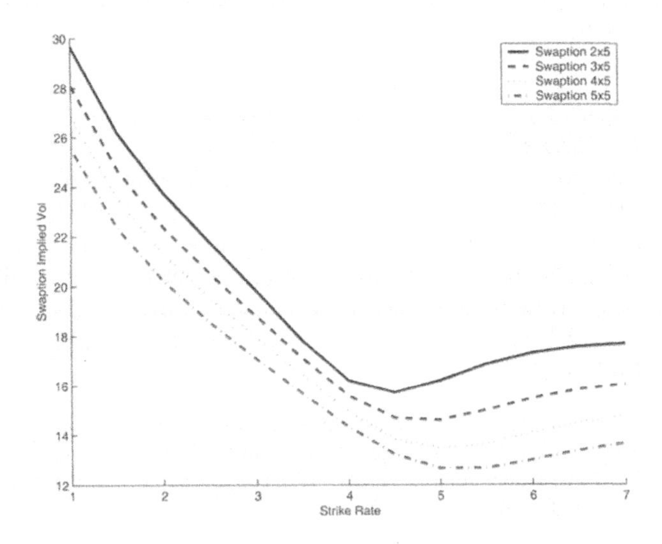

Fig. 12.8. The implied-volatility smile/skew for the 2 x 5, 3 x 5, 4 x 5 and 5 x 5 swaptions.

12.9 Joint Calibration to Caps and Swaptions

The previous example of joint calibration to caps and swaptions is not quite satisfactory from a practical point of view, due to the reduced number of swaptions we included in the calibration set. We thus conclude this section

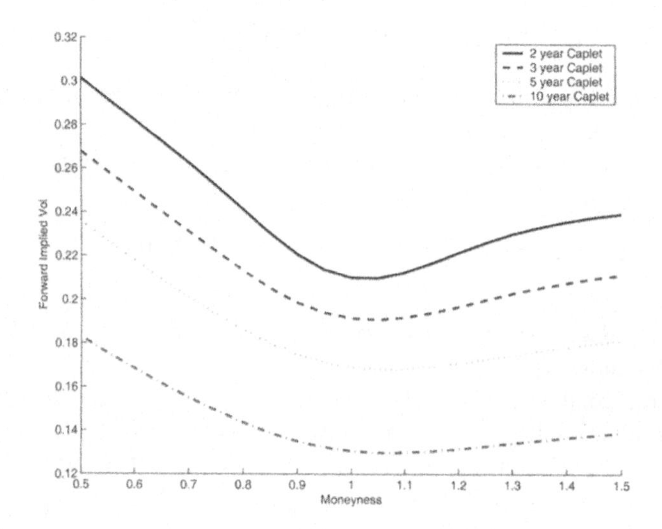

Fig. 12.9. One-year forward volatility smiles of caplets setting after two, three, five and ten years.

by describing a more realistic case of simultaneous calibration to the two interest-rate markets.[8]

To this end, we consider EUR data as of December 30, 2004. The cap and swaption volatilities we use are reported in Tables 12.14, 12.15 and in Table 12.16, respectively. Cap maturities range from three to twenty years, whereas swaption maturities and tenors vary from one to ten years. Accordingly, the relevant forward rates in our data set are all semiannual, with expiries from 0.5 to 19.5 years.[9] The discount factors for the relevant maturities are reported in Table 12.13.

[8] We thank Andrea Pallavicini for his fundamental contribution in the model implementation and for this empirical example in particular.

[9] When using semiannual forward rates, we have to change the definition of swap rate in (12.7) as follows

$$S_{\alpha,\beta}(t) = \frac{\sum_{h=\alpha'+1}^{\beta'} \tau'_h P(t,T'_h) F_h(t)}{\sum_{k=\alpha+1}^{\beta} \tau_k P(t,T_k)} = \sum_{h=\alpha'+1}^{\beta'} \omega'_h(t) F_h(t)$$

$$\omega'_h(t) := \frac{\tau'_h P(t,T'_h) F_h(t)}{\sum_{k=\alpha+1}^{\beta} \tau_k P(t,T_k)}$$

where times T' and T are, respectively, semiannually and annually spaced, and τ' and τ are the respective year fractions. The swaption price (12.8) is then replaced by the one obtained by re-writing the procedure in Section 12.3, accounting for the new weights ω'.

We then specify an instantaneous covariation structure in (12.4), which we regard as flexible enough to accommodate the given 442 market volatilities. Setting again $N = 3$, we assume that:

- the forward-rate volatilities σ_j^i, in each of the three scenarios, are parameterized according to the formulation (6.11);
- the shift parameters are the same for all forward rates, namely $\alpha_j^i = \alpha_k^i$ for each j, k, but can be different in different scenarios;
- the correlations between forward rates are given by the following generalization of (12.10)

$$\rho_{i,j} = a + (1 - a)\exp[-|i - j|(b - c\max(i,j) + d\max(i,j)^2)], \qquad (12.13)$$

where a, b, c and d are positive constants.

The model parameters to be used in the calibration process are therefore 243, namely 234 (=39x3x2) volatility parameters, 3 shift parameters, 4 correlation parameters and 2 probabilities λ's.

The calibration is performed by minimizing the sum of squared percentage differences between model and market prices over all caps and swaptions in our data set. Given the high number of optimization parameters, the fitting procedure can be rather cumbersome and time consuming, not to talk about the concrete possibility to get stuck in some local minimum. For this reason, we start with a fast sequential calibration to the cap volatilities, obtaining some preliminary values for the covariance parameters. Using these values as initial guess, we then run a global optimization including the swaption volatilities.

The results of our fitting procedure are reported in Tables 12.17, 12.18 and 12.19, where we show the percentage differences between model and market volatilities. We can see that calibration errors are usually rather small, especially in the caps matrix. Some large differences are found for swaptions with long maturity and short tenor. This, however, could be expected and simply reflects the classical difficulty to accommodate caplets and swaptions with one-year tenor simultaneously.

T	$P(0,T)$	T	$P(0,T)$
0.5y	0.989000	10.5y	0.669968
1.0y	0.976796	11.0y	0.654140
1.5y	0.963574	11.5y	0.639033
2.0y	0.949249	12.0y	0.623980
2.5y	0.934460	12.5y	0.609400
3.0y	0.918808	13.0y	0.594737
3.5y	0.903221	13.5y	0.580627
4.0y	0.886782	14.0y	0.566456
4.5y	0.870740	14.5y	0.552794
5.0y	0.853781	15.0y	0.539091
5.5y	0.836993	15.5y	0.526013
6.0y	0.819938	16.0y	0.512916
6.5y	0.802873	16.5y	0.500369
7.0y	0.785538	17.0y	0.488020
7.5y	0.768596	17.5y	0.476035
8.0y	0.751387	18.0y	0.464187
8.5y	0.734803	18.5y	0.452806
9.0y	0.717940	19.0y	0.441439
9.5y	0.701854	19.5y	0.430530
10.0y	0.685384	20.0y	0.419701

Table 12.13. Discount factors for the relevant maturities in the EUR market as of December 30, 2004.

	Maturities								
Strikes	3	4	5	6	7	8	9	10	11
2.50%	24.30%	24.10%	23.80%	23.50%	23.20%	23.00%	22.70%	22.40%	22.14%
2.75%	23.75%	23.33%	22.91%	22.52%	22.24%	21.99%	21.70%	21.39%	21.13%
3.00%	23.80%	23.00%	22.30%	21.80%	21.50%	21.10%	20.80%	20.50%	20.24%
3.25%	23.74%	22.64%	21.76%	21.20%	20.82%	20.33%	19.99%	19.70%	19.45%
3.50%	23.60%	22.30%	21.30%	20.70%	20.20%	19.70%	19.30%	19.00%	18.75%
3.75%	23.58%	22.15%	21.00%	20.31%	19.71%	19.22%	18.76%	18.41%	18.16%
4.00%	23.70%	22.20%	20.90%	20.10%	19.40%	18.90%	18.40%	18.00%	17.74%
4.25%	23.95%	22.40%	21.01%	20.09%	19.31%	18.75%	18.22%	17.79%	17.49%
4.50%	24.25%	22.65%	21.20%	20.15%	19.30%	18.65%	18.10%	17.65%	17.31%
4.75%	24.52%	22.87%	21.35%	20.16%	19.24%	18.51%	17.94%	17.47%	17.08%
5.00%	24.80%	23.10%	21.50%	20.20%	19.20%	18.40%	17.80%	17.30%	16.88%
5.25%	25.13%	23.41%	21.73%	20.35%	19.27%	18.41%	17.76%	17.23%	16.80%
5.50%	25.50%	23.79%	22.02%	20.60%	19.43%	18.52%	17.82%	17.26%	16.82%
5.75%	25.90%	24.20%	22.36%	20.90%	19.66%	18.69%	17.94%	17.36%	16.93%
6.00%	26.30%	24.60%	22.70%	21.20%	19.90%	18.90%	18.10%	17.50%	17.07%
6.25%	26.69%	24.97%	23.03%	21.47%	20.13%	19.10%	18.26%	17.66%	17.23%
6.50%	27.07%	25.32%	23.35%	21.71%	20.36%	19.30%	18.43%	17.84%	17.40%
6.75%	27.44%	25.66%	23.67%	21.95%	20.58%	19.50%	18.61%	18.02%	17.57%
7.00%	27.80%	26.00%	24.00%	22.20%	20.80%	19.70%	18.80%	18.20%	17.74%

Table 12.14. Cap implied volatilities for maturities up to eleven years in the EUR market as of December 30, 2004.

	Maturities								
Strikes	12	13	14	15	16	17	18	19	20
2.50%	21.90%	21.66%	21.42%	21.20%	21.00%	20.81%	20.63%	20.46%	20.30%
2.75%	20.89%	20.64%	20.39%	20.15%	19.94%	19.75%	19.58%	19.42%	19.26%
3.00%	20.00%	19.73%	19.46%	19.20%	18.98%	18.80%	18.65%	18.52%	18.40%
3.25%	19.20%	18.93%	18.66%	18.40%	18.18%	17.99%	17.82%	17.67%	17.53%
3.50%	18.50%	18.24%	17.96%	17.70%	17.46%	17.25%	17.06%	16.88%	16.70%
3.75%	17.93%	17.64%	17.33%	17.03%	16.77%	16.55%	16.36%	16.18%	16.02%
4.00%	17.50%	17.19%	16.84%	16.50%	16.22%	15.99%	15.80%	15.65%	15.50%
4.25%	17.23%	16.90%	16.54%	16.18%	15.89%	15.65%	15.46%	15.29%	15.13%
4.50%	17.00%	16.66%	16.29%	15.95%	15.65%	15.40%	15.18%	14.99%	14.80%
4.75%	16.73%	16.37%	16.01%	15.66%	15.36%	15.10%	14.86%	14.64%	14.43%
5.00%	16.50%	16.12%	15.75%	15.40%	15.09%	14.81%	14.56%	14.33%	14.10%
5.25%	16.41%	16.02%	15.63%	15.26%	14.94%	14.65%	14.39%	14.15%	13.91%
5.50%	16.44%	16.04%	15.63%	15.24%	14.90%	14.60%	14.34%	14.09%	13.86%
5.75%	16.55%	16.14%	15.71%	15.30%	14.95%	14.64%	14.37%	14.13%	13.90%
6.00%	16.70%	16.28%	15.83%	15.40%	15.03%	14.72%	14.46%	14.22%	14.00%
6.25%	16.85%	16.41%	15.95%	15.51%	15.14%	14.83%	14.57%	14.34%	14.13%
6.50%	16.99%	16.54%	16.07%	15.63%	15.26%	14.95%	14.70%	14.48%	14.27%
6.75%	17.14%	16.68%	16.20%	15.76%	15.39%	15.09%	14.84%	14.63%	14.43%
7.00%	17.30%	16.83%	16.34%	15.90%	15.53%	15.24%	15.00%	14.79%	14.60%

Table 12.15. Cap implied volatilities for maturities from twelve to twenty years in the EUR market as of December 30, 2004.

	Tenors									
Maturities	1	2	3	4	5	6	7	8	9	10
1	24.30%	23.70%	22.50%	21.60%	20.50%	19.10%	18.20%	17.50%	16.90%	16.50%
2	22.90%	21.50%	20.30%	19.30%	18.20%	17.40%	16.60%	16.10%	15.80%	15.50%
3	20.80%	19.70%	18.50%	17.50%	16.70%	16.10%	15.60%	15.20%	15.00%	14.80%
4	19.20%	18.20%	17.10%	16.30%	15.70%	15.10%	14.80%	14.50%	14.30%	14.20%
5	17.80%	16.90%	16.00%	15.30%	14.80%	14.40%	14.10%	13.90%	13.80%	13.80%
6	16.68%	15.89%	15.08%	14.47%	14.03%	13.77%	13.51%	13.42%	13.35%	13.32%
7	15.80%	15.10%	14.30%	13.80%	13.40%	13.20%	13.00%	13.00%	12.90%	12.80%
8	15.06%	14.44%	13.66%	13.24%	12.90%	12.71%	12.54%	12.58%	12.46%	12.35%
9	14.41%	13.85%	13.10%	12.75%	12.48%	12.29%	12.11%	12.14%	12.03%	11.96%
10	13.80%	13.30%	12.60%	12.30%	12.10%	11.90%	11.70%	11.70%	11.60%	11.60%

Table 12.16. Swaption implied volatilities in the EUR market as of December 30, 2004.

	Maturities								
Strikes	3	4	5	6	7	8	9	10	11
2.50%	0.05%	1.31%	1.86%	2.60%	2.66%	2.47%	2.43%	2.23%	1.61%
2.75%	0.89%	1.32%	1.56%	2.39%	2.27%	2.31%	2.23%	2.14%	1.69%
3.00%	-0.45%	-0.12%	0.55%	1.48%	1.26%	1.94%	1.94%	1.85%	1.48%
3.25%	-1.34%	-1.13%	-0.44%	0.35%	0.29%	1.31%	1.50%	1.37%	1.05%
3.50%	-1.68%	-1.70%	-1.19%	-0.63%	-0.44%	0.54%	0.95%	0.83%	0.55%
3.75%	-1.84%	-2.22%	-1.70%	-1.22%	-0.92%	-0.21%	0.32%	0.34%	0.06%
4.00%	-1.90%	-2.68%	-2.14%	-1.64%	-1.39%	-0.96%	-0.43%	-0.31%	-0.58%
4.25%	-1.93%	-3.05%	-2.69%	-2.13%	-2.05%	-1.67%	-1.30%	-1.26%	-1.44%
4.50%	-1.79%	-3.12%	-2.95%	-2.26%	-2.33%	-1.89%	-1.71%	-1.83%	-1.87%
4.75%	-1.37%	-2.73%	-2.62%	-1.67%	-1.78%	-1.24%	-1.22%	-1.45%	-1.35%
5.00%	-0.94%	-2.28%	-2.10%	-0.90%	-0.93%	-0.29%	-0.33%	-0.60%	-0.41%
5.25%	-0.74%	-2.12%	-1.85%	-0.57%	-0.43%	0.32%	0.34%	0.10%	0.27%
5.50%	-0.80%	-2.27%	-1.91%	-0.67%	-0.34%	0.52%	0.67%	0.49%	0.57%
5.75%	-1.06%	-2.61%	-2.18%	-1.03%	-0.53%	0.41%	0.74%	0.59%	0.56%
6.00%	-1.44%	-3.01%	-2.56%	-1.44%	-0.86%	0.14%	0.64%	0.50%	0.37%
6.25%	-1.88%	-3.37%	-2.95%	-1.75%	-1.18%	-0.16%	0.47%	0.27%	0.11%
6.50%	-2.37%	-3.73%	-3.37%	-1.99%	-1.49%	-0.46%	0.23%	-0.05%	-0.20%
6.75%	-2.89%	-4.13%	-3.84%	-2.29%	-1.84%	-0.81%	-0.10%	-0.45%	-0.56%
7.00%	-3.44%	-4.62%	-4.40%	-2.69%	-2.25%	-1.21%	-0.51%	-0.89%	-0.98%

Table 12.17. Calibration result: percentage differences between SLMUP and market cap volatilities for maturities up to eleven years.

	Maturities								
Strikes	3	4	5	6	7	8	9	10	11
2.50%	1.34%	0.95%	0.50%	-0.05%	-0.38%	-0.83%	-1.33%	-1.87%	-2.40%
2.75%	1.44%	1.18%	0.92%	0.57%	0.33%	-0.05%	-0.51%	-1.03%	-1.55%
3.00%	1.24%	1.14%	1.11%	0.97%	0.83%	0.44%	-0.09%	-0.72%	-1.39%
3.25%	0.85%	0.81%	0.85%	0.80%	0.74%	0.46%	0.06%	-0.44%	-0.96%
3.50%	0.36%	0.34%	0.44%	0.48%	0.53%	0.39%	0.16%	-0.13%	-0.45%
3.75%	-0.22%	-0.14%	0.18%	0.46%	0.66%	0.60%	0.39%	0.06%	-0.32%
4.00%	-0.90%	-0.74%	-0.22%	0.28%	0.60%	0.56%	0.28%	-0.18%	-0.71%
4.25%	-1.68%	-1.51%	-1.00%	-0.53%	-0.20%	-0.23%	-0.50%	-0.93%	-1.43%
4.50%	-1.96%	-1.81%	-1.44%	-1.14%	-0.88%	-0.89%	-1.09%	-1.40%	-1.78%
4.75%	-1.30%	-1.14%	-0.89%	-0.72%	-0.50%	-0.50%	-0.65%	-0.87%	-1.14%
5.00%	-0.28%	-0.11%	0.11%	0.20%	0.41%	0.43%	0.32%	0.14%	-0.07%
5.25%	0.34%	0.51%	0.74%	0.85%	1.06%	1.10%	1.00%	0.81%	0.61%
5.50%	0.50%	0.66%	0.93%	1.12%	1.37%	1.41%	1.29%	1.07%	0.82%
5.75%	0.36%	0.50%	0.84%	1.11%	1.40%	1.43%	1.27%	0.99%	0.66%
6.00%	0.10%	0.24%	0.63%	0.97%	1.27%	1.28%	1.07%	0.71%	0.30%
6.25%	-0.14%	0.04%	0.47%	0.83%	1.10%	1.07%	0.79%	0.35%	-0.15%
6.50%	-0.35%	-0.13%	0.30%	0.63%	0.87%	0.78%	0.43%	-0.08%	-0.67%
6.75%	-0.61%	-0.34%	0.07%	0.34%	0.53%	0.38%	-0.03%	-0.62%	-1.26%
7.00%	-0.94%	-0.66%	-0.29%	-0.09%	0.06%	-0.13%	-0.59%	-1.22%	-1.92%

Table 12.18. Calibration result: percentage differences between SLMUP and market cap volatilities for maturities from twelve to twenty years.

	Tenors									
Maturities	1	2	3	4	5	6	7	8	9	10
1	-0.01%	-0.08%	-0.94%	-2.97%	-3.21%	-1.86%	-1.38%	-1.42%	-1.44%	-0.56%
2	1.60%	0.08%	-0.13%	-0.31%	-0.09%	0.08%	0.84%	0.43%	1.03%	0.55%
3	-1.94%	-1.62%	-0.10%	0.05%	0.54%	0.28%	-0.12%	1.53%	0.47%	0.38%
4	-2.59%	-0.63%	0.07%	0.39%	-0.04%	0.05%	1.69%	1.38%	1.38%	0.27%
5	2.31%	-0.31%	0.43%	0.60%	-0.11%	2.76%	2.33%	2.27%	1.03%	-0.57%
6	-1.78%	-0.40%	0.40%	0.00%	3.74%	2.73%	3.05%	1.59%	0.46%	-1.63%
7	2.74%	0.97%	0.73%	5.23%	4.63%	4.26%	3.43%	1.66%	-0.09%	-1.21%
8	2.24%	-1.11%	5.88%	4.51%	5.09%	4.07%	3.60%	0.50%	-0.57%	-1.04%
9	-1.08%	5.36%	5.03%	5.55%	4.98%	4.60%	3.00%	0.49%	-0.08%	-0.71%
10	11.06%	4.83%	7.94%	7.11%	6.44%	4.44%	3.51%	1.76%	1.22%	0.06%

Table 12.19. Calibration result: percentage differences between SLMUP and market swaption volatilities.

Part V

EXAMPLES OF MARKET PAYOFFS

13. Pricing Derivatives on a Single Interest-Rate Curve

Increasingly, problems do not rule out practice, but support it. Instead of finding that practice is too difficult, that we have too many problems, we see that the problems themselves are the jewels, and we devote ourselves to be with them in a way we never dreamt of before.

Charlotte Joko Beck, "Nothing Special: Living Zen", 1995, HarperCollins.

In this chapter, we present a sample of financial products we believe to be representative of a large portion of the interest-rate market. We will use different models (mostly the LFM and the G2++ model) for different problems, and try to clarify the advantages of each model. All the discounted payoffs will be calculated at time $t = 0$.

Before starting, we remark upon the possible use of an approximated LIBOR market model (LFM) for pricing some of the products we will consider.

It is possible to freeze part of the drift of the LFM dynamics so as to obtain a geometric Brownian motion. This is what was done for example in Section 6.16 to derive approximated formulas for terminal correlations. In that section, we derived such a dynamics under the T_γ-forward-adjusted measure Q^γ:

$$dF_k(t) = \bar{\mu}_{\gamma,k}(t)F_k(t)\,dt + \sigma_k(t)F_k(t)\,dZ_k(t)\,, \tag{13.1}$$

where

$$\begin{aligned}
\mu_{\gamma,k}(t) &:= -\sum_{j=k+1}^{\gamma} \frac{\rho_{k,j}\tau_j\sigma_j(t)F_j(0)}{1+\tau_jF_j(0)}\,, \quad k<\gamma\,,\\
\mu_{\gamma,\gamma}(t) &:= 0\,, \quad k=\gamma\,,\\
\mu_{\gamma,k}(t) &:= \sum_{j=\gamma+1}^{k} \frac{\rho_{k,j}\tau_j\sigma_j(t)F_j(0)}{1+\tau_jF_j(0)}\,, \quad k>\gamma\,,\\
\bar{\mu}_{\gamma,k}(t) &:= \sigma_k(t)\mu_{\gamma,k}(t).
\end{aligned}$$

This dynamics gives access, in some cases, to a number of techniques which have been developed for the basic Black and Scholes setup, for example, in equity and FX markets. Moreover, this "freezing-part-of-the-drift" technique can be combined with drift interpolation so as to allow for rates that are not in the fundamental (spanning) family corresponding to the particular LFM being implemented.

We detail this possible "interpolate and freeze (part of the) drift" approach in case of accrual swaps in Section 13.13.1, but the method is rather general and, when used in combination with other possible approximations, can be used for other products.

Finally, even if one keeps on using Monte Carlo evaluation, the frozen-drift approximation leads to a process (geometric Brownian motion) that is much easier to propagate in time and requires no small discretization step in the propagation, allowing instead for "one-shot" simulation also over long periods of time.

In the following, we assume we are given a set of dates $T_\alpha, \ldots, T_i \ldots, T_{\beta+1}$ with associated year fractions $\tau_\alpha, \ldots, \tau_i \ldots, \tau_{\beta+1}$.

13.1 In-Arrears Swaps

An in-arrears swap is an IRS that resets at dates $T_{\alpha+1}, \ldots, T_\beta$ and pays at the same dates, with unit notional amount and with fixed-leg rate K. More precisely, the discounted payoff of an in-arrears swap (of "payer" type) can be expressed via

$$\sum_{i=\alpha+1}^{\beta} D(0, T_i)\tau_{i+1}(F_{i+1}(T_i) - K).$$

The value of such a contract is, therefore,

$$\mathbf{IAS} = E\left[\sum_{i=\alpha+1}^{\beta} D(0, T_i)\tau_{i+1}(F_{i+1}(T_i) - K)\right],$$

where we omit arguments in the "IAS" notation for brevity.

Before calculating the expectations, it is convenient to make some adjustments. We shall use the following identity (obtained easily via iterated conditioning, as seen in Proposition 2.8.1):

$$E[XD(0,T)] = E\left[\frac{XD(0,S)}{P(T,S)}\right] \quad \text{for all } 0 < T < S, \tag{13.2}$$

where X is a T-measurable random variable.

To value the above contract, notice that

$$
\begin{aligned}
& E\left\{\sum_{i=\alpha+1}^{\beta} D(0,T_i)\tau_{i+1}(F_{i+1}(T_i)-K)\right\} \\
&= E\left\{\sum_{i=\alpha+1}^{\beta} D(0,T_i)\left[\frac{1}{P(T_i,T_{i+1})}-(1+\tau_{i+1}K)\right]\right\} \\
&= E\left\{\sum_{i=\alpha+1}^{\beta}\left[\frac{D(0,T_{i+1})}{P(T_i,T_{i+1})^2}-D(0,T_i)(1+\tau_{i+1}K)\right]\right\}
\end{aligned}
$$

$$
\begin{aligned}
&= \sum_{i=\alpha+1}^{\beta}\left\{P(0,T_{i+1})E^{i+1}\left[\frac{1}{P(T_i,T_{i+1})^2}\right]-P(0,T_i)(1+\tau_{i+1}K)\right\} \\
&= \sum_{i=\alpha+1}^{\beta}\left\{P(0,T_{i+1})E^{i+1}\left[(1+\tau_{i+1}F_{i+1}(T_i))^2\right]-P(0,T_i)(1+\tau_{i+1}K)\right\}.
\end{aligned}
$$

Computing the expected value is an easy task, since we know that, under Q^{i+1}, F_{i+1} has the driftless (martingale) lognormal dynamics

$$dF_{i+1}(t)=\sigma_{i+1}(t)F_{i+1}(t)dZ_{i+1}(t)\ ,$$

so that, remembering the resulting lognormal distribution of $F_{i+1}^2(T_i)$, one has

$$E^{i+1}\left(F_{i+1}^2(T_i)\right)=F_{i+1}^2(0)\exp\left[\int_0^{T_i}\sigma_{i+1}^2(t)dt\right]=F_{i+1}^2(0)\exp(v_{i+1}^2)$$

where the v's have been defined in (6.19) and are deduced from cap prices. We obtain

$$
\begin{aligned}
\mathbf{IAS} = & \sum_{i=\alpha+1}^{\beta}\{P(0,T_{i+1})\left[1+2\tau_{i+1}F_{i+1}(0)+\tau_{i+1}^2F_{i+1}^2(0)\exp(v_{i+1}^2)\right] \\
& -(1+\tau_{i+1}K)P(0,T_i)\}. \qquad (13.3)
\end{aligned}
$$

Contrary to the plain-vanilla case, this price depends on the volatility of forward rates through the caplet volatilities v. Notice however that correlations between different rates are not involved in this product, as one expects from the additive and "one-rate-per-time" nature of the payoff.

Remark 13.1.1. If caplet prices are available in the market for each maturity T_i and strike K, we can price an in-arrears swap consistently with the observed caplet smile by noting that we can write

$$F^2=2\int_0^{+\infty}(F-K)^+dK,$$

so that

$$P(0,T_{i+1})E^{i+1}\left[F_{i+1}^2(T_i)\right] = 2\int_0^{+\infty} P(0,T_{i+1})E^{i+1}[(F_{i+1}(T_i)-K)^+]\,dK$$
$$= 2\int_0^{+\infty} \frac{1}{\tau_{i+1}}\mathbf{Cpl}^{\mathrm{Mkt}}(0,T_i,T_{i+1},\tau_{i+1},K)\,dK.$$

Equation (13.3) can then be modified accordingly.

13.2 In-Arrears Caps

An in-arrears cap is composed by caplets resetting at dates $T_{\alpha+1},\ldots,T_\beta$ and paying at the same dates, with unit notional amount and strike rate K. More precisely, the discounted payoff of an in-arrears cap can be expressed via

$$\sum_{i=\alpha+1}^{\beta} D(0,T_i)\tau_{i+1}(F_{i+1}(T_i)-K)^+.$$

The value of such a contract is, therefore,

$$\mathbf{IAC} = E\left[\sum_{i=\alpha+1}^{\beta} D(0,T_i)\tau_{i+1}(F_{i+1}(T_i)-K)^+\right].$$

The payoff is the same as in the case of in-arrears swaps, except for the positive-part operator.

13.2.1 A First Analytical Formula (LFM)

We apply the same reasoning we used for in-arrears swaps, obtaining:

$$\mathbf{IAC} = \sum_{i=\alpha+1}^{\beta} \tau_{i+1}P(0,T_{i+1})E^{i+1}\left[(1+\tau_{i+1}F_{i+1}(T_i))(F_{i+1}(T_i)-K)^+\right]$$
$$= \sum_{i=\alpha+1}^{\beta} \tau_{i+1}P(0,T_{i+1})\left(E^{i+1}[(F_{i+1}(T_i)-K)^+]\right.$$
$$\left. +\tau_{i+1}E^{i+1}\left[F_{i+1}(T_i)(F_{i+1}(T_i)-K)^+\right]\right)$$
$$= \sum_{i=\alpha+1}^{\beta} \tau_{i+1}P(0,T_{i+1})\left[\mathrm{Bl}(K,F_{i+1}(0),v_{i+1}) + \tau_{i+1}g(K,F_{i+1}(0),v_{i+1})\right],$$

$$g(K,F,v) := F^2\exp[v^2]\,\Phi\left(\frac{3v}{2}-\frac{1}{v}\ln\frac{K}{F}\right) - FK\,\Phi\left(\frac{v}{2}-\frac{1}{v}\ln\frac{K}{F}\right),$$

where "Bl" and v have been defined in (1.26), (6.19) and above. In-arrears caps do not depend on the correlation of different rates but just on the caplet volatilities v, as one expects again from the additive and "one-rate-per-time" nature of the payoff.

13.2.2 A Second Analytical Formula (G2++)

The above expectations can also be easily computed under the Gaussian G2++ model, by exploiting the lognormal distribution of bond prices. After lengthy but straightforward calculations we obtain:

$$\mathbf{IAC} = \sum_{i=\alpha+1}^{\beta} P(0,T_i)\left[\frac{P(0,T_i)}{P(0,T_{i+1})} e^{\Sigma(T_i,T_{i+1})^2} \Phi\left(\frac{\ln\frac{P(0,T_i)}{\widetilde{K}_i P(0,T_{i+1})} + \frac{3}{2}\Sigma(T_i,T_{i+1})^2}{\Sigma(T_i,T_{i+1})}\right)\right.$$
$$\left. - \widetilde{K}_i \Phi\left(\frac{\ln\frac{P(0,T_i)}{\widetilde{K}_i P(0,T_{i+1})} + \frac{1}{2}\Sigma(T_i,T_{i+1})^2}{\Sigma(T_i,T_{i+1})}\right)\right] \tau_{i+1},$$

where $\widetilde{K}_i = 1 + K\tau_i$ and

$$\Sigma(T,S)^2 = \frac{\sigma^2}{2a^3}\left[1 - e^{-a(S-T)}\right]^2\left[1 - e^{-2aT}\right] + \frac{\eta^2}{2b^3}\left[1 - e^{-b(S-T)}\right]^2\left[1 - e^{-2bT}\right]$$
$$+ 2\rho\frac{\sigma\eta}{ab(a+b)}\left[1 - e^{-a(S-T)}\right]\left[1 - e^{-b(S-T)}\right]\left[1 - e^{-(a+b)T}\right].$$

13.3 Autocaps

We adopt the same notation, terminology and conventions as in Section 6.4, and take $\alpha = 0$. An autocap is similar to a cap, but at most $\gamma \leq \beta$ caplets can be exercised, and they *have* to be automatically exercised when in the money. Therefore, the discounted payoff can be written as

$$\sum_{i=1}^{\beta} \tau_i \left[F(T_{i-1}; T_{i-1}, T_i) - K\right]^+ D(0,T_i)\, 1\{A_i\},$$

$$A_i = \{\text{at most } \gamma \text{ among } F_1(T_0), \ldots, F_i(T_{i-1}) \text{ are larger than } K\},$$

where $1\{A\}$ denotes the indicator function for the set A.

The pricing of this contract can be obtained by considering the risk-neutral expectation E of its discounted payoff:

$$E\left[\sum_{i=1}^{\beta} \tau_i (F_i(T_{i-1}) - K)^+ D(0,T_i)\, 1\{A_i\}\right]$$
$$= P(0,T_\beta) \sum_{i=1}^{\beta} \tau_i E^\beta \left[\frac{(F_i(T_{i-1}) - K)^+ \, 1\{A_i\}}{P(T_i,T_\beta)}\right],$$

where we have used (13.2) (equivalently, the remarks of Section 2.8).

Notice that the A_i term depends not only on the forward rate $F_i(T_{i-1})$, but also on $F_1(T_0), ..., F_{i-1}(T_{i-2})$. Therefore, a "path-dependent" feature is introduced in the contract. If we attempt to price this contract by a Monte Carlo method, in order to compute the discounted payoff we need to generate paths under Q^β for the vector (whose dimension decreases over time)

$$F_{\beta(t)}(t), \ldots, F_\beta(t),$$

where we recall that $t \in (T_{\beta(t)-2}, T_{\beta(t)-1}]$. These paths can be deduced from discretizing the dynamics (6.14). In our setting, such dynamics reads $(k = \beta(t), \beta(t)+1, \ldots, \beta)$

$$dF_k(t) = -\sigma_k(t)F_k(t) \sum_{j=k+1}^{\beta} \frac{\rho_{k,j}\tau_j\sigma_j(t)F_j(t)}{1+\tau_jF_j(t)} dt + \sigma_k(t)F_k(t)dZ_k^\beta(t). \quad (13.4)$$

Taking logs and using the Milstein scheme, analogously to what was done for swaptions in Section 6.10, yields the desired simulated paths:

$$\ln F_k^{\Delta t}(t+\Delta t) = \ln F_k^{\Delta t}(t) - \sigma_k(t) \sum_{j=k+1}^{\beta} \frac{\rho_{k,j}\tau_j\sigma_j(t)F_j^{\Delta t}(t)}{1+\tau_jF_j^{\Delta t}(t)} \Delta t - \frac{\sigma_k(t)^2}{2}\Delta t$$
$$+\sigma_k(t)(Z_k^\beta(t+\Delta t) - Z_k^\beta(t)). \quad (13.5)$$

Actually, here too, one can improve the scheme by resorting to more refined shocks, in the spirit of Remark 6.10.1.

13.4 Caps with Deferred Caplets

These are caps for which all caplets payments occur at the final time T_β. The discounted payoff is, therefore,

$$\sum_{i=1}^{\beta} \tau_i(F_i(T_{i-1}) - K)^+ D(0, T_\beta).$$

The pricing of this "deferred" cap can be obtained by considering the risk-neutral expectation E of its discounted payoff:

$$E\left[\sum_{i=1}^{\beta} \tau_i(F_i(T_{i-1}) - K)^+ D(0, T_\beta)\right] = P(0, T_\beta) \sum_{i=1}^{\beta} \tau_i E^\beta\left[(F_i(T_{i-1}) - K)^+\right].$$

The expected value can be computed through a Monte Carlo method based on the discretized Q^β dynamics (13.5).

13.4.1 A First Analytical Formula (LFM)

The above formula requires to compute the expected values

$$E^{\beta}\left[(F_i(T_{i-1}) - K)^+\right].$$

This is a case where the LFM with partially frozen drift can be of help in deriving analytical approximations. Indeed, consider the approximate LFM dynamics (13.1) with $\gamma = \beta$. Then, the above expectation is easily computed as a Black and Scholes call-option price (see Appendix D):

$$\exp\left(\int_0^{T_{i-1}} \bar{\mu}_{\beta,i}(t)dt\right) F_i(0)\,\Phi\left(\frac{\ln\frac{F_i(0)}{K} + \int_0^{T_{i-1}}\left[\bar{\mu}_{\beta,i}(t) + \frac{\sigma_i^2(t)}{2}\right]dt}{\sqrt{\int_0^{T_{i-1}}\sigma_i^2(t)dt}}\right)$$
$$-K\Phi\left(\frac{\ln\frac{F_i(0)}{K} + \int_0^{T_{i-1}}\left[\bar{\mu}_{\beta,i}(t) - \frac{\sigma_i^2(t)}{2}\right]dt}{\sqrt{\int_0^{T_{i-1}}\sigma_i^2(t)dt}}\right).$$

Replacing the expectation with such expression in the above summation, we obtain an analytical formula for the price of the cap with deferred caplets.

13.4.2 A Second Analytical Formula (G2++)

The expectations

$$E^{\beta}\left[(F_i(T_{i-1}) - K)^+\right]$$

can also be easily computed under the Gaussian G2++ model, by again exploiting the lognormal distribution of bond prices. After lengthy but straightforward calculations we obtain:

$$\frac{1}{\tau_i}\left[\frac{P(0,T_{i-1})}{P(0,T_i)}e^{\psi(0,T_{i-1},T_i,T_\beta,1)}\Phi\left(\frac{\ln\frac{P(0,T_{i-1})}{\widetilde{K}P(0,T_i)} + \psi(0,T_{i-1},T_i,T_\beta,\frac{3}{2})}{\sqrt{\psi(T_{i-1},T_i,T_i,2)}}\right)\right.$$
$$\left. - \widetilde{K}\Phi\left(\frac{\ln\frac{P(0,T_{i-1})}{\widetilde{K}P(0,T_i)} + \psi(0,T_{i-1},T_i,T_\beta,\frac{1}{2})}{\sqrt{\psi(T_{i-1},T_i,T_i,2)}}\right)\right],$$

where $\widetilde{K} := 1 + \tau_i K$ and

$$\begin{aligned}
&\psi(T,S,\tau,\lambda)\\
&:= \frac{\sigma^2}{2a^3}\left[1 - e^{-a(S-T)}\right]\left[1 - e^{-2aT}\right]\left[e^{-a(\tau-T)} - 1 + \lambda - \lambda e^{-a(S-T)}\right]\\
&+ \frac{\eta^2}{2b^3}\left[1 - e^{-b(S-T)}\right]\left[1 - e^{-2bT}\right]\left[e^{-b(\tau-T)} - 1 + \lambda - \lambda e^{-b(S-T)}\right]\\
&+ \frac{\rho\sigma\eta}{ab(a+b)}\left[1 - e^{-a(S-T)}\right]\left[1 - e^{-(a+b)T}\right]\left[e^{-b(\tau-T)} - 1 + \lambda - \lambda e^{-b(S-T)}\right]\\
&+ \frac{\rho\sigma\eta}{ab(a+b)}\left[1 - e^{-b(S-T)}\right]\left[1 - e^{-(a+b)T}\right]\left[e^{-a(\tau-T)} - 1 + \lambda - \lambda e^{-a(S-T)}\right].
\end{aligned}$$

Notice that, using the previous notation, we can write $\psi(T_{i-1}, T_i, T_i, 2) = \Sigma(T_{i-1}, T_i)^2$.

13.5 Ratchet Caps and Floors

A ratchet cap is a cap where the strike is updated at each caplet, based on the previous realization of the relevant interest rate. More specifically, we have typically that the simplest ratchet contract has the following payoff. Consider as usual a tenor structure $\mathcal{T} = \{T_1, \ldots, T_\beta\}$ for spanning forward rates. A ratchet cap first resetting at T_α and paying at $T_{\alpha+1}, \ldots, T_\beta$ pays the following payoff:

$$\sum_{i=\alpha+1}^{\beta} D(0, T_i)\tau_i \left[L(T_{i-1}, T_i) - (L(T_{i-2}, T_{i-1}) + X)\right]^+ , \tag{13.6}$$

Notice that if we set $K_i := L(T_{i-2}, T_{i-1}) + X$ for all i's this is a set of caplets with (random) strikes K_i.

The quantity X is a margin, which can be either positive or negative.

A sticky ratchet cap is instead given by the following discounted payoff:

$$\sum_{i=\alpha+1}^{\beta} D(0, T_i)\tau_i \left[L(T_{i-1}, T_i) - X_i\right]^+ , \tag{13.7}$$

$$X_i = \max\left(L(T_{i-2}, T_{i-1}) \pm \bar{X}, X_{i-1} \pm \bar{X}\right), \; X_\alpha := L(T_{\alpha-1}, T_\alpha).$$

There are versions with "min" replacing "max" in the X_i's definition. The quantity $\bar{X}$ is a spread that can be positive or negative.

In general a sticky ratchet cap has to be valued through Monte Carlo simulation. Indeed, the risk-neutral expectation of the discounted payoff (13.7) leads to

$$E\left\{\sum_{i=\alpha+1}^{\beta} D(0, T_i)\tau_i \left[L(T_{i-1}, T_i) - X_i\right]^+\right\} \tag{13.8}$$

$$= P(0, T_\beta) \sum_{i=\alpha+1}^{\beta} \tau_i E^\beta \left\{\frac{\left[L(T_{i-1}, T_i) - X_i\right]^+}{P(T_i, T_\beta)}\right\}.$$

Since the forward-rate dynamics under Q^β of

$$F_{\beta(t)}(t), \ldots, F_\beta(t)$$

is known as from (13.4), a Monte Carlo pricing can be carried out in the usual manner.

However, for the non-sticky ratchet cap payoff (13.6) we may investigate possible analytical approximations based on the usual "freezing the drift" technique for the LIBOR market model.

13.5.1 Analytical Approximation for Ratchet Caps with the LFM

We may study an analytical formula based on a drift approximation in the LIBOR market model, similar to the one used in deriving approximated swaptions volatilities and terminal correlations in Chapter 6.

We proceed as follows. Consider the discounted payoff (13.6). All we need to compute is the expectation

$$E\{D(0,T_i)\,[L(T_{i-1},T_i) - (L(T_{i-2},T_{i-1}) + X)]^+\}$$
$$= P(0,T_i)E^i\{[F_i(T_{i-1}) - F_{i-1}(T_{i-2}) - X]^+\} =: P(0,T_i)m_i$$

and then add terms. In the above expectation, the key rates evolve as follows under the measure Q^i:

$$dF_i(t) = \sigma_i(t)F_i(t)dZ_i(t),$$
$$dF_{i-1}(t) = -\frac{\rho_{i-1,i}\tau_i\sigma_i(t)F_i(t)}{1+\tau_iF_i(t)}F_{i-1}(t)dt + \sigma_{i-1}(t)F_{i-1}(t)dZ_{i-1}(t).$$

As usual, in such dynamics we do not know the distribution of $F_{i-1}(t)$. But, since F_{i-1} and F_i's reset times are adjacent, we may freeze the drift in F_{i-1} and be rather confident on the resulting approximations. We thus replace the second SDE by

$$dF_{i-1}(t) = \bar{\mu}(t)F_{i-1}(t)dt + \sigma_{i-1}(t)F_{i-1}(t)dZ_{i-1}(t),$$
$$\bar{\mu}(t) := -\frac{\rho_{i-1,i}\tau_i\sigma_i(t)F_i(0)}{1+\tau_iF_i(0)}.$$

Now both F_{i-1} and F_i follow (correlated) geometric Brownian motions as in the Black and Scholes model.

Now consider the case $X > 0$.

If we set $S_1 := F_i$, $S_2 := F_{i-1}$, $r - q_1 := 0$, $r - q_2 := \bar{\mu}(t)1\{t < T_{i-2}\}$, $\sigma_1 := \sigma_i(t)$, $\sigma_2 := \sigma_{i-1}(t)1\{t < T_{i-2}\}$, $a = 1$, $b = -1$, and $\omega = 1$, we may consider Appendix E and view our dynamics as (E.2) and our payoff as (E.3), by slightly adjusting to the fact that no discounting should occur in our case.

If $X < 0$, instead, we just switch the definitions of S_1 and S_2 above, $S_1 := F_{i-1}$, $S_2 = F_i$ etc., and then take $\omega = -1$. In the calculations below we assume $X > 0$.

As a matter of fact, our coefficients here are time-dependent, but this does not change substantially the derivation. It follows that our expected value

$$m_i := E^i\{[F_i(T_{i-1}) - F_{i-1}(T_{i-2}) - X]^+\}$$

is given by formula (E.5) when taking into account the above substitutions, i.e. one needs to apply said formula with

$$a = 1,\ \omega = 1,\ t = 0,\ S_1(t) = F_i(0),\ q_1\mathcal{T} = 0,\ q_2\mathcal{T} = \int_0^{T_{i-2}} \bar{\mu}(u)du,$$
$$r = 0,\ \sigma_1^2\mathcal{T} = \int_0^{T_{i-1}} \sigma_i^2(u)du,\ \sigma_2^2\mathcal{T} = \int_0^{T_{i-2}} \sigma_{i-1}^2(u)du,$$
$$\rho = \rho_{i-1,i},\ \ K = X.$$

Once we have the m_i's, our ratchet price is given by

$$\sum_{i=\alpha+1}^{\beta} P(0,T_i)\tau_i m_i.$$

The above price depends on a one-dimensional numerical integration. There is a case, though, where this is not necessary. Indeed, if $X = 0$, we obtain a special ratchet cap that, under the lognormal assumption, we may value analytically through the Margrabe formula for the option exchanging one asset for another. This formula may be derived by applying our formula to the case $X = 0$, or for example by applying the Formula at page 309 of Hunt and Kennedy (2000). The final formula is as follows:

$$E\left\{\sum_{i=\alpha+1}^{\beta} D(0,T_i)\tau_i\,[L(T_{i-1},T_i) - L(T_{i-2},T_{i-1})]^+\right\}$$
$$= E\left\{\sum_{i=\alpha+1}^{\beta} D(0,T_i)\tau_i\,[F_i(T_{i-1}) - F_{i-1}(T_{i-2})]^+\right\}$$
$$\approx \sum_{i=\alpha+1}^{\beta} \tau_i\ P(0,T_i)\left[F_i(0)\Phi(d_1^i) - F_{i-1}(0)\exp\left(\int_0^{T_{i-2}} \bar{\mu}(u)du\right)\Phi(d_2^i)\right],$$
$$d_{1,2}^i = \frac{\ln(F_i(0)/F_{i-1}(0)) - \int_0^{T_{i-2}} \bar{\mu}(u)du}{R_i} \pm \tfrac{1}{2}R_i,$$
$$R_i = \left(\int_0^{T_{i-1}} \sigma_i^2(u)du + \int_0^{T_{i-2}} (\sigma_{i-1}^2(u) - 2\rho_{i-1,i}\sigma_{i-1}(u)\sigma_i(u))du\right)^{\frac{1}{2}}$$

In this section we dealt with ratchet caps. The treatment of ratchet floors is analogous.

13.6 Ratchets (One-Way Floaters)

We give a short description of one-way floaters in the following. We assume a unit nominal amount.

- Institution A pays to B (a percentage γ of) a reference floating rate (plus a constant spread S) at dates $\mathcal{T} = \{T_1, \ldots, T_\beta\}$. Formally, at time T_i institution A pays to B

$$(\gamma F_i(T_{i-1}) + S)\tau_i .$$

- Institution B pays to A a coupon that is given by the reference rate plus a spread X at dates $\mathcal{T}$, floored and capped respectively by the previous coupon and by the previous coupon plus an increment Y. Formally, at time T_i with $i > 1$, institution B pays to A the coupon

$$c_i = \begin{cases} (F_i(T_{i-1}) + X)\tau_i & \text{if } c_{i-1} \leq (F_i(T_{i-1}) + X)\tau_i \leq c_{i-1} + Y, \\ c_{i-1} & \text{if } (F_i(T_{i-1}) + X)\tau_i < c_{i-1}, \\ c_{i-1} + Y & \text{if } (F_i(T_{i-1}) + X)\tau_i > c_{i-1} + Y, \end{cases}$$

At the first payment time T_1, institution B pays to A the coupon

$$(F_1(T_0) + X)\tau_1.$$

The discounted payoff as seen from institution A is

$$\sum_{i=1}^{\beta} D(0, T_i) \left[c_i - (\gamma F_i(T_{i-1}) + S)\tau_i\right]$$

and the value to A of the contract is the risk-neutral expectation

$$E\left\{\sum_{i=1}^{\beta} D(0, T_i) \left[c_i - (\gamma F_i(T_{i-1}) + S)\tau_i\right]\right\}$$
$$= P(0, T_\beta) \sum_{i=1}^{\beta} E^\beta \left[\frac{c_i - (\gamma F_i(T_{i-1}) + S)\tau_i}{P(T_i, T_\beta)}\right].$$

Since the forward-rate dynamics under Q^β of

$$F_{\beta(t)}(t), \ldots, F_\beta(t)$$

is known as from (13.4), a Monte Carlo pricing can be carried out in the usual manner.

13.7 Constant-Maturity Swaps (CMS)

13.7.1 CMS with the LFM

A constant-maturity swap is a financial product structured as follows. We assume a unit nominal amount. Let us denote by $\mathcal{T} = \{T_0, \ldots, T_n\}$ a set of payment dates at which coupons are to be paid. We assume, for simplicity, such dates to be one-year spaced.

- At time T_{i-1} (in some variants at time T_i), $i \geq 1$, institution A pays to B the c-year swap rate resetting at time T_{i-1}. Formally, at time T_{i-1} institution A pays to B

$$S_{i-1,i-1+c}(T_{i-1})\,\tau_i\,,$$

where, as usual,

$$S_{i-1,i-1+c}(t) = \frac{P(t,T_{i-1}) - P(t,T_{i-1+c})}{\sum_{k=i}^{i-1+c} \tau_k P(t,T_k)}\,. \tag{13.9}$$

- Institution B pays to A a fixed rate K.

The net value of the contract to B at time 0 is

$$\begin{aligned} &E\left(\sum_{i=1}^{n} D(0,T_{i-1})(S_{i-1,i-1+c}(T_{i-1}) - K)\tau_i\right) \\ &= \sum_{i=1}^{n} \tau_i P(0,T_{i-1})\left[E^{i-1}\left(S_{i-1,i-1+c}(T_{i-1})\right) - K\right] \\ &= \sum_{i=1}^{n} \tau_i \left(P(0,T_n)\,E^n\left(\frac{S_{i-1,i-1+c}(T_{i-1})}{P(T_{i-1},T_n)}\right) - KP(0,T_{i-1})\right) \end{aligned} \tag{13.10}$$

We need only compute either

$$E^{i-1}\left[S_{i-1,i-1+c}(T_{i-1})\right] \quad \text{or} \quad E^n[S_{i-1,i-1+c}(T_{i-1})/P(T_{i-1},T_n)] \tag{13.11}$$

for all i's. At first sight, one might think to discretize equation (6.39) for the dynamics of the forward swap rate and compute the required expectation through a Monte Carlo simulation. However, notice that forward rates appear in the drift m^α of such equation, so that we are forced to evolve forward rates anyway. As a consequence, we can use equation (13.9) jointly with Monte Carlo simulated forward-rate dynamics, and do away with the dynamics (6.39), thus directly recovering the swap rate $S_{i-1,i-1+c}(T_{i-1})$ from the T_{i-1} values of the (Monte Carlo generated) spanning forward rates

$$F_i(T_{i-1}), F_{i+1}(T_{i-1}), \ldots, F_{i-1+c}(T_{i-1}).$$

Analogously to the autocaps case, such forward rates can be generated according to the usual discretized (Milstein) dynamics (13.5) based on Gaussian shocks and under the unique measure Q^n.

Finally, an analytical approximation is possible if adopting the usual drift freezing procedure. Indeed, resort to the usual formula expressing swap rates as a weighted average of forward rates: $S_{\alpha,\beta}(T_\alpha) = \sum_{i=\alpha+1}^{\beta} w_i(T_\alpha)F_i(T_\alpha) \approx \sum_{i=\alpha+1}^{\beta} w_i(0)F_i(T_\alpha)$ and compute

$$E^\alpha S_{\alpha,\beta}(T_\alpha) \approx \sum_{j=\alpha+1}^{\beta} w_j(0) E^\alpha F_j(T_\alpha) \approx \sum_{j=\alpha+1}^{\beta} w_j(0) e^{\int_0^{T_\alpha} \bar{\mu}_{\alpha,j}(t)dt} F_j(0)$$

We have frozen again the drift in the dynamics of the F's under Q^α. By applying this formula to the case $\alpha = i-1$ and $\beta = i-1+c$ we obtain the first of the terms (13.11) (the second one can be obtained similarly) and have a very fast approximated formula for CMS in that case. This can be compared with classical market convexity adjustments (below). The two methods give similar results when volatilities are not too high.

13.7.2 CMS with the G2++ Model

It is possible to price a CMS with the G2++ model (4.4). See the related Section 14.2.2 on quanto CMS's in the next chapter. We do not repeat things here, since the CMS pricing procedure can be easily deduced from the procedure for the more general quanto-CMS case.

13.8 The Convexity Adjustment and Applications to CMS

13.8.1 Natural and Unnatural Time Lags

As with so many things, it was simply a matter of time.
The Time Trapper, Zero Hour – End of an Era, LSH 61, 1994, DC Comics

To appropriately introduce the convexity-adjustment technique, we quickly recall the pricing formulas for swaps. We begin by a plain-vanilla swap with natural time lag.

Consider an IRS that resets at dates $T_\alpha, T_{\alpha+1}, \dots, T_{\beta-1}$ and pays at dates $T_{\alpha+1}, \dots, T_\beta$, with unit notional amount. The fact that the payment indexed by the LIBOR rate resetting at time T_i for the maturity T_{i+1} occurs precisely at time T_{i+1} is referred to as a "natural time lag". This renders the swap price independent of the volatility of rates. Indeed, let us consider only the variable swap leg. The discounted value of this leg can be expressed either via the swap rate or via forward rates. In effect, the discounted payoff is given by

$$D(0,T_\alpha) S_{\alpha,\beta}(T_\alpha) \sum_{i=\alpha+1}^{\beta} \tau_i P(T_\alpha, T_i),$$

which is equivalent to

$$\sum_{i=\alpha+1}^{\beta} D(0,T_i) \tau_i F_i(T_{i-1}).$$

The value of such a leg is easily computed in both cases as

$$
\begin{aligned}
E\left[\sum_{i=\alpha+1}^{\beta} D(0,T_i)\tau_i F_i(T_{i-1})\right] &= \sum_{i=\alpha+1}^{\beta} P(0,T_i)\tau_i E^i\left[F_i(T_{i-1})\right] \\
= \sum_{i=\alpha+1}^{\beta} P(0,T_i)\tau_i F_i(0) &= \sum_{i=\alpha+1}^{\beta} \left[P(0,T_{i-1}) - P(0,T_i)\right] \\
&= P(0,T_\alpha) - P(0,T_\beta).
\end{aligned}
$$

From the last formula notice that, as is well known, neither volatility nor correlation of rates affect this financial product.

Now, let us reconsider in-arrears swaps. Consider the variable leg of an IRS that resets at dates $T_{\alpha+1},\ldots,T_\beta$ and pays *at the same dates*, with unit notional amount. We say this swap has an "unnatural time lag". This term is justified by seeing that the price of such a leg depends on volatility. Indeed, see formula (13.3) with $K = 0$.

Contrary to the plain-vanilla case, the in-arrears-swap price depends on the volatility of forward rates through their average volatilities v, which are usually deduced inverting cap prices through Black's formula.

The "natural/unnatural" terminology reflects the above calculations. A *natural time lag* for the variable leg of a swap makes the value of such a leg *independent of the rates volatility.* On the contrary, an *unnatural time lag* makes the value of the variable leg volatility dependent.

As a corollary, we can derive the corresponding formulas for forward-rate agreements. Suppose we are now at time 0, and at time T_2 the contract pays the LIBOR rate resetting at time $T_1 < T_2$ and maturing at T_2. As usual, we denote this rate by $L(T_1,T_2) = F_2(T_1)$ and we denote by τ the year fraction between T_1 and T_2. The contract value is therefore, consistently with the general FRA notation previously established,

$$
\begin{aligned}
-\mathbf{FRA}(0,T_1,T_2,0) &= E[D(0,T_2)\tau F_2(T_1)] \\
&= P(0,T_2)\tau E^2[F_2(T_1)] = P(0,T_2)\tau F_2(0) \\
&= P(0,T_1) - P(0,T_2).
\end{aligned}
$$

If we have an in-arrears FRA, this time the contract pays at time T_1 the LIBOR rate resetting at the same time $T_1 < T_2$ and maturing at T_2. By reasoning in an analogous way to the case of in-arrears swaps, we obtain

$$
\begin{aligned}
\mathbf{IAFRA} &= P(0,T_2)\left[1 + 2\tau F_2(0) + \tau^2 F_2^2(0)\exp(v_2^2(T_1))\right] - P(0,T_1) \\
&= P(0,T_1)\left[1 + \frac{\tau\ F_2(0) + \tau^2 F_2^2(0)\exp(v_2^2(T_1))}{1+\tau F_2(0)}\right] - P(0,T_1) \\
&= P(0,T_2)\tau F_2(0)\left(1 + \tau F_2(0)\exp(v_2^2(T_1))\right) \\
&\approx P(0,T_2)\tau F_2(0)\left(1 + \tau F_2(0) + \tau v_2^2(T_1)F_2(0)\right) \\
&= P(0,T_1)\tau F_2(0) + P(0,T_2)\tau^2 F_2^2(0)v_2^2(T_1) \qquad (13.12)
\end{aligned}
$$

13.8.2 The Convexity-Adjustment Technique

The time is out of joint. O cursed spite,
That ever I was born to set it right
Hamlet, I.5

The convexity-adjustment technique can be attempted any time there is an unnatural time lag. We consider its application to a single payment.

Assume a swap rate is involved, and that the payment $\tau_i S_{\alpha,\beta}(T_{i-k})$ is due at time T_i, $i - k \leq i \leq \alpha < \beta$.

Remark 13.8.1. **(CMS)** This is typical of constant-maturity swaps (CMS) where we have $i = \alpha$ and $k = 1$ or $k = 0$. This is the case where the convexity adjustment works well and is also supported by the output of more sophisticated models like, for instance, the G2++ model. If k is large, the correction can be quite wrong. Therefore, in such cases, the correction discussed here should be considered with care.

We are far from the "usual" IRS case, because the rate being exchanged at each payment instant is a *swap* rate rather than a *LIBOR* rate.

A first adjustment

The forward swap rate $S_{\alpha,\beta}$ is originally defined as related to an IRS that pays at times $\alpha+1, \ldots, \beta$: $S_{\alpha,\beta}(T)$ at time T, $T \leq T_\alpha$, is the fixed rate such that the fixed leg of the above IRS has value equal to that of the floating leg. In case of reimbursement of the notional amount, such a value at time T is always $P(T, T_\alpha)$ (see Definition 1.5.2 and the subsequent comments), so that we can write ("FL" stands for "floating leg")

$$\mathrm{FL}_{\alpha,\beta}(T) = P(T, T_\alpha) = S_{\alpha,\beta}(T) \sum_{i=\alpha+1}^{\beta} \tau_i P(T, T_i) + P(T, T_\beta) .$$

Now rewrite the same expression with the discount factors coming from a flat yield curve fixed at a level y (annually compounded) at time T_α, ("FFL" stands for Flat Floating Leg):

$$\mathrm{FFL}_{\alpha,\beta}(T; y) = S_{\alpha,\beta}(T) \sum_{i=\alpha+1}^{\beta} \tau_i \frac{P(T, T_\alpha)}{(1+y)^{\tau_{\alpha,i}}} + \frac{P(T, T_\alpha)}{(1+y)^{\tau_{\alpha,\beta}}} ,$$

where $\tau_{\alpha,i}$ denotes the year fraction between T_α and T_i.

If one allows for the first-order expansion

$$\delta S_{\alpha,\beta}(T) = (1 + S_{\alpha,\beta}(T))^\delta - 1 ,$$

and takes $T_i = i\delta$, $\tau_{\alpha,i} = (i - \alpha)\delta$ and $\tau_i = \delta$, it is easy to see that the above flat floating leg coincides with $P(T, T_\alpha)$ only for $y = S_{\alpha,\beta}(T)$,

$$\mathrm{FFL}_{\alpha,\beta}(T; S_{\alpha,\beta}(T)) = P(T, T_\alpha).$$

Therefore, the value of y around which the flat-curve approximation is to be considered is the forward swap rate $S_{\alpha,\beta}(T)$, in that it is the flat rate that agrees with the non-flat case as far as the price of the floating leg is concerned:

$$\mathrm{FFL}_{\alpha,\beta}(T; S_{\alpha,\beta}(T)) = \mathrm{FL}_{\alpha,\beta}(T).$$

Consider now the expectation

$$\begin{aligned} &E_0^T \left[P(0,T)\mathrm{FFL}_{\alpha,\beta}(T; S_{\alpha,\beta}(T)) - \mathrm{FFL}_{\alpha,\beta}(0; S_{\alpha,\beta}(0))\right] \\ &\quad = E_0^T \left[P(0,T)\frac{P(T,T_\alpha)}{P(T,T)} - P(0,T_\alpha)\right] = 0\,. \end{aligned} \tag{13.13}$$

At this point, we proceed by defining the following quantity $\Phi_{\alpha,\beta}(y)$ through a slight approximation of the argument of the above expectation, where we assume $P(0,T)P(T,T_\alpha) \approx P(0,T_\alpha)$:

$$P(0,T)\mathrm{FFL}_{\alpha,\beta}(T,y) \approx S_{\alpha,\beta}(T) \sum_{i=\alpha+1}^{\beta} \tau_i \frac{P(0,T_\alpha)}{(1+y)^{\tau_{\alpha,i}}} + \frac{P(0,T_\alpha)}{(1+y)^{\tau_{\alpha,\beta}}} =: \Phi_{\alpha,\beta}(T,y).$$

We expand Φ through a second-order Taylor expansion around $y = S_{\alpha,\beta}(0)$, we evaluate the resulting expression at $y = S_{\alpha,\beta}(T)$, we then solve for $S_{\alpha,\beta}(T) - S_{\alpha,\beta}(0)$ and introduce a further approximation:

$$\begin{aligned} S_{\alpha,\beta}(T) - S_{\alpha,\beta}(0) &\approx \frac{\Phi_{\alpha,\beta}(T, S_{\alpha,\beta}(T)) - \Phi_{\alpha,\beta}(T, S_{\alpha,\beta}(0))}{\Phi'_{\alpha,\beta}(T, S_{\alpha,\beta}(0))} \\ &\quad - \frac{(S_{\alpha,\beta}(T) - S_{\alpha,\beta}(0))^2}{2} \frac{\Phi''_{\alpha,\beta}(T, S_{\alpha,\beta}(0))}{\Phi'_{\alpha,\beta}(T, S_{\alpha,\beta}(0))} \\ &\approx \frac{\Phi_{\alpha,\beta}(T, S_{\alpha,\beta}(T)) - \Phi_{\alpha,\beta}(0, S_{\alpha,\beta}(0))}{\Phi'_{\alpha,\beta}(0, S_{\alpha,\beta}(0))} \\ &\quad - \frac{(S_{\alpha,\beta}(T) - S_{\alpha,\beta}(0))^2}{2} \frac{\Phi''_{\alpha,\beta}(0, S_{\alpha,\beta}(0))}{\Phi'_{\alpha,\beta}(0, S_{\alpha,\beta}(0))}, \end{aligned} \tag{13.14}$$

where the superscript $'$ denotes partial derivative with respect to y.

Now take expectation on both sides under the measure Q^T. The first term on the right-hand side has expectation zero, due to equation (13.13).

We further assume that we can approximate the true Q^T-dynamics of $S_{\alpha,\beta}$ by its lognormal $Q^{\alpha,\beta}$-dynamics (6.37), an approximation that has been shown to work well in most situations for the LFM when T is close to T_α (see the tests on the distributions of the swap rate described at the end of Section 8.3 and the related results in Section 8.4). We obtain:

$$\begin{aligned} E_0^T \left[(S_{\alpha,\beta}(T) - S_{\alpha,\beta}(0))^2\right] &\approx E_0^{\alpha,\beta} \left[(S_{\alpha,\beta}(T) - S_{\alpha,\beta}(0))^2\right] \\ &= S_{\alpha,\beta}(0)^2 (e^{v^2_{\alpha,\beta}(T)} - 1) \approx S^2_{\alpha,\beta}(0)\, v^2_{\alpha,\beta}(T), \end{aligned}$$

where

$$v_{\alpha,\beta}^2(T) = \int_0^T (\sigma^{(\alpha,\beta)}(t))^2 \, dt$$

is the average variance of the forward swap rate in the interval $[0, T]$ times the interval length. Now, we can evaluate (13.14) by taking expectation on both sides:

$$E_0^T [S_{\alpha,\beta}(T)] \approx S_{\alpha,\beta}(0) - \tfrac{1}{2} S_{\alpha,\beta}^2(0) v_{\alpha,\beta}^2(T) \frac{\Phi''_{\alpha,\beta}(0, S_{\alpha,\beta}(0))}{\Phi'_{\alpha,\beta}(0, S_{\alpha,\beta}(0))} \,. \quad (13.15)$$

A second adjustment

A second adjustment we can consider is based on neglecting the final reimbursement of the notional amount in the above IRS. We thus define Ψ as Φ without notional reimbursement,

$$\Psi_{\alpha,\beta}(y) := S_{\alpha,\beta}(T) \sum_{i=\alpha+1}^{\beta} \tau_i \frac{P(0, T_\alpha)}{(1+y)^{\tau_{\alpha,i}}} \,.$$

Assuming that also

$$E_0^T [\Psi_{\alpha,\beta}(S_{\alpha,\beta}(T)) - \Psi_{\alpha,\beta}(S_{\alpha,\beta}(0))] \approx 0,$$

as for Φ when taking expectations on both sides of (13.14), and using again a second-order expansion, it follows that

$$\boxed{E_0^T [S_{\alpha,\beta}(T)] \approx S_{\alpha,\beta}(0) - \tfrac{1}{2} S_{\alpha,\beta}^2(0) v_{\alpha,\beta}^2(T) \frac{\Psi''_{\alpha,\beta}(S_{\alpha,\beta}(0))}{\Psi'_{\alpha,\beta}(S_{\alpha,\beta}(0))}} \,, \quad (13.16)$$

where the ratio $\Psi''_{\alpha,\beta}(S_{\alpha,\beta}(0))/\Psi'_{\alpha,\beta}(S_{\alpha,\beta}(0))$ is independent of T.

This is the formula that is usually considered in the market for convexity adjustments (especially for CMS), see for example Hull (1997), in particular formula (16.13) and the related Example 16.8. The approximation works well when T is not too far away from T_α, as implied by the "Q^T vs $Q^{\alpha,\beta}$" dynamics approximation for the forward swap rate.

Let us now apply this formula to specific situations.

Floating leg with swap-rate-indexed payments

Suppose we need to compute the present value of our generic payment,

$$E[\tau_i D(0, T_i) S_{\alpha,\beta}(T_{i-k})] \,.$$

Move under the T_i-forward measure, to obtain

$$P(0, T_i)E^i[\tau_i S_{\alpha,\beta}(T_{i-k})].$$

The first rougher approximation is to treat the measure Q^i as if it were the swap measure $Q^{\alpha,\beta}$, under which S can be modeled through the lognormal martingale

$$dS_{\alpha,\beta}(t) = \sigma^{(\alpha,\beta)}(t) S_{\alpha,\beta}(t)\, dW_t .$$

Under this approximation, we would then have

$$E[\tau_i D(0, T_i) S_{\alpha,\beta}(T_{i-k})] \approx \tau_i P(0, T_i) S_{\alpha,\beta}(0).$$

The convexity adjustment (13.16) leads to the following modification of this last formula:

$$\begin{aligned}
&E[\tau_i D(0, T_i) S_{\alpha,\beta}(T_{i-k})] \\
&\quad = \tau_i P(0, T_i) E^i[S_{\alpha,\beta}(T_{i-k})] \approx \tau_i P(0, T_i) E^{i-k}[S_{\alpha,\beta}(T_{i-k})] \\
&\quad \approx \tau_i P(0, T_i) \left[S_{\alpha,\beta}(0) - \tfrac{1}{2} S^2_{\alpha,\beta}(0) v^2_{\alpha,\beta}(T_{i-k}) \frac{\Psi''_{\alpha,\beta}(S_{\alpha,\beta}(0))}{\Psi'_{\alpha,\beta}(S_{\alpha,\beta}(0))} \right].
\end{aligned}$$

As anticipated in Remark 13.8.1, this approximation turns out to work well only for small values of k. Therefore, if k is large, the correction should be considered with due care.

We now check that, in case of an in-arrears FRA, this formula is consistent with the value found earlier by exact evaluation. We take $\alpha = i = 1$, $k = 0$, $\beta = 2$ and $\tau_{1,2} = \tau$, so that $S_{\alpha,\beta}(t) = F(t; T_1, T_2)$ and

$$\Psi_{1,2}(y) = \frac{C}{(1 + \tau y)},$$

with C a suitable constant, and where we have used simple compounding instead of annual compounding. Notice that

$$\frac{\Psi''_{1,2}(y)}{\Psi'_{1,2}(y)} = \frac{-2\tau}{(1 + \tau y)},$$

so that the convexity-adjustment formula (13.16) yields

$$\tau P(0, T_1) \left[F_2(0) + \tau \frac{F_2^2(0) v_2^2(T_1)}{1 + \tau F_2(0)} \right]$$

$$= \tau P(0, T_1) F_2(0) + \tau^2 P(0, T_2) F_2^2(0) v_2^2(T_1),$$

which is the same result found, at first order in v_2^2, by exact evaluation in (13.12).

13.8.3 Deducing a Simple Lognormal Dynamics from the Adjustment

We can easily adjust the approximate driftless dynamics

$$dS_{\alpha,\beta}(t) = \sigma^{(\alpha,\beta)}(t) S_{\alpha,\beta}(t)\, dW_t\,,$$

for which

$$E_0^T\left[S_{\alpha,\beta}(T_{i-k})\right] = S_{\alpha,\beta}(0)\,,$$

to a new dynamics

$$dS_{\alpha,\beta}(t) = \mu^{\alpha,\beta} S_{\alpha,\beta}(t)\, dt + \sigma^{(\alpha,\beta)}(t) S_{\alpha,\beta}(t)\, dW_t\,, \tag{13.17}$$

for which

$$E_0^T\left[S_{\alpha,\beta}(T_{i-k})\right] = S_{\alpha,\beta}(0) - \tfrac{1}{2} S^2_{\alpha,\beta}(0) v^2_{\alpha,\beta}(T_{i-k}) \frac{\Psi''_{\alpha,\beta}(S_{\alpha,\beta}(0))}{\Psi'_{\alpha,\beta}(S_{\alpha,\beta}(0))}\,,$$

consistently with the convexity-adjustment evaluation. Since the dynamics (13.17) produces

$$E_0^T\left[S_{\alpha,\beta}(T_{i-k})\right] = S_{\alpha,\beta}(0)\exp(\mu^{\alpha,\beta} T_{i-k}) \approx S_{\alpha,\beta}(0)(1 + \mu^{\alpha,\beta} T_{i-k}),$$

at first order in $\mu^{\alpha,\beta} T_{i-k}$, it suffices to set

$$\mu^{\alpha,\beta} = -\tfrac{1}{2} S_{\alpha,\beta}(0) \frac{v^2_{\alpha,\beta}(T_{i-k})}{T_{i-k}} \frac{\Psi''_{\alpha,\beta}(S_{\alpha,\beta}(0))}{\Psi'_{\alpha,\beta}(S_{\alpha,\beta}(0))}.$$

Notice that in case the instantaneous forward-swap-rate volatility $\sigma^{\alpha,\beta}$ is assumed to be constant, we have

$$\frac{v^2_{\alpha,\beta}(T_{i-k})}{T_{i-k}} = (\sigma^{\alpha,\beta})^2.$$

This approximation is however rather rough and should not be used to evaluate nonlinear payoffs, unless a considerable amount of testing has been performed and acceptable errors are found.

13.8.4 Application to CMS

We have seen before that a constant-maturity swap has a floating leg that pays at times $T_{\alpha+1}, \ldots, T_\beta$ the swap rates

$$S_{\alpha,\alpha+c}(T_\alpha), S_{\alpha+1,\alpha+1+c}(T_{\alpha+1}), \ldots, S_{\beta-1,\beta-1+c}(T_{\beta-1})\,.$$

Therefore, at each payment instant $T_{\alpha+k+1}$, such leg pays a certain prespecified c-year swap rate resetting at the previous instant $T_{\alpha+k}$. In some variants, instead, it pays at $T_{\alpha+k+1}$ a certain pre-specified swap rate resetting at the same instant. We will consider here the first version.

The value of the generic CMS payment is given by

$$E\left[D(0,T_{i+1})\tau_{i+1}S_{i,i+c}(T_i)\right] = \tau_{i+1}P(0,T_{i+1})E^{i+1}S_{i,i+c}(T_i)$$
$$\approx \tau_{i+1}P(0,T_{i+1})\left[S_{i,i+c}(0) - \tfrac{1}{2}S^2_{i,i+c}(0)v^2_{i,i+c}(T_i)\frac{\Psi''_{i,i+c}(S_{i,i+c}(0))}{\Psi'_{i,i+c}(S_{i,i+c}(0))}\right],$$

see also Example 16.8 in Hull (1997). The CMS price is then obtained by adding terms for i ranging from α to $\beta-1$. Recall that the adjustment used here has been derived under a number of approximations. As such, it can be improved. Indeed, the classical adjustment has been found to be not completely satisfactory by some traders, especially in some market situations involving volatility smiles. For other works on CMS adjustments see for example Pugachevsky (2001).

13.8.5 Forward Rate Resetting Unnaturally and Average-Rate Swaps

We consider now the following problem, which can have several applications. Consider two time instants s,u and a payment date T, $s < u < T$. Assume we have a contract that pays at time T the spot LIBOR rate resetting at time s for the maturity u:

$$L(s,u) = F(s;s,u).$$

In case $T = u$ we have a natural time lag. Indeed, the contract value at time 0 is the risk-neutral expectation of the discounted payoff

$$E_0[D(0,T)F(s;s,T)] = P(0,T)E_0^T[F(s;s,T)] = P(0,T)F(0;s,T)$$

and does not depend on volatility specifications.

If $T > u$, the above formula no longer holds. However, we can still evaluate the contract as follows.

Consider the no-arbitrage forward-rate dynamics for $F(t) = F(t;s,u)$ under the T-forward-adjusted measure Q^T:

$$dF(t) = -\sigma_{s,u}\sigma_{u,T}\tau(u,T)F(t)\frac{F(t;u,T)}{1+\tau(u,T)F(t;u,T)}\,dt + \sigma_{s,u}F(t)\,dW_t^T,$$

where $\sigma_{s,u}$ is the instantaneous volatility of $F(t) = F(t;s,u)$ and $\sigma_{u,T}$ is the instantaneous volatility of $F(t;u,T)$, and both are assumed to be constant (otherwise they can be replaced with the square roots of the average variances of $F(t;s,u)$ and $F(t;u,T)$, respectively, over $[0,\ s]$). The quantity $\tau(a,b)$ denotes in general the time between dates a and b in years.

We assume unit correlation between $F(t; s, u)$ and $F(t; u, T)$, since usually T and u are close. If this is not the case, a ρ parameter can be included in the drift of the above process.

With the usual deterministic-percentage-drift approximation we can write

$$dF(t) = -\sigma_{s,u}\sigma_{u,T}\tau(u,T)F(t)\frac{F(0;u,T)}{1+\tau(u,T)F(0;u,T)}\,dt + \sigma_{s,u}F(t)\,dW_t^T.$$

This new process has lognormal distribution under the T-forward measure and it can be easily seen that its expected value, conditional on the information available at time 0, under the T-forward measure, is

$$E_0^T[F(s;s,u)] = F(0;s,u)\exp\left(-\tau(u,T)\sigma_{s,u}\sigma_{u,T}\,s\,\frac{F(0;u,T)}{1+\tau(u,T)F(0;u,T)}\right).$$

We are now able to price the discounted payoff

$$E_0[D(0,T)L(s,u)] = P(0,T)E_0^T[F(s;s,u)]$$

$$= P(0,T)F(0;s,u)\exp\left(-\tau(u,T)\sigma_{s,u}\sigma_{u,T}\,s\,\frac{F(0;u,T)}{1+\tau(u,T)F(0;u,T)}\right).$$

As an example, consider a contract that pays at a future time T the average value of the 3-month ($3m$) LIBOR rates in the days $t_1 < t_2 < \ldots < t_n$, $t_n < T$, with δ_i denoting the year fraction between t_i and $t_i + 3m$. This is a possible example of a leg of an average-rate swap.

If the notional is N, the contract price is

$$E_0\left[D(0,T)\frac{\sum_{i=1}^n \delta_i N L(t_i, t_i+3m)}{n}\right]$$

$$= \frac{P(0,T)}{n}N\sum_{i=1}^n \delta_i E_0^T\left[F(t_i;t_i,t_i+3m)\right]$$

and is given by

$$\frac{P(0,T)}{n}N\sum_{i=1}^n \delta_i F(0;t_i,t_i+3m)$$

$$\cdot\exp\left[-\sigma_{t_i,t_i+3m}\sigma_{t_i+3m,T}\,t_i\,\frac{\tau(t_i+3m,T)F(0;t_i+3m,T)}{1+\tau(t_i+3m,T)F(0;t_i+3m,T)}\right].$$

Notice that the correction to the "brute-force" formula

$$\frac{P(0,T)}{n}N\sum_{i=1}^n \delta_i F(0;t_i,t_i+3m)$$

is multiplicative for each term and is given by the exponentials. The correction effect is to (slightly) reduce the "brute-force" value, since the exponents are negative. The correction might be not negligible for large values of the volatilities.

The difficulty in applying the above formula lies in the fact that the forward rate $F(0; t_i + 3m, T)$ can be rather atypical as for expiry or maturity dates. Therefore, apart from few exceptions, its volatility cannot be recovered exactly from market cap prices. However, a synthetic volatility deduced from volatilities of "smaller" forward rates nested in $F(0; t_i + 3m, T)$ can be used for this purpose, or arguments similar to those of Section 6.20 can be employed.

At a first stage, the above formula can be used to have a feeling on the order of magnitude of the adjustment due to second-order effects, and to decide whether these should be taken into account or not.

13.9 Average Rate Caps

An average-rate cap is a stream of average-rate caplets. An average rate caplet is a contract that pays the average of a given LIBOR rate over a pre-defined period, capped at a given strike K.

Assume that the LIBOR rate is a six month LIBOR rate, detected at dates $t_1, \ldots, t_j, \ldots, t_{n_i}$, all such dates preceding time T_i. Assume moreover that the payoff is to be valued at time T_{i+1}. Typically the t_j's are roughly thirty and range over one month in the semester preceding T_i, with a one-day time step. The T_i's are semiannually spaced.

The discounted payoff of the average-rate caplet at time T_{i+1} can be written formally as

$$D(0, T_{i+1})\tau_{i+1} \left(\frac{\sum_{j=1}^{n_i} L(t_j, t_j + 6m)}{n_i} - K \right)^+ .$$

The first attempt at valuing this product may be a Monte Carlo valuation involving interpolation, either of rates directly or of their dynamics, similarly to what we have seen in average-rate swaps in Section 13.8.5. The problem in this approach is that, if we apply it naively, we need simulating a lot of in-between rates that are not in the canonical family of our LIBOR model. This can be done along the lines of what we will describe later on in Section 13.13.

Notice that, given the high correlation that is expected between one-month adjacent forward rates, it may be pointless to simulate jointly all thirty forward rates in a month concurring to the average rate. A subset of such rates can be simulated and the average of the rates in this subset may suffice in standard market situations for practical purposes. Numerical investigations for assessing the size of this subset may be needed.

Here, however, we present an analytical approximation that is deduced from basket options. The method is particularly fortunate here because semi-annual forward LIBOR rates are quite homogeneous "assets" and are likely to have high correlation when taken within a one-month period.

Recall the dynamics of $F(t; u, u + \delta)$, where we set $\delta = 6m$. Since $F(\cdot; u, u + \delta)$ is outside our fundamental family of forward rates, we use the drift-interpolation technique of Section 6.21.1. Let $F_u(t) := F(t; u, u+\delta)$ and use (6.78) plus (6.16) (the second to move from Z^i to Z^{i+1}) to obtain, through partially frozen coefficients and a few rearrangements ($T_i \leq u + \delta < T_{i+1}$):

$$\begin{aligned} dF_u(t) &= \mu_u(t)F_u(t)\,dt + \sigma_u(t)F_u(t)\,dZ^{i+1}(t), \\ \mu_u(t) &:= -\frac{(T_{i+1} - (u + \delta))F_{i+1}(0)}{1 + \tau_{i+1}F_{i+1}(0)}\sigma_u(t)\rho(u, T_{i,i+1})\sigma_{i+1}(t) \;. \end{aligned} \quad (13.18)$$

The quantity $\sigma_u(t) := \sigma(t; u, u + \delta)$ is the instantaneous percentage volatility of the related forward rate, and is usually obtained by some kind of interpolation from the "standard rates" volatilities σ_k's. The quantity $\rho(u, T_{i,i+1})$ is the instantaneous correlation between the forward LIBOR rate F_u and F_{i+1}, and is typically assumed to be equal to $\rho_{i,i+1}$.

We may consider the mean and second moment of the average LIBOR rate associated with the above dynamics (13.18) under the T_{i+1} forward measure:

$$\begin{aligned} &E^{i+1}\sum_{j=1}^{n_i} L(t_j, t_j + 6m)/n_i = \\ &= \sum_{j=1}^{n_i} E^{i+1}F(t_j; t_j, t_j + 6m)/n_i \\ &= \sum_{j=1}^{n_i} F_{t_j}(0)\exp\left(\int_0^{t_j} \mu_{t_j}(s)ds\right)/n_i =: \boxed{M_i}, \end{aligned}$$

$$\begin{aligned} &E^{i+1}\left[\left(\sum_{j=1}^{n_i} L(t_j, t_j + 6m)/n_i\right)^2\right] = \\ &= \sum_{j,h=1}^{n_i} E^{i+1}(L(t_j, t_j + 6m), L(t_h, t_h + 6m))/(n_i)^2 \\ &= \sum_{j,h=1}^{n_i} F_{t_j}(0)F_{t_h}(0)\exp\left(\int_0^{t_j} \mu_{t_j}(s)ds + \int_0^{t_h} \mu_{t_h}(s)ds + \right. \\ &\qquad \left. +\rho(t_j, t_h)\int_0^{\min(t_j,t_h)} \sigma_{t_j}(s)\sigma_{t_h}(s)ds\right)/(n_i)^2 =: \boxed{(V_i)^2} \end{aligned}$$

where $\rho(t_j, t_h)$ is set to one when $j = h$ and is set to a common correlation value $\bar{\rho}$ for $j \neq h$. One may take $\bar{\rho} = \rho_{i,i+1}$, typically equal or quite close to one, or else one may choose some functional forms for this quantity.

The approximate average rate caplet price is then, assuming the average to be lognormally distributed with the above mean and variance, given by a Black-Scholes formula. Recall that if X_t is an underlying geometric Brownian motion with "risk free rate" set to zero, with a time-varying dividend yield responsible for a nonzero drift in the dynamics, and with a time-varying volatility, the Black-Scholes formula for the payoff $(X_T - K)^+$ can be written directly in terms of the ("risk neutral") mean and variance of the terminal X_T as

$$E_0(X_T)\Phi(d_1(K, E_0(X_T), (\ln(E_0(X_T^2)/E_0(X_T)^2))^{1/2})) - K\Phi(d_2),$$

see also Appendix D, so that we obtain, in our case,

$$\mathbf{ARCpl}(0, [t_1, t_{n_i}], T_i, K) = P(0, T_{i+1})\tau_{i+1}\mathrm{Bl}\left(K, M_i, (\ln((V_i/M_i)^2))^{1/2}\right).$$

A final remark is in order in case the contract is re-valued some time after the beginning, say at time t_a, with $t_1 < t_a < t_{n_i}$. In this case part of the average is already known and the discounted payoff becomes

$$D(t_a, T_{i+1})\tau_{i+1}\left(\frac{\sum_{j=a+1}^{n_i} L(t_j, t_j + 6m)}{n_i} - \left[K - \frac{\sum_{j=1}^{a} L(t_j, t_j + 6m)}{n_i}\right]\right)^+.$$

In this case one has to be careful and check that the realized part of the mean does not offset completely the strike K: indeed, if the quantity inside square brackets is negative, the payoff loses its optionality and becomes analogous to an average rate swap, and as such is to be valued.

13.10 Captions and Floortions

A caption is an option that gives its holder the right to enter at a future time T_γ a cap whose first caplet resets at date $T_\alpha \geq T_\gamma$ and whose subsequent caplets reset at times $T_{\alpha+1}, \ldots, T_{\beta-1}$ with T_β the last payment date. The strike rate for this cap will be denoted by K. The price the holder of the caption will pay for this future cap is fixed as the caption strike and will be denoted by X. We can therefore express the caption payoff as a call payoff on the underlying cap.

We assume a unit notional amount. The T_γ value of the underlying cap described above is given by the usual Black formula (see for instance Section 6.4.3), computed at time T_γ instead of time 0,

$$\sum_{i=\alpha+1}^{\beta} \tau_i P(T_\gamma, T_i)\,\mathrm{Bl}(K, F_i(T_\gamma), \sqrt{T_{i-1} - T_\gamma}\; V(T_\gamma, T_{i-1})),$$

where the average volatility $V(\cdot, \cdot)$ was defined in Section 6.5.

The caption discounted payoff, expressed as a call payoff, can be written as

$$D(0,T_\gamma)\left\{\sum_{i=\alpha+1}^{\beta}\tau_i P(T_\gamma,T_i)\,\mathrm{Bl}(K,F_i(T_\gamma),\sqrt{T_{i-1}-T_\gamma}\;V(T_\gamma,T_{i-1}))-X\right\}^+.$$

The caption value is given by the risk-neutral expectation of this payoff, which in turn is given by

$$P(0,T_\gamma)E^\gamma\left\{\sum_{i=\alpha+1}^{\beta}\tau_i P(T_\gamma,T_i)\,\mathrm{Bl}(K,F_i(T_\gamma),\sqrt{T_{i-1}-T_\gamma}\,V(T_\gamma,T_{i-1}))-X\right\}^+.$$

Once again, the expected value can be computed through a Monte Carlo method, given the simulated values of

$$F_{\gamma+1}(T_\gamma),F_{\gamma+2}(T_\gamma),\ldots,F_\beta(T_\gamma)$$

under Q^γ, obtained through the usual discretized Milstein dynamics (13.5).

13.11 Zero-Coupon Swaptions

In this section we introduce zero-coupon swaptions and explain an approximated analytical method to price them. A payer (receiver) zero-coupon swaption is a contract giving the right to enter a payer (receiver) zero-coupon IRS at a future time. A zero-coupon IRS is an IRS where a single fixed payment is due at the unique (final) payment date T_β for the fixed leg in exchange for a stream of usual floating payments $\tau_i L(T_{i-1},T_i)$ at times T_i in $T_{\alpha+1},T_{\alpha+2},\ldots,T_\beta$ (usual floating leg). In formulas, the discounted payoff of a payer zero-coupon IRS is, at time $t\le T_\alpha$:

$$D(t,T_\alpha)\left[\sum_{i=\alpha+1}^{\beta}P(T_\alpha,T_i)\tau_i F_i(T_\alpha)-P(T_\alpha,T_\beta)\tau_{\alpha,\beta}K\right],$$

where $\tau_{\alpha,\beta}$ is the year fraction between T_α and T_β. The analogous payoff for a receiver zero-coupon IRS is obviously given by the opposite quantity.

Taking risk-neutral expectation, we obtain easily the contract value as

$$P(t,T_\alpha)-P(t,T_\beta)-\tau_{\alpha,\beta}KP(t,T_\beta),$$

which is the typical value of a floating leg minus the value of a fixed leg with a single final payment.

The value of the strike rate K that renders the contract fair is obtained by equating to zero the above value and solving in K. One obtains $K=F(t;T_\alpha,T_\beta)$. Indeed, we could have reasoned as follows. The value of the swap is independent of the number of payments on the floating leg, since the floating leg always values at par, no matter the number of payments (see Section 1.5.2 and the related remarks). Therefore, we might as well have taken a floating leg paying only in T_β the amount $\tau_{\alpha,\beta}L(T_\alpha, T_\beta)$. This would have given us again a standard swaption, standard in the sense that the two legs of the underlying IRS have the same payment dates (collapsing to T_β) and the unique reset date T_α. In such a one-payment case, the swap rate collapses to a forward rate, so that we should not be surprised to find out that the forward swap rate in this particular case is simply a forward rate.

An option to enter a payer zero-coupon IRS is a payer zero-coupon swaption, and the related payoff is

$$D(t, T_\alpha) \left[\sum_{i=\alpha+1}^{\beta} P(T_\alpha, T_i)\tau_i F_i(T_\alpha) - P(T_\alpha, T_\beta)\tau_{\alpha,\beta} K \right]^+ ,$$

or, equivalently, by expressing the F's in terms of discount factors,

$$D(t, T_\alpha) \left[1 - P(T_\alpha, T_\beta) - P(T_\alpha, T_\beta)\tau_{\alpha,\beta} K\right]^+ ,$$

which in turn can be written as

$$D(t, T_\alpha)\tau_{\alpha,\beta} P(T_\alpha, T_\beta) \left[F(T_\alpha; T_\alpha, T_\beta) - K\right]^+ .$$

Notice that, from the point of view of the payoff structure, this is merely a caplet. As such, it can be priced easily through Black's formula for caplets. The problem, however, is that such a formula requires the integrated percentage volatility of the forward rate $F(\cdot; T_\alpha, T_\beta)$, which is a forward rate over a non-standard period. Indeed, $F(\cdot; T_\alpha, T_\beta)$ is not in our usual family of spanning forward rates, unless we are in the trivial case $\beta = \alpha+1$. Therefore, since the market provides us (through standard caps and swaptions) with volatility data for standard forward rates, we need a formula for deriving the integrated percentage volatility of the forward rate $F(\cdot; T_\alpha, T_\beta)$ from volatility data of the standard forward rates $F_{\alpha+1}, \ldots, F_\beta$. The reasoning is once again based on the "freezing the drift" procedure, leading to an approximately lognormal dynamics for our standard forward rates.

Denote for simplicity $F(t) := F(t; T_\alpha, T_\beta)$ and $\tau := \tau_{\alpha,\beta}$.

We begin by noticing that, through straightforward algebra, we have (write everything in terms of discount factors to check)

$$1 + \tau F(t) = \prod_{j=\alpha+1}^{\beta} (1 + \tau_j F_j(t)).$$

It follows that

$$\ln(1 + \tau F(t)) = \sum_{j=\alpha+1}^{\beta} \ln(1 + \tau_j F_j(t)),$$

so that

$$d\ln(1+\tau F(t)) = \sum_{j=\alpha+1}^{\beta} d\ln(1+\tau_j F_j(t)) = \sum_{j=\alpha+1}^{\beta} \frac{\tau_j dF_j(t)}{1+\tau_j F_j(t)} + (\ldots)dt.$$

Now, since

$$dF(t) = \frac{1+\tau F(t)}{\tau} d\ln(1+\tau F(t)) + (\ldots)dt,$$

we obtain from the above expression

$$dF(t) = \frac{1+\tau F(t)}{\tau} \sum_{j=\alpha+1}^{\beta} \frac{\tau_j dF_j(t)}{1+\tau_j F_j(t)} + (\ldots)dt.$$

Take variance (conditional on the information up to time t) on both sides:

$$\mathrm{Var}\left(\frac{dF(t)}{F(t)}\right) = \left[\frac{1+\tau F(t)}{\tau F(t)}\right]^2 \sum_{i,j=\alpha+1}^{\beta} \frac{\tau_i \tau_j \rho_{i,j} \sigma_i(t) \sigma_j(t) F_i(t) F_j(t)}{(1+\tau_i F_i(t))(1+\tau_j F_j(t))} dt.$$

Now freeze all $t's$ to zero except for the σ's, and integrate over $[0, T_\alpha]$:

$$(v_{\alpha,\beta}^{zc})^2 := \left[\frac{1+\tau F(0)}{\tau F(0)}\right]^2 \sum_{i,j=\alpha+1}^{\beta} \frac{\tau_i \tau_j \rho_{i,j} F_i(0) F_j(0)}{(1+\tau_i F_i(0))(1+\tau_j F_j(0))} \int_0^{T_\alpha} \sigma_i(t)\sigma_j(t)dt.$$

To price the zero-coupon swaption it is then enough to put this quantity into the related Black's formula:

$$\mathbf{ZCPS} = \tau P(0, T_\beta)\mathrm{Bl}(K, F(0), v_{\alpha,\beta}^{zc}).$$

We can check the accuracy of this formula against the usual Monte Carlo pricing based on the exact dynamics of the forward rates. In the tests all swaptions are at-the-money. We have done this under the data of case (3.a) of the volatility tests of Section 8.3, and in other situations. Under the data of Section 8.3, we considered first the case $T_\alpha = 2y$, $T_\beta = 19y$. We obtained the implied volatility $v_{\alpha,\beta}^{zcMC}/\sqrt{T_\alpha}$ by inverting the Monte Carlo price through Black's formula:

$$\mathbf{MCZCPS} = \tau P(0, T_\beta)\mathrm{Bl}(F(0), F(0), v_{\alpha,\beta}^{zcMC}).$$

We found, in this case:

$$\frac{v_{\alpha,\beta}^{zcMC}}{\sqrt{T_\alpha}} = 0.1410, \quad \frac{v_{\alpha,\beta}^{zc}}{\sqrt{T_\alpha}} = 0.1455.$$

A two-side 98% window for the Monte Carlo volatility defined as in Section 8.3 is in this case $[0.1404 \quad 0.1416]$. Our algebraic approximation falls out of the 98% window, but of a small amount if compared with the distance from the volatility of the corresponding plain-vanilla European swaption. In fact, the standard at-the-money plain-vanilla swaption with the same initial reset date and final payment date, whose algebraic approximation has been found to be accurate in Section 8.3, has volatility

$$\frac{v_{\alpha,\beta}^{\text{LFM}}}{\sqrt{T_\alpha}} = 0.0997.$$

We have also considered the case $T_\alpha = 10y$, $T_\beta = 19y$. We obtained

$$\frac{v_{\alpha,\beta}^{zcMC}}{\sqrt{T_\alpha}} = 0.1081, \quad \frac{v_{\alpha,\beta}^{zc}}{\sqrt{T_\alpha}} = 0.1114.$$

Now a two-side 98% window for the Monte Carlo volatility defined as in Section 8.3 is [0.1076 0.1086]. Again, our algebraic approximation falls out of the 98% window of a small amount when compared with the discrepancy with respect to the corresponding standard swaption, resulting in a volatility

$$\frac{v_{\alpha,\beta}^{\text{LFM}}}{\sqrt{T_\alpha}} = 0.0897.$$

In the two examples above we notice that the at-the-money standard swaption has always a lower volatility (and hence price) than the corresponding at-the-money zero-coupon swaption. We may wonder whether this is a general feature. Indeed, we have the following.

Remark 13.11.1. **(Comparison between zero-coupon swaptions and corresponding standard swaptions).** A first remark is due for a comparison between the zero-coupon swaption volatility $v_{\alpha,\beta}^{zc}$ and the corresponding European-swaption approximation $v_{\alpha,\beta}^{\text{LFM}}$. If we rewrite the latter as

$$(v_{\alpha,\beta}^{\text{LFM}})^2 = \sum_{i,j=\alpha+1}^{\beta} \rho_{i,j}\lambda_i\lambda_j \int_0^{T_\alpha} \sigma_i(t)\sigma_j(t)dt, \quad \lambda_i = \frac{w_i(0)F_i(0)}{S_{\alpha,\beta}(0)},$$

it is easy to check that

$$(v_{\alpha,\beta}^{zc})^2 = \sum_{i,j=\alpha+1}^{\beta} \rho_{i,j}\mu_i\mu_j \int_0^{T_\alpha} \sigma_i(t)\sigma_j(t)dt,$$

where

$$\mu_i = \frac{P(0,T_\alpha)}{P(0,T_{i-1})}\lambda_i \geq \lambda_i,$$

the discrepancy increasing with the payment index i. It follows that, for positive correlations, the zero-coupon swaption volatility is always larger than the corresponding plain vanilla swaption volatility, the difference increasing with the tenor $T_\beta - T_\alpha$, for each given T_α.

A final remark concerns the possibility to price zero-coupon swaptions with other models.

Remark 13.11.2. **(Pricing zero-coupon swaptions with other models).** Zero-coupon swaptions can be priced analytically under all short-rate models admitting explicit formulas for European options on zero-coupon bonds and, accordingly, for caplets. For instance, under the CIR++ model (3.76) we can use formula (3.79), whereas under the G2++ model (4.4) we can resort to formula (4.28).

13.12 Eurodollar Futures

A Eurodollar-futures contract gives its owner the payoff

$$X\,(1 - L(S_1, S_2))$$

at the future time $S_1 < S_2$, where X is a notional amount, and the year fraction between S_1 and S_2 is denoted by τ. The fair price of this contract at time t is

$$\begin{aligned} V_t = E_t[X\,(1 - L(S_1, S_2))] &= X\,(1 - E_t[L(S_1, S_2)]) \\ &= X\left(1 + \frac{1}{\tau} - \frac{1}{\tau} E_t\left[\frac{1}{P(S_1, S_2)}\right]\right), \end{aligned} \tag{13.19}$$

and takes into account continuous rebalancing (see for example Sandmann and Sondermann (1997) and their reference to the related work of Cox Ingersoll and Ross).

The problem is computing the expectation

$$E_t\left[\frac{1}{P(S_1, S_2)}\right] = E_t\left[\frac{P(S_1, S_1)}{P(S_1, S_2)}\right].$$

If we were under the S_2-forward-adjusted measure this would be simply

$$\frac{P(0, S_1)}{P(0, S_2)} = 1 + \tau F(0; S_1, S_2)\,,$$

and the price would reduce to

$$X(1 - F(0; S_1, S_2)).$$

Instead, we need the expectation under the risk-neutral measure.

Since we need to compute

$$E_t[L(S_1, S_2)] = E_t[F(S_1; S_1, S_2)],$$

the result will depend on the interest-rate model we are using.

13.12.1 The Shifted Two-Factor Vasicek G2++ Model

We can use the two-additive-factor Gaussian model described in Chapter 4. Consistently with the notation adopted there, recall that

$$P(t,T) = \frac{P^M(0,T)}{P^M(0,t)} \exp\{\mathcal{A}(t,T)\},$$

$$\mathcal{A}(t,T) = \frac{1}{2}[V(t,T) - V(0,T) + V(0,t)] - \frac{1-e^{-a(T-t)}}{a} x(t) - \frac{1-e^{-b(T-t)}}{b} y(t),$$

with V defined as in(4.10), so that

$$E_t\left[\frac{1}{P(T_1,T_2)}\right] = \frac{P^M(0,T_1)}{P^M(0,T_2)} E_t\left[\exp\{-\mathcal{A}(T_1,T_2)\}\right]$$

$$= \frac{P^M(0,T_1)}{P^M(0,T_2)} \exp\left\{-\frac{1}{2}[V(T_1,T_2) - V(0,T_2) + V(0,T_1)] + \frac{1-e^{-a(T_2-T_1)}}{a} x(t) e^{-a(T_1-t)} + \frac{1-e^{-b(T_2-T_1)}}{b} y(t) e^{-b(T_1-t)}\right.$$

$$+ \left(\frac{1-e^{-a(T_2-T_1)}}{a}\right)^2 \frac{\sigma^2}{4a}\left[1-e^{-2a(T_1-t)}\right] + \left(\frac{1-e^{-b(T_2-T_1)}}{b}\right)^2 \frac{\eta^2}{4b}\left[1-e^{-2b(T_1-t)}\right]$$

$$\left. + \frac{(1-e^{-a(T_2-T_1)})(1-e^{-b(T_2-T_1)})}{ab} \cdot \rho \frac{\sigma\eta}{a+b}\left[1-e^{-(a+b)(T_1-t)}\right]\right\}.$$

By substituting this algebraic formula in (13.19) one has the value of the Eurodollar-futures contract. Notice that if this is evaluated at time 0, since $x_0 = y_0 = t = 0$ the above formula simplifies a little. Typically $X = 100$ and $\tau = 0.25$.

We have then considered a set of parameters coming from a typical calibration of the G2++ model to swaptions volatilities and to the zero-coupon curve of the Euro market. The values of these parameters are: $a = 0.0234$; $b = 0.0015$; $\sigma = 0.0081429$; $\eta = 0.0020949$; $\rho = -0.2536$.

We have finally computed prices for increasing maturities T_1 (from three months to ten years), while keeping $T_2 = T_1 + 0.25$, and we considered the differences

$$\text{Spread}(T_1) := E_0[L(T_1, T_1 + 0.25)] - F(0; T_1, T_1 + 0.25)$$

as T_1 increases. Such differences, in basis points (hundredths of a percentage point), are shown in Figure 13.1 below.

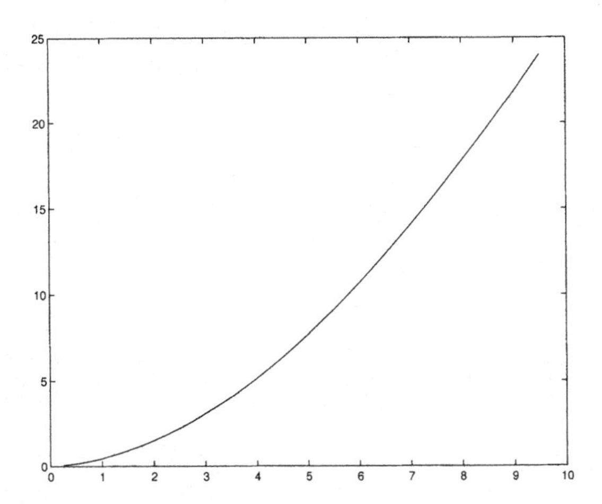

Fig. 13.1. Spread(T_1) in basis points plotted against T_1

The (upward concave) qualitative behaviour of the correction agrees with what is usually experienced in the market.

13.12.2 Eurodollar Futures with the LFM

Since we need to compute

$$E_t[L(S_1, S_2)] = E_t[F(S_1; S_1, S_2)],$$

we need the dynamics of the forward rate $F(\cdot; S_1, S_2)$ under the risk-neutral measure. This can be obtained starting from the martingale dynamics under the numeraire $P(\cdot, S_2)$ and moving to the bank-account numeraire via the "change of numeraire toolkit". This new dynamics involves the bond-price dynamics of $P(\cdot, S_2)$, which in turn can be expressed in terms of spanning forward rates. Therefore, in the relevant dynamics, correlation and volatilities of all spanning forward rates are involved. Subsequently, the forward-rates dynamics need be discretized and a Monte Carlo method can be applied to compute the relevant expectation under the risk-neutral measure.

In detail, assume we have a set of expiry/maturity dates $\{T_0, T_1, \ldots, T_M\}$ for a family of spanning forward rates, with $T_{M-1} = S_1$ and $T_M = S_2$. As we explained in Section 6.3, the forward-rate dynamics under the risk-neutral measure is given by (6.17).

Here, we would need to model the instantaneous forward rate f in the initial interval $(t, T_{\beta(t)-1}]$ to close the equations, but if we discretize these equations (for the logarithm of F's) with a Milstein scheme *exactly at the time instants* $\{T_0, T_1, \ldots, T_M\}$, we are in no need to model f. One sees easily that this is the same as discretizing the LFM dynamics (6.18) under the spot LIBOR measure whose numeraire is the discretely-rebalanced bank account. As usual, a Monte Carlo method, based on the jointly Gaussian distributions of the shocks for different components, can be applied to propagate all $F's$ up to time $T_{M-1} = S_1$ in order to evaluate the final expectation

$$E_0[F_M(T_{M-1})] = E_0[F(S_1; S_1, S_2)] .$$

Again, we can freeze part of the drift in the Spot-LIBOR-measure dynamics (6.18) thus obtaining

$$dF_k(t) = \sigma_k(t)\, F_k(t) \sum_{j=\beta(t)}^{k} \frac{\tau_j\, \rho_{j,k}\, \sigma_j(t)\, F_j(0)}{1 + \tau_j F_j(0)}\, dt + \sigma_k(t)\, F_k(t)\, dZ_k^d(t) ,$$

which is a geometric Brownian motion. Under this dynamics, the above expected value is easily computed in terms of the now deterministic percentage drift.

13.13 LFM Pricing with "In-Between" Spot Rates

Let us assume the current time to be $t = 0$, and let us denote by $\mathcal{T} = \{T_0, \ldots, T_n\}$ a set of payment dates, at which coupons of a certain financial instrument are to be paid. Such dates are assumed to be equally δ-spaced for simplicity. We also denote by $\mathcal{G} = \{g_1, \ldots, g_m\}$ the set of future dates at which a reference rate (typically the six-month LIBOR rate) is quoted in the market up to time T_n.

We denote by $L(t)$ the relevant reference rate at time t with maturity $t + \delta$. Forward-rate dynamics will be considered under the forward-adjusted measure Q^n corresponding to the final payment time T_n.

Consider a financial product whose payoff depends on the (for example daily) evolution of the reference rate L in each reset/payment interval.

In order to Monte Carlo price this product based on the forward-rate dynamics of the LFM, we need to recover at any time t the reference rate, which we assume to be the δ spot rate $L(t) = F(t; t, t + \delta)$, from the family of spanning forward rates at our disposal at times $T_{\beta(t)-2}$ and $T_{\beta(t)-1}$, i.e. at the dates in $\mathcal{T}$ that are closest to the current time t: $T_{\beta(t)-2} < t \leq T_{\beta(t)-1}$. In particular, we have both

$$L(T_{\beta(t)-2}) = F_{\beta(t)-1}(T_{\beta(t)-2})$$

and

$$L(T_{\beta(t)-1}) = F_{\beta(t)}(T_{\beta(t)-1}) \ .$$

How do we obtain $L(t)$ from $L(T_{\beta(t)-2})$ and $L(T_{\beta(t)-1})$? We have faced this problem earlier in Sections 6.21.1 and 6.21.2, proposing both a "drift interpolation" and a "bridging" technique.

We now present some particular products depending on "in-between" rates, and we will tacitly assume that "in-between" rates have been obtained through one of these methods.

13.13.1 Accrual Swaps

We give a short description of accrual swaps in the following. We assume a unit nominal amount.

- Institution A pays to B (a percentage γ of) the reference rate L (plus a spread S) at dates $\mathcal{T}$. Formally, at time T_i institution A pays to B

$$(\gamma L(T_{i-1}) + S)\tau_i \, ,$$

  where τ_i is the year fraction between the payment dates T_{i-1} and T_i.
- Institution B pays to A, at time T_i, a percentage α of the reference rate plus a spread Q, times the relative number of days between T_{i-1} and T_i where the reference rate L was in the corridor $L_1 \leq L \leq L_2$. Formally, at time T_i, institution B pays to A the coupon

$$c(T_i) = (\alpha L(T_{i-1}) + Q)\, \tau_i \, \frac{\sum_{g \in \mathcal{G} \cap [T_{i-1}, T_i)} 1\{L_1 \leq L(g) \leq L_2\}}{\#\{\mathcal{G} \bigcap [T_{i-1}, T_i)\}} \, , \tag{13.20}$$

  where, as usual, # denotes the number of elements of a set (cardinality).

When the simulated paths for L are available, we are able to evaluate the accrual swap. The discounted payoff as seen from institution A is

$$\sum_{i=1}^{n} D(0,T_i)\tau_i \Bigg[(\alpha L(T_{i-1}) + Q) \frac{\sum_{g \in \mathcal{G} \cap [T_{i-1}, T_i)} 1\{L_1 \leq L(g) \leq L_2\}}{\#\{\mathcal{G} \bigcap [T_{i-1}, T_i)\}} - (\gamma L(T_{i-1}) + S) \Bigg] \, ,$$

so that the value to A of the accrual swap is the risk-neutral expectation

$$E\left\{\sum_{i=1}^{n} D(0,T_i)\tau_i\left[(\alpha\ L(T_{i-1})+Q)\ \frac{\sum_{g\in\mathcal{G}\cap[T_{i-1},T_i)} 1\{L_1 \le L(g) \le L_2\}}{\#\{\mathcal{G}\bigcap[T_{i-1},T_i)\}}\right.\right.$$

$$\left.\left.-(\gamma L(T_{i-1})+S)\right]\right\}$$

$$= P(0,T_n)\sum_{i=1}^{n}\tau_i E^n\left\{\frac{1}{P(T_i,T_n)}\right.$$

$$\left.\cdot\left[(\alpha\ L(T_{i-1})+Q)\frac{\sum_{g\in\mathcal{G}\cap[T_{i-1},T_i)} 1\{L_1 \le L(g) \le L_2\}}{\#\{\mathcal{G}\bigcap[T_{i-1},T_i)\}} - (\gamma L(T_{i-1})+S)\right]\right\}$$

Both the forward-rate dynamics and the related approximated L dynamics under Q^n are known and a Monte Carlo pricing can be carried out.

Analytical Formula for Accrual Swaps

Alternatively, we may study an analytical formula based on a drift approximation in the LIBOR market model, similar to the one used in deriving approximated swaptions volatilities and terminal correlations in Chapter 6. We proceed as follows.

We concentrate on the non-trivial leg, paid by B to A. Let us focus on the single discounted payment occurring at time T_i. It will suffice to add up all contributions after each one has been priced. We have seen above the payment at time T_i to be given by (13.20). Instead of expressing every coupon under the terminal measure, let us write

$$E_0[D(0,T_i)c(T_i)] = P(0,T_i)E_0^i[c(T_i)].$$

Our task is then reduced to computing the expected value $E_0^i[c(T_i)]$, which, in turn, amounts to computing, by additive decomposition, expected values such as

$$E_0^i[L(T_{i-1})\,1\{L_1 \le L(u) \le L_2\}],\quad E_0^i[1\{L_1 \le L(u) \le L_2\}]$$

under the T_i-forward-adjusted measure Q^i and for $T_{i-1} \le u < T_i$. These may be rewritten in terms of forward rates as

$$E_0^i[F_i(T_{i-1})\,1\{L_1 \le F(u;u,u+\delta) \le L_2\}],\quad Q^i\{L_1 \le F(u;u,u+\delta) \le L_2\},$$

where we have expressed the expected value of an indicator function directly as a probability.

Now, in order to handle such expressions we consider approximated forward-rate dynamics. Actually, no approximation is needed for F_i under Q^i, since its drift is zero and we have a nice geometric Brownian motion. Instead, we act on the dynamics of $F(t;u,u+\delta)$. Since $F(\cdot;u,u+\delta)$ is not in our fundamental family of forward rates, we use the drift-interpolation technique seen in Section 6.21.1. If we set $F_u(t) = F(t; u, u+\delta)$ for brevity, by applying formula (6.78), with partially frozen coefficients and a few rearrangements, we obtain (notice that $T_i \leq u + \delta < T_{i+1}$)

$$\begin{aligned} dF_u(t) &= \mu(t) F_u(t)\, dt + \sigma(t; u, u+\delta) F_u(t)\, dZ^i(t), \\ \mu(t) &:= \frac{(u+\delta-T_i)F_{i+1}(0)}{1+\tau_{i+1}F_{i+1}(0)}\, \sigma(t; u, u+\delta)\rho_{i,i+1}\sigma_{i+1}(t) \end{aligned} \tag{13.21}$$

where $\sigma(t; u, u+\delta)$ is the instantaneous volatility of the related forward rate, and is usually obtained by some kind of interpolation from the "standard rates" volatilities σ_k's.

Our approximation has produced a fundamental effect. The process (13.21) is now a geometric Brownian motion, and we can apply a standard "Black and Scholes technology" to our pricing problem.

Let us recall the following Black and Scholes fundamental setup. Assume we are given two asset prices following correlated geometric Brownian motions under the relevant measure,

$$\begin{aligned} dS_t &= \mu_1(t) S_t\, dt + v_1(t) S_t\, dZ_1(t), \\ dA_t &= \mu_2(t) A_t\, dt + v_2(t) A_t\, dZ_2(t), \quad dZ_1\, dZ_2 = \rho\, dt \end{aligned}$$

(all coefficients being deterministic). We can easily calculate, through laborious but straightforward computations, for $T < u$,

$$\begin{aligned} E_0[S(T)\, 1\{L_1 \leq A(u) \leq L_2\}] &= S(0) \exp\left(\int_0^T \mu_1(s) ds \right) \\ \cdot \Bigg[\Phi\left(\frac{\ln(L_2/A(0)) - \int_0^u \left(\mu_2(s) - \frac{v_2(s)^2}{2}\right) ds - \rho \int_0^T v_1(s) v_2(s) ds}{\sqrt{\int_0^u v_2(s)^2 ds}} \right) \\ - \Phi\left(\frac{\ln(L_1/A(0)) - \int_0^u \left(\mu_2(s) - \frac{v_2(s)^2}{2}\right) ds - \rho \int_0^T v_1(s) v_2(s) ds}{\sqrt{\int_0^u v_2(s)^2 ds}} \right) \Bigg]. \end{aligned}$$

We can apply this formula to our case, by setting $T = T_{i-1}$, $S(t) = F_i(t)$, $A(t) = F_u(t)$, $\mu_1(t) = 0$, $\mu_2(t) = \mu(t)$, $v_1(t) = \sigma_{i+1}(t)$, $v_2(t) = \sigma(t; u, u+\delta)$, $\rho = \rho_{i,i+1}$. This provides us with the terms

$$E_0^i[F_i(T_{i-1})\, 1\{L_1 \leq F(u; u, u+\delta) \leq L_2\}],$$

where the impact of correlation is evident. On the other hand, we may compute

$$E_0[1\{L_1 \le A(u) \le L_2\}] = \left[\Phi\left(\frac{\ln(L_2/A(0)) - \int_0^u \left(\mu_2(s) - \frac{v_2(s)^2}{2}\right)ds}{\sqrt{\int_0^u v_2(s)^2 ds}}\right)\right.$$
$$\left. - \Phi\left(\frac{\ln(L_1/A(0)) - \int_0^u \left(\mu_2(s) - \frac{v_2(s)^2}{2}\right)ds}{\sqrt{\int_0^u v_2(s)^2 ds}}\right)\right],$$

from which terms

$$Q^i\{L_1 \le F(u; u, u+\delta) \le L_2\}$$

are readily computed. Now, putting all the pieces together, we obtain the accrual-swap price. The "frozen drift" approximation guarantees us that for short maturities this formula should work well. However, we have seen that the drift "freezing approximation" above usually does not take us far away from the lognormal distribution even for large maturities, as we have observed in the density plots given in Chapter 8. Problems might only occur with pathological or very large volatilities.

Finally, we would like to point out that the "freezing part of the drift" method can usually be used to transform the distributionally unknown LFM dynamics into the geometric-Brownian-motion dynamics of the basic Black and Scholes lognormal setup. As a consequence, this method can be of help in all cases where forward rates play the role of underlying assets under the relevant measure and where the basic Black and Scholes setup leads to analytical formulas. Before adopting the thus derived approximated formulas, however, one should test them against a Monte Carlo pricing, carried out through the true LFM dynamics, in a sufficiently large number of market situations.

13.13.2 Trigger Swaps

A trigger swap is an interest-rate swap periodically paying a certain *reference* rate against a fixed payment. This swap "comes to life" or "terminates" when a certain *index* rate hits a prespecified level. It is somehow similar to barrier options in the FX or equity markets. Usually, the two rates coincide, but the index rate is observed at a higher frequency than the payment frequency. For example, the index rate and the reference rate can both coincide with the six-month LIBOR rate, which can be observed daily for the indexing and every six months for the payments.

There are four standard basic types of trigger swaps: Down and Out (DO), Up and Out (UO), Down and In (DI), Up and In (UI). Let the prespecified level be H.

- DO: The initial index rate is above H. The swap terminates its payments ("goes OUT") as soon as the index rate hits the level H (from above, i.e. going "DOWN").

- UO: The initial index rate is below H. The swap terminates its payments ("goes OUT") as soon as the index rate hits the level H (from below, i.e. going "UP").
- DI: The initial index rate is above H. The swap starts its payments ("goes IN") as soon as the index rate hits the level H (from above, i.e. going "DOWN").
- UI: The initial index rate is below H. The swap starts its payments ("goes IN") as soon as the index rate hits the level H (from below, i.e. going "UP").

The payoff from a DO trigger swap can be expressed formally as follows. As for accrual swaps, we assume the current time to be $t = 0$, and we denote by $\mathcal{T} = \{T_0, \ldots, T_n\}$ a set of payment dates, at which payments occur. Such dates are assumed to be equally δ-spaced. We also denote by $\mathcal{G} = \{g_1, \ldots, g_m\}$ the set of future dates at which the reference rate (typically the six-month LIBOR rate) is quoted in the market up to time T_n.

We assume the index rate and reference rate to coincide. We denote by $L(t)$ the reference rate at the generic time instant t with maturity $t + \delta$. Forward-rate dynamics will be considered under the forward-adjusted measure Q^n corresponding to the final payment time T_n.

We assume unit nominal amount. If the swap is still alive at time $t = T_{i-1}$, then at time T_i the following will occur:

- Institution A pays to B the fixed rate K at time T_i *if* at all previous instants in the interval $(T_{i-1}, T_i]$ the index rate L is above the triggering barrier H. Formally, if the swap is still alive at time T_{i-1}, at time T_i institution A pays to B

$$\begin{aligned} &K\tau_i \prod_{g \in \mathcal{G} \cap (T_{i-1}, T_i]} 1\{L(g) > H\} \\ &= K\tau_i 1\big\{ \min\{L(g),\ g \in \mathcal{G} \cap (T_{i-1}, T_i]\} > H \big\}, \end{aligned}$$

  where τ_i is the year fraction between the payment dates T_{i-1} and T_i.
- Institution B pays to A (a percentage α of) the reference rate L at the last reset date T_{i-1} (plus a spread Q) *if* at all previous instants of the interval $(T_{i-1}, T_i]$ the index rate L is above the triggering barrier H. Formally, at time T_i institution B pays to A

$$(\alpha L(T_{i-1}) + Q)\, \tau_i\, 1\{\min\{L(g),\ g \in \mathcal{G} \cap (T_{i-1}, T_i]\} > H\}$$

The complete discounted payoff as seen from institution A can be expressed as

$$\sum_{i=1}^{n} D(0, T_i)\, (\alpha L(T_{i-1}) + Q - K)\, \tau_i\, 1\big\{ \min\{L(g),\ g \in \mathcal{G} \cap (T_0, T_i]\} > H \big\}$$

and the contract value to institution A is

$$E\left[\sum_{i=1}^{n} D(0,T_i)\,(\alpha L(T_{i-1}) + Q - K)\;\tau_i\, 1\{\min\{L(g),\ g \in \mathcal{G} \cap (T_0, T_i]\} > H\}\right]$$

$$= P(0,T_n) \sum_{i=1}^{n} \tau_i\, E^n\left[\frac{(\alpha L(T_{i-1}) + Q - K)\; 1\{\min\{L(g),\ g \in \ldots\} > H\}}{P(T_i, T_n)}\right].$$

Once again, it is enough to recover spot rates $L(T_i) = F_{i+1}(T_i)$ and discount factors $P(T_i, T_n)$ by generating for all i's spanning forward rates

$$F_{i+1}(T_i), F_{i+2}(T_i), \ldots, F_n(T_i)$$

under Q^n according to the usual discretized (Milstein) dynamics (analogously to the autocaps case (13.5)), and apply either the "drift interpolation" or the "bridging" technique of Sections 6.21.1 and 6.21.2 to recover in-between rates $L(g)$.

13.14 LFM Pricing with Early Exercise and Possible Path Dependence

Here we shortly present the "Least Squared" Monte Carlo method for pricing early-exercise (and possibly path-dependent) products through Monte Carlo simulation in the LFM. This idea first appeared for general American options in Tilley (1993), followed by Carrière (1996) and then Longstaff and Schwartz (2001), who apply the apparatus to swaptions, as pointed out to us by Lacey (2002).

What we will present here can be intended as a solution of the following two different and yet related problems.

1. How can we use Monte Carlo for early-exercise (non path-dependent) products? This can be necessary when in presence of non-Markovian dynamics or of large dimensionality of the underlying process, as we shall see in a moment.
2. How can we price derivatives that show at the same time path dependence and early-exercise features, even in the favorable cases of low dimensionality and Markovian dynamics?

In Chapter 3 we observed that the pricing of early-exercise products can be carried out through binomial/trinomial trees, and that Monte Carlo is instead suited to treat path-dependent products. Here, before proposing a promising extension of the Monte Carlo method, we shortly recall what was already remarked there, in the beginning of Sections 3.11.2 and 3.11.3.

Trees can be used for early-exercise products when the fundamental underlying variable is low-dimensional (say one or two-dimensional), as happens typically with short-rate models. In such cases, the tree is the ideal instrument, given its "backward-in-time" nature. We know the value of the payoff in each final node, and move backward in time, by updating the value of continuation through discounting. At each node of the tree we can compare the backwardly propagated value of continuation with the payoff evaluated at that node, and decide whether exercise is to be considered or not at that point. After the exercise decision has been taken, the backward induction restarts and we continue to propagate backwards the updated value. When we reach the initial node of the tree, at time 0, we have (an approximation of) the price of our early-exercise product. Thus trees are ideally suited to "travel backward in time".

The other family of products that is usually considered is the family of "path-dependent" payoffs. Such products can be exercised only at a final date, but their final payoffs depend on the history of the underlying variable up to the final time, and not only on the value of the underlying variable at maturity. For path-dependent products, the Monte Carlo method is ideally suited, since it works through forward propagation in time of the underlying variable, by simulating its transition density between dates where the underlying-variable history matters to the final payoff. Monte Carlo is thus ideally suited to "travel forward in time".

In principle, trees have problems mainly in two situations. The first case concerns high dimensionality. If the underlying variable follows a high-dimensional process (in practice with dimension larger than two or three, as in case of the LFM, for example), the tree is practically impossible to consider, since the computational time grows roughly exponentially with the dimension. Moreover, there are also difficulties in handling correlations and other aspects, so that trees become extremely difficult to use.

The second case where trees have major problems is with path-dependent products. When we try and propagate backwards the contract value from the final nodes we are immediately in trouble, since to value the payoff at a given node (and at any final node in particular) we need to know the past history of the underlying variable. But this past history is not determined yet, since we move backward in time. This method, therefore, is not applicable in a standard way.

Actually, there are ad-hoc procedures to render trees able to price particular path-dependent products in the basic Black and Scholes setting, for example barrier and lookback options. However, in general there is no consolidated and realistic recipe on how using a tree for path-dependent payoffs, and moreover, when dealing with interest-rate derivatives models, we are usually outside the Black and Scholes framework.

As for the Monte Carlo method, it does better with respect to high dimensionality, in that computational time grows roughly *linearly* with the dimension, and it is also suited to parallel computing. However, there are problems with early exercise. Since we propagate trajectories forward in time, we have no means to know whether it is optimal to continue or to exercise at a certain time. Therefore, Monte Carlo cannot be used, in its original formulation, for the large range of products involving early exercise features.

However, Carrière (1996) and Longstaff and Schwartz (2001) have proposed an approximated method to make Monte Carlo techniques work also in presence of early-exercise features. The resulting method is very promising, since it allows for the pricing of instruments with high-dimensional underlying variables, path dependence and early exercise at the same time.

Clearly, the method needs further testing beyond what shown in Longstaff and Schwartz (2001), especially on practical cases concerning interest-rate models. Still, test results in Longstaff and Schwartz (2001) look rather encouraging so as to justify a general exposition of the method and of its possible developments even before extensive testing has been carried out. This generality, together with the potential of the method of not exceeding the true value of the early-exercise contract, could result in a supremacy of Monte Carlo over trees and finite-difference methods in general, especially if numerically efficient Monte Carlo methods are brought into play.

Research on improvements of the basic Monte Carlo setup, based on weighted paths and other techniques have received considerable attention in the literature, thus further strengthening the interest in the Least Squared method.

We now review this method for a generic product whose final payoff is nonnegative and dependent on a (possibly multi-dimensional) underlying variable X.

Assume we have a product that can be exercised at times $0 = t_1, \dots, t_N$, whose immediate-exercise value at each time t_k depends on part of the history of an underlying process $X(t)$ up to time t_k itself. Typically, the value can depend on $X(s_1), \dots, X(s_{j_k})$, where the times $s_1 < s_2 < \dots < s_{j_k} \leq t_k$ are the ones contributing to the immediate-exercise payoff at time t_k. In detail, we assume that, if exercised at time t_k, the product pays immediately the Cash flow from Exercise (CE) given by

$$\mathrm{CE}(t_k) := \mathrm{CE}(t_k; X(s_1), \dots, X(s_{j_k})).$$

This value has to be compared with the backwardly Cumulated discounted cash flows from Continuation (CC) at the same time, namely the value of the contract at t_k when this has not been exercised before or at t_k itself,

$$\mathrm{CC}(t_k) := \mathrm{CC}(t_k; X(s_1), \dots, X(s_{j_k})).$$

We assume we are computing prices under a generic numeraire asset $U(t)$. In their paper, Longstaff and Schwartz (2001) take the bank account as fundamental numeraire, and work under the risk-neutral measure.

The method can be summarized through the following scheme:

1. Choose a number of paths, np.

2. (Choice of the basis functions). For each time t_k, choose i_k basis functions

$$\phi_1(t_k, x_1, \ldots, x_{j_k}), \ldots, \phi_{i_k}(t_k, x_1, \ldots, x_{j_k})$$

that will be used in approximating the continuation value as a function of the past and present values $X(s_1), X(s_2), \ldots, X(s_{j_k})$ of the underlying variable up to time t_k (see step 7 below).
3. (Simulating the underlying variables). Simulate np paths for both the underlying variable X and the numeraire U from time t_1 to time t_n. Make sure of including the reset times $s_1, \ldots, s_{j_n}$ among the dates at which X and U are simulated. Typically, this simulation is "exact" if the transition distributions of X and U are known, like, for example, in the case of geometric Brownian motion or linear-Gaussian processes as in Hull and White's models. Alternatively, a numerical discretization scheme for SDEs such as the Euler or Milstein schemes can be employed if this transition density is not known. In any case, denote by

$$X^j(t_k), U^j(t_k)$$

the simulated values of X and U respectively under the j-th scenario at time t_k. More generally, the superscript on a stochastic quantity will denote the quantity itself under the scenario given by the superscript index.
4. (Computing the payoff at final time). Set

$$\mathrm{CC}^j(t_n) := \mathrm{CE}(t_n; X^j(s_1), \ldots, X^j(s_{j_n})).$$

(The backwardly Cumulated discounted cash flow from Continuation at final time is simply the exercise value at that time).
5. (Positioning the initial step at final time). Set $k = n$. We position ourselves at the final exercise time. Now the iterative part of the scheme begins.
6. (Consider only scenarios where the immediate-exercise value of the contract is positive). Set

$$I_{k-1} := \{j \in \{1, 2, \ldots, np\} : \mathrm{CE}(t_{k-1}; X^j(s_1), \ldots, X^j(s_{j_{k-1}})) > 0\}.$$

We thus focus only on scenarios where the exercise value is strictly positive at the current evaluation time t_{k-1};
7. (Regressing the discounted continuation value on the chosen basis functions). In this step, we aim at approximating the discounted continuation value at current time t_{k-1} as a linear combination of the basis functions

$$\phi_1(t_{k-1}, x_1, \ldots, x_{j_{k-1}}), \ldots, \phi_{i_{k-1}}(t_{k-1}, x_1, \ldots, x_{j_{k-1}})$$

through a regression, so as to estimate the combinators λ in

$$\frac{U^j(t_{k-1})}{U^j(t_k)}\mathrm{CC}^j(t_k) = \sum_{h=1}^{i_{k-1}} \lambda_h(t_{k-1})\, \phi_h(t_{k-1}, X^j(s_1), \ldots, X^j(s_{j_{k-1}})),$$

where $j \in I_{k-1}$.
On the left-hand side of the above equation, we have the continuation value an instant later discounted back at current time t_{k-1} through the chosen numeraire U. Notice that if the numeraire is the bank account $B(t) = \exp(rt)$, with deterministic constant r, as in Longstaff and Schwartz (2001), then the U's ratio reduces to $\exp(-r(t_k - t_{k-1}))$.
On the right-hand side of the same equation, we have a linear combination of the chosen basis functions, corresponding ideally to a truncated L^2 expansion. The step could be made exact with an infinite expansion ($i_{k-1} = \infty$), when the conditional expectation defining the actual continuation value above behaves nicely in an L^2 sense. See Longstaff and Schwartz (2001) for further details.

8. Store the exercise flag (EF) over scenarios at time t_{k-1}:

$$\mathrm{EF}(j, t_{k-1}) := 1\Bigg\{ \mathrm{CE}(t_{k-1}; X^j(s_1), \ldots, X^j(s_{j_{k-1}})) > \sum_{h=1}^{i_{k-1}} \lambda_h(t_{k-1})\, \phi_h(t_{k-1}, X^j(s_1), \ldots, X^j(s_{j_{k-1}})) \Bigg\}.$$

This flag is set to one when exercise is the convenient choice, and to zero when continuation is in order. Again, $1\{\cdots\}$ denotes the indicator function of the set between curly brackets.
When $\mathrm{EF}(j, t_{k-1})$ is one, set all its subsequent values to zero, $\mathrm{EF}(j, t_h) := 0$ for all $h > k-1$.

9. Set

$$\mathrm{CC}^j(t_{k-1}) := \frac{U^j(t_{k-1})}{U^j(t_k)}\mathrm{CC}^j(t_k) \text{ if } \mathrm{EF}(j, t_{k-1}) = 0 \text{ (continuation)},$$

and set

$$\mathrm{CC}^j(t_{k-1}) := \mathrm{CE}(t_{k-1}; X^j(s_1), \ldots, X^j(s_{j_{k-1}})) \text{ if } \mathrm{EF}(j, t_{k-1}) = 1 \text{ (exercise)}$$

10. If $k = 2$, *i.e.* $t_{k-1} = 0$, stop, otherwise replace k with $k-1$ and restart from point 6.

13.15 LFM: Pricing Bermudan Swaptions

Bermudan swaptions are options to enter an IRS not only at its first reset date, but also at subsequent reset dates of the underlying IRS, at least in some of the simplest formulations.

Let again $\mathcal{T} = \{T_1, \ldots, T_n\}$ be a set of reset and payment dates. Recall that we denote by $\mathbf{PS}(T_i, T_k, \{T_k, \ldots, T_n\}, K)$ the price at time T_i of a (payer) swaption maturing at time T_k, which gives its holder the right to enter at time T_k an interest-rate swap with first reset date T_k and payment dates $T_{k+1}, \ldots, T_n$ at the fixed strike rate K. We will abbreviate this price by $\mathbf{PS}_{k,n}(T_i)$. This price is known as a function of the present value for basis point $C_{k,n}(T_i)$ and of the forward swap rate $S_{k,n}(T_i)$ through Black's formula for swaptions.

Definition 13.15.1. (Bermudan Swaption). *A (payer) Bermudan swaption is a swaption characterized by three dates $T_k < T_h < T_n$, giving its holder the right to enter* at any time T_l in-between T_k and T_h *(included) into an interest-rate swap with first reset in T_l, last payment in T_n and fixed rate K. Thus, the swap start and length depend on the instant T_l when the option is exercised. We denote by $\mathbf{PBS}_{k,h,n}(T_i)$ the value of such a Bermudan swaption at time T_i, with $T_i \leq T_k$.*[1]

Pricing Bermudan swaptions with the LFM has to be handled through tailor-made methods, since the model is not ideally suited for the implementation of recombining lattices. A possible alternative to the tailor-made techniques is the general Least Squared "Monte Carlo Regression" (LSMC) approach reviewed in Section (13.14), which is indeed quite general and usually results in good approximations. However, the method itself has to be tailored (choice of the basis functions, ...) when applied to Bermudan swaptions in the LFM.

13.15.1 Least Squared Monte Carlo Approach

As we just noticed, the LSMC method can be used to price Bermudan swaptions in the LFM. Longstaff and Schwartz (2001), however, tested the LSMC method (in the section "valuing swaptions in a string model" of their paper) by actually considering a version of the so called string model. In practice, when working in a finite set of expiries/maturities, string models are often equivalent to the LIBOR market model (LFM). For more details on string models, see for example Santa Clara and Sornette (2001) or Longstaff, Santa Clara and Schwartz (2001). In the specific application of string models we are considering here, Longstaff and Schwartz (2001) used directly bond-prices dynamics and bond-prices volatilities instead of forward-rates volatilities. For completeness, we here illustrate their procedure.

A Bermudan swaption is considered, where the underlying swap starts at the initial time with given reset and payment dates. The swaption's holder has the right to exercise the option at some fixed dates and enter the swap, whose life span decreases as time moves forward.

[1] There are other types of Bermudan swaptions, but for our purposes the type described here suffices.

The underlying swap, with a ten-year maturity, resets semi-annually, and exercise can occur at any reset date after one year, one year included and ten years excluded. There are therefore nineteen exercise dates. In propagating the zero-coupon-bond prices,

$$P(\cdot,\ 0.5y), P(\cdot, 1y), P(\cdot,\ 1.5y), \ldots, P(\cdot, 10y),$$

the LSMC method starts from the twenty-dimensional vector above, and the dimension decreases by one each six months. The bond-price dynamics has as percentage risk-neutral drift the risk-free rate, which is approximated with the corresponding six-month continuously-compounded rate

$$r(t) \approx -2 \ln P(t, t + 0.5y),$$

thus closing the set of equations in P for the discretized approximate dynamics once the volatility has been assigned. Indeed, the approximate dynamics reads now

$$dP(t, T_i) = -2 \ln P(t, t + 0.5y) P(t, T_i) dt + \sigma_{P_i}(t) P(t, T_i) dZ_i(t), \quad T_i = 0.5i,$$

$i = 1, \ldots, 20$. When these equations are discretized at times T_i, we obtain a closed set of equations, since the drift rate now involves a bond price in the family.

The Z_i's are correlated Brownian motions under the risk-neutral measure. In the simulation it is assumed that

$$dZ_i dZ_j = \exp(-k|i - j|) dt,$$

where k is a positive constant, and the Z vector is kept twenty-dimensional.

At the exercise time T_i, the basis functions of the algorithm are selected as:

$$1, P(\cdot, T_i), \ldots, P(\cdot, T_{20}), \frac{1 - P(T_i, T_{20})}{\sum_{j=i+1}^{20} 0.5 P(T_i, T_j)}, \left[\frac{1 - P(T_i, T_{20})}{\sum_{j=i+1}^{20} 0.5 P(T_i, T_j)}\right]^2,$$
$$\left[\frac{1 - P(T_i, T_{20})}{\sum_{j=i+1}^{20} 0.5 P(T_i, T_j)}\right]^3,$$

where the last three terms are simply the underlying swap rate $S_{i,20}(T_i)$ and its second and third powers.

At the first exercise time ($i = 3$), there are 22 basis functions, their number decreasing as time goes by. Longstaff and Schwartz state that adding further functions does not change the option value, so that one can infer the valuation to be correct, given that the approximated value never exceeds the real value. See also the related discussion in Longstaff and Schwartz (2001).

Notice that Longstaff and Schwartz have assumed a deterministic bond-price percentage volatility. This is not really consistent with the LFM distribution for lognormal forward rates. Therefore, as already mentioned above, the model analyzed in this section is not a LFM, from a theoretical point of view.

13.15.2 Carr and Yang's Approach

Carr and Yang (1997) use simulations to develop a Markov-chain approximation for the valuation of Bermudan swaptions in the LFM. Their method stems from the observation that, given the tenor structure

$$T_1, \dots, T_n,$$

one can represent the whole yield curve along the structure by just knowing the evolution of a chosen numeraire. Take for example the numeraire $P(\cdot, T_n)$, associated with the terminal measure Q^n. At a time T_i in the tenor structure, the whole (Zero-bond) curve

$$P(T_i, T_{i+1}), P(T_i, T_{i+2}), \dots, P(T_i, T_n)$$

can be obtained as follows. Recall that by definition of numeraire we have

$$\frac{P(T_i, T_j)}{P(T_i, T_n)} = E^n_{T_i}\left[\frac{P(T_j, T_j)}{P(T_j, T_n)}\right],$$

or

$$P(T_i, T_j) = P(T_i, T_n)\, E^n_{T_i}\left[\frac{1}{P(T_j, T_n)}\right], \tag{13.22}$$

so that we can compute each $P(T_i, T_j)$ by knowing the current value of the numeraire $P(\cdot, T_n)$ and its distribution under its own measure Q^n. The exercise decision, at any instant, can thus be reduced to knowledge of the distributional properties of the single process $P(\cdot, T_n)$.

Based on this observation, Carr and Yang (1997) found a way to construct a Markov chain approximating the migration of $P(\cdot, T_n)$ in between areas of a selected partition of $[0, 1]$. Partitioning $[0, 1]$ in $I_1(t), I_2(t), \dots, I_{l(t)}(t)$, so that $[0, 1]$ is given by the disjoint union of the sets I, the Markov chain is constructed as follows.

First, simulate spanning forward-rate dynamics $F_{i+1}(t)^j, \dots, F_n(t)^j$ under several scenarios, each scenario denoted by a superscript j, up to a generic time $t = T_i$. Second, obtain the numeraire bond price $P(t, T_n)^j$ from these simulations under each scenario j. Third, define the transition matrix between "state" h at time $t = T_i$ and "state" k at time $t + \Delta = T_{i+1}$ as

$$p_{h,k}(t) := \frac{\#\{j : \; P(t, T_n)^j \in I_h(t) \;\text{ and }\; P(t + \Delta, T_n)^j \in I_k(t + \Delta)\}}{\#\{j : \; P(t, T_n)^j \in I_h(t)\}}.$$

Then one defines $\bar{P}_h(t, T_n)$ as the average of the $P(t, T_n)^j$'s in $I_h(t)$,

$$\bar{P}_h(t, T_n) := \frac{\sum_{j:\; P(t,T_n)^j \in I_h(t)} P(t, T_n)^j}{\#\{j : \; P(t, T_n)^j \in I_h(t)\}}.$$

Consider the chain $X(t)$ with states $\{1, 2, \dots, l(t)\}$ and probability $p_{h,k}(t)$ of going from $X(t) = h$ to $X(t + \Delta) = k$. Our $\bar{P}_h(t, T_n)$ can be considered as a discrete-space approximation of the numeraire $P(t, T_n)$ when $X(t) = h$.

The chain X summarizes the true dynamics of $P(t, T_n)$ into a Markov process that can be used for approximately simulating $P(t, T_n)$. We can therefore simulate the whole yield curve in the spirit of the relationship (13.22). Then backward induction becomes possible by using the Markov chain instead of the original paths for the numeraire.

We move backwards in time by means of the process $\bar{P}_{X(t)}(t, T_n)$ in place of the process $P(t, T_n)$, with $\bar{P}_{X(t)}(t, T_n)$ that assumes only a finite set of possible values at each instant. The transition probabilities allow us to roll back the relevant expectations and the Bermudan swaption can be easily priced through backward induction, see Carr and Yang (1997) for the details and for numerical tests.

A similar approach has been suggested in Clewlow and Strickland (1998) for a Gaussian multi-factor Heath-Jarrow-Morton model (and not the LFM), where again the early-exercise opportunity is evaluated in terms of a single variable. This variable is taken to be the fixed leg of the underlying interest-rate swap. Since the floating leg is always valued on par at reset dates, this choice amounts roughly to considering the value of the underlying interest-rate swap as fundamental single process at the reset dates.

The approximate specification of the early-exercise region as a function of the underlying variable is found by using a single-factor extended Vasicek (Hull and White) approximation of the multi-factor model. With the one-factor model one obtains the approximate early-exercise region via a recombining tree for the short rate, by determining the critical values of the underlying interest-rate swap at the early-exercise dates through backward induction on the tree. Choosing only one factor allows for a richer discretization in time and this yields an accurate exercise region.

Once the exercise decision has been estimated as a function of the underlying swap through the tree, one runs a Monte Carlo simulation for the original multi-factor model, where each early-exercise opportunity, when encountered, is evaluated as the (known) approximate function of the underlying swap.

This method seems to be robust. It provides one with a lower bound for the Bermudan swaption price, due to the sub-optimal exercise region, as in the LSMC method. A similar method for the LFM has been proposed by Andersen (1999), and we review it in the following.

13.15.3 Andersen's Approach

Andersen (1999) proposed a method similar to that of Clewlow and Strickland (1998). Again, the early-exercise region is extracted by a low-dimensional parameterization, consisting of a small number of key variables (these including the underlying interest-rate swap as in Clewlow and Strickland), but the approximated early-exercise region, as a function of these variables, is not determined through a *one-factor* model. Rather, an optimization on a separate simulation *for the whole multi-factor model* is considered in order to determine this function. The method can be summarized as follows. We adopt the notation introduced earlier in Definition 13.15.1.

We now provide a scheme summarizing a possible formulation of Andersen's method for approximately computing $\mathbf{PBS}_{k,h,n}(T_k)$.

1) Choose a function f approximating for each T_l the optimal exercise flag $\mathcal{I}(T_l)$, depending for example on the nested European swaptions and on a function $H = H(T_l)$ to be determined,

$$\mathcal{I}(T_l) \approx f(\mathbf{PS}_{l,n}(T_l), \mathbf{PS}_{l+1,n}(T_l), \ldots, \mathbf{PS}_{h,n}(T_l), H(T_l)).$$

The optimal exercise flag $\mathcal{I}(T_l)(\omega)$ at time T_l, under the path ω, is defined to be one when exercise is optimal at T_l along the trajectory ω and 0 when the continuation value at T_l is larger than the exercise value along ω. As usual, ω is omitted in the notation.

2) Simulate, through the LFM dynamics for the forward LIBOR rates, in a set of scenarios indexed by j, all the variables

$$\mathbf{PS}^j_{l,n}(T_l),\ \mathbf{PS}^j_{l+1,n}(T_l), \ldots,\ \mathbf{PS}^j_{h,n}(T_l),\ B^j_d(T_l)$$

entering in f's expression above, for all $l = k, k+1, \ldots, h$. The last quantity is the discrete-bank-account numeraire that is used for discounting, i.e.

$$B_d(T_l) = \prod_{m=1}^{l} [1 + \tau_m F_m(T_{m-1})],$$

which is determined by the simulated forward-rate dynamics of the LFM, with T_0 denoting the initial time. Notice that the first variable $\mathbf{PS}^j_{l,n}(T_l)$ involves the interest-rate swap whose swap rate is $S_{l,n}(T_l)$, which was the (unique) "early-exercise flag" variable in the Clewlow and Strickland method.

3) Compute by backward induction all values of $H(T_l)$ from $T_l = T_h$ to $T_l = T_k$ as follows:

- 3.a) The final $H(T_h)$ has to be known from the requirement

$$f(\mathbf{PS}_{h,n}(T_h), H(T_h)) = 1\{\mathbf{PS}_{h,n}(T_h) > 0\}.$$

This is to say that at the last possible exercise date we simply exercise if the underlying European swaption has strictly positive value, as should be. Set $m = h$.

- 3.b) Find $H(T_{m-1})$ as follows. For each simulated path j, solve the optimization problem

$$H^j(T_{m-1}) = \arg\sup_H \Big\{ f(\mathbf{PS}^j_{m-1,n}(T_{m-1}), \mathbf{PS}^j_{m,n}(T_{m-1}), \ldots \\ \ldots, \mathbf{PS}^j_{h,n}(T_{m-1}), H)\ \mathbf{PS}^j_{m-1,n}(T_{m-1}) \\ + \frac{B^j_d(T_{m-1})}{B^j_d(T_m)} (1 - f)\ \mathbf{PBS}^j_{m,h,n}(T_m) \Big\},$$

where we omit f's arguments in the second half of the expression for brevity, and where the expression between curly brackets basically reads as:
if (exercise(H)) then (current underlying European swaption)
else (present value of one-period ahead Bermudan swaption).
We thus look for the value of H in the exercise strategy that maximizes the option value in each scenario.
Notice also that we can write

$$\frac{B_d(T_{m-1})}{B_d(T_m)} = \frac{1}{1 + \tau_m F_m(T_{m-1})}.$$

Let $\mathbf{PBS}^j_{m-1,h,n}(T_{m-1})$ be the supremum corresponding to the above $H^j(T_{m-1})$.
Average over all scenarios j and find $H(T_{m-1})$ from the $H^j(T_{m-1})$'s.

- 3.c) If $m-1$ equals k then move to point 4), otherwise decrease m by one and restart from point 3.b).

4) Now that H is known at all times, compute the Bermudan-swaption price $\mathbf{PBS}_{k,h,n}(T_k)$ through a new simulation with a larger number of paths and with the approximated exercise function given by f.

Andersen (1999) proposed as possible examples of approximate early-exercise function f two possibilities. First, one can set

$$\mathcal{I}(T_l) = 1\{\mathbf{PS}_{l,n}(T_l) > H(T_l)\}.$$

With this choice we say that early exercise will depend on the longest nested European swaption exceeding a level H. A second possibility is setting

$$\mathcal{I}(T_l) = 1\Big\{\mathbf{PS}_{l,n}(T_l) > H(T_l) \text{ and } \max_{p=l+1,\dots,h} \mathbf{PS}_{p,n}(T_l) \leq \mathbf{PS}_{l,n}(T_l)\Big\}.$$

This choice is more refined than the previous one and amounts to adding the requirement that all the other nested future European swaptions, when valued at T_l, have a lower value than the current longest one. This intuitively amounts to saying that, in the context of European swaptions evaluated now, the most convenient is the current longest one. Then, as before, the option is to be exercised if this longest swaption exceeds a level $H(T_l)$.

The second choice is more refined but also more computationally demanding. Indeed, with the first choice, f depends only on the present value per basis point C and on the underlying swap rate (both defining the relevant European swaption), so that backward induction concerns only these two variables and memory requirements are not a problem.

Andersen (1999) also made several considerations on the possible computational efficiency of the method and on low memory requirements. The first Monte Carlo simulation involved in steps 1)-3) usually requires a low number of paths, whereas the evaluation in step 4) requires usually a higher number of scenarios. For other considerations and numerical results, see Andersen (1999). We also mention that Pedersen (1999), among several other issues, considers a comparison of the Andersen method with the Least Squared Monte Carlo method summarized in Section 13.15.1.

13.15.4 Numerical Example

We now develop a pricing example of a Bermudan swaption with the LFM. Precisely, we consider a y-non-call-x Bermudan swaption, which matures in year y and can be exercised at the beginning of any year, starting from year x. By exercising the option, the holder enters a swap starting at the time of exercise and ending at year $y + 1$.[2]

Bermudan swaptions are typically hedged with the corresponding co-terminal ATM European swaptions.For this reason, we devise a pricing procedure that is based on calibrating exactly the co-terminal ATM swaptions, while achieving a satisfactory fit to the upper triangular portion of the volatility matrix at hand. The calibration is based on a recursive algorithm, which is referred to as "diagonal recursive calibration" (DRC). Further details can be found in Lvov (2005), where an extensive treatment of pricing and hedging issues is provided.

In our example, we consider $M = y$ forward rates and impose the following structures of instantaneous covariations between them:

- The instantaneous volatility of each forward rate F_k is assumed to be given by
$$\sigma_k(t) = \psi_k v(T_{k-1} - t; \alpha), \tag{13.23}$$
where v is a deterministic function, $\alpha = \{\alpha_1, \alpha_2, \ldots\}$ is a vector of parameters and ψ_k's are rate-specific coefficients. Precisely, we test two different structures. The first one is based on the assumption that the function v is constant (identically equal to one), with the related LFM calibration being referred to as diagonal recursive calibration with constant volatilities (DRC-CV). The second instantaneous volatility is instead assumed to be time-varying as in (6.13):
$$v(\tau; \alpha) = (\tau\alpha_1 + \alpha_2)\, e^{-\tau\alpha_3} + \alpha_4, \tag{13.24}$$
where α_1, α_2, α_3 and α_4 are real constants. The related LFM calibration is referred to as diagonal recursive calibration with time-varying volatilities (DRC-TVV).

[2] All swaps that can be potentially entered into by exercising the Bermudan swaption are referred to as co-terminal since they terminate on the same date. European swaptions that have co-terminal swaps for underlying are referred to as co-terminal swaptions.

- The instantaneous correlation between rates F_i and F_j is assumed to be given by the Schoenmakers and Coffey (2002) parametrization, which produces a full-rank positive-definite correlation matrix:

$$\rho_{i,j} = \exp\left[-\frac{|j-i|}{M-1} \cdot \left(-\ln\beta_3 + \beta_1 \frac{i^2+j^2+ij-3Mi-3Mj+3i+3j+2M^2-M-4}{(M-2)(M-3)} - \beta_2 \frac{i^2+j^2+ij-Mi-Mj-3i-3j+3M^2-2}{(M-2)(M-3)} \right) \right],$$

where $i,j=1,\ldots,M$, $3\beta_1 \geq \beta_2 \geq 0$, $0 \leq \beta_1+\beta_2 \leq -\ln\beta_3$.

In the DRC-CV, the optimization is performed over parameters in the structure of instantaneous correlations only. For the DRC-TVV, optimization is over both volatility and correlation parameters.

We then consider two more LFM calibrations, based on the cascade algorithm of Section 7.4, which, for a given instantaneous correlation matrix, allows to achieve an exact fit to the upper triangle of the swaption matrix, as long as the method does not fail. The first cascade calibration (CC-1) uses the correlation matrix produced by the DRC-CV, whereas the second one (CC-2) uses that implied by the DRC-TVV.

To price European swaptions, we resort to the approximation (6.67). Therefore, the Black implied volatility of an ATM European swaption maturing at time T_a, and whose underlying swap starts at time T_a and pays at times $T_{a+1},\ldots,T_b$, is approximated by

$$\begin{aligned} \nu_{a,b}^2 &= \sum_{i,j=a+1}^{b} \frac{w_i(0)w_j(0)F_i(0)F_j(0)}{T_a S_{a,b}(0)^2} \rho_{i,j} \int_0^{T_a} \sigma_i(t)\sigma_j(t)\,dt \\ &= \sum_{i,j=a+1}^{b} \frac{F_i^* F_j^* \rho_{i,j}}{T_a S_{a,b}(0)^2} \psi_i \psi_j \int_0^{T_a} v(T_{i-1}-t;\alpha) v(T_{j-1}-t;\alpha)\,dt \end{aligned} \qquad (13.25)$$

where

$$S_{a,b}(0) = \sum_{i=a+1}^{b} w_i(0)F_i(0), \quad \omega_i(0) = \frac{\tau_i P(0,T_i)}{\sum_{j=a+1}^{b} \tau_j P(0,T_j)},$$

and we set, for each k,

$$F_k^* := w_k(0)F_k(0).$$

In order to price the above Bermudan swaption, we start by calibrating the LFM parameter vectors α, $\beta=(\beta_1,\beta_2,\beta_3)$, and $\psi=(\psi_1,\ldots,\psi_y)$, with the family of forward rates spanning the period from time zero to time T_y, being times T_i one-year spaced and $T_0 = 1$ year. In the following, we describe the analytical recursive procedure that allows for an exact calibration to all co-terminal swaptions.[3]

First, we consider the co-terminal swaption maturing at the same time, $T_{y-1} = y$ years, as the Bermudan swaption. This is the last co-terminal swaption, and its underlying is a one-period swap. Its implied volatility $\nu_{y-1,y}$ satisfies

$$\nu_{y-1,y}^2 = \frac{1}{T_{y-1}}\psi_y^2 \int_0^{T_{y-1}} v(T_{y-1} - t; \alpha)^2 \, dt.$$

Therefore, for a given vector of parameters α, an exact calibration to the last co-terminal swaption can be achieved by setting

$$\psi_y = \sqrt{\frac{\hat{\nu}_{y-1,y}^2 T_{y-1}}{\int_0^{T_{y-1}} v(T_{y-1} - t; \alpha)^2 \, dt}}, \tag{13.26}$$

where $\hat{\nu}_{y-1,y}$ denotes the market volatility of the last co-terminal swaption.

We consider next a co-terminal swaption maturing at time T_{y-2}. For a given set of volatility and correlation parameters, the implied volatility of this swaption can be approximated, using (13.25), by

$$\nu_{y-2,y}^2 = \sum_{i,j=y-1}^{y} \frac{F_i^* F_j^* \rho_{i,j}}{T_{y-2} S_{y-2,y}(0)^2} \psi_i \psi_j \int_0^{T_{y-2}} v(T_{i-1} - t; \alpha) v(T_{j-1} - t; \alpha) \, dt.$$

Defining

$$I_{k,m}^x := \int_0^{T_x} v(T_{k-1} - t; \alpha) v(T_{m-1} - t; \alpha) \, dt,$$

the implied volatility approximation can then be written as

$$T_{y-2} \left(\nu_{y-2,y} S_{y-2,y}(0)\right)^2 = \sum_{i,j=y-1}^{y} F_i^* F_j^* \psi_i \psi_j \rho_{i,j} I_{i,j}^{y-2},$$

which yields

$$\begin{aligned} &T_{y-2} \left(\nu_{y-2,y} S_{y-2,y}(0)\right)^2 \\ &= F_{y-1}^{*2} \psi_{y-1}^2 I_{y-1,y-1}^{y-2} + 2F_{y-1}^* F_y^* \psi_{y-1} \psi_y \rho_{y-1,y} I_{y-1,y}^{y-2} + F_y^{*2} \psi_y^2 I_{y,y}^{y-2}. \end{aligned}$$

Assuming that the last co-terminal swaption has been calibrated exactly by setting ψ_y as in (13.26), an exact calibration to next co-terminal swaption

[3] The calibration is performed on implied volatilities. Since formula (13.25) typically yields a very accurate approximation, our calibration to co-terminal swaptions is referred to as "exact" even though we do not match prices exactly.

can be achieved by setting ψ_{y-1} equal to the higher positive solution to the following second-order algebraic equation[4]

$$c_1\hat{\psi}^2 + c_2\hat{\psi} + c_3 = 0 \tag{13.27}$$

where

$$\begin{aligned}
c_1 &= F_{y-1}^{*2} I_{y-1,y-1}^{y-2} \\
c_2 &= 2F_{y-1}^{*} F_y^{*} \psi_y \rho_{y-1,y} I_{y-1,y}^{y-2} \\
c_3 &= F_y^{*2} \psi_y^2 I_{y,y}^{y-2} - T_{y-2}\left(\hat{\nu}_{y-2,y} S_{y-2,y}(0)\right)^2,
\end{aligned}$$

with $\hat{\nu}_{y-2,y}$ denoting the market volatility of the swaption.

Similarly, an exact calibration to the co-terminal swaption maturing at time T_{y-3} is obtained by setting ψ_{y-2} equal to the higher positive solution to the quadratic equation (13.27) with coefficients given by

$$\begin{aligned}
c_1 &= F_{y-2}^{*2} I_{y-2,y-2}^{y-3} \\
c_2 &= 2F_{y-2}^{*} F_{y-1}^{*} \psi_{y-1} \rho_{y-2,y-1} I_{y-2,y-1}^{y-3} + 2F_{y-2}^{*} F_y^{*} \psi_y \rho_{y-2,y} I_{y-2,y}^{y-3} \\
c_3 &= F_{y-1}^{*2} \psi_{y-1}^2 I_{y-1,y-1}^{y-3} + F_y^{*2} \psi_y^2 I_{y,y}^{y-3} + 2F_{y-1}^{*} F_y^{*} \rho_{y-1,y} \psi_{y-1} \psi_y I_{y-1,y}^{y-3} \\
&\quad - T_{y-3}\left(\hat{\nu}_{y-3,y} S_{y-3,y}(0)\right)^2,
\end{aligned}$$

where $\hat{\nu}_{y-3,y}^2$ is the market volatility of the co-terminal swaption maturing in year T_{y-3}.

In general, an exact calibration to a co-terminal swaption maturing at time T_{y-n} , $0 < n \leq y$, is achieved recursively by setting ψ_{y-n+1} equal to the higher positive solution to equation (13.27) with coefficients

$$\begin{aligned}
c_1 &= F_{y-n+1}^{*2} I_{y-n+1,y-n+1}^{y-n} \\
c_2 &= 2F_{y-n+1}^{*} \sum_{i=0}^{n-2} F_{y-i}^{*} \psi_{y-i} \rho_{y-n+1,y-i} I_{y-n+1,y-i}^{y-n} \\
c_3 &= \sum_{i,j=0}^{n-2} F_{y-i}^{*} F_{y-j}^{*} \rho_{y-i,y-j} \psi_{y-i} \psi_{y-j} I_{y-i,y-j}^{y-n} \\
&\quad - T_{y-n}\left(\hat{\nu}_{y-n,y} S_{y-n,y}(0)\right)^2
\end{aligned}$$

where $\hat{\nu}_{y-n,y}$ is the market volatility of the co-terminal swaption maturing in year T_{y-n}.

The vector $(\hat{\psi}_1, ..., \hat{\psi}_y)$ formed by the above recursive routine, will be denoted by $\psi_{\alpha,\beta}$. It follows that if $\psi = \psi_{\alpha,\beta}$ an exact calibration to all co-terminal swaptions has been achieved for a given set of volatility parameters α and correlation parameters β.

[4] We assume that all the algebraic equations we have to solve in this procedure possess at least one positive real root. In practice, this appears to be the case even for stressed input data.

Calibration to the rest of the data is done numerically as follows. For a given set of parameters α, β and $\psi = \psi_{\alpha,\beta}$, we use formula (13.25) to form a vector $\theta_{\alpha,\beta,\psi}$ of model implied volatilities for all swaptions in the data set, excluding the co-terminal ones. Denoting by $\hat{\theta}$ the vector of corresponding market volatilities, we then solve (numerically) the following minimization problem

$$\min_{\alpha,\beta} \left[|\theta_{\alpha,\beta,\psi} - \hat{\theta}|^2 : \psi = \psi_{\alpha,\beta} \right], \tag{13.28}$$

where the exact calibration to co-terminal swaptions is performed on every iteration in the numerical algorithm, before the target function is evaluated. This allows to achieve an exact calibration to co-terminal swaptions analytically, without imposing additional constraints on the minimization.

Once optimal parameters α, β and $\psi = \psi_{\alpha,\beta}$ are found, we can use them to compute volatilities and correlations of the forward rates involved in the Bermudan swaption evaluation.[5]

Pricing Results. Our numerical test is conducted on a 10-non-call-1 Bermudan swaptions. The market data we use comprises implied volatilities of ATM European swaptions and annual forward rates as observed on the 11-th of August 2004 in the Euro market. Swaption implied volatilities are reported in Table 13.1, whereas the initial forward rates are shown in Table 13.2. The missing entries in the swaption matrix have been filled in by means of a log-linear interpolation.

Since our objective is to price a ten-year Bermudan swaption, only the upper triangle of the interpolated swaption matrix will be used for the LFM calibration.

	Maturities						
Tenors	1y	2y	3y	4y	5y	7y	10y
1y	25.2	21.8	19.1	17.3	15.9	13.9	12.4
2y	23.5	20.1	17.9	16.3	15.0	13.3	11.8
3y	21.4	18.7	16.8	15.3	14.2	12.6	11.3
4y	19.4	17.4	15.7	14.4	13.4	12.0	10.8
5y	18.0	16.3	14.7	13.5	12.7	11.4	10.3
6y	16.8	15.3	13.8	13.0	12.2	11.2	10.1
7y	15.9	14.6	13.4	12.6	12.0	11.1	10.1
8y	15.1	14.0	13.0	12.4	11.8	10.9	10.0
9y	14.5	13.5	12.8	12.1	11.6	10.8	9.9
10y	13.9	13.2	12.5	11.9	11.5	10.8	9.9

Table 13.1. Implied volatilities, in percentage points, of ATM European Swaptions as observed on the 11-th of August 2004 in the Euro market.

[5] For a detailed description of the calibration algorithm we refer to Lvov (2005).

Expiry	F(0)
1y	0.0298658
2y	0.0366475
3y	0.0409215
4y	0.0443771
5y	0.0474538
6y	0.0497249
7y	0.0513826
8y	0.052244
9y	0.0529868
10y	0.0539636

Table 13.2. Initial annual forward rates as observed on the 11-th of August 2004 in the Euro market.

The purpose of the calibration is to find the LFM parameters for the family of annual LIBOR rates underlying a 10-non-call-1 Bermudan swaption. This family spans the period from year one to year eleven and has ten forward rates that are "alive" at time zero. The set of instruments available for calibration comprises fifty-five ATM European swaptions, ten of which are co-terminal to the Bermudan swaption in question and shall be fitted exactly.

The number of model parameters is 13 for the DRC-CV, 16 for the DRC-TVV and 55 for both the CC-1 and the CC-2. Both the DRC-CV and the DRC-TVV use 10 parameters for the exact calibration to the co-terminal swaptions, leaving the remaining ones (respectively 3 and 6) the task of fitting the other 45 swaptions. Both the CC-1 and the CC-2, instead, use all their parameters for an exact calibration to all swaptions in the upper triangular matrix, including the co-terminal ones.

After performing our four LFM calibrations, we then price three 10-non-call-1 payer's Bermudan swaptions, with strikes of 4.5%, which equals the forward swap rate starting in year one and terminating in year eleven (ATM), of 3.5% (ITM) and 5.5% (OTM). All prices are obtained with 10,000 paths generated using Sobol random numbers. The early-exercise boundary is constructed with the least-squares regression method explained in Section 13.14. The set of basis functions comprises all underlying forward swap rates, their squares and a constant. Prices and standard deviations are reported in Table 13.3.[6]

It can be seen from Table 13.3 that, for all considered strikes, Bermudan swaption prices obtained with differently calibrated LFMs are very similar. In particular, the maximum absolute relative difference between prices associated to different calibrations is less than 1% for ATM swaptions, less than

[6] Standard deviations are computed without accounting for the fact that the simulation employs Sobol numbers. Therefore, they can be viewed as a conservative estimate of price accuracy.

Strike	DRC-CV	DRC-TVV	CC-1	CC-2
4.5% (ATM)	51.51 (0.55)	51.31 (0.53)	51.62 (0.55)	51.73 (0.53)
3.5% (ITM)	94.87 (0.62)	94.67 (0.61)	94.74 (0.62)	94.85 (0.61)
5.5% (OTM)	26.19 (0.41)	26.24 (0.40)	26.42 (0.41)	26.71 (0.40)

Table 13.3. Monte-Carlo prices and standard errors for a 10-non-call-1 payer's Bermudan swaption with 1000 notional, computed with differently calibrated LFMs.

0.25% for ITM swaptions and is within 2% in the OTM case.[7] No calibration appears to lead systematically to the lowest or highest option price.

The above results seem to indicate that Bermudan swaption prices are not very sensitive to the specific choice of volatility and correlation structures, as long as the corresponding LFM is calibrated exactly to co-terminal swaptions along with a best fit to the upper triangular matrix. Furthermore, since both DRCs lead to Bermudan swaption prices that are consistent with those obtained with the two cascade calibrations, an under-fitting of above-diagonal elements does not seem to lead to a significant miss-pricing of a Bermudan swaption.[8]

13.16 New Generation of Contracts

In their continuous struggle to attract institutional, retail and corporate clients, financial institutions have been proposing more and more innovative interest-rate derivatives. The competition is based on designing increasingly sophisticated payoffs, depending on both LIBOR and swap rates and possibly involving path-dependent and callability features at the same time. A typical example is a callable swap where payments at LIBOR plus a spread are exchanged for amounts depending on one or more CMS rates.

The pricing, and especially hedging, of this new generation of contracts can be rather involved and problematic. Using for instance a LIBOR model, the valuation procedure should be ideally based on a joint calibration to the cap and swaption markets, possibly accounting for the respective smiles and skews. Moreover, the Monte Carlo simulation of forward rates and resulting payoffs must be coupled with a procedure for estimating the early-exercise region, like for instance the least-squares method of Section 13.14.

Unfortunately, no universal procedure is readily available for pricing and hedging general payoffs. In fact, tailor-made methods accounting for the specific nature of the contract in question, are usually necessary for a faster and more robust implementation. However, as a general recipe for the LFM, one

[7] We also notice that part of the price discrepancies is due to numerical noise.

[8] Many other considerations and empirical studies on the calibration, pricing and hedging of Bermudan swaptions can be found in Lvov (2005).

can tackle various forms of path dependence as in Sections 13.5, 13.9 and 13.13, whereas early-exercise features can be addressed by following the procedures in Sections 13.14 and 13.15. The implementation of clever tricks and shortcuts, anyway, is often indispensable for a pricing model to be efficiently applied in practice.

Among the variety of contracts that have recently appeared in the interest-rate market, we have here selected two derivatives, which are both extremely popular and successful and can be taken as main representatives of LIBOR-based and CMS-based derivatives, respectively. Their payoffs are described in the following sections.

13.16.1 Target Redemption Notes

A Target Redemption Note, often referred to by its acronym TARN, is a note paying coupons whose overall amount is initially set by the contract and must be fully paid within the note's maturity. Coupon payments are subject to trigger conditions, so that the note may fail to pay on some of the contract's payment times. Moreover, as soon as the sum of already-paid coupons equals the value set by the contract, the note automatically expires paying the notional.

With respect to a classical coupon-bearing bond, a TARN presents two major differences: i) a coupon may not be paid on a (possible) payment time; ii) the actual maturity of the note can be shorter than the nominal maturity.

The attractive features of a TARN are: i) the usually-large coupon that is paid on the first years, and ii) the certainty of receiving a given overall coupon within a given maturity.

A TARN is characterized by a set of payment times $T_1, T_2, \ldots, T_M$, where $T_i = i$ years, a fixed coupon c, a trigger level X and an overall sum of coupons S. The actual coupon $C(T_i)$ paid at time T_i is

$$C(T_i) = \begin{cases} c & i = 1, \ldots, k \\ \min\left[c, \left(S - \sum_{j=1}^{i-1} C(T_j)\right)^+\right] 1\{\omega L(T_{i-1}, T_i) \le \omega X\} & k < i < M \\ \left(S - \sum_{j=1}^{M-1} C(T_j)\right)^+ & i = M \end{cases} \tag{13.29}$$

where year fractions are all set equal to one, $\omega = 1$ (-1) for an upper (lower) trigger level, and where no trigger condition is present at maturity T_M so as to grant that

$$\sum_{j=1}^{M} C(T_j) = S.$$

The last non-zero coupon payment occurs at time T_{i^*},

$$i^* := \min\left\{ i : \sum_{j=1}^{i} C(T_j) = S \right\},$$

where the (unit) nominal value is also paid. The random time T_{i^*} is therefore the actual maturity of the note.

The no-arbitrage value of the above TARN is given by

$$E\left\{ \sum_{i=1}^{i^*} D(0,T_i)C(T_i) + D(0,T_{i^*}) \right\}$$
$$= P(0,T_M)E^M \left[\sum_{i=1}^{i^*} \frac{C(T_i)}{P(T_i,T_M)} + \frac{1}{P(T_{i^*},T_M)} \right],$$

which can be calculated, using a LIBOR model, by Monte Carlo simulation exactly in the same way as explained in Section 13.3.

Remark 13.16.1. **(TARN's and swaps).** Target redemption notes are often swapped at LIBOR plus a spread. The resulting swap has a standard floating leg, paying LIBOR plus a margin, and a "fixed" leg paying coupons given by (13.29). In this case, the whole swap terminates at T_{i^*}.

Remark 13.16.2. **(Alternative TARN's).** Several versions of coupons (13.29) can be considered and some have also been proposed in the market. For instance, one can replace constant payments c with time-dependent ones, c_i. One can also impose that a low coupon is paid whenever the trigger condition is not fulfilled. Moreover, the trigger variable (the LIBOR rate in the above case) can be replaced by a CMS rate, a CMS spread or something more involved. The trigger condition itself can be eliminated and coupons c_i can be made stochastic. In principle, any payment structure can be "tarned" by imposing that the overall sum of effectively-paid coupons equals an a-priori given quantity.

13.16.2 CMS Spread Options

CMS spread options are options on the spread between two different swap rates. The first is typically a long-maturity swap rate (from ten years onwards), the second a short-maturity one (one or two years).

Considering the usual time structure $T_1, T_2, \ldots$, and given $\alpha < \beta < \gamma$, a CMS spread option pay off at time T_α

$$[S_{\alpha,\gamma}(T_\alpha) - S_{\alpha,\beta}(T_\alpha) - X]^+, \tag{13.30}$$

where X is the option's strike. CMS spread options are thus traded as bets or hedges against a steepening or a flattening of the zero-coupon curve.

The no-arbitrage value of the payoff (13.30) at time zero is given by

$$P(0,T_\alpha)E^\alpha\{[S_{\alpha,\gamma}(T_\alpha) - S_{\alpha,\beta}(T_\alpha) - X]^+\}, \tag{13.31}$$

and can be calculated as soon as we know the joint distribution of the pair $(S_{\alpha,\gamma}(T_\alpha), S_{\alpha,\beta}(T_\alpha))$ under the T_α-forward measure.

Such a distribution can be generated, for instance, in the G2++ model by Monte Carlo simulation of factors x and y according to Lemma 4.2.2, or in a LFM by simulating trajectories of the relevant forward rates.

The simplest valuation procedure, however, is based on assuming that the logarithms of the swap rates are jointly normally distributed as in a Black and Scholes (1973) model with two underlying assets, see also Appendix E. A formal justification of this approach is given by resorting to a LSM and to suitable approximations. In fact, let us assume that

$$dS_{\alpha,x}(t) = \sigma_{\alpha,x}S_{\alpha,x}(t)\,dZ_{\alpha,x}(t), \quad \text{under } Q^{\alpha,x}, \; x \in \{\beta,\gamma\}, \tag{13.32}$$

where $\sigma_{\alpha,\beta}$ and $\sigma_{\alpha,\gamma}$ are positive constants, and $Z_{\alpha,x}$ is a standard Brownian motion under $Q^{\alpha,x}$. The dynamics of both swap rates under the T_α-forward measure, which are needed to calculate (13.31), are given by

$$\begin{aligned} dS_{\alpha,\beta}(t) &= \mu_{\alpha,\beta}(t)S_{\alpha,\beta}(t) + \sigma_{\alpha,\beta}S_{\alpha,\beta}(t)\,dZ_\beta(t), \\ dS_{\alpha,\gamma}(t) &= \mu_{\alpha,\gamma}(t)S_{\alpha,\gamma}(t) + \sigma_{\alpha,\gamma}S_{\alpha,\gamma}(t)\,dZ_\gamma(t), \end{aligned} \tag{13.33}$$

where $dZ_\beta(t)\,dZ_\gamma(t) = \rho\,dt$, and the drift terms $\mu_{\alpha,\beta}$ and $\mu_{\alpha,\gamma}$ are implicitly defined in (6.39). Setting $\bar{\mu}_x := \mu_{\alpha,x}(0)$, we then resort to the classical freezing technique and replace $\mu_{\alpha,x}(t)$ with $\bar{\mu}_x$, so that the dynamics of both swap rates under Q^α are approximately lognormal and

$$E^\alpha[S_{\alpha,x}(T_\alpha)] \approx S_{\alpha,x}(0)\,e^{\bar{\mu}_x T_\alpha}, \quad x \in \{\beta,\gamma\}.$$

Applying the result of Appendix E, we finally have that the price at time zero of the above CMS spread option is

$$\mathbf{CMSSO}(0,T_\alpha,X;\beta,\gamma) = P(0,T_\alpha)\int_{-\infty}^{+\infty} \frac{1}{\sqrt{2\pi}}e^{-\frac{1}{2}v^2}f(v)\,dv, \tag{13.34}$$

where

$$\begin{aligned} f(v) =& S_{\alpha,\gamma}(0)\exp\left[\bar{\mu}_\gamma T_\alpha - \frac{1}{2}\rho^2\sigma^2_{\alpha,\gamma}T_\alpha + \rho\sigma_{\alpha,\gamma}\sqrt{T_\alpha}v\right] \\ &\cdot \Phi\left(\frac{\ln\frac{S_{\alpha,\gamma}(0)}{h(v)} + [\bar{\mu}_\gamma + (\frac{1}{2} - \rho^2)\sigma^2_{\alpha,\gamma}]T_\alpha + \rho\sigma_{\alpha,\gamma}\sqrt{T_\alpha}v}{\sigma_{\alpha,\gamma}\sqrt{T_\alpha}\sqrt{1-\rho^2}}\right) \\ &- h(v)\Phi\left(\frac{\ln\frac{S_{\alpha,\gamma}(0)}{h(v)} + (\bar{\mu}_\gamma - \frac{1}{2}\sigma^2_{\alpha,\gamma})T_\alpha + \rho\sigma_{\alpha,\gamma}\sqrt{T_\alpha}v}{\sigma_{\alpha,\gamma}\sqrt{T_\alpha}\sqrt{1-\rho^2}}\right) \end{aligned}$$

and

$$h(v) = X + S_{\alpha,\beta}(0)e^{(\bar{\mu}_\beta - \frac{1}{2}\sigma_{\alpha,\beta}^2)T_\alpha + \sigma_{\alpha,\beta}\sqrt{T_\alpha}v}.$$

Price (13.34) can easily be calculated by taking the swap-rate volatilities $\sigma_{\alpha,\gamma}$ and $\sigma_{\alpha,\beta}$ from the swaption market and by inferring the constants $\bar{\mu}_\gamma$ and $\bar{\mu}_\beta$ from the respective convexity adjustments (13.16). The correlation parameter ρ can be estimated historically.

Remark 13.16.3. **(Callable CMS spread options).** Payments (13.30) are typically swapped in the market at LIBOR plus a spread. The value of the resulting structure can be simply calculated by pricing the CMS-based payments according to formula (13.34). Such a swap is often callable. This, however, complicates considerably the valuation procedure. In this case, the simple LSM above can not be used in a straightforward manner, and classical solutions like the G2++ model or the LIBOR market model are actually preferable.

14. Pricing Derivatives on Two Interest-Rate Curves

So curiosity is in a sense the heart of practice.
Charlotte Joko Beck, "Nothing Special: Living Zen", 1995, HarperCollins.

In this chapter, we explain how one can model both a first (domestic) and a second (foreign) interest-rate curve, each by a two-factor additive Gaussian short-rate model, in order to Monte Carlo price a quanto constant-maturity swap and similar contracts, which we will present in the following sections.

We need to compute the payoff of a contract of this kind: party A pays party B a certain rate associated with a first currency "1", say Euro to fix ideas. Party B pays A the currency "1" amount expressed by a certain rate associated with a second currency "2", say British Pounds (GBP). To compute the expectations involved in pricing this kind of contracts, a good strategy can be to use a short-rate model with analytical formulas for zero-coupon bond prices in terms of the factors concurring to the short-rate. In such cases, we carry out the different rates involved in the contract in terms of the discount factors, so that the above expectations are easily computed once the distribution of the short-rate factors is known.

At the same time, the chosen model has to allow for a realistic volatility structure for each curve, and also correlation between different rates has to be modeled realistically, since usually swap rates are involved. A good choice, therefore, can be given by selecting a two-factor Gaussian short-rate model for each curve.

The modeling of stochastic interest rates in a multi-currency setting has been tackled by many authors under different assumptions and frameworks. We quote, among others, the works of Andreasen (1995), Frey and Sommer (1996), Flesaker and Hughston (1996), Rogers (1997), Mikkelsen (2001), Schlögl (2002), Pelsser (2003) and Amin (2003).

14.1 The Attractive Features of G2++ for Multi-Curve Payoffs

14.1.1 The Model

The basic model we adopt for each curve is described in detail in Chapter 4. Here, we briefly reintroduce it, and slightly adapt the notation to the present context.

We assume that the dynamics of the instantaneous short-rate process for curve "1", under the associated risk-adjusted measure Q_1, is given by

$$\text{Curve "1" (Euro): } r_1(t) = x_1(t) + y_1(t) + \varphi_1(t), \quad r_1(0) = r_0^1, \tag{14.1}$$

where the processes $\{x_1(t) : t \geq 0\}$ and $\{y_1(t) : t \geq 0\}$ satisfy

$$\begin{aligned} dx_1(t) &= -a_1 x_1(t)dt + \sigma_1 dW_1^x(t), \quad x_1(0) = 0, \\ dy_1(t) &= -b_1 y_1(t)dt + \eta_1 dW_1^y(t), \quad y_1(0) = 0, \end{aligned} \tag{14.2}$$

where W_1^x and W_1^y are two correlated Brownian motions, $dW_1^x(t)dW_1^y(t) = \rho_1 dt$, and r_0^1, a_1, b_1, σ_1, η_1 and ρ_1 are suitable constants. The function φ_1 is deterministic and well defined in the time interval $[0, T_n]$, with T_n a given time horizon, for example 10, 30 or 50 (years). In particular, $\varphi_1(0) = r_0^1$. We denote by $\mathcal{F}_t^1$ the sigma-field generated by the pair (x_1, y_1) up to time t.

Simple integration of equations (14.2) implies that for each $s < t$

$$\begin{aligned} r_1(t) = x_1(s)e^{-a_1(t-s)} + y_1(s)e^{-b_1(t-s)} + \sigma_1 \int_s^t e^{-a_1(t-u)} dW_1^x(u) \\ + \eta_1 \int_s^t e^{-b_1(t-u)} dW_1^y(u) + \varphi_1(t), \end{aligned}$$

meaning that $r_1(t)$ conditional on $\mathcal{F}_s^1$ is normally distributed with mean and variance given respectively by

$$\begin{aligned} E\{r_1(t)|\mathcal{F}_s^1\} &= x_1(s)e^{-a_1(t-s)} + y_1(s)e^{-b_1(t-s)} + \varphi_1(t), \\ \text{Var}\{r_1(t)|\mathcal{F}_s^1\} &= \frac{\sigma_1^2}{2a_1}\left[1 - e^{-2a_1(t-s)}\right] + \frac{\eta_1^2}{2b_1}\left[1 - e^{-2b_1(t-s)}\right] \\ &\quad + 2\rho_1 \frac{\sigma_1 \eta_1}{a_1 + b_1}\left[1 - e^{-(a_1+b_1)(t-s)}\right]. \end{aligned} \tag{14.3}$$

This was already noticed in (4.6) of Chapter 4.

The second curve is modeled analogously. We in fact assume that the dynamics of the instantaneous short-rate process for curve "2", under the associated risk-adjusted measure Q_2, is given by

$$\text{Curve "2" (GBP): } r_2(t) = x_2(t) + y_2(t) + \varphi_2(t), \quad r_2(0) = r_0^2, \tag{14.4}$$

where the processes $\{x_2(t) : t \geq 0\}$ and $\{y_2(t) : t \geq 0\}$ satisfy

$$dx_2(t) = -a_2 x_2(t)dt + \sigma_2 dW_2^x(t), \quad x_2(0) = 0, \tag{14.5}$$
$$dy_2(t) = -b_2 y_2(t)dt + \eta_2 dW_2^y(t), \quad y_2(0) = 0,$$

where W_2^x and W_2^y are two correlated Brownian motions, $dW_2^x(t)dW_2^y(t) = \rho_2 dt$, and r_0^2, a_2, b_2, σ_2, η_2 and ρ_2 are suitable constants. The function φ_2 is deterministic and well defined in the time interval $[0, T_n]$. In particular, $\varphi_2(0) = r_0^2$. We denote by $\mathcal{F}_t^2$ the sigma-field generated by the pair (x_2, y_2) up to time t. The explicit expression and transition densities of r_2 are completely analogous to those for curve "1". It is indeed sufficient to replace subscripts and superscripts "1" with "2".

Now consider the market instantaneous forward rates for the two curves "1" and "2", respectively, at the initial time 0:

$$f_1^M(0,T) = -\frac{\partial \ln P_1^M(0,T)}{\partial T},$$
$$f_2^M(0,T) = -\frac{\partial \ln P_2^M(0,T)}{\partial T},$$

where the superscript "M" denotes financial quantities as observed in the market.

It is shown in Section 4.2.2 that by choosing

$$\varphi_1(T) = f_1^M(0,T) + \frac{\sigma_1^2}{2a_1^2}\left(1 - e^{-a_1 T}\right)^2 + \frac{\eta_1^2}{2b_1^2}\left(1 - e^{-b_1 T}\right)^2$$
$$+\rho_1 \frac{\sigma_1 \eta_1}{a_1 b_1}\left(1 - e^{-a_1 T}\right)\left(1 - e^{-b_1 T}\right),$$
$$\varphi_2(T) = f_2^M(0,T) + \frac{\sigma_2^2}{2a_2^2}\left(1 - e^{-a_2 T}\right)^2 + \frac{\eta_2^2}{2b_2^2}\left(1 - e^{-b_2 T}\right)^2$$
$$+\rho_2 \frac{\sigma_2 \eta_2}{a_2 b_2}\left(1 - e^{-a_2 T}\right)\left(1 - e^{-b_2 T}\right),$$

as in (4.12) of Chapter 4, the initial term structures of discount factors $T \mapsto P_1^M(0,T)$ and $T \mapsto P_2^M(0,T)$ for the curves "1" and "2" are perfectly reproduced by the above models for r_1 and r_2, respectively.

An equivalent condition is (we just write it for curve "1")

$$\exp\left\{-\int_t^T \varphi_1(u)du\right\} = \frac{P_1^M(0,T)}{P_1^M(0,t)} \exp\left\{-\frac{1}{2}[V_1(0,T) - V_1(0,t)]\right\}, \tag{14.6}$$

where in general

$$
\begin{aligned}
V_1(t,T) = &\frac{\sigma_1^2}{a_1^2}\left[T-t+\frac{2}{a_1}e^{-a_1(T-t)}-\frac{1}{2a_1}e^{-2a_1(T-t)}-\frac{3}{2a_1}\right]\\
&+\frac{\eta_1^2}{b_1^2}\left[T-t+\frac{2}{b_1}e^{-b_1(T-t)}-\frac{1}{2b_1}e^{-2b_1(T-t)}-\frac{3}{2b_1}\right]\\
&+2\rho_1\frac{\sigma_1\eta_1}{a_1b_1}\left[T-t+\frac{e^{-a_1(T-t)}-1}{a_1}+\frac{e^{-b_1(T-t)}-1}{b_1}\right.\\
&\left.-\frac{e^{-(a_1+b_1)(T-t)}-1}{a_1+b_1}\right],
\end{aligned}
\tag{14.7}
$$

as in (4.10) of Chapter 4.

The corresponding condition and definition for curve "2" are analogous, with the subscript "2" replacing the subscript "1".

14.1.2 Interaction Between Models of the Two Curves "1" and "2"

So far we have worked on the models for the curves "1" (Euro) and "2" (GBP) separately. We now consider quantities describing the interaction of the two curves. We assume the following instantaneous correlations between the factors of the two curves:

$$
d\begin{bmatrix} W_1^x \\ W_1^y \\ W_2^x \\ W_2^y \end{bmatrix} d\left[W_1^x\ W_1^y\ W_2^x\ W_2^y\right] = \begin{bmatrix} 1 & \rho_1 & \gamma_{x1,x2} & \gamma_{x1,y2} \\ . & 1 & \gamma_{y1,x2} & \gamma_{y1,y2} \\ . & . & 1 & \rho_2 \\ . & . & . & 1 \end{bmatrix} dt
$$

where the entries that are not specified are determined by symmetry. We are therefore assuming that this matrix is positive semidefinite and:

- The instantaneous correlation between shocks in the first factor of the first curve and the second factor of the first curve is the previously introduced ρ_1.
- The instantaneous correlation between shocks in the first factor of the first curve and the first factor of the second curve is the new parameter $\gamma_{x1,x2}$.
- The instantaneous correlation between shocks in the first factor of the first curve and the second factor of the second curve is the new parameter $\gamma_{x1,y2}$.
- The instantaneous correlation between shocks in the second factor of the first curve and the second factor of the second curve is the new parameter $\gamma_{y1,y2}$.
- The instantaneous correlation between shocks in the second factor of the first curve and the first factor of the second curve is the new parameter $\gamma_{y1,x2}$.
- The instantaneous correlation between shocks in the first factor of the second curve and the second factor of the second curve is the previously introduced ρ_2.

However, a trader may find it difficult to express views on correlations between single factors. Indeed, it would be preferable to express views on the instantaneous correlations between the two rates r_1 and r_2 themselves. This can be obtained by observing that

$$\mathrm{Corr}\{dr_1, dr_2\} = \frac{\sigma_1\sigma_2\gamma_{x1,x2} + \eta_1\sigma_2\gamma_{y1,x2} + \sigma_1\eta_2\gamma_{x1,y2} + \eta_1\eta_2\gamma_{y1,y2}}{\sqrt{\sigma_1^2 + \eta_1^2 + 2\rho_1\sigma_1\eta_1}\,\sqrt{\sigma_2^2 + \eta_2^2 + 2\rho_2\sigma_2\eta_2}}\,. \tag{14.8}$$

Now notice that the parameters σ_1, η_1, ρ_1 can be determined by calibration of r_1 to the cap or swaption markets related to curve "1" (Euro), whereas σ_2, η_2, ρ_2 can be determined by calibration of r_2 to the cap or swaption markets related to curve "2" (GBP). At this point, after these two separate calibrations, the trader may be willing to express a view on the instantaneous correlation $\mathrm{Corr}\{dr_1, dr_2\}$ between the two curves. However, we still have four unknowns $\gamma_{x,y}$ and just one equation. A simplifying assumption that can be made is that

$$\gamma_{x1,x2} = \gamma_{x1,y2} = \gamma_{y1,x2} = \gamma_{y1,y2} =: \gamma.$$

Now, when the trader expresses her view on $\mathrm{Corr}\{dr_1, dr_2\}$, equation (14.8) can be solved in the unique unknown γ:

$$\gamma = \mathrm{Corr}\{dr_1, dr_2\}\frac{\sqrt{\sigma_1^2 + \eta_1^2 + 2\rho_1\sigma_1\eta_1}\,\sqrt{\sigma_2^2 + \eta_2^2 + 2\rho_2\sigma_2\eta_2}}{(\sigma_1 + \eta_1)(\sigma_2 + \eta_2)}. \tag{14.9}$$

It is immediate to see that γ is always a number whose absolute value is smaller than one, and therefore a "legal" correlation parameter:

$$-1 \le -|\mathrm{Corr}\{dr_1, dr_2\}| \le \gamma \le |\mathrm{Corr}\{dr_1, dr_2\}| \le 1\,.$$

Notice that extreme cases are allowed: $\gamma = 0$ translates the view of no instantaneous correlation between curves "1" and "2", $\mathrm{Corr}\{dr_1, dr_2\} = 0$, whereas perfect positive or negative correlations, $\mathrm{Corr}\{dr_1, dr_2\} = \pm 1$, can be attained respectively with

$$\gamma = \pm\frac{\sqrt{\sigma_1^2 + \eta_1^2 + 2\rho_1\sigma_1\eta_1}\,\sqrt{\sigma_2^2 + \eta_2^2 + 2\rho_2\sigma_2\eta_2}}{(\sigma_1 + \eta_1)(\sigma_2 + \eta_2)}\,.$$

We remind that the above dynamics for x and y are given under the respective risk-neutral measures: x_1 and y_1 are modeled under the risk-neutral measure Q_1 for curve "1", whereas x_2 and y_2 are modeled under the risk-neutral measure Q_2 for curve "2".

14.1.3 The Two-Models Dynamics under a Unique Convenient Forward Measure

We need to express the equations for the factors x_2 and y_2 describing curve "2" (GBP) under the risk-neutral measure Q_1, related to curve "1" (Euro).

This in turn requires modeling the exchange rate between markets "1" and "2" (GBP). Let $X(t)$ denote the amount of currency "2" (GBP) needed to buy one unit of currency "1" (Euro). We assume the following no-arbitrage dynamics for X under Q_2:

$$dX(t) = ((r_2(t) - r_1(t))X(t)\,dt + \nu X(t)\,dW^X(t)\,.$$

We need also to model the instantaneous correlations between the exchange rate X and the two factors of curve "2" (GBP), x_2 and y_2. We assume

$$d\begin{bmatrix} W^X \\ W_2^x \\ W_2^y \end{bmatrix} d\left[W^X\ W_2^x\ W_2^y\right] = \begin{bmatrix} 1 & c_{x2,X} & c_{y2,X} \\ . & 1 & \rho_2 \\ . & . & 1 \end{bmatrix} dt$$

where the entries that are not specified are determined by symmetry. We are therefore assuming that:

- The instantaneous correlation between shocks in the first factor of the second curve "2" and the exchange rate between markets "1" and "2" is the new parameter $c_{x2,X}$.
- The instantaneous correlation between shocks in the second factor of the second curve "2" and the exchange rate between markets "1" and "2" is the new parameter $c_{y2,X}$.
- The instantaneous correlation between shocks in the first factor of the second curve and the second factor of the second curve is the previously introduced ρ_2.

The trader may find it difficult to express views or to estimate correlation between single factors of curve "2" and the exchange rate. What is reasonable to expect, instead, is some view on the correlation between the instantaneous rate of curve "2" as a whole and the exchange rate. Notice that such a correlation is given by

$$\mathrm{Corr}\{dX, dr_2\} = \frac{\sigma_2 c_{x2,X} + \eta_2 c_{y2,X}}{\sqrt{\sigma_2^2 + \eta_2^2 + 2\rho_2\sigma_2\eta_2}}\,, \tag{14.10}$$

so that by inverting this formula one can translate the trader's view into model parameters. However, equation (14.10) cannot be inverted as it stands, since there are two unknowns. A simplifying assumption that can be made at this point is

$$c_{x2,X} = c_{y2,X} =: c_X\,.$$

Following this assumption (14.10) can now be inverted so as to yield

$$c_X = \mathrm{Corr}\{dX, dr_2\}\frac{\sqrt{\sigma_2^2 + \eta_2^2 + 2\rho_2\sigma_2\eta_2}}{\sigma_2 + \eta_2}\,, \tag{14.11}$$

so that, by means of this formula, one can translate the trader's view into the model parameter c_X. Notice that the fraction in the above formula is a positive number smaller or equal than one. As a consequence,

$$-1 \le -|\mathrm{Corr}\{dX, dr_2\}| \le c_X \le |\mathrm{Corr}\{dX, dr_2\}| \le 1,$$

i.e. c_X is always a viable correlation. Notice also that extreme cases are allowed: $c_X = 0$ translates the view of no instantaneous correlation between curve "2" and the exchange rate, $\mathrm{Corr}\{dX, dr_2\} = 0$, whereas perfect positive or negative correlation, $\mathrm{Corr}\{dX, dr_2\} = \pm 1$, can be attained respectively with

$$c_X = \pm\sqrt{\sigma_2^2 + \eta_2^2 + 2\rho_2\sigma_2\eta_2}/(\sigma_2 + \eta_2).$$

We can now express the equations for the factors x_2 and y_2, describing curve "2" (GBP) under the risk-neutral measure Q_1 associated with market "1". We use the change-of-numeraire technique and move from the bank-account numeraire for market "2" to the numeraire

$$X(t) \times (\text{bank account of market "1"}) .$$

This is the right change of numeraire for moving from measure Q_2 to measure Q_1 as explained in Section 2.9 on numeraire changes between domestic and foreign markets. Beware that, in Section 2.9, our exchange rate X was $1/\mathcal{Q}$, while X was an asset price. Moreover B^f denoted the bank account of market "2".

We can apply the change-of-numeraire toolkit (as described in Section 2.3) obtaining

$$\begin{aligned} dx_2(t) &= [-a_2 x_2(t) + \sigma_2 \nu c_{x2,X}]\, dt + \sigma_2 dU_2^x(t), \\ dy_2(t) &= [-b_2 y_2(t) + \eta_2 \nu c_{y2,X}]\, dt + \eta_2 dU_2^y(t), \end{aligned} \tag{14.12}$$

where U_2^x and U_2^y are Brownian motions under Q_1, whose correlation structure is the same as that of W_2^x and W_2^y.

14.2 Quanto Constant-Maturity Swaps

A quanto CMS is a financial product involving both multi-currency issues and unnatural time lags for rate payments. Therefore, we will develop this example to a certain degree, up to a scheme for implementing the G2++ model for quanto CMS.

14.2.1 Quanto CMS: The Contract

Consider the following contract. Party A pays to party B an amount in a first currency "1" (Euro) expressed by the c-year swap rate associated with a second currency "2" (GBP). Payments occur once every δ years (typically $\delta = 0.5$ years $= 6$ months). Notice that here we slightly generalize the CMS case of Section 13.7 to payment intervals different from one year. On the same payment dates, party B pays to party A the δ-year simply-compounded LIBOR rate associated with the currency "1". Formally, the payoff can be expressed as follows.

We assume unit nominal amount. Let us assume the current time to be $t = 0$, and let us denote by $\mathcal{T} = \{T_1, \ldots, T_n\}$ the set of payment dates at which the flows of the two legs are to be exchanged. We denote by τ_i the year fraction between T_{i-1} and T_i, which will be close in general to the single value δ. We take $T_0 = 0$.

- At time T_i, $i \geq 1$, party A pays to B, in currency "1", the amount expressed by the c-year swap rate $S_{i-1,i-1+c/\delta}(T_{i-1})$ for the curve "2" as reset in time T_{i-1}:

$$S_{i-1,i-1+c/\delta}(T_{i-1})\tau_i \text{ units of currency "1".}$$

  In general, the (forward) swap rate at time $t \leq T_{j-1}$ for a swap whose payments occur at times $T_j, T_{j+1}, \ldots, T_{m-1}$ and with first reset date at T_{j-1} is defined as

$$S_{j-1,m-1}(t) = \frac{P_2(t, T_{j-1}) - P_2(t, T_{m-1})}{\sum_{k=j}^{m-1} \tau_k P_2(t, T_k)} \tag{14.13}$$

  (where indices are shifted by one for later notation convenience). In general, $P_2(t, T)$ denotes the discount factor for the curve "2" at time t for maturity T. An analogous definition is given for curve "1".
- At time T_i, $i \geq 1$, Institution B pays to A the LIBOR rate for curve "1":

$$\tau_i L_1(T_{i-1}, T_i) \text{ units of currency "1".}$$

  The quantity

$$L_1(T_{i-1}, T_i) = \frac{1}{\tau_i}\left(\frac{1}{P_1(T_{i-1}, T_i)} - 1\right)$$

  denotes the LIBOR rate at time T_{i-1} for maturity T_i for curve "1".

The net value of the contract as seen by B at time 0 is

$$\begin{aligned} &E_1\left\{\sum_{i=1}^{n} \exp\left(-\int_0^{T_i} r_1(s)\,ds\right)[S_{i-1,i-1+c/\delta}(T_{i-1}) - L_1(T_{i-1}, T_i)]\tau_i\right\} \\ &= P_1(0, T_n)\sum_{i=1}^{n} \tau_i E_1^n\left\{\frac{S_{i-1,i-1+c/\delta}(T_{i-1})}{P_1(T_i, T_n)}\right\} - (1 - P_1(0, T_n)), \end{aligned} \tag{14.14}$$

where E_1 denotes the risk-neutral expectation for curve "1", and E_1^i is in general expectation with respect to the T_i-forward measure $Q_1^{T_i}$ for curve "1".

We need to compute

$$E_1^n[S_{i-1,i-1+c/\delta}(T_{i-1})/P_1(T_i,T_n)] \tag{14.15}$$

for all i.

To this purpose, we remember the considerations made in the introduction of this chapter. We can use a short-rate model with analytical formulas for zero-coupon bond prices in terms of the factors concurring to the short rate. In such a case, (14.13) provides an analytical formula for swap rates in terms of such factors, and it is enough to simulate paths for the factors of the short rates up to T_n under the measure Q_1^n. If such a dynamics is linear, the transition densities are Gaussian. The transition of the factors of the short rate between two instants T_{i-1}, T_i can thus be simulated one-shot from the such Gaussian densities, without resorting to a time-discretization of the dynamics between T_{i-1} and T_i.

At the same time, the chosen model has to allow for a realistic volatility structure for each curve, and moreover correlation between different rates has to be modeled in curve "2", since swap rates are involved. A good choice can therefore be given by selecting a two-factor Gaussian short rate model for each curve.

14.2.2 Quanto CMS: The G2++ Model

Starting from the models described in Section 14.1, we deduce the analytical formulas needed for the quanto-CMS payoff.

The model allows for the following bond-price formula for curve "1":

$$\begin{aligned} P_1(t,T;x_1(t),y_1(t)) = &\frac{P_1^M(0,T)}{P_1^M(0,t)} \exp\left\{\frac{1}{2}[V_1(t,T) - V_1(0,T) + V_1(0,t)]\right. \\ &\left. - \frac{1-e^{-a_1(T-t)}}{a_1} x_1(t) - \frac{1-e^{-b_1(T-t)}}{b_1} y_1(t)\right\}, \end{aligned}$$

where $V_1(t,T)$ is defined in (14.7) (see also (4.14) in Chapter 4).

The corresponding bond-price formula for curve "2" is analogous, with the subscript "2" replacing the subscript "1".

Swap rates can be computed analytically in terms of x and y from the above formula for discount factors. For the (forward) swap rate of curve "2" we can rewrite (14.13) in terms of model quantities:

$$S_{j-1,m-1}(t;x_2(t),y_2(t)) = \frac{P_2(t,T_{j-1};x_2(t),y_2(t)) - P_2(t,T_{m-1};x_2(t),y_2(t))}{\sum_{k=j}^{m-1} \tau_k P_2(t,T_k;x_2(t),y_2(t))}. \tag{14.17}$$

In order to compute (14.15), we need the dynamics of x_1, y_1, x_2 and y_2 under the T_n-forward measure Q_1^n for curve "1" (Euro).

To this purpose, recall that in Section 14.1.3 we derived the equations for the factors x_2 and y_2 of curve "2" (GBP) under the risk-neutral measure Q_1 associated with market "1" (Euro). After the exchange rate between markets

"1" and "2", we modeled the instantaneous correlations between the exchange rate "X" and the two factors, x_2 and y_2, of curve "2" (GBP), thus obtaining the dynamics (14.12).

Consider now the dynamics for x_1, y_1, x_2 and y_2, as from (14.2) and (14.12). A second change of numeraire for market "1" can now be performed, moving from the (market-"1") bank-account to the (market-"1") T_n-maturity bond price. We set $T_n = T$. By applying again the change-of-numeraire toolkit, and taking into account the equivalence shown in Section 2.9 or at the end of Section 14.1.3, we obtain the following dynamics:

$$
\begin{aligned}
dx_1(t) &= \left[-a_1x_1(t) - \frac{\sigma_1^2}{a_1}(1-e^{-a_1(T-t)}) - \rho_1\frac{\sigma_1\eta_1}{b_1}(1-e^{-b_1(T-t)})\right]dt \\
&\quad + \sigma_1 dZ_1^x(t),
\end{aligned}
$$

$$
\begin{aligned}
dy_1(t) &= \left[-b_1y_1(t) - \frac{\eta_1^2}{b_1}(1-e^{-b_1(T-t)}) - \rho_1\frac{\sigma_1\eta_1}{a_1}(1-e^{-a_1(T-t)})\right]dt \\
&\quad + \eta_1 dZ_1^y(t), \\
dx_2(t) &= \left[-a_2x_2(t) - \sigma_2\left(\gamma_{x1,x2}\frac{\sigma_1}{a_1}(1-e^{-a_1(T-t)})\right.\right. \\
&\quad \left.\left. + \gamma_{x2,y1}\frac{\eta_1}{b_1}(1-e^{-b_1(T-t)}) - \nu c_{x2,X}\right)\right] dt + \sigma_2 dZ_2^x(t), \\
dy_2(t) &= \left[-b_2y_2(t) - \eta_2\left(\gamma_{y1,y2}\frac{\eta_1}{b_1}(1-e^{-b_1(T-t)})\right.\right. \\
&\quad \left.\left. + \gamma_{x1,y2}\frac{\sigma_1}{a_1}(1-e^{-a_1(T-t)}) - \nu c_{y2,X}\right)\right] dt + \eta_2 dZ_2^y(t),
\end{aligned}
$$

where the Z's are Brownian motions under Q_1^n with the same correlation structure as the W's.

By integrating the four-dimensional Gaussian process above and using a four-dimensional version of Ito's isometry, one obtains the exact transition density as follows:

$$
\begin{aligned}
x_1(t) &= e^{-a_1(t-s)}x_1(s) - M_{x1}^T(s,t) + N_1(t-s), \\
y_1(t) &= e^{-b_1(t-s)}y_1(s) - M_{y1}^T(s,t) + N_2(t-s), \\
x_2(t) &= e^{-a_2(t-s)}x_2(s) - M_{x2}^T(s,t) + N_3(t-s), \\
y_2(t) &= e^{-b_2(t-s)}y_2(s) - M_{y2}^T(s,t) + N_4(t-s),
\end{aligned}
\tag{14.18}
$$

for $s \le t \le T$, where

$$\begin{aligned}
M_{x1}^T(s,t) &= \left(\frac{\sigma_1^2}{a_1^2} + \rho_1 \frac{\sigma_1 \eta_1}{a_1 b_1}\right) \left[1 - e^{-a_1(t-s)}\right] \\
&\quad - \frac{\sigma_1^2}{2a_1^2} \left[e^{-a_1(T-t)} - e^{-a_1(T+t-2s)}\right] \\
&\quad - \frac{\rho_1 \sigma_1 \eta_1}{b_1(a_1+b_1)} \left[e^{-b_1(T-t)} - e^{-b_1 T - a_1 t + (a_1+b_1)s}\right], \\
M_{y1}^T(s,t) &= \left(\frac{\eta_1^2}{b_1^2} + \rho_1 \frac{\sigma_1 \eta_1}{a_1 b_1}\right) \left[1 - e^{-b_1(t-s)}\right] \\
&\quad - \frac{\eta_1^2}{2b_1^2} \left[e^{-b_1(T-t)} - e^{-b_1(T+t-2s)}\right] \\
&\quad - \frac{\rho_1 \sigma_1 \eta_1}{a_1(a_1+b_1)} \left[e^{-a_1(T-t)} - e^{-a_1 T - b_1 t + (a_1+b_1)s}\right],
\end{aligned}$$

and

$$\begin{aligned}
M_{x2}^T(s,t) &= \frac{\sigma_1 \sigma_2 \gamma_{x1,x2}}{a_1} \left(\frac{1 - e^{-a_2(t-s)}}{a_2} - \frac{e^{-a_1(T-t)} - e^{-a_1(T-s) - a_2(t-s)}}{a_1 + a_2}\right) \\
&\quad + \frac{\eta_1 \sigma_2 \gamma_{x2,y1}}{b_1} \left(\frac{1 - e^{-a_2(t-s)}}{a_2} - \frac{e^{-b_1(T-t)} - e^{-b_1(T-s) - a_2(t-s)}}{b_1 + a_2}\right) \\
&\quad - \sigma_2 \nu c_{X,x2} \frac{1 - e^{-a_2(t-s)}}{a_2} \\
M_{y2}^T(s,t) &= \frac{\eta_1 \eta_2 \gamma_{y1,y2}}{b_1} \left(\frac{1 - e^{-b_2(t-s)}}{b_2} - \frac{e^{-b_1(T-t)} - e^{-b_1(T-s) - b_2(t-s)}}{b_1 + b_2}\right) \\
&\quad + \frac{\sigma_1 \eta_2 \gamma_{y2,x1}}{a_1} \left(\frac{1 - e^{-b_2(t-s)}}{b_2} - \frac{e^{-a_1(T-t)} - e^{-a_1(T-s) - b_2(t-s)}}{a_1 + b_2}\right) \\
&\quad - \eta_2 \nu c_{X,y2} \frac{1 - e^{-b_2(t-s)}}{b_2}
\end{aligned}$$

and where, finally, $N(t-s)$ is a four-dimensional Gaussian random vector with zero mean and covariance matrix $C(t-s)$ given by

$$\begin{bmatrix}
\sigma_1^2 \frac{1-e^{-2a_1(t-s)}}{2a_1} & \cdot & \cdot & \cdot \\
\sigma_1 \eta_1 \frac{1-e^{-(a_1+b_1)(t-s)}}{a_1+b_1} \rho_1 & \eta_1^2 \frac{1-e^{-2b_1(t-s)}}{2b_1} & \cdot & \cdot \\
\sigma_1 \sigma_2 \frac{1-e^{-(a_1+a_2)(t-s)}}{a_1+a_2} \gamma_{x1,x2} & \eta_1 \sigma_2 \frac{1-e^{-(b_1+a_2)(t-s)}}{b_1+a_2} \gamma_{y1,x2} & (***) & \cdot \\
\sigma_1 \eta_2 \frac{1-e^{-(a_1+b_2)(t-s)}}{a_1+b_2} \gamma_{x1,y2} & \eta_1 \eta_2 \frac{1-e^{-(b_1+b_2)(t-s)}}{b_1+b_2} \gamma_{y1,y2} & (**) & (*)
\end{bmatrix}$$

$$(***) = \sigma_2^2 \frac{1 - e^{-2a_2(t-s)}}{2a_2},$$

$$(**) = \eta_2 \sigma_2 \frac{1 - e^{-(b_2+a_2)(t-s)}}{b_2 + a_2} \rho_2, \quad (*) = \eta_2^2 \frac{1 - e^{-2b_2(t-s)}}{2b_2}.$$

Quanto CMS: Monte Carlo Pricing and Examples. We now explain how one can price the contract described in Subsection 14.2.1. Recall that the net value of the contract as seen by B at time 0 is

$$E_1\left\{\sum_{i=1}^{n}\exp\left(-\int_0^{T_i} r_1(s)\,ds\right)[S_{i-1,i-1+c/\delta}(T_{i-1})-L_1(T_{i-1},T_i)]\tau_i\right\}$$
$$=P_1^M(0,T_n)\sum_{i=1}^{n}\tau_i\,E_1^n\left[\frac{S_{i-1,i-1+c/\delta}(T_{i-1};x_2(T_{i-1}),y_2(T_{i-1}))}{P_1(T_i,T_n;x_1(T_i),y_1(T_i))}\right]$$
$$-(1-P_1^M(0,T_n)). \tag{14.19}$$

We need therefore to compute

$$E_1^n\left[\frac{S_{i-1,i-1+c/\delta}(T_{i-1};x_2(T_{i-1}),y_2(T_{i-1}))}{P_1(T_i,T_n;x_1(T_i),y_1(T_i))}\right] \tag{14.20}$$

for all i's and then substitute back in the above formula. Proceed as follows.

Inputs. We first describe the inputs necessary for the evaluation of the contract.

1. The number np of scenarios for the Monte Carlo evaluation.
2. The initial curves "1" (Euro) and "2" (GBP) at time 0 (as curves of discount factors, possibly interpolated): $T \mapsto P_1^M(0,T)$ and $T \mapsto P_2^M(0,T)$.
3. The parameters $a_1, b_1, \sigma_1, \eta_1$ and ρ_1, obtained by calibrating model (14.1) to the caps or swaptions markets "1" (Euro).
4. The parameters $a_2, b_2, \sigma_2, \eta_2$ and ρ_2, obtained by calibrating model (14.4) to the caps or swaptions markets "2" (GBP).
5. The instantaneous correlation between the two curves, $\text{Corr}\{dr_1, dr_2\}$, from which the corresponding model parameter γ can be computed via (14.9).
6. The instantaneous correlation between curve "2" (GBP) and the exchange rate Euro/GBP, $\text{Corr}\{dX, dr_2\}$, from which the corresponding model parameter c_X can be computed via (14.11).
7. The percentage annualized volatility ν of the exchange rate Euro/GBP.

Scheme. We can now give a schematic description of the use of the G2++ model for Monte Carlo pricing a quanto CMS, i.e. for computing (14.19).

1. Set the current time t to $t = T_0 = 0$. Accordingly, set $x_1(T_0) = y_1(T_0) = x_2(T_0) = y_2(T_0) = 0$. Set $i = 1$.
2. If $i \geq 2$, compute the following (curve-"2") discount factors at time $t = T_{i-1}$, in each scenario "p", by using formula (14.16) for curve "2":

$$P_2(T_{i-1},T_i;x_2^p(T_{i-1}),y_2^p(T_{i-1})),$$
$$P_2(T_{i-1},T_{i+1};x_2^p(T_{i-1}),y_2^p(T_{i-1})),$$
$$\ldots,P_2(T_{i-1},T_{i-1+c/\delta};x_2^p(T_{i-1}),y_2^p(T_{i-1})).$$

   Else, if $i = 1$, the above discount factors at time $T_{i-1} = 0$ are known as an input.

3. If $i \geq 2$, compute the model swap rate at time $t = T_{i-1}$ in each scenario "p", $S^p_{i-1,i-1+c/\delta} = S_{i-1,i-1+c/\delta}(T_{i-1}; x^p_2(T_{i-1}), y^p_2(T_{i-1}))$, via formula (14.17) with $j = i$, $m = i + c/\delta$ and $t = T_{i-1}$. Again, if $i = 1$, such swap rates at time $T_{i-1} = 0$ are known from the input.
4. Use formula (14.18) to generate np realizations

$$x^p_1(T_i),\ y^p_1(T_i),\ x^p_2(T_i),\ y^p_2(T_i),\ \ p = 1, 2, \ldots, \text{np}$$

   of $x_1(T_i)$, $y_1(T_i)$, $x_2(T_i)$, $y_2(T_i)$ starting from the previously generated $x_1(T_{i-1})$, $y_1(T_{i-1})$, $x_2(T_{i-1})$, $y_2(T_{i-1})$. Formula (14.18) is to be applied with $s = T_{i-1}$, $t = T_i$, and with the np new realizations of $N_1, \ldots, N_4$ generated from a four–dimensional Gaussian variable with mean zero and covariance matrix $C(T_i - T_{i-1})$.
5. Compute the T_n-discount factor for curve "1" at time $t = T_i$ in each scenario "p" by using formula (14.16):

$$P_1(T_i, T_n; x^p_1(T_i), y^p_1(T_i)).$$

6. Compute in each scenario "p" the ratio

$$\frac{S^p_{i-1,i-1+c/\delta}(T_{i-1})}{P_1(T_i, T_n; x^p_1(T_i), y^p_1(T_i))}$$

7. Average the above quantities over all scenarios. We obtain the Monte Carlo evaluation of (14.20). Store this value.
8. Increase i by one.
9. If $i \leq n$ then go back to step 2, otherwise
10. Compute the final price by adding all the terms computed and stored at step 7, according to formula (14.19).

The above scheme can be easily generalized to cases where optional features are added to either leg of the contract.

Analysis of a specific contract. We now present a specific example that is based on a concrete problem we solved via the above method. Precisely, we consider the following contract, which is specified in two parts.

First Part. First, it is given a fixed leg switching, at a given future instant, to a "floored" CMS fraction, which is exchanged for a canonical floating leg. Both legs are based on rates from market "1". Formally, maintaining notation of Subsection 14.2.1,

- party A pays to B a fixed-rate payment

$$\tau_i R_F \text{ units of currency "1"},\quad i = 1, 2, \ldots, f,$$

  at instants T_i for $i \leq f$, and pays the "floored" fraction of the swap rate

$$\max[Y S^1_{i-1,i-1+c/\delta}(T_{i-1}), R_c]\tau_i \text{ units of currency "1"},$$

at later instants $T_{f+1}, \ldots, T_n$, where Y is a fraction-parameter to be determined. Notice that here S^1 is the swap rate for curve "1".

- At time T_i, $i \geq 1$, Institution B pays to A the LIBOR rate for curve "1":

$$\tau_i L_1(T_{i-1}, T_i) \text{ units of currency "1"}.$$

One needs to find the value of Y such that the present value of this first CMS-like contract is zero. The net value of the contract as seen by B at time 0 is

$$\begin{aligned}
&E_1\left(\sum_{i=1}^{f} \exp\left(-\int_0^{T_i} r_1(s)\,ds\right)(R_F - L_1(T_{i-1}, T_i))\tau_i\right)\\
&+ E_1\left[\sum_{i=f+1}^{n} \exp\left(-\int_0^{T_i} r_1(s)\,ds\right)(\max[Y S^1_{i-1,i-1+c/\delta}(T_{i-1}), R_c]\right.\\
&\qquad\qquad\qquad\qquad \left. - L_1(T_{i-1}, T_i))\tau_i\right]\\
&= \sum_{i=1}^{f} R_F \tau_i P_1(0, T_i) + P_1(0, T_n) \sum_{i=f+1}^{n} \tau_i E_1^n\left[\frac{\max[Y S^1_{i-1,i-1+c/\delta}(T_{i-1}), R_c]}{P_1(T_i, T_n)}\right]\\
&\quad - (1 - P_1(0, T_n)).
\end{aligned}$$

Therefore, if such a value has to be zero, we need to solve the following nonlinear equation in Y:

$$\begin{aligned}
&\sum_{i=1}^{f} R_F \tau_i P_1(0, T_i) \ + P_1(0, T_n) \sum_{i=f+1}^{n} \tau_i E_1^n\left[\frac{\max[Y S^1_{i-1,i-1+c/\delta}(T_{i-1}), R_c]}{P_1(T_i, T_n)}\right]\\
&\quad -(1 - P_1(0, T_n)) = 0\ .
\end{aligned}$$

The difficulty is in computing the expected values. These can be evaluated through a Monte Carlo method, which will be iterated in order to solve for the unknown Y. Fortunately, the Gaussian G2++ model allows for a quick Monte Carlo evaluation, so as to render the task feasible.

Second Part. Now that the "fair" swap-fraction Y has been determined, we consider the following quanto "floored" CMS.

We are given the currency-"1" amount paid by a fixed leg switching, at a given future instant, to a "floored" CMS fraction for curve "2", which is exchanged for a canonical floating leg from curve "1". Formally, maintaining notation of Subsection 14.2.1,

- party A pays a fixed rate payment

$$\tau_i R_F \text{ units of currency "1"}, \quad i = 1, 2, \ldots, f,$$

at instants T_i for $i \leq f$, and pays the "floored" fraction of the swap rate

$$\max[Y S_{i-1,i-1+c/\delta}(T_{i-1}), R_c]\tau_i \text{ units of currency "1"},$$

at later instants $T_{f+1}, \ldots, T_n$, where Y is the fraction parameter determined before. Notice that here S is the swap rate for curve "2" (thus introducing the quanto feature of the contract).
- At time T_i, $i \geq 1$, Institution B pays to A the LIBOR rate for curve "1":

$$\tau_i L_1(T_{i-1}, T_i) \text{ units of currency "1"}.$$

The net value of the contract as seen by B at time 0 is

$$E_1\left(\sum_{i=1}^{f} \exp\left(-\int_0^{T_i} r_1(s)ds\right)(R_F - L_1(T_{i-1}, T_i))\tau_i\right) \tag{14.22}$$

$$+E_1\left[\sum_{i=f+1}^{n} \exp\left(-\int_0^{T_i} r_1(s)ds\right)(\max[Y S_{i-1,i-1+c/\delta}(T_{i-1}), R_c] - L_1(T_{i-1}, T_i))\tau_i\right]$$

$$= \sum_{i=1}^{f} R_F \tau_i P_1(0, T_i) + P_1(0, T_n) \sum_{i=f+1}^{n} \tau_i E_1^n\left[\frac{\max[Y S_{i-1,i-1+c/\delta}(T_{i-1}), R_c]}{P_1(T_i, T_n)}\right] - (1 - P_1(0, T_n)).$$

Again, the only difficulty is in evaluating the expected values. However, a Monte Carlo method and related scheme can be applied in a completely analogous way to the one presented earlier for standard quanto CMS.

14.2.3 Quanto CMS: Quanto Adjustment

Suppose we are again pricing a payoff that involves swap rates *in a foreign currency*. Denote quantities related to the foreign currency by the superscript or subscript "(2)". We need to compute

$$E[\tau_i D(0, T_i) S_{\alpha,\beta}^{(2)}(T_{i-k})] ,$$

where E denotes expectation under the domestic risk-neutral measure.

As in the original derivation of Black's formula for caps, we now assume that, in both markets, discounting occurs with deterministic rates, thus implicitly assuming equivalence between every T_i-forward measure and the risk-neutral measure for the same market.

Consider the foreign-market convexity-adjusted dynamics

$$dS^{(2)}_{\alpha,\beta}(t) = \mu^{\alpha,\beta}_{(2)} S^{(2)}_{\alpha,\beta}(t)\,dt + \sigma^{(\alpha,\beta)}_{(2)}(t) S^{(2)}_{\alpha,\beta}(t)\,dW^{(2)}_S(t).$$

This is basically the dynamics (13.17) rewritten for the foreign market, i.e. under Q_2. However, to evaluate the above expectation, we need the dynamics of the foreign swap rate under the *domestic* risk-neutral measure. This in turn requires modeling the exchange rate between the domestic and foreign markets.

Let $X(t)$ denote, as before, the amount of foreign currency needed to buy one unit of domestic currency. Assume the following no-arbitrage dynamics for X under Q_2:

$$dX(t) = ((r_2(t) - r(t))X(t)dt + \nu X(t)dW^{(2)}_X(t).$$

We need also to model the instantaneous correlations between the exchange rate X and the foreign swap rate:

$$dW^{(2)}_S(t)dW^{(2)}_X(t) = cdt.$$

We use the change-of-numeraire technique, described in Section 2.3, and move from the bank-account numeraire for the foreign market (2) to the numeraire

$$X(t) \times (\text{bank account of domestic market}).$$

This is the right change of numeraire for moving from measure Q_2 to the domestic measure Q, as explained in Section 2.9. We obtain the following dynamics under Q:

$$dS^{(2)}_{\alpha,\beta}(t) = \left[\mu^{\alpha,\beta}_{(2)} + \sigma^{(\alpha,\beta)}_{(2)}(t)\nu c\right] S^{(2)}_{\alpha,\beta}(t)dt + \sigma^{(\alpha,\beta)}_{(2)}(t) S^{(2)}_{\alpha,\beta}(t)dW_S(t).$$

With this last dynamics it is immediate to check that

$$\begin{aligned}
E\left[\tau_i D(0,T_i) S^{(2)}_{\alpha,\beta}(T_{i-k})\right] &= \tau_i P(0,T_i)\, E^i\left[S^{(2)}_{\alpha,\beta}(T_{i-k})\right] \\
&\approx \exp\left[\nu c \int_0^{T_{i-k}} \sigma^{(\alpha,\beta)}_{(2)}(t)dt\right] \tau_i P(0,T_i) \\
&\quad \cdot \left[S^{(2)}_{\alpha,\beta}(0) - \tfrac{1}{2}(S^{(2)}_{\alpha,\beta}(0))^2 (v^{(2)}_{\alpha,\beta}(T_{i-k}))^2 \frac{\Psi''_{\alpha,\beta}(S^{(2)}_{\alpha,\beta}(0))}{\Psi'_{\alpha,\beta}(S^{(2)}_{\alpha,\beta}(0))}\right].
\end{aligned}$$

If the instantaneous volatility of the foreign swap rate is assumed to be constant,

$$\sigma^{(\alpha,\beta)}_{(2)}(t) = \sigma^{(\alpha,\beta)}_{(2)},$$

then the above formula reduces to

$$\approx \exp\left[\sigma_{(2)}^{(\alpha,\beta)} T_{i-k} \nu c\right] \tau_i P(0, T_i)$$

$$\cdot \left[S_{\alpha,\beta}^{(2)}(0) - \tfrac{1}{2}(S_{\alpha,\beta}^{(2)}(0))^2 (\sigma_{(2)}^{(\alpha,\beta)})^2 T_{i-k} \frac{\Psi''_{\alpha,\beta}(S_{\alpha,\beta}^{(2)}(0))}{\Psi'_{\alpha,\beta}(S_{\alpha,\beta}^{(2)}(0))} \right].$$

Notice that the quantity inside square brackets is the convexity-adjusted expectation for the foreign market alone. One can compute it as shown in Chapter 13 (for the domestic market), and subsequently multiply by the domestic-market discount factor $\tau_i P(0, T_i)$. The further correction due to the quanto feature is given by the factor

$$\exp\left[\sigma_{(2)}^{(\alpha,\beta)} T_{i-k} \nu c\right].$$

If either the correlation c or the volatilities are small, the effect of this correction is negligible.

14.3 Differential Swaps

We now introduce differential swaps, which are also called quanto swaps. In order to maintain this section as self-contained as possible, we will reintroduce some notation used in earlier sections. The main difference with quanto CMS, of which differential swaps are a particular case, is that a closed-form formula is available within the G2++ model. A market-like closed-form formula is also available for differential swaps, and we will derive it in the later Section 14.4, in the context of general quanto derivatives.

14.3.1 The Contract

Consider the following contract. Party A pays to party B in a first (domestic) currency "1" (Euro) an amount expressed by the δ-simply-compounded LIBOR rate associated with a second (foreign) currency "2" (GBP). Payments occur once every δ years (typically $\delta = 0.5$ years $= 6$ months). On the same payment dates, party B pays to party A the δ-simply-compounded LIBOR rate associated with currency "1". Formally, the payoff can be expressed as follows.

We assume unit nominal amount. Let us assume the current time to be $t = 0$, and let us denote by $\mathcal{T} = \{T_1, \ldots, T_n\}$ a set of payment dates at which the flows of the two legs are to be exchanged. We denote by τ_i the year fraction between T_{i-1} and T_i, which will be close in general to the single value δ. We take $T_0 = 0$.

- At time T_i, $i \geq 1$, party A pays to B, in currency "1", the amount expressed by the δ-simply-compounded LIBOR rate for curve "2":

$$L_2(T_{i-1}, T_i)\tau_i \text{ units of currency "1".}$$

where the quantity

$$L_2(T_{i-1}, T_i) = \frac{1}{\tau_i}\left(\frac{1}{P_2(T_{i-1}, T_i)} - 1\right) \tag{14.23}$$

denotes the LIBOR rate at time T_{i-1} for maturity T_i for curve "2", and where in general $P_2(t, T)$ denotes the discount factor, for curve "2", at time t for maturity T. Analogous definitions of LIBOR rates and discount factors are given for curve "1".

- At time T_i, $i \geq 1$, institution B pays to A the LIBOR rate for curve "1" plus a spread K:

$$\tau_i(L_1(T_{i-1}, T_i) + K) \text{ units of currency "1".}$$

The net value of the contract as seen by B at time 0 is

$$E_1\left(\sum_{i=1}^{n} \exp\left(-\int_0^{T_i} r_1(s)ds\right)(L_2(T_{i-1}, T_i) - L_1(T_{i-1}, T_i) - K)\tau_i\right)$$
$$= \sum_{i=1}^{n} \tau_i P_1(0, T_i)[E_1^i(L_2(T_{i-1}, T_i)) - K] - (1 - P_1(0, T_n)), \tag{14.24}$$

where E_1^i denotes the T_i-forward-measure expectation for curve "1". The related measure is denoted by Q_1^i.

We need to compute

$$E_1^i[L_2(T_{i-1}, T_i)] \tag{14.25}$$

for all i's.

To this purpose, as in the case of a quanto CMS, a good strategy can be to use a short-rate model with analytical formulas for zero-coupon rates in terms of the factors concurring to the short rate. In such a case, the above expectations are easily computed once the distribution of the short-rate factors is known.

At the same time, the chosen model has to allow for a realistic volatility structure for each curve. A good choice can again be given by selecting the two-factor Gaussian G2++ short-rate model for each curve.

14.3.2 Differential Swaps with the G2++ Model

Starting from the models described in Subsection 14.1.1, we deduce the analytical formulas needed for the differential-swap payoff.

The model allows for the following bond-price formula for curve "2":

$$P_2(t,T;x_2(t),y_2(t)) = \frac{P_2^M(0,T)}{P_2^M(0,t)} \exp\left\{\frac{1}{2}[V_2(t,T) - V_2(0,T) + V_2(0,t)] - \frac{1-e^{-a_2(T-t)}}{a_2}x_2(t) - \frac{1-e^{-b_2(T-t)}}{b_2}y_2(t)\right\}, \quad (14.26)$$

where $V_2(t,T)$ is defined analogously to $V_1(t,T)$ given in (14.7).

We now determine the dynamics of x_2 and y_2 under Q_1^i.

As shown in Subsection 14.2.2, using the change-of-numeraire technique we can express the equations for the factors x_2 and y_2 describing curve "2" (GBP) under the measure Q_1^i associated with market "1", obtaining the transition equation between $s = 0$ and any instant t, with $T = T_i$. In particular, we obtain

$$\begin{aligned} x_2(t) &= -M_{x2}^T(0,t) + N_x(t), \\ y_2(t) &= -M_{y2}^T(0,t) + N_y(t), \end{aligned} \quad (14.27)$$

where $(N_x(t), N_y(t))$ is a two-dimensional Gaussian random vector with zero mean and covariance matrix $C(t)$ given by

$$\begin{bmatrix} \sigma_2^2 \frac{1-e^{-2a_2t}}{2a_2} & \eta_2\sigma_2 \frac{1-e^{-(b_2+a_2)t}}{b_2+a_2}\rho_2 \\ \cdot & \eta_2^2 \frac{1-e^{-2b_2t}}{2b_2} \end{bmatrix}.$$

It is now possible to compute the following expected value through formula (14.26) combined with the just derived distribution of $x_2(t)$ and $y_2(t)$, starting from $x_2(0)$ and $y_2(0)$, under Q_1^i:

$$E_1^i\left[\frac{1}{P_2(t,T;x_2(t);y_2(t))}\right] = \frac{P_2^M(0,t)}{P_2^M(0,T)} \exp\{-\tfrac{1}{2}[V_2(t,T) - V_2(0,T) + V_2(0,t)]\} \cdot E_1^i\left\{\exp\left[\frac{1-e^{-a_2(T-t)}}{a_2}x_2(t) + \frac{1-e^{-b_2(T-t)}}{b_2}y_2(t)\right]\right\}.$$

Now use the joint distribution of $x_2(t)$ and $y_2(t)$ under Q_1^i as from (14.27) to obtain

$$\begin{aligned} \psi_2(t,T) &:= E_1^T\left[\frac{1}{P_2(t,T;x_2(t);y_2(t))}\right] \\ &= \frac{P_2^M(0,t)}{P_2^M(0,T)} \exp\left\{-\frac{1}{2}[V_2(t,T) - V_2(0,T) + V_2(0,t)]\right. \\ &\quad - \frac{1-e^{-a_2(T-t)}}{a_2}M_{x2}^T(0,t) - \frac{1-e^{-b_2(T-t)}}{b_2}M_{y2}^T(0,t) \\ &\quad + \frac{1}{2}\left[\sigma_2^2\left(\frac{1-e^{-a_2(T-t)}}{a_2}\right)^2 \frac{1-e^{-2a_2t}}{2a_2}\right. \end{aligned}$$

$$+2\eta_2\sigma_2\rho_2 \frac{(1-e^{-a_2(T-t)})(1-e^{-b_2(T-t)})(1-e^{-(a_2+b_2)(T-t)})}{a_2 b_2(a_2+b_2)}$$

$$+\eta_2^2 \left(\frac{1-e^{-b_2(T-t)}}{b_2}\right)^2 \frac{1-e^{-2b_2 t}}{2b_2}\Bigg]\Bigg\}$$

Therefore, by taking into account definition (14.23), we have that

$$E_1^T[L_2(t,T)] = \frac{1}{\tau}(\psi_2(t,T)-1) \ ,$$

where τ denotes the year fraction between t and T. We can then compute the contract value (14.24) as follows:

$$\sum_{i=1}^{n} \tau_i P_1(0,T_i) E_1^i[L_2(T_{i-1},T_i)] - (1-P_1(0,T_n))$$

$$= \sum_{i=1}^{n} P_1(0,T_i)\,(\psi_2(T_{i-1},T_i) - 1 - \tau_i K) - (1-P_1(0,T_n))$$

There is no need for a numerical scheme here, since we have found a closed-form formula. This formula requires the following inputs:

1. The initial bond prices $P_1(0,T_1),\ldots,P_1(0,T_n)$ for the zero curve "1" (Euro) and the analogous quantities for the initial zero curve "2" (GBP).
2. The parameters $a_2, b_2, \sigma_2, \eta_2$ and ρ_2, obtained by calibration of model (14.4, 14.5) to caps or swaptions in market "2" (GBP).
3. The instantaneous correlation between curve "1" (Euro) and curve "2" (GBP), $\mathrm{Corr}\{dr_1, dr_2\}$, from which the corresponding model parameter γ can be computed via (14.9).
4. The instantaneous correlation between curve "2" (GBP) and the exchange rate Euro/GBP, $\mathrm{Corr}\{dX, dr_2\}$, from which the corresponding model parameter c_X can be computed via (14.11).
5. The percentage annualized volatility ν of the exchange rate Euro/GBP.

14.3.3 A Market-Like Formula

In Subsection 14.4.3 we will present a market-like closed formula for differential swaps, as an easy consequence of the general formulas for quanto caps and floors.

14.4 Market Formulas for Basic Quanto Derivatives

This section is self-contained and can be read independently of the previous sections in this chapter. Here we derive the arbitrage-free price of some fundamental derivatives with multi-currency features. We start with the pricing of a quanto caplet by modeling the relevant quantities through lognormal martingales under a given forward measure. Such assumptions immediately lead to a Black-like pricing formula with coefficients that can be all expressed in terms of relevant financial quantities. Analogously, we also derive the arbitrage-free price of a quanto floorlet and, by extension, of a quanto cap, of a quanto floor and of a quanto swap.

14.4.1 The Pricing of Quanto Caplets/Floorlets

Given a domestic market and a foreign market, let us assume that the term structures of discount factors that are observed in the domestic and foreign markets at time t are respectively given by $T \mapsto P(t,T)$ and $T \mapsto P^f(t,T)$ for $T \geq t$. Let us denote by $\mathcal{X}(t)$ the exchange rate at time t between the currencies in the two markets, in that 1 unit of the foreign currency equals $\mathcal{X}(t)$ units of the domestic currency. Notice that $\mathcal{X}(t)$ is the reciprocal of the exchange rate $X(t)$ defined in Subsection 14.1.3.

Given the future times T_1 and T_2, a quanto caplet pays off at time T_2

$$\left[F^f(T_1;T_1,T_2) - K\right]^+ \tau_{1,2} N \quad \text{in domestic currency,} \tag{14.28}$$

where N is the nominal value, K is the caplet rate (strike), $\tau_{1,2}$ is the year fraction between times T_1 and T_2 and $F^f(t;T_1,T_2)$ is the forward rate in the foreign market at time t for the interval $[T_1,T_2]$, i.e.,

$$F^f(t;T_1,T_2) = \frac{P^f(t,T_1) - P^f(t,T_2)}{\tau_{1,2} P^f(t,T_2)},$$

where the year fraction is assumed to be the same in both markets. The no-arbitrage value at time t of the payoff (14.28) is then given by

$$\mathbf{QCpl}(t,T_1,T_2,N,K) = \tau_{1,2} N P(t,T_2) E^2 \left\{ \left[F^f(T_1;T_1,T_2) - K\right]^+ | \mathcal{F}_t \right\},$$

where E^2 denotes the expectation under the domestic forward measure Q^2 induced by the numeraire $P(t,T_2)$.

In order to compute this expectation we must know the distribution of $F^f(T_1;T_1,T_2)$ under the measure Q^2. Notice that, under the foreign forward measure associated with the numeraire $P^f(t,T_2)$, $F^f(t;T_1,T_2)$ is a martingale, which is here assumed to be a (driftless) geometric Brownian motion:

$$dF^f(t;T_1,T_2) = \sigma F^f(t;T_1,T_2) dW^f(t).$$

However, under the domestic forward measure Q^2, $F^f(t;T_1,T_2)$ displays a drift that we shall derive as follows.

Let us define the forward exchange rate at time t maturing at time T_2 as

$$F_{\mathcal{X}}(t,T_2) = \mathcal{X}(t) \frac{P^f(t,T_2)}{P(t,T_2)},$$

which is a martingale under Q^2 since $\mathcal{X}(t)P^f(t,T_2)$ is the price of a tradable asset in the domestic market, and assume that

$$dF_{\mathcal{X}}(t,T_2) = \sigma_{F\mathcal{X}} F_{\mathcal{X}}(t,T_2)\, dW_{\mathcal{X}}(t),$$

where $W_{\mathcal{X}}$ is a standard Brownian motion under Q^2, with $dW_{\mathcal{X}}(t)\, dW^f(t) = \rho\, dt$.

Then, applying the change-of-numeraire-toolkit formula (2.14) with $S = P^f(\cdot,T_2)$, $U = P(\cdot,T_2)/\mathcal{X}(\cdot)$ and $\mu_t^S \equiv 0$, we have that the dynamics of $F^f(t;T_1,T_2)$ under the domestic forward measure Q^2 is

$$dF^f(t;T_1,T_2) = F^f(t;T_1,T_2)\big[-\rho\sigma_{F\mathcal{X}}\sigma\, dt + \sigma\, dW_2^f(t)\big],$$

so that, under Q^2, $F^f(T_1;T_1,T_2)$ is lognormally distributed with

$$E^2\left\{\ln \frac{F^f(T_1;T_1,T_2)}{F^f(t;T_1,T_2)}\bigg|\mathcal{F}_t\right\} = \left(\mu - \frac{1}{2}\sigma^2\right)(T_1-t),$$
$$\mathrm{Var}^2\left\{\ln \frac{F^f(T_1;T_1,T_2)}{F^f(t;T_1,T_2)}\bigg|\mathcal{F}_t\right\} = \sigma^2(T_1-t),$$

where we set

$$\mu := -\rho\sigma_{F\mathcal{X}}\sigma.$$

This immediately implies, see also (D.2) in Appendix D, that

$$\mathbf{QCpl}(t,T_1,T_2,N,K) = \tau_{1,2}NP(t,T_2)\left[F^f(t;T_1,T_2)e^{\mu(T_1-t)}\Phi(d_1) - K\Phi(d_2)\right]$$
$$d_1 = \frac{\ln\frac{F^f(t;T_1,T_2)}{K} + (\mu+\frac{1}{2}\sigma^2)(T_1-t)}{\sigma\sqrt{T_1-t}}$$
$$d_2 = d_1 - \sigma\sqrt{T_1-t}.$$

Analogously, the arbitrage-free price of a quanto floorlet that pays off at time T_2

$$\left[K - F^f(T_1;T_1,T_2)\right]^+ \tau_{1,2}N \quad \text{in domestic currency,}$$

is

$$\mathbf{QFll}(t,T_1,T_2,N,K) = \tau_{1,2}NP(t,T_2) \cdot \left[-F^f(t;T_1,T_2)e^{\mu(T_1-t)}\Phi(-d_1) + K\Phi(-d_2)\right].$$

14.4.2 The Pricing of Quanto Caps/Floors

As to the pricing of caps and floors, we denote by $D = \{d_1, d_2, \ldots, d_n\}$ the set of the cap/floor payment dates and by $\mathcal{T} = \{T_0, T_1, \ldots, T_n\}$ the set of the corresponding times, meaning that T_i is the difference in years between d_i and the settlement date t, and where T_0 is the first reset time. Moreover, we denote by τ_i the year fraction for the time interval $(T_{i-1}, T_i]$, $i = 1, \dots, n$, by σ_i the (proportional) volatility of $F^f(t; T_{i-1}, T_i)$, by ρ_i the instantaneous correlation between $F_\mathcal{X}(t, T_i)$ and $F^f(t; T_{i-1}, T_i)$, by $\sigma^i_{F\mathcal{X}}$ the (proportional) volatility of the forward exchange rate $F_\mathcal{X}(t, T_i)$. We set $\tau := \{\tau_1, \dots, \tau_n\}$ and

$$\mu_i = -\rho_i \sigma^i_{F\mathcal{X}} \sigma_i.$$

Since the price of a cap (floor) is the sum of the prices of the underlying caplets (floorlets),[1] the price at time t of a cap with cap rate (strike) K, nominal value N and set of times $\mathcal{T}$ is then given by

$$\begin{aligned}
&\mathbf{QCap}(t, \mathcal{T}, \tau, N, K)\\
&\quad = \sum_{i=1}^{n} \tau_i N P(t, T_i) \Bigg[F^f(t; T_{i-1}, T_i) e^{\mu_i (T_{i-1} - t)} \\
&\qquad \cdot \Phi\left(\frac{\ln \frac{F^f(t; T_{i-1}, T_i)}{K} + \left(\mu_i + \frac{1}{2}\sigma_i^2\right)(T_{i-1} - t)}{\sigma_i \sqrt{T_{i-1} - t}} \right) \\
&\qquad - K \Phi\left(\frac{\ln \frac{F^f(t; T_{i-1}, T_i)}{K} + \left(\mu_i - \frac{1}{2}\sigma_i^2\right)(T_{i-1} - t)}{\sigma_i \sqrt{T_{i-1} - t}} \right) \Bigg],
\end{aligned} \tag{14.29}$$

and the price of the corresponding floor is

$$\begin{aligned}
&\mathbf{QFlr}(t, \mathcal{T}, \tau, N, K)\\
&\quad = \sum_{i=1}^{n} \tau_i N P(t, T_i) \Bigg[- F^f(t; T_{i-1}, T_i) e^{\mu_i (T_{i-1} - t)} \\
&\qquad \cdot \Phi\left(\frac{\ln \frac{K}{F^f(t; T_{i-1}, T_i)} - \left(\mu_i + \frac{1}{2}\sigma_i^2\right)(T_{i-1} - t)}{\sigma_i \sqrt{T_{i-1} - t}} \right) \\
&\qquad + K \Phi\left(\frac{\ln \frac{K}{F^f(t; T_{i-1}, T_i)} - \left(\mu_i - \frac{1}{2}\sigma_i^2\right)(T_{i-1} - t)}{\sigma_i \sqrt{T_{i-1} - t}} \right) \Bigg].
\end{aligned} \tag{14.30}$$

14.4.3 The Pricing of Differential Swaps

A differential swap, or quanto swap, has been defined in the previous Section 14.3. Suppose for a moment that we take away the LIBOR payment in the leg paid by institution B to institution A. Then the price at time t of the corresponding quanto swap where we receive the floating rate and we pay the fixed rate K is simply the difference

[1] Some care is actually needed since the previous valuation of time-T_2 options has been dealt with independently from possible preceding or subsequent payoffs, thus practically considering different models for different payment times. For a clearer explanation of this fact and a more consistent pricing we refer to Section 14.5.4 below.

$$\mathbf{QS}(t, \mathcal{T}, \tau, N, K) = \mathbf{QCap}(t, \mathcal{T}, \tau, N, K) - \mathbf{QFlr}(t, \mathcal{T}, \tau, N, K),$$

since $(F - K)^+ - (K - F)^+ = F - K$. Hence,[2]

$$\mathbf{QS}(t, \mathcal{T}, \tau, N, K) = \sum_{i=1}^{n} \tau_i N P(t, T_i) \left[F^f(t; T_{i-1}, T_i) e^{\mu_i (T_{i-1} - t)} - K \right].$$

If we put back the domestic LIBOR rate together with the fixed-rate payment K in the leg from B to A, we have to subtract the corresponding value to the above formula, thus obtaining

$$\sum_{i=1}^{n} \tau_i N P(t, T_i) \left[F^f(t; T_{i-1}, T_i) e^{\mu_i (T_{i-1} - t)} - K \right] - N(1 - P(t, T_n)).$$

14.4.4 The Pricing of Quanto Swaptions

We now consider a further multi-currency derivative, namely a European-style quanto swaption, and show how to price it by means of the general formula for spread options reported in Appendix E.[3]

Let $\mathcal{T} = \{T = T_0, T_1, T_2, \ldots, T_n\}$ be a set of $n + 1$ times with $0 < T < T_1 < T_2 < \cdots < T_n$. Assume we are given a domestic and a foreign financial markets, where discount factors and forward swap rates corresponding to $\mathcal{T}$ are defined. As before, we denote respectively by $P(t, T_i)$ and $P^f(t, T_i)$ the domestic and foreign discount factors at time t for maturity T_i, $i = 0, 1, \ldots, n$. The exchange rate between the two market currencies is again assumed to evolve according to the above process $\mathcal{X}$, meaning that, at any time t, one unit of foreign currency is worth $\mathcal{X}(t)$ units of domestic currency.

The domestic forward swap rate at time $0 < t < T$ corresponding to the swap starting at time T and with payment times T_i, $i = 1, \ldots, n$, is given by

$$S(t) = \frac{P(t, T) - P(t, T_n)}{C(t)}, \tag{14.31}$$

where

$$C(t) = \sum_{i=1}^{n} \tau_i P(t, T_i), \tag{14.32}$$

---

[2] We again refer to Section 14.5.4 below for a more correct management of multiple payment times.

[3] Such a formula is based on the assumption that the two underlying assets evolve according to geometric Brownian motions with nonzero instantaneous correlation. The financial market constituted by these assets and a deterministic bond is complete, meaning that arbitrage-free prices are given by risk-neutral valuation. However, no closed formula is obtainable for the price of the spread option. We will have instead to resort to numerical integration of a function involving both the cumulative distribution and probability density functions of a standard normal random variable.

and τ_i is again the year fraction over the period from time T_{i-1} to T_i, and $\tau := \{\tau_1, \dots, \tau_n\}$. Analogously, the foreign forward swap-rate at time $0 < t < T$ corresponding to the swap starting at time T and with payment times T_i, $i = 1, \dots, n$, is given by

$$S^f(t) = \frac{P^f(t,T) - P^f(t,T_n)}{C^f(t)}, \tag{14.33}$$

where

$$C^f(t) = \sum_{i=1}^{n} \tau_i P^f(t,T_i), \tag{14.34}$$

and, for simplicity, it is assumed that the year fractions are the same in both markets.

A (European-style) quanto swaption is, roughly speaking, an option on the difference between these two swap rates, with this difference being denominated in units of domestic currency. In formulas, given the swap rates (14.31) and (14.33), the strike $K > 0$ and assuming that the option maturity is T, the quanto-swaption payoff at time T is

$$C(T)[wS^f(T) - wS(T) - wK]^+, \tag{14.35}$$

where either $w = 1$ or $w = -1$. Choosing $C(t)$ as numeraire, the arbitrage-free price of (14.35) at time t can be written as

$$\mathbf{QES}(t, \mathcal{T}, \tau, K) = C(t)E^C\left\{[wS^f(T) - wS(T) - wK]^+ | \mathcal{F}_t\right\}, \tag{14.36}$$

where E^C denotes expectation under the domestic forward-swap measure Q^C induced by $C(t)$ and $\mathcal{F}_t$ is the sigma-field generated by (S^f, S) up to time t.

By definition of domestic forward-swap measure, the forward-swap-rate process S is a martingale under such a measure. Assuming lognormal dynamics, the domestic forward-swap process under Q^C is then described by

$$dS(t) = \sigma S(t)dW(t), \tag{14.37}$$

where σ is a positive real number and W is a Q^C-Brownian motion.

The forward-swap-rate process S^f is a martingale under the foreign forward-swap measure, but under Q^C displays the drift that has been calculated by Hunt and Pelsser (1998). Precisely, they assumed that under Q^C the two following martingales

$$M_1(t) = \frac{\mathcal{X}(t)(P^f(t,T) - P^f(t,T_n))}{C(t)}$$

$$M_2(t) = \frac{\mathcal{X}(t)C^f(t)}{C(t)}$$

are lognormally distributed, so that also $S^f(t) = \frac{M_1(t)}{M_2(t)}$ is lognormally distributed with dynamics given by

$$dS^f(t) = S^f(t)[\mu^f dt + \sigma^f dW^f(t)],$$

where μ^f and σ^f are real constants and W^f is a Brownian motion under Q^C. They also showed that the drift rate $\mu^f dt$ is equal to minus the instantaneous covariance of $\ln M_2(t)$ and $\ln S^f(t)$ and stated that, for practical purposes,[4] $M_2(t)$ can be replaced by the time-T forward exchange rate

$$F_{\mathcal{X}}(t,T) = \frac{\mathcal{X}(t)P^f(t,T)}{P(t,T)}.$$

By doing so, we can get the approximated equality

$$\mu^f = -\rho_{F,f}\sigma^f\sigma_F,$$

where σ_F is the (assumed constant) proportional volatility of the process $F_{\mathcal{X}}$ and $\rho_{F,f}$ is the instantaneous correlation between $F_{\mathcal{X}}$ and S^f, i.e. between the forward exchange rate and the foreign swap rate.

Knowing the distribution of S and S^f under the measure Q^C, we can now calculate the expectation in (14.36) by means of the results in Appendix E. Notice, in fact, that under Q^C the joint distribution of

$$\left(\ln\frac{S^f(T)}{S^f(t)}, \ln\frac{S(T)}{S(t)}\right) \text{ conditional on } \mathcal{F}_t$$

is bivariate normal with with mean vector

$$\begin{bmatrix}\mu_x \\ \mu_y\end{bmatrix} = \begin{bmatrix}(\mu^f - \frac{1}{2}(\sigma^f)^2) \\ -\frac{1}{2}\sigma^2\end{bmatrix}\tau$$

and covariance matrix

$$\begin{bmatrix}\sigma_x^2 & \rho\sigma_x\sigma_y \\ \rho\sigma_x\sigma_y & \sigma_y^2\end{bmatrix} = \begin{bmatrix}(\sigma^f)^2 & \rho\sigma^f\sigma \\ \rho\sigma^f\sigma & \sigma^2\end{bmatrix}\tau$$

where $\tau = T - t$ and ρ is the instantaneous correlation between the two forward swap-rates S^f and S, i.e., $dW(t)dW^f(t) = \rho\, dt$. Therefore, we can apply the same decomposition as in (E.6) to obtain

$$\mathbf{QES}(t) = C(t)\int_{-\infty}^{+\infty}\left[\int_{-\infty}^{+\infty}\left(wS^f(t)e^x - wS(t)e^y - wK\right)^+ f_{X|Y}(x,y)dx\right] \cdot f_Y(y)dy, \tag{14.38}$$

with $f_{X|Y}$ and f_Y defined in (14.47), which, by formula (E.7), becomes

[4] That is, to be able to explicitly calculate this covariance from market data.

$$
\begin{aligned}
C(t) \int_{-\infty}^{+\infty} \Bigg\{ & w S^f(t) \exp\left[\mu_x + \rho\sigma_x \frac{y-\mu_y}{\sigma_y} + \frac{1}{2}\sigma_x^2(1-\rho^2)\right] \cdot \\
& \cdot \Phi\left(w \frac{\mu_x + \rho\sigma_x \frac{y-\mu_y}{\sigma_y} - \ln\frac{K+S(t)e^y}{S^f(t)} + \sigma_x^2(1-\rho^2)}{\sigma_x\sqrt{1-\rho^2}} \right) \\
& - w\left(K + S(t)e^y\right)\Phi\left(w \frac{\mu_x + \rho\sigma_x \frac{y-\mu_y}{\sigma_y} - \ln\frac{K+S(t)e^y}{S^f(t)}}{\sigma_x\sqrt{1-\rho^2}} \right) \Bigg\} f_Y(y)\,dy.
\end{aligned}
\tag{14.39}
$$

We are then ready to state the following.

Proposition 14.4.1. *The unique arbitrage-free price of the quanto swaption described by the payoff (14.35) is*

$$
\mathbf{QES}(t) = C(t)\int_{-\infty}^{+\infty} \frac{1}{\sqrt{2\pi}} e^{-\frac{1}{2}v^2} f(v)dv, \tag{14.40}
$$

where

$$
\begin{aligned}
f(v) = w S^f(t) \exp & \left[(\mu^f - \frac{1}{2}\rho^2(\sigma^f)^2)\tau + \rho\sigma^f\sqrt{\tau}v\right] \\
& \cdot \Phi\left(w \frac{\ln\frac{S^f(t)}{h(v)} + [\mu^f + (\frac{1}{2}-\rho^2)(\sigma^f)^2]\tau + \rho\sigma^f\sqrt{\tau}v}{\sigma^f\sqrt{\tau}\sqrt{1-\rho^2}} \right) \\
& - w h(v) \Phi\left(w \frac{\ln\frac{S^f(t)}{h(v)} + [\mu^f - \frac{1}{2}(\sigma^f)^2]\tau + \rho\sigma^f\sqrt{\tau}v}{\sigma^f\sqrt{\tau}\sqrt{1-\rho^2}} \right)
\end{aligned}
$$

and

$$
h(v) = K + S(t)e^{-\frac{1}{2}\sigma^2\tau + \sigma\sqrt{\tau}v}
$$

Proof. Formula (14.40) immediately follows from (14.39) by remembering the definition of μ_x, μ_y, σ_x and σ_y and performing the variable change $v = \frac{y-\mu_y}{\sigma_y}$. □

The formula (14.40) can be computed through an easy numerical integration. As to the parameters being involved in it, besides the obvious inputs $S^f(t)$, $S(t)$, T and K, we have to remark that σ^f, σ and σ_F can be immediately obtained from market data of implied volatilities, whereas ρ and $\rho_{F,f}$ can be estimated historically.

14.5 Pricing of Options on two Currency LIBOR Rates

We finally show how to price options that explicitly depend on two LIBOR rates belonging to two different currencies (the former is domestic, the latter foreign). To this end, we apply the same apparatus based on market models, which we have developed in the previous section for quanto options. More precisely, we assume the following.

Given the future times T_{i-1} and T_i, $i = 1, \dots, M$ and $T_0 > 0$, the domestic and foreign forward rates at time t for the interval $[T_{i-1}, T_i]$ are, respectively,

$$F_i(t) = F(t; T_{i-1}, T_i) = \frac{P(t, T_{i-1}) - P(t, T_i)}{\tau_i P(t, T_i)}$$

$$F_i^f(t) = F^f(t; T_{i-1}, T_i) = \frac{P^f(t, T_{i-1}) - P^f(t, T_i)}{\tau_i P^f(t, T_i)}$$

where τ_i is the year fraction between times T_{i-1} and T_i, which is assumed to be the same in both markets.

Denoting by $F_{\mathcal{X}}(t, T_i)$ the forward exchange rate at time t for maturity T_i,

$$F_{\mathcal{X}}(t, T_i) = \mathcal{X}(t) \frac{P^f(t, T_i)}{P(t, T_i)},$$

and assuming constant (proportional) volatilities, the two forward rates evolve under the domestic forward measure Q^i according to

$$\begin{aligned} dF_i(t) &= \sigma_i F_i(t)\, dW_i(t), \\ dF_i^f(t) &= F_i^f(t)\big[-\rho\sigma_{F\mathcal{X}}\sigma_i^f\, dt + \sigma_i^f\, dW_i^f(t)\big], \end{aligned}$$

where W_i and W_i^f are two standard Brownian motions with instantaneous correlation ρ_i, $\rho = \rho_{\mathcal{X},i}$ is the instantaneous correlation between $F_{\mathcal{X}}(\cdot, T_i)$ and $F_i^f(\cdot)$, and $\sigma_{F\mathcal{X}} = \sigma_{F\mathcal{X}}^i$ is the assumed constant (proportional) volatility of the forward exchange rate $F_{\mathcal{X}}(t, T_i)$:

$$dF_{\mathcal{X}}(t, T_i) = \sigma_{F\mathcal{X}} F_{\mathcal{X}}(t, T_i)\, dW_{\mathcal{X}}(t),$$

where $W_{\mathcal{X}}$ is a standard Brownian motion under Q^i, with $dW_{\mathcal{X}}(t)\, dW_i^f(t) = \rho\, dt$, and where, for ease of notation, we omit the superscript i in $\sigma_{F\mathcal{X}}^i$ and the subscripts in $\rho_{\mathcal{X},i}$ when dealing only with maturity T_i.

Let us then consider a derivative whose payoff at time T_i is a function $g(F_i(T_{i-1}), F_i^f(T_{i-1}))$. By formula (2.24), the no-arbitrage value at time t of such a payoff is

$$P(t, T_i) E^i \big\{ g(F_i(T_{i-1}), F_i^f(T_{i-1})) | \mathcal{F}_t \big\}. \tag{14.41}$$

In the following, we will calculate this expectation for a few specific choices of the payoff function g. In particular, we will analyze the fundamental case of an option written on the spread between the two LIBOR rates and derive closed-form formulas for both the "up-front" and the "in-arrears" cases. Explicit formulas will also be derived for options on the product of the two rates as well as for trigger swaps.

14.5.1 Spread Options

A spread option on the two LIBOR rates $L(T_{i-1}, T_i)$ and $L^f(T_{i-1}, T_i)$ is a derivative paying off at time T_i, in domestic currency,

$$\begin{aligned} &\tau_i N \left[\omega\big(L(T_{i-1}, T_i) - L^f(T_{i-1}, T_i) + K\big)\right]^+ \\ &= \tau_i N \left[\omega\big(F_i(T_{i-1}) - F_i^f(T_{i-1}) + K\big)\right]^+, \end{aligned} \tag{14.42}$$

where N is the nominal value, K is the contract margin and $\omega = 1$ for a call and $\omega = -1$ for a put.

An "in-arrears" spread option pays off the same quantity at time T_{i-1}. By Proposition 2.8.1, see also Section 13.1, this is equivalent to paying off at time T_i

$$\tau_i N \left[\omega\big(F_i(T_{i-1}) - F_i^f(T_{i-1}) + K\big)\right]^+ \big(1 + \tau_i F_i(T_{i-1})\big). \tag{14.43}$$

The two payoffs (14.42) and (14.43) can be summarized into

$$\tau_i N \left[\omega\big(F_i(T_{i-1}) - F_i^f(T_{i-1}) + K\big)\right]^+ \big(1 + \psi \tau_i F_i(T_{i-1})\big), \tag{14.44}$$

where $\psi = 1$ for the "in-arrears" case and $\psi = 0$ otherwise.

Proposition 14.5.1. *The no-arbitrage value at time t of the payoff (14.44) is given by*

$$\begin{aligned} &\mathbf{LSO}(t, T_{i-1}, T_i, \tau_i, N, K, \omega, \psi) \\ &= \tau_i N P(t, T_i) \int_{-\infty}^{+\infty} \frac{1}{\sqrt{2\pi}} e^{-\frac{1}{2}v^2} \big[1 + \psi \tau_i \big(h(v) - K\big)\big] f(v)\, dv, \end{aligned} \tag{14.45}$$

where

$$\begin{aligned} f(v) = &\Bigg[-\omega F_i^f(t) e^{\mu_y + \rho_i \sigma_y v + \frac{1}{2}\sigma_y^2(1-\rho_i^2)} \\ &\cdot \Phi\left(-\omega \frac{\ln \frac{F_i^f(t)}{h(v)} + \mu_y + \rho_i \sigma_y v + \sigma_y^2(1-\rho_i^2)}{\sigma_y \sqrt{1-\rho_i^2}}\right) \\ &+ \omega h(v) \Phi\left(-\omega \frac{\ln \frac{F_i^f(t)}{h(v)} + \mu_y + \rho_i \sigma_y v}{\sigma_y \sqrt{1-\rho_i^2}}\right)\Bigg] 1_{\{h(v)>0\}} \\ &+ \tfrac{1}{2}(1-\omega) 1_{\{h(v)\le 0\}} \left[-h(v) + F_i^f(t) e^{\mu_y + \rho_i \sigma_y v + \frac{1}{2}\sigma_y^2(1-\rho_i)^2}\right] \end{aligned}$$

and

$$h(v) = K + F_i(t)e^{\mu_x + \sigma_x v}$$
$$\mu_x = -\tfrac{1}{2}\sigma_x^2$$
$$\mu_y = -\rho\sigma_{FX}\sigma_i^f \tau - \tfrac{1}{2}\sigma_y^2$$
$$\sigma_x = \sigma_i\sqrt{\tau}$$
$$\sigma_y = \sigma_i^f\sqrt{\tau}$$
$$\tau = T_{i-1} - t$$

Proof. By formula (14.41), the no-arbitrage value at time t of the payoff (14.44) is

$$\tau_i NP(t, T_i)E^i\left\{\left[\omega\big(F_i(T_{i-1}) - F_i^f(T_{i-1}) + K\big)\right]^+ (1 + \psi\tau_i F_i(T_{i-1}))|\mathcal{F}_t\right\}. \tag{14.46}$$

Defining, see also Appendix E,

$$X := \ln\frac{F_i(T_{i-1})}{F_i(t)},$$
$$Y := \ln\frac{F_i^f(T_{i-1})}{F_i^f(t)},$$

the joint density function $f_{X,Y}$ of (X, Y) under the measure Q^i is bivariate normal with mean vector and variance-covariance matrix respectively given by

$$M_{X,Y} = \begin{bmatrix}\mu_x \\ \mu_y\end{bmatrix}, \quad V_{X,Y} = \begin{bmatrix}\sigma_x^2 & \rho_i\sigma_x\sigma_y \\ \rho_i\sigma_x\sigma_y & \sigma_y^2\end{bmatrix}$$

that is

$$f_{X,Y}(x,y) = \frac{1}{2\pi\sigma_x\sigma_y\sqrt{1-\rho_i^2}}\exp\left[-\frac{\left(\frac{x-\mu_x}{\sigma_x}\right)^2 - 2\rho_i\frac{x-\mu_x}{\sigma_x}\frac{y-\mu_y}{\sigma_y} + \left(\frac{y-\mu_y}{\sigma_y}\right)^2}{2(1-\rho_i^2)}\right].$$

Since

$$f_{X,Y}(x,y) = f_{Y|X}(x,y)f_X(x),$$

where

$$f_{Y|X}(x,y) = \frac{1}{\sigma_y\sqrt{2\pi}\sqrt{1-\rho_i^2}}\exp\left[-\frac{\left(\frac{y-\mu_y}{\sigma_y} - \rho_i\frac{x-\mu_x}{\sigma_x}\right)^2}{2(1-\rho_i^2)}\right] \tag{14.47}$$

$$f_X(x) = \frac{1}{\sigma_x\sqrt{2\pi}}\exp\left[-\frac{1}{2}\left(\frac{x-\mu_x}{\sigma_x}\right)^2\right],$$

the expectation in (14.46) can be written as

$$\int_{-\infty}^{+\infty} (1 + \psi\tau_i F_i(t)e^x) \cdot \left[\int_{-\infty}^{+\infty} \left(\omega F_i(t)e^x - \omega F_i^f(t)e^y + \omega K \right)^+ f_{Y|X}(x,y)\,dy \right] f_X(x)\,dx$$

The expression between square brackets can be calculated analytically by distinguishing two cases:

1. $F_i(t)e^x + K \leq 0$.
   If $\omega = 1$, the expression is equal to 0 (the positive part of a negative number is zero). If $\omega = -1$, instead,

$$\left[\cdots\right] = -F_i(t)e^x - K + F_i^f(t) \int_{-\infty}^{+\infty} e^y f_{Y|X}(x,y)\,dy$$
$$= -F_i(t)e^x - K + F_i^f(t)e^{\mu_y + \rho_i\sigma_y \frac{x-\mu_x}{\sigma_x} + \frac{1}{2}\sigma_y^2(1-\rho_i^2)}$$

2. $F_i(t)e^x + K > 0$.
   Set $\bar{K} := F_i(t)e^x + K$ and $\bar{\omega} := -\omega$. Then

$$\left[\cdots\right] = \int_{-\infty}^{+\infty} \left(\bar{\omega} F_i^f(t)e^y - \bar{\omega}K \right)^+ f_{Y|X}(x,y)\,dy$$
$$= \bar{\omega} F_i^f(t)e^{\mu_y + \rho_i\sigma_y \frac{x-\mu_x}{\sigma_x} + \frac{1}{2}\sigma_y^2(1-\rho_i^2)} \cdot \Phi\left(\bar{\omega} \frac{\ln \frac{F_i^f(t)}{F_i(t)e^x + K} + \mu_y + \rho_i\sigma_y \frac{x-\mu_x}{\sigma_x} + \sigma_y^2(1-\rho_i^2)}{\sigma_y\sqrt{1-\rho_i^2}} \right)$$
$$- \bar{\omega}(F_i(t)e^x + K)\Phi\left(\bar{\omega} \frac{\ln \frac{F_i^f(t)}{F_i(t)e^x + K} + \mu_y + \rho_i\sigma_y \frac{x-\mu_x}{\sigma_x}}{\sigma_y\sqrt{1-\rho_i^2}} \right)$$

   by formula (D.1) in Appendix D.

Finally, to obtain (14.45), we simply have to set $v := (x - \mu_x)/\sigma_x$. □

14.5.2 Options on the Product

The second example we consider in this section is that of an option written on the product of the two LIBOR rates $L(T_{i-1}, T_i)$ and $L^f(T_{i-1}, T_i)$, whose payoff at time T_i, in domestic currency, is

$$\tau_i N \left[\omega(L(T_{i-1}, T_i)L^f(T_{i-1}, T_i) - K)\right]^+ = \tau_i N \left[\omega\left(F_i(T_{i-1})F_i^f(T_{i-1}) - K\right)\right]^+, \tag{14.48}$$

where N is the nominal value, K is the strike price and $\omega = 1$ for a call and $\omega = -1$ for a put.

Proposition 14.5.2. *The no-arbitrage value at time* t *of the payoff (14.48) is given by*

$$\begin{aligned}
&\mathbf{LP}(t, T_{i-1}, T_i, \tau_i, N, K, \omega) = \tau_i N P(t, T_i) \\
&\cdot \Big[\omega F_i(t) F_i^f(t) e^{[-\rho\sigma_F\chi\sigma_i^f + \rho_i\sigma_i\sigma_i^f]\tau} \\
&\quad \cdot \Phi\left(\omega \frac{\ln \frac{F_i(t)F_i^f(t)}{K} + [2\rho_i\sigma_i\sigma_i^f - \rho\sigma_F\chi\sigma_i^f + \frac{1}{2}\sigma_i^2 + \frac{1}{2}(\sigma_i^f)^2]\tau}{\sqrt{[\sigma_i^2 + (\sigma_i^f)^2 + 2\rho_i\sigma_i\sigma_i^f]\tau}}\right) \qquad (14.49)\\
&\quad -\omega K \Phi\left(\omega \frac{\ln \frac{F_i(t)F_i^f(t)}{K} - [\rho\sigma_F\chi\sigma_i^f + \frac{1}{2}\sigma_i^2 + \frac{1}{2}(\sigma_i^f)^2]\tau}{\sqrt{[\sigma_i^2 + (\sigma_i^f)^2 + 2\rho_i\sigma_i\sigma_i^f]\tau}}\right)\Big]
\end{aligned}$$

Proof. Since

$$\begin{aligned}
&F_i(T_{i-1}) F_i^f(T_{i-1}) \\
&= F_i(t) F_i^f(t) e^{-[\rho\sigma_F\chi\sigma_i^f + \frac{1}{2}\sigma_i^2 + \frac{1}{2}(\sigma_i^f)^2]\tau + \sigma_i[W_i(T_{i-1}) - W_i(t)] + \sigma_i^f[W_i^f(T_{i-1}) - W_i^f(t)]},
\end{aligned}$$

we have that, under Q^i,

$$\begin{aligned}
&\ln\big[F_i(T_{i-1}) F_i^f(T_{i-1}) | \mathcal{F}_t\big] \sim \mathcal{N}(M, V^2), \\
&M = \ln\big[F_i(t) F_i^f(t)\big] - [\rho\sigma_F\chi\sigma_i^f + \tfrac{1}{2}\sigma_i^2 + \tfrac{1}{2}(\sigma_i^f)^2]\tau, \\
&V = \sqrt{[\sigma_i^2 + (\sigma_i^f)^2 + 2\rho_i\sigma_i\sigma_i^f]\tau}.
\end{aligned}$$

To obtain (14.49), we simply have to remember (14.41) and apply again formula (D.1) in Appendix D. □

14.5.3 Trigger Swaps

We here consider a swap where, in one leg, different payments are triggered by different levels of either the domestic LIBOR rate or the foreign one.

In formulas, this leg of the trigger swap pays off at time T_i, in domestic currency, either

$$\tau_i N \left[\left(a F_i(T_{i-1}) + b F_i^f(T_{i-1}) + c\right) 1_{\{\omega F_i(T_{i-1}) \geq \omega K\}}\right] \left(1 + \psi \tau_i F_i(T_{i-1})\right), \qquad (14.50)$$

or, in case the payment is triggered by the foreign rate,

$$\tau_i N \left[\left(a F_i(T_{i-1}) + b F_i^f(T_{i-1}) + c\right) 1_{\{\omega F_i^f(T_{i-1}) \geq \omega K\}}\right] \left(1 + \psi \tau_i F_i(T_{i-1})\right), \qquad (14.51)$$

where N is the nominal value, a, b, c are real constants specified by the contract, ω is either 1 or -1, $\psi = 1$ for the "in-arrears" case and $\psi = 0$ otherwise.

Proposition 14.5.3. *The no-arbitrage value at time t of the payoff (14.50) is given by*

$$\begin{aligned}
&\mathbf{TSD}(t,T_{i-1},T_i,\tau_i,N,K,\omega,\psi)\\
&= \tau_i N P(t,T_i)\left[(a+c\psi\tau_i)F_i(t)\Phi\left(\omega\frac{\ln\frac{F_i(t)}{K}+\frac{1}{2}\sigma_i^2\tau}{\sigma_i\sqrt{\tau}}\right)\right.\\
&+a\psi\tau_i F_i^2(t)e^{\sigma_i^2\tau}\Phi\left(\omega\frac{\ln\frac{F_i(t)}{K}+\frac{3}{2}\sigma_i^2\tau}{\sigma_i\sqrt{\tau}}\right)+c\Phi\left(\omega\frac{\ln\frac{F_i(t)}{K}-\frac{1}{2}\sigma_i^2\tau}{\sigma_i\sqrt{\tau}}\right)\\
&+bF_i^f(t)e^{-\rho\sigma_{FX}\sigma_i^f\tau}\Phi\left(\omega\frac{\ln\frac{F_i(t)}{K}+\left[\rho_i\sigma_i\sigma_i^f-\frac{1}{2}\sigma_i^2\right]\tau}{\sigma_i\sqrt{\tau}}\right)\\
&\left.+b\psi\tau_i F_i(t)F_i^f(t)e^{[-\rho\sigma_{FX}\sigma_i^f+\rho_i\sigma_i\sigma_i^f]\tau}\Phi\left(\omega\frac{\ln\frac{F_i(t)}{K}+\left[\rho_i\sigma_i\sigma_i^f+\frac{1}{2}\sigma_i^2\right]\tau}{\sigma_i\sqrt{\tau}}\right)\right].
\end{aligned}\tag{14.52}$$

The no-arbitrage value at time t of the payoff (14.51) is instead given by

$$\begin{aligned}
&\mathbf{TSF}(t,T_{i-1},T_i,\tau_i,N,K,\omega,\psi)\\
&= \tau_i N P(t,T_i)\left[c\Phi\left(\omega\frac{\ln\frac{F_i^f(t)}{K}-\left[\rho\sigma_{FX}+\frac{1}{2}\sigma_i^f\right]\sigma_i^f\tau}{\sigma_i^f\sqrt{\tau}}\right)\right.\\
&+(a+c\psi\tau_i)F_i(t)\Phi\left(\omega\frac{\ln\frac{F_i^f(t)}{K}-\left[\rho\sigma_{FX}+\frac{1}{2}\sigma_i^f-\rho_i\sigma_i\right]\sigma_i^f\tau}{\sigma_i^f\sqrt{\tau}}\right)\\
&+a\psi\tau_i F_i^2(t)e^{\sigma_i^2\tau}\Phi\left(\omega\frac{\ln\frac{F_i^f(t)}{K}-\left[\rho\sigma_{FX}+\frac{1}{2}\sigma_i^f-2\rho_i\sigma_i\right]\sigma_i^f\tau}{\sigma_i^f\sqrt{\tau}}\right)\\
&+bF_i^f(t)e^{-\rho\sigma_{FX}\sigma_i^f\tau}\Phi\left(\omega\frac{\ln\frac{F_i^f(t)}{K}+\left[-\rho\sigma_{FX}+\frac{1}{2}\sigma_i^f\right]\sigma_i^f\tau}{\sigma_i^f\sqrt{\tau}}\right)\\
&+b\psi\tau_i F_i(t)F_i^f(t)e^{[-\rho\sigma_{FX}\sigma_i^f+\rho_i\sigma_i\sigma_i^f]\tau}\\
&\left.\cdot\Phi\left(\omega\frac{\ln\frac{F_i^f(t)}{K}+\left[-\rho\sigma_{FX}+\frac{1}{2}\sigma_i^f+\rho_i\sigma_i\right]\sigma_i^f\tau}{\sigma_i^f\sqrt{\tau}}\right)\right].
\end{aligned}\tag{14.53}$$

Proof. The proof is quite similar in spirit to that of Proposition 14.5.1 and is therefore omitted. The only difference is that here the outer integral, in both cases, can be explicitly calculated, too. □

14.5.4 Dealing with Multiple Dates

A final fundamental remark is in order. When dealing with several payment times simultaneously, Schlögl (2002) noticed that we cannot assume that the volatilities $\sigma^i_{F\mathcal{X}}$, σ_i and σ^f_i are positive constants for all i. In fact, any two consecutive forward exchange rates and the corresponding domestic and foreign forward rates are constrained by the following relation

$$\frac{F_\mathcal{X}(t,T_i)}{F_\mathcal{X}(t,T_{i-1})} = \frac{1+\tau_i F_i(t)}{1+\tau_i F^f_i(t)}. \tag{14.54}$$

Clearly, if we assume that $\sigma^i_{F\mathcal{X}}$, σ_i and σ^f_i are positive constants, $\sigma^{i-1}_{F\mathcal{X}}$ cannot be constant as well, and its admissible values are obtained by equating the (instantaneous) quadratic variations on both sides of (14.54).

However, resorting to the classical trick of freezing the forward rates at their time 0 value in the diffusion coefficients of the right-hand-side of (14.54), we can still deal with forward exchange-rate volatilities that are approximately constant. For instance, assuming perfect correlation in both sides,

$$\begin{aligned}
\sigma^{i-1}_{F\mathcal{X}} &= \sigma^i_{F\mathcal{X}} + \sigma^f_i \frac{\tau_i F^f_i(t)}{1+\tau_i F^f_i(t)} - \sigma_i \frac{\tau_i F_i(t)}{1+\tau_i F_i(t)} \\
&\approx \sigma^i_{F\mathcal{X}} + \sigma^f_i \frac{\tau_i F^f_i(0)}{1+\tau_i F^f_i(0)} - \sigma_i \frac{\tau_i F_i(0)}{1+\tau_i F_i(0)}.
\end{aligned}$$

Therefore, applying this "freezing" procedure for each $i < M$ starting from $\sigma^M_{F\mathcal{X}}$, or equivalently for each $i > 0$ starting from $\sigma^0_{F\mathcal{X}}$, we can still assume that the volatilities $\sigma^i_{F\mathcal{X}}$ are all constant and set to one of their admissible values.

Part VI

INFLATION

15. Pricing of Inflation-Indexed Derivatives

I wasn't affected by inflation - I had nothing to inflate.
Gerald Barzan, humorist.

European governments have been issuing inflation-indexed bonds since the beginning of the 80's, but it is only in the very last years that these bonds, and inflation-indexed derivatives in general, have become more and more popular.[1]

Inflation is defined in terms of the percentage increments of a reference index, the Consumer Price Index (CPI), which is a representative basket of goods and services. The evolution of two major European and American inflation indices from September 2001 to July 2004 is shown in Figure 15.1.

In theory, and also in practice, inflation can become negative, so that, to preserve positivity of coupons, the inflation rate is typically floored at zero, thus implicitly offering a zero-strike floor in conjunction with the "pure" inflation-linked bond.[2]

Floors with low strikes are the most actively traded options on inflation rates. Other extremely popular derivatives are inflation-indexed swaps, where the inflation rate is either payed on an annual basis or with a single amount at the swap maturity.

All these inflation-indexed derivatives require a specific model to be valued. Their pricing has been tackled by, among others, Barone and Castagna (1997) and Jarrow and Yildirim (2003) (JY), who proposed similar frameworks based on a foreign-currency analogy.[3] In both articles, what is modelled is the evolution of the instantaneous nominal and real rates and of the CPI, which is interpreted as the "exchange rate" between the nominal and real economies. In this setting, the valuation of an inflation-indexed payoff becomes equivalent to that of a cross-currency interest rate derivative.

[1] A thorough reference to inflation-indexed securities is the book by Deacon, Derry and Mirfendereski (2004).

[2] A comprehensive guide to inflation-indexed derivatives is that of The Royal Bank of Scotland (2003).

[3] Other references for the pricing of inflation-indexed derivatives are van Bezooyen et al. (1997), Hughston (1998), Cairns (2000), Jamshidian (2002a, 2002b) and Beletski and Korn (2004).

Exploiting this foreign-currency analogy, Mercurio (2005) introduced two LFMs to model interest rates in the nominal and real economies, respectively. He then derived closed-form formulas for both inflation swaps and caps.[4] A different approach has been independently proposed by Kazziha (1999), Belgrade, Benhamou and Koehler (2004) and Mercurio (2005), who considered a market model based on forward indices. They all noticed that a forward CPI is a martingale under the corresponding (nominal) forward measure and, by assuming driftless lognormal dynamics, derived explicit formulas for swap and caps that are much simpler than in the LFM case.

The purpose of this chapter is to define the main types of inflation-indexed swaps and caps present in the market and price them analytically and consistently with no arbitrage. To this end we will review and use the JY model, the Mercurio (2005) application of the LFM, and the market models of Kazziha (1999), Belgrade, Benhamou and Koehler (2004) and Mercurio (2005).

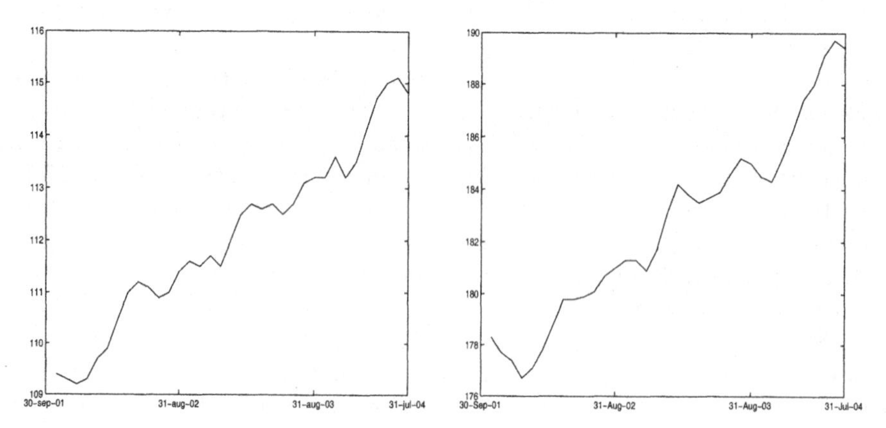

Fig. 15.1. Left: EUR CPI Unrevised Ex-Tobacco. Right: USD CPI Urban Consumers NSA. Monthly closing values from 30-Sep-01 to 21-Jul-04.

15.1 The Foreign-Currency Analogy

Many of the articles on inflation modeling are based on the so called *foreign-currency analogy*, according to which real rates are viewed as interest rates in the real (i.e. foreign) economy, and the CPI is interpreted as the exchange rate between the nominal (i.e. domestic) and real "currencies".

This analogy is perfectly motivated by the same definition of CPI. In fact, denoting by $I(t)$ the value of the CPI at time t, the reference basket of good and services can be bought with $I(0)$ units of currency at time $t = 0$, whereas at time $t = T$ one needs to spend $I(T)$. Equivalently, compared to

[4] Cap prices are actually derived under the assumption of deterministic real rates.

our purchasing power at time zero, one unit of currency at time T is worth $I(0)/I(T)$ in real terms. Therefore, setting the normalizing constant $I(0)$ to one, we can convert a nominal value into the corresponding real value by simply dividing by the CPI's value at that time, pretty much as we do when converting amounts in one currency into another.

The interest rate that is associated with the real economy is simply referred to as real rate, even though, prima facie, this may create some bewilderment. This rate, however, is to be intended as the "expected real rate" for the future interval it applies to, or better as the real rate we can lock in by suitably trading in inflation swaps. The true real rate will be only known at the end of the corresponding period, as soon as the value of the CPI at that time is known. This is why there is no redundancy in modeling both this real rate and the inflation index together with the nominal rate.

15.2 Definitions and Notation

We generally denote quantities related to the nominal and real economies with the subscripts n and r, respectively.

The term structures of discount factors, at time t, for the nominal and real economies are respectively given by $T \mapsto P_n(t,T)$ and $T \mapsto P_r(t,T)$ for $T \geq t$.

Given the future times T_{i-1} and T_i, the related forward LIBOR rates, at time t, are defined by (1.20), namely

$$F_x(t;T_{i-1},T_i) = \frac{P_x(t,T_{i-1}) - P_x(t,T_i)}{\tau_i P_x(t,T_i)}, \quad x \in \{n,r\},$$

where τ_i is the year fraction for the interval $[T_{i-1},T_i]$, which is assumed to be the same for both nominal and real rates.

The nominal and real instantaneous forward rates at time t for maturity T are defined by (1.23), namely

$$f_x(t,T) = -\frac{\partial \ln P_x(t,T)}{\partial T}, \quad x \in \{n,r\}.$$

We then denote the nominal and real instantaneous short rates, respectively, by

$$\begin{aligned} n(t) &= f_n(t,t) \\ r(t) &= f_r(t,t). \end{aligned}$$

We denote by Q_n and Q_r the nominal and real risk-neutral measures, respectively, and by E_x the expectation associated with Q_x, $x \in \{n,r\}$.

Finally, the forward CPI at time t for maturity T_i is denoted by $\mathcal{I}_i(t)$ and defined by

$$\mathcal{I}_i(t) := I(t)\frac{P_r(t,T_i)}{P_n(t,T_i)}. \tag{15.1}$$

15.3 The JY Model

We briefly review here the approach proposed by Jarrow and Yildirim (2003) for modeling inflation and interest rates. We will restrict our analysis to a specific parametrization for the forward rates volatilities.

Under the real-world probability space $(\Omega, \mathcal{F}, P)$, with associated filtration $\mathcal{F}_t$, Jarrow and Yildirim considered the following evolution for the nominal and real instantaneous forward rates and for the CPI

$$\begin{aligned} df_n(t,T) &= \alpha_n(t,T)\,dt + \varsigma_n(t,T)\,dW_n^P(t) \\ df_r(t,T) &= \alpha_r(t,T)\,dt + \varsigma_r(t,T)\,dW_r^P(t) \\ dI(t) &= I(t)\mu(t)\,dt + \sigma_I I(t)\,dW_I^P(t) \end{aligned}$$

with $I(0) = I_0 > 0$, and

$$f_x(0,T) = f_x^M(0,T), \quad x \in \{n, r\},$$

where

- (W_n^P, W_r^P, W_I^P) is a Brownian motion with correlations $\rho_{n,r}$, $\rho_{n,I}$ and $\rho_{r,I}$;
- α_n, α_r and μ are adapted processes;
- ς_n and ς_r are deterministic functions;
- σ_I is a positive constant;
- $f_n^M(0,T)$ and $f_r^M(0,T)$ are, respectively, the nominal and real instantaneous forward rates observed in the market at time 0 for maturity T.

To ease the calculation of the derivatives' prices in the next sections, we choose to model the forward rate volatilities as

$$\begin{aligned} \varsigma_n(t,T) &= \sigma_n\, e^{-a_n(T-t)}, \\ \varsigma_r(t,T) &= \sigma_r\, e^{-a_r(T-t)}, \end{aligned}$$

where σ_n σ_r, a_n and a_r are positive constants. Using this parametrization, we can resort to the equivalent formulation in terms of instantaneous short rates, as explained in Section 5.2, and rephrase the main result by Jarrow and Yildirim (2003) as follows.

Proposition 15.3.1. *The Q_n-dynamics of the instantaneous nominal rate, the instantaneous real rate and the CPI are*

$$\begin{aligned} dn(t) &= [\vartheta_n(t) - a_n n(t)]\,dt + \sigma_n\,dW_n(t) \\ dr(t) &= [\vartheta_r(t) - \rho_{r,I}\sigma_I\sigma_r - a_r r(t)]\,dt + \sigma_r\,dW_r(t) \\ dI(t) &= I(t)[n(t) - r(t)]\,dt + \sigma_I I(t)\,dW_I(t), \end{aligned} \tag{15.2}$$

where (W_n, W_r, W_I) is a Brownian motion with correlations $\rho_{n,r}$, $\rho_{n,I}$ and $\rho_{r,I}$, and $\vartheta_n(t)$ and $\vartheta_r(t)$ are deterministic functions to be used to exactly fit the term structures of nominal and real rates, respectively, i.e.*:*

$$\vartheta_x(t) = \frac{\partial f_x(0,t)}{\partial T} + a_x f_x(0,t) + \frac{\sigma_x^2}{2a_x}(1 - e^{-2a_x t}), \quad x \in \{n, r\},$$

where $\frac{\partial f_x}{\partial T}$ *denotes partial derivative of* f_x *with respect to its second argument.*

Jarrow and Yildirim thus assumed that both nominal and real (instantaneous) rates are normally distributed under their respective risk-neutral measures. They then proved that the real rate r is still an Ornstein-Uhlenbeck process under the nominal measure Q_n, and that the inflation index $I(t)$, at each time t, is lognormally distributed under Q_n, since we can write, for each $t < T$,

$$I(T) = I(t)\, e^{\int_t^T [n(u)-r(u)]\, du - \frac{1}{2}\sigma_I^2(T-t) + \sigma_I (W_I(T) - W_I(t))}. \tag{15.3}$$

In the following sections, we will apply the JY model to the valuation of inflation-indexed swaps and caps.

16. Inflation-Indexed Swaps

Given a set of dates $T_1, \ldots, T_M$, an Inflation-Indexed Swap (IIS) is a swap where, on each payment date, Party A pays Party B the inflation rate over a predefined period, while Party B pays Party A a fixed rate. The inflation rate is calculated as the percentage return of the CPI index over the time interval it applies to. Two are the main IIS traded in the market: the zero coupon (ZC) swap and the year-on-year (YY) swap.

In a ZCIIS, at the final time T_M, assuming $T_M = M$ years, Party B pays Party A the fixed amount

$$N[(1+K)^M - 1], \tag{16.1}$$

where K and N are, respectively, the contract fixed rate and nominal value. In exchange for this fixed payment, Party A pays Party B, at the final time T_M, the floating amount

$$N\left[\frac{I(T_M)}{I_0} - 1\right]. \tag{16.2}$$

In a YYIIS, at each time T_i, Party B pays Party A the fixed amount

$$N\varphi_i K,$$

where φ_i is the contract fixed-leg year fraction for the interval $[T_{i-1}, T_i]$, while Party A pays Party B the (floating) amount

$$N\psi_i\left[\frac{I(T_i)}{I(T_{i-1})} - 1\right], \tag{16.3}$$

where ψ_i is the floating-leg year fraction for the interval $[T_{i-1}, T_i]$, $T_0 := 0$ and N is again the contract nominal value.

Both ZC and YY swaps are quoted, in the market, in terms of the corresponding fixed rate K. The ZCIIS and YYIIS (mid) fixed-rate quotes in the Euro market on October 7th 2004 are shown in Figure 16.1, for maturities up to twenty years. The reference CPI is the Euro-zone ex-tobacco index.

16.1 Pricing of a ZCIIS

Standard no-arbitrage pricing theory implies that the value at time t, $0 \leq t < T_M$, of the inflation-indexed leg of the ZCIIS is

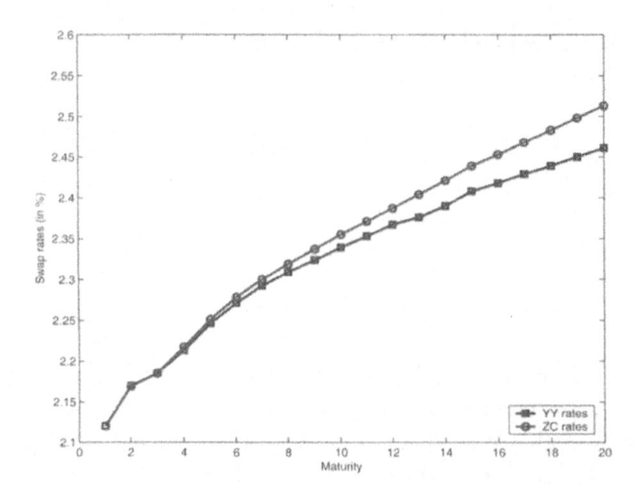

Fig. 16.1. Euro inflation swap rates as of October 7, 2004.

$$\mathbf{ZCIIS}(t, T_M, I_0, N) = NE_n \left\{ e^{-\int_t^{T_M} n(u)\,du} \left[\frac{I(T_M)}{I_0} - 1 \right] \Big| \mathcal{F}_t \right\}, \tag{16.4}$$

where $\mathcal{F}_t$ denotes the σ-algebra generated by the relevant underlying processes up to time t.

By the foreign-currency analogy, the nominal price of a real zero-coupon bond equals the nominal price of the contract paying off one unit of the CPI index at bond maturity, see also the general formula (2.31). In formulas, for each $t < T$:

$$I(t)P_r(t,T) = I(t)E_r \left\{ e^{-\int_t^T r(u)\,du} \big| \mathcal{F}_t \right\} = E_n \left\{ e^{-\int_t^T n(u)\,du} I(T) \big| \mathcal{F}_t \right\}. \tag{16.5}$$

Therefore, (16.4) becomes

$$\mathbf{ZCIIS}(t, T_M, I_0, N) = N \left[\frac{I(t)}{I_0} P_r(t, T_M) - P_n(t, T_M) \right], \tag{16.6}$$

which at time $t = 0$ simplifies to

$$\mathbf{ZCIIS}(0, T_M, N) = N[P_r(0, T_M) - P_n(0, T_M)]. \tag{16.7}$$

Formulas (16.6) and (16.7) yield model-independent prices, which are not based on specific assumptions on the evolution of the interest rate market, but simply follow from the absence of arbitrage. This result is extremely important since it enables us to strip, with no ambiguity, real zero-coupon bond prices from the quoted prices of zero-coupon inflation-indexed swaps.

In fact, the market quotes values of $K = K(T_M)$ for some given maturities T_M, so that equating (16.7) with the (nominal) present value of (16.1), and getting the discount factor $P_n(0, T_M)$ from the current (nominal) zero-coupon curve, we can solve for the unknown $P_r(0,T_M)$. We thus obtain the discount factor for maturity T_M in the real economy:[1]

$$P_r(0,T_M) = P_n(0,T_M)(1 + K(T_M))^M. \tag{16.8}$$

Remark 16.1.1. **(ZCIIS and Forward CPI).** Kazziha (1999) defines the T-forward CPI at time t as the fixed amount X to be exchanged at time T for the CPI $I(T)$, for which such a swap has zero value at time t, in analogy with the definition of a forward LIBOR rate we gave in Chapter 1. From formula (16.5), we immediately obtain

$$I(t)P_r(t,T) = XP_n(t,T).$$

This is consistent with definition (15.1), which was directly based on the foreign-currency analogy.

The advantage of Kazziha's approach is that no foreign-currency analogy is required for the definition of the forward CPI's $\mathcal{I}_i$, and the pricing system she defines is only based on nominal zero-coupon bonds and forward CPI's. In her setting, the value at time zero of a T_M-forward CPI can be obtained from the market quote $K(T_M)$ by applying this simple formula

$$\mathcal{I}_M(0) = I(0)(1 + K(T_M))^M,$$

which is perfectly equivalent to (16.8).

16.2 Pricing of a YYIIS

Compared to that of a ZCIIS, the valuation of a YYIIS is more involved. Notice, in fact, that the value at time $t < T_i$ of the payoff (16.3) at time T_i is

$$\mathbf{YYIIS}(t, T_{i-1}, T_i, \psi_i, N) = N\psi_i E_n \left\{ e^{-\int_t^{T_i} n(u)\,du} \left[\frac{I(T_i)}{I(T_{i-1})} - 1 \right] \Big| \mathcal{F}_t \right\}, \tag{16.9}$$

which, assuming $t < T_{i-1}$ (otherwise we fall back to the previous case), can be calculated as

$$N\psi_i E_n \left\{ e^{-\int_t^{T_{i-1}} n(u)\,du} E_n \left[e^{-\int_{T_{i-1}}^{T_i} n(u)\,du} \left(\frac{I(T_i)}{I(T_{i-1})} - 1 \right) \Big| \mathcal{F}_{T_{i-1}} \right] \Big| \mathcal{F}_t \right\}. \tag{16.10}$$

The inner expectation is nothing but $\mathbf{ZCIIS}(T_{i-1}, T_i, I(T_{i-1}), 1)$, so that we obtain

$$\begin{aligned} & N\psi_i E_n \left\{ e^{-\int_t^{T_{i-1}} n(u)\,du} [P_r(T_{i-1}, T_i) - P_n(T_{i-1}, T_i)] \Big| \mathcal{F}_t \right\} \\ &= N\psi_i E_n \left\{ e^{-\int_t^{T_{i-1}} n(u)\,du} P_r(T_{i-1}, T_i) \Big| \mathcal{F}_t \right\} - N\psi_i P_n(t, T_i). \end{aligned} \tag{16.11}$$

[1] The real discount factors for intermediate maturities can be inferred by taking into account the typical seasonality effects in inflation.

The last expectation can be viewed as the nominal price of a derivative paying off, in nominal units, the real zero-coupon bond price $P_r(T_{i-1}, T_i)$ at time T_{i-1}. If real rates were deterministic, then this price would simply be the present value, in nominal terms, of the forward price of the real bond. In this case, in fact, we would have:

$$\begin{aligned} E_n\left\{e^{-\int_t^{T_{i-1}} n(u)\,du} P_r(T_{i-1}, T_i)\big|\mathcal{F}_t\right\} &= P_r(T_{i-1}, T_i) P_n(t, T_{i-1}) \\ &= \frac{P_r(t, T_i)}{P_r(t, T_{i-1})} P_n(t, T_{i-1}). \end{aligned}$$

In practice, however, real rates are stochastic and the expected value in (16.11) is model dependent. For instance, under dynamics (15.2), the forward price of the real bond must be corrected by a factor depending on both the nominal and real interest rates volatilities and on the respective correlation. This is explained in the following.

16.3 Pricing of a YYIIS with the JY Model

Denoting by Q_n^T the nominal T-forward measure for a general maturity T and by E_n^T the associated expectation, we can write:

$$\begin{aligned} &\mathbf{YYIIS}(t, T_{i-1}, T_i, \psi_i, N) \\ &= N\psi_i P_n(t, T_{i-1}) E_n^{T_{i-1}}\left\{P_r(T_{i-1}, T_i)\big|\mathcal{F}_t\right\} - N\psi_i P_n(t, T_i). \end{aligned} \tag{16.12}$$

Remembering formula (3.39) for the zero-coupon bond price in the Hull and White (1994b) model:

$$\begin{aligned} P_r(t,T) &= A_r(t,T) e^{-B_r(t,T) r(t)}, \\ B_r(t,T) &= \frac{1}{a_r}\left[1 - e^{-a_r(T-t)}\right], \\ A_r(t,T) &= \frac{P_r^M(0,T)}{P_r^M(0,t)} \exp\left\{B_r(t,T) f_r^M(0,t) - \frac{\sigma_r^2}{4a_r}(1 - e^{-2a_r t}) B_r(t,T)^2\right\}, \end{aligned} \tag{16.13}$$

and noting that, by the change-of-numeraire toolkit in Section 2.3, and formula (2.12) in particular, the real instantaneous rate evolves under $Q_n^{T_{i-1}}$ according to

$$dr(t) = \left[-\rho_{n,r}\sigma_n\sigma_r B_n(t, T_{i-1}) + \vartheta_r(t) - \rho_{r,I}\sigma_I\sigma_r - a_r r(t)\right] dt + \sigma_r\, dW_r^{T_{i-1}}(t) \tag{16.14}$$

with $W_r^{T_{i-1}}$ a $Q_n^{T_{i-1}}$-Brownian motion, we have that the real bond price $P_r(T_{i-1}, T_i)$ is lognormally distributed under $Q_n^{T_{i-1}}$, since $r(T_{i-1})$ is still a normal random variable under this (nominal) forward measure. After some tedious, but straightforward, algebra we finally obtain

$$\begin{aligned}&\mathbf{YYIIS}(t,T_{i-1},T_i,\psi_i,N)\\&=N\psi_iP_n(t,T_{i-1})\frac{P_r(t,T_i)}{P_r(t,T_{i-1})}e^{C(t,T_{i-1},T_i)}-N\psi_iP_n(t,T_i),\end{aligned}\tag{16.15}$$

where

$$\begin{aligned}C(t,T_{i-1},T_i)=&\sigma_rB_r(T_{i-1},T_i)\Big[B_r(t,T_{i-1})\Big(\rho_{r,I}\sigma_I-\tfrac{1}{2}\sigma_rB_r(t,T_{i-1})\\&+\frac{\rho_{n,r}\sigma_n}{a_n+a_r}(1+a_rB_n(t,T_{i-1}))\Big)-\frac{\rho_{n,r}\sigma_n}{a_n+a_r}B_n(t,T_{i-1})\Big].\end{aligned}$$

The expectation of a real zero-coupon bond price under a nominal forward measure, in the JY model, is thus equal to the current forward price of the real bond multiplied by a correction factor, which depends on the (instantaneous) volatilities of the nominal rate, the real rate and the CPI, on the (instantaneous) correlation between nominal and real rates, and on the (instantaneous) correlation between the real rate and the CPI.

The exponential of C is the correction term we mentioned above. This term accounts for the stochasticity of real rates and, indeed, vanishes for $\sigma_r = 0$.

The value at time t of the inflation-indexed leg of the swap is simply obtained by summing up the values of all floating payments. We thus get

$$\begin{aligned}\mathbf{YYIIS}(t,\mathcal{T},\Psi,N)=&N\psi_{\iota(t)}\left[\frac{I(t)}{I(T_{\iota(t)-1})}P_r(t,T_{\iota(t)})-P_n(t,T_{\iota(t)})\right]\\+N\sum_{i=\iota(t)+1}^{M}&\psi_i\left[P_n(t,T_{i-1})\frac{P_r(t,T_i)}{P_r(t,T_{i-1})}e^{C(t,T_{i-1},T_i)}-P_n(t,T_i)\right],\end{aligned}\tag{16.16}$$

where we set $\mathcal{T} := \{T_1, \dots, T_M\}$, $\Psi := \{\psi_1, \dots, \psi_M\}$ and $\iota(t) = \min\{i : T_i > t\}$,[2] and where the first payment after time t has been priced according to (16.6). In particular at $t = 0$,

$$\begin{aligned}&\mathbf{YYIIS}(0,\mathcal{T},\Psi,N)=N\psi_1[P_r(0,T_1)-P_n(0,T_1)]\\&+N\sum_{i=2}^{M}\psi_i\left[P_n(0,T_{i-1})\frac{P_r(0,T_i)}{P_r(0,T_{i-1})}e^{C(0,T_{i-1},T_i)}-P_n(0,T_i)\right]\\&=N\sum_{i=1}^{M}\psi_iP_n(0,T_i)\left[\frac{1+\tau_iF_n(0;T_{i-1},T_i)}{1+\tau_iF_r(0;T_{i-1},T_i)}e^{C(0,T_{i-1},T_i)}-1\right].\end{aligned}\tag{16.17}$$

The advantage of using Gaussian models for nominal and real rates is clear as far as analytical tractability is concerned. However, the possibility of negative rates and the difficulty in estimating historically the real rate parameters led to alternative approaches. We now illustrate two different market models that have been proposed for alternative valuations of a YYIIS and other inflation-indexed derivatives.

[2] By definition, $T_{\iota(t)-1} \le t < T_{\iota(t)}$.

16.4 Pricing of a YYIIS with a First Market Model

For an alternative pricing of the above YYIIS, we notice that we can change measure and, as explained in Section 2.8, re-write the expectation in (16.12) as

$$\begin{aligned} P_n(t,T_{i-1})E_n^{T_{i-1}}\left\{P_r(T_{i-1},T_i)|\mathcal{F}_t\right\} &= P_n(t,T_i)E_n^{T_i}\left\{\frac{P_r(T_{i-1},T_i)}{P_n(T_{i-1},T_i)}|\mathcal{F}_t\right\} \\ &= P_n(t,T_i)E_n^{T_i}\left\{\frac{1+\tau_i F_n(T_{i-1};T_{i-1},T_i)}{1+\tau_i F_r(T_{i-1};T_{i-1},T_i)}|\mathcal{F}_t\right\}, \end{aligned} \tag{16.18}$$

which can be calculated as soon as we specify the distribution of both forward rates under the nominal T_i-forward measure.

It seems natural, therefore, to resort to a LFM, which postulates the evolution of simply-compounded forward rates, namely the variables that explicitly enter the last expectation, see Section 6.3. This approach, followed by Mercurio (2005), is detailed in the following.

Since $I(t)P_r(t,T_i)$ is the price of an asset in the nominal economy, we have that the forward CPI

$$\mathcal{I}_i(t) = I(t)\frac{P_r(t,T_i)}{P_n(t,T_i)}$$

is a martingale under $Q_n^{T_i}$ by the definition itself of $Q_n^{T_i}$. Assuming lognormal dynamics for $\mathcal{I}_i$,

$$d\mathcal{I}_i(t) = \sigma_{I,i}\mathcal{I}_i(t)\,dW_i^I(t), \tag{16.19}$$

where $\sigma_{I,i}$ is a positive constant and W_i^I is a $Q_n^{T_i}$-Brownian motion, and assuming also that both nominal and real forward rates follow a LFM, the analogy with cross-currency derivatives pricing implies that the dynamics of $F_n(\cdot;T_{i-1},T_i)$ and $F_r(\cdot;T_{i-1},T_i)$ under $Q_n^{T_i}$ are given by (see Section 14.4)

$$\begin{aligned} dF_n(t;T_{i-1},T_i) &= \sigma_{n,i}F_n(t;T_{i-1},T_i)\,dW_i^n(t), \\ dF_r(t;T_{i-1},T_i) &= F_r(t;T_{i-1},T_i)\left[-\rho_{I,r,i}\sigma_{I,i}\sigma_{r,i}\,dt + \sigma_{r,i}\,dW_i^r(t)\right], \end{aligned} \tag{16.20}$$

where $\sigma_{n,i}$ and $\sigma_{r,i}$ are positive constants, W_i^n and W_i^r are two Brownian motions with instantaneous correlation ρ_i, and $\rho_{I,r,i}$ is the instantaneous correlation between $\mathcal{I}_i(\cdot)$ and $F_r(\cdot;T_{i-1},T_i)$, i.e. $dW_i^I(t)\,dW_i^r(t) = \rho_{I,r,i}\,dt$.

Allowing $\sigma_{I,i}$, $\sigma_{n,i}$ and $\sigma_{r,i}$ to be deterministic functions of time does not complicate the calculations below. We assume hereafter that such volatilities are constant for ease of notation only. In practice, however, the implications of using constant or time-dependent coefficients should be carefully analyzed. See also Chapter 7 and Remark 18.0.1 below.

The expectation in (16.18) can then be easily calculated with a numerical integration by noting that, under $Q_n^{T_i}$ and conditional on $\mathcal{F}_t$, the pair[3]

[3] To lighten the notation, we simply write (X_i, Y_i) instead of $(X_i(t), Y_i(t))$.

$$(X_i, Y_i) = \left(\ln \frac{F_n(T_{i-1}; T_{i-1}, T_i)}{F_n(t; T_{i-1}, T_i)}, \ln \frac{F_r(T_{i-1}; T_{i-1}, T_i)}{F_r(t; T_{i-1}, T_i)} \right) \tag{16.21}$$

is distributed as a bivariate normal random variable with mean vector and variance-covariance matrix, respectively, given by

$$M_{X_i,Y_i} = \begin{bmatrix} \mu_{x,i}(t) \\ \mu_{y,i}(t) \end{bmatrix}, \quad V_{X_i,Y_i} = \begin{bmatrix} \sigma_{x,i}^2(t) & \rho_i \sigma_{x,i}(t)\sigma_{y,i}(t) \\ \rho_i \sigma_{x,i}(t)\sigma_{y,i}(t) & \sigma_{y,i}^2(t) \end{bmatrix}, \tag{16.22}$$

where

$$\mu_{x,i}(t) = -\tfrac{1}{2}\sigma_{n,i}^2(T_{i-1} - t), \quad \sigma_{x,i}(t) = \sigma_{n,i}\sqrt{T_{i-1} - t},$$
$$\mu_{y,i}(t) = \left[-\tfrac{1}{2}\sigma_{r,i}^2 - \rho_{I,r,i}\sigma_{I,i}\sigma_{r,i} \right](T_{i-1} - t), \quad \sigma_{y,i}(t) = \sigma_{r,i}\sqrt{T_{i-1} - t}.$$

It is well known that the density $f_{X_i,Y_i}(x, y)$ of (X_i, Y_i) can be decomposed as[4]

$$f_{X_i,Y_i}(x, y) = f_{X_i|Y_i}(x, y) f_{Y_i}(y),$$

where

$$f_{X_i|Y_i}(x, y) = \frac{1}{\sigma_{x,i}(t)\sqrt{2\pi}\sqrt{1 - \rho_i^2}} \exp\left[-\frac{\left(\frac{x - \mu_{x,i}(t)}{\sigma_{x,i}(t)} - \rho_i \frac{y - \mu_{y,i}(t)}{\sigma_{y,i}(t)} \right)^2}{2(1 - \rho_i^2)} \right]$$

$$f_{Y_i}(y) = \frac{1}{\sigma_{y,i}(t)\sqrt{2\pi}} \exp\left[-\frac{1}{2} \left(\frac{y - \mu_{y,i}(t)}{\sigma_{y,i}(t)} \right)^2 \right]. \tag{16.23}$$

The last expectation in (16.18) can thus be calculated as

$$\int_{-\infty}^{+\infty} \frac{\int_{-\infty}^{+\infty} (1 + \tau_i F_n(t; T_{i-1}, T_i) e^x) f_{X_i|Y_i}(x, y)\, dx}{1 + \tau_i F_r(t; T_{i-1}, T_i)\, e^y} f_{Y_i}(y)\, dy$$
$$= \int_{-\infty}^{+\infty} \frac{1 + \tau_i F_n(t; T_{i-1}, T_i)\, e^{\mu_{x,i}(t) + \rho_i \sigma_{x,i}(t) \frac{y - \mu_{y,i}(t)}{\sigma_{y,i}(t)} + \frac{1}{2}\sigma_{x,i}^2(t)(1 - \rho_i^2)}}{1 + \tau_i F_r(t; T_{i-1}, T_i)\, e^y} f_{Y_i}(y)\, dy$$
$$= \int_{-\infty}^{+\infty} \frac{1 + \tau_i F_n(t; T_{i-1}, T_i)\, e^{\rho_i \sigma_{x,i}(t) z - \frac{1}{2}\sigma_{x,i}^2(t)\rho_i^2}}{1 + \tau_i F_r(t; T_{i-1}, T_i)\, e^{\mu_{y,i}(t) + \sigma_{y,i}(t) z}} \frac{1}{\sqrt{2\pi}} e^{-\frac{1}{2}z^2}\, dz,$$

yielding:

$$\mathbf{YYIIS}(t, T_{i-1}, T_i, \psi_i, N)$$
$$= N\psi_i P_n(t, T_i) \int_{-\infty}^{+\infty} \frac{1 + \tau_i F_n(t; T_{i-1}, T_i)\, e^{\rho_i \sigma_{x,i}(t) z - \frac{1}{2}\sigma_{x,i}^2(t)\rho_i^2}}{1 + \tau_i F_r(t; T_{i-1}, T_i)\, e^{\mu_{y,i}(t) + \sigma_{y,i}(t) z}} \frac{1}{\sqrt{2\pi}} e^{-\frac{1}{2}z^2}\, dz$$
$$- N\psi_i P_n(t, T_i). \tag{16.24}$$

[4] See also Appendix E for a similar calculation.

To value the whole inflation-indexed leg of the swap some care is needed, since we cannot simply sum up the values (16.24) of the single floating payments. In fact, as noted by Schlögl (2002) in a multi-currency version of the LFM,[5] we cannot assume that the volatilities $\sigma_{I,i}$, $\sigma_{n,i}$ and $\sigma_{r,i}$ are positive constants for all i, because there exists a precise relation between two consecutive forward CPIs and the corresponding nominal and real forward rates, namely:

$$\frac{\mathcal{I}_i(t)}{\mathcal{I}_{i-1}(t)} = \frac{1 + \tau_i F_n(t; T_{i-1}, T_i)}{1 + \tau_i F_r(t; T_{i-1}, T_i)}. \tag{16.25}$$

Clearly, if we assume that $\sigma_{I,i}$, $\sigma_{n,i}$ and $\sigma_{r,i}$ are positive constants, $\sigma_{I,i-1}$ cannot be constant as well, and its admissible values are obtained by equating the (instantaneous) quadratic variations on both sides of (16.25).

However, by freezing the forward rates at their time 0 value in the diffusion coefficients of the right-hand-side of (16.25), we can still get forward CPI volatilities that are approximately constant. For instance, in the one-factor model case,

$$\begin{aligned}\sigma_{I,i-1} &= \sigma_{I,i} + \sigma_{r,i}\frac{\tau_i F_r(t; T_{i-1}, T_i)}{1 + \tau_i F_r(t; T_{i-1}, T_i)} - \sigma_{n,i}\frac{\tau_i F_n(t; T_{i-1}, T_i)}{1 + \tau_i F_n(t; T_{i-1}, T_i)} \\ &\approx \sigma_{I,i} + \sigma_{r,i}\frac{\tau_i F_r(0; T_{i-1}, T_i)}{1 + \tau_i F_r(0; T_{i-1}, T_i)} - \sigma_{n,i}\frac{\tau_i F_n(0; T_{i-1}, T_i)}{1 + \tau_i F_n(0; T_{i-1}, T_i)}.\end{aligned}$$

Therefore, applying this "freezing" procedure for each $i < M$ starting from $\sigma_{I,M}$, or equivalently for each $i > 2$ starting from $\sigma_{I,1}$, we can still assume that the volatilities $\sigma_{I,i}$ are all constant and set to one of their admissible values. The value at time t of the inflation-indexed leg of the swap is thus given by

$$\begin{aligned}\mathbf{YYIIS}(t, \mathcal{T}, \Psi, N) &= N\psi_{\iota(t)}\left[\frac{I(t)}{I(T_{\iota(t)-1})}P_r(t, T_{\iota(t)}) - P_n(t, T_{\iota(t)})\right] \\ &+ N\sum_{i=\iota(t)+1}^{M} \psi_i P_n(t, T_i) \\ &\cdot \left[\int_{-\infty}^{+\infty} \frac{1 + \tau_i F_n(t; T_{i-1}, T_i)\, e^{\rho_i \sigma_{x,i}(t) z - \frac{1}{2}\sigma_{x,i}^2(t)\rho_i^2}}{1 + \tau_i F_r(t; T_{i-1}, T_i)\, e^{\mu_{y,i}(t) + \sigma_{y,i}(t) z}} \frac{1}{\sqrt{2\pi}} e^{-\frac{1}{2}z^2}\, dz - 1\right].\end{aligned} \tag{16.26}$$

In particular at $t = 0$,

[5] See also Section 14.5.4.

$$\mathbf{YYIIS}(0,\mathcal{T},\Psi,N) = N\psi_1[P_r(0,T_1) - P_n(0,T_1)] + N\sum_{i=2}^{M}\psi_i P_n(0,T_i)$$
$$\cdot\left[\int_{-\infty}^{+\infty}\frac{1+\tau_i F_n(0;T_{i-1},T_i)\,e^{\rho_i\sigma_{x,i}(0)z-\frac{1}{2}\sigma_{x,i}^2(0)\rho_i^2}}{1+\tau_i F_r(0;T_{i-1},T_i)\,e^{\mu_{y,i}(0)+\sigma_{y,i}(0)z}}\frac{1}{\sqrt{2\pi}}e^{-\frac{1}{2}z^2}\,dz - 1\right]$$
$$= N\sum_{i=1}^{M}\psi_i P_n(0,T_i)$$
$$\cdot\left[\int_{-\infty}^{+\infty}\frac{1+\tau_i F_n(0;T_{i-1},T_i)\,e^{\rho_i\sigma_{x,i}(0)z-\frac{1}{2}\sigma_{x,i}^2(0)\rho_i^2}}{1+\tau_i F_r(0;T_{i-1},T_i)\,e^{\mu_{y,i}(0)+\sigma_{y,i}(0)z}}\frac{1}{\sqrt{2\pi}}e^{-\frac{1}{2}z^2}\,dz - 1\right]. \tag{16.27}$$

This YYIIS price depends on the following parameters: the (instantaneous) volatilities of nominal and real forward rates and their correlations, for each payment time T_i, $i = 2,\dots,M$; the (instantaneous) volatilities of forward inflation indices and their correlations with real forward rates, again for each $i = 2,\dots,M$.

Compared with expression (16.17), formula (16.27) looks more complicated both in terms of input parameters and in terms of the calculations involved. However, one-dimensional numerical integrations are not so cumbersome and time consuming. Moreover, as is typical in a market model, the input parameters can be determined more easily than those coming from the previous short-rate approach. In this respect, formula (16.27) is preferable to (16.17).

As in the JY case, valuing a YYIIS with a LFM has the drawback that the volatility of real rates may be hard to estimate, especially when resorting to a historical calibration. This is why, in the literature, a second market model has been proposed, which enables us to overcome this estimation issue. In the following section we will review this approach, which has been independently developed by Kazziha (1999), Belgrade, Benhamou and Koehler (2004) and Mercurio (2005).

16.5 Pricing of a YYIIS with a Second Market Model

Applying the definition of forward CPI and using the fact that $\mathcal{I}_i$ is a martingale under $Q_n^{T_i}$, we can also write, for $t < T_{i-1}$,

$$\begin{aligned}\mathbf{YYIIS}(t,T_{i-1},T_i,\psi_i,N) &= N\psi_i P(t,T_i)E_n^{T_i}\left\{\frac{I(T_i)}{I(T_{i-1})} - 1\middle|\mathcal{F}_t\right\}\\ &= N\psi_i P(t,T_i)E_n^{T_i}\left\{\frac{\mathcal{I}_i(T_i)}{\mathcal{I}_{i-1}(T_{i-1})} - 1\middle|\mathcal{F}_t\right\}\\ &= N\psi_i P(t,T_i)E_n^{T_i}\left\{\frac{\mathcal{I}_i(T_{i-1})}{\mathcal{I}_{i-1}(T_{i-1})} - 1\middle|\mathcal{F}_t\right\}.\end{aligned} \tag{16.28}$$

The dynamics of $\mathcal{I}_i$ under $Q_n^{T_i}$ is given by (16.19) and an analogous evolution holds for $\mathcal{I}_{i-1}$ under $Q_n^{T_{i-1}}$. The dynamics of $\mathcal{I}_{i-1}$ under $Q_n^{T_i}$ can be derived by applying the change-of-numeraire toolkit in Section 2.3. We get:

$$d\mathcal{I}_{i-1}(t) = \mathcal{I}_{i-1}(t)\sigma_{I,i-1}\left[-\frac{\tau_i\sigma_{n,i}F_n(t;T_{i-1},T_i)}{1+\tau_iF_n(t;T_{i-1},T_i)}\rho_{I,n,i}\,dt + dW_{i-1}^I(t)\right], \tag{16.29}$$

where $\sigma_{I,i-1}$ is a positive constant, W_{i-1}^I is a $Q_n^{T_i}$-Brownian motion with $dW_{i-1}^I(t)\,dW_i^I(t) = \rho_{I,i}\,dt$, and $\rho_{I,n,i}$ is the instantaneous correlation between $\mathcal{I}_{i-1}(\cdot)$ and $F_n(\cdot;T_{i-1},T_i)$.

The evolution of $\mathcal{I}_{i-1}$, under $Q_n^{T_i}$, depends on the nominal forward rate $F_n(\cdot;T_{i-1},T_i)$, so that the calculation of (16.28) is rather involved in general. To avoid unpleasant complications, like those induced by higher-dimensional integrations, we freeze the drift in (16.29) at its current time-t value, so that $\mathcal{I}_{i-1}(T_{i-1})$ conditional on $\mathcal{F}_t$ is lognormally distributed also under $Q_n^{T_i}$. This leads to

$$E_n^{T_i}\left\{\frac{\mathcal{I}_i(T_{i-1})}{\mathcal{I}_{i-1}(T_{i-1})}\Big|\mathcal{F}_t\right\} = \frac{\mathcal{I}_i(t)}{\mathcal{I}_{i-1}(t)}e^{D_i(t)},$$

where

$$D_i(t) = \sigma_{I,i-1}\left[\frac{\tau_i\sigma_{n,i}F_n(t;T_{i-1},T_i)}{1+\tau_iF_n(t;T_{i-1},T_i)}\rho_{I,n,i} - \rho_{I,i}\sigma_{I,i} + \sigma_{I,i-1}\right](T_{i-1}-t),$$

so that

$$\begin{aligned}\mathbf{YYIIS}(t,T_{i-1},T_i,\psi_i,N) &= N\psi_iP_n(t,T_i)\left[\frac{\mathcal{I}_i(t)}{\mathcal{I}_{i-1}(t)}e^{D_i(t)} - 1\right]\\ &= N\psi_iP_n(t,T_i)\left[\frac{P_n(t,T_{i-1})P_r(t,T_i)}{P_n(t,T_i)P_r(t,T_{i-1})}e^{D_i(t)} - 1\right].\end{aligned} \tag{16.30}$$

Finally, the value at time t of the inflation-indexed leg of the swap is

$$\begin{aligned}\mathbf{YYIIS}(t,\mathcal{T},\Psi,N) &= N\psi_{\iota(t)}P_n(t,T_{\iota(t)})\left[\frac{\mathcal{I}_{\iota(t)}(t)}{I(T_{\iota(t)-1})} - 1\right]\\ &\quad + N\sum_{i=\iota(t)+1}^{M}\psi_iP_n(t,T_i)\left[\frac{\mathcal{I}_i(t)}{\mathcal{I}_{i-1}(t)}e^{D_i(t)} - 1\right]\\ &= N\psi_{\iota(t)}\left[\frac{I(t)}{I(T_{\iota(t)-1})}P_r(t,T_{\iota(t)}) - P_n(t,T_{\iota(t)})\right]\\ &\quad + N\sum_{i=\iota(t)+1}^{M}\psi_i\left[P_n(t,T_{i-1})\frac{P_r(t,T_i)}{P_r(t,T_{i-1})}e^{D_i(t)} - P_n(t,T_i)\right].\end{aligned} \tag{16.31}$$

In particular at $t=0$,

$$\begin{aligned}
\mathbf{YYIIS}(0,\mathcal{T},\Psi,N) &= N\sum_{i=1}^{M}\psi_i P_n(0,T_i)\left[\frac{\mathcal{I}_i(0)}{\mathcal{I}_{i-1}(0)}\,e^{D_i(0)}-1\right]\\
&= N\psi_1[P_r(0,T_1)-P_n(0,T_1)]\\
&\quad + N\sum_{i=2}^{M}\psi_i\left[P_n(0,T_{i-1})\frac{P_r(0,T_i)}{P_r(0,T_{i-1})}\,e^{D_i(0)}-P_n(0,T_i)\right]\\
&= N\sum_{i=1}^{M}\psi_i P_n(0,T_i)\left[\frac{1+\tau_i F_n(0;T_{i-1},T_i)}{1+\tau_i F_r(0;T_{i-1},T_i)}\,e^{D_i(0)}-1\right].
\end{aligned} \tag{16.32}$$

This YYIIS price depends on the following parameters: the (instantaneous) volatilities of forward inflation indices and their correlations; the (instantaneous) volatilities of nominal forward rates; the instantaneous correlations between forward inflation indices and nominal forward rates.

Expression (16.32) looks pretty similar to (16.17) and may be preferred to (16.27) since it combines the advantage of a fully-analytical formula with that of a market-model approach. Moreover, contrary to (16.27), the correction term D does not depend on the volatility of real rates.

A drawback of formula (16.32) is that the approximation it is based on may be rough for long maturities T_i. In fact, such a formula is exact when the correlations $\rho_{I,n,i}$ are set to zero and the terms D_i are simplified accordingly. In general, however, such correlations can have a non-negligible impact on the D_i, and non-zero values can be found when calibrating the model to YYIIS market data.

To visualize the magnitude of the correction terms D_i in the pricing formula (16.32), we plot in Figure 16.2 the values of $D_i(0)$ corresponding to setting $T_i = i$ years, $i = 2, 3, \ldots, 20$, $\sigma_{I,i} = 0.006$, $\sigma_{n,i} = 0.22$, $\rho_{I,n,i} = 0.2$, $\rho_{I,i} = 0.6$, for each i, and where the forward rates $F_n(0; T_{i-1}, T_i)$ are stripped from the Euro nominal zero-coupon curve as of 7 October 2004.

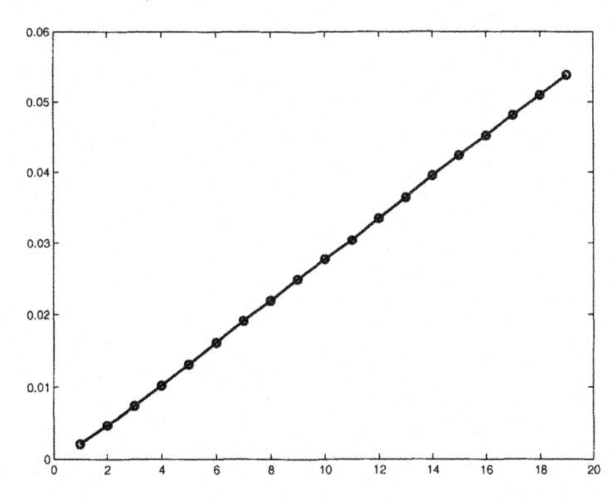

Fig. 16.2. Plot of values $D_i(0)$, in percentage points, for $i = 2, 3, \ldots, 20$.

17. Inflation-Indexed Caplets/Floorlets

An Inflation-Indexed Caplet (IIC) is a call option on the inflation rate implied by the CPI index. Analogously, an Inflation-Indexed Floorlet (IIF) is a put option on the same inflation rate. In formulas, at time T_i, the IICF payoff is

$$N\psi_i \left[\omega\left(\frac{I(T_i)}{I(T_{i-1})} - 1 - \kappa\right)\right]^+, \tag{17.1}$$

where κ is the IICF strike, ψ_i is the contract year fraction for the interval $[T_{i-1}, T_i]$, N is the contract nominal value, and $\omega = 1$ for a caplet and $\omega = -1$ for a floorlet.

Setting $K := 1 + \kappa$, standard no-arbitrage pricing theory implies that the value at time $t \leq T_{i-1}$ of the payoff (17.1) at time T_i is

$$\begin{aligned}
&\mathbf{IICplt}(t, T_{i-1}, T_i, \psi_i, K, N, \omega) \\
&= N\psi_i E_n \left\{ e^{-\int_t^{T_i} n(u)\,du} \left[\omega\left(\frac{I(T_i)}{I(T_{i-1})} - K\right)\right]^+ |\mathcal{F}_t \right\} \\
&= N\psi_i P_n(t, T_i) E_n^{T_i} \left\{ \left[\omega\left(\frac{I(T_i)}{I(T_{i-1})} - K\right)\right]^+ |\mathcal{F}_t \right\}.
\end{aligned} \tag{17.2}$$

The pricing of an IICF is therefore similar to that of a forward-start (cliquet) option. We now derive analytical formulas for (17.2) both under the JY model and under the second market model.

17.1 Pricing with the JY Model

As previously mentioned, assuming Gaussian nominal and real rates leads to a CPI that is lognormally distributed under Q_n. Under this assumption, the type of distribution of the CPI is preserved when we move to a nominal forward measure. Precisely, the ratio $I(T_i)/I(T_{i-1})$ conditional on $\mathcal{F}_t$ is lognormally distributed also under $Q_n^{T_i}$. This implies that (17.2) can be calculated as soon as we know the expectation of this ratio and the variance of its logarithm. Notice, in fact, that if X is a lognormal random variable with $E(X) = m$ and $\mathrm{Std}[\ln(X)] = v$, we then have, from formula (D.2) in Appendix D,

$$E\left\{[\omega(X-K)]^+\right\} = \omega m \Phi\left(\omega \frac{\ln \frac{m}{K} + \frac{1}{2}v^2}{v}\right) - \omega K \Phi\left(\omega \frac{\ln \frac{m}{K} - \frac{1}{2}v^2}{v}\right). \quad (17.3)$$

The conditional expectation of $I(T_i)/I(T_{i-1})$ is immediately obtained through the price of a YYIIS, and formula (16.15) in particular:

$$E_n^{T_i}\left\{\frac{I(T_i)}{I(T_{i-1})}\Big|\mathcal{F}_t\right\} = \frac{P_n(t,T_{i-1})}{P_n(t,T_i)}\frac{P_r(t,T_i)}{P_r(t,T_{i-1})}e^{C(t,T_{i-1},T_i)}.$$

The variance of the logarithm of the ratio can be equivalently calculated under the (nominal) risk-neutral measure.[1] After tedious, but straightforward, calculations we get

$$\mathrm{Var}_n^{T_i}\left\{\ln\frac{I(T_i)}{I(T_{i-1})}\Big|\mathcal{F}_t\right\} = V^2(t,T_{i-1},T_i),$$

where

$$\begin{aligned}
V^2(t,T_{i-1},T_i) &= \frac{\sigma_n^2}{2a_n^3}(1-e^{-a_n(T_i-T_{i-1})})^2[1-e^{-2a_n(T_{i-1}-t)}] + \sigma_I^2(T_i-T_{i-1}) \\
&+ \frac{\sigma_r^2}{2a_r^3}(1-e^{-a_r(T_i-T_{i-1})})^2[1-e^{-2a_r(T_{i-1}-t)}] - 2\rho_{n,r}\frac{\sigma_n\sigma_r}{a_n a_r(a_n+a_r)} \\
&\cdot (1-e^{-a_n(T_i-T_{i-1})})(1-e^{-a_r(T_i-T_{i-1})})[1-e^{-(a_n+a_r)(T_{i-1}-t)}] \\
&+ \frac{\sigma_n^2}{a_n^2}\left[T_i - T_{i-1} + \frac{2}{a_n}e^{-a_n(T_i-T_{i-1})} - \frac{1}{2a_n}e^{-2a_n(T_i-T_{i-1})} - \frac{3}{2a_n}\right] \\
&+ \frac{\sigma_r^2}{a_r^2}\left[T_i - T_{i-1} + \frac{2}{a_r}e^{-a_r(T_i-T_{i-1})} - \frac{1}{2a_r}e^{-2a_r(T_i-T_{i-1})} - \frac{3}{2a_r}\right] \\
&- 2\rho_{n,r}\frac{\sigma_n\sigma_r}{a_n a_r}\left[T_i - T_{i-1} - \frac{1-e^{-a_n(T_i-T_{i-1})}}{a_n} - \frac{1-e^{-a_r(T_i-T_{i-1})}}{a_r}\right. \\
&\left. + \frac{1-e^{-(a_n+a_r)(T_i-T_{i-1})}}{a_n+a_r}\right] + 2\rho_{n,I}\frac{\sigma_n\sigma_I}{a_n}\left[T_i - T_{i-1} - \frac{1-e^{-a_n(T_i-T_{i-1})}}{a_n}\right] \\
&-2\rho_{r,I}\frac{\sigma_r\sigma_I}{a_r}\left[T_i - T_{i-1} - \frac{1-e^{-a_r(T_i-T_{i-1})}}{a_r}\right].
\end{aligned}$$

By (17.3), we thus have[2]

[1] This is because the change of measure produces only a deterministic additive term, which has no impact in the variance calculation.

[2] A similar formula is in the guide of The Royal Bank of Scotland (2003).

$$\mathbf{IICplt}(t, T_{i-1}, T_i, \psi_i, K, N, \omega) = \omega N \psi_i P_n(t, T_i) \left[\frac{P_n(t, T_{i-1})}{P_n(t, T_i)} \frac{P_r(t, T_i)}{P_r(t, T_{i-1})} \right.$$

$$\cdot e^{C(t,T_{i-1},T_i)} \Phi \left(\omega \frac{\ln \frac{P_n(t,T_{i-1})P_r(t,T_i)}{K P_n(t,T_i) P_r(t,T_{i-1})} + C(t, T_{i-1}, T_i) + \frac{1}{2} V^2(t, T_{i-1}, T_i)}{V(t, T_{i-1}, T_i)} \right)$$

$$\left. - K \Phi \left(\omega \frac{\ln \frac{P_n(t,T_{i-1})P_r(t,T_i)}{K P_n(t,T_i) P_r(t,T_{i-1})} + C(t, T_{i-1}, T_i) - \frac{1}{2} V^2(t, T_{i-1}, T_i)}{V(t, T_{i-1}, T_i)} \right) \right]. \tag{17.4}$$

17.2 Pricing with the Second Market Model

To calculate (17.2) under a market model, one first applies the tower property of conditional expectations to get

$$\mathbf{IICplt}(t, T_{i-1}, T_i, \psi_i, K, N, \omega)$$
$$= N \psi_i P_n(t, T_i) E_n^{T_i} \left\{ \frac{E_n^{T_i} \left\{ [\omega(I(T_i) - K I(T_{i-1}))]^+ | \mathcal{F}_{T_{i-1}} \right\}}{I(T_{i-1})} \Big| \mathcal{F}_t \right\}, \tag{17.5}$$

where it is assumed that $I(T_{i-1}) > 0$.

Sticking to a market-model approach, the calculation of the outer expectation in (17.5) depends on whether one models forward rates or directly the forward CPI. We here follow the approach of Section 16.5 since it leads to a simpler formula with less input parameters. For completeness, the former case is dealt with in the appendix of this chapter.

Assuming that (16.19) holds and remembering that $I(T_i) = \mathcal{I}_i(T_i)$, we have

$$E_n^{T_i} \{ [\omega(I(T_i) - K I(T_{i-1}))]^+ | \mathcal{F}_{T_{i-1}} \}$$
$$= E_n^{T_i} \{ [\omega(\mathcal{I}_i(T_i) - K I(T_{i-1}))]^+ | \mathcal{F}_{T_{i-1}} \}$$
$$= \omega \mathcal{I}_i(T_{i-1}) \Phi \left(\omega \frac{\ln \frac{\mathcal{I}_i(T_{i-1})}{K I(T_{i-1})} + \frac{1}{2} \sigma_{I,i}^2 (T_i - T_{i-1})}{\sigma_{I,i} \sqrt{T_i - T_{i-1}}} \right)$$
$$- \omega K I(T_{i-1}) \Phi \left(\omega \frac{\ln \frac{\mathcal{I}_i(T_{i-1})}{K I(T_{i-1})} - \frac{1}{2} \sigma_{I,i}^2 (T_i - T_{i-1})}{\sigma_{I,i} \sqrt{T_i - T_{i-1}}} \right),$$

so that, by (17.5) and the definition of $\mathcal{I}_{i-1}$,

$$\begin{aligned}
&\mathbf{IICplt}(t, T_{i-1}, T_i, \psi_i, K, N, \omega) = \omega N \psi_i P_n(t, T_i) \\
&\cdot E_n^{T_i}\left\{\frac{\mathcal{I}_i(T_{i-1})}{\mathcal{I}_{i-1}(T_{i-1})}\Phi\left(\omega\frac{\ln\frac{\mathcal{I}_i(T_{i-1})}{K\mathcal{I}_{i-1}(T_{i-1})} + \frac{1}{2}\sigma_{I,i}^2(T_i - T_{i-1})}{\sigma_{I,i}\sqrt{T_i - T_{i-1}}}\right)\right. \\
&\left. -K\Phi\left(\omega\frac{\ln\frac{\mathcal{I}_i(T_{i-1})}{K\mathcal{I}_{i-1}(T_{i-1})} - \frac{1}{2}\sigma_{I,i}^2(T_i - T_{i-1})}{\sigma_{I,i}\sqrt{T_i - T_{i-1}}}\right)|\mathcal{F}_t\right\}.
\end{aligned} \tag{17.6}$$

Remembering (16.29) and freezing again the drift at its time-t value, we have that under $Q_n^{T_i}$ the ratio $\mathcal{I}_i(T_{i-1})/\mathcal{I}_{i-1}(T_{i-1})$, conditional on $\mathcal{F}_t$, is lognormally distributed. Precisely, setting

$$V_i(t) := \sqrt{(\sigma_{I,i-1}^2 + \sigma_{I,i}^2 - 2\rho_{I,i}\sigma_{I,i-1}\sigma_{I,i})(T_{i-1} - t)}$$

we have that

$$\ln\frac{\mathcal{I}_i(T_{i-1})}{\mathcal{I}_{i-1}(T_{i-1})}|\mathcal{F}_t \sim \mathcal{N}\left(\ln\frac{\mathcal{I}_i(t)}{\mathcal{I}_{i-1}(t)} + D_i(t) - \frac{1}{2}V_i^2(t), V_i^2(t)\right).$$

Straightforward algebra then leads to

$$\begin{aligned}
&\mathbf{IICplt}(t, T_{i-1}, T_i, \psi_i, K, N, \omega) \\
&= \omega N \psi_i P_n(t, T_i)\left[\frac{\mathcal{I}_i(t)}{\mathcal{I}_{i-1}(t)} e^{D_i(t)}\Phi\left(\omega\frac{\ln\frac{\mathcal{I}_i(t)}{K\mathcal{I}_{i-1}(t)} + D_i(t) + \frac{1}{2}\mathcal{V}_i^2(t)}{\mathcal{V}_i(t)}\right)\right. \\
&\left. - K\Phi\left(\omega\frac{\ln\frac{\mathcal{I}_i(t)}{K\mathcal{I}_{i-1}(t)} + D_i(t) - \frac{1}{2}\mathcal{V}_i^2(t)}{\mathcal{V}_i(t)}\right)\right] \\
&= \omega N \psi_i P_n(t, T_i)\left[\frac{1 + \tau_i F_n(t; T_{i-1}, T_i)}{1 + \tau_i F_r(t; T_{i-1}, T_i)} e^{D_i(t)}\right. \\
&\cdot \Phi\left(\omega\frac{\ln\frac{1+\tau_i F_n(t;T_{i-1},T_i)}{K[1+\tau_i F_r(t;T_{i-1},T_i)]} + D_i(t) + \frac{1}{2}\mathcal{V}_i^2(t)}{\mathcal{V}_i(t)}\right) \\
&\left. - K\Phi\left(\omega\frac{\ln\frac{1+\tau_i F_n(t;T_{i-1},T_i)}{K[1+\tau_i F_r(t;T_{i-1},T_i)]} + D_i(t) - \frac{1}{2}\mathcal{V}_i^2(t)}{\mathcal{V}_i(t)}\right)\right],
\end{aligned} \tag{17.7}$$

where we set

$$\mathcal{V}_i(t) = \sqrt{V_i^2(t) + \sigma_{I,i}^2(T_i - T_{i-1})}.$$

Analogously to the YYIIS prices (16.30) and (16.31), this caplet price depends on the (instantaneous) volatilities of forward inflation indices and their correlations, the (instantaneous) volatilities of nominal forward rates, and the instantaneous correlations between forward inflation indices and nominal forward rates. Therefore, formula (17.7) has, in terms of input parameters, all the advantages and drawbacks of the swap price (16.31).

The analogy with the Black and Scholes (1973) formula renders (17.7) quite appealing from a practical point of view, and provides a further support for the modeling of forward CPIs as geometric Brownian motions under their associated measures.

17.3 Inflation-Indexed Caps

An Inflation-Indexed Cap (IICap) is a stream of inflation-indexed caplets. An analogous definition holds for an Inflation-Indexed Floor (IIFloor). Given the set of dates $T_0, T_1, \dots, T_M$, with $T_0 = 0$, an IICapFloor pays off, at each time T_i, $1, \dots, M$,

$$N\psi_i \left[\omega\left(\frac{I(T_i)}{I(T_{i-1})} - 1 - \kappa\right)\right]^+, \tag{17.8}$$

where κ is the IICapFloor strike.

Again setting $K := 1 + \kappa$, standard no-arbitrage pricing theory implies that the value at time 0 of the above IICapFloor is

$$\begin{aligned}
&\mathbf{IICapFloor}(0, \mathcal{T}, \Psi, K, N, \omega)\\
&= N\sum_{i=1}^{M} \psi_i E_n\left\{e^{-\int_0^{T_i} n(u)\,du}\left[\omega\left(\frac{I(T_i)}{I(T_{i-1})} - K\right)\right]^+\right\}\\
&= N\sum_{i=1}^{M} P_n(0, T_i)\psi_i E_n^{T_i}\left\{\left[\omega\left(\frac{I(T_i)}{I(T_{i-1})} - K\right)\right]^+\right\},
\end{aligned}$$

where we again set $\mathcal{T} := \{T_1, \dots, T_M\}$ and $\Psi := \{\psi_1, \dots, \psi_M\}$.

Sticking to the second market model, from (17.7), we immediately get

$$\begin{aligned}
\mathbf{IICapFloor}(0, \mathcal{T}, \Psi, K, N, \omega) &= \omega N \sum_{i=1}^{M} \psi_i P_n(0, T_i)\left[\frac{1 + \tau_i F_n(0; T_{i-1}, T_i)}{1 + \tau_i F_r(0; T_{i-1}, T_i)}\right.\\
&\quad \cdot e^{D_i(0)}\Phi\left(\omega\frac{\ln\frac{1+\tau_i F_n(0;T_{i-1},T_i)}{K[1+\tau_i F_r(0;T_{i-1},T_i)]} + D_i(0) + \frac{1}{2}\mathcal{V}_i^2(0)}{\mathcal{V}_i(0)}\right)\\
&\quad \left. - K\Phi\left(\omega\frac{\ln\frac{1+\tau_i F_n(0;T_{i-1},T_i)}{K[1+\tau_i F_r(0;T_{i-1},T_i)]} + D_i(0) - \frac{1}{2}\mathcal{V}_i^2(0)}{\mathcal{V}_i(0)}\right)\right].
\end{aligned} \tag{17.9}$$

Appendix: IICapFloor Pricing with the LFM

We here show how to price a IICapFloor with the LFM of Section 16.4. To this end, we notice that, by (17.6) and the definition of $\mathcal{I}_i$

$$\begin{aligned}&\mathbf{IICplt}(t)\\&= \omega N \psi_i P_n(t, T_i) E_n^{T_i}\left\{ \frac{P_r(T_{i-1}, T_i)}{P_n(T_{i-1}, T_i)} \Phi\left(\omega \frac{\ln \frac{P_r(T_{i-1},T_i)}{K P_n(T_{i-1},T_i)} + \frac{1}{2}\sigma_{I,i}^2 (T_i - T_{i-1})}{\sigma_{I,i}\sqrt{T_i - T_{i-1}}}\right)\right.\\&\left. -K\Phi\left(\omega \frac{\ln \frac{P_r(T_{i-1},T_i)}{K P_n(T_{i-1},T_i)} - \frac{1}{2}\sigma_{I,i}^2 (T_i - T_{i-1})}{\sigma_{I,i}\sqrt{T_i - T_{i-1}}}\right) \Big| \mathcal{F}_t \right\},\end{aligned} \tag{17.10}$$

and that the expectation in (17.10) can be rewritten as

$$\begin{aligned}&E_n^{T_i}\left\{ \frac{1 + \tau_i F_n(T_{i-1}; T_{i-1}, T_i)}{1 + \tau_i F_r(T_{i-1}; T_{i-1}, T_i)} \right.\\&\cdot \Phi\left(\omega \frac{\ln \frac{1+\tau_i F_n(T_{i-1};T_{i-1},T_i)}{K[1+\tau_i F_r(T_{i-1};T_{i-1},T_i)]} + \frac{1}{2}\sigma_{I,i}^2 (T_i - T_{i-1})}{\sigma_{I,i}\sqrt{T_i - T_{i-1}}}\right)\\&\left. - K\Phi\left(\omega \frac{\ln \frac{1+\tau_i F_n(T_{i-1};T_{i-1},T_i)}{K[1+\tau_i F_r(T_{i-1};T_{i-1},T_i)]} - \frac{1}{2}\sigma_{I,i}^2 (T_i - T_{i-1})}{\sigma_{I,i}\sqrt{T_i - T_{i-1}}}\right) \Big| \mathcal{F}_t \right\}.\end{aligned} \tag{17.11}$$

Following the approach of Section 16.4, we assume that nominal and real forward rates evolve according to (16.20) and impose that the forward CPI volatilities are constant by the freezing procedure at the end of the section. The expectation in (17.11) can then be easily calculated with a numerical integration by noting again that the pair (16.21) is distributed as a bivariate normal random variable with mean vector and variance-covariance matrix given by (16.22).

The dimensionality of the problem to solve can, however, be reduced by assuming deterministic real rates, as we do in the following.[3]

Under deterministic real rates, the future LIBOR value $F_r(T_{i-1}; T_{i-1}, T_i)$ is simply equal to the current forward rate $F_r(0; T_{i-1}, T_i)$, so that we can write

$$\begin{aligned}&\mathbf{IICap}(0, \mathcal{T}, \Psi, K, N, \omega) = \omega N \sum_{i=1}^{M} \psi_i P_n(0, T_i) E_n^{T_i} \left\{ \frac{1 + \tau_i F_n(T_{i-1}; T_{i-1}, T_i)}{1 + \tau_i F_r(0; T_{i-1}, T_i)} \right.\\&\cdot \Phi\left(\omega \frac{\ln \frac{1+\tau_i F_n(T_{i-1};T_{i-1},T_i)}{K[1+\tau_i F_r(0;T_{i-1},T_i)]} + \frac{1}{2}\sigma_{I,i}^2 (T_i - T_{i-1})}{\sigma_{I,i}\sqrt{T_i - T_{i-1}}}\right)\end{aligned}$$

[3] We have previously noticed that both the volatility of real rates and their correlation with nominal rates can play an important role in the pricing of inflation-linked derivatives. We here assume deterministic real rates, for simplicity, even though these parameters should be explicitly taken into account, in general.

$$-K\Phi\left(\omega\frac{\ln\frac{1+\tau_iF_n(T_{i-1};T_{i-1},T_i)}{K[1+\tau_iF_r(0;T_{i-1},T_i)]}-\frac{1}{2}\sigma_{I,i}^2(T_i-T_{i-1})}{\sigma_{I,i}\sqrt{T_i-T_{i-1}}}\right)\Bigg\}. \tag{17.12}$$

Since the nominal forward rate $F_n(\cdot;T_{i-1},T_i)$ follows the LFM (16.20), we finally obtain:

$$\begin{aligned}
\mathbf{IICap}(0,\mathcal{T},\Psi,K,N,\omega) &= \omega N\psi_1\left[P_r(0,T_1)\Phi\left(\omega\frac{\ln\frac{P_r(0,T_1)}{KP_n(0,T_1)}+\frac{1}{2}\sigma_{I,1}^2T_1}{\sigma_{I,1}\sqrt{T_1}}\right)\right.\\
&\left.-KP_n(0,T_1)\Phi\left(\omega\frac{\ln\frac{P_r(0,T_1)}{KP_n(0,T_1)}-\frac{1}{2}\sigma_{I,1}^2T_1}{\sigma_{I,1}\sqrt{T_1}}\right)\right]+\omega N\sum_{i=2}^{M}\psi_iP_n(0,T_i)\\
&\cdot\int_{-\infty}^{+\infty}J(x)\frac{1}{\sigma_{n,i}\sqrt{2\pi T_{i-1}}}e^{-\frac{1}{2}\left(\frac{x+\frac{1}{2}\sigma_{n,i}^2T_{i-1}}{\sigma_{n,i}\sqrt{T_{i-1}}}\right)^2}dx,
\end{aligned} \tag{17.13}$$

where

$$\begin{aligned}
J(x) :=& \frac{1+\tau_iF_n(0;T_{i-1},T_i)e^x}{1+\tau_iF_r(0;T_{i-1},T_i)}\Phi\left(\omega\frac{\ln\frac{1+\tau_iF_n(0;T_{i-1},T_i)e^x}{K[1+\tau_iF_r(0;T_{i-1},T_i)]}+\frac{1}{2}\sigma_{I,i}^2(T_i-T_{i-1})}{\sigma_{I,i}\sqrt{T_i-T_{i-1}}}\right)\\
&-K\Phi\left(\omega\frac{\ln\frac{1+\tau_iF_n(0;T_{i-1},T_i)e^x}{K[1+\tau_iF_r(0;T_{i-1},T_i)]}-\frac{1}{2}\sigma_{I,i}^2(T_i-T_{i-1})}{\sigma_{I,i}\sqrt{T_i-T_{i-1}}}\right)
\end{aligned}$$

18. Calibration to market data

In this section, we consider an example of calibration to Euro market data as of October 7, 2004. Precisely, we test the performance of the JY model and the two market models as far as the calibration to inflation-indexed swaps is concerned, with some model parameters being previously fitted to at-the-money (nominal) cap volatilities. The zero-coupon and year-on-year swap rates we consider are plotted in Figure 16.1.

As explained in Section 16, we use the zero-coupon rates to strip the current real discount factors for the relevant maturities. From these discount factors, we then derive the real forward rates that enter the pricing functions (16.17), (16.27) and (16.32) for the JY model, the first and the second market models, respectively.

The model parameters that best fit the given set of market data are found by minimizing the square absolute difference between model and market YYIIS fixed rates, under some constraints we introduce to avoid over-parametrization. For a given vector of model parameters $\underline{p}$, the model YYIIS fixed rate is defined as the rate $K = K(\underline{p})$ that renders the corresponding YYIIS a zero-value contract at the current time.

The JY formula (16.17) involves seven parameters: a_n, σ_n, a_r, σ_r, σ_I, $\rho_{n,r}$ and $\rho_{r,I}$. We reduce, however, this number to five, by finding a_n and σ_n through a previous calibration to the at-the-money (nominal) caps market.

The first market model formula (16.27) involves five parameters for each payment time from the second year onwards: $\sigma_{n,i}$, $\sigma_{r,i}$, $\sigma_{I,i}$, ρ_i, and $\rho_{I,r,i}$, for $i = 2, \dots, M = 20$. In this case, we reduce the optimization parameters to five, $c_1, c_2, \dots, c_5$, since we automatically calibrate each $\sigma_{n,i}$ to the at-the-money (nominal) cap volatilities, and set $\sigma_{r,i} = c_1/[1+c_2(T_i-T_2)]$, $\sigma_{I,1} = c_3$, with the subsequent $\sigma_{I,i}$'s that are computed remembering (16.25), $\rho_i = c_4$ and $\rho_{I,r,i} = c_5$, for each $i > 1$.

Also the second market model formula (16.32) involves five parameters for each payment time from the second year onwards: $\sigma_{I,i-1}$, $\sigma_{n,i}$, $\rho_{I,i}$, $\rho_{I,n,i}$, and $\sigma_{I,i}$, for $i = 2, \dots, M = 20$. In this last case, we reduce the optimization parameters to four: $c_1, \dots, c_4$. In fact, each $\sigma_{n,i}$ is again calibrated automatically to the at-the-money (nominal) cap volatilities. We then set $\rho_{I,i} = 1-(1-c_1)\exp(-c_2 T_{i-1})$, $\rho_{I,n,i} = c_3$ and $\sigma_{I,i} = c_4$, for each $i > 1$.

Our calibration results are shown in Table 18.1, where the three model swap rates (in percentage points) are compared with the market ones. The

Maturity	Market	JY	MM1	MM2
1	2.120	2.120	2.120	2.120
2	2.170	2.169	2.168	2.168
3	2.185	2.186	2.186	2.184
4	2.213	2.217	2.218	2.215
5	2.246	2.250	2.250	2.247
6	2.271	2.276	2.275	2.272
7	2.292	2.296	2.295	2.293
8	2.309	2.314	2.312	2.310
9	2.324	2.324	2.322	2.320
10	2.339	2.345	2.343	2.341
11	2.353	2.358	2.356	2.355
12	2.367	2.371	2.369	2.369
13	2.383	2.385	2.383	2.383
14	2.390	2.397	2.396	2.396
15	2.408	2.410	2.410	2.410
16	2.418	2.420	2.421	2.421
17	2.429	2.430	2.431	2.432
18	2.439	2.439	2.442	2.443
19	2.450	2.448	2.453	2.454
20	2.461	2.457	2.463	2.465

Table 18.1. Comparison between market YYIIS fixed rates (in percentage points) and those implied by the JY model, the first market model (MM1) and the second one (MM2).

performance of these models is quite satisfactory. In fact, the largest difference between a model swap rate and the corresponding market value is 0.7 basis points, which is negligible also because typical bid-ask spreads in the market are between five and ten basis points.

Even though the three models are equivalent in terms of calibration to market YYIIS rates, they can however imply quite different prices when away-from-the money derivatives are considered. As an example, we price zero-strike floors, for maturities from four to twenty years, with the JY model and the second market model, using the parameters coming from the previous calibration. The result of this test is shown in Figure 18.1.

Looking at Figure 18.1, one may suspect the existence of a systematic bias between the JY and second market models. However, this is not true in general, and the two models are closer than one may think at a first sight. In fact, both the JY model and the second market model are based on forward CPIs that are lognormally distributed or, equivalently, on inflation rates that have a shifted lognormal density. Therefore, from a distributional point of view, the two models are rather similar.

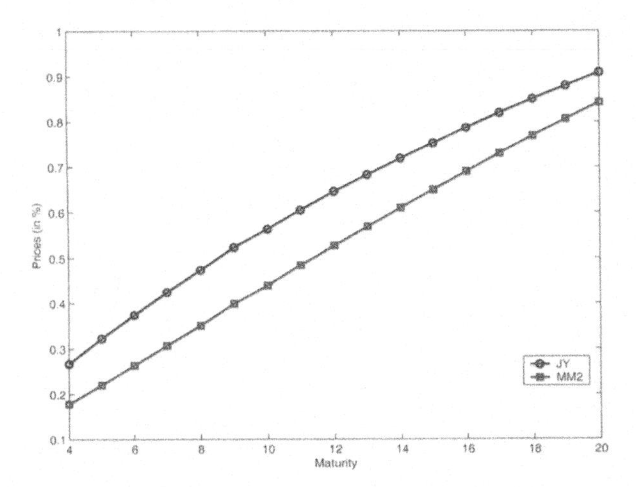

Fig. 18.1. Comparison of zero-strike floors prices implied by the JY and second market models, for different maturities.

The price differences in the figure can then be explained by the fact that even large changes in the indices volatilities can have a negligible effect on the YYIIS price (16.32), whereas the same changes can impact sensibly on the corresponding floor price (17.9). Some care in the calibration to the market of inflation derivatives is therefore required. In this respect, our experience is that one should avoid considering swap prices alone and at least include some option prices in the fitting procedure so as to perform a more sensible calibration. By doing so, one can still achieve a very good fit to the swap and zero-strike floor markets simultaneously, as we show in Figure 18.2, where we compare market floor prices with the calibrated ones (the calibration to YYIIS rates is as good as that of Table 18.1).

Remark 18.0.1. **(Correlation between forward inflation rates)**. Using constant CPI volatilities has the drawback that the implied terminal correlation between forward inflation rates may be too low, whatever the choice of instantaneous correlations $\rho_{I,i}$. This typically leads to calibrated values of $\sigma_{I,i}$, $i = 2, \dots, M$, that are too low compared with those implied by the prices of ZC caps (*i.e.* options on the CPI), when traded in the market. Higher values for the terminal correlation between inflation rates and for the CPI integrated volatilities can be obtained by introducing a suitable time-dependent parametrization for the CPI instantaneous volatility $\sigma_{I,i}$.

Remark 18.0.2. **(Smile Effect)**. One may argue that a smile effect should be taken into account as soon as we calibrate to out-of-the-money options like zero-strike floors. To this end, one may extend either market models in this section by allowing for an uncertain volatility in the spirit of Gatarek (2003) and Brigo, Mercurio and Rapisarda (2004), see also Section 12, or may

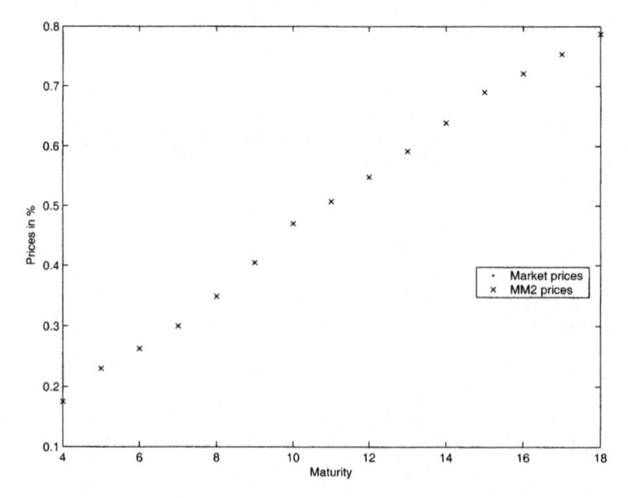

Fig. 18.2. Zero-strike floor prices implied by the second market model, after calibration to market quotes (both swaps and floors) as of October 7, 2004.

directly resort to the tailor-made approach of Mercurio and Moreni (2006), which is reviewed in the following chapter. However, the "flat-smile" market models of this section are good enough for many purposes, also considering that reliable prices for a wide range of options may be hard to get in the market.

19. Introducing Stochastic Volatility

We have seen in the previous chapters that the pricing of II derivatives is typically addressed by resorting to a foreign-currency analogy. In fact, one can convert nominal values into real ones simply by dividing by the current value of the reference CPI. The most relevant application of the foreign-currency analogy is due to Jarrow and Yildrim (2003), who modelled the dynamics of the CPI along with nominal and real rates, both assumed to follow a one-factor Gaussian process in the HJM framework.

We have also seen that an alternative approach, independently proposed by Kazziha (1999), Belgrade et al. (2004) and Mercurio (2005), is based on a market model whose underlying variables are forward CPI's evolving as driftless geometric Brownian motions under associated forward measures. As a consequence, each II caplet can be priced with a Black-like formula, which can be safely used to price caps with different maturities but same strike. When simultaneously pricing caps with different strikes, however, the natural question arises whether one should include some kind of smile effect, since this is a consolidated practice in all developed options markets.

To address such an issue, let us focus on the first maturity T_1. As at time T_0, $\mathcal{I}_0$ is known and equal to $I(0)$, the first caplet is in fact a plain-vanilla call on $\mathcal{I}_1$ which, by definition, is a martingale under the T_1-forward measure. Assuming the following lognormal-type dynamics for $\mathcal{I}_1$

$$d\mathcal{I}_1(t) = \sigma_1 \mathcal{I}_1(t)\, dZ_1^I(t),$$

where σ_1 is a positive constant and Z_1^I is a standard Brownian motion, the price at time 0 of the first caplet is thus given by, see (17.7),

$$\begin{aligned} P(0,T_1)&\left[\frac{\mathcal{I}_1(0)}{I(0)}\Phi\left(\frac{\ln\frac{\mathcal{I}_1(0)}{(1+K)I(0)} + \frac{1}{2}\sigma_1^2 T_1}{\sigma_1\sqrt{T_1}}\right)\right.\\ &\left. - (1+K)\Phi\left(\frac{\ln\frac{\mathcal{I}_1(0)}{(1+K)I(0)} - \frac{1}{2}\sigma_1^2 T_1}{\sigma_1\sqrt{T_1}}\right)\right]. \end{aligned} \tag{19.1}$$

With caplet quotes available for a number of strikes, it is straightforward to test whether the II caplet market is flat-smiled or not. To this end, we proceed as follows. We first take the ATM strike,[1] and calculate the ATM implied volatility by inversion of formula (19.1). We then use this same volatility to price a set of caplets with different strikes. The results we find are reported in Figure 19.1, where we consider USD market data as of November 3th, 2004. It is clear that the ATM implied volatility underestimates OTM prices and overestimate ITM ones, revealing an implicit skew in the market prices. This simple test, therefore, suggests the need for a model that goes beyond

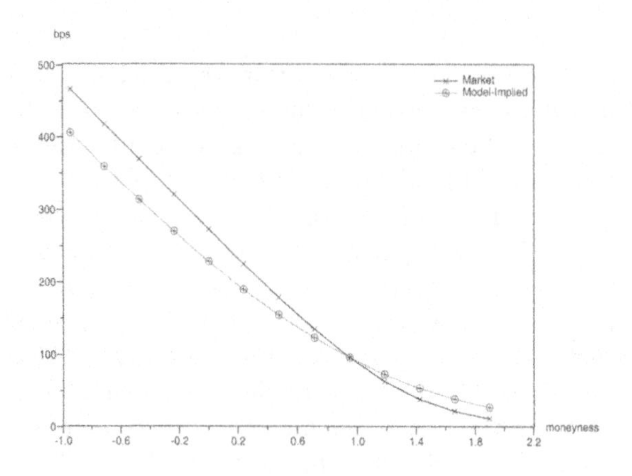

Fig. 19.1. Market prices versus model prices with ATM implied volatility for one-year caplets. Moneyness is defined as: $K/(\mathcal{I}_1(0)/I(0) - 1)$.

the geometric Brownian motion. This is also motivated by the dependence of forward CPI's on forward rates and the empirical evidence of a stochastic volatility for interest rates.

An extension of the above lognormal market model has been proposed by Mercurio and Moreni (2006), who introduced stochastic volatility as in Heston (1993). In this chapter, we will review their main results.

Hereafter, we will drop the subscript n when denoting nominal quantities, for ease of notation.

19.1 Modeling Forward CPI's with Stochastic Volatility

Mercurio and Moreni (2006) assume that nominal forward rates

$$F_i(t) = [P(t, T_{i-1})/P(t, T_i) - 1]/\tau_i,$$

are lognormally distributed according to an LFM with constant volatilities, and that forward CPI's, $\mathcal{I}_i$, have a Heston-like evolution with a common

[1] Or the closest possible one in case this is not immediately available.

volatility $V(t)$ that follows a mean-reverting square-root process, see Heston (1993), under a given *reference measure* $\mathbb{Q}$:

$$\begin{aligned} dF_i(t)/F_i(t) &= (\dots)\,dt + \sigma_i^F\, dZ_i^{\mathbb{Q},F}(t) \\ d\mathcal{I}_i(t)/\mathcal{I}_i(t) &= (\dots)\,dt + \sigma_i^I \sqrt{V(t)}\, dZ_i^{\mathbb{Q},I}(t) \\ dV(t) &= \alpha(\theta - V(t))\,dt + \epsilon\sqrt{V(t)}\, dW^{\mathbb{Q}}(t), \end{aligned} \tag{19.2}$$

where σ_i^F's, σ_i^I's, α, θ and ϵ are positive constants, $2\alpha\theta > \epsilon$ to ensure positiveness of V, and where we allow for correlations between Brownian motions $Z_i^{\mathbb{Q},F}, Z_i^{\mathbb{Q},I}, W^{\mathbb{Q}}$.

Instead of defining the volatility's dynamics, and parameters α, θ, ϵ, under a risk-neutral measure or under the terminal forward measure Q^M, Mercurio and Moreni choose $\mathbb{Q}$ to be equal to the spot LIBOR measure Q^d, which is, to many extents, payoff independent, allowing the valuation of a single payoff

$$\left(\frac{I(T_i) - I(T_{i-1})}{I(T_{i-1})} - K\right)^+ = \left(\frac{\mathcal{I}_i(T_i)}{\mathcal{I}_{i-1}(T_{i-1})} - K - 1\right)^+$$

irrespectively of the number of caplets that follow. In fact, taking Q^M as reference measure, the volatility's evolution would depend on the choice of the last maturity T_M. Choosing instead a particular risk-free measure, would hide the evaluation issue of the market price of volatility risk.

The classical change of measure technique, see Section 2.3, implies that, under Q^d, the forward rates and forward CPI's dynamics are given by

$$\begin{aligned} dF_i(t)/F_i(t) &= \sigma_i^F \left[\sum_{l=\beta(t)+1}^{i} \sigma_l^F \rho_{i,l}^F \frac{\tau_l F_l(t)}{1 + \tau_l F_l(t)} dt + dZ_i^{0,F}(t) \right] \\ d\mathcal{I}_i(t)/\mathcal{I}_i(t) &= \sqrt{V(t)}\,\sigma_i^I \left[\sum_{l=\beta(t)+1}^{i} \sigma_l^F \rho_{l,i}^{F,I} \frac{\tau_l F_l(t)}{1 + \tau_l F_l(t)} dt + dZ_i^{0,I}(t) \right] \\ dZ_i^{0,F}(t)\, dZ_l^{0,F}(t) &= \rho_{i,l}^F\, dt \\ dZ_i^{0,I}(t)\, dZ_l^{0,F}(t) &= \rho_{l,i}^{F,I}\, dt \end{aligned} \tag{19.3}$$

while $V(t)$ evolves as in (19.2), and where the last two equations define the correlation parameters $\rho_{i,l}^F$ and $\rho_{l,i}^{F,I}$.

II caplets and floorlets paying at time T_j are derivatives that depend on $\mathcal{I}_j$ and $\mathcal{I}_{j-1}$. Therefore, it is convenient to express dynamics under the forward measure Q^j, under which, by definition, both F_j and $\mathcal{I}_j$ are martingales.

When we take $P(t, T_j)$ as numeraire, the relevant quantities $\mathcal{I}_j(\cdot)$, $\mathcal{I}_{j-1}(\cdot)$ and $X_j(\cdot) := \ln(\mathcal{I}_j(\cdot)/\mathcal{I}_{j-1}(\cdot))$ satisfy the following SDE's

$$\begin{aligned}
d\mathcal{I}_j(t)/\mathcal{I}_j(t) &= \sqrt{V(t)}\,\sigma_j^I\,dZ_j^I(t)\\
d\mathcal{I}_{j-1}(t)/\mathcal{I}_{j-1}(t) &= \sqrt{V(t)}\,\sigma_{j-1}^I\left[-\frac{\tau_j F_j(t)}{1+\tau_j F_j(t)}\,\sigma_j^F\rho_{j,j-1}^{F,I}dt + dZ_{j-1}^I(t)\right]\\
dX_j(t) &= \left[\frac{V(t)}{2}\left((\sigma_{j-1}^I)^2-(\sigma_j^I)^2\right)+\sqrt{V(t)}\sigma_{j-1}^I\sigma_j^F\rho_{j,j-1}^{F,I}\frac{\tau_j F_j(t)}{1+\tau_j F_j(t)}\right]dt\\
&\quad+\sqrt{V(t)}[\sigma_j^I\,dZ_j^I(t)-\sigma_{j-1}^I\,dZ_{j-1}^I(t)]
\end{aligned} \tag{19.4}$$

while the volatility evolves according to

$$\begin{aligned}
dV(t) &= \left[\alpha\theta-\epsilon m_j(t)\sqrt{V(t)}-\alpha V(t)\right]dt+\epsilon\sqrt{V(t)}\,dW(t)\\
m_j(t) &= \sum_{l=\beta(t)+1}^{j}\frac{\tau_l F_l(t)}{1+\tau_l F_l(t)}\sigma_l^F\rho_l^{F,V}
\end{aligned} \tag{19.5}$$

where $dZ_l^F(t)\,dW(t)=\rho_l^{F,V}dt$, for each l.

We denote by $\rho_{j,l}^I$ the correlation between forward CPI's $\mathcal{I}_j$ and $\mathcal{I}_l$,

$$\rho_{j,l}^I\,dt = dZ_j^I(t)\,dZ_l^I(t),$$

and by $\rho_i^{I,V}$ the correlation between the forward CPI $\mathcal{I}_i$ and the volatility,

$$\rho_i^{I,V}\,dt := dZ_i^I(t)\,dW(t).$$

19.2 Pricing Formulae

Regardless of the model chosen for the CPI's evolution, the price at time $t\le T_j$ of the j-th caplet, is, under measure Q^j,

$$\begin{aligned}
\mathbf{IICplt}(t,T_{j-1},T_j,K,1) &= P(t,T_j)E_t^j\left(\frac{\mathcal{I}_j(T_j)}{\mathcal{I}_{j-1}(T_{j-1})}-(K+1)\right)^+\\
&= P(t,T_j)\int_{-\infty}^{+\infty}(\mathrm{e}^s-\mathrm{e}^k)^+q_t^j(s)\,ds
\end{aligned} \tag{19.6}$$

where $k=\ln(K+1)$, $q_t^j(s)ds := Q^j\left\{\ln\left(\mathcal{I}_j(T_j)/\mathcal{I}_{j-1}(T_{j-1})\right)\in[s,s+ds)|\mathcal{F}_t\right\}$, and E_t^j denotes expectation under Q^j conditional on the σ-algebra $\mathcal{F}_t$.

The difficulty in the calculation of (19.6) is that, instead of having a payoff depending on a single asset as in case of standard or forward-start (cliquet) options, here the payoff depends on the ratio of *two different assets, $\mathcal{I}_j$ and $\mathcal{I}_{j-1}$ at two different times*, T_j and T_{j-1}.

Following Carr and Madan (1998), the caplet price (19.6) can be rewritten in terms of its (renormalized) Fourier Transform (FT), obtaining

$$\mathbf{IICplt}(t,T_{j-1},T_j,K,1) = P(t,T_j)\frac{\mathrm{e}^{-\eta k}}{2\pi}\int_{-\infty}^{+\infty}\mathrm{e}^{-iuk}\frac{\phi_t^j(u-(\eta+1)i)}{(\eta+iu)(\eta+1+iu)}du$$
$$= P(t,T_j)\frac{\mathrm{e}^{-\eta k}}{\pi}\mathrm{Re}\int_0^{+\infty}\mathrm{e}^{-iuk}\frac{\phi_t^j(u-(\eta+1)i)}{(\eta+iu)(\eta+1+iu)}du \tag{19.7}$$

where the only unknown function is the conditional characteristic function $\phi_t^j(\cdot)$ of $\ln(\mathcal{I}_j(T_j)/\mathcal{I}_{j-1}(T_{j-1}))$,

$$\phi_t^j(u) = E_t^j\left[\mathrm{e}^{iu\ln(\mathcal{I}_j(T_j)/\mathcal{I}_{j-1}(T_{j-1}))}\right],$$

and where $\eta \in \mathbb{R}^+$ is used to ensure L^2-integrability when $k \to -\infty$.

Analogously, the price of a II floorlet is given by, see Appendix B of this chapter,

$$\mathbf{IICplt}(t,T_{j-1},T_j,K,-1) = P(t,T_j)E_t^j\left((K+1)-\frac{\mathcal{I}_j(T_j)}{\mathcal{I}_{j-1}(T_{j-1})}\right)^+$$
$$= P(t,T_j)\frac{\mathrm{e}^{\tilde{\eta}k}}{\pi}\mathrm{Re}\int_0^{+\infty}\mathrm{e}^{-iuk}\frac{\phi_t^j(u+i(\tilde{\eta}-1))}{(iu-\tilde{\eta}+1)(iu-\tilde{\eta})}du$$

where $\tilde{\eta} \in (1,+\infty)$ has been introduced, like η in (19.7), for FT regularization. As explained in Appendix B, it is also possible to compute floorlet prices by a put-call parity that holds between II caplets and floorlets.

The pricing of II caplets and floorlets reduces, therefore, to the calculation of a conditional characteristic function. With an explicit formula available for ϕ_t^j, pricing can be quite fast thanks to well established FT techniques, thus allowing a not too expensive calibration procedure.

By definition of characteristic function and the Markov property, we can write

$$\phi_t^j(u) = H\left(V(t), \ln(\mathcal{I}_j(t)), \ln(\mathcal{I}_{j-1}(t)), F_1(t), \ldots, F_j(t); u\right), \tag{19.8}$$

where H is the solution of a PDE that can be found through Feynman-Kac's theorem, see Appendix C of the book.

It is clear that in the general framework of (19.4), due to the unpleasant presence of drift terms like $\sqrt{V(t)}F_l(t)/(1+\tau_l F_l(t))$, the PDE satisfied by H is rather involved and does not seem to be explicitly solvable but under particular conditions or approximations. As a consequence, Mercurio and Moreni (2006) restrict their attention to particular cases where the dependence on forward rates can be eliminated so as to yield an explicit expression for $\phi_t^j(u)$, which is obtained by solving iteratively two Heston PDE's with different diffusion coefficients and boundary conditions.

19.2.1 Exact Solution for the Uncorrelated Case

The forward CPI's and volatility's dynamics (19.4) and (19.5) can be considerably simplified by assuming that

$$\rho_{i,l}^{F,I} = 0, \text{ for each } i, l = 1, \ldots, M, \tag{19.9}$$

so that each $\mathcal{I}_j$ is a driftless geometric Brownian motion under every forward measure, and

$$\rho_i^{F,V} = 0, \text{ for each } i = 1, \ldots, M, \tag{19.10}$$

so as to completely separate the SDE's of forward rates and volatility.

Introducing (19.9) and (19.10) is extremely convenient from a computational point of view, since it allows the derivation of an explicit expression for the function H in (19.8).

Assumption (19.9) has also an empirical justification. In fact, under the lognormal market model of Section 16.5, the prices of both year-on-year swaps and caplets are not very sensitive to changes in the correlation between forward rates and forward CPI's. Calibration to cross-sectional data thus likely provides no empirical evidence that the *implied* correlation's value is necessarily different than zero.

Under (19.9) and (19.10), the cumbersome terms depending on forward rates disappear from the drifts of SDE's (19.4) and (19.5), thus leading to

$$\begin{aligned}
dY_j(t) &= -\tfrac{1}{2}V(t)(\sigma_j^I)^2 dt + \sqrt{V(t)}\,\sigma_j^I(t)\,dZ_j^I(t)\\
dX_j(t) &= \frac{V(t)}{2}\left((\sigma_{j-1}^I)^2 - (\sigma_j^I)^2\right)dt + \sqrt{V(t)}[\sigma_j^I\,dZ_j^I(t) - \sigma_{j-1}^I\,dZ_{j-1}^I(t)]\\
dV(t) &= [\alpha\theta - \alpha V(t)]\,dt + \epsilon\sqrt{V(t)}\,dW(t)
\end{aligned} \tag{19.11}$$

where we set $Y_j(t) := \ln(\mathcal{I}_j(t))$.

To produce dynamics (19.11), all correlations that explicitly enter the previous SDEs are set to zero. However, one still allows for non-zero values for $\rho_{j,l}^I$ and for $\rho_i^{I,V}$, namely for the correlation between different forward CPI's and for the correlation between forward CPI's and the volatility, respectively. Assumptions (19.9) and (19.10), therefore, are not so restrictive as far as the pricing of II derivatives is concerned.[2]

To make ϕ_t^j explicit, let us rewrite (19.8) as

$$\phi_t^j(u) = E_t^j\left[e^{iu\left(Y_j(T_j) - Y_{j-1}(T_{j-1})\right)}\right] = E_t^j\left[e^{-iuY_{j-1}(T_{j-1})}E_{T_{j-1}}^j\left(e^{iuY_j(T_j)}\right)\right] \tag{19.12}$$

and notice that $E_{T_{j-1}}^j\left(e^{iuY_j(T_j)}\right)$ is nothing but the characteristic function of $\ln(\mathcal{I}_j(T_j))$ conditional on $\mathcal{F}_{T_{j-1}}$. From the first and third equations of (19.11) and solving a Heston-like PDE, as detailed in Appendix A of this chapter, we have that

$$E_{T_{j-1}}^j\left(e^{iuY_j(T_j)}\right) = \exp\left\{A_Y(\tau_j, u) + B_Y(\tau_j, u)V(T_{j-1}) + iuY_j(T_{j-1})\right\}$$

[2] When pricing hybrid products that are based on inflation and interest rates, we should allow for general correlations and resort to some *freezing* approximations as explained in Section 19.2.2.

where

$$
\begin{aligned}
B_Y(s,u) &= \frac{\gamma - b}{2a}\left[\frac{1-\mathrm{e}^{\gamma s}}{1-\frac{b-\gamma}{b+\gamma}\mathrm{e}^{\gamma s}}\right] \\
A_Y(s,u) &= \frac{\alpha\theta(\gamma - b)}{2a}s - \frac{\alpha\theta}{a}\ln\left[\frac{1-\frac{b-\gamma}{b+\gamma}\mathrm{e}^{\gamma s}}{1-\frac{b-\gamma}{b+\gamma}}\right]
\end{aligned}
\tag{19.13}
$$

and

$$
\begin{aligned}
a &= \epsilon^2/2, & c &= -\, iu(\sigma_j^I)^2/2 - (\sigma_j^I)^2 u^2/2, \\
b &= iu\sigma_j^I \epsilon \rho_j^{I,V} - \alpha, & \gamma &= \sqrt{b^2 - 4ac}.
\end{aligned}
$$

As a consequence, since $X_j(T_{j-1}) = Y_j(T_{j-1}) - Y_{j-1}(T_{j-1})$, we can write

$$
\phi_t^j(u) = \mathrm{e}^{A_Y(\tau_j,u)} E_t^j\left[\mathrm{e}^{iuX_j(T_{j-1}) + B_Y(\tau_j,u)V(T_{j-1})}\right]. \tag{19.14}
$$

The residual expectation in (19.14), which coincides with its argument if $t \in [T_{j-1}, T_j)$, is the characteristic function of the couple $(X_j(T_{j-1}), V(T_{j-1}))$ evaluated at point $(u, -iB_Y(\tau_j, u))$, and conditional on $\mathcal{F}_t$. An explicit formula for $\phi_t^j(u)$ can then be obtained by solving a new PDE, which is determined by the second and third equation of (19.11). Proceeding as before, we finally have

$$
\phi_t^j(u) = \exp\left\{A_Y(\tau_j,u) + A_X(T_{j-1}-t,u)B_X(T_{j-1}-t,u)V(t) + iuX_j(t)\right\} \tag{19.15}
$$

where

$$
\begin{aligned}
B_X(\tau,u) =& B_Y(\tau_j,u) + \frac{\bar\gamma - \bar b - 2\bar a B_Y(\tau_j,u)}{2\bar a}\left[\frac{1-\mathrm{e}^{\bar\gamma\tau}}{1-\frac{2\bar a B_Y(\tau_j,u)+\bar b-\bar\gamma}{2\bar a B_Y(\tau_j,u)+\bar b+\bar\gamma}\mathrm{e}^{\bar\gamma\tau}}\right] \\
A_X(\tau,u) =& \frac{\alpha\theta(\bar\gamma - \bar b)}{2\bar a}\tau - \frac{\alpha\theta}{\bar a}\ln\left[\frac{1-\frac{2\bar a B_Y(\tau_j,u)+\bar b-\bar\gamma}{2\bar a B_Y(\tau_j,u)+\bar b+\bar\gamma}\mathrm{e}^{\bar\gamma\tau}}{1-\frac{2\bar a B_Y(\tau_j,u)+\bar b-\bar\gamma}{2\bar a B_Y(\tau_j,u)+\bar b+\bar\gamma}}\right]
\end{aligned}
\tag{19.16}
$$

and

$$
\begin{aligned}
&\bar a = \epsilon^2/2, \qquad \bar b = iu\epsilon(\sigma_j^I\rho_j^{I,V} - \sigma_{j-1}^I\rho_{j-1}^{I,V}) - \alpha \\
&\bar c = iu\big((\sigma_{j-1}^I)^2 - (\sigma_j^I)^2\big)/2 - \big((\sigma_{j-1}^I)^2 + (\sigma_j^I)^2 - 2\sigma_j^I\sigma_{j-1}^I\rho_{j,j-1}^I\big)u^2/2 \\
&\bar\gamma = \sqrt{\bar b^2 - 4\bar a\bar c},
\end{aligned}
\tag{19.17}
$$

where the case $\sigma_j^I = \sigma_{j-1}^I$ with $\rho_{j,j-1}^I = 1$ has been discarded because it leads to a degenerate deterministic evolution for X_j.

Function (19.15) gives the exact solution to the above pricing problem under (19.9) and (19.10). Under non-zero correlation structures, the pricing of caplets is less straightforward. However, as shown by Mercurio and Moreni (2006), the general case of dynamics (19.4) and (19.5) can be addressed by resorting to suitable approximations. This is illustrated in the following section.

19.2.2 Approximated Dynamics for Non-zero Correlations

The classical way to handle unpleasant drift terms in the LIBOR market model is by freezing them at their time-0 value, see Chapter 6. Here, we deal with the forward-rates dependent ratios in the drifts of (19.4) by using a similar technique.

The drift terms that involve forward rates are

$$D_l(t) := \sqrt{V(t)}\frac{F_l(t)}{1+\tau_l F_l(t)}$$

and depend on the volatility, too.

A first way to freeze these terms consists of setting

$$D_l(t) \approx D_l(0),$$

and changing the asymptotic volatility value from θ to

$$\tilde{\theta} := \theta - \frac{\epsilon}{\alpha}\sum_{l=1}^{j} D_l(0)\tau_l \sigma_l^F \rho_l^{F,V}, \tag{19.18}$$

which leads to the following (approximated) SDEs for X_j and V

$$\begin{aligned} dX_j(t) \approx & \left[\frac{V(t)}{2}((\sigma_{j-1}^I)^2 - (\sigma_j^I)^2) + D_j(0)\tau_j \sigma_{j-1}^I \sigma_j^F \rho_{j,j-1}^{F,I}\right] dt \\ & + \sqrt{V(t)}[\sigma_j^I \, dZ_j^I(t) - \sigma_{j-1}^I \, dZ_{j-1}^I(t)] \\ dV(t) \approx & \alpha(\tilde{\theta} - V(t))\, dt + \epsilon\sqrt{V(t)}\, dW(t). \end{aligned} \tag{19.19}$$

The dynamics of X_j in (19.19) differs from that in equation (19.11) for a constant drift term. Such a term, however, is innocuous and the relevant characteristic functions can still be calculated by means of Appendix A in this chapter.

A second possibility for a tractable approximation is to set

$$D_l(t) \approx \frac{F_l(t)}{1+\tau_l F_l(t)}\frac{V(t)}{\sqrt{V(t)}} \approx \frac{F_l(0)}{1+\tau_l F_l(0)}\frac{V(t)}{\sqrt{V(0)}} = D_l(0)\frac{V(t)}{V(0)}, \tag{19.20}$$

where the freezing is done with the purpose of producing a linear term in $V(t)$. This leads to the following (approximated) SDEs for X_j and V

$$\begin{aligned} dX_j(t) \approx & V(t)\left[\frac{1}{2}((\sigma_{j-1}^I)^2 - (\sigma_j^I)^2) + \frac{D_j(0)}{V(0)}\tau_j \sigma_{j-1}^I \sigma_j^F \rho_{j,j-1}^{F,I}\right] dt \\ & + \sqrt{V(t)}[\sigma_j^I \, dZ_j^I(t) - \sigma_{j-1}^I \, dZ_{j-1}^I(t)] \\ dV(t) \approx & \bar{\alpha}(\bar{\theta} - V(t))\, dt + \epsilon\sqrt{V(t)}\, dW(t), \end{aligned} \tag{19.21}$$

where

$$\bar{\alpha} := \alpha + \frac{\epsilon}{V(0)} \sum_{l=1}^{j} D_l(0)\tau_l \sigma_l^F \rho_l^{F,V} \tag{19.22}$$
$$\bar{\theta} := \alpha\theta/\bar{\alpha}$$

Also in this second case, the relevant characteristic functions can be calculated according to Appendix A.

Both approximations (19.19) and (19.21) lead to characteristic functions that can be explicitly calculated by applying the result in Appendix A, and analogously to the "uncorrelated" case. Given the similarity with (19.15), the explicit solutions in these two cases are not reported here for brevity.

Remark 19.2.1. For the approximated processes $V(t)$ to be meaningful, we must require $\tilde{\theta} > 0$ in (19.19) and $\bar{\alpha} > 0$ in (19.21). Moreover, for the origin to be inaccessible, conditions $2\tilde{\theta}\alpha > \epsilon$ and $2\bar{\theta}\bar{\alpha} > \epsilon$ must be imposed, with the latter that is automatically satisfied since $2\theta\alpha > \epsilon$ by assumption.

To test the goodness of the above approximations (19.19) and (19.21), one should perform a Monte Carlo simulation of processes (19.4) and (19.5) and compare the analytical caplet prices coming from the approximations with the corresponding Monte Carlo price windows calculated numerically. This procedure, however, is rather involved, since it also requires the joint simulation of all forward rates.

Mercurio and Moreni (2006) conduct a much quicker test by simply comparing the caplet prices implied by the two approximations. In fact, the two freezings can lead to quite different terms when the volatility $V(t)$ deviates from its initial value $V(0)$. Therefore, if the approximations are quite close to each other, one may conclude that they should also be very close to the true model's value.

Mercurio and Moreni also compare the two approximations with the exact price obtained under assumptions (19.9) and (19.10). Using EUR market data as of October 7, 2004, and setting $\rho_i^{F,V} = \rho_{i,l}^{F,I} = \rho_i^{I,V} = -0.2$, $\rho_{i,i-1}^{I} = 1 - 1.5\mathrm{e}^{-0.08(i-2)}$, $\alpha = 0.2$, $\theta = 0.001$, $V(0) = 0.001$, $\epsilon = 0.01$ and $\sigma_i^I = 1 - 0.05(i-1)$, for $i, l = 1, \ldots, 5$, they found that the price percentage differences for caplets with maturities from 1 to 5 years and for a wide range of strikes, are of less than 0.2%, thus being negligible if compared with typical bid-ask spreads in the market. Their result can also be interpreted by stating that already the exact price (19.15) can be considered a good approximation of the true model's price.

19.3 Example of Calibration

We now illustrate an example of calibration to II options, based on USD market data as of November 3, 2004. The stochastic-volatility model (19.4) with (19.5) is tested under the assumption that $\rho_{i,l}^{F,I} = 0$ and $\rho_i^{F,V} = 0$ for each i, l. The quoted cap prices are reported on Table 19.1 for a range of strikes going from -1.5% to 0.5%.

T_i	K				
	-1.5%	-1.0%	-0.5%	0%	0.5%
1y	416.7	368.2	319.8	271.8	224.3
2y	822.4	727.6	633.3	540.1	448.7
3y	1212.7	1074.0	936.5	800.7	667.8
4y	1588.4	1408.0	1229.2	1052.8	880.5
5y	1952.9	1732.8	1514.6	1299.4	1089.1
6y	2288.2	2030.7	1775.6	1524.1	1278.3
7y	2612.4	2319.7	2029.8	1744.2	1465.2
8y	2911.5	2585.9	2263.6	1946.2	1636.5
9y	3197.2	2840.7	2488.0	2140.9	1802.5
10y	3467.7	3082.3	2701.1	2326.3	1961.1

Table 19.1. II Cap prices (in bps) for different strikes and maturities as of November 3, 2004, in the USD market.

The calibration is performed on caplet prices, which are stripped from the cap ones by simply taking differences between consecutive maturities, see Table 19.2.

T_i	K				
	-1.5%	-1.0%	-0.5%	0%	0.5%
1	416.7	368.2	319.8	271.8	224.3
2	405.7	359.4	313.5	268.3	224.4
3	390.3	346.4	303.2	260.6	219.1
4	375.7	334.0	292.7	252.1	212.7
5	364.5	324.8	285.4	246.6	208.6
6	335.3	297.9	261.0	224.7	189.2
7	324.2	289.0	254.2	220.1	186.9
8	299.1	266.2	233.8	202.0	171.3
9	285.7	254.8	224.4	194.7	166.0
10	270.5	241.6	213.1	185.4	158.6

Table 19.2. II Caplet prices (in bps) for different strikes and maturities as of November 3, 2004, in the USD market.

The model's price at time zero of a caplet with maturity T_i is given by formula (19.15) with $t = 0$, and depends on both the T_i-forward and the T_{i-1}-forward CPI's. The value at time zero of a T_i-forward CPI, where $T_i = i$ years, is obtained from the market quote $S(T_i)$ of the corresponding ZCIIS by applying the following relation

$$\mathcal{I}_i(0) = I(0)(1 + S(T_i))^i.$$

The CPI's value on November 3, 2004 was 190.91. The USD discount factors and the ZC swap rates observed on that date are reported in Table 19.3, together with the implied forward CPI's.

T_i	$P(0,T_i)$	ZC rates	$\mathcal{I}_i(0)$
1y	0.97701	2.111%	194.94
2y	0.94982	2.188%	199.35
3y	0.91835	2.240%	204.03
4y	0.88433	2.278%	208.91
5y	0.84862	2.293%	213.82
6y	0.81179	2.300%	218.82
7y	0.77460	2.310%	224.00
8y	0.73785	2.320%	229.36
9y	0.70218	2.325%	234.78
10y	0.66773	2.335%	240.48

Table 19.3. USD discount factors, ZC swap rates and implied forward CPI's, on November 3, 2004.

The model parameters to be calibrated are:

- The volatility parameters ϵ, α, θ, $V(0)$;
- The forward CPI's volatility coefficients σ_i^I, $i = 1, \dots, M$;
- The correlations $\rho_{i-1,i}^I$ between consecutive forward CPI's, $i = 2, \dots, M$;
- The correlations $\rho_i^{I,V}$ between forward CPI's and the volatility, $i = 1, \dots, M$;

We stress that, under assumptions (19.9) and (19.10), no forward-rate parameter enters the pricing formula.

To reduce the degrees of freedom at hand, we then parameterize the CPI's correlations as

$$\rho_{i,i-1}^I = 1 - (1 - \rho_0)\mathrm{e}^{-\lambda T_{i-2}}, \quad i = 2, \dots, M, \tag{19.23}$$

where ρ_0 is a correlation parameter and λ is a positive number. Moreover, we notice that for any $\kappa > 0$, one obtains equivalent SDE's by scaling the model parameters in (19.4) and (19.5) as follows:

$$\theta \to \kappa\theta, \quad \epsilon \to \sqrt{\kappa}\epsilon, \quad \sigma_i^I \to \frac{\sigma_i^I}{\sqrt{\kappa}}, \quad V(0) \to \kappa V(0).$$

To avoid redundancies, therefore, we set $\sigma_1^I = 1$, which corresponds to choosing $\kappa = (\sigma_1^I)^2$ as scaling factor.

Given the parameterizations and simplifications above, the calibration is eventually performed on the following $2M - 5$ parameters:

$$\epsilon, \alpha, \theta, V(0), \rho_0, \lambda, \sigma_2^I, \dots, \sigma_M^I, \rho_1^{I,V}, \dots, \rho_M^{I,V}.$$

The first caplet price depends on five parameters, the second on nine, and from the third on, to obtain the overall number of parameters at hand, we simply have to add two to the previous maturity's number.

Aiming to reproduce caplets prices for a wide range of strikes, we solve the calibration problem by minimizing the sum of squared percentage differences between model and market prices.

In Tables 19.4 and 19.5, we report the result of our minimization by listing, respectively, the best-fit prices and calibration errors for a reduced range of maturities.

			K		
T_i	-1.5%	-1.0%	-0.5%	0%	0.5%
1	412.0	366.3	319.8	272.3	225.9
2	401.4	356.2	312.1	268.6	226.5
3	390.8	347.1	303.4	260.5	218.6

Table 19.4. Model caplet prices after calibration.

			K		
T_i	-1.5%	-1.0%	-0.5%	0%	0.5%
1	-1.12	-0.51	-0.0033	0.17	0.72
2	-1.05	-0.88	-0.46	0.093	0.95
3	0.14	0.20	0.073	-0.031	-0.20

Table 19.5. Percentage errors on caplet prices after calibration.

Prices of short-maturity caplets are recovered with a very good precision. When we increase the caplet maturity, however, the fitting quality slightly worsen and the improvement induced by our stochastic volatility model over the deterministic volatility LFCPIM tends to reduce.

Postulating a unique volatility process for all forward CPI's seems too restrictive if we aim at calibrating many maturities simultaneously. One may then consider a different stochastic volatility process for each forward CPI, introducing new volatility and correlation parameters. In this case, however, caplet prices may be difficult to derive in closed form, since the pricing of each caplet (but the first one) would involve an extra volatility process.

Appendix A: Heston PDE

Consider the following Heston-like SDE for a stochastic process X:

$$
\begin{aligned}
dX(t) =& (\omega + \mu V(t))\, dt + \sigma \sqrt{V(t)}\, dZ(t) \\
dV(t) =& \alpha(\theta - V(t))\, dt + \epsilon \sqrt{V(t)}\, dW(t) \\
dZ(t)\, dW(t) =& \rho\, dt
\end{aligned}
\tag{19.24}
$$

Given times t and T, $t \leq T$, we are interested in evaluating the characteristic function of the couple $(X(T), V(T))$, conditional on $\mathcal{F}_t$,

$$
\phi(X(t), V(t), t; T, u_1, u_2) = E_t[\mathrm{e}^{iu_1 X(T) + iu_2 V(T)}]. \tag{19.25}
$$

The Feynman-Kac formula allows us to write a PDE satisfied by ϕ:

$$
\begin{cases}
\dfrac{\partial \phi}{\partial t} + [\alpha\theta - \alpha v]\dfrac{\partial \phi}{\partial v} + [\omega + v\mu]\dfrac{\partial \phi}{\partial x} + \frac{1}{2}\dfrac{\partial^2 \phi}{\partial x^2} v\sigma^2 + \frac{1}{2}\epsilon^2 v \dfrac{\partial^2 \phi}{\partial v^2} + \sigma\epsilon\rho v \dfrac{\partial^2 \phi}{\partial v \partial x} = 0 \\
\phi(x, v, t; T, u_1, u_2) = \mathrm{e}^{iu_1 x + iu_2 v}
\end{cases}
\tag{19.26}
$$

Following Heston (1993), we guess a solution like

$$
\begin{aligned}
\phi(x, v, t; T, u_1, u_2) &= \exp[A(T - t, u_1, u_2) + B(T - t, u_1, u_2)v + iu_1 x] \\
A(0, u_1, u_2) &= 0 \\
B(0, u_1, u_2) &= iu_2,
\end{aligned}
\tag{19.27}
$$

which is suggested by the linearity in v of the PDE's coefficients. The PDE in Heston (1993) is recovered by setting $u_2 = 0$.

Let us set $\tau := T - t$. Once we plug our *ansatz* solution into (19.26), we have (for simplicity, we drop dependence on u_1, u_2)

$$
\begin{aligned}
\frac{\partial A(\tau)}{\partial \tau} =& B(\tau)\alpha\theta + iu_1\omega \\
\frac{\partial B(\tau)}{\partial \tau} =& \frac{\epsilon^2}{2} B^2(\tau) + B(\tau)(iu_1\sigma\epsilon\rho - \alpha) + iu_1\mu - \tfrac{1}{2}\sigma^2 u_1^2.
\end{aligned}
\tag{19.28}
$$

The second equation is in the form

$$
\frac{\partial B(\tau)}{\partial \tau} = aB^2(\tau) + bB(\tau) + c, \qquad a, b, c \in \mathbb{C} \tag{19.29}
$$

and can be solved, for instance,[3] by setting $B_{\pm} = \frac{1}{2a}\left[-b \pm \sqrt{b^2 - 4ac}\right] := \frac{1}{2a}(-b \pm \gamma)$, $\gamma = \sqrt{b^2 - 4ac}$ and by rewriting (19.28) as

$$
\left(\frac{1}{B - B_+} - \frac{1}{B - B_-}\right)\partial B = \gamma \partial \tau. \tag{19.30}
$$

After some algebra, we have, for each $\tau > 0$,

[3] See Wu and Zhang (2002) for an equivalent derivation.

$$
\begin{aligned}
B(\tau) &= \frac{(\gamma - b)}{2a} \frac{1 - \frac{b+\gamma}{b-\gamma} \cdot \frac{2aB(0)+b-\gamma}{2aB(0)+b+\gamma} e^{\gamma\tau}}{1 - \frac{2aB(0)+b-\gamma}{2aB(0)+b+\gamma} e^{\gamma\tau}} \\
&= B(0) + \frac{\gamma - b - 2aB(0)}{2a} \left[\frac{1 - e^{\gamma\tau}}{1 - \frac{2aB(0)+b-\gamma}{2aB(0)+b+\gamma} e^{\gamma\tau}} \right]
\end{aligned}
\tag{19.31}
$$

which coincides with the corresponding Heston's formula. Function A is then obtained by integration

$$
A(\tau) - A(0) = iu_1\omega\tau + \frac{\alpha\theta(\gamma - \beta)}{2a\gamma} \int_1^{e^{\gamma\tau}} \left[\frac{1}{y} + \frac{f - d}{1 - fy} \right] dy
$$

with $f = (2aB(0) + b - \gamma)/(2aB(0) + b + \gamma)$ and $d = f(b + \gamma)/(b - \gamma)$, so that, finally,

$$
A(\tau) = A(0) + \left[iu_1\omega + \frac{\alpha\theta(\gamma - b)}{2a} \right] \tau - \frac{\alpha\theta}{a} \ln \left[\frac{1 - \frac{2aB(0)+b-\gamma}{2aB(0)+b+\gamma} e^{\gamma\tau}}{1 - \frac{2aB(0)+b-\gamma}{2aB(0)+b+\gamma}} \right] \tag{19.32}
$$

with

$$
\begin{aligned}
a &= \epsilon^2/2 \\
b &= iu_1\sigma\epsilon\rho - \alpha \\
c &= iu_1\mu - \sigma^2 u_1^2/2 \\
A(0) &= 0 \\
B(0) &= iu_2.
\end{aligned}
$$

Appendix B: Floorlet Pricing

Let us consider a T_j-maturity floorlet, whose time-t value is given by

$$
\begin{aligned}
\mathbf{IICplt}(t, T_{j-1}, T_j, K, -1) &= P(t, T_j) \int_{-\infty}^{+\infty} (e^k - e^s)^+ q_t^j(s) ds \\
&= P(t, T_j) \int_{-\infty}^{k} (e^k - e^s) q_t^j(s) ds
\end{aligned}
$$

where $k = \ln(K + 1)$ and q_t^j is defined in Section 19.2.

To price a call option with an FT approach, Carr and Madan (1998) tackled the non-integrability issue of the call price for $k \to -\infty$, by introducing a regularizing factor leading to a modified caplet price that tends to zero whenever $k \to -\infty$. Here, instead, we have that the floorlet price, as function of k, is not square integrable on the positive semi-axis, when $k \to +\infty$. We thus choose $\tilde{\eta} > 1$ and introduce an exponential dumping factor by setting

$$\mathbf{IICplt}(t, T_{j-1}, T_j, K, -1) = P(t, T_j) \mathrm{e}^{\tilde{\eta} k} H(k) \\ := P(t, T_j) \mathrm{e}^{\tilde{\eta} k} \int_{-\infty}^{k} \mathrm{e}^{-\tilde{\eta} k} (\mathrm{e}^k - \mathrm{e}^s) q_t^j(s) ds. \tag{19.33}$$

It is now easy to compute the FT $\hat{H}(u)$ of $H(k)$ in terms of the characteristic function ϕ_t^j as

$$\begin{aligned} \hat{H}(u) &= \int_{-\infty}^{+\infty} dk\, \mathrm{e}^{iku} H(k) = \int_{-\infty}^{+\infty} dk \int_{-\infty}^{k} ds\, \mathrm{e}^{(iu - \tilde{\eta})k} (\mathrm{e}^k - \mathrm{e}^s) q_t^j(s) \\ &= \int_{-\infty}^{+\infty} ds\, q_t^j(s) \int_{s}^{+\infty} dk\, \mathrm{e}^{(iu-\tilde{\eta})k} (\mathrm{e}^k - \mathrm{e}^s) \\ &= \int_{-\infty}^{+\infty} \frac{\mathrm{e}^{i[u + i(\tilde{\eta} - 1)]s}}{(iu - \tilde{\eta} + 1)(iu - \tilde{\eta})} q_t^j(s)\, ds = \frac{\phi_t^j(u + i(\tilde{\eta} - 1))}{(iu - \tilde{\eta} + 1)(iu - \tilde{\eta})}. \end{aligned} \tag{19.34}$$

The floorlet price is then calculated by means of the inverse FT:

$$\begin{aligned} &\mathbf{IICplt}(t, T_{j-1}, T_j, K, -1) \\ &= P(t, T_j) \frac{\mathrm{e}^{\tilde{\eta} k}}{\pi} \mathrm{Re} \int_{0}^{+\infty} \mathrm{e}^{-iuk} \frac{\phi_t^j(u + i(\tilde{\eta} - 1))}{(iu - \tilde{\eta} + 1)(iu - \tilde{\eta})} du. \end{aligned} \tag{19.35}$$

We finally remark that, also for II options, it is possible to derive an explicit put-call parity. In fact, since $(x - y)^+ - (y - x)^+ = x - y$, we have that,

$$\begin{aligned} &\mathbf{IICplt}(t, T_{j-1}, T_j, K, 1) - \mathbf{IICplt}(t, T_{j-1}, T_j, K, -1) \\ &= P(t, T_j) E_t^j \left\{ \frac{\mathcal{I}_j(T_j)}{\mathcal{I}_{j-1}(T_{j-1})} - (K+1) \right\} \\ &= P(t, T_j) \left\{ E_t^j \left[\frac{\mathcal{I}_j(T_{j-1})}{\mathcal{I}_{j-1}(T_{j-1})} \right] - (K+1) \right\} \\ &= P(t, T_j) \left\{ E_t^j \left[\mathrm{e}^{X_j(T_{j-1})} \right] - (K+1) \right\}, \end{aligned} \tag{19.36}$$

where the second equality holds since $\mathcal{I}_j$ is a martingale under $\mathbb{Q}^j$. The residual expectation in (19.36) is the characteristic function of $X_j(T_{j-1})$ evaluated at point $-i$, and can be easily computed as explained in Appendix A and in Section 19.2.1.

20. Pricing Hybrids with an Inflation Component

In the recent years, there has been an increasing interest for hybrid structures whose payoff is based on assets belonging to different markets. Among them, derivatives with an inflation component are getting more and more popular. In this chapter, we tackle the pricing issue of a specific hybrid payoff when no smile effects are taken into account. The valuation of more general structures is to be dealt with on a case by case basis and is likely to involve numerical routines as Monte Carlo.

20.1 A Simple Hybrid Payoff

Let us consider a contract that pays off at maturity T the maximum between the return on an equity index and the return on the CPI. In formulas, the time-T payoff is

$$N \max\left\{\frac{I(T)}{I_0}, \frac{S(T)}{S_0}\right\}, \tag{20.1}$$

where S denotes the price of the equity index, S_0 its value at time 0 and N is the contract's nominal amount.

Pricing with the JY Model

Standard risk-neutral pricing implies that the value at time t, $0 \leq t < T$, of the contract is

$$N E_n\left\{e^{-\int_t^T n(u)\,du} \max\left[\frac{I(T)}{I_0}, \frac{S(T)}{S_0}\right] |\mathcal{F}_t\right\}, \tag{20.2}$$

where $\mathcal{F}_t$ denotes the σ-algebra generated by the (I, S) up to time t.

Let us assume that the asset price S evolves, under the risk-adjusted measure Q_n, according to

$$dS(t) = S(t)[(n(t) - y)dt + \eta\, dZ(t)], \tag{20.3}$$

where Z is a Q_n-Brownian motion and $y > 0$ is the asset dividend yield, and that the CPI follows the JY dynamics (15.2), namely

$$dI(t) = I(t)[n(t) - r(t)]\,dt + \sigma_I I(t)\,dW_I(t),$$
$$dn(t) = [\vartheta_n(t) - a_n n(t)]\,dt + \sigma_n\,dW_n(t)$$
$$dr(t) = [\vartheta_r(t) - \rho_{r,I}\sigma_I\sigma_r - a_r r(t)]\,dt + \sigma_r\,dW_r(t)$$

where (W_n, W_r, W_I) is a Brownian motion with correlations $\rho_{n,r}$, $\rho_{n,I}$ and $\rho_{r,I}$, $\vartheta_n(t)$ and $\vartheta_r(t)$ are deterministic functions, and σ_n σ_r, a_n, a_r and σ_I are positive constants.

To calculate the expectation in (20.2), it is more convenient to apply a measure change and move to the measure $Q^{\bar{S}}$ associated to the numeraire $\bar{S}(t) := S(t)\exp(yt)$, which satisfies the SDE

$$d\bar{S}(t) = \bar{S}(t)[n(t)dt + \eta\,dZ(t)].$$

In fact

$$\begin{aligned} &NE_n\left\{e^{-\int_t^T n(u)\,du}\max\left[\frac{I(T)}{I_0}, \frac{S(T)}{S_0}\right]|\mathcal{F}_t\right\} \\ &= N\bar{S}(t)E^{\bar{S}}\left\{\max\left[\frac{I(T)}{I_0\bar{S}(T)}, \frac{S(T)}{S_0\bar{S}(T)}\right]|\mathcal{F}_t\right\} \\ &= NS(t)\,e^{yt}E^{\bar{S}}\left\{\max\left[\frac{I(T)P_r(T,T)}{I_0\bar{S}(T)}, \frac{e^{-yT}}{S_0}\right]|\mathcal{F}_t\right\}. \end{aligned} \tag{20.4}$$

Being $I(t)P_r(t,T)$ the price of a tradable asset, the process

$$X(t) := \frac{I(t)P_r(t,T)}{\bar{S}(t)}$$

is a martingale under $Q^{\bar{S}}$. Precisely, X evolves under $Q^{\bar{S}}$ according to the (driftless) geometric Brownian motion

$$dX(t) = X(t)\varsigma(t,T)\,dW(t),$$

where W is a standard Brownian motion under $Q^{\bar{S}}$,

$$\begin{aligned} \varsigma(t,T)^2 := \sigma_I^2 + \sigma_r^2 B_r(t,T)^2 + \eta^2 - 2\rho_{r,I}\sigma_I\sigma_r B_r(t,T) \\ - 2\rho_{S,I}\sigma_I\eta + 2\rho_{r,S}\eta\sigma_r B_r(t,T), \end{aligned}$$

$B_r(t,T)$ is defined in (16.13), and the instantaneous correlations $\rho_{S,I}$ and $\rho_{r,S}$ are defined by

$$dZ(t)\,dW_I(t) = \rho_{S,I}\,dt,$$
$$dZ(t)\,dW_r(t) = \rho_{r,S}\,dt.$$

The last expectation in (20.4) can then be calculated as in the Black and Scholes (1973) model with time-dependent volatility and zero drift. We thus obtain

$$
N\left[\frac{I(t)P_r(t,T)}{I_0}\Phi\left(\frac{\ln\frac{S_0 I(t)P_r(t,T)}{I_0 S(t)e^{-y(T-t)}}+\frac{1}{2}\Sigma^2(t,T)}{\Sigma(t,T)}\right)\right.
$$
$$
\left.+\frac{S(t)e^{-y(T-t)}}{S_0}\Phi\left(\frac{\ln\frac{I_0 S(t)e^{-y(T-t)}}{S_0 I(t)P_r(t,T)}+\frac{1}{2}\Sigma^2(t,T)}{\Sigma(t,T)}\right)\right], \tag{20.5}
$$

where

$$
\Sigma^2(t,T)=\int_t^T \varsigma^2(u,T)\,du.
$$

Pricing with a Market Model

The payoff (20.1) can also be priced by using the market model of Section 16.5, where the forward CPI's are assumed to evolve according to geometric Brownian motions under the respective forward measure. To this end, we can conveniently resort to the T-forward measure Q_n^T, under which the price of the contract can be written as follows:

$$
NP_n(t,T)E_n^T\left\{\max\left\{\frac{I(T)}{I_0},\frac{S(T)}{S_0}\right\}|\mathcal{F}_t\right\}. \tag{20.6}
$$

Defining the forward assets

$$
\begin{aligned}
\mathcal{I}(t) &:= I(t)\frac{P_r(t,T)}{P_n(t,T)},\\
\mathcal{S}(t) &:= S(t)\frac{e^{-y(T-t)}}{P_n(t,T)},
\end{aligned} \tag{20.7}
$$

which are both martingales under Q_n^T, and re-writing (20.1) as

$$
N\max\left\{\frac{\mathcal{I}(T)}{I_0},\frac{\mathcal{S}(T)}{S_0}\right\}, \tag{20.8}
$$

the price (20.6) becomes

$$
NP_n(t,T)E_n^T\left\{\max\left\{\frac{\mathcal{I}(T)}{I_0},\frac{\mathcal{S}(T)}{S_0}\right\}|\mathcal{F}_t\right\}. \tag{20.9}
$$

Assuming the following lognormal-type dynamics under Q_n^T:

$$
\begin{aligned}
d\mathcal{I}(t) &= \mathcal{I}(t)\sigma_I\,dW_I(t),\\
d\mathcal{S}(t) &= \mathcal{S}(t)\sigma_S\,dW_S(t),
\end{aligned}
$$

where σ_I and σ_S are positive constants and W_I and W_S are two Brownian motions with instantaneous correlation ρ, the expectation in (20.9) can then be calculated as in a two-dimensional Black and Scholes (1973) model. We thus obtain:

$$NP_n(t,T)\left[\frac{\mathcal{I}(t)}{I_0}\Phi\left(\frac{\ln\frac{\mathcal{I}(t)S_0}{I_0 S(t)}+\frac{1}{2}\sigma^2(T-t)}{\sigma\sqrt{T-t}}\right)\right.$$
$$\left.+\frac{S(t)}{S_0}\Phi\left(\frac{\ln\frac{I_0 S(t)}{\mathcal{I}(t)S_0}+\frac{1}{2}\sigma^2(T-t)}{\sigma\sqrt{T-t}}\right)\right], \qquad (20.10)$$

where

$$\sigma=\sqrt{\sigma_I^2+\sigma_S^2-2\rho\sigma_I\sigma_S}.$$

Remembering the definitions (20.7), we see that the claim price (20.10) is perfectly analogous to (20.5). In fact, for a single payoff occurring at time T, we obtain exactly the same formula at time t if (and only if)

$$\sigma\sqrt{T-t}=\Sigma(t,T),$$

so that we may be indifferent between modeling spot or forward quantities (both CPI and equity index), when smile effects on inflation are neglected. The choice between either approach is usually driven, besides personal taste, by the payoff structure to price and also by the calibration instruments one can rely on.

Part VII

CREDIT

21. Introduction and Pricing under Counterparty Risk

"Failure is not an option"

Apollo XIII Mission to the Moon Rescue Motto

"Duke of Alencon, this was your default,
That, being captain of the watch to-night,
Did look no better to that weighty charge."
King Henry VI, Part 1, Act 2, Scene I.

This chapter starts the credit part of the book. In this first chapter we are going to introduce the financial payoffs and the families of rates we are dealing with in the following. But before doing so, we present a guided tour to give some orientation and general feeling for this credit part of the book. The guided tour is given in Section 21.1. Then we introduce as first credit payoffs the prototypical defaultable bonds in Section 21.2. Credit Default Swaps (CDS) payoffs and defaultable floaters, including a relationship between the two, are presented in Section 21.3. In particular, in Section 21.3.5 we explain that CDS' allow in principle to strip default or survival probabilities in a model independent way, and in Section 21.3.7 we consider some different definitions of CDS forward rates, with analogies with earlier parts of the book, and with LIBOR vs swap rates in particular. In Section 21.3.8 instead we explore in detail possible equivalence between CDS payoffs and rates and defaultable floaters payoffs and rates. We anticipate some facts on intensity models that will be clarified completely later on in Chapter 22, where we also present more detailed results on CDS calibration.

The remaining part of this first chapter on credit describes some other single name credit derivatives. In Section 21.4 we introduce CDS options payoffs, pointing out some formal analogies with the swaption payoff encountered earlier in the book. In Section 21.5 we introduce constant maturity CDS, a product that has grown in popularity in recent times. This product presents analogies with constant maturity swaps in the default free market. Finally, we close the chapter with counterparty risk pricing in interest rate derivatives. We show how to include the event that the counterparty may default in the risk neutral valuation of the financial payoff. This is particularly important after the recent regulatory directions given by the Basel II agreement and subsequent amendments, and also by the "IAS 39" (international accounting standard) system requirements. This part is introduced in Section 21.6.

21.1 Introduction and Guided Tour

Green Lantern? Kyle Rayner? Introductions: I became self aware in the year 850000330. I am a diamond-generation intelligent machine colony, DNA-programmed with tyler miraclo gene biosoftware... I am Hourman– also known as "the master of time"
JLA, Rock of Ages part 3, 1997

These final chapters face the challenge of dealing with credit derivatives. Is there a reason for treating credit derivatives in a book on interest-rate derivatives? We believe the answer is in the affirmative, and for a number of reasons. Let us see a few of them.

Counterparty Risk

Default is now permeating the valuation of derivatives in any area. Whenever a derivative is traded over the counter, the default risk of the counterparty should, in principle, enter the valuation. Recently, regulatory institutions insisted on the need to include this kind of risk in the pricing paradigm, as for example pointed out in the Basel II framework and also in the "IAS 39" (international accounting standard) system. What does it mean to include counterparty risk? Counterparty risk is the risk that the counterparty owing us some payments (according to an agreed derivative transaction) defaults. If this happens, instead of said payments we receive just a fraction (recovery) of their value at the instant of default. This is a problem to us, and we wish to include this default risk into the risk neutral valuation paradigm. In a sense, when we do so we obtain a payoff depending on the underlying of the basic derivative and on the default of the counterparty. The latter feature transforms our payoff in a credit (actually hybrid) derivative, and default modeling is needed to carry out the risk neutral pricing. This applies to all derivatives, and in particular to interest rate derivatives, the subject of this book.

Technical analogies

We will see below that there are two main kinds of credit models. One of them, the so called reduced form models framework, uses techniques that are very similar to interest rate models, so that we may (and will) use the earlier results in the book to obtain some credit derivatives models almost "for free".

Formal analogies

We will see that the basic credit derivatives, Credit Default Swaps, and the associated forward rates, bear some formal analogies to interest rate swaps as we have seen them earlier in this book. Even the market models for CDS (Credit Default Swap) options, although being complicated by technicalities that are absent in the earlier parts of the book, bear astonishing resemblances with the LIBOR and swap models seen earlier.

Credit Derivatives are interesting in their own right

As we will see credit derivatives pose problems that are absent in all the theory and practice seen before in the book, so that this is an interesting opportunity to examine different market products and different theoretical apparatus from probability, statistic, stochastic calculus etc.

Now that we have motivated the credit part of the book, we proceed with the guided tour on credit derivatives in general and on what we treat here in particular.

Our discussion has to begin with the default event. A default event is an event where a firm cannot face its obligation on the payments owed to some entity.

Mathematically, default is represented by means of the default time. The default time, typically denoted by τ, is a random time that can be modeled in several ways. There are essentially two paradigms that emerged over the years.

21.1.1 Reduced form (Intensity) models

Reduced form models (also called *intensity models* when a suitable context is possible) describe default by means of an exogenous jump process; more precisely, the default time τ is the first jump time of an important kind of stochastic process, the Poisson process. The Poisson process can have deterministic or stochastic (Cox process) intensity. With these models default is not triggered by basic market observables but has an exogenous component that is independent of all the default free market information. Monitoring the default free market (interest rates, exchange rates, etc) does not give complete information on the default process, and there is no economic rationale behind default. This family of models is particularly suited to model credit spreads and in its basic formulation is easy to calibrate to Credit Default Swap (CDS) or corporate bond data. See also references at the beginning of Chapter 8 in Bielecki and Rutkowski (2001) for a summary of the literature on intensity models. We cite here Duffie and Singleton (1999) and Lando (1998), and Brigo and Alfonsi (2003, 2005) as a reference for explicit calibration of a tractable stochastic intensity model to CDS data. But one of the crucial points we would like to clarify in this guided tour is: why are these models well suited for credit spreads?

In basic reduced form or intensity models, the default time τ is the first jump of a Poisson process. The simplest Poisson process, i.e. the time-homogeneous Poisson process, is the purely-jumps analogous of Brownian motion (that is instead continuous), in that it is a process with stationary independent increments. Moving from this basic process one can, through a time-change, obtain time-inhomogeneous Poisson processes. We are not going into details here, but recall that the first jump time of a (time-inhomogeneous) Poisson process obeys roughly the following:

Having not defaulted (jumped) before t, the (risk neutral) probability of defaulting (jumping) in the next dt instants is

$$\mathbb{Q}(\tau \in [t, t + dt) | \tau > t, \text{market info up to } t) = \lambda(t) dt$$

where the "probability" dt factor λ is assumed here for simplicity to be strictly positive and is in general called *intensity* or *hazard rate.*

Define also the further quantity

$$\Lambda(t) := \int_0^t \lambda(u) du,$$

i.e. the *cumulated intensity, cumulated hazard rate,* or also *Hazard function.* Now, assume for simplicity λ to be deterministic; since it is positive, its integral Λ will be strictly increasing. One of the important facts about Poisson processes is a property of the transformation of the jump time τ according to its own cumulated intensity Λ. We have

$$\Lambda(\tau) =: \xi \sim \text{exponential standard random variable}$$

and this exponential random variable is independent of all other variables (interest rates, equities, intensities themselves in case these are stochastic, etc.). By inverting this last equation we have that

$$\tau = \Lambda^{-1}(\xi).$$

But if we recall for a second the cumulative distribution function of an exponential random variable, i.e. $\mathbb{Q}(\xi \leq x) = 1 - e^{-x}$, we can show immediately that

$$\mathbb{Q}\{\tau > t\} = \mathbb{Q}\{\Lambda(\tau) > \Lambda(t)\} = \mathbb{Q}\{\xi > \Lambda(t)\} = e^{-\int_0^t \lambda(u) du},$$

where the first equality comes from Λ being strictly increasing.

At this point of the book it should be obvious what the last term looks like: It is a discount factor. Even more, if intensity is stochastic the corresponding expression, i.e. the probability to default after t, or to survive t, or *survival probability* at t, is

$$\mathbb{Q}\{\tau > t\} = \mathbb{E}[e^{-\int_0^t \lambda(u)du}],$$

but this is just the price of a zero coupon bond in an interest rate model with short rate r replaced by λ. This is why survival probabilities are interpreted as zero coupon bonds and intensities λ as instantaneous credit spreads. Thus one can choose any (positive!) interest short-rate model seen earlier in the book for λ. We will see in particular what happens when choosing the CIR++ model as a possible model for λ. Necessity to have positive intensity will eliminate the rich family of Gaussian models, leaving us with the choice between lognormal or CIR processes. Given the need for analytical tractability to speed up calibration, among other considerations, we will resort indeed to CIR++.

What is more, we will see in Section 22.4 that if we actually compute the price of a defaultable zero coupon bond, i.e. of a contract paying one unit of currency at final maturity T when the underlying name has not defaulted by T (meaning $\tau > T$), in a stochastic intensity setting, we have as price

$$\mathbb{E}[D(0,T)1_{\{\tau>T\}}] = \mathbb{E}[e^{-\int_0^T (r(u)+\lambda(u))du}]$$

from which we see that the interpretation of the intensity λ as credit spread is once again confirmed, in that it is added to the short rate r to obtain the relevant bond price.

An important remark that should not go unnoticed is that, as hinted at above, ξ *is independent of all default free market quantities and represents an external source of randomness that makes reduced form models incomplete* (more on this later).

Incompleteness is not a cost we face without reward. The exogenous component, besides making the model incomplete, makes default unpredictable in an important technical sense, and allows for non-null instantaneous credit spreads, contrary to the other important family of (basic) firm value (or structural) models. These are usually considered to be positive and realistic features.

To sum up: Intensity models allow us to translate our earlier interest rate models technology into default modeling, even if at the cost of an exogenous term entering the model and of the lack of an economic interpretation of the default event (contrary to the firm value models we see below). This translation makes these credit derivatives models extremely natural to study in a book about interest rate models. This is not the case for structural models, though, as we are going to explain below. But before doing so, our tour moves to examine market models for CDS options, that remain in a way close to reduced form models.

21.1.2 CDS Options Market Models

One may notice from the above discussion that we did not introduce the key ingredients of intensity models by referring directly to market payoffs. In a sense, what happened is similar to what we have done with short-rate models. Since every payoff, by no arbitrage reasons, is the discounted expectation of minus the integral of the short rate times something (this something being "one" in case of zero coupon bonds), it seems to make sense to model the short rate as the variable that explains the whole curve. Indeed, the short rate is the core of the discounting mechanism:

$$\text{Price}_0 = \mathbb{E}\left[\exp\left(-\int_0^{\text{Maturity}} \boxed{r_t}\, dt\right) \text{Payoff(Maturity)}\right].$$

As such, it is one of the first variables one would choose to model. We have done exactly so in the earlier Chapters 3 and 4 on short rate models. However, as the book progressed, for a number of reasons we have seen in Chapter 6 that modern interest rate models, such as the market models, resort to variables that have a more direct link with market payoffs. The LIBOR (LFM) model for example models forward LIBOR rates, i.e. the rates that make Forward Rate Agreements (FRA) contracts fair. The swap models (LSM) consider the rates that make interest rate swaps contracts fair, called forward swap rates. In different terms, the market models model rates that are obtained by setting to zero the values of some reference financial contracts. In the Chapter on the LIBOR market model we have seen that this allows for much more general and realistic volatility and correlation structures, and for much more powerful calibration and diagnostics than with simple short rate models.

However, no arbitrage valuation allows us to write both forward LIBOR rates and forward swap rates in terms of zero coupon bonds, that in turn can be obtained as suitable expectations of functionals of the short rate r:

$$F(t,T,U) = \frac{1}{U-T}\left(\frac{\mathbb{E}_t[\exp(-\int_t^T \boxed{r_t}\, dt)]}{\mathbb{E}_t[\exp(-\int_t^U \boxed{r_t}\, dt)]} - 1\right). \tag{21.1}$$

This means that in principle market models could be described by means of a sophisticated enough model for r. This model however is of no interest, since the direct modeling of forward rates (once no arbitrage is ensured by defining suitable drifts in their dynamics through changes of numeraire) is preferable and gives the advantages described above.

Market models for CDS options are similar in spirit, and in a way they are to intensity models what the swap models are to short rate models in the light of what we just said above.

$$\begin{array}{ccccc} \text{Fixed Income} & r & \text{vs} & S_{\alpha,\beta} \\ \updownarrow & \updownarrow & & \updownarrow \\ \text{Credit} & \lambda & \text{vs} & R_{a,b} \end{array}$$

But to explain this more clearly, we need to define CDS forward rates $R_{a,b}$ and thus give an idea of what CDS's themselves (Credit Default Swaps) are.

Credit Default Swaps are basic protection contracts that became quite liquid in the last few years. CDS's are now actively traded and have become a sort of basic product of the single-name credit derivatives area, analogously to interest-rate swaps and FRA's being basic products in the interest-rate derivatives world.

As a consequence, the need is no longer to have a model to be used to value CDS's, but rather to consider a model that can be *calibrated* to CDS's, i.e. to take CDS's as inputs, in order to price more complex credit derivatives.

CDS options on single names are not liquid yet, but the interest for these products is growing in the market. We may expect models will have to incorporate CDS options rather than price them in a near future, similarly to what happened to CDS themselves.

A CDS contract ensures protection against default. Two companies "A" (Protection buyer) and "B" (Protection seller) agree on the following.

If a third company "C" (Reference Credit) defaults at time $\tau = \tau_C$, with $T_a < \tau < T_b$, "B" pays to "A" a certain (deterministic) cash amount LGD (Loss Given Default of the reference credit "C"). In turn, "A" pays to "B" a rate R at times $T_{a+1}, \ldots, T_b$ or until default. Set $\alpha_i = T_i - T_{i-1}$ and $T_0 = 0$.

Protection Seller B				Protection Buyer A
	$\rightarrow$	protection LGD at default τ_C if $T_a < \tau_C \leq T_b$	$\rightarrow$	
	$\leftarrow$	rate R at $T_{a+1}, \ldots, T_b$ or until default τ_C	$\leftarrow$	

(protection leg and premium leg respectively). The cash amount LGD is a *protection* for "A" in case "C" defaults. Typically LGD = notional, or "notional - recovery" $= 1 -$ REC.

A typical stylized case occurs when "A" has bought a corporate bond issued by "C" and is waiting for the coupons and final notional payment from "C": If "C" defaults before the corporate bond maturity, "A" does not receive such payments. "A" then goes to "B" and buys some protection against this risk, asking "B" a payment that roughly amounts to the loss on the bond (e.g. notional minus deterministic recovery) that A would face in case "C" defaults.

Usually, at evaluation time (say t) the amount R is set at a value $R_{a,b}(t)$ that makes the contract fair, i.e. such that the present value of the two exchanged flows is zero. This is how the market quotes CDS's: CDS are quoted via their fair R's (Bid and Ask). This is analogous to the forward swap rate $S_{\alpha,\beta}(t)$ being the rate that sets to zero the value at time t of an interest rate swap first resetting in T_α and exchanging payments in $T_{\alpha+1}, \ldots, T_\beta$.

Market models then consider the joint dynamics of CDS forward rates $dR_{a,b}(t)$ with different a, b directly, without bothering with the underlying stochastic intensity model $d\lambda_t$ that could be responsible for such dynamics. No arbitrage conditions are satisfied if these dynamics are derived according to suitable change of numeraire techniques, as we will do. This is similar to modeling $dS_{\alpha,\beta}(t)$ without bothering about the dr_t model that could lead to such a dynamics.

$$\begin{array}{rllll} \text{Fixed Income:} & \text{Modeling} & dS_{\alpha,\beta} & \text{without} & dr \\ & & \updownarrow & & \updownarrow \\ \text{Credit:} & \text{Modeling} & dR_{a,b} & \text{without} & d\lambda \end{array}$$

We will see that if we choose again a geometric Brownian motion for dR under a suitable (and tricky non-vanishing) numeraire we obtain Black's formula for CDS options. Then we will see a few examples of related implied volatilities, and they will turn out to be huge when compared to swaptions volatilities. Finally, given the generality of the martingale dynamics, smile models will be easily introduced when needed. We also point out that these market models allow for the valuation of Constant Maturity CDS in a rather natural way that bears a lot of resemblance to constant maturity swap valuation under the LIBOR market model, seen in Section 13.7.

21.1.3 Firm Value (or Structural) Models

Structural models are based on the work by Merton (1974), in which a firm life is linked to its ability to pay back its debt. Let us suppose that a firm issues a bond to finance its activities and also that this bond has maturity T. At final time T, if the firm is not able to reimburse all the bondholders we can say that there has been a default event. In this context default may occur only at final time T and is triggered by the value of the firm being below the debt level. In a more realistic and sophisticated structural model (Black and Cox (BC) (1976), part of the family of *first passage time models*) default can happen also before maturity T. In first passage time models the default time is the first instant where the firm value hits from above either a deterministic (possibly time varying) or a stochastic barrier, ideally associated with safety covenants forcing the firm to early bankruptcy in case of important credit deterioration. In this sense the firm value is seen as a generic asset and these models use the same mathematics of barrier options pricing models.

More in detail, the fundamental hypothesis of the standard structural models is that the underlying process is a Geometric Brownian Motion (GBM), which is also the kind of process commonly used for equity stocks in the Black Scholes model. Classical structural models (Merton, Black Cox) postulate a GBM (Black and Scholes) lognormal dynamics for the value of the firm V. This lognormality assumption is considered to be acceptable. Crouhy et al (2000) report that "this assumption [lognormal V] is quite robust and, according to KMVs own empirical studies, actual data conform quite well to this hypothesis.".

In these models the value of the firm V is the sum of the firm equity value S and of the firm debt value D. The firm equity value S, in particular, can be seen as a kind of (vanilla or barrier-like) option on the value of the firm V. Merton typically assumes a zero-coupon debt at a terminal maturity T. Black Cox assume, besides a possible zero coupon debt, safety covenants forcing the firm to declare bankruptcy and pay back its debt with what is left as soon as the value of the firm itself goes below a "safety level" barrier. This is what introduces the barrier option technology in structural models for default.

In Merton's model there is a debt maturity T, a debt face value L and the company defaults at final maturity (and only then) if the value of the firm V_T is below the debt L to be paid.

The debt value at time $t < T$ is thus

$$D_t = \mathbb{E}_t[D(t,T)\min(V_T, L)] = \mathbb{E}_t[D(t,T)[V_T - (V_T - L)^+]] =$$

$$= \mathbb{E}_t[D(t,T)[L - (L - V_T)^+]] = P(t,T)L - \text{Put}(t,T;V_t,L)$$

where Put(time, maturity, underlying, strike) is a put option price and one assumes deterministic interest rates ($D(t,T) = P(t,T)$).

The equity value can be derived as a difference between the value of the firm and the debt:

$$S_t = V_t - D_t = V_t - P(t,T)L + \text{Put}(t,T;V_t,L) = \text{Call}(t,T;V_t,L)$$

(by put-call parity) so that, as is well known, in Merton's model the equity can be interpreted as a call option on the value of the firm.

Let us now move to the Black Cox (BC) model. In this model we have safety covenants in place, in that the firm is forced to reimburse its debt as soon as its value V_t hits a low enough "safety level" $H(t)$. Assuming a debt face value of L at final maturity T as before, an obvious candidate for this "safety level" is the final debt present value discounted back at time t, i.e. $LP(t,T)$. However, one may want to cut some slack to the counterparty, giving it some time to recover even if the level goes below $LP(t,T)$, and the "safety level" can be chosen to be lower than $LP(t,T)$.

In any case, once the barrier is chosen, the price of a zero coupon corporate bond with maturity $T_1 < T$ is the risk neutral expectation of a final payoff at T_1 that is one in all scenarios where the barrier has not been touched (no early default) and zero (or a recovery amount) in all scenarios where the barrier has been touched. Clearly, as said before, this is a barrier option pricing problem. First passage time models make use of barrier options techniques. Notice also the different nature of the default time in a stylized case: Now τ can be defined as

$$\inf\{t \geq 0 : V_t \leq H(t)\}$$

if this quantity is smaller than the final debt maturity T, and by T in the other case if further $V_T < L$. In all other cases there is no default. Notice that the "inf" quantity is the first time V hits the barrier H, hence the term "first passage models".

For a summary of the literature on structural models, possibly with stochastic interest rates and default barriers, we refer for example to Chapter 3 of Bielecki and Rutkowski (2001). It is important to notice that structural models make some implicit but important assumptions: As we have just seen, they assume that the firm value follows a random process similar to the one used to describe generic stocks in equity markets, and that it is possible to observe this value at any time. This assumption is often debated, but in basic structural models it is usually maintained. Therefore, unlike intensity models, here the default process can be completely monitored based on default free market information and comes less as a surprise. Also, structural models in their basic formulations and with standard barriers (Merton, BC) have few parameters in their dynamics and cannot be calibrated exactly to structured data such as CDS quotes along different maturities. Brigo and Tarenghi (2004, 2005) have extended the Black-Cox first passage model first by means of a time-varying volatility and curved barrier (2004), and then further by random barrier and volatility scenarios (2005). Their approach maintains tractability and they calibrate the term structure of CDS rates, showing also a calibration case study based on Parmalat and Vodafone data. Results are refined with different parameterizations in Brigo and Morini (2006).

These developments and a detailed study of structural models would almost make a book in themselves, especially if considering the comparison with intensity models. We do not treat structural models in this book, since these models depart from the interest rate analogies we have illustrated for intensity models and their presence would be somehow less natural. The cover of the book owes a little to structural models, though.

21.1.4 Further Models

We should also say that the picture given here is a simple version of the credit models area: Hybrid models mixing structural and intensity characteristics have been introduced, although we will not go through them here.

Jump diffusion models are also proposed in lieu of the diffusions we used above. A jump diffusion dynamics may be considered both in the structural and reduced form frameworks. We will present an example of jump diffusion intensity models in the next chapter.

21.1.5 The Multi-name picture: FtD, CDO and Copula Functions

More importantly, when dealing in multi name situations with credit derivatives, the notion of dependence needs to be generalized, simple correlation being not suited to model dependence among variables that are not jointly instantaneous Gaussian shocks (Brownian motions). Copula functions are then used in conjunction with intensity models to deal with this task, and that is another area that departs considerably from what we studied in this book and that we will not address in detail here. We only hint at copula functions in this guided tour as a matter of "general culture", since they became such a popular topic in the last years that some readers might find themselves at a loss if knowing nothing about copulas.

The reader who is not interested in multi-name credit derivatives or in copula functions in particular may skip the remaining part of this section.

Before delving into the modeling issues and in copula functions, let us see at least two possible multi-name credit derivative payoffs requiring multi-name credit models.

We may start with a first to default basket. This contract is similar to a CDS but involves more names.

21.1.6 First to Default (FtD) Basket.

Consider a First to Default contract. There are n reference credits. Let τ_i be the default time of reference credit i, and let τ^i be the default time of the i-th name that defaults.

τ^i depends on the trajectories. If in a trajectory ω first defaults name 3, second name 5 and third name 2, we have

$$\tau^1(\omega) = \tau_3(\omega), \quad \tau^2(\omega) = \tau_5(\omega), \quad \tau^3(\omega) = \tau_2(\omega).$$

In a different trajectory the order can be different.

A first to default (FtD) is similar to a CDS on a single name but this time the default that calls for the payment of protection and for the end of the premium leg is the first default in the underlying basket of reference credits.

Two companies "A" (Protection buyer) and "B" (Protection seller) agree on the following. If the first defaulting company among the underlying names $1, 2, ..., n$ (reference credits) defaults at time $\tau^1(\omega) = \tau_{i_1(\omega)}(\omega) = \min(\tau_1(\omega), ..., \tau_n(\omega))$, with $T_a < \tau^1 < T_b$, "B" pays to "A" a certain cash amount $\text{L}_{\text{GD}_{i_1}}$. In turn, "A" pays to "B" a rate R at times $T_{a+1}, \ldots, T_b$ or until default τ^1.

Protection Seller B	$\rightarrow$ $\text{L}_{\text{GD}_{i_1}}$ at 1st default $\tau^1 = \tau_{i_1}$ if in $[T_a, T_b]$ $\rightarrow$ $\leftarrow$ rate R at $T_{a+1}, \ldots, T_b$ or until 1st default τ^1 $\leftarrow$	Protection Buyer A

(protection leg and premium leg respectively). The cash amount LGD_{i_1} is a *protection* for "A" in case the first name defaults before T_b and is name i_1. Typically LGD = notional, or "notional - recovery" $= 1 - \text{REC}$. Notice that i_1 is a random variable. In standard FtD, written by sectors, all names have the same REC (0.5 for financial and 0.4 for telecoms for example) and LGD is always the same.

Usually the FtD's are composed by a small number of reference entities: A typical size for a FtD is of 5-6 underlying names.

As we have already seen, a FtD contract works similarly to the standard CDS, with the main difference that the protection is paid against the first reference entity in the basket experiencing default. This way the protection seller assumes more risk with respect to selling protection on a single name.

This fact leads to higher rates paid from the protection buyer. The protection seller is attracted by the leverage obtained investing in such a structure. In case of a default, the protection seller is due to make a single payment relative to a single reference entity, but it receives a larger rate before default than with a single CDS, due to the higher riskiness of the strategy.

From the protection buyer viewpoint, a FtD is seen as a lower cost method of partially hedging multiple credits. However, since the protection is paid only in the case of one default (the first-to-default), the buyer keeps the risk of multiple defaults.

The last considerations lead to the following: The premium paid by the protection buyer depends not only on the probability of default of each name in the basket but also on the **default dependence** among these names.

The premium paid is essentially the premium paid for a single name protection **plus** a part due to the likelihood of multiple defaults.

In case of perfect dependence, i.e. all reference credits defaulting under the same triggering ξ, obviously the premium paid for the FtD basket would be equal to the premium paid for the riskiest name (the correlation is usually meant not directly on the default times but on the variables ξ triggering default in the reduced form model; we will go deeply into this topic later, when discussing copula functions. Now we only give a feeling for this).

Indeed, assume positive deterministic intensities λ_i, cumulated intensities Λ_i and defaults of each name i, τ_i, as first jump of a related Poisson process:

$$\tau_1 = \Lambda_1^{-1}(\xi_1), \ldots, \tau_n = \Lambda_n^{-1}(\xi_n).$$

If the ξ are all equal, the smallest τ_i (i.e. the first default) is the one corresponding to the largest Λ_i, i.e. to the riskiest name as far as credit spreads are concerned. Then the related premium rate R will be the rate of the corresponding CDS.

At the same time the premium paid for a FtD has to be lower than the sum of the single different premia of the basket components, since the protection is paid in case of a single default. Indeed, the sum of the CDS rates R_i corresponds to protection against all of the names. As such, this includes in particular protection against the first name, and then the premium rate of the sum of CDS has to be larger than the premium rate in the FtD.

Notice that the payoff depends on the joint distribution of default times (and then on their dependence) through the following term:

$$\tau^1 = \min(\tau_1, \tau_2, \ldots, \tau_n).$$

As before, the price of this product, in principle, can be computed by risk neutral expectation, requiring the joint distribution of the default times (and thus their dependence).

We now move to the second example of fundamental multiname credit derivative.

21.1.7 Collateralized Debt Obligation (CDO) Tranches.

Collateralized Debt Obligations (CDO) are products that are related to the loss distribution of a pool of names. Synthetic CDO's with maturity T are obtained by putting together a collection of CDS with the same maturity on different names, $1, 2, ..., n$, and then "tranching" the total loss

$$\text{Loss}(T) = \sum_{i=1}^{n} \text{Lgd}_i 1_{\{\tau_i \leq T\}} = \sum_{i=1}^{n} (1 - \text{Rec}_i) 1_{\{\tau_i \leq T\}}$$

associated with the pool along two **attachment points** A and B $(A < B)$.

The indicator $1_{\{\tau_i \leq T\}}$ is 1 in scenarios where the i-th name has defaulted before T and 0 otherwise, so that $\text{Loss}(T)$ is the sum of the losses due to all names that have defaulted before or at T.

In formula, the tranched loss at time t is

$$\text{Loss}^{tr}_{A,B}(t) := \frac{1}{B-A} \left[(\text{Loss}(t) - A) 1_{\{A < \text{Loss}(t) \leq B\}} + (B - A) 1_{\{\text{Loss}(t) > B\}} \right]$$

If the total loss is below A, there is no loss in the tranche. If the total loss is between A and B, then the total loss minus A (re-scaled by the tranche thickness $B - A$) is the loss of the tranche. If the total loss exceeds B, the tranche loss stops at $B - A$ re-scaled by the thickness, i.e. to one.

The process is actually a dynamic one. Once enough names have defaulted and the loss has reached A, each time the tranched loss increases the corresponding partial loss payment is made to the protection buyer, until maturity arrives or until the total loss exceeds B (protection leg). In the other leg, a premium rate is paid periodically on the outstanding notional of the tranche, that is exhausted after the total loss exceeds B.

More specifically, the general flows of payment between two counterparties entering a CDO tranche contract is represented as follows

Prot Seller	$\rightarrow$ Prot. $d\text{Loss}^{tr}_{A,B}(\tau^i)$ at the i-th default τ^i if $T_a < \tau^i \le T_b$ $\rightarrow$ $\leftarrow$ rate R at $T_{a+1},\dots,T_b$ on the outstanding notional $\leftarrow$	Prot Buyer

where

$$d\text{Loss}^{tr}_{A,B}(\tau^i) = \frac{1}{B-A}\left[\min(B,\text{Loss}(\tau^i)) - \max(A,\text{Loss}(\tau^{i-1}))\right]^+$$

is the discrete protection payment made only in correspondence of some default (this is the *protection leg*). Notice: $d\text{Loss}^{tr}_{A,B}(t) = 0$ if $t \notin \{\tau_1,\tau_2,\dots,\tau_n\}$. In this sense a better notation would be $\Delta\text{Loss}^{tr}_{A,B}(t)$, since $\text{Loss}^{tr}_{A,B}$ changes only by finite jumps, although in the literature the "d" notation is used in connection with integrals in the Loss.

On the other side the protection buyer pays a premium rate R on the following outstanding notional of the tranche (*premium leg*),

$$\text{Outst}^{tr}_{A,B}(t) = 1 - \text{Loss}^{tr}_{A,B}(t)$$

This time the rate is paid **on the "survived" positive (re-scaled) notional at the relevant payment time** and not on the whole notional.

Clearly the payoff is completely determined once the default times

$$\tau_1,\tau_2,\dots,\tau_n$$

of all names, determining their "order statistics" $\tau^1,\dots,\tau^n$, are given. Since dependence on these times $\tau_1,\tau_2,\dots,\tau_n$ through their "order statistics" is nonlinear, their joint distribution, and thus their dependence, is needed for risk-neutral valuation of the CDO tranche.

21.1.8 Where can we introduce dependence?

Where does dependence enter these FtD and CDO payoffs exactly? Dependence is introduced across the default times $\tau_1,\tau_2,\tau_3...$ of different names entering the above payoffs as follows.

We have seen above that in reduced form models, transforming the default time τ by its cumulated intensity $\Lambda(t) = \int_0^t \lambda(s)ds$ leads to an exponential random variable independent of any default-free quantity:

$$\Lambda(\tau) = \xi \sim \text{exponential},\ \ \mathcal{F}-\text{independent}.$$

If we assume λ to be positive, we may define τ as

$$\tau = \Lambda^{-1}(\xi)$$

If we have several names $1,2,...,n$, we may define dependence between the default times

$$\tau_1 = \Lambda_1^{-1}(\xi_1),\dots,\tau_n = \Lambda_n^{-1}(\xi_n)$$

essentially in three ways.

1. Put dependence in (stochastic) intensities of the different names and keep the ξ of the different names independent;
2. Put dependence among the ξ of the different names and keep the (stochastic or trivially deterministic) intensities λ_i independent;
3. Put dependence both among (stochastic) intensities λ of the different names and among the ξ of the different names;

Let us give a look at these three possibilities with more detail.

With choice 1), one may induce dependence among the $\lambda_i(t)$ by taking diffusion dynamics for each of them and correlating the Brownian motions:

$$d\lambda_i(t) = \mu_i(t, \lambda_i(t))dt + \sigma_i(t, \lambda_i(t))dW_i(t),$$

$$d\lambda_j(t) = \mu_j(t, \lambda_j(t))dt + \sigma_j(t, \lambda_j(t))dW_j(t),$$

$$dW_i dW_j = \rho_{i,j}dt, \quad \xi_i, \xi_j \text{independent}$$

The advantages with this choice are possible partial tractability and ease of implementation. Also, the default of one name does not affect the intensity of other names. The correlation can be estimated historically from time series of credit spreads and inserted into the model. Also, with stochastic intensity we may model correlation between interest rates and credit spreads, that is considered to be an important feature in some situations.

The disadvantages consist in a non realistic (too low) level of dependence across default events $1_{\{\tau_i<T\}}, 1_{\{\tau_j<T\}}$. See for example Jouanin et al. (2001).

With choice 2) we meet the framework that is currently used for quoting implied correlation (in index tranches) in the market.

Advantages: We can take even deterministic intensities, which makes life easier for stripping single name default probabilities. We can reproduce realistically large levels of dependence across default times of different names by putting dependence structures (called "copula functions", more on this later) on the ξ's.

Disadvantages: There is no natural and feasible historical source for estimating the copula, that is often calibrated by means of dubious considerations. Furthermore, default of one name affects the intensity of other names through partial derivatives of the imposed copula, see for example Schönbucher and Schubert (2001). In case of deterministic intensities this approach ignores credit spreads volatilities, which can be rather large (besides ignoring credit spread correlations in general).

Choice 3) leads to the most complicated framework.

The advantages of this third solution are that it takes into account possible credit spread volatility, and can produce a sufficient amount of dependence among default times.

Disadvantages: there is no natural and feasible historical source for estimating the copula, that is often calibrated by means of dubious considerations. Moreover, as before, default of one name affects the intensity of other names through partial derivatives of the copula. Calculations are quite complicated, due to the presence of stochasticity both in the intensities λ_i's and in the ξ's.

21.1.9 Copula Functions.

Let us focus now on choice 2, involving copula functions. What are copula functions and why are they introduced?

It is well known that linear correlation is not enough to express the dependence between two random variables in an efficient way, in general.

Example: take X standard Gaussian and take $Y = X^3$. Y is a deterministic one-to-one transformation of X, so that the two variables give exactly the same information and should have maximum dependence. However, if we take the linear correlation between X and Y we easily get

$$(E(X^3X) - E(X^3)E(X))/(\text{Std}(X^3)\text{Std}(X)) = 3/\sqrt{15} = \sqrt{3}/\sqrt{5} < 1.$$

We obtain a dependence measure that is smaller than 1 (1 corresponds to maximum dependence).

So *correlation is not a good measure of dependence in this case.*

In standard financial models this problem with correlation as a dependence measure is usually absent because in standard modeling we are concerned with dependence between instantaneous Brownian shocks, which are jointly Gaussian. *Correlation works well for jointly Gaussian variables*, so that as long as we are concerned with instantaneous correlations in jointly Gaussian shocks we do not need to generalize our notion of dependence.

In credit derivatives with intensity models we may find ourselves in the situation where we need to introduce dependence between the *exponential* components $\xi = \Lambda(\tau)$ of Poisson processes for different names. This is usually done by means of copula functions.

Let us see how we arrive at the definition of copula function. A fundamental fact about transformation of random variables is the following. Given a random variable X, we may transform it in several ways through a deterministic function: $2X$, X^5, $\exp(X)$,... A particularly interesting transformation function is the cumulative distribution function F_X of X. Set $U = F_X(X)$ and assume for simplicity F_X to be invertible (say strictly increasing and continuous). Let us compute F_U in a generic point $u \in [0, 1]$.

$$\begin{aligned} F_U(u) &= \mathbb{Q}(U \leq u) = \mathbb{Q}(F_X(X) \leq u) = \mathbb{Q}(F_X(X) \leq F_X(F_X^{-1}(u))) \\ &= \mathbb{Q}(X \leq F_X^{-1}(u))) = F_X(F_X^{-1}(u)) = u. \end{aligned}$$

However, the identity distribution function $F_U(u) = u$ is characteristic of a uniform random variable in $[0\ 1]$. This means that $U = F_X(X)$ is a **uniform** random variable. Notice also that since F_X is one to one, U contains the same information as X.

The idea then is to transform all random variables X by their F_X obtaining all uniform variables that contain the same information as the starting X. This way we rid ourselves of marginal distributions, obtaining only uniform random variables, and can concentrate on introducing dependence directly for these standardized uniforms.

Indeed, let $(U_1, ..., U_n)$ be a random vector with uniform margins and joint distribution $C(u_1, ..., u_n)$. $C(u_1, ..., u_n)$ is the *copula* of the random vector. It can be characterized by a number of properties that we do not repeat here. See for example Nelsen (1999), Joe (1997), Cherubini et al (2004), or Embrechts et al (2001).

An important result is given by *Sklar's theorem*: Let H be an n-dimensional distribution function with margins $F_1, ..., F_n$. Then there exists an n-copula C such that for all $\mathbf{x}$ in $\bar{\mathbb{R}}^n$,

$$H(x_1, ..., x_n) = C(F_1(x_1), ..., F_n(x_n)). \tag{21.2}$$

This result tells us that what we hinted at previously works. Indeed, one would intuitively write

$$\begin{aligned} H(x_1, ..., x_n) &= \mathbb{Q}(X_1 \le x_1, \dots, X_n \le x_n) \\ &= \mathbb{Q}(F_1(X_1) \le F_1(x_1), ..., F_n(X_n) \le F_n(x_n)) \\ &= \mathbb{Q}(U_1 \le F_1(x_1), ..., U_n \le F_n(x_n)) = C(F_1(x_1), ..., F_n(x_n)) \end{aligned}$$

where C is the joint distribution function of uniforms $U_1, ..., U_n$.

Sklar's theorem: For any joint distribution function $H(x_1, ..., x_n)$ with margins $F_1, ..., F_n$ there exists a copula function $C(u_1, ..., u_n)$ (i.e. a joint distribution function on n uniforms) such that

$$H(x_1, ..., x_n) = C(F_1(x_1), ..., F_n(x_n)).$$

Notice that C contains the pure dependence information. Notice the important point: Correlation between two variables is just a number, whereas a copula function between two variables is a two dimensional function.

We may also write

$$C(u_1, ..., u_n) = H(F_1^{-1}(u_1), ..., F_n^{-1}(u_n)). \tag{21.3}$$

from which we see that we may use any known joint distribution H function to define a copula C.

Consider again our example with X standard Gaussian and $Y = X^3$, $Z = X$. The copula between X and Z is the copula expressing maximum dependence (also correlation works in this case: $\mathrm{Corr}(X, Z) = 1$). This copula is the joint distribution of $U_1 = F_X(X)$ and $U_2 = F_Z(Z) = F_X(X) = U_1$,

$$\mathbb{Q}(U_1 < u_1, U_1 < u_2) = \mathbb{Q}(U_1 < \min(u_1, u_2)) = \min(u_1, u_2).$$

So this "min" copula corresponds to maximum dependence. Now consider $Y = X^3$ and the dependence between X and Y. We have seen above that linear correlation fails to be an efficient measure of dependence in this case. Call $U_3 = F_Y(Y)$. Notice that

$$F_{X^3}(x^3) = \mathbb{Q}(X^3 \leq x^3) = \mathbb{Q}(X \leq x) = F_X(x) \quad \text{for all} \quad x,$$

so that in particular

$$U_3 = F_Y(Y) = \boxed{F_{X^3}(X^3) = F_X(X)} = U_1.$$

Consider the copula between X and Y. Since $U_3 = U_1$, this copula is

$$\mathbb{Q}(U_1 < u_1, U_3 < u_2) = \mathbb{Q}(U_1 < u_1, U_1 < u_2) = \min(u_1, u_2),$$

the same as before. So with copulas also X and X^3 get maximum dependence, as should be.

This example actually has a more general version: if $g_1, ..., g_n$ are (say strictly increasing) one-to-one transformations, then the copula of some given $X_1, ..., X_n$ is the same as the copula for $g_1(X_1), ..., g_n(X_n)$ (not so for correlation). So *the copula is invariant for deterministic transformations that preserve the information.* This tells us again that copulas are really expressing the core of dependence.

Also, it can be proved that every copula C is bounded between two functions C^+ and C^-, which are known as the Fréchet-Hoeffding bounds:

$$C(u_1, u_2, \ldots, u_n)^- \leq C(u_1, u_2, \ldots, u_n) \leq C(u_1, u_2, \ldots, u_n)^+$$

where

$$C(u_1, u_2, \ldots, u_n)^- = \max(u_1 + u_2 + \ldots + u_n + 1 - n, 0)$$

and

$$C(u_1, u_2, \ldots, u_n)^+ = \min(u_1, u_2, \ldots, u_n)$$

as in our example above. While C^+ is a copula in general, C^- is a copula only in dimension 2. We can define also an "orthogonal" copula $C^\perp$ corresponding to independent variables:

$$C(u_1, u_2, \ldots, u_n)^\perp = u_1 \cdot u_2 \cdot \ldots \cdot u_n$$

Then C^+ is the copula corresponding to the maximum dependence and C^- is (in dimension 2) the copula corresponding to the maximum negative dependence. $C^\perp$ corresponds to perfect independence between two variables.

We also recall the notion of *survival copula*: this is defined as

$$\mathbb{P}[X_1 > x_1, \ldots, X_n > x_n] = \breve{C}(\bar{F}_1(x_1), \ldots, \bar{F}_n(x_n))$$

where the $\bar{F}$'s are the margins survival functions (i.e. for example $\bar{F}_1(x_1) = \mathbb{Q}(X_1 > x_1) = 1 - F_1(x_1)$). The survival copula is not linked to the copula in a simple way: It can be proved that in two dimensions the following relation holds:

$$\breve{C}(u,v) = u + v - 1 + C(1-u, 1-v).$$

In general, if one is able to compute survival copulas from the original copula, one obtains yet one more family of copulas for each given copula family (except with the Frank family in dimension two, as we hint at below).

Before starting to introduce the most important families of copulas, let us define the concept of tail dependence.

The concept of tail dependence relates to the amount of dependence in the upper-right quadrant tail or lower-left-quadrant tail of a bivariate distribution. It is a concept that is relevant for the study of dependence between extreme values. Roughly speaking, it is the idea of "fat tails" for the dependence structure.

It turns out that tail dependence between two continuous random variables X and Y is a copula property and hence the amount of tail dependence is invariant under strictly increasing transformations of X and Y.

Let (X,Y) be a pair of continuous random variables with marginal distribution functions F_X and F_Y. The coefficient of upper tail dependence of (X,Y) is

$$\lim_{u \uparrow 1} \mathbb{Q}\{Y > F_Y^{-1}(u) | X > F_X^{-1}(u)\} = \lambda_U$$

provided that the limit $\lambda_U \in [0,1]$ exists. If $\lambda_U \in (0,1]$, X and Y are said to be asymptotically dependent in the upper tail; if $\lambda_U = 0$, X and Y are said to be asymptotically independent in the upper tail.

Since $\mathbb{Q}\{Y > F_Y^{-1}(u) | X > F_X^{-1}(u)\}$ can be rewritten as

$$\frac{1 - \mathbb{Q}\{X \le F_X^{-1}(u)\} - \mathbb{Q}\{Y \le F_Y^{-1}(u)\} + \mathbb{Q}\{X \le F_X^{-1}(u), Y \le F_Y^{-1}(u)\}}{1 - \mathbb{Q}\{X \le F_X^{-1}(u)\}}$$

an alternative and equivalent definition (for continuous random variables), from which it is seen that the concept of tail dependence is indeed a copula property, is the following:

$$\lim_{u \uparrow 1} (1 - 2u + C(u,u))/(1-u) = \lambda_U.$$

A more compact characterization of upper tail dependence can be given in terms of the survival copula:

$$\lim_{v \downarrow 0} \breve{C}(v,v)/v = \lambda_U.$$

The concept of lower tail dependence can be defined in a similar way. If the limit

$$\lim_{u \downarrow 0} \mathbb{Q}\{Y \le F_Y^{-1}(u) | X \le F_X^{-1}(u)\} = \lim_{u \downarrow 0} C(u,u)/u = \lambda_L$$

exists, then C has lower tail dependence if $\lambda_L \in (0,1]$, and lower tail independence if $\lambda_L = 0$.

The canonical copula is the Gaussian (or Normal) copula.

"È normale..." [It's normal...]

Aurélien Alfonsi, commenting on his derivation of a particularly difficult and counterintuitive result, after two beers at "la Scrofa Semi-lanuta", Milan, December 2002.

The Gaussian copula is obtained by using a multivariate n-dimensional normal distribution Φ_R^n with standard Gaussian margins and correlation matrix R as multivariate distribution H:

$$C_{\mathcal{N}(R)}(u_1,...,u_n) = \Phi_R^n(\Phi^{-1}(u_1),...,\Phi^{-1}(u_n)) \tag{21.4}$$

where Φ^{-1} is the inverse of the usual standard normal cumulative distribution function. Unfortunately this copula cannot be expressed in closed form. Indeed, in the 2-dimensional case we have:

$$C_{\mathcal{N}(R)}(u,v) = \int_{-\infty}^{\Phi^{-1}(u)} \int_{-\infty}^{\Phi^{-1}(v)} \frac{1}{2\pi(1-\rho^2)^{1/2}} \exp\left\{-\frac{s^2 - 2\rho st + t^2}{2(1-\rho^2)}\right\} ds\, dt, \tag{21.5}$$

ρ being the (only) correlation parameter in the 2×2 matrix R. Notice that in case we are modeling the dependence among n names, the correlation matrix R in principle has $n(n-1)/2$ free parameters.

Now some properties of the Gaussian copula for $\rho \in (-1, 1)$:

- Neither upper nor lower tail dependence;
- $C(u,v) = C(v,u)$ i.e. *exchangeable copula.*

Let us now move to **Archimedean copulas**. These are an important class of copulas with an important feature: they can be expressed in closed form. In general Archimedean copulas arise from a particular function φ called the *generator* of the copula. In particular, if $\varphi : [0,1] \to [0,\infty)$ is a continuous, strictly decreasing function such that $\varphi(1) = 0$, then

$$C(u,v) = \varphi^{[-1]}(\varphi(u) + \varphi(v)) \tag{21.6}$$

is a copula if and only if φ is convex (in other words it must be $\varphi' < 0$ and $\varphi'' > 0$ under differentiability). We recall that $\varphi^{[-1]}$ is the *pseudo-inverse* of φ defined as:

$$\varphi^{[-1]}(t) = \begin{cases} \varphi^{-1}(t) & 0 \le t \le \varphi(0) \\ 0 & t > \varphi(0) \end{cases}.$$

If $\lim_{t\to 0} \varphi(t) = +\infty$ we say that φ is a *strict* generator and the copula is said to be a *strict* copula.

It happens that even if Archimedean copulas are known in closed form, they are difficult to simulate. On the contrary Gaussian copulas are not known in closed form but are easier to simulate.

According to the particular generator used, we have different families of copula functions. We give examples in dimension 2 but they can be easily generalized. Now let us see a few particular cases of archimedean copulas.

Clayton family. Let us choose $\varphi(t) = (t^{-\theta}-1)/\theta$ where $\theta \in [-1,\infty)\backslash\{0\}$. Then the Clayton family is:

$$C_\theta(u,v) = \max([u^{-\theta} + v^{-\theta} - 1], 0)^{-1/\theta}. \tag{21.7}$$

If $\theta > 0$ the copulas are strict and the copula expression simplifies to

$$C_\theta(u,v) = (u^{-\theta} + v^{-\theta} - 1)^{-1/\theta}. \tag{21.8}$$

The Clayton copula has lower tail dependence for $\theta > 0$, and $C_{-1} = C^-$, $\lim_{\theta\to 0} C_\theta = C^\perp$ and $\lim_{\theta\to\infty} C_\theta = C^+$.

Frank family.

Let us choose $\varphi(t) = -\ln \frac{e^{-\theta t}-1}{e^{-\theta}-1}$, where $\theta \in \mathbb{R}\backslash\{0\}$. This originates the Frank family

$$C_\theta(u,v) = -\frac{1}{\theta}\ln\left(1 + \frac{(e^{-\theta u}-1)(e^{-\theta v}-1)}{e^{-\theta}-1}\right). \tag{21.9}$$

The Frank copulas are strict Archimedean copulas. Furthermore

$$\lim_{\theta\to-\infty} C_\theta = C^-, \quad \lim_{\theta\to\infty} C_\theta = C^+, \quad \text{and} \lim_{\theta\to 0} C_\theta = C^\perp.$$

The members of the Frank family have no tail dependence (neither upper nor lower).

The members of the Frank family are the only Archimedean copulas which satisfy the equation $C(u,v) = \breve{C}(u,v)$ (see for example Embrechts et al (2001)).

Gumbel family. Let us choose $\varphi(t) = (-\ln t)^\theta$, where $\theta \geq 1$. This gives the Gumbel family

$$C_\theta(u,v) = \exp(-[(-\ln u)^\theta + (-\ln v)^\theta]^{1/\theta}). \tag{21.10}$$

The Gumbel copulas are strict Archimedean copulas. Furthermore

$$\lim_{\theta\to\infty} C_\theta = C^+, \; C_1 = C^\perp.$$

Gumbel copulas describe **only positive dependence** between random variables; moreover they feature upper tail dependence.

We now leave the realm of Archimedean copulas and move to some important families of non-archimedean copulas.

t-Copulas. If the vector $\mathbf{X}$ of random variables has the stochastic representation $\mathbf{X} \sim \mu + \frac{\sqrt{\nu}}{\sqrt{S}}\mathbf{Z}$ where $\mu \in \mathbb{R}^n$, ν is a positive integer, $S \simeq \chi^2_\nu$ and $\mathbf{Z} \simeq \mathcal{N}(\mathbf{0}, \Sigma)$ are independent, where Σ is an $n \times n$ covariance matrix, then

$\mathbf{X}$ has an n-variate t_ν-distribution with mean μ (for $\nu > 1$) and covariance matrix $\frac{\nu}{\nu-2}\Sigma$ (for $\nu > 2$). If $\nu \leq 2$ then $\mathrm{Cov}(\mathbf{X})$ is not defined. In this case we just interpret Σ as being the shape parameter of the distribution of $\mathbf{X}$.

The copula of $\mathbf{X}$ defined above can be written as

$$C^t_{\nu,R}(\mathbf{u}) = t^n_{\nu,R}(t_\nu^{-1}(u_1), \ldots, t_\nu^{-1}(u_n))$$

where $R_{ij} = \Sigma_{ij}/\sqrt{\Sigma_{ii}\Sigma_{jj}}$ for $i, j \in \{1, \ldots, n\}$ and where $t^n_{\nu,R}$ denotes the distribution function of $\sqrt{\nu}\mathbf{Y}/\sqrt{S}$ where $S \simeq \chi^2_\nu$ and $\mathbf{Y} \simeq \mathcal{N}(\mathbf{0}, R)$ are independent. Here t_ν denotes the (equal) margins of $t^n_{\nu,R}$, i.e. the distribution function of $\sqrt{\nu}Y_1/\sqrt{S}$.

In the bivariate case the copula expression can be written as

$$C^t_{\nu,R}(u, v) = \int_{-\infty}^{t_\nu^{-1}(u)} \int_{-\infty}^{t_\nu^{-1}(v)} \frac{1}{2\pi(1 - R_{12}^2)^{1/2}} \left\{1 + \frac{s^2 - 2R_{12}st + t^2}{\nu(1 - R_{12}^2)}\right\}^{-\frac{\nu+2}{2}} ds\, dt$$

Note that R_{12} is simply the usual linear correlation coefficient of the corresponding bivariate t_ν-distribution if $\nu > 2$.

Summary on copula properties

In the following table we collect the properties of the different copulas considered so far.

Copula	*Positive Dependence*	*Independ*	*Negative Dependence*	*Upper Tail Dep*	*Lower Tail Dep*
Clayton $\theta \in [-1, +\infty)$ $\theta \neq 0$	$C \to C^+$ $\theta \to +\infty$	$C \to C^\perp$ $\theta \to 0$	$C = C^-$ $\theta = -1$	no	only for $\theta > 0$
Frank $\theta \in \mathbb{R}\backslash\{0\}$	$C \to C^+$ $\theta \to +\infty$	$C \to C^\perp$ $\theta \to 0$	$C \to C^-$ $\theta \to -\infty$	no	no
Gumbel $\theta \in [1, +\infty)$	$C \to C^+$ $\theta \to +\infty$	$C = C^\perp$ $\theta = 1$	no negative dependence	yes	no
Gaussian $\rho \in (-1, 1)$	$C \to C^+$ $\rho \to +1$	$C = C^\perp$ $\rho = 0$	$C \to C^-$ $\rho \to -1$	no	no
t-Copula $R_{12} \in (-1, 1)$	$C \to C^+$ $R_{12} \to +1$	$C = C^\perp$ $R_{12} = 0$	$C \to C^-$ $R_{12} \to -1$	yes	yes

We add that all the copulas we have introduced above are exchangeable. In some situation it is appropriated to have non-exchangeable copulas. We introduce one such family below. Indeed, before closing this crash course on copula functions, we introduce a recent development.

The Alfonsi-Brigo periodic copulas. This family has been introduced recently in Alfonsi and Brigo (2005). We say in the following that a copula admits a *density* when $\frac{\partial^2 C}{\partial u_1 \partial u_2} = c(u_1, u_2)$ exists in the ordinary sense, so that typically

$$C(u_1, u_2) = \int_0^{u_1} \int_0^{u_2} c(x_1, x_2) dx_1 \, dx_2.$$

Periodic copulas have a density that can be written in the form

$$c(u_1, u_2) = \tilde{c}(u_1 + u_2)$$
$$(\text{resp. } c(u_1, u_2) = \tilde{c}(u_1 - u_2))$$

for a function $\tilde{c} : \mathbb{R} \to \mathbb{R}$. To produce a copula, $\tilde{c}$ must be nonnegative and verify:

$$\int_0^{u_1} \int_0^1 \tilde{c}(x_1 \pm x_2) dx_1 dx_2 = u_1, \ \forall u_1 \in [0, 1],$$
$$\int_0^1 \int_0^{u_2} \tilde{c}(x_1 \pm x_2) dx_1 dx_2 = u_2, \ \forall u_2 \in [0, 1].$$

Differentiating with respect to u_1 and u_2 respectively, we obtain

$$\int_0^1 \tilde{c}(u_1 \pm x_2) dx_2 = 1, \quad \forall u_1 \in [0, 1],$$
$$\int_0^1 \tilde{c}(x_1 \pm u_2) dx_1 = 1, \quad \forall u_2 \in [0, 1].$$

Differentiating further the first relation, since $\int_0^1 \tilde{c}(u_1 + x_2) dx_2 = \int_{u_1}^{u_1+1} \tilde{c}(x_2) dx_2$ (resp. $\int_0^1 \tilde{c}(u_1 - x_2) dx_2 = \int_{u_1-1}^{u_1} \tilde{c}(x_2) dx_2$) , we obtain:

$$\tilde{c}(u_1 + 1) = \tilde{c}(u_1) \ \forall u_1 \in [0, 1], \quad (\text{resp. } \tilde{c}(u_1 - 1) = \tilde{c}(u_1) \ \forall u_1 \in [0, 1]).$$

Thus, a consequence of requiring $c(u_1, u_2) := \tilde{c}(u_1 \pm u_2)$ to be the density of a copula is that $\tilde{c}$ has to be 1-periodic (at least on $[-1, 2]$, but its value outside this interval is irrelevant) and that $\int_0^1 \tilde{c}(u) du = 1$. Conversely, it is easy to see that if $\tilde{c}$ is a nonnegative 1-periodic function such that $\int_0^1 \tilde{c}(u) du = 1$, then

$$\widetilde{C}^-(u_1, u_2) := \int_0^{u_1} \int_0^{u_2} \tilde{c}(x_1 + x_2) dx_1 dx_2$$
$$(\text{resp. } \widetilde{C}^+(u_1, u_2) := \int_0^{u_1} \int_0^{u_2} \tilde{c}(x_1 - x_2) dx_1 dx_2)$$

satisfies conditions leading to a copula function. The related copula is called, with a slight abuse of language, *periodic copula.* We note here that copulas obtained with these densities form a convex set since a convex combination of 1-periodic nonnegative functions satisfying $\int_0^1 \tilde{c}(u) du = 1$ is also a 1-periodic nonnegative function with integral 1 on a period. Notice further that the use of the "-" and "+" signs appears to be counterintuitive (one would exchange the above signs), but there is a reason for this that is clarified in Alfonsi and Brigo (2005).

At times, rather than characterizing copulas through their densities, it is preferable to have a direct characterization of the copula itself. To characterize periodic copulas without explicitly referring to their densities, denote by φ the primitive of $\tilde{c}$ that vanishes at 0, and set $\Phi(x) := \int_0^x \varphi(u)du$, so that Φ is a double primitive of $\tilde{c}$. We can then rewrite the above periodic copula as follows:

$$\widetilde{C}^-(u_1, u_2) = \int_0^{u_1}\int_0^{u_2} \tilde{c}(x_1 + x_2)dx_1 dx_2 = \Phi(u_1 + u_2) - \Phi(u_1) - \Phi(u_2),$$
$$\widetilde{C}^+(u_1, u_2) = \int_0^{u_1}\int_0^{u_2} \tilde{c}(x_1 - x_2)dx_1 dx_2 = \Phi(u_1) + \Phi(-u_2) - \Phi(u_1 - u_2)$$

and we see that the first copula is always exchangeable (symmetric), in that $\widetilde{C}^-(u_1, u_2) = \widetilde{C}^-(u_2, u_1)$, whereas the second one can be non symmetric if Φ is not par, i.e. if $\Phi(-x) \neq \Phi(x)$ for some x. We have thus characterized our periodic copulas in terms of double primitives Φ of periodic functions $\tilde{c}$.

The resulting copulas do not feature tail dependence, can easily range from C^- to C^+, are relatively easy to simulate and can be extended to any dimension beyond two. For further information see Alfonsi and Brigo (2005).

21.1.10 Dynamic Loss models.

To close this tour on multi-name credit derivatives and models, we notice that copula models are *static* models of dependence. There is no dynamics in the copula, and the search for dynamical dependence (or "correlation") modeling has just begun. It is important to have dependence dynamics to price, for example, CDO tranche *options.* Copula functions are just the first step in dependence modeling. An attempt in dynamical default dependence modeling is hinted at in Section 22.8.7 below, through correlated jump-diffusion intensities and independent ξ's.

More promising and direct approaches resorting to direct modeling of the loss dynamics are available. This approach is pursued in Bennani (2005), in Schönbucher (2005), in Sidenius, Piterbarg and Andersen (2005) and in Di Graziano and Rogers (2005). The general tendency considers general frameworks for modeling quantities directly related to the loss distribution of a pool of names, rather than zooming on the single defaults of the names in the pool.

A recent concrete approach that is analytically tractable and with realistic and concrete examples of simultaneous calibration to several quoted index tranches of different maturities is given for example in Brigo, Pallavicini and Torresetti (2006).

21.1.11 What data are available in the market?

It is a capital mistake to theorize before one has data. Insensibly one begins to twist facts to suit theories, instead of theories to suit facts.
Sherlock Holmes, "A Scandal in Bohemia". Quoted by Jeffrey R. Bohn in "A Survey of Contingent-Claims Approaches to Risky Debt Valuation".

After this long tour intermixing models and products, the reader might benefit from a summary on what is directly available in the market in terms of market quotes.

Concerning single-name products, involving default of single companies, the main data are defaultable (or corporate) bond prices and CDS rates. We will see both zero-coupon and coupon-bearing corporate bonds as well as corporate floaters in their prototypical forms in Sections 21.2 and 21.3.8.

However, these corporate bonds will be used more as building blocks than as concrete market instruments. Market actual corporate bonds involve a number of special features departing from the prototypical products we introduce below.

The situation is different as far as CDS are concerned. CDS are now highly standardized products focusing on default of single names. When available, liquid CDS data are preferred to bond data for this reason. The payoffs for CDS we introduce are representing the actual payoffs one finds in the market. This consistency and absence of "special features" in CDS is the reason why we will privilege CDS as sources of single-name credit data over corporate bonds. The tools we give here allow the reader to work with bonds as well, but the reader will have to go patiently through the bond details and special features himself in order to do so.

CDS are quoted through the rates (or "spreads") R in their premium legs that render them fair at inception. CDS rates on one name are usually quoted for protection extending up to one to ten years. Mostly, maturities T_b of 1, 3, 5, 7 and 10 years are quoted. These data may be used to calibrate either intensity or structural models to credit data, by finding the model parameters matching the default probabilities implicit in CDS prices to the default probabilities implied by the models themselves.

A few prices of CDS options on single names are quoted in the market as well, but bid-offer spreads are usually large and these quotes are not liquid, as we see in particular in Chapter 23. For the time being it may be safe not to include single-name CDS option data in model calibration, although this might rapidly change.

As far as multi-name credit derivatives are concerned, the quoting mechanisms are similar. The premium rate R_{index} on a standardized pool of CDS on different names (the DJiTRAXX index, involving 125 European names, and the DJCDX index, involving 125 US names are the main examples) is quoted when protecting against the whole loss of the pool of names, although some particular conventions on outstanding notionals and recovery rates are taken, which we partly illustrate with the trade example below.

But also standardized CDO tranche quotes are available in the market. Premium rates $R_{\text{index}}^{A_i,B_i}$ in the premium legs rendering the different tranches fair at inception are considered, for CDO's on standardized pools of names (again DJiTRAXX and DJCDX) with standardized attachment points (A_i, B_i) and standardized maturities T_b. The protection maturities T_b are 5y and 10y. The standardized attachment/detachment points of the tranches are, as percentages:

$$A_1(=0) - B_1, \quad A_2(=B_1) - B_2, \quad A_3(=B_2) - B_3, \quad \ldots, A_5(=B_4) - B_5,$$

where for the i-TRAXX index we have

$$A_1 = 0,\ A_2 = 3\%,\ A_3 = 6\%,\ A_4 = 9\%,\ A_5 = 12\%,\ A_6 = 22\%$$

while for the CDX index we have

$$A_1 = 0,\ A_2 = 3\%,\ A_3 = 7\%,\ A_4 = 10\%,\ A_5 = 15\%,\ A_6 = 30\%.$$

Just to illustrate this important family of products, we consider an example concerning an iTRAXX tranche trade.

Begin(example of i-TRAXX tranche trade)

An investor sells EUR 10mn protection on the 3%-6% tranche with a 5y maturity. We assume a fair tanche-credit-spread (or rate) $R_{\text{index}}^{3,6}$ of 135bp. Therefore, the market maker pays the investor 135bp per annum quarterly on a notional amount of EUR 10mn. We assume that each underlying name has the same recovery $\text{Rec} = 40\%$.

- Each single name in the portfolio has a credit position in the index of $1/125 = 0.8\%$ and participates to the aggregate loss in terms of $0.8\% \times \text{Lgd} = 0.8\% \times 0.6 = 0.48\%$, since each time a default occurs the recovery is saved, leading to a loss of $\text{Lgd} = 1 - \text{Rec} = 1 - 0.4 = 0.6$.
- This means that each default corresponds to a loss of 0.48% in the global portfolio.
- After 6 defaults, the total loss in the portfolio is EUR $0.48\% \times 6 = 2.88\%$, and the tranche buyer is still protected since the tranche starts at 3% and we are at $2.88 < 3$.
- When the 7th name in the pool defaults the total loss amounts to 3.36% and the lower attachment point (that is 3%) of the tranche is reached.
- To compute the loss of the tranche we have to normalize the total loss with respect to the tranche size: The net loss in the tranche is then (3.36%-3%)/3% $\times$ 10mn = EUR 1.2mn which is immediately paid by the protection seller to the protection buyer.
- The notional amount on which the premium is paid reduces to 10mn - 1.2mn = EUR 8.8mn, and the investor receives every month a premium of 135bp on EUR 8.8mn until maturity or until the next default.
- Each following default leads to change in the tranche loss (paid by the protection seller) of $0.48\%/3\% \times 10\text{mn} =$ EUR 1.6mn, and the tranche notional decreases correspondingly.

- After the 13th default the total loss exceeds 6% ($13 \times 0.48\% = 6.24\%$) and the tranche is completely wiped out.
- In this case one last payment is made of (6%-5.76%)/3% $\times$ 10mn = EUR 0.8mn to the protection buyer, which in turn stops paying the premium since the outstanding notional has reduced to zero.

The different tranches offer different kinds of protection to the investor. An equity tranche (0-3%) buyer suffers of every default in the portfolio, which leads to a decrease of tranche notional on which the periodic premium is paid and conversely to contingent protection payments.

On the other side, the buyer of more senior tranches (e.g. 9%-12%) is more protected against few defaults: Indeed these tranches are affected only in case of a large number of defaults.

This difference leads to different premia paid to buy protection. It is natural to see that the periodic premium paid to buy protection is inversely proportional to the tranche seniority: The higher A_i, B_i, the lower $R_{\text{index}}^{A_i,B_i}$.

A technical note: The premium for the equity tranche is usually very large, so it is market practice to pay it as a fixed running premium of $R_{\text{index}}^{0,3} = 500$bps plus an upfront payment (computed in a way such that the total value is zero at inception).

End(example of i-TRAXX tranche trade)

However, iTRAXX tanches are now quoted in the market through a mechanism involving copula functions. The fair tranche premium-rate $R_{\text{index}}^{A_i,B_i}$ is translated into an implied correlation parameter as follows.

Consider a single standardized CDO tranche on (A_i, B_i) and the related fair premium rate $R_{\text{index}}^{A_i,B_i}(0)$ at inception time $t = 0$. One may postulate a Gaussian copula linking the ξ_i's of different names defaults τ_i, $i = 1, 2, \ldots, 125$, with all parameters in the copula correlation matrix set to a common value $\rho_{h,k} = \bar{\rho}$. Then the equation obtained by equating the two legs, with the market fair CDO premium rate $R_{\text{index}}^{A_i,B_i}(0)$ inserted in the premium leg, can be solved numerically in $\bar{\rho} = \bar{\rho}_{A_i,B_i}$. This parameter $\bar{\rho}_{A_i,B_i}$ is then called "implied correlation" for the considered attachment points A_i, B_i. Actually there are two kinds of implied correlation. When we re-map quotes by referring to nested tranches $0, B_i$,

$$0, B_1, \quad 0, B_2, \quad 0, B_3, \ldots, \quad 0, B_5,$$

all having zero as lower attachment point and increasing standardized detachment points B, then the relevant correlations $\bar{\rho}_{0,B_i}$ are called "base correlations". Instead, if we keep the adjacent sequence of attachment-detachment intervals, partitioning the loss distribution domain, then we have the so called "compound correlation" $\bar{\rho}_{A_i,B_i}$. The market often quotes iTRAXX tranche prices through implied correlation for the different standardized attachment-detachment points above. The plot of implied correlation with respect to such standardized points is the so called correlation smile or skew (according to whether one uses compound $i \mapsto \bar{\rho}_{A_i,B_i}$ or base $B_i \mapsto \bar{\rho}_{0,B_i}$ correlations). Since we do not pursue multi-name credit derivatives further in this book, we stop here as far as market quotes for multi-name payoffs are concerned.

We just give an example of market correlation smile and skew in Figures 21.1 and 21.2.

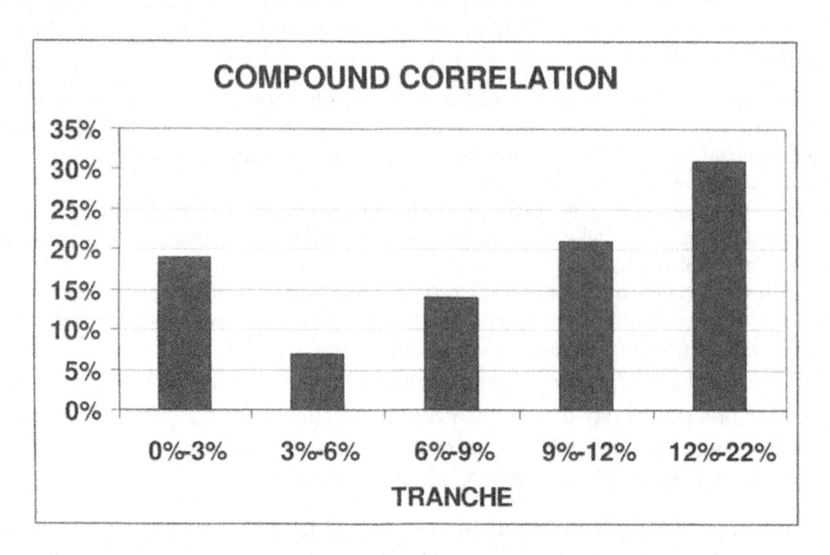

Fig. 21.1. Example of compound correlation structure for the DJ iTraxx (correlation smile $(A_i, B_i) \mapsto \bar{\rho}_{A_i,B_i}$). Correlations are expressed as percentages: 5% = 0.05 etc

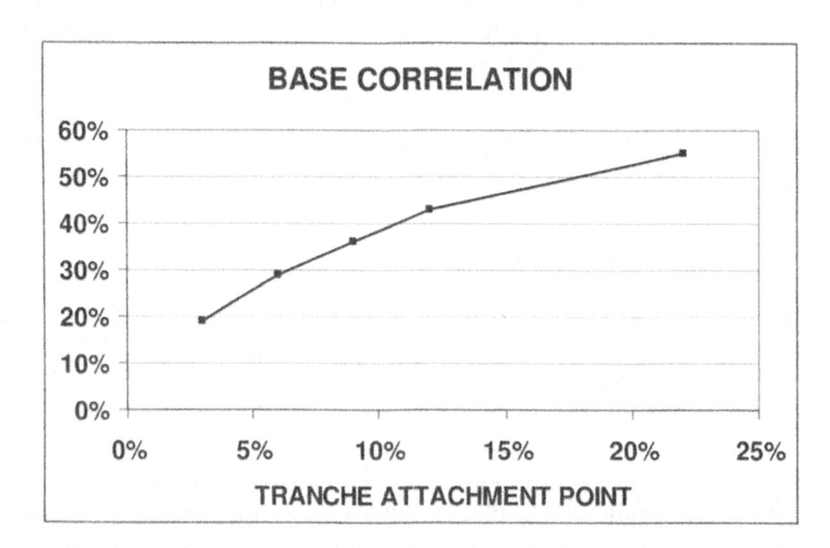

Fig. 21.2. Example of base correlation structure for the DJ iTraxx (Correlation skew $B_i \mapsto \bar{\rho}_{0,B_i}$).

The above quoting mechanism assumes Gaussian copulas (and deterministic credit spreads on each name) to characterize defaults, and collapses $125 \times 124/2$ dependence parameters $\rho_{h,k}$ to one parameter $\bar{\rho}$. The reader may find this machinery to be a little hazardous, and she might have a point. One should keep in mind, though, that this is meant as a quoting mechanism and that one does not really believe defaults to be characterized by such assumptions. Besides, much as in the quoting mechanism for implied volatilities with respect to strike and maturities, in the volatility smile modeling framework, here we do not have a real model consistently underlying all the quotes. Indeed, we use a Gaussian copula model but we change the parameter $\bar{\rho}_{A_i,B_i}$ (and thus the particular model) every time we change the tranche, similarly to how we change the volatility in the Black Scholes formula to match the market option price when the strike changes in the volatility smile market. This is not a single model that is consistent at the same time with all tranches, in the same way as the Black and Scholes model is not a single model consistent with the whole volatility smile at a single time. One needs a more sophisticated model, in a similar way to how one needs a volatility smile model to account for the market volatility smile. Such models are starting to appear recently and have been investigated in the last few years for the first time.

21.2 Defaultable (corporate) zero coupon bonds

After the guided tour, we begin the description of some credit derivatives payoffs that will play a fundamental role in the following.

Similarly to the zero coupon bond $P(t,T)$ being one of the possible fundamental quantities for describing the interest-rate curve, the defaultable zero coupon bond $\bar{P}(t,T)$ is one of the fundamental objects used to describe credit curves.

If we denote by τ the default time of the reference company, the value of a bond issued by the company and promising the payment of 1 unit of currency at time T, as seen from time t, is

$$\mathbf{1}_{\{\tau>t\}}\bar{P}(t,T) := \mathbb{E}\{D(t,T)\mathbf{1}_{\{\tau>T\}}|\mathcal{G}_t\} \tag{21.11}$$

where $\mathcal{G}_t = \mathcal{F}_t \vee \sigma(\{\tau < u\}, u \leq t)$ represents the flow of information on whether default occurred before t and if so at what time exactly (sigma-field $\sigma(\{\tau < u\}, u \leq t)$), and on the default free market variables up to t (sigma-field $\mathcal{F}_t$). $\mathbb{E}$ denotes the risk-neutral expectation in the enlarged probability space supporting τ.

The "indicator" function $\mathbf{1}_{\text{condition}}$ is 1 if "condition" is satisfied and 0 otherwise. In particular, $\mathbf{1}_{\{\tau>T\}}$ reads 1 if default τ did not occur before T, and 0 in the other case.

We understand then that (ignoring recovery) $\mathbf{1}_{\{\tau>T\}}$ is the correct payoff for a corporate bond at time T: the contract pays 1 if the company has not defaulted, and 0 if it defaulted before T.

If we include a recovery rate Rec paid at default τ, we have as discounted payoff

$$D(t,T)\mathbf{1}_{\{\tau>T\}} + \text{Rec} D(t,\tau)\mathbf{1}_{\{\tau\le T\}}$$

If we include a recovery rate Rec paid at maturity T, we have as discounted payoff

$$D(t,T)\mathbf{1}_{\{\tau>T\}}+\text{Rec}D(t,T)\mathbf{1}_{\{\tau\le T\}} = D(t,T)(1-\mathbf{1}_{\{\tau\le T\}})+\text{Rec}D(t,T)\mathbf{1}_{\{\tau\le T\}} =$$

$$= D(t,T)1 - \text{Lgd}D(t,T)\mathbf{1}_{\{\tau\le T\}} = \text{Rec}D(t,T) + \text{Lgd}D(t,T)\mathbf{1}_{\{\tau>T\}}$$

and the price is

$$\mathbf{1}_{\{\tau>t\}}\bar{P}^{\text{REC}}(t,T) = \text{Rec}P(t,T) + \mathbf{1}_{\{\tau>t\}}\text{Lgd}\ \bar{P}(t,T)$$

where $\text{Lgd} = 1 - \text{Rec}$ denotes the Loss Given Default on a unit notional. However, we will use defaultable zero coupon bond prices more as building blocks than as financial products in themselves; as such, we will assume them to have zero recovery.

21.2.1 Defaultable (corporate) coupon bonds

We consider a bond issued by counterparty "C" with default time τ paying coupons $c_{a+1}, c_{a+2}, \ldots, c_b$ at times $T_{a+1}, \ldots, T_b$, with recovery rate Rec. The discounted payoff at time t of this bond is (assuming a constant absolute recovery cash flow Rec at the first T_i following default):

$$\sum_{i=a+1}^{b} c_i D(t,T_i)\mathbf{1}_{\{\tau>T_i\}} + D(t,T_b)\mathbf{1}_{\{\tau>T_b\}} + \text{Rec} \sum_{i=a+1}^{b} D(t,T_i)\mathbf{1}_{\{\tau\in(T_{i-1},T_i]\}}$$

This can be read as
"Coupons if no default" + "Notional reimbursement if no default" + "recovery if early default".

21.3 Credit Default Swaps and Defaultable Floaters

We have already encountered CDS's payoffs in the guided tour above, to which we refer for some colloquial considerations on them. Here we present the mathematical formulation of their payoffs, with possible variants.

More in detail, we consider some alternative expressions for CDS payoffs, stemming from different conventions on the payment flows and on the protection leg for these contracts. We consider standard running CDS (RCDS), postponed payments running CDS (PRCDS), and briefly upfront CDS (UCDS). Each different running CDS definition implies a different definition of forward CDS rate, which we consider with some detail.

Following Brigo (2004, 2005) we introduce defaultable floating rate notes (DFRN's). We point out which kind of CDS payoff produces a forward CDS rate that is equal to the fair spread in the considered DFRN. An approximated equivalence between CDS's and DFRN's is established. Equivalence of CDS and DFRN's has been known for a while in the market, but usually in the simple and stylized case with continuous flows of payments. Here we consider a discrete set of flows, as in real market contracts, and find that the equivalence holds only after postponing or anticipating some relevant default indicators or discount factors.

We briefly investigate the possibility to express forward CDS rates in terms of some basic rates and discuss a possible analogy with the LIBOR and swap default-free models.

We will use this setup in a later chapter to discuss the change of numeraire approach to deriving a Black-like formula for CDS options, allowing us to quote CDS options through their implied volatilities.

The CDS and defaultable floaters part is structured as follows: Section 21.3.1 introduces notation, different kinds of CDS discounted payoffs, and in Section 21.3.2 we see the main definition of CDS forward rate. The notion of CDS implied hazard function and its possible use as quoting mechanism is recalled in Section 21.3.5, where we explain also that CDS' allow in principle to strip default or survival probabilities in a model independent way. Upfront CDS's are hinted at. *For some issues we will have to anticipate facts on intensity models. We ask the reader to trust us with these anticipations, full light will be given in Chapter 22.*

Section 21.3.7 examines some possible variant definitions of CDS rates. Furthermore, we examine the relationship between CDS rates on different periods and point out some parallels with the default free LIBOR and swap market rates.

Section 21.3.8 introduces defaultable floating rate notes and explores their relationship with CDS payoffs, finding equivalence under some payment schedules.

21.3.1 CDS payoffs: Different Formulations

We recall briefly some basic definitions for CDS's. Consider a CDS where we exchange protection payment rates R at times $T_{a+1}, \ldots, T_b$ or until default (the "premium leg") in exchange for a single protection payment $\mathrm{L{\scriptsize GD}}$ (loss given default, the "protection leg") at the default time τ of a reference entity

"C", provided that $T_a < \tau \leq T_b$. This is called a "running CDS" (RCDS) discounted payoff.

Protection Seller B	$\rightarrow$ protection LGD at default τ_C if $T_a < \tau_C \leq T_b$ $\rightarrow$ $\leftarrow$ rate R at $T_{a+1}, \ldots, T_b$ or until default τ_C $\leftarrow$	Protection Buyer A

The first line is called "protection leg", whereas the second line is called "premium leg". Formally, we may write the RCDS discounted value at time t as seen from "B" as

$$\Pi_{\mathrm{RCDS}a,b}(t) := D(t,\tau)(\tau - T_{\beta(\tau)-1})R\mathbf{1}_{\{T_a<\tau<T_b\}} \tag{21.12}$$
$$+ \sum_{i=a+1}^{b} D(t,T_i)\alpha_i R\mathbf{1}_{\{\tau \geq T_i\}} - \mathbf{1}_{\{T_a<\tau\leq T_b\}} D(t,\tau)\,\mathrm{L{\scriptstyle GD}}$$

where $t \in [T_{\beta(t)-1}, T_{\beta(t)})$, i.e. $T_{\beta(t)}$ is the first date among the T_i's that follows t, and where α_i is the year fraction between T_{i-1} and T_i. As elsewhere in the book, the stochastic discount factor at time t for maturity T is denoted by $D(t,T) = B(t)/B(T)$, where $B(t) = \exp(\int_0^t r_u du)$ denotes the bank-account numeraire, r being the instantaneous short interest rate.

We explicitly point out that we are assuming the offered protection amount LGD to be deterministic. Typically LGD $= 1 -$ REC, where the recovery rate REC is assumed to be deterministic and the notional is set to one.

Sometimes a slightly different payoff is considered for RCDS contracts. Instead of considering the exact default time τ, the protection payment LGD is postponed to the first time T_i following default, i.e. to $T_{\beta(\tau)}$. If the grid is three-or six months spaced, this postponement consists in a few months at worst. With this formulation, the CDS discounted payoff as seen from "B" can be written as

$$\Pi_{\mathrm{PRCDS}a,b}(t) := \sum_{i=a+1}^{b} D(t,T_i)\alpha_i R\mathbf{1}_{\{\tau \geq T_i\}} \tag{21.13}$$
$$- \sum_{i=a+1}^{b} \mathbf{1}_{\{T_{i-1}<\tau\leq T_i\}} D(t,T_i)\,\mathrm{L{\scriptstyle GD}},$$

which we term "Postponed payoffs Running CDS" (PRCDS) discounted payoff. Compare with the earlier discounted payout (21.12) where the protection payment occurs exactly at τ. The advantage of the postponed protection payment is that no accrued-interest term in $(\tau - T_{\beta(\tau)-1})$ is necessary, and also that all payments occur at the canonical grid of the T_i's. The postponed payout is better for deriving market models of CDS rates dynamics, as we shall see later on.

A slightly different postponed discounted payoff is given by

$$\Pi_{\text{PR2CDS}a,b}(t) := \sum_{i=a+1}^{b} D(t,T_i)\alpha_i R \mathbf{1}_{\{\tau > T_{i-1}\}} - \sum_{i=a+1}^{b} \mathbf{1}_{\{T_{i-1} < \tau \leq T_i\}} D(t,T_i) \text{ L}_{\text{GD}} \tag{21.14}$$

(notice the T_{i-1} in the indicators of the first summation). We see that we are including one more R-payment with respect to the earlier postponed case. In a way this is appropriate, since by pretending default is occurring at $T_{\beta(\tau)}$ instead of τ we are in fact introducing one more whole interval in the "premium leg", and we have to account for this interval.

From a different point of view, and since the protection leg, even if postponed, is discounted with the appropriate discount factor taking into account postponement, notice that in cases where τ is slightly larger than T_i then the first postponed payoff (21.13) is a better approximation of the actual one. Instead, in cases where τ is slightly smaller than T_i, the postponed payoff (21.14) represents a better approximation. We will see the different implications of these two payoffs.

Recently, there has been some interest in "upfront CDS" contracts (Veronesi (2003)). In this version, the present value of the protection leg is paid upfront by the party that is buying protection. In other terms, instead of exchanging a possible protection payment for some coupons, one exchanges it with an upfront payment.

The discounted payoff of the protection leg is simply

$$\Pi_{\text{UCDS}a,b}(t) := \mathbf{1}_{\{T_a < \tau \leq T_b\}} D(t,\tau) \text{ L}_{\text{GD}} = \sum_{i=a+1}^{b} \mathbf{1}_{\{T_{i-1} < \tau \leq T_i\}} D(t,\tau) \text{ L}_{\text{GD}}. \tag{21.15}$$

Alternatively, one can approximate this leg by a "postponed payment" version, where we postpone the protection payment until the first T_i following default τ:

$$\Pi_{\text{UPCDS}a,b}(t) := \sum_{i=a+1}^{b} \mathbf{1}_{\{T_{i-1} < \tau \leq T_i\}} D(t,T_i) \text{ L}_{\text{GD}}. \tag{21.16}$$

21.3.2 CDS pricing formulas

We denote by $\text{CDS}(t, [T_{a+1}, \ldots, T_b], T_a, T_b, R, \text{L}_{\text{GD}})$ the price at time t of the above standard running CDS. At times some terms are omitted, such as for example the list of payment dates $[T_{a+1}, \ldots, T_b]$, and to shorten notation further we may write $\text{CDS}_{a,b}(t, R, \text{L}_{\text{GD}})$. We add the prefixes "PR1" or "PR2" to denote, respectively, the analogous prices for the postponed payoffs (21.13) and (21.14). We add the prefix "U" (upfront) to denote the present value at t of the protection leg (21.15) of the CDS, and "UP" (upfront postponed) in case we are considering the present value of (21.16).

The pricing formulas for these payoffs depend on the assumptions on interest-rate dynamics and on the default time τ. If τ is assumed to be independent of interest rates, then model independent valuation formulas for CDS's involving directly default (or survival) probabilities and default free zero coupon bonds are available, see the calculation of the two CDS legs given later on in 21.22 and 21.23.

Most of times, in the remaining part of this section we place ourselves in a stochastic intensity framework, where the intensity will turn out to be an $\mathcal{F}_t$-adapted continuous positive process, $\mathcal{F}_t$ denoting as before the basic filtration without default, typically representing the information flow of interest rates, intensities and possibly other default-free market quantities. Indeed, most of times we assume default to be modeled as the first jump time of a Cox process (see also Appendix C.6). More specifically, in the Cox process setting we have $\tau = \Lambda^{-1}(\xi)$, where Λ is a stochastic hazard function which we assume to be $\mathcal{F}_t$ adapted, absolutely continuous and strictly increasing, and ξ is standard exponentially distributed and independent of $\{\mathcal{F}_t,\ t \geq 0\}$. These assumptions imply the existence of a positive adapted process λ, which we assume also to be right continuous and limited on the left, such that $\Lambda(t) = \int_0^t \lambda_s ds$ for all t. We will see some credit models that model directly and explicitly the stochastic intensity in Chapter 22, and also different credit models that consider models for some market quantities embedding the impact of the relevant intensity model that is consistent with them, without considering the intensity explicitly (Chapter 23). Or even more generally, we may have market models where one directly models market-related rates without presuming existence of a default intensity, see for example Jamshidian (2004) and Brigo and Morini (2005).

In general, whichever the model, we can compute the CDS price according to risk-neutral valuation (see for example Bielecki and Rutkowski (2001)):

$$\mathrm{CDS}_{a,b}(t, R, \mathrm{L{\scriptsize GD}}) = \mathbb{E}\left\{\Pi_{\mathrm{RCDS}_{a,b}}(t)|\mathcal{G}_t\right\}. \tag{21.17}$$

With the aim of developing a general definition of CDS forward rates with far-reaching implications, we now consider whether we may replace the $\mathcal{G}_t$ conditioning in the above expectation with a milder conditioning.

21.3.3 Changing filtration: $\mathcal{F}_t$ without default VS complete $\mathcal{G}_t$

The above expected value is with respect to the filtration $\mathcal{G}_t$ including default monitoring, since at the moment where we compute the price we do know if the underlying name has defaulted or not. Given that the default status at a given present instant is known, when we price with a risk neutral expectation we need to condition on this information, besides the default free one. In other terms, we need to add to the default free market filtration $\mathcal{F}_t$ (the one we have used in the whole book before the credit part) the information on whether default occurred so far, and if yes when exactly (to which we refer as "default monitoring", and which we denote by $\sigma(\{\tau < u\}, u \leq t)$ if t is the present time). In formula: $\mathcal{G}_t = \mathcal{F}_t \vee \sigma(\{\tau < u\}, u \leq t)$. However, in some cases it would be better to keep on computing prices as expectations with respect to the old default free filtration $\mathcal{F}_t$. This is possible. Indeed, the expected value (21.17) can also be written as

$$\mathrm{CDS}_{a,b}(t, R, \mathrm{L}_{\mathrm{GD}}) = \frac{\mathbf{1}_{\{\tau>t\}}}{\mathbb{Q}(\tau > t|\mathcal{F}_t)} \mathbb{E}\left\{\Pi_{\mathrm{RCDS}_{a,b}}(t)|\mathcal{F}_t\right\}, \qquad (21.18)$$

for a sketchy proof see Section 22.5 later on, where we introduce and prove the related filtration switching formula. This second expression, and the analogous definitions with postponed payoffs, will be fundamental for introducing the market model for CDS options in a rigorous way. Looking at this last formula we see that we can change filtration when pricing, provided we scale the expectation by a normalizing term. There is need also to include explicitly the condition that the reference credit has not defaulted before valuation time, as implied by the indicator $\mathbf{1}_{\{\tau>t\}}$ in the numerator. This is not necessary for CDS payoffs when conditioning on $\mathcal{G}_t$, since CDS payoffs have zero value in paths with early default and the filtration $\mathcal{G}_t$ "knows" whether there has been an early default or not. In fact, in general for $T > t$ and for a general payoff X we have

$$\mathbb{E}[\mathbf{1}_{\{\tau>T\}}X|\mathcal{G}_t] = \mathbb{E}[\mathbf{1}_{\{\tau>t\}}\mathbf{1}_{\{\tau>T\}}X|\mathcal{G}_t] = \mathbf{1}_{\{\tau>t\}}\mathbb{E}[\mathbf{1}_{\{\tau>T\}}X|\mathcal{G}_t]$$

where the first equality follows from the fact that if the company is still going at time T then it is still going also at time t for sure, and the second equality follows from the fact that the information $\mathbf{1}_{\{\tau>t\}}$ is available in $\mathcal{G}_t$, so that it can be carried out from the expectation. This information is not available in $\mathcal{F}_t$ though, so that when we switch to the $\mathcal{F}_t$ expectation it has to be included explicitly.

The possibility to compute risk neutral prices as expectations conditional on the default free filtration $\mathcal{F}_t$ rather than conditional on the larger $\mathcal{G}_t$ (that includes also explicit default monitoring) is an important aspect whose usefulness will be soon clear.

Remembering the detailed expression of the CDS payoff given in (21.12), which we substitute in (21.18), we can write the price to the protection seller as

$$\mathrm{CDS}_{a,b}(t, R, \mathrm{L_{GD}}) = \frac{\mathbf{1}_{\{\tau>t\}}}{\mathbb{Q}(\tau > t|\mathcal{F}_t)} \cdot \tag{21.19}$$

$$\cdot\Big\{R\,\mathbb{E}[D(t,\tau)(\tau - T_{\beta(\tau)-1})\mathbf{1}_{\{T_a<\tau<T_b\}}|\mathcal{F}_t]$$

$$+\sum_{i=a+1}^{b} \alpha_i R\,\mathbb{E}[D(t,T_i)\mathbf{1}_{\{\tau\geq T_i\}}|\mathcal{F}_t]$$

$$-\mathrm{L_{GD}}\,\mathbb{E}[\mathbf{1}_{\{T_a<\tau\leq T_b\}}D(t,\tau)|\mathcal{F}_t]\Big\}.$$

We can apply the same change of filtration also to the price of a defaultable zero coupon bond. This price would be, in principle, given by (21.11). However, using the filtration switching formula, we can write under very general assumptions

$$\mathbb{E}[D(t,T)\mathbf{1}_{\{\tau>T\}}|\mathcal{G}_t] = \frac{\mathbf{1}_{\{\tau>t\}}}{\mathbb{Q}(\tau > t|\mathcal{F}_t)}\mathbb{E}[D(t,T)\mathbf{1}_{\{\tau>T\}}|\mathcal{F}_t]$$

$$= \mathbf{1}_{\{\tau>t\}}\bar{P}(t,T)$$

If we substitute this last equation into (21.19) we obtain

$$\mathrm{CDS}_{a,b}(t, R, \mathrm{L_{GD}}) = \frac{\mathbf{1}_{\{\tau>t\}}}{\mathbb{Q}(\tau > t|\mathcal{F}_t)} \cdot \tag{21.20}$$

$$\cdot\Big\{R\,\mathbb{E}[D(t,\tau)(\tau - T_{\beta(\tau)-1})\mathbf{1}_{\{T_a<\tau<T_b\}}|\mathcal{F}_t]$$

$$+\sum_{i=a+1}^{b} \alpha_i R\,\mathbb{Q}(\tau > t|\mathcal{F}_t)\bar{P}(t,T_i)$$

$$-\mathrm{L_{GD}}\,\mathbb{E}[\mathbf{1}_{\{T_a<\tau\leq T_b\}}D(t,\tau)|\mathcal{F}_t]\Big\}$$

21.3.4 CDS forward rates: The first definition

Let us go back now to defining CDS forward rates. For the time being, let us deal with the definition of (running) CDS forward rate $R_{a,b}(t)$. This can be defined as that R that makes the CDS value equal to zero at time t, so that

$$\mathrm{CDS}_{a,b}(t, R_{a,b}(t), \mathrm{L_{GD}}) = 0.$$

The idea is then solving this equation in $R_{a,b}(t)$. In doing this one has to be careful. It is best to use the expression coming from (21.18) rather than (21.17). Equate thus (21.19) to zero and derive R correspondingly. Strictly speaking, the resulting R would be defined on $\{\tau > t\}$ only, since elsewhere the equation is satisfied automatically thanks to the indicator in front of the expression, regardless of R. Since the value of R does not matter when $\{\tau < t\}$, the equation being satisfied automatically, we need not worry about $\{\tau < t\}$ and may define, in general,

$$R_{a,b}(t) = \frac{\text{L}_{\text{GD}}\ \mathbb{E}[D(t,\tau)\mathbf{1}_{\{T_a<\tau\le T_b\}}|\mathcal{F}_t]}{\sum_{i=a+1}^{b}\alpha_i\mathbb{Q}(\tau > t|\mathcal{F}_t)\bar{P}(t,T_i) + \text{accrual}_t}, \tag{21.21}$$

where accrual_t is the accrual term

$$\text{accrual}_t := \mathbb{E}\left\{D(t,\tau)(\tau - T_{\beta(\tau)-1})\mathbf{1}_{\{T_a<\tau<T_b\}}|\mathcal{F}_t\right\}.$$

This approach to defining $R_{a,b}$ amounts to equating to zero only the expected value part in (21.18), i.e. the terms in-between large curly brackets in (21.19) or (21.20), and in a sense is a way of privileging $\mathcal{F}_t$ expected values to $\mathcal{G}_t$ ones. The technical tool that would allow us to do this is the filtration switching formula. This formula holds also under assumptions more general than those adopted here, as shown in the above-mentioned Jeanblanc Rutkowski (2000) result, and this is the spirit of part of the work in Jamshidian (2004). In particular, the assumption that could generalize our Cox process setting is "Conditional independence for subfiltrations" (Jamshidian (2004)), called "martingale invariance property" in Jeanblanc and Rutkowski (2000). See also Brigo and Morini (2005).

21.3.5 Market quotes, model independent implied survival probabilities and implied hazard functions

Now we explain shortly how the market quotes running and upfront CDS prices. First we notice that typically the T's are three- months spaced. Let us begin with running CDS's. Usually at time $t = 0$, provided default has not yet occurred, the market sets R to a value $R^{\text{MID}}_{a,b}(0)$ that makes the CDS fair at time 0, i.e. such that $\text{CDS}_{a,b}(0, R^{\text{MID}}_{a,b}(0), \text{L}_{\text{GD}}) = 0$. In fact, in the market running CDS's used to be quoted at a time 0 through a bid and an ask value for this "fair" $R^{\text{MID}}_{a,b}(0)$, for CDS's with $T_a = 0$ and with T_b spanning a set of canonical final maturities, $T_b = 1y$ up to $T_b = 10y$. As time moved on of, say, $\Delta t = 1$ day, the market shifted the T's of Δt, setting $T_a = 0 + \Delta t, \ldots, T_b = 10y + \Delta t$, and then quoted $R^{\text{MID}}_{a,b}(\Delta t)$ satisfying $\text{CDS}_{a,b}(\Delta t, R^{\text{MID}}_{a,b}(\Delta t), \text{L}_{\text{GD}}) = 0$. This means that as time moved on, the maturities increased and the times to maturity remained constant.

Recently, the quoting mechanism has changed and has become more similar to the mechanism of the futures markets. Let 0 be the current time. Maturities $T_a, \ldots, T_b$ are fixed at the original time 0 to some values such as 1y, 2y, 3y etc and then, as time moves for example to 1 day, the CDS maturities are not shifted correspondingly of 1 day as before but remain 1y,2y etc from the original time 0. This means that the times to maturity of the quoted CDS's decrease as time passes. When the quoting time approaches maturity, a new set of maturities is fixed and so on. A detail concerning the "fixed maturities" paradigm is that when the first maturity T_a is less than one month away from the quoting time (say t), the payoff two terms

$$(T_a - t)D(t, T_a)R\mathbf{1}_{\{\tau > T_a\}} + (T_{a+1} - T_a)D(t, T_{a+1})R\mathbf{1}_{\{\tau > T_{a+1}\}}$$

are replaced by

$$(T_{a+1} - t)D(t, T_{a+1})R\mathbf{1}_{\{\tau > T_{a+1}\}}$$

in determining the "fair" R. If we neglect this last convention, once we fix the quoting time (say to t) the method to strip implied survival probabilities is the same. And even with the last convention in place, the two mechanisms coincide at the exact valuation dates $t = T_i$.

We now present a model independent valuation formula for CDS that assumes independence between interest rates and the default time.

Assume the stochastic discount factors $D(s, t)$ to be independent of the default time τ for all possible $0 < s < t$.

The premium leg of the CDS at time 0 can be valued as follows:

$$\begin{aligned}
\text{PremiumLeg}_{a,b}(R) &= \mathbb{E}[D(0,\tau)(\tau - T_{\beta(\tau)-1})R\mathbf{1}_{\{T_a<\tau<T_b\}}] + \\
&\quad + \sum_{i=a+1}^{b} \mathbb{E}[D(0,T_i)\alpha_i R\mathbf{1}_{\{\tau \geq T_i\}}] \\
&= \mathbb{E}\left[\int_{t=0}^{\infty} D(0,t)(t - T_{\beta(t)-1})R\mathbf{1}_{\{T_a<t<T_b\}}\mathbf{1}_{\{\tau\in[t,t+dt]\}}\right] + \\
&\quad + \sum_{i=a+1}^{b} \mathbb{E}[D(0,T_i)]\alpha_i R\, \mathbb{E}[\mathbf{1}_{\{\tau \geq T_i\}}] \\
&= \int_{t=T_a}^{T_b} \mathbb{E}[D(0,t)(t - T_{\beta(t)-1})R\, \mathbf{1}_{\{\tau\in[t,t+dt)\}}] + \\
&\quad + \sum_{i=a+1}^{b} P(0,T_i)\alpha_i R\, \mathbb{Q}(\tau \geq T_i) \\
&= \int_{t=T_a}^{T_b} \mathbb{E}[D(0,t)](t - T_{\beta(t)-1})R\, \mathbb{E}[\mathbf{1}_{\{\tau\in[t,t+dt)\}}] + \\
&\quad + \sum_{i=a+1}^{b} P(0,T_i)\alpha_i R\, \mathbb{Q}(\tau \geq T_i) \\
&= R\int_{t=T_a}^{T_b} P(0,t)(t - T_{\beta(t)-1})\mathbb{Q}(\tau\in[t,t+dt)) + \\
&\quad + R\sum_{i=a+1}^{b} P(0,T_i)\alpha_i \mathbb{Q}(\tau \geq T_i),
\end{aligned}$$

where we have used independence in factoring the above expectations. We have thus, by rearranging terms and introducing a "unit-premium" premium leg:

$$\text{PremiumLeg}_{a,b}(R; P(0,\cdot), \mathbb{Q}(\tau > \cdot)) = R \ \text{PremiumLeg1}_{a,b}(P(0,\cdot), \mathbb{Q}(\tau > \cdot)),$$

$$\text{PremiumLeg1}_{a,b}(P(0,\cdot), \mathbb{Q}(\tau > \cdot)) := -\int_{T_a}^{T_b} P(0,t)(t - T_{\beta(t)-1}) d_t \boxed{\mathbb{Q}(\tau \geq t)}$$
$$+ \sum_{i=a+1}^{b} P(0,T_i)\alpha_i \boxed{\mathbb{Q}(\tau \geq T_i)} \qquad (21.22)$$

This formula is indeed model independent given the initial zero coupon curve (bonds) at time 0 observed in the market (i.e. $P(0,\cdot)$) and given the survival probabilities $\mathbb{Q}(\tau \geq \cdot)$ at time 0 (terms in the boxes).

A similar formula holds for the protection leg, again under independence between default τ and interest rates.

$$\text{ProtecLeg}_{a,b}(\text{L}_{\text{GD}}) = \mathbb{E}[\mathbf{1}_{\{T_a < \tau \leq T_b\}} D(0,\tau) \ \text{L}_{\text{GD}}]$$
$$= \text{L}_{\text{GD}} \ \mathbb{E}\left[\int_{t=0}^{\infty} \mathbf{1}_{\{T_a < t \leq T_b\}} D(0,t) \mathbf{1}_{\{\tau \in [t,t+dt)\}}\right]$$
$$= \text{L}_{\text{GD}} \left[\int_{t=T_a}^{T_b} \mathbb{E}[D(0,t)\mathbf{1}_{\{\tau \in [t,t+dt)\}}]\right]$$
$$= \text{L}_{\text{GD}} \int_{t=T_a}^{T_b} \mathbb{E}[D(0,t)]\mathbb{E}[\mathbf{1}_{\{\tau \in [t,t+dt)\}}]$$
$$= \text{L}_{\text{GD}} \int_{t=T_a}^{T_b} P(0,t)\mathbb{Q}(\tau \in [t,t+dt))$$

so that we have, by introducing a "unit-notional" protection leg:

$$\text{ProtecLeg}_{a,b}(\text{L}_{\text{GD}}; P(0,\cdot), \mathbb{Q}(\tau > \cdot)) = \text{L}_{\text{GD}} \ \text{ProtecLeg1}_{a,b}(P(0,\cdot), \mathbb{Q}(\tau > \cdot)),$$

$$\text{ProtecLeg1}_{a,b}(P(0,\cdot), \mathbb{Q}(\tau > \cdot)) := -\int_{T_a}^{T_b} P(0,t) \ d_t \boxed{\mathbb{Q}(\tau \geq t)} \qquad (21.23)$$

This formula too is model independent given the initial zero coupon curve (bonds) at time 0 observed in the market and given the survival probabilities at time 0 (term in the box).

The integrals in the survival probabilities given in the above formulas can be valued as Stieltjes integrals in the survival probabilities themselves, and can easily be approximated numerically by summations through Riemann-Stieltjes sums, considering a low enough discretization time step.

Now recall that the market quotes, at time 0, the fair $R = R_{0,b}^{\text{mkt MID}}(0)$ (actually bid and ask quotes are available for this fair R) equating the two legs for a set of CDS with initial protection time $T_a = 0$ and final protection time $T_b \in \{1y, 2y, 3y, 4y, 5y, 6y.7y, 8y, 9y, 10y\}$, although often only a subset of the maturities $\{1y, 3y, 5y, 7y, 10y\}$ is available. Solve then

$$\text{ProtLeg}_{0,b}(\text{L}_{\text{GD}}; P(0,\cdot), \mathbb{Q}(\tau > \cdot))) = \text{PremLeg}_{0,b}(R_{0,b}^{\text{mktMID}}(0); P(0,\cdot), \mathbb{Q}(\tau > \cdot))$$

in portions of $\mathbb{Q}(\tau > \cdot)$ starting from $T_b = 1y$, finding the market implied survival $\{\mathbb{Q}(\tau \geq t), t \leq 1y\}$; plugging this into the $T_b = 2y$ CDS legs formulas, and then solving the same equation with $T_b = 2y$, we find the market implied survival $\{\mathbb{Q}(\tau \geq t), t \in (1y, 2y]\}$, and so on up to $T_b = 10y$.

This is a way to strip survival probabilities from CDS quotes in a model independent way. No need to assume an intensity or a structural model for default here.

However, the market in doing the above stripping typically resorts to hazard functions, assuming existence of hazard functions associated with the default time. We now assume existence of a deterministic intensity, as in deterministic intensity models, and briefly illustrate the notion of implied deterministic cumulated intensity (hazard function), satisfying

$$\mathbb{Q}\{\tau \geq t\} = \exp(-\Gamma(t)), \quad \mathbb{Q}\{s < \tau \leq t\} = \exp(-\Gamma(s)) - \exp(-\Gamma(t)).$$

More details and examples will be given in Chapter 22, fully devoted to intensity models. We anticipate now a few elements that are useful for the discussion. The market Γ's are obtained by inverting a pricing formula based on the assumption that τ is the first jump time of a Poisson process with deterministic intensity $\lambda_t = \gamma(t) = d\Gamma(t)/dt$. The interpretation of this function is: probability of defaulting in $[t, t+dt)$ having not defaulted before t is $\gamma(t)dt$:

$$\mathbb{Q}(\tau \in [t, t+dt) | \tau > t, \mathcal{F}_t) = \gamma(t)dt.$$

In this case one can derive a formula for CDS prices based on integrals of γ, and on the initial interest-rate curve, resulting from the above expectation:

$$\begin{aligned} \text{CDS}_{a,b}(t, R, \text{L}_{\text{GD}}; \Gamma(\cdot)) &= \qquad (21.24) \\ = \mathbf{1}_{\{\tau > t\}} \Bigg[-R \int_{T_a}^{T_b} & P(t,u)(u - T_{\beta(u)-1}) d_u(e^{-(\Gamma(u)-\Gamma(t))}) \\ &+ \sum_{i=a+1}^{b} P(t,T_i) R \alpha_i e^{\Gamma(t)-\Gamma(T_i)} \\ &+ \text{L}_{\text{GD}} \int_{T_a}^{T_b} P(t,u) d_u(e^{-(\Gamma(u)-\Gamma(t))}) \Bigg]. \end{aligned}$$

By equating to zero the above expression in γ for $t = 0, T_a = 0$, after plugging in the relevant market quotes for R, one can extract the γ's corresponding to CDS market quotes for increasing maturities T_b and obtain market implied γ^{mkt} and Γ^{mkt}'s.

More in detail, one finds the Γ^{mkt}'s solving

$$\text{CDS}_{0,b}(0, R_{0,b}^{\text{mkt MID}}(0), \text{L}_{\text{GD}}; \Gamma^{\text{mkt}}(0 \div T_b)) = 0, \quad T_b = 1y, 2y, 3y, 5y, 7y, 10y.$$

If we are given $R_{0,b}^{\text{mkt MID}}(0)$ for different maturities T_b, we can assume a *piecewise linear* (or at times constant) γ, and invert prices in an iterative way as T_b increases, deriving each time the new part of γ that is consistent with the R for the new increased maturity.

It is important to point out that usually the actual model one assumes for τ is more complex and may involve stochastic intensity either directly or through stochastic modeling of the R dynamics itself. Even so, the γ^{mkt}'s are retained as a mere quoting mechanism for CDS rate market quotes, and may be taken as inputs in the calibration of more complex models, as we will do in particular with the SSRD stochastic intensity model in Chapter 22.

Upfront CDS are simply quoted through the present value of the protection leg. Under deterministic hazard rates γ, we have

$$\text{UCDS}(t, T_a, T_b, R, \text{L}_{\text{GD}}; \Gamma(\cdot)) = -\mathbf{1}_{\{\tau > t\}} \text{L}_{\text{GD}} \int_{T_a}^{T_b} P(t,u) d_u(e^{-(\Gamma(u)-\Gamma(t))}).$$

As before, by equating to the corresponding upfront market quote the above expression in γ, one can extract the γ's corresponding to UCDS market quotes for increasing maturities and obtain again market implied γ^{mkt} and Γ^{mkt}'s.

Once the implied γ are estimated, it is easy to switch from the "running CDS quote" R to the "upfront CDS quote" UCDS, or vice versa. Indeed, we see that, without postponed payments, the two quotes are linked by

$$\begin{aligned} &\text{UCDS}(t, T_a, T_b, R, \text{L}_{\text{GD}}; \Gamma^{\text{mkt}}(\cdot)) = \\ &= R_{a,b}(t) \Bigg[- \int_{T_a}^{T_b} P(t,u)(u - T_{\beta(u)-1}) d_u(e^{-(\Gamma^{\text{mkt}}(u)-\Gamma^{\text{mkt}}(t))}) + \\ &\quad + \sum_{i=a+1}^{n} P(t,T_i)\alpha_i e^{\Gamma^{\text{mkt}}(t)-\Gamma^{\text{mkt}}(T_i)} \Bigg] \end{aligned}$$

We present some concrete examples of calibrated hazard rates γ in Section 22.3

21.3.6 A simpler formula for calibrating intensity to a single CDS

The market makes intensive use of a simpler formula for calibrating a constant intensity (and thus hazard rate) $\gamma(t) = \gamma$ to a single CDS, say $\text{CDS}_{0,b}$. The formula is the following:

$$\boxed{\gamma = \frac{R_{0,b}(0)}{\text{L}_{\text{GD}}}}. \tag{21.25}$$

This formula is very handy: one does not need the interest rate curve to apply it. Also, if we recall what anticipated in Section 21.1.1, i.e. that the intensity $\gamma = \lambda$ can be interpreted as an instantaneous credit spread, then the interpretation as credit spread extends to R.

In the present context, this simple formula shows us that, given a constant hazard rate (and subsequent independence between the default time and interest rates), *the CDS premium rate R can really be interpreted as a credit spread, or a default probability.* We derive this formula now.

Assume we have a stylized CDS contract for protection in $[0, T]$ under independence between interest rates ($D(0,t)$'s) and the default time τ. The premium leg pays continuously until default the premium rate R of the CDS: this means that in the interval $[t,\ t+dt]$ the premium leg pays "$R\ dt$". By discounting each premium flow "$R\ dt$" from the time t where it occurs to time 0 we obtain $D(0,t)R\ dt$, and by adding up all the premiums in different instants of the period $[0,T]$ where default has not yet occurred ($\tau > t$) we get

$$\int_0^T D(0,t)1_{\{\tau>t\}} R\ dt.$$

The protection leg is as usual. We can then write

$$\text{PremiumLeg} = \mathbb{E}\left[\int_0^T D(0,t)1_{\{\tau>t\}} R dt\right] = R\int_0^T \mathbb{E}[D(0,t)1_{\{\tau>t\}}]dt =$$

$$= R\int_0^T \mathbb{E}[D(0,t)]\mathbb{E}[1_{\{\tau>t\}}]dt = R\int_0^T P(0,t)\mathbb{Q}(\tau>t)dt$$

and

$$\text{ProtectionLeg} = \mathbb{E}[\text{L}_{\text{GD}} D(0,\tau)1_{\{\tau\le T\}}]$$

$$= \text{L}_{\text{GD}}\int_0^T \mathbb{E}[D(0,t)1_{\{\tau\in[t,t+dt)\}}] = \text{L}_{\text{GD}}\int_0^T \mathbb{E}[D(0,t)]\mathbb{E}[1_{\{\tau\in[t,t+dt)\}}] =$$

$$= \text{L}_{\text{GD}}\int_0^T P(0,t)\mathbb{Q}(\tau\in[t,t+dt)) = -\text{L}_{\text{GD}}\int_0^T P(0,t)d_t\mathbb{Q}(\tau>t).$$

Assume that the default curve comes from a constant intensity model, where default is the first jump of a time homogeneous Poisson process: $\mathbb{Q}(\tau > t) = e^{-\gamma t}$. Substitute

$$\mathbb{Q}(\tau>t) = e^{-\gamma t},\quad d\,\mathbb{Q}(\tau>t) = -\gamma e^{-\gamma t}\,dt = -\gamma\mathbb{Q}(\tau>t)\,dt$$

to obtain

$$\text{ProtectionLeg} = -\text{L}_{\text{GD}}\int_0^T P(0,t)\ d_t\mathbb{Q}(\tau>t) = \gamma\text{L}_{\text{GD}}\int_0^T P(0,t)\mathbb{Q}(\tau>t)dt$$

Now recall that the market quotes the fair R equating the two legs. Solve then

$$\text{ProtectionLeg} = \text{PremiumLeg}$$

i.e.

$$\gamma \text{L}_{\text{GD}} \int_0^T P(0,t)\mathbb{Q}(\tau > t)dt = R \int_0^T P(0,t)\mathbb{Q}(\tau > t)dt$$

to obtain our initial formula above.

Clearly this formula is only approximated, due to the assumptions of continuous payments in the premium leg, and it does not take into account the term structure of CDS, since it is based on a single quote for R; however, it can be used in any situation where one needs a quick calibration of the default intensity or probability to a single (say for example the 5y) CDS quote.

21.3.7 Different Definitions of CDS Forward Rates and Analogies with the LIBOR and SWAP rates

The procedure of equating to 0 the current price of a contract to derive a sensible definition of forward rate is rather common. For example, we have seen in Chapter 1 that the default free forward LIBOR rate $F(t,S,T)$ is obtained as the rate at time t that makes the time-t price of a Forward Rate Agreement contract (FRA) with expiry S and maturity T vanish. In the same chapter we introduced an analogous definition of forward swap rate at time t as the rate in the fixed leg of the swap that makes the swap value at time t equal to 0.

In the current context, we can set a CDS price to zero to derive a forward CDS rate. Clearly, the obtained rate changes according to the different running CDS payoff we consider. For example, by equating to 0 expression (21.18) and solving in R, we have the standard running CDS forward rate given in (21.21). We may wonder about what we would have obtained as definition of forward CDS rates when considering CDS payoffs PRCDS with postponed protection payments (21.13) or even PR2CDS (21.14). By straightforwardly adapting the above derivation, we would have obtained a CDS forward rate defined as

$$R_{a,b}^{\text{PR}}(t) = \frac{\text{L}_{\text{GD}} \sum_{i=a+1}^{b} \mathbb{E}[D(t,T_i)\mathbf{1}_{\{T_{i-1}<\tau\le T_i\}}|\mathcal{F}_t]}{\sum_{i=a+1}^{b} \alpha_i \mathbb{E}[D(t,T_i)\mathbf{1}_{\{\tau>T_i\}}|\mathcal{F}_t]} =$$

$$= \frac{\text{L}_{\text{GD}} \sum_{i=a+1}^{b} \mathbb{E}[D(t,T_i)\mathbf{1}_{\{T_{i-1}<\tau\le T_i\}}|\mathcal{F}_t]}{\sum_{i=a+1}^{b} \alpha_i \mathbb{Q}(\tau > t|\mathcal{F}_t)\bar{P}(t,T_i)},$$

and

$$R_{a,b}^{\text{PR2}}(t) = \frac{\text{L}_{\text{GD}} \sum_{i=a+1}^{b} \mathbb{E}[D(t,T_i)\mathbf{1}_{\{T_{i-1}<\tau\le T_i\}}|\mathcal{F}_t]}{\sum_{i=a+1}^{b} \alpha_i \mathbb{E}[D(t,T_i)\mathbf{1}_{\{\tau>T_{i-1}\}}|\mathcal{F}_t]}$$

(where "PR" and "PR2" stand for "postponed-running" payoffs of the first and second kind, respectively).

Can we use the forward CDS rate definition, limited to a one-period interval, to introduce defaultable one-period forward rates? A straightforward generalization of the definition of forward LIBOR rates to the defaultable case is given for example in Schönbucher (2000). This definition mimics the definition in the default free case, in that from zero-coupon bonds one builds a "defaultable forward LIBOR rate"

$$\bar{F}(t;T_{j-1},T_j) := (1/\alpha_j)\left(\frac{1_{\{\tau>t\}}\bar{P}(t,T_{j-1})}{1_{\{\tau>t\}}\bar{P}(t,T_j)} - 1\right)$$

on $\tau > t$. However, as noticed earlier, the default free F is obtained as the *fair rate* at time t for a Forward Rate Agreement contract (FRA). Can we see $\bar{F}$ as the fair rate for a sort of defaultable FRA? Since the most liquid credit instruments are CDS's, consider a running postponed CDS on a one-period interval, with $T_a = T_{j-1}$ and $T_b = T_j$. We obtain (take $\text{L}_{\text{GD}} = 1$)

$$R_j^{PR}(t) := \frac{\mathbb{E}[D(t,T_j)\mathbf{1}_{\{T_{j-1}<\tau\le T_j\}}|\mathcal{F}_t]}{\alpha_j \mathbb{Q}(\tau>t|\mathcal{F}_t)\bar{P}(t,T_j)} = \tag{21.26}$$

$$= \frac{\mathbb{E}[D(t,T_j)\mathbf{1}_{\{\tau>T_{j-1}\}}|\mathcal{F}_t] - \mathbb{E}[D(t,T_j)\mathbf{1}_{\{\tau>T_j\}}|\mathcal{F}_t]}{\alpha_j \mathbb{Q}(\tau>t|\mathcal{F}_t)\bar{P}(t,T_j)}$$

where we have set $R_j^{PR} := R_{j-1,j}^{PR}$. The analogous part of $\bar{F}_j = \bar{F}(\cdot,T_{j-1},T_j)$ would be, after adjusting the conditioning to $\mathcal{F}_t$ ($\hat{F}_j(t) = \bar{F}_j(t)$ on $\tau > t$ but $\hat{F}$ is defined also on $\tau \le t$)

$$\hat{F}_j(t) = \frac{\mathbb{E}[D(t,\mathbf{T_{j-1}})\mathbf{1}_{\{\tau>T_{j-1}\}}|\mathcal{F}_t] - \mathbb{E}[D(t,T_j)\mathbf{1}_{\{\tau>T_j\}}|\mathcal{F}_t]}{\alpha_j \mathbb{Q}(\tau>t|\mathcal{F}_t)\bar{P}(t,T_j)}.$$

The difference is that in R_j^{PR}'s numerator we are taking expectation of a quantity that vanishes in all paths where $\tau > T_j$, whereas in $\bar{F}$ the corresponding quantity does not vanish necessarily in paths with $\tau > T_j$. Moreover, R_j comes from requiring a zero initial value at inception of a particular financial contract, consistently with earlier definitions of forward LIBOR and swap rates.

Schönbucher (2000) defines the discrete tenor credit spread, in general, to be

$$H_j(t) := \frac{1}{\alpha_j}\left(\frac{1_{\{\tau>t\}}\bar{P}(t,T_{j-1})/P(t,T_{j-1})}{1_{\{\tau>t\}}\bar{P}(t,T_j)/P(t,T_j)} - 1\right)$$

(in $\tau > t$), and it is easy to see that we get, in $\tau > t$:

$$H_j(t) = R_j^{PR}(t),$$

but under independence of the default intensity and the interest rates, and not in general. Again, in general R_j comes from imposing a one-period CDS to be fair whereas H_j does not.

A last remark concerns an analogy with the default-free LIBOR market model, where we have Formula (6.34) linking swap rates to forward rates through a weighted average:

$$S_{a,b}(t) = \frac{\sum_{i=a+1}^{b} \alpha_i P(t,T_i) F_i(t)}{\sum_{k=a+1}^{b} \alpha_k P(t,T_k)} = \sum_{i=a+1}^{b} w_i(t; F(t))\; F_i(t).$$

This is useful since it leads to an approximated formula for swaptions in the LIBOR LFM model, as we have seen in the related chapter. A similar approach can be obtained for CDS forward rates. It is easy to check that

$$R_{a,b}^{PR}(t) = \frac{\sum_{i=a+1}^{b} \alpha_i \bar{P}(t,T_i) R_i^{PR}(t)}{\sum_{i=a+1}^{b} \alpha_i \bar{P}(t,T_i)} \tag{21.27}$$

$$= \sum_{i=a+1}^{b} \bar{w}_i(t) R_i^{PR}(t) \approx \sum_{i=a+1}^{b} \bar{w}_i(0) R_i^{PR}(t).$$

A similar relationship for $R_{a,b}^{PR2}$ involving a weighted average of one-period rates is obtained when resorting to the second type of postponed payoff.

A possible lack of analogy with the swap rates is that the $\bar{w}$'s cannot be expressed as functions of the R_i's only, unless we make some particular assumptions on the correlation between default intensities and interest rates. However, if we freeze the $\bar{w}$'s to time 0, which we have seen to work in the default-free LIBOR model, we obtain easily a useful approximate expression for $R_{a,b}$ and its volatility in terms of R_i's and their volatilities/correlations.

More generally, when not freezing, the presence of stochastic intensities besides stochastic interest rates adds degrees of freedom. Now the $\bar{P}$'s (and thus the $\bar{w}$'s) can be determined as functions for example of one- and two-period rates. Indeed, it is easy to show that

$$\bar{P}(t,T_i) = \bar{P}(t,T_{i-1}) \frac{\alpha_{i-1}(R_{i-1}^{PR}(t) - R_{i-2,i}^{PR}(t))}{\alpha_i (R_{i-2,i}^{PR}(t) - R_i^{PR}(t))}. \tag{21.28}$$

With this formulation one has to assume $R_{i-2,i}^{PR}(t) - R_i^{PR}(t) \neq 0$. We will discuss later the dynamics of forward CDS rates. For the time being let us keep in mind that the exact weights $\bar{w}(t)$ in (21.27) are completely specified in terms of $R_i(t)$'s and $R_{i-2,i}(t)$'s, so that if we include these two families of rates in our dynamics the "system" is closed in that we also know all the relevant $\bar{P}$'s.

21.3.8 Defaultable Floater and CDS

Consider a prototypical defaultable floating rate note (FRN).

Definition 21.3.1. Prototypical defaultable floating-rate note. *A prototypical defaultable floating-rate note is a contract ensuring the payment at future times $T_{a+1}, \ldots, T_b$ of the LIBOR rates that reset at the previous instants $T_a, \ldots, T_{b-1}$ plus a spread X, each payment conditional on the issuer having not defaulted before the relevant previous instant. Moreover, the note pays a last cash flow consisting of the reimbursement of the notional value of the note at final time T_b if the issuer has not defaulted earlier. We assume a deterministic recovery value R_{EC} to be paid at the first T_i following default if default occurs before T_b. The note is said to quote at par if its value is equivalent to the value of the notional paid at the first reset time T_a in case default has not occurred before T_a.*

Recall that in a default free world the fair spread making the FRN quote at par is 0, since a prototypical FRN quotes already at par without spreads, see the discussion following Definition 1.5.2.

When in presence of Default, the note discounted payoff, including the initial cash flow on 1 paid in T_a, is

$$\Pi_{\mathrm{DFRN}a,b} = -D(t,T_a)1_{\{\tau>T_a\}} + \sum_{i=a+1}^{b} \alpha_i D(t,T_i)(L(T_{i-1},T_i) + X)\mathbf{1}_{\{\tau>T_i\}}$$

$$+D(t,T_b)1_{\{\tau>T_b\}} + \mathrm{R}_{\mathrm{EC}} \sum_{i=a+1}^{b} D(t,T_i)\mathbf{1}_{\{T_{i-1}<\tau\leq T_i\}},$$

where R_{EC} is the recovery rate, i.e. the percentage of the notional that is paid in replacement of the notional in case of default, and it is paid at the first instant among $T_{a+1}, \ldots, T_b$ following default. This is the correct definition of DFRN, in that the default monitoring is made at the payment time T_i of each LIBOR flow $L(T_{i-1}, T_i)$, and not at its reset T_{i-1}. The problem with such definition is that it has no equivalent in terms of approximated CDS payoff. This is due to the fact that it is difficult to disentangle the LIBOR rate L from the indicator and stochastic discount factor in such a way to obtain expectations of pure stochastic discount factors times default indicators. This becomes possible if we replace $1_{\{\tau>T_i\}}$ in the first summation with $1_{\{\tau>T_{i-1}\}}$, as one can see from computations (21.29) below. This amounts to monitor default at the reset times of LIBOR flows rather than at the payment times, in that the payment is made only if the company has not defaulted at the time of reset of the relevant LIBOR rate.

The same computations, in case we keep $1_{\{\tau>T_i\}}$ in the first summation, even in the simplified case where $\mathrm{L}_{\mathrm{GD}} = 1$ and interest rates are independent of default intensities, would lead us to a corresponding definition of CDS forward rate where protection is paid at the last instant T_i *before* default, which is not natural since one should anticipate default.

A first approximated DFRN payoff

We thus consider two alternative definitions of DFRN. The first one is obtained by moving the default indicator of $L(T_{i-1}, T_i) + X$ from T_i to T_{i-1}. The related FRN discounted payoff is defined as follows:

$$\Pi_{\text{DFRN2}a,b} = -D(t, T_a)1_{\{\tau > T_a\}} + \sum_{i=a+1}^{b} \alpha_i D(t, T_i)(L(T_{i-1}, T_i) + X) \boxed{\mathbf{1}_{\{\tau > T_{i-1}\}}}$$

$$+D(t, T_b)1_{\{\tau > T_b\}} + \text{Rec} \sum_{i=a+1}^{b} D(t, T_i)\mathbf{1}_{\{T_{i-1} \le \tau < T_i\}},$$

Recall that, in the CDS payoff, $\text{Lgd} = 1 - \text{Rec}$. We may now value the above discounted payoff at time t and derive the value of X that makes it 0. Define

$$\text{DFRN2}_{a,b}(t, X, \text{Rec}) = \mathbb{E}\{\Pi_{\text{DFRN2}a,b} | \mathcal{G}_t\}$$

$$= \mathbf{1}_{\{\tau > t\}} \mathbb{E}\{\Pi_{\text{DFRN2}a,b} | \mathcal{F}_t\} / \mathbb{Q}(\tau > t | \mathcal{F}_t)$$

and solve $\mathbb{E}\{\Pi_{\text{DFRN2}a,b} | \mathcal{F}_t\} = 0$ in X. The only nontrivial part is computing

$$\alpha_i \mathbb{E}[D(t, T_i)L(T_{i-1}, T_i)\mathbf{1}_{\{\tau > T_{i-1}\}} | \mathcal{F}_t]$$

$$= \alpha_i \mathbb{E}[\mathbb{E}[D(t, T_i)L(T_{i-1}, T_i)\mathbf{1}_{\{\tau > T_{i-1}\}} | \mathcal{F}_{T_{i-1}}] | \mathcal{F}_t] = \ldots$$

Under a Cox process setting for τ for example, we can write

$$\ldots = \alpha_i \mathbb{E}[\mathbb{E}[D(t, T_i)L(T_{i-1}, T_i)\mathbf{1}_{\{\xi > \Lambda(T_{i-1})\}} | \mathcal{F}_{T_{i-1}}] | \mathcal{F}_t] = \tag{21.29}$$

$$= \alpha_i \mathbb{E}[D(t, T_{i-1})L(T_{i-1}, T_i) \exp(-\Lambda(T_{i-1})) \mathbb{E}[D(T_{i-1}, T_i) | \mathcal{F}_{T_{i-1}}] | \mathcal{F}_t] =$$

$$= \alpha_i \mathbb{E}[\exp(-\Lambda(T_{i-1})) D(t, T_{i-1}) L(T_{i-1}, T_i) P(T_{i-1}, T_i) | \mathcal{F}_t] =$$

$$= \mathbb{E}[\exp(-\Lambda(T_{i-1})) D(t, T_{i-1})(1 - P(T_{i-1}, T_i)) | \mathcal{F}_t] =$$

$$= \mathbb{E}[D(t, T_{i-1})(1 - P(T_{i-1}, T_i))\mathbf{1}_{\{\tau > T_{i-1}\}} | \mathcal{F}_t] =$$

$$= \mathbb{E}[D(t, T_{i-1})\mathbf{1}_{\{\tau > T_{i-1}\}} | \mathcal{F}_t] - \mathbb{E}[D(t, T_i)\mathbf{1}_{\{\tau > T_{i-1}\}} | \mathcal{F}_t]$$

Now the LIBOR flow has vanished from the above payoff and we have expressed everything in terms of pure discount factors and default indicators. Some of these computations could have been performed under more general assumptions simply by means of standard and model independent arguments. Indeed, it would suffice to assume Jamshidian's (2004) conditional independence. We carried out the calculations in the Cox process setting with explicit intensity so that the reader may try them when not replacing $1_{\{\tau > T_i\}}$ to see what goes wrong.

For brevity now we denote by $\mathbb{E}_t$ the risk neutral expectation conditional on $\mathcal{F}_t$.

We may write also

$$\mathrm{DFRN2}_{a,b}(t, X, \mathrm{R_{EC}}) =$$

$$= (1_{\{\tau>t\}}/\mathbb{Q}(\tau > t|\mathcal{F}_t))\Big[- \mathbb{E}_t[D(t,T_a)1_{\{\tau>T_a\}}] + \mathbb{E}_t[D(t,T_b)1_{\{\tau>T_b\}}]$$

$$\sum_{i=a+1}^{b} \mathbb{E}_t[(D(t,T_i) - D(t,T_{i-1}))\mathbf{1}_{\{\tau>T_{i-1}\}}] + X \sum_{i=a+1}^{b} \alpha_i \mathbb{E}_t[D(t,T_i)\mathbf{1}_{\{\tau>T_{i-1}\}}]$$

$$+\mathrm{R_{EC}} \sum_{i=a+1}^{b} \mathbb{E}_t[D(t,T_i)\mathbf{1}_{\{T_{i-1}<\tau\leq T_i\}}]\Big].$$

We may simplify terms in the summations and obtain

$$\mathrm{DFRN2}_{a,b}(t, X, \mathrm{R_{EC}}) = \frac{1_{\{\tau>t\}}}{\mathbb{Q}(\tau > t|\mathcal{F}_t)}\Big[- \mathrm{L_{GD}} \sum_{i=a+1}^{b} \mathbb{E}_t[D(t,T_i)\mathbf{1}_{\{T_{i-1}<\tau\leq T_i\}}]$$

$$+X \sum_{i=a+1}^{b} \alpha_i \mathbb{E}_t[D(t,T_i)\mathbf{1}_{\{\tau>T_{i-1}\}}]\Big],$$

from which we notice en passant that

$$\mathrm{DFRN2}_{a,b}(t, X, \mathrm{R_{EC}}) = \mathrm{PR2CDS}_{a,b}(t, X, 1 - \mathrm{R_{EC}}). \tag{21.30}$$

By taking into account this result, the expression for X that makes the DFRN quote at par is clearly the running "postponed of the second kind" CDS forward rate

$$X^{(2)}_{a,b}(t) = R^{PR2}_{a,b}(t),$$

i.e. *the fair spread in a defualtable floating rate note is equal to the running postponed CDS forward rate.*

A second approximated DFRN payoff

The second alternative definition of DFRN, leading to a useful relationship with approximated CDS payoffs, is obtained by moving the default indicator of $L(T_{i-1}, T_i) + X$ from T_i to T_{i-1} but only for the LIBOR flow, not for the spread X. This payoff is closer to the original Π_{DFRN} payoff than the approximated Π_{DFRN2} payoff considered above. Set

$$\Pi_{\mathrm{DFRN1}a,b} = -D(t,T_a)1_{\{\tau>T_a\}}$$

$$+ \sum_{i=a+1}^{b} \alpha_i D(t,T_i)(L(T_{i-1},T_i) \boxed{\mathbf{1}_{\{\tau>T_{i-1}\}}} + X\mathbf{1}_{\{\tau>T_i\}})$$

$$+D(t,T_b)1_{\{\tau>T_b\}} + \text{R}_{\text{EC}} \sum_{i=a+1}^{b} D(t,T_i)\mathbf{1}_{\{T_{i-1}<\tau\leq T_i\}}.$$

By calling $\text{DFRN1}_{a,b}(t,X,\text{R}_{\text{EC}})$ the t-value of the above payoff and by going through the computations we can see easily that this time

$$\text{DFRN1}_{a,b}(t,X,\text{R}_{\text{EC}}) = \text{PRCDS}_{a,b}(t,X,1-\text{R}_{\text{EC}}), \tag{21.31}$$

and that, as far as fair spreads are concerned,

$$X_{a,b}^{(1)}(t) = R_{a,b}^{PR}(t).$$

21.4 CDS Options and Callable Defaultable Floaters

Consider the option for a protection buyer to enter a CDS at a future time $T_a > 0$, $T_a < T_b$, paying a fixed premium rate K at times $T_{a+1}, \ldots, T_b$ or until default of the reference credit, in exchange for a protection payment L_{GD} against possible default in $[T_a, T_b]$ of the reference credit (payer CDS option). If the option is exercised and default occurs in $[T_a, T_b]$ then L_{GD} is received by the protection buyer. By noticing that the market CDS rate $R_{a,b}(T_a)$ will set the CDS value in T_a to 0, the payoff can be written as the discounted difference between said CDS and the corresponding CDS with CDS premium rate K. We will see below that this is equivalent to a call option on the future CDS fair rate $R_{a,b}(T_a)$. The discounted CDS option payoff reads, at time t,

$$\begin{aligned}\Pi_{\text{CallCDS}_{a,b}}(t;K) = D(t,T_a)[&\text{CDS}_{a,b}(T_a, R_{a,b}(T_a), \text{L}_{\text{GD}}) \\ &-\text{CDS}_{a,b}(T_a, K, \text{L}_{\text{GD}})]^+,\end{aligned}$$

so that, by writing the CDS expressions explicitly (as from (21.20) with $t = T_a$) we have

$$\begin{aligned}\Pi_{\text{CallCDS}_{a,b}}(t;K) = \frac{1_{\{\tau>T_a\}}}{\mathbb{Q}(\tau > T_a|\mathcal{F}_{T_a})} D(t,T_a) \Bigg[&\sum_{i=a+1}^{b} \alpha_i \mathbb{Q}(\tau > T_a|\mathcal{F}_{T_a})\bar{P}(T_a,T_i) + \\ &+ \mathbb{E}\left\{D(T_a,\tau)(\tau - T_{\beta(\tau)-1})\mathbf{1}_{\{\tau<T_b\}}|\mathcal{F}_{T_a}\right\} \Bigg] (R_{a,b}(T_a) - K)^+\end{aligned}$$

These options can be introduced also for postponed CDS. When keeping the original running payoff, if we neglect the accrued interest term, that is known to be small compared to the remaining terms, the payoff simplifies to the following *approximated one*

$$1_{\{\tau>T_a\}} D(t,T_a) \left[\sum_{i=a+1}^{b} \alpha_i \bar{P}(T_a,T_i) \right] (R_{a,b}(T_a) - K)^+$$

This is also the *exact* payoff if we use as underlying CDS price the expectation of the first type of postponed CDS, given in (21.13), and if we replace R by R^{PR}. The term between square brackets is often called defaultable *Annuity* or defaultable *PVBP* (Present Value per Basis Point). We thus have, for the payer CDS option:

Survival at maturity * Discount * [Annuity] * (Call payoff on CDS Rate)

Notice the analogies with the swaption payoff in the interest rate market:

Discount * [Annuity] * (Call payoff on Swap Rate)

Indeed, remember the swaption payoff (6.36) and notice that if we take away the "bar" from the bonds, replace R by S and forget about the default indicator then we have the same structure as with the swaption payoff.

This analogy leads to several ideas and to a completely analogous pricing formula and notion of implied volatility. We will discuss this in Chapter 23, where we will also see some numerical examples of implied volatilities from single name CDS options.

Finally we point out that, given the equivalence between CDS and defaultable floating rate notes, CDS options can be viewed as the option component of callable defaultable floating rate notes.

21.5 Constant Maturity CDS

Consider a contract protecting in $[T_a, T_b]$ against default of a reference credit "C".

If default occurs in $[T_a, T_b]$, a protection payment $\mathrm{L{\scriptstyle GD}}$ is made from the protection seller "B" to the protection buyer "A" at the first T_j following the default time.

This is called, as before, "protection leg".

In exchange for this protection "A" pays to "B" at each T_j before default the "$c+1$–long" (constant maturity) CDS rate $R_{j-1,j+c}(T_{j-1})$ prevailing at time T_{j-1}, with "c" integer > 0.

Notice that a product with $c = 0$ would be fair (initial price $= 0$) and equivalent to a standard CDS (more on this in Chapter 23). It would be a sort of "floating rate CDS", see also Calamaro and Nassar (2004).

We can summarize the situation as follows:

Standard CDS: (initial present value is 0 if $R = R_{a,b}(t)$)

Protection	$\rightarrow$ protection $\mathrm{L{\scriptsize GD}}$ at default τ_C if $T_a < \tau_C \le T_b$ $\rightarrow$	Protection
Seller B	$\leftarrow$ rate R at $T_{a+1}, \ldots, T_b$ or until default τ_C $\leftarrow$	Buyer A

"Floating Rate Equivalent" of Standard CDS: (initial present value is 0)

B	$\rightarrow$ protection $\mathrm{L{\scriptsize GD}}$ at default τ_C if $T_a < \tau_C \le T_b$ $\rightarrow$	
	$\leftarrow$ rate $R_{i-1,i}(T_{i-1})$ at $T_i = T_{a+1}, \ldots, T_b$ or until τ_C $\leftarrow$	A

CMCDS: (initial present value?? Unknown, we need a model)

B	$\rightarrow$ protection $\mathrm{L{\scriptsize GD}}$ at default τ_C if $T_a < \tau_C \le T_b$ $\rightarrow$	
	$\leftarrow$ $R_{i-1,i+c}(T_{i-1})$ at $T_i = T_{a+1}, \ldots, T_b$ or until τ_C $\leftarrow$	A

Similarly to constant maturity swaps, we will see in Section 23.5 that CMCDS bring about a convexity adjustment depending on volatility and correlations among CDS forward rates with different tenors.

Finally, in the market CMCDS are often capped, in that a limit is set to the premium rate $R_{i-1,i+c}(T_{i-1})$ that is paid at each period. In this case it is still possible to find a closed form approximated formula, by resorting to the change of numeraire and drift freezing, plus moment matching techniques for finding approximate (possibly shifted) lognormal distributions for the relevant CDS rates under the suitable measures. We will present a detailed formula with Monte Carlo tests in future work.

For the time being, we give a little market feeling for this product, leaving the detailed treatment to Section 23.5. CDS rates R are usually postponed CDS rates R^{PR} in the following subsection.

21.5.1 Some interesting Financial features of CMCDS

CMCDS have been signalled to us by Salcoacci (2003) as a possibly interesting product. As pointed out for example in Kakodkar and Galiani (2004), or in Due and Ahluwalia (2004), CMCDS are attractive for market investors for the following reasons. Market participants who are investing by selling protection and are optimistic about credit fundamentals future patterns might be concerned about CDS-rates (or spread) widening (increase) risk. Suppose that an investor, a protection seller, has entered a (receiver) CDS position at time 0 when the CDS premium rate is $R(0)$. If at a later time t CDS rates increase (widen) to $R(t) > R(0)$, this means protection has become more expensive. Our investor, with his CDS position set in the past, still receives the $R(0)$ rate, i.e. less than he would receive according to the current market conditions, corresponding to the larger $R(t)$ CDS rate. This is a risk that the holder of a receiver CDS (protection seller) may be willing to cover in some way.

Given formula (21.25) seen earlier, CDS-rates widening (increase) can indeed be interpreted as a widening of the (possibly deterministic) intensity (hazard rate) $\lambda(=\gamma)$ in an intensity setting. Widening of R is thus linked to widening of γ. Since the intensity γ is in turn essentially a "credit spread", as anticipated in Section 21.1.1, we have the term "spread widening".

Since it is notoriously difficult to predict the time of spread widening, and since traditional defensive approaches, when timed incorrectly, could prove to be of little use and also expensive, could an investor resort to CMCDS against this worry? This could be a reasonable solution for the following reasons:

- CMCDS' allow protection selling investors to take floating spread exposure to a credit (see the above formal definition of CMCDS as generalization of "floating rate CDS").
- From the previous point, CMCDS' serve as an effective hedge against spread widening for protection selling investors.
- Again thanks to their "floating rate" feature, CMCDS' have a significantly lower mark to market than a similar plain CDS in the event of parallel spread widening.
- CMCDS' benefit from curve steepening environment: even the simplified Formula

$$\mathrm{CDS}_{\mathrm{CM}_{a,b,c}}(0, \mathrm{L}_{\mathrm{GD}}; \rho = 0) = \sum_{j=a+1}^{b} \alpha_j \bar{P}(0, T_j)(R_{j-1,j+c}(0) \ - R_{j-1,j}(0))$$

  (see the discussion preceeding (23.24) later on) shows that an increase in CDS one-period forward rates implies an increase in the CMCDS position, since in this case longer forward rates $R_{j-1,j+c}$, being an average of increasing one-period rates, get larger than the first one-period rate R_j.
- From the shape of the payoffs we have defined above it is clear that CMCDS' are useful in splitting default risk from spread risk. Spread risk is the risk of an unfavourable change in value of the position due to changes in the R's but not to an actual default event. Default risk is the risk of the actual default event. A short (sell) CMCDS long (buy) CDS position isolates spread risk while hedging default risk (since the protection leg is the same and the default event is the same).

For further discussion on market features the reader is referred to the technical documents by Kakodkar and Galiani (2004), Due and Ahluwalia (2004), Calamaro and Nassar (2004).

21.6 Interest-Rate Payoffs with Counterparty Risk

In this section we show how to handle counterparty risk when pricing some basic interest-rate payoffs. In particular we are going to analyze in detail *counterparty-risk* (or shortly "risky") *Interest Rate Swaps*. The reason to introduce counterparty risk when evaluating a contract is linked to the fact that many financial contracts are traded over the counter (OTC), so that the credit quality of the counterparty can be important. This is particularly appropriated especially when thinking of the different defaults experienced by some important companies in recent years. A further motivation comes from the recommendations of regulatory institutions and of the Basel II framework and "IAS 39" (international accounting standard) system in particular.

We are going to face the problem from the viewpoint of a safe (default-free) counterparty entering a financial contract with another counterparty which has a positive probability of defaulting before the maturity of the contract itself. In general when investing in risky assets we require a *risk premium* which in some sense is a reward for assuming the default risk: If we think, for example, of a corporate bond, we know that the yield is higher than the corresponding yield of an equivalent treasury bond, and this difference is usually called *credit spread*. The positive credit spread implies a lower price for the bond when compared to default free bonds. This is a typical feature of every asset we are going to prove below in general: *The value of a generic claim subject to counterparty risk is always smaller than the value of a similar claim traded with a counterparty having null default probability.*

When evaluating default risky assets one has to introduce the default probabilities in the pricing models. We consider Credit Default Swaps as liquid sources of market default (or, equivalently, survival) probabilities. Different models can be used to calibrate CDS data and obtain survival probabilities: In Brigo and Tarenghi (2004, 2005) and Brigo and Morini (2006) for example *firm value models (or structural models)* are used, whereas in Brigo and Alfonsi (2003, 2005) (more on this later) a stochastic intensity model is used.

Remark 21.6.1. **(Model-independent formulas).** *As we were suggested to point out in Brigo and Masetti (2005b) by Pykhtin (2005), the formulas we develop in this section can be interpreted as model independent, with the only assumption of independence between the default time τ of the counterparty and default free interest rates. The spirit is the same as we have seen for CDS valuation in Section 21.3.5. Indeed, recall that according to Formulas (21.22,21.23) a term structure of CDS's can be calibrated through model independent survival probabilities. Although in this section, for simplicity, we will assume existence of a deterministic intensity consistent with such survival probabilities, this is not necessary and all the formulas given here stand as model-independent, with the only assumption of independence between τ*

and $D(s,t)$ for all $0 < s < t$. This is the real assumption we make throughout this section and the related subsections.

Indeed, we stick to this convention by adopting intensity models in their simplest formulation (deterministic intensity), which we briefly introduced earlier in this chapter and will develop further in Chapter 22.

Our discussion of counterparty risk pricing starts in Section 21.6.1 with a general formula for counterparty risk valuation in a derivative transaction. We will show that the derivative price in presence of counterparty risk is just the default free price minus a discounted option term in scenarios of early default times the loss given default. The option is on the residual present value at time of default. In Section 21.6.2 we apply this formula to a single interest rate swap (IRS). We find the already known result (see also Arvanitis and Gregory (2001) Chapter 6, Bielecki and Rutkowski (2001) Chapter 14, among other references) that the IRS price under counterparty risk is the sum of swaption prices with different maturities, each weighted with the probability of defaulting around that maturity. Things become more interesting when we consider a portfolio of IRS's towards a single counterparty in presence of netting agreements. Roughly speaking, a netting agreement implies that when a counterparty defaults, all positions towards that counterparty are netted, i.e. are valued as a portfolio. If the portfolio value is positive to the non-defaulted company, then this company receives only the recovery fraction of it. If the value is negative, the non-defaulted company pays it in full to the defaulted company. This is different from applying the recovery mechanism to single positions, since an option on a portfolio is smaller than the corresponding portfolio of options. Indeed, a netting agreement may be helpful in diminishing the counterparty risk price component, since an option on a portfolio is smaller than the portfolio of options (similarly to how a swaptions is smaller than a cap, as noticed in Section 1.6). Therefore, under netting agreements, when default occurs, we need to consider the option on the residual present value of the whole portfolio. This option cannot be valued as a standard swaption, and we need either to resort to Monte Carlo simulation (under the LIBOR model, or alternatively the swap model) or to derive analytical approximations. We derive an analytical approximation based on the standard "drift freezing" technique for swaptions pricing in the LIBOR model in Brigo and Masetti (2005, 2005b). The approximated formula is well suited to risk management, where the computational time under each risk factors scenario is crucial and an analytical approximation keeps it small.

21.6.1 General Valuation of Counterparty Risk

Let us call T the final maturity of the payoff we are going to evaluate. If $\tau > T$ there is no default of the counterparty during the life of the product and the counterparty has no problems in repaying the investors. On the contrary, if

$\tau \le T$ the counterparty cannot fulfill its obligations any longer and the following happens. At τ the Net Present Value (NPV) of the residual payoff until maturity is computed: If this NPV is negative (respectively positive) for the investor (defaulted counterparty), it is completely paid (received) by the investor (counterparty) itself. If the NPV is positive (negative) for the investor (counterparty), only a recovery fraction Rec of the NPV is exchanged. In this section all the expectations $\mathbb{E}_t$ are taken under the measure $\mathbb{Q}$ and with respect to the filtration $\mathcal{G}_t$.

Let us call $\Pi^D(t)$ the payoff of a generic defaultable claim at t and $\text{Cashflow}(u,s)$ the net cash flows of the claim between time u and time s, discounted back at u, all payoffs seen from the point of view of the company facing counterparty risk. Then we have $\text{NPV}(\tau) = \mathbb{E}_\tau\{\text{Cashflow}(\tau, T)\}$ and

$$\begin{aligned} \Pi^D(t) &= \mathbf{1}_{\{\tau > T\}} \text{Cashflow}(t,T) + \\ &+ \mathbf{1}_{\{\tau \le T\}} \left[\text{Cashflow}(t,\tau) + D(t,\tau) \left(\text{Rec}\,(\text{NPV}(\tau))^+ - (-\text{NPV}(\tau))^+ \right) \right]. \end{aligned} \tag{21.32}$$

The expected value of this last expression is the general price under counterparty risk. Indeed, if there is no early default this expression reduces to risk neutral valuation of the payoff (first term in the right hand side); in case of early default, the payments due before default occurs are received (second term), and then if the residual net present value is positive only a recovery of it is received (third term), whereas if it is negative it is paid in full (fourth term).

Calling $\Pi(t)$ the payoff for an equivalent claim with a default-free counterparty, it is possible to prove the following

Proposition 21.6.1. (General counterparty risk pricing formula). *At valuation time t, and on $\{\tau > t\}$, the price of our payoff under counterparty risk is*

$$\mathbb{E}_t\{\Pi^D(t)\} = \mathbb{E}_t\{\Pi(t)\} - \text{Lgd}\,\mathbb{E}_t\{\mathbf{1}_{\{\tau \le T\}} D(t,\tau)\,(NPV(\tau))^+\} \tag{21.33}$$

where $\text{Lgd} = 1 - \text{Rec}$ is the Loss Given Default *and the recovery fraction Rec is assumed to be deterministic. It is clear that the value of a defaultable claim is the sum of the value of the corresponding default-free claim minus an option part, in the specific a call option (with zero strike) on the residual NPV giving nonzero contribution only in scenarios where $\tau \le T$. Counterparty risk thus adds an optionality level to the original payoff, and in particular payoffs whose default-free valuation is model independent become model dependent.*

From this proposition it is clear that, as anticipated earlier, the value of a generic claim subject to counterparty risk is always smaller than the value of a similar claim traded with a counterparty having null default probability.

Finally notice that, as is clear from the proof, the proposition holds under more general assumptions than those adopted in this section (in particular, one does not need independence of τ from interest rates).

We now prove the proposition. The reader that is not interested in technicalities may go directly to the next section, where we start to apply this formula to particular financial contracts, in the specific to interest rate swaps.

Proof. Since

$$\Pi(t) = \text{Cashflow}(t,T) = \mathbf{1}_{\{\tau>T\}}\text{Cashflow}(t,T) + \mathbf{1}_{\{\tau\le T\}}\text{Cashflow}(t,T) \tag{21.34}$$

we can rewrite the terms inside the expectations in the right hand side of (21.33) as

$$\begin{aligned}
&\mathbf{1}_{\{\tau>T\}}\text{Cashflow}(t,T) + \mathbf{1}_{\{\tau\le T\}}\text{Cashflow}(t,T)\\
&+ \{(\text{Rec}-1)[\mathbf{1}_{\{\tau\le T\}}D(t,\tau)(\text{NPV}(\tau))^+]\}\\
&= \mathbf{1}_{\{\tau>T\}}\text{Cashflow}(t,T) + \mathbf{1}_{\{\tau\le T\}}\text{Cashflow}(t,T)\\
&+ \text{Rec}\ \mathbf{1}_{\{\tau\le T\}}D(t,\tau)(\text{NPV}(\tau))^+ - \mathbf{1}_{\{\tau\le T\}}D(t,\tau)(\text{NPV}(\tau))^+
\end{aligned} \tag{21.35}$$

The second and the fourth terms have expectation equal to

$$\begin{aligned}
&\mathbb{E}_t[\mathbf{1}_{\{\tau\le T\}}\text{Cashflow}(t,T) - \mathbf{1}_{\{\tau\le T\}}D(t,\tau)(\text{NPV}(\tau))^+] =\\
&= \mathbb{E}_t\{\mathbf{1}_{\{\tau\le T\}}[\text{Cashflow}(t,\tau) + D(t,\tau)E_\tau[\text{Cashflow}(\tau,T)]\\
&\quad -D(t,\tau)(E_\tau[\text{Cashflow}(\tau,T)])^+]\}\\
&= \mathbb{E}_t\{\mathbf{1}_{\{\tau\le T\}}[\text{Cashflow}(t,\tau) - D(t,\tau)(E_\tau[\text{Cashflow}(\tau,T)])^-]\}\\
&= \mathbb{E}\{\mathbf{1}_{\{\tau\le T\}}[\text{Cashflow}(t,\tau) - D(t,\tau)(E_\tau[-\text{Cashflow}(\tau,T)])^+]\}\\
&= \mathbb{E}\{\mathbf{1}_{\{\tau\le T\}}[\text{Cashflow}(t,\tau) - D(t,\tau)(-\text{NPV}(\tau))^+]\}
\end{aligned} \tag{21.36}$$

(where we used the tower property of conditional expectation) since by no-arbitrage

$$\begin{aligned}
\mathbb{E}_\tau[\mathbf{1}_{\{\tau\le T\}}\text{Cashflow}(t,T)] = [\mathbf{1}_{\{\tau\le T\}}\{&\text{Cashflow}(t,\tau)\\
&+D(t,\tau)E_\tau[\text{Cashflow}(\tau,T)]\}
\end{aligned} \tag{21.37}$$

and $f = f^+ - f^- = f^+ - (-f)^+$. Substituting (21.36) into (21.35) and then the result inside the expectation in the right hand side of (21.33), we have just the price of (21.32).

□

21.6.2 Counterparty Risk in single Interest Rate Swaps (IRS)

The original result for standard IRS with counterparty risk is due to Soerensen and Bollier (1994), as pointed out by Cherubini (2005). Here we derive the result from scratch, consistently with the notation of our book.

Let us suppose that we are a default-free counterparty "A" entering a payer swap with a defaultable counterparty "B", exchanging fixed for floating

payments at times $T_{a+1}, \ldots, T_b$.
Denote by α_i the year fraction between T_{i-1} and T_i

Of course, if we consider the possibility that "B" may default, the correct spread to pay in the fixed leg to make the contract fair at inception is lower than the forward swap rate $S_{a,b}(t)$, since we are willing to be rewarded for bearing this default risk. In particular, using the previous formula (21.33) we find the price of a payer (forward) swap under default risk of the counterparty (for the derivation see also Bielecki and Rutkowski (2001), the receiver IRS case is analogous)

$$\mathbf{PFS}^D(t) = \mathbf{PFS}(t) - \text{L}_{\text{GD}} \cdot \text{OP}(t) \tag{21.38}$$

where OP is the option part

$$\begin{aligned}\text{OP}(t) &= \mathbb{E}_t\{\mathbf{1}_{\{\tau \le T_b\}} D(t,\tau)(\text{NPV}(\tau))^+\} = \\ &= \int_{T_a}^{T_b} \mathbf{PS}(t; s, T_b, K, S(t; s, T_b), \sigma_{s,T_b}) d_s \mathbb{Q}_t(\tau \le s) \end{aligned} \tag{21.39}$$

being $\mathbf{PS}(t; s, T_b, K, S(t; s, T_b), \sigma_{s,T_b})$ the price in t of a payer swaption (it would be a receiver if the basic IRS were a receiver IRS) with maturity s, strike K, underlying forward swap rate $S(t; s, T_b)$, volatility σ_{s,T_b} and underlying swap with maturity T_b. When $s = T_j$ for some j we replace the arguments s, T_b by indices j, b, consistently with our earlier notation on forward swap rates in the book. From now on in this section $\mathbb{Q}_t$ denotes expectation conditional on $\tau > t$ and on $\mathcal{F}_t$, both information including $\mathcal{G}_t$ when put together. The proof is simple, here we only sketch it. The only nontrivial term can be written as

$$\begin{aligned}\mathbb{E}_t\{\mathbf{1}_{\{\tau \le T_b\}} D(t,\tau)(\text{NPV}(\tau))^+\} &= \mathbb{E}_t\left[\int_t^{T_b} D(t,s)(\text{NPV}(s))^+ 1_{\{\tau \in [s,s+ds)\}}\right] \\ &= \int_t^{T_b} \mathbb{E}_t[D(t,s)(\text{NPV}(s))^+ 1_{\{\tau \in [s,s+ds)\}}] \\ &= \int_t^{T_b} \mathbb{E}_t[D(t,s)(\text{NPV}(s))^+]\mathbb{E}_t[1_{\{\tau \in [s,s+ds)\}}] \\ &= \int_t^{T_b} \mathbb{E}_t[D(t,s)(\text{NPV}(s))^+]\mathbb{Q}_t\{\tau \in [s,s+ds)\} \\ &= \int_t^{T_b} \mathbb{E}_t[D(t,s)(\text{NPV}(s))^+] d_s \mathbb{Q}_t\{\tau \le s\}\end{aligned}$$

and since the residual NPV of an IRS is a forward start IRS, the positive part of this turns out to be a swaption discounted payoff, whose expectation $\mathbb{E}_t[D(t,u)(\text{NPV}(u))^+]$ gives the swaption price in the integrand. In the above steps we have first used Fubini's theorem to switch the time integral with the expectation and then the independence of τ from all interest rate quantities, allowing us to factor the expectation. Notice that formally we should have written $\delta(\tau - s)ds$ (with δ the Dirac mass centered in 0) rather than $1_{\{\tau\in[s,s+ds)\}}$; as in other parts of the book, we privilege our more intuitive if less rigorous notation.

We can simplify (21.39) through some assumptions: We allow the default to happen only at points T_i of the grid of payments of the fixed leg. In particular two different specifications could be applied: One for which the default is anticipated to the first T_i preceding τ and one for which it is postponed to the first T_i following τ. In this way the option part in (21.39) is simplified. Indeed, in the case of the postponed ("P") payoff we obtain

$$\begin{aligned}\mathrm{OP}^P(t) &= \sum_{i=a+1}^{b-1} \mathbb{Q}_t\{\tau \in (T_{i-1}, T_i]\}\, \mathbf{PS}_{i,b}(t; K, S_{i,b}(t), \sigma_{i,b}) \qquad (21.40)\\ &= \sum_{i=a+1}^{b-1} (\mathbb{Q}_t(\tau > T_{i-1}) - \mathbb{Q}_t(\tau > T_i))\, \mathbf{PS}_{i,b}(t; K, S_{i,b}(t), \sigma_{i,b})\end{aligned}$$

and this can be easily computed summing across the T_i's and using the default probabilities implicitly given in market CDS prices.
A similar result can be obtained considering the anticipated ("A") default

$$\begin{aligned}\mathrm{OP}^A(t) &= \sum_{i=a+1}^{b} \mathbb{Q}_t\{\tau \in (T_{i-1}, T_i]\}\, \mathbf{PS}_{i-1,b}(t; K, S_{i-1,b}(t), \sigma_{i-1,b})\\ &= \sum_{i=a+1}^{b} (\mathbb{Q}_t(\tau > T_{i-1}) - \mathbb{Q}_t(\tau > T_i))\, \mathbf{PS}_{i-1,b}(t; K, S_{i-1,b}(t), \sigma_{i-1,b})\end{aligned} \qquad (21.41)$$

We carried out some numerical experiments to analyze the impact of counterparty risk on the fair rate of the swap. For the discounts and the swap rates we used the data of March 10th, 2004. The volatility matrix of the swaptions has been chosen arbitrarily and in particular we kept a fix value of 15%. We also considered different default-risk profiles for the counterparty, studying stylized cases of high default risk, medium default risk and low default risk. Also, we choose a piecewise constant intensity γ. The choice of the shape of γ poses some problems. Mainly, one might face the problem of evaluating a 30 years swap when the market quotes CDS' only up to a 10 years maturity. In this case we need to strip the intensities from the available CDS' and then we have to extrapolate the intensity values for the longer maturities, or perhaps use bond information if available. If we use piecewise linear intensity then, when extrapolating up to 20 years by prolonging we could find strange results (in principle also with negative probabilities). The use of a piecewise constant intensity has the drawback of not being continuous but generally provides less dramatic results (at least granting positive default probabilities after 10 years). In Table 21.1 we report the survival probabilities (that, remember, in principle could be derived as model independent) and the intensities in the three cases with different credit quality for the counterparty while in Table 21.2 we report the risk-free swap rates for different maturities together with the spread that has to be subtracted in the defalt risky cases to make the swap fair when including counterparty risk.

Date	Intensity	Survival	Intensity	Survival	Intensity	Survival
10-mar-04	0.0036	100.00%	0.0202	100.00%	0.0534	100.00%
12-mar-05	0.0036	99.64%	0.0202	97.96%	0.0534	94.70%
12-mar-07	0.0065	98.34%	0.0231	93.48%	0.0564	84.47%
12-mar-09	0.0099	96.38%	0.0266	88.57%	0.0600	74.78%
12-mar-11	0.0111	94.24%	0.0278	83.71%	0.0614	66.03%
12-mar-14	0.0177	89.31%	0.0349	75.27%	0.0696	53.42%
12-mar-19	0.0177	81.64%	0.0349	63.05%	0.0696	37.53%
12-mar-24	0.0177	74.63%	0.0349	52.80%	0.0696	26.36%
12-mar-29	0.0177	68.22%	0.0349	44.23%	0.0696	18.51%
12-mar-34	0.0177	62.36%	0.0349	37.05%	0.0696	13.01%

Table 21.1. Intensities and related survival probabilities in three different cases (from left to right: low, medium and high default risk) for the credit quality of the counterparty.

Maturity (yrs)	Risk-free swap rate	Antic.	Postp.	Antic.	Postp.	Antic.	Postp.
5	3.249%	0.64	0.51	1.91	1.80	4.27	4.25
10	4.074%	2.52	2.16	6.09	5.8	12.28	12.26
15	4.463%	4.92	4.48	10.52	10.2	19.55	19.68
20	4.675%	7.24	6.77	14.51	14.21	25.45	25.77
25	4.775%	9.1	8.64	17.53	17.28	29.46	29.93
30	4.811%	10.51	10.06	19.67	19.46	31.97	32.54

Table 21.2. Risk-free implied swap rate and related counterparty risk spread negative adjustment (under low , medium and high default risk as given in Table 21.1). We report the spread to be subtracted (in basis points) from the risk free interest rate for both anticipated and postponed default approximations.

We see that, as expected, the spread adjustment (to be subtracted from the implied risk-free swap rate) grows together with the default riskiness of the counterparty, and also with the increasing maturity of the underlying swap. Also we see that the difference between the two approximations is very low (most of times it is smaller than 0.5 bps). One could decide to use an average of the two values, just to reach a better proxy for the exact correction in (21.39), but in any case the error would be negligible for most practical purposes.

As a further remark we mention the fact that in case we enter a *receiver* IRS and still consider that only our counterparty can default, a similar procedure can be applied, but in that case we have a higher value for the swap rate K than in the default free case (it is intuitive since we are receiving the fixed leg and we want a premium to bear the default risk of "B").

So far we considered only the case of counterparty risk for a single interest rate swap. What happens, however, in case of a portfolio of several IRS towards a single counterparty, in presence of netting agreements? At the default time of the counterparty, the portfolio value is netted, and if the total portfolio value is positive to us, we only receive a recovery of it, whereas if it is negative we pay the whole related amount to the counterparty. The difference with the earlier case is that now we have options on portfolios of IRS and not on single standard IRS. As a consequence, we no longer get a stream of standard swaptions. Brigo and Masetti (2005, 2005b), however, derive approximated formulas based on the usual drift freezing technique for the LIBOR model plus moment matching solutions, also in case of netting agreements or non-standard IRS (amortizing, zero-coupon, bullet, etc.), and test it with Monte Carlo simulations.

We now go back to single payoffs without netting and present a summary of basic interest rate products valuation formulas under counterparty risk. The derivation is analogous or even simpler than that of the interest rate swaps. Default is taken in the anticipated formulation. The postponed formulation is analogous, and one could also take an average of the two formulations outputs as a final approximation.

Swaptions (payer and receiver)

$$\boxed{\mathbf{PS}^D_{a,b}(t;K) = \mathbf{PS}_{a,b}(t;K)(1 - \text{L}_{\text{GD}}\,\mathbb{Q}_t\{\tau \le T_a\})}$$

$$\boxed{\mathbf{RS}^D_{a,b}(t;K) = \mathbf{RS}_{a,b}(t;K)(1 - \text{L}_{\text{GD}}\,\mathbb{Q}_t\{\tau \le T_a\})}$$

Caplets e Floorlets

$$\boxed{\mathbf{Cpl}^D(t,K) = \mathbf{Cpl}(t,K)(1 - \text{L}_{\text{GD}}\mathbb{Q}_t\{\tau \le T\})}$$

$$\boxed{\mathbf{Fll}^D(t,K) = \mathbf{Fll}(t,K)(1 - \text{L}_{\text{GD}}\mathbb{Q}_t\{\tau \le T\})}$$

Interest Rate Swaps (payer and receiver)

$$\boxed{\mathbf{PFS}^D_{a,b}(t,K) = \mathbf{PFS}_{a,b}(t,K) - \text{L}_{\text{GD}} \sum_{j=a+1}^{b} \mathbb{Q}_t\{\tau \in (T_{j-1},T_j]\}\mathbf{PS}_{j-1,b}(t;K)}$$

$$\mathbf{RFS}^D_{a,b}(t,K) = \mathbf{RFS}_{a,b}(t,K) - \text{L}_{\text{GD}} \sum_{j=a+1}^{b} \mathbb{Q}_t\{\tau \in (T_{j-1}, T_j]\} \mathbf{RS}_{j-1,b}(t;K)$$

Caps e Floors

$$\mathbf{Cap}^D_{a,b}(t,K) = \mathbf{Cap}_{a,b}(t,K) - \text{L}_{\text{GD}} \sum_{j=a+1}^{b} \mathbb{Q}_t\{\tau \in (T_{j-1}, T_j]\} \mathbf{Cap}_{j-1,b}(t,K)$$

$$\mathbf{Flr}^D_{a,b}(t,K) = \mathbf{Flr}_{a,b}(t,K) - \text{L}_{\text{GD}} \sum_{j=a+1}^{b} \mathbb{Q}_t\{\tau \in (T_{j-1}, T_j]\} \mathbf{Flr}_{j-1,b}(t,K)$$

22. Intensity Models

This... fog has been theotropically engineered with Apokolips technology ... super-stochastic effects are creating local zones of increased... confusion...

"JLA: Rock of Ages", DC Comics

In this chapter we focus completely on intensity models, exploring in details also the issues we have anticipated in the earlier chapter in order to be able to deal with Credit Default Swap (CDS) and notions of implied hazard rates and functions. Before proceeding further with the chapter we advise the reader that has not done so yet to have a look at the guided tour in Section 21.1, especially at the subsection on intensity models, just to put the intensity models we are going to examine here into perspective. In the introductory section we recall some of the aspects seen earlier on intensity models in general and also present the detailed structure of the chapter.

22.1 Introduction and Chapter Description

Intensity models, part of the family of reduced form models, all move from the basic idea of describing the default time τ as the first jump time of a Poisson process. Default is not induced by basic market observables and/or economic fundamentals, but has an exogenous component that is independent of all the default free market information. Monitoring the default free market (interest rates, exchange rates, etc) does not give complete information on the default process, and there is no economic rationale behind default. This family of models is particularly suited to model credit spreads and in its basic formulation is easy to calibrate to Credit Default Swap (CDS) or corporate bond data. We have presented a sketch of the reason why this happens already in Section 21.1.1.

Here we are going into more details. The basic facts from probability are essentially the theory of Poisson and Cox processes. The reader who is not familiar with such notions is encouraged to have a look at Appendix C.6. This is not strictly necessary for a first, more colloquial understanding of the subject, since in Section 22.2 of this chapter we are going to review the main facts on Poisson processes that will be useful in the following. We start from the simplest, constant intensity Poisson process and explain the interpretation of the intensity as a probability of first jumping (defaulting) per unit of time. We then move to time-inhomogeneous Poisson processes (time varying deterministic intensity), that allow to model credit spreads without volatility. Here the probability of first jumping (defaulting) is a function of time. Further, we move to stochastic intensity Poisson processes, where the probability of first jumping (defaulting) is itself random and follows a stochastic process of a certain kind. This last case is referred to as "Cox process" approach, or "doubly stochastic Poisson process", since we have stochasticity both in the jump variable ξ and in the intensity λ. This approach allows us to take into account credit spread volatility. In all three cases of constant, deterministic-time-varying and stochastic intensity we point out how the Poisson process structure allows to view survival probabilities as discount factors, the intensity as credit spread, and how this helps us in recycling the interest-rate technology for default modeling.

In Section 22.3 we analyze in detail the CDS calibration with deterministic intensity models, illustrating the notion of implied hazard function with a case study based on Parmalat CDS data.

Section 22.4 illustrates how the only hope of inducing dependence between the default event and interest rates in a diffusion setting is through a stochastic intensity correlated with the interest rate. If we keep deterministic intensities we cannot really have dependence between the default time and the short rate when we adopt a diffusion model for the short rate r.

Section 22.5 illustrates the fundamental idea of conditioning only to the partial information of the default free market when pricing credit derivatives. The related fundamental filtration switching formula is introduced and proven. This is possible following a fundamental technical result concerning filtrations. Although at first sight this result appears to be a mere technicality, it has fundamental consequences in that it will allow us later to define the CDS market model under a measure that is equivalent to the risk neutral one. Also, our definition of forward CDS rate itself owes much to this result, as pointed out earlier in Section 21.3.3.

Section 22.6 explains how to simulate the default time, illustrating the notion of standard error and presenting suggestions on how to keep the number of paths under control. These suggestions take into account peculiarities of default modeling that make the variance reduction more difficult than in the default free market case.

Section 22.7 introduces our choice for the stochastic intensity in a diffusion setting, following Brigo and Alfonsi (2003, 2005) and Brigo and Cousot (2004). We term the stochastic intensity and interest rate model SSRD: Shifted Square Root Diffusion model. This is a CIR++ model for the intensity λ correlated with a CIR++ model for the short rate r. We argue this choice is the only reasonable one in a diffusion setting for the intensity given that one wishes analytical tractability for survival probabilities (CDS calibration) and positivity of the intensity process. We show how to calibrate the SSRD model to CDS quotes and interest rate data in a separable way, and argue that the instantaneous correlation ρ between r and λ has a negligible impact on the CDS price, allowing us to maintain the separability of the λ and r calibration in practice even when $\rho \neq 0$.

We present some original numerical schemes for the simulation of the SSRD model that preserve positivity of the discretized process and analyze the convergence of such schemes. We also introduce a Gaussian mapping technique that maps the model into a two factor Gaussian model, where calculations in presence of correlation are much easier. We analyze the mapping procedure and its accuracy by means of Monte Carlo tests. We also analyze the impact of the correlation ρ on some prototypical payoffs, finding cases where ρ does have an impact on the payoff, contrary to the CDS case. As an exercise we price a cancellable structure with the stochastic intensity model. We also introduce a CDS option closed form formula under deterministic r and CIR++ stochastic λ, a particular case of the SSRD model. We analyze implied CDS volatilities patterns in the full SSRD case by means of Monte Carlo simulation. Finally, we explain why the CIR++ model for λ cannot attain large levels (such as 50%) of implied option volatilities for CDS rates, and introduce jumps in the CIR++ model, hinting at the JCIR and JCIR++ models and at their possible calibration to both CDS and options. This model is developed with some detail, is still affine and the related survival probability formulas are re-cast in a notation resembling the classic CIR bond price formula, as from Brigo and El-Bachir (2005). In a multi-name situation, one might even attempt to put a dependence structure across the jumps in the different jump-diffusion intensities to induce a possibly strong dependence in the default times of the different names themselves, while keeping the ξ variables of different names independent. This would allow us to preserve all the benefits of keeping independent ξ's while allowing us perhaps to attain high levels of dependence for the default indicators of different names, solving the problems of solution 1) in Section 21.1.8. This is currently under investigation.

22.2 Poisson processes

"I need strategy. I need tricks. I need to remember my science or I'm history."

Flash/Wally West, from "JLA: New World Order", DC Comics.

Poisson processes are described into detail in Appendix C.6. Here we recall some important facts.

22.2.1 Time homogeneous Poisson processes

A "time homogeneous Poisson process" $\{M_t,\ t \geq 0\}$ is a *unit-jump increasing, right continuous process* with *stationary independent increments* and null initial condition, $M_0 = 0$.

It turns out these seemingly general requirements imply strong facts on the process distribution and on its jump times. Indeed, let $\tau^1, \tau^2, \ldots, \tau^m, \ldots$ be the first, second etc. jump times of M. There exists a positive constant $\bar{\gamma}$ such that

$$\mathbb{Q}\{M_t = 0\} = \mathbb{Q}\{\tau^1 > t\} = \exp(-\bar{\gamma}t)$$

for all t. Also,

$$\lim_{t \to 0} \mathbb{Q}\{M_t \geq 2\}/t = 0$$

(zero probability of having more than one jump in arbitrarily small intervals) and

$$\lim_{t \to 0} \mathbb{Q}\{M_t = 1\}/t = \bar{\gamma}.$$

The constant $\bar{\gamma}$ further obeys

$$\bar{\gamma} = \mathbb{E}(M_t)/t = \text{Var}(M_t)/t$$

(average arrival rate and variance per unit of time).

The distribution of the process itself turns out to be the Poisson law, in that

$$\mathbb{Q}\{M_t - M_s = k\} = e^{-\bar{\gamma}(t-s)}(\bar{\gamma}(t-s))^k/k!$$

i.e. $M_t - M_s \sim$ PoissonRandomVariable($(t-s)\bar{\gamma}$), with $M_t - M_s$ independent of $\sigma(\{M_u, u \leq s\})$, i.e. of the history of the process itself up to time s.

As for the jump times, their distribution is interesting.
Indeed, $\tau^1, \tau^2 - \tau^1, \tau^3 - \tau^2, \ldots$ i.e. the times between one jump and the next one, are independent and identically distributed (iid) as an exponential random variable of parameter $\bar{\gamma}$ (with mean $1/\bar{\gamma}$). In particular, $\bar{\gamma}\ \tau^1$ is a standard exponential random variable. It is this exponential distribution that gives us the interpretation of the intensity $\bar{\gamma}\, dt$ as the probability of defaulting (first jumping) in $[t, t+dt)$ having not defaulted (jumped) before: If we define the default time as the first jump, $\bar{\tau} := \tau^1$, then $\mathbb{Q}\{\bar{\tau} \in [t, t+dt) | \bar{\tau} \geq t\} = \bar{\gamma}\, dt$

In the simplest intensity model the default time is indeed modelled as τ^1.

Summing up, if we define the default time as $\bar{\tau} := \tau^1$, the first jump time of M_t, then $\bar{\tau}$ has the following properties: $\bar{\gamma}\ \bar{\tau} \sim$ exponentialRandVar(1), independent of $\mathcal{F}$, and we have $\mathbb{Q}\{\bar{\tau} > t\} =$

$$= \mathbb{Q}\{\bar{\gamma}\bar{\tau} > \bar{\gamma}t\} = \mathbb{Q}\{\text{exponentialRV}(1) > \bar{\gamma}t\} = e^{-\bar{\gamma}t} \Rightarrow$$

$$\Rightarrow \mathbb{Q}\{\bar{\tau} \in [t, t+dt) | \bar{\tau} \geq t\} = \frac{\mathbb{Q}\{\bar{\tau} \in [t, t+dt) \bigcap \bar{\tau} \geq t\}}{\mathbb{Q}\{\bar{\tau} > t\}} = \frac{\mathbb{Q}\{\bar{\tau} \in [t, t+dt)\}}{\mathbb{Q}\{\bar{\tau} > t\}}$$

$$= \frac{\mathbb{Q}\{\bar{\tau} > t\} - \mathbb{Q}\{\bar{\tau} > t + dt\}}{\mathbb{Q}\{\bar{\tau} > t\}} = \frac{e^{-\bar{\gamma}t} - e^{-\bar{\gamma}(t+dt)}}{e^{-\bar{\gamma}t}} \approx \bar{\gamma}\, dt$$

for small dt; *"probability that company defaults in (arbitrarily small) "dt" years given that it has not defaulted so far is* $\bar{\gamma}\, dt$." Also, the probability of defaulting between s and t is

$$\mathbb{Q}\{s < \bar{\tau} \leq t\} = \exp(-\bar{\gamma}s) - \exp(-\bar{\gamma}t) \approx \bar{\gamma}(t - s),$$

where the approximation holds only when t is close to s.

In particular, the formula

$$\mathbb{Q}\{\bar{\tau} > t\} = e^{-\bar{\gamma}t}$$

is very important. It tells us that *survival probabilities have the same structure as discount factors, with the default intensity playing the role of interest rates.* Indeed, in a world with deterministic and constant short rate r we have

$$P(0, t) = e^{-r\, t}.$$

This analogy survival probability/zero coupon bond is an important consequence of the exponential distribution for times between jumps.

It is this fundamental property of jumps of Poisson processes that allows us to see survival probabilities as discount factors, and thus default intensities as credit spreads. This allows us to use much of the interest-rate technology in default modeling under this kind of reduced form models.

22.2.2 Time inhomogeneous Poisson Processes

We consider now *deterministic time-varying* intensity $\gamma(t)$, which we assume to be a positive and piecewise (right-) continuous function. We define

$$\Gamma(t) := \int_0^t \gamma(u)du,$$

the *cumulated intensity, cumulated hazard rate,* or also *Hazard function.*

If M_t is a Standard Poisson Process, i.e. a Poisson Process with intensity one, then a *time-inhomogeneous Poisson Process* N_t with intensity γ is defined as

$$N_t = M_{\Gamma(t)}.$$

So a time inhomogeneous Poisson process is just a time-changed standard Poisson process.

The process N_t is still increasing by jumps of size 1, its increments are still independent, but they are *no longer identically distributed* due to the "time distortion" introduced by Γ.

From $N_t = M_{\Gamma(t)}$ we have obviously

N jumps the first time at $\tau \iff M$ jumps the first time at $\Gamma(\tau)$.

But since we know that M is standard Poisson Process for which the first jump time is an exponential random variable, then

$$\Gamma(\tau) =: \xi \sim \text{exponentialRandomVariable}(1).$$

This gives us a fundamental well known fact on Poisson processes: *If we transform the first jump time of a Poisson process according to its cumulated intensity we obtain a standard exponential random variable independent of all previous processes in the given probability space (independent of $\mathcal{F}$).*

By inverting this last equation we have that

$$\tau = \Gamma^{-1}(\xi),$$

with ξ standard exponential random variable, which implicitly suggests us how to simulate the default time. Also, we have easily

$$\mathbb{Q}\{s < \tau \leq t\} = \mathbb{Q}\{\Gamma(s) < \Gamma(\tau) \leq \Gamma(t)\} = \mathbb{Q}\{\Gamma(s) < \xi \leq \Gamma(t)\} =$$

$$= \mathbb{Q}\{\xi > \Gamma(s)\} - \mathbb{Q}\{\xi > \Gamma(t)\} = \exp(-\Gamma(s)) - \exp(-\Gamma(t)) \text{ i.e.}$$

"prob of default between s and t is "$e^{-\int_0^s \gamma(u)du} - e^{-\int_0^t \gamma(u)du} \approx \int_s^t \gamma(u)du$" (where the final approximation is good for small exponents). It is easy to show, along the same lines, that

$$\mathbb{Q}\{\tau \in [t, t+dt)\} = \mathbb{Q}\{\tau \geq t\} - \mathbb{Q}\{\tau \geq t + dt\} = e^{-\int_0^t \gamma(u)du}\gamma(t)\, dt. \quad (22.1)$$

However, conditional on survival at time t, we may compute

$$\mathbb{Q}\{\tau \in [t, t+dt) | \tau \geq t\} = \frac{\mathbb{Q}\{\tau \in [t, t+dt)\}}{\mathbb{Q}\{\tau \geq t\}} = \gamma(t)\, dt.$$

"probability that company defaults in (arbitrarily small) "dt" years given that it has not defaulted so far is $\gamma(t)\, dt$." In particular, the formula

$$\mathbb{Q}\{\tau > t\} = \exp(-\Gamma(t)) = \exp\left(-\int_0^t \gamma(s)ds\right)$$

tells us again that the Poisson Process core structure with exponentially distributed in-between-jump times is allowing us to see survival probabilities as discount factors, and thus default intensities as credit spreads.

Indeed, in an interest rate world with deterministic short interest rate $r(t)$ we have

$$P(0,t) = \exp\left(-\int_0^t r(s)ds\right).$$

Again, this analogy allows us to translate much of the interest rate ideas into the credit modeling world.

However, ξ *is independent of all default free market quantities and represents an external source of randomness that makes reduced form models incomplete* (more on this later).
Here the time-varying nature of γ allows us to take into account a possible term structure of credit spreads. It is actually this model that is used to strip default probabilities from CDS quotes, for example.

This formulation does not take into account credit spread volatility. For this, we need to move to the next section.

22.2.3 Cox Processes

Intensity can be also time-varying and *stochastic*: in that case it is assumed to be at least a $\mathcal{F}_t$-adapted and right continuous (and thus progressive) process and is denoted by λ_t and the *cumulated intensity* or *hazard process* is the random variable $\Lambda(T) = \int_0^T \lambda_t dt$. We assume $\lambda_t > 0$. Remember that the requirement to be "$\mathcal{F}_t$-adapted" means that given $\mathcal{F}_t$, i.e. the default-free market information up to time t, we know λ from 0 to t. This intuitively says that the randomness we allow into the intensity is induced by the default free market. In a *Cox process* with stochastic intensity λ, conditional on $\mathcal{F}_t$ (or just on $\mathcal{F}_t^\lambda = \sigma(\{\lambda_s : s \leq t\})$, i.e. just on the paths of λ), we still have a Poisson process structure, and all facts we have seen for the case with $\gamma(t)$ still hold, conditional on λ and with λ replacing γ. In particular, we have again that the first jump time of the process, transformed through its cumulated intensity, is an exponential random variable independent of $\mathcal{F}_t$:

$$\Lambda(\tau) = \xi$$

with ξ standard exponential independent of $\mathcal{F}_t$. Then we have that default can be defined as

$$\tau := \Lambda^{-1}(\xi),$$

that as before provides us with suggestions on how to simulate the default time in this setting.

Notice that in this setting not only ξ is random (and still independent of anything else, included λ), but λ itself is stochastic. This is why Cox processes are at times called "doubly stochastic Poisson processes".

With default coming from a Cox process we still have

$$\mathbb{Q}\{\tau \in [t, t + dt) | \tau \geq t, \mathcal{F}_t\} = \lambda_t \; dt.$$

This reads, if "t=now":
"probability that company defaults in (small) "dt" years given that it has not defaulted so far and given the default-free-market information so far is λ_t dt."

For the survival probability we have

$$\mathbb{Q}\{\tau \geq t\} = \mathbb{Q}\{\Lambda(\tau) \geq \Lambda(t)\} = \mathbb{Q}\left\{\xi \geq \int_0^t \lambda(u) du\right\}$$

$$= \mathbb{E}\left[\mathbb{Q}\left\{\xi \geq \int_0^t \lambda(u)du \middle| \mathcal{F}_t^\lambda\right\}\right] = \mathbb{E}\left[e^{-\int_0^t \lambda(u)du}\right]$$

which is completely analogous to the bond price formula in a short rate model with interest rate λ. Indeed, we have seen many times in the book that under a stochastic short rate r the zero coupon bond price formula is

$$P(0,t) = \mathbb{E}\left[e^{-\int_0^t r(u)du}\right].$$

Cox processes thus allow to drag the stochastic interest-rates technology and paradigms into default modeling. But again ξ *is independent of all default free market quantities (of $\mathcal{F}$, of λ, of r...) and represents an external source of randomness that makes reduced form models incomplete.*

Now the time varying nature of λ may account for the term structure of credit spreads, while the stochasticity of λ can be used to introduce credit spread volatility. For example, in a diffusion setting, we set

$$d\lambda_t = b(t, \lambda_t)dt + \sigma(t, \lambda_t)\, dW_t.$$

We will see some explicit examples like the SSRD and CIR++ models later on. We will also consider the possibility to add jumps to the stochastic intensity, leading to the JCIR and JCIR++ jump-diffusion stochastic intensity models.

Summing up:

- Standard Poisson Processes (with unit constant intensity, i.e. instantaneous jump probability) are the probabilistic basis; *The intensity* (or instantaneous credit spread) *is constant* and set to one. Time-homogeneous Poisson processes are completely analogous but the intensity is a constant possibly different from 1.
- Time inhomogeneous Poisson processes can be built based on standard Poisson processes and on a given *deterministic time-varying intensity*; these are often used as a quoting mechanism for credit spreads in *CDS and Corporate Bond* contracts, the intensity being also interpreted as an *instantaneous credit spread*;
- Cox processes: If the intensity is Stochastic, conditional on the intensity filtration we have a time-inhomogeneous Poisson process. These models can be used for more sophisticated credit derivatives and take into account also *credit spread volatility.*

22.3 CDS Calibration and Implied Hazard Rates/ Intensities

Reduced form models are the models that are most commonly used in the market to infer implied default probabilities from market quotes.

Market instruments from which these probabilities are drawn are especially CDS and Bonds.

We will see in some detail the procedure concerning CDS.

The reduced-form model used for this purpose is the time-inhomogeneous Poisson Process, with time varying intensity $\gamma(t)$ and cumulated intensity/hazard function $\Gamma(t) = \int_0^t \gamma(u)du$ seen in Section 22.2.2.

As a fundamental example we may take the hazard rate γ to be deterministic and *piecewise constant*:

$$\gamma(t) = \gamma_i \text{ for } t \in [T_{i-1}, T_i), \quad (\gamma_1, \gamma_2, \ldots, \gamma_i, \ldots)$$

where the T_i's span the relevant maturities.

Notice that $\Gamma(t) = \int_0^t \gamma(s)ds = \sum_{i=1}^{\beta(t)-1}(T_{i+1} - T_i)\gamma_i + (t - T_{\beta(t)-1})\gamma_{\beta(t)}$
Recall that $\beta(t)$ is the index of the first T_i following t. Set

$$\Gamma_j := \int_0^{T_j} \gamma(s)ds = \sum_{i=1}^{j}(T_i - T_{i-1})\gamma_i.$$

In this context, compute for example the protection leg term of a CDS:

$$\text{L}_{\text{GD}}\, \mathbb{E}[D(0,\tau)\mathbf{1}_{\{T_a<\tau<T_b\}}|\mathcal{F}] = \text{L}_{\text{GD}} \int_0^\infty \mathbb{E}[D(0,u)\mathbf{1}_{\{T_a<u<T_b\}}]\mathbb{Q}(\tau \in [u,\, u+du))$$

$$= \text{L}_{\text{GD}} \int_{T_a}^{T_b} \mathbb{E}[D(0,u)]\mathbb{Q}(\tau \in [u, u+du))$$

$$= \text{L}_{\text{GD}} \int_{T_a}^{T_b} P(0,u)\gamma(u)\exp\left(-\int_0^u \gamma(s)ds\right)du$$

$$= \text{L}_{\text{GD}} \sum_{i=a+1}^{b} \gamma_i \int_{T_{i-1}}^{T_i} \exp(-\Gamma_{i-1} - \gamma_i(u - T_{i-1}))P(0,u)du$$

where we have used (22.1).

With similar computations for the other CDS terms, it can be shown that under this formulation we have

$$\text{CDS}_{a,b}(t, R, \text{L}_{\text{GD}}; \Gamma(\cdot)) = \mathbf{1}_{\{\tau>t\}}\left[R\int_{T_a}^{T_b} P(t,u)(T_{\beta(u)-1} - u)d(e^{-(\Gamma(u)-\Gamma(t))})\right.$$

$$\left.+ \sum_{i=a+1}^{b} P(t,T_i)R\alpha_i e^{-(\Gamma(T_i)-\Gamma(t))} + \text{L}_{\text{GD}}\int_{T_a}^{T_b} P(t,u)d(e^{-(\Gamma(u)-\Gamma(t))})\right]$$

and in particular

$$\mathrm{CDS}_{a,b}(0,R,\mathrm{L_{GD}};\Gamma(\cdot)) = \left[R\int_{T_a}^{T_b} P(0,u)(T_{\beta(u)-1}-u)d(e^{-\Gamma(u)})\right.$$

$$\left. + \sum_{i=a+1}^{b} P(0,T_i)R\alpha_i e^{-\Gamma(T_i)} + \mathrm{L_{GD}}\int_{T_a}^{T_b} P(0,u)d(e^{-\Gamma(u)})\right] \tag{22.2}$$

so that, with our piecewise constant example for γ we obtain

$$\mathrm{CDS}_{a,b}(0,R,\mathrm{L_{GD}};\Gamma(\cdot)) =$$

$$= R\sum_{i=a+1}^{b}\gamma_i\int_{T_{i-1}}^{T_i}\exp(-\Gamma_{i-1}-\gamma_i(u-T_{i-1}))P(0,u)(u-T_{i-1})du$$

$$+R\sum_{i=a+1}^{b} P(0,T_i)\alpha_i e^{-\Gamma(T_i)}$$

$$-\mathrm{L_{GD}}\sum_{i=a+1}^{b}\gamma_i\int_{T_{i-1}}^{T_i}\exp(-\Gamma_{i-1}-\gamma_i(u-T_{i-1}))P(0,u)du.$$

Now in the market $T_a = 0$ and we have fair R quotes for $T_b = 1y, 2y, 3y, \ldots, 10y$, with T_i's resetting quarterly. We solve

$$\mathrm{CDS}_{0,1y}(0,R_{0,1y}^{MKT},\mathrm{L_{GD}};\gamma_1=\gamma_2=\gamma_3=\gamma_4=:\gamma^1)=0;$$

$$\mathrm{CDS}_{0,2y}(0,R_{0,2y}^{MKT},\mathrm{L_{GD}};\gamma^1;\gamma_5=\gamma_6=\gamma_7=\gamma_8=:\gamma^2)=0;\ldots.$$

and so on, each time finding the new four (or more in case some quotes are missing) intensity parameters.

At this point we present some numerical examples, based on Parmalat CDS data across some dates. We start by September 10, 2003, and finish with December 10, 2003, when Parmalat entered a deep crisis.

Parmalat CDS calibration, September 10th, 2003

Recovery Rate = 40%.

Maturity T_b (yr)	Maturity (dates)	$R_{0,b}$
1	20-Sep-04	192.5
3	20-Sep-06	215
5	20-Sep-08	225
7	20-Sep-10	235
10	20-Sep-13	235

Table 22.1. Maturity dates and corresponding CDS quotes in bps for T_0 = September 10th, 2003.

date	intensity γ	survival pr $\exp(-\Gamma)$
10-Sep-03	3.199%	100.000%
20-Sep-04	3.199%	96.714%
20-Sep-06	4.388%	89.552%
22-Sep-08	3.659%	82.508%
20-Sep-10	5.308%	75.357%
20-Sep-13	2.338%	67.078%

Table 22.2. Calibration with piecewise linear intensity on September 10th, 2003.

date	intensity γ	survival pr $\exp(-\Gamma)$
10-Sep-03	3.199%	100.000%
20-Sep-04	3.199%	96.714%
20-Sep-06	3.780%	89.578%
22-Sep-08	4.033%	82.516%
20-Sep-10	4.458%	75.402%
20-Sep-13	3.891%	66.978%

Table 22.3. Calibration with piecewise constant intensity on September 10th, 2003.

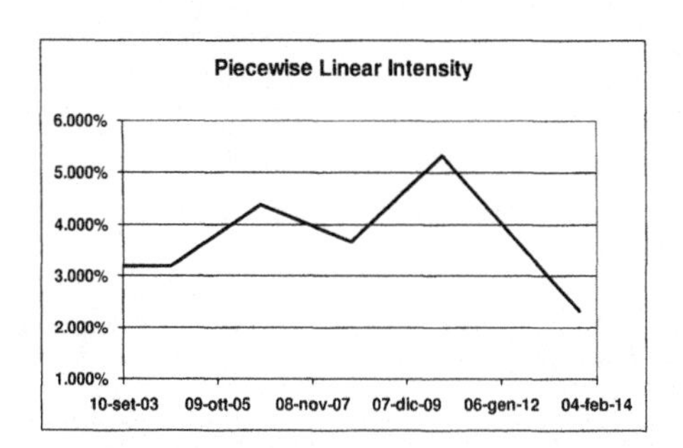

Fig. 22.1. Piecewise linear intensity γ calibrated on CDS quotes on September 10th, 2003.

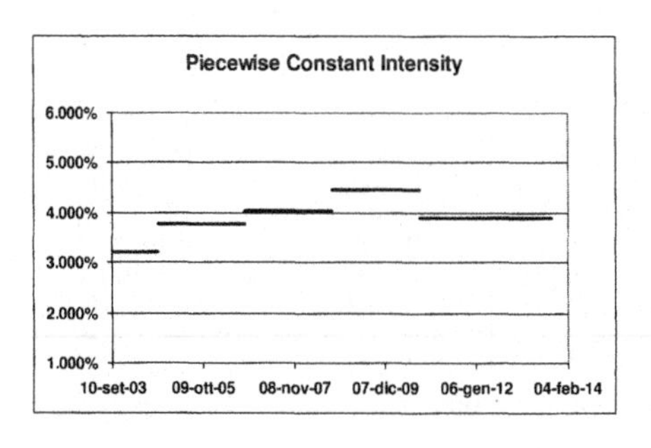

Fig. 22.2. Piecewise constant intensity γ calibrated on CDS quotes on September 10th, 2003.

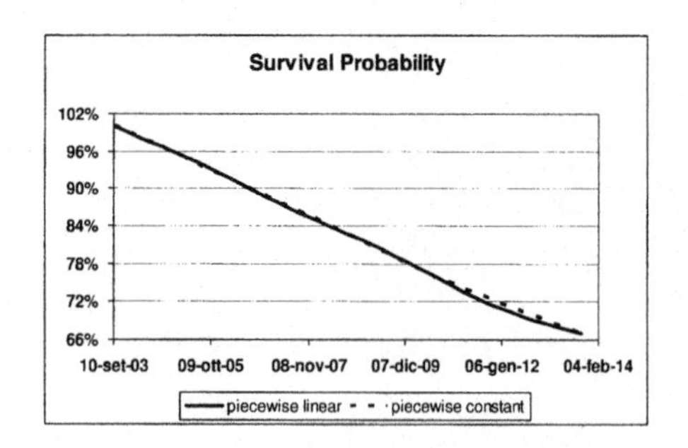

Fig. 22.3. survival probability $\exp(-\Gamma)$ resulting from calibration on CDS quotes on September 10th, 2003.

Parmalat CDS calibration, November 28th, 2003

Recovery Rate = 40%.

Maturity T_b (yr)	Maturity (dates)	$R_{0,b}$
1	20-Dec-04	725
3	20-Dec-06	630
5	20-Dec-08	570
7	20-Dec-10	570
10	20-Dec-13	570

Table 22.4. Maturity dates and corresponding CDS quotes in bps relative to $T_0 =$ November 28th, 2003.

date	intensity γ	survival pr $\exp(-\Gamma)$
28-Nov-03	12.047%	100.000%
20-Dec-04	12.047%	87.824%
20-Dec-06	6.545%	72.736%
22-Dec-08	8.226%	62.581%
20-Dec-10	10.779%	51.640%
20-Dec-13	7.880%	38.872%

Table 22.5. Calibration with piecewise linear intensity on November 28th, 2003.

date	intensity γ	survival pr $\exp(-\Gamma)$
28-Nov-03	12.047%	100.000%
20-Dec-04	12.047%	87.824%
20-Dec-06	9.426%	72.545%
22-Dec-08	7.331%	62.486%
20-Dec-10	9.441%	51.626%
20-Dec-13	9.437%	38.734%

Table 22.6. Calibration with piecewise constant intensity on November 28th, 2003.

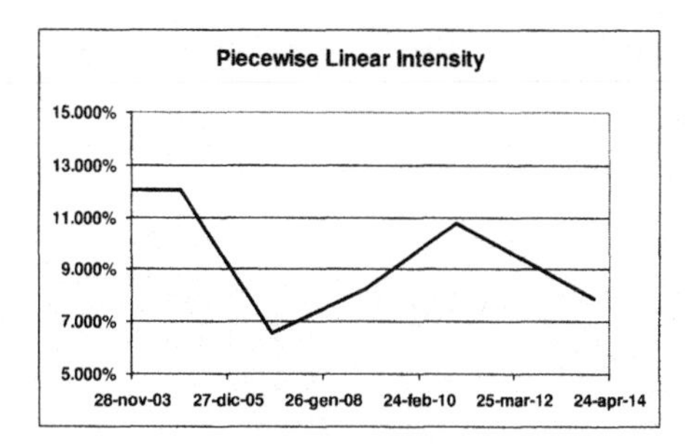

Fig. 22.4. Piecewise linear intensity γ calibrated on CDS quotes on November 28th, 2003.

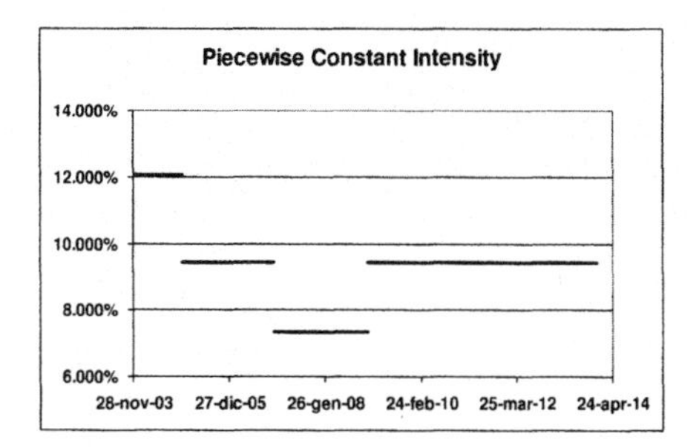

Fig. 22.5. Piecewise constant intensity γ calibrated on CDS quotes on November 28th, 2003.

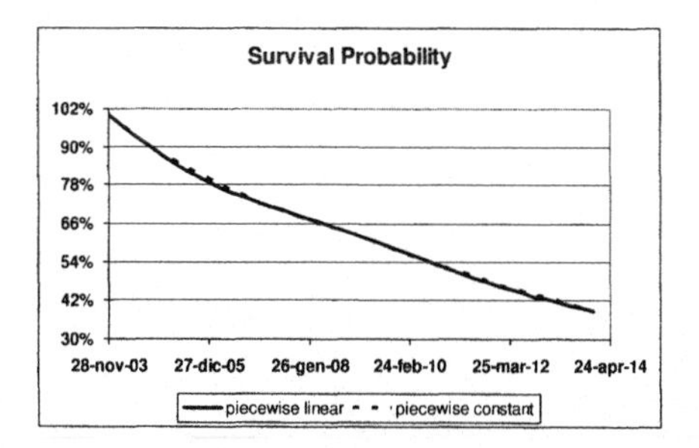

Fig. 22.6. Survival probability $\exp(-\Gamma)$ resulting from calibration on CDS quotes on November 28th, 2003.

Parmalat CDS calibration, December 8th, 2003

Recovery Rate = 25%.

Maturity T_b (yr)	Maturity (dates)	$R_{0,b}$
1	20-Dec-04	1450
3	20-Dec-06	1200
5	20-Dec-08	940
7	20-Dec-10	850
10	20-Dec-13	850

Table 22.7. Maturity Dates and corresponding CDS quotes in bps relative to T_0 = December 8th, 2003.

date	intensity γ	survival pr $\exp(-\Gamma)$
08-Dec-03	19.272%	100.000%
20-Dec-04	19.272%	81.680%
20-Dec-06	7.263%	62.413%
22-Dec-08	2.393%	56.570%
20-Dec-10	11.205%	49.303%
20-Dec-13	11.318%	34.993%

Table 22.8. Calibration with piecewise linear intensity on December 8th, 2003.

date	intensity γ	survival pr $\exp(-\Gamma)$
08-Dec-03	19.272%	100.000%
20-Dec-04	19.272%	81.680%
20-Dec-06	13.650%	61.931%
22-Dec-08	4.834%	56.126%
20-Dec-10	6.500%	49.213%
20-Dec-13	11.256%	34.934%

Table 22.9. Calibration with piecewise constant intensity on December 8th, 2003.

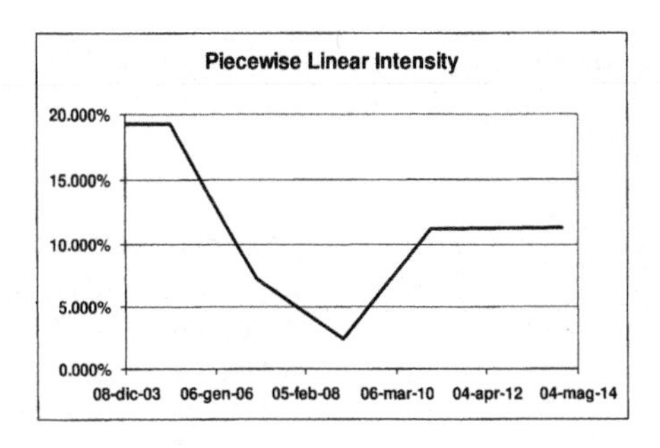

Fig. 22.7. Piecewise linear intensity γ calibrated on CDS quotes on December 8th, 2003.

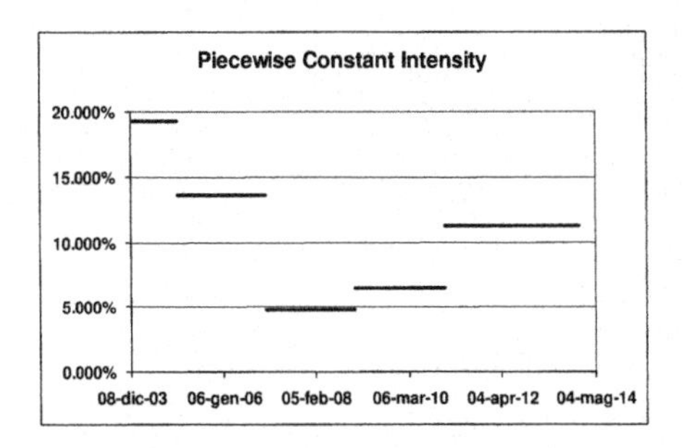

Fig. 22.8. Piecewise constant intensity γ calibrated on CDS quotes on December 8th, 2003.

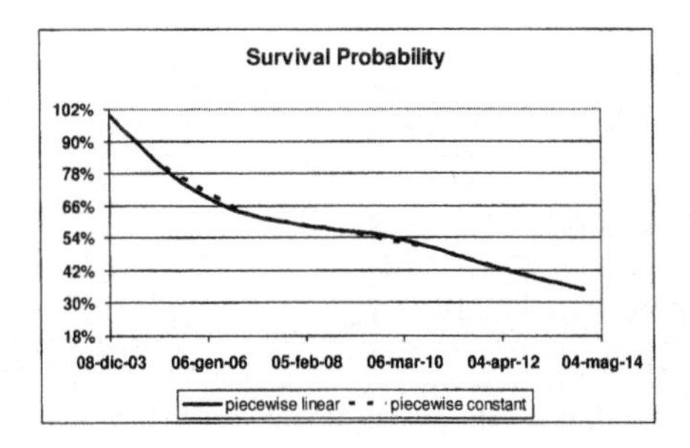

Fig. 22.9. Survival probability $\exp(-\Gamma)$ resulting from calibration on CDS quotes on December 8th, 2003.

Parmalat CDS calibration, December 10th, 2003

Recovery Rate = 15%.

Maturity T_b (yr)	Maturity (dates)	$R_{0,b}$
1	20-Dec-04	5050
3	20-Dec-06	2100
5	20-Dec-08	1500
7	20-Dec-10	1250
10	20-Dec-13	1100

Table 22.10. Maturity Dates and corresponding CDS quotes in bps relative to T_0 = December 10th, 2003.

date	intensity γ	survival pr $\exp(-\Gamma)$
10-Dec-03	55.483%	100.000%
20-Dec-04	55.483%	56.018%
20-Dec-06	-61.665%	59.642%
22-Dec-08	84.397%	47.321%
20-Dec-10	-86.408%	48.293%
20-Dec-13	123.208%	27.581%

Table 22.11. Calibration with piecewise linear intensity and postponed payoff 1 ("PR") on Dec 10, 2003.

date	intensity γ	survival pr $\exp(-\Gamma)$
10-Dec-03	55.483%	100.000%
20-Dec-04	55.483%	56.018%
20-Dec-06	0.807%	55.109%
22-Dec-08	4.017%	50.780%
20-Dec-10	4.292%	46.559%
20-Dec-13	5.980%	38.809%

Table 22.12. Calibration with piecewise const intensity and postponed payoff 1 ("PR") on Dec 10, 2003.

date	intensity γ	survival pr $\exp(-\Gamma)$
10-Dec-03	64.188%	100.000%
20-Dec-04	64.188%	51.150%
20-Dec-06	-80.163%	60.144%
22-Dec-08	108.007%	45.299%
20-Dec-10	-113.620%	47.944%
20-Dec-13	162.645%	22.732%

Table 22.13. Calibration with pwise linear intensity and postponed payoff 2 ("PR2").

date	intensity γ	survival pr $\exp(-\Gamma)$
10-Dec-03	64.188%	100.000%
20-Dec-04	64.188%	51.150%
20-Dec-06	-3.270%	54.657%
22-Dec-08	3.900%	50.484%
20-Dec-10	4.282%	46.297%
20-Dec-13	6.065%	38.491%

Table 22.14. Calibration with piecewise constant intensity and postponed payoff of kind 2 ("PR2").

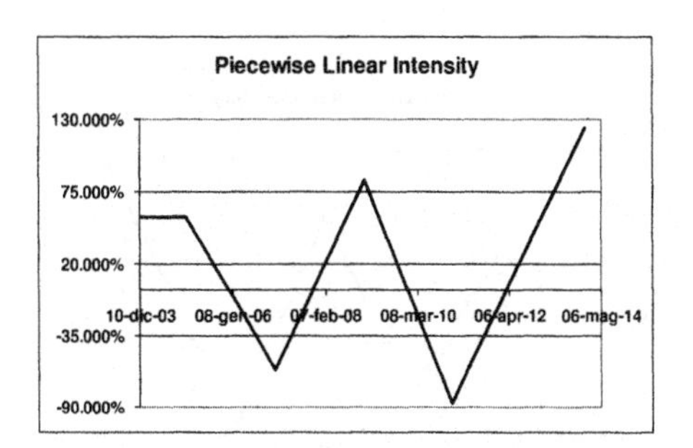

Fig. 22.10. Piecewise linear intensity γ calibrated on CDS quotes on December 10th, 2003, with postponed payoff of kind 1 ("PR").

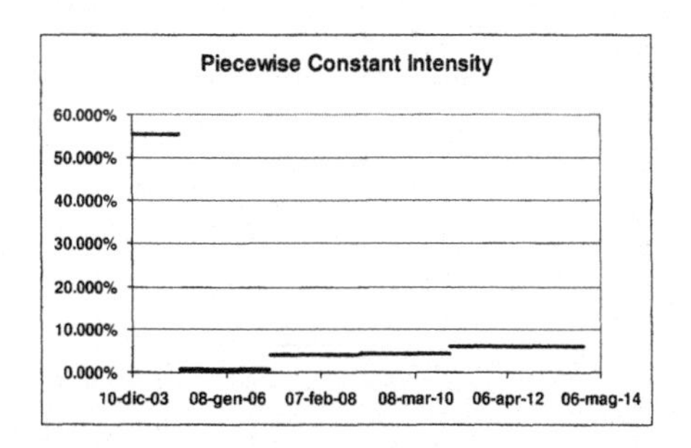

Fig. 22.11. Piecewise constant intensity γ calibrated on CDS quotes on December 10th, 2003, with postponed payoff of kind 1 ("PR").

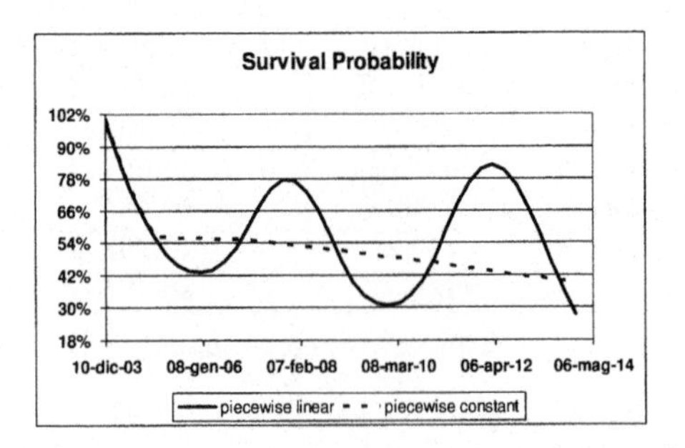

Fig. 22.12. Survival probability $\exp(-\Gamma)$ resulting from calibration on CDS quotes on December 10th, 2003, with postponed payoff of kind 1 ("PR").

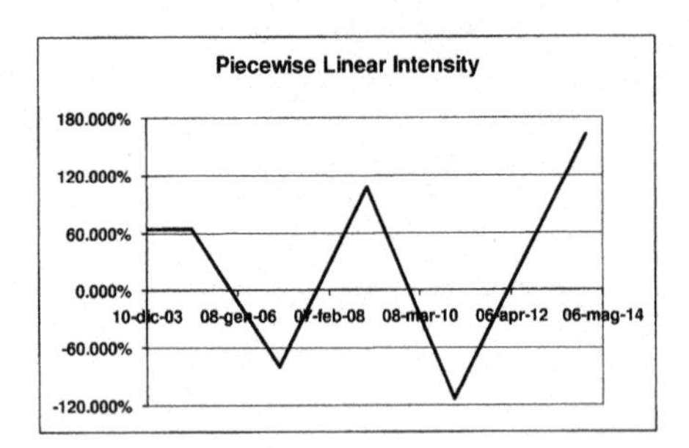

Fig. 22.13. Piecewise linear intensity γ calibrated on CDS quotes on December 10th, 2003, with postponed payoff of kind 2 ("PR2").

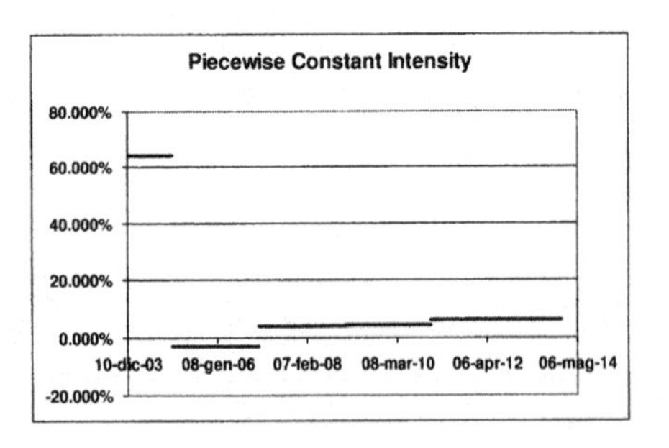

Fig. 22.14. Piecewise constant intensity γ calibrated on CDS quotes on December 10th, 2003, with postponed payoff of kind 2 ("PR2").

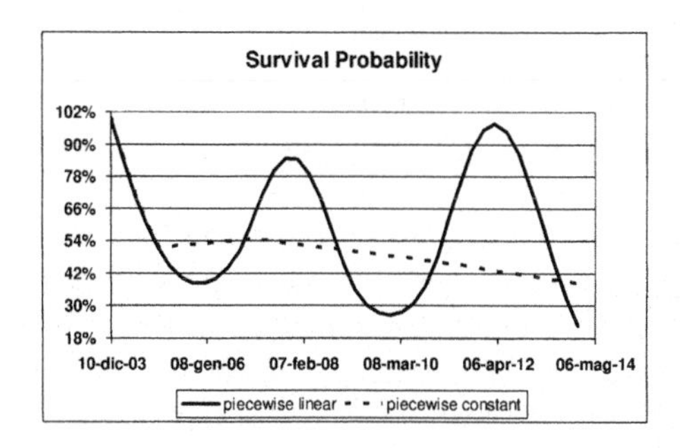

Fig. 22.15. Survival probability $\exp(-\Gamma)$ resulting from calibration on CDS quotes on December 10th, 2003, with postponed payoff of kind 2 ("PR2").

Comments on the calibration examples

The above examples lead us to the following considerations. In all the calibrated intensity graphs but the first one (September 10, 2003) the highest intensity is on the first period. This means that the market is perceiving the first interval as the most risky. Given the lower intensities in the following periods, the market perception is that if the firm survives the first period then the situation will improve considerably. This feature looks extreme in the last date (December 10, 2003).

Also, we can see that in general intensities raise as the calibration date approaches December 10, 2003. This is expected, since CDS rates are raising as well due to the coming crisis.

One particular feature deserves attention: the extremely negative intensity in both Figures 22.10 and 22.13, and the slightly negative intensity in Figure 22.14 (second segment).

The intensity is a probability per unit of time, so that it has to be non-negative. Negative intensities violate this condition and may give survival probabilities that increase with the maturity. Notice that these figures refer to the calibration with postponed CDS payoffs of kind 1 and 2 (PR and PR2). If we keep the real CDS payoff including the accrual term, we have results similar to case PR2, i.e. with negative intensities. The only case where the intensity does not go negative is with the PRCDS payout, as shown in Figure 22.11, although the intensity gets very close to zero. We deduce that

- When we calibrate an intensity model to very distressed names, we may occasionally find negative intensities; this is a clear signal saying that the situation is pathological and that we should move carefully;
- In general the piecewise constant intensity formulation, although being discontinuous, looks more solid than the piecewise linear (continuous) one. Indeed, while the piecewise linear intensity in the last date goes extremely negative, in the PRCDS case the piecewise constant intensity remains even positive. And when, like in the PR2CDS case, the piecewise constant in-

tensity goes negative it does so only of a small amount. A possible interpretation for this is the above-mentioned market belief that if Parmalat endures the sudden crisis, corresponding to large short-term intensity, it is going to do well afterwards, corresponding to lower subsequent intensities. This extreme change in perception along the term structure is best rendered with a piecewise constant intensity, since forcing continuity like in the piecewise linear case clearly induces an unstable behaviour.
- We have the possibility of reversing part of the folklore on intensity VS structural models. Structural models are appreciated for their economic foundations. Yet they are deemed to be less flexible and to have less calibration power than intensity models. However, in Brigo and Tarenghi (2004) and Brigo and Morini (2006) we calibrate the same Parmalat CDS data we have used here and show that the AT1P structural model calibrates exactly the CDS term structure with "legal" (i.e. positive and real volatility) parameters even when the intensity models may degenerate to negative intensities.

22.4 Inducing dependence between Interest-rates and the default event

If we assume interest rates to be stochastic and to be driven by Brownian motions as sources of randomness, since a Poisson process and a Brownian motion defined on the same probability space are independent (see for example Bielecki and Rutkowski (2001), p. 188), the processes N and r are independent. *We can thus assume the stochastic discount factor* $D(s,t) = \exp(-\int_s^t r_u du)$, *and the default time* τ *to be independent under deterministic intensities* for τ. This is implicit in what we have seen before: if

$$\tau = \Gamma^{-1}(\xi)$$

with Γ strictly increasing deterministic Hazard function and ξ a standard exponential random variable *independent of anything else* (and of interest rates in particular), clearly τ will be *independent of interest rates.*

This indirectly tells us that our only hope for inducing dependence between interest rates and the default event is through the intensity, since ξ will always be independent of interest rates. The only possibility is taking stochastic intensities $\lambda_t > 0$ correlated with interest rates; this will induce a dependence between

$$\tau = \Lambda^{-1}(\xi) \quad \text{and} \quad r$$

coming from λ. We will do so explicitly under the SSRD stochastic intensity model. If we retain deterministic intensities, the bond price simplifies. By independence

$$\bar{P}(0,T) = \mathbb{E}[D(0,T)1_{\{\tau>T\}}] = \mathbb{E}[D(0,T)]\mathbb{E}[1_{\{\tau>T\}}] = P(0,T)\mathbb{Q}(\tau > T)$$

so that we see that
*defaultable zero-coupon bond = zero-coupon bond * survival probability.*

This only holds under independence. Indeed, with stochastic intensity

$$\bar{P}(0,T) = \mathbb{E}[D(0,T)1_{\{\tau>T\}}] = \mathbb{E}[D(0,T)1_{\{\Lambda(\tau)>\Lambda(T)\}}] = \mathbb{E}[D(0,T)1_{\{\xi>\Lambda(T)\}}] =$$

$$= \mathbb{E}[\mathbb{E}(D(0,T)1_{\{\xi>\Lambda(T)\}}|\mathcal{F}_T)] = \mathbb{E}[D(0,T)\mathbb{E}(1_{\{\xi>\Lambda(T)\}}|\mathcal{F}_T)] =$$

$$= \mathbb{E}[D(0,T)\exp(-\Lambda(T))] = \mathbb{E}[e^{-\int_0^T(r_s+\lambda_s)ds}].$$

This formula is general, and if λ and r are correlated, this correlation should impact this expectation and thus the defaultable bond $\bar{P}$. However, if λ and r are independent we can write the same equality as in the deterministic intensity case:

$$\bar{P}(0,T) = \mathbb{E}[D(0,T)\exp(-\Lambda(T))] = \mathbb{E}[e^{-\int_0^T(r_s+\lambda_s)ds}] = \mathbb{E}[e^{-\int_0^T r_s ds}\, e^{-\int_0^T \lambda_s ds}]$$

$$= \mathbb{E}[e^{-\int_0^T r_s ds}]\mathbb{E}[e^{-\int_0^T \lambda_s ds}] = P(0,T)\mathbb{Q}(\tau > T).$$

22.5 The Filtration Switching Formula: Pricing under partial information

Here we consider again the result we have used in defining CDS forward rates as in Section 21.3.4, a result based on the earlier Formula (21.18) in particular.

Under very general measurability conditions for the payoff (typically the payoff is assumed to be $\mathcal{G}_\infty$-measurable, i.e. known at time infinity) and for $t < T$ we have

The Filtration Switching Formula:

$$\boxed{\mathbb{E}(1_{\{\tau>T\}}\text{Payoff}|\mathcal{G}_t) = \frac{1_{\{\tau>t\}}}{\mathbb{Q}\{\tau > t|\mathcal{F}_t\}}\mathbb{E}(1_{\{\tau>T\}}\text{Payoff}|\mathcal{F}_t)} \tag{22.3}$$

where we recall that $\mathcal{G}_t = \mathcal{F}_t \vee \sigma(\{\tau < u\}, u \leq t)$,
$\mathcal{F}_t$="info-on-default-free-markets-up-to-t";
$\sigma(\{\tau < u\}, u \leq t)$ ="info if default before t, and, if so, when exactly".

The most general form of this result is given in Bielecki and Rutkowski (2001) formula (5.1) p. 143, or more in particular in Jeanblanc and Rutkowski (2000).

Switching from $\mathcal{G}$ expectations to $\mathcal{F}$ expectations is important because for some variables the $\mathcal{F}$ conditional expectations are easier to compute. Also, a careful use of conditioning to partial information $\mathcal{F}$ instead of complete one $\mathcal{G}$ will enable us to avoid problems concerning non equivalence of pricing measures later on, when we will derive a market model for CDS options in Chapter 23.

Given its importance, we present a sketchy proof of this result now.

Sketch of the proof of the Filtration Switching Formula

Proof. We start by recalling that

$$\mathcal{G}_t = \mathcal{F}_t \vee \sigma(\{\tau \leq u\}, u \leq t).$$

Now let X be a $\mathcal{G}_\infty$ measurable payoff. We can write

$$\mathbb{E}[\mathbf{1}_{\{\tau>T\}} X | \mathcal{G}_t] = \mathbb{E}[\mathbf{1}_{\{\tau>t\}} \mathbf{1}_{\{\tau>T\}} X | \mathcal{G}_t] =$$

$$= \mathbf{1}_{\{\tau>t\}} \mathbb{E}[\mathbf{1}_{\{\tau>T\}} X | \mathcal{G}_t] = \mathbf{1}_{\{\tau>t\}} \mathbb{E}[\mathbf{1}_{\{\tau>T\}} X | \mathcal{F}_t, \sigma((\tau \leq u),\ u \leq t)] =$$

where we have used the above definition of $\mathcal{G}_t$. Then

$$= \mathbf{1}_{\{\tau>t\}} \mathbb{E}[\mathbf{1}_{\{\tau>T\}} X | \mathcal{F}_t, \{\tau \geq t\}] =$$

since the expected value of $X 1_{\{\tau>T\}}$ gives no contribution in scenarios with $\tau < t$, so that the only useful information in $\sigma((\tau \leq u),\ u \leq t)$ is whether $\tau \geq t$. Also, since for any expectation E, any random variable Y and event A we have

$$E(Y|A) = E(Y 1_{\{A\}})/\text{Prob}(A),$$

we may write, taking $E = \mathbb{E}[\cdot|\mathcal{F}_t]$,

$$= \frac{\mathbf{1}_{\{\tau>t\}}}{\mathbb{Q}(\tau > t|\mathcal{F}_t)} \mathbb{E}[\mathbf{1}_{\{\tau>t\}} \mathbf{1}_{\{\tau>T\}} X | \mathcal{F}_t] = \frac{\mathbf{1}_{\{\tau>t\}}}{\mathbb{Q}(\tau > t|\mathcal{F}_t)} \mathbb{E}[\mathbf{1}_{\{\tau>T\}} X | \mathcal{F}_t].$$

□

The reader may easily realize that the proof does not rely on any particular intensity-model setup but is rather general.

22.6 Default Simulation in reduced form models

I saw an infinity of paths, endless changing...

Metron, "JLA: Rock of Ages", DC Comics

Assume we have obtained the (positive deterministic) hazard rates γ's from CDS market quotes with the usual procedure.

How do we simulate the default time τ? Recall that $\tau = \Gamma^{-1}(\xi)$. Then we can adopt the following scheme.

a) Simulate a standard exponential random variable ξ. This can be done for example by simulating samples $u^1, \ldots, u^m$ from a uniform variable U in $[0, 1]$ and then taking $\xi^1 := -\ln(1 - u^1), \ldots, \xi^m = -\ln(1 - u^m)$.
b) For each sample ξ^j, solve $\Gamma(\tau^j) = \xi^j$ in τ^j; the obtained solutions are the simulated samples $\tau^1, \ldots, \tau^m$ of the default time τ.

Possible problem: if some obtained ξ's are very large, we may not have Γ going far enough in time to solve the equation, because our CDS data from which we deduced Γ stopped before. As an example, consider the Hazard function of Figure 22.16

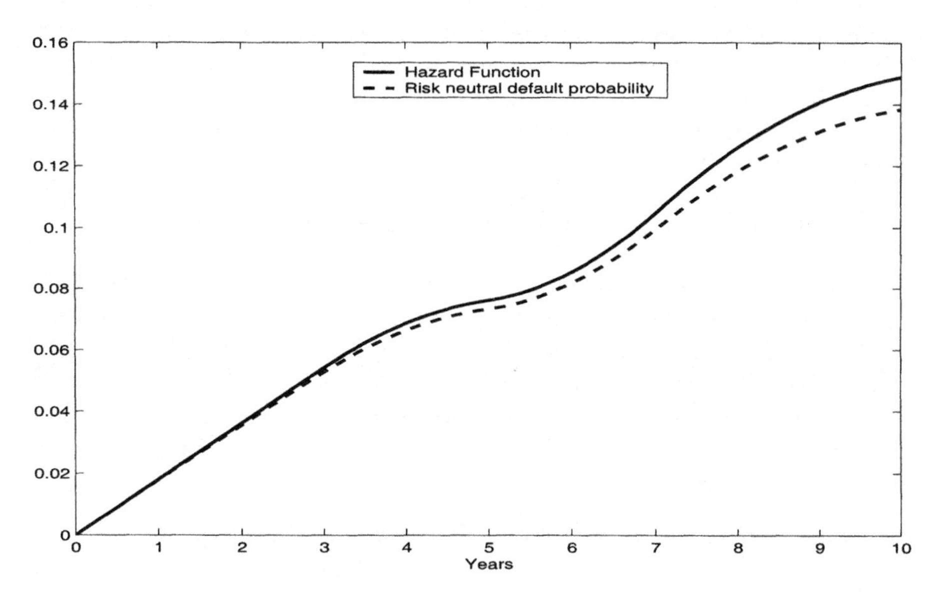

Fig. 22.16. Hazard function Γ and risk-neutral default probability $1 - e^{-\Gamma}$ for Merrill-Lynch CDS's, October 25, 2002

If in this case we have obtained a $\xi_j = 0.3$, we cannot solve $\Gamma(\tau^j) = 0.3$ from our data, unless we extend in some way the graph of Γ beyond its maturity. Of course the probability of ξ being larger than 0.14, our final Γ, is large: $\mathbb{Q}(\xi > 0.14) = \exp(-0.14) = 0.87$, so that this is an important problem we can find very often from the simulated ξ_i (in our example in 87% of simulations)

A second problem of the simulation is that in presence of default, to obtain an acceptable precision with a Monte-Carlo algorithm, it is unfortunately necessary to simulate a quite large number of scenarios.

Indeed, variances are quite large in relative terms, due essentially to indicator terms in the payoffs, such as for example $1_{\{\tau<T\}}$ in $1_{\{\tau<T\}}\text{L}_{\text{GD}}D(0,\tau)$. A quick example can help us to clarify this important second point.

Assume we want a MC simulation for $\mathbb{E}[1_{\{\tau<T\}}]$. Compute the variance

$$\text{Var}(1_{\{\tau<T\}}) = \mathbb{E}1^2_{\{\tau<T\}} - (\mathbb{E}1_{\{\tau<T\}})^2 = \mathbb{E}1_{\{\tau<T\}} - (\mathbb{E}1_{\{\tau<T\}})^2.$$

Consider for example the data given in the Figure 22.16 above and take $T = 5y$. Notice that $\mathbb{E}1_{\{\tau<T\}} = \mathbb{Q}(\tau < T)$ is the risk neutral probability to default in 5y for ML. From the graph we see that this is about 0.07. Then the above variance is about $0.07 - 0.07^2 = 0.0651$, and the standard deviation is $\sqrt{0.0651} = 0.2551$.

We know that the standard error in the Monte Carlo method is given by the standard deviation of the object we are simulating divided by the square root of the number of paths (this is reviewed below, in Section 22.6.1). So we have that the standard error is about $0.2551/\sqrt{m}$ (where m, as before, is the number of paths or scenarios). Now, we are estimating a quantity that is about 0.07 and we would like to have a standard error below, say, one basis point. But if we wish our standard error to be below one basis point (i.e. 1/10000) we need to set

$$0.2551/\sqrt{m} < 1/10000 \quad \Rightarrow m > (10000 * 0.2551)^2 = 6507601.$$

We may slightly improve the situation by setting a threshold barrier $\bar{B}$ such that we will be interested only in Γ's with $\Gamma(T) < \bar{B}$. This means we will be able to retrieve τ's with $\Gamma(\tau) < \bar{B}$, i.e. $\tau < \Gamma^{-1}(\bar{B})$. Default times larger than this will not be simulated.

This is natural since most payoffs become known when the default time exceeds a given final maturity T_b, so that there is no need to simulate them.

Indeed, if Π is a payoff we can write the price as

$$\begin{aligned}\mathbb{E}\,\Pi &= \mathbb{E}[\Pi|\Gamma(\tau) < \bar{B}]\mathbb{Q}(\Gamma(\tau) < \bar{B}) + \mathbb{E}[\Pi|\Gamma(\tau) > \bar{B}]\mathbb{Q}(\Gamma(\tau) > \bar{B})\\ &= \mathbb{E}[\Pi|\xi < \bar{B}]\mathbb{Q}(\xi < \bar{B}) + \mathbb{E}[\Pi|\xi > \bar{B}]\mathbb{Q}(\xi > \bar{B})\\ &= \mathbb{E}[\Pi|\Gamma(\tau) < \bar{B}](1 - e^{-\bar{B}}) + \mathbb{E}[\Pi|\Gamma(\tau) \geq \bar{B}]e^{-\bar{B}}.\end{aligned}$$

Now, in payoffs that become trivial for large enough default times (like Bonds or CDS for example) the second expected value on the right hand side need not be computed. Since the second term is usually known without simulation, the idea is then to simulate default times $\tau = \Gamma^{-1}(\xi)$ *conditional on* $\xi < \bar{B}$.

So, if ξ is an exponential random variable with parameter one we just simulate $\xi|\xi < \bar{B}$, whose density is easily seen to be

$$p_{\xi|\xi<\bar{B}}(u) = 1_{\{u<\bar{B}\}}e^{-u}/(1 - e^{-\bar{B}}).$$

From the exponential distribution we see that simulating m scenarios for ξ amounts to simulate $m(1-e^{-\bar{B}})$ scenarios with $\xi < \bar{B}$ and $me^{-\bar{B}}$ with $\xi \geq \bar{B}$. So in turn simulating $M = m(1 - e^{-\bar{B}})$ scenarios for $\xi < \bar{B}$, as we will do, amounts to simulate in total $m = M/(1-e^{-\bar{B}})$ scenarios, the extra scenarios corresponding to the known value of the payoff when $\tau > \Gamma^{-1}(\bar{B})$.

Dividing by $1 - e^{-\bar{B}}$ may help us increase efficiency (in our examples typically it increases the number of scenarios by a factor 10), but a large amount of scenarios remains to be generated, and the time needed for Monte Carlo simulation remains large.

For example, in Figure 22.16, if the payoff becomes trivial after 10y, we may consider only $\tau < 10$, i.e. $\Gamma(\tau) < \Gamma(10) = 0.15 =: \bar{B}$.

Then $1/(1 - e^{-\bar{B}}) = 7.18$. If the payoff "stops" earlier after 5y, since $\Gamma(5y) = 0.08$, we have $1/(1 - e^{-\bar{B}}) = 13$.

In presence of much larger default probabilities, like for Parmalat on December 8, 2003, where we had a 35% survival probability in 10y, i.e. $\Gamma(10y) = 1.0518 =: \bar{B}$, we would obtain a much poorer reduction: $1/(1 - e^{-\bar{B}}) = 1.54$. In all cases the above trick can be combined with a control variate technique to reduce the variance of the Monte Carlo estimate (and thus the standard error). We address this below, after reviewing the notion of standard error.

22.6.1 Standard error

"Of course it does. A conservative estimate puts material reality as ten thousand million light years across. Even using the combined computational function of all the processors in the united planets capable of dealing with numerical scale, it would take at least ninety years to..."

Brainiac 5, "Legion Lost", 2000, DC Comics

First some notation. Suppose we have to value a financial payoff depending on the default time τ. Let the discounted payoff be

$$\Pi := X(\tau)$$

i.e. a given function of the default time of the underlying credit.

Assume we generate m scenarios for the default time, each scenario denoted by an upper index: τ^j denotes the j-th scenario of the default time.

Let Π^j be the corresponding j-th scenario for the discounted payoff. The Monte Carlo price of our payoff is computed, based on the simulated paths, as

$$\mathbb{E}[\Pi(T)] = \sum_{j=1}^{m} \Pi^j/m$$

where the default time τ in Π has been simulated under the risk neutral measure.

We wish to have an estimate of the error we have when approximating the true expectation $\mathbb{E}(\Pi)$ by its Monte Carlo estimate $\sum_{j=1}^{m} \Pi^j/m$. To do so, the classic reasoning is as follows. We have already seen a similar argument in the chapter on the LIBOR model. We repeat the reasoning here to keep the chapter more self contained.

Let us view $(\Pi^j)_j$ as a sequence of independent identically distributed (iid) random variables, distributed as Π. By the central limit theorem, we know that under suitable assumptions one has

$$\frac{\sum_{j=1}^{m}(\Pi^j - \mathbb{E}(\Pi))}{\sqrt{m}\,\mathrm{Std}(\Pi)} \to \mathcal{N}(0,1),$$

in law, as $m \to \infty$, from which we may write, approximately and for large m:

$$\frac{\sum_{j=1}^{m} \Pi^j}{m} - \mathbb{E}(\Pi) \sim \frac{\text{Std}(\Pi)}{\sqrt{m}} \mathcal{N}(0,1).$$

It follows that

$$\mathbb{Q}\left\{\left|\frac{\sum_{j=1}^{m} \Pi^j}{m} - \mathbb{E}(\Pi)\right| < \epsilon\right\} = \mathbb{Q}\left\{|\mathcal{N}(0,1)| < \epsilon \frac{\sqrt{m}}{\text{Std}(\Pi)}\right\}$$

$$= 2\Phi\left(\epsilon \frac{\sqrt{m}}{\text{Std}(\Pi)}\right) - 1,$$

where as usual Φ denotes the cumulative distribution function of the standard Gaussian random variable.

The above equation gives the probability that our Monte Carlo estimate $\sum_{j=1}^{m} \Pi^j/m$ is not farther than ϵ from the true expectation $\mathbb{E}(\Pi)$ we wish to estimate. Typically, one sets a desired value for this probability, say 0.98, and derives ϵ by solving

$$2\Phi\left(\epsilon \frac{\sqrt{m}}{\text{Std}(\Pi)}\right) - 1 = 0.98.$$

For example, since we know from the Φ tables that

$$2\Phi(z) - 1 = 0.98 \iff \Phi(z) = 0.99 \iff z \approx 2.33,$$

we have that

$$\epsilon = 2.33 \frac{\text{Std}(\Pi)}{\sqrt{m}}.$$

The true value of $\mathbb{E}(\Pi)$ is thus inside the "window"

$$\left[\frac{\sum_{j=1}^{m} \Pi^j}{m} - 2.33 \frac{\text{Std}(\Pi)}{\sqrt{m}}, \quad \frac{\sum_{j=1}^{m} \Pi^j}{m} + 2.33 \frac{\text{Std}(\Pi)}{\sqrt{m}}\right]$$

with a 98% probability. This is called a 98% confidence interval for $\mathbb{E}(\Pi)$. Other typical confidence levels are given in Table 6.1, which we report here:

$2\Phi(z) - 1$	$z \approx$
99%	2.58
98%	2.33
95.45%	2
95%	1.96
90%	1.65
68.27%	1

We can see that, all things being equal, as m increases, the window shrinks as $1/\sqrt{m}$, which is worse than $1/m$. If we need to reduce the window size to one tenth, we have to increase the number of scenarios by a factor 100. Sometimes, to reach a chosen accuracy (a small enough window), we need to take a huge number of scenarios m. When this is too time-consuming, there are "variance-reduction" techniques that may be used to reduce the above window size.

A more fundamental problem with the above window is that the true standard deviation $\mathrm{Std}(\Pi)$ of the payoff is usually unknown. This is typically replaced by the known sample standard deviation obtained by the simulated paths,

$$(\widehat{\mathrm{Std}}(\Pi;m))^2 := \sum_{j=1}^{m}(\Pi^j)^2/m - (\sum_{j=1}^{m}\Pi^j/m)^2$$

and the actual 98% Monte Carlo window we compute is

$$\left[\frac{\sum_{j=1}^m \Pi^j}{m} - 2.33\,\frac{\widehat{\mathrm{Std}}(\Pi;m)}{\sqrt{m}},\quad \frac{\sum_{j=1}^m \Pi^j}{m} + 2.33\,\frac{\widehat{\mathrm{Std}}(\Pi;m)}{\sqrt{m}}\right]. \tag{22.4}$$

To obtain a 95% (narrower) window it is enough to replace 2.33 by 1.96, and to obtain a (still narrower) 90% window it is enough to replace 2.33 by 1.65. All other sizes may be derived by the Φ tables.

We know that in some cases, to obtain a 98% window whose (half-) width $2.33\,\widehat{\mathrm{Std}}(\Pi;m)/\sqrt{m}$ is small enough, we are forced to take a huge number of paths m. This can be a problem for computational time. A way to reduce the impact of this problem is, for a given m that we deem to be large enough, to find alternatives that reduce the variance $(\widehat{\mathrm{Std}}(\Pi;m))^2$, thus narrowing the above window without increasing m.

One of the most effective methods to do this is the control variate technique.

22.6.2 Variance Reduction with Control Variate

"I have been mathematically simulating [in my head] several hundred possible scenarios to account for the corruption of his programming..." "Several hundred, Superman? Two days ago you could have calculated several bilion simulations simultaneously"

Superman One Million and Flash One Million, DC One Million (1999)

We begin by selecting an alternative payoff Π^{an} which we know how to evaluate analytically, in that

$$\mathbb{E}(\Pi^{\mathrm{an}}) = \pi^{\mathrm{an}}$$

is known. When we simulate our original payoff Π we now simulate also the analytical payoff Π^{an} as a function of the same scenarios for the underlying variables. We define a new control-variate estimator for $\mathbb{E}\Pi$ as

$$\widehat{\Pi}_c(k;m) := \frac{\sum_{j=1}^m \Pi^j}{m} + k\left(\frac{\sum_{j=1}^m \Pi^{\mathrm{an},j}}{m} - \pi^{\mathrm{an}}\right),$$

with k a constant to be determined. When viewing Π^j as independent identically distributed (iid) copies of Π and $\Pi^{\mathrm{an},j}$ as iid copies of Π^{an}, the above estimator remains unbiased, since we are subtracting the true known mean π^{an} from the correction term in k. So, once we have found that the estimator has not been biased by our correction, we may wonder whether our correction can be used to lower the variance.

Consider the random variable

$$\Pi_c(k) := \Pi + k(\Pi^{\mathrm{an}} - \pi^{\mathrm{an}})$$

whose expectation is the $\mathbb{E}(\Pi)$ we are estimating, and compute

$$\mathrm{Var}(\Pi_c(k)) = \mathrm{Var}(\Pi) + k^2\mathrm{Var}(\Pi^{\mathrm{an}}) + 2k\mathrm{Corr}(\Pi, \Pi^{\mathrm{an}})\mathrm{Std}(\Pi)\mathrm{Std}(\Pi^{\mathrm{an}}).$$

We may minimize this function of k by differentiating and setting the first derivative to zero.

We obtain easily that the variance is minimized by the following value of k:

$$k^* := -\mathrm{Corr}(\Pi, \Pi^{\mathrm{an}})\mathrm{Std}(\Pi)\ /\mathrm{Std}(\Pi^{\mathrm{an}}).$$

By plugging $k = k^*$ into the above expression, we obtain easily

$$\mathrm{Var}(\Pi_c(k^*)) = \mathrm{Var}(\Pi)(1 - \mathrm{Corr}(\Pi, \Pi^{\mathrm{an}})^2),$$

from which we see that $\Pi_c(k^*)$ has a smaller variance than our original Π, the smaller this variance the larger (in absolute value) the correlation between Π and Π^{an}. Accordingly, when moving to simulated quantities, we set

$$\widehat{\mathrm{Std}}(\Pi_c(k^*);m) = \widehat{\mathrm{Std}}(\Pi;m)(1 - \widehat{\mathrm{Corr}}(\Pi, \Pi^{\mathrm{an}};m)^2)^{1/2},$$

where $\widehat{\mathrm{Corr}}(\Pi, \Pi^{\mathrm{an}};m)$ is the sample correlation

$$\widehat{\mathrm{Corr}}(\Pi, \Pi^{\mathrm{an}};m) = \frac{\widehat{\mathrm{Cov}}(\Pi, \Pi^{\mathrm{an}};m)}{\widehat{\mathrm{Std}}(\Pi;m)\ \widehat{\mathrm{Std}}(\Pi^{\mathrm{an}};m)}$$

and the sample covariance is

$$\widehat{\mathrm{Cov}}(\Pi, \Pi^{\mathrm{an}};m) = \sum_{j=1}^m \Pi^j\Pi^{\mathrm{an},j}/m - (\sum_{j=1}^m \Pi^j)(\sum_{j=1}^m \Pi^{\mathrm{an},j})/(m^2)$$

and

$$(\widehat{\mathrm{Std}}(\Pi^{\mathrm{an}}; m))^2 := \sum_{j=1}^{m} (\Pi^{\mathrm{an},j})^2/m - (\sum_{j=1}^{m} \Pi^{\mathrm{an},j}/m)^2.$$

One may include the correction factor $m/(m-1)$ to correct for the bias of the variance estimator, although the correction is irrelevant for large m.

We see from

$$\widehat{\mathrm{Std}}(\Pi_c(k^*); m) = \widehat{\mathrm{Std}}(\Pi; m)(1 - \widehat{\mathrm{Corr}}(\Pi, \Pi^{\mathrm{an}}; m)^2)^{1/2},$$

that for the variance reduction to be relevant, we need to choose the analytical payoff Π^{an} to be as (positively or negatively) correlated as possible with the original payoff Π. Notice that in the limit case of correlation equal to one the variance shrinks to zero.

The window for our control-variate Monte Carlo estimate $\widehat{\Pi}_c(k; m)$ of $\mathbb{E}(\Pi)$ is now:

$$\left[\widehat{\Pi}_c(k; m) - 2.33\, \frac{\widehat{\mathrm{Std}}(\Pi_c(k^*); m)}{\sqrt{m}},\quad \widehat{\Pi}_c(k; m) + 2.33\, \frac{\widehat{\mathrm{Std}}(\Pi_c(k^*); m)}{\sqrt{m}}\right],$$

This window is narrower than the corresponding simple Monte Carlo one by a factor $(1 - \widehat{\mathrm{Corr}}(\Pi, \Pi^{\mathrm{an}}; m)^2)^{1/2}$.

We may wonder about a good possible Π^{an}. We may select as Π^{an} the simplest payoff equal to the underlying default indicator:

$$\Pi^{\mathrm{an}} := 1_{\{\tau < T\}},$$

whose expectation is known analytically and is simply the probability of defaulting by time T (as is for example stripped from CDS contracts). Alternatively, we may take sums or differences of default indicators for different T's if different maturities are involved:

$$\Pi^{\mathrm{an}} := 1_{\{\tau < T_1\}} + 1_{\{\tau < T_2\}} + \ldots + 1_{\{\tau < T_n\}},$$

or

$$\Pi^{\mathrm{an}} := 1_{\{\tau < T_2\}} - 1_{\{\tau < T_1\}} = 1_{\{\tau \in [T_1, T_2)\}},$$

and so on. We may also let Π^{an} zoom on the time intervals $[T_1, T_2)$ where the original payoff most insists.

22.7 Stochastic Intensity: The SSRD model

In this section we consider a model with stochastic intensity and interest rates. The default is then the first jump of a Cox process, that is a Poisson process with stochastic intensity.

As we have explained earlier in the guided tour in Section 21.1.1 and more in detail in Section 22.2.3, in this kind of models the intensity λ is a stochastic process but, conditional on the filtration generated by λ itself, the process N whose first jump represents default remains a time-inhomogeneous Poisson process with intensity $\gamma = \lambda$.

Our assumptions on the short-rate process r and on the intensity dynamics come from the earlier part of the book, and this is where we actually use the interest rate models technology also for the default event, almost for free.

In particular, the model we consider is the CIR++ model introduced in Section 3.9. Although we will recall here some of the features of this model, we advise the reader who is not familiar with the CIR (Section 3.2.3) and CIR++ (Section 3.9) models to have a look at the earlier sections and related references to other parts of the book before proceeding further. We give a summary anyway, with a slightly generalized notation, so as to be able to distinguish the CIR++ component for interest rates from the CIR++ component for stochastic intensities. The model we introduce is called "Shifted Square Root Diffusion" (SSRD) model, and is based on Brigo and Alfonsi (2003, 2005).

22.7.1 A two-factor shifted square-root diffusion model for intensity and interest rates (Brigo and Alfonsi (2003))

We begin by illustrating the interest-rate part of the model.

CIR++ short-rate model (Brigo and Mercurio (2001a))

We write the short-rate r_t as the sum of a deterministic function φ and of a Markovian process x_t^α:

$$r_t = x_t^\alpha + \varphi(t;\alpha)\ , \quad t \geq 0, \tag{22.5}$$

where φ depends on the parameter vector α (which includes x_0^α) and is integrable on closed intervals. Notice that x_0^α is indeed one more parameter at our disposal: we are free to select its value as long as

$$\varphi(0;\alpha) = r_0 - x_0\ .$$

We take as reference model for x the Cox-Ingersoll-Ross (1985) process:

$$dx_t^\alpha = k(\theta - x_t^\alpha)dt + \sigma\sqrt{x_t^\alpha}dW_t,$$

where the parameter vector is $\alpha = (k, \theta, \sigma, x_0^\alpha)$, with k, θ, σ, x_0^α positive deterministic constants. As usual, W is a standard Brownian motion process under the risk neutral measure, representing the stochastic shock in our dynamics. The condition

$$2k\theta > \sigma^2$$

ensures that the origin is inaccessible to the reference model, so that the process x^α remains positive. As is well known, this process x^α features a noncentral *chi-square* distribution, and yields an affine term-structure of interest rates. Accordingly, analytical formulas for prices of zero-coupon bond options, caps and floors, and, through Jamshidian's decomposition, coupon-bearing bond options and swaptions, can be derived (see also Section 3.2.3). We can therefore consider the CIR++ model, consisting of our extension (22.5), and calculate the analytical formulas implied by such a model, by simply adapting the analogous explicit expressions for the reference CIR model as given in Cox et al. (1985). Denote by f instantaneous forward rates, i.e. $f(t,T) = -\partial \ln P(t,T)/\partial T$.

The initial market zero-coupon interest-rate curve $T \mapsto P^{\text{mkt}}(0,T)$ is automatically calibrated by our model if we set $\varphi(t;\alpha) = \varphi^{\text{CIR}}(t;\alpha)$ where

$$\begin{aligned}\varphi^{\text{CIR}}(t;\alpha) &= f^{\text{mkt}}(0,t) - f^{\text{CIR}}(0,t;\alpha), \qquad (22.6)\\ f^{\text{CIR}}(0,t;\alpha) &= 2k\theta \frac{(\exp\{th\}-1)}{2h+(k+h)(\exp\{th\}-1)}\\ &\quad + x_0 \frac{4h^2\exp\{th\}}{[2h+(k+h)(\exp\{th\}-1)]^2}\end{aligned}$$

with

$$h = \sqrt{k^2 + 2\sigma^2}.$$

For restrictions on the α's that keep r positive see Section 3.9.3.

The price at time t of a zero-coupon bond maturing at time T is

$$\begin{aligned}P(t,T) = &\frac{P^{\text{mkt}}(0,T)A(0,t;\alpha)\exp\{-B(0,t;\alpha)x_0\}}{P^{\text{mkt}}(0,t)A(0,T;\alpha)\exp\{-B(0,T;\alpha)x_0\}}\cdot \qquad (22.7)\\ &\cdot P^{\text{CIR}}(t,T,r_t-\varphi^{\text{CIR}}(t;\alpha);\alpha)\end{aligned}$$

where

$$P^{\text{CIR}}(t,T,x_t;\alpha) = \mathbb{E}_t(e^{-\int_t^T x^\alpha(u)du}) = A(t,T;\alpha)\exp\{-B(t,T;\alpha)x_t\}$$

is the bond price formula for the basic CIR model given in Section 3.2.3; from this bond formula continuously compound spot rates $R(t,T)$ (still affine in r_t), the spot LIBOR rates $L(t,T)$, forward LIBOR rates $F(t,T,S)$ and all other kind of rates can be easily computed as explicit functions of r_t. We omit the argument α when clear from the context.

The cap option price formula for the CIR++ model can be derived easily in closed form from the corresponding formula for the basic CIR model. This formula is a function of the parameters α. In our application we will calibrate the parameters α to cap prices, by inverting the analytical CIR++ formula, so that our interest rate model is calibrated to the initial zero coupon curve through φ and to the cap market through α. See also the calibration example in Section 3.14.

After summarizing the interest-rate part of the model, we now move to the intensity part of the model.

CIR++ intensity model (Brigo and Alfonsi (2003, 2005))

For the intensity model we adopt a similar approach, in that we set

$$\lambda_t = y_t^\beta + \psi(t;\beta)\ , \quad t \geq 0, \tag{22.8}$$

where ψ is a deterministic function, depending on the parameter vector β (which includes y_0^β), that is integrable on closed intervals. As before, y_0^β is indeed one more parameter at our disposal: We are free to select its value as long as

$$\psi(0;\beta) = \lambda_0 - y_0\ .$$

We take y again of the form:

$$dy_t^\beta = \kappa(\mu - y_t^\beta)dt + \nu\sqrt{y_t^\beta}dZ_t,$$

where the parameter vector is $\beta = (\kappa, \mu, \nu, y_0^\beta)$, with κ, μ, ν, y_0^β positive deterministic constants. As usual, Z is a standard Brownian motion process under the risk neutral measure, representing the stochastic shock in our dynamics. Again we assume the origin to be inaccessible, i.e.

$$2\kappa\mu > \nu^2.$$

For restrictions on the β's that keep λ positive, as is required in intensity models, see again the suggestions given in Section 3.9.3.

We will often use the integrated quantities

$$\Lambda(t) = \int_0^t \lambda_s ds, \quad Y^\beta(t) = \int_0^t y_s^\beta ds, \quad \text{and } \Psi(t,\beta) = \int_0^t \psi(s,\beta)ds.$$

We take the short interest-rate and the intensity processes to be correlated, by assuming the driving Brownian motions W and Z to be instantaneously correlated according to

$$dW_t\ dZ_t = \rho\ dt.$$

This way to model the intensity and the short interest rate can be viewed as a generalization of a particular case of the Lando's (1998) approach, and can also be seen as a generalization of a particular case of the Duffie and Singleton (1997, 1999) square-root diffusion model (see for example Bielecki and Rutkowski (2001), pp 253-258). In both cases we add a non homogeneous term to recover exactly fundamental market data in the spirit of our earlier deterministic shift extension of Section 3.8 (see also Brigo and Mercurio (2001a)).

22.7.2 Calibrating the joint stochastic model to CDS: Separability

With the above choice for λ, in the credit derivatives world we have formulas that are analogous to the ones for interest-rate derivatives products. Consider for example the risk-neutral survival probability. We have easily, as we have seen several times, that

$$\mathbb{Q}(\tau > t) = \mathbb{E}[1_{\{\tau>t\}}] = \mathbb{E}[\mathbb{E}(1_{\{\tau>t\}}|\mathcal{F}^\lambda)] =$$
$$= \mathbb{E}[\mathbb{E}(1_{\{\Lambda(\tau)>\Lambda(t)\}}|\mathcal{F}^\lambda)] = \mathbb{E}e^{-\Lambda(t)} = \mathbb{E}(e^{-\int_0^t \lambda(u)du}),$$

since, conditional on λ, $\Lambda(\tau)$ is an exponential random variable ξ with parameter one. Notice that, if λ were a short-rate process, the last expectation of the "stochastic discount factor" would simply be the zero-coupon bond price in our interest-rate model. In other terms, notice the analogy we already pointed out:

$$P(0,t) = \mathbb{E}(e^{-\int_0^t r(u)du}), \quad \mathbb{Q}(\tau > t) = \mathbb{E}[1_{\{\tau>t\}}] = \mathbb{E}(e^{-\int_0^t \lambda(u)du}).$$

So we see that survival probabilities for the λ model are the analogous of zero-coupon bond prices P in the r model. Thus if we choose for λ a CIR++ process, survival probabilities will be given by the CIR++ model bond price formula.

In particular, we can express credit default swaps data through the implied hazard function Γ according to the method described in Section 21.3.5. The related implied Γ, stripped from CDS data, will be denoted by $\Gamma = \Gamma^{\text{mkt}}$, while the associated hazard rate will be denoted by $\gamma = \gamma^{\text{mkt}}$.

Now assume $\rho = 0$. This implies that the intensity λ and the cumulated intensity Λ are independent of the short rate r, and of interest rates in general. Since $\tau = \Lambda^{-1}(\xi)$ and ξ is also independent of interest rates, in this $\rho = 0$ case the default time τ and interest rate quantities $r, D(s,t), ...$ are independent. It follows that everything we stated on CDS valuation in Section 21.3.5 holds, and in particular the model independent formulas (21.22) and (21.23) hold. In other terms, a CDS valuation (say at time 0 and seen from the protection seller) when $\rho = 0$ becomes model independent and is given by the formula

$$\text{CDS}_{a,b}(0, R; P(0,\cdot), \mathbb{Q}(\tau > \cdot)) = R\Big[-\int_{T_a}^{T_b} P(0,t)(t - T_{\beta(t)-1})d_t\mathbb{Q}(\tau \geq t)$$
$$+ \sum_{i=a+1}^{b} P(0,T_i)\alpha_i\mathbb{Q}(\tau \geq T_i)\Big] + \text{L}_{\text{GD}}\int_{T_a}^{T_b} P(0,t)\, d_t\mathbb{Q}(\tau \geq t) \quad (22.9)$$

This formula is indeed model independent given the initial zero coupon curve (bonds) at time 0 observed in the market (i.e. $P(0,\cdot)$) and given the survival probabilities $\mathbb{Q}(\tau \geq \cdot)$ at time 0. However, this is also the formula for the SSRD model corresponding to the case $\rho = 0$. This means that if we strip survival probabilities from CDS in a model independent way, to calibrate the market CDS quotes we just need to make sure that the survival probabilities we strip from CDS are correctly reproduced by the SSRD model. Since the survival probabilities in the SSRD model are given by

$$\mathbb{Q}(\tau > t)_{model} = \mathbb{E}(e^{-\Lambda(t)}) = \mathbb{E}\exp\left(-\Psi(t,\beta) - Y^\beta(t)\right) \tag{22.10}$$

we just need to make sure

$$\mathbb{E}\exp\left(-\Psi(t,\beta) - Y^\beta(t)\right) = \mathbb{Q}(\tau > t)_{market}.$$

If we agree to express survival probabilities through implied hazard rates and functions, so that $\mathbb{Q}(\tau > t)_{market} = e^{-\Gamma^{\text{mkt}}(t)}$, then Equation (22.10) reads

$$\mathbb{E}\exp\left(-\Psi(t,\beta) - Y^\beta(t)\right) = e^{-\Gamma^{\text{mkt}}(t)} \tag{22.11}$$

from which

$$\Psi(t,\beta) = \Gamma^{\text{mkt}}(t) + \ln(\mathbb{E}(e^{-Y^\beta(t)})) = \Gamma^{\text{mkt}}(t) + \ln(P^{\text{CIR}}(0,t,y_0;\beta)), \tag{22.12}$$

where we choose the parameters β in order to have a positive function ψ (i.e. an increasing Ψ). Thus, if ψ is selected according to this last formula, as we will assume from now on, the model is calibrated to the market implied hazard function Γ^{mkt}, i.e. to CDS data.

The whole chain of reasoning relies on the possibility to disentangle survival probabilities from discount factors, in a way that leads to the model independent formula (22.9) for CDS prices. This allows us to calibrate directly the survival probabilities of the model to the survival probabilities from the market. The condition of independence between τ and r is fundamental for this disentangling result to hold, and is satisfied only when $\rho = 0$.

Indeed, if $\rho \neq 0$ then Formula (22.9) no longer holds and to calibrate cds data we need to solve

$$\text{CDS}_{0,b}^{\text{SSRD}}(0, R_{0,b}^{\text{MKT}}(0); \alpha, \beta, \varphi(\cdot;\alpha), \psi(\cdot,\beta), \rho) = 0$$

for a set of canonical maturities T_b, typically ranging from one year to ten years, jointly with analogous equations for interest rate derivatives involving α and φ. This involves the joint parameters of the r and λ models and the correlation ρ to be calibrated all at the same time.

Let us now compute the CDS price in the SSRD model in general.

The $\text{CDS}_{a,b}(t, R, \text{L}_{\text{GD}})$ price in a stochastic intensity model is given by

$$\begin{aligned}\text{CDS}_{a,b}(t, R, \text{L}_{\text{GD}}) &= \mathbf{1}_{\{\tau>t\}}\mathbb{E}\Big\{D(t,\tau)(\tau - T_{\beta(\tau)-1})R\mathbf{1}_{\{T_a<\tau<T_b\}} \\ &+ \sum_{i=a+1}^{b} D(t,T_i)\alpha_i R\mathbf{1}_{\{\tau>T_i\}} - \mathbf{1}_{\{T_a<\tau<T_b\}}D(t,\tau)\,\text{L}_{\text{GD}}|\mathcal{G}_t\Big\}.\end{aligned}$$

Let us expand the first term. We obtain

$$
\begin{aligned}
&1_{\{\tau>t\}}\mathbb{E}\left\{D(t,\tau)(\tau-T_{\beta(\tau)-1})R1_{\{T_a<\tau<T_b\}}\middle|\mathcal{G}_t\right\}\\
&=\frac{1_{\{\tau>t\}}}{\mathbb{Q}(\tau>t|\mathcal{F}_t)}\mathbb{E}\left\{D(t,\tau)(\tau-T_{\beta(\tau)-1})R1_{\{T_a<\tau<T_b\}}\middle|\mathcal{F}_t\right\}\\
&=\frac{1_{\{\tau>t\}}}{\exp(-\int_0^t\lambda_s ds)}\mathbb{E}\left\{\int_t^\infty D(t,s)(s-T_{\beta(s)-1})R1_{\{T_a<s<T_b\}}1_{\{\tau\in[s,s+ds)\}}\middle|\mathcal{F}_t\right\}=\\
&=\frac{1_{\{\tau>t\}}}{\exp(-\int_0^t\lambda_s ds)}\mathbb{E}\left[\mathbb{E}\left\{\int_{T_a}^{T_b} D(t,s)(s-T_{\beta(s)-1})R1_{\{\tau\in[s,s+ds)\}}\middle|\mathcal{F}_{T_b}\right\}\middle|\mathcal{F}_t\right]\\
&=\frac{1_{\{\tau>t\}}}{\exp(-\int_0^t\lambda_s ds)}\mathbb{E}\left[\int_{T_a}^{T_b} D(t,s)(s-T_{\beta(s)-1})R\mathbb{E}\left\{1_{\{\tau\in[s,s+ds)\}}\middle|\mathcal{F}_{T_b}\right\}\middle|\mathcal{F}_t\right]\\
&=\frac{1_{\{\tau>t\}}}{\exp(-\int_0^t\lambda_s ds)}\mathbb{E}\left[\int_{T_a}^{T_b} D(t,s)(s-T_{\beta(s)-1})R\,\mathbb{Q}\left\{\tau\in[s,s+ds)\middle|\mathcal{F}_{T_b}\right\}\middle|\mathcal{F}_t\right]\\
&=\frac{1_{\{\tau>t\}}}{\exp(-\int_0^t\lambda_s ds)}\mathbb{E}\left[\int_{T_a}^{T_b} D(t,s)(s-T_{\beta(s)-1})R\exp\left(-\int_0^s\lambda_u du\right)\lambda_s ds\middle|\mathcal{F}_t\right]\\
&=1_{\{\tau>t\}}\mathbb{E}\left[\int_{T_a}^{T_b} D(t,s)(s-T_{\beta(s)-1})R\exp\left(-\int_t^s\lambda_u du\right)\lambda_s ds\middle|\mathcal{F}_t\right]
\end{aligned}
$$

where we have used the filtration switching formula (22.3) in the first equality and equation (22.1) in the last two steps, since we are conditioning on $\mathcal{F}_{T_b}$, which makes intensity deterministic and thus formula (22.1) applicable, provided we substitute γ with λ (and t with s). By recalling that $D(t,s)=\exp(-\int_t^s r_u du)$ and carrying out analogous calculations for the remaining CDS terms, we can write eventually, under any stochastic intensity model:

$$
\begin{aligned}
\mathrm{CDS}_{a,b}(t,R,\mathrm{L{\scriptsize GD}}) &= 1_{\{\tau>t\}}\left\{R\sum_{i=a+1}^{b}\alpha_i\mathbb{E}\left[\exp\left(-\int_t^{T_i}(r_s+\lambda_s)ds\right)\middle|\mathcal{F}_t\right]\right.\\
&\quad +R\int_{T_a}^{T_b}\mathbb{E}\left[\exp\left(-\int_t^u(r_s+\lambda_s)ds\right)\lambda_u\middle|\mathcal{F}_t\right](u-T_{\beta(u)-1})du\\
&\quad \left.-\mathrm{L{\scriptsize GD}}\int_{T_a}^{T_b}\mathbb{E}\left[\exp\left(-\int_t^u(r_s+\lambda_s)ds\right)\lambda_u\middle|\mathcal{F}_t\right]du\right\} \qquad (22.13)
\end{aligned}
$$

When considering the above expected values in the SSRD model with $\rho\neq 0$, given the particular λ and r processes and the parameters in their dynamics, we obtain a formula $\mathrm{CDS}_{a,b}^{\mathrm{SSRD}}(0,R;\alpha,\beta,\varphi(\cdot;\alpha),\psi(\cdot,\beta),\rho)$ for CDS valuation in the SSRD model. However, this formula admits no closed form. One is then forced in principle to solve

$$\text{CDS}_{0,b}^{\text{SSRD}}(0, R_{0,b}^{\text{MKT}}(0); \alpha, \beta, \varphi(\cdot; \alpha), \psi(\cdot, \beta), \rho) = 0 \tag{22.14}$$

numerically, together with the equations for interest rate derivatives calibration, in the unknowns $\alpha, \beta, \varphi(\cdot; \alpha), \psi(\cdot, \beta), \rho$. This may be a rather dramatic task. None of this problems is present when $\rho = 0$. Indeed, as a double-check we now compute the above general expression under the assumption that $\rho = 0$. The reader not interested in the double check may skip directly to the end of the double check part.

Begin double-check: Assuming $\rho = 0$ in the general CDS expression (22.13).

Assuming $\rho = 0$, the expectations appearing in the above expression can be computed as follows:

$$\begin{aligned}
&\mathbb{E}\left[\exp\left(-\int_t^{T_i}(r_s+\lambda_s)ds\right)|\mathcal{F}_t\right] = \\
&= \mathbb{E}\left[\exp\left(-\int_t^{T_i} r_s ds\right)\exp\left(-\int_t^{T_i}\lambda_s ds\right)\Big|\mathcal{F}_t\right] \\
&= \mathbb{E}\left[\exp\left(-\int_t^{T_i} r_s ds\right)\Big|\mathcal{F}_t\right]\mathbb{E}\left[\exp\left(-\int_t^{T_i}\lambda_s ds\right)\Big|\mathcal{F}_t\right] \\
&= \exp(\Psi(t,\beta)-\Psi(T_i,\beta))P^{\text{CIR}}(t,T_i;y_t,\beta)\times \\
&\quad\times\exp(\Phi(t,\alpha)-\Phi(T_i,\alpha))P^{\text{CIR}}(t,T_i;x_t,\alpha).
\end{aligned} \tag{22.15}$$

Further, again with $\rho = 0$, we may compute

$$\mathbb{E}\left[\exp\left(-\int_t^{u}(r_s+\lambda_s)ds\right)\lambda_u\Big|\mathcal{F}_t\right] = \tag{22.16}$$

$$\begin{aligned}
&= \mathbb{E}\left[\exp\left(-\int_t^{u} r_s ds\right)\Big|\mathcal{F}_t\right]\mathbb{E}\left[\exp\left(-\int_t^{u}\lambda_s ds\right)\lambda_u\Big|\mathcal{F}_t\right] = \\
&= \mathbb{E}\left[\exp\left(-\int_t^{u} r_s ds\right)\Big|\mathcal{F}_t\right]\left(-\frac{d}{du}\mathbb{E}\left[\exp\left(-\int_t^{u}\lambda_s ds\right)\Big|\mathcal{F}_t\right]\right) = \\
&= -\exp(\Phi(t,\alpha)-\Phi(u,\alpha))P^{\text{CIR}}(t,u;x_t,\alpha)\times \\
&\quad\times\frac{d}{du}\left[\exp(\Psi(t,\beta)-\Psi(u,\beta))P^{\text{CIR}}(t,u;y_t,\beta)\right]
\end{aligned}$$

so that all terms are known analytically given x_t and y_t.

Now, if we are using the deterministic shifts φ and ψ coming from market data at time t given by P^{mkt} (or equivalently f^{mkt}) and Γ^{mkt}, according to the procedures outlined above (in particular see (22.6) and (22.12)), then by construction all the above terms give a CDS value that is consistent with the model-independent CDS value where survival probabilities are re-cast in terms of hazard functions, according to formula (22.2) with $\Gamma = \Gamma^{\mathrm{mkt}}$.

Indeed, with φ coming from (22.6) and ψ chosen via (22.12), bonds and survival probabilities are by construction consistent with model-independent market bonds and with model-independent survival probabilities, to which we associated the relevant market implied hazard functions, and expression (22.15) reduces to

$$P^{\mathrm{mkt}}(t,T_i)\left(e^{-(\Gamma^{\mathrm{mkt}}(T_i)-\Gamma^{\mathrm{mkt}}(t))}\right)$$

while expression (22.16) reduces to

$$-P^{\mathrm{mkt}}(t,u)\frac{d}{du}\left(e^{-(\Gamma^{\mathrm{mkt}}(u)-\Gamma^{\mathrm{mkt}}(t))}\right)$$
$$= P^{\mathrm{mkt}}(t,u)\gamma^{\mathrm{mkt}}(u)\left(e^{-(\Gamma^{\mathrm{mkt}}(u)-\Gamma^{\mathrm{mkt}}(t))}\right).$$

If for simplicity we take the initial time $t = 0$ the two terms reduce to

$$P^{\mathrm{mkt}}(0,u)\gamma^{\mathrm{mkt}}(u)e^{-\Gamma^{\mathrm{mkt}}(u)},\quad P^{\mathrm{mkt}}(0,T_i)e^{-\Gamma^{\mathrm{mkt}}(T_i)}$$

and, when we substitute them in the CDS-price formula (22.13) valued at $t = 0$ we find the same expression as in Equation (22.2). For this to hold, i.e. for the SSRD model price to be consistent with the implied hazard rate price we needed to assume $\rho = 0$, leading to the factorizations given in formulas (22.15) and (22.16).

End of double-check.

To sum up, we have shown that in case $\rho = 0$,

i) Calibrating the model survival probabilities to the CDS-implied hazard function Γ^{mkt} by choosing ψ according to (22.12)

is equivalent to

ii) directly calibrate the (r,λ)-model by setting to zero the model CDS prices with the market quoted $R_{0,b}$'s in the premium legs.

More precisely, we have shown that if $\rho = 0$ and $\psi(\cdot;\beta)$ is selected according to (22.12), then the price ii) of the $\mathrm{CDS}_{0,b}$ payoff under the stochastic intensity model λ is the same price i) obtained under hazard rate γ^{mkt} and is given by (22.2). So, as we knew already, when $\rho = 0$ the CDS price does not depend on the dynamics of (λ, r), and in particular it does not depend on $k, \theta, \sigma, \kappa, \nu$ and μ. We will verify this also numerically in Table 22.22: by amplifying intensity parameters through an increase of κ, ν and μ we do not substantially affect the CDS price in case $\rho = 0$.

However, if $\rho \neq 0$, the factorizations (22.15) and (22.16) on which the equivalence with an implied hazard function is based do not hold. As we have seen in (22.14), the CDS price under the SSRD model becomes different and is in principle dependent on the dynamics; now the two procedures i) and ii) are not equivalent: The correct one would be to equate to zero the model CDS prices and solve Equation (22.14) (together with the interest derivatives equations), to obtain the model parameters, including in principle ρ.

This is annoying, since the attractive feature of the model is the separate and semi-automatic calibration of the interest-rate part to interest-rate data and of the intensity part to credit market data. Indeed, in the separable case the credit derivatives desk might ask for the α parameters and the $\varphi(\cdot;\alpha)$ curve to the interest-rate derivatives desk, and then proceed with finding β and $\psi(\cdot;\beta)$ from CDS data. This ensures also a consistency of the interest rate model that is used in credit derivatives evaluation with the interest rate model that is used for default-free derivatives. This separate automatic calibration no longer holds if we introduce ρ, since now the dynamics of interest rates is also affecting the CDS price.

However, we will see below in Table 22.22 that the impact of ρ is typically negligible on CDSs, even in case intensity parameters are increased by a factor from 3 to 5.

We can thus calibrate CDS data with $\rho = 0$, using the separate calibration procedure outlined above, and then set ρ to a desired value, based perhaps on historical estimation or on market views.

Once we have done this and calibrated CDS data through $\psi(\cdot,\beta)$, we are left with the parameters β, which can be used to calibrate further products, similarly to the way the α parameters of the r model are used to calibrate cap prices after calibration of the zero-coupon curve in the interest rate market. However, this will be interesting when option data on the credit derivatives market will become more liquid. Currently the bid-ask spreads for single name CDS options are large and suggest to consider these first quotes with caution. At the moment we content ourselves of calibrating only CDS's for the credit part. To help specifying β without further data we impose a constraint on the calibration of CDS's. We require the β's to be found to keep Ψ positive and increasing and to minimize $\int_0^T \psi(s,\beta)^2 ds$. This minimization amounts to contain the departure of λ from its time-homogeneous component y^β as much as possible. Indeed, if we take as criterion the integrated squared difference between "instantaneous forward rates" γ^{mkt} in the market and $f^{\text{CIR}}(\cdot;\beta)$ in our homogeneous CIR model with β parameters, constraining these differences to be positive at all points, the related minimization gives us the time-homogeneous CIR model β that is closest to market data under the given constraints.

We calibrated the same CDS data used to obtain the implied hazard function Γ given in Figure 22.16 up to a ten years maturity.

In Figure 22.17 we give the related piecewise linear hazard rate $\gamma^{\text{mkt}}(t)$ obtained by calibrating the 1y, 3y, 5y, 7y and 10y CDS's on Merrill-Lynch on October 2002. In Figure 22.16 the related risk-neutral default probabilities

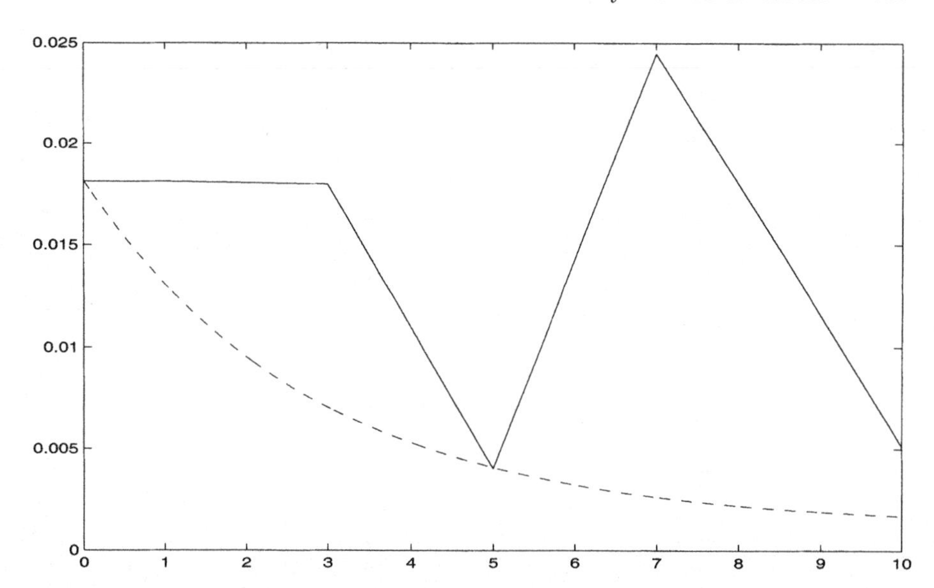

Fig. 22.17. Graph of the implied deterministic intensity $t \mapsto \gamma^{\text{mkt}}(t)$ for Merrill-Lynch CDS's of several maturities on October 25, 2002 (continuous line) and the best approximating hazard rate coming from a time-homogeneous CIR model (dashed line) satisfying the relevant constraints. We will extend this CIR model to CIR++ to recover exactly γ^{mkt}

are given. These are equal, first order in the hazard function, to the hazard function $\Gamma(t)$ itself, since $\mathbb{Q}\{\tau < t\} = 1 - \exp(-\Gamma(t)) \approx \Gamma(t)$ for small Γ.

Our calibration of the SSRD model produced the following results

$$\beta: \quad \kappa = 0.354201, \ \mu = 0.00121853, \ \nu = 0.0238186; \ y_0 = 0.0181,$$

with the ψ function plotted in Fig 22.18. The interest-rate model part has been calibrated to the initial zero curve and to cap prices. The parameters are

$$\alpha: \quad k = 0.528905, \ \theta = 0.0319904, \quad \sigma = 0.130035, \ x_0 = 8.32349 \times 10^{-5}.$$

To check that, as anticipated above, the impact of the correlation ρ is negligible on CDS's we reprice the 5y CDS we used in the above calibration with $\rho = 0$, ceteris paribus, by setting first $\rho = -1$ and then $\rho = 1$. As usual, the amount R renders the CDS fair at time 0, thus giving $\text{CDS}_{0,b}(0, R, \text{L}_{\text{GD}}) = 0$ with the deterministic model or with the stochastic model when $\rho = 0$. In our case (market data of October 25, 2002) the MID value R corresponding to $R^{\text{BID}} = 0.009$ and $R^{\text{ASK}} = 0.0098$ is $R = 0.0094$, while $\text{L}_{\text{GD}} = 0.593$, corresponding to a recovery rate of $\text{R}_{\text{EC}} = .407$. With this R and the above (r, λ) model calibrated with $\rho = 0$ we now set ρ to different values and, by the "Gaussian mapping" approximation technique described below to model (r, λ), we obtain the results given in Table 22.21. With a Monte Carlo method

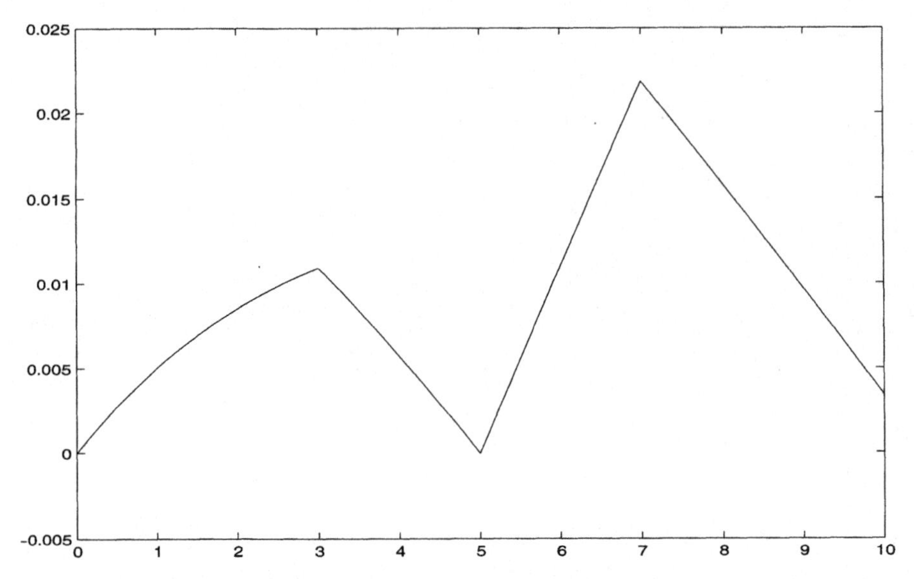

Fig. 22.18. ψ function for the CIR++ model for λ calibrated to Merrill-Lynch CDS's of maturities up to 10y on October 25, 2002

for the true dynamics we obtain substantially the same conclusions, as illustrated in Table 22.22. It is evident that the impact of rates/intensities correlation is almost negligible on CDS's, and typically well within a small fraction of the bid-ask spread. Indeed, with the above market quotes, in the case $\rho = 0$, we have

$$\mathrm{CDS}_{0,5y}(0, R^{\mathrm{BID}}_{0,5y}, \mathrm{L_{GD}}) = -17.14E-4, \quad \mathrm{CDS}_{0,5y}(0, R^{\mathrm{ASK}}_{0,5y}, \mathrm{L_{GD}}) = 17.16E-4. \tag{22.17}$$

So we see that the possible excursion of the CDS value due to correlation as from Table 22.21 is less than one tenth of the CDS excursion corresponding to the market bid-ask spread induced by bid-ask quotes in the premium rate. This excursion is basically negligible. This is further confirmed when Monte Carlo valuation replaces the Gaussian dependence mapping approximation, as one can see from Table 22.22.

There is one possible objection to our numerical investigation of the correlation impact. The fact is that the ν parameter is quite small. Indeed, with the β parameters above we would have the following initial "percentage instantaneous volatility" for the time-homogeneous part of the intensity, i.e. for y:

$$\frac{\nu\sqrt{y_0}}{y_0} = \frac{\nu}{\sqrt{y_0}} \approx 0.18.$$

Thus the percentage volatility has an order of magnitude about 18%. This value is small compared to typical CDS rate implied volatilities (see Section 23.2.3). We will consider further tests on the correlation impact with

CDS options later in the chapter, including the possibility of adding jumps in the model in order to increase the volatilities implied by the model for CDS options.

22.7.3 Discretization schemes for simulating (λ, r)

The SSRD model allows for known non-central chi-squared transition densities in the case with 0 correlation. Indeed, in this case we have two independent CIR processes, and to have their joint distribution it is enough to couple the single known distributions of the two processes (noncentral chi-squared distributions). However, when ρ is not zero we are not able to characterize the joint distribution of the two processes and we need to resort to numerical methods to obtain the joint distribution of r and λ and of their functionals needed for discounting and evaluating payoffs. This is the reason why one assumes $\rho = 0$ in the CIR2 and CIR2++ models seen in Section 4.3. Here, however, we remove this assumption, and our findings may be attempted also on the CIR2 and CIR2++ models.

The typical technique when $\rho \neq 0$ consists in adopting a discretization scheme for the relevant SDEs and then to simulate the Gaussian shocks corresponding to the joint Brownian motions increments in the discretized dynamics. Let us mention here that even in the case $\rho = 0$, discretization schemes may be faster than a simulation based on the known distribution when we have to compute path-dependent payoffs.

Euler and Milstein explicit schemes for simulating (λ, r). The easiest choice is given by the Euler scheme we have already used earlier in the book, see also Appendix C . Let $t_0 = 0 < t_1 < ... < t_n = T$ be a discretization of the interval $[0, T]$. We write the second Brownian motion Z, correlated with the first Brownian motion W, by means of a third Brownian motion V independent of W. We write Z as $Z_t = \rho W_t + \sqrt{1-\rho^2} V_t$ (Cholesky decomposition), where V_t is a Brownian motion independent of W, and we obtain the increments of (W, Z) between t_i and t_{i+1} through simulation of the increments of W and V (independent, centered Gaussian variables with variance $t_{i+1} - t_i$). So from now on when we mention simulation of (W, Z) this is obtained through a simulation based on independent (W, V) and the above formula.

The Euler scheme reads:

$$\tilde{x}^{\alpha}_{t_{i+1}} = \tilde{x}^{\alpha}_{t_i} + k(\theta - \tilde{x}^{\alpha}_{t_i})(t_{i+1} - t_i) + \sigma\sqrt{\tilde{x}^{\alpha}_{t_i}}(W_{t_{i+1}} - W_{t_i})$$

$$\tilde{y}^{\beta}_{t_{i+1}} = \tilde{y}^{\beta}_{t_i} + \kappa(\mu - \tilde{y}^{\beta}_{t_i})(t_{i+1} - t_i) + \nu\sqrt{\tilde{y}^{\beta}_{t_i}}(Z_{t_{i+1}} - Z_{t_i}),$$

starting from $(\tilde{x}^{\alpha}_{t_0}, \tilde{y}^{\beta}_{t_0}) = (x^{\alpha}_0, y^{\beta}_0)$. Although the regularity conditions that ensure a better convergence for the Milstein scheme are not satisfied here (the diffusion coefficient is not Lipschitz), one may try to apply it anyway. The related equation for $\tilde{x}^{\alpha}_{t_i}$ is as follows:

$$\tilde{x}^\alpha_{t_{i+1}} = \tilde{x}^\alpha_{t_i} + k(\theta - \tilde{x}^\alpha_{t_i})(t_{i+1} - t_i) + \sigma\sqrt{\tilde{x}^\alpha_{t_i}}(W_{t_{i+1}} - W_{t_i}) +$$
$$+\frac{1}{4}\sigma^2[(W_{t_{i+1}} - W_{t_i})^2 - (t_{i+1} - t_i)]$$

with an analogous expression for $\tilde{y}^\beta_{t_i}$, replacing $\alpha = (k, \theta, \sigma, x^\alpha_0)$ by $\beta = (\kappa, \mu, \nu, y^\beta_0)$ and $(W_{t_{i+1}} - W_{t_i})$ by $(Z_{t_{i+1}} - Z_{t_i})$ (from now, we will only write the schemes for x^α). The major drawback of the previous explicit schemes is that they do not ensure positivity of $\tilde{x}^\alpha_{t_i}$ (resp. $\tilde{y}^\beta_{t_i}$) and are thus not well defined since the square-root of a negative number would move the domain to complex numbers. It is possible to correct the above problem as follows: when we obtain a negative value at t_{i+1}, we go back to t_i and we can simulate a Brownian bridge on $[t_i, t_{i+1}]$, with a time step small enough to retrieve the positivity which is ensured in the continuous case when $2k\theta > \sigma^2$. However, this solution presents the drawback to require other simulations of random variables.

Two "modified versions" of the Euler scheme have been proposed by Deelstra and Delbaen (1998) and Diop (2003). Deelstra and Delbaen consider:

$$\tilde{x}^\alpha_{t_{i+1}} = \tilde{x}^\alpha_{t_i} + k(\theta - \tilde{x}^\alpha_{t_i})(t_{i+1} - t_i) + \sigma\sqrt{\tilde{x}^\alpha_{t_i}\mathbf{1}_{\{\tilde{x}^\alpha_{t_i} > 0\}}}(W_{t_{i+1}} - W_{t_i}).$$

This scheme allows $\tilde{x}^\alpha_{t_i}$ to take negative values, but the positive part (indicator) in the square-root makes it well defined by zeroing the diffusion coefficient in cases where the argument of the square root would go negative. Diop (2003) studies the reflected Euler scheme:

$$\tilde{x}^\alpha_{t_{i+1}} = |\tilde{x}^\alpha_{t_i} + k(\theta - \tilde{x}^\alpha_{t_i})(t_{i+1} - t_i) + \sigma\sqrt{\tilde{x}^\alpha_{t_i}}(W_{t_{i+1}} - W_{t_i})|.$$

However, for CIR processes we may also obtain ad-hoc schemes as follows.

Implicit Euler schemes and derived Explicit schemes (Brigo and Alfonsi (2003, 2005), Alfonsi (2005)). Let us remark that, for a sufficiently regular partition of $[0, T]$, when $\max\{t_{i+1} - t_i, 0 \le i < n\} \to 0$ we have

$$x^\alpha_t = x^\alpha_0 + \int_0^t k(\theta - x^\alpha_s)ds + \sigma\int_0^t \sqrt{x^\alpha_s}dW_s$$
$$= \lim\left[x^\alpha_0 + \sum_{i;t_i<t} k(\theta - x^\alpha_{t_i})(t_{i+1} - t_i) + \sigma\sum_{i;t_i<t}\sqrt{x^\alpha_{t_i}}(W_{t_{i+1}} - W_{t_i})\right]$$

$$= \lim\left[x^\alpha_0 + \sum_{i;t_i<t} k(\theta - x^\alpha_{t_{i+1}})(t_{i+1} - t_i) + \sigma\sum_{i;t_i<t}\sqrt{x^\alpha_{t_{i+1}}}(W_{t_{i+1}} - W_{t_i})\right.$$

$$-\sigma \sum_{i;t_i<t} (\sqrt{x^\alpha_{t_{i+1}}} - \sqrt{x^\alpha_{t_i}})(W_{t_{i+1}} - W_{t_i})\Bigg]$$

$$= \lim \left[x^\alpha_0 + \sum_{i;t_i<t} (k\theta - kx^\alpha_{t_{i+1}})(t_{i+1} - t_i) + \sigma \sum_{i;t_i<t} \sqrt{x^\alpha_{t_{i+1}}}(W_{t_{i+1}} - W_{t_i})\right] +$$

$$- \lim \sum_{i;t_i<t} \frac{\sigma^2}{2}(t_{i+1} - t_i)$$

$$= \lim \left[x^\alpha_0 + \sum_{i;t_i<t} (k\theta - \frac{\sigma^2}{2} - kx^\alpha_{t_{i+1}})(t_{i+1} - t_i) + \sigma \sum_{i;t_i<t} \sqrt{x^\alpha_{t_{i+1}}}(W_{t_{i+1}} - W_{t_i})\right],$$

since $d\langle\sqrt{x^\alpha_t}, W_t\rangle = d\sqrt{x^\alpha_t}\, dW_t = \sigma\, dt/2$. It is then natural to introduce the following implicit scheme (Brigo and Alfonsi (2003, 2005)):

$$\tilde{x}^\alpha_{t_{i+1}} = \tilde{x}^\alpha_{t_i} + (k\theta - \frac{\sigma^2}{2} - k\tilde{x}^\alpha_{t_{i+1}})(t_{i+1} - t_i) + \sigma\sqrt{\tilde{x}^\alpha_{t_{i+1}}}(W_{t_{i+1}} - W_{t_i}).$$

It follows that $\sqrt{\tilde{x}^\alpha_{t_{i+1}}}$ is the unique positive root (when $2k\theta > \sigma^2$ and $k > 0$) of the second-degree polynomial

$$P(X) = (1 + k(t_{i+1} - t_i))X^2 - \sigma(W_{t_{i+1}} - W_{t_i})X - (\tilde{x}^\alpha_{t_i} + (k\theta - \frac{\sigma^2}{2})(t_{i+1} - t_i)),$$

and we get

$$\tilde{x}^\alpha_{t_{i+1}} = \left(\frac{\sigma(W_{t_{i+1}} - W_{t_i}) + \sqrt{\Delta_{t_i}}}{2(1 + k(t_{i+1} - t_i))}\right)^2, \qquad (22.18)$$

where

$$\Delta_{t_i} = \sigma^2(W_{t_{i+1}} - W_{t_i})^2 + 4(\tilde{x}^\alpha_{t_i} + (k\theta - \frac{\sigma^2}{2})(t_{i+1} - t_i))(1 + k(t_{i+1} - t_i)),$$

with a similar formula for $\tilde{y}^\beta_{t_{i+1}}$. The positivity of $\tilde{x}^\alpha_{t_{i+1}}$ is guaranteed by construction. Since this expression is clearly increasing in $\tilde{x}^\alpha_{t_i}$, we obtain also the monotonicity property: "For a given path $(W_{t_i}(\omega))_i$, $x_0 \le \bar{x}_0$ implies $\tilde{x}^\alpha_{t_i}(\omega) \le \tilde{x}^{\bar{\alpha}}_{t_i}(\omega)$ for all t_i's". This property is important, since it holds for the original process in continuous time [1]. In the same spirit, one can look at the diffusion of the square-root process:

$$d\sqrt{\tilde{x}^\alpha_t} = \frac{k\theta - \sigma^2/4}{2\sqrt{\tilde{x}^\alpha_t}}dt - \frac{k}{2}\sqrt{\tilde{x}^\alpha_t}dt + \frac{\sigma}{2}dW_t$$

[1] Indeed, if we set $\delta_t = x^{\bar{\alpha}}_t - x^\alpha_t$ with $\bar{x}_0 > x_0$, we have $d\delta_t = \delta_t(-kdt + \sigma/(\sqrt{x^\alpha_t} + \sqrt{x^{\bar{\alpha}}_t})dW_t)$. Thus, δ_t appears as a Doleans exponential process and remains positive for all t.

and write the implicit scheme. This leads to another second-degree equation in $\sqrt{\tilde{x}^\alpha_{t_{i+1}}}$ (Alfonsi (2005)):

$$\left(1+\frac{k}{2}(t_{i+1}-t_i)\right)\tilde{x}^\alpha_{t_{i+1}} - \left[\frac{\sigma}{2}(W_{t_{i+1}}-W_{t_i})+\sqrt{\tilde{x}^\alpha_{t_i}}\right]\sqrt{\tilde{x}^\alpha_{t_{i+1}}}+$$

$$-\frac{k\theta-\sigma^2/4}{2}(t_{i+1}-t_i)=0$$

that has also only one positive root when $\sigma^2 < 4k\theta$ and $k>0$, and it gives:

$$\tilde{x}^\alpha_{t_{i+1}} = \left(\frac{\frac{\sigma}{2}(W_{t_{i+1}}-W_{t_i})+\sqrt{\tilde{x}^\alpha_{t_i}}+\sqrt{\Delta^{\surd}_{t_i}}}{2(1+\frac{k}{2}(t_{i+1}-t_i))}\right)^2 \tag{22.19}$$

where:

$$\Delta^{\surd}_{t_i} = (\frac{\sigma}{2}(W_{t_{i+1}}-W_{t_i})+\sqrt{\tilde{x}^\alpha_{t_i}})^2+2\left(1+\frac{k}{2}(t_{i+1}-t_i)\right)(k\theta-\sigma^2/4)(t_{i+1}-t_i).$$

This implicit scheme on the square-root ensures also the positivity and the property of monotonicity since once again $\tilde{x}^\alpha_{t_{i+1}}$ is an increasing function of $\tilde{x}^\alpha_{t_i}$. It is then interesting to make a Taylor expansion of both schemes up to order one in $(t_{i+1}-t_i)$ and order two in $(W_{t_{i+1}}-W_{t_i})$. This gives for the implicit scheme:

$$\begin{aligned}\tilde{x}^\alpha_{t_{i+1}} &\approx \tilde{x}^\alpha_{t_i}(1-k(t_{i+1}-t_i))+\sigma\sqrt{\tilde{x}^\alpha_{t_i}}(W_{t_{i+1}}-W_{t_i})\\ &+\sigma^2/2(W_{t_{i+1}}-W_{t_i})^2+(k\theta-\sigma^2/2)(t_{i+1}-t_i)\end{aligned}$$

and for the implicit scheme on the square-root:

$$\begin{aligned}\tilde{x}^\alpha_{t_{i+1}} &\approx \tilde{x}^\alpha_{t_i}(1-k(t_{i+1}-t_i))+\sigma\sqrt{\tilde{x}^\alpha_{t_i}}(W_{t_{i+1}}-W_{t_i})\\ &+\sigma^2/4(W_{t_{i+1}}-W_{t_i})^2+(k\theta-\sigma^2/4)(t_{i+1}-t_i).\end{aligned}$$

We remark that this last expansion is the one of the Milstein scheme. These expansions suggest us a family of explicit schemes $E(\lambda)$ for $0\le\lambda\le k\theta-\sigma^2/4$ ensuring nonnegative values but not the property of monotonicity (Alfonsi (2005)) **(Discretization schemes** $E(\lambda)$, $0\le\lambda\le k\theta-\sigma^2/4$**)**:

$$\begin{aligned}\tilde{x}^\alpha_{t_{i+1}} &= \left(\left(1-\frac{k}{2}(t_{i+1}-t_i)\right)\sqrt{\tilde{x}^\alpha_{t_i}}+\frac{\sigma(W_{t_{i+1}}-W_{t_i})}{2(1-\frac{k}{2}(t_{i+1}-t_i))}\right)^2 \\ &+(k\theta-\sigma^2/4)(t_{i+1}-t_i)+\lambda[(W_{t_{i+1}}-W_{t_i})^2-(t_{i+1}-t_i)].\end{aligned} \tag{22.20}$$

The scheme with $\lambda=0$ corresponds to the expansion of the implicit scheme on the square-root (and thus of Milstein) while the one with $\lambda=\sigma^2/4$ has the same expansion as the implicit scheme. Let us finally mention that we can extend the schemes presented in this subsection to parameters that do not fulfill the standard CIR conditions as follows:

- For the implicit schemes which are defined with second-degree polynomials, we will set $\tilde{x}^{\alpha}_{t_{i+1}} = 0$ when the discriminant is negative and else use formulas (22.18) or (22.19).
- For the explicit schemes $E(\lambda)$, we simply define $\tilde{x}^{\alpha}_{t_{i+1}}$ as the positive part of the right-hand side of (22.20).

We will use these extensions, when needed, for the simulations in the next section.

22.7.4 Study of the convergence of the discretization schemes for simulating CIR processes (Alfonsi (2005))

From now on, we will only consider the regular grid $\{t_i^n = iT/n,\ 0 \le i \le n\}$ and we will denote by $\tilde{x}^{\alpha,n}$ a discretization scheme on this time-grid. We have at our disposal several schemes, and we would like to know which one to choose in a general situation. We will follow the paper of Alfonsi (2005) and compare the convergence of the different schemes presented in the previous section, both theoretically and numerically. There are two main kinds of convergence we may study. We will first focus, in the next subsection, on the strong convergence, that is on the pathwise convergence, and then we will examine the weak convergence (i.e. the convergence in law at a fixed time). This will allow us to outline one preferred scheme, $E(0)$, that gathers most of the properties of convergence one may wish when using a a Monte-Carlo algorithm. For our study, we focus on the one-dimensional process (x_t^{α}).

Strong convergence. A payoff on x^{α} can always be seen as a function of the whole path, $F(x_t^{\alpha}, 0 \le t \le T)$. In general, we can approximate this function by a function $\tilde{F}^n(x^{\alpha}_{t_i^n}, 0 \le i \le n)$. As a standard example, if

$$F(x_t^{\alpha}, 0 \le t \le T) = \int_0^T x_t^{\alpha} dt,$$

we would naturally take

$$\tilde{F}^n(x^{\alpha}_{t_i^n}, 0 \le i \le n) = \frac{T}{n} \sum_{i=0}^{n-1} x^{\alpha}_{t_i^n}.$$

Thus, using a Monte-Carlo algorithm, we will calculate $\mathbb{E}\left[\tilde{F}^n(\tilde{x}^{\alpha,n}_{t_i^n}, 0 \le i \le n)\right]$ in order to approximate $\mathbb{E}\left[F(x_t^{\alpha}, 0 \le t \le T)\right]$. It follows that the approximation error we face here,

$$Err^n := \mathbb{E}\left[\tilde{F}^n(\tilde{x}^{\alpha,n}_{t_i^n}, 0 \le i \le n)\right] - \mathbb{E}\left[F(x_t^{\alpha}, 0 \le t \le T)\right] \qquad (22.21)$$

can be written as the sum of the error due to the discretization of F,

$$\mathbb{E}\left[\tilde{F}^n(x^{\alpha}_{t_i^n}, 0 \le i \le n)\right] - \mathbb{E}\left[F(x_t^{\alpha}, 0 \le t \le T)\right]$$

which is typically of order C/n for our example, and the error due to the discretization scheme for the "underlying" process x:

$$\mathbb{E}\left[\tilde{F}^n(\tilde{x}^{\alpha,n}_{t^n_i}, 0 \le i \le n)\right] - \mathbb{E}\left[\tilde{F}^n(x^{\alpha}_{t^n_i}, 0 \le i \le n)\right].$$

When $\tilde{F}^n$ is Lipschiz with respect to $(x^{\alpha}_{t^n_i}, 0 \le i \le n)$, which is the case in our example, we can bound the absolute value of this error by

$$K\mathbb{E}\left[\sup_{0\le i\le n} |\tilde{x}^{\alpha,n}_{t^n_i} - x^{\alpha}_{t^n_i}|\right].$$

Thus, it is interesting to study the convergence of

$$\mathbb{E}\left[\sup_{0\le i\le n} |\tilde{x}^{\alpha,n}_{t^n_i} - x^{\alpha}_{t^n_i}|\right]$$

towards zero. If this convergence holds, we say that there is strong convergence of the discretization scheme towards the continuous process $(x^{\alpha}_t,\ 0 \le t \le T)$. Deelstra and Delbaen (1998) have established the strong convergence of their scheme. Their proof mainly relies on two points: the construction of an extension of the scheme to all time instants $t \in [0, T]$ that is adapted, and the use of Yamada's functions that are smooth approximations of the absolute value function. Using the same technique, Alfonsi (2005) has shown the convergence of the implicit scheme for $2k\theta > \sigma^2$ and of the explicit schemes for $0 \le \lambda \le k\theta - \sigma^2/4$. Even if the implicit scheme on the square-root does not enter in the framework of this proof, numerical investigations suggest this scheme to converge strongly. Diop (2003) uses a nice argument to obtain the speed of strong convergence for the reflected scheme. Under some technical assumptions on parameters, she proves that

$$\mathbb{E}\left[\sup_{0\le i\le n} |\tilde{x}^{\alpha,n}_{t^n_i} - x^{\alpha}_{t^n_i}|\right] \le C/\sqrt{n}.$$

The theoretical study of the speed of the strong convergence is in general tough, mainly because the diffusion coefficient is not Lipschitz near the origin. We can try to assess the speed of strong convergence numerically, but the problem is that it is not obvious how to simulate the discretization scheme and the original CIR process on the same probability space. Thus, we cannot calculate directly

$$\mathbb{E}\left[\sup_{0\le i\le n} |\tilde{x}^{\alpha,n}_{t^n_i} - x^{\alpha}_{t^n_i}|\right]$$

by a Monte-Carlo method. That is why we need the following lemma stating that, to get an idea of the speed of the strong convergence, it is sufficient to compute the expectation of the sup-distance between the scheme and the same scheme with half the time step (Alfonsi (2005)).

Lemma 22.7.1. *Let us consider a scheme* $(\tilde{x}^{\alpha,n}_{t^n_i})$ *that converges towards a continuous process* x^α_t *in the following strong sense:*

$$\mathbb{E}\left[\sup_{0\le i\le n} |\tilde{x}^{\alpha,n}_{t^n_i} - x^\alpha_{t^n_i}|\right] \underset{n\to\infty}{\longrightarrow} 0. \tag{22.22}$$

Then, for any $a > 0$ *and* $b \ge 0$,

$$\mathbb{E}\left[\sup_{0\le i\le n} |\tilde{x}^{\alpha,n}_{t^n_i} - x^\alpha_{t^n_i}|\right] = O\left(\frac{(\ln n)^b}{n^a}\right) \text{ if and only if}$$

$$\text{if and only if } S_n := \mathbb{E}\left[\sup_{0\le i\le n} |\tilde{x}^{\alpha,n}_{t^n_i} - \tilde{x}^{\alpha,2n}_{t^{2n}_{2i}}|\right] = O\left(\frac{(\ln n)^b}{n^a}\right).$$

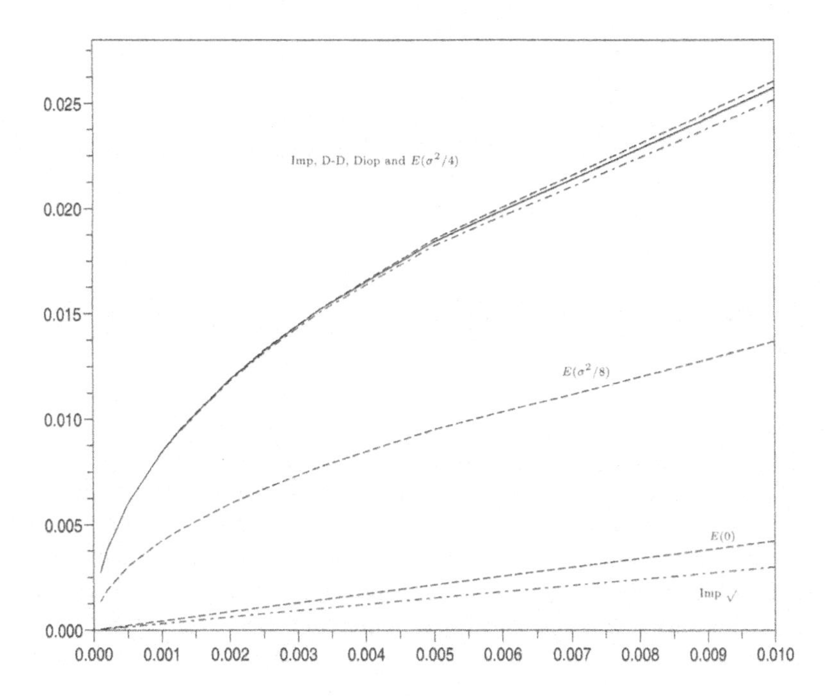

Fig. 22.19. S_n as a function of the time-step $1/n$ for $x_0 = 1$, $k = 1$, $\theta = 1$ with $\sigma = 1$.

For $T = 1$ we have plotted in Figure 22.19 S_n as a function of the time step $1/n$ for parameters with $\sigma^2 < 2k\theta$. In Figure 22.20 we do the same in

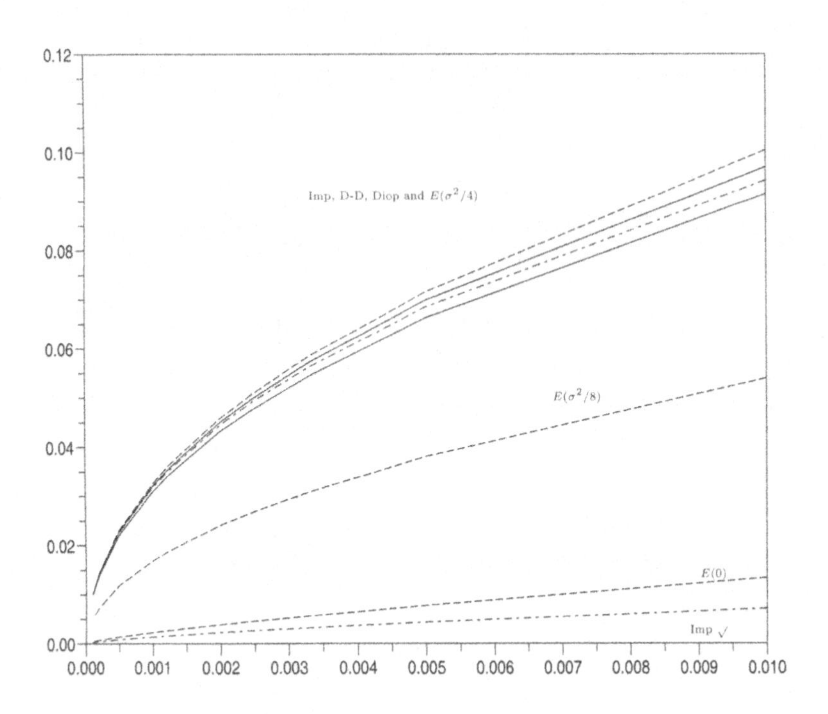

Fig. 22.20. S_n in function of the time-step $1/n$ for $x_0 = 1$, $k = 1$, $\theta = 1$ with $\sigma = \sqrt{3}$.

a case where $\sigma^2 > 2k\theta$. It seems that the schemes $E(0)$ and "implicit on the square-root" converge strongly in a much better way than the others. This is not fully surprising if we remember that these schemes have the same expansion as the Milstein scheme (indeed, we know that for SDEs driven by Lipschitz coefficients, the strong convergence of the Milstein scheme is of order C/n while it is of order $C/\sqrt{n}$ for the Euler scheme). To get an idea of the exact order of convergence, let us postulate that $S_n \sim C/n^a$ with $a > 0$. Thanks to the above lemma, this implies a speed of strong convergence of order $1/n^a$. To estimate a, we remark that

$$\log_{10}(S_n) - \log_{10}(S_{10n}) \underset{n \to +\infty}{\longrightarrow} a,$$

and we have reported $\log_{10}(S_n) - \log_{10}(S_{10n})$ for $n = 200$ in Figure 22.21. We have plotted the result as a function of the parameter $\sigma^2/(2k\theta)$ since this is the quantity playing a key role in the CIR diffusion. This can be understood easily as a time-scaling. For the "implicit scheme on the square root" and the

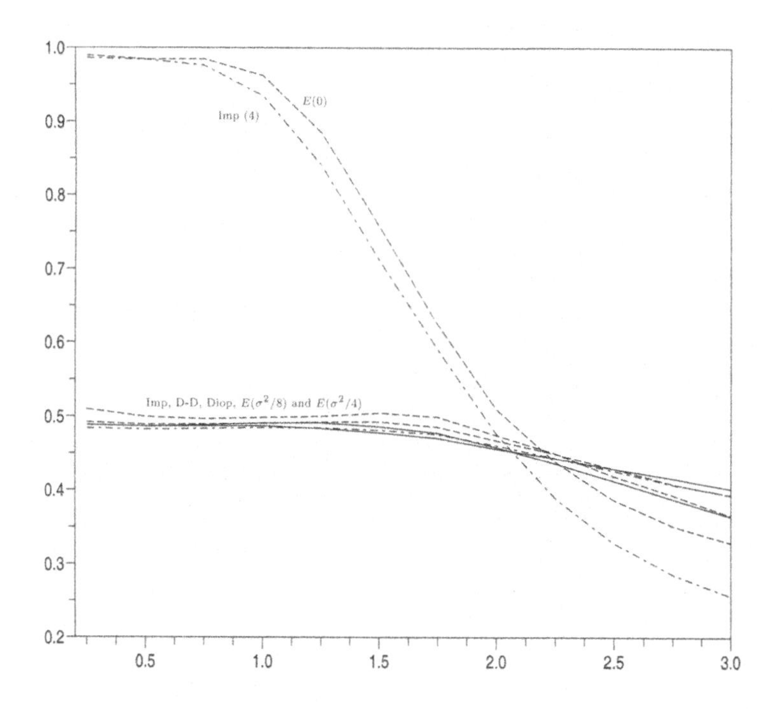

Fig. 22.21. Convergence speed of S_n: estimation of the a parameter as a function of $\sigma^2/(2k\theta)$ for $x_0 = 1$, $k = 1$ and $\theta = 1$.

scheme $E(0)$, the estimated a is close to 1 for $\sigma^2 < 2k\theta$ and decreases from 1 to 1/2 for $2k\theta < \sigma^2 < 4k\theta$. For the other schemes, the estimated value of a is close to 1/2 for $\sigma^2 < 4k\theta$. The decreasing pattern we observe is due to the fact that the process spends more time in the neighborhood of 0 where the square-root is not Lipschitz. Obviously, the speed of convergence may be more complex than the one we postulated, but this investigation can give us a good idea of the actual behaviour.

Weak convergence and weak error expansion. The speed of the strong convergence, as we have seen, allows us to estimate the approximation error by an upper bound for a quite general class of payoffs. However, it may happen that, for particular payoffs, the actual error is much smaller than the bound given by the strong convergence estimate. In this subsection, we are interested in the convergence of $f(\tilde{x}_T^{\alpha,n})$ towards $f(x_T^\alpha)$ for some functions $f : \mathbb{R}_+ \to \mathbb{R}$. Indeed, this is the approximation error (22.21) when $F(x_t^\alpha, 0 \le t \le T) = f(x_T^\alpha)$. In that case the discretized function $\tilde{F}^n(x_{t_i^n}^\alpha, 0 \le i \le n) = f(x_T^\alpha)$ is equal to F, so that the error is only due to the discretization scheme, and is equal to

$$Err^n = \mathbb{E}[f(\tilde{x}_T^{\alpha,n})] - \mathbb{E}[f(x_T^\alpha)].$$

Diop (2003) shows that for her scheme, if f is a four times continuously differentiable and bounded function with derivatives that are all bounded, then there is a constant that does not depend on n such that

$$|Err^n| \leq C/n^{\min(1,k\theta/\sigma^2)}.$$

Alfonsi (2005) shows the following proposition.

Proposition 22.7.1. *Let $f : \mathbb{R}_+ \to \mathbb{R}$ be a four times continuously differentiable function such that $\exists A, m > 0$ such that $\forall x \geq 0, |f^{(4)}(x)| \leq A(1 + x^m)$, where $f^{(4)}$ denotes the fourth derivative of f. Let $(\tilde{x}^{\alpha,n})$ be either the implicit scheme with $\sigma^2 < 2k\theta$ or the explicit scheme $E(\lambda)$ with $0 \leq \lambda \leq k\theta - \sigma^2/4$. Then, the weak error is of order $1/n$:*

$$|Err^n| \leq C/n$$

for a constant C that depends on α, T and f, but not on n.

The proof uses the argument that was introduced by Talay and Tubaro (1990) to prove the analogous result for the Euler scheme when the coefficients of the SDE are Lipschitz. To illustrate this convergence, we have plotted, with $T = 1$, $\mathbb{E}(f(\tilde{x}_1^{\alpha,2n}))$ as a function of $1/n$ for different parameters ($2k\theta > \sigma^2$ for Fig. 22.22 and $2k\theta < \sigma^2$ for Fig. 22.23) for a chosen function f. We see that for small values of σ (Fig. 22.22) all the schemes seem to have an error Err^n of order $O(1/n)$, whereas for large values of σ (Fig. 22.23) only the Explicit schemes and the Deelstra-Delbaen scheme show patterns that are compatible with an error of order $O(1/n)$.

It is also interesting to have a further expansion of the error. Indeed, if we know for example that $Err^n = c_1/n + O(1/n^2)$, using the Romberg method (see for example Klöden and Platen (1995)), we see that

$$\begin{aligned} 2\mathbb{E}[f(\tilde{x}_T^{\alpha,2n})] - \mathbb{E}[f(\tilde{x}_T^{\alpha,n})] &= \mathbb{E}[f(x_T^\alpha)] + 2Err^{2n} - Err^n \\ &= \mathbb{E}[f(x_T^\alpha)] + O(1/n^2). \end{aligned}$$

Thus,

$$2\mathbb{E}[f(\tilde{x}_T^{\alpha,2n})] - \mathbb{E}[f(\tilde{x}_T^{\alpha,n})]$$

converges with a quadratic speed towards the desired expectation. The following proposition (Alfonsi (2005)) states that we have the required expansion of the weak error, if f is regular enough, for the Explicit schemes $E(\lambda)$.

Proposition 22.7.2. *Let ν be a positive integer and let $f : \mathbb{R}_+ \to \mathbb{R}$ be infinitely continuously differentiable and such that for all nonnegative integers $q, \exists A_q > 0, m_q \in \mathbb{N} : \ |f^{(q)}(x)| \leq A_q(1 + x^{m_q})$. Let $(\tilde{x}^{\alpha,n})$ be the explicit scheme $E(\lambda)$ with $0 \leq \lambda \leq k\theta - \sigma^2/4$. Then, the weak error has an expansion up to order ν:*

$$Err^n = c_1/n + c_2/n^2 + .. + c_{\nu-1}/n^{\nu-1} + O(1/n^\nu).$$

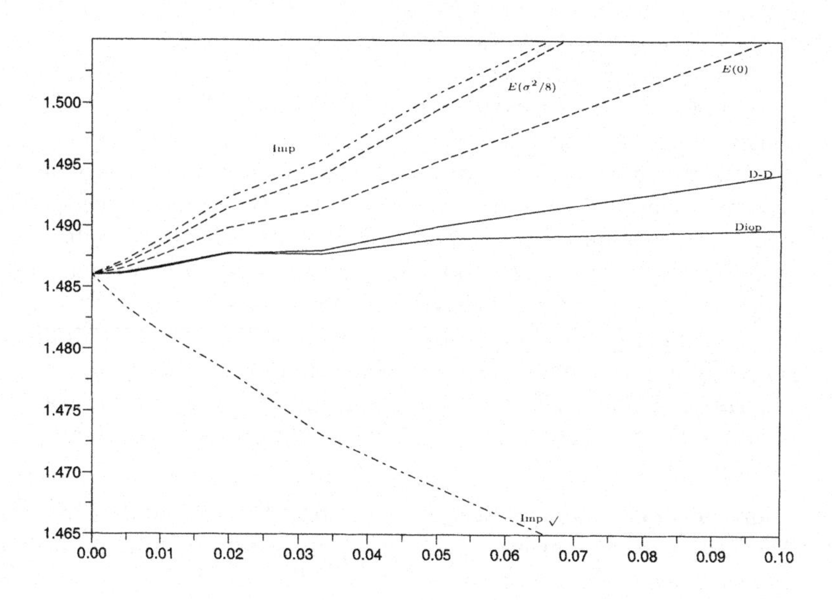

Fig. 22.22. $\mathbb{E}(f(\tilde{x}_1^{\alpha,2n}))$ as a function of $1/n$ with $f(x) = \frac{5+3x^4}{2+5x}$ for $x_0 = 0$, $k = 1$, $\theta = 1$ and $\sigma = 1$.

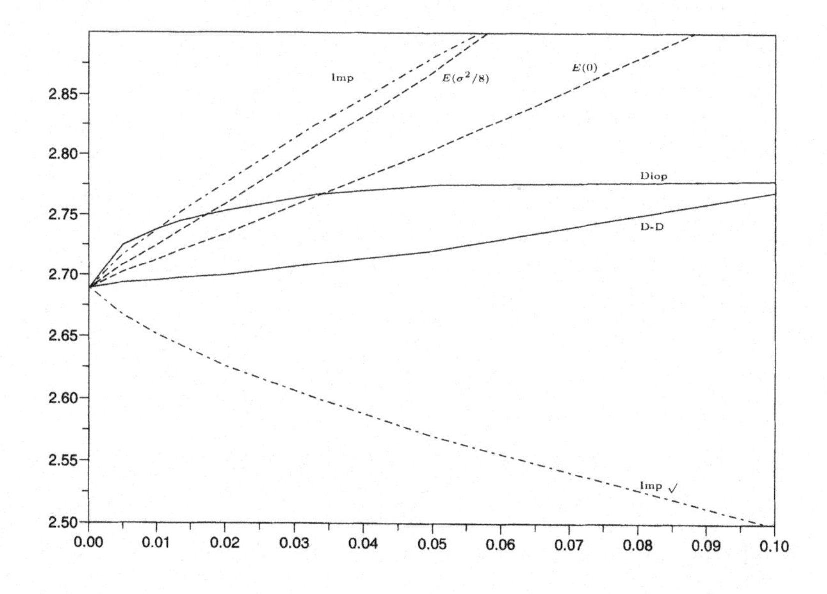

Fig. 22.23. $\mathbb{E}(f(\tilde{x}_1^{\alpha,2n}))$ as a function of $1/n$ with $f(x) = \frac{5+3x^4}{2+5x}$ for $x_0 = 0$, $k = 1$, $\theta = 1$ and $\sigma = \sqrt{3}$.

We have calculated the same expectations (Fig. 22.24 and 22.25), using now the Romberg approximation $2\mathbb{E}[f(\tilde{x}_1^{\alpha,2n})] - \mathbb{E}[f(\tilde{x}_1^{\alpha,n})]$ to estimate $\mathbb{E}[f(x_1^\alpha)]$. Let us mention here that the time required is of the same order as the time required by $\mathbb{E}[f(\tilde{x}_1^{\alpha,2n})]$. Indeed, this time is mainly proportional to n, and the extra cost due to the calculation of $\mathbb{E}[f(\tilde{x}_1^{\alpha,n})]$ multiplies the computation time by $3/2$. The figure 22.24 shows that in both the cases $\sigma^2 \leq 2k\theta$ and $\sigma^2 > 2k\theta$, Diop's scheme and the implicit schemes do not present a quadratic convergence. As expected, the explicit schemes $E(0)$ and $E(\sigma^2/8)$ have a quadratic shape. As concerns the Deelstra-Delbaen scheme, let us first say that for large time-steps, negative values may occur frequently, which explains the strange behaviour we observe. However, for time-steps that are small enough, the convergence seems compatible with a quadratic convergence. To appreciate the efficiency of the Romberg method, we can compare Figures 22.24 and 22.25 to Figures 22.22 and 22.23, and see for example that with $n = 50$ ($1/n = 0.02$), we are already quite close to the right value with the Explicit schemes $E(\lambda)$.

Convergence towards zero-coupon bond prices and other payoffs. Another kind of pathwise function that arises naturally when dealing with interest-rates or default intensities is the following one:

$$F(x_t^\alpha, 0 \leq t \leq T) = \exp\left(-\int_0^T x_t^\alpha dt\right).$$

Once again, one may wonder whether the error (22.21) is not better than the one suggested by the speed of strong convergence. We take in this case the approximated function

$$\tilde{F}^n(x_{t_i^n}^\alpha, 0 \leq i \leq n) = \exp\left(-\frac{T}{n}\sum_{i=0}^{n-1} x_{t_i^n}^\alpha\right).$$

On some numerical examples, we see that we have good convergence for all the schemes. Indeed, with $T = 1$ and with a time step of $1/n$ corresponding to $n = 20$, with parameters $(k, \theta, \sigma, x_0) = (0.1, 0.1, 0.1, 0.1)$ and with a 95% precision of 5×10^{-6}, we get for

$$\mathbb{E}\left[\exp\left(-\frac{1}{n}\sum_{i=0}^{n-1} \tilde{x}_{t_i^n}^{\alpha,n}\right)\right]$$

Implicit	Implicit $\sqrt{}$	Diop	Deelstra-Delb.	$E(0)$	$E(\sigma^2/8)$
0.904993	0.904981	0.904982	0.904982	0.904971	0.904971

while the value calculated with the analytical formula is

$$\mathbb{E}\left[\exp\left(-\int_0^1 x_t^\alpha dt\right)\right] = 0.904977.$$

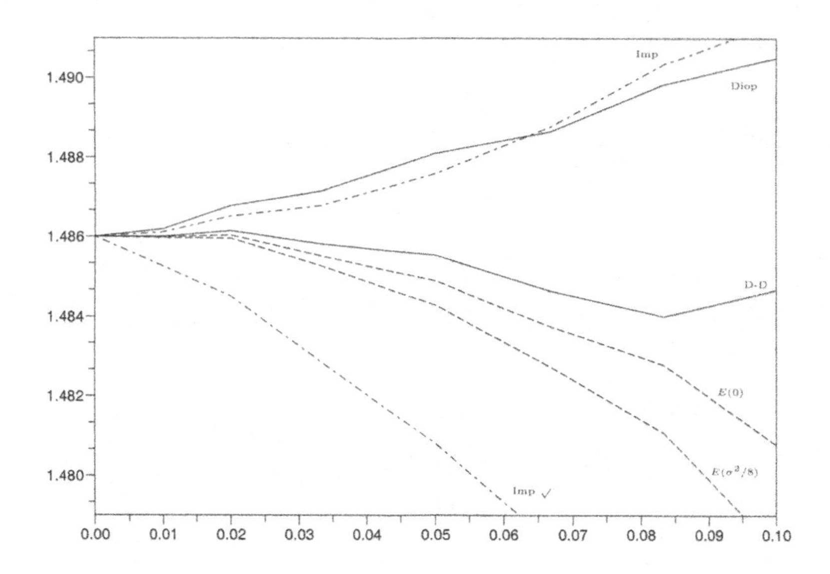

Fig. 22.24. $2\mathbb{E}[f(\tilde{x}_1^{\alpha,2n})]-\mathbb{E}[f(\tilde{x}_1^{\alpha,n})]$ as a function of $1/n$ with $f(x) = (5+3x^4)/(2+5x)$ for $x_0 = 0$, $k = 1$, $\theta = 1$ and $\sigma = 1$.

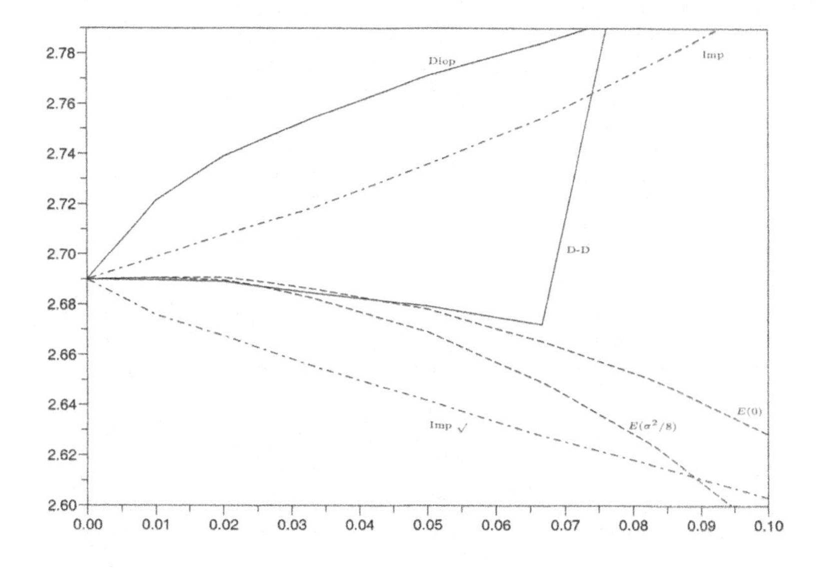

Fig. 22.25. $2\mathbb{E}[f(\tilde{x}_1^{\alpha,2n})]-\mathbb{E}[f(\tilde{x}_1^{\alpha,n})]$ as a function of $1/n$ with $f(x) = (5+3x^4)/(2+5x)$ for $x_0 = 0$, $k = 1$, $\theta = 1$ and $\sigma = \sqrt{3}$.

To consider the difference between the schemes with a time horizon fixed to $T = 1$, we have to increase the parameters, even if they are not realistic from a financial point of view.

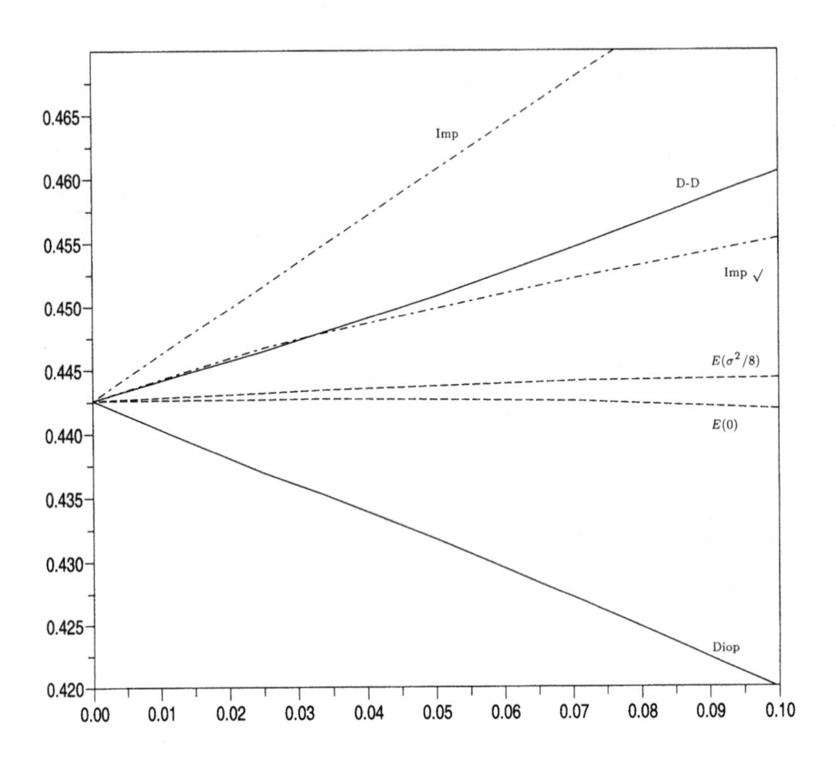

Fig. 22.26. $\mathbb{E}\left[\exp\left(-\frac{1}{n}\sum_{i=0}^{n-1}\tilde{x}_{t_i^n}^{\alpha,n}\right)\right]$ as a function of $1/n$ for $x_0 = 1$, $k = 1$, $\theta = 1$ and $\sigma = \sqrt{3}$.

Thus, we have plotted in Figure 22.26 the convergence of

$$\mathbb{E}\left[\exp\left(-\frac{1}{n}\sum_{i=0}^{n-1}\tilde{x}_{t_i^n}^{\alpha,n}\right)\right]$$

towards

$$\mathbb{E}\left[\exp\left(-\int_0^1 x_t^\alpha dt\right)\right].$$

For the chosen parameters, we see that the Explicit schemes $E(0)$ and $E(\sigma^2/8)$ are quite close to the right value with $n = 10$. Most of the schemes present a linear convergence telling us that in practice the Romberg method can be used successfully.

Obviously, when we use a Monte-Carlo algorithm, this is in general to calculate payoff values

$$\mathbb{E}\left[F(x_t^\alpha, 0 \leq t \leq T)\right]$$

whose expression is far more complicated than the one of the zero-coupon bond here or a European plain vanilla option (with no path dependence, studied with the weak convergence). However, considering the convergence obtained in these simple cases, the general situation is not hopeless, especially when using the explicit schemes $E(\lambda)$, and the error (22.21) can have the expansion $Err^n = c_1/n + O(1/n^a)$ with $a > 1$. If it appears so (numerically), the Romberg approximation

$$\mathbb{E}\left[\tilde{F}^{2n}(\tilde{x}_{t_i^{2n}}^{\alpha,2n}, 0 \leq i \leq 2n)\right] - \mathbb{E}\left[\tilde{F}^{n}(\tilde{x}_{t_i^{n}}^{\alpha,n}, 0 \leq i \leq n)\right]$$
$$= \mathbb{E}\left[F(x_t^\alpha, 0 \leq t \leq T)\right] + O(1/n^a)$$

converges faster than

$$\mathbb{E}\left[\tilde{F}^{n}(\tilde{x}_{t_i^{n}}^{\alpha,n}, 0 \leq i \leq n)\right]$$

towards the desired expectation.

A comparative study of the schemes. We have summarized in Table 22.15 the theoretical results obtained for each scheme, listing on one hand their algebraic properties of positivity and monotonicity, and on the other hand the convergence results we already obtained. The star in Y* means that the result has been established under some assumption on the parameters while the question mark indicates that no result has been shown yet. Table 22.16 presents the results of the numerical tests presented here. All these results tend to show that the explicit scheme $E(0)$ is the one that gathers the most interesting properties of convergence for the purpose of a Monte-Carlo algorithm. This scheme is moreover really easy to implement. That is why in the general case, we will prefer this scheme, at least for $\sigma^2 \leq 4k\theta$.

	I	$\mathrm{I}_{\surd}$	Diop	Deel-Delb.	$E(0)$	$E(\lambda)$
Nonnegativity	Y	Y	Y	N	Y	Y
Monotonicity	Y	Y	N	N	N	N
Strong CV	Y	?	Y*	Y	Y	Y
Weak CV rate in $1/n$	Y	?	Y*	?	Y	Y
Weak error expansion	?	?	?	?	Y	Y

Table 22.15. Theoretical results. 'I": Implicit ($\sigma^2 \leq 2k\theta$), "$\mathrm{I}_{\surd}$": Implicit square root$_{\surd}$ ($\sigma^2 \leq 4k\theta$); $E(0)$ assumes $\sigma^2 \leq 4k\theta$; $E(\lambda)$ assumes $0 < \lambda \leq k\theta - \sigma^2/4$

	I	I $\sqrt{}$	Diop	Deel-Delb	$E(0)$	$E(\lambda)$
Strong CV order	$\approx 1/2$	≈ 1	$\approx 1/2$	$\approx 1/2$	≈ 1	$\approx 1/2$
Weak CV rate $(1/n)$	Y	Y	Y	Y	Y	Y
Romberg $(1/n^2)$	N	N	N	Y	Y	Y
Strong CV order	$\approx 1/2$	$>\approx 1/2$	$\approx 1/2$	$\approx 1/2$	$>\approx 1/2$	$\approx 1/2$
Weak CV rate $(1/n)$	?	?	N	Y	Y	Y
Romberg $(1/n^2)$	N	N	N	Y?	Y	Y

Table 22.16. Numerical results, the first sub-table (first three rows) refers to the case $\sigma^2 \in [0, 2k\theta]$, whereas the second sub-table (last three rows) refers to the case $\sigma^2 \in [2k\theta, 4k\theta]$.

22.7.5 Gaussian dependence mapping: A tractable approximated SSRD

To obtain an acceptable precision with a Monte-Carlo algorithm, it is unfortunately necessary to simulate a quite large number of scenarios. We have given a detailed discussion on why this happens in the case with deterministic intensity, in Section 22.6. We have seen that the variance of the CDS simulated net present value is quite large in relative terms, due essentially to the indicator term in $1_{\{\tau<T\}}\text{L}_{\text{GD}}D(0,\tau)$. A quick example considering the variance

$$\mathrm{Var}(1_{\{\tau<T\}}) = \mathbb{E}1^2_{\{\tau<T\}} - (\mathbb{E}1_{\{\tau<T\}})^2 = \mathbb{E}1_{\{\tau<T\}} - (\mathbb{E}1_{\{\tau<T\}})^2$$

helped us clarify this point with some concrete numbers.

Here the situation is analogous, the stochasticity of intensity does not change things substantially. We can use the same tricks we devised in Section 22.6. Without going through the control variate technique again, that remains essentially unaltered, we re-explain the threshold technique for the case with stochastic intensity.

As before, we may improve the situation by using a threshold barrier $\bar{B}$ such that $\mathbb{Q}(\Lambda(T) < \bar{B}) \simeq 1$. We thus assume that default may occur only when $\Lambda(\tau) < \bar{B}$. The idea is then to simulate default times conditional on $\xi := \Lambda(\tau) < \bar{B}$. Indeed, we see that if Π_{RCDS} is any CDS discounted payoff, recalling that $\Lambda(\tau)$ is exponential with parameter 1 independent of $\mathcal{F}$, we have that

$$\mathbb{E}\ \Pi_{\text{RCDS}} = \mathbb{E}[\Pi_{\text{RCDS}}|\Lambda(\tau) < \bar{B}](1 - e^{-\bar{B}}) + \mathbb{E}[\Pi_{\text{RCDS}}|\Lambda(\tau) \geq \bar{B}]e^{-\bar{B}}.$$

The CDS value is known in case $\xi > \bar{B}$, since in this case default has not occurred before final maturity and the CDS price is $R_{0,b}(0)\sum_{i=1}^{b} P(0,T_i)\alpha_i$. Our simulations then need concern only the first term, so if ξ is an exponential random variable with parameter one we just simulate $\xi|\xi < \bar{B}$, whose density is easily seen to be

$$p_{\xi|\xi<\bar{B}}(u) = 1_{\{u<\bar{B}\}}e^{-u}/(1 - e^{-\bar{B}}).$$

From the exponential distribution we see that simulating m scenarios for ξ amounts to simulate $m(1-e^{-\bar{B}})$ scenarios with $\xi < \bar{B}$ and $me^{-\bar{B}}$ with $\xi \geq \bar{B}$. So in turn simulating $M = m(1-e^{-\bar{B}})$ scenarios for $\xi < \bar{B}$, as we will do, amounts to simulate in total $m = M/(1-e^{-\bar{B}})$ scenarios, the extra scenarios corresponding to the known value $R_{0,b}\sum_{i=1}^{b} P(0,T_i)\alpha_i$ of the CDS in case of default. Dividing by $1-e^{-\bar{B}}$ may help us increase efficiency (in our examples typically it increases the number of scenarios by a factor 10), but a large amount of scenarios remains to be generated, and the time needed for Monte Carlo simulation remains large.

With the SSRD model, using the independence of $\xi = \Lambda(\tau)$ from $\mathcal{F}$ (and thus λ and r), the value of the CDS at time 0 can be written, by simple passages, as expression (22.13), that we repeat here in the particular case $t = 0$ and $T_a = 0$:

$$\begin{aligned}
\mathrm{CDS}_{0,b}(0,R,\mathrm{L_{GD}}) &= \Big\{ R\sum_{i=1}^{b}\alpha_i \mathbb{E}\left[\exp\left(-\int_0^{T_i}(r_s+\lambda_s)ds\right)\right] \\
&\quad + R\int_0^{T_b}\mathbb{E}\left[\exp\left(-\int_0^{u}(r_s+\lambda_s)ds\right)\lambda_u\right](u-T_{\beta(u)-1})du \\
&\quad - \mathrm{L_{GD}}\int_0^{T_b}\mathbb{E}\left[\exp\left(-\int_0^{u}(r_s+\lambda_s)ds\right)\lambda_u\right]du \Big\}
\end{aligned}$$

We did the above computations earlier when $\rho = 0$, and we know that in that case we obtain a formula equivalent to the model independent valuation formula (22.9). But what happens when $\rho \neq 0$? This is an important question, since the terms in λ and r appearing in the above formula are quite common in credit derivatives evaluation and it would be a good idea to have an approximated formula to compute them when $\rho \neq 0$.

Our idea is to "map" the two-dimensional CIR dynamics in an analogous tractable two-dimensional Gaussian dynamics that preserves as much as possible of the original CIR structure, and then do calculations with the Gaussian model. Recall that the CIR process and the Vasicek process for interest rates give both affine models. The first one is more convenient because it ensures positive values while the second one is more analytically tractable. Indeed, in the SSRD we have no formula for $\mathbb{E}[\exp(-\int_0^T(x_s^\alpha + y_s^\beta)ds)]$ and $\mathbb{E}[\exp(-\int_0^T(x_s^\alpha + y_s^\beta)ds)y_T^\beta]$ when $\rho \neq 0$, while in the Vasicek case, we can easily derive such formulas from the following

Lemma 22.7.2. *Let $A = m_A + \sigma_A N_A$ and $B = m_B + \sigma_B N_B$ be two random variables such that N_A and N_B are two correlated standard Gaussian random variables with $[N_A, N_B]$ jointly Gaussian vector with correlation $\bar{\rho}$. Then,*

$$\mathbb{E}(e^{-A}B) = m_B e^{-m_A+\frac{1}{2}\sigma_A^2} - \bar{\rho}\sigma_A\sigma_B e^{-m_A+\frac{1-\bar{\rho}^2}{2}\sigma_A^2} \tag{22.23}$$

Lemma 22.7.3. *Let $x_t^{\alpha,V}$ and $y_t^{\beta,V}$ be two Vasicek processes as follows:*

$$\begin{aligned} dy_t^{\beta,V} &= \kappa(\mu - y_t^{\beta,V})dt + \nu dZ_t, \\ dx_t^{\alpha,V} &= k(\theta - x_t^{\alpha,V})dt + \sigma dW_t \end{aligned} \tag{22.24}$$

with $dW_t\ dZ_t = \rho\ dt$. Then $A = \int_0^T (x_t^{\alpha,V} + y_t^{\beta,V})dt$ and $B = y_T^{\beta,V}$ are Gaussian random variables with respective means:

$$\begin{aligned} m_A &= (\mu + \theta)T - [(\theta - x_0)g(k,T) + (\mu - y_0)g(\kappa,T)] \\ m_B &= \mu - (\mu - y_0)e^{-\kappa T} \end{aligned}$$

respective variances:

$$\begin{aligned} \sigma_A^2 &= \left(\frac{\nu}{\kappa}\right)^2 (T - 2g(\kappa,T) + g(2\kappa,T)) + \left(\frac{\sigma}{k}\right)^2 (T - 2g(k,T) + g(2k,T)) \\ &\quad + \frac{2\rho\nu\sigma}{k\kappa}(T - g(\kappa,T) - g(k,T) + g(\kappa + k,T)) \\ \sigma_B^2 &= \nu^2 g(2\kappa,T) \end{aligned}$$

and correlation:

$$\bar{\rho} = \frac{1}{\sigma_A \sigma_B}\left[\frac{\nu^2}{\kappa}(g(\kappa,T) - g(2\kappa,T)) + \frac{\rho\sigma\nu}{k}(g(\kappa,T) - g(\kappa + k,T))\right]$$

where $g(k,T) = (1 - e^{-kT})/k$.

Thus, we are able to calculate $\mathbb{E}[\exp(-\int_0^T (x_t^{\alpha,V} + y_t^{\beta,V})dt)y_T^{\beta,V}]$ and $\mathbb{E}[\exp(-\int_0^T (x_t^{\alpha,V} + y_t^{\beta,V})dt)]$ (taking $m_B = 1$ and $\sigma_B = 0$); and taking for y^V a degenerated case ($\mu = \kappa = y_0 = 1, \nu = 0$), we obtain the well known formula for the bond price in the Vasicek model, which in our notation reads

$$\begin{aligned} \mathbb{E}\left[\exp\left(-\int_0^T x_s^{\alpha,V} ds\right)\right] &= A^V(0,T;\alpha)\exp(-B^V(0,T;\alpha)x_0) \qquad (22.25) \\ &= \exp\left(-\theta t + (\theta - x_0)g(k,t) + \frac{1}{2}\left(\frac{\sigma}{k}\right)^2 (t - 2g(k,t) + g(2k,t))\right). \end{aligned}$$

The idea is then to approximate the expectation by these formulas. More precisely, on $[0,T]$ we consider a particular Vasicek volatility in the dynamics (22.24), corresponding to taking $\alpha_T := (x_0, k, \theta, \sigma^{V,T})$ (resp. $\beta_T = (y_0, \kappa, \mu, \nu^{V,T})$) such that

$$\mathbb{E}\left[\exp\left(-\int_0^T x_s^{\alpha_T,V} ds\right)\right] = \mathbb{E}\left[\exp\left(-\int_0^T x_s^{\alpha} ds\right)\right]$$

$$\left(\text{resp. } \mathbb{E}\left[\exp\left(-\int_0^T y_s^{\beta_T,V} ds\right)\right] = \mathbb{E}\left[\exp\left(-\int_0^T y_s^{\beta} ds\right)\right]\right)$$

where on the right hand sides we have the CIR processes. In the above equations expectations on both sides are analytically known, being bond price formulas for the Vasicek and CIR models respectively, and the inversions needed to retrieve $\sigma^{V,T}$ and $\nu^{V,T}$ are quite easy since the expression (22.25) is monotone with respect to σ. In practical cases, these volatilities exist, and can be seen as some sort of means of time-averages of $\sigma\sqrt{x_s^\alpha}$ (resp. $\nu\sqrt{y_s^\beta}$) on $[0,T]$. We then adopt the following approximations to estimate the impact of correlation:

$$\mathbb{E}\left[\exp\left(-\int_0^T (x_s^\alpha + y_s^\beta)ds\right)\right] \approx \mathbb{E}\left[\exp\left(-\int_0^T (x_s^{\alpha_T,V} + y_s^{\beta_T,V})ds\right)\right], \quad (22.26)$$

$$\mathbb{E}\left[\exp\left(-\int_0^T (x_s^\alpha + y_s^\beta)ds\right) y_T^\beta\right] \approx \mathbb{E}\left[\exp\left(-\int_0^T (x_s^{\alpha_T,V} + y_s^{\beta_T,V})ds\right) y_T^{\beta_T,V}\right] + \Delta \quad (22.27)$$

where

$$\Delta = \mathbb{E}\left[\exp\left(-\int_0^T x_s^\alpha ds\right)\right]\mathbb{E}\left[\exp\left(-\int_0^T y_s^\beta ds\right) y_T^\beta\right]$$
$$-\mathbb{E}\left[\exp\left(-\int_0^T x_s^{\alpha,T,V} ds\right)\right]\mathbb{E}\left[\exp\left(-\int_0^T y_s^{\beta,T,V} ds\right) y_T^{\beta,T,V}\right]$$

and where we use the known analytical expressions for the right-hand sides.

22.7.6 Numerical Tests: Gaussian Mapping and Correlation Impact

We perform numerical tests for formulas (22.26) and (22.27) and for the related CDS prices, based on Monte Carlo simulations of the left-hand sides. We take the α and β parameters as from the final part of Section 22.7.2, and assume $T = 5y$. We obtain the results of Tables 22.17 and 22.18. The Vasicek mapped volatilities are $\sigma^{V,5y} = 0.016580$ and $\nu^{V,5y} = 0.0025675$.

To check the quality of the approximation under stress, we multiply all parameters k, θ, σ and κ, μ, ν by three and check again the approximation. We obtain the results shown in Tables 22.19 and 22.20, and now the Vasicek mapped volatilities are $\sigma^{V,5y} = 0.108596$ and $\nu^{V,5y} = 0.0060675$.

If the values in Table 22.17 were interpreted as bond prices, the corresponding continuously compounded spot rates would be $-\ln(0.86191)/5 = 0.02972$ and $-\ln(0.861762)/5 = 0.029755$, respectively, giving a small difference.

If the values in Table 22.19 were interpreted as bond prices, the corresponding continuously compounded spot rates would be instead

	ρ = -1	ρ =1
LHS of (22.26)	0.86191 (0.861815 0.862004)	0.8624 (0.862272 0.862529)
RHS of (22.26)	0.861762,	0.862554

Table 22.17. MC simulation for the quality of the approximation (22.26)

	ρ = -1	ρ =1
LHS of (22.27)	3.5848E-3 (3.57946 3.59014)	3.44852E-3 (3.44408 3.45295)
RHS of (22.27)	3.59831E-3	3.43174E-3

Table 22.18. MC simulation for the quality of the approximation (22.27)

	ρ = -1	ρ =1
LHS of (22.26)	0.64232 (0.642106 0.642534)	0.644151 (0.643909 0.644393)
RHS of (22.26)	0.641989	0.643904

Table 22.19. MC simulation for the quality of the approximation (22.26) under stress

$-\ln(0.64232)/5 = 0.088534$ and $-\ln(0.641989)/5 = 0.088637$, so that we see a larger difference than before, ranging around 1 basis point, which is however still contained.

	ρ = -1	ρ =1
LHS of (22.27)	2.4757E-3 (2.46991 2.48149)	2.27465E-3 (2.27018 2.27913)
RHS of (22.27)	2.53527	2.24435

Table 22.20. MC simulation for the quality of the approximation (22.27) under stress

So we may trust the approximation to work well within the typical market bid-ask spreads for CDS's. Indeed, we consider the valuation of CDS's both by Monte Carlo simulation and by the Gaussian dependence mapped model, where we apply formulas (22.26) and (22.27) each time with the most convenient maturity T for that part of the CDS payoff we are evaluating.

In Table 22.21 we give the results of the application of the approximations (22.26) and (22.27) to CDS valuation in presence of correlation $\rho \neq 0$ under the parameters given in Section 22.7.2.

In Table 22.22 we give instead the corresponding Monte Carlo simulation for the extreme cases $\rho = -1$ and $\rho = 1$ and the known case $\rho = 0$, based on 140.000 paths with control variate variance reduction technique, both under the usual parameters of Section 22.7.2 and under some amplified λ parameters. The Gaussian mapping approximation, even in the case of increased parameters, remains well within a small fraction of the CDS bid-ask spread (22.17).

ρ	-1	-0.5	0	0.5	1
cds	-1.12E-4	-0.554E-4	0.012E-4	0.578E-4	1.14E-4

Table 22.21. 5y CDS price as a function of ρ with Gaussian mapping

CDS prices	Gaussian Mapping	Monte Carlo value and 95% window
$\rho = -1$	-1.12E-4	-1.48625E-4 (-1.79586 -1.17664)
$\rho = 0$	0.012E-4	0.17708E-4 (-0.142444 0.496605)
$\rho = 1$	1.14E-4	1.25475E-4 (0.922997 1.5865)

Same run with κ, ν increased by a factor 5 and μ by a factor 3 :

CDS prices	Gaussian Mapping	Monte Carlo value and 95% window
$\rho = -1$	-1.03E-4	-1.77E-4 (-2.02 -1.51)
$\rho = 0$	0.021E-4	0.143E-4 (-0.138 0.424)
$\rho = 1$	1.07E-4	1.08E-4 (0.78 1.37)

Table 22.22. 5y CDS prices as a function of ρ with MC simulation

22.7.7 The impact of correlation on a few "test payoffs"

It can be interesting to study the main terms that appear in basic payoffs of the credit derivatives world from the point of view of the impact of the correlation ρ between interest rates r and stochastic default intensities λ. Precisely, we will study here the influence of the correlation ρ in the following payoffs

$$A = L(T - 1y, T)D(0, T)1_{\{\tau < T\}}, \quad B = D(0, \tau)1_{\{\tau < T\}} \qquad (22.28)$$
$$C = D(0, \tau \wedge T), \quad D = D(0, T)L(T - 1y, T)1_{\{\tau \in [T-1y, T]\}},$$

under the SSRD correlated model. We will see that in all cases even high correlations between r and λ induce a small effect on the particular functional forms of $D(0, \cdot)$ in r and of indicators of the default times τ in λ. Higher effects are observed, in relative terms, when terms such as $L(T - 1y, T)$ and $1_{\{\tau \in [T-1y, T]\}}$ are included in the payoff. Indeed, the indicator isolates λ between $T - 1y$ and T, while L isolates r between $T - 1y$ and T. Thus we have a sort of more direct correlation between r and λ in the same interval, and this explains the highest percentage influcence of correlation observed in this case. Results are summarized in Table 22.23. As expected, D is the case where the correlation influence is most visible in relative terms. We have used the same paths for W and Z when changing ρ from -1 to 1, and we have taken $T = 5y$ and the same parameters in the dynamics as in Section 22.7.2.

To check that indeed it is the "localization" of λ and r in the same interval $[T - 1y, T] = [4y, 5y]$ that generates the high relative influence of ρ, we consider also the terms

$$E = D(0, 5)L(4, 5)1_{\{\tau \in [3,4]\}}, \; F = D(0, 5)L(4, 5)1_{\{\tau \in [2,3]\}}, \qquad (22.29)$$
$$G = D(0, 5)L(4, 5)1_{\{\tau \in [1,2]\}}, \; H = D(0, 5)L(4, 5)1_{\{\tau \in [0,1]\}}$$

and check that the correlation decreases as τ gets far from the 4y LIBOR reset date. This is indeed the case, as one can see from Table 22.24.

	$\rho = -1$	$\rho = 1$	relative variation	absolute variation
A	30.3672E-4	31.1962	+2.73%	+0.829E-4
B	679.197E-4	676.208	-0.44%	-2.989E-4
C	8207.23E-4	8209.61	+0.03%	+2.38E-4
D	2.77376E-4	3.10889	+12.08%	+0.34E-4

Table 22.23. Influence of ρ on the terms A,B,C and D defined in (22.28)

	$\rho = -1$	$\rho = 1$	relative variation	absolute variation
E	5.6E-4	5.88E-4	+5.010%	+0.281E-4
F	7.16E-4	7.31 E-4	+2.09%	+0.149E-4
G	7.41E-4	7.44E-4	+0.36%	2.66E-6
H	7.55E-4	7.56E-4	+0.056%	4.26 E-7

Table 22.24. Influence of ρ on the terms E,F,G and H defined in (22.29)

22.7.8 A pricing example: A Cancellable Structure

In this section we present an example of a payoff that can be valued with the calibrated (λ, r) model. We consider a sort of cancellable swap with a recovery value. This is similar to a payoff we actually priced with the model.

A first company "A" owns a bond issued by name "C", and receives from "C" once an year at time T_i a payment consisting of $L(T_i - 1, T_i) + s$, where s is a spread ($s = 50$ basis points), up to a final date $T = T_n = 5y$. We assume unit year fractions for simplicity.

$$\text{“C” (until possible default)} \rightarrow L(T_i - 1y, T_i) + s \rightarrow \text{“A”},$$

In turn, "A" has a swap with a bank "B", where "A" turns the payment $L(T_i - 1y, T_i) + s$ to "B",

$$\text{“A”} \rightarrow L(T_i - 1, T_i) + s \rightarrow \text{“B”},$$

and, in exchange for this, the bank "A" receives from "B" some fixed payments that we express as the percentages of the unit nominal value given in (22.30) below.

$$\text{“A”} \leftarrow \begin{array}{|c|c|}\hline \text{Year} & \% \\ \hline T_1 = 1 & \alpha_1 = 4.20 \\ T_2 = 2 & \alpha_2 = 3.75 \\ T_3 = 3 & \alpha_3 = 3.25 \\ T_4 = 4 & \alpha_4 = 0.50 \\ T_5 = T_n = T = 5 & \alpha_5 = 0.50 \\ \hline \end{array} \leftarrow \text{“B”} \qquad (22.30)$$

However, if "C" defaults, "A" receives a recovery rate R_{EC} from "C" (typically one recovers from $\text{R}_{\text{EC}} = 0$ to 0.5 out of 1), and still has to pay the remaining payments $L(T_i - 1, T_i) + s$ to "B".

"A" wishes to have the possibility to cancel the swap with "B" in case both "C" defaults and the recovery rate R_{EC} is not enough to close the swap with "B" without incurring in a loss.

Continuing the swap after the default τ implies for "A" to pay cash flows whose total discounted value at time τ is (including the recovery rate R_{EC}):

$$-\text{R}_{\text{EC}} + \sum_{i=\beta(\tau)}^{n} P(\tau, T_i)\left(-\alpha_i + s + F(\tau; T_{i-1}, T_i)\right) \qquad (22.31)$$

where as usual $F(\tau; T_{i-1}, T_i) = (P(\tau, T_{i-1})/P(\tau, T_i) - 1)/(T_i - T_{i-1})$ is the forward LIBOR rate at time τ between T_{i-1} and T_i. "A" wishes to cancel this payment when it is positive. By simple algebra, and substituting the definition of F, this cancellation has the following value at time τ:

$$\left[\sum_{i=\beta(\tau)}^{5} \left(P(\tau, T_i)(s - \alpha_i) + P(\tau, T_{i-1}) - P(\tau, T_i)\right) - \text{R}_{\text{EC}}\right]^+ .$$

Thus we need computing

$$\mathbb{E}\left\{D(0,\tau)1_{\{\tau < T_n\}}\left[\sum_{i=\beta(\tau)}^{5} \left(P(\tau, T_i)(s - \alpha_i) + P(\tau, T_{i-1}) - P(\tau, T_i)\right) - \text{R}_{\text{EC}}\right]^+\right\}. \qquad (22.32)$$

By a joint simulation of (λ, r) this payoff can be easily valued. Indeed, from the simulation of Λ and $\xi = \Lambda(\tau)$ one obtains a simulation of τ, and thus, through the joint simulation of r, is able to build scenarios of r_τ. Since all bonds $P(\tau, T)$ are known functions of r_τ in the SSRD CIR++ model, we simply have to discount these scenarios from τ to 0 and then average along scenarios.

Our results, with the same interest-rate and default-intensity dynamics (r, λ) as in Section 22.7.2 are reported in Tables 22.25 (recovery $\text{R}_{\text{EC}} = 0.1$), 22.26 (recovery $\text{R}_{\text{EC}} = 0$) and 22.27 (recovery $\text{R}_{\text{EC}} = 0$ and stressed parameters, κ and ν increased by a factor 5 and μ by a factor 3).

Results show that for this nonlinear payoff correlation may have a relevant impact. It is interesting to notice that the correlation pattern is inverted when randomness increases as in the last table, since the value decreases as the correlation increases, contrary to the earlier cases. This may be explained qualitatively as follows. The indicator term $1_{\{\tau<T_5\}}$ selects relatively high values of λ. In case of positive correlation ρ, high λ's correspond to high r's (and thus a low discount factor $D(0,\tau)$). So in (22.31) the F term is "dominating" the remaining terms and selects a high value for the inner payoff in (22.32). In turn, $D(0,\tau)$ is low, and the combined effect depends on the dynamic parameters of the model, which is what we observe in our examples.

Again in the case with amplified parameters in intensities, in Table 22.27, we observe possible excursions of about 15 basis points due to correlation. So cancellable swaps turn out to be more sensitive to correlation than the almost insensitive CDS's.

$s \downarrow$	$\rho \to$ -1	0	1	Det
-100	0.59 (0.56, 0.62)	0.78 (0.74, 0.82)	1.09 (1.05, 1.12)	0
-50	1.075 (1.03, 1.12)	1.45 (1.40, 1.50)	1.92 (1.86, 1.98)	0
0	2.1 (2.04, 2.17)	2.68 (2.61, 2.75)	3.40 (3.31, 3.48)	0
+50	4.56 (4.47, 4.65)	5.53 (5.43, 5.63)	6.63 (6.52, 6.75)	2.35
+100	11.61 (11.47, 11.75)	12.92 (12.77 13.07)	14.45 (14.28, 14.62)	11.87

Table 22.25. Cancellable swap price in basis points (10^{-4}) as a function of ρ and s with MC simulation, REC = 0.1, "Det" for deterministic model

32.56 (32.15, 32.97)	34.26 (33.83, 34.69)	36.24 (35.78, 36.70)	34.38
43.48 (42.96, 44.00)	45.19 (44.65, 45.74)	47.03 (46.46, 47.59)	45.08
54.351 (53.71, 54.99)	55.59 (54.94, 56.25)	57.40 (56.72, 58.08)	55.79
64.91 (64.15, 65.67)	66.26 (65.48, 67.04)	68.25 (67.45, 69.05)	66.49
75.64 (74.76, 76.53)	76.78 (75.88 77.68)	78.81 (77.89, 79.73)	77.20

Table 22.26. Same table as before with REC = 0: Cancellable swap price in basis points (10^{-4}) as a function of ρ and s with MC simulation, REC = 0, "Det" for deterministic model

22.7.9 CDS Options and Jamshidian's Decomposition

We consider now CDS options. We will deal with these options, their market model and some examples of implied volatility in Chapter 23. Also, a description of these contracts has been given in Section 21.4.

59.06 (58.63, 59.49)	50.23 (49.86, 50.60)	44.92 (44.58, 45.26)	34.38
74.11 (73.59, 74.63)	65.58 (65.12, 66.03)	60.17 (59.75, 60.60)	45.08
89.60 (88.99, 90.22)	80.97 (80.41, 81.52)	75.56 (75.04, 76.08)	55.79
104.76 (104.04, 105.48)	96.55 (95.89, 97.20)	91.21 (90.58, 91.83)	66.49
119.99 (119.18, 120.81)	111.50 (110.75 112.26)	106.40 (105.68, 107.13)	77.20

Table 22.27. Same table as before with stressed SSRD parameters: Cancellable swap price in basis points (10^{-4}) as a function of ρ and s under stressed parameters with MC simulation, $\text{R}\textsc{ec} = 0$, "Det" for deterministic model

Here we address first the pricing of these products with the CDS-calibrated SSRD model under deterministic interest rates. In this case we can find a closed form formula.

The case with deterministic interest rates (Brigo 2004, Brigo and Alfonsi 2003). We developed this formula by an initial hint of Ouyang (2003). Consider the option to enter a CDS at a future time $T_a > 0$, $T_a < T_b$, receiving protection $\text{L}\textsc{gd}$ against default up to time T_b, in exchange for a fixed protection rate K (payer CDS option). At T_a there is the option of entering a CDS paying a fixed rate K at times $T_{a+1}, \ldots, T_b$ or until default, in exchange for protection against a possible default in $[T_a, T_b]$. If default occurs a protection payment $\text{L}\textsc{gd}$ is received. By noticing that the market CDS rate $R_{a,b}(T_a)$ at T_a will set the CDS value in T_a to 0, the payoff can be written as the discounted difference between said CDS and the corresponding CDS with rate K. We have that the payoff at T_a reads

$$\Pi_a := [\overbrace{\text{CDS}_{a,b}(T_a, R_{a,b}(T_a), \text{L}\textsc{gd})}^{0} - \text{CDS}_{a,b}(T_a, K, \text{L}\textsc{gd})]^+$$

$$= [-\text{CDS}_{a,b}(T_a, K, \text{L}\textsc{gd})]^+$$

$$= 1_{\{\tau > T_a\}} \Bigg(\mathbb{E}\Bigg[- D(T_a, \tau)(\tau - T_{\beta(\tau)-1}) K 1_{\{\tau < T_b\}}$$

$$- \sum_{i=a+1}^{b} D(T_a, T_i)\alpha_i K 1_{\{\tau > T_i\}} + 1_{\{\tau < T_b\}} D(T_a, \tau)\, \text{L}\textsc{gd} | \mathcal{G}_{T_a} \Bigg] \Bigg)^+$$

$$= 1_{\{\tau > T_a\}} \Bigg\{ -K \int_{T_a}^{T_b} \mathbb{E}\left[\exp\left(-\int_{T_a}^{u} (r_s + \lambda_s) ds \right) \lambda_u | \mathcal{F}_{T_a} \right] (u - T_{\beta(u)-1}) du$$

$$-K \sum_{i=a+1}^{b} \alpha_i \mathbb{E}\left[\exp\left(-\int_{T_a}^{T_i} (r_s + \lambda_s) ds \right) | \mathcal{F}_{T_a} \right]$$

$$+\text{L}\textsc{gd} \int_{T_a}^{T_b} \mathbb{E}\left[\exp\left(-\int_{T_a}^{u} (r_s + \lambda_s) ds \right) \lambda_u | \mathcal{F}_{T_a} \right] du \Bigg\}^+$$

$$=: 1_{\{\tau > T_a\}} \left[-\text{CDS}^{\mathcal{F}}_{a,b}(T_a, K, \text{L}\textsc{gd}; x_{T_a}, y_{T_a}) \right]^+ \qquad (22.33)$$

where $\text{CDS}^{\mathcal{F}}_{a,b}(T_a, K, \text{L}_{\text{GD}}; x_{T_a}, y_{T_a})$ is the quantity inside curly brackets. If we take deterministic interest rates r this reads

$$\Pi_a = 1_{\{\tau > T_a\}} \Big\{ -K \sum_{i=a+1}^{b} \alpha_i P(T_a, T_i) \mathbb{E}\left[\exp\left(-\int_{T_a}^{T_i} \lambda_s ds \right) \Big| \mathcal{F}_{T_a} \right]$$

$$-K \int_{T_a}^{T_b} \mathbb{E}\left[\exp\left(-\int_{T_a}^{u} \lambda_s ds \right) \lambda_u \Big| \mathcal{F}_{T_a} \right] P(T_a, u)(u - T_{\beta(u)-1}) du$$

$$+\text{L}_{\text{GD}} \int_{T_a}^{T_b} P(T_a, u) \mathbb{E}\left[\exp\left(-\int_{T_a}^{u} \lambda_s ds \right) \lambda_u \Big| \mathcal{F}_{T_a} \right] du \Big\}^{+}.$$

Define

$$H(t, T; y_t^{\beta}) := \mathbb{E}\left[\exp\left(-\int_t^T \lambda_s ds \right) \Big| \mathcal{F}_t \right]$$

and notice that

$$\mathbb{E}\left[\exp\left(-\int_t^T \lambda_s ds \right) \lambda_T \Big| \mathcal{F}_t \right] = -\frac{d}{dT} \mathbb{E}\left[\exp\left(-\int_t^T \lambda_s ds \right) \Big| \mathcal{F}_t \right]$$

$$= -\frac{d}{dT} H(t, T)$$

Write then

$$\Pi_a = 1_{\{\tau > T_a\}} \Big\{ K \int_{T_a}^{T_b} P(T_a, u)(u - T_{\beta(u)-1}) \frac{d}{du} H(T_a, u) du$$

$$-K \sum_{i=a+1}^{b} \alpha_i P(T_a, T_i) H(T_a, T_i) - \text{L}_{\text{GD}} \int_{T_a}^{T_b} P(T_a, u) \frac{d}{du} H(T_a, u) du \Big\}^{+}.$$

Note that the first two summations add up to a positive quantity, since they are expectations of positive terms.

By integrating by parts in the first and third integral, we obtain, by defining $q(u) := -dP(T_a, u)/du$,

$$\Pi_a = 1_{\{\tau > T_a\}} \Big\{ \text{L}_{\text{GD}} - \int_{T_a}^{T_b} \big[\text{L}_{\text{GD}} q(u) + K P(T_a, T_{\beta(u)}) \delta_{T_{\beta(u)}}(u)$$

$$-K(u - T_{\beta(u)-1}) q(u) - K P(T_a, T_{\beta(u)}) \delta_{T_{\beta(u)}}(u)$$

$$+\text{L}_{\text{GD}} \delta_{T_b}(u) P(T_a, u) + K P(T_a, u) \big] H(T_a, u) du \Big\}^{+}.$$

Define

$$h(u) := \text{L}_{\text{GD}}\, q(u) - K(u - T_{\beta(u)-1}) q(u) + \text{L}_{\text{GD}}\, \delta_{T_b}(u) P(T_a, u) + K P(T_a, u),$$

where $\delta_x(y)dy = \delta_0(y-x)dy$ is the Dirac delta function centered in x and valued at y, which we occasionally write as $1_{\{x\in[y,y+dy)\}}$.

as that can be written also as

so that

$$\Pi_a = 1_{\{\tau>T_a\}}\left\{\text{LGD} - \int_{T_a}^{T_b} h(u)H(T_a,u;y^{\beta}_{T_a})du\right\}^{+} \tag{22.34}$$

It is easy to check, by remembering the signs of the terms of which the above coefficients are expectations, that

$$h(u) > 0 \ \text{ for all } \ u.$$

Now we look for a term y^* such that

$$\int_{T_a}^{T_b} h(u)H(T_a,u;y^*)du = \text{LGD}. \tag{22.35}$$

It is easy to see that in general $H(t,T;y)$ is decreasing in y for all t,T. This equation can be solved, since $h(u)$ is known and deterministic and since H is completely given in terms of the CIR bond price formula and is deterministic once the argument y is selected. Furthermore, either a solution exists or the option valuation is not necessary. Indeed, consider first the limit of the left hand side for $y^*\to\infty$. We have

$$\lim_{y^*\to\infty}\int_{T_a}^{T_b} h(u)H(T_a,u;y^*)du = 0 < \text{LGD},$$

which shows that for y^* large enough we always go below the value LGD. Then consider the limit of the left hand side for $y^*\to 0$:

$$\lim_{y^*\to 0+}\int_{T_a}^{T_b} h(u)H(T_a,u;y^*)du =$$

$$= \text{LGD} + \int_{T_a}^{T_b}\left[\text{LGD}P(T_a,u)\frac{\partial H(T_a,u;0)}{\partial u} + (K(u-T_{\beta(u)-1})q(u)+ \right.$$

$$\left. +KP(T_a,u))H(T_a,u;0)\right]du.$$

Now if the integral in the last expression is positive then we have that the limit is larger than LGD and by continuity and monotonicity there is always a solution y^* giving LGD. If instead the integral in the last expression is negative, then the limit is smaller than LGD and we have that (22.35) admits no solution, in that its left hand side is always smaller than the right hand side. However, this implies in turn that the expression inside curly brackets in the payoff (22.34) is always positive and thus the contract loses its optionality and can be valued by taking the expectation without positive part, giving as option price simply $-\mathrm{CDS}_{a,b}(t, K, \mathrm{L}_{\mathrm{GD}}) > 0$, a forward start payer CDS. In case y^* exists, instead, we may rewrite our discounted payoff as

$$\Pi_a = 1_{\{\tau > T_a\}} \left\{ \int_{T_a}^{T_b} h(u)(H(T_a, u; y^*) - H(T_a, u; y^\beta_{T_a}))du \right\}^+$$

Since $H(t, T; y)$ is decreasing in y for all t, T, all terms $(H(T_a, u; y^*) - H(T_a, u; y^\beta_{T_a}))$ have the same sign, which will be positive if $y^\beta_{T_a} > y^*$ or negative otherwise. Since all such terms have the same sign, we may write

$$\Pi_a = 1_{\{\tau > T_a\}} \left\{ \int_{T_a}^{T_b} h(u)(H(T_a, u; y^*) - H(T_a, u; y^\beta_{T_a}))^+ du \right\} =: 1_{\{\tau > T_a\}} Q_a.$$

Now compute the price as

$$\mathbb{E}[D(0, T_a)\Pi_a] = P(0, T_a)\mathbb{E}[1_{\{\tau > T_a\}} Q_a] = P(0, T_a)\mathbb{E}\left[\exp\left(-\int_0^{T_a} \lambda_s ds\right) Q_a\right] =$$

$$= \int_{T_a}^{T_b} h(u)\mathbb{E}\left[\exp\left(-\int_0^{T_a} \lambda_s ds\right)(H(T_a, u; y^*) - H(T_a, u; y^\beta_{T_a}))^+\right] du$$

From a structural point of view, the $H(T_a, u; y^\beta_{T_a})$'s are like zero coupon bond prices in a CIR++ model with short term interest rate λ, for maturity T_a on bonds maturing at u. Thus, each term in the summation is $h(u)$ times a zero-coupon bond like call option with strike $K^*_u = H(T_a, u; y^*)$. A formula for such options is given for example in (3.78).

If one maintains stochastic interest rates with possibly non-null ρ, then a possibility is to use the Gaussian mapped processes x^V and y^V introduced earlier and to reason as for pricing swaptions with the G2++ model through Jamshidian's decomposition and one-dimensional Gaussian numerical integration, along the lines of the procedures leading to (4.31). Clearly the resulting formula has to be tested against Monte Carlo simulation.

The full SSRD model: stochastic interest rates. Consider again the general payoff expression (22.33) for CDS options in the SSRD model but now do not assume interest rates to be deterministic. In particular, let the r CIR++ parameters k, θ, σ, x_0 and the instantaneous correlation ρ between r and λ be non-trivial.

Assuming $\rho = 0$ from T_a on, the expectations appearing in the above expression (22.33) can be computed according to formulas (22.15) and (22.16) with $t = T_a$. From these terms valued in $t = T_a$ it follows immediately that all terms are known analytically given the simulated paths of x_{T_a} and y_{T_a}, which are to be simulated with nonzero ρ from time 0 to time T_a. Putting all pieces together, without forgetting the indicator $1_{\{\tau > T_a\}}$, we may value the CDS option payoff in (22.33) by simulation.

$$\mathbb{E}\left\{D(t,T_a)[-\text{CDS}_{a,b}(T_a,K,\text{L}_{\text{GD}})]^+|\mathcal{G}_t\right\}$$

$$= \mathbb{E}\left\{D(t,T_a)1_{\{\tau>T_a\}}[-\text{CDS}^{\mathcal{F}}_{a,b}(T_a,K,\text{L}_{\text{GD}};x_{T_a},y_{T_a})]^+|\mathcal{G}_t\right\}$$

$$= \frac{1_{\{\tau>t\}}}{\exp(-\Lambda(t))}\mathbb{E}\left\{D(t,T_a)1_{\{\tau>T_a\}}[-\text{CDS}^{\mathcal{F}}_{a,b}(T_a,K,\text{L}_{\text{GD}};x_{T_a},y_{T_a})]^+|\mathcal{F}_t\right\}$$

$$= \frac{1_{\{\tau>t\}}}{\exp(-\Lambda(t))}\mathbb{E}\left\{\mathbb{E}\left[D(t,T_a)1_{\{\tau>T_a\}}[-\text{CDS}^{\mathcal{F}}_{a,b}(T_a,K,\text{L}_{\text{GD}};x_{T_a},y_{T_a})]^+|\mathcal{F}_{T_a}\right]|\mathcal{F}_t\right\}$$

$$= \frac{1_{\{\tau>t\}}}{\exp(-\Lambda(t))}\mathbb{E}\left\{D(t,T_a)[-\text{CDS}^{\mathcal{F}}_{a,b}(T_a,K,\text{L}_{\text{GD}};x_{T_a},y_{T_a})]^+\mathbb{E}\left[1_{\{\tau>T_a\}}|\mathcal{F}_{T_a}\right]|\mathcal{F}_t\right\}$$

$$= 1_{\{\tau>t\}}\mathbb{E}\left\{D(t,T_a)\exp(-\Lambda(T_a)+\Lambda(t))[-\text{CDS}^{\mathcal{F}}_{a,b}(T_a,K,\text{L}_{\text{GD}};x_{T_a},y_{T_a})]^+|\mathcal{F}_t\right\}$$

$$= \boxed{1_{\{\tau>t\}}\mathbb{E}\left\{\exp\left(-\int_t^{T_a}(r_s+\lambda_s)ds\right)[-\text{CDS}^{\mathcal{F}}_{a,b}(T_a,K,\text{L}_{\text{GD}};x_{T_a},y_{T_a})]^+|\mathcal{F}_t\right\}}, \tag{22.36}$$

where we have used the filtration switching formula (22.3), iterated conditioning, and the fact that in general

$$\mathbb{Q}(\tau > T|\mathcal{F}_T) = \mathbb{Q}(\Lambda(\tau) > \Lambda(T)|\mathcal{F}_T) = \mathbb{Q}(\xi > \Lambda(T)|\mathcal{F}_T) = e^{-\Lambda(T)}.$$

The assumption above that $\rho = 0$ from T_a on allows us to compute the $\mathcal{F}$-measurable part of the CDS payoff, i.e. $\text{CDS}^{\mathcal{F}}$, as a function of the simulated x_{T_a} and y_{T_a} without further simulation from T_a to T_b. It suffices to use formulas (22.15) and (22.16) with $t = T_a$ in the payoff terms. However, we have to check that we can set $\rho = 0$ from T_a on without harm. We know from the discussion at the end of Section 22.7.2 and from Brigo and Alfonsi (2003, 2005) that ρ has little impact on "at the money" CDS contracts valued at time 0. We plan to check whether this is the case with the option payoff from T_a on.

We will thus compute the option price both by taking $\rho = 0$ from T_a on and by keeping the nonzero ρ also in $[T_a, T_b]$. In the latter case we can resort to the "sub-path" method. We simulate n paths of λ and r from 0 to T_a, and then for each T_a realization we sub-simulate m paths up to T_b to compute the inner discounted payoff at T_a conditional on the T_a scenario. We need a way to estimate the standard error of the Monte Carlo method in a sub-path setup.

SSRD standard error upper bound under nonzero correlation. Write the above option payoff at maturity T_a by collecting the expected values as

$$(-\text{CDS}_{a,b}(T_a, K, \text{L}_{\text{GD}}))^+ =$$

$$= 1_{\{\tau > T_a\}} \Bigg\{ \mathbb{E}\Bigg[-K \int_{T_a}^{T_b} \exp\left(-\int_{T_a}^{u}(r_s+\lambda_s)ds\right)\lambda_u(u - T_{\beta(u)-1})du$$

$$-K \sum_{i=a+1}^{b} \alpha_i \exp\left(-\int_{T_a}^{T_i}(r_s+\lambda_s)ds\right)$$

$$+\text{L}_{\text{GD}} \int_{T_a}^{T_b} \exp\left(-\int_{T_a}^{u}(r_s+\lambda_s)ds\right)\lambda_u du \Bigg| \mathcal{F}_{T_a}\Bigg]\Bigg\}^+ .$$

Call X the part of this expression inside the expectation after the indicator, i.e. the part inside square brackets. The CDS option price can be written as

$$\mathbb{E}\left[\exp\left(-\int_0^{T_a}(r_s+\lambda_s)ds\right)(\mathbb{E}_{T_a}X)^+\right] = \mathbb{E}\left[Y \cdot (\mathbb{E}_{T_a}X)^+\right]$$

where Y denotes the exponential term and $\mathbb{E}_T[\cdot] = \mathbb{E}[\cdot|\mathcal{F}_T]$. The method we use is generate some scenarios ω_i, $i = 1, \dots, n$ for r and λ up to T_a. Then, conditional on each such ω_i, we generate m sub-paths $\omega_{i,j}$, $j = 1, \dots, m$ for r and λ from T_a to T_b. We call Y^i the realization of Y corresponding to ω_i and $X^{i,j}$ the realization of X corresponding to $\omega_{i,j}$.

Our Monte Carlo estimate for the above price will then be

$$\Pi_{MC} = \frac{1}{n}\sum_{i=1}^{n} Y^i \left(\frac{1}{m}\sum_{j=1}^{m} X^{i,j}\right)^+$$

To estimate the MC error, we calculate

$$V_{n,m} := \mathbb{E}\left[\left(\frac{1}{n}\sum_{i=1}^{n} Y^i \left(\frac{1}{m}\sum_{j=1}^{m} X^{i,j}\right)^+ - \mathbb{E}[Y^1 \mathbb{E}_{\mathcal{F}^1_{T_a}}(X^{1,1})^+]\right)^2\right],$$

with $\mathcal{F}^1_{T_a}$ the default free filtration at time T_a given that further $Y = Y^1$.

This is the good measure of the error, since it is telling us how far we are from the true price, given by the expected value in the expression. By expanding and through straightforward calculations, we obtain

$$V_{n,m} = \mathbb{E}\left[\frac{1}{n^2}\sum_{i_1=1}^{n}\sum_{i_2=1}^{n} Y^{i_1}Y^{i_2}\left(\frac{1}{m}\sum_{j=1}^{m} X^{i_1,j}\right)^+ \left(\frac{1}{m}\sum_{j=1}^{m} X^{i_2,j}\right)^+\right.$$

$$\left. -\frac{2}{n}\sum_{i=1}^{n} Y^i \left(\frac{1}{m}\sum_{j=1}^{m} X^{i,j}\right)^+ \mathbb{E}[Y^1 \mathbb{E}_{\mathcal{F}^1_{T_a}}(X^{1,1})^+] + \mathbb{E}[Y^1 \mathbb{E}_{\mathcal{F}^1_{T_a}}(X^{1,1})^+]^2\right]$$

$$
= \frac{1}{n}\mathbb{E}\left[\left(Y^1\left(\frac{1}{m}\sum_{j=1}^{m}X^{1,j}\right)^+\right)^2\right] + \left(1-\frac{1}{n}\right)\mathbb{E}\left[Y^1\left(\frac{1}{m}\sum_{j=1}^{m}X^{1,j}\right)^+\right]^2
$$

$$
-2\mathbb{E}\left[Y^1\left(\frac{1}{m}\sum_{j=1}^{m}X^{1,j}\right)^+\right]\mathbb{E}[Y^1\mathbb{E}_{\mathcal{F}^1_{T_a}}(X^{1,1})^+] + \mathbb{E}[Y^1\mathbb{E}_{\mathcal{F}^1_{T_a}}(X^{1,1})^+]^2
$$

$$
= \frac{1}{n}\mathrm{Var}\left[Y^1\left(\frac{1}{m}\sum_{j=1}^{m}X^{1,j}\right)^+\right]
$$

$$
+\left(\mathbb{E}\left[Y^1\left(\frac{1}{m}\sum_{j=1}^{m}X^{1,j}\right)^+\right] - \mathbb{E}[Y^1\mathbb{E}_{\mathcal{F}^1_{T_a}}(X^{1,1})^+]\right)^2
$$

$$
= \frac{1}{n}\mathrm{Var}\left[Y^1\left(\frac{1}{m}\sum_{j=1}^{m}X^{1,j}\right)^+\right]
$$

$$
+\left(\mathbb{E}\left[Y^1\left(\left(\frac{1}{m}\sum_{j=1}^{m}X^{1,j}\right)^+ - \mathbb{E}_{\mathcal{F}^1_{T_a}}(X^{1,1})^+\right)\right]\right)^2
$$

$$
\overset{Schwarz}{\leq} \frac{1}{n}\mathrm{Var}\left[Y^1\left(\frac{1}{m}\sum_{j=1}^{m}X^{1,j}\right)^+\right]
$$

$$
+\mathbb{E}((Y^1)^2)\mathbb{E}\left[\left(\left(\frac{1}{m}\sum_{j=1}^{m}X^{1,j}\right)^+ - \mathbb{E}_{\mathcal{F}^1_{T_a}}(X^{1,1})^+\right)^2\right]
$$

$$
\overset{|x^+-y^+|\leq|x-y|}{\leq} \frac{1}{n}\mathrm{Var}\left[Y^1\left(\frac{1}{m}\sum_{j=1}^{m}X^{1,j}\right)^+\right]
$$

$$
+\mathbb{E}((Y^1)^2)\mathbb{E}\left[\left(\frac{1}{m}\sum_{j=1}^{m}X^{1,j} - \mathbb{E}_{\mathcal{F}^1_{T_a}}(X^{1,1})\right)^2\right]
$$

$$
= \frac{1}{n}\mathrm{Var}\left[Y^1\left(\frac{1}{m}\sum_{j=1}^{m}X^{1,j}\right)^+\right] + \frac{1}{m}\mathbb{E}((Y^1)^2)\mathbb{E}\left[\mathrm{Var}[X^{1,j}|\mathcal{F}_{T_a}]\right]
$$

In the last expression we obtained, each part can be estimated respectively by its sample counterparts, leading to the following estimate of the Monte Carlo standard error based on simulated paths:

$$
\hat{V}_{n,m} = \frac{1}{n}\left[\frac{1}{n-1}\sum_{i=1}^{n}\left(y_i\left(\frac{1}{m}\sum_{j=1}^{m}x^{i,j}\right)^+\right)^2 - \left(\frac{1}{n}\sum_{i=1}^{n}y_i\left(\frac{1}{m}\sum_{j=1}^{m}x^{i,j}\right)^+\right)^2\right] +
$$

$$
+\frac{1}{m}\left(\frac{1}{n}\sum_{i=1}^{n}y_i^2\right)\left(\frac{1}{n}\sum_{i=1}^{n}\left(\frac{1}{m-1}\sum_{j=1}^{m}(x^{i,j})^2 - \left(\frac{1}{m}\sum_{j=1}^{m}x^{i,j}\right)^2\right)\right),
$$

and we can mimic the standard case for a 95% standard error by multiplying the square-root of this upper bound by 1.96.

Alfonsi (2005b) considers some cases based on Parmalat CDS data on December 8, 2003 (a deep crisis date). The CDS data are the same we have seen in Section (22.3). He takes parameters

$$k = 0.528905, \theta = 0.0319904, \sigma = 0.130035, x_0 = 8.32349E - 05$$

(cap calibration) and

$$\beta : \kappa = 0.583307, \mu = 0.0149846, \nu = 0.0479776, y_0 = 0.192973$$

coming from minimizing $\int_0^{10y} \psi^2(t;\beta)dt$, as we did in Section 22.7.2. He adds shifts φ and ψ to match exactly the zero coupon interest rate curve and the CDS term structure. The discretization step in the Monte Carlo explicit scheme $E(0)$ seen in (22.20) is 1/100 year.

Alfonsi (2005b) considers the following option:

First option.

- Option with maturity $T_a = 1y$, final CDS date $T_b = 5y$, and strike at the money forward: $K = R_{a,b}(0) = 0.0748048$.
- Option price with $\rho = 0$: 0.00438654;
- 95% standard-error estimate: 0.000926927.
- Number of paths: $n = 10000, m = 1000000$.
- Option price with deterministic interest rates and stochastic intensity (CIR++ for λ) : 0.00444254;
- 95% MC standard error in this case: 0.000472689.
- Number of paths in this case: $n = 10000$.

From this first experiment we notice that the impact of stochastic r, in case $\rho = 0$, is quite small in the CDS option price, as we expect from the model independent nature of underlying CDS prices when τ is independent of r. Almost all the value comes from the intensity volatility. Also, the standard error in the deterministic r case is a real standard error, since no sub-paths method and subsequent error estimate is needed.

Consider now a second option:

Second option.

- Option with maturity $T_a = 4y$, final CDS date $T_b = 5y$, and strike $K = R_{a,b}(0) = 0.0280203$;
- Option price under the full SSRD model with $\rho = 0$ is 0.000845052;
- The 95% standard-error estimate is 0.000756794;
- Number of paths: $n = 1000, m = 100000$.
- Option price under the model with deterministic r: 0.000845232;
- 95% MC standard error in this case: 0.000135247;
- Number of paths in this case: $n = 1000$.

Notice that the error measure in case of full SSRD is very large, but keep in mind this is an upper bound of the real MC error. The price itself looks again very similar in the two cases of full SSRD and deterministic r.

Now let us move to correlation $\rho = 1$. Consider again the

Second option (with $\rho = 1$ everywhere).

- We keep $\rho = 1$ up to time T_b (sub-paths MC method);
- The option value is 0.00100042;
- The 95% MC error estimate is 0.000760244.

Now we check what happens when setting correlation to zero from T_a on.

Second option (with $\rho = 1$ in $[0, T_a]$ and $\rho = 0$ in $[T_a, T_b]$).

- We keep $\rho = 1$ up to time T_a and use deterministic r from T_a on;
- The option value under deterministic r from T_a on is 0.0010523;
- 95% MC standard error 0.000156311.

These tests are rather partial but they point out that, at first sight, killing the correlation in the CDS option payoff from T_a on does not affect the option price much, at least for options on CDS that are not too long. We are planning further tests in the future to check this is true. If these tests confirm the possibility to neglect ρ from T_a on, the patterns for implied volatilities as functions of the SSRD model parameters found by Brigo and Cousot (2004) hold. Indeed, in the general SSRD model, one may compute the CDS option price by means of Monte Carlo simulations, equate this Monte Carlo price to the market model Formula (23.13) we are going to derive in the next chapter, applied to the same CDS option at $t = 0$, and solve in $\sigma_{a,b}$. This $\sigma_{a,b}$ is then the implied volatility coming from the SSRD price. The first numerical results we found in a number of cases point out the following patterns of $\sigma_{a,b}$ in terms of SSRD model parameters, when we set $\rho = 0$ from T_a on.:

Param :	$\kappa \uparrow$	$\mu \uparrow$	$\nu \uparrow$	$y_0 \uparrow$	$\rho\uparrow$
$\sigma_{a,b}^{imp}$:	$\downarrow$	$\uparrow$	$\uparrow$	$\uparrow$	$\downarrow$

Table 22.28. Implied volatility patterns for prices generated by the SSRD model with $\rho = 0$ from T_a on.

For more details see also Brigo and Cousot (2004). The patterns are reasonable. When κ increases (all other things being equal) the time-homogeneous core of the stochastic intensity has a higher speed of mean reversion and then randomness reduces more quickly in time, so that the implied volatility reduces. When μ increases, the asymptotic mean of the homogeneous part of the intensity increases, so that we have higher intensity and thus, since instantaneous volatility is proportional to the increased $\sqrt{y}$, more randomness.

When ν increases, clearly randomness of λ increases so that it is natural for the implied volatility to increase. We also find that increasing y_0 (the initial point of the time-homogeneous part of the intensity) increases the implied volatility, while increasing the correlation ρ decreases the implied volatility.

22.7.10 Bermudan CDS Options

Bermudan CDS options do not exist yet in the market. However, anticipating possible market developments, Ben Ameur, Brigo and Errais (2005) define the Credit Default Swap (CDS) Bermudan option, and devise a numerical method to price it. The method proposed is general and could be used to price any early exercise single name CDS-based contingent claim. The authors, according to what we have seen earlier in this chapter, model the default event within the doubly stochastic intensity (or Cox process) framework where the intensity is defined by the CIR++ model. As we have seen earlier, the model is calibrated exactly to the market term structure of CDS, as in Brigo and Alfonsi (2003, 2005). The numerical method developed in Ben Ameur, Brigo and Errais (2005) is based on dynamic programming with finite elements and piecewise linear approximation.

The CIR++ intensity models suffers the drawback of not being able to reproduce high implied volatilities for vanilla CDS options, as we will point out in the next section. However, we will suggest a remedy to this situation. Basically one may add jumps to the CIR++ model and obtain a model that is still in the affine family, retaining analytical tractability, and that can be used in conjunction with dynamic programming to price Bermudan CDS options. The reader can adapt the results in Ben Ameur, Brigo and Errais (2005) to price Bermudan CDS options under the jump extended CIR model. This way, the model used to price the Bermudan option will be consistent with the high implied volatilities observed in the vanilla CDS options market. We now introduce the jump-extended CIR model for stochastic intensity.

22.8 Stochastic diffusion intensity is not enough: Adding jumps. The JCIR(++) Model

There is a problem we should mention in closing this chapter. The SSRD stochastic intensity model we introduced is not always capable of generating high levels of implied volatility. Indeed, the numerical experiments in Brigo and Cousot (2004) point out that it is quite difficult to find implied volatilities with an order of magnitude of 30% with the SSRD model. This is due essentially to two reasons. First, we need to keep the shift ψ positive and this limits the configurations of parameters in a way that renders high implied volatilities hard to attain. The second problem is more fundamental and is related to the structure of the square root diffusion dynamics. Assume for a moment that we give up the shift and work with a time homogenous CIR model for λ, or in other terms $\lambda = y$, $\psi = 0$, so that

$$d\lambda_t = \kappa(\mu - \lambda_t)dt + \nu\sqrt{\lambda_t}dZ_t.$$

Intuitively, high implied volatility for the option prices generated by this model corresponds to a high volatility parameter ν in the intensity dynamics. However, to restrict the values attainable by λ in the positive domain, we need to ensure that the following condition is satisfied

$$2\kappa\mu > \nu^2.$$

This condition implies that if ν is large, κ and/or μ are forced to assume large values as well. None of these possibilities is really desirable though. Actually, drastically increasing μ means increasing the mean reversion level of the intensity process, so that λ is supposed to tend to possibly very high values. Alternatively, increasing κ drastically can counter the increase in ν as far as the SSRD implied volatility is concerned. Indeed, as we commented already after Table 22.28, large κ means a large speed of mean reversion, which means that the trajectories will tend to regroup around μ faster, so that the system will have less stochasticity, for a given value of ν. To sum up, we may increase ν to increase the implied volatility but this will force us, due to the positivity condition, to increase κ, whose effect will counter the initial increase in ν. In practice, we have been able in realistic situations to go up to a 30% implied volatility. However, we will see in some examples in the next chapter that implied volatilities in CDS markets may easily exceed 50%. This means that we can easily find situations where no realistic configuration of the parameters κ, ν, μ, y_0 of the SSRD model can generate market implied volatilities.

This is particularly annoying as the square root diffusion is convenient to work with, since it restricts the intensity to positive values only and at the same time is relatively tractable. Other known positive diffusion models that are tractable include the squared gaussian process studied by Leippold and Wu (2002) among others, and time-changed square root diffusions as suggested by Musiela (2005) for volatility modeling. The logarithm time changed square root diffusion proposed by Musiela (2005) is particularly attractive here since it allows to increase κ without introducing such a drastic increase in the speed of mean reversion as it transforms the exponential mean reversion of the square root diffusion to a milder polynomial mean reversion.

For a more detailed account of these issues, see the discussion in Brigo and El Bachir (2005), where it is suggested to introduce a jump component in the intensity process, as we do now.

22.8.1 The jump-diffusion CIR model (JCIR)

We consider a special case of the class of Affine Jump Diffusions (AJD) (see for example Duffie, Pan and Singleton (2000), and Duffie and Garleanu

(2001)). The dynamics of λ would then satisfy

$$d\lambda_t = \kappa(\mu - \lambda_t)dt + \nu\sqrt{\lambda_t}dZ_t + dJ_t$$

where J is a pure jump process with jumps arrival rate $\alpha > 0$ and jump sizes distribution π on $\mathbb{R}^+$.

Notice an important point: the jump process J we just introduced is a jump in the stochastic intensity dynamics, not the already introduced fundamental jump in the default process. Recall indeed that τ is the first jump of a suitable Cox process. This first jump is not a jump of J. When J alone jumps the intensity is affected by an increase but there is no default.

Notice also that we restrict the jumps to be positive, preserving the attractive feature of positive default intensity implied by the basic CIR dynamics. Further, assume that π is an exponential distribution with mean $\gamma > 0$, and that

$$J_t = \sum_{i=1}^{M_t} Y_i$$

where M is a time-homogeneous Poisson process with intensity α, the Ys being exponentially distributed with parameter γ. The larger α, the more frequent the jumps, and the larger γ, the larger the sizes of the occurring jumps. We denote the resulting jump process by $J^{\alpha,\gamma}$, to point out the parameters influencing its dynamics. We write

$$d\lambda_t = \kappa(\mu - \lambda_t)dt + \nu\sqrt{\lambda_t}dZ_t + dJ_t^{\alpha,\gamma}, \tag{22.37}$$

so that we can see all the parameters in the dynamics.

22.8.2 Bond (or Survival Probability) Formula.

Since this model belongs to the tractable affine jump diffusion (AJD) class of models, the survival probability has the typical "log-affine" shape

$$\mathbb{Q}\{\tau > T|\mathcal{G}_t\} = \mathbf{1}_{\{\tau>t\}}\bar{\alpha}(t,T)\exp(-\bar{\beta}(t,T)\lambda_t) =: \mathbf{1}_{\{\tau>t\}}P^{JCIR}(t,T,\lambda_t)$$

where the functional forms of the terms $\bar{\alpha}$ and $\bar{\beta}$ with respect to the parameters $\kappa, \mu, \nu, \alpha, \gamma$ are given in the appendix of Duffie and Garleanu (2001), or Christensen (2002) whose results are summarized in the appendix of Lando (2004) and are obtained by solving the usual Riccati equations. These expressions for $\bar{\alpha}$ and $\bar{\beta}$ can be recast in a form that is similar to the classical terms A and B in the bond price formula for CIR as in Brigo and El Bachir (2005):

$$\bar{\alpha}(t,T) = A(t,T)\left(\frac{2h\exp\left(\frac{h+\kappa+2\gamma}{2}(T-t)\right)}{2h+(\kappa+h+2\gamma)(\exp^{h(T-t)}-1)}\right)^{\frac{2\alpha\gamma}{\nu^2-2\kappa\gamma-2\gamma^2}} \tag{22.38}$$

$$\bar{\beta}(t,T) = B(t,T) \tag{22.39}$$

where $A(t,T)$, $B(t,T)$ are the terms from the CIR model, and similarly $h = \sqrt{\kappa^2 + 2\nu^2}$.

In this expression one has to be careful. Given the denominator in the exponent of the large round brackets, one sees that this denominator can be rewritten as

$$\nu^2 - 2\kappa\gamma - 2\gamma^2 = \tfrac{1}{2}[h^2 - (\kappa + 2\gamma)^2],$$

and is zero if $h = \kappa + 2\gamma$. One can see through a limit, that when this happens the expression above for $\bar{\alpha}$ has to be substituted by the following one:

$$\bar{\alpha}(t,T) = A(t,T)\exp\left(-2\alpha\gamma\left[\frac{T-t}{\kappa+h+2\gamma} + \frac{e^{-h(T-t)}-1}{h(\kappa+h+2\gamma)}\right]\right).$$

22.8.3 Exact calibration of CDS: The JCIR++ model

In general, our jump-diffusion square root process above could be shifted again according to the usual trick to obtain an exact calibration to credit default swaps. Indeed all we need to compute the shift is the bond price formula for the homogenous model as given above. The shift reproducing exactly CDS quotes would then be the following generalization of (22.12):

$$\Psi^J(t,\beta) = \Gamma^{mkt}(t) + \ln\left(\mathbb{E}\left[\exp^{-\int_0^t \lambda_s ds}\right]\right) = \Gamma^{mkt}(t) + \ln\left(P^{JCIR}(0,t,\lambda_0;\beta)\right) \tag{22.40}$$

This leads to the Jump-diffusion CIR++ model (JCIR++). Still, the addition of the jump component makes it more difficult to find conditions guaranteeing the shift ψ^J to be positive (or, equivalently, its integral Ψ^J increasing). At the same time, it is good to notice that now the basic model without shift has six parameters

$$\kappa, \mu, \nu, \lambda_0, \alpha, \gamma$$

that we might try to use to calibrate 5 CDS quotes plus one option volatility. This is currently under investigation.

22.8.4 Simulation

Simulating the (possibly shifted) J-CIR process is no more difficult than simulating the diffusion part only. Indeed, since the Brownian Z and the compound Poisson process $J^{\alpha,\gamma}$ are independent, Mikulevicius and Platen (1988) propose to generate jump times and jump amplitudes, then proceed with the diffusion discretization schemes adding the jumps at the times when they occur. To apply this method here, one only needs to be able to generate Poisson jump times and exponentially distributed jump sizes, in addition to one of the schemes discussed for CIR.

More specifically, for each path one can generate a Poisson random variable with parameter $\alpha(T-t)$ giving the number of jumps to be simulated in the period of length $[t,T]$. Then simulate all the jump times as independently distributed uniforms on $[t,T]$, and generate the jump magnitude at each jump time as an independently distributed exponential random variable. At this stage, one has all the jump times and the corresponding jump magnitudes for the path under simulation, and needs to proceed with the diffusion part. For that purpose, it is enough to augment the discretization $t_0 = 0 < t_1 < \cdots < t_n = T$ by all the simulated jump times, and proceed with the schemes already discussed, except that for a jump time t_j, we will need to add the jump amplitude:

$$\widetilde{y}^{\beta}_{t_j} = g(\widetilde{y}^{\beta}_{t_{j-1}}) + \widetilde{I}_{t_j}$$

where g is the functional form used to simulate $\widetilde{y}^{\beta}_{t_j}$ given $\widetilde{y}^{\beta}_{t_{j-1}}$ for the chosen scheme, and $\widetilde{I}_{t_j}$ is the size of the simulated jump at time t_j.

22.8.5 Jamshidian's Decomposition.

"This one's tricky. You have to use imaginary numbers, like eleventeen".

Hobbes, helping Calvin with Maths

Another convenient property of this extended jump-diffusion model is the fact that the Jamshidian's decomposition based formula introduced earlier in Section 22.7.9 still holds for the jump-diffusion case. Indeed, the same arguments used to derive that formula still hold in the presence of jumps. Consider the same notation as in Section 22.7.9. The formula is:

$$\mathbb{E}[D^{\mathrm{JCIR}}(0,T_a)\Pi_a] = \tag{22.41}$$
$$= \int_{T_a}^{T_b} h^{JCIR}(u)\mathbb{E}\left[e^{-\int_0^{T_a}\lambda_s ds}\left(H^{\mathrm{JCIR}}(T_a,u;y^*) - H^{\mathrm{JCIR}}(T_a,u;y^{\beta}_{T_a})\right)^+\right]du$$

where the superscript JCIR underlines the need to replace the CIR formulas with their jump diffusion equivalents.

As already observed in Section 22.7.9, each term in the summation is $h^{\mathrm{JCIR}}(u)$ times a zero-coupon bond like put option with strike $K_u^* = H^{\mathrm{JCIR}}(T_a,u;y^*)$. Without jumps, this put option price is given in (3.78). However, since we introduced jumps in the model, we need a new formula for this put option. Thanks to the relative tractability of the AJD framework, this formula can be obtained explicitly. For the general theory, the reader is referred to Duffie, Pan and Singleton (2000), while important elements of the solution for the specific process considered here have been derived by

Christensen (2002) whose results are summarized in the appendix of Lando (2004). Denoting $\bar{\alpha}_a^u := \bar{\alpha}(T_a, u)$, $\bar{\beta}_a^u := \bar{\beta}(T_a, u)$ and $c^u := \ln(\frac{K_u^*}{\bar{\alpha}_a^u})$:

$$\mathbb{E}\left[e^{-\int_0^{T_a} \lambda_s ds}\left(K_u^* - H^{\text{JCIR}}(T_a, u; y_{T_a}^{\beta})\right)^+\right] =$$

$$= K_u^* \mathbb{E}\left[e^{-\int_0^{T_a} \lambda_s ds} \mathbf{1}_{\{c^u \geq -\bar{\beta}_a^u \lambda_{T_a}\}}\right] + \tag{22.42}$$

$$-\bar{\alpha}_a^u \mathbb{E}\left[e^{-\int_0^{T_a} \lambda_s ds} e^{-\bar{\beta}_a^u \lambda_{T_a}} \mathbf{1}_{\{c^u \geq -\bar{\beta}_a^u \lambda_{T_a}\}}\right] = \ldots$$

which admits the following semi-explicit solution:

$$\ldots = K_u^* \Pi_1(T_a, u, \lambda_0, c^u) - \bar{\alpha}_a^u \Pi_2(T_a, u, \lambda_0, c^u) \tag{22.43}$$

where Π_1 and Π_2 can be obtained from the solution to the "extended transform" of the affine jump diffusion (AJD) model at hand. The "extended transform" refers to the following expectation:

$$\psi(z, T_a, \lambda_0) = \mathbb{E}\left[\exp\left(z\lambda_{T_a} - \int_0^{T_a} \lambda_s ds\right)\right]$$

with $z \in \mathbb{C}$.

Again, a solution can be found in Christensen (2002). The expression obtained is analytic up to an integral. We present below a version that is consistent with our previous notation:

$$\psi(z, T_a, \lambda_0) = \bar{\alpha}_\psi(T_a) e^{-\bar{\beta}_\psi(T_a)\lambda_0} \tag{22.44}$$

where

$$\bar{\beta}_\psi(T_a) = \frac{-2zh + (2 + z(\kappa - h))(e^{hT_a} - 1)}{2h + (h + \kappa - z\nu^2)(e^{hT_a} - 1)} \tag{22.45}$$

$$\bar{\alpha}_\psi(T_a) = \left[\frac{2he^{\frac{(\kappa+h)}{2}T_a}}{2h + (h + \kappa - z\nu^2)(e^{hT_a} - 1)}\right]^{\frac{2\kappa\mu}{\nu^2}}$$

$$\times \left[\frac{2h(1 - z\gamma)e^{\left(\frac{h+\kappa+2\gamma}{2}T_a\right)}}{2h(1 - z\gamma) + [h + \kappa - z\nu^2 + \gamma(2 - z(h - \kappa))](e^{hT_a} - 1)}\right]^{\frac{2\alpha\gamma}{\nu^2 - 2\kappa\gamma - 2\gamma^2}} \tag{22.46}$$

Now, define the following quantity:

$$\Pi(x, T_a, u, \lambda_0, c^u) = \frac{1}{2}\psi(x, T_a, \lambda_0) - \frac{1}{\pi}\int_0^\infty \frac{\text{Im}\left[e^{-ivc^u}\psi(x - iv\bar{\beta}_a^u, T_a, \lambda_0)\right]}{v} dv$$

where i is the imaginary unit of complex numbers. A cumbersome but explicit expression for the imaginary part in the above expression can be found in Christensen (2002) or Lando (2004).

Π_1 and Π_2 can be rewritten as:

$$\Pi_1(T_a, u, \lambda_0, c^u) = \Pi(0, T_a, u, \lambda_0, c^u) \tag{22.47}$$

$$\Pi_2(T_a, u, \lambda_0, c^u) = \Pi(-\bar{\beta}_a^u, T_a, u, \lambda_0, c^u) \tag{22.48}$$

allowing us to write the CDS option price as:

$$\int_{T_a}^{T_b} h^{JCIR}(u) \left[K_u^* \Pi(0, T_a, u, \lambda_0, c^u) du - \bar{\alpha}_u^* \Pi(-\bar{\beta}_a^u, T_a, u, \lambda_0, c^u)\right] du \tag{22.49}$$

Hence, pricing CDS options using Jamshidian decomposition in this AJD model requires numerically computing double integrals. For more details and the use of less computationally intensive methods, the reader can refer to Brigo and El Bachir (2005) for some approximations in the same spirit as the ones used in Glasserman and Merener (2002), and Schrager and Pelsser (2004) for swaption pricing.

22.8.6 Attaining high levels of CDS implied volatility

Finally, we check with at least one example that the JCIR model can attain high implied volatilities. Consider then the JCIR process (22.37) without shift and set the diffusion-part intensity parameters to

$$\kappa = 0.35, \quad \mu = 0.045, \quad \nu = 0.15, \quad \lambda_0 = 0.035,$$

with deterministic interest rate curve. Consider then different possibilities for the values of the jump parameters α and γ. Call again $\sigma_{a,b}^{imp}$ the implied volatility associated to the JCIR model. In the JCIR model, one may compute the CDS option price by means of the methods we just described, or through Monte Carlo simulation. Once the price has been computed, we equate the Monte Carlo price to the market model Formula (23.13) we are going to derive in the next chapter, applied to the same CDS option at $t = 0$, and solve in $\sigma_{a,b}$. This $\sigma_{a,b}$ is then the implied volatility coming from the JCIR price. Then, for the chosen diffusion dynamics parameters and for different values of α and γ we check both the implied volatility $\sigma_{a,b}$ and the CDS rate $R_{a,b}(0)$. We need to check the CDS rate because we are not using the shift ψ^J. The model is not reproducing the market CDS rates, so that we check that the spreads produced by the model are not unrealistic. We take the option maturity $T_a = 1y$ and the underlying CDS tenor $T_b = 5y$.

- $\alpha = 0, \gamma = 0$ (no jumps): We obtain $\sigma_{a,b}^{imp} = 0.28$;
  $R_{0,1y}(0) = 0.02587, \quad R_{0,3y} = 0.0271, \quad R_{0,5y} = 0.0277, \quad R_{0,10y} = 0.0284$;
- $\alpha = 0.1, \gamma = 0.1$: $\sigma_{a,b}^{imp} = 0.4$;
  $R_{0,1y} = 0.02886, \quad R_{0,3y} = 0.0337, \quad R_{0,5y} = 0.0364, \quad R_{0,10y} = 0.0393$;
- $\alpha = 0.15, \gamma = 0.15$: $\sigma_{a,b}^{imp} = 0.57$;
  $R_{0,1y} = .03254, \quad R_{0,3y} = 0.0412, \quad R_{0,5y} = 0.0458, \quad R_{0,10y} = 0.0507$;

As we see, in principle high implied volatilities are possible. In the last case we attain a 57% implied volatility, even though the underlying 1y CDS rate is about 325 basis points (0.03254). The problem the model may have in its time homogeneous formulation is "decoupling" high implied volatilities from high CDS rates. Can we get high implied volatilities with relatively low underlying CDS rates? The shift cannot help much there, since typically it becomes negative in cases with low market CDS rates an inputs. This problem is currently under investigation.

22.8.7 JCIR(++) models as a multi-name possibility

Finally, as regards multi-name situations, for a more detailed discussion on the simulation of similar jump diffusion processes in a general multidimensional context (using Levy Copulas first introduced by Tankov (2003) to model the dependence between the jump components), the reader can refer to El Bachir (2005). Here we comment briefly on the possible benefits of the jump diffusion setting in multi-name situations.

Recall that, in a multi-name situation with names $1, 2, \dots, n$ that may default, in an intensity setting the default time of names $1, 2, \dots, n$ can be expressed as

$$\tau_1 = \Lambda_1^{-1}(\xi_1), \dots, \tau_n = \Lambda_n^{-1}(\xi_n).$$

We need to induce dependence (or "correlation") between these default times of different names.

With choice 1) in Section 21.1.8, one may induce dependence among the $\lambda_i(t)$ by taking diffusion dynamics for each of them and correlating the Brownian motions:

$$d\lambda_i(t) = \mu_i(t, \lambda_i(t))dt + \sigma_i(t, \lambda_i(t))dW_i(t),$$

$$d\lambda_j(t) = \mu_j(t, \lambda_j(t))dt + \sigma_j(t, \lambda_j(t))dW_j(t),$$

$$dW_i dW_j = \rho_{i,j} dt, \quad \xi_i, \xi_j \text{ independent.}$$

As explained earlier, the advantages with this choice are possible tractability and ease of implementation. Also, the default of one name does not affect the intensity of others. The intensity correlation across names can be estimated historically from time series of credit spreads and inserted into the model. Also, with stochastic intensity we may model correlation between interest rates and credit spreads, that is considered to be an important feature in some situations.

The disadvantages consist in a non realistic (too low) level of dependence across default events $1_{\{\tau_i < T\}}, 1_{\{\tau_j < T\}}$. See for example Jouanin et al. (2001).

But now, suppose we replace the above intensity processes for different names with jump-diffusion versions:

$$d\lambda_i(t) = \mu_i(t, \lambda_i(t))dt + \sigma_i(t, \lambda_i(t))dW_i(t) + dJ_i(t),$$

$$d\lambda_j(t) = \mu_j(t, \lambda_j(t))dt + \sigma_j(t, \lambda_j(t))dW_j(t) + dJ_j(t),$$

$$\xi_i, \xi_j \text{ independent,}$$

where now we correlate the jumps J in the diffusion jump processes. The fact that we keep the ξ's still independent makes life easier, tractability and single name calibration are retained, but now we need to check whether the correlation induced across indicators $1_{\{\tau_i<T\}}, 1_{\{\tau_j<T\}}$ is high enough, and in particular higher than in the pure diffusion setting above. We will investigate this in further work. This could lead to a realistic "dynamical dependence (correlation) model", and could be an attempt at solving the problem pointed out at the end of Section 21.1.9. Some preliminary tests of ours (Pallavicini (2006)) suggest that one may need to adopt very large intensity jump sizes (with order of magnitude $\gamma = 0.5$ or even 1) and very low frequencies for the intensity jumps themselves (with order of magnitude $\alpha = 0.01$), with perfect correlation in the jump processes J (all the intensities have the same driving jump process $J^{\alpha,\gamma}$). This may allow one to attain a high enough correlation between default indicators, for practical purposes. The cost seems to be an extreme behaviour for the intensities, they jump very rarely but when they do jump they do so all together and by a huge amount. This forces a large enough correlation in the default indicators even with independent ξ's, but looks like a possibly unrealistic feature for the intensity dynamics. More investigations are in order before deciding whether to adopt or reject this framework.

22.9 Conclusions and further research

J'onn: "I fear the Justice League's greatest challenge lies just ahead..."
Kal: "Doesn't it always, J'onn?"

"Death Star", DC One Million 4, 1998, DC Comics

In this Chapter we have presented a partial view across the field of intensity models. This part of the book (together with the previous and subsequent chapters) has been written originally with the purpose of illustrating the possible use of previously developed interest rate modeling tools in the credit derivatives area. This, however, brought a life of their own to credit models and forced us to think of extensions (like the Jump-CIR++ model) that we never considered for the interest rate market.

Of course there are many more intensity and more generally reduced form models beyond our chosen CIR++, SSRD, JCIR(++) specifications. Since this is a book mostly devoted to interest rate models, we do not list further models here. We only hint at the fact that the reader may try and adapt positive interest rate models seen earlier in the book to the stochastic intensity setting.

Similarly to the evolution leading from short-rate models to the HJM framework, one could model instantaneous forward rates associated with the stochastic intensity and adopt a HJM approach to intensity models. This is what is illustrated at the beginning of Chapter 5 for interest rate models. A similar transition is possible with intensity models. However, we will not pursue this approach here, since the positivity constraint greatly limits the useful models coming out of an explicit HJM approach. As with short rate models, much of the effort is saved and little advantages are lost by modeling directly the instantaneous intensity λ as we did in this chapter.

Finally, as we have seen in the guided tour at beginning of Chapter 6, a more important transition occurred when moving to market models. In several ways, the decision to model directly market "observable" rates having discrete tenors has turned out to be fundamental. The modelled rates are rates that render financial contracts such as forward rate agreements and interest rate swaps fair. These models turn out to be consistent with market quoting mechanisms of volatilities by construction, and embed correlation and volatility information in quite a natural way.

In the next chapter we try a similar transition for intensity models: we drop specific, explicit modeling of the stochastic intensity and resort to models for rates that make credit default swap contracts fair. However, the dynamics for a "CDS market model" poses some subtleties on the pricing measures; furthermore, we may have to model two families of rates rather than one. We attack these problems in the next chapter.

23. CDS Options Market Models

"I have had my results for a long time, but I do not yet know how I am to arrive at them".

Gauss, quoted in A. Arber, "The Mind and the Eye", 1954.

"If I only had the theorems!
Then I should find the proofs easily enough". Riemann

In this final chapter devoted to credit we close an ideal path going through the following steps:

- Illustrate the most representative market payoffs that motivate the adoption of stochastic dynamical models;
- Start with models for theoretical and not-directly-observable market variables that are at the core of the financial phenomenon influencing the market under examination; in the interest rate market the core variable is the short rate r. The variable r is indeed at the core of the basic engine for discounting:

$$D(t,T) = \exp\left(-\int_t^T \boxed{r_s} ds\right).$$

For default models, we consider the default intensity λ whose interpretation as local probability of default per unit of time helps intuition:

$$\mathbb{Q}(\tau \in [t, t+dt) | \tau > t, \text{market info up to } t) = \lambda(t)dt.$$

The variable λ is the not-directly-observable theoretical variable at the core of the default model thanks to its role in defining the default time: The time-integral Λ of λ characterizes the default time τ through

$$\tau = \boxed{\Lambda}^{-1}(\text{exponentialRandomVariable}), \quad \Lambda(T) = \int_0^T \boxed{\lambda_s}\, ds.$$

- For the interest rate part we have seen a long path motivating the abandoning of r for the direct modeling of observable market (forward LIBOR) F_i or (forward swap) $S_{\alpha,\beta}$ rates.

A similar path for credit derivatives has just started, with the CDS forward rates $R_{a,b}$ as underlying variables, as explained for example in Brigo (2004, 2005).

$$\begin{array}{llcl} \text{Fixed Income:} & \not{r} & \longrightarrow & F_i \text{ or } S_{\alpha,\beta} \\ & \updownarrow & & \updownarrow \\ \text{Credit:} & \not{\lambda} & \longrightarrow & R_i(?) \text{ or } R_{a,b} \end{array} \tag{23.1}$$

This will be the subject of this chapter.
A good idea at this point is also re-reading Section 21.1.2 illustrating the parallels with interest rate modeling evolution.

- Probably, this "modern" path to credit modeling is not going to be as strong as the earlier analogous path in the default free interest rate market. Indeed, in the interest rate market through drift interpolation and other techniques we could still find a complete characterization of the market curve in terms of our LIBOR model. In the credit world this is not possible given the fact that the default time comes as

$$\tau = \Lambda^{-1}(\text{exponentialRandomVariable}).$$

Even after replacing the intensity λ with observable CDS forward rates R's, we would still need λ and an exogenous exponential random variable independent of everything else to obtain the default time. Sure, we might deduce some kind of intensity λ from the dynamics of the discrete tenor $R_{a,b}$'s of CDS, but this would be complicated; notice that indeed in the interest rate world we never derived the r dynamics that is consistent with LIBOR (or a family of swap) models (plus some interpolated rates). We never went from the F_i's (or $S_{\alpha,\beta}$'s) back to r. Doing so involves choices on the interpolation techniques and other aspects that we need to control in order to maintain the model arbitrage free (and the "short rate" positive in case of intensity models). If on one hand the discrete-tenor credit model for observable CDS forward rates $R_{a,b}$'s can be a good tool for CDS options and other payoffs (such as constant maturity CDS for example), on the other hand it is difficult to see this as a total substitute for explicit intensity λ models. For the same reasons one may doubt that this model will attain the same level of success as the LIBOR model in replacing short rate r models in the whole single-name credit derivatives market. In other terms, the lower horizontal arrow in (23.1) is probably not going to be as important as the upper one.
- Nonetheless, it may be useful to have a market Black-like formula for CDS options, since this would allow us at least to quote options as implied volatilities, a quoting mechanism traders are familiar with. And if we can derive this formula in a fully rigorous arbitrage free context, then all the better for us. This is the task we address in this chapter.

- There is more: not only will we give an arbitrage free derivation of the CDS option market formula, based on a direct dynamics for the relevant underlying CDS rate $R_{a,b}$, but we will also consider the joint dynamics of a family of CDS forward rates R_i under a single pricing measure. From this family all other CDS rates can be built. This is similar to modeling a family of forward LIBOR rates F_i from which all other (swap) rates $S_{\alpha,\beta}$ can be built.
- This market model for the family of CDS rates will be helpful in deriving the price of other products such as constant maturity CDS.

The chapter is ideally linked with sections of earlier chapters. In particular, we recall that in Sections from 21.3.1 to 21.3.6 we introduced notation, different kinds of CDS discounted payoffs, and the main definition of CDS forward rate. The notion of CDS implied hazard function and its possible use as quoting mechanism were recalled. Upfront CDS's were hinted at.

Section 21.3.7 examined some possible variant definitions of CDS rates. Furthermore, we examined the relationship between CDS rates on different periods and pointed out some parallels with the default free LIBOR and swap market rates.

Section 21.3.8 introduced defaultable floating rate notes and explored their relationship with CDS payoffs, finding equivalence under some payment schedules.

In this chapter we start with Section 23.1. This section describes the payoffs and structural analogies between CDS options and callable defaultable floating rate notes (DFRN).

Section 23.2 introduces the market formula for CDS options and callable DFRN, based on a rigorous change of numeraire technique. Numerical examples of implied volatilities from CDS option quotes are given, and are found to be rather high, in agreement with previous studies dealing with historical CDS rate volatilities.

Section 23.3 discusses possible developments towards a complete specifications of the vector dynamics of CDS forward rates under a single pricing measure, based on one-period CDS rates.

Section 23.4 gives some brief hints on modeling of the volatility smile for CDS options, based on the general framework introduced earlier.

Section 23.5 illustrates how to use the market model to derive an approximated formula for CMCDS that is based on a sort of convexity adjustment and bears resemblance to the formula for valuing constant maturity swaps with the LIBOR model, as seen at the end of Section 13.7.1. The adjustment is illustrated with several numerical examples. We also present some stylized facts and market sentiment on the CMCDS products.

23.1 CDS Options and Callable Defaultable Floaters

We have already introduced CDS options in Section 21.4. We recall the notation and the basic payoffs here.

Consider the option for a protection buyer to enter a CDS at a future time $T_a > 0$, $T_a < T_b$, paying a fixed premium rate K at times $T_{a+1}, \ldots, T_b$ or until default, in exchange for a protection payment L_{GD} against possible default in $[T_a, T_b]$ (payer CDS option). If the option is exercised and default occurs in $[T_a, T_b]$ then L_{GD} is received by the protection buyer. By noticing that the market CDS rate $R_{a,b}(T_a)$ will set the underlying CDS value in T_a to 0, the payoff can be written as the discounted difference between said CDS and the corresponding CDS with CDS premium rate K. We will see below that this is equivalent to a call option on the future CDS fair rate $R_{a,b}(T_a)$. The discounted CDS option payoff reads, at time t,

$$\begin{aligned}\Pi_{\text{CallCDS}_{a,b}}(t;K) &= D(t,T_a)\ \left[-\text{CDS}_{a,b}(T_a,K,\text{L}_{\text{GD}})\right]^+ \\ &= D(t,T_a)[\underbrace{\text{CDS}_{a,b}(T_a,R_{a,b}(T_a),\text{L}_{\text{GD}})}_{0} - \text{CDS}_{a,b}(T_a,K,\text{L}_{\text{GD}})]^+,\end{aligned}$$

so that, by writing the CDS expressions explicitly (as from (21.20) with $t = T_a$) we have

$$\begin{aligned}\Pi_{\text{CallCDS}_{a,b}}(t;K) = \frac{1_{\{\tau>T_a\}}}{\mathbb{Q}(\tau > T_a|\mathcal{F}_{T_a})} D(t,T_a)\Bigg[\sum_{i=a+1}^{b} \alpha_i \mathbb{Q}(\tau > T_a|\mathcal{F}_{T_a})\bar{P}(T_a,T_i) + \\ + \mathbb{E}\left\{D(T_a,\tau)(\tau - T_{\beta(\tau)-1})\mathbf{1}_{\{\tau<T_b\}}|\mathcal{F}_{T_a}\right\}\Bigg](R_{a,b}(T_a) - K)^+. \quad (23.2)\end{aligned}$$

We have an expression that looks like a numeraire times a call option on the CDS rate in T_a. These options can be introduced for every type of underlying CDS. We have illustrated the standard CDS case, and we will consider the postponed CDS cases below.

The quantity inside square brackets in (23.2) will play a key role in the following. We will often neglect the accrued interest term in $(\tau - T_{\beta(\tau)-1})$ and consider the related simplified payoff: in such a case the quantity between square brackets is denoted by $\widehat{C}_{a,b}(T_a)$ and is called ("no survival-indicator"-) "defaultable present value per basis point (DPVBP) numeraire" (and sometimes "defaultable annuity"). Actually the real DPVBP or defaultable annuity would have a default indicator $1_{\{\tau> \cdot\}}$ term in front of the summation, but by a slight abuse of language we call DPVBP the expression without indicator. More generally, at time t, we set

$$\widehat{C}_{a,b}(t) := \mathbb{Q}(\tau > t|\mathcal{F}_t)\bar{C}_{a,b}(t), \quad \bar{C}_{a,b}(t) := \sum_{i=a+1}^{b} \alpha_i \bar{P}(t,T_i).$$

When including as a factor the indicator $1_{\{\tau>t\}}$, this quantity is the price, at time t, of a portfolio of defaultable zero-coupon bonds with zero recovery and with different maturities, and as such it is the price of a tradable asset. The original work of Schönbucher (2000, 2004) with the so called "survival measure" is in this spirit, in that the "numeraire" is taken with the indicator, so that it may vanish and the measure it defines is not equivalent to the risk neutral measure. If we keep the indicator away, following in spirit part of the work in Jamshidian (2004), Brigo (2004, 2005) and Brigo and Morini (2005), this quantity maintains a link with said price and is always strictly positive, so that we are allowed to take it as numeraire. In the book we follow this path rather than the survival measure approach. This path maintains the theory more similar to the default free interest rate theory and avoids situations with non-equivalent pricing measures. The related probability measure, equivalent to the risk neutral measure, is denoted by $\widehat{\mathbb{Q}}^{a,b}$ and the related expectation by $\widehat{\mathbb{E}}^{a,b}$. Incidentally, we notice that our assumptions on recovery rates are rather simple. A more advanced approach, in this respect, is pursued in Wu (2006).

By neglecting the accrued interest term, the option discounted payoff simplifies to

$$1_{\{\tau>T_a\}} D(t,T_a) \left[\sum_{i=a+1}^{b} \alpha_i \bar{P}(T_a,T_i) \right] (R_{a,b}(T_a) - K)^+ \tag{23.3}$$

but this is only an approximated payoff and not the exact one.

Let us follow the same derivation under the postponed CDS payoff of the first kind. Consider thus

$$\begin{aligned} \Pi_{\text{CallPRCDS}_{a,b}}(t;K) &= D(t,T_a)[\overbrace{\text{PRCDS}_{a,b}(T_a, R^{PR}_{a,b}(T_a), \text{L}_{\text{GD}})}^{0} + \\ &\qquad -\text{PRCDS}_{a,b}(T_a, K, \text{L}_{\text{GD}}) \quad]^+ \\ &= D(t,T_a) \quad [-\text{PRCDS}_{a,b}(T_a, K, \text{L}_{\text{GD}})]^+ \quad . \end{aligned} \tag{23.4}$$

Incidentally, we notice that by (21.31) this is equivalent to

$$D(t,T_a)[-\text{DFRN1}_{a,b}(T_a, K, \text{R}_{\text{EC}})]^+, \tag{23.5}$$

with $\text{L}_{\text{GD}} = 1 - \text{R}_{\text{EC}}$, or to

$$D(t,T_a)[\underbrace{\text{DFRN1}_{a,b}(T_a, X_{a,b}(T_a), \text{R}_{\text{EC}})}_{0} - \text{DFRN1}_{a,b}(T_a, K, \text{R}_{\text{EC}})]^+.$$

By expanding the expression of PRCDS we obtain as *exact* discounted payoff the quantity $\Pi_{\text{CallPRCDS}_{a,b}}(t,K) =$

$$= 1_{\{\tau > T_a\}} D(t, T_a) \left[\sum_{i=a+1}^{b} \alpha_i \bar{P}(T_a, T_i) \right] (R_{a,b}^{PR}(T_a) - K)^+, \tag{23.6}$$

which is structurally identical to the approximated *payoff (23.3) for the standard CDS case.* En passant we have also created a link between the first postponed CDS option payoff and an option on an approximated DFRN. Notice in particular that the quantity in front of the optional part is the same as in the earlier standard approximated discounted payoff, i.e. the DPVBP.

We may also consider the postponed running CDS of the second kind. The related discounted CDS option payoff reads, at time t,

$$\Pi_{\text{CallPR2CDS}_{a,b}}(t, K) = D(t, T_a)[\text{PR2CDS}_{a,b}(T_a, R_{a,b}^{PR2}(T_a), \text{L}_{\text{GD}}) - \text{PR2CDS}_{a,b}(T_a, K, \text{L}_{\text{GD}})]^+, \tag{23.7}$$

and incidentally, given (21.30), this is equivalent to

$$D(t, T_a)[\text{DFRN2}_{a,b}(T_a, X_{a,b}(T_a), \text{R}_{\text{EC}}) - \text{DFRN2}_{a,b}(T_a, K, \text{R}_{\text{EC}})]^+, \tag{23.8}$$

with $\text{R}_{\text{EC}} = 1 - \text{L}_{\text{GD}}$, or, by expanding the expression for PR2CDS, as

$$\frac{1_{\{\tau > T_a\}} D(t, T_a)}{\mathbb{Q}(\tau > T_a | \mathcal{F}_{T_a})} \sum_{i=a+1}^{b} \alpha_i \mathbb{E}[D(T_a, T_i) 1_{\{\tau > T_{i-1}\}} | \mathcal{F}_{T_a}] (R_{a,b}^{PR2}(T_a) - K)^+ \tag{23.9}$$

Again we have equivalence between CDS options and options on the defaultable floater.

23.1.1 Once-callable defaultable floaters

The option on the floater can be seen as the optional component of a callable DFRN. A DFRN with final maturity T_b is issued at time 0 with a fair rate $X_{0,b}(0)$ in such a way that $\text{DFRN}_{0,b}(0, X_{0,b}(0), \text{R}_{\text{EC}}) = 0$. Suppose that this FRN includes a callability feature: at time T_a the issuer has the right to take back the subsequent FRN flows and replace them with the notional 1. The issuer will do so only if the present value in T_a of the subsequent FRN flows is larger than 1 in T_a. This is equivalent, for the note holder, to receive $1_{\{\tau > T_a\}} + \min(\text{DFRN}_{a,b}(T_a, X_{0,b}(0), \text{R}_{\text{EC}}), 0) = 1_{\{\tau > T_a\}} + \text{DFRN}_{a,b}(T_a, X_{0,b}(0), \text{R}_{\text{EC}}) - (\text{DFRN}_{a,b}(T_a, X_{0,b}(0), \text{R}_{\text{EC}}))^+$ at time T_a if no default has occurred by then (recall that in our notation $\text{DFRN}_{a,b}$ includes a negative cash flow of 1 at time T_a).

The component $1_{\{\tau > T_a\}} + \text{DFRN}_{a,b}(T_a, X_{0,b}(0), \text{R}_{\text{EC}})$ when valued at time 0 is simply the residual part of the original DFRN without callability features from T_a on, so that when added to the previous payments in $[0\ T_a]$ its present value is 0. This happens because $X_{0,b}(0)$ is the fair rate for the total DFRN at time 0. The component $(\text{DFRN}_{a,b}(T_a, X_{0,b}(0), \text{R}_{\text{EC}}))^+ = (\text{DFRN}_{a,b}(T_a, X_{0,b}(0), \text{R}_{\text{EC}}) - \text{DFRN}_{a,b}(T_a, X_{a,b}(T_a), \text{R}_{\text{EC}}))^+$ is structurally equivalent to a CDS option, provided we approximate its payoff with DFRN1 or DFRN2, as we have seen earlier, and we may value it if we have a model for CDS options. We are deriving a market formula for such options in the next section, so that we will be implicitly deriving an approximate formula for once-callable defaultable floaters.

23.2 A market formula for CDS options and callable defaultable floaters

As in more traditional option markets, one may wish to introduce a notion of implied volatility for CDS options. This would be a volatility associated to the relevant underlying CDS rate R. In order to do so rigorously, one has to come up with an appropriate dynamics for $R_{a,b}$ directly, rather than modeling instantaneous default intensities explicitly. This somehow parallels what we find in the default-free interest rate market when we resort to the swap market model for $S_{\alpha,\beta}$ as opposed for example to a one-factor short-rate model for r for pricing swaptions. In a one-factor short-rate model the dynamics of the forward swap rate is a byproduct of the short-rate dynamics itself, through Ito's formula: Indeed, $S_{\alpha,\beta}(t)$ can be computed in terms of $P(t,T_i)$'s, and $P(t,T_i) = P(t,T_i;r_t)$ is a function of r_t in one-factor short rate models, so that also $S_{\alpha,\beta}(t) = S_{\alpha,\beta}(t,r_t)$ is a function of r_t, and its dynamics can be obtained through Ito's formula given the dynamics of r. On the contrary, the market model for swaptions directly postulates, under the relevant numeraire a (lognormal) dynamics for the forward swap rate $S_{\alpha,\beta}(t)$, with no explicit reference to r.

The framework we develop in this chapter is similar, with λ replacing r and $R_{a,b}$ replacing $S_{\alpha,\beta}$.

$$\begin{array}{lccc} \text{Fixed Income:} & \text{Modeling } dS_{\alpha,\beta} & \text{without} & dr \\ & \updownarrow & & \updownarrow \\ \text{Credit:} & \text{Modeling } dR_{a,b} & \text{without} & d\lambda \end{array}$$

23.2.1 Market formulas for CDS Options

In the case of CDS options the market model is derived as follows. First, let us ignore the accruing term in $(\tau - T_{\beta(\tau)-1})$, by replacing it with zero. It can be seen that typically the order of magnitude of this term is negligible with respect to the remaining terms in the payoff. Failing this negligibility, one may reformulate the payoff by postponing the default payment to the first date among the T_i's following τ, i.e. to $T_{\beta(\tau)}$, as we shall see in Section 23.2.2 below. This amounts to considering as underlying a payoff corresponding to (21.13) and eliminates the accruing term altogether, even though it slightly modifies the option payoff.

In any case, take as numeraire the DPVBP or annuity $\widehat{C}_{a,b}$, so that

$$R_{a,b}(t) = \frac{\text{LGD}\ \mathbb{E}[D(t,\tau)1_{\{T_a<\tau\leq T_b\}}|\mathcal{F}_t]}{\sum_{i=a+1}^{b}\alpha_i \mathbb{Q}(\tau>t|\mathcal{F}_t)\bar{P}(t,T_i)} = \frac{\text{LGD}\ \mathbb{E}[D(t,\tau)1_{\{T_a<\tau\leq T_b\}}|\mathcal{F}_t]}{\widehat{C}_{a,b}(t)}, \tag{23.10}$$

$t \leq T_a$. $R_{a,b}$ has as numerator the price of an upfront CDS, and can be interpreted as the ratio between a tradable asset and our numeraire. As such, it is a martingale under this numeraire's measure and can be modeled as a Black-Scholes driftless geometric Brownian motion, leading to a Black and Scholes formula for CDS options. As for the initial condition, notice that, when $R_{a,b}(0)$ is not quoted directly by the market, we may infer it for example by the market implied γ^{mkt} according to

$$R_{a,b}(0) = \frac{-\text{LGD}\int_{T_a}^{T_b} P(0,u)d(e^{-\Gamma^{\text{mkt}}(u)})}{\sum_{i=a+1}^{b}\alpha_i P(0,T_i)e^{-\Gamma^{\text{mkt}}(T_i)}} .$$

This formula presumes independence between the default time and interest rates, and as we have seen in Section 22.7.6 this is not a problem for CDS and their forward rates, since said correlation has a low impact on them. We may occasionally maintain this assumption then in computing $R_{a,b}(0)$ but we drop it in general when pricing the option, where correlation has a more relevant impact.

By resorting to the change of numeraire starting from (23.3) (thus ignoring the accruing term or working with the postponed payoff), and by using the filtration switching formula, we see that

$$\mathbb{E}\{1_{\{\tau>T_a\}}D(t,T_a)\sum_{i=a+1}^{b}\alpha_i\bar{P}(T_a,T_i)(R_{a,b}(T_a)-K)^+|\mathcal{G}_t\} \tag{23.11}$$

$$= \frac{1_{\{\tau>t\}}}{\mathbb{Q}(\tau>t|\mathcal{F}_t)}\mathbb{E}\{1_{\{\tau>T_a\}}D(t,T_a)\sum_{i=a+1}^{b}\alpha_i\bar{P}(T_a,T_i)(R_{a,b}(T_a)-K)^+|\mathcal{F}_t\}$$

$$= \frac{1_{\{\tau>t\}}}{\mathbb{Q}(\tau>t|\mathcal{F}_t)}\mathbb{E}\big[\mathbb{E}\{1_{\{\tau>T_a\}}D(t,T_a)\sum_{i=a+1}^{b}\alpha_i\bar{P}(T_a,T_i)(R_{a,b}(T_a)-K)^+|\mathcal{F}_{T_a}\}|\mathcal{F}_t\big]$$

$$= \frac{1_{\{\tau>t\}}}{\mathbb{Q}(\tau > t|\mathcal{F}_t)}\mathbb{E}\big[D(t,T_a)\sum_{i=a+1}^{b}\alpha_i\bar{P}(T_a,T_i)(R_{a,b}(T_a)-K)^+\mathbb{E}\{1_{\{\tau>T_a\}}|\mathcal{F}_{T_a}\}|\mathcal{F}_t\big]$$

$$= \frac{1_{\{\tau>t\}}}{\mathbb{Q}(\tau > t|\mathcal{F}_t)}\mathbb{E}\big[D(t,T_a)\sum_{i=a+1}^{b}\mathbb{Q}(\tau > T_a|\mathcal{F}_{T_a})\alpha_i\bar{P}(T_a,T_i)(R_{a,b}(T_a)-K)^+|\mathcal{F}_t\big]$$

$$= \frac{1_{\{\tau>t\}}}{\mathbb{Q}(\tau > t|\mathcal{F}_t)}\mathbb{E}\big[D(t,T_a)\widehat{C}_{a,b}(T_a)(R_{a,b}(T_a)-K)^+|\mathcal{F}_t\big]$$

$$= \frac{1_{\{\tau>t\}}}{\mathbb{Q}(\tau > t|\mathcal{F}_t)}\mathbb{E}^{\boxed{B}}\left[\frac{\boxed{B(t)}}{\boxed{B(T_a)}}\widehat{C}_{a,b}(T_a)(R_{a,b}(T_a)-K)^+|\mathcal{F}_t\right]$$

$$= \frac{1_{\{\tau>t\}}}{\mathbb{Q}(\tau > t|\mathcal{F}_t)}\widehat{\mathbb{E}}^{\boxed{a,b}}\left[\frac{\boxed{\widehat{C}_{a,b}(t)}}{\boxed{\widehat{C}_{a,b}(T_a)}}\widehat{C}_{a,b}(T_a)(R_{a,b}(T_a)-K)^+|\mathcal{F}_t\right]$$

$$= \frac{1_{\{\tau>t\}}}{\mathbb{Q}(\tau > t|\mathcal{F}_t)}\widehat{C}_{a,b}(t)\widehat{\mathbb{E}}^{a,b}[(R_{a,b}(T_a)-K)^+|\mathcal{F}_t]$$

$$= 1_{\{\tau>t\}}\bar{C}_{a,b}(t)\widehat{\mathbb{E}}^{a,b}[(R_{a,b}(T_a)-K)^+|\mathcal{F}_t]$$

(the use of boxes refers to the change of numeraire invariance/ "Fact Two" property seen in Section 2.3 and used many times in the book) and we may take

$$dR_{a,b}(t) = \sigma_{a,b}R_{a,b}(t)dW^{a,b}(t), \tag{23.12}$$

where $W^{a,b}$ is a Brownian motion under $\widehat{\mathbb{Q}}^{a,b}$, leading to a market formula for the CDS option. We have ("call" refers to the call structure in R, the option is a "payer CDS option")

$$\begin{aligned} \text{CallCDS}_{a,b}(t,K,\text{L}_{\text{GD}}) &= \mathbb{E}[\Pi_{\text{CallCDS}_{a,b}}(t,K;\text{L}_{\text{GD}})|\mathcal{G}_t] \qquad (23.13)\\ &= \mathbb{E}\{1_{\{\tau>T_a\}}D(t,T_a)\bar{C}_{a,b}(T_a)(R_{a,b}(T_a)-K)^+|\mathcal{G}_t\} = \\ &= 1_{\{\tau>t\}}\bar{C}_{a,b}(t)[R_{a,b}(t)N(d_1(t)) - KN(d_2(t))] \end{aligned}$$

$$d_{1,2} = \left(\ln(R_{a,b}(t)/K) \pm (T_a - t)\sigma^2_{a,b}/2\right)/(\sigma_{a,b}\sqrt{T_a-t}).$$

As happens in most markets, this formula could be used as a quoting mechanism rather than as a real model formula. That is, the market price can be converted into its implied volatility matching the given price when substituted in the above formula.

23.2.2 Market Formula for callable DFRN

Since we are also interested in the parallel with DFRN's, let us derive the analogous market model formula under running CDS's postponed payoffs of the first and second kind.

Let us begin with the postponed CDS payoff of the first kind, PRCDS. The derivation goes trough as above and we obtain easily the same model as in (23.12) and (23.13) with R^{PR} replacing R everywhere. This time the formula is exact since we have discarded no accruing term.

If we consider the second kind of approximation for FRN's, the option price is obtained as the price of a CDS option, where the CDS is a postponed CDS of the second kind. Compute then

$$\mathbb{E}\Big\{D(t,T_a)[\mathrm{PR2CDS}_{a,b}(T_a, R^{PR2}_{a,b}(T_a), \mathrm{L}_{\mathrm{GD}})$$

$$-\mathrm{PR2CDS}_{a,b}(T_a, K, \mathrm{L}_{\mathrm{GD}})]^+|\mathcal{G}_t\Big\}$$

$$= \mathbb{E}\Big\{\frac{1_{\{\tau>T_a\}}}{\mathbb{Q}(\tau > T_a|\mathcal{F}_{T_a})}D(t,T_a)\sum_{i=a+1}^{b}\alpha_i\mathbb{E}_{T_a}[D(T_a,T_i)1_{\{\tau>T_{i-1}\}}]\cdot$$

$$\cdot(R^{PR2}_{a,b}(T_a) - K)^+|\mathcal{G}_t\Big\}$$

$$= \mathbb{E}\Big\{1_{\{\tau>T_a\}}D(t,T_a)\sum_{i=a+1}^{b}\alpha_i\frac{\mathbb{E}_{T_a}[D(T_a,T_i)1_{\{\tau>T_{i-1}\}}]}{\mathbb{Q}(\tau > T_a|\mathcal{F}_{T_a})}\cdot$$

$$\cdot(R^{PR2}_{a,b}(T_a) - K)^+|\mathcal{G}_t\Big\} = \dots$$

This time let us take as numeraire

$$\check{C}_{a,b}(t) := \sum_{i=a+1}^{b}\alpha_i\mathbb{E}[D(t,T_i)1_{\{\tau>T_{i-1}\}}|\mathcal{F}_t],$$

$$(\text{notice}\quad \widehat{C}_{a,b}(t) = \sum_{i=a+1}^{b}\alpha_i\mathbb{E}[D(t,T_i)1_{\{\tau>T_i\}}|\mathcal{F}_t]).$$

This quantity is positive, and when including the indicator $1_{\{\tau>t\}}$ this is, not surprisingly, a multiple of the premium leg of a PR2CDS at time t. We may also view it as a ("no survival-indicator"-) portfolio of defaultable bonds where the default maturity is one-period-displaced with respect to the payment maturity. Thus this quantity is only approximately a numeraire. To shorten notation, denote $\mathbb{E}[\cdot|\mathcal{F}_t]$ by $\mathbb{E}_t$. Compute, using the filtration switching formula, iterated conditioning and the change of numeraire:

$$\dots = \mathbb{E}\left\{1_{\{\tau > T_a\}} D(t,T_a) \frac{\check{C}_{a,b}(T_a)}{\mathbb{Q}(\tau > T_a | \mathcal{F}_{T_a})} (R^{PR2}_{a,b}(T_a) - K)^+ | \mathcal{G}_t \right\}$$

$$= \frac{1_{\{\tau > t\}}}{\mathbb{Q}(\tau > t|\mathcal{F}_t)} \mathbb{E}_t \left\{1_{\{\tau > T_a\}} D(t,T_a) \frac{\check{C}_{a,b}(T_a)}{\mathbb{Q}(\tau > T_a | \mathcal{F}_{T_a})} (R^{PR2}_{a,b}(T_a) - K)^+ \right\}$$

$$= \frac{1_{\{\tau > t\}}}{\mathbb{Q}(\tau > t|\mathcal{F}_t)} \mathbb{E}_t \left\{\mathbb{E}_{T_a}\left[1_{\{\tau > T_a\}} D(t,T_a) \frac{\check{C}_{a,b}(T_a)}{\mathbb{Q}(\tau > T_a | \mathcal{F}_{T_a})} (R^{PR2}_{a,b}(T_a) - K)^+ \right]\right\}$$

$$= \frac{1_{\{\tau > t\}}}{\mathbb{Q}(\tau > t|\mathcal{F}_t)} \mathbb{E}_t \left\{D(t,T_a) \frac{\check{C}_{a,b}(T_a)}{\mathbb{Q}(\tau > T_a | \mathcal{F}_{T_a})} (R^{PR2}_{a,b}(T_a) - K)^+ \mathbb{E}_{T_a}\left[1_{\{\tau > T_a\}}\right]\right\}$$

$$= \frac{1_{\{\tau > t\}}}{\mathbb{Q}(\tau > t|\mathcal{F}_t)} \mathbb{E}_t \left\{D(t,T_a) \check{C}_{a,b}(T_a) (R^{PR2}_{a,b}(T_a) - K)^+ \right\}$$

$$= \frac{1_{\{\tau > t\}}}{\mathbb{Q}(\tau > t|\mathcal{F}_t)} \mathbb{E}_t^{\boxed{B}} \left\{\frac{\boxed{B(t)}}{\boxed{B(T_a)}} \check{C}_{a,b}(T_a) (R^{PR2}_{a,b}(T_a) - K)^+ \right\}$$

$$= \frac{1_{\{\tau > t\}}}{\mathbb{Q}(\tau > t|\mathcal{F}_t)} \check{\mathbb{E}}_t^{\boxed{a,b}} \left\{\frac{\boxed{\check{C}_{a,b}(t)}}{\boxed{\check{C}_{a,b}(T_a)}} \check{C}_{a,b}(T_a) (R^{PR2}_{a,b}(T_a) - K)^+ \right\}$$

$$= \frac{1_{\{\tau > t\}}}{\mathbb{Q}(\tau > t|\mathcal{F}_t)} \check{C}_{a,b}(t) \check{\mathbb{E}}_t^{a,b} \left\{(R^{PR2}_{a,b}(T_a) - K)^+ \right\}.$$

Now notice that R^{PR2} can be also written as

$$R^{PR2}_{a,b}(t) = \frac{\sum_{i=a+1}^{b} \mathbb{E}[D(t,T_i) 1_{\{T_{i-1} < \tau \le T_i\}} | \mathcal{F}_t]}{\check{C}_{a,b}(t)},$$

so that it is a martingale under $\check{\mathbb{Q}}^{a,b}$. As such, we may model it as

$$dR^{PR2}_{a,b}(t) = \sigma_{a,b} R^{PR2}_{a,b}(t) d\check{W}^{a,b}(t) \tag{23.14}$$

and compute the above expectation accordingly. We obtain, as price of the payer CDS option,

$$\mathbb{E}\{1_{\{\tau > T_a\}} D(t,T_a) \frac{\check{C}_{a,b}(T_a)}{\mathbb{Q}(\tau > T_a | \mathcal{F}_{T_a})} (R^{PR2}_{a,b}(T_a) - K)^+ | \mathcal{G}_t\} = \tag{23.15}$$

$$= 1_{\{\tau > t\}} \frac{\check{C}_{a,b}(t)}{\mathbb{Q}(\tau > t|\mathcal{F}_t)} [R^{PR2}_{a,b}(t) N(d_1) - K N(d_2)]$$

where d_1 and d_2 are defined as usual in terms of $R^{PR2}_{a,b}(t)$, K and σ.

Which model should one use between DFRN1 and DFRN2 when dealing with DFRN options? DFRN1 has the advantage of better approximating the real DFRN; further, the related market model is derived under a quantity resembling more a numeraire than in the DFRN2 case. DFRN2 is derived only under a quantity that is more approximately a numeraire and is a worse approximation of the real DFRN, but the related CDS payoff PR2CDS is in some cases a better approximation of a real CDS than PRCDS.

23.2.3 Examples of Implied Volatilities from the Market

We present now some CDS options implied volatilities obtained with the postponed payoff of the first and second kind. We consider three companies C1, C2 and C3 on the Euro market and the related CDS options quotes as of March 26, 2004; the recovery is $R_{EC} = 0.4$; C1 and C3 are in the telephonic sector, whereas C2 is a car industry. C1 = Deutsche Telecom; C2 = Daimler Chrysler; C3 = France Telecom. We have $L_{GD} = 1 - 0.4 = 0.6$; $T_0 =$ March 26 2004 (0); We consider two possible maturities T_a =June 20 2004 (86d≈3m) and T'_a =Dec 20 2004 (269d≈9m); T_b = june 20 2009 (5y87d); we consider receiver option quotes (puts on R) in basis points (i.e. 1E-4 units on a notional of 1). We obtain the results presented in Table 23.1.

	bid	mid	ask	$R_{0,b}(0)$	$R^{PR}_{a,b}(0)$	$R^{PR2}_{a,b}(0)$	K	$\sigma^{PR}_{a,b}$	$\sigma^{PR(2)}_{a,b}$
C1(T_a)	14	24	34	60	61.497	61.495	60	50.31	50.18
C2	32	39	46	94.5	97.326	97.319	94	54.68	54.48
C3	18	25	32	61	62.697	62.694	61	52.01	51.88
C1(T'_a)	28	35	42	60	65.352	65.344	61	51.45	51.32

Table 23.1. bid, mid and ask option prices, spot CDS rates, CDS forward rates, strikes and implied volatilities on three companies on March 26, 2004. Rates are in basis points and volatilities are percentages.

Implied volatilities are rather high when compared with typical interest-rate default free swaption volatilities. However, the values we find have the same order of magnitude as some of the values found by Hull and White (2003) via historical estimation. Further, we see that while the option prices differ considerably, the related implied volatilities are rather similar. This shows the usefulness of a rigorous model for implied volatilities. The mere price quotes could have left one uncertain on whether the credit spread variabilities implicit in the different companies were quite different from each other or similar.

We analyze also the implied volatilities and CDS forward rates under different payoff formulations and under stress. Table 23.1 shows that the impact of changing postponement from PR to PR2 (maintaining the same $R_{0,b}(0)$'s and re-stripping Γ's to get $R_{a,b}(0)$'s) leaves both CDS forward rates and implied volatilities almost unchanged.

In Table 23.2 we check the impact of the recovery rate on implied volatilities and CDS forward rates. Every time we change the recovery we re-calibrate the Γ's, since the only direct market quotes are the $R_{0,b}(0)$'s, which we cannot change, and our uncertainty is on the recovery rate that might change. As we can see from the table the impact of the recovery rate is rather small, but we have to keep in mind that the CDS option payoff is built in such a way that the recovery direct flow in L_{GD} cancels and the recovery remains only implicitly inside the initial condition $R_{a,b}(0)$ for the dynamics of the forward

	REC = 0.2	REC = 0.3	REC = 0.4	REC = 0.5	REC = 0.6
$\sigma_{a,b}^{PR}$:					
C1(T_a)	50.02	50.14	50.31	50.54	50.90
C2	54.22	54.42	54.68	55.05	55.62
C3	51.71	51.83	52.01	52.25	52.61
C1(T_a')	51.13	51.27	51.45	51.71	52.10
$R_{a,b}^{PR}$:					
C1(T_a)	61.488	61.492	61.497	61.504	61.514
C2	97.303	97.313	97.326	97.346	97.374
C3	62.687	62.691	62.697	62.704	62.716
C1(T_a')	65.320	65.334	65.352	65.377	65.415

Table 23.2. Impact of recovery rates on the implied volatility and on the CDS forward rates for the PR payoff. Volatilities are expressed as percentages and rates as basis points

$R_{a,b}$, as one can see for example from the payoff (23.6), where LGD does not appear explicitly. It is $R_{a,b}(0)$ that depends on the stripped Γ's which, in turn, depend on the recovery (and on $R_{0,b}(0)$'s).

	shift −0.5%	0	+0.5%
C1(T_a)	49.68	50.31	50.93
C2	54.02	54.68	55.34
C3	51.36	52.01	52.65

shift −0.5%	0	+0.5%
61.480	61.497	61.514
97.294	97.326	97.358
62.677	62.697	62.716

Table 23.3. Implied volatilities $\sigma_{a,b}$ (left, as percentages) and forward CDS rates $R_{a,b}^{PR}$ (right, as basis points) as the simply compounded rates of the zero coupon interest rate curve are shifted uniformly for all maturities.

In Table 23.3 we check the impact of a shift in the simply compounded rates of the zero coupon interest rate curve on CDS forward rates and implied volatilities. Every time we shift the curve we recalibrate the Γ's, while maintaining the same input $R_{0,b}(0)$'s. We see that the shift has a more relevant impact than the recovery rate, an impact that remains small.

We also include the zero coupon curve we used, in Table 23.4, and the CDS market quotes we used, in Table 23.5.

T	$P(0,T)$	T	$P(0,T)$	T	$P(0,T)$
26-mar-04	1	30-dec-04	0.985454616	28-mar-13	0.701853679
29-mar-04	0.999829196	30-mar-06	0.956335676	31-mar-14	0.665778313
31-mar-04	0.9997158	30-mar-07	0.9261161	30-mar-15	0.630686684
06-apr-04	0.999372341	31-mar-08	0.891575268	30-mar-16	0.597987523
30-apr-04	0.99806645	30-mar-09	0.85486229	30-mar-17	0.566052224
31-may-04	0.996398755	30-mar-10	0.816705705	29-mar-18	0.535085529
30-jun-04	0.994847843	30-mar-11	0.777867013	29-mar-19	0.505632535
30-sep-04	0.99014189	30-mar-12	0.739273058		

Table 23.4. Euro curve for Zero coupon bonds $P(0,T)$ for several maturities T as of march 26, 2004.

Maturity T_b	$R_{0,b}$ (C1)	$R_{0,b}$ (C2)	$R_{0,b}$ (C3)
1y	30	38.5	27
3y	49	72.5	49
5y	60	94.5	61
7y	69	104.5	73

Table 23.5. Quoted CDS spot rates for four maturities for the three names in basis points as of march 26, 2004

23.3 Towards a Completely Specified Market Model

"I saw the possibility of it long ago, and I explored it into existence". He sounded apologetic. "I would never have used it except under extreme circumstances".

Patricia McKillip, Harpist in the wind (1979)

So far we have been able to rigorously justify the market CDS option formula by directly modeling the single underlying CDS dynamics. However, to completely specify the market model we need to show how the dynamics of $R_{a,b}$ (or, better, of a family of rates generating all CDS rates) changes when changing numeraire. We refer to the PRCDS payoff, and all CDS rates in this section and related subsections, unless differently specified, are "PR" rates, even if for brevity we omit the "PR" notation. Following Brigo (2004, 2005, 2005b) we present three possible choices of families of rates dynamics completely specifying the CDS term structure dynamics (a further interesting choice not included here is in Brigo and Morini (2005)). The first two choices are still sketchy and incomplete in details. They are not fully operational yet. The reader who is not interested in speculation and on a first attack to this problem may skip directly to our third choice, given in Section 23.3.3, where we introduce a family of approximated rates allowing one to price Constant Maturity CDS and other possible products depending on the CDS term structure dynamics in general.

23.3.1 First Choice. One-period and two-period rates

The first case we address is a family of one-period rates. This is to say that we are trying to build a sort of forward LIBOR model for CDS rates. As the LIBOR model is based on one-period forward rates, our first choice of a market model for CDS options will be based on one-period rates. In other terms, we try and close the loop:

$$\begin{array}{lcl} \text{Fixed Income: } F_i = F(\cdot, T_{i-1}, T_i), \ _{i=\alpha+1,\dots,\beta} & \longrightarrow & S_{\alpha,\beta} \\ \updownarrow & & \updownarrow \\ \text{Credit:} \quad R_i = R_{i-1,i}(\cdot), \quad _{i=a+1,\dots,b} & ? \longrightarrow ? & R_{a,b} \end{array}$$

Since Formula (21.28) will be fundamental in the following, we report it here:

$$\bar{P}(t,T_i) = \bar{P}(t,T_{i-1}) \frac{\alpha_{i-1}(R_{i-1}^{PR}(t) - R_{i-2,i}^{PR}(t))}{\alpha_i(R_{i-2,i}^{PR}(t) - R_i^{PR}(t))}. \tag{23.16}$$

Notice that this formula assumes

$$R_{i-2,i}^{PR}(t) - R_i^{PR}(t) \neq 0.$$

This condition can be enforced at time zero, while for the future times t it can be checked a posteriori once the dynamics of one- and two- period rates has been selected.

The fundamental components of our numeraires $\widehat{C}$ are the $\bar{P}$'s. The $\bar{P}$'s, through (23.16), can be reduced to a function of a common initial $\bar{P}$ (that cancels when considering the relevant ratios) and of one- and two-period rates R_k, $R_{k-2,k}$ in the relevant range. We start then by writing the (martingale) dynamics of one- and two-period rates R_k and $R_{k-2,k}$ each under its canonical numeraire $\widehat{\mathbb{Q}}^{k-1,k}$ and $\widehat{\mathbb{Q}}^{k-2,k}$. At this point we use the change of numeraire technique on each of this rates to write their dynamics under a single preferred $\widehat{\mathbb{Q}}^{\cdot,\cdot}$. This is possible in terms of quadratic covariations between the one- and two-period rates being modeled and the analogous rates concurring to form the $\bar{P}$'s entering the relevant $\hat{C}$. We detail this scheme in the following part of this subsection below.

Now if we have the dynamics for all the relevant R_k, $R_{k-2,k}$'s under a common measure $\widehat{\mathbb{Q}}^{\cdot,\cdot}$, we proceed as follows. $R_{a,b}$ is completely specified in terms of one and two-period rates through our earlier formula (21.27) which we report here:

$$\begin{aligned} R_{a,b}^{PR}(t) &= \frac{\sum_{i=a+1}^{b} \alpha_i R_i^{PR}(t) \bar{P}(t,T_i)}{\sum_{i=a+1}^{b} \alpha_i \bar{P}(t,T_i)} \\ &= \sum_{i=a+1}^{b} \bar{w}_i(t) R_i^{PR}(t) \approx \sum_{i=a+1}^{b} \bar{w}_i(0) R_i^{PR}(t), \end{aligned} \tag{23.17}$$

and through (23.16), that provides the $\bar{P}$'s. Then through our rates R_k and $R_{k-2,k}$ dynamics we have indirectly also $R_{a,b}$'s dynamics.

Notice that if first we assign the dynamics of one period rates, then the dynamics of the two-period rates has to be selected carefully. For example, two-period rates will have to be selected into a range determined by one-period rates to avoid $\bar{P}(t,T_k)/\bar{P}(t,T_{k-1})$ to be negative or larger than one. The use of suitable martingale dynamics for each $R_{k-2,k}$ under $\widehat{\mathbb{Q}}^{k-2,k}$ ensuring this property is currently under investigation.

If we are concerned about lognormality of R's, leading to Black-like formulas for CDS options, one of the possible choices is to impose one-period rates R_k to have a lognormal distribution under their canonical measures. It suffices to postulate a driftless geometric Brownian motion dynamics for each such rate under its associated measure. The resulting $R_{a,b}$ will only be approximately lognormal, especially under the freezing approximation for the weights $\bar{w}$, but this is the case also with LIBOR vs SWAP models, since lognormal one-period swap rates (i.e. forward LIBOR rates) and multi period swap rates cannot be all lognormal (each under its canonical measure), as we have pointed out in Section 6.8. In Section 8.1 we have seen that the distance of the swap rate distribution in the LIBOR model from the lognormal family is small, so that the LIBOR and SWAP market models are practically consistent in most situations. The important difference between the CDS market models we are trying to develop here and the LIBOR model is that here we need also two-period rates to close the system. The need for two-period rates stems from the additional degrees of freedom coming from stochastic intensity whose "maturities", in rates like R's, are not always temporally aligned with the stochastic interest rates maturities. More precisely, notice that in the numerator of the last term in (21.26), which we report here:

$$R_j^{PR}(t) = \frac{\mathbb{E}[D(t,T_j)\mathbf{1}_{\{\tau>T_{j-1}\}}|\mathcal{F}_t] - \mathbb{E}[D(t,T_j)\mathbf{1}_{\{\tau>T_j\}}|\mathcal{F}_t]}{\alpha_j\mathbb{Q}(\tau>t|\mathcal{F}_t)\bar{P}(t,T_j)} \quad (23.18)$$

we have not only $\bar{P}$ (second term in the numerator) but also a term in $D(t,T_j)\mathbf{1}_{\{\tau>T_{j-1}\}}$ (first term in the numerator). Notice the misaligned T_{j-1} and T_j. This adds degrees of freedom that are accounted for by considering two-period rates.

A final remark is that the freezing approximation is typically questionable when volatilities are very large. Since, as we have seen in Section 23.2.3, at the moment implied volatilities in the CDS option market are rather large, the freezing approximation has to be considered with care.

Detailed scheme for the change of numeraire technique. Let us postulate the following dynamics for one- and two- period CDS forward rates. Recall that $R_j = R_{j-1,j}$.

$$dR_j(t) = \sigma_j(t)R_j(t)dZ_j^j(t)$$

$$dR_{j-2,j}(t) = \nu_j(t;R)R_{j-2,j}(t)dV_j^{j-2,j}(t)$$

In the Brownian shocks Z and V the upper index denotes the measure (i.e. the measure associated with the numeraires $\widehat{C}_{j-1,j}, \widehat{C}_{j-2,j}$ in the above case) and the lower index denotes to which component of the one- and two-period rate vectors the shock refers. The volatilities σ are deterministic, whereas the ν's depend on the one-period R's. We assume correlations

$$dZ_i dZ_j = \rho_{i,j}dt, \quad dV_i dV_j = \eta_{i,j}dt, \quad dZ_i dV_j = \theta_{i,j}dt$$

and

$$R_{i-2,i}(t) \in (\min(R_{i-1}(t), [R_{i-1}(t)+R_i(t)]/2), \max(R_{i-1}(t), [R_{i-1}(t)+R_i(t)]/2)).$$

This latter condition ensures that the resulting $\bar{P}$ from formula (23.16) be positive and decreasing with respect to the maturity, i.e.

$$0 < \bar{P}(t,T_i)/\bar{P}(t,T_{i-1}) < 1.$$

The specific definition of ν ensuring this property is currently under investigation.

We aim at finding the drift of a generic R_j under the measure associated with $\widehat{C}_{i-1,i}$, let us say for $j \geq i$.

The change of numeraire toolkit provides the formula relating shocks under $\widehat{C}_{i-1,i}$ to shocks under $\widehat{C}_{j-2,j}$, see for example Formula 2.13.

We can write, by slightly rephrasing said formula:

$$d\begin{bmatrix} Z^{j-2,j} \\ V^{j-2,j} \end{bmatrix} = d\begin{bmatrix} Z^i \\ V^i \end{bmatrix} - \text{CorrMatrix} \times \text{VectDiffusionCoefficient}\left(\ln\left(\frac{\widehat{C}_{j-2,j}}{\widehat{C}_{i-1,i}}\right)\right)' dt$$

Let us abbreviate "Vector Diffusion Coefficient" by "DC".

This is actually a sort of operator for diffusion processes that works as follows. $\text{DC}(X_t)$ is the row vector $\mathbf{v}$ in

$$dX_t = (...)dt + \mathbf{v}\; d\begin{bmatrix} Z_t \\ V_t \end{bmatrix}$$

for diffusion processes X with Z and V column vectors Brownian motions common to all relevant diffusion processes. This is to say that if for example $dR_1 = \sigma_1 R_1 dZ_1^1$, then

$$\text{DC}(R_1) = [\sigma_1 R_1,\ 0,\ 0, \ldots,\ 0].$$

Let us call Q the total correlation matrix including ρ, η and θ. We have

$$d\begin{bmatrix} Z^{j-2,j} \\ V^{j-2,j} \end{bmatrix} = d\begin{bmatrix} Z^i \\ V^i \end{bmatrix} - Q\ \text{DC}\left(\ln\left(\frac{\widehat{C}_{j-2,j}}{\widehat{C}_{i-1,i}}\right)\right) dt.$$

Now we need to compute

$$\mathrm{DC}\left(\ln\left(\frac{\widehat{C}_{j-2,j}}{\widehat{C}_{i-1,i}}\right)\right) = \mathrm{DC}\left(\ln\left(\frac{\alpha_{j-1}\bar{P}(t,T_{j-1}) + \alpha_j \bar{P}(t,T_j)}{\alpha_i \bar{P}(t,T_i)}\right)\right) =$$

$$= \mathrm{DC}\left(\ln\left(\frac{\alpha_{j-1}}{\alpha_i}\frac{\alpha_i}{\alpha_{j-1}}\prod_{k=i+1}^{j-1}\frac{R_{k-1}-R_{k-2,k}}{R_{k-2,k}-R_k} + \frac{\alpha_j}{\alpha_i}\frac{\alpha_i}{\alpha_j}\prod_{k=i+1}^{j}\frac{R_{k-1}-R_{k-2,k}}{R_{k-2,k}-R_k}\right)\right)$$

$$= \mathrm{DC}\left(\ln\left(\left[\prod_{k=i+1}^{j-1}\frac{R_{k-1}-R_{k-2,k}}{R_{k-2,k}-R_k}\right]\left[1+\frac{R_{j-1}-R_{j-2,j}}{R_{j-2,j}-R_j}\right]\right)\right)$$

$$= \mathrm{DC}\left(\sum_{k=i+1}^{j-1}\ln\left(\frac{R_{k-1}-R_{k-2,k}}{R_{k-2,k}-R_k}\right)\right) + \mathrm{DC}\left(\ln\left(\frac{R_{j-1}-R_j}{R_{j-2,j}-R_j}\right)\right)$$

$$= \sum_{k=i+1}^{j-1}\mathrm{DC}\left(\ln\left(\frac{R_{k-1}-R_{k-2,k}}{R_{k-2,k}-R_k}\right)\right) + \mathrm{DC}\left(\ln\left(\frac{R_{j-1}-R_j}{R_{j-2,j}-R_j}\right)\right) =$$

$$= \sum_{k=i+1}^{j-1}[\mathrm{DC}(\ln(R_{k-1}-R_{k-2,k})) - \mathrm{DC}(\ln(R_{k-2,k}-R_k))] +$$

$$+\mathrm{DC}(\ln(R_{j-1}-R_j)) - \mathrm{DC}(\ln(R_{j-2,j}-R_j))$$

$$= \sum_{k=i+1}^{j-1}\frac{\mathrm{DC}(\,R_{k-1}-R_{k-2,k})}{R_{k-1}-R_{k-2,k}} - \sum_{k=i+1}^{j-1}\frac{\mathrm{DC}(\,R_{k-2,k}-R_k)}{R_{k-2,k}-R_k} +$$

$$+\frac{\mathrm{DC}(\,R_{j-1}-R_j)}{R_{j-1}-R_j} - \frac{\mathrm{DC}(\,R_{j-2,j}-R_j)}{R_{j-2,j}-R_j} =$$

$$= \sum_{k=i+1}^{j-1}\frac{(\,\mathrm{DC}(R_{k-1})-\mathrm{DC}(R_{k-2,k}))}{R_{k-1}-R_{k-2,k}} - \sum_{k=i+1}^{j-1}\frac{(\,\mathrm{DC}(R_{k-2,k})-\mathrm{DC}(R_k))}{R_{k-2,k}-R_k}$$

$$+\frac{\mathrm{DC}(R_{j-1})-\mathrm{DC}(R_j)}{R_{j-1}-R_j} - \frac{\mathrm{DC}(R_{j-2,j})-\mathrm{DC}(R_j)}{R_{j-2,j}-R_j}.$$

It follows that

$$dZ_m^{j-2,j} - dZ_m^i = -\sum_{k=i+1}^{j-1}\frac{(\rho_{k-1,m}\sigma_{k-1}R_{k-1} - \theta_{m,k}\nu_k R_{k-2,k})}{R_{k-1}-R_{k-2,k}}dt$$

$$+\sum_{k=i+1}^{j-1}\frac{(\theta_{m,k}\nu_k R_{k-2,k} - \rho_{k,m}\sigma_k R_k)}{R_{k-2,k}-R_k}dt - \frac{\rho_{j-1,m}\sigma_{j-1}R_{j-1} - \rho_{j,m}\sigma_j R_j}{R_{j-1}-R_j}dt$$

$$+\frac{\theta_{m,j}\nu_j R_{j-2,j} - \rho_{j,m}\sigma_j R_j}{R_{j-2,j}-R_j}\,dt$$

and

$$dV_m^{j-2,j} - dV_m^i = -\sum_{k=i+1}^{j-1} \frac{(\theta_{k-1,m}\sigma_{k-1}R_{k-1} - \eta_{m,k}\nu_k R_{k-2,k})}{R_{k-1} - R_{k-2,k}} dt$$

$$+\sum_{k=i+1}^{j-1} \frac{(\eta_{m,k}\nu_k R_{k-2,k} - \theta_{k,m}\sigma_k R_k)}{R_{k-2,k} - R_k} dt - \frac{\theta_{j-1,m}\sigma_{j-1}R_{j-1} - \theta_{j,m}\sigma_j R_j}{R_{j-1} - R_j} dt$$

$$+\frac{\eta_{j,m}\nu_j R_{j-2,j} - \theta_{j,m}\sigma_j R_j}{R_{j-2,j} - R_j} dt =: \bar{\phi}_m^{i,j}\ dt.$$

Therefore, by subtracting from the first equation, taking $h > i$:

$$dZ_m^h - dZ_m^i = dZ_m^{j-2,j} - dZ_m^i - (dZ_m^{j-2,j} - dZ_m^h) =$$

$$= -\sum_{k=i+1}^{h} \frac{(\rho_{k-1,m}\sigma_{k-1}R_{k-1} - \theta_{m,k}\nu_k R_{k-2,k})}{R_{k-1} - R_{k-2,k}} dt$$

$$+\sum_{k=i+1}^{h} \frac{(\theta_{m,k}\nu_k R_{k-2,k} - \rho_{k,m}\sigma_k R_k)}{R_{k-2,k} - R_k} dt =: \bar{\mu}_m^{i,h} dt$$

so that we finally obtain (taking $h = j$)

$$dR_j(t) = \sigma_j R_j(t)(\bar{\mu}_j^{i,j} dt + dZ_j^i(t))$$

$$dR_{j-2,j}(t) = \nu_j R_{j-2,j}(t))(\bar{\phi}_j^{i,j} dt + dV_j^i(t)),$$

or, by setting

$$\mu_j^i := \bar{\mu}_j^{i,j}\ \sigma_j,\quad \phi_j^i := \bar{\phi}_j^{i,j}\ \nu_j,$$

we have

$$dR_j(t) = R_j(t)(\mu_j^i dt + \sigma_j dZ_j^i(t)),\quad dR_{j-2,j}(t) = R_{j-2,j}(t)(\phi_j^i dt + \nu_j dV_j^i(t)),$$

and since μ and ϕ are completely determined by one- and two- period rates vectors $R = [R_{i-1,i}]_i$ and $R^{(2)} = [R_{i-2,i}]_i$, the system is closed. We can write a vector SDE which is a vector diffusion for all the one- and two-period rates under any of the $\widehat{C}_{i-1,i}$ measures:

$$d\begin{bmatrix} R \\ R^{(2)} \end{bmatrix} = \mathrm{diag}(\mu(R, R^{(2)}), \phi(R, R^{(2)})) \begin{bmatrix} R \\ R^{(2)} \end{bmatrix} dt + \mathrm{diag}(\sigma, \nu) \begin{bmatrix} R \\ R^{(2)} \end{bmatrix} d\begin{bmatrix} Z^i \\ V^i \end{bmatrix}.$$

Notation $\mathrm{diag}(x)$ in meant to denote a diagonal matrix whose diagonal entries are given in the vector x.

At this point a Monte Carlo simulation of the process, based on a discretization scheme for the above vector SDE is possible. One only needs to know the initial CDS rates $R(0), R^{(2)}(0)$, which if not directly available one can build by suitably stripping spot CDS rates. Given the volatilities and correlations, one can easily simulate the scheme by means of standard Gaussian shocks.

If C is the Cholesky decomposition of the correlation Q ($Q = CC'$ with C lower triangular matrix) and W is a standard Brownian motion under $\widehat{C}_{i-1,i}$, we can write

$$d\begin{bmatrix} R \\ R^{(2)} \end{bmatrix} = \text{diag}(\mu(R, R^{(2)}), \phi(R, R^{(2)})) \begin{bmatrix} R \\ R^{(2)} \end{bmatrix} dt \qquad (23.19)$$
$$+\text{diag}(\sigma, \nu) \begin{bmatrix} R \\ R^{(2)} \end{bmatrix} C\, dW$$

The log process can be easily simulated with a Milstein scheme.

23.3.2 Second Choice: Co-terminal and one-period CDS rates market model

Our second choice is based on co-terminal CDS rates. Indeed, let us take a family of CDS rates $R_{a,b}, R_{a+1,b}, \ldots, R_{i,b}, \ldots, R_{b-1,b}$. Keep in mind that we are always referring to PR rates. Can we write the dynamics of all such rates under (say) $\widehat{\mathbb{Q}}^{a,b}$? The answer is affirmative if we take into account the following equality, which is not difficult to prove with some basic algebra:

$$\widehat{C}_{i,b}(t) = \mathbb{Q}(\tau > t|\mathcal{F}_t)\bar{P}(t, T_b) \prod_{k=i+1}^{b-1} \frac{R_{k,b}(t) - R_k(t)}{R_{k-1,b}(t) - R_k(t)}, \qquad i = a, \ldots, b-2. \qquad (23.20)$$

Notice that the term in front of the product is just $\widehat{C}_{b-1,b}(t)$. Notice also that the one-period rates canonical numeraires $\bar{P}$ can be obtained from the above numeraires via $\widehat{C}_{i,b} - \widehat{C}_{i-1,b}$'s. Take into account that we need to assume $R_{k-1,b}(t) \neq R_k(t)$. Analogously to what seen previously for the one- and two- period rates case, we can assume this to hold at time 0 and then we can check a posteriori whether the probability that this condition is violated at future times is zero under the selected dynamics.

As before, the set of rates and of different-numeraires ratios is not a closed system. To close the system we need to include one-period rates $R_{a+1}, \ldots, R_b$. In this framework we may derive the joint dynamics of

$$R_{a,b}, R_{a+1,b}, R_{a+2,b}, \ldots, R_{b-1,b}; \quad R_{a+1}, R_{a+2}, \ldots, R_{b-1}$$

under a common measure (say $\widehat{\mathbb{Q}}^{a,b}$) as follows. First assume a lognormal driftless geometric Brownian motion dynamics for $R_{a,b}$ under $\widehat{\mathbb{Q}}^{a,b}$ and suitable martingale dynamics for every other rate under its canonical measure.

These different dynamics have to be chosen so as to enforce the needed constraints on the $\widehat{C}_{k,b}(t)$, such as for example $\widehat{C}_{k-1,b}(t) > \widehat{C}_{k,b}(t)$ and similar inequalities implying the correct behavior of the embedded $\bar{P}$'s. Take then a generic rate in the family and write its dynamics under $\widehat{\mathbb{Q}}^{a,b}$ with the following method. Thanks to the change of numeraire technique, the drift of this rate dynamics under $\widehat{\mathbb{Q}}^{a,b}$ will be a function of the quadratic covariation between the rate being modeled and the ratio of $\widehat{C}_{a,b}(t)$ with the canonical numeraire of the selected rate itself. Thanks to (23.20), this ratio is a function of the rates in the family and therefore the relevant quadratic covariation can be expressed simply as a suitable function of the volatilities and correlation ("diffusion coefficients" and "instantaneous Brownian covariations" are more precise terms) of the one- and multi-period rates in our family. As before no inconsistency is introduced, thanks to the additional degrees of freedom stemming from stochastic intensity. Under this second "co-terminal" formulation we can obtain the Black-like market formula above for the $T_a - T_b$ tenor in the context of a consistent and "closed" market model. The definition of suitable martingale dynamics for the above one-period CDS rates and for CDS forward rates on other tenors in general is under investigation.

23.3.3 Third choice. Approximation: One-period CDS rates dynamics

Consider now the one-period CDS rates $R_j(t) = R_{j-1,j}(t)$. Go through the following approximation:

$$R_j(t) := R_{j-1,j}(t) = \text{L}_{\text{GD}} \frac{\mathbb{E}[D(t,T_j)1_{\{T_{j-1}<\tau\le T_j\}}|\mathcal{F}_t]}{\alpha_j \mathbb{Q}(\tau > t|\mathcal{F}_t)\bar{P}(t,T_j)}$$

$$= \text{L}_{\text{GD}} \frac{\mathbb{E}[D(t,T_j)1_{\{\tau>T_{j-1}\}}|\mathcal{F}_t] - \mathbb{E}[D(t,T_j)1_{\{\tau>T_j\}}|\mathcal{F}_t]}{\alpha_j \mathbb{Q}(\tau > t|\mathcal{F}_t)\bar{P}(t,T_j)} =$$

$$= \text{L}_{\text{GD}} \frac{\mathbb{E}[D(t,T_{j-1})1_{\{\tau>T_{j-1}\}} \boxed{D(t,T_j)/D(t,T_{j-1})} |\mathcal{F}_t] - \mathbb{E}[D(t,T_j)1_{\{\tau>T_j\}}|\mathcal{F}_t]}{\alpha_j \mathbb{Q}(\tau > t|\mathcal{F}_t)\bar{P}(t,T_j)} = \dots$$

At this point we approximate the boxed ratio of stochastic discount factors with the related zero-coupon bonds, obtaining

$$\approx \text{L}_{\text{GD}} \frac{\mathbb{E}[D(t,T_{j-1})1_{\{\tau>T_{j-1}\}}|\mathcal{F}_t] \boxed{P(t,T_j)/P(t,T_{j-1})} - \mathbb{E}[D(t,T_j)1_{\{\tau>T_j\}}|\mathcal{F}_t]}{\alpha_j \mathbb{Q}(\tau > t|\mathcal{F}_t)\bar{P}(t,T_j)}$$

$$= \text{L}_{\text{GD}} \frac{\bar{P}(t,T_{j-1})P(t,T_j)/P(t,T_{j-1}) - \bar{P}(t,T_j)}{\alpha_j \bar{P}(t,T_j)}$$

$$= \frac{\text{L}_{\text{GD}}}{\alpha_j}\left(\frac{\bar{P}(t,T_{j-1})}{(1+\alpha_j F_j(t))\bar{P}(t,T_j)} - 1\right) = \dots$$

At this point, by inserting a LIBOR model for the forward LIBOR rates F_j we could proceed towards a joint dynamics for CDS rates and interest rates, both stochastic. However, let us simplify further the rates by freezing F_j to time zero:

$$\ldots \approx \frac{\text{L}_{\text{GD}}}{\alpha_j}\left(\frac{\bar{P}(t,T_{j-1})}{(1+\alpha_j F_j(0))\bar{P}(t,T_j)} - 1\right) =: \widetilde{R}_j(t).$$

This last definition can be inverted so as to have

$$\frac{\bar{P}(t,T_{j-1})}{\bar{P}(t,T_j)} = \left(\frac{\alpha_j}{\text{L}_{\text{GD}}}\widetilde{R}_j + 1\right)(1+\alpha_j F_j(0)) > 1 \qquad (23.21)$$

as long as $\widetilde{R} > 0$, provided that $F_j(0) > 0$ as should be. This means that we are free to select any martingale dynamics for $\widetilde{R}_j$ under $\widehat{\mathbb{Q}}^{j-1,j}$, as long as $\widetilde{R}_j$ remains positive. Choose then such a family of $\widetilde{R}$ as building blocks

$$d\widetilde{R}_i(t) = \sigma_i(t)\widetilde{R}_i(t)dZ_i^i(t), \quad \text{for all } i$$

and define the $\bar{P}$ by using (23.21) to obtain inductively $\bar{P}(t,T_j)$ from $\bar{P}(t,T_{j-1})$ and from $\widetilde{R}_j$. *This way, the numeraires $\bar{P}$ become functions only of the $\widetilde{R}$'s, so that now the system is closed and all one has to model is the one-period rates $\widetilde{R}$ vector. No need to model auxiliary rates in order to be able to close the system.*

In this context the change of numeraire becomes

$$\begin{aligned}
dZ^j &= dZ^i - \rho \text{DC}\left(\ln\left(\frac{\widehat{C}_{j-1,j}}{\widehat{C}_{i-1,i}}\right)\right)' dt \\
&= dZ^i - \rho \text{DC}\left(\ln\left(\frac{\bar{P}(t,T_j)}{\bar{P}(t,T_i)}\right)\right)' dt = \\
&= dZ^i - \rho \text{DC} \ln\left[\left(\prod_{h=j+1}^{i}\left(\frac{\alpha_h}{\text{L}_{\text{GD}}}\widetilde{R}_h + 1\right)(1+\alpha_h F_h(0))\right)\right]' dt \\
&= dZ^i - \rho \sum_{h=j+1}^{i} \text{DC} \ln\left(\left(\frac{\alpha_h}{\text{L}_{\text{GD}}}\widetilde{R}_h + 1\right)(1+\alpha_h F_h(0))\right)' dt = \\
&= dZ^i - \rho \sum_{h=j+1}^{i} \text{DC} \ln\left(\left(\frac{\alpha_h}{\text{L}_{\text{GD}}}\widetilde{R}_h + 1\right)\right)' dt \\
&= dZ^i - \rho \sum_{h=j+1}^{i} \frac{1}{\widetilde{R}_h + \frac{\text{L}_{\text{GD}}}{\alpha_h}} \text{DC}(\widetilde{R}_h)' dt
\end{aligned}$$

so that we can write

$$dZ_k^j = dZ_k^i - \sum_{h=j+1}^{i} \rho_{k,h} \frac{\sigma_h(t)\widetilde{R}_h}{\widetilde{R}_h + \frac{\mathrm{L_{GD}}}{\alpha_h}} dt$$

from which we have the dynamics of $\widetilde{R}_i$ under $\widehat{\mathbb{Q}}^j$:

$$d\widetilde{R}_i = \sigma_i \widetilde{R}_i dZ_i^i = \sigma_i \widetilde{R}_i \left(dZ_i^j + \sum_{h=j+1}^{i} \rho_{j,h} \frac{\sigma_h \widetilde{R}_h}{\widetilde{R}_h + \frac{\mathrm{L_{GD}}}{\alpha_h}} dt \right) \tag{23.22}$$

$$=: \widetilde{R}_i(\tilde{\mu}_i^j(\widetilde{R})dt + \sigma_i dZ_i^j).$$

In the above derivation, as explained in the previous subsections, we abbreviated "Vector Diffusion Coefficient" by "DC", a different notation to express Formula 2.13.

At this point a Monte Carlo simulation of the vector process R, based on a discretization scheme for the above vector SDE is possible. One only needs to know the initial CDS rates $\widetilde{R}(0)$, which if not directly available one can build by suitably stripping spot CDS rates. Given the volatilities and correlations, one can easily simulate the scheme by means of standard Gaussian shocks.

A final remark is in order: although our approximated family of rates $\widetilde{R}$ is based on partial freezing of the F's, this does not mean that we are assuming deterministic interest rates (or independence between interest rates and the default time) altogether. Indeed, the building blocks for our $\widetilde{R}$ are the $\bar{P}$'s that embed possibly correlated stochastic interest rates and default events.

23.4 Hints at Smile Modeling

None of our equipment is even registering his presence; according to the machines, he's completely abstract.

J'onn J'onzz, "JLA: Rock of Ages", DC Comics

Finally, we consider the possibility of including a volatility smile in our CDS options model. Since the derivation is general, we may replace the dynamics (23.12) or (23.14) by a different "local volatility" dynamics

$$dR_{a,b}^{PR}(t) = \nu_{a,b}(t, R_{a,b}^{PR}(t))R_{a,b}^{PR}(t)dW^{a,b}(t)$$

with ν a suitable deterministic function of time and state. We might choose the CEV dynamics, a displaced diffusion dynamics, an hyperbolic sine densities mixture dynamics or a lognormal mixture dynamics. Several tractable choices are possible already in the local volatility diffusion setup, and one may select a smile dynamics for the LIBOR or swap model and use it to model R. There are several possible choices. For example, one may select $\nu_{a,b}$ from Brigo and Mercurio (2003) or Brigo Mercurio and Sartorelli (2003). Also, the uncertain volatility dynamics from Brigo, Mercurio and Rapisarda (2004) can be adapted to this context. More generally, see Chapters 9-12 for several candidate smile dynamics.

23.5 Constant Maturity Credit Default Swaps (CMCDS) with the market model

Constant Maturity Credit Default Swaps (CMCDS) are receiving increasing attention in the financial community. In this section, we aim at deriving an approximated no-arbitrage market valuation formula for CMCDS. We move from the CDS options market model in Brigo (2004, 2005), previously described in this chapter, and derive a formula for CMCDS that is the analogous of the formula for constant maturity swaps in the default free swap market under the LIBOR market model, seen at the end of Section 13.7.1. For some market feeling on CMCDS we invite the reader who jumped directly here to have a look at Section 21.5 before proceeding further.

This section is based on Brigo (2005b).

We define CMCDS and give immediately the main result of the section, the approximated pricing formula, in terms of CDS forward rates and of their volatilities and correlations. We point out some analogies with constant maturity swaps, showing that a "convexity adjustment"-like correction is present. Without such correction, or with zero correlations, the formula returns an obvious deterministic-credit-spread expression for the CMCDS price.

Once the main result has been described, we move to proving it. Through the one-period rates dynamics introduced in Section 23.3.3 and a drift-freezing approximation we then prove the formula for CMCDS pricing and give some numerical examples highlighting the role of the participation rate and of the convexity adjustment.

In the following subsections, R is the postponed CDS rate R^{PR}.

23.5.1 CDS and Constant Maturity CDS

In a Constant Maturity CDS (CMCDS) with first reset in T_a and with final maturity T_b, protection L_{GD} on a reference credit "C" against default in $[T_a, T_b]$ is given from a protection seller "B" to a protection buyer "A". In exchange for this protection, a "constant maturity" CDS rate is paid.

We know by definition that the fair rate to be paid at T_i for protection against default in $[T_{i-1}, T_i]$ would be $R_{i-1,i}(T_{i-1}) = R_i(T_{i-1})$. This leads us to recall the following

Remark 23.5.1. **(A "floating-rate" CDS).** A contract that protects in T_a, T_b can be in principle decomposed into a stream of contracts, each single contract protecting in $[T_{j-1}, T_j]$, for $j = a+1, \ldots, b$, say with protection payment L_{GD} postponed to T_j if default occurs in $[T_{j-1}, T_j]$. In each single period, the rate $R_j(T_{j-1})$ paid at T_j makes the exchange fair, so that in total a contract offering protection L_{GD} on a reference credit "C" in $[T_a, T_b]$ in exchange for payment of rates $R_{a+1}(T_a), \ldots, R_j(T_{j-1}), \ldots, R_b(T_{b-1})$ at times $T_{a+1}, \ldots, T_j, \ldots, T_b$ is fair, i.e. has zero initial present value. This product can be seen as a sort of floating rate CDS.

However, in CMCDS's the rate that is paid at each period for protection is not the related one-period CDS rate, as would be natural from the above remark, but a longer period CDS rate. See also Calamaro and Nassar (2004). Consider indeed the following

Definition 23.5.1. (Constant Maturity CDS). *Consider a contract protecting in $[T_a, T_b]$ against default of a reference credit "C". If default occurs in $[T_a, T_b]$, a protection payment L_{GD} is made from the protection seller "B" to the protection buyer "A" at the first T_j following the default time. This is called "protection leg". In exchange for this protection "A" pays to "B" at each T_j before default a "$c+1$–long" (constant maturity) CDS rate $R_{j-1,j+c}(T_{j-1})$ (times a year fraction $\alpha_j = T_j - T_{j-1}$), with "c" an integer larger than zero. Notice that for $c = 0$ we would obtain the fair "floating rate" CDS above, whose initial value would be zero.*

Compare:
"Floating Rate Equivalent" of Standard CDS: (initial present value is 0)

Prot. Seller B	→ protection L_{GD} at default τ_C if $T_a < \tau_C \le T_b$ → ← rate $R_{i-1,i}(T_{i-1})$ at $T_i = T_{a+1}, \ldots, T_b$ or until τ_C ←	Prot. Buyer A

CMCDS: (initial present value?? Unknown, we need a model)

B	→ protection L_{GD} at default τ_C if $T_a < \tau_C \le T_b$ → ← $R_{i-1,i+c}(T_{i-1})$ at $T_i = T_{a+1}, \ldots, T_b$ or until τ_C ←	A

As explained before, in the market CMCDS are often capped, in that a limit is set to the premium rate $R_{i-1,i+c}(T_{i-1})$ that is paid at each period. In this case it is still possible to find a closed form approximated formula, by resorting to the change of numeraire and drift freezing, plus moment matching techniques for finding approximate (possibly shifted) lognormal distributions for the relevant CDS rates under the suitable measures. We will present a detailed formula with Monte Carlo tests in future work.

Going back to the basic case above, given that $c > 0$ in the definition of CMCDS, the value of the contract will be nonzero in general, so that we have to find this value at the initial time 0 if we are to price this kind of transaction. We face this task by resorting to the CDS market model derived earlier.

The value of the CMCDS to "B" is the value of the premium leg minus the value of the protection leg. The protection leg valuation is trivial, since this is the same leg as in a standard forward start $[T_a, T_b]$ CDS. As such, it is for example equal to

$$R_{a,b}(0) \sum_{j=a+1}^{b} \alpha_j \bar{P}(0,T_j) = \sum_{j=a+1}^{b} \alpha_j R_j(0) \bar{P}(0,T_j).$$

This value has to be subtracted to the premium leg. The non-trivial part is indeed computing the premium leg value at initial time 0.

Notice that the final formula we give for this leg can be implemented easily on a spreadsheet, requiring no numerical apparatus. Here it is:

Proposition 23.5.1. (Main Result: An approximated formula for CMCDS (Brigo 2005b)) *Consider the Constant Maturity CDS defined in Definition 23.5.1. The present value at initial time 0 of the CMCDS to the protection seller "B" is*

$$CDS_{CM_{a,b,c}}(0, L_{GD}) = \sum_{j=a+1}^{b} \alpha_j \bar{P}(0,T_j) \left\{ \sum_{i=j}^{j+c} \frac{\alpha_i \bar{P}(0,T_i)}{\sum_{h=j}^{j+c} \alpha_h \bar{P}(0,T_h)} \cdot \right. \tag{23.23}$$

$$\left. \widetilde{R}_i(0) \exp\left[T_{j-1} \sigma_i \cdot \left(\sum_{k=j+1}^{i} \rho_{j,k} \frac{\sigma_k \widetilde{R}_k(0)}{\widetilde{R}_k(0) + L_{GD}/\alpha_k} \right) \right] - R_j(0) \right\}$$

where $R_k(0)$ are the one-period CDS forward rates for protection in $[T_{k-1}, T_k]$. These CDS forward rates can be computed from quoted spot CDS rates $R_{0,k}(0)$ and corporate zero coupon bonds $\bar{P}(0,T_k)$ via

$$R_k(0) = \frac{R_{0,k}(0) \sum_{h=1}^{k} \alpha_h \bar{P}(0,T_h) - R_{0,k-1}(0) \sum_{h=1}^{k-1} \alpha_h \bar{P}(0,T_h)}{\alpha_k \bar{P}(0,T_k)}$$

while $\widetilde{R}_k(0)$ are approximations of the $R_k(0)$ (equal in case of independence of interest rates and credit spreads) in terms of corporate $\bar{P}$ and default free P zero coupon bonds given by

$$R_k(0) \approx \widetilde{R}_k(0) = L_{GD} \frac{\bar{P}(0,T_{k-1}) P(0,T_k)/P(0,T_{k-1}) - \bar{P}(0,T_k)}{\alpha_k \bar{P}(0,T_k)}$$

and where: σ_k *is the volatility of* $R_k(t)$*, assumed constant (we deal with the time-varying volatility in the proof below);*

$\rho_{i,j}$ *is the instantaneous correlation between* R_i *and* R_j*;*

One-period forward CDS rates volatilities σ_k *can in principle be stripped from longer period CDS volatilities, similarly to how forward LIBOR rates volatilities can be stripped from swaptions volatilities in the LIBOR model. This stripping is made possible from an approximated volatility formula based on drift freezing (formula (6.67) for the LIBOR case). Cascade methods are also available for this (as in Brigo and Morini (2003) or in Sections 7.4-7.7), although for the time being the only available CDS options all have short maturities and the lack of a liquid market discourages this kind of approach. For the time being the above formula can be employed with stylized values of volatilities to have an idea of the impact of the "convexity adjustments". Finally, one may consider using historical volatilities and correlations in the formula as first guesses.*

As a further remark we notice that, if not for the exponential term (which vanishes for example when ρ*'s are set to zero) this expression would be, not surprisingly,*

$$CDS_{CM_{a,b,c}}(0, L_{GD}; \rho = 0) = \sum_{j=a+1}^{b} \alpha_j \bar{P}(0,T_j)(R_{j-1,j+c}(0) - R_{j-1,j}(0)) \quad (23.24)$$

The exponential term in (23.23), i.e.

$$\exp\left[T_{j-1}\sigma_i \cdot \left(\sum_{k=j+1}^{i} \rho_{j,k} \frac{\sigma_k \widetilde{R}_k(0)}{\widetilde{R}_k(0) + L_{GD}/\alpha_k}\right)\right],$$

can be considered indeed to be a sort of "convexity adjustment" similar in spirit to the convexity adjustment needed to value constant maturity swaps with the LIBOR model in the default-free market.

Finally, this formula should be tested against prices obtained via Monte Carlo simulation of the dynamics (23.19) before being employed massively. One should make sure that for the order of magnitude of volatilities, correlations and initial CDS rates present in the market at a given time the freezing approximation works well. We plan to analyze this approximation against Monte Carlo simulation in further work. This future work is the reason why we derived some possible candidate exact rates dynamics above, although the general framework is still incomplete.

23.5.2 Proof of the main result

Consider the drift term in formula (23.22). If we compute $\widehat{\mathbb{E}}^{j-1,j}[\widetilde{R}_i(T_{j-1})]$ we obtain

$$\widehat{\mathbb{E}}^{j-1,j}[\widetilde{R}_i(T_{j-1})] \approx \widetilde{R}_i(0)\exp\left\{\int_0^{T_{j-1}} \widetilde{\mu}_i^j(\widetilde{R}(0))du\right\}$$

$$= \widetilde{R}_i(0)\exp\left\{\sum_{k=j+1}^{i} \frac{\widetilde{R}_k(0)}{\widetilde{R}_k(0)+\mathrm{L{\scriptstyle GD}}/\alpha_k}\rho_{j,k}\int_0^{T_{j-1}} \sigma_i(u)\sigma_k(u)du\right\} \quad (23.25)$$

and, if we take volatilities σ to be constant, we have

$$\approx \widetilde{R}_i(0)\exp\left\{T_{j-1}\sigma_i\cdot\left(\sum_{k=j+1}^{i}\rho_{j,k}\frac{\sigma_k\widetilde{R}_k(0)}{\widetilde{R}_k(0)+\mathrm{L{\scriptstyle GD}}/\alpha_k}\right)\right\}. \quad (23.26)$$

Under independence between intensities and interest rates, a harmless assumption when dealing with plain CDS, by definition of R_j it is easy to show that at time 0, both the original $R_j(0)$ and the approximated $\widetilde{R}_j(0)$ are given in terms of the survival probabilities as

$$R_j(0) = \widetilde{R}_j(0) = \mathrm{L{\scriptstyle GD}}/\alpha_j\left(\frac{\mathbb{Q}(\tau > T_{j-1})}{\mathbb{Q}(\tau > T_j)} - 1\right) \quad (23.27)$$

and hence (23.25) is reduced to

$$\widehat{\mathbb{E}}^{j-1,j}[\widetilde{R}_i(T_{j-1})] = \frac{\mathrm{L{\scriptstyle GD}}}{\alpha_i}\left(\frac{\mathbb{Q}(\tau > T_{i-1})}{\mathbb{Q}(\tau > T_i)} - 1\right)\cdot \quad (23.28)$$

$$\cdot\exp\left\{T_{j-1}\sigma_i\sum_{k=j+1}^{i}\rho_{j,k}\sigma_k\left(1-\frac{\mathbb{Q}(\tau > T_k)}{\mathbb{Q}(\tau > T_{k-1})}\right)\right\}.$$

However, there is no need to assume independence between interest rates and default in our approach, or even deterministic interest rates. Indeed, our general definition of $\widetilde{R}$ involves defaultable bonds $\bar{P}$ embedding both interest-rate and default risk, with possibly correlated interest rates and credit spreads. We only make an approximation partly resembling independence "locally" in defining the rates $\widetilde{R}$ from the R's, and an independence assumption as an approximation when stripping initial CDS forward curves (initial conditions for our dynamics). But our stochastic dynamics for the $\widetilde{R}$'s may well embed stochasticity of interest rates.

Now, based on the approximated dynamics (23.22) and the related expectation above, we prove the main result of the section, i.e. Proposition 23.5.1.

Proof. To prove the proposition, we compute the price of the premium leg as

$$\sum_{j=a+1}^{b}\alpha_j\mathbb{E}_0[D(0,T_j)1_{\{\tau>T_j\}}R_{j-1,j+c}(T_{j-1})] = \ldots$$

The first approximation we consider is (23.17) applied to $R_{j-1,j+c}(T_{j-1})$, so that

$$R_{j-1,j+c}(T_{j-1}) \approx \sum_{i=j}^{j+c} \bar{w}_i^j(0) R_i(T_{j-1}), \quad \bar{w}_i^j(0) = \frac{\alpha_i \bar{P}(0,T_i)}{\sum_{h=j}^{j+c} \alpha_h \bar{P}(0,T_h)}.$$

Then by substituting this in the premium leg expression we have

$$\ldots \approx \sum_{j=a+1}^{b} \sum_{i=j}^{j+c} \alpha_j \bar{w}_i^j(0) \mathbb{E}_0[D(0,T_j) 1_{\{\tau > T_j\}} R_i(T_{j-1})] =$$

$$= \sum_{j=a+1}^{b} \sum_{i=j}^{j+c} \alpha_j \bar{w}_i^j(0) \mathbb{E}_0[D(0,T_j) R_i(T_{j-1}) \mathbb{E}(1_{\{\tau > T_j\}} | \mathcal{F}_{T_j})]$$

$$= \sum_{j=a+1}^{b} \sum_{i=j}^{j+c} \bar{w}_i^j(0) \mathbb{E}_0^{\boxed{B}} \left[\frac{\boxed{B(0)}}{\boxed{B(T_j)}} (R_i(T_{j-1})\, \widehat{C}_{j-1,j}(T_j)) \right]$$

$$= \sum_{j=a+1}^{b} \sum_{i=j}^{j+c} \bar{w}_i^j(0) \widehat{\mathbb{E}}_0^{\boxed{j-1,j}} \left[\frac{\boxed{\widehat{C}_{j-1,j}(0)}}{\boxed{\widehat{C}_{j-1,j}(T_j)}} (R_i(T_{j-1})\, \widehat{C}_{j-1,j}(T_j)) \right]$$

$$= \sum_{j=a+1}^{b} \sum_{i=j}^{j+c} \bar{w}_i^j(0) \widehat{C}_{j-1,j}(0) \widehat{\mathbb{E}}_0^{j-1,j}[R_i(T_{j-1})]$$

$$= \sum_{j=a+1}^{b} \sum_{i=j}^{j+c} \alpha_j \bar{w}_i^j(0) \bar{P}(0,T_j) \widehat{\mathbb{E}}_0^{j-1,j}[R_i(T_{j-1})] = \ldots$$

where we have applied the change of numeraire, moving from the risk neutral numeraire B to the numeraires $\widehat{C}_{j-1,j}$'s (again the use of boxes refers to the change of numeraire invariance/ "Fact Two" property seen in Section 2.3). The last expected value can be computed based on (23.26). By substituting the expected value expression, we obtain the final formula. □

23.5.3 A few numerical examples

We report input data and outputs for a name with relatively large CDS forward rates. We consider the FIAT car company CDS market quotes as of December 20, 2004. Since in Brigo and Alfonsi (2003), or in Section 22.7.6, we have some evidence on the fact that CDS prices depend very little on the correlation between interest rates and credit spreads, when stripping credit spreads from CDS data we may assume independence between interest rates and credit spreads. This leads to a model where it is easy to strip default probabilities (hazard rates) from CDS prices, as hinted at again in Brigo and Alfonsi (2003) and in Sections 22.3 and 22.7. Using this independence assumption, we strip default (or survival) probabilities (actually hazard functions Γ^{mkt}) from CDS quotes with increasing maturities.

Inputs. We take as inputs the following Fiat CDS rates and use mid quotes

T_b	$R_{0,b}^{BID}(bps)$	$R_{0,b}^{ASK}$
1Y	99.9	175.57
2Y	172.5	231.38
3Y	243.73	286.13
5Y	348.85	366.54
7Y	380	410
10Y	395.16	412.73

We take $\text{Rec} = 0.4$ (so that $\text{Lgd} = 0.6$). The input zero coupon curve, and the survival risk-neutral probabilities stripped from FIAT CDS quotes are given at quarterly intervals in Table 23.6.

Outputs. We start by giving a table for

$$\text{Conv}(\sigma,\rho) := \text{CDS}_{\text{CM}a,b,c}(0, \text{Lgd}, \sigma, \rho) - \text{CDS}_{\text{CM}a,b,c}(0, \text{Lgd}; \rho = 0).$$

The first term is computed by assuming the volatilities σ_i of forward one-period CDS rates R_i to have a common value σ and the pairwise correlations $\rho_{i,j}$ to have a common value ρ. This first term is then given by formula (23.23). The second term is the simpler value (23.24) where no correction due to CDS forward rate dynamics is accounted for. This difference then gives us the impact of volatilities and correlations of CDS rates on the CMCDS price. The difference is always positive, similarly to what happens to analogous constant maturity swaps in default free markets under similar conditions on volatilities and correlation. It is the impact of "convexity" on the CMCDS valuation. We take $a = 0$, $b = 20$ (5y final maturity) and $c = 21$ (which means we are considering non-standard 5y6m CDS rates in the CMCDS premium leg, $c = 19$ would amount to a 5y CDS rate).

We obtain

$\text{Conv}(\sigma,\rho)$	ρ: 0.7	0.8	0.9	0.99
σ: 0.1	0.000659	0.000754	0.000848	0.000933
0.2	0.002662	0.003047	0.003435	0.003784
0.4	0.011066	0.012742	0.014442	0.015995
0.6	0.026619	0.030964	0.035464	0.039652

The "convexity difference" increases with respect both to correlation and volatility, as expected.

α_i	T_i	$P(0,T_i)$	$\mathbb{Q}(\tau > T_i)$
0	0	0.99994	0.99994
0.24444	0.24444	0.99459	0.99429
0.25556	0.5	0.989	0.98856
0.25556	0.75556	0.98309	0.98279
0.25278	1.0083	0.97709	0.97712
0.25	1.2583	0.97098	0.97155
0.25556	1.5139	0.96458	0.96433
0.25556	1.7694	0.958	0.95409
0.25278	2.0222	0.95133	0.94108
0.25	2.2722	0.94448	0.92552
0.25556	2.5278	0.9373	0.9086
0.25556	2.7833	0.92995	0.89227
0.25278	3.0361	0.92251	0.87669
0.25278	3.2889	0.91502	0.86165
0.25556	3.5444	0.90731	0.84618
0.25556	3.8	0.89929	0.82931
0.25278	4.0528	0.89139	0.81203
0.25	4.3028	0.88373	0.79449
0.25556	4.5583	0.87544	0.77495
0.25556	4.8139	0.8673	0.75531
0.25278	5.0667	0.85906	0.73503
0.25	5.3167	0.85085	0.7142
0.25556	5.5722	0.84255	0.69403
0.25556	5.8278	0.83417	0.67559
0.25278	6.0806	0.82572	0.65879
0.25	6.3306	0.81735	0.64351
0.25556	6.5861	0.80892	0.62968
0.25556	6.8417	0.80035	0.61709
0.25278	7.0944	0.79182	0.60594
0.25278	7.3472	0.78344	0.59601
0.25556	7.6028	0.77494	0.58651
0.25556	7.8583	0.76641	0.57685
0.25278	8.1111	0.75794	0.56715
0.25	8.3611	0.74977	0.55744
0.25556	8.6167	0.74141	0.5474
0.25556	8.8722	0.73303	0.53726
0.25278	9.125	0.72474	0.52713
0.25	9.375	0.7168	0.51705
0.25556	9.6306	0.70869	0.50667
0.25556	9.8861	0.70041	0.49601
0.25278	10.139	0.69241	0.48565
0.25	10.389	0.6849	0.4756

Table 23.6. FIAT CDS Survival Probabilities and default free-zero coupon bond prices as of December 20, 2004.

The next table reports the so called "participation rate" $\phi_{a,b,c}(\sigma,\rho)$ for a CMCDS with final $T_b = 5y$ ($a = 0, b = 20$, recalling that resets occur quarterly), with $5y6m$ constant maturity CDS rates ($c = 21$),

$$\phi_{0,20,21}(\sigma,\rho) = \frac{\text{“premium leg CDS”}}{\text{“premium leg CMCDS”}}$$

$$= \frac{\sum_{j=1}^{20} \alpha_j \bar{P}(0,T_j) R_{0,20}(0)}{\sum_{j=1}^{20} \alpha_j \widehat{\mathbb{E}}_0^{j-1,j}[D(0,T_j) 1_{\{\tau>T_j\}} R_{j-1,j+21}(T_{j-1})]},$$

The CMCDS premium leg is computed with our approximated market model based on one-period rates $\widetilde{R}$. As we see from the outputs, the participation rate decreases with volatility and correlation, as is expected from the "convexity adjustment" effect.

$\phi_{0,20,21}(\sigma,\rho)$	ρ: 0.7	0.8	0.9	0.99
σ: 0.1	0.71358	0.71325	0.71292	0.71262
0.2	0.70664	0.70532	0.704	0.70281
0.4	0.67894	0.67368	0.66842	0.66368
0.6	0.63302	0.62128	0.60957	0.59907

Finally, we fix volatilities and correlations and check how the patterns change when changing final maturity $T_b = T_i$. We consider the following quantities at time 0 and with $T_a = 0$:

$$x_i = \frac{\text{“Constant maturity rate”}}{\text{“standard rate”}} = \frac{R_{i-1,i+c}(0)}{R_{0,b}(0)}, \quad i = 1,\dots,b$$

$$y_i = \frac{\widehat{\mathbb{E}}_0^{i-1,i}[D(0,T_i) 1_{\{\tau>T_i\}} R_{i-1,i+c}(T_{i-1})]}{\bar{P}(0,T_i) R_{0,b}(0)}, \quad i = 1,\dots,b$$

$$z_i = \frac{\widehat{\mathbb{E}}_0^{i-1,i}[D(0,T_i) 1_{\{\tau>T_i\}} R_{i-1,i+c}(T_{i-1})]}{\bar{P}(0,T_i) R_{i-1,i+c}(0)}, \quad i = 1,\dots,b$$

$$\psi_i = \frac{\text{“premium leg CDS”}}{\text{“premium leg CMCDS”}} = \frac{\sum_{j=1}^{i} \alpha_j \bar{P}(0,T_j) R_{0,i}(0)}{\sum_{j=1}^{i} \alpha_j \bar{P}(0,T_j) R_{j-1,j+c}(0)}, \quad i = 1,\dots,b$$

$$\phi_i = \frac{\text{“premium leg CDS”}}{\text{“premium leg CMCDS with convexity”}}$$

$$= \frac{\sum_{j=1}^{i} \alpha_j \bar{P}(0,T_j) R_{0,i}(0)}{\sum_{j=1}^{i} \alpha_j \widehat{\mathbb{E}}_0^{j-1,j}[D(0,T_j) 1_{\{\tau>T_j\}} R_{j-1,j+c}(T_{j-1})]}.$$

The x_i's measure how the constant maturity CDS rate differs multiplicatively from the standard CDS rate, so they are a measure of how the constant maturity CDS differs from a standard CDS in the premium rate paid at each period. We find an increasing pattern in T_i as partly expected from the fact that the input CDS rates $R_{0,b}^{BID,ASK}$ are increasing with respect to maturity T_b.

x_i	y_i	z_i	ψ_i	ϕ_i	
					$\sigma = 0.4$;
					$\rho = 0.9$;
1.0668	1.0668	1	0.37773	0.37773	
1.1288	1.1359	1.0063	0.36281	0.36162	REC = 0.4;
1.1914	1.2075	1.0135	0.35281	0.35039	a=0;
1.2525	1.2792	1.0214	0.34359	0.33993	c = 20;
1.3107	1.3495	1.0297	0.33512	0.33024	b = 20;
1.3673	1.4193	1.038	0.34187	0.33548	
1.4171	1.4826	1.0462	0.36905	0.36064	
1.4515	1.53	1.0541	0.40755	0.39664	
1.4716	1.5622	1.0616	0.45262	0.43881	
1.4798	1.5818	1.0689	0.49477	0.47785	
1.4837	1.5979	1.0769	0.52661	0.50671	
1.4905	1.6175	1.0852	0.55072	0.52799	
1.4999	1.6403	1.0936	0.56931	0.54384	
1.5122	1.666	1.1018	0.58674	0.55846	
1.5236	1.69	1.1092	0.60704	0.57574	
1.5275	1.706	1.1168	0.62715	0.5928	
1.5274	1.7174	1.1244	0.64681	0.60938	
1.5249	1.7236	1.1303	0.67017	0.62939	
1.5106	1.7173	1.1368	0.69254	0.64843	
1.4924	1.7047	1.1422	0.71589	0.66842	

Table 23.7. Outputs for a range of terminal dates $T_b = T_i$ spanning five years at quarterly intervals

The y_i's measure the same effect while taking into account "convexity", i.e. future randomness of the payoff and correlation. The y_i's would reduce to the x_i's if correlations ρ were taken equal to 0. The y maintain the increasing pattern with respect to T_i.

The z_i's measure the multiplicative impact of "convexity" at each payment time in the premium leg of the CMCDS, in that they are due to contributions stemming from volatilities σ and correlations ρ of CDS rates. The impact is increasing with maturity T_i, as expected from the sign in the exponent of the convexity adjustments and from the positive signs of correlations (and volatilities).

Finally, as seen above, the ψ_i's are the so called "participation rates" for different terminal maturities $T_b = T_i$. They give the ratio between the premium leg in a standard CDS protecting in $[0, T_i]$ and the premium leg in CMCDS for the same protection interval when ignoring the convexity adjustment due to correlation and volatilities. The ϕ_i's are the participation rates computed when taking into account convexity due to volatilities and correlations. We have seen a particular participation rate ϕ earlier. Numerical results are given in Table 23.7.

In the table above for ϕ_i we obtain an initially decreasing pattern followed by a longer increasing pattern for both ψ and ϕ. Notice that, on the longest participation rate, in the last row of the related table, convexity has an impact moving from a 71.59% participation rate when not including "convexity" (ignoring correlations and volatilities) to a 66.84% participation rate when including convexity. There is a 4.8% difference in the participation rate of this FIAT 5y-5y6m CMCDS due to "convexity", when correlations are set to 0.9 and volatilities at 40%.

Part VIII

APPENDICES

A. Other Interest-Rate Models

In this appendix we introduce brief sketches of some of the models that are known in the literature and that have not been included in the previous chapters. All models are arbitrage free, and we will not discuss no-arbitrage implications further. Instead, we synthetically explain in what these models differ from the previous models and what are their original features. We also give references for the readers who might wish to deepen their knowledge of a specific approach. Clearly, presenting all the models that have been proposed in the literature is a huge task. We only present a few, without any claim to completeness of the treatment. Indeed, there are certainly several other relevant and worthy models that have not been included here, and we make the excuse that a choice is necessary since it is impossible to do justice to all the models appeared over the years. The reader interested in models that have not appeared in this book can also check other books on interest rate models such as for example James and Webber (2000).

A.1 Brennan and Schwartz's Model

Brennan and Schwartz (1979, 1982) consider a model based on modeling both the short and consol rates. The consol bond is, roughly speaking, a claim paying perpetually a constant dividend rate. If this constant dividend is assumed to be q, the payoff contribution in the infinitesimal time interval $(s, s+ds]$ is the dividend q per unit of time times the time amount ds discounted at time t:

$$D(t,s)qds \ .$$

By adding all these infinitesimal discounted contributions we obtain

$$\int_t^{+\infty} D(t,s)qds$$

whose risk neutral expectation gives the consol bond value at time t:

$$E_t\left[\int_t^{+\infty} D(t,s)qds\right] = q\int_t^{+\infty} P(t,s)ds \ .$$

Now notice that if at time t all rates were equal to a single rate $L(t)$ (continuous compounding) the above quantity would reduce to

$$\int_t^{+\infty} q e^{-L(t)(s-t)} ds = \frac{q}{L(t)} .$$

By solving

$$q \int_t^{+\infty} P(t,s) ds = \frac{q}{L(t)}$$

we find the *consol rate* L at time t. The consol rate at time t is that unique rate at time t discounting at which one recovers the correct price at time t of a consol bond.

$L(t)$ is a synthesis of the whole term structure up to infinity, and can be shown to incorporate information on the *steepness* of the yield curve at time t, since it involves discount functions for very large maturities and thus provides an indication of the "height" of the extreme far part of the yield curve at time t.

Brennan and Schwartz model the joint evolution of r and L with two particular diffusion processes and allow for correlation between the two. The dynamics of r and L are given under the objective measure, and market prices of risk are introduced to move to the risk neutral measure and price bonds. It is clear that at a given time t one can interpret the short rate $r(t)$ as the level of the curve at time t. Therefore, since $L(t)$ is related to the steepness of the curve, the Brennan and Schwartz model can be seen as a model incorporating level and steepness in the yield-curve evolution. The advantage of the model is that the two "factors" r and L have a clear and immediate financial interpretation.

Brennan and Schwartz (1979) analyze government bonds of the Canadian market with their proposed model, obtaining satisfactory results. However, besides these attractive features, there are problems concerning analytical tractability and other aspects.

The interested reader is referred also to Chapter 15 of Rebonato (1998).

A.2 Balduzzi, Das, Foresi and Sundaram's Model

Balduzzi, Das, Foresi and Sundaram (1996) propose a general framework to three-factor short-rate models leading to affine term structures. As a fundamental example of their framework they consider the following three-factor model under the objective measure:

$$\begin{aligned}
dr(t) &= k(\theta(t) - r(t))dt + \sqrt{v(t)}dZ(t), \\
d\theta(t) &= \alpha(\beta - \theta(t))dt + \eta dW(t), \\
dv(t) &= a(b - v(t))dt + \phi\sqrt{v(t)}dV(t),
\end{aligned}$$

where Z, W and V are Brownian motions such that

$$dZ\, dV = \rho\, dt, \quad dZ\, dW = 0, \quad dZ\, dV = 0,$$

and k, α,β, η, a, b, ϕ and ρ are constants. The short-rate equation features a time-varying "mean reversion" θ that follows a Vasicek-like process, while the instantaneous absolute volatility v follows a CIR-like process.

This model has been often mentioned by our traders in relationship to factor analysis of the term structure of interest rates under the objective measure.

To move under the risk-neutral measure, particular forms of market price of risk have to be chosen for the three processes.

By doing so, one obtains an affine term-structure model, in the sense of the following multidimensional generalization of the formulation given in Section 3.2.4:

$$P(t,T) = A(T-t) \exp[-r(t)B(T-t) - \theta(t)C(T-t) - v(t)D(T-t)] \; .$$

While B and C can be computed analytically, A and D have to be found through numerical solutions.

Balduzzi, Das, Foresi and Sundaram (1996) examine the possible shapes of the zero-coupon interest-rate curves implied by the above three-factor dynamics. That is, for a fixed t they consider the τ-"curve" $\tau \mapsto R(t, t+\tau)$, and they try to see how this whole map changes according to changes in the factors r, v, θ. They thus find that the three proposed factors account for classical features of the evolution of the term-structure "curve": r provides the *level* of the curve, v is strictly related to the *curvature* and centers its influence at medium-term maturities, while finally θ dictates the steepness of the term structure. This work should be checked by researchers interested in the dynamics of the whole yield curve under the objective measure.

A.3 Flesaker and Hughston's Model

After the short-rate setup, Flesaker and Hughston (1996) (FH) were among the first to propose an entirely new approach to interest-rate modeling resulting in concrete models that are not part of the short-rate world. They in fact model quantities related directly to the *state-price density* (known also as *pricing kernel, pricing operator*). In our context the state-price density can also be viewed as the reciprocal $1/U$ of a chosen numeraire U.

The general framework of FH consists of modeling directly the bank-account numeraire as follows. Consider a family $(M(t,T))_{T \geq t}$ of positive martingales such that

$$M(t,T) \geq M(t,T+\tau) \;\; \text{for all} \;\; t \geq T, \tau \geq 0 \; .$$

The M's are a specification of

$$M(t,T) = \frac{P(t,T)}{B(t)} = E_t[D(0,T)] \ .$$

For those willing to relate our brief exposition to the Flesaker and Hughston (1996) notation, we specify that our $M(t,T)$ is Flesaker and Hughston's $\Delta_{tT}/\rho_t = \Delta_t P(t,T)/\rho_t$, where ρ is the Radon-Nikodym derivative dQ/dQ^U, Q^U being the particular pricing measure under which we work, and Q being the risk-neutral measure.

You may notice that modeling the M's amounts to modeling the ratio between the two privileged numeraires $P(t,T)$ (T-forward measure) and $B(t)$ (risk-neutral measure).

The model allows for positive interest rates in quite a natural way. Roughly, the fact that the M's are taken decreasing in T makes the resulting zero-coupon-bond prices $P(t,T) = B(t)M(t,T)$ decreasing in T, thus ensuring positive rates.

Flesaker and Hughston develop a nice framework starting from particular specifications of M (or Δ). A particularly interesting model is obtained by them when setting $\Delta_{tT} = A_T + B_T S_t$, where A and B are decreasing and positive deterministic functions and S is a unit-initialized (possibly time-changed) geometric Brownian motion. By this approach, bond prices are obtained as

$$P(t,T) = \frac{A_T + B_T S_t}{A_t + B_t S_t},$$

i.e. as a rational function of a lognormal variable S. The same consideration applies to other classical quantities in the interest-rate world, see for example Rapisarda and Silvotti (2001). This is the reason why this particular version of FH's model is called the *rational lognormal model.*

This model features some appealing properties, which follow easily from the rational bond-price formula above. The model prices caps and swaptions analytically with formulas similar to but different from the corresponding market Black formulas. Moreover, this model interacts well with exchange rates, and is particularly suited to be used in situations involving interest-rate curves of different currencies.

However, as a general framework, the model has to be dealt with carefully. Rogers (1997) shows a possible problem: Not all choices of M as above correspond to interest-rate models. Rogers, in fact, presents the example $M(t,T) = \exp(W_t - t/2 - T)$. This definition of M satisfies the conditions above, but $M(t,t) = \exp(W_t - 3t/2)$ cannot bear the interpretation $P(t,t)/B(t) = 1/B(t)$ as requested, since it is not decreasing in t. We notice that the rational lognormal specification avoids this problem, but in general one has to beware this point. Still, when possible, one can check the models resulting from the FH framework on a case by case basis.

For an empirical investigation on the calibration performances of the FH model to market data, also in comparison with other classical one-factor models seen in Chapter 3, see Rapisarda and Silvotti (2001).

A.4 Rogers's Potential Approach

After pointing out that, notwithstanding all its merits, FH's framework has to be handled with care, Rogers (1997) presents a similar framework, which is more rooted in the classical theory of Markov processes. This approach follows the ideas of Constantinides (1992). The model is basically obtained by specifying two objects: A Markov process X, and a positive function f. Then one defines the state-price density ζ_t in terms of X and f. Recall that the state-price density can be essentially interpreted also as the reciprocal of a chosen numeraire, so that $1/\zeta_t$ can be viewed as our basic numeraire under which the model will work and under which discounting will take place. Rogers sets $\zeta_t = e^{-\alpha t} R_\alpha(\alpha - G)f$ and obtains the short rate as $r_t = [(\alpha - G)f(X_t)]/f(X_t)$, where R_α is the resolvent of the process X and G its generator, see Rogers (1997) for the details. Then different models based on different X and f are derived, most of which are based on taking X as Gaussian. As for several other models, empirical investigations on the calibration to caps or swaptions, on the implied evolution for the term structure of volatilities, on terminal correlations between rates and other features are needed. A final note on this approach is that it is suited to model curves in different currencies in quite an elegant and natural way.

A.5 Markov Functional Models

Hunt, Kennedy and Pelsser (1998b) proposed the Markov functional models. Roughly, in this setup it is assumed that the zero-coupon-bond prices P are functionals of some low-dimensional Markov process x, $P(t,T) = \Pi(t,T;x_t)$ for $t \in [0, \mathcal{D}(T)]$, with $\mathcal{D}(T) \leq T$. $\mathcal{D}$ is the "boundary curve for time" and is given as a function of maturities, and is typically $\mathcal{D}(T) = T$ if $T \leq \bar{T}$, or $\mathcal{D}(T) = \bar{T}$ if $T > \bar{T}$, where $\bar{T}$ is some terminal maturity. The process x is assumed to be Markov under a numeraire U, which is in turn a function N of x: $U_t = N(t, x_t)$.

The model is actually specified by assigning the law of x under U, the functional form Π of the discount factors on the boundary, $\Pi(\mathcal{D}(T), T; \cdot)$, and finally the functional form $N(t, \cdot)$.

In other terms, one chooses a fundamental low-dimensional Markov process x, and the model specification requires knowledge of the functional form Π in the chosen process of the zero-coupon-bond price on some boundary $\mathcal{D}$, and the law of a numeraire N. The functional form Π can be chosen so as to calibrate the model to relevant market prices, while the remaining freedom on the choice of the law of the Markov process x is what can allow one to make the model realistic. For example, the freedom in choosing the functional form N can be used to reproduce the marginal distributions of swap rates that are particularly relevant for the calibration considered.

From a methodological point of view, given that the notion of numeraire is close to that of state-price density, and that here the numeraire is taken to depend on a low-dimensional driving Markov process, the reader will notice that this approach looks similar, in some aspects, to the FH and the potential approaches. The main difference is the possibility to choose functional forms so as to match market prices of interest-rate options. Indeed, market distributions of rates at the relevant times are considered and matched as much as possible. Hunt, Kennedy and Pelsser (1998b) note the similarities with the FH and potential models, but observe that both these approaches fail to calibrate well to the distributions of market rates. They point out that the unique feature of their approach is the possibility to recover the correct distributions of the relevant market rates (as in the market models) while keeping low dimensionality at the same time (contrary to the market models). For further details, the interested reader is obviously referred to Hunt, Kennedy and Pelsser (1998b).

"I trust your studies will not have been in vain."

Metron, JLA 15, 1997, DC Comics.

B. Pricing Equity Derivatives under Stochastic Rates

Rates are so low and I've got my eye on a hot stock.
John Grisham, "The Broker", 2005.

The well consolidated theory for pricing equity derivatives under the Black and Scholes (1973) model is based on the assumption of deterministic interest rates. Such an assumption is harmless in most situations since the interest-rates variability is usually negligible if compared to the variability observed in equity markets. When pricing a long-maturity option, however, the stochastic feature of interest rates has a stronger impact on the option price. In such cases it is therefore advisable to relax the assumption of deterministic rates.

The general interest rate theory developed in the first part of this book can be applied to the pricing of equity derivatives under the assumption of stochastic interest rates. We then do so and consider a continuous-time economy where interest rates are stochastic and asset prices evolve according to a geometric Brownian motion. Explicit formulas for European options on a given asset are provided when the instantaneous spot rate follows the Hull and White (1994a) process (3.33). Our pricing procedure is based on the derivation of the asset and rate dynamics under the forward measure whose numeraire is the bond price with the same maturity as the option's. The option prices are then obtained by computing the expectation (2.24).

If the instantaneous spot rate follows a lognormal process like that of Black and Karasinski (1991), as in (3.54), it is not possible to derive closed form formulas for the above European option prices. We can anyway construct an approximating tree, which can also be used for pricing more complex derivatives.

B.1 The Short Rate and Asset-Price Dynamics

We consider a continuous-time economy where interest rates are stochastic and the price of a given tradable asset evolves according to a geometric Brownian motion. We assume that under the risk adjusted measure Q, the dynamics of the instantaneous short rate is given by the Hull and White (1994a) process, see (3.33),

$$dr(t) = [\theta(t) - ar(t)]dt + \sigma dW(t), \quad r(0) = r_0, \tag{B.1}$$

and that the asset price evolves according to

$$dS(t) = S(t)[(r(t) - y)dt + \eta dZ(t)], \quad S(0) = S_0, \tag{B.2}$$

where W and Z are two correlated Brownian motions with

$$dZ(t)\, dW(t) = \rho\, dt,$$

ρ being the instantaneous-correlation parameter between the asset price and the short interest rate, and where $y > 0$ is the asset dividend yield, r_0, a, σ, η and S_0 are positive constants and θ is a deterministic function that is well defined in the time interval $[0, T^*]$, with T^* a given time horizon. We denote by $\mathcal{F}_t$ the sigma-field generated by (r, S) up to time t.

We know from Section 3.3 that we can write

$$r(t) = x(t) + \varphi(t), \quad r(0) = r_0, \tag{B.3}$$

where the process x satisfies

$$dx(t) = -ax(t)dt + \sigma dW(t), \quad x(0) = 0, \tag{B.4}$$

and the function φ is deterministic and well defined in the time interval $[0, T^*]$. In particular, $\varphi(0) = r_0$.

Let us now assume that the term structure of discount factors that is currently observed in the market is given by the sufficiently smooth function $T \mapsto P^M(0, T)$.

We remember from Section 3.3 that, denoting by $f^M(0, T)$ the instantaneous forward rate at time 0 for a maturity T implied by the term structure $T \mapsto P^M(0, T)$, i.e.,

$$f^M(0, T) = -\frac{\partial \ln P^M(0, T)}{\partial T},$$

in order to exactly fit the observed term structure, we must have that, for each T, see (3.36),

$$\varphi(T) = f^M(0, T) + \frac{\sigma^2}{2a^2}\left(1 - e^{-aT}\right)^2. \tag{B.5}$$

Simple integration of the previous SDE's implies that, for each $s < t$,

$$\begin{aligned} r(t) &= r(s)e^{-a(t-s)} + \int_s^t e^{-a(t-u)}\theta(u)du + \sigma \int_s^t e^{-a(t-u)}dW(u) \\ &= x(s)e^{-a(t-s)} + \sigma \int_s^t e^{-a(t-u)}dW(u) + \varphi(t) \end{aligned}$$

and

$$S(t) = S(s) \exp \left\{ \int_s^t r(u)du - y(t-s) - \tfrac{1}{2}\eta^2(t-s) + \eta(Z(t) - Z(s)) \right\}.$$

This means that, conditional on $\mathcal{F}_s$, $r(t)$ is normally distributed with mean and variance given respectively by

$$\begin{aligned} E\{r(t)|\mathcal{F}_s\} &= x(s)e^{-a(t-s)} + \varphi(t), \\ \mathrm{Var}\{r(t)|\mathcal{F}_s\} &= \frac{\sigma^2}{2a}\left[1 - e^{-2a(t-s)}\right]. \end{aligned} \tag{B.6}$$

(see (3.37)). Moreover, also $\ln S(t)$ conditional on $\mathcal{F}_s$ is normally distributed. Its mean and variance are calculated as follows.

By using the following equality, see also Appendix A at the end of Chapter 4,

$$\int_t^T x(u)du = \frac{1 - e^{-a(T-t)}}{a} x(t) + \frac{\sigma}{a} \int_t^T \left[1 - e^{-a(T-u)}\right] dW(u), \tag{B.7}$$

we have that

$$\begin{aligned} \ln \frac{S(t)}{S(s)} &= \int_s^t r(u)du - y(t-s) - \tfrac{1}{2}\eta^2(t-s) + \eta(Z(t) - Z(s)) \\ &= \frac{1 - e^{-a(t-s)}}{a} x(s) + \frac{\sigma}{a} \int_s^t (1 - e^{-a(t-u)})dW(u) + \int_s^t \varphi(u)du \\ &\quad - y(t-s) - \tfrac{1}{2}\eta^2(t-s) + \eta \int_s^t dZ(u), \end{aligned}$$

so that

$$\begin{aligned} E\left\{\ln \frac{S(t)}{S(s)} | \mathcal{F}_s \right\} &= \frac{1 - e^{-a(t-s)}}{a} \left[r(s) - f^M(0,s) - \frac{\sigma^2}{2a^2}\left(1 - e^{-as}\right)^2 \right] \\ &\quad - y(t-s) - \tfrac{1}{2}\eta^2(t-s) + \ln \frac{P^M(0,s)}{P^M(0,t)} \\ &\quad + \frac{\sigma^2}{2a^2}\left[t - s + \frac{2}{a}\left(e^{-at} - e^{-as}\right) - \frac{1}{2a}\left(e^{-2at} - e^{-2as}\right) \right] \end{aligned}$$

and

$$\begin{aligned} \mathrm{Var}\left\{\ln \frac{S(t)}{S(s)} | \mathcal{F}_s \right\} &= \frac{\sigma^2}{a^2} \int_s^t (1 - e^{-a(t-u)})^2 du + \eta^2(t-s) \\ &\quad + 2\rho \frac{\sigma\eta}{a} \int_s^t (1 - e^{-a(t-u)})du \\ &= \frac{\sigma^2}{a^2}\left[t - s - \frac{2}{a}\left(1 - e^{-a(t-s)}\right) + \frac{1}{2a}\left(1 - e^{-2a(t-s)}\right) \right] \\ &\quad + \eta^2(t-s) + 2\rho\frac{\sigma\eta}{a}\left[t - s - \frac{1}{a}\left(1 - e^{-a(t-s)}\right) \right]. \end{aligned}$$

B.1.1 The Dynamics under the Forward Measure

The dynamics of the processes r and S can be also expressed in terms of two independent Brownian motions $\widetilde{W}$ and $\widetilde{Z}$ as follows (Cholesky decomposition):

$$\begin{aligned} dr(t) &= [\theta(t) - ar(t)]dt + \sigma d\widetilde{W}(t), \\ dS(t) &= S(t)[(r(t) - y)dt + \eta\rho d\widetilde{W}(t) + \eta\sqrt{1-\rho^2}d\widetilde{Z}(t)], \end{aligned} \tag{B.8}$$

where

$$\begin{aligned} dW(t) &= d\widetilde{W}(t), \\ dZ(t) &= \rho d\widetilde{W}(t) + \sqrt{1-\rho^2}d\widetilde{Z}(t). \end{aligned}$$

This decomposition makes it easier to perform a measure transformation. In fact, for any fixed maturity T, let us denote by Q^T the T-forward (risk-adjusted) measure, i.e., the probability measure that is defined by the Radon-Nikodym derivative, see Chapter 2,

$$\begin{aligned} \frac{dQ^T}{dQ} &= \frac{\exp\left\{-\int_0^T r(u)du\right\}}{P(0,T)} \\ &= \frac{\exp\left\{-\int_0^T x(u)du - \int_0^T \varphi(u)du\right\}}{P(0,T)} \\ &= \frac{\exp\left\{-\frac{\sigma}{a}\int_0^T \left[1-e^{-a(T-u)}\right]d\widetilde{W}(u) - \int_0^T f^M(0,u)du - \int_0^T \frac{\sigma^2}{2a^2}\left(1-e^{-au}\right)^2 du\right\}}{P(0,T)} \\ &= \exp\left\{-\frac{\sigma}{a}\int_0^T \left[1-e^{-a(T-u)}\right]d\widetilde{W}(u) - \int_0^T \frac{\sigma^2}{2a^2}\left[1-e^{-a(T-u)}\right]^2 du\right\}, \end{aligned} \tag{B.9}$$

where equalities (B.5) and (B.7) have been taken into account.

The Girsanov theorem then implies that the two processes $\widetilde{W}^T$ and $\widetilde{Z}^T$ defined by

$$\begin{aligned} d\widetilde{W}^T(t) &= d\widetilde{W}(t) + \frac{\sigma}{a}\left[1-e^{-a(T-t)}\right]dt \\ d\widetilde{Z}^T(t) &= d\widetilde{Z}(t) \end{aligned} \tag{B.10}$$

are two independent Brownian motions under the measure Q^T. Therefore the (joint) dynamics of r and S under Q^T are given by

$$\begin{aligned} dr(t) &= \left[\theta(t) - \frac{\sigma^2}{a}\left(1-e^{-a(T-t)}\right) - ar(t)\right]dt + \sigma d\widetilde{W}^T(t), \\ dS(t) &= S(t)\left[(r(t)-y) - \rho\frac{\sigma\eta}{a}\left(1-e^{-a(T-t)}\right)\right]dt \\ &\quad + S(t)\left[\eta\rho d\widetilde{W}^T(t) + \eta\sqrt{1-\rho^2}d\widetilde{Z}^T(t)\right]. \end{aligned} \tag{B.11}$$

Integrating equation (B.11) yields, for each $t < T$,

$$\begin{aligned}
r(t) &= r(s)e^{-a(t-s)} + \int_s^t e^{-a(t-u)}\theta(u)du - \frac{\sigma^2}{a}\int_s^t e^{-a(t-u)}\left[1 - e^{-a(T-u)}\right] du \\
&\quad + \sigma \int_s^t e^{-a(t-u)} d\widetilde{W}^T(u) \\
&= x(s)\, e^{-a(t-s)} - \frac{\sigma^2}{a}\int_s^t e^{-a(t-u)}\left[1 - e^{-a(T-u)}\right] du \\
&\quad + \sigma \int_s^t e^{-a(t-u)} d\widetilde{W}^T(u) + \varphi(t)
\end{aligned}$$

and

$$\begin{aligned}
S(T) &= S(t)\exp\Big\{\int_t^T r(u)du - y(T-t) - \rho\frac{\sigma\eta}{a}\int_t^T \left(1 - e^{-a(T-u)}\right) du \\
&\quad - \tfrac{1}{2}\eta^2(T-t) + \eta\rho(\widetilde{W}^T(T) - \widetilde{W}^T(t)) + \eta\sqrt{1-\rho^2}(\widetilde{Z}^T(T) - \widetilde{Z}^T(t))\Big\} \\
&= S(t)\exp\Big\{\frac{1 - e^{-a(T-t)}}{a}x(t) + \frac{\sigma}{a}\int_t^T \left[1 - e^{-a(T-u)}\right] d\widetilde{W}^T(u) \\
&\quad - \frac{\sigma^2}{a}\int_t^T\int_t^u e^{-a(u-s)}\left[1 - e^{-a(T-s)}\right] ds\, du + \int_t^T f^M(0,u)du \\
&\quad + \frac{\sigma^2}{2a^2}\int_t^T \left(1 - e^{-au}\right)^2 du - y(T-t) - \rho\frac{\sigma\eta}{a}\int_t^T \left(1 - e^{-a(T-u)}\right) du \\
&\quad - \tfrac{1}{2}\eta^2(T-t) + \eta\rho(\widetilde{W}^T(T) - \widetilde{W}^T(t)) + \eta\sqrt{1-\rho^2}(\widetilde{Z}^T(T) - \widetilde{Z}^T(t))\Big\}.
\end{aligned}$$

Straightforward calculations lead to

$$\begin{aligned}
&E^T\left\{\ln\frac{S(T)}{S(t)}|\mathcal{F}_t\right\} \\
&= \frac{1 - e^{-a(T-t)}}{a}x(t) - \frac{\sigma^2}{a^2}\left[T - t + \frac{2}{a}e^{-a(T-t)} - \frac{1}{2a}e^{-2a(T-t)} - \frac{3}{2a}\right] \\
&\quad + \ln\frac{P^M(0,t)}{P^M(0,T)} + \frac{\sigma^2}{2a^2}\left[T - t + \frac{2}{a}\left(e^{-aT} - e^{-at}\right) - \frac{1}{2a}\left(e^{-2aT} - e^{-2at}\right)\right] \\
&\quad - y(T-t) - \rho\frac{\sigma\eta}{a}\left[T - t - \frac{1}{a}\left(1 - e^{-a(T-t)}\right)\right] - \tfrac{1}{2}\eta^2(T-t) \\
&= \frac{1 - e^{-a(T-t)}}{a}x(t) - V(t,T) + \ln\frac{P^M(0,t)}{P^M(0,T)} + \tfrac{1}{2}\left[V(0,T) - V(0,t)\right] \\
&\quad - y(T-t) - \rho\frac{\sigma\eta}{a}\left[T - t - \frac{1}{a}\left(1 - e^{-a(T-t)}\right)\right] - \tfrac{1}{2}\eta^2(T-t) \\
&= -\ln(P(t,T)) - \rho\frac{\sigma\eta}{a}\left[T - t - \frac{1}{a}\left(1 - e^{-a(T-t)}\right)\right] \\
&\quad - (y + \tfrac{1}{2}\eta^2)(T-t) - \tfrac{1}{2}V(t,T),
\end{aligned}$$

where

$$V(t,T) := \frac{\sigma^2}{a^2}\left[T - t + \frac{2}{a}e^{-a(T-t)} - \frac{1}{2a}e^{-2a(T-t)} - \frac{3}{2a}\right],$$

and

$$\mathrm{Var}^T\left\{\ln\frac{S(T)}{S(t)}|\mathcal{F}_t\right\} = V(t,T) + \eta^2(T-t)$$
$$+ 2\rho\frac{\sigma\eta}{a}\left[T - t - \frac{1}{a}\left(1 - e^{-a(T-t)}\right)\right].$$

B.2 The Pricing of a European Option on the Given Asset

The results of the previous section allow us to explicitly calculate the price of a European option when the asset price is modeled by a geometric Brownian motion and interest rates are stochastic and evolve according to the Hull and White (1994a) process.

From the general results and assumptions of Chapter 2, we know that the arbitrage-free option price, for a strike K and a maturity T, is

$$\mathcal{O}(t,T,K) = P(t,T)E^{Q^T}\left\{[\psi(S(T)-K)]^+|\mathcal{F}_t\right\},$$

with $\psi \in \{-1,1\}$, and can be calculated by means of formula (D.2). Indeed, we have just to replace M and V^2 in such a formula respectively with $m(t,T)$ and $v^2(t,T)$:

$$m(t,T) := \ln\frac{S(t)}{P(t,T)} - \rho\frac{\sigma\eta}{a}\left[T - t - \frac{1}{a}\left(1 - e^{-a(T-t)}\right)\right]$$
$$- (y + \tfrac{1}{2}\eta^2)(T-t) - \tfrac{1}{2}V(t,T),$$
$$v^2(t,T) := V(t,T) + \eta^2(T-t) + 2\rho\frac{\sigma\eta}{a}\left[T - t - \frac{1}{a}\left(1 - e^{-a(T-t)}\right)\right].$$

We therefore have the following.

Proposition B.2.1. *The price at time t of a European option with maturity T, strike K, and written on the asset S is given by*

$$\mathcal{O}(t,T,K) = \psi S(t)e^{-y(T-t)}\Phi\left(\psi\frac{\ln\frac{S(t)}{KP(t,T)} - y(T-t) + \frac{1}{2}v^2(t,T)}{v(t,T)}\right)$$
$$- \psi KP(t,T)\Phi\left(\psi\frac{\ln\frac{S(t)}{KP(t,T)} - y(T-t) - \frac{1}{2}v^2(t,T)}{v(t,T)}\right). \tag{B.12}$$

where $\psi = 1$ for a call and $\psi = -1$ for a put.

Formula (B.12) is in agreement with the result of Merton (1973), who derived a closed-form solution for the price of a European call option on a risky asset when interest rates are stochastic. He assumed the existence of a zero-coupon-bond price following a diffusion process where the variance rate of the instantaneous bond's return is deterministic and vanishes at the bond's maturity. Merton's model gives the following formula for the price of the above European call option ($\psi = 1$)

$$S(t)e^{-y(T-t)}\Phi\left(\frac{\ln\frac{S(t)}{KP(t,T)} - y(T-t) + \frac{1}{2}u_t^2}{u_t}\right)$$
$$-KP(t,T)\Phi\left(\frac{\ln\frac{S(t)}{KP(t,T)} - y(T-t) - \frac{1}{2}u_t^2}{u_t}\right),$$

where u_t^2 satisfies

$$u_t^2 = \int_t^T \mathrm{Var}\left\{d\ln\frac{S(\tau)}{P(\tau,T)}\right\},$$

meaning that u_t^2 is the integrated variance of the instantaneous return of $S(t)/P(t,T)$. This formula holds in general for Gaussian short-rate models, not necessarily one-factor. In particular, one can use this expression under a multi-factor Gaussian model such as G2++.

Remark B.2.1. The difference $v^2(t,T) - \eta^2(T-t)$ is an increasing function of the time to maturity $T-t$. Therefore, as expected, the larger the option maturity, the larger the impact of the stochastic behaviour of interest rates on the option price.

B.3 A More General Model

We now assume that the dynamics of the instantaneous short rate under the risk-adjusted measure Q is given by the more general process

$$\begin{cases} dx(t) = -ax(t)dt + \sigma dW(t), \;\; x(0) = 0, \\ r(t) = f(x(t) + \alpha(t)), \end{cases} \tag{B.13}$$

where f is a deterministic real function with inverse g (i.e., $g(f(x)) = x$) and α is a deterministic function that is properly chosen so as to exactly fit the current term structure of spot rates. The asset price is still assumed to evolve according to (B.2), i.e.,

$$dS(t) = S(t)[(r(t) - y)dt + \eta dZ(t)], \;\; S(0) = S_0,$$

where W and Z are again two Brownian motions with $dZ(t)dW(t) = \rho dt$. Putting $\bar{S}(t) = \ln(S(t)/S_0)$, we denote by $\mathcal{F}_t$ the sigma-field generated by $(x, \bar{S})$ up to time t.

Assuming enough regularity of the deterministic functions above, by Ito's lemma, the dynamics of r is given by

$$dr(t) = \left[f'(g(r(t)))\alpha'(t) - af'(g(r(t)))(g(r(t)) - \alpha(t)) + \tfrac{1}{2}f''(g(r(t)))\sigma^2\right] dt \\ + \sigma f'(g(r(t)))dW(t), \tag{B.14}$$

with $'$ denoting a derivative, so that r's absolute volatility is $\sigma f'(g(r(t)))$ and the instantaneous correlation between r and S is ρ.

For example, we can retrieve the Black and Karasinski (1991) model, see (3.54), by setting $f(x) = \exp(x)$, so that $g(x) = \ln(x)$. In this case, (B.14) becomes

$$dr(t) = \left[r(t)(\alpha'(t) + a\alpha(t)) - ar(t)\ln(r(t)) + \tfrac{1}{2}\sigma^2 r(t)\right] dt + \sigma r(t)dW(t).$$

The model (B.13) is not analytically tractable in general. We have, therefore, to resort to numerical procedures even to price the simplest derivatives. Indeed, even the fundamental quantities of interest-rate modeling, i.e. bond prices, do not have a closed-form expression with this model.

To this end, we first construct two trinomial trees, one for the process r and the other for the process S, and then we merge the two trees accounting for the proper correlation. The construction of the trinomial trees for r and S is reviewed in the following.

B.3.1 The Construction of an Approximating Tree for r

We first construct a trinomial tree for the process x based on a generalization of the Hull and White (1993) procedure so as to allow for a variable time step. See also Appendix F.

Let us assume that we are given a sequence of times $t_0, t_1, \ldots, t_m$. We set $\Delta t_i = t_{i+1} - t_i$. We denote the value of x on the j-th node at time t_i by $x_{i,j}$ and we set $x_{i,j} := j\Delta x_i$, where the vertical step Δx_i is constant at each time t_i. The mean and standard deviation of $x(t_{i+1})$ conditional on $x = x_{i,j}$ are $x_{i,j} + M_i x_{i,j}$ and $\sqrt{V_i}$, respectively, where

$$M_i := e^{-a\Delta t_i} - 1,$$
$$V_i := \frac{\sigma^2}{2a}\left(1 - e^{-2a\Delta t_i}\right).$$

The branching procedure is established as follows. From $x_{i,j}$, the variable x can move to $x_{i+1,k-1}$, to $x_{i+1,k}$ or to $x_{i+1,k+1}$, where k is chosen so that $x_{i+1,k}$ is as close as possible to $x_{i,j} + M_i x_{i,j}$. The associated probabilities are calculated by matching the mean and variance of the increments of the original continuous-time process. We obtain

$$\begin{cases} p_u(i,j) = \dfrac{\sigma^2 \Delta t_i}{2\Delta x_{i+1}^2} + \dfrac{\eta^2(i,j)}{2\Delta x_{i+1}^2} + \dfrac{\eta(i,j)}{2\Delta x_{i+1}}, \\ p_m(i,j) = 1 - \dfrac{\sigma^2 \Delta t_i}{\Delta x_{i+1}^2} - \dfrac{\eta^2(i,j)}{\Delta x_{i+1}^2}, \\ p_d(i,j) = \dfrac{\sigma^2 \Delta t_i}{2\Delta x_{i+1}^2} + \dfrac{\eta^2(i,j)}{2\Delta x_{i+1}^2} - \dfrac{\eta(i,j)}{2\Delta x_{i+1}}, \end{cases}$$

where $p_u(i,j)$, $p_m(i,j)$ and $p_d(i,j)$ denote the up, middle and down branching probabilities on the j-th node at time t_i, respectively, and where

$$\eta(i,j) = x_{i,j} M_i + j\Delta x_i - k\Delta x_{i+1}, \tag{B.15}$$

with k to be determined at each node in the tree as previously explained. A portion of this trinomial tree for x is displayed in Figure B.1. The branching

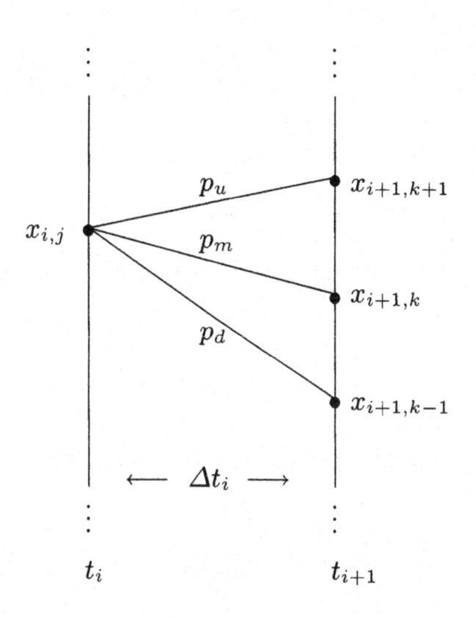

Fig. B.1. Evolution of the process x at node (i,j).

probabilities should always lie in the interval $[0,1]$. This can be ensured by choosing Δx_{i+1} in such a way that, at each node in the tree, $x_{i+1,k-1}$ and $x_{i+1,k+1}$ bracket the expected value of x in the next time interval conditional to $x_{i,j}$. Hull and White (1993) suggested $\Delta x_{i+1} = \sqrt{3V_i}$ as an appropriate choice. We denote by $\underline{j}_i$ and $\bar{j}_i$ the minimum and the maximum levels at each time step i.

Proceeding in the tree construction, the following step is the displacement of the tree nodes according to the function α. To this end, we denote by $Q(i,j)$ the present value of an instrument paying 1 if node (i,j) is reached and zero otherwise. The values of the displacement $\alpha_i := \alpha(t_i)$ at period i and of $Q(i,j)$ are calculated recursively from $\alpha(0) = g(r_0)$. Precisely, as soon as the value of α_i has been determined, the values $Q(i+1,j)$, $j = \underline{j}_i, \ldots, \bar{j}_i$, are calculated through

$$Q(i+1,j) = \sum_k Q(i,k)p(k,j)\exp(-f(\alpha_i + k\Delta x_i)\Delta t_i),$$

where $p(k,j)$ is the probability of moving from node (i,k) to node $(i+1,j)$ and the sum is over all values of k for which such probability is non-zero.

After deriving the value of $Q(i,j)$, for each j, the value of α_i is calculated by numerically solving

$$\sum_{j=-n_i}^{n_i} Q(i,j)\exp\left[-f(\alpha_i + j\Delta x_i)\Delta t_i\right] - P^M(0,t_{i+1}) = 0$$

where $P^M(0,t_{i+1})$ denotes (as before) the market price at the initial time of a discount bond with maturity t_{i+1} and n_i is the number of nodes on each side of the central node at the i-th time step in the tree.

To obtain the approximating tree for r we finally have to apply the function f to each node value.

B.3.2 The Approximating Tree for S

We first build an approximating tree for $\bar{S}$ assuming that interest rates are constant and equal to $\bar{r}$.

Setting $\Delta\bar{S}_{i+1} := \eta\sqrt{3\Delta t_i}$, for each i, we denote by $q_u(i,l)$, $q_m(i,l)$ and $q_d(i,l)$ the probabilities at period i of moving from $\bar{S}_{i,l} := l\Delta\bar{S}_i$ to $\bar{S}_{i+1,h+1} = \bar{S}_{i+1,h} + \Delta\bar{S}_{i+1}$, $\bar{S}_{i+1,h}$ and $\bar{S}_{i+1,h-1} = \bar{S}_{i+1,h} - \Delta\bar{S}_{i+1}$, respectively. Applying the same "first-two-moments-matching" procedure as before, we obtain

$$\begin{cases} q_u(i,l) = \dfrac{1}{6} + \dfrac{\xi_{l,h}^2}{6\eta^2\Delta t_i} + \dfrac{\xi_{l,h}}{2\sqrt{3}\eta\sqrt{\Delta t_i}}, \\ q_m(i,l) = \dfrac{2}{3} - \dfrac{\xi_{l,h}^2}{3\eta^2\Delta t_i}, \\ q_d(i,l) = \dfrac{1}{6} + \dfrac{\xi_{l,h}^2}{6\eta^2\Delta t_i} - \dfrac{\xi_{l,h}}{2\sqrt{3}\eta\sqrt{\Delta t_i}}, \end{cases} \tag{B.16}$$

where

$$h = \text{round}\left(\frac{\bar{S}_{i,l} + (\bar{r} - y - \frac{1}{2}\eta^2)\Delta t_i}{\Delta\bar{S}_{i+1}}\right),$$
$$\xi_{l,h} = \bar{S}_{i,l} + (\bar{r} - y - \tfrac{1}{2}\eta^2)\Delta t_i - \bar{S}_{i+1,h}.$$

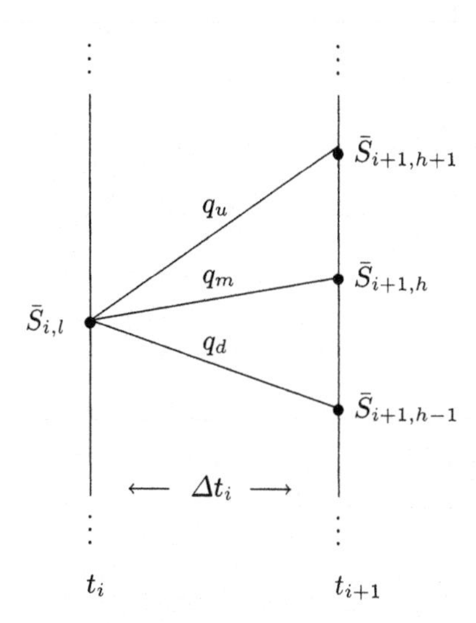

Fig. B.2. Evolution of the process $\bar{S}$ starting from the value $\bar{S}_{i,l}$ at time t_i.

The resulting tree geometry is displayed in Figure B.2. To finally obtain the approximating tree for S we have simply to transform each node value by applying the relation $S(t) = S_0 \exp(\bar{S}(t))$.

B.3.3 The Two-Dimensional Tree

In the construction of the joint two-dimensional tree, we have to remember that the marginal probabilities (B.16) explicitly depend on the spot rate r, so that they vary as soon as r varies in the joint tree.

Along the procedure suggested by Hull and White (1994b), we build a preliminary tree by assuming zero correlation. See again Appendix F.

Denoting by (i, j, l) the tree node at time t_i where $x = x_{i,j} = j\Delta x_i$ and $\bar{S} = \bar{S}_{i,l} = l\Delta\bar{S}_i$, if k and h are chosen as before, we denote by

- π_{uu} the probability of moving from (i, j, l) to $(i+1, k+1, h+1)$;
- π_{um} the probability of moving from (i, j, l) to $(i+1, k+1, h)$;
- π_{ud} the probability of moving from (i, j, l) to $(i+1, k+1, h-1)$;
- π_{mu} the probability of moving from (i, j, l) to $(i+1, k, h+1)$;
- π_{mm} the probability of moving from (i, j, l) to $(i+1, k, h)$;
- π_{md} the probability of moving from (i, j, l) to $(i+1, k, h-1)$;
- π_{du} the probability of moving from (i, j, l) to $(i+1, k-1, h+1)$;
- π_{dm} the probability of moving from (i, j, l) to $(i+1, k-1, h)$;
- π_{dd} the probability of moving from (i, j, l) to $(i+1, k-1, h-1)$;

Under zero correlation, these probabilities are simply the product of the corresponding marginal probabilities, where in (B.16) we replace $\bar{r}$ with

$f(x_{i,j} + \alpha_i)$. For example,

$$\pi_{ud} = \left[\frac{\sigma^2 \Delta t_i}{2\Delta x_{i+1}^2} + \frac{\eta^2(i,j)}{2\Delta x_{i+1}^2} + \frac{\eta(i,j)}{2\Delta x_{i+1}}\right]\left[\frac{1}{6} + \frac{\xi_{l,h}^2}{6\eta^2 \Delta t_i} - \frac{\xi_{l,h}}{2\sqrt{3}\eta\sqrt{\Delta t_i}}\right].$$

Let us now define Π_0 as the matrix of these probabilities, i.e.,

$$\Pi_0 := \begin{pmatrix} \pi_{ud} & \pi_{um} & \pi_{uu} \\ \pi_{md} & \pi_{mm} & \pi_{mu} \\ \pi_{dd} & \pi_{dm} & \pi_{du} \end{pmatrix}.$$

To account for the proper correlation ρ we have to shift each probability in Π_0 in such a way that the sum of the shifts in each row and each column is zero (so that the marginal distributions are maintained) and the following is verified:

$$\begin{aligned} &\Delta x_{i+1}\Delta \bar{S}_{i+1}\left(-\pi_{ud} - \varepsilon_{ud} + \pi_{uu} + \varepsilon_{uu} + \pi_{dd} + \varepsilon_{dd} - \pi_{du} - \varepsilon_{du}\right) \\ &\quad = \mathrm{Cov}\{x(t_{i+1}), \bar{S}(t_{i+1})|\mathcal{F}_{t_i}\} \\ &\quad = \sigma\eta\rho\Delta t_i, \end{aligned}$$

at first order in Δt_i. Clearly, no middle "m" term appears, since such terms correspond to at least one of the two components being displaced by 0, thus giving no contribution to the above covariance. Now, recalling that

$$\Delta x_{i+1} = \sqrt{3V_i} = \sqrt{3\frac{\sigma^2}{2a}(1 - e^{-2a\Delta t_i})} \approx \sigma\sqrt{3\Delta t_i}$$

and that

$$\Delta \bar{S}_{i+1} = \eta\sqrt{3\Delta t_i},$$

we obtain that the above equation becomes, in the limit,

$$-\varepsilon_{ud} + \varepsilon_{uu} + \varepsilon_{dd} - \varepsilon_{du} = \frac{\rho}{3},$$

where ε_{ud} denotes the shift for π_{ud} (the definition of the other shifts is analogous).

Solving for the previous constraints, a possible solution, which is similar to that proposed by Hull and White (1994b), is

$$\begin{aligned} \Pi_\rho &= \Pi_0 + \rho(\Pi_1^l - \Pi_0^l), \text{ if } \rho > 0 \\ \Pi_\rho &= \Pi_0 - \rho(\Pi_{-1}^l - \Pi_0^l), \text{ if } \rho < 0 \end{aligned}$$

where Π_ρ is the probability matrix that accounts for the right correlation ρ and Π_1^l, Π_0^l and Π_{-1}^l are the probability matrices in the limit ($\Delta t_i = 0$) respectively for $\rho = 1$, $\rho = 0$ and $\rho = -1$. More explicitly,

$$\Pi_\rho = \Pi_0 + \frac{\rho}{36}\begin{pmatrix} -1 & -4 & 5 \\ -4 & 8 & -4 \\ 5 & -4 & -1 \end{pmatrix} \text{ if } \rho > 0$$

and

$$\Pi_\rho = \Pi_0 - \frac{\rho}{36}\begin{pmatrix} 5 & -4 & -1 \\ -4 & 8 & -4 \\ -1 & -4 & 5 \end{pmatrix} \text{ if } \rho < 0.$$

However, it may happen that one or more entries in Π_ρ are negative. To overcome this drawback, we are compelled to modify the correlation on node (i, j, l), for example substituting Π_ρ with Π_0. Assuming a different correlation on such node, and on all the tree nodes where some probabilities are negative, has a negligible impact on a derivative price as long as the mesh of $\{t_0, t_1, \ldots, t_m\}$ is sufficiently small. Notice, in fact, that in the limit for such a mesh going to zero, the elements of Π_ρ are all positive, being Π_ρ a convex linear combination of two matrices with positive entries.

C. A Crash Intro to Stochastic Differential Equations and Poisson Processes

The principle of generating small amounts of finite improbability by simply hooking the logic circuits of a Bambleweeny 57 Sub-Meson Brain to an atomic vector plotter suspended in a strong Brownian motion producer (say a nice hot cup of tea) were of course well understood [...]
Douglas Adams, "The Hitch-Hiker's Guide to the Galaxy"

This book uses continuous time stochastic calculus as a mathematical tool for financial modeling. In this appendix we plan to give a quick (informal) introduction to stochastic differential equations (SDEs) for the reader who is not familiar with this field. In order to have stochastic differential equations defined we will illustrate the Brownian motion process, a fundamental continuous-paths process. Given its importance in default modeling, we also introduce the Poisson process, to some extent the purely jump analogous of Brownian motion. Brownian motions and Poisson processes are among the most important random processes of probability.

These notes are far from being complete or fully rigorous, in that we privilege the intuitive aspect, but we give references for the reader who is willing to deepen her knowledge on such matters.

We note that the understanding, and subsequent implementation, of most of the essential and important issues in interest rate modeling do not require excessively-exotic tools of stochastic calculus. The basic paradigms, risk neutral valuation and change of numeraire, in fact, essentially involve Ito's formula and the Girsanov theorem. We therefore introduce quickly and intuitively such results, which are used especially in Chapter 2. The fact that we do not insist upon more advanced tools of stochastic calculus is mainly due to our belief that the fundamental questions to address in practice can be very often solved with the basic tools above.

C.1 From Deterministic to Stochastic Differential Equations

Here we present a quick and informal introduction to SDEs. We consider the scalar case to simplify exposition.

We consider a probability space $(\Omega, \mathcal{F}, (\mathcal{F}_t)_t, \mathbb{P})$. The usual interpretation of this space as an experiment can help intuition. The generic experiment result is denoted by $\omega \in \Omega$; Ω represents the set of all possible outcomes of the random experiment, and the σ-field $\mathcal{F}$ represents the set of events $A \subset \Omega$ with which we shall work. The σ-field $\mathcal{F}_t$ represents the information available up to time t. We have $\mathcal{F}_t \subseteq \mathcal{F}_u \subseteq \mathcal{F}$ for all $t \leq u$, meaning that "the information increases in time", never exceeding the whole set of events $\mathcal{F}$. The family of σ-fields $(\mathcal{F}_t)_{t\geq 0}$ is called filtration.

If the experiment result is ω and $\omega \in A \in \mathcal{F}$, we say that the event A occurred. If $\omega \in A \in \mathcal{F}_t$, we say that the event A occurred at a time smaller or equal to t.

We use the symbol $\mathbb{E}$ to denote expectation, and $\mathbb{E}[\cdot|\mathcal{F}]$ denotes expectation conditional on the information contained in $\mathcal{F}$.

We begin by a simple example. Consider a population growth model. Let $x(t) = x_t \in \mathbb{R}$, $x_t \geq 0$, be the population at time $t \geq 0$. The simplest model for the population growth is obtained by assuming that the growth rate dx_t/dt is proportional to the current population. This can be translated into the differential equation:

$$dx_t = Kx_t dt, \quad x_0,$$

where K is a real constant. Now suppose that, due to some complications, it is no longer realistic to assume the initial condition x_0 to be a deterministic constant. Then we may decide to let x_0 be a random variable $X_0(\omega)$, and to model the population growth by the differential equation:

$$dX_t(\omega) = KX_t(\omega)dt, \quad X_0(\omega).$$

The solution of this last equation is $X_t(\omega) = X_0(\omega)\exp[Kt]$. Note that $X_t(\omega)$ is a random variable, but all its randomness comes from the initial condition $X_0(\omega)$. For each experiment result ω, the map $t \mapsto X_t(\omega)$ is called the path of X associated to ω.

As a further step, suppose that not even K is known for certain, but that also our knowledge of K is perturbed by some randomness, which we model as the "increment" of a stochastic process $\{W_t(\omega),\ t \geq 0\}$, so that

$$dX_t(\omega) = (Kdt + dW_t(\omega))\, X_t(\omega)\ , \quad X_0(\omega), \quad K \geq 0. \tag{C.1}$$

Here, $dW_t(\omega)$ represents a noise process that adds randomness to K.

Equation (C.1) is an example of stochastic differential equation (SDE). More generally, a SDE is written as

$$dX_t(\omega) = f_t(X_t(\omega))dt + \sigma_t(X_t(\omega))dW_t(\omega), \quad X_0(\omega)\,. \tag{C.2}$$

The function f, corresponding to the deterministic part of the SDE, is called the *drift*. The function σ_t (or sometimes its square $a_t := \sigma_t^2$) is called the *diffusion coefficient*. Note that the randomness enters the differential equation from two sources: The "noise term" $\sigma_t(\cdot)dW_t(\omega)$ and the initial condition $X_0(\omega)$.

Usually, the solution X of the SDE is called also a *diffusion process*, because of the fact that some particular SDE's can be used to arrive at a model of physical diffusion. In general the paths $t \mapsto X_t(\omega)$ of a diffusion process X are continuous.

Brownian Motion

The process whose "increments" $dW_t(\omega)$ are candidate for representing the noise process in (C.2) is the Brownian motion. This process has important properties: It has stationary and independent Gaussian increments "$dW_t(\omega)$", or more precisely for any $0 < s < t < u$ and any $h > 0$:

$$W_u(\omega) - W_t(\omega) \text{ independent of } W_t(\omega) - W_s(\omega) \quad \text{(indep. increments)} \tag{C.3}$$

$$W_{t+h}(\omega) - W_{s+h}(\omega) \sim W_t(\omega) - W_s(\omega) \quad \text{(stationary increments)} \tag{C.4}$$

$$W_t(\omega) - W_s(\omega) \sim \mathcal{N}(0, t-s) \quad \text{(Gaussian increments).} \tag{C.5}$$

The above assumptions imply intuitively that, for example, $W_{t+h}(\omega) - W_t(\omega)$ is independent of the history of W up to time t. Therefore, $W_{t+h}(\omega) - W_t(\omega)$ can assume *any* value independently of $\{W_s(\omega), s \leq t\}$.

The definition of Brownian motion requires also the paths $t \mapsto W_t(\omega)$ to be continuous. It turns out that the properties listed above imply that although the path be continuous, they are (almost surely) nowhere differentiable. In fact, the paths have unbounded variation, and hence

$$\dot{W}_t(\omega) = \frac{d}{dt} W_t(\omega)$$

does not exist.

Stochastic Integrals

Since $\dot{W}_t(\omega)$ does not exist, what meaning can we give to equation (C.2)? The answer relies on rewriting (C.2) in integral form:

$$X_t(\omega) = X_0(\omega) + \int_0^t f_s(X_s(\omega))\, ds + \int_0^t \sigma_s(X_s(\omega))\, dW_s(\omega), \tag{C.6}$$

so that, from now on, all differential equations involving terms like dW are meant as integral equations, in the same way as (C.2) will be an abbreviation for (C.6).

However, we are not done yet, since we have to deal with the new problem of defining an integral like $\int_0^t \sigma_s(X_s(\omega))dW_s(\omega)$. A priori it is not possible to define it as a Stieltjes integral on the paths, since they have unbounded variation. Nonetheless, under some "reasonable" assumptions that we do not mention (see for example Chapter 3 of Øksendal (1992)), it is still possible to define such integrals a la Stieltjes. The price to be paid is that the resulting integral will depend on the chosen points of the sub-partitions (whose mesh tends to zero) used in the limit that defines the integral. More specifically, consider the following definition. Take an interval $[0, T]$ and consider the following dyadic partition of $[0, T]$ depending on an integer n,

$$T_i^n = \min\left(T, \frac{i}{2^n}\right), \quad i = 0, 1, \ldots, \infty.$$

Notice that from a certain i on all terms collapse to T, i.e. $T_i^n = T$ for all $i > 2^n T$. For each n we have such a partition, and when n increases the partition contains more elements, giving a better discrete approximation of the continuous interval $[0, T]$. Then define the integral as

$$\int_0^T \phi_s(\omega) dW_s(\omega) = \lim_{n\to\infty} \sum_{i=0}^{\infty} \phi_{t_i^n}(\omega)[W_{T_{i+1}^n}(\omega) - W_{T_i^n}(\omega)]$$

where t_i^n is any point in the interval $[T_i^n, T_{i+1}^n)$. Now, by choosing $t_i^n := T_i^n$ (initial point of the subinterval) we have the definition of the Ito integral, whereas by taking $t_i^n := (T_i^n + T_{i+1}^n)/2$ (middle point) we obtain a different result, the Stratonovich integral.

The Ito integral has interesting probabilistic properties (for example, it is a *martingale*, an important type of stochastic process that will be briefly defined below), but leads to a calculus where the standard chain rule is not preserved since there is a non-zero contribution of the second order terms. On the contrary, although probabilistically less interesting, the Stratonovich integral does preserve the ordinary chain rule, and is preferable from the viewpoint of properties of the paths.

To better understand the difference between these two definitions, we can resort to the following classical example of stochastic integral computed both with the Ito calculus and the Stratonovich calculus:

$$\text{Ito} \Rightarrow \int_0^t W_s(\omega) dW_s(\omega) = \frac{W_t(\omega)^2}{2} - \frac{1}{2}t\,,$$
$$\text{Stratonovich} \Rightarrow \int_0^t W_s(\omega) dW_s(\omega) = \frac{W_t(\omega)^2}{2}\,.$$

To distinguish between the two definition, a symbol "$\circ$" is often introduced to denote the Stratonovich version as follows:

$$\int_0^t W_s(\omega) \circ dW_s(\omega).$$

In differential notation, one then has

$$\text{Ito} \Rightarrow d(W_t(\omega)^2) = dt + 2W_t(\omega)dW_t(\omega)\,, \tag{C.7}$$
$$\text{Stratonovich} \Rightarrow d(W_t(\omega)^2) = 2W_t(\omega) \circ dW_t(\omega)\,, \tag{C.8}$$

In the Ito version, the "dt" term originates from second order effects, which are not negligible like in ordinary calculus. Note that the first integral is a martingale (so that, for example, it has constant expected value equal to zero, which is an important probabilistic property), but does not satisfy formal rules of calculus, as instead does the second one (which is not a martingale).[1]

[1] We must point out, anyway, that it is possible to transform an equation written in the Ito form into an equation in the Stratonovich form with the same solution by altering the drift, and vice-versa. This is referred to as the Ito-Stratonovich transformation, see Øksendal (1992), Chapter 3.

In general, stochastic integrals are defined a la Lebesgue rather than a la Riemann-Stieltjes. One defines the stochastic integral for increasingly more sophisticated integrands (indicators, simple functions...), and then takes the limit in some sense. The reader interested in the deeper mathematical aspects of stochastic integration in connection with SDEs may consult books such as, for example, Øksendal (1992) or, for a more advanced treatment, Rogers and Williams (1987).

Martingales, Driftless SDEs and Semimartingales

In our discussion above, we have mentioned the concept of martingale. To give a quick idea, consider a process X satisfying the following measurability and integrability conditions.

Measurability: $\mathcal{F}_t$ includes all the information on X up to time t, usually expressed in the literature by saying that $(X_t)_t$ is *adapted* to $(\mathcal{F}_t)_t$;

Integrability: the relevant expected values exist.

A martingale is a process satisfying these two conditions and such that the following property holds for each $t \leq T$:

$$\mathbb{E}[X_T|\mathcal{F}_t] = X_t.$$

This definition states that, if we consider t as the present time, the expected value at a future time T given the current information is equal to the current value. This is, among other things, a picture of a "fair game", where it is not possible to gain or lose on average. It turns out that the martingale property is also suited to model the absence of arbitrage in mathematical finance. To avoid arbitrage, one requires that certain fundamental processes of the economy be martingales, so that there are no "safe" ways to make money from nothing out of them.

Consider an SDE admitting a unique solution (resulting in a diffusion process, as stated above). This solution is a martingale when the equation has zero drift. In other terms, the solution of the SDE (C.2) is a martingale when $f_t(\cdot) = 0$ for all t:

$$dX_t(\omega) = \sigma_t(X_t(\omega))dW_t(\omega), \quad X_0(\omega).$$

Therefore, in diffusion-processes language, martingale means driftless diffusion process.

A submartingale is a similar process X satisfying instead

$$\mathbb{E}[X_T|\mathcal{F}_t] \geq X_t.$$

This means that the expected value of the process grows in time, and that averages of future values of the process given the current information always exceed (or at least are equal to) the current value.

Similarly, a supermartingale satisfies

$$\mathbb{E}[X_T|\mathcal{F}_t] \leq X_t,$$

and the expected value of the process decreases in time, so that averages of future values of the process given the current information are always smaller than (or at most are equal to) the current value.

A process X that is either a supermartingale or a submartingale is usually termed semimartingale.

Quadratic Variation

The quadratic variation of a stochastic process Y_t with continuous paths $t \mapsto Y_t(\omega)$ is defined as follows:

$$\langle Y \rangle_T = \lim_{n \to \infty} \sum_{i=1}^{\infty} (Y_{T_i^n}(\omega) - Y_{T_{i-1}^n}(\omega))^2.$$

Intuitively this could be written as a "second order" integral:

$$\langle Y \rangle_T = \text{"} \int_0^T (dY_s(\omega))^2 \text{"} ,$$

or, even more intuitively, in the differential form

$$\text{"}d\langle Y \rangle_t = dY_t(\omega) dY_t(\omega)\text{"} .$$

It is easy to check that a process Y whose paths $t \mapsto Y_t(\omega)$ are differentiable for almost all ω satisfies $\langle Y \rangle_t = 0$. In case Y is a Brownian motion, it can be proved, instead, that

$$\langle W \rangle_T = T, \quad \text{for each } T,$$

which can be written in a more informal way as

$$\boxed{dW_t(\omega)\, dW_t(\omega) = dt} .$$

Again, this comes from the fact that the Brownian motion moves so quickly that second order effects are not negligible. Instead, a process whose trajectories are differentiable cannot move so quickly, and therefore its second order effects do not contribute.

In case the process Y is equal to the deterministic process $t \mapsto t$, so that $dY_t = dt$, we immediately retrieve the classical result from (deterministic) calculus:

$$\boxed{dt\, dt = 0} .$$

Quadratic Covariation

One can also define the quadratic covariation of two processes Y and Z with continuous paths as

$$\langle Y, Z\rangle_T = \lim_{n\to\infty} \sum_{i=1}^{\infty} (Y_{T_i^n}(\omega) - Y_{T_{i-1}^n}(\omega))(Z_{T_i^n}(\omega) - Z_{T_{i-1}^n}(\omega)).$$

Intuitively this could be written as a "second order" integral:

$$\langle Y, Z\rangle_T = \text{"} \int_0^T dY_s(\omega) dZ_s(\omega) \text{"},$$

or, in differential form,

$$\text{"}d\langle Y, Z\rangle_t = dY_t(\omega) dZ_t(\omega)\text{"} .$$

It is then easy to check that, denoting again by t the deterministic process $t \mapsto t$,

$$\langle W, t\rangle_T = 0, \quad \text{for each } T,$$

which can be informally written as

$$\boxed{dW_t(\omega)\, dt = 0} .$$

Solution to a General SDE

Let us go back to our general SDE, and let us take time-homogeneous coefficients for simplicity:

$$dX_t(\omega) = f(X_t(\omega))dt + \sigma(X_t(\omega))dW_t(\omega), \quad X_0(\omega) . \tag{C.9}$$

Under which conditions does it admit a unique solution in the Ito sense? Standard theory tells us that it is enough to have both the f and σ coefficients satisfying Lipschitz continuity (and linear growth, which does not follow automatically in the time-inhomogeneous case or with *local* Lipschitz continuity only). These *sufficient* conditions are valid for deterministic differential equations as well, and can be weakened, especially in dimension one. Typical examples showing how, without Lipschitz continuity or linear growth, existence and uniqueness of solutions can fail are the following:

$$dx_t = x_t^2\, dt, \quad x_0 = 1 \quad \Rightarrow x_t = \frac{1}{1-t}, \quad t \in [0, 1),$$

(explosion in finite time) and

$$dx_t = 3x_t^{2/3}\, dt, \quad x_0 = 0 \quad \Rightarrow x_t = 1_{(a,\infty)}(t)\; (t-a)^3, \quad t \in [0, +\infty)$$

for any positive a (no uniqueness).

The proof of the fact that the existence and uniqueness of a solution to a SDE is guaranteed by Lipschitz continuity and linear growth of the coefficients, is similar in spirit to the proof for deterministic equations. See again Øksendal (1992) for the details.

Interpretation of the Coefficients of the SDE

We conclude by presenting an interpretation of the drift and diffusion coefficient for a SDE. For deterministic differential equations such as

$$dx_t = f(x_t)dt\,,$$

with a smooth function f, one clearly has

$$\lim_{h\to 0} \frac{x_{t+h} - x_t}{h}\bigg|_{x_t=y} = f(y)\,,$$

$$\lim_{h\to 0} \frac{(x_{t+h} - x_t)^2}{h}\bigg|_{x_t=y} = 0.$$

The "analogous" relations for the SDE

$$dX_t(\omega) = f(X_t(\omega))dt + \sigma(X_t(\omega))dW_t(\omega),$$

are the following:

$$\lim_{h\to 0} \mathbb{E}\left\{ \frac{X_{t+h}(\omega) - X_t(\omega)}{h} \bigg| X_t = y \right\} = f(y)$$

$$\lim_{h\to 0} \mathbb{E}\left\{ \frac{[X_{t+h}(\omega) - X_t(\omega)]^2}{h} \bigg| X_t = y \right\} = \sigma^2(y)\,.$$

The second limit is non-zero because of the "infinite velocity" of Brownian motion, while the first limit is the analogous of the deterministic case.

C.2 Ito's Formula

Now we are ready to introduce the famous Ito formula, which gives the "chain rule" for differentials in a stochastic context.

For deterministic differential equations such as

$$dx_t = f(x_t)dt,$$

given a smooth transformation $\phi(t,x)$, one can write the evolution of $\phi(t,x_t)$ via the chain rule:

$$d\phi(t,x_t) = \frac{\partial\phi}{\partial t}(t,x_t)dt + \frac{\partial\phi}{\partial x}(t,x_t)dx_t\,. \tag{C.10}$$

We already observed in (C.7) that whenever a Brownian motion is involved, such a fundamental rule of calculus needs to be modified. The general formulation of the chain rule for stochastic differential equations is the following.

Let $\phi(t,x)$ be a smooth function and let $X_t(\omega)$ be the unique solution of the stochastic differential equation (C.9). Then, Ito's formula reads as

$$d\phi(t,X_t(\omega)) = \frac{\partial\phi}{\partial t}(t,X_t(\omega))dt + \frac{\partial\phi}{\partial x}(t,X_t(\omega))dX_t(\omega) + \tfrac{1}{2}\frac{\partial^2\phi}{\partial x^2}(t,X_t(\omega))dX_t(\omega)dX_t(\omega), \tag{C.11}$$

or, in a more compact notation,

$$\boxed{d\phi(t,X_t) = \frac{\partial\phi}{\partial t}(t,X_t)dt + \frac{\partial\phi}{\partial x}(t,X_t)dX_t + \tfrac{1}{2}\frac{\partial^2\phi}{\partial x^2}(t,X_t)d\langle X\rangle_t}\,.$$

Comparing equation (C.11) with its "deterministic" counterpart (C.10), we notice that the extra term

$$\tfrac{1}{2}\frac{\partial^2\phi}{\partial x^2}(t,X_t(\omega))dX_t(\omega)dX_t(\omega)$$

appears in our stochastic context, and this is the term due to the Ito integral.[2]

The term $dX_t(\omega)dX_t(\omega)$ can be developed algebraically by taking into account the rules on the quadratic variation and covariation seen above:

$$dW_t(\omega)\,dW_t(\omega) = dt, \quad dW_t(\omega)\,dt = 0, \quad dt\,dt = 0\,.$$

We thus obtain

$$d\phi(t,X_t(\omega)) = \frac{\partial\phi}{\partial t}(t,X_t(\omega))dt + \frac{\partial\phi}{\partial x}(t,X_t(\omega))f(X_t(\omega))dt$$
$$+\tfrac{1}{2}\frac{\partial^2\phi}{\partial x^2}(t,X_t(\omega))\sigma^2(X_t)dt + \frac{\partial\phi}{\partial x}(t,X_t(\omega))\sigma(X_t(\omega))dW_t(\omega).$$

Stochastic Leibnitz Rule

Also the classical Leibnitz rule for differentiation of a product of functions is modified, analogously to the chain rule. The related formula can be derived as a corollary of Ito's formula in two dimensions, and is reported below.

For deterministic and differentiable functions x and y we have the deterministic Leibnitz rule

$$d(x_t\,y_t) = x_t\,dy_t + y_t\,dx_t.$$

[2] Notice that, when writing instead the equation for $d\phi(t,X_t(\omega))$ in the Stratonovich sense, we would re-obtain

$$d\phi(t,X_t(\omega)) = \frac{\partial\phi}{\partial t}(t,X_t(\omega))dt + \frac{\partial\phi}{\partial x}(t,X_t(\omega))\circ dX_t(\omega)\,,$$

since formal rules of calculus are preserved.

For two diffusion processes (and more generally semimartingales) $X_t(\omega)$ and $Y_t(\omega)$ we have instead[3]

$$d(X_t(\omega)\,Y_t(\omega)) = X_t(\omega)\,dY_t(\omega) + Y_t(\omega)\,dX_t(\omega) + dX_t(\omega)\,dY_t(\omega),$$

or, in more compact notation,

$$\boxed{d(X_t\,Y_t) = X_t\,dY_t + Y_t\,dX_t + d\langle X,Y\rangle_t}\,.$$

C.3 Discretizing SDEs for Monte Carlo: Euler and Milstein Schemes

When one cannot solve an SDE explicitly, it is possible to simulate its trajectories through a discretization scheme. Here we briefly review the Euler scheme. For a more detailed treatment the reader is referred to Klöden and Platen (1995).

Consider again the SDE

$$dX_t(\omega) = f(X_t(\omega))dt + \sigma(X_t(\omega))dW_t(\omega), \quad x_0$$

where for simplicity we took a deterministic initial condition. Let us integrate this equation between s and $s+\Delta s$:

$$X_{s+\Delta s}(\omega) = X_s(\omega) + \int_s^{s+\Delta s} f(X_t(\omega))dt + \int_s^{s+\Delta s} \sigma(X_t(\omega))dW_t(\omega).$$

The Euler scheme consists of approximating this integral equation by

$$\bar{X}_{s+\Delta s}(\omega) = \bar{X}_s(\omega) + f(\bar{X}_s(\omega))\Delta s + \sigma(\bar{X}_s(\omega))(W_{s+\Delta s}(\omega) - W_s(\omega)) \quad \text{(C.12)}$$

with $\bar{X}_0(\omega) = x_0$. If we apply this formula iteratively for a given set of s's, say

$$s = s_1, s_2, \ldots, s_m, \quad s_1 = 0, \quad s_m = T,$$

we obtain a discretized approximation $\bar{X}$ of the solution X of the above SDE. For a definition of the order of convergence of this and other schemes see Klöden and Platen (1995).

A stronger convergence is attained with a more refined scheme, called Milstein scheme. We do not review the Milstein scheme here. We only hint at the fact that when the diffusion coefficient is deterministic, say $\sigma(X_t(\omega)) = \sigma(t)$, with $t \mapsto \sigma(t)$ a deterministic function of time, the Euler and Milstein schemes coincide. Therefore it is preferable, when possible, to apply the Euler scheme to SDEs with deterministic diffusion coefficients, since this ensures the same stronger convergence of the Milstein scheme.

[3] Clearly, using the Stratonovich calculus, we would still have $d(X_t\,Y_t) = X_t \circ dY_t + Y_t \circ dX_t$.

These discretization schemes can be useful for Monte Carlo simulation. Indeed, suppose we need to compute the expected value of a functional of the solution X of the above SDE, say for simplicity

$$\mathbb{E}_0[\phi(X_{s_1}(\omega), \ldots, X_{s_m}(\omega))], \quad s_1 = 0, \quad s_m = T$$

(this is a typical pricing problem for path-dependent payoffs in mathematical finance). Assume also that the times s are close to each other.

We may decide to compute an approximation of this expectation as follows.

1. Select the number N of scenarios for the Monte Carlo method.
2. Set the initial value to $\bar{X}_0^j = x_0$ for all scenarios $j = 1, \ldots, N$.
3. Set $k = 1$.
4. Set $s = s_k$ and $\Delta s = s_{k+1} - s_k$ (so that $s + \Delta s = s_{k+1}$).
5. Generate N new realizations ΔW^j ($j = 1, \ldots, N$) of a standard Gaussian distribution $\mathcal{N}(0, 1)$ multiplied by $\sqrt{\Delta s}$, thus simulating the distribution of
$$(W_{s+\Delta s}(\omega) - W_s(\omega)).$$
6. Apply formula (C.12) for each scenario $j = 1, \ldots, N$ with the generated shocks:
$$\bar{X}_{s+\Delta s}^j = \bar{X}_s^j + f(\bar{X}_s^j)\Delta s + \sigma(\bar{X}_s^j)\Delta W^j.$$
7. Store $\bar{X}_{s+\Delta s}^j$ for all j.
8. If $s + \Delta s = s_m$ then stop, otherwise increase k by one and start again from point 4.
9. Approximate the expected value by
$$\frac{\sum_{j=1}^N \phi(\bar{X}_{s_1}^j, \ldots, \bar{X}_{s_m}^j)}{N}.$$

Notice that the increments can be generated as new independent draws from a Gaussian distribution at each iteration because the Brownian motion has independent (stationary Gaussian) increments.

Notice also that in case the s's are not close to each other, one needs to infra-discretize the equation between two of the s's, and apply several times the basic scheme (C.12) at the infra-instants. For simplicity, we assumed above that the times s's in the expectation are close enough for the scheme to be applied directly without further discretization.

Discretization schemes are useful and powerful because they allow us to replace, in small intervals, the unknown exact distribution for the transition $X_s \to X_{s+\Delta s}$ with the known (and easy to simulate) Gaussian distribution in the shocks of the transition $\bar{X}_s \to \bar{X}_{s+\Delta s}$.

C.4 Examples

We now present some relevant examples of SDEs that are often encountered throughout the book.

Linear SDEs with Deterministic Diffusion Coefficient

A SDE is said to be linear if both its drift and diffusion coefficients are first order polynomials (or affine functions) in the state variable. We here consider the particular case:

$$dX_t(\omega) = (\beta_t X_t(\omega) + \alpha_t)dt + v_t dW_t(\omega), \; X_0(\omega) = x_0, \tag{C.13}$$

where α, β, v are deterministic functions of time that are regular enough to ensure existence and uniqueness of a solution.

It can be shown that a stochastic integral of a deterministic function is the same both in the Stratonovich and in the Ito sense. As a consequence, by writing (C.13) in integral form we see that the same equation holds in the Stratonovich sense:

$$dX_t(\omega) = (\beta_t X_t(\omega) + \alpha_t)dt + v_t \circ dW_t(\omega), \; X_0(\omega) = x_0,$$

so that we can solve it by ordinary calculus for linear differential equations: We obtain

$$\begin{aligned} X_t(\omega) &= e^{\int_0^t \beta_s ds}\left[x_0 + \int_0^t e^{-\int_0^s \beta_u du}\alpha_s ds + \int_0^t e^{-\int_0^s \beta_u du} v_s dW_s(\omega)\right] \\ &= x_0 e^{\int_0^t \beta_s ds} + \int_0^t e^{\int_s^t \beta_u du}\alpha_s ds + \int_0^t e^{\int_s^t \beta_u du} v_s dW_s(\omega). \end{aligned}$$

A remarkable fact is that the distribution of the solution X_t is normal at each time t. Intuitively, this holds since the last stochastic integral is a limit of a sum of independent normal random variables. Indeed, we have

$$X_t \sim \mathcal{N}\left(x_0 e^{\int_0^t \beta_s ds} + \int_0^t e^{\int_s^t \beta_u du}\alpha_s ds, \int_0^t e^{2\int_s^t \beta_u du} v_s^2 ds\right).$$

The major examples of models based on a SDE like (C.13) are that of Vasicek (1978) and that of Hull and White (1990). See also equations (3.5) and (3.32).

Lognormal Linear SDEs

Another interesting example of linear SDE is that where the diffusion coefficient is a first order homogeneous polynomial in the underlying variable. This SDE can be obtained as an exponential of a linear equation with deterministic diffusion coefficient. Indeed, let us take $Y_t = \exp(X_t)$, where X_t evolves according to (C.13), and write by Ito's formula

$$dY_t(\omega) = de^{X_t(\omega)} = e^{X_t(\omega)}dX_t(\omega) + \tfrac{1}{2}e^{X_t(\omega)}dX_t(\omega)\,dX_t(\omega)$$
$$= Y_t(\omega)[\alpha_t + \beta_t \ln Y_t(\omega) + \tfrac{1}{2}v_t^2]dt + \; v_t Y_t(\omega) dW_t(\omega).$$

As a consequence, the process Y has a lognormal marginal density. A major example of model based on such a SDE is the Black and Karasinski (1991) model. See also (3.53).

Geometric Brownian Motion

The geometric Brownian motion is a particular case of a process satisfying a lognormal linear SDE. Its evolution is defined according to

$$dX_t(\omega) = \mu X_t(\omega)dt + \sigma X_t(\omega)dW_t(\omega),$$

where μ and σ are positive constants. To check that X is indeed a lognormal process, one can compute $d\ln(X_t)$ via Ito's formula and obtain

$$X_t(\omega) = X_0 \exp\left\{\left(\mu - \frac{1}{2}\sigma^2\right)t + \sigma W_t(\omega)\right\}.$$

From the seminal work of Black and Scholes (1973) on, processes of this type are frequently used in option pricing theory to model general asset price dynamics. Notice that this process is a submartingale, in that clearly

$$\mathbb{E}[X_T|\mathcal{F}_t] = e^{\mu(T-t)}X_t \geq X_t.$$

Finally, notice also that by setting $Y_t(\omega) = e^{-\mu t}X_t(\omega)$, we obtain

$$dY_t(\omega) = \sigma Y_t(\omega)dW_t(\omega).$$

Therefore, since the drift of this last SDE is zero, $e^{-\mu t}X_t$ is a martingale.

Square-Root Processes

An interesting case of non-linear SDE is given by

$$dX_t(\omega) = (\beta_t X_t(\omega) + \alpha_t)dt + v_t\sqrt{X_t(\omega)}\,dW_t(\omega), \quad X_0(\omega) = x_0. \tag{C.14}$$

A process following such dynamics is commonly referred to as square-root process. Major examples of models based on this dynamics are the Cox, Ingerssoll and Ross (1985) instantaneous interest rate model, see also (3.21), and a particular case of the constant-elasticity of variance (CEV) model for stock prices:

$$dX_t(\omega) = \mu X_t(\omega)dt + \sigma\sqrt{X_t(\omega)}\,dW_t(\omega).$$

In general, square-root processes are naturally linked to non-central χ-square distributions.[4] In particular, there are simplified versions of (C.14) for which the resulting process X is strictly positive and analytically tractable, like in the case of the Cox, Ingerssoll and Ross (1985) model.

[4] They may also display exit boundaries, for example the CEV process features absorption in the origin.

C.5 Two Important Theorems

We now introduce (informally) two important results. See Øksendal (1992) for more details.

The Feynman-Kac Theorem

The Feynman-Kac theorem, under certain assumptions, allows us to express the solution of a given partial differential equation (PDE) as the expected value of a function of a suitable diffusion process whose drift and diffusion coefficient are defined in terms of the PDE coefficients.

Theorem C.5.1. [The Feynman-Kac Theorem] *Given Lipschitz continuous $f(x)$ and $\sigma(x)$ and a smooth ϕ, the solution of the PDE*

$$\frac{\partial V}{\partial t}(t,x) + \frac{\partial V}{\partial x}(t,x)f(x) + \frac{1}{2}\frac{\partial^2 V}{\partial x^2}(t,x)\sigma^2(x) = rV(t,x) \tag{C.15}$$

with terminal boundary condition

$$V(T,x) = \phi(x) \tag{C.16}$$

can be expressed as the following expected value

$$V(t,x) = e^{-r(T-t)}\ \widetilde{\mathbb{E}}\{\phi(X_T)|X_t = x\}, \tag{C.17}$$

where the diffusion process X has dynamics, starting from x at time t, given by

$$dX_s(\omega) = f(X_s(\omega))ds + \sigma(X_s(\omega))d\widetilde{W}_s(\omega), \quad s \geq t,\ X_t(\omega) = x \tag{C.18}$$

under the probability measure $\widetilde{\mathbb{P}}$ under which the expectation $\widetilde{\mathbb{E}}\{\cdot\}$ is taken. The process $\widetilde{W}$ is a standard Brownian motion under $\widetilde{\mathbb{P}}$.

Notice that the terminal condition determines the function ϕ of the diffusion process whose expectation is relevant, whereas the PDE coefficients determine the dynamics of the diffusion process.

This theorem is important because it establishes a link between the PDE's of traditional analysis and physics and diffusion processes in stochastic calculus. Solutions of PDE's can be interpreted as expectations of suitable transformations of solutions of stochastic differential equations and vice versa.

The Girsanov Theorem

The Girsanov theorem shows how a SDE changes due to changes in the underlying probability measure. It is based on the fact that the SDE drift depends on the particular probability measure $\mathbb{P}$ in our probability space

$(\Omega, \mathcal{F}, (\mathcal{F}_t)_t, \mathbb{P})$, and that, if we change the probability measure in a "regular" way, the drift of the equation changes while the diffusion coefficient remains the same. The Girsanov theorem can be thus useful when we want to modify the drift coefficient of a SDE. Indeed, suppose that we are given two measures $\mathbb{P}^*$ and $\mathbb{P}$ on the space $(\Omega, \mathcal{F}, (\mathcal{F}_t)_t)$. Two such measures are said to be equivalent, written $\mathbb{P}^* \sim \mathbb{P}$, if they share the same sets of null probability (or of probability one, which is equivalent). Therefore two measures are equivalent when they agree on which events of $\mathcal{F}$ hold almost surely. Accordingly, a proposition holds almost surely under $\mathbb{P}$ if and only if it holds almost surely under $\mathbb{P}^*$. Similar definitions apply also for the measures restriction to $\mathcal{F}_t$, thus expressing equivalence of the two measures up to time t.

When two measures are equivalent, it is possible to express the first in terms of the second through the Radon-Nikodym derivative. Indeed, there exists a martingale ρ_t on $(\Omega, \mathcal{F}, (\mathcal{F}_t)_t, \mathbb{P})$ such that

$$\mathbb{P}^*(A) = \int_A \rho_t(\omega) d\mathbb{P}(\omega), \quad A \in \mathcal{F}_t,$$

which can be written in a more concise form as

$$\left. \frac{d\mathbb{P}^*}{d\mathbb{P}} \right|_{\mathcal{F}_t} = \rho_t.$$

The process ρ_t is called the Radon-Nikodym derivative of $\mathbb{P}^*$ with respect to $\mathbb{P}$ restricted to $\mathcal{F}_t$.

When in need of computing the expected value of an integrable random variable X, it may be useful to switch from one measure to another equivalent one. Indeed, it is possible to prove that the following equivalence holds:

$$\mathbb{E}^*[X] = \int_\Omega X(\omega) d\mathbb{P}^*(\omega) = \int_\Omega X(\omega) \frac{d\mathbb{P}^*}{d\mathbb{P}}(\omega) d\mathbb{P}(\omega) = \mathbb{E}\left[X \frac{d\mathbb{P}^*}{d\mathbb{P}}\right],$$

where $\mathbb{E}^*$ and $\mathbb{E}$ denote expected values with respect to the probability measures $\mathbb{P}^*$ and $\mathbb{P}$, respectively. More generally, when dealing with conditional expectations, we can prove that

$$\mathbb{E}^*[X|\mathcal{F}_t] = \frac{\mathbb{E}\left[X \frac{d\mathbb{P}^*}{d\mathbb{P}} |\mathcal{F}_t\right]}{\rho_t}.$$

Theorem C.5.2. [The Girsanov theorem] *Consider again the stochastic differential equation, with Lipschitz coefficients,*

$$dX_t(\omega) = f(X_t(\omega))dt + \sigma(X_t(\omega))dW_t(\omega), \quad x_0,$$

under $\mathbb{P}$. Let be given a new drift $f^*(x)$ and assume $(f^*(x) - f(x))/\sigma(x)$ to be bounded. Define the measure $\mathbb{P}^*$ by

$$\frac{d\mathbb{P}^*}{d\mathbb{P}}(\omega)\bigg|_{\mathcal{F}_t} = \exp\left\{ -\frac{1}{2}\int_0^t \left(\frac{f^*(X_s(\omega)) - f(X_s(\omega))}{\sigma(X_s(\omega))}\right)^2 ds + \int_0^t \frac{f^*(X_s(\omega)) - f(X_s(\omega))}{\sigma(X_s(\omega))} dW_s(\omega) \right\}.$$

Then $\mathbb{P}^*$ is equivalent to $\mathbb{P}$. Moreover, the process W^* defined by

$$dW_t^*(\omega) = -\left[\frac{f^*(X_t(\omega)) - f(X_t(\omega))}{\sigma(X_t(\omega))}\right] dt + dW_t(\omega)$$

is a Brownian motion under $\mathbb{P}^*$, and

$$dX_t(\omega) = f^*(X_t(\omega))dt + \sigma(X_t(\omega))dW_t^*(\omega), \quad x_0.$$

As already noticed, this theorem is fundamental when we wish to change the drift of a SDE. It is now clear that we can do this by defining a new probability measure $\mathbb{P}^*$, via a suitable Radon-Nikodym derivative, in terms of the difference "desired drift - given drift".

In mathematical finance, a classical example of application of the Girsanov theorem is when one moves from the "real-world" asset price dynamics

$$dS_t(\omega) = \mu S_t(\omega)dt + \sigma S_t(\omega)dW_t(\omega)$$

to the risk-neutral ones

$$dS_t(\omega) = rS_t(\omega)dt + \sigma S_t(\omega)dW_t^*(\omega).$$

This is accomplished by setting

$$\frac{d\mathbb{P}^*}{d\mathbb{P}}(\omega)\bigg|_{\mathcal{F}_t} = \exp\left\{-\frac{1}{2}\left(\frac{\mu - r}{\sigma}\right)^2 t - \frac{\mu - r}{\sigma}W_t(\omega)\right\}. \tag{C.19}$$

We finally stress that above we assumed boundedness for simplicity, but less stringent assumptions are possible for the theorem to hold. See for example Øksendal (1992).

C.6 A Crash Intro to Poisson Processes

Given their importance in default modeling, and their growing interest to the financial community in addressing jump-diffusion models, we cannot close the book without mentioning Poisson processes, that are the purely jump analogous of the Brownian motion with which we started the appendix.

A time homogeneous Poisson process is a unit-jump increasing, right continuous process M_t with stationary independent increments and $M_0 = 0$.

One notices immediately that the Poisson process shares the properties expressed by (C.3) and (C.4) with the Brownian motion process. Actually, if we substituted "unit jump increasing, right continuous" by "continuous" we would obtain a characterization of Brownian motion with time-linear drift. Indeed, an important theorem of stochastic calculus due to Lévy states that a *continuous* process with independent and stationary increments and null initial condition is necessarily of the form

$$bt + \sigma W_t$$

with b and σ deterministic constants and W a Brownian motion (see for example Rogers and Williams (1987), Theorem I.28.12). This shows that Poisson Processes and Brownian motions are analogous processes in the purely jump processes family and in the continuous processes family respectively. In particular, they are both particular cases of the larger family of Levy processes, i.e. particular cases of processes with stationary independent increments and with right continuous and left limit paths.

The first results on Poisson processes are given by the following facts:

First properties of Poisson Processes. Let M be a time homogeneous Poisson process. Then

1) There exists a positive real number $\bar{\gamma}$ such that $\mathbb{P}\{M_t = 0\} = \exp(-\bar{\gamma}t)$ for all t.

2) $\lim_{t\to 0} \mathbb{P}\{M_t \geq 2\}/t = 0$.

3) $\lim_{t\to 0} \mathbb{P}\{M_t = 1\}/t = \bar{\gamma}$.

The first point states that the probability of having no jumps up to some given time is an exponential function of minus that (possibly re-scaled) time. The second point tells us that the probability of having more than one jump in an arbitrarily small time going to zero goes to zero faster than the time itself. So, roughly speaking, in small intervals we can have at most one jump. The third point tells us that the probability of having exactly one jump in a small time, re-scaled by the time itself, is the constant we find in the exponent of the exponential function found in the first point.

Also, in classical Poisson process theory, starting from the above first results, one proves the following

Further properties of Poisson Processes. Let M be a time homogeneous Poisson process. Then

$$\mathbb{P}\{M_t - M_s = k\} = e^{-\bar{\gamma}(t-s)}(\bar{\gamma}(t-s))^k/k!$$

i.e. $M_t - M_s \sim \text{PoissonLaw}((t-s)\bar{\gamma})$, with $M_t - M_s$ independent of $\sigma(\{M_u, u \leq s\})$.

This second set of properties tells us that the number of jumps of a Poisson process follows the Poisson law (hence the name of the process).

What is amazing at first sight is that a requirement on the properties of trajectories (unit jumps and right continuity) plus a requirement on the increments, i.e. independence and stationarity, completely determines the law of the process. This happens with Brownian motion (with time-linear drift) too, but the different requirement on the paths (they are to be continuous with Brownian motion) implies a Gaussian law for the process, contrary to the Poisson case where we get the Poisson law. Thus seemingly general requirements on the increments (stationarity and independence) determine completely the law of the process in case paths are required to be continuous (Brownian motion, Gaussian law) or unit-jump increasing and right continuous (Poisson process, Poisson law).

A different characterization of the Poisson process (that could also be used as a definition) is given now:

A different characterization of Poisson processes.

M is a time-homogeneous Poisson process with parameter $\bar{\gamma}$
if and only if
$\mathbb{E}[M_{t+s} - M_t | M_u : u \leq t] = \bar{\gamma}s$ for all $t, s > 0$ and M is unit-jump increasing, right continuous with $M_0 = 0$ (see for example Cinlar (1975)).

This characterization is remarkably mild at first sight, compared to the standard definition.

At this point we may wonder whether we can have some further intuition for the parameter $\bar{\gamma}$. Actually, from the above properties we have easily the following

Interpretation of $\bar{\gamma}$ as average arrival rate for unit of time. Let M be a time-homogeneous Poisson process. Then

$$\bar{\gamma} = \mathbb{E}(M_t)/t = \text{Var}(M_t)/t.$$

A fundamental result (also for financial applications) concerns the distribution of the intervals of time between two jumps of the process.

Exponential distribution for the time between two jumps. Let M be a time homogeneous Poisson process. Let $\tau^1, \tau^2, \ldots, \tau^m, \ldots$ be the first, second etc. jump times of M. Then $\tau^1, \tau^2 - \tau^1, \tau^3 - \tau^2, \ldots$, i.e. the times between any jump and the subsequent one, are i.i.d. $\sim$ exponential($\bar{\gamma}$) (or, equivalently, the random variables $\bar{\gamma}\tau^1, \bar{\gamma}(\tau^2 - \tau^1), \bar{\gamma}(\tau^3 - \tau^2)$.. are i.i.d. $\sim$ exponential(1))

In the simplest intensity models for credit derivatives the default time is modeled as τ^1, so that the time of the first jump becomes particularly important. An immediate important consequence of the last property is that the probability of having the first jump in a small time interval $[t, \quad t+dt]$ given that this first jump did not occur before t is $\bar{\gamma}dt$, so that $\bar{\gamma}$ bears also the interpretation of probability of having a new jump about t given that we have not had it before t. In formula

$$\mathbb{P}\{\tau^1 \in [t, t+dt) | \tau^1 \geq t\} = \bar{\gamma}\, dt.$$

C.6.1 Time inhomogeneous Poisson Processes

We consider now a deterministic time-varying intensity $\gamma(t)$ (called also hazard rate), which we assume to be a strictly positive and piecewise continuous function. We define

$$\Gamma(t) := \int_0^t \gamma(u)du,$$

the cumulated intensity, cumulated hazard rate, or also Hazard function.

If M_t is a Standard Poisson process, i.e. a Poisson process with intensity one, than a time-inhomogeneous Poisson process N_t with intensity $\gamma(t)$ is defined as

$$N_t = M_{\Gamma(t)}.$$

So a time inhomogeneous Poisson process is just a time-changed standard Poisson process.

N_t is still increasing by jumps of size 1, its increments are still independent, but they are *no longer identically distributed* (stationary) due to the "time distortion" introduced by the possibly nonlinear Γ.

From $N_t = M_{\Gamma(t)}$ we have obviously that N jumps the first time at τ if and only if M jumps the first time at $\Gamma(\tau)$.

But since we know that M is a standard Poisson Process for which the first jump time is exponentially distributed, then we have

$$\Gamma(\tau) =: \xi \sim \text{exponential}(1).$$

By inverting this last equation we have that

$$\boxed{\tau = \Gamma^{-1}(\xi)}\ ,$$

with ξ standard exponential random variable. Also, we have easily

$$\mathbb{P}\{s < \tau < t\} = \mathbb{P}\{\Gamma(s) < \Gamma(\tau) < \Gamma(t)\} = \mathbb{P}\{\Gamma(s) < \xi < \Gamma(t)\} =$$

$$= \mathbb{P}\{\xi > \Gamma(s)\} - \mathbb{P}\{\xi > \Gamma(t)\} = \exp(-\Gamma(s)) - \exp(-\Gamma(t))$$

i.e. "Probability of first jumping between s and t" is

$$e^{-\int_0^s \gamma(u)du} - e^{-\int_0^t \gamma(u)du} = e^{-\int_0^s \gamma(u)du}(1 - e^{-\int_s^t \gamma(u)du}) \approx e^{-\int_0^s \gamma(u)du} \int_s^t \gamma(u)du$$

(where the final approximation is good for small exponents). Following this, we have that the "probability of first jumping between s and t given that one has not jumped before s" is

$$\mathbb{P}\{s < \tau < t | \tau > s\} = \frac{\mathbb{P}\{\Gamma(s) < \Gamma(\tau) < \Gamma(t)\}}{\mathbb{P}\{\Gamma(\tau) > \Gamma(s)\ \}} = \frac{e^{-\int_0^s \gamma(u)du} - e^{-\int_0^t \gamma(u)du}}{e^{-\int_0^s \gamma(u)du}}$$

$$= 1 - e^{-\int_s^t \gamma(u)du} \approx \int_s^t \gamma(u)du = (t-s) \boxed{\frac{1}{t-s}\int_s^t \gamma(u)du},$$

(where, again, the final approximation is good for small exponents). The boxed term is a sort of time-averaged intensity between s and t.

It is easy to show, along the same lines, that

$$\mathbb{P}\{\tau \in [t, t+dt) | \tau \geq t\} = \gamma(t)\ dt.$$

"Probability that first jump occurs in the (arbitrarily small) next "dt" instants given that we had no jump so far is $\gamma(t)\ dt$."

Notice that a fundamental fact from probability tells us that ξ is independent of all possible Brownian motions in the same probability space where the Poisson process is defined, and also of the intensity itself when this is assumed to be stochastic, as we are going to assume now.

C.6.2 Doubly Stochastic Poisson Processes (or Cox Processes)

Intensity, besides being time varying, can also be stochastic: in that case it is assumed to be at least a $\mathcal{F}_t$-adapted and right continuous (and thus progressive) process and is denoted by λ_t and the cumulated intensity or hazard process is the random variable $\Lambda(T) = \int_0^T \lambda_t dt$. We assume $\lambda_t > 0$.

We recall again that "$\mathcal{F}_t$-adapted" means essentially that given the information $\mathcal{F}_t$ we know λ from 0 to t.

A Poisson process with stochastic intensity λ is called a doubly stochastic Poisson process, or Cox process. The term doubly stochastic is due to the fact that besides having stochasticity in the jump component ξ, we have also stochasticity in the probability of jumping, i.e. on the intensity. We observe that in the definition of Cox process as a Poisson process with stochastic intensity it is implicit that conditional on $\mathcal{F}^\lambda$ (i.e. on λ), we still have a Poisson process structure and all facts we have seen for the case with deterministic intensity $\gamma(t)$ still hold, conditional on λ and replacing γ with λ.

We have that, for Cox processes, the first jump time can be represented as $\tau := \Lambda^{-1}(\xi)$.

Notice once again that here not only ξ is random (and still independent of anything else, included λ), but λ itself is stochastic. With Cox processes we have $\mathbb{P}\{\tau \in [t, t+dt) | \tau \geq t, \mathcal{F}_t\} = \lambda_t\ dt$. This reads, if "t=now":

"The probability that the process first jumps in (a small) time "dt" given that it has not jumped so far and given the $\mathcal{F}_t$ information is $\lambda_t\, dt$."

Under standard assumptions one can show that

$$\mathbb{P}\{\tau \geq s\} = \mathbb{P}\{\Lambda(\tau) \geq \Lambda(s)\} = \mathbb{P}\left\{\xi \geq \int_0^s \lambda(u)du\right\} =$$

$$= \mathbb{E}\left[\mathbb{P}\left\{\xi \geq \int_0^s \lambda(u)du \middle| \mathcal{F}^\lambda\right\}\right] = \mathbb{E}\left[e^{-\int_0^s \lambda(u)du}\right]$$

which, in a financial context, where τ is typically a default time, is completely analogous to the bond price formula in a short rate model with interest rate λ replacing r.

Cox processes thus allow to drag the interest-rate technology and paradigms into default modeling. But again ξ is independent of all default free market quantities (of $\mathcal{F}$, of λ, of r...) and represents an external source of randomness that makes reduced form models incomplete.

C.6.3 Compound Poisson processes

A generalization of the time-homogeneous Poisson process different from the time-inhomogeneous Poisson process and from the Cox process is the compound Poisson process.

A compound Poisson process is obtained by taking a time-homogeneous Poisson process and by replacing the jumps of unit size 1 with jumps of size distributed according to i.i.d. random variables $Y_1, Y_2, \ldots, Y_n, \ldots$, all independent of the basic Poisson process.

Indeed, a time-homogeneous Poisson process M with intensity $\bar{\gamma} > 0$ can be trivially written as

$$M_t = \sum_{i=1}^{M_t} \boxed{1}.$$

Now replace the boxed "1" by i.i.d. non-negative random variables

$$Y_1, Y_2, \ldots, Y_n, \ldots,$$

all distributed according to a distribution function F and independent of M. We obtain

$$J_t^{F,\bar{\gamma}} = \sum_{i=1}^{M_t} \boxed{Y_i}.$$

The process J is a compound Poisson process.

C.6.4 Jump-diffusion Processes

The interest in compound Poisson processes is given by their possible use as "jumpy shocks" processes in jump diffusion models, as opposed to the continuous shock process given by Brownian motion. We have seen an example of such a situation when introducing the JCIR jump diffusion model in Section 22.8, where the jumpy shocks associated to a compound Poisson process have been added to the usual Brownian shocks.

In general, a candidate jump-diffusion process is written as

$$dX_t(\omega) = f_t(X_t(\omega))dt + \sigma_t(X_t(\omega))dW_t(\omega) + \eta_t(X_t(\omega))dJ_t^{F,\bar{\gamma}}(\omega).$$

Here dJ denotes the jump-increments in the compound Poisson process J. If we call $\tau^1(\omega), \tau^2(\omega), \tau^3(\omega) \ldots$ the first, second, third... jump times of the basic process M, then

$$d\ J_t^{F,\bar{\gamma}}(\omega) = Y_i(\omega) \ \text{ if } \ t = \tau^i(\omega), \ \text{ and } \ 0 \ \text{ otherwise.}$$

We see that the shock dJ in $[t, t + dt)$ is always finite (rather than infinitesimal/small of order "dt" or "$\sqrt{dt}$") or null.

Clearly, the more we increase $\bar{\gamma}$, the more frequent the jumps in the system. Also, the larger the values implied by the distribution F, the larger the jump sizes. These degrees of freedom allow for a large variability of situations.

Finally, we notice that the above-mentioned *Lévy processes* have been characterized as limits of compositions of independent families of *compound Poisson processes* and *Brownian motions* (see for example Karlin and Taylor (1981)). The basic mathematical framework for reaching Lévy processes thus includes the above compound Poisson process and the Brownian motion. Obviously in their basic formulation Lévy processes incorporate the Brownian motion and the (compound) Poisson process as particular cases, but not the jump diffusion and Cox processes in general. The financial community is now considering processes with Lévy *shocks*, or Lévy processes under stochastic time-changes (see for example Carr and Wu (2004)). These processes encompass a large family of earlier models based on jump-diffusions and stochastic volatility. For further references on Lévy processes in finance see also Eberlein (2001), Eberlein and Özkan (2005), and Eberlein, Jacod and Raible (2005).

D. A Useful Calculation

Lemma. *Let M, V and K be real numbers with V and K positive. Then, for $\omega \in \{-1, 1\}$,*

$$\int_{-\infty}^{+\infty} \frac{1}{\sqrt{2\pi}V}[\omega(e^y - K)]^+ e^{-\frac{1}{2}\frac{(y-M)^2}{V^2}} dy$$
$$= \omega e^{M+\frac{1}{2}V^2} \Phi\left(\omega \frac{M - \ln(K) + V^2}{V}\right) - \omega K \Phi\left(\omega \frac{M - \ln(K)}{V}\right). \tag{D.1}$$

Proof. The above integral

$$\int_{-\infty}^{+\infty} \frac{1}{\sqrt{2\pi}V}[\omega(e^y - K)]^+ e^{-\frac{1}{2}\frac{(y-M)^2}{V^2}} dy$$

is calculated as follows:

$$= \int_{\ln(K)}^{+\infty\cdot\omega} \frac{1}{\sqrt{2\pi}V}(e^y - K) e^{-\frac{1}{2}\frac{(y-M)^2}{V^2}} dy$$
$$= \int_{\frac{\ln(K)-M}{V}}^{+\infty\cdot\omega} \frac{1}{\sqrt{2\pi}}(e^{M+Vz} - K) e^{-\frac{1}{2}z^2} dz$$
$$= e^{M+\frac{1}{2}V^2} \int_{\frac{\ln(K)-M}{V}}^{+\infty\cdot\omega} \frac{1}{\sqrt{2\pi}} e^{-\frac{1}{2}(z-V)^2} dz - K \int_{\frac{\ln(K)-M}{V}}^{+\infty\cdot\omega} \frac{1}{\sqrt{2\pi}} e^{-\frac{1}{2}z^2} dz$$
$$= e^{M+\frac{1}{2}V^2} \left[\Phi(+\infty\cdot\omega) - \Phi\left(\frac{\ln(K) - M - V^2}{V}\right)\right]$$
$$- K \left[\Phi(+\infty\cdot\omega) - \Phi\left(\frac{\ln(K) - M}{V}\right)\right]$$
$$= e^{M+\frac{1}{2}V^2} \omega \Phi\left(-\omega \frac{\ln(K) - M - V^2}{V}\right) - K\omega \Phi\left(-\omega \frac{\ln(K) - M}{V}\right).$$

□

Proposition. *Let X be random variable that is lognormally distributed, and denote by M and V the mean and standard deviation of $Y := \ln(X)$. Then*

$$
\begin{aligned}
E\left\{[\omega(X-K)]^+\right\} =& \omega e^{M+\frac{1}{2}V^2}\Phi\left(\omega\frac{M-\ln(K)+V^2}{V}\right) \\
&- \omega K \Phi\left(\omega\frac{M-\ln(K)}{V}\right),
\end{aligned}
\tag{D.2}
$$

for each $K > 0$, $\omega \in \{-1, 1\}$, where E denotes expectation with respect to X's distribution and Φ denotes the cumulative standard normal distribution function.

Proof. We just have to notice that

$$
E\left\{[\omega(X-K)]^+\right\} = \int_{-\infty}^{+\infty} \frac{1}{\sqrt{2\pi}V}[\omega(e^y - K)]^+ e^{-\frac{1}{2}\frac{(y-M)^2}{V^2}}\,dy,
$$

and apply the previous lemma. □

E. A Second Useful Calculation

The Pricing of a Spread Option

Consider two assets whose prices S_1 and S_2 evolve, under the real-world measure, according to

$$\begin{aligned} dS_1(t) &= S_1(t)[m_1 dt + \sigma_1 dW_1(t)], \quad S_1(0) = s_1, \\ dS_2(t) &= S_2(t)[m_2 dt + \sigma_2 dW_2(t)], \quad S_2(0) = s_2, \\ dW_1(t)dW_2(t) &= \rho \, dt, \end{aligned} \tag{E.1}$$

where m_1, m_2, σ_1, σ_2, s_1 and s_2 are positive real numbers and W_1 and W_2 are Brownian motions with instantaneous correlation ρ.

Assume that the assets pay continuous dividend yields q_1 and q_2, with q_1 and q_2 positive real numbers, and that interest rates are constant for all maturities and equal to the positive real number r.

The dynamics (E.1) implies there exists a unique equivalent martingale measure Q under which S_1 and S_2 evolve according to

$$\begin{aligned} dS_1(t) &= S_1(t)[(r - q_1)dt + \sigma_1 dW_1^Q(t)], \quad S_1(0) = s_1, \\ dS_2(t) &= S_2(t)[(r - q_2)dt + \sigma_2 dW_2^Q(t)], \quad S_2(0) = s_2, \end{aligned} \tag{E.2}$$

where W_1^Q and W_2^Q are Brownian motions under Q with instantaneous correlation ρ.

Consider now an option on the spread between the two assets. Precisely, fix a maturity T, a positive real number a, a negative real number b, a strike price $K > 0$. The spread-option payoff at time T is then defined by

$$H = (awS_1(T) + bwS_2(T) - wK)^+, \tag{E.3}$$

where $w = 1$ for a call and $w = -1$ for a put.

The existence of a unique equivalent martingale measure implies that there exists a unique arbitrage-free price for the spread option at any time $t \in [0, T]$. Such a price is given by, see also (2.2),

$$\pi_t = e^{-r(T-t)} E^Q \left\{ (awS_1(T) + bwS_2(T) - wK)^+ | \mathcal{F}_t \right\}, \tag{E.4}$$

where E^Q denotes expectation under Q and $\mathcal{F}_t$ is the sigma-field generated by (S_1, S_2) up to time t.

For a nonzero strike price, the expectation in (E.4) cannot be computed in an explicit fashion. However, a pseudo-analytical formula can be derived in terms of improper integrals. This is explained in the following.

Proposition. The unique arbitrage-free price of the payoff (E.3) at maturity T is

$$\pi_t = \int_{-\infty}^{+\infty} \frac{1}{\sqrt{2\pi}} e^{-\frac{1}{2}v^2} f(v)dv, \tag{E.5}$$

where

$$\begin{aligned} f(v) = {} & awS_1(t) \exp\left[-q_1\tau - \frac{1}{2}\rho^2\sigma_1^2\tau + \rho\sigma_1\sqrt{\tau}v\right] \\ & \cdot \Phi\left(w \frac{\ln \frac{aS_1(t)}{h(v)} + [\mu_1 + (\frac{1}{2} - \rho^2)\sigma_1^2]\tau + \rho\sigma_1\sqrt{\tau}v}{\sigma_1\sqrt{\tau}\sqrt{1-\rho^2}} \right) \\ & - wh(v)e^{-r\tau}\Phi\left(w \frac{\ln \frac{aS_1(t)}{h(v)} + (\mu_1 - \frac{1}{2}\sigma_1^2)\tau + \rho\sigma_1\sqrt{\tau}v}{\sigma_1\sqrt{\tau}\sqrt{1-\rho^2}} \right) \end{aligned}$$

and

$$\begin{aligned} h(v) &= K - bS_2(t)e^{(\mu_2 - \frac{1}{2}\sigma_2^2)\tau + \sigma_2\sqrt{\tau}v} \\ \mu_1 &= r - q_1 \\ \mu_2 &= r - q_2 \\ \tau &= T - t \end{aligned}$$

with $\Phi(\cdot)$ denoting the standard normal cumulative distribution function.

Proof. Defining

$$\begin{aligned} X &:= \ln \frac{S_1(T)}{S_1(t)}, \\ Y &:= \ln \frac{S_2(T)}{S_2(t)}, \end{aligned}$$

the joint density function $f_{X,Y}$ of (X, Y) under the measure Q is bivariate normal with mean vector

$$M_{X,Y} = \begin{bmatrix} \mu_x \\ \mu_y \end{bmatrix} := \begin{bmatrix} (\mu_1 - \frac{1}{2}\sigma_1^2)\tau \\ (\mu_2 - \frac{1}{2}\sigma_2^2)\tau \end{bmatrix}$$

and covariance matrix

$$V_{X,Y} = \begin{bmatrix} \sigma_x^2 & \rho\sigma_x\sigma_y \\ \rho\sigma_x\sigma_y & \sigma_y^2 \end{bmatrix} := \begin{bmatrix} \sigma_1^2\tau & \rho\sigma_1\sigma_2\tau \\ \rho\sigma_1\sigma_2\tau & \sigma_2^2\tau \end{bmatrix},$$

that is

$$f_{X,Y}(x,y) = \frac{1}{2\pi\sigma_x\sigma_y\sqrt{1-\rho^2}} \exp\left[-\frac{\left(\frac{x-\mu_x}{\sigma_x}\right)^2 - 2\rho\frac{x-\mu_x}{\sigma_x}\frac{y-\mu_y}{\sigma_y} + \left(\frac{y-\mu_y}{\sigma_y}\right)^2}{2(1-\rho^2)}\right].$$

It is well known that

$$f_{X,Y}(x,y) = f_{X|Y}(x,y) f_Y(y),$$

where

$$f_{X|Y}(x,y) = \frac{1}{\sigma_x\sqrt{2\pi}\sqrt{1-\rho^2}} \exp\left[-\frac{\left(\frac{x-\mu_x}{\sigma_x} - \rho\frac{y-\mu_y}{\sigma_y}\right)^2}{2(1-\rho^2)}\right]$$

$$f_Y(y) = \frac{1}{\sigma_y\sqrt{2\pi}} \exp\left[-\frac{1}{2}\left(\frac{y-\mu_y}{\sigma_y}\right)^2\right].$$

This implies that

$$\pi_t = e^{-r\tau}\int_{-\infty}^{+\infty}\int_{-\infty}^{+\infty} (awS_1(t)e^x + bwS_2(t)e^y - wK)^+ f_{X,Y}(x,y)dxdy$$
$$= e^{-r\tau}\int_{-\infty}^{+\infty}\int_{-\infty}^{+\infty} (awS_1(t)e^x + bwS_2(t)e^y - wK)^+ f_{X|Y}(x,y)f_Y(y)dxdy$$
$$= e^{-r\tau}\int_{-\infty}^{+\infty}\left[\int_{-\infty}^{+\infty} (awS_1(t)e^x + bwS_2(t)e^y - wK)^+ f_{X|Y}(x,y)dx\right] f_Y(y)dy. \tag{E.6}$$

The inner integral can be explicitly computed, since the term $bwS_2(t)e^y$ can be viewed as a constant when integrating with respect to x, and turns out to be of the Black-Scholes type since $f_{X|Y}(x,y)$, for a fixed y, is itself the density function of a Gaussian random variable. More precisely, one applies the following formula, see also (D.1) in Appendix D, where $w \in \{-1, 1\}$ and $A > 0$, $K > 0$, M and $V > 0$ are real numbers,

$$\int_{-\infty}^{+\infty} \frac{1}{\sqrt{2\pi}V}(wAe^z - wK)^+ e^{-\frac{1}{2}\frac{(z-M)^2}{V^2}} dz$$
$$= wAe^{M+\frac{1}{2}V^2}\Phi\left(w\frac{M - \ln\frac{K}{A} + V^2}{V}\right) - wK\Phi\left(w\frac{M - \ln\frac{K}{A}}{V}\right),$$

to get that

$$\int_{-\infty}^{+\infty} \left(awS_1(t)e^x + bwS_2(t)e^y - wK\right)^+ f_{X|Y}(x,y)dx$$
$$= awS_1(t)\exp\left[\mu_x + \rho\sigma_x\frac{y-\mu_y}{\sigma_y} + \frac{1}{2}\sigma_x^2(1-\rho^2)\right]$$
$$\cdot \Phi\left(w\frac{\mu_x + \rho\sigma_x\frac{y-\mu_y}{\sigma_y} - \ln\frac{K-bS_2(t)e^y}{aS_1(t)} + \sigma_x^2(1-\rho^2)}{\sigma_x\sqrt{1-\rho^2}}\right) \tag{E.7}$$
$$- w\left(K - bS_2(t)e^y\right)\Phi\left(w\frac{\mu_x + \rho\sigma_x\frac{y-\mu_y}{\sigma_y} - \ln\frac{K-bS_2(t)e^y}{aS_1(t)}}{\sigma_x\sqrt{1-\rho^2}}\right).$$

The definition of μ_x, μ_y, σ_x and σ_y and the variable change $v = \frac{y-\mu_y}{\sigma_y}$ then lead to (E.5). □

F. Approximating Diffusions with Trees

The Holy One directed his steps to that blessed Bodhi-tree beneath whose shade he was to accomplish his search.
Paul Carus, "The Gospel of Buddha", 1894.

In this appendix, we show how to approximate a diffusion process with a tree. The general procedure we outline is used throughout the book in the tree construction for both one-factor and two-factor short-rate models. In the one-factor case, the tree is constructed by imposing that the conditional local mean and variances at each node are equal to those of the basic continuous-time process. The geometry of the tree is then designed so as to ensure the positivity of all branching probabilities. In the two-factor case, instead, we first construct the trees for the two factors along the procedure that applies to one-factor diffusions. We then construct a two-dimensional tree by imposing that the tree marginal distributions match those of the two factors' trees and by imposing the correct local correlation structure so as to preserve the positivity of all branching probabilities as well.

The tree construction procedure we propose in this appendix is mostly based on heuristic arguments. Indeed, the proof of formal convergence results is beyond the scope of this section. When deriving an approximating tree, we simply have in mind that, at first order in the amplitude of the time step, the diffusion transition density is normal and hence completely specified by its mean and variance.

Approximating a one-factor diffusion

Let us consider the diffusion process X that evolves according to

$$dX_t = \mu(t, X_t)dt + \sigma(t, X_t)dW_t, \tag{F.1}$$

where μ and σ are smooth scalar real functions and W is a scalar standard Brownian motion.

We want to discretize the dynamics (F.1) both in time and in space. Precisely, we want to construct a trinomial tree that suitably approximates the evolution of the process X.

To this end, we fix a finite set of times $0 = t_0 < t_1 < \cdots < t_n = T$ and we set $\Delta t_i = t_{i+1} - t_i$. At each time t_i, we have a finite number of equispaced states, with constant vertical step Δx_i to be suitably determined. We set $x_{i,j} = j\Delta x_i$.

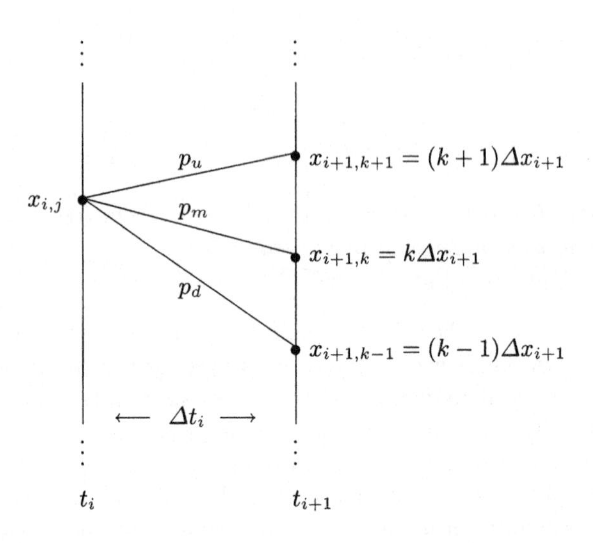

Fig. F.1. Evolution of the process x starting from $x_{i,j}$ at time t_i and moving to $x_{i+1,k+1}$, $x_{i+1,k}$ or $x_{i+1,k-1}$ at time t_{i+1} with probabilities p_u, p_m and p_d, respectively.

The tree geometry is displayed in Figure F.1. Assuming that at time t_i we are on the j-th node with associated value $x_{i,j}$, the process can move to $x_{i+1,k+1}$, $x_{i+1,k}$ or $x_{i+1,k-1}$ at time t_{i+1} with probabilities p_u, p_m and p_d, respectively. The central node is therefore the k-th node at time t_{i+1}, where also the level k is to be suitably determined.

Denoting by $M_{i,j}$ and $V_{i,j}^2$ the mean and the variance of X at time t_{i+1} conditional on $X(t_i) = x_{i,j}$, i.e.,

$$\begin{aligned} E\{X(t_{i+1})|X(t_i) = x_{i,j}\} &= M_{i,j} \\ \mathrm{Var}\{X(t_{i+1})|X(t_i) = x_{i,j}\} &= V_{i,j}^2, \end{aligned}$$

we want to find p_u, p_m and p_d such that these conditional mean and variance match those in the tree. Precisely, noting that $x_{i+1,k+1} = x_{i+1,k} + \Delta x_{i+1}$ and $x_{i+1,k-1} = x_{i+1,k} - \Delta x_{i+1}$, we look for positive constants p_u, p_m and p_d summing up to one and satisfying

$$\begin{cases} p_u(x_{i+1,k} + \Delta x_{i+1}) + p_m x_{i+1,k} + p_d(x_{i+1,k} - \Delta x_{i+1}) = M_{i,j} \\ p_u(x_{i+1,k} + \Delta x_{i+1})^2 + p_m x_{i+1,k}^2 + p_d(x_{i+1,k} - \Delta x_{i+1})^2 = V_{i,j}^2 + M_{i,j}^2. \end{cases}$$

Simple algebra leads to

$$\begin{cases} x_{i+1,k} + (p_u - p_d)\Delta x_{i+1} = M_{i,j} \\ x_{i+1,k}^2 + 2x_{i+1,k}\Delta x_{i+1}(p_u - p_d) + \Delta x_{i+1}^2(p_u + p_d) = V_{i,j}^2 + M_{i,j}^2. \end{cases}$$

Setting $\eta_{j,k} = M_{i,j} - x_{i+1,k}$,[1] we finally obtain

$$\begin{cases} (p_u - p_d)\Delta x_{i+1} = \eta_{j,k} \\ (p_u + p_d)\Delta x_{i+1}^2 = V_{i,j}^2 + \eta_{j,k}^2, \end{cases}$$

so that, remembering that $p_m = 1 - p_u - p_d$, the candidate probabilities are

$$\begin{cases} p_u = \frac{V_{i,j}^2}{2\Delta x_{i+1}^2} + \frac{\eta_{j,k}^2}{2\Delta x_{i+1}^2} + \frac{\eta_{j,k}}{2\Delta x_{i+1}}, \\ p_m = 1 - \frac{V_{i,j}^2}{\Delta x_{i+1}^2} - \frac{\eta_{j,k}^2}{\Delta x_{i+1}^2}, \\ p_d = \frac{V_{i,j}^2}{2\Delta x_{i+1}^2} + \frac{\eta_{j,k}^2}{2\Delta x_{i+1}^2} - \frac{\eta_{j,k}}{2\Delta x_{i+1}}. \end{cases}$$

In general, there is no guarantee that p_u, p_m and p_d are actual probabilities, because the expressions defining them could be negative. We then have to exploit the available degrees of freedom in order to obtain quantities that are always positive. To this end, we make the assumption that $V_{i,j}$ is independent of j,[2] so that from now on we simply write V_i instead of $V_{i,j}$. We then set $\Delta x_{i+1} = V_i\sqrt{3}$,[3] and we choose the level k, and hence $\eta_{j,k}$, in such a way that $x_{i+1,k}$ is as close as possible to $M_{i,j}$. As a consequence,

$$k = \text{round}\left(\frac{M_{i,j}}{\Delta x_{i+1}}\right), \tag{F.2}$$

where round(x) is the closest integer to the real number x. Moreover,

$$\begin{cases} p_u = \frac{1}{6} + \frac{\eta_{j,k}^2}{6V_i^2} + \frac{\eta_{j,k}}{2\sqrt{3}V_i}, \\ p_m = \frac{2}{3} - \frac{\eta_{j,k}^2}{3V_i^2}, \\ p_d = \frac{1}{6} + \frac{\eta_{j,k}^2}{6V_i^2} - \frac{\eta_{j,k}}{2\sqrt{3}V_i}. \end{cases} \tag{F.3}$$

It is easily seen that both p_u and p_d are positive for every value of $\eta_{j,k}$, whereas p_m is positive if and only if $|\eta_{j,k}| \leq V_i\sqrt{2}$. However, defining k as in (F.2) implies that $|\eta_{j,k}| \leq V_i\sqrt{3}/2$, hence the condition for the positivity of p_m is satisfied, too.

As a conclusion, (F.3), under the condition (F.2), are actual probabilities such that the corresponding trinomial tree has conditional (local) mean and variance that match those of the continuous-time process X.

[1] We omit to express the dependence on the index i to lighten the notation.

[2] This assumption indeed holds true in all the specific examples we consider in this book. In general, however, one has to look for a suitable transformation leading to a deterministic-volatility process, as in Sections 3.9.1 and 5.3.

[3] This choice, motivated by convergence purposes, is a standard one. See for instance Hull and White (1993, 1994).

Approximating a two-factor diffusion

Consider a process Z that is the sum of two diffusion processes X and Y, i.e.,

$$Z_t = X_t + Y_t,$$

where

$$dX_t = \mu^X(t, X_t)dt + \sigma^X(t, X_t)dW_t^X,$$
$$dY_t = \mu^Y(t, Y_t)dt + \sigma^Y(t, Y_t)dW_t^Y,$$

with μ^X, μ^Y, σ^X and σ^Y smooth real functions and W^X and W^Y two correlated standard Brownian motions with $dW_t^X dW_t^Y = \rho dt$, $\rho \in [-1, 1]$.

To construct an approximating tree for Z, we first construct the two trees approximating the processes X and Y along the procedure being previously illustrated. We then build a two-dimensional tree by locally imposing the right marginal distributions and the right correlation. To this end, we follow a similar procedure to that suggested by Hull and White (1994c), thus building a preliminary tree based on the assumption of zero correlation.

We denote by $M_{i,j}^X$ and $(V_i^X)^2$ the mean and the variance of X at time t_{i+1} conditional on $X(t_i) = x_{i,j}$, and by $M_{i,l}^Y$ and $(V_i^Y)^2$ the mean and the variance of Y at time t_{i+1} conditional on $Y(t_i) = y_{i,l}$, i.e.,

$$E\{X(t_{i+1})|X(t_i) = x_{i,j}\} = M_{i,j}^X,$$
$$\text{Var}\{X(t_{i+1})|X(t_i) = x_{i,j}\} = (V_i^X)^2,$$
$$E\{Y(t_{i+1})|Y(t_i) = y_{i,l}\} = M_{i,l}^Y,$$
$$\text{Var}\{Y(t_{i+1})|Y(t_i) = y_{i,l}\} = (V_i^Y)^2,$$

where, analogously to the one-factor case, we assume that the two variances are independent of j and l, respectively. We then denote by (i, j, l) the tree node at time t_i where X equals $x_{i,j} = j\Delta x_i$ and $Y = y_{i,l} = l\Delta y_i$, and by

- π_{uu} the probability of moving from (i, j, l) to $(i + 1, k + 1, h + 1)$,
- π_{um} the probability of moving from (i, j, l) to $(i + 1, k + 1, h)$,
- π_{ud} the probability of moving from (i, j, l) to $(i + 1, k + 1, h - 1)$,
- π_{mu} the probability of moving from (i, j, l) to $(i + 1, k, h + 1)$,
- π_{mm} the probability of moving from (i, j, l) to $(i + 1, k, h)$,
- π_{md} the probability of moving from (i, j, l) to $(i + 1, k, h - 1)$,
- π_{du} the probability of moving from (i, j, l) to $(i + 1, k - 1, h + 1)$,
- π_{dm} the probability of moving from (i, j, l) to $(i + 1, k - 1, h)$,
- π_{dd} the probability of moving from (i, j, l) to $(i + 1, k - 1, h - 1)$,

where k and h are chosen so that $x_{i+1,k} = k\Delta x_{i+1}$ is as close as possible to $M_{i,j}^X$ and $y_{i+1,h} = h\Delta y_{i+1}$ is as close as possible to $M_{i,l}^Y$.

Under zero correlation, these probabilities are simply the product of the corresponding marginal probabilities. For instance,

$$\pi_{ud} = \left(\frac{1}{6} + \frac{(\eta^X_{j,k})^2}{6(V^X_i)^2} + \frac{\eta^X_{j,k}}{2\sqrt{3}V^X_i}\right)\left(\frac{1}{6} + \frac{(\eta^Y_{l,h})^2}{6(V^Y_i)^2} - \frac{\eta^Y_{l,h}}{2\sqrt{3}V^Y_i}\right),$$

where $\eta^X_{j,k} = M^X_{i,j} - x_{i+1,k}$ and $\eta^Y_{l,h} = M^Y_{i,l} - y_{i+1,h}$.

Let us now define Π_0 as the matrix of the "zero-correlation" probabilities, i.e.,

$$\Pi_0 = \begin{pmatrix} \pi_{ud} & \pi_{um} & \pi_{uu} \\ \pi_{md} & \pi_{mm} & \pi_{mu} \\ \pi_{dd} & \pi_{dm} & \pi_{du} \end{pmatrix}$$

To account for the proper correlation ρ we have to shift each probability in Π_0 in such a way that the sum of the shifts in each row and each column is zero (so that the marginal distributions are maintained) and the following is verified:

$$\frac{\text{Cov}\{X(t_{i+1}), Y(t_{i+1})|X(t_i), Y(t_i)\}}{V^X_i V^Y_i}$$
$$= \frac{\Delta x_{i+1}\Delta y_{i+1}\left(-\pi_{ud} - \varepsilon_{ud} + \pi_{uu} + \varepsilon_{uu} + \pi_{dd} + \varepsilon_{dd} - \pi_{du} - \varepsilon_{du}\right)}{V^X_i V^Y_i} = \rho,$$

at first order in Δt_i, where ε_{ud} denotes the shift for π_{ud} (the definition of the other shifts is analogous). Since $\Delta x_{i+1} = V^X_i\sqrt{3}$ and $\Delta y_{i+1} = V^Y_i\sqrt{3}$, this condition can be written as

$$-\pi_{ud} - \varepsilon_{ud} + \pi_{uu} + \varepsilon_{uu} + \pi_{dd} + \varepsilon_{dd} - \pi_{du} - \varepsilon_{du} = \frac{\rho}{3},$$

becoming, in the limit for $\Delta t_i \to 0$,

$$-\varepsilon_{ud} + \varepsilon_{uu} + \varepsilon_{dd} - \varepsilon_{du} = \frac{\rho}{3}.$$

Solving for the previous constraints, a possible solution, that is similar to that proposed by Hull and White (1994c), is

$$\Pi_\rho = \Pi_0 + \rho(\Pi^l_1 - \Pi^l_0), \text{ if } \rho > 0$$
$$\Pi_\rho = \Pi_0 - \rho(\Pi^l_{-1} - \Pi^l_0), \text{ if } \rho < 0$$

where Π_ρ is the probability matrix that accounts for the right correlation ρ and Π^l_1, Π^l_0 and Π^l_{-1} are the probability matrices in the limit, respectively for $\rho = 1$, $\rho = 0$ and $\rho = -1$. More explicitly,

$$\Pi_\rho = \Pi_0 + \frac{\rho}{36}\begin{pmatrix} -1 & -4 & 5 \\ -4 & 8 & -4 \\ 5 & -4 & -1 \end{pmatrix} \text{ if } \rho > 0$$

and

$$\Pi_\rho = \Pi_0 - \frac{\rho}{36}\begin{pmatrix} 5 & -4 & -1 \\ -4 & 8 & -4 \\ -1 & -4 & 5 \end{pmatrix} \text{ if } \rho < 0.$$

However, it may happen that one or more entries in Π_ρ are negative. To overcome this drawback, we are compelled to modify the correlation on node (i, j, l), for example substituting Π_ρ with Π_0. Assuming a different correlation on such node, and on all the tree nodes where some probabilities are negative, has a negligible impact on a derivative price as long as the mesh of $\{t_0, t_1, \ldots, t_m\}$ is sufficiently small. Notice, in fact, that in the limit for such mesh going to zero, the elements of Π_ρ are all positive, being Π_ρ a convex linear combination of two matrices with positive entries.

G. Trivia and Frequently Asked Questions

"I don't want to hear about Green Arrow's love life...
This is the JLA, Kyle"
Flash/Wally West to Green Lantern/Kyle Rayner
in JLA 12, 1997, DC Comics

"Not by birth is one an outcast; not by birth is one a brahman. By deed one becomes an outcast, by deed one becomes a brahman."
Gotama Buddha, Vasala Sutta, 21

Many readers still confuse the two authors and keep on asking questions on the book, on the quotes, on who does what etc. We answer some of the most frequent questions here. This appendix is meant as a relaxing break the reader may visit any time.

Question: Why the delay and the long wait for this second edition?
Answer: It had to do with all other tasks one is supposed to attend to inside an investment bank, especially with external university and training course teaching activities, conferences, scientific committees etc taking place all the time. The authors apologize for the delay.

Q: Damiano... Fabio... but who is Brigo and who is Mercurio of the two??
A: Damiano is the guy with short balding hair, pale skin and light blue eyes, with a craze for motorbikes, animation, comics, Shakespeare, religion, general science, cosmology, fiction and quotes; Fabio is the guy with curly hair and hazel eyes, a craze for sports and art (from prehistory to present), and a passion for travelling, learning new cultures and religion.
If still in doubt and curious, one may check the authors web pages, both contain pictures:
`http://www.damianobrigo.it/`   `http://www.fabiomercurio.it/`

Q: How old are the two authors?

A: When the first edition came out the authors were about 35 years old, by the time you are reading this second edition the two authors should be about forty years old.

Q: Since when have the two authors been working together?

A: The two authors have been in touch since 1985, when they both attended Mathematics in Padua. Then they both studied for a PhD in the Netherlands (Damiano in Stochastic Systems and Nonlinear Filtering with differential geometry in Amsterdam/Rennes and Fabio in Mathematical Finance in Rotterdam). Fabio entered an Italian bank in October 1996 and Damiano joined him in February 1997. They both moved to Banca IMI in 1998 and have been colleagues since then.

Q: What are their current roles?

A: Currently, they are both employed in the Financial Engineering department of Banca IMI. Damiano is head of the Credit Models team, while Fabio is head of the Financial Models team. In the last three years Damiano has been in charge for the modeling, pricing and hedging of Credit Derivatives and Counterparty risk, while formerly he worked on interest rate derivatives, smile modeling, risk measurement and quantitative analysis in the bank. Now hybrid products including credit features (and thus virtually any derivative product, given counterparty risk) have entered his field of research. Damiano basically worked in every field of derivatives pricing, if with different extents, including obviously credit.

Fabio, instead, has been in charge for the modeling, pricing and hedging of Equity, Foreign Exchange and Interest-Rate derivatives in the bank for the last five years. Now, he is also in charge of the modeling of Commodity, Inflation and Hybrid (all but credit) derivatives.

Q: Who had the idea of writing this book first?

A: It looks like there has been an event at the dawn of time that changed history and edited memories. Different hypertime fluxes branched together and now the two authors have different memories as to who thought about the book first. Just for the record, Damiano is convinced he mentioned the idea during one of the late 1997 evening informal scientific meetings at Cariplo Bank, inspired by a draft of Björk (1997) and while working on the CIR++ model, while Fabio is sure he suggested it first in 1998 after the first technical reports on interest rate models in Banca IMI had been assembled in response to the first tasks in this second bank.

Since time machines are not available yet and given that navigating hypertime can be rather hazardous anyway[1] it is difficult to know what actually transpired. But at least the book is solid and in your hands anyway...

[1] see for example Morrison and Semekis (85271) and Kesel and Grummett (1999)

Q: What about all these quotes? Who put them in the book and why?

A: All quotes (Shakespeare, Douglas Adams, Comics, Babbage, Guth, Buddhism, Christianity, Tolkien, etc) in the first edition but a few (Einstein and Grisham) are Damiano's. He thought they could give a nice break from long hours of studying. Besides, he thinks they are fun. Fabio joined the game in this second edition and the new quotes you find in the inflation and smile chapters are his.

Q: Are the quotes placed randomly?

A: No, they are placed in places where they have some link with the matter under discussion. The link may be ironical or serious, but most of times it is there. In some cases this is obvious, in some other cases it in not so obvious, but the reader is clearly welcome to be creative also in this respect.

Q: How did readers react to quotes?

A: Reactions have been extremely positive, much to the authors' surprise. Especially successful have been the quote of Babbage on inputs and outputs, the Batman and Superman closing quote after "talking to the traders", Tolkien based quotes (well in advance the popular movies on Tolkien's trilogy), and the opening Kara's quote (now closing this second edition), a warning and an example, if fictional, the authors can never match. Also Douglas Adams' brownian motion quote at the beginning of the SDE's appendix has been much appreciated.

Q: Why "Don't Panic"? Now even in the back cover...

A: Because, especially now, it is a quite large and heavy looking book. One can really be scared at first sight. The book is friendly, but one may not guess so from outside. Besides, the authors themselves may benefit from this warning when keeping the book on their desks :-)

Last but not least, it is a homage to Douglas Adams who passed away in 2001, slightly after the first edition of this book, including his quotes, had appeared.

Q: How on Earth did you think of this "talking to the traders" appendix?

A: The "talking to the traders" appendix was born out of a dream of Damiano, during one night's sleep in the last weeks of work on the first edition.

Q: Who designed the cover? What does it mean?

A: The initial cover sketch of this new edition has been drawn by Damiano. It represents an interest rate volatility smile surface on which default arrives: the idea is having smile and credit, that are among the new features, on the cover. A second cover was proposed by Damiano later, consisting of a spiral galaxy with two jets exploding from its center orthogonally to the disc, one jet with an inflation equation, a second jet with a smile equation, and a "τ" in the center. Fabio even had the brilliant idea of replacing the horse-head logo of the publisher with the horse-head nebula. This was however refused by the publisher for technical and management reasons, and time constraints.

Q: Why not splitting the book into two volumes?

A: The authors did not really expect to exceed one thousand pages, and when they did it was too late to change plans with the publisher.

Q: Is there a sort of web site for the book?

A: The following page includes the table of contents, a book description, and downloads for readers owing the first edition and not willing to buy the second one. The address is

`http://www.damianobrigo.it/book.html`

Then of course there is the publisher's page.

Q: Can software codes or software source files be available upon request?

A: Unfortunately this is not possible, the codes (even prototypes) are reserved for the Bank activities and cannot be disclosed, due to our bank internal regulations.

Q: Is a soft-cover, paperback or textbook version going to appear?

A: Not in the short term, but it is a possibility the authors are considering.

Q: Do the authors have plans for a third edition?

A: - Beep - Wrong question :-)

[This book] has already supplanted the great Encyclopedia Galactica as the standard repository of all knowledge and wisdom, for though it has many omissions and contains much that is apocryphal, or at least wildly inaccurate, it scores over the older, more pedestrian work in two important respects. First, it is slightly cheaper; and secondly it has the words "Don't Panic" inscribed in large friendly letters on its cover.

Douglas Adams (1952-2001)

H. Talking to the Traders

'A logical division of Labor.
Interviews require a certain facility for interpersonal interaction, sometimes referred to as "people skills" '
J'onn J'onnz in reply to Batman in "Martian Manhunter annual" 2, 1999, DC Comics.

"My body aches, Superman, and I'm hungry. That's all I'm sure of. But I'll call this a learning experience if you do."
Batman, DC One Million, DC Comics, 1999.

In this appendix, we would like to reproduce an hypothetical conversation between a trader in interest rate derivatives and a quantitative analyst eager to get acquainted with some specific market practice. This virtual interview reflects our personal experience of interaction with traders, considering some of the traders' opinions we have collected over the years.

We would also like to warn the reader that he/she must not be deceived by the smooth flowing of this interview. Traders and quants tend to speak different languages and sentences or reasoning must be often rephrased for a reciprocal understanding. However, after an unavoidable initial effort, and many imprecations that (hopefully) fade away on the tip of both tongues, traders and quants get to know each other, becoming (more and more) aware of how important can be maintaining a reciprocal cooperation. Thus, in the interview below we have tried to reproduce this atmosphere at times. We took the symbol ":-)" from the internet jargon to mean that a humorous remark needs not to be taken as offence or hostility.

Short rate models, HJM and market models: Where now?

1. Quant: *Among all the short rate models we implemented for you in the software unit you call "big toy", namely* HW, CIR++, BK, EEV, G2++, *which ones do you use most frequently and why?*
   Trader: It depends. For our purposes, we mostly use BK (or alternatively EEV) and G2++ (equivalent to two factor Hull and White model, HW2).
2. Q: *Why this choice?*
   T: Well, first off, these are among the most popular models in the short rate world, and as such we tend to trust them more. Next, they involve kind of opposite assumptions on the short rate distribution. One model takes it to be Gaussian, the other lognormal.

3. Q: *So what's your problem with CIR++ for example?*
   T: The problem is that this noncentral chi square distribution stays in-between the other two cases as far as tails are concerned, and so we prefer ranging a wider interval by considering the lognormal and normal models. Not to talk about the numerical problems one is likely to face when dealing with chi square distribution functions.
4. Q: *Why considering two so different models?*
   T: Their difference is exactly the reason why we take them both. We expect, and have actually experienced, that when calibrated on the same data and used for managing products that are *at the money* (ATM), the two models yield roughly the same answers. Instead, as soon as moneyness moves, the tails of the rates distributions become more relevant, so that one model tends to amplify prices while the other underestimates them. Therefore, when dealing with *in the money* (ITM) or *out of the money* (OTM) options, we take as final solution a (somehow weighted) average of the answers provided by the two models.
5. Q: *Can you give me some further insight? I mean, why is moneyness so important?*
   T: Ok, it's very simple. In the market, we usually observe what is called the "smile effect". This basically means that implied volatilities are not all traded at a same level. It usually happens that OTM volatilities are higher than the corresponding ITM ones, and this can have a serious impact in the pricing of derivatives.
6. Q: *Can you be more explicit?*
   T: Yes, just to fix ideas, let's say you must price an OTM receiver swaption of Bermudan type. If you calibrate your favorite interest rate model to ATM volatilities (they are the only ones to be actively quoted in the market), you miss the fact that an OTM receiver European swaption is priced with a higher volatility than the corresponding ATM one. Assuming a normal distribution of rates can then be helpful, since it assigns a higher probability to low rates, thus rendering the OTM swaption more valuable. At the same time, however, the market implicitly assumes that rates are lognormally distributed. This is why the correct price is somehow in between the Gaussian and lognormal prices.
7. Q: *Do you think short rate models still have a future?*
   T: Who knows. The HJM developments did not threaten seriously the short-rate world but rather incorporated it. However, the recent growth of market models certainly threatens the short-rate model survival.
8. Q: *What can be impressing is the calibrating capability: The LIBOR market model can calibrate a lot of swaptions and caps at the same time, at least in principle...*
   T: Sure, but that is not necessarily a uniform advantage. Remember what I call the "uncertainty principle of modeling", the more a model fits, the less it explains.

9. Q: *What do you mean exactly?*
   T: I mean, if a model recovers a huge number of financial observables by construction, it cannot be able to explain well what is happening, exactly because it "eats" everything you feed it on. On the contrary, a model that cannot calibrate a huge number of prices will signal problems. Then, I can watch the data and decide whether the problem is with the model limitations or there are some pathologies with the market structures given as input. So the "poor" model warns me, while a too rich model is always happy and does not help me that much to sense danger.
10. Q: *I think I read something similar on Rebonato's book on interest rate models*
    T: Yes, I agree completely with Rebonato on this issue, as well as on so many other things.
11. Q: *This looks like a point in favor of short rate models.*
    T: Yes. Let me ask you, are there reasons to prefer short rate models as far as implementation issues are concerned?
12. Q: *Well, a lot of features are simplified with short rate models. You can build trees and price early exercise products easily enough, but only in case of low dimension, say two or at most three. Notice also that, in case of correlated factors, the tree construction can be tricky already in the two-factor case. With high dimensional trees, moreover, you easily have problems concerning the speed of convergence and the computer memory required to run your pricing routines. However, the tree construction for the short-rate models we have implemented is rather straightforward.*
    T: Which is not the case with the market model. What about Monte Carlo?
13. Q: *Well, with tractable models such as HW2 or CIR++, from the short rate factors all kind of rates (forward and swap rates for any expiry/maturity/payment dates) can be recovered easily. Instead, in the market model you need to interpolate when leaving the preassigned dates, since you simulate finite rates with preassigned expiries and maturities. Also, with HW2 and CIR++, Monte Carlo simulation is easy, since the factors dynamics under the forward measure are known explicitly as well as their transition densities. In fact, "one-shot" simulations are possible and no discretizations of the factors equations are required, contrary to the market models. These can be attractive features especially for risk management purposes.*
    T: So you see some further advantages of short rate models.

Some hedging issues

1. Q: *Well, I was also wondering about hedging. What about hedging?*
   T: Hedging is a different matter. Hedging can be done naively on the basis of shifting the required market observable, recalibrate, and compute the

difference in prices divided by the shift amount. This is good to compute "global" sensitivities, i.e. sensitivities with respect to uniform shifts of input market structures, but...

2. Q: *Yes?*
   T: But we have problems with breakdowns of sensitivities. When we need sensitivities to single inputs, your big toy©doesn't help that much, you see... [gets excited, moves his hands frantically]... these @##!@%%&@!! volatility surfaces move like this [traces surfaces contours and tangent planes in the air]... How the @##!@%%&@!! can I compute sensitivities say to single swaption prices? And I need them...
3. Q: *Ahem, ever the diplomat :-) ...let me rephrase it... are you saying that the influence of a local shift in a market observable is distributed globally on the parameters by the calibration, so that the local effects of the variations are lost? I see... Hedging seems possible when shifting uniformly market curves and surfaces in input, but when shifting single points, the effect is probably lost and possibly confused with other possible causes.*
   T: Right, that's what I said, but you took all the colour out of it :-). A short rate model with only one time-dependent function (used for exact calibration to the zero-coupon curve) has too few parameters to appreciate the influence of local changes in the input volatility structures. Shifting two rather different points can cause the same change in the parameters, due to the flattening of the information implied by the low number of parameters. A market model, instead, can often appreciate such a change by distinguishing the two cases. In the book you even described a volatility parameterization in the LIBOR market model (LFM) where you have a one to one correspondence between forward rates volatilities and swaption prices. This can be incredibly helpful for hedging.
4. Q: *Can you elaborate more on this?*
   T: What I wanted to say is that this can allow for computation of sensitivities with respect to a single swaption price (volatility) used in the calibration. This is important for Vega breakdown analysis and sensitivity to volatility in general. Indeed, it helps us understand what portion of the (implied) volatility surface has a significant effect on the price of the considered derivative. We can then construct a hedging strategy based on the European swaptions corresponding to that volatility surface portion.
5. Q: *I understand. But don't you think you can achieve similar results by introducing further time-dependent coefficients in the short rate dynamics? This is actually the original approach followed by Hull and White, Black, Derman and Toy or Black and Karasinski.*
   T: You're right. However, too many time-dependent parameters can be dangerous. And not only due to the overparameterization problems I mentioned before. Indeed, the price to pay for a better, possibly exact

fitting of market data, is an unplausible evolution of the future term structure of volatility, in that the implied future volatility surfaces can have unrealistic shapes.

6. Q: *Uhm, [an amused look appears on the quant's face...] how comes that traders give so much importance to aesthetics? :-) ...*
   T: Heh, of course it's deeper than that! See, an unplausible future volatility structure can badly affect the pricing of instruments that implicitly depend on such a volatility structure, like a Bermudan swaption for instance. To fix ideas, suppose you have to price a Bermudan swaption, and you decide to calibrate to the underlying European swaptions. Using the general formulation of the Hull and White one-factor model, you can find infinitely many specifications of the parameters leading to an exact fitting of these European swaptions. This is not harmless. You'll indeed find that the corresponding prices wanders freely in such a wide range that you can't give any meaningful answer.
7. Q: *So there are both advantages and disadvantages with short rate models. Which do you think are dominant?*
   T: Disadvantages. Probably, in due time, market models will finally replace short rate models completely. We will see. But let me give you a warning. When the HJM theory came out, a lot of people thought that the interest rate theory was dead and everything had been done. Yet, we have seen how many things happened since then. So we should be careful in thinking market models are the final and complete solution to all the problems in interest rate models... and who knows, maybe short rate models will come back one day, after they have vanished.

Are market models completely developed?

1. Q: *What are - in your opinion - the areas where market models are not yet completely effective?*
   T: Let me think... I guess that multi-currency products are still an issue. In the book you pointed out that products such as quanto CMS with optional features are easily priced with two correlated two-factor extended Vasicek models. You were also able to translate the relevant model correlations into correlations I would be able to express a view upon. However, pricing similar products with correlated LIBOR market models for each curve can be more complicated and computationally demanding.
2. Q: *Actually, I have recently seen some interesting work on the multi-currency extensions of the LIBOR market model, so it seems some research is being done in that direction, too. What else?*
   T: Well, market models are relatively young and there is still so much to learn about them. You know, models such as HW or BK have been on the market for years, have been also implemented by several commercial softwares, and their limits are well known. I feel market models are yet

at a preliminary stage in that respect, although more and more studies on their practical implementation are being done and about to become public domain.

3. *Q: So you feel that LIBOR market models are yet to be fully tested?*
   T: I mean, just think of how many commercial software companies implemented a comprehensive version of the LIBOR market model, compared to older short rate models. I think there are not those many. At the same time, however, I must say that the big market players have their own knowledge and expertise. Just think that the top financial institutions were using some kind of market models well before they appeared in the literature.
4. Q: *I heard several commercial software companies are working on the LIBOR market model, and probably two or three of them already have a version of it. However, I think it is difficult to check whether the proposed implementations are satisfactory in the sense we outlined in the chapter we devoted to market models.*
   T: In some cases the possibility for a full investigation would allow the client to deduce key features of the model implementation, of the covariance structure, of the method for computing sensitivities, and so on. And we know that the added value of a model often lies in an efficient implementation rather than on nice "mathematical features". A company may not be willing to give that away, not even to a good client.
5. Q: *Does that mean that you don't care that much about mathematical rigor?*
   T: Wait, don't put words in my mouth. But consider an example: Many financial institutions have been pricing products with BK's model for years. This is a model where the average bank account is infinite after an infinitesimal time...
6. Q: *And yet BK was and, to some extent, still is quite a successful model...*
   T: Indeed. Now let me ask you: If you put your money in a bank account, and your model tells you that the expected value of the money you'll have in say one second is infinite, would you consider such a model satisfactory?
7. Q: *Mmmmhhh... as a mathematician I would probably have perplexities about this infinite expected future bank account.*
   T: Right. However, when you implement this model you usually resort to a tree, and with trees the danger of infinite expectations is avoided. The tree has a finite number of states, and thus avoids this undesirable feature. So, as a practitioner, I would not worry too much about this explosion issue, given that, in change for this, we get a lognormal distribution for the short rate and usually a good fitting to a low number of selected swaption prices. To sum up: Mathematical rigor is important, but it is not everything, and at times it is a good idea to put the accent on different aspects.

Modeling the smile in market models

1. Q: *All right. Back to market models: Other open issues?*
   T: Of course, smile modeling. Andersen and Andreasen had the idea of applying the CEV structure to the LIBOR market model and have developed interesting approximations. Yet, the CEV structure does not have enough flexibility for many practical purposes.
2. Q: *What do you think of our "shifted lognormal mixture dynamics"?*
   T: I would like to see some approximations leading to a quick pricing of swaptions, something similar to the tricks leading to Rebonato's or Hull and White's formulas for swaption volatilities in the LIBOR market model (LFM). Then with enough parameters in your mixture dynamics you might try a calibration to part of the swaptions smile. And... well...
3. Q: *Yes?*
   T: It would be interesting to see something on the transition densities implied by your dynamics, not merely on the marginal distribution. This could give information on the volatility structures implied by your dynamics at future times.
4. Q: *You know, I almost feel we have exchanged places, with you discussing such technical matters. I mean, you are asking quite technical questions. Ok, I'll put that on schedule. Not that we hadn't thought about it...*
   T: Don't worry. This is a kind of aspect that has been ignored in several models I have seen around. However, you should not be too surprised by my questions. Quantitative traders often have quite some experience, and can thus appreciate the models implications and limitations in connections with their market applications.
5. Q: *I have no doubts about it. It's just that in other banks we have had more difficulties in interacting with traders. Thanks for your time. Even if this "interview" has been slightly unfocused and informal, I think our readers might appreciate it.*
   T: My pleasure. It's not so many times that a trader can give an opinion or some teachings in a book about the doctrine. I am glad I have been given this possibility, and let me wish good luck to all your readers in their efforts either in the market or in quantitative research.
   Q: *Good idea: Good luck!*

Talking to the Traders... FIVE YEARS LATER.

> *"I thought of lots of things. It got darker. The glare of the red neon sign spread farther and farther across the ceiling. [...] After a little while I felt a little better, but very little. I needed a drink, I needed a lot of life insurance, I needed a vacation, I needed a home in the country. What I had was a coat, a hat and a gun. I put them on and went out of the room."*
> Philip Marlowe, "Farewell, my lovely", 1940

> *Instead of trench coats, stubble, and plotless angst in the alleyways and avenues, I went for bright costumes, aspirational nobility, and widescreen imaginative romps through space and time*
> Grant Morrison on his JLA run (1997-2000), DC Comics.

What's new?

1. T: Hey, "long time no see", as they say. How's been your work in these last five years? Did you have any luck?
2. Q: *There have been a lot of new issues in the last years, and it was fun working on them.*
3. T: Such as?
4. Q: *Smile modeling grew more and more important, to the point that in many situations one cannot afford to ignore it.*
5. T: Why, was there a time when you could ignore it?
6. Q: *Right, models should always be perfect and should also have always been perfect in the past, is that what you think? Remember you have been pricing products with one factor short rate models up to some time ago...*
7. T: Well, relax, you know we aim at the best...
8. Q: *Fair enough. Now smile seems to be a big issue. Brokers quote swaptions smiles with the SABR model. If even brokers resort to a model to quote the smile it means that the smile has entered the heart of the market to some extent.*
9. T: It should have happened before, but better late than never.
10. Q: *Can you explain to me why the SABR model, of all existing models?*
11. T: I think it's used for its simplicity. It is a compromise between richer and more realistic models such as Heston's and analytical tractability/speed of execution. And it's "Black-Scholes-like" in many ways.
12. Q: *Yes, and it's not that bad as concerns other models. Since this model is only used as a quoting mechanism, we could simply calibrate the swaption prices underlying it to a more realistic model.*
13. T: Mmmhhhh... but if traders believe in this models, price and hedge according to it, then the model becomes "true" by definition...
14. Q: *Dangerous reasoning...*
15. T: Nonetheless, one has to be careful in dismissing the model.
16. Q: *Correct. Is there any other model you like?*

17. T: Well, the SLMUP looks promising. I like the idea of mixing LIBOR models, thinking of different volatility and shift scenarios. I'm a just concerned about the unrealistic dynamics it assumes. There's so much criticism about it.
18. Q: *I understand. However, if you were using a stochastic volatility model, wouldn't you calibrate and change the model parameters every day?*
19. T: Of course. My option book is marked to market daily.
20. Q: *Exactly! So, your model tomorrow is different from the one you are assuming today.*
21. T: In practice yes. Always.
22. Q: *And what about the story that Delta-hedging with the SLMUP gives you a systematically wrong hedge?*
23. T: I guess they mean you can only hedge an average scenario, so that, whatever scenario occurs, you are never hedged in practice.
24. Q: *Hum, sounds no good ...*
25. T: Wait, wait! We option traders are not so crazy to forget about volatility risk. Don't you know that Vega-hedging is essential for us?
26. Q: *[Grumbling...] Go to the point, please.*
27. T: It's pretty simple. Combining Delta-hedging with Vega-hedging reduces inconsistencies and P&L swings. At least this is my practical experience.
28. Q: *Great! Nice to hear. And, is there any other model you like?*
29. T: Well, hard to say. Did I ever mention to you my list of wishes?
30. Q: *Hum, refresh my memory ...*
31. T: Ahah, check this out! 1) As simple as possible; 2) Easy to understand; 3) Fast; 4) Tractable; 5) Accommodating market data (possibly caps and swaptions simultaneously); 6) Allowing for calculation of bucketed sensitivities; 7) Stable and robust ...
32. Q: *Hey mate, stop please, I've heard enough!*
33. T: Come on, don't tell me you are scared, you brave man!
34. Q: *Yeah, I like challenges, but this is pure utopia. Maybe in the future, who knows ...*
35. T: Ok, I understand. What is your opinion then?
36. Q: *Well, a universal model would be ideal. But I'm afraid that we have to resort to different solutions for different products. Are you fine with that?*
37. T: Hum, my first requirement is to be aligned with the market. For example, if I have to price a Bermudan swaption I must necessarily calibrate my model to the underlying vanillas. As a consequence, different Bermudans may be priced with different model parameters. Going further, if I change contract, say a ratchet, I'm also willing to change model, as long as it properly accounts for the quoted cap smile.
38. Q: *Yes, but aren't you afraid of inconsistencies? I mean, let's say that you use a G2++ model for pricing Bermudans and a LIBOR model to*

price a ratchet. Can you freely add the sensitivities coming from the two models? Isn't it like adding apples and oranges?

39. T: Of course, but you know [shrugging], as long as I can make money out of it ...
40. Q: *Ok man, fair enough!*

Inflation

1. T: And besides smile modeling?
2. Q: *There's inflation, for example.*
3. T: Oh yeah, sure! There's been an increasing demand for inflation-linked products.
4. Q: *Ahah, interesting! And what type of derivatives is typically traded?*
5. T: In the government market, bonds paying you some kind of average inflation rate, with an embedded floor at zero.
6. Q: *Of course! Inflation can go negative, can't it?*
7. T: Not very likely on annual basis, but monthly inflation is a different story. Think for instance to the sales effect in January.
8. Q: *Hey, that's great! It means that I can model inflation with a Gaussian process and no one will ever criticize me!*
9. T: Wait! Wait! Inflation rates can't be smaller than -1! Yet, a normal density is fair enough, I guess.
10. Q: *Normal densities rule!*
11. T: Ok, ok. Does it mean that you would model inflation as an Ornstein-Uhlenbeck process? Yes, but what about its drift? And what pricing measure would you choose?
12. Q: *Good points. The original models for pricing inflation-indexed derivatives were based on the so called "foreign currency analogy".*
13. T: Hum, sounds cryptical ...
14. Q: *In fact, it's rather easy. The idea is to imagine that we have two distinct economies, a nominal and a real. And the inflation index, whose percentage returns define inflation rates, is viewed as the exchange rate between the two economies.*
15. T: You're telling me that pricing inflation-indexed derivatives is like pricing cross-currency derivatives?
16. Q: *You're fast, dude! But, let me also tell you there exists a different approach, based on the philosophy of market models.*
17. T: Sounds interesting.
18. Q: *Indeed. You model forward inflation indices and you need no foreign-currency analogy, which is to some extents unnatural and not everybody likes it.*
19. T: Great! And I imagine this market model has a great deal of tractability.
20. Q: *You got it!*

21. T: There's just one problem ...
22. Q: *[Frowning] Name it!*
23. T: The inflation derivatives market is still rather undeveloped, especially the interbank one.
24. Q: *And what are the traded instruments?*
25. T: Swaps and a few options, but we are far from having an active market with reliable quotes.
26. Q: *Any hope for the future?*
27. T: Who knows! It looks like the next semester may be the good one, but we've said that other times.
28. Q: *Ok, I see. In the meanwhile, I keep on modeling new features. You never know. If the inflation market takes off, I'm pretty sure you'll soon start asking me for the pricing of caps and swaption under smile effects. I must be ready by that time.*
29. T: Man, that's music to my ears!

Counterparty Risk

> *The recent tendency is to assume this embarrassment can be resolved by inflation or some other adjustment of the physics of the very early universe.*
> Peebles, P.J.E., Principles of Physical Cosmology, Princeton University Press, 1993.

> *"The air tastes rotten, my gravity rod is registering toxic levels of background radiation and viral pollution. Everything seems so squeezed together... when does tesseract technology get here?"*
> Starman One Million, in: DC One Million (1999), DC Comics

1. T: What else?
2. Q: *Counterparty risk has taken a relevant role, especially after the Basel II framework took off*
3. T: Is it really relevant? I seem to remember that the impact of this risk on the NPV of a swap, for example, is very small.
4. Q: *Of course it depends on the riskiness of the counterparty. In general the component is small in quiet situations, but for very long swaps with counterparties having high default probabilities the situation can be different, with a quite relevant impact.*
5. T: Interesting... is the inclusion of counterparty risk painless in general or what?
6. Q: *No, it's rather "bloody painful", as you would put it. Even in simple counterparty risk, when there are no guarantees such as posted collateral and you charge the whole risk upfront, as a component of the price! This*

is going to affect your trades soon!! From now on you may sleep in a preoccupied state :-)

7. T: Whoa, gimme a break. What's so "bloody" about it? Man, can't you come up with an add-on or something easy? It would be cool!! You would simply adjust the earlier counterparty risk-free price with this add-on, depending perhaps almost only on the credit quality of the counterparty, on its implied default probabilities. No need to reprice anything. Keep the old prices and adjust them. And we all sleep well!!
8. Q: *Sure, that would be cool... unfortunately it's not possible. See, it's not hard to show that the price in presence of counterparty risk is just the default free earlier price minus a discounted option term in scenarios of early default.*
9. T: Oh @##!@%%&@!! I see... an option.. that would make the price model dependent even if the earlier default-free payoff was model independent... boy o boy...
10. Q: *The option is on the residual present value at time of default. You had it right: even payoffs whose valuation is model independent become model dependent due to counterparty risk.*
11. T: That means that a swap...
12. Q: *... has to be priced as a portfolio of swaptions when we include counterparty risk, so that you need an interest rate model even if before introducing counterparty risk you did not need it. Indeed, to value a swap, even amortizing or zero coupon, you don't need a model. To value the same swap under counterparty risk you not only need a default model for the counterparty...*
13. T: ... but also an interest rate model. I see.
14. Q: *Also, the option maturity is random because it is related to the default time. Here correlation between the default time and the underlying of the default-free contract can be important. Sometimes this correlation is ignored by taking deterministic default intensity to avoid complications.*
15. T: Too bad...
16. Q: *In general counterparty risk introduces the need of default modeling and adds one level of "optionality" to the default-free payoff. Yeah, a complicated story ...*

Credit Derivatives

Relativistic effects take over as a body approaches light-speed. Visual input will begin to blue-shift and my body's mass will increase towards infinity
Grant Morrison's take on Flash/Wally West, JLA: New World Order

1. T: And credit derivatives? Have they become relevant for your job?

2. Q: *Yes, going back to the counterparty risk problem, we need default models; the models have to be calibrated to liquid credit derivatives to get market-implied risk-neutral default probabilities.*
3. T: What kind of products can you use to deduce implied default probabilities?
4. Q: *Typically defaultable (corporate) bonds or credit default swaps (CDS). CDS are often more liquid and more standardized, so that when available they are preferred.*
5. T: But when corporate bonds and CDS are both available for the given counterparty, are they consistent with each other? I think there is a difference between the default probabilities and hazard rates implied by the two products... there should be a "basis"...
6. Q: *Indeed there is. But let's say we stick to CDS for simplicity*
7. T: Good. And how is the default modeling and calibration accomplished?
8. Q: *We used mostly reduced form models (also called intensity models in some situations...)*
9. T: Oh yeah, that French-fish stuff, right?[1]
10. Q: *Man, you think you are funny, don't you? :-) Leave the cabaret to me, and let me speak!! Intensity models describe the default time as the first jump of a Poisson process with some intensity. The intensity is related to the "local" default probability, i.e. to the probability of defaulting around each time instant.*
11. T: What is so good about these models?
12. Q: *Several things. Mostly, the fact that default (actually survival) probabilities are analogous to bond prices in the interest rate market if the intensity replaces the short rate. The intensity can thus be interpreted as an instantaneous credit spread.*
13. T: You mean that from the modeling point of view, modeling intensities = modeling short rates?? Whoa!!!
14. Q: *Exactly so. The only thing is that intensity (being related to a probability) needs to be positive... no Gaussian models, then. Hull and White for example is not ok, but CIR++ could work.*
15. T: From this discussion I infer that the intensity is always stochastic, like the short rate in a short-rate model?
16. Q: *Not necessarily... A Poisson process can have deterministic or stochastic intensity. In the latter case it is called a Cox process.*
17. T: But the Poisson structure is still there even under stochastic intensity?
18. Q: *Yes, conditional on the intensity itself. Also, you should know that with these models default is not triggered by basic market observables but has an exogenous component that is independent of all the default free market information. Monitoring the default free market (interest rates, exchange rates, etc) does not give complete information on the default process, and there is no economic rationale behind default.*

[1] Fish = "Poisson" in French

19. T: So in this sense intensity models are not "complete"?
20. Q: *Indeed... but even so this family of models is particularly suited to model credit spreads and in its basic formulation is easy to calibrate to Credit Default Swap or corporate bond data. Hence its popularity.*
21. T: Ok, so intensity models are the credit analogous of short rate models in the interest rate market. Is there also an analogous for the LIBOR or Swap market models? Are there impliedvolatilties? whatkindofmarketquantitiesdoyoumodel?
22. Q: *Hey, take it easy... come back from superluminal speed and calm down even if the market is rallying... stop looking at that screen and look at me... heed me...*
23. T: Ok, ok.. I slow down
24. Q: *Ok, so... we did not introduce the key ingredients of intensity models by referring directly to market payoffs. This is similar to what we have done with short-rate models. However, modern interest rate models, such as the market models, resort to variables that have a more direct link with market payoffs. The LIBOR model for example models forward rates that make Forward Rate Agreements (FRA) contracts fair.*
25. T: and the swap model modelsratesthatmake...
26. Q: *slow down... The swap models consider the rates that make interest rate swaps contracts fair. In different terms, the market models consider rates that are obtained by setting to zero the values of some financial contracts. This allows for much more general and realistic volatility and correlation structures, and for much more powerful calibration and diagnostics than with simple short rate models.*
27. T: ... and sometimes puts you in a lot of troubles as well...
28. Q: *Think positive (and I can't believe I'm saying this to you!!)... However, no arbitrage valuation allows us to write both forward LIBOR rates and forward swap rates in terms of zero coupon bonds, that in turn can be obtained as suitable expectations of functionals of the short rate r. This means that in principle market models could be described by means of a sophisticated enough model for r.*
29. T: ... but these models would not look like our old short rate models...
30. Q: *Indeed, and this r model is of no interest, since the direct modeling of forward rates is better and gives the advantages above.*
31. T: Ok, so to make it simple the r model equivalent to market models would look like a mess... so I guess you forget about it. I guess you do the same with intensity models, you find some market rates and model them, forgetting about the λ model that would be equivalent with these rates... WHOA look at the market NOOOOOO @##!@%%&@!!
32. Q: *I told you not to watch that screen... your colleague is monitoring the market, don't worry and talk to me.*
33. T: "Talk to me" your b#tt... you're not the one whose trading book is experiencing intense gravity @##!@%%&@!!

34. Q: *Breath slowly and deeply... ok... I'll try to hypnotize you... your eyes are heavy... do not watch the screen... your limbs are heavy and you feel relaxed... you feel an intense need to talk about modeling....*
35. T: Yes, let us talk about modeling. I am heavy and comfortably warm...
36. Q: *Again, before looking at that monitor you had it right. Market models for CDS options are similar in spirit, and in a way they are to intensity models what the swap models are to short rate models. You model rates that make CDS fair directly, and forget about stochastic intensity. If you consider LIBOR rates = one-period CDS rates (i.e. rates of CDS protecting on a single period of the tenor structure), and Swap rates = CDS rates, you find several analogies between a "CDS options market model" and the swap/LIBOR market models.*
37. T: But isn't default complicating life with the pricing measure and all that technobabble? At least a colleague of the credit desk once told me so and hinted at a "survival measure" or something...
38. Q: *Wait, one thing at a time... If you take a defaultable asset as numeraire, in the paths where it vanishes because of default you get zero as numeraire, which is not allowed. So you have two choices: either you do a change on measure only on the set of paths where the numeraire is still 'alive', but in doing so you lose the equivalence of pricing measures...*
39. T: Or?
40. Q: *Or you price under partial information, defining rates only with respect to the default-free market filtration: this is made possible by a technical result called martingale invariance property or co-adaptedness, and you are fine with an equivalent pricing measure.*
41. T: Interesting... but are CDS options traded that much?
42. Q: *My head is spinning... can't you stay on a topic for more than 20 seconds? Ok, yes, they are not traded that much, especially on single names they are rather illiquid. Also, implied volatilities are quite large, much larger than swaption volatilities. Eventually we will also need a smile model there too*
43. T: So we closed the loop, going back to the smile...
44. Q: *Don't make me start again... it has been like a ride in a roller coaster :-)*
45. T: Ok, no kidding... nice talking to you, though. I hope for the next time we don't have to wait five years :-)
46. Q: *I hope we are talking again earlier, but in the meantime...*
47. T: ...let's do it again:
48. Q and T: To all of our readers: **Don't Panic and Good Luck!!**

'It looks like our troubles are over for the time being, Batman'
'Our troubles are just beginning, Superman'
Day of Judgment 5, 1999, DC Comics

Mō kaeru koto nai ⟨We shall not return⟩
uchū no hoshi yo ⟨star of the universe!⟩
saraba saraba ⟨Farewell farewell⟩
mō aewashinai ⟨we will never meet again.⟩
keredo namida wa iranai ⟨But tears are not appropriated⟩
bokura ni wa haha ga iru ⟨[since] we have a mother⟩
chichi ga iru ⟨[and] a father,⟩
heiwa de yutakana ⟨we have a prosperous⟩
chikyū ga aru ⟨[and finally] peaceful Earth.⟩
kagayake! ⟨Shine!⟩
bokura no hoshi yo ⟨Star of us all!⟩
eien ni kagayake! ⟨Forever shine!⟩

Mō tatakai wa nai ⟨The war has ended⟩
midori no daichi yo ⟨green mother Earth!⟩
Hashire hashire ⟨Run run⟩
Kaze yori mo hayaku ⟨faster than even wind.⟩
Soshite hohoemi wasurezu ⟨[And do] not forget to smile⟩
Bokura ni wa ai ga aru. ⟨[since] we have love⟩
tomo ga aru ⟨and [we have] friends,⟩
Seigi to yūki no ⟨justice and courage⟩
nakama ga aru ⟨are our companions.⟩
kagayake! ⟨Shine!⟩
bokura no hoshi yo ⟨Star of us all!⟩
eien ni kagayake! ⟨Forever shine!⟩

"Uchū no hoshi yo eien ni" ⟨Star of the universe - forever⟩, closing theme of "Muteki Choujin Zanbot 3", Music by Takeo Watanabe, 1977.

Damiano is grateful to dottoressa Cristina Dalle Vedove for the Romaji transcription and the Italian translation.

"We don't do it for the glory.
We don't do it for the recognition...
We do it because it needs to be done.
Because if we don't, no one else will.
And we do it even if no one knows what we've done.
Even if no one knows we exist.
Even if no one remembers we ever existed."

Kara, "Christmas with the Superheroes" 2,
"Should Auld Acquaintance Be Forgot", 1989, DC Comics

References

1. Aït-Sahalia, Y. (1996) Testing Continuous-Time Models of the Spot Rate. *The Review of Financial Studies* 9, 385-426.
2. Alexander, C. (2001). Market Models: A Guide to Financial Data Analysis. Wiley.
3. Alexander, C. (2002). Common Correlation Structures for Calibrating the Libor Model, University of Reading. ISMA Center Preprint.
4. Alexander, C., Brintalos, G., and Nogueira, L. (2003) Short and Long Term Smile Effects: The Binomial Normal Mixture Diffusion Model. ISMA Centre working paper.
5. Alfonsi, A. (2005) On the discretization schemes for the CIR (and Bessel squared) processes. *CERMICS internal report*, May 2005. Available at `http://cermics.enpc.fr/reports/CERMICS-2005/CERMICS-2005-279.pdf`
6. Alfonsi, A. (2005b) Private communication.
7. Alfonsi, A., and Brigo, D. (2005). New Families of Copulas Based on Periodic Functions. *Communications in Statistics: Theory and Methods*, Vol. 34, N. 7.
8. Amari, S-I.(1985). *Differential Geometric Methods in Statistics.* Lecture Notes in Statistics, 28. Springer, Berlin.
9. Amin, A. (2003) Multi-Factor Cross Currency LIBOR Market Models: Implementation, Calibration and Examples. Available online at: `http://www.geocities.com/anan2999/files/cross40.pdf`
10. Amin, K., and Morton, A. (1994) Implied Volatility Functions in Arbitrage Free Term Structure Models. *Journal of Financial Economics* 35, 141-180.
11. Andersen, L. (1999). A Simple Approach to the Pricing of Bermudan Swaptions in the Multi–Factor LIBOR Market Model, preprint.
12. Andersen, L., and Andreasen, J. (2000). Volatility Skews and Extensions of the LIBOR Market Model. *Applied Mathematical Finance* 7, 1-32.
13. Andersen, L., and Brotherton-Ratcliffe, R. (2001) Extended Libor Market Models with Stochastic Volatility. Working Paper. Available online at: `http://papers.ssrn.com/sol3/papers.cfm?abstract_id=294853`
14. Andreasen, J.F. (1995) Pricing by Arbitrage in an International Economy. *Research in International Business and Finance* 12, 93116.
15. Arvanitis, A., and Gregory, J. (2001). Credit: The Complete Guide to Pricing, Hedging and Risk Management. Risk Books, London.
16. Avellaneda, M., Newman, J. (1998) Positive Interest Rates and Nonlinear Term Structure Models. Preprint, Courant Institute of Mathematical Sciences, New York University.
17. Balduzzi, P., Das, S.R., Foresi, S., Sundaram, R. (1996) A Simple Approach to Three-Factor Term Structure Models. *The Journal of Fixed Income* 6, 43-53.
18. Balland, P. and Hughston, L.P. (2000) Markov Market Model Consistent with Cap Smile. *International Journal of Theoretical and Applied Finance* 3(2), 161-181.

19. Barone, E., and Castagna, A. (1997) The Information Content of Tips. Internal report. SanPaolo IMI, Turin, and Banca IMI, Milan.
20. Baxter, M.W. (1997) General Interest-Rate Models and the Universality of HJM. *Mathematics of Derivative Securities*, M.A.H. Dempster, S.R. Pliska, eds. Cambridge University Press (1997), Cambridge, pp. 315-335.
21. Belgrade, N., Benhamou, E., and Koehler, E. (2004) A Market Model for Inflation. Preprint, CDC Ixis Capital Markets. Available online at: `http://papers.ssrn.com/sol3/papers.cfm?abstract_id=576081`
22. Ben Ameur, H., Brigo, D., and Errais, E. (2005). Pricing Credit Default Swaps Bermudan Options: An Approximate Dynamic Programming Approach. *Working paper*, available at `http://ssrn.com/abstract=715801`
23. Bennani, N. (2005). The forward loss model: a dynamic term structure approach for the pricing of portfolio credit derivatives. `http://www.defaultrisk.com/pp_crdrv_95.htm`
24. Bielecki T., Rutkowski M. (2001). Credit risk: Modeling, Valuation and Hedging. Springer
25. Björk, T. (1997) Interest rate Theory. In *Financial Mathematics, Bressanone 1996*, W. Runggaldier, ed. *Lecture Notes in Math.* 1656, Springer, Berlin Heidelberg New York, pp. 53-122.
26. Björk, T. (1998) Arbitrage Theory in Continuous Time. Oxford University Press.
27. Black, F. (1976). The pricing of commodity contracts. *Journal of Financial Economics* 3, 167-179.
28. Black, F., and Cox, (1976). Valuing corporate securities: Some effects of bond indenture provisions. Journal of Finance 31, 351-367
29. Black, F., Derman, E., Toy, W. (1990) A One-Factor Model of Interest Rates and its Application to Treasury Bond Options. *Financial Analysts Journal* 46, 33-39.
30. Black, F., Karasinski, P. (1991) Bond and Option Pricing when Short Rates are Lognormal. *Financial Analysts Journal* 47, 52-59.
31. Black, F., Scholes, M. (1973) The Pricing of Options and Corporate Liabilities. *Journal of Political Economy* 81, 637-654.
32. Bliss, R., Ritchken, P. (1996) Empirical Tests of Two State-Variable Heath, Jarrow and Morton Models. *Journal of Money, Credit and Banking* 18, 426-447.
33. Brace, A. (1996) Dual Swap and Swaption Formulae in Forward Models. *FMMA notes* working paper.
34. Brace, A. (1997) Rank–2 Swaption Formulae. *UNSW* preprint.
35. Brace, A. (1998) Simulation in the GHJM and LFM models. *FMMA notes* working paper.
36. Brace, A., Dun, T., and Barton, G. (1998) Towards a Central Interest Rate Model. *FMMA notes* working paper. Also published in *Handbooks in Mathematical Finance: Topics in Option Pricing, Interest Rates and Risk Management* (2001), Cambridge University Press, Cambridge.
37. Brace, A., Goldys, B., Klebaner, F., and Womersley, R. (2001) Market Model of Stochastic Implied Volatility with application to the BGM Model. Working Paper S01-1, Department of Statistics, University of New South Wales, Sydney.
38. Brace, A., Gatarek D., Musiela, M. (1997) The Market Model of Interest Rate Dynamics. *Mathematical Finance* 7, 127-155.
39. Brace, A., Musiela, M. (1994) A Multifactor Gauss Markov Implementation of Heath, Jarrow, Morton. *Mathematical Finance* 4, 259-283.

40. Brace, A., Musiela, M. (1997) Swap Derivatives in a Gaussian HJM Framework. *Mathematics of Derivative Securities*, M.A.H. Dempster, S.R. Pliska, eds. Cambridge University Press, Cambridge, pp. 336-368.
41. Brace, A., Musiela, M., and Schlögl, E. (1998) A Simulation Algorithm Based on Measure Relationships in the Lognormal Market Models. Preprint.
42. Breeden, D.T. and Litzenberger, R.H. (1978) Prices of State-Contingent Claims Implicit in Option Prices. *Journal of Business* 51, 621-651.
43. Brenner, R.J., Harjes, R.H., Kroner, K.F. (1996) Another Look at Models of the Short-Term Interest Rate. *Journal of Financial and Quantitative Analysis* 31, 85-107.
44. Brennan, M. J., and Schwartz, E. (1979). A Continuous Time Approach to the Pricing of Bonds. *Journal of Banking and Finance* 3 , 133-155.
45. Brennan, M. J., and Schwartz, E. (1982) An Equilibrium Model of Bond Prices and a Test of Market Efficiency. *Journal of Financial and Quantitative Analysis* 17, 301-329.
46. Brigo, D. (1999). Diffusion Processes, Manifolds of Exponential Densities, and Nonlinear Filtering. in: O.E. Barndorff-Nielsen and E. B. Vedel Jensen (Editors), *Geometry in Present Day Science*, World Scientific.
47. Brigo, D. (2002). A Note on Correlation and Rank Reduction. Working paper. Available at `http://www.damianobrigo.it/correl.pdf`
48. Brigo, D. (2002b). The general mixture-diffusion SDE and its relationship with an uncertain-volatility option model with volatility-asset decorrelation. Posted on the SSRN network.
49. Brigo, D. (2004). Candidate Market Models and the Calibrated CIR++ Stochastic Intensity Model for Credit Default Swap Options and Callable Floaters. In: *Proceedings of the 4-th ICS Conference*, Tokyo, March 18-19, 2004. Extended version available at `http://www.damianobrigo.it/cdsmm.pdf`
50. Brigo, D. (2005). Market Models for CDS Options and Callable Floaters, *Risk*, January issue. Also in: *Derivatives Trading and Option Pricing*, Dunbar N. (Editor), Risk Books, 2005.
51. Brigo, D. (2005b). Constant Maturity Credit Default Swap Pricing with Market Models. Working paper. Available at ssrn.com and at `http://www.damianobrigo.it/cmcdsweb.pdf`. Updated version to appear in *Risk Magazine*, June 2006 issue.
52. Brigo, D. (2005c). Fixed Income Course Lecture Notes for Bocconi University CLEFIN Students. Available at `http://www.damianobrigo.it/bocconi.html`
53. Brigo, D., and Alfonsi, A. (2003). A two-dimensional CIR++ shifted diffusion model with automatic calibration to credit defalult swaps and interest rate derivatives data. In: *Proceedings of the 6-th Columbia=JAFEE International Conference*, Tokyo, March 15-16, 2003, 563-585. *Extended internal report* available at `http://www.damianobrigo.it/cirppcredit.pdf`.
54. Brigo, D., and Alfonsi, A. (2005). Credit Default Swap Calibration and Derivatives Pricing with the SSRD Stochastic Intensity Model. *Finance and Stochastics*, Vol 9, n 1.
55. Brigo, D., Capitani, C., and Mercurio, F. (2000) On the Joint Calibration of the LIBOR Market Model to Caps and Swaptions Data. Internal Report, Banca IMI, Milan.
56. Brigo, D., and Cousot, L. (2004) A Comparison between the stochastic intensity SSRD Model and the Market Model for CDS Options Pricing. Working paper available at `http://www.defaultrisk.com`, presented at the Third Bachelier Conference on Mathematical Finance, Chicago, July 21-24, 2004. To appear in the *International Journal of Theoretical and Applied Finance*.

57. Brigo, D., and El Bachir, N. (2005). CDS options with an affine jump diffusion model and efficient numerical approximations, with general implied volatility and positive intensity issues. Working paper.
58. Brigo, D., and Hanzon, B. (1998). On some filtering problems arising in mathematical finance. *Insurance: Mathematics and Economics*, 22 (1) pp. 53-64.
59. Brigo, D., Hanzon, B., and Le Gland, F. (1999). Approximate Nonlinear Filtering by Projection on Exponential Manifolds of Densities. *Bernoulli*, Vol. 5, N. 3 (1999), pp. 495–534.
60. Brigo D., and Liinev J. (2002). On the distributional difference between the Libor and the Swap market models. Available at `http://www.damianobrigo.it/liborswapdistance.pdf`. Updated version published in *Quantitative Finance*, vol. 5, N. 5 (2005), pp. 433-442.
61. Brigo, D., and Masetti, M. (2005). A Formula for Interest Rate Swaps Valuation under Counterparty Risk in presence of Netting Agreements. Working paper available at `http://www.damianobrigo.it/IRSnetcounternew.pdf`
62. Brigo, D., and Masetti, M. (2005b). Risk Neutral Pricing of Counterparty Risk. In: Pykhtin, M. (Editor), *Counterparty Credit Risk Modeling: Risk Management, Pricing and Regulation.* Risk Books, London.
63. Brigo, D., Mercurio, F. (1998) On Deterministic Shift Extensions of Short-Rate Models. Internal Report, Banca IMI, Milan. Available at `http://www.damianobrigo.it` and at `http://www.fabiomercurio.it`.
64. Brigo, D., Mercurio, F. (2000a) Fitting Volatility Smiles with Analytically Tractable Asset Price Models. Internal Report, Banca IMI, Milan. Available at `http://www.damianobrigo.it` and at `http://www.fabiomercurio.it`.
65. Brigo, D., Mercurio, F. (2000b) A Mixed-up Smile. *Risk*, September, 123-126.
66. Brigo, D., Mercurio, F. (2000c) The CIR++ Model and Other Deterministic-Shift Extensions of Short-Rate Models, in: *Proceedings of the Columbia-JAFEE International Conference* held in Tokyo on December 16-17, 2000, pp. 563-584.
67. Brigo, D., Mercurio, F. (2001a) A Deterministic-Shift Extension of Analytically-Tractable and Time-Homogenous Short-Rate Models. *Finance & Stochastics* 5, 369-388.
68. Brigo, D., Mercurio, F. (2001b) Displaced and Mixture Diffusions for Analytically-Tractable Smile Models. In *Mathematical Finance - Bachelier Congress 2000*, Geman, H., Madan, D.B., Pliska, S.R., Vorst, A.C.F., eds. *Springer Finance*, Springer, Berlin Heidelberg New York.
69. Brigo, D., Mercurio, F. (2001c) Interest Rate Models: Theory and Practice. First Edition. Springer, Berlin Heidelberg New York.
70. Brigo, D., and Mercurio, F. (2002a). Calibrating LIBOR. *Risk*, 117-121.
71. Brigo, D., Mercurio, F. (2002b). Lognormal-Mixture Dynamics and Calibration to Market Volatility Smiles. *International Journal of Theoretical & Applied Finance* 5(4), 427-446.
72. Brigo, D., Mercurio, F. (2003). Analytical Pricing of the Smile in a Forward LIBOR Market Model. *Quantitative Finance* 3(1), 15-27.
73. Brigo, D., Mercurio, F., and Morini, M. (2002). Different covariance parameterizations of the Libor market model and joint caps/swaptions calibration. Available at `www.exoticderivatives.com/Files/Papers/brigomercuriomorini.pdf` Working paper version of the following reference.
74. Brigo, D., Mercurio, F., and Morini, M. (2005). The Libor Model Dynamics: Approximations, Calibration and Diagnostics. *European Journal of Operation Research* 163 (2005), pp 30-41.

75. Brigo, D., Mercurio, F., and Rapisarda, F. (2002). Parameterizing Correlations: a Geometric Interpretation. Preprint, available at http://it.geocities.com/rapix/Correlations.pdf
76. Brigo, D., Mercurio, F., and Rapisarda, F. (2004) Smile at the uncertainty. *Risk* 17(5), 97-101.
77. Brigo, D., Mercurio, F., and Sartorelli, G. (2003). Alternative Asset-price Dynamics and Volatility Smile". *Quantitative Finance* 3(3), 173-183.
78. Brigo, D., and Morini, M. (2002). New Developments on the Analytical Cascade Swaption Calibration of the LIBOR market model. Paper presented at the Fourth Italian Workshop on Mathematical Finance, ICER, Turin, January 30-31, 2003.
79. Brigo, D., and Morini, M. (2003). An empirically efficient analytical cascade calibration of the LIBOR model based only on directly quoted swaptions data. Reduced version in *Proceedings of the FEA 2004 Conference at MIT, Cambridge, Massachusetts, November 8-10* and forthcoming in the *Journal of Derivatives.* Available at http://ssrn.com/abstract=552581
80. Brigo, D., and Morini, M. (2005). CDS Market Formulas and Models. In: *Proceedings of the 18th Annual Warwick Options Conference*, September 30 2005, Warwick, UK.
81. Brigo, D., and Morini, M. (2006). Structural credit calibration. *Risk Magazine*, April issue.
82. Brigo, D., Pallavicini, A., and Torresetti, R. (2006). Calibration of CDO Tranches with the dynamical Generalized-Poisson Loss model. Working paper, available at ssrn.com
83. Brigo, D. and Tarenghi, M. (2004). Credit Default Swap Calibration and Equity Swap Valuation under Counterparty risk with a Tractable Structural Model. Working Paper, available at www.damianobrigo.it/cdsstructural.pdf. Reduced version in *Proceedings of the FEA 2004 Conference at MIT, Cambridge, Massachusetts, November 8-10* and in *Proceedings of the Counterparty Credit Risk 2005 C.R.E.D.I.T. conference*, Venice, Sept 22-23, Vol 1.
84. Brigo, D. and Tarenghi, M. (2005). Credit Default Swap Calibration and Counterparty Risk Valuation with a Scenario based First Passage Model. Working Paper, available at www.damianobrigo.it/cdsscenario1p.pdf Also in: *Proceedings of the Counterparty Credit Risk 2005 C.R.E.D.I.T. conference*, Venice, Sept 22-23, Vol 1.
85. Bühler, W., Uhrig-Homburg, M., Walter, U., Weber, T. (1999) An Empirical Comparison of Forward-Rate and Spot-Rate Models for Valuing Interest-Rate Options. *The Journal of Finance* 54, 269-305.
86. Cairns, A.J.G. (2000) A multifactor model for the term structure and inflation for long-term risk management with an extension to the equities market. Preprint. Heriot-Watt University, Edinburgh Available online at: http://www.ma.hw.ac.uk/~andrewc/papers/ajgc25.pdf
87. Calamaro, J.P., and Nassar, T. (2004). CMCDS: The Path to Floating Credit Spread Products. Deutsche Bank technical document.
88. Cannabero, E. (1995) Where Do One-Factor Interest Rate Models Fail? *The Journal of Fixed Income* 5, 31-52.
89. Carr, P., and Madan, D. (1998) Option Valuation Using the Fast Fourier Transform. Working Paper, Morgan Stanley and University of Maryland.
90. Carr, P., Tari, M. and Zariphopoulou, T. (1999). Closed Form Option Valuation with Smiles. Preprint. NationsBanc Montgomery Securities.
91. Carr, P., Yang, G. (1997). Simulating Bermudan Interest Rate Derivatives, Courant Institute at New York University, Preprint.

92. Carr, P., and Wu, L. (2004). Time-changed Levy processes and option pricing. *Journal of Financial Economics*, vol. 71(1), pages 113-141.
93. Carrière, J. (1996). Valuation of the Early-Exercise Price for Options using Simulations and Nonparametric Regression. *Insurance: Mathematics and Economics*, Vol.19, No. 1, pp 19-30.
94. Carverhill, A. (1994) When is the Short Rate Markovian? *Mathematical Finance* 4, 305-312.
95. Castagna, A. (2001) Private Communication.
96. Chan, K.C., Karolyi, G.A., Longstaff, F.S., Sanders, A.B. (1992) An Emprical Comparison of Alternative Models of the Term Structure of Interest Rates. *The Journal of Finance* 47, 1209-1228.
97. Chapman, D.A., Long, A.B., Pearson, N.D. (1999) Using Proxies for the Short Rate: When Are There Months Like an Instant? *The Review of Financial Studies* 12, 763-806.
98. Chen, R., Scott, L. (1992) Pricing Interest Rate Options in a Two Factor Cox-Ingersoll-Ross Model of the Term Structure. *The Review of Financial Studies* 5, 613-636.
99. Chen, R., Scott, L. (1995) Interest Rate Options in Multifactor Cox-Ingersoll-Ross Models of the Term Structure. *The Journal of Derivatives* 3, 53-72.
100. Cherubini, U. (2005). Counterparty Risk in Derivatives and Collateral Policies: The Replicating Portfolio Approach. In: *Proceedings of the Counterparty Credit Risk 2005 C.R.E.D.I.T. conference*, Venice, Sept 22-23, Vol 1.
101. Cherubini, U., Luciano, E., and Vecchiato, W. (2004). Copula Methods in Finance. Wiley.
102. Christensen, J. H. (2002) Kreditderivater og deres prisfastsaettelse. Ph.D. Thesis, Institute of Economics, University of Copenhagen
103. Cinlar, E. (1975). Introduction to Stochastic Processes. Prentice Hall, Englewood Cliffs, N.J.
104. Clewlow, L., Strickland, C. (1997) Monte Carlo Valuation of Interest Rate Derivatives Under Stochastic Volatility. *The Journal of Fixed Income* 7, 35-45.
105. Clewlow, L., Strickland, C. (1998) Pricing Interest Rate Exotics in Multi-Factor Gaussian Interest Rate Models, working paper.
106. Constantinides, G. (1992) A Theory of the Nominal Term Structure of Interest Rates. *The Review of Financial Studies* 5 , 531-552.
107. Cox, J.C. (1975) Notes on Option Pricing I: Constant Elasticity of Variance Diffusions. Working paper. Stanford University.
108. Cox, J.C., and Ross S. (1976) The Valuation of Options for Alternative Stochastic Processes. *Journal of Financial Economics* 3, 145-166.
109. Cox, J.C., Ingersoll, J.E., and Ross, S.A. (1985) A Theory of the Term Structure of Interest Rates. *Econometrica 53*, 385-407.
110. Cox, J.C., Ross, S.A., and Rubinstein, M. (1979) Option Pricing: A Simplified Approach. *Journal of Financial Economics* 7, 229-263.
111. Crouhy M., Galai, D., and Mark, R. (2000). A comparative analysis of current credit risk models. *Journal of Banking and Finance* 24, 59-117
112. Davis, M. (1998) A Note on the Forward Measure. *Finance and Stochastics*, 2, pp. 19-28.
113. Deacon, M., Derry, A., and Mirfendereski, D. (2004) Inflation-indexed Securities. Wiley. Chichester.
114. De Jong, F., Driessen, J., and Pelsser, A. (1999). LIBOR and Swap Market Models for the Pricing of Interest Rate Derivatives: An Empirical Analysis. Preprint.
115. De Jong, F., Driessen, J., Pelsser, A. (2001). LIBOR versus swap market models: an empirical comparison. *European Finance Review* 5, 201-237.

116. Deelstra, G. and Delbaen, F (1998). Convergence of Discretized Stochastic (Interest Rate) Processes with Stochastic Drift Term. Appl. Stochastic Models Data Anal. 14, 77-84
117. Derman, E., and Kani, I. (1994) Riding on a Smile. *Risk* February, 32-39.
118. Derman, E., and Kani, I. (1998) Stochastic Implied Trees: Arbitrage Pricing with Stochastic Term and Strike Structure of Volatility. *International Journal of Theoretical and Applied Finance* 1, 61-110.
119. Di Graziano, G., and Rogers, C. (2005). A New Approach to the Modeling and Pricing of Correlation Credit Derivatives. Working paper.
120. Diop, A. (2003). Sur la discrétisation et le comportement à petit bruit d'EDS multidimensionnelles dont les coefficients sont à dérivées singulières, Ph.D Thesis, INRIA. *(available at http://www.inria.fr/rrrt/tu-0785.html)*
121. Dominici, L. (2002) Private Communication, Interest-Rate Derivatives Desk, Banca IMI.
122. Dothan, L.U. (1978) On the Term Structure of Interest Rates. *Journal of Financial Economics* 6, 59-69.
123. Dudewicz, E.J., and Van der Meulen, E.C. (1987). The empiric entropy, a new approach to nonparametric entropy estimation, in Puri M.L. et al. eds., *New Perspectives in Theoretical and Applied Statistics*, Wiley, New York, 207-227
124. Due, J., and Ahluwalia, R. (2004). Introduction to Constant Maturity CDS and CDO's. JPMorgan technical document.
125. Duffie, D. (1996) Dynamic Asset Pricing Theory, 2d. ed. Princeton: Princeton University Press.
126. Duffie, D., and Garleanu, N. (2001). Risk and Valuation of Collateralized Debt Obligations. Financial Analysts Journal, Vol. 57, n. 1, 41-59
127. Duffie, D., Kan, R. (1996) A Yield-Factor Model of Interest Rates. *Mathematical Finance* 64, 379-406.
128. Duffie, D., Pan, J. and Singleton, K. (2000). Transform Analysis and Asset Pricing for Affine Jump Diffusions. Econometrica, Vol. 68, pp. 1343-1376
129. Duffie, D., and Singleton, K.J. (1997) An Econometric Model of the Term Structure of Interest-Rate Swap Yields. The Journal of Finance 52, 1287-1321.
130. Duffie, D., and Singleton, K.J. (1999). Modeling Term Structures of Defaultable Bonds. The Review of Financial Studies 12, 687-719.
131. Dun, T., Schlögl, E., and Barton, G. (1999) Simulated Swaption Hedging in the Lognormal Forward LIBOR Model. Preprint.
132. Dupire, B. (1994) Pricing with a Smile. *Risk* January, 18-20.
133. Dupire, B. (1997) Pricing and Hedging with Smiles. *Mathematics of Derivative Securities*, edited by M.A.H. Dempster and S.R. Pliska, Cambridge University Press, Cambridge, 103-111.
134. Dybvig, P.H. (1988) Bond and Bond Option Pricing Based on the Current Term Structure. Working Paper, Washington University.
135. Dybvig, P.H. (1997) Bond and Bond Option Pricing Based on the Current Term Structure. *Mathematics of Derivative Securities*, Michael A. H. Dempster and Stanley R. Pliska, eds. Cambridge: Cambridge University Press, pp. 271-293.
136. Eberlein. E. (2001) Application of generalized hyperbolic Lévy motions to finance. In Lévy Processes: Theory and Applications, O. E. Barndorff-Nielsen, T. Mikosch, and S. Resnick (Eds.), Birkhäuser (2001), 319337.
137. Eberlein, E., Jacod, J. and Raible, S. (2005). Lévy term structure models: No-arbitrage and completeness. *Finance and Stochastics* 9, 6788.
138. Eberlein, E., and Keller, U. (1995) Hyperbolic Distributions in Finance. *Bernoulli* 1, 281- 299.

139. Eberlein, E., and Özkan, F. (2005) The Levy LIBOR model. *Finance and Stochastics* 9, 327-348.
140. El Bachir, N. (2005). Levy copulas for dependent jump diffusion intensities and credit dependence pricing. Working paper.
141. Embrechts, P., Lindskog, F., and McNeil, A. (2001). Modeling dependence with copulas and applications to risk management. Working paper.
142. Errais, E., Mauri, G., and Mercurio, F. (2004). Capturing the Skew in Interest Rate Derivatives: A Shifted Lognormal LIBOR Model with Uncertain Parameters. Internal Report. Banca IMI.
143. Flesaker, B. (1993) Testing the Heath-Jarrow-Morton/Ho-Lee Model of Interest Rate Contingent Claims Pricing. *Journal of Financial and Quantitative Analysis* 28, 483-496.
144. Flesaker, B. (1996) Exotic Interest Rate Options. *Exotic Options: The State of the Art*, L. Clewlow and C. Strickland, eds. London: Chapman and Hall, Ch. 6.
145. Flesaker, B., Hughston, L. (1996) Positive Interest. *Risk* 9, 46-49.
146. Flesaker, B., Hughston, L. (1997) Dynamic Models for Yield Curve Evolution. *Mathematics of Derivative Securities*, M.A.H. Dempster, S.R. Pliska, eds. Cambridge University Press, Cambridge, pp. 294-314.
147. Frey, R., and Sommer, D. (1996) A Systematic Approach to Pricing and Hedging International Derivatives with Interest Rate Risk. *Applied Mathematical Finance* 3(4), 295317.
148. Gatarek, D. (2003) LIBOR market model with stochastic volatility. Deloitte&Touche. Available online at:
`http://papers.ssrn.com/sol3/papers.cfm?abstract_id=359001`
149. Geman, H., El Karoui, N., Rochet, J.C. (1995) Changes of Numeraire, Changes of Probability Measures and Pricing of Options. *Journal of Applied Probability* 32, 443-458.
150. Geyer, A.L.J., Pichler, S. (1999) A State-Space Approach to Estimate and Test Multifactor Cox-Ingersoll-Ross Models of the Term Structure. *The Journal of Financial Research* 22, 107-130.
151. Gibbons, M., Ramaswamy, K. (1993) A Test of the Cox, Ingersoll and Ross Model of the Term Structure. *The Review of Financial Studies* 6, 619-632.
152. Glasserman, P., and Kou, S.G. (2003) The Term Structure of Simple Forward Rates with Jump Risk. *Mathematical Finance* 13(3), 383-410.
153. Glasserman, P., and Merener, N. (2001) Numerical Solutions of Jump-Diffusion LIBOR Market Models. Working paper, Columbia University.
154. Glasserman, P. and Merener, N. (2003). Cap and Swaption Approximations in LIBOR Market Models with Jumps. Journal of Computational Finance, Vol. 7, n 1.
155. Glasserman, P., and Zhao, X. (1999) Fast Greeks in Forward LIBOR Models. Preprint.
156. Glasserman, P., and Zhao, X. (2000). Arbitrage-Free Discretization of Lognormal Forward LIBOR and Swap Rate Models. *Finance and Stochastics*, 4
157. Grant, D., Vora, G. (1999) Implementing No-Arbitrage Term Structure of Interest Rate Models in Discrete Time When Interest Rates are Normally Distributed. *The Journal of Fixed Income* 8, 85-98.
158. Hagan, P.S., Kumar, D., Lesniewski, A.S., Woodward, D.E. (2002) Managing Smile Risk. *Wilmott magazine*, September, 84-108.
159. Harrison, J.M., Kreps, D.M. (1979) Martingales and Arbitrage in Multiperiod Securities Markets, *Journal of Economic Theory* 20, 381-408.

160. Harrison, J.M., Pliska, S.R. (1981) Martingales and Stochastic Integrals in the Theory of Continuous Trading, *Stochastic Processes and their Applications* 11, 215-260.
161. Harrison, J.M., Pliska, S.R. (1983) A Stochastic Calculus Model of Continuous Trading: Complete Markets, *Stochastic Processes and their Applications* 15, 313-316.
162. Heath, D., Jarrow, R., Morton, A. (1990a) Bond Pricing and the Term Structure of Interest Rates: A Discrete Time Approximation. *Journal of Financial Quantitative Analysis* 25, 419-440.
163. Heath, D., Jarrow, R., Morton, A. (1990b) Contingent Claim Valuation with a Random Evolution of Interest Rates. *Review of Futures Markets* 9, 54-76.
164. Heath, D., Jarrow, R., Morton, A. (1992) Bond Pricing and the Term Structure of Interest Rates: A New Methodology. *Econometrica* 60, 77-105.
165. Heston, S.L. (1993) A Closed-Form Solution for Options with Stochastic Volatility with Applications to Bond and Currency Options. *The Review of Financial Studies* 6, 327-343.
166. Ho, T.S.Y., Lee, S.-B. (1986) Term Structure Movements and the Pricing of Interest Rate Contingent Claims. *The Journal of Finance* 41, 1011-1029.
167. Hughston, L.P. (1998) Inflation Derivatives. Working paper. Merrill Lynch.
168. Hull, J. (1997) Options, Futures, and Other Derivatives, 3rd. edition. Upper Saddle River, New Jersey: Prentice-Hall.
169. Hull, J., White, A. (1987) The Pricing of Options on Assets with Stochastic Volatilities. *Journal of Financial and Quantitative Analysis* 3, 281-300.
170. Hull, J., White, A. (1990a) Valuing Derivative Securities Using the Explicit Finite Difference Method. *Journal of Financial and Quantitative Analysis* 25, 87-100.
171. Hull, J., White, A. (1990b) Pricing Interest Rate Derivative Securities. *The Review of Financial Studies* 3, 573-592.
172. Hull, J., White, A. (1993a) Bond Option Pricing Based on a Model for the Evolution of Bond Prices. *Advances in Futures and Options Research* 6, 1-13.
173. Hull, J., White, A. (1993b) Efficient Procedures for Valuing European and American Path-Dependent Options. *The Journal of Derivatives* 1, 21-31.
174. Hull, J., White, A. (1993c) The Pricing of Options on Interest-Rate Caps and Floors Using the Hull-White Model. *The Journal of Financial Engineering* 2, 287-296.
175. Hull, J., White, A. (1993d) One-Factor Interest Rate Models and the Valuation of Interest Rate Derivative Securities. *Journal of Financial and Quantitative Analysis* 28, 235-254.
176. Hull, J., White, A. (1994a) Branching Out. *Risk* 7, 34-37.
177. Hull, J., White, A. (1994b) Numerical Procedures for Implementing Term Structure Models I: Single-Factor Models. *The Journal of Derivatives* 2, 7-16.
178. Hull, J., White, A. (1994c) Numerical Procedures for Implementing Term Structure Models II: Two-Factor Models. *The Journal of Derivatives* 2, 37-47.
179. Hull, J., White, A. (1995a) Hull-White on Derivatives. London: Risk.
180. Hull, J., White, A. (1995b) A Note on the Models of Hull and White for Pricing Options on the Term Structure: Response. *The Journal of Fixed Income* 5 (September, 1995), 97-102.
181. Hull, J., White, A. (1996) Using Hull-White Interest Rate Trees. *The Journal of Derivatives* 3, 26-36.
182. Hull, J., White, A. (1997) Taking Rates to the Limits. *Risk* 10, 168-169.
183. Hull, J. White, A. (1999). Forward Rate Volatilities, Swap Rate Volatilities, and the Implementation of the LIBOR Market Model. Preprint.

184. Hull, J. White, A. (2003). The Valuation of Credit Default Swap Options. *Journal of Derivatives*, Vol 10, n 3, pp. 40-50
185. Hunt, P., Kennedy, J. (2000) Financial Derivatives in Theory and Practice. Wiley. Chichester.
186. Hunt, P., Pelsser, A. (1998) Arbitrage-Free Pricing of Quanto-Swaptions *The Journal of Financial Engineering* 7, 25-33.
187. Hunt, P., Kennedy, J., Pelsser, A. (1998a) Fit and Run. *Risk* 11, pp. 65-67.
188. Hunt, P., Kennedy, J., Pelsser, A. (1998b) Markov–Functional Interest Rate Models. Working Paper, University of Warwick.
189. Inui, K., Masaaki, K. (1998) A Markovian Framework in Multi-Factor Heath-Jarrow-Morton Models. *Journal of Financial and Quantitative Analysis* 33, 423-440.
190. Ivanov, A.V., and Rozhkova, M.N. (1981). Properties of the statistical estimate of the entropy of a random vector with a probability density. *Problems of Information Transmission*, 17, 171-178.
191. Jäckel, P., and Rebonato, R. (2000). Linking Caplet and Swaption Volatilities in a BGM/J Framework: Approximate Solutions, *QUARC* preprint.
192. James, J., Webber, N. (2000) Interest Rate Modeling. Wiley. Chichester.
193. Jamshidian, F. (1988) The One-Factor Gaussian Interest Rate Model: Theory and Implementation. Working Paper, Merril Lynch Capital Markets.
194. Jamshidian, F. (1989) An Exact Bond Option Pricing Formula. *The Journal of Finance* 44, 205-209.
195. Jamshidian, F. (1991) Bond and Option Evaluation in the Gaussian Interest Rate Model. *Research in Finance* 9, 131-170.
196. Jamshidian, F. (1995) A Simple Class of Square-Root Interest Rate Models. *Applied Mathematical Finance* 2, 61-72.
197. Jamshidian, F. (1996) Sorting out Swaptions. *Risk*, March, 59-60.
198. Jamshidian, F. (1997) LIBOR and Swap Market Models and Measures. *Finance and Stochastics* 1, 293-330.
199. Jamshidian, F. (2002a) Pricing and Partially Hedging Inflation-Index Bonds. Working paper. NIB Capital, The Hague.
200. Jamshidian, F. (2002b) Report on Calibration of Gaussian Real Term Structure Model. Working paper. NIB Capital, The Hague.
201. Jamshidian, F. (2004) Valuation of Credit Default Swaps and Swaptions. *Finance and Stochastics* 8, 343-371.
202. Jamshidian, F., and Zhu, Y. (1997) Scenario Simulation: Theory and methodology. *Finance and Stochastics* 1, 43–67.
203. Jarrow, R.A. (1996) Modeling Fixed Income Securities and Interest Rate Options. New York: McGraw-Hill.
204. Jarrow, R.A. (1997) The HJM Model: Its Past, Present, and Future. *The Journal of Financial Engineering* 6, 269-279.
205. Jarrow, R.A., Madan, D. (1995) Option Pricing Using the Term Structure of Interest Rates to Hedge Systematic Discontinuities in Asset Returns. *Mathematical Finance* 5, 311-336.
206. Jarrow, R.A., Turnbull, S.M. (1994) Delta, Gamma, and Bucket Hedging of Interest Rate Derivatives. *Applied Mathematical Finance* 1, 21-48.
207. Jarrow, R., and Yildirim, Y. (2003) Pricing Treasury Inflation Protected Securities and Related Derivatives using an HJM Model. *Journal of Financial and Quantitative Analysis* 38(2), 409-430.
208. Jeanblanc, M., and Rutkowski, M. (2000). Default Risk and Hazard Process. In: Geman, Madan, Pliska and Vorst (eds), *Mathematical Finance Bachelier Congress 2000*, Springer .

209. Joe, H. (1997). Multivariate Models and Dependence Concepts. Chapman & Hall, London.
210. Joshi, M. (2003). The Concepts and Practice of Mathematical Finance. Cambridge University Press.
211. Joshi, M., and Rebonato, R. (2003) A stochastic-volatility, displaced-diffusion extension of the LIBOR market model. *Quantitative Finance* 3, 458-469.
212. Jouanin J.-F., Rapuch G., Riboulet G., Roncalli T. (2001). Modeling dependence for credit derivatives with copulas, Groupe de Recherche Opérationnelle, Crédit Lyonnais, France.
213. Jourdain, B. (2004) Loss of Martingality in Asset Price Models with Lognormal Stochastic Volatility. Working paper. ENPC-CERMICS.
214. Kagan, A.M. , Linnik, Y.V., and Rao, C.R. (1973). *Characterization problems in Mathematical Statistics.* John Wiley and Sons, New York.
215. Kakodkar, A., and Galiani, S. (2004). Constant Maturity CDS: Bullish on Credit, Bearish on Spreads. Merrill Lynch Technical Document.
216. Karlin, S., and Taylor, H.M. (1981). A Second Course in Stochastic Processes. Academic Press, New York.
217. Kazziha, S. (1999). Interest Rate Models, Inflation-based Derivatives, Trigger Notes And Cross-Currency Swaptions. PhD Thesis, Imperial College of Science, Technology and Medicine. London.
218. Kennedy, D.P. (1997) Characterizing Gaussian Models of the Term Structure of Interest Rates. *Mathematical Finance* 2, 107-118.
219. Kesel, K., and Grummett, T. (1999). Hypertension. Superboy 60-66, DC Comics, New York.
220. Kijima, M., Nagayama, I. (1994) Efficient Numerical Procedures for the Hull-White Extended Vasicek Model. *The Journal of Financial Engineering* 3, 275-292.
221. Kijima, M., Nagayama, I. (1996) A Numerical Procedure for the General One-Factor Interest Rate Model. *The Journal of Financial Engineering* 5, 317-337.
222. Klöden, P.E., Platen, E. (1995) Numerical Solutions of Stochastic Differential Equations. Springer, Berlin, Heidelberg, New York.
223. Lacey, Raymond (2002). Private Communication.
224. Lando, D. (1998). On Cox processes and credit risky securities. Review of Derivatives Research 2, 99-120
225. Lando, D. (2004). Credit risk modeling: theory and applications. Princeton University Press
226. Ledoit, O., and Santa-Clara, P. (1998) Relative Pricing of Options with Stochastic Volatility. Working paper, Anderson Graduate School of Management, University of California, Los Angeles.
227. Leippold, M. and Wu, L. (2002) Asset Pricing under the Quadratic Class. Journal of Financial and Quantitative Analysis, Vol. 37, n. 2
228. Li, A., Ritchken, P., Sankarasubramanian, L. (1995a) Lattice Models for Pricing American Interest Rate Claims. *The Journal of Finance* 50, 719-737.
229. Li, A., Ritchken, P., Sankarasubramanian, L. (1995b) Lattice Works. *Risk* 8, 65-69.
230. Liinev J., Eberlein E. (2004) Forward swap market models with jumps. Handelingen Contactforum 2nd Actuarial and Financial Mathematics Day, 6 February 2004, Vanmaele, M. et al. (Eds.), Koninklijke Vlaamse Academie van België voor Wetenschappen en Kunsten, Brussel, 83-94.
231. Longstaff, F.A., Santa Clara, P., and Schwartz, E.S. (2001) Throwing Away a Billion Dollars: The Cost of Suboptimal Exercise Strategies in the Swaptions Market. *Journal of Financial Economics* 62, 39-66.

232. Longstaff, F.A., Schwartz, E.S. (1992a) Interest Rate Volatility and the Term Structure: A Two-Factor General Equilibrium Model. *The Journal of Finance* 47, 1259-1282.
233. Longstaff, F.A., Schwartz, E.S. (1992b) A Two-Factor Interest Rate Model and Contingent Claims Valuation. *The Journal of Fixed Income* 3, 16-23.
234. Longstaff, F.A., Schwartz, E.S. (1993) Implementation of the Longstaff-Schwartz Interest Rate Model. *The Journal of Fixed Income* 3, 7-14.
235. Longstaff, F. A., and Schwartz, E.S. (2001). Valuing American Options by Simulation: A Simple Least-Squares Approach. *The Review of Financial Studies* 14, 113-147.
236. Lvov, D. (2005). Monte Carlo Methods for Pricing and Hedging: Applications to Bermudans Swaptions and Convertible Bonds. PhD. thesis. ISMA Centre. University of Reading.
237. Madan, D. B., and E. Seneta (1990) The Variance Gamma Model for Share Market Returns. *Journal of Business* 63, 511524.
238. Maghsoodi, Y. (1996) Solution of the Extended CIR Term Structure and Bond Option Valuation. *Mathematical Finance* 6, 89-109.
239. Marris, D. (1999). Financial Option Pricing and Skewed Volatility. M.Phil Thesis. Statistical Laboratory, University of Cambridge.
240. Mauri, G. (2001) Private Communication.
241. Meister, M. (2004) Smile Modeling in the LIBOR Market Model. Diploma Thesis, University of Karlsruhe.
242. Mercurio, F. (2002) A multi-stage uncertain-volatility model. Banca IMI internal report. Available online at `http://www.fabiomercurio.it/UncertainVol.pdf`
243. Mercurio, F. (2005) Pricing Inflation-Indexed Derivatives, *Quantitative Finance*, 5(3), 289-302.
244. Mercurio, F., Moraleda, J.M. (2000) An Analytically Tractable Interest Rate Model with Humped Volatility. *European Journal of Operational Research* 120, 205-214.
245. Mercurio, F., Moraleda, J.M. (2001) A Family of Humped Volatility Models. *The European Journal of Finance* 7, 93-116.
246. Mercurio, F., and Moreni, N. (2006) Inflation with a smile, *Risk* 19(3), 70-75.
247. Mercurio, F., and Morini, M. (2005) Uncertain Shifts: a Tractable Libor Model for Cap and Swaption Smile. Internal Report, Banca IMI, Milan.
248. Mercurio, F., and Pallavicini, A. (2005) Swaption Skews and Convexity Adjustments. Internal Report, Banca IMI, Milan.
249. Merton R. (1973) Theory of Rational Option Pricing, *Bell Journal of Economics and Management Science* 4, 141-183.
250. Merton R. (1974) On the Pricing of Corporate Debt: The Risk Structure of Interest Rates. The Journal of Finance 29, 449-470.
251. Merton, R.C. (1976) Option Pricing When Underlying Stock Returns Are Discontinuous. *Journal of Financial Economics* 3, 125-144.
252. Mikkelsen, P. (2001) Cross-Currency LIBOR Market Models. CAF working paper 85.
253. Mikulevicius, R., and Platen, E. (1988). Time discrete Taylor approximations for Ito processes with jump component. Mathematische Nachrichten 138, 93-104
254. Miller, E. (2003). A New Class of Entropy Estimators for Multi-Dimensional Densities. In: *International Conference on Acoustics, Speech, and Signal Processing*, proceedings.
255. Miltersen, K.R., Sandmann K., Sondermann D. (1997) Closed Form Solutions for Term Structure Derivatives with Log-Normal Interest Rates. *The Journal of Finance* 52, 409-430.

256. Miron, P., Swannel, P. (1991) Pricing and Hedging Swaps. Euromoney Books, London.
257. Miyazaki, K., Toshihiro, Y. (1998) Valuation Model of Yield-Spread Options in the HJM Framework. *The Journal of Financial Engineering* 7, 98-107.
258. Moraleda, J.M., Vorst, A.C.F. (1997) Pricing American Interest Rate Claims with Humped Volatility Models. *The Journal of Banking and Finance* 21, 1131-1157.
259. Morini, M., (2002). Calibrazione e strutture di covarianza del LIBOR Market Model. Teoria, software e applicazioni a dati di mercato. Master thesis, University of Pavia.
260. Morini, M., (2003). An analytical cascade calibration algorithm for the LIBOR model based only on directly quoted swaptions volatilities. Working paper.
261. Morini, M., and Webber, N. (2003). An EZI method to reduce the rank of a correlation matrix. FORC Preprint, Warwick Business School.
262. Morrison, G., and Semekis, V. (85271). DC One million 1-4, DC comics, New York.
263. Muck, M. (2003). On the Similarity between Displaced Diffusion and Constant Elasticity of Variance Market Models of the Term Structure. Working paper.
264. Musiela, M. (2005). Mathematical issues with volatility modeling. Presentated at the Conference *Quantitative Finance: Developments, Applications and Problems*. Newton Institute, Cambridge.
265. Musiela, M., and Rutkowski, M. (1997). Continuous-Time Term Structure Models: Forward Measure Approach. *Finance and Stochastics*, 4, pp. 261–292.
266. Musiela, M. and Rutkowski, M. (1998) Martingale Methods in Financial Modeling. Springer. Berlin.
267. Nelsen, R. (1999). An Introduction to Copulas. Springer, New York.
268. Nelson, D.B., Ramaswamy, K. (1990) Simple Binomial Processes as Diffusion Approximations in Financial Models. *The Review of Financial Studies* 3, 393-430.
269. Øksendal, B. (1992). Stochastic Differential Equations: An Introduction with Applications. Springer. Berlin.
270. Ouyang, Wei (2003). Private Communication.
271. Pallavicini, A. (2006). Private communication, Credit Models department of Banca IMI.
272. Pedersen, M.B. (1999) Bermudan Swaptions in the LIBOR Market Model, SimCorp Financial Research Working Paper.
273. Pelsser, A. (1996) Efficient Methods for Valuing and Managing Interest Rate and other Derivative Securities. PhD Dissertation, Erasmus University Rotterdam, The Netherlands.
274. Pelsser, A. (2000) Efficient Methods for Valuing Interest Rate Derivatives. Springer. Heidelberg.
275. Pelsser, A. (2003) Mathematical Foundation of Convexity Correction. *Quantitative Finance* 3(1), 59-65
276. Pelsser, A., Vorst, A.C.F. (1998) Pricing of Flexible and Limit Caps. Report 9809, Erasmus University Rotterdam.
277. Piterbarg, V. (2003) A Stochastic Volatility Forward LIBOR Model with a Term Structure of Volatility Smiles. Working Paper. Bank of America.
278. Pugachevsky, D. (2001) Forward CMS Rate Adjustment. *Risk* 14 (March, 2001), 125-128.
279. Pykhtin, M. (2005). Private communication concerning Brigo and Masetti (2005b).

280. Rapisarda, F., Silvotti, R. (2001) Implementation and Performance of Various Stochastic Models for Interest-Rate Derivatives. *Applied Stochastic Models in Business and Industry* 17, 109-120.
281. Rebonato, R. (1998) Interest Rate Option Models. Second Edition. Wiley, Chichester.
282. Rebonato, R. (1999a) Calibrating the BGM Model. *Risk* 12 (March, 1999), 74-79.
283. Rebonato, R. (1999b) On the Pricing Implications of the Joint Lognormal Assumption for the Swaption and Cap Market. *The Journal of Computational Finance* 2, 57-76.
284. Rebonato, R. (1999c) On the Simultaneous Calibration of Multifactor Lognormal Interest Rate Models to Black Volatilities and to the Correlation Matrix. *The Journal of Computational Finance* 2, 5-27.
285. Rebonato, R. (1999d) Volatility and Correlation (First Edition). Wiley, Chichester.
286. Rebonato, R. (2001) Points of Interest: Review of "Interest Rate Models: Theory and Practice", *Risk Magazine.*
287. Rebonato, R. (2002) Modern Pricing of Interest Rate Derivatives: The LIBOR Market Model and Beyond. Princeton University Press.
288. Rebonato, R., and Jäckel, P. (1999). The most general methodology to create a valid correlation matrix for risk management and option pricing purposes, QUARC preprint.
289. Rebonato, R., and Joshi, M. (2001). A Joint Empirical and Theoretical Investigation of the Modes of Deformation of Swaption Matrices: Implications for Model Choice, *QUARC* preprint.
290. Renault, E., Touzi, N. (1996). Option Hedging and Implied Volatilities in a Stochastic Volatility Model. *Mathematical Finance* 6, 279-302.
291. Rendleman, R.J., Bartter, B.J. (1980) The Pricing of Options on Debt Securities. *Journal of Financial and Quantitative Analysis* 15, 11-24.
292. Ritchken, P., Sankarasubramanian, L. (1995) Volatility Structures of Forward Rates and the Dynamics of the Term Structure. *Mathematical Finance* 5, 55-72.
293. Rogers, L.C.G. (1995) Which Model for Term-Structure of Interest Rates Should One Use? in M. Davis, D. Duffie, W. Fleming and S. Shreve, eds. Mathematical Finance. IMA Vol. Math. Appl. 65, New York: Springer.
294. Rogers, L.C.G. (1996) Gaussian Errors. *Risk* 9, 42-45.
295. Rogers, L.C.G. (1997) The Potential Approach to the Term Structure of Interest Rates and Foreign Exchange Rates. *Mathematical Finance* 7, 157-176.
296. Rogers, L.C.G., and Williams, D. (1987) Diffusions, Markov Processes and Martingales, Vol. II, Wiley and Sons, New York.
297. Rubinstein, M. (1983) Displaced Diffusion Option Pricing. *Journal of Finance* 38, 213-217.
298. Rubinstein, M. (1991) Pay Now, Choose Later. *Risk* 4, 13.
299. Rutkowski, M. (1996) On Continuous-Time Models of Term Structure of Interest Rates. In Stochastic Processes and Related Topics. H. J. Englebert, H. Föllmer, and J. Zabczyk, eds. New York: Gordon and Beach.
300. Rutkowski, M. (1999) Models of Forward LIBOR and Swap Rates. Preprint.
301. Salcoacci, M. (2003). Private Communication.
302. Sandmann, K., Sondermann, D. (1993) A Term Structure Model and the Pricing of Interest Rate Derivatives. *The Review of Futures Markets* 12, 391-423.
303. Sandmann, K., Sondermann, D. (1997) A Note on the Stability of Lognormal Interest Rate Models and the Pricing of Eurodollar Futures. *Mathematical Finance* 7, 119-128.

304. Santa Clara, P, and Sornette, D. (2001) The Dynamics of the Forward Interest Rate Curve with Stochastic String Shocks. *The Review of Financial Studies* 14, 149-185.
305. Schlögl, E. (2002) A Multicurrency Extension of the Lognormal Interest Rate Market Models. *Finance and Stochastics* 6(2), 173196.
306. Schlögl, E., Sommer, D. (1998) Factor Models and the Shape of the Term Structure. *The Journal of Financial Engineering* 7, 79-88.
307. Schmidt, W.M. (1997) On A General Class of One-Factor Models for the Term Structure of Interest Rates. *Finance and Stochastics* 1, 3-24.
308. Schönbucher, P. (1999) A Market Model of Stochastic Implied Volatility. *Philosophical Transactions of the Royal Society*, Series A, Vol. 357, No. 1758, pp. 2071-2092.
309. Schönbucher, P. (2000) A Libor Market Model with Default Risk. Working paper.
310. Schönbucher, P. (2004) A Measure of Survival. *Risk*, August issue, 79-85
311. Schönbucher, P. (2005). Portfolio Losses and the Term Structure of Loss Transition Rates: A new methodology for the pricing of portfolio credit derivatives. `http://www.defaultrisk.com/pp_model_74.htm`
312. Schönbucher, P., and Schubert, D. (2001). Copula-Dependent Default Risk in Intensity Models, working paper.
313. Schoenmakers, J. (2005). *Robust Libor Modeling and Pricing of Derivative Products*, Chapman and Hall, Boca Raton.
314. Schoenmakers, J. , and Coffey, C. (2000). Stable implied calibration of a multifactor Libor model via a semi-parametric correlation structure, Weierstrass Institute Preprint n. 611.
315. Schoenmakers, J., and Coffey, C. (2002). Systematic generation of parametric correlation structures for the Libor market model, Preprint. Updated version in *International Journal of Theoretical and Applied Finance* 6(4), 1-13.
316. Schrager, D.F., and Pelsser, A.A.J. (2004). Pricing Swaptions in Affine Term Structure Models. Working paper.
317. Scott, L. (1995) The Valuation of Interest Rate Derivatives in a Multi-Factor Term-Structure Model with Deterministic Components. University of Georgia. Working Paper.
318. Sidenius, J., Piterbarg, V., Andersen, L. (2005). A New Framework for Dynamic Credit Portfolio Loss Modeling. Available at `http://defaultrisk.com/pp_model_83.htm`
319. Sorensen, E.H., and Bollier, T. F. (1994). Pricing Swap Default Risk. *Financial Analysts Journal*, 50. 23–33.
320. Talay, D. and Tubaro, L. (1990). Expansion of the global error for numerical schemes solving stochastic differential equations. Stochastic Analysis and Applications, Vol. 8 n 4, 94-120.
321. Tankov, P. (2003). Dependence structure of Levy processes with applications in risk management. Rapport interne 502, CMAP, Ecole Polytechnique.
322. The Royal Bank of Scotland (2003) Guide to Inflation-Linked Products. Risk.
323. Tilley, J. (1993). Valuing American options in a path simulation model. *Trans. Soc. Actuaries* 45, 83-104.
324. van Bezooyen, J.T.S., Exley, C.J. and Smith, A.D. (1997) A market-based approach to valuing LPI liabilities. Available online at: `http://www.gemstudy.com/DefinedBenefitPensionsDownloads`
325. Vasicek, O. (1976). A test for normality based on sample entropy. *J. R. Statist. Soc. B*, 38, pp. 54–59.
326. Vasicek, O. (1977) An Equilibrium Characterization of the Term Structure. *Journal of Financial Economics* 5, 177-188.

327. Veronesi, F. (2003). Private Communication. Credit Derivatives Desk of Banca IMI.
328. Wu, L. (2006). Arbitrage Pricing of single-name Credit Derivatives. Working paper.
329. Wu, L., and Zhang, F. (2002) Libor Market Model: from Deterministic to Stochastic Volatility. Working Paper. Claremont Graduated University and Hong Kong University of Science and Technology.
330. Zhang, Z.-T., and Wu, L. (2001). Optimal Low-rank approximation of Correlation Matrices. Working paper available at `http://www.math.ust.hk/~malwu/`

Index

Printing: Krips bv, Meppel
Binding: Stürtz, Würzburg